THE ENUMERATION OF LICHENIZED FUNGI IN CHINA

WEI Jiangchun

中国林業出版社
CFPH China Forestry Publishing House

图书在版编目(CIP)数据

中国地衣型真菌综览 = The Enumeration of Lichenized Fungi in China : 英文 / 魏江春著. -- 北京 : 中国林业出版社, 2020.8
ISBN 978-7-5219-0770-4

Ⅰ. ①中… Ⅱ. ①魏… Ⅲ. ①地衣类－真菌－中国－英文 Ⅳ. ①Q949.34

中国版本图书馆CIP数据核字(2020)第166266号

内容简介

《中国地衣型真菌综览》，简称“综览”，收录了3085种，其中地衣型真菌3050种，包括子囊菌3041种，担子菌9种；地衣外生真菌以星号“*” 为标记35种，包括子囊菌31种，担子菌4种。每一分类群均附有学名正名及其汉名、拉丁基原异名以及出现在有关中国地衣型和地衣外生真菌文献中的学名异名。同时，每一分类群均附有省、自治区及直辖市一级的产地及其作者与文献。该“综览”是中国地衣型真菌资源及生命科学研究的信息存取系统。它在自然界地衣多样性与资源研发二者之间起着不可缺少的桥梁作用，也是《中国地衣志》编研的重要参考文献。

该“综览”可供地衣学、真菌学、微生物学、植物学、生物多样性、资源生物学以及环境生物学领域的科技工作者以及高等院校生命科学领域的师生参考。

中国林业出版社·林业分社
策划、责任编辑：于界芬

出版发行 中国林业出版社
(100009 北京西城区德内大街刘海胡同 7 号)
网　　址 http://www.forestry.gov.cn/lycb.html
电　　话 (010) 83143542
印　　刷 河北京平诚乾印刷有限公司
版　　次 2020 年 12 月第 1 版
印　　次 2020 年 12 月第 1 次
开　　本 787mm × 1092mm　1/16
印　　张 38.5　**彩插** 16 面
字　　数 1380 千字
定　　价 273.00 元

Wei Jiangchun
魏江春

Wei Jiangchun, Dr. BioSci., CAS member, Honorary Professor of UCAS, Research Professor of IMCAS.

In order to fill in the blank subject in China, he was sent to the Graduate School of the USSR Academy of Sciences to study the lichenlology from 1958 to 1962.

After his return to China in 1962, he carried out a research on the flora of lichens of Mt. Taibai shan in Shaanxi, Mt. Fanjing shan in Guizhou, Mt. Emei shan in Sichuan, Mt.Hengshan in Hunan, Mt.Lushan in Jiangxi, and Mt. Qomolangma peak in Xizang (Tibet), Northeast and Northwest China. The first Fungarium Lichenum (Herbarium Mycologicum Academiae Sinicae:HMAS-L) in China were established by him.

In 1972, at his initiative and with the co-organization of the relevant experts, a pre-compilation research and compilation of "Flora Cryptogams" including *Flora Algarum Marinarum Sinicarum, Flora Algarum Sinicarum Aquae Dulcis*, *Flora Fungorum Sinicorum*, *Flora Lichenum Sinicorum* and *Flora*

Bryophytorum Sinicorum were carried out by national algologists, mycologists, lichenologists and bryologists under the head of the Editorial Committee of Cryptogamic Flora, CAS since 1973. Through nearly half a century of pre-compilation research and compilation of 5 Cryptogamic Florae, including 27 volumes of *Flora Lichenum Sinicorum* will be completed. During this period, more than 50 young lichenologists with master's or doctor's degrees engaged in lichen research were trained. All of this will be used as a gift to celebrate the 100th anniversary of the Communist Party of China.

On the basis of the above achievements, he moved the Chinese lichenology from phenotype research to a comprehensive study of phenotypic and genotypic combination. A new order Umbilicariales was established by him in 2007. His team completed the whole genome sequencing and analysis of the world's first lichenized fungus-*Endocarpon pusillum* Hedwig. Fourteen PKS (polyketide synthase) genes, 2 NRPS(non-ribosomal peptide synthesase) genes and a large number of drought resistance genes were found in the full genome of *End. pusillum* that never produces any secondary metabolites. The findings provide a scientific basis for the development of secondary metabolite resources in the lichenized fungi that have never produced and less secondary metabolites in the way of triggering the silent gene pathway. Aiming at "Desert Biological Carpet Project" and "Vision of Desert changing Field", the drought resistance gene of lichen fungi is being used to study perennial drought resistant transgenic sand fixing turf plants, drought resistant transgenic wheat and rice. Thus, it is in the first and leading position in the field of lichenized mycology of the world.

As one of the founders of the *Acta Mycologica Sinica* and *Mycosystema*, the first chairman of the Mycological Society of China, and the first Director of the Systematic Mycology and

Lichenology in the Chinese Academy of Sciences, he has made an important contribution to the unity of the national mycologists for development of mycology in China.

So far, 135 scientific papers,4 monographs and 3 compiles have been published. In addition, *Flora Lichenum Sinicorum* volume I has been completed; volumes 9 and 26 are in the process of being finalized. The monograph of the Umbilicariaceae in the world is being written.

He served as deputy director of the institute of microbiology, Chinese Academy of Sciences (CAS), and the director of its degree Committee, Member of the Asian Committee of the International Society of Mycology; Current Editor-in-Chief of "Editorial Committee of Cryptogamic Flora of China, CAS", Honorary President of the Mycological Society of China, Member of the Scientific Committee on Endangered Species of the People's Republic of China, Deputy Director of the editorial Committee of *Catalogue of Life China*, Member of the Science Consultant Committee for CNC-DIVERSITAS, etc.

魏江春，字青川，生物科学博士，中国科学院院士，中国科学院大学荣誉讲席教授，中国科学院微生物研究所研究员。

为了填补我国空白学科，中国科学院应用真菌学研究所于 1958 年派他前往苏联科学院研究生院攻读地衣学专业。他于 1962 年学成回国后，对秦岭太白山、贵州梵净山以及四川、云南、湖南、江西直到西藏珠穆朗玛峰以及东北和西北地区进行了地衣区系考察研究，建立了我国第一个地衣标本室。1972 年，在他的倡议和相关专家的共同组织下，启动了由全国非维管束孢子植物学家参加的包括《中国海藻志》《中国淡水藻志》《中国真菌志》《中国地衣志》和《中国苔藓志》在内的中国科学院中国孢子植物志的编前研究和在研究基础上的编写工作。编研成果将于 2021 年向第一个百年献礼。

在上述成绩的基础上，他将我国地衣学从表型研究推向表型与基因型相结合的综合研究阶段。完成了世界上首个地衣型真菌-石果衣真菌全基因组的测序和分析。为从不产生任何次级代谢产物的石果衣真菌全基因组中率先发现了14个沉默的聚酮合成酶(polyketide syntheases，PKS）基因和2个非核糖体肽合成酶(non ribosomal peptide synthetase，NRPS）基因，为激活沉默基因途径从不产生和少产生次级代谢产物的地衣型真菌中开发次级代谢产物资源提供了科学依据；从中发现了大量抗干旱基因，以“沙漠生物地毯工程”和“沙漠变良田愿景”为目标，正在利用地衣型真菌抗旱基因进行多年生抗旱转基因固沙草皮植物以及抗旱转基因小麦和水稻的研究，从而在地衣型真菌学领域处于世界首创和领跑地位。

作为《真菌学报》和 *Mycosystema*（《菌物学报》）的创始人之一和中国菌物学会的首届理事长，以及中国科学院真菌地衣系统学开放研究实验室的首届主任，在团结全国菌物学家为我国菌物学的发展做出了重要贡献。

迄今已发表论文135篇，专著4册，编译3册，共计142 篇(册)。此外，已经完成了《中国地衣志》第一卷的编研任务；正在进行《中国地衣志》第九卷和第二十六卷的定稿工作，以及世界范围石耳科的专著撰写工作。

历任中国科学院微生物所副所长及所学位委员会主任、国际菌物学会亚洲委员会委员；现任中国科学院中国孢子植物志编委会主编、中国菌物学会名誉理事长、中华人民共和国濒危物种科学委员会委员、国际生物多样性中国委员会顾问委员、《中国生物物种名录》编委会副主任等。

Fig. 1 The coastal rocks of Hainan Island. A dark lichen below J.C.Wei's left hand on rock, i.e. Roccella sinensis Nyl. (See Fig.2)

Fig. 2 **Roccella sinensis** Nyl.
on the coastal rocks of Hainan Island

Fig. 3 " Shihai ", the rock-block field of Wudalianchi, a rare wonder of volcanic lava in the world. It's covered with lichens. Heilongjiang

Fig. 4 **Rhizoplaca chrysoleuca** (Sm.) Zopf
on rock-block of Wudalianchi, Heilongjiang

Fig. 5 **Acarospora gobiensis** H. Magn.
on calciferous stone in Xinjiang

Fig. 6 **Lecanora alpigena** (Ach.) Cl. Roux
on rock in Xinjiang

Fig. 7 **Zeroviella mandschurica** (Zahlbr.) S.Y. Kondr. & Hur
on rock in Mt. Wuling shan, Hebei

Fig. 8 **Rhizoplaca melanophthalma** (DC.) Leuckert
on rock in Xinjiang

Fig. 9 **Protoparmeliopsis peltata** (Ramond) Arup, Zhao Xin & Lumbsch
on rock in Xinjiang

Fig. 10 **Protoparmeliopsis muralis** (Schreb.) M. Choisy
on rock in Mt. Fanjing shan, Guizhou

Fig. 11 **Boreoplaca ultrafrigida** Timdal
on rock in Mt. Tualopodingzi, Jilin

Fig. 12 **Umbilicaria esculenta** (Miyoshi) Minks
on rock in Mt. Tulaopodingzi, Jilin

Fig. 13 **Umbilicaria caroliniana** Tuck.
on rock in Mangui, Neimenggu

Fig. 14 **Umbilicaria hypococcinea** (Jatta) Llano
on rocks in Mt. Guangtou shan, Shaanxi

Fig. 15 **Umbilicaria hypococcinea** (Jatta) Llano
on rocks in Mt. Guangtou shan, Shaanxi

Fig. 16 **Lasallia pertusa** (Rassad.) Llano
on rocks in Xinjiang

Fig. 17 **Stereocaulon sorediiferum** Hue
on rocks in Guangxi

Fig. 18 **Psora decipiens** (Hedw.) Hoffm.
on desert sandy soil in Ningxia

Fig. 19 **Toninia tristis** (Th. Fr.) Th. Fr.
on desert sandy soil in Ningxia

Fig. 20 **Endocarpon crystallinum** J.C. Wei & Jun Yang
on desert sandy soil in Yanggao county, Shanxi

Fig. 21 **Gyalidea asteriscus** subsp. **gracilispora** Jun Yang & J.C. Wei
on desert sandy soil in Mt. Xiaowutai shan, Hebei

Fig. 22 **Cladonia arbuscula** (Wallr.) Flot.
on the ground in forest of Mt. Changbai shan, Jilin

Fig. 23 **Cladonia borealis** S. Stenroos
on the deadwood in the forest of Xinjiang

Fig. 24 **Cladonia macilenta** Hoffm.
on the ground in the forest of Mt. Changbai shan, Jilin

Fig.25 **Cladonia pleurota** (Flörke) Schaer.
on the ground in the forest in Mt. Fanjing shan, Guizhou

Fig. 26 **Cladonia alinii** Trass
on the ground in the forest in Mt. Changbai shan, Jilin

Fig. 27 **Cladonia mongolica** Ahti
on the ground in the forest of Mt. Changbai shan, Jilin

Fig. 28 **Cladonia ramulosa** (With.) J.R. Laundon
on the ground in Mt. Wuyi shan of Fujian

Fig. 29 **Cladonia botrytes** (K.G. Hagen) Willd
on the rotten wood in the Shuanghe National Nature Reserve, Helongjiang

Fig. 30 **Allocetraria subteres** (Asahina) J.C.Wei
on the meadow steppe in Xizang. Scale 1mm

Fig.31 **Thamnolia vermicularis (Sw.) Schaer**.
on the ground in the Tundra zone of Mt. Changbai shan, Jilin

Fig. 32 **Lobaria scrobiculata** (Scop.) P. Gaertn.
on rocks among mosses in the forest of Xinjiang

Fig. 33 **Flavoparmelia caperata** (L.) Hale
on tree bark in Wudalianchi, Heilongjiang

Fig. 34 **Flavopunctelia soredica** (Nyl.) Hale
on tree bark in Wudalianchi, Heilongjiang

Fig. 35 **Nephromopsis komarovii** (Elenk.) J.C.Wei
on tree bark in the forest of Mt. Changbai shan, Jilin

Fig. 36 **Parmotrema tinctorum** (Dilese ex Nyl.) Hale
on the tree bark in Hainan island

Fig. 37 **Menegazzia asahinae** (Yasuda ex Asahina) R. Sant.
on tree bark in Mt. Changbai shan, Jilin

Fig. 38 **Menegazzia terebrata** (Hoffm.) A. Massal.
on tree bark in the forest of Mt. Changbai shan, Jilin

Fig. 39 **Solorina crocea** (L.) Ach.
on the rock among mosses in the Mt. Fanjing shan, Guizhou

Fig. 40 **Ramalina sinensis** Jatta
on tree bark in Wudalianchi, Heilongjiang

Fig. 41 **Heterodermia boryi** (Fée) Kr.P. Singh & S.R.Singh
on trees in Mt. Wuyi shan, Fujian

Fig. 42 **Pseudocyphellaria aurata** (Ach.) Vain.
on tree bark in Mt. Wuyi shan, Fujian

Fig. 43 **Lasallia mayebarae** (M.Sato) Asahina
on trunk of Pinus densata, Yunnan

Fig. 44 **Oxneria fallax** (Arnold) S.Y. Kondr. & Kärnefelt
on tree trunk in the forest of Mt. Changbai shan, Jilin

Fig. 45 **Roccella montagnei** Bél.
on tree trunk in Hainan island

Fig. 39 **Evernia mesomorpha** Nyl.
on tree bark in the forest of Xinjiang

Fig. 47 **Dolichousnea diffracta** (Vain.) Articus
on tree trunk in the forest of Mt. Changbai shan, Jilin

Preface

Lichens were recognized autonomous organisms, coequal with mosses before 1866. De Bary in 1866, followed by Schwendener in 1867, finally came to the realization that lichens are a union of two unrelated organisms.

In fact, "A lichen is a stable self-supporting association of a fungus (mycobiont) and an alga or cyanobacterium (photobiont). More precisely, a lichen is an ecologically obligate, stable mutualism between an exhabitant fungal partner and an inhabitant population of extracellularly located unicellular or filamentous algal or cyanobacterial cells" (Hawksworth & Honegger, in Williams (Eds.) (Plant galls: 77, 1994 and Kirk et al. 378, 2008).

In other words, a lichen is a stable fungal-algal or fungal-cyanobacterial extracellularly mutualistic community in the ecosystem of the Earth's biosphere. Each mutualistic community is composed of a lichenized fungus as a constructive species together with a corresponding alga or cyanobactreium as a companion species. Sometimes a lichenicolous and/or endolichenic fungus, including Basidiomycete yeasts embedded in the cortex of some macrolichens (Spribille et al., 2016), cyanobacterium within cephalodium, and epilichenic or so called lichenicolous lichens are also participated as accidental species in the mutualistic community.

Each species in the fungal-algal mutualistic community has the scientific name and systematic position of its own. So called the scientific name of lichen and its systematic position, in fact, is the scientific name of lichenized fungus, and its systematic position is in the kingdom Fungi. The scientific name of the alga and cyanobacterium in the mutualistic community is that of the alga and cyanobacterium, and their systematic position is in the algae and bacteria respectively.

In the fungal-algal or fungal-cyanobacterial mutualistic community, the alga or cyanobacterium provides the source of carbon for the lichenized fungus, and the lichenized fungus provides anti adversity protection for the alga or cyanobacterium. In rare cases, the lichenized fungus does not provide anti adversity protection for the alga, such as some species in *Coenogonium*.

An Enumeration of Lichens in China (1991) was published 29 years ago. Science and technology have been developed rapidly during this period. Systematic mycology has entered from the classic period of phenotype to the modern period of genophenotypically comprehensive analysis, which as bridge between fungal diversity in the nature and R & D of fungal resources has promoted the development of lichenized fungal species and genetic resource biology. In the same time, a number of young Chinese lichenologists have grown up. It is imperative to compile the revised Secondary Edition of the *An Enumeration of Lichens in China* and renamed as *The Enumeration of Lichenized Fungi in China* faced with the presnt situation.

The Enumeration of Lichenized Fungi in China (abbreviation: "Enumeration") contains 3,085 species belonging to 445 genera, 99 families, 28 orders, 10 classes, and 2 phyla in the kingdom Fungi. Among them 3,050 species are lichenized fungi, including 3041 species of Ascomycota, 9 species of Basidiomycota; 35 species belonging to 26 genera with an asterisk " * " are lichenicolous fungi, included 31 species of Ascomycota belonging to 23 genera, and 4 species of Basidiomycota belonging to 3 genera.

The classification system in the present "Enumeration" is in the literatures of Hibbett et al. (2007), Lumbsch et al. (2007), and Miadlikowska J. et al. (2014) for reference. An index to the correct Latin & Chinises names of listed taxa, their Latin basionyms & synonyms in the taxonomic literature relating to China lichens is provided.

The localities of every taxa with their corresponding references and their authors in the present "Enumeration" are cited. The arrangement order of the localities in the provinces, autonomous regions and municipalities are arranged by their capitals given by the CCTV1 during the weather forecast every evening as follows:

Beijing, Heilongjiang (Harbin), Jilin (Changchun), Liaoning (Shenyang), Tianjin, Neimenggu (Huhehaote), Xinjiang (Wurumuqi), Ningxia (Yinchuan), Qinghai (Xining), Gansu (Lanzhou), Shaanxi (Xian), Xizang (Lasa), Sichuan (Chengdu), Chongqing, Guizhou (Guiyang), Yunnan (Kunming), Shanxi (Taiyuan), Hebei (Shijiazhuang), Shandong (Jinan), Henan (Zhengzhou), Anhui (Hefei), Jiangsu (Nanjing), Shanghai, Hubei (Wuhan), Hunan (Changsha), Jiangxi (Nanchang), Zhejiang (Hangzhou), Fujian (Fuzhou), Taiwan (Taibei), Guangxi (Nanning), Hainan (Haikou), Guangdong (Guangzhou), Hong Kong (= Xianggang), and Macao (= Aomen).

The fungal diversity is the second only to insects in the earth biosphere. A world total of 270,000 species of vascular plants has been assumed. The ratio between the number of vascular plants and fungi is about 1:6, the number of the fungi in the world is conservatively estimated at 1.5 million (Hawksworth, 1991). So, the total of 30.000 species of vascular plants has been assumed in China, the number of the fungi in China is conservatively estimated at 180,000. Around 20% of all fungi are lichenized (Kirk et al., 2008), the number of the lichenized fungi in China is conservatively estimated at 36,000 species.

Each species contains one or several genophenotypically similar populations, which is consisting of genophenotypically similar individuals. Each species is a gene pool. The species diversity is the diversity of gene pool. The gene itself, however, is not valuable in vitro (Zhang, 2000). So, the ecosystem diversity is the life's cradle of species diversity. Therefore, the biodiversity, in fact, is the species diversity in the ecosystem diversity of the earth biosphere.

The first duty of the systematic biology of lichenized fungi is to distinguish, comb and arrange the species diversity from the nature in rank system of species, genus, family, order, class, and phylum according to their relationship. The bioinformation of the rank system, fungarium, and culture collections are the bases of the three systems for storage and retrieval (Wei, 2010). So, the three systems can play an indispensable role of bridge between lichenized fungal diversity in the nature and R & D of their resources, including discovery for life sciences (Zhang & Wei, 2017). On the one hand, owing to the existence of PKS genes silencing in lichenized fungi, on the other hand, lichenized fungi are inherently most resistant to drought, low temperature, salt, heavy metal, and radiation. So, the researches of species and genetic resource biology of the lichenized fungi are pregnant with tremendous potentialities for scienece and practical application.

The present "Enumeration" as a system for storage and retrieval of bioinformation will play an indispensable role of bridge between the lichenized fungal diversity in the nature and R & D of their resources during the sustainable development of human beings as well.

According to the Darwin's theory of common descent, the organisms in the terrestrial biosphere are all descended from common parents (Darwin 1872). Therefore, all living organisms have some of the characteristics of their common ancestor. So, it may be feasible to classify by both their genotypic and phenotypic features of the common ancestor that they retain. The result of phenogenotypical analyses may reflect the genetic relationship among taxa from their common ancestor. The phenogenotypical analyses not only can distinguish the different normal taxa, but also the complicated ones, such as the marginal taxa (Zhang & Wei, 2017), which can not be clustered correctly by the soft ware in single or multigene phylogenetic analyses (Davydov et al., 2010; Miadlikowska J. et al., 2014).

On the basis of biodiversity and its systematics, the species and genetic resource biology would be important area of the biology of the lichenized fungi. The genomic sequencing and analysis of the lichenized fungus *Endocarpon pusillum* (Wang Y. Y. et al., 2014) and its comparative transcriptome analysis have promoted the researches of the species and genetic resource biology for lichenized fungi (Li H, Wei J. C., 2016; Li H. et al. 2017; Zhang Y. L. et al., 2017).

The present "Enumeration" is a collection of the lichenized and lichenicolous fungi in China, based on the original research papers concerning the taxonomy of lichenized fungi, and as a reference for the compilation of the *Flora Lichenum Sinicorum*. The changes of some taxa of Basidiomycota remain to be revised in the future. Something missing or improper in the "Enumeration" is unavoidable. A revised study on the taxa in the "Enumeration", and the discovery of new taxa in the nature remain to be carried out by the authors and editors of the *Flora Lichenum Sinicorum* in the future. A pre-compilation research and the compilation based on the research of the *Flora Lichenum Sinicorum*, first has to play its role as a bridge between lichenized fungal diversity in the nature and R & D of their resources, the next will enrich the knowledge of the diversity of lichenized fungi and accumulate scientific data for the origin and evolution of life.

During the work, Prof. Abdulla Abbas enabled the literatures on the lichenized fungi in Xinjiang region, Prof. Jia Ze-Feng, Prof. Ren Qiang, and Ms Guo Wei helped to collect some literatures on lichenized fungi from different regions in China after 1991, to whom all I express my sincere thanks.

I would be grateful to Professors Chen Jian-Bin, Abdulla Abbas, Guo Shou-Yu, Wang Li-Song, Jia Ze-Feng, Deng Hong and Dr. Hurnisa Xahidin and Wang Ke et al. for their valuable comments and suggestions during the compilation of the "Enumeration".

I want to thank my wife and my children's care, help and support in my life and work.

Wei Jiang-chun

June 26, 2019, Beijing

前　言

在1866年以前，地衣曾被认为是等同于绿色苔藓植物那样单一的自养植物。De Bary (1866) 和 Schwendener (1867)率先认识到地衣是两种不同生物的联合体。

实际上，地衣是稳定的菌藻自我支撑联合体。确切地说，地衣是真菌和单细胞或丝状藻类或蓝细菌细胞之间生态学专化而稳定的胞外互惠共生(Hawksworth & Honegger, in Williams (Eds.) (Plant galls: 77, 1994 and Kirk et al. p. 378, 2008)。

换言之，地衣是地球生物圈内生态系统中稳定的菌藻胞外互惠共生群落。每一个共生群落都是由一种地衣型真菌作为建群种和一种相应的藻类或蓝细菌作为伴生种所组成。除此而外，菌藻共生群落中有时还伴有生长在地衣体外表的外生真菌，生长在地衣体内的内生真菌，包括一些大型地衣皮层中的担子菌纲酵母菌(Spribille et al., 2016)，衣瘿蓝细菌，以及地衣体表的附生地衣等作为偶见种。

菌藻共生群落中的每一个种均具有各自的科学名称及其系统地位。所谓地衣的科学名称及其系统地位，实际上，正是地衣型真菌的科学名称及其在真菌界的系统地位。而菌藻共生群落中的藻类和蓝细菌的科学名称及其系统地位则分别为藻类和细菌各自的科学名称及其系统地位。

《中国地衣综览》(1991)问世以来已经历了29年的科技快速发展时期。系统真菌学已经由传统的表型分类学时期进入表型与基因型相结合的现代分类学时期。地衣型真菌的分类研究获得了长足进展，并作为自然界物种多样性及其物种与基因资源研发之间的桥梁将我国地衣型真菌生物学推向崭新的时代，新一代年轻的中国地衣型真菌学家也在茁壮成长。为了进一步发挥地衣型真菌分类学在自然界物种多样性与系统性在物种与基因资源研发之间的桥梁作用，《中国地衣综览》的再版修订并更名为《中国地衣型真菌综览》的编撰势在必行。

《中国地衣型真菌综览》，简称"综览"，收录了3085种，分隶于真菌界的445属，99科，28目，10纲，2门。其中3050种为地衣型真菌，包括子囊菌3041种，担子菌9种；以星号"*"为标记的26属35种为地衣外生真菌，其中子囊菌23属31种，担子菌3属4种。

该"综览"中的分类系统是以Hibbett et al. (2007), Lumbsch et al. (2007) 及 Miadlikowska J. et al. (2014)的著作为参考。其分类单元的拉丁正名及其基原异名以及出现在涉及中国地衣型真菌的国内外分类学文献中的异名及其有关文献则可通过文献引证查询。"综览"中的地衣型和地衣外生真菌产地引证到省、自治区及直辖市一级。省、自治区、直辖市的排列顺序按照中央电视台综合频道每晚新闻联播后天气预报中各省及自治区的首府与直辖市的汉语拼音名称顺序排列如下：

Beijing, Heilongjiang (Harbin), Jilin (Changchun), Liaoning (Shenyang), Tianjin, Neimenggu (Huhehaote), Xinjiang (Wurumuqi), Ningxia (Yinchuan), Qinghai (Xining), Gansu (Lanzhou), Shaanxi (Xian), Xizang (Lasa), Sichuan (Chengdu), Chongqing, Guizhou (Guiyang), Yunnan (Kunming), Shanxi (Taiyuan), Hebei (Shijiazhuang), Shandong (Jinan), Henan (Zhengzhou), Anhui (Hefei), Jiangsu (Nanjing), Shanghai, Hubei (Wuhan), Hunan (Changsha), Jiangxi (Nanchang), Zhejiang (Hangzhou), Fujian (Fuzhou), Taiwan (Taibei), Guangxi (Nanning), Hainan (Haikou), Guangdong (Guangzhou), Hong Kong (= Xianggang), and Macao (= Aomen).

地球生物圈内真菌的多样性仅次于动物界的昆虫多样性。在对地球生物圈内真菌物种数量的估计中,以每一种维管束植物有 6 种真菌估计,全世界 27 万种维管束植物至少有 150 万种真菌(Hawksworth,1991)。以同样的方法估计,我国已知维管束植物按 3 万种计,至少有 18 万种真菌;其中地衣型真菌占真菌界的 20% 计(Kirk et al. ,2008),我国至少应有地衣型真菌 3.6 万种。

由表基相似的若干生物个体组成居群;由一至若干表基相似的居群组成生物种。每一个生物种都是一个独特的基因库。然而,基因本身在生物个体之外是没有生存价值的(张玉静,2000)。因此,物种多样性亦即基因库的多样性。而生态系统多样性则是生物物种多样性,亦即基因库多样性的生存摇篮。简言之,生物多样性为生存于地球生物圈内生态系统多样性中的生物物种多样性。

地衣型真菌分类学的首要任务就在于对地球生物圈内丰富多样而杂乱无章又不断演化中的物种多样性,通过表型与基因型相结合的综合分析,对其进行识别并梳理成种、属、科、目、纲、门、界的有序等级分类系统。这些等级分类系统作为信息存取系统,连同物种原型标本存取系统以及菌、藻活体培养物存取系统一起(魏,2010)的三大存取系统,在自然界生物多样性与生物资源研发以及生命科学研究之间发挥着不可缺少的桥梁作用(Zhang & Wei, 2017)。一方面,由于地衣型真菌中的聚酮合酶(PKS)基因往往处于沉默状态;另方面,地衣型真菌对干旱、低温、盐碱、重金属以及辐射具有极强的抗性;因此,对于地衣型真菌物种和基因资源生物学的研究在科学与实际应用中具有巨大潜力。《中国地衣型真菌综览》作为信息存取系统将在自然界地衣型真菌物种多样性及其物种与基因资源研发之间发挥必要的桥梁作用。

根据达尔文进化论的共同祖先学说,每一类生物都来自一个共同的祖先(Darwin C, 1872)。当今地球生物圈如此丰富多彩的生物多样性,都是由一个共同祖先通过其基因突变或重组,引起基因组结构的变化,进而致使蛋白质中氨基酸序列的改变,最终导致表型性状的差异,从而在漫长的生物演化过程中,通过自然选择逐步分化而形成的。如此丰富多样的后代生物或多或少都保留其共同祖先的某些基因型和表现型性状,即表基共同祖征。表基共同祖征分析不仅能够识别不同的正常类群,而且还能分辨复杂的对象,如边缘种(Zhang & Wei,2017);而面对后者的复杂类群,通过软件进行的单基因或多基因片段的系统发育分析则难以分辨(Davydov et al. 2010; Miadlikowska et al. ,2014)。

以生物多样性及其系统性为基础,进行地衣物种与基因资源生物学的研究是地衣生物学研究的主要方向和领域。地衣型真菌石果衣全基因组的测序和分析(Wang Y. Y. et al. , 2014)及其转录组分析(Wang Y. Y. et al. ,2015)已经促进了地衣物种资源与基因资源生物学的研究进展(Li H. & Wei J. C. ,2016; Wang J. H. et al. ,2017; Jiang S. H. ,2017; Li H. et al. ,2017; Zhang Y. L. et al. ,2017)。

本"综览"是国内外有关中国地衣型及地衣外生真菌分类学研究文献中的种类总汇,以供《中国地衣志》编研的参考。至于担子菌门部分类群的变动,有待今后订正。“综览”内容有所遗漏或不妥之处在所难免,有待读者及《中国地衣志》的作者和编者予以补充和修订。《中国地衣志》的编前研究和在研究基础上的编写,首先,将在自然界地衣型真菌物种和基因多样性及其研究与开发之间发挥桥梁作用;其次,将丰富人类关于生物多样性及生物演化的科学知识。

在本“综览”修订和编写期间,阿不都拉·阿巴斯教授提供新疆大学作者们的有关地衣分类学文献;贾泽峰教授、任强教授和郭威女士在文献收集方面给与很大的帮助。在此谨对他(她)们表示衷心感谢!对陈健斌、阿不都拉·阿巴斯、郭守玉、王立松、贾泽峰、邓红诸教授先生以及吾尔尼莎博士和王科等在“综览”编撰中提供的宝贵意见和建议表示衷心感谢!最后,我想感谢我的妻子儿女们在我的生活与工作中对我一直的关心、帮助和支持!

魏江春

2019 年 6 月 26 日于北京

CONTENTS

Preface

I Ascomycota

II Basidiomycota

Fungi R. T. Moore 真菌界

Bot. Mar. 23: 371(1980).

Fungi T. L. Jahn & F. F. Jahn, How to Know the Protozoa: 7(1949), nomen invalidum

Dikarya Hibbett, T. Y. James & Vilgalys in Hibbett et al. 双核菌亚界

Mycol. Res. 111(5): 518(2007).

Neomycota Caval. -Sm., Rev. Biol. 73: 209(1998). —*Carpomycetaceae* Bessey, Univ. Studies, Univ. Nebr. 7: 294(1907).

I: Ascomycota Caval. -Sm. 子囊菌门

Biol. Rev. 73: 247(1998).

Pezizomycotina O. E. Erikss. & Winka 果囊菌亚门

Myconet 1(1): 9(1997).

1. Arthoniomycetes O. E. Erikss. & Winka 斑衣菌纲

Myconet 1(1): 4(1997).

[1] Arthoniales Henssen ex D. Hawksw. & O. E. Erikss. 斑衣菌目

Syst. Ascom. 5(1): 177(1986).

Based on Henssen in Henssen & Jahns, Lichenes: 393, 1973, no Latin diagnosis

(1) Arthoniaceae Rchb. 斑衣菌科

Deut. Bot. Herb. -Buch: 13(1841).

See *Reichenbach, Handb. nat. Pfl. -Syst.*: 131(1837), nom. inval., Art. 32. 1.

Allarthonia(Nyl.) Zahlbr. 星衣属

in Engler & Prantl, *Nat. Pflanzenfam.*, Teil. I(Leipzig) 1: 91(1903).

≡*Arthonia* subgen. *Allarthonia* Nyl., *Flora*, Regensburg 61: 246(1878).

Allarthonia bohlinii H. Magn. 包氏星衣

Lich. Centr. Asia 1: 32(1940). Lamb, *Index Nom. Lich.* 1963: 21.

Type: Gansu, at about 4, 100 m, 12/I/1932, Bohlin no. 77a in S.

On calcareous rock.

Allarthonia yunnana Zahlbr. 云南星衣

in Handel-Mazzetti, *Symb. Sin.* 3: 36(1930) & *Cat. Lich. Univ.* 8: 182(1932).

Type: Yunnan, at about 1, 300 m, 7/III/1915, Handel-Mazzetti no. 13091.

On tree trunk.

Allarthothelium(Vain.) Zahlbr. 荞麦皮衣属

in Engler & Prantl, *Nat. Pflanzenfam.*, Teil. I(Leipzig) 1: 241(1908).

≡*Arthonia* subgen. *Allarthothelium* Vain., *J. Bot.*, Lond. 34: 263(1896).

Allarthothelium sparsum Zahlbr. 散荞麦皮衣

in Handel-Mazzetti, *Symb. Sin.* 3: 37(1930) & *Cat. Lich. Univ.* 8: 184(1932).

Type: Yunnan, at about 3, 100 m, 22/VI/1915, Handel-Mazzetti no. 6890.

On bark.

Arthonia Ach. 斑衣属

Neues J. Bot. 1(3): 3(1806).

Arthonia antillarum(Fée) Nyl. 枯草斑衣

Flora, Regensburg 50: 7(1867). f. antillarum 原变型

≡*Coniocarpon antillarum* Fée, *Suppl. Cryptog. Écorc. Officin.*, 94(1837).

On bark of tree.

Shanghai(Xu B. S., 1989, p. 167), Guangdong(Zahlbr., 1930 b, p. 36), Hongkong (Thrower, 1988, p. 50).

Notes: the citation of Zahlbr. for Guangdong is according to Rabh., who is based on Kremp.

1873, p. 467, where is only quoted *Arthonia antillarum* f. *spermogonifera* Rabh.

f. spermogonifera Rabh. 精孢变型

On bark.

Guangdong(Kremp. , 1873, p. 467 as Wampoa) .

Arthonia apotheciorum(A. Massal.) Almq. 囊盘斑衣

K. svenska Vetensk-Akad. Handl. , ser. 2 17(no. 6) : 58(1880) .

≡*Sphaeria apotheciorum* A. Massal. , *Ric. auton. lich. crost.* (Verona) : 26(1852) .

Guangdong(L. Rabh. 1873, p. 286; Kremp. 1874, p. 34) .

Arthonia astropica Kremp. 星斑衣

Chinesische Flechten. Flora 56(30) : 466(1873) .

Misapplied name:

Arthonia cf. *schoepfiae* auct. non Zahlbr. in Thrower, Hong Kong lichens, 1988, pp. 15 & 56, revised by Aptroot & Seaward, 1999, p. 65) .

On exposed trees. Tropical Asiar emains uncertain.

Hongkong(Thrower, 1988, pp. 15, 51, 56; Aptroot & Seaward, 1999, p. 65) .

Arthonia biseptella Nyl. 双隔斑衣

Sertum Lichen, Trop. Labuan et Singapore, 35(1892) .

Hongkong(Seaward and Aptroot 2005, p. 283 according to Pfister 1978: 123) .

Arthonia cinereopruinosa Schaer. 灰霜斑衣

Enum. critic. lich. europ. (Bern) : 243(1850) .

On bark of tree.

Fujian(Wu J. N. et al. 1981, p. 2) .

Arthonia cinnabarina(DC.) Wallr. 朱砂斑衣

Fl. crypt. Germ. (Norimbergae) 3: 320(1831) . var. cinnabarina 原变种

≡*Coniocarpon cinnabarinum* DC. , in Lamarck & de Candolle, Fl. franç. , Edn 3(Paris) 2: 323(1805) .

On exposed base of *Phoenix* palm. Cosmopolitan.

Xizang(Obermayer, 2004, p. 487) , Hongkong(Aptroot & Seaward, 1999, p. 65) .

=*Arthonia cinnabarina* var. *anerythraea* Nyl. , *Lich. Scand.* (Helsinki) : 257(1861) ; Zahlbr. in Feddes, *Repertorium sp. nov.* 31: 206(1933) .

=*Arthonia tumidula*(Ach.) Ach. , *Neues J. Bot.* 1(3) : 11(1806) ; Wu J. N. & Qian Z. G. in Xu B. S. *Cryptogamic flora of the Yangtze Delta and adjecent regions* 168(1989) .

≡*Spiloma tumidula* Ach. , *Methodus*, Sectio prior: 11(1803) .

On trees

Shanghai(Xu B. S. , 1989, p. 168, as *A. tumidula*) , Taiwan(Zahlbr. , 1933 b, p. 206 as *A. cinnabarina* var. *anerythraea*; Wang & Lai, 1973, p. 88, as *A. cinnabarina* var. *anerythraea*) .

var. coccinea(Flörke) Zahlbr. 红色变种

Cat. Lich. Univ. 2: 24(1922) [1924] ; Zahlbr. in Handel-Mazzetti, *Symb. Sin.* 3: 36(1930) .

≡*Conioloma coccineum* Flörke, *Deutsche Lich.* 2: no. 21(1815) .

=*Arthonia cinnabarina* var. *adspersa*(Mont.) Nyl. , *Mém. Soc. Imp. Sci. Nat. Cherbourg* 5: 132(1857) ; Zahlbr. in Handel-Mazzetti, *Symb. Sin.* 3: 36(1930) .

≡*Ustalia adspersa* Mont. , *Annls Sci. Nat.* , *Bot.* , sér. 2 18: 278(1842) .

On bark of trees.

Yunnan (Zahlbr. , 1930b, p. 36) , Guangdong (Kremp. 1873, p. 467 as *Arthonia cinnabarina* var. *adspersa* (Mont.) Nyl. & 1874 a, p. 34; Zahlbr. , 1930b, p. 36) , Shanghai(Kremp. 1873, p. 467; Nyl. & Cromb. 1883, p. 65; Zahlbr. 1930b, p. 36) , China(prov. not given: Hue, 1891, p. 177) .

Arthonia clemens(Tul.) Th. Fr. 静斑衣

K. svenska Vetensk-Akad. Handl., ny följd 7(no. 2): 46(1867).

≡*Phacopsis clemens* Tul., *Annls Sci. Nat.*, *Bot.*, sér. 3 17: 124(1852).

On *Lecidea ochrorufa* (Magnusson, 1940, pp. 57 – 8).

Gansu(Magnusson, 1940, pp. 57 – 8; Hawksw. & M. Cole, 2003, p. 360 as lichenicolous fungus, and noted: " Magnusson was uncertain as to the specific identification, and the name has often been misapllied".

Arthonia complanata Fée 扁平斑衣

Essai Crypt. Exot. (Paris): 54(1825)[1824].

Taiwan(Aptroot, and Sparrius, 2003, p. 3).

Arthonia elegans(Ach.) Almq. 丽斑衣

K. svenska Vetensk-Akad. Handl., ser. 2 17(no. 6): 19(1880).

≡*Spiloma elegans* Ach., *Lich. Univ.*: 1 – 696(1810)

On *Roystonea regia*. New to Asia, so far known from Europe.

Taiwan(Aptroot, and Sparrius, 2003, p. 3).

Arthonia glebosa Tuck. 小块斑衣

Gen. lich. (Amherst): 221(1872).

On ground.

Xizang(Obermayer, 2004, p. 487 – 488).

Arthonia ilicina Taylor 圣栎斑衣

in Mackay, *Fl. Hibern.* 2: 105(1836).

On *Ficus beeshiana*. New to Asia, but furthermore cosmopolitan.

Taiwan(Aptroot, and Sparrius, 2003, p. 3).

Arthonia leioplacella Zahlbr. 花面斑衣

in Handel-Mazzetti, *Symb. Sin.* 3: 35(1930) & *Cat. Lich. Univ.* 8: 178(1932).

Type: Yunnan, Handel-Mazzetti no. 11425.

On exposed trees.

Hongkong(Thrower, 1988, p. 54; Aptroot & Seaward, 1999, p. 65).

Arthonia lividula Vain. 肝色斑衣

Ann. Acad. Sci. fenn., Ser. A 15(no. 6): 309(1921).

≡*Arthoniopsis lividula*(Vain.) Zahlbr., *Cat. Lich. Univers.* 2: 140(1922)[1924].

On Pandanus leaves.

Hongkong(Aptroot et al., 2001, p. 321).

Arthonia lopingensis Zahlbr. 罗平斑衣

in Handel-Mazzetti, *Symb. Sin.* 3: 36(1930) & *Cat. Lich. Univ.* 8: 178(1932).

Misapplied name:

Arthonia albofuscescens auct. non Tuck. in Thrower, Hong Kong lichens, 1988, pp. 15 & 49, revised by Aptroot & Seaward, 1999, . 65 – 66).

Arthonia cf. *antillarum* auct. non(Fée) Nyl. in Thrower, Hong Kong lichens, 1988, pp. 15 & 50, revised by Aptroot & Seaward, 1999, p. 65 – 66).

On exposed trees, especially branches. Tropical Asia remains uncertain.

Type locality: Yunnan, Handel-Mazzetti no. 10107.

Hongkong(Thrower, 1988, pp. 15, 49, 50, 55; Aptroot & Seaward, 1999, p. 65 – 66; Aptroot et al., 2001, p. 321).

Arthonia ochropallens Zahlbr. 黄白斑衣

in Fedde, *Repertorium*, 31: 206(1933).

Type: Taiwan(Asahina no. 317).

On bark of trees.

Taiwan(M. Lamb, 1963, p. 38; Wang & Lai, 1973, p. 88).

Arthonia palmulacea(Müll. Arg.) R. Sant. 掌斑衣

Symb. bot. upsal. 12(1): 87(1952).

≡*Arthoniopsis palmulacea* Müll. Arg., *J. Linn. Soc., Bot.* 29: 238(1892).

≡*Eremothecella palmulacea*(Müll. Arg.) Sérus., *Syst. Ascom.* 11(1): 43(1992).

On bamboo stems.

Hongkong(Aptroot et al., 2001, p. 321).

Arthonia parantillarum Aptroot 准枯草斑衣

Fungal Diversity 14: 3 – 4(2003).

Type: Taiwan, Pingtung County: Kenting Forest Recreation Area, near guesthouse, 200 m alt., 51QTE743300, on *Roystonea regia*. Aptroot 43224(B-holotype, ABL, TUNG-isotypes) and Sparrius 5323, 5342(hb Sparrius-isotypes), 17 October 2001.

Arthonia radiata(Pers.) Ach. 斑衣

K. Vetensk-Acad. Nya Handl. 29: 131(1808). f. radiata 原变型

≡*Opegrapha radiata* Pers., Ann. Bot. (Usteri) 1: 29(1794)

On trees.

Shanghai, Zhejiang(Xu B. S., 1989, p. 168), Taiwan(Aptroot and Sparrius, 2003, p. 4).

f. astroidea Ach. (Ach.) Ach. 星状变型

Lich. Univ.: 1 – 696(1810).

≡*Arthonia astroidea* Ach., *Neues J. Bot.* 1(3): 17(1806).

On bark of tree.

Shaanxi(Jatta, 1902, p. 479; Zahlbr., 1930 b, p. 35), Yunnan(Zahlbr., 1930 b, p. 35).

Arthonia reniformis(Pers.) Röhl. 肾斑衣

Deutschl. Fl. (Frankfurt) 3(2): 29(1813).

≡*Opegrapha reniformis* Pers. (1794); Zahlbr., *Cat. Lich. Univ.* 2191922), p. 94.

On bark of Schoepfia jasminodora.

Yunnan(Zahlbr., 1930 b, p. 35).

Arthonia schoepfiae Zahlbr. 铁青树斑衣

in Handel-Mazzetti, *Symb. Sin.* 3: 35(1930) & *Cat. Lich. Univ.* 8: 180(1932).

Type: Yunnan, at about 1, 600m, 12/VI/1917, Handel-Mazzetti no. 13094.

On bark of *Schoepfia jasminodora*.

Arthonia spadicea Leight. 枣红斑衣

Ann. Mag. nat. Hist., Ser. 2 13: 442(1854).

Misapplied name:

Arthonia complanata auct. non Fée in Thrower, *Hong Kong Lichens*, 1988, pp. 15 & 52, revised by Aptroot & Seaward, 1999, p. 66).

On shaded trees in forests. Noto-boreal(distributed in warm temperate areas of both hemispheres.

Taiwan(Aptroot and Sparrius, 2003, p. 4), Hongkong(Thrower, 1988, pp. 15, 52 as *Arthonia complanata* auct. non Fée; Aptroot & Seaward, 1999, p. 66; Aptroot et al., 2001, p. 321).

Arthonia swartziana Ach. 类星斑衣

Neues J. Bot. 1(3): 13(1806).

Hongkong(Seaward and Aptroot 2005, p. 283 and mentioned as *Arthonia astroidea* var. *swartziana* Nyl. cited by Pfister 1978: 123).

Arthonia trilocularis Müll. Arg. 三腔斑衣

Flora, Regensburg 64(15): 233(1881).

On living leaves of ferns, Pandanus and trees in forests. Pantropical.

Hongkong(Aptroot & Seaward, 1999, p. 66; Aptroot et al. , 2001, p. 321) . Living culture CBS – 101371.

Arthonia varia(Ach.) Nyl. 异斑衣

Mém. Soc. Imp. Sci. Nat. Cherbourg 5: 132(1857) . var. varia 原变种

≡*Opegrapha abnormis* var. *varia* Ach. , *Lich. univ.* : 1 – 696(1810) .

Hongkong(Form indetermined in Herb. Tuck. , Hue, 1891, p. 179) .

var. stenographella(Nyl.) Vain. 狭文变种

In *Annal. Acad. Scient. Fennic.* , Ser. A, Vol. XV, no. 6, 303(1921) .

≡*Arthonia stenographella* Nyl. in *Acta Soc. Scient. Fennic.* , vol. VII, 1863, p. 496.

On bark of bushes.

Hongkong(Seaward and Aptroot 2005, p. 283 and mentioned as *Arthonia stenographella* Nyl. cited by Pfister 1978: 123,) .

Arthonia vinosa Leight. 酒红斑衣

Ann. Mag. nat. Hist. , Ser. 2 18: 331(1856)

Xizang(Tibet, Obermayer, 2004, p. 488) .

Arthothelium A. Massal. **芝麻粒衣属**

Ric. auton. lich. crost. (Verona) : 54(1852) .

Arthothelium chiodectoides(Nyl.) Zahlbr. 座盘芝麻粒衣

Cat. Lich. Univ. 2: 122(1922) [1924] .

≡*Arthonia chiodectoides* Nyl. , *Flora*, Regensburg 52: 72(1869) .

On *Castanopsis*. Aptroot 52508(ABL) and Sparrius 6107(Herb. Sparrius) , 11 October 2001.

Taiwan(Aptroot and Sparrius, 2003, p. 4) .

Arthothelium dispersum Mudd 散芝麻粒衣

Man. Brit. Lich. : 252(1861) .

Corticolous.

Jiangsu(Wu J. L. , 1987, p. 39; Xu B. S. , 1989, p. 168) .

Arthothelium fecundum Zahlbr. 繁育芝麻粒衣

in Handel-Mazzetti, *Symb. Sin.* 3: 37(1930) & *Cat. Lich. Univ.* 8: 183(1932) .

Type: Yunnan, Handel-Mazzetti no. 6027.

Arthothelium ruanum(A. Massal.) Körber 无花果芝麻粒衣

Parerga lichenol. (Breslau) 3: 263(1861) .

≡*Arthonia ruana* A. Massal. , *Ric. auton. lich. crost.* (Verona) : 47(1852) .

On *Ficus microcarpa*. Aptroot 53205(ABL) , 17 October 2001.

Taiwan(Aptroot and Sparrius, 2003, p. 4) .

Arthothelium spectabile Flot. ex Massal. 芝麻粒衣

Ric. auton. lich. crost. (Verona) : 54(1852) .

Hunan(Zahlbr. , 1930 b, p. 37) .

Cryptothecia Stirt. **隐囊衣属**

Proc. Roy. phil. Soc. Glasgow 10: 164(1876) [1877] .

= *Herpothallon* Tobler, *Flora*, Regensburg 131: 446(1937) .

= *Herpothallonomyces* Cif. & Tomas. , *Atti Ist. bot. Univ. Lab. crittog. Pavia*, Ser. 4 10(2) : 292(1954) .

Cryptothecia aleurella (Nyl.) Makhija & Patwardhan 粉隐囊衣

Biovigyanam 11(1) : 3(1985) .

≡*Arthonia aleurella* Nyl. , *Sert. Lich. Trop. Labuan Singapore*: 22(1891) .

On *Cryptomeria japonica*.

Taiwan(Aptroot and Sparrius, 2003, p. 15) .

Cryptothecia aleurocarpa (Nyl.) Makhija & Patwardhan 粉果隐囊衣

Biovigyanam 11(1):3(1985).

≡*Arthonia aleurocarpa* Nyl. in *Annal. Scienc. Nat.*, *Botan.*, ser. 5., vol. VII, 1867, p. 340, Zahlbr., *Cat. Lich. Univ.* 2: 120.

On *Lonicera cumingiana*.

Taiwan(Aptroot and Sparrius, 2003, p. 15).

Cryptothecia candida(Kremp.) R. Sant. 白色隐囊衣

Symb. bot. upsal. 12(no. 1):65(1952).

≡*Myriostigma candidum* Kremp. [as '*candicum*'], Lich. Foliic. Leg. Beccari: 22(1874) & Nuov. Giorn. Botan. Ital., vol. VII, 1875, p. 45. tab. I, fig. 29 in Zahlbr., *Cat. Lich. Univ.* 2: 122.

Yunnan(Aptroot et al., 2003, p. 44), Hongkong(Aptroot et al., 2001, p. 325).

Cryptothecia polymorpha Makhija & Patw. 多形隐囊衣

Biovigyanam 13(2):47(1987).

On tree trunk in low secondary forest.

Hongkong(Aptroot et al., 2001, p. 325).

Cryptothecia subnidulans Stirton 隐囊衣

Proc. Roy. phil. Soc. Glasgow 10: 164(1876)[1877].

On *Zelkova serrata*.

Taiwan(Aptroot and Sparrius, 2003, p. 15).

Cryptothecia subtecta Stirt. 亚藏隐囊衣

Proc. Roy. phil. Soc. Glasgow 11: 320(1879)[1878].

On trees, e. g. *Kandelia*, in forests. Tropical Asia.

Hongkong(Aptroot & Seaward, 1999, p. 73).

On *Acacia confusa*.

Taiwan(Aptroot and Sparrius, 2003, p. 15).

Herpothallon Tobler **腹枝衣属**

Flora, Regensburg 131: 446(1937).

Notes: Since the genus *Cryptothecia* Stirt. is a current name, *Herpothallon* Tobler as its synonym, then the species under the genus *Herpothallon* remain to be revised.

Herpothallon rubrocinctum (Ehrenb.) Aptroot, Lücking & G. Thor 红带腹枝衣

in Aptroot, Thor, Lücking, Elix & Chaves, *Biblthca Lichenol.* 99: 61(2009).

≡*Hypochnus rubrocinctus* Ehrenb., in Nees von Esenbeck(ed.), *Horae Phys. Berol.*: 84, tab. 17, fig. 3(1820); Zahlbr. in Handel-Mazzetti, *Symb. Sin.* 3: 63(1930).

≡*Cryptothecia rubrocincta*(Ehrenb.) G. Thor, *Bryologist* 94: 280(1991).

= *Chiodecton sanguineum* Vain., *Acta Soc. Fauna Flora Fenn.* 7(no. 2): 143(1890); Zahlbr. *in* Handel-Mazzetti, *Symb. Sin.* 3: 63(1930).

Guizhou(Zahlbr., 1930b, p. 63).

Herpothallon weii Y. L. Cheng & H. Y. Wang 魏氏腹枝衣

in Cheng, Ning, Xu, Zhang, Wang & Zhao, *Mycotaxon* 119: 440(2012).

Type: China. Guizhou province, Mt. Leigong, on bark, alt. 1000 m, 2 Nov. 2009, Q. Tian 20103366. (Holotype in SDNU).

Stirtonia A. L. Sm. **小斑衣属**

Trans. *Br. mycol. Soc.* 11(3-4): 195(1926).

Stirtonia indica Makhija & Patw. 印度小斑衣

Mycotaxon 67: 300(1998).

On *Lonicera cumingiana*, and *Castanopsis* sp.

Taiwan(Aptroot & Sparrius 2003, p. 46).

Tylophoron Nyl. ex Stizenb. **载瘤衣属**

Ber. Tät. St Gall. naturw. Ges. : 117(1862)[1861 –62].

Tylophoron moderatum Nyl. 轻载瘤衣

Bot. Ztg. 20: 279(1862).

ssp. orientale Zahlbr. 东方亚种

in Fedde, *Repertorium* 31: 206(1933).

Type: Taiwan, Mt. Ali shan, Asahina no. 380.

On bark of tree.

Taiwan(Wang & Lai, 1973, p. 97).

(2) Chrysotrichaceae Zahlbr. **金絮衣科**

[as 'Chrysothricaceae'] in Engler & Prantl, *Nat. Pflanzenfam.*, Teil. I(Leipzig) 1*: 117(1905).

Chrysothrix Mont. **金絮衣属**

Annls Sci. Nat., Bot., sér. 3 18: 312(1852).

Chrysothrix candelaris(L.) J. R. Laundon 烛金絮衣

Lichenologist 13(2): 110(1981).

≡*Byssus candelaris* L., *Sp. pl.* 2: 1169(1753).

On trees in open areas and in forests. Comsopolitan.

Zhejiang(D. Hawksw. & Weng, 1990, p. 515), Hongkong(Thrower, 1988, p. 74; Aptroot & Seaward, 1999, p. 71; Aptroot et al., 2001, p. 324), Taiwan(Aptroot and Sparrius, 2003, p. 14).

Chemical race I(pinastric acid): Xizang(W. Obermayer, 2004, p. 494).

Chemical race II(calycin): Sichuan(W. Obermayer, 2004, p. 494).

Chemical race III(pinastric acid and calycin): Xizang(W. Obermayer, 2004, p. 494).

Chrysothrix chlorina(Ach.) J. R. Laundon 金絮衣

Lichenologist 13(2): 106(1981).

≡*Lichen chlorinus* Ach., *Lich. suec. prodr.* (Linköping): 6(1799)[1798].

≡*Lepra chlorina*(Ach.) DC., in Lamarck & de Candolle, *Fl. franç.*, Edn 3(Paris) 2: 68(1805); Zahlbr. in Handel-Mazzetti, *Symb. Sin.* 3: 244(1930).

On the ground.

Shaanxi(Jatta, 1902, p. 480; Zahlbr., 1930b, p. 244), Xizang(W. Obermayer, 2004, p. 494).

Chrysothrix xanthina(Vain.) Kalb 黄金絮衣

Biblthca Lichenol. 78: 144(2001).

≡*Lepraria xanthina* Vain., in Hiern, *Cat. Welwitsch Afric. Pl.* 2(2): 463(1901).

On *Delonix* p., and *Mangifera indica*.

Taiwan(Aptroot and Sparrius, 2003, p. 14).

(3) Melaspileaceae Walt. Watson **黑斑衣科**

New Phytol. 28: 94(1929).

Melaspilea Nyl. **黑斑衣属**

Act. Soc. linn. Bordeaux 21: 416(1857).

Melaspilea diplasiospora(Nyl.) Müll. Arg. 黑斑衣

Mém. Soc. Phys. Hist. nat. Genève 29(no. 8): 22(1887).

≡*Opegrapha diplasiospora* Nyl., *Acta Soc. Sci. fenn.* 7(2): 476(1863).

Misapplied name:

Melaspilea sp. in Thrower, 1988, pp. 16, 121, = *Opegrapha* sp., revised by Aptroot & Seaward, 1999, p. 82.

On exposed trees. Known from neotropics, Australia, and Asia.

Hongkong(Aptroot & Seaward, 1999, p. 82).

Melaspilea urceolata(Fr.) Almb. 平黑斑衣

in Grummann, *Cat. Lich. Germ.* (Stuttgart): 20(1963).

≡*Lecanactis urceolata* Fr., *Syst. orb. veg.* (Lundae) 1: 288(1825)

On the walls of Macao Fort.

Guangdong(Nyl. & Cromb. 1883, p. 65; Hue, 1891, p. 126).

(4) Roccellaceae Chevall. **染料衣科**

[as '*Rocellaceae*'] *Fl. gén. env. Paris*(Paris) 1: 604(1826).

Bactrospora A. Massal. **菌孢衣属**

Ric. auton. lich. crost. (Verona): 133(1852).

Bactrospora myriadea(Fée) Egea & Torrente 多杆菌孢衣

Lichenologist 25(3): 245(1993).

≡*Coniocarpon myriadeum* Fée, *Essai Cryptog. écorc. Officin.*, 99(1824).

On *Delonix regia*, *Mangifera indica*, *Acacia confusa*, and eutrophicated bark.

Taiwan(Aptroot and Sparrius, 2003, p. 8), Hongkong(Aptroot et al., 2001, p. 321).

Chiodecton Ach. **座盘衣属**

Syn. meth. lich. (Lund): 108(1814).

Chiodecton aurantiacoflavum B. de Lesd. 橘黄座盘衣

Rev. Bryol. Lichénol., N. S. 7: 61(1934).

On shale.

Taiwan(Aptroot and Sparrius, 2003, p. 13).

Chiodecton congestulum Nyl. 乳酪座盘衣

Bull. Soc. linn. Normandie, sér. 2 2: 106(1868).

= *Chiodecton stromaticum* C. Knight, *Trans. Linn. Soc. London*, Bot. 2: 42(1882)

= *Chiodecton butyraceum* Zahlbr. in Fedde, *Repertorium* 31: 222(1933) & *Cat. Lich. Univ.* 10: 199(1940), synonymized by Thor, 1990 The lichen genus Chiodecton and five allied genera. -Opers Bot. 103: 37, 41(1990) [1991].

Type: Taiwan, Rengechi, 1925 Asahina F 319(W, lectotype).

Corticolous.

Taiwan(Lamb, 1963, p. 164; Wang & Lai, 1973, p. 89; Thor, 1990, 37, 41), Hongkong(Aptroot et al., 2001, p. 324, on rock).

Chiodecton mucorinum Zahlbr. 白霉座盘衣

in Handel-Mazzetti, *Symb. Sin.* 3: 63(1930) & *Cat. Lich. Univ.* 8: 222(1932).

Type: Fujian, Kushan near Fuzhou(Fuchou), alt. 500 – 600m, 1926 Chung no. 593(W, holotype, FH isotype, Thor, 1990, p. 53).

Corticolous.

Cresponea Egea & Torrente **鸡冠衣属**

Mycotaxon 48: 302(1993).

Cresponea leprieurii(Mont.) Egea & Torrente 勒氏鸡冠衣

Mycotaxon 48: 317(1993).

≡*Lecidea leprieurii* Mont., *Annls Sci. Nat.*, *Bot.*, sér. 4 16: 56(1851).

On trees, e. g. *Callitris*, in forest. Known from neotropics and Australsia.

Hongkong(Aptroot & Seaward, 1999, p. 73).

Cresponea premnea(Ach.) Egea & Torrente 鸡冠衣

Mycotaxon 48: 324(1993).

≡*Lecidea premnea* Ach., *Lich. univ.*: 173(1810).

On shale.

Taiwan(Aptroot and Sparrius, 2003, p. 14).

Cresponea proximata(Nyl.) Egea & Torrente 近鸡冠衣

Mycotaxon 48: 328(1993) .

≡*Lecidea proximata* Nyl. , *Annls Sci. Nat.* , *Bot.* , sér. 4 19: 356(1863) .

On trees in forests. Known from tropical Asia, the neotropics and Australasia.

Hongkong(Aptroot & Seaward, 1999, p. 73; Seaward & Aptroot, 2005, p. 284, Pfister 1978: 120, as *Lecanactis premnea*; Aptroot et al. , 2001, p. 325) , Taiwan(Aptroot and Sparrius, 2003, p. 14) .

Dichosporidium Pat. **双子衣属**

in Duss, *Enum. Champ.* Guadeloupe(Lons-le-Saunier) : 20(1903) .

Dichosporidium boschianum(Mont.) G. Thor 柱双子衣

Op. bot. 103: 64(1991) .

≡*Chiodecton boschianum* Mont. , *Syll. gen. sp. crypt.* (Paris) : 356(1856) .

On trees, e. g. *Liquidambar* sp. , in forest. Known from tropical Asia and Australasia.

Hongkong(Aptroot & Seaward, 1999, p. 73) , Taiwan(Aptroot and Sparrius, 2003, p. 15) .

Dichosporidium nigrocinctum(Ehrenb.) G. Thor 黑双子衣

Op. bot. 103: 71(1991) .

≡*Hypochnus nigrocinctus* Ehrenb. , *Enum. fung. Chamisso coll.* : 85(1820) .

On trees in forest. Known from the neotropics.

Hongkong(Aptroot & Seaward, 1999, p. 73) .

Enterographa Fée **全缘衣属**

Essai Crypt. Exot. (Paris) : xxxii, cx, 57(1825) [1824] .

Enterographa hainanensis B. Gao & J. C. Wei 海南全缘衣

[as ' *hainanesis*'] , *Mycosystema* 28(2) : 176(2009) .

Typus: China, Hainan, Bawangling, alt. 670m, in foliis *Salicis* sp. , Leg. Zhang Tao HN333 – 1(HMAS-L) .

Enterographa pallidella(Nyl.) Redinger 白全缘衣

Reprium nov. Spec. Regni veg. 43: 61(1938) .

≡*Platygrapha pallidella* Nyl. , *Flora*, Regensburg 50: 6(1867) .

Misapplied name:

Arthonia cf. *fissurina auct. non* Nyl. in Thrower, 1988, pp. 15, 53.

On exposed trees and branches. Pantropical.

Hongkong(Thrower, 1988, pp. 15, 53; Aptroot & Seward, 1999, p. 74; Aptroot et al. , 2001, p. 326) .

Enterographa praepallens(Nyl.) Redinger 灰全缘衣

Reprium nov. Spec. Regni veg. 43: 66(1938) .

≡*Stigmatidium praepallens* Nyl. , Lich. Japon. : 84(1890) .

On sheltered, overhanging, coastal granite rocks. Known from tropical Asia and Ausralasia.

Hongkong(Aptroot & Seward, 1999, p. 74 – 75; Aptroot et al. , 2001, p. 326) .

Haplodina Zahlbr. **单碗衣属**

in Handel-Mazzetti, *Symb. Sin.* 3: 65(1930) & *Cat. Lich. Univ.* 8: 228(1932) .

Haplodina alutacea Zahlbr. 淡棕单碗衣

in Handel-Mazzetti, *Symb. Sin.* 3: 66(1930) & *Cat. Lich. Univ.* 8: 228(1932) .

Type: Yunnan, Handel-Mazzetti no. 6595.

On rocks.

Haplodina corticola Zahlbr. 树皮单碗衣

in Handel-Mazzetti, *Symb. Sin.* 3: 65(1930) & *Cat. Lich. Univ.* 8: 228(1932) .

Type: Sichuan, Handel-Mazzetti no. 1380.

On bark of Alnus ferdinandi-coburgii.

Haplodina microcarpa Zahlbr. 微果单碗衣

in Handel-Mazzetti, *Symb. Sin.* 3: 65(1930) & *Cat. Lich. Univ.* 8: 228(1932).

Type: Yunnan, Handel-Mazzetti no. 5722.

On twigs of *Ternstroemis japonica*

Lecanactis Körb. **碗衣属**

Syst. lich. germ. (Breslau): 275(1855).

Lecanactis limosescens Zahlbr. 泥碗衣

in Fedde, *Repertorium*, 31: 224(1933).

Type: Taiwan, Asahina no. 174.

On bark of trees.

Taiwan(Wang & Lai, 1973, p. 91).

Lecanactis macrocarpoides Zahlbr. 类大果碗衣

in Feddes, Lichenes in Feddes, *Repertorium sp. nov.* 31: 224(1933).

Type locality: Taiwan, Asahina no. 169.

On bark of trees.

Taiwan(Lamb, 1963, p. 286; Wang & Lai, 1973, p. 91).

Lecanactis quassiae(Fée) Zahlbr. 碗衣

Cat. Lich. Univ. 2: 544(1923) [1924]; Zahlbr. in Handel-Mazzetti, *Symb. Sin.* 3: 66(1930) as *L. auassiae*(Fée) Zahlbr.

≡*Lecidea quassiae* Fée, *Suppl. Écorc. Officin.*, 1837, p. 104.

On bark of tree.

Fujian(Zahlbr., 1930 b, p. 66).

Lecanactis submorosa Zahlbr. 台湾碗衣

in Feddes, Lichenes in Feddes, *Repertorium sp. nov.* 33: 22(1933). var. submorosa 原变种

Type: Taiwan, Asahian nos. 161, 187(syntypes).

On rocks.

Taiwan(Wang & Lai, 1973, p. 91).

var. laetior Zahlbr. 光亮变种

in Feddes, Lichenes in Feddes, *Repertorium sp. nov.* 33: 23(1933).

Type: Taiwan, Asahina no. 188.

Taiwan(Wang & Lai, 1973, p. 91).

Mazosia A. Massal. **叶睛衣属**

Neagenea Lich: 9(1854).

Mazosia dispersa(J. Hedrick) R. Sant. 散叶睛衣

Symb. bot. upsal. 12(no. 1): 120(1952).

≡*Calenia dispersa* J. Hedrick, *Publ. Carnegie Inst. Wash.* 461: 111(1935).

On living leaves of *Pandanus* in forests. Pantropical.

Hongkong(Aptroot & Seaward, 1999, p. 82).

Mazosia melanophthalma(Müll. Arg.) R. Sant. 小疣叶睛衣

Symb. bot. upsal. 12(no. 1): 117(1952)

≡*Opegrapha melanophthalma* Müll. Arg., *Flora*, Regensburg 66: 34(no. 684)(1883).

On living leaves of *Pandanus* sp. in forests.

Foliicolous. Pantropical.

Yunnan(J. C. Wei & Y. M. Jiang, 1991, p. 208, on leaves of *Knema furfuracea* and of *Cetis giganticarpa*; Aptroot et al., 2003, pp. 44), Guangdong(J. C. Wei &Y. M. Jiang, 1991, p. 208,

On leaves of *Tectaria* sp.), Hongkong(Thrower, 1988, pp. 16, 119; Aptroot & Seaward, 1999, p. 82; Aptroot &

Sipman 2001, p. 331).

On leaves of *Diospyros maritime* and *Asplenium nidus*.

Taiwan(Aptroot & Sparrius, 2003, p. 28; Aptroot et al., 2003, pp. 42 – 43).

Mazosia ocellata(Nyl.) R. C. Harris 单眼叶睛衣

Some Florida Lichens (New York): 43(1990).

≡*Platygrapha ocellata* Nyl., *Acta Soc. Sci. fenn.* 7(2): 478(1863).

On *Erythrina variegata*.

Taiwan(Aptroot & Sparrius, 2003, p. 28), Hongkong(Aptroot & Sipman 2001, p. 331).

Mazosia paupercula(Müll. Arg.) R. Sant. 光滑叶睛衣

Symb. bot. upsal. 12(no. 1): 128(1952).

≡*Opegrapha paupercula* Müll. Arg., *Hedwigia* 30: 241(1891).

On leaves of *Knema furfuracea*

Yunnan(J. C. Wei & Y. M. Jiang, 1991, p. 210).

Mazosia phyllosema(Nyl.) Zahlbr. 无疣叶睛衣

Cat. Lich. Univ. 2: 503(1923)

≡*Platygrapha phyllosema* Nyl., *Bull. Soc. linn. Normandie*, sér. 2 7: 171(1873).

On leaves of *Knema furfuracea*, and on living leaves of Pandanus in forest. Pantropical.

Yunnan(J. C. Wei & Y. M. Jiang, 1991, p. 211; Aptroot et al., 2003, pp. 44), Hongkong(Aptroot & Seaward, 1999, p. 82; Aptroot & Sipman 2001, p. 331).

Mazosia rotula(Mont.) A. Massal. 叶睛衣

Neagenea Lich: 9(1854).

≡*Strigula rotula* Mont., in Sagra, *Historia física, polirica y nayturál de la islea de Cuba* 9: 2(1845).

Foliicolous.

Sichuan(Zahlbr., 1930 b, p. 64; R. Sant. 1952, p. 115), Yunnan (Zahlbr., 1930 b, p. 64; R. Sant. 1952, p. 115; J. C. Wei & Y. M. Jiang, 1991, p. 211, on leaves of *Knema furfuracea*).

Note: This species is placed under "Unverified records" in Santensson's book, and is probably *Mazosia phyllosema* rather than *M. rotula* according to Santesson(1952, p. 115). It has been already verified by Wei & Jiang(1991, p. 211).

Opegrapha Ach. **孔文衣属**

K. Vetensk-Acad. Nya Handl. 30: 97(1809).

Opegrapha calcarea Turner ex Sm. 钙孔文衣

in Smith & Sowerby, *Engl. Bot.* 25: tab. 1790(1807)

On raised coral reef.

Taiwan(Aptroot & Sparrius, 2003, p. 31).

Opegrapha ectolechiacearum Matzer & R. Sant. 外孔文衣

in *Matzer, Mycol. Pap.* 171: 66(1996).

Yunnan(Aptroot et al., 2003, p. 44).

Opegrapha filicina Mont. 线孔文衣

in Sagra, *Historia física, polirica y nayturál de la islea de Cuba* 2: 184(1842) [1838 – 1842]

On *Knema furfuracea*

Yunnan(J. C. Wei & Y. M. Jiang, 1991, p. 207).

Opegrapha herbarum Mont. 草孔文衣

in Guillemin, *Arch. Bot.* (Forlì) 2: 302(1833).

On *Pemphis maritima* branches.

Taiwan(Aptroot & Sparrius, 2003, p. 31).

Opegrapha melanospila Müll. Arg. 黑孔文衣

Flora, Regensburg 60(30):474(1877).

On conglomeratic rock.

Taiwan(Aptroot & Sparrius, 2003, p. 31).

Opegrapha rufescens Pers. 微红孔文衣

Ann. Bot. (Usteri) 1:29(1794).

On bark of trees.

Jiangsu(Wu J. L. 1987, p. 39; Xu B. S., 1989, p. 169).

Opegrapha sp. 孔文衣未定种

Misappliname(revised by Aptroo& Seaward in *Tropical Bryology* 17:82(1999)):

Melaspilea sp. in Thrower, *Hong Kong Lichens*, 1988:16, 121.

Hongkong(Thrower, 1988, pp. 16 &121; Aptroo& Seaward, 1999, p. 82).

Opegrapha subclausa Zahlbr. 亚裂孔文衣

in Handel-Mazzetti, *Symb. Sin.* 3:38(1930).

Type: Hunan, Handel-Mazzetti no. 11023.

On bark of *Pistacia* sp.

Opegrapha varia Pers. 多变孔文衣

Ann. Bot. (Usteri) 1:30(1794).

On shaded trees. Cosmopolitan.

Hongkong(Aptroot & Seaward, 1999, p. 83; Aptroot et al., 2001, p. 335).

Opegrapha vulgata(Ach.) Ach. 孔文衣

Methodus, Sectio prior: 20(1803).

≡*Lichen vulgatus* Ach., *Lich. suec. prodr.* (Linköping):21(1799)[1798].

var. subsiderella Nyl. 亚星变种

Lich. Scand. (Helsinki):255(1861).

On bark of *Salix* sp. and of *Liquidambar* sp.

Hunan(Zahlbr., 1930 b, p. 38), Shanghai(Nyl. & Cromb. 1883, p. 65; Hue, 1891, p. 172; Zahlbr., 1930 b, p. 38).

Roccella DC. **染料衣属**

in Lamarck & de Candolle, *Fl. franç.*, Edn 3(Paris) 2:334(1805).

Roccella montagnei Bél. 莽氏染料衣

Voy. Indes Or., *Bot.* 2:117(1834).

On trees, rarely on rocks.

Zhejiang(North latitude 30 X 120 east longitude: Darbishire, 1898, p. 945: < i-iv >, < Taf. 2 – 7, fig. 13 – 26 >, & 1928, map 5), Hainan(JCWei et al., 2013, p. 238).

Roccella sinensis Nyl. 中华染料衣

Synops. Lich. 1:261(1860).

Type: Hongkong , Macao.

Misapplied name:

Roccella tinctoria f. *hypomecha* sensu Mey. et Flot. Verh. der Kaiserl. Leopold-Carol Academie der Naturforschen, vol. 19, Suppl. 1. Breslau und Bonn 1843, p. 214, non Ach.,

On marine rocks.

Hongkong(including Macao: Mey. & Flot. 1843, p. 214; Darbishire, 1898, p. 39 & 1928, map. 4; Thrower, 1988, p. 165), China(prov. not indicated: Nyl. 1860, p. 261; Hue, 1890, p. 269; Zahlbr., 1924, p. 522 & 1930 b, p. 64).

On supra-littoral granitic rock. Endemic to China.

Darbishire, 1898, p. 39 & 1928, map. 4- probably erroneously based on Meyen & Flotow, 1843, p. 214, which is actually a record of *R. tinctoria* f. *hypomecha* Ach. From Macao(J. C. Wei, 1991, p. 228).

Hongkong(Thrower, 1988, pp. 17, 165; Chu, 1997, p. 48; Aptroot & Seaward, 1999, p. 93 – 94), Hainan(J. C. Wei et al., 2013, p. 238).

Roccellina Darb. **小染衣属**

Ber. dt. bot. Ges. 16: 11, 14(1898).

Roccellina nipponica(Nyl.) Tehler 日本小染衣

[as ' niponica'], *Op. bot.* 70: 70(1983).

≡*Dirina nipponica* Nyl., *Lich. Japon.* : 50(1890).

On overhanging supra-littoral granitic rock. Endemic to China and Japna.

Taiwan(Aptroot and Sparrius, 2003, p. 45), Hongkong(Aptroot & Seaward, 1999, p. 94; Aptroot & Sipman, 2001, p. 340),

Ropalospora A. Massal. **锥形孢属**

Atti Inst. Veneto Sci. lett., ed Arti, Sér. 3 5: 263(1860) [1859 – 1860].

Ropalospora chlorantha(Tuck.) S. Ekman 绿锥形孢

Bryologist 96(4): 586(1993).

≡*Biatora chlorantha* Tuck., *Proc. Amer. Acad. Arts & Sci.* 1: 252(1848).

On tree bark.

Yunnan(Hu L. et al., 2013, pp. 440 – 442).

Ropalospora phaeoplaca(Zahlbr.) S. Ekman 黑锥形孢

Op. bot. 127: 128(1996).

≡*Bacidia phaeoplaca* Zahlbr., *Bot. Mag.*, Tokyo 41: 333(1927).

On tree branches, *Cryptomeria japonica*, *Shiia carlesii*, *Pseudotsuga* sp. and *Pisana* sp.

Yunnan, Guizhou, Fujian(Hu L. et al., 2013, pp. 442 – 444), TAIWAN(Aptroot and Sparrius, 2003, p. 45).

Schismatomma Flot. & Körb. ex A. Massal. **裂盘衣属**

Ric. auton. lich. crost. (Verona): 55(1852).

Schismatomma margaritaceum Zahlbr. 珍珠裂盘衣

Flechten der Insel Formosa in Feddes, *Repertorium sp. nov.* 33: 23(1933).

Syntype localities: Taiwan, Asahina nos. 323, 325(US, W).

On bark of *Cinnamon* sp.

Taiwan(Lamb, 1963, p. 659; Wang & Lai, 1973, p. 96; Tehler, 1993, p. 30).

Sclerophyton Eschw. **硬衣属**

Syst. Lich. : 14, 25(1824).

Sclerophyton actinoboloides Zahlbr. 辐硬衣

in Feddes, *Repertorium sp. nov.* 31: 223(1933).

Type: Taiwan, Asahina no. 320.

On bark of trees.

Taiwan(Zahlbr., 1940, p. 198; Lamb, 1963, p. 660; Wang & Lai, 1973, p. 96).

Zwackhia Körb. **拟孔衣属**

Syst. lich. germ. (Breslau): 285(1855).

Zwackhia prosodea(Afzel.) Ertz 拟孔衣

in Diederich, Ertz, Eichler, Cezanne, Boom, Fischer, Killmann, Van den Broeck & Sérusiaux, *Bull. Soc. Nat. luxemb.* 113: 106(2012).

≡*Opegrapha prosodea* Ach., *Methodus, Sectio prior*: 22(1803).

var. *quassiicola*(Fée) 苦木变种

≡*Opegrapha bonplandii* var. *quassiicola* Fée [as ' *quassiaecola*'], Essai Crypt. Exot. (Paris): 26, tab. V, fig. 5

(1824).

≡*Opegrapha prosodea* var. *quassiicola*(Fée) Vain. [as '*quassiaecola*'], *Ann. Acad. Sci. fenn.*, Ser. A 15(no. 6): 268(1921), J. C. Wei, *Enum. Lich. China*, 1991, p. 161.

On bark of trees.

Fujian(Zahlbr. 1934 b, p. 196).

var. *sclerocarpa*(Mey. et Flot.) 硬果变种

≡*Graphis sclerocarpa* Mey. et Flot. in *Nova Actaq Acaqd. Leopold. -Carolin.*, vol. XIX, Suppl. 229(1843).

≡*Opegrapha prosodea* var. *sclerocarpa*(Mey. et Flot.) Vain., *Ann. Acad. Sci. fenn.*, Ser. A15(no. 6): 269 (1921), J. C. Wei, *Enum. Lich. China*, 1991, p. 161.

On bark of trees.

Taiwan(Zahlbr., 1933 b, p. 207; Wang & Lai, 1973, p. 92).

Zwackhia viridis(Ach.) Poetsch & Schied. 绿色拟孔衣

Syst. Aufz. Krypt. Pfl.: 186(1872).

≡*Opegrapha rubella* var. *viridis* Ach., Methodus, Sectio prior: 22(1803).

≡*Opegrapha viridis*(Ach.) Ach., Syn. meth. lich. (Lund): 73(1814).

= *Opegrapha multiseptata* var. *plicatula* Redinger in *Ark. f. Bot.* XXIXA, 19: 36(1940); Aptroot & Seaward, 83 (1999).

On trees, e. g. *Kandelia*. Cosmopolitan.

Hongkong(Thrower, 1988, pp. 16, 173; Aptroot & Seaward, 1999, p. 83).

On *Mangifera indica*.

Taiwan(Aptroot & Sparrius, 2003, p. 31, as *Opegrapha viridis*).

2. Dothideomycetes O. E. Erikss. & Winka **座囊菌纲**

Myconet 1(1): 5(1997).

1) Dothideomycetidae P. M. Kirk, P. F. Cannon, J. C. David & Stalpers **座囊菌亚纲**

ex C. L. Schoch Spatafora, Crous & Shoemaker, in Schoch, Shoemaker, Seifert, Hambleton, Spatafora & Crous, *Mycologia* 98(6): 1045(2007) [2006].

≡*Dothideomycetidae* P. M. Kirk, P. F. Cannon, J. C. David & Stalpers, Ainsworth & Bisby's *Dictionary of the Fungi*, Edn 9(Wallingford): 165(2001), Nom. inval., Art. 39. 1(Melbourne).

[2] Capnodiales Woron. **煤炱目**

Annls mycol. 23(1/2): 177(1925).

* **Endococcus** Nyl. **内球菌属**

Mém. Soc. Sci. nat. Cherbourg 3: 193(1855).

* Endococcus propinquus(Körb.) D. Hawksw. 邻内球菌

Bot. Notiser 132(3): 287(1979).

≡*Microthelia propinqua* Körb., *Syst. lich. germ.* (Breslau): 374(1855).

Syn. *Discothecium gemmiferum* auct., non(Tayl.) Vouaux 1913).

On *Aspicilia asiatica*(as *Lecanora asiatica*, H. Magn. 1940, p. 90).

Gansu(D. Hawksw. & M. Cole, 2003, p. 360).

(5) Mycosphaerellaceae Lindau **球囊菌科**

in Engler & Prantl *Nat. Pflanzenfam.*, Teil. I(Leipzig) 1(1): 421(1897).

* **Stigmidium** Trevis. **斑点菌属**

Conspect. Verruc.: 17(1860).

* Stigmidium arthrorhaphidis Hafellner & Obermayer 节瘤斑点菌

Cryptogamie, Bryol. Lichénol. 16(3): 186(1995).

Typus: China, Tibet(= prov. Xizang), Himalaya Range, 280 km SSE of Lhasa, 40 km SW of Mainling, 29°

03'N/93°58'E, 4400 - 4500 m alt., *Rhododendron* shrubs, 12. VIII. 1994, W. Obermayer(3014) (GZU); auf *Arthrorhaphis citrinella* (GZUHolotypus; UPS-Isotypus).

Xizang, sichuan(J. Hafellner & W. Obermayer, 1995, pp. 186 – 187; W. Obermayer, 2004, pp. 510 – 511).

* Stigmidium calopadiae Matzer 丽斑点菌

Mycol. Pap. 171: 157(1996).

Yunnan(Aptroot et al., 2003, p. 45).

* Stigmidium cupulare(Pat.) D. Hawksw. 勺斑点菌

in D. Hawksw. & Cole, *Mycosystema* 22(3): 362(2003).

≡*Pharcidia cupularis* Pat., *Bull. Soc. mycol. Fr.* 18: 303(1902).

On *Sticta nylanderiana*(as *S. platyphylla*)

Yunnan (D. Hawksw. & M. Cole, 2003, p. 362).

* Stigmidium epiphyllum Matzer 叶面斑点菌

Mycol. Pap. 171: 159(1996).

Yunnan(Aptroot et al., 2003, p. 45).

* Stigmidium trichotheliorum Matzer 毛斑点菌

Mycol. Pap. 171: 164(1996).

Yunnan(Aptroot et al., 2003, p. 45).

2) **Pleosporomycetidae** C. L. Schoch, Spatafora, Crous & Shoemaker **格孢腔菌亚纲**

in Schoch, Shoemaker, Seifert, Hambleton, Spatafora & Crous *Mycologia* 98(6): 1048(2007)[2006].

[3] **Pleosporales** Luttr. ex M. E. Barr **格孢腔菌目**

Prodr. Cl. Loculoasc. (Amherst): 67(1987)

(6) **Didymosphaeriaceae** Munk **双球菌科**

Dansk bot. Ark. 15(no. 2): 128(1953).

* **Apiosporella** Höhn. **蜜孢菌属**

Sber. Akad. Wiss. Wien, Math. -naturw. Kl., Abt. 1 118: 1215 [59 repr.] (1909).

* Apiosporella caudata(Kernst.) Keissl. 尾蜜孢菌

Beih. bot. Zbl., Abt. 2 37: 266(1920).

≡*Cercidospora caudata* Kernst., *Verh. zool. -bot. Ges.* Wien 44: 212(1894).

≡*Didymosphaeria caudata*(Kernst.) Magnus in Dalla Torre & Sarnth(anamorphic).

On *Collema* sp., *Lecanora* sp., and *Lecidea tessellata*(as *L. pavimentans*, 1940, pp. 50 – 51.).

(Magn. 1940, pp. 50 – 51).

Gansu(Hawksw. & M. Cole, 2003, p. 360).

(7) **Polycoccaceae** Ertz, Hafellner & Diederich **弯衣菌科**

in Ertz, Diederich, Lawrey, Franz Berger, Freebury, Coppins, Gardiennet & Hafellner

Fungal Diversity 74: 82(2015).

* **Polycoccum** Saut. ex Körb. **多果菌属**

Parerga lichenol. (Breslau) 5: 470(1865).

* Polycoccum bryonthae(Arnold) Vězda 苔多果菌

Česká Mykol. 23(2): 109(1969).

≡*Endococcus* bryonthae *Arnold in* Lichenol. Fragment. *XVI, p.* 26, *tab. II, fig.* 15, *in Flora,* 1874.

Syn. Didymosphaeria bryonthae*(Arnold) G. Winter* 1885

On Acarospora cervina(= A. cinereoalbida)*(Magn.* 1944*)*.

Neimenggu(Hawksw. & M. Cole, 2003, *p.* 362*)*.

[4] **Strigulales** Lücking, M. P. Nelsen & K. D. Hyde **叶上衣目**

in Hyde et al., *Fungal Diversity* 63: 10(2013).

(8) Strigulaceae Zahlbr. 叶上衣科

in Engler, *Syllabus*, Edn 2(Berlin) : 46(1898) .

See Fries, Syst. orb. veg. 1: 110(1825) , not published at family rank, (as ' subordo') .

Strigula Fr. 叶上衣属

Syst. mycol. (Lundae) 2(2) : 535(1823) .

Strigula acuticonidiarumm S. H. Jiang, X. L. Wei & J. C. Wei 尖分孢叶上衣

MycoKeys 19: 34(2017) .

Type. China, Guangxi: Nanning City, Long' an County, Longhu mountain natural reserve. 22°57′42″N, 107°37′40″E, 150 m alt. , on living leaves, 1 Dec 2015, S. H. Jiang GX201511085 (HMAS-L 0138045-holotype) .

Strigula africana Vain. 非洲叶上衣

in Hiern, *Cat. Welwitsch Afric*. Pl. 2(2) : 461(1901) ; Zahlbr. , 1930 b, p. 31.

On leaves of evergreen trees.

Yunnan, Fujian(Zahlbr. , 1930b, p. 31) .

Strigula antillarum(Fée) Müll. Arg. 安地叶上衣

Bot. Jahrb. Syst. 6: 379(1885) .

Yunnan, Guangxi, Hainan(Jiang et al. , 2016, p. 800) .

Strigula concreta(Fée) R. Sant. 硬叶上衣

Symb. bot. upsal. 12(no. 1) : 177(1952) .

≡ *Craspedon concretum* Fée, *Essai Crypt. Exot*. (Paris) : xciv, c(1825) [1824] .

On leaves of *Machilus thunbergii*.

Yunnan(Aptroot et al. , 2003, p. 45) , Taiwan(Aptroot 2003, p. 168; Aptroot et al. , 2003, pp. 42 – 43) .

Strigula delicata Serus. 花朵叶上衣

Bryologist 101(1) : 147(1998) .

On living leaves.

Xizang(Jiang S. H. et al. , 2017, p. 395) .

Strigula guangxiensis S. H. Jiang, X. L. Wei & J. C. Wei 广西叶上衣

MycoKeys 19: 37(2017) .

Type. China, Guangxi: Nanning, Long' an County, Longhu mountain natural reserve. 22°57′42″N, 107°37′40″E, 150 m alt. , on living leaves, 1 Dec 2015, S. H. Jiang GX201511127 (HMAS-L 0138040-holotype) .

Strigula jamesii(Swinscow) R. C. Harris 詹氏叶上衣

in D. Hawksw. , James & Coppins, *Lichenologist* 12(1) : 107(1980) .

≡ *Geisleria jamesii* Swinscow, *Lichenologist* 3: 420(1967) .

On *Podocarpus* sp. New to Asia.

Taiwan(Aptroot 2003, p. 168) .

Strigula janeirensis(Müll. Arg.) Lücking 尾孢叶上衣

Trop. Bryol. 15: 65(1998) .

≡ *Phylloporina janeirensis* Müll. *Arg.* , *Flora*, Rengensburg 73: 198(no. 1563) (1890) .

on living leaves.

Hainan(Jiang S. H. et al. , 2017, p. 396) .

Strigula laureriformis Aptroot & Lücking 月桂叶上衣

Biblioth. Lichenol. 97: 131(2008) .

Xizang(Jiang et al. , 2016, p. 800) .

Strigula macrocarpa Vain. 大果叶上衣

Ann. Acad. Sci. fenn. , Ser. A 19(no. 15) : 20(1923) .

On leaves of *Amaplelopris* sp. and of *Prosartema stellaria*, etc.

Yunnan(Wei & Jiang, 1991, p. 213; Aptroot et al. , 2003, p. 45) .

Strigula maculata(Cooke & Massee) R. Sant. 斑点叶上衣

Symb. bot. upsal. 12(no. 1) : 186(1952) .

≡*Micropeltis maculata* Cooke & Massee, in Cooke, *Grevillea* 18(no. 86) : 35(1889) .

Yunnan(Aptroot et al. , 2003, p. 45) .

Strigula melanobapha(Kremp.) R. Sant. 绵纹叶上衣

Symb. bot. upsal. 12(no. 1) : 188(1952) .

≡*Verrucaria melanobapha* Kremp. , *Lich. Foliic. Leg. Beccari*: 18(1874) .

= *Strigula fibrillosa* Zahlbr. in Handel-Mazzetti, *Symb. Sin.* 3: 31(1930) , revised by R. Sant. 1952, p. 188.

Type locality: Fujian, Chung nos. 319, 487b, 488.

On living leaves.

Yunnan(J. C. Wei & Y. M. Jiang, 1991, p. 213, on leaves of *Phoebe lanceolata* and of *Drypetes indica*, etc.), Fujian(R. Sant. 1952, p. 188) .

Strigula minor(Vězda) Cl. Roux & Sérus. 小叶上衣

in Puntillo, Bricaud & Sérusiaux, *Cryptog. Mycol.* 21(3) : 175(2000) .

≡*Raciborskiella minor* Vězda, *Folia geobot. phytotax.* 18(1) : 49(1983) .

Yunnan(Aptroot et al. , 2003, p. 45) .

Strigula muriformis Aptroot & Diederich 砖壁叶上衣

In Aptroot, Diederich, Sérusiaux & Sipman, *Biblthca Lichenol.* 64: 188(1997) .

On *Castonopsis* sp. New to northern hemisphere.

Taiwan(Aptroot 2003, pp. 168 – 169) .

Strigula nemathora Mont. , in Sagra 丝状叶上衣

Historia física, polirica y nayturál de la islea de Cuba 2: 139(1842) [1838 – 1842]

On leaves of *Camellia* sp. and *Turpina* sp.

Yunnan(Aptroot et al. , 2003, p. 45) , Taiwan(Aptroot 2003, p. 169; Aptroot et al. , 2003, pp. 42 – 43) .

Strigula nemathora f. hypothelia(Nyl.) Lücking 腹生乳头变型

Fl. Neotrop. , Monogr. 103: 248(2008) .

≡*Strigula hypothelia* Nyl. , *Bull. Soc. linn. Normandie,* sér. 2 2: 520(1868) .

Yunnan(Aptroot et al. , 2003, p. 45) .

Strigula nitidula Mont. 健美叶上衣

In Sagra, *Historia física, polirica y nayturál de la islea de Cuba* 2: 139(1842) [1838 – 1842] .

Yunnan(Aptroot et al. , 2003, p. 45) .

Strigula phaea(Ach.) R. C. Harris 黑色叶上衣

in Tucker & Harris, *Bryologist* 83(1) : 18(1980) .

≡*Verrucaria phaea* Ach. , *Syn. meth. lich.* (Lund) : 88(1814) .

On granite boulder.

Hongkong(Aptroot & Sipman, 2001, p. 340) .

Strigula phyllogena(Müll. Arg.) R. C. Harris 原生叶上衣

More Florida Lichens, Incl. 10 Cent Tour Pyrenol. (New York) : 159(1995) .

≡*Porina phyllogena* Müll. Arg. , *Flora*, Regensburg 66: 27(no. 663) (1883) .

Yunnan(Aptroot et al. , 2003, p. 45) .

Strigula prasina Müll. Arg. 亮叶上衣

Flora 68: 343(1885) .

Hainan(Jiang et al. , 2016, p. 802) .

Strigula schizospora R. Sant. 裂孢叶上衣

Symb. bot. upsal. 12(no. 1) : 175(1952) .

Misapplied name:

Porina phyllogena auct. non Müll. Arg. in Thrower, 1988, pp. 17, 150, revised by Aptroot & Seaward, 1999, p. 95.

On living leaves of trees. Pantropical.

Hongkong(Thrower, 1988, pp. 17, 150; Aptroot & Seaward, 1999, p. 95).

Strigula sinoaustralis S. H. Jiang, X. L. Wei & J. C. Wei 华南叶上衣

Mycotaxon 131: 797(2016).

Type: China. Guangxi Province, Shangsi County, Shiwan mountain forest park, along the way to Yingkesong and Jiulongsong, 21°54′16″N 107°54′09″E, alt. 271 m, on living leaves, 26/V/2015, Xin-Li Wei & Jiao-Hong Wang GX20150342(Holotype, HMAS-L 0137203; GenBank KX216699).

Strigula smaragdula Fr. 叶上衣

Linnaea 5: 550(1830); Aptroot 2003: 169.

= *Strigula elegans*(Fée) Müll. Arg., *Flora, Regensburg* 63(3): 41(1880); R. Sant. 1952, p. 160; Lai M. J. 2000: 85. Revised by Aptroot 2003: 169.

= *Strigula elegans* var. *nematora* Müll. Arg. in Engler, *Botan. Jahrbuch* 6: 380(1885); Zahlbr., *Flechten der Insel Formosa*, 31: 205(1933), revised by Aptroot 2003: 169.

On leaves of *Machilus thunbergii* and other leaves.

Guizhou(R. Sant. 1952, p. 169 ; Zhang T & J. C. Wei, 2006, p. 10), Yunnan(R. Sant. 1952, p. 169; J. C. Wei & Y. M. Jiang, 1991, p. 213; Aptroot et al., 2003, p. 45), Hubei, Hunan(R. Sant. 1952, p. 169), Fujian(R. Sant. 1952, p. 169), Taiwan(Zahlbr., 1933 b, p. 205; Wang & Lai, 1973, p. 97; R. Sant. 1952, p. 169; Aptroot 2003, p. 169; Aptroot et al., 2003, pp. 42 – 43, On *Trpinia* sp.).

Strigula subelegans Vain. 亚丽叶上衣

Ann. Acad. Sci. fenn., Ser. A 19(no. 15): 23(1923).

On leaves of broad-leaved trees

Yunnan(J. C. Wei & Y. M. Jiang, 1991, p. 214).

Strigula submuriformis(R. C. Harris) R. C. Harris 亚砖壁叶上衣

In Egan, *Bryologist* 90(2): 164(1987).

≡ *Arthopyrenia submuriformis* R. C. Harris, Michigan Bot. 12(1): 15(1973)

On bark of trees. Known from neotropics and Asia.

Hongkong(Aptroot & Seaward, 1999, p. 95).

Strigula subtilissima(Fée) Müll. Arg. 精细叶上衣

Flora, Regensburg 66: 32(no. 678) (1883).

≡ *Racoplaca subtilissima* Fée, *Essai Crypt. Exot.* (Paris): xciv, xcix(1825) [1824].

On living leaves of trees. Pantropical.

Yunnan(Aptroot et al., 2003, p. 45), Hainan(Aptroot et al., 2003, p. 45), Hongkong (Aptroot & Seaward, 1999, p. 95).

Strigula univelbiserialis S. H. Jiang, X. L. Wei & J. C. Wei 单双列孢叶上衣

Mycoscience 58: 393(2017).

Type: China, Guangxi Province, Yulin City, Beiliu County, Darongshan nature reserve. 225104600 N, 1101603100 E, alt. 495 m, on living leaves of shrubs, 7 Dec 2015, S. H. Jiang GX201511004(holotype HMAS-L 0137656).

Strigula viridiseda(Nyl.) R. C. Harris 绿色叶上衣

In Tucker & Harris, *Bryologist* 83(1): 18(1980).

≡ *Verrucaria viridiseda* Nyl., *Expos. Synopt. Pyrenocarp.*: 55(1858).

On *Debeyraecea edulis*, *Erythrina variegate*, *Ficus nervosa* and *Liquidamber* sp.

Taiwan(Aptroot 2003, p. 169).

Dothideomycetes, families incertae sedis 座囊菌纲未定位科

(9) Arthopyreniaceae Walt. Watson **星核衣科**

New Phytol. 28: 107(1929).

Arthopyrenia A. Massal. **星核衣属**

Ric. auton. lich. crost. (Verona): 165(1852).

Arthopyrenia amaura Zahlbr. 暗星核衣

in Handel-Mazzetti, *Symb. Sin.* 3: 19(1930) & *Cat. Lich. Univ.* 8: 81(1932).

Type locality: Hunan, Mt. Yuelu shan near Changsha city, at about 200 m, 10/XII/1917, Handel-Mazzetti no. 11426.

On trunk of *Randia henryi.*

Arthopyrenia analepta(Ach.) A. Massal. 假星核衣

Ric. auton. lich. crost. (Verona): 165(1852).

≡*Lichen analeptus* Ach., *Lich. suec. prodr.* (Linköping): 15(1799) [1798].

=*Arthopyrenia lapponina* Anzi, *Comm. Soc. crittog. Ital.* 2(1): 25(1864), J. C. Wei, *Enum. Lich. China*, 1991, p. 29.

=*Arthopyrenia fallax*(Nyl.) Arnold, *Verh. zool. -bot. Ges. Wien* 23: 505(1873), J. C. Wei, *Enum. Lich. China*, 1991, p. 29.

≡*Verrucaria epidermidis* var. *fallax* Nyl., *Bot. Notiser*: 178(1852).

On trees.

Jiangsu(Xu B. S., 1989, p. 170).

Arthopyrenia antecellens(Nyl.) Arnold [as 'anticellens'] 异星核衣

Flora, Regensburg 53: 485(1870).

≡*Verrucaria antecellens* Nyl., *Flora*, Regensburg 49: 86(1866).

On *Rhododendron* wood.

Taiwan(Aptroot 2003, p. 158).

Arthopyrenia cinchonae(Ach.) Müll. Arg. 疑星核衣

Flora, Regensburg 66: 289(1883).

≡*Verrucaria cinchonae* Ach., Syn. meth. lich. (Lund): 90(1814).

On *Tsuga formosana.*

Taiwan(Aptroot 2003, p. 158, & 2004, p. 34).

=*Microthelia ambigua* Zahlbr., in Fedde, *Repertorium* 31: 199(1933), synonymized by Aptroot, 2004, p. 34.

Type: Taiwan, Asahina F 314(W holotype n. v.).

≡*Arthopyrenia ambigua*(Zahlbr.) D. Hawksw., *Bull. Brit. Mus. (Nat. Hist.) Bot.* 14(2): 133(1985), J. C. Enum. Lich. China, 1991, p. 25.

Taiwan(Zahlbr., 1940, p. 70; Lamb, 1963, p. 417; Wang & Lai, 1973, p. 92).

Arthopyrenia cinereopruinosa(Schaer.) A. Massal. 灰霜星核衣

Symmict. Lich: 117(1855).

≡*Verrucaria cinereopruinosa* Schaer., *Lich. helv. spicil.* 6: 343(1833).

On *Populus* sp.

Taiwan(Aptroot 2003, p. 158).

Arthopyrenia extensa Zahlbr. 展星核衣

in Handel-Mazzetti, *Symb. Sin.* 3: 19(1930) &*Cat. Lich. Univ.* 8: 85(1932).

Syntype localities: Yunnan, Handel-Mazzetti nos. 8331, 8455, 9260.

On bark of *Sorbus* sp. and of *Betula* sp.

Arthopyrenia punctiformis A. Massal. 斑点星核衣

Ric. auton. lich. crost. (Verona): 168, fig. 335(1852).

On *Pemphis maritima*.

Taiwan(Aptroot 2003, p. 158).

Arthopyrenia saxicola A. Massal. 岩生星核衣

Symmict. Lich: 107(1855).

On rocks.

Notes. This is an inconspicuous, possibly non-lichenized, saxicolous species(Aptroot et al., 2001, p. 121).

Hongkong(Aptroot et al., 2001, p. 321).

Arthopyrenia subantecellens Müll. Arg. 亚突星核衣

Bot. Jb. 6: 403(1885).

On bark of tree.

Shaanxi(Jatta, 1902, p. 480; Zahlbr., 1930 b, p. 19).

(10) Monoblastiaceae Walt. Watson **单芽菌科**

New Phytol. 28: 106(1929).

Anisomeridium(Müll. Arg.) M. Choisy **异形菌属**

Icon. Lich. Univ. 1: 24(1928).

≡*Arthopyrenia* sect. *Anisomeridium* Müll. Arg., *Flora*, Regensburg 66: 290(1883).

Anisomeridium albisedum(Nyl.) R. C. Harris 白景天异形菌

in Egan, *Bryologist* 90(2): 163(1987).

≡*Verrucaria viridiseda* f. *albiseda* Nyl., *Lich. Japon*.: 108(1890).

On a tree.

Hongkong(Aptroot et al., 2001, p. 319).

Anisomeridium americanum(A. Massal.) R. C. Harris 美洲异形菌

More Florida Lichens, Incl. 10 Cent Tour Pyrenol. (New York): 144(1995).

≡*Arthopyrenia americana* A. Massal., *Ric. auton. lich. crost.* (Verona) (1852).

On *Ehretia* sp. and *Ficus microcarpa*.

Taiwan(Aptroot, 2003, p. 157).

Anisomeridium anisolobum(Müll. Arg.) Aptroot 不等芽异形菌

in Aptroot, Diederich, Sérusiaux & Sipman, *Biblthca Lichenol*. 57: 21(1995).

≡*Arthopyrenia anisoloba* Müll. Arg., *Flora*, Regensburg 66(19): 305(1883)

= *Anisomeridium feeanum*(Müll. Arg.) R. C. Harris, in Thrower, 1988, p. 15, revised by Aptroot & Seaward, 1999, p. 62).

On shaded trees, but also on shaded rock, especially in forestgs. Pantropical.

Taiwan(Aptroot, 2003, p. 157), Hongkong(Thrower, 1988, p. 15; Aptroot & Seaward, 1999, p. 62; Aptroot et al., 2001, p. 319; Seaward and Aptroot 2005, p. 283, Pfister 1978: 126, as *Verrucaria biformis*).

Anisomeridium biforme(Borrer) R. C. Harris 双异形菌

in Vězda, *Folia geobot. phytotax*. 20: 207(1978).

≡*Verrucaria biformis* Borrer, *Suppl. Engl. Bot*. 1: tab. 2617, fig. 1(1831).

On a tree.

Hongkong(Aptroot et al., 2001, p. 319).

Anisomeridium carinthiacum(J. Steiner) R. C. Harris 龙骨瓣异形菌

in Egan, *Bryologist* 90(2): 163(1987).

≡*Arthopyrenia carinthiaca* J. Steiner, *Öst. bot. Z*. 63: 335(1913).

On shale.

Taiwan(Aptroot, 2003, p. 157).

Anisomeridium conorostratum Aptroot 球果异形菌

in Aptroot & Seaward, *Trop. Bryol*. 17: 62(1999).

Typus: Hongkong, Hongkong island, Lung Fu Shan, mountain slope, alt. 200m(HKU(M) —holotypus, ABL—isotypus).

Anisomeridium consobrinum(Nyl.) Aptroot 联异形菌

in Aptroot, Diederich, Sérusiaux & Sipman, *Biblthca Lichenol.* 57:21(1995).

≡*Verrucaria consobrina* Nyl., *Annls Sci. Nat.*, *Bot.*, sér. 4 15:53(1861).

On shaded trees. Tropical Asia and Oceania.

Taiwan(Aptroot, 2003, p. 157), Hongkong(Aptroot & Seaward, 1999, p. 63; Seaward and Aptroot 2005, p. 283, (a) Pfister 1978: 126, as *Verrucaria insulata*. On various bushes.

Mountains. 23. 3. 1854. C. Wright. (b) Pfister 1978: 126, as *Verrucaria* sp. On bark of Rhus. Mountains. n. d. C. Wright. (c) Not cited in Pfister 1978 Mountains. n. d. C. Wright. [Label as *Verrucaria septemseptata* Nyl.].

Anisomeridium hydei Aptroot 海德异形菌

in Aptroot & Seaward 1999 *Tropical Bryology* 17:62 –64 –65.

Typus: Hongkong, Pok Fu Lam, University Campus, alt. 150m, Aptroot 43026[HKU(M) —holotypus, ABL—isotypus].

Anisomeridium polypori(Ellis & Everh.) M. E. Barr 聚孔异形菌

in Barr, Huhndorf & Rogerson, *Mem. N. Y. bot. Gdn* 79:76(1996).

≡*Apiospora polypori* Ellis & Everh., *N. Amer. Pyren.* (Newfield):311(1892).

Pok Fu Lam, University Campus, 1998 Aptroot 43016; Lung Fu Shan, 1998 Aptroot 43054 &43100.

On *Machilus thunbergii*. Cosmopolitan.

Taiwan(Aptroot, 2003, p. 157), Hongkong(Aptroot & Seaward, 1999, p. 65; Aptroot et al., 2001, p. 319).

Anisomeridium subnexum(Nyl.) R. C. Harris 亚织异形菌

More Florida Lichens, Incl. 10 Cent Tour Pyrenol. (New York):150(1995).

≡*Verrucaria subnexa* Nyl., *Bull. Soc. linn. Normandie*, sér. 2 7:181(1873).

Misapplied name:

Arthopyrenia annulata auct. non R. C. Harris *ined.* in Thrower, 1988, pp. 15, 57, revised by Aptroot & Seaward, 1999, p. 65).

On shaded trees and branches, mostly in forests. Tropical Asia and Oceania.

Hongkong(Thrower, 1988, pp. 15, 57; Aptroot & Seaward, 1999, p. 65; Aptroot et al., 2001, p. 319), Taiwan (Aptroot, and Sparrius, 2003, p. 3).

Anisomeridium tamarindi(Fée) R. C. Harris 特码拉异形菌

in Tucker & Harris, *Bryologist* 83(1):4(1980).

≡*Verrucaria tamarindi* Fée Suppl. *Essai Cryptog. écorc.* Officin., 1837, p. 85.

On shaded trees in forest. Pantropical.

Taiwan(Aptroot, and Sparrius, 2003, p. 3), Hongkong(Aptroot & Seaward, 1999, p. 65).

Anisomeridium terminatum(Nyl.) R. C. Harris [as ‘terminata’] 顶生异形菌

More Florida Lichens, Incl. 10 Cent Tour Pyrenol. (New York):130(1995).

≡*Verrucaria terminata* Nyl., *Bull. Soc. linn. Normandie*, sér. 2 7:180(1873).

On shaded trees in forest. Pantropical.

Hongkong(Aptroot & Seaward, 1999, p. 65; Aptroot et al. 2001, p. 319).

Anisomeridium tetrasporum Aptroot & Sipman 四孢异形菌

J. Hattori bot. Lab. 91:319(2001).

Hongkong(Aptroot et al., 2001, p. 319).

Anisomeridium throwerae R. C. Harris 斯柔氏异形菌

More Florida Lichens, Incl. 10 Cent Tour Pyrenol. (New York):150(1995).

On shaded trees in forest, so far only known from Hong Kong, where it is common(Aptroot & Sipman 2002).

Taiwan(Aptroot, 2003, pp. 157 – 158), Hongkong(Aptroot & Seaward, 1999, p. 65; Aptroot et al., 2001, p. 320).

Caprettia Bat. & H. Maia **卷须菌属**

Atas Inst. Micol. Univ. Pernambuco 2: 377(1965).

Caprettia sp. 卷须菌未定种

On *Arenga* sp.

Yunnan(Aptroot et al., 2003, pp. 44), Taiwan(Aptroot et al., 2003, pp. 42 – 43).

Distothelia Aptroot **散乳头菌属**

In Seaward and Aptroot. *The Bryologist*, 108(2): 284(2005).

Distothelia isthmospora Aptroot 散乳头菌

in Seaward & Aptroot, *Bryologist* 108(2): 284(2005).

Type: China. Hong Kong Island: Hillside; on bark of Rhus [and other bushes]; 21 April 1854, C. Wright(FH, holotype, sub *Verrucaria thelena* Ach.). Pfister 1978: 126, as *Verrucaria thelena* var. *albidior* Nyl. (incorrect identification). See also D. Hawksw. (1985: 30 – 132).

Hongkong(Seaward & Aptroot, 2005, p. 284).

(11) **Mycoporaceae** Zahlbr. **菌孔科**

in Engler & Prantl, *Nat. Pflanzenfam.*, Teil. I(Leipzig) 1 *: 77(1903).

Mycoporum Flot. ex Nyl. **菌孔属**

Mém. Soc. Imp. Sci. Nat. Cherbourg 3: 186(1855).

Mycoporum eschweileri(Müll. Arg.) R. C. Harris 艾氏菌孔

More Florida Lichens, Incl. 10 *Cent Tour Pyrenol.* (New York): 69(1995).

≡*Mycoporellum eschweileri* Müll. Arg., *Flora*, Regensburg 72: 526(1889).

≡*Tomasellia eschweileri* (Müll. Arg.) R. C. Harris, in Tucker & Harris, *Bryologist* 83(1): 19(1980).

=*Arthonia punctiformis* Eschw., in Martius, *Fl. Bras.* 1: 110(1833) [1829 – 1833].

On trees. Pantropical.

Hongkong(Thrower, 1988, p. 17; Aptroot & Seaward, 1999, p. 95), Taiwan(Aptroot 2003, p. 170).

(12) **Parmulariaceae** E. Müll. & Arx ex M. E. Barr **小梅科**

Mycologia 71: 944(1979).

Parmularia Lév. **小梅属**

Annls Sci. Nat., Bot., sér. 3 5: 286(1846).

Parmularia melanophthalma(DC.) Räsänen 垫脐小梅

Ann. bot. Soc. Zool. -Bot. fenn. Vanamo 12(no. 1): 63(1939).

≡*Squamaria melanophthalma* DC., in Lamarck & de Candolle, *Fl. franç.*, Edn 3(Paris) 2: 376(1805).

≡*Rhizoplaca melanophthalma*(DC.) Leuckert, *Nova Hedwigia* 28: 72(1977).

≡*Lecanora chrysoleuca* var. *melanophthalma*(DC.) Th. Fr., Lich. Scand. (Upsaliae) 1(1): 225(1871).

On rocks.

Xinjiang(Abbas et al., 1993, p. 80), Xizang(J. C. Wei, 1984, p. 211; J. C. Wei & Jiang, 1986, p. 33).

(13) **Trypetheliaceae** Eschw. **乳嘴衣科**

In Goebel & Kunze, *Syst. Lich.*: 17(1824). Published as ' Cohors'; see Fée, Essai crypt. écorc.: xxxvi(1824), nom. inval., Art. 32. 1(b); see Art. 18. 4.

≡ *Trypetheliaceae* Zenker, in Goebel & Kunze, Pharmaceutische Waarenkunde(Eisenach) 1(3): 123(1827) [1827 – 1829].

Astrothelium Eschw. **星果衣属**

Syst. Lich.: 18, 26(1824).

Astrothelium cinnamomeum(Eschw.) Müll. Arg. 肉桂星果衣

Flora, Regensburg 67(35) : 670(1884).

≡*Pyrenastrum cinnamomeum* Eschw., in Martius, *Icon. Plant. Cryptog.* 2: 18(1828) [1828 – 32]

On tree in forest. Pantropical.

Hongkong(Aptroot & Seaward, 1999, p. 66).

Astrothelium sp. 星果衣未定种

On bark.

Hainan(J. C. Wei et al., 2013, p. 227).

Astrothelium speciosum Zahlbr. 丽星果衣

Lichenes Rariores et Critici Exsiccati: no. 252(1927).

China(prov. not indicated: Zahlbr., 1932 a, p. 132).

Astrothelium variolosum(Ach.) Müll. Arg. 瘤星果衣

Flora, Regensburg 68(12) : 255(1885).

≡*Trypethelium variolosum* Ach., Syn. meth. lich. (Lund) : 104(1814).

On bark of trees.

Taiwan(Zahlbr., 1933b, p. 203).

Laurera Rchb. **桂冠衣属**

Deut. Bot. Herb. -Buch: 15(1841).

Laurera megasperma(Mont.) Riddle 巨子桂冠衣

Bull. Torrey bot. Club 44(7) : 323(1917).

≡*Trypethelium megaspermum* Mont., Annls Sci. Nat., Bot., sér. 2 19: 68(1843).

On *Abies kawakamii* and *Pseudotsuga wilsoniana*.

= *Polyblastiopsis pertusarina* Zahlbr. *Repert. Spec. Nov. Regni Veg.* 31: 199(1933).

Type: Taiwan, Prov. Chiai, Mt. Arisan, Toroyen, 24 December 1925, Asahina F 299(TNS holotype).

Taiwan(Zahlbr., 1940, p. 85; Wang & Lai, 1973, p. 95; Aptroot 2003, p. 161; revised by Aptroot, 2004, p. 35).

Megalotremis Aptroot **巨杨衣属**

Biblthca Lichenol. 44: 124(1991).

Megalotremis endobrya(Döbbeler & Poelt) Aptroot 内巨杨衣

in Aptroot, Lücking, Sipman, Umaña & Chaves, *Biblthca Lichenol.* 97: 68(2008).

≡*Anisomeridium endobryum*(Döbbeler & Poelt) R. C. Harris, *More Florida Lichens, Incl.* 10 *Cent Tour Pyrenol.* (New York) : 146(1995).

≡*Arthopyrenia endobrya* Döbbeler & Poelt, *Pl. Syst. Evol.* 138(3 – 4) : 276(1981).

* **Musaespora** Aptroot & Sipman **曲孢衣属**

Lichenologist 25(2) : 123(1993).

* Musaespora coccinea Aptroot & Sipman 丹曲孢衣

Lichenologist 25(2) : 124(1993).

Hainan(Aptroot et al., 2003, p. 46).

Mycomicrothelia Keissl. **疣核衣属**

Rabh. Krypt. -Fl., Edn 2(Leipzig) 9(1. 2) : 7(1936).

Mycomicrothelia fumosula(Zahlbr.) D. Hawksw. 疣核衣

Bull. Br. Mus. nat. Hist., Bot. 14(2) : 86(1985).

≡*Microthelia fumosula* Zahlbr., in Handel-Mazzetti, Symb. Sinic. 3: 18(1930).

Type locality: Guizhou, Handel-Mazzetti no. 10728.

On twigs of *Pittosporum floribundum*.

Mycomicrothelia subfallens(Müll. Arg.) D. Hawksw. 亚落疣核衣

Bull. Br. Mus. nat. Hist., Bot. 14(2) : 111(1985).

≡*Microthelia subfallens* Müll. Arg. , in Engler, *Bot. Jb.* 6: 416(1885).

On *Ficus* sp. and *Lonicera cumingiana*.

Taiwan(Aptroot 2003, p. 161).

Polymeridium(Müll. Arg.) R. C. Harris **聚扇衣属**

In Tucker & Harris, *Bryologist* 83(1): 12(1980).

Polymeridium campylothelioides Aptroot & Sipman 弯曲聚扇衣

In Aptroot, Diederich, Sérusiaux & Sipman, *Biblthca Lichenol.* 57: 37(1995)

On Semecarpus. New to northern hemisphere.

Taiwan(Aptroot 2003, p. 163).

Polymeridium proponens(Nyl.) R. C. Harris 前聚扇衣

Bolm Mus. paraense 'Emílio Goeldi', sér. bot. 7(2): 637(1993)[1991].

≡*Verrucaria proponens* Nyl. , *Bull. Soc. linn. Normandie*, sér. 2 2: 130(1868)

On sheltered trees in forests. Pantropical.

Hongkong(Thrower, 1988, pp. 17, 149 as *Polymeridium* sp.; Aptroot & Seaward, 1999, p. 88).

Pseudopyrenula Müll. Arg. **拟核衣属**

Flora, Regensburg 66: 247(1883).

Pseudopyrenula bicincta Zahlbr. 双围拟核衣

Feddes Repert. 31: 200(1933)

Type: Taiwan, Asahina no. 304.

Taiwan(Zahlbr. , 1940, p. 85; Lamb, 1963, p. 601; Wang & Lai, 1973, p. 96).

Pseudopyrenula quintaria Zahlbr. 五胞拟核衣

In Handel-Mazzetti, *Symb. Sin.* 3: 20(1930) & *Cat. Lich.* Univ. 8: 97(1932).

Type locality: Guizhou, Handel-Mazzetti no. 10427.

On bark of *Antidesma microphyllum*

Pseudopyrenula subgregaria Müll. Arg. 群生拟核衣

Bot. Jb. 6: 408(1885).

On bark.

Hongkong(Thrower, 1988, p. 155).

Pseudopyrenula subnudata Müll. Arg. 亚曲拟核衣

In *Flora* LXVI, 1883, pp. 240 &272.

= *Pseudopyrenula subgregaria* Müll. Arg. in Thrower, 1988, pp. 17, 155, mentioned by Aptroot & Seaward, 1999, p. 90.

On exposed or sheltered trees. Pantropical.

Hongkong(Thrower, 1988, pp. 17, 155, as *P. subgregaria*; Aptroot & Seaward, 1999, p. 89 – 90; Aptoot & Sipman, 2001, p. 338).

Pseudopyrenula tropica(Ach.) Müll. Arg. 热带拟核衣

Flora, Regensburg 66: 248(1883).

≡*Verrucaria tropica* Ach. , *Lich. univ.* : 1 – 696(1810).

On bark of trees.

Guangdong(Krphbr. 1873, p. 467; Zahlbr. , 1930 b, p. 20).

Trypetheliopsis Asahina **红乳嘴衣属**

J. Jap. Bot. 13: 319(1937).

Trypetheliopsis boninensis Asahina 红乳嘴衣

J. Jap. Bot. 13(5): 319(1937).

On bark of *Cinnamoum randaiense*.

Taiwan(Kurok. 1978, p. 29).

Trypetheliopsis gigas(Zahlbr.) Aptroot 巨红乳嘴衣

in Kashiwadani, Aptroot & Moon, *Biblthca Lichenol.* 99: 249(2009) .

≡*Mycoporellum gigas* Zahlbr. , *Annals Cryptog. Exot.* 1(2) : 117(1928) .

≡*Musaespora gigas* (Zahlbr.) R. C. Harris, More Florida Lichens, Incl. 10 Cent Tour Pyrenol. (New York) : 133(1995) .

On a tree. Aptroot 48726(CBS culture109004) .

Hongkong(Aptroot et al. , 2001, p. 333, as *Musaespora gigas*) .

Trypethelium Spreng. **乳嘴衣属**

Anleit. Kennt. Gew. 3: 350(1804) .

Trypethelium eluteriae Spreng. 乳嘴衣

Anleit. Kennt. Gew. 3: 350(1804) . var. eluteriae

On bark of tree. Pantropical.

Jiangsu, Shanghai, Zhejiang(Xu B. S. , 1989, p. 174) , Taiwan(Zahlbr. , 1933b, p. 204; Asahina & M. Sato in Asahina, 1939, p. 617) , Hongkong(Thrower, 1988, pp. 17, 177; Aptroot & Seaward, 1999, p. 96; Aptroot & Sipman, 2001, p. 341) , Taiwan(Aptroot 2003, p. 170) .

var. citrinum(Eschw.) Müll. Arg. 柠檬变种

Bot. Jb. 6: 393(1885) .

≡*Astrothelium varium* var. *citrinum* Eschw. , in von Martius, *Fl. Bras.* 1(1) : 162(1833) [1829 – 1833] .

= *Trypethelium sprengelii* Ach. , *Lich. univ.* : 306(1810) , Zahlbr. , 29(1930) .

On bark of tree.

Guangdong(Kremp. 1874b, p. 60; Zahlbr. , 29(1930) .

var. conglobatum(Ach.) Zahlbr. 聚球变种

Cat. Lich. Univ. 1: 491(1922) .

≡*Trypethelium conglobatum* Ach. *Synops. Lich.* 1814, p. 105.

On bank of tree.

Taiwan(Zahlbr. , 1933 b, p. 204; Wang & Lai, 1973, p. 97) .

Trypethelium epileucodes Nyl. 白色乳嘴衣

Lich. Japon. : 116(1890) .

On sheltered trees in forests. Known from Malaysia, Australia, Papua New Guinea and China. Living culture CBS 101279.

Hongkong(Aptroot & Seaward, 1999, p. 96) .

Trypethelium leprosulum Zahlbr. 斑点乳嘴衣

Annal. Mycolog. 30: 429(1932) .

Syntype: Fujian, near Fuzhou, Chung nos. 371, 376.

On bark of trees in forest.

Fujian(Zahlbr. , 1934 b, p. 196 & 1940, p. 102; Lamb, 1963, p. 723) .

Trypethelium nitidiusculum(Nyl.) R. C. Harris 光乳嘴衣

Acta Amazon. , Supl. 14(1 – 2) : 75(1986) [1984] .

≡*Verrucaria nitidiuscula* Nyl. , *Acta Soc. Sci. fenn.* 7(2) : 491(1863) .

On exposed trees, e. g. *Callitris* sp. , in parks and open forests. Pantropical.

Hongkong(Aptroot & Seaward, 1999, p. 96; Aptroot & Sipman, 2001, p. 341) .

= *Trypethelium formosanum* Zahlbr. Repert. *Spec. Nov. Regni Veg.* 31: 204(1933) .

Type: Taiwan, Prov. Taipei, Taihoku, 19 December 1925, Asahina F 311(TNS holotype) .

Revised by Aptroot, 2004, p. 36.

Taiwan(Zahlbr. , 1940, p. 102; Lamb, 1963, p. 723; Wang & Lai, 1973, p. 97) .

Trypethelium subnitidiusculum Makhija & Patw. 亚光乳嘴衣

J. Hattori bot. Lab. 73: 207(1993).

On tree branches. India & China Taiwan.

Taiwan(Aptroot 2003, p. 170).

Trypethelium tropicum(Ach.) Müll. Arg. 热带乳嘴衣

Bot. Jb. 6: 393(1885).

≡ *Verrucaria tropica* Ach., *Lich. Univ.*: 1 – 696(1810).

= *Pseudopyrenula bicincta* Zahlbr. *Repert. Spec. Nov. Regni Veg.* 31: 200(1933).

Type: Taiwan, Prov. Changhua, Rengechi, 30 December 1925, Asahina F 304(W holotype n. v., TNS isotype).

Taiwan(Zahlbr., 1940, p. 85; Lamb, 1963, p. 601; Wang & Lai, 1973, p. 96; Aptroot 2003, pp. 170 – 171, on *Delonix regia*; revised by Aptroot, 2004, p. 35).

On exposed trees of *Callitris* sp. Pantropical.

Hongkong(Aptroot & Seaward, 1999, pp. 96 – 97; Aptroot & Sipman, 2001, p. 341; Seaward & Aptroot, 2005, p. 286, Pfister 1978: 126, as *Verrucaria tropica*).

Dothideomycetes, genera incertae sedis **座囊菌纲未定位属**

* **Cercidospora** Körb. **杆孢属**

Parerga lichenol. (Breslau) 5: 465(1865).

* Cercidospora soror Obermayer & Triebel 群杆孢

in Hafellner & Obermayer, *Cryptogamie, Bryol. Lichénol.* 16(3): 182(1995).

On *Arthrorhaphis alpina* var. *jungens* on soil above red sandstone, and *A. vacillans*.

Sichuan, Xizang(J. Hafellner & W. Obermayer 1995, pp. 182 – 185).

* Cercidospora trypetheliza (Nyl.) Hafellner & Obermayer 边杆孢

Cryptogamie, Bryol. Lichénol. 1995, 16(3): 180.

≡ *Lecidea trypetheliza* Nylander, *Acta Soc. Sci. Fenn.* 7: 402(1863).

= *Cercidospora arthrorhaphidicola* Alstrup, in Alstrup et al.

Xizang(J. Hafellner & W. Obermayer pp. 180 – 182).

* **Karschia** Körb. **隔孢黑盘菌属**

Parerga lichenol. (Breslau) 5: 459(1865).

* Karschia talcophila(Ach.) Körb. 隔孢黑盘菌

Parerga lichenol. (Breslau) 5: 460(1865).

≡ *Lecidea talcophila* Ach., *Lich. Univ.*: 183(1810).

On *Peltula zabolotnoji*(as *Heppia kansuensis* by H. Magn. in 1940, p. 43).

Gansu(Hawksw. & M. Cole, 2003, p. 361).

Kirschsteiniothelia D. Hawksw. **半球果衣属**

J. Linn. Soc., Bot. 91: 182(1985).

Kirschsteiniothelia aethiops(Sacc.) D. Hawksw. 半球果衣

J. Linn. Soc., Bot. 91(1 – 2): 185(1985).

On *Shiia calesii*.

Taiwan(Aptroot 2003, p. 161).

* **Lichenostigma** Hafellner **柱头菌属**

Herzogia 6(1 – 2): 301(1983).

* Lichenostigma cosmopolites Hafellner & Calat. 廣柱头菌

Mycotaxon 72: 108(1999).

On *Xanthoparmelia scabrosa* (Aptroot & Sipman 2001).

Hongkong(Hawksw. & M. Cole, 2003, p. 361; Aptroot & Sipman 2001, p. 331).

Mycoporellum Müll. Arg. **乳头衣属**

Revue mycol., Toulouse 6(no. 21): 14(1884)

Mycoporellum leucoplacellum Zahlbr. 白乳头衣

Feddes Repert. 31: 205(1933).

Type: Taiwan, Asahina no. 215.

On bark of trees.

Taiwan(Zahlbr., 1940, p. 109; Wang & Lai, 1973, p. 92).

Mycoporellum subpomaceum Zahlbr. 鸭梨乳头衣

in Handel-Mazzetti, *Symb. Sin.* 3: 32(1930).

On bark of *Jasminum* sp.

Type: Yunnan, Handel-Mazzetti no. 5830.

3. Eurotiomycetes O. E. Erikss. & Winka **散囊菌纲**

Myconet 1(1): 6(1997).

1) Mycocaliciomycitidae Tibell **粉衣亚纲**

Mycological Research 111: 528(2007).

[5] Mycocaliciales Tibell & Wedin **粉衣目**

Mycologia 92(3): 579(2000).

(14) Mycocaliciaceae A. F. W. Schmidt **粉衣科**

Mitt. Staatsinst. Allg. Bot. Hamburg 13: 127(1970).

Chaenothecopsis Vain. **类口果粉衣属**

Acta Soc. Fauna Flora Fenn. 57(no. 1): 70(1927).

Chaenothecopsis pusiola(Ach.) Vain. 类口果粉衣

Acta Soc. Fauna Flora Fenn. 57(no. 1): 70(1927).

≡ *Calicium pusiolum* Ach., *K. Vetensk-Acad. Nya Handl.*: 231(1817).

= *Chaenothecopsis lignicola*(Nádv.) A. F. W. Schmidt, *Mitt. Staatsinst. Allg. Bot. Hamburg* 13: 153(1970).

≡ *Calicium lignicola* Nádv. [as '*lignicolum*'], *Preslia* 18 – 19: 128(1940).

On wood.

Taiwan(Sparrius et al., 2002, 359).

Mycocalicium Vain. **粉菌属**

Acta Soc. Fauna Flora Fenn. 7(no. 2): 182(1890).

Mycocalicium subtile(Pers.) Szatala 粉菌衣

Magy. Bot. Lapok 24: 47(1925).

≡ *Calicium subtile* Pers., *Tent. disp. meth. fung.* (Lipsiae): 60(1797).

On wood.

Taiwan(Sparrius et al., 2002, 359).

[6] Pyrenulales Fink ex D. Hawksw. & O. E. Erikss. **小核衣目**

Syst. Ascom. 5(1): 182(1986).

Lyromma Bat. & H. Maia **里拉菌属**

Atas Inst. Micol. Univ. Pernambuco 2: 359(1965).

Lyromma nectandrae Bat. & H. Maia (Anamorphic Pyrelulales) 里拉菌

Atas Inst. Micol. Univ. Recife 2: 360(1965).

On *Hydrangea* sp.

Yunnan(Aptroot et al., 2003, p. 44), taiwan(Aptroot et al., 2003, pp. 42 – 43).

(15) Massariaceae Nitschke **粉团科**

Verh. naturh. Ver. preuss. Rheinl. 26: 73(1869).

Navicella Fabre **舟型菌属**

Annls Sci. Nat., *Bot.*, sér. 6 9: 96(1879)[1878].

Navicella diabola Aptroot 诡舟型菌

J. Hattori bot. Lab. 93: 161 – 162(2003).

Type: Nantou County: 44km WNW of Hulen, Meifeng, broadleaf forest remnant in valley, 2050m. alt., 51RUG146655, on living bark of Castanopsis sp., Sparrius 5992 (ABLO, holotype-hb Sparrius, isotype), 10Oct. 2001.

Taiwan(Aptroot 2003, p. 161).

(16) Pyrenulaceae Rabenh. **小核衣科**

Krypt. -Fl. Sachsen, Abth. 2(Breslau): 42(1870).

Anthracothecium Hampe ex A. Massal. **炭壳衣属**

Atti Inst. Veneto Sci. lett., ed Arti, Sér. 3 5: 330(1860)[1859 – 1860].

Anthracothecium chrysophorum Zahlbr. 金叶炭壳衣

In Handel-Mazzetti, *Symb. Sin.* 3: 28(1930) & *Cat. Lich. Univ.* 8: 121(1932).

Type: Fujian, Chung no. 392.

Anthracothecium columellatum(Vain.) Zahlbr. 囊轴炭壳衣

Cat. Lich. Univ. 8: 121(1931)[1932].

≡*Bottaria columellata* Vain., *Ann. Acad. Sci. Fenn.*, Ser. A 15(no. 6): 325(1921).

On *Ficus* sp.

Taiwan(Aptroot, 2003, p. 158).

Anthracothecium libricola(Fée) Müll. Arg. 内皮炭壳衣

Linnaea 43: 44(1880).

≡*Pyrenula libricola* Fée, *Essai Crypt. Exot.*, *Suppl. Révis.* (Paris): 82(1837).

On bark of trees.

Fujian(Zahlbr., 1930 b, p. 27).

Anthracothecium majus(Zahlbr.) Kashiw. & Kurok. 大炭壳衣

Journ. Jap. Bot. 56(11): 348(1981).

≡*Anthracothecium angulatum* Zahlbr. var. *majus* Zahlbr. in Fedde, *Repertorium*, 31: 203(1933).

Type: Taiwan, Mt. Alishan, Y. Asahina F – 326, holotype in W, isotype in TNS(not seen).

Taiwan(Zahlbr., 1940, p. 98; M. Lamb, 1963, p. 31; Wang & Lai, 1973, p. 88; Kashiw. & Kurok. 1981, p. 348).

Anthracothecium nanum(Zahlbr.) R. C. Harris 矮小炭壳衣

Mem. N. Y. bot. Gdn 49: 79(1989).

≡*Pleurotheliopsis nana* Zahlbr., *Annls Mycol.* 33(1/2): 40(1935).

On *Aphananthe* sp. and *Ardisia sieboldii*.

Taiwan(Aptroot, 2003, p. 158).

Anthracothecium oculatum Müll. Arg. 眼点炭壳衣

Nuovo G. bot. ital. 23: 404(1891).

= *Pyrenula neoculata* Aptroot, *Australas. Lichenol.* 60: 38(2007).

= *Anthracothecium angulatum* Zahlbr. in Feddes, *Repertorium sp. nov.* 31: 202(1933).

Type: Taiwan, Mt. Alishan, Y. Asahina 298(holotype) in G & isotype in TNS(not seen).

Taiwan(Zahlbr., 1940, p. 98; M. Lamb, 1963, p. 31; Wang & Lai, 1973, p. 88; Kashiw. & Kurok. 1981, p. 350).

Anthracothecium pachycheilum(Tuck.) Zahlbr. 厚唇炭壳衣

Cat. Lich. Univ. 1: 465(1922).

≡*Pyrenula pachycheila* Tuck., *Gen. lich.* (Amherst): 274(1872).

= *Anthracothecium fraternale* Zahlbr., in Handel-Mazzetti, *Symb. Sin.* 3: 27(1930) & *Cat. Lich. Univ.* 8: 121

(1932).

Type: Fujian, Mt. Gushan, at about 500 – 600 m, Chung no. 445 b, holotype in W(not seen).

"F" is an abbreviation from "Fukien" in Zahlbr.'s work(1930 b, p. 27), but not that for Taiwan (as "Formosa") as miscited in Kashiw. & Kurok.'s paper(1981, p. 35).

Anthracothecium prasinum(Eschw.) R. C. Harris 韭绿炭壳衣

In Egan, *Bryologist* 90(2): 163(1987).

≡ *Verrucaria prasina* Eschw., in von Martius, *Fl. Bras.* 1(1): 124(1833)[1829 – 1833].

= *Anthracothecium angulatum* Zahlbr. in Fedde, *Repertorium*, 31: 202(1933), synonymized by Aptroot, 2004: 32.

Type: Taiwan, Mt. Arisan, Toroyen, December 1925, Y. Asahina 298(TNS lectotype).

Taiwan(Zahlbr., 1940, p. 98; M. Lamb, 1963, p. 31; Wang & Lai, 1973, p. 88; Kashiw. & Kurok. 1981, p. 350; Aptroot, 2004, p. 32).

= *Anthracothecium angulatum* var. *majus* Zahlbr. in Fedde, *Repertorium*, 31: 203(1933), synonymized by Aptroot, 2004: 32.

Type: Taiwan, Mt. Alishan, Y. Asahina F-326, holotype in W, isotype in TNS.

Taiwan(Zahlbr., 1940, p. 98; M. Lamb, 1963, p. 31; Wang & Lai, 1973, p. 88; Kashiw. & Kurok. 1981, p. 348; Aptroot, 2004, p. 32).

On *Machilus thunbergii*.

Taiwan(Aptroot 2003, p. 158).

Anthracothecium speciosum Zahlbr. 丽炭壳衣

In Handel-Mazzetti, *Symb. Sin.* 3: 27(1930) & *Cat. Lich. Univ.* 8: 123(1932).

Syntype: Hunan, Handel-Mazzetti no. 11193 & Fujian, Chung nos. 395, 396, 397 a, 468 a, 595.

Lithothelium Müll. Arg. **石乳头衣属**

Bot. Jb. 6: 386(1885).

Lithothelium illotum(Nyl.) Aptroot 污石乳头衣

Biblthca Lichenol. 44: 60(1991).

≡ *Verrucaria illota* Nyl., *Flora*, Regensburg 59: 364(1876).

On *Zelkova serrata*, *Ficus microcarpa* and *Ardisia sieboldii*. Pantropical species.

Taiwan(Aptroot 2003, p. 161).

Lithothelium obtectum(Müll. Arg.) Aptroot 覆石乳头衣

Biblthca Lichenol. 44: 62(1991).

≡ *Sagedia obtecta* Müll. Arg., *Linnaea* 63: 42(1880).

On *Machlus thungbergii*. Pantropical species.

Taiwan(Aptroot 2003, p. 161).

Lithothelium submuriforme R. C. Harris & Aptroot 亚砖壁石乳头衣

Biblthca Lichenol. 44: 69(1991).

On *Machlus thungbergii*. New to northern hemisphere.

Taiwan(Aptroot 2003, p. 161).

Lithothelium triseptatum(Nyl.) Aptroot 三隔石乳头衣

Biblthca Lichenol. 44: 70(1991).

≡ *Verrucaria conoidea* var. *triseptata* Nyl., *Act. Soc. linn. Bordeaux* 21: 435(1856).

On raised coral reef. New to Asia.

Taiwan(Aptroot 2003, p. 161).

Pyrenula Ach. **小核衣属**

K. Vetensk-Acad. Nya Handl. 30: 160(1809).

Pyrenula acutalis R. C. Harris 尖小核衣

Mem. N. Y. bot. Gdn 49: 85(1989).

On *Semecarpus*.

Taiwan(Aptroot 2003, p. 165).

Pyrenula albidopunctata Zahlbr. 白点小核衣

in Fedde, *Repertorium*, 31: 201(1933).

Type: Taiwan, Faurie no. 308.

On bark of trees.

Taiwan(Zahlbr., 1940, p. 95; Wang & Lai, 1973, p. 96).

Pyrenula anomala(Ach.) Vain. 畸小核衣

Ann. Acad. Sci. fenn., Ser. A 6(no. 7): 189(1915).

≡*Trypethelium anomalum* Ach., *Syn. meth. lich.* (Lund): 105(1814).

≡*Melanotheca anomala* (Ach.) A. Massal., *Atti Inst. Veneto Sci. lett., ed Arti*, Sér. 3 5: 334(1860) [1859 – 1860] in Thrower, 1988, p. 16, mentioned by Aptroot & Seaward, 1999, p. 90.

Misapplied name:

Parathelium microcarpum auct. non Riddle in Thrower, 1988, p. 16.

On sheltered trees in forests. Pantropical.

Hongkong(Thrower, 1988, pp. 16, 120; Aptroot & Seaward, 1999, p. 90).

Pyrenula aspistea(Afzel. ex Ach.) Ach. 盾小核衣

Mag. Gesell. naturf. Freunde, Berlin 6: 17(1814)

≡*Verrucaria aspistea* Afzel. ex Ach., *Methodus*, Sectio prior: 121(1803)

=*Pyrenula bonplandiae* Fée, *Ess. Cryptog. Ecerc. Offic.* 1: 74, tab. XXI, fig. 3(1824), Zahlbr.: 201(1933).

On bark of trees. Pantropical.

Taiwan(Zahlbr., 1933 b, p. 201; Wang & Lai, 1973, p. 96), Hongkong(Aptroot & Seaward, 1999, p. 90; Aptroot & Sipman, 2001, p. 338).

Pyrenula astroidea(Fée) R. C. Harris 星小核衣

Mem. N. Y. bot. Gdn 49: 87(1989).

≡*Parmentaria astroidea* Fée, *Essai Crypt. Exot.* (Paris): xci, 70(1825)[1824].

=*Anthracothecium* sp. in Thrower, 1988, p. 15, mentioned by Aptroot & Seaward, 1999, p. 90.

On sheltered trees in forest. Pantropical.

Hongkong(Aptroot & Seaward, 1999, p. 90), Taiwan(Aptroot 2003, p. 166, on *Lonicera cumingiana*).

=*Pleurotheliopsis asahinae* Zahlbr. Repert. Spec. Nov. Regni Veg. 31: 204 – 205(1933), revised by Aptroot, 35 (2004).

Type: Taiwan, Taipei, Taihoku, 19 December 1925, Asahina F 297(TNS holotype).

Corticola.

Pyrenula chondriana Zahlbr. 软骨小核衣

In Handel-Mazzerri, *Symb. Sin.* 3: 24(1930) & *Cat. Lich. Univ.* 8: 114(1932).

Type: Fujian, Mt. Gu shan, Cung no. 591 c.

On bark of trees.

Pyrenula chungii Zahlbr. 钟氏小核衣

In Handel-Mazzerri, *Symb. Sin.* 3: 25(1930) & *Cat. Lich. Univ.* 8: 114(1932).

Syntype: Fujian, Mt. Gu shan, Chung nos. 396 a, 580 b.

On bark of trees.

Pyrenula concatervans(Nyl.) R. C. Harris 联小核衣

In Tucker & Harris, *Bryologist* 83(1): 15(1980).

≡*Verrucaria concatervans* Nyl., *Lich. Japon.*: 109(1890).

On *Zelkova serrata*.

Taiwan(Aptroot 2003, p. 166).

Pyrenula confinis(Nyl.) R. C. Harris 邻小核衣

More Florida Lichens, Incl. 10 Cent Tour Pyrenol. (New York) : 109(1995).

≡*Verrucaria confinis* Nyl. , *Annls Sci. Nat.* , *Bot.* , sér. 4 3: 173(1855).

On *Ficus* sp. and *Roystonea regia*.

Taiwan(Aptroot 2003, p. 166).

Pyrenula cuprescens Zahlbr. 铜色小核衣

In Handel-Mazzerri, *Symb. Sin.* 3: 26(1930) & *Cat. Lich. Univ.* 8: 115(1932).

Type: Hunan, Mt. Yuelu shan, Handel-Mazzetti no. 11423.

On bark of *Eurya nitida*.

Pyrenula dermatodes(Borrer) Schaerer 皮小核衣

Enum. critic. lich. europ. (Bern) : 213(1850).

≡*Verrucaria dermatodes* Borrer, *Suppl. Engl. Bot.* 1: tab. 2607, fig. 2(1831).

On tree, on *Shiia calesii* & *Pseudotsuga wilsoniana*.

Taiwan(Aptroot 2003, p. 166).

Pyrenula falsaria(Zahlbr.) R. C. Harris 拟小核衣

Mem. N. Y. bot. Gdn 49: 92(1989).

≡*Anthracothecium falsarium* Zahlbr. , *Annls mycol.* 33: 37(1935).

On sheltered trees in forests. Pantropical.

Hongkong(Aptroot & Seaward, 1999, p. 90).

Pyrenula hunana Zahlbr. 湖南小核衣

In Handel-Mazzerri, *Symb. Sin.* 3: 26(1930) & *Cat. Lich. Univ.* 8: 116(1932).

Type: Hunan, Brauer no. 12769.

On twigs of *Pseudolarix fortunei*.

Pyrenula japonica Kurok. 日本小核衣

In Kurok. & Nakanishi, *Mem. Natn Sci. Mus*, Tokyo 4: 66(1971).

On trees.

Anhui, Shanghai, Zhejiang(Xu B. S. , 1989, p. 173).

Pyrenula kelungana Zahlbr. 台湾小核衣

In Fedde, *Repertorium*, 31: 202(1933).

Type: Taiwan, Faurie no. 241.

On twigs.

Taiwan(Zahlbr. , 1940, p. 96; Lamb, 1963, p. 609; Wang & Lai, 1973, p. 96).

Pyrenula laii Aptroot 赖氏小核衣

J. Hattori bot. Lab. 93: 166(2003).

Type: Taiwan , Hualien County. 43km WNW of Hualien, 2100m alt. , 51RUG1566, on *Machilus thunbergii*, Aptroot 52296(ABL, holotype, 11Oct. 2001).

Pyrenula leucostoma Ach. 白气孔小核衣

Syn. meth. lich. (Lund) : 124(1814).

On *Ficus* sp.

Taiwan(Aptroot 2003, p. 167), Hongkong(Aptroot & Sipman, 2001, p. 338).

Pyrenula leucotrypa(Nyl.) Upreti 白穴小核衣

Nova Hedwigia 66(3 – 4) : 570(1998).

≡*Trypethelium leucotrypum* Nyl. , *Flora*, Regensburg 50: 9(1867).

On sheltered trees in forest remnant. Known from India, including the Andaman Islands.

Hongkong(Aptroot & Seaward, 1999, p. 90) .

Pyrenula macrocarpa Massal. 巨果小核衣

Ricerch. Auton. Lich. 164(1852) , Zahlbr. *Cat. Lich. Univ.* 1: 517(1922) .

On *Aphananthe* sp. , and *Machilus thunbergii*.

Taiwan(Aptroot 2003, p. 167) .

Pyrenula macularis(Zahlbr.) R. C. Harris 斑点小核衣

Mem. N. Y. bot. Gdn 49: 94(1989) .

≡*Anthracothecium maculare* Zahlbr. , *Mycologia* 22(2) : 70(1930) .

On exposed trees, e. g. *Bombax* sp. in forest. Pantropical.

Hongkong(Aptroot & Seaward, 1999, p. 90) , Taiwan(Aptroot 2003, p. 167) .

Pyrenula mamillana(Ach.) Trevis. 乳头小核衣

Conspect. Verruc. : 13(1860) .

≡*Verrucaria mamillana* Ach. , *Methodus, Sectio prior:* 120(1803) .

=*Pyrenuia kunthii* Fee, *Suppl. Essai Cryptog. Eoorc. Offic.* : 80(1837) ; Zahlbr. , Catal. Lich. 1: 434. & in Fedde, *Repertorium, Flechten der Insel Formosa* 31: 201(1933) .

Soozan prope Teihoku, ad corticem arboris(Fauricno. 332) .

In regionibus tropicis frequens.

≡*Verrucaria kunthii* Fée, *Essai Crypt. Exot.* (Paris) : 88(1825) [1824] .

Type: Taiwan, Fauricno. 332

On bark of trees.

Taiwan(Zahlbr. , 1933, p. 201; Wang & Lai, 1973, p. 96) .

=*Pyrenula marginata* Hook. , in Kunth, *Syn. pl. (Paris)* 1: 20(1822) .

On bark of trees.

Yunnan(Zahlbr. , 1930 b, p. 22) .

=*Pyrenula santensis*(Nyl.) Müll. Arg. , *Flora*, Regensburg 65(25) : 400(1882) .

≡*Verrucaria marginata* var. *santensis* Nyl. , Expos. *Synopt. Pyrenocarp.* : 45(1858) .

On bark of *Quercus gilliana*. On sheltered rees in forests. Pantropical.

Sichuan(Zahlbr. , 1930 b, p. 24) , Shanghai(Nyl. & Cromb. 1883, p. 65; Zahlbr. , 1930 b, p. 24) , Taiwan(Wang & Lai, 1973, p. 96; Zahlbr. , 1933, p. 201) , Hongkong(Aptroot & Seaward, 1999, pp. 90 – 91; Aptroot & Sipman, 2001, p. 338; Seaward & Aptroot, 2005, p. 285, Pfister 1978: 125, as *Verrucaria marginata* var. *diminuens* Nyl.) .

Pyrenula manhaviensis Zahlbr. 曼耗小核衣

In Handel-Mazzetti, *Symb. Sin.* 3: 23(1930) & *Cat. Lich. Univ.* 8: 116(1932) .

Type: Yunnan, Manhao, Handel-Mazzetti no. 5911.

On tree trunk.

Pyrenula nebulosa Zahlbr. 云小核衣

In Handel-Mazzerri, *Symb. Sin.* 3: 23(1930) &*Cat. Lich. Univ.* 8: 117(1932) .

Type: Fujian, Mt. Gu shan, Chung no 445 c.

On bark of trees.

Pyrenula nitida(Weigel) Ach. 小核衣

Syn. meth. lich. (Lund) : 125(1814) .

≡*Sphaeria nitida* Weigel, *Observ. Bot.* : 45(1772) .

≡*Verrucaria nitida*(Weigel) Schrad. , *J. Bot.* (Schrader) 1: 79(1801) , Zahlbr. , 26(1930) .

On bark of trees.

Shanghai(Nyl. & Cromb. 1883, p. 66; Zahlbr. , 1930 b, p. 26; Asahina & M. Sato in Asahina, 1939, p. 615) .

Pyrenula nitidula(Bres.) R. C. Harris 亮小核衣

In Aptroot, Diederich, Sérusiaux & Sipman, *Biblthca Lichenol*. 64: 164(1997).

≡*Melanomma nitidulum* Bres. , in Bresadola, Hennings & Magnus, *Bot. Jb*. 17: 500(1893).

On *Delonix regia*.

Taiwan(Aptroot 2003, p. 167), Hongkong(Aptroot & Sipman, 2001, p. 338).

Pyrenula ochraceoflava(Nyl.) R. C. Harris 黄褐小核衣

Mem. N. Y. bot. Gdn 49: 96(1989).

≡*Verrucaria ochraceoflava* Nyl. , *Acta Soc. Sci. fenn*. 7(2): 491(1863).

On exposed *Pandunus* sp. in coastal area. Pantropical.

Hongkong(Aptroot & Seaward, 1999, p. 91), TAIWAN(Aptroot 2003, p. 167).

Pyrenula parvinuclea(Meyen & Flot.) Aptroot 微小核衣

In Aptroot, Diederich, Sérusiaux & Sipman, *Biblthca Lichenol*. 64: 165(1997).

≡*Verrucaria parvinuclea* Meyen & Flot. , *Nova Acta Acad. Caes. Leop. -Carol. Nat. Cur.* , Suppl. 1 19: 231 (1843).

On exposed trees in coastal area. Pantropical.

Hongkong(Aptroot & Seaward, 1999, p. 91), Taiwan(Aptroot 2003, pp. 167 – 168).

Pyrenula pertusarina Zahlbr. 孔小核衣

In Handel-Mazzerri, *Symb. Sin*. 3: 24(1930) & *Cat. Lich. Univ*. 8: 119(1932).

Syntype: Fujian, Chung nos. 347 a, b, 397, 468 b, 603 a.

On twigs.

Pyrenula pileata Vain. 帽状小核衣

Ann. Acad. Sci. fenn. , Ser. A 15(no. 6): 336(1921).

On sheltered trees in forests. Tropical Asian.

Hongkong(Aptroot & Seaward, 1999, p. 91; Aptroot & Sipman, 2001, p. 338), Taiwan(Aptroot 2003, p. 168).

Pyrenula pseudobufonia(Rehm) R. C. Harris 略小核衣

Evansia 2(3): 46(1985).

≡*Clypeosphaeria pseudobufonia* Rehm, *Annls mycol*. 2(2): 176(1904).

=*Pyrenula neglecta* R. C. Harris in Thrower, 1988, pp. 17, 156, mentioned by Aptroot & Seaward, 1999, p. 91.

On bark. On exposed and sheltered trees. Known from East Asia and North America.

Hongkong(Thrower, 1988, pp. 17, 156; Aptroot & Seaward, 1999, p. 91; Aptroot & Sipman, 2001, p. 338), Taiwan(Aptroot 2003, p. 168).

Pyrenula pyrenuloides(Mont.) R. C. Harris 类小核衣

Mem. N. Y. bot. Gdn 49: 99(1989).

≡*Trypethelium pyrenuloides* Mont. , *Annls Sci. Nat.* , *Bot.* , sér. 2 19: 69(1843).

On exposed and sheltered trees in forests. Pantropical.

Hongkong(Aptroot & Seaward, 1999, p. 91).

Pyrenula quassiicola Fée [as ' quassiaecola'] 枯木小核衣

Essai Crypt. Exot. , *Suppl. Révis.* (Paris): 79(1837).

On *Machilus thunbergii* and *Ardisia sieboldii*.

Taiwan(Aptroot 2003, p. 168).

Pyrenula quercuum Zahlbr. 栎小核衣

In Handel-Mazzerri, *Symb. Sin*. 3: 27(1930) & *Cat. Lich. Univ*. 8: 119(1932).

Type: Sichuan, Handel-Mazzetti no. 2448.

On twigs.

Pyrenula rubromaculata Vain. 红点小核衣

Bot. Mag., Tokyo 35: 77(1921).

≡*Melanotheca rubromaculata*(Vain.) Zahlbr.

Cat. Lich. Univers. 8: 126(1931).

Corticolous.

Taiwan(Zahlbr., 1933 b, p. 203; Wang & Lai, 1973, p. 92).

Pyrenula schutschensis Zahlbr. 云南小核衣

In Handel-Mazzerri, *Symb. Sin.* 3: 22(1930) & *Cat. Lich. Univ.* 8: 119(1932).

Type: Yunnan, Handel-Mazzetti no. 9488.

On bark of *Prunus cornuta*.

Pyrenula subferruginea(Malme) R. C. Harris 锈红小核衣

Mem. N. Y. bot. Gdn 49: 101(1989)

≡*Parathelium subferrugineum* Malme, *Ark. Bot.* 19(no. 8): 17(1924)

On *Pemphis maritima*.

Taiwan(Aptroot 2003, p. 168).

Pyrenula tunicata Zahlbr. 膜小核衣

In Handel-Mazzerri, *Symb. Sin.* 3: 25(1930) & *Cat. Lich. Univ.* 8: 120(1932).

Type: Yunnan, Handel-Mazzetti no. 6750.

On pine tree.

(17) Requienellaceae Boise **粒核衣科**

Mycologia 78(1): 37(1986).

Granulopyrenis Aptroot **粒核衣属**

Biblthca Lichenol. 44: 91(1991).

Granulopyrenis seawardii Aptroot & Sipman 西沃氏粒核衣

J. Hattori bot. Lab. 91: 326(2001).

Type: China, Hongkong, Hongkong island, Sipman45175 = Aptroot48920 (B-holotype; HKU (M), ABL-isotypes); culture CBS109025.

On weathered wood of a tree trunk and decorticated branches on the rocky slope at the coast.

(18) Xanthopyreniaceae Zahlbr. **黄核衣科**

In Engler & Prantl, *Nat. Pflanzenfam.*, Edn 2(Leipzig) 8: 91(1926). Nom. rejic. prop., see Hawksworth & Eriksson, *Taxon* 37: 190(1988).

Pyrenocollema Reinke **核胶衣属**

In Pringsheim, *Jb. wiss. Bot.* 28: 463(1895).

Pyrenocollema halodytes(Nyl.) R. C. Harris 嗜盐核胶衣

In Egan, *Bryologist* 90(2): 164(1987).

≡*Verrucaria halodytes* Nyl., *Mém. Soc. Imp. Sci. Nat.* Cherbourg 5: 142(1857).

=*Arthopyrenia lithoralis* (Leighton) Arnold(nom. illegit.)

=*Arthopyrenia sublithoralis* (Leighton) Arnold(Santesson 1939, p. 63 & Aptroot & Seaward, 1999, p. 90).

At the castle, on *Tetraclita squamosa* & Fk *Balanus* sp. On maritime rocks, including granite, but characteristically on calcareous shells of barnacles in the littoral zone. Cosmopolitan.

Hongkong(R. Sant. 1939, p. 63; Aptroot & Seaward, 1999, p. 90).

* **Zwackhiomyces** Grube & Hafellner **兹瓦克菌属**

Nova Hedwigia 51(3 - 4): 305(1990).

* Zwackhiomyces dispersus(J. Lahm ex Körb.) Triebel & Grube 散兹瓦克菌

In Grube & Hafellner, *Nova Hedwigia* 51(3 - 4): 314(1990).

≡*Arthopyrenia dispersa* J. Lahm ex Körb. *Parerga lichenol.* (Breslau) 5: 388(1865).

On *Lecanora subisabellina*(Magnusson 1944).

Neimenggu(Hawksw. & M. Cole, 2003, p. 362).

* Zwackhiomyces sphinctrinoides(Zwackh) Grube & Hafellner 斯芬兹瓦克菌

Nova Hedwigia 51(3 –4):327(1990).

≡*Endococcus sphinctrinoides* Zwackh, *Flora*, Regensburg 47: 88(1864).

≡ *Didymella sphinctrinoides*(Zwackh) Berl. & Voglino, in Saccardo, Syll. fung., Addit. I-IV(Abellini): 89 (1886).

On *Xanthoria elegans*(as *Caloplaca elegans*) (Magnusson 1940), On *Aspicilia maculata*(syn. *Lecanora maculata*) (Magn., 1944).

Neimenggu, Gansu(D. Hawksw. & M. Cole, 2003, p. 362).

(19) Celotheliaceae Lücking, Aptroot & Sipman, **干瘤菌科**

In Aptroot, Lücking, Sipman, Umaña & Chaves, *Biblthca Lichenol.* 97: 12(2008).

Celothelium A. Massal. **干瘤菌属**

Atti Inst. Veneto Sci. lett., ed Arti, Sér. 35: 332(1860)[1859 –1860].

Celothelium aciculiferum(Nyl.) Vain. 小针干瘤菌

Ann. Acad. Sci. fenn., Ser. A 6(7):209(1915).

≡*Melanotheca aciculifera* Nyl., *Expos. Synopt. Pyrenocarp.*: 71(1858).

Misapplied name:

Celothelium dominicanum(Vain.) M. B. Aguirre, *Bull. Br. Mus. nat. Hist.*, *Bot.* 21(2): 139(1991), Aptroot & Seaward, 1999, p. 71.

On trees in open forest. Pantropical.

Hongkong(Aptroot & Seaward, 1999, p. 71 as *C. dominicanum*; Aptroot et al., 2001, p. 324).

[7] Verrucariales Mattick ex D. Hawksw. & O. E. Erikss. **瓶口衣目**

Syst. Ascom. 5(1):183(1986).

(20) Verrucariaceae Eschw. **瓶口衣科**

In Goebel & Kunze, *Syst. Lich.*: 15(1824).

Agonimia Zahlbr. **鳞砖孢属**

Öst. bot. Z. 59: 350(1909).

Agonimia opuntiella(Buschardt & Poelt) Vězda 工鳞砖孢

Lichenes Rariores Exsiccati 33(nos 321 –330):4, no. 330(1997).

≡*Physcia opuntiella* Buschardt & Poelt, in Poelt, *Flora*, Regensburg 169: 24(1980).

On trees, soil, and sandstone bouders, Cosmopolitan.

Taiwan(Aptroot, 2003, p. 156).

Agonimia pacifica(H. Harada) Diederich 太平洋鳞砖孢

In Aptroot, Diederich, Sérusiaux & Sipman, *Biblthca Lichenol.* 64: 12(1997).

≡*Agonimiella pacifica* H. Harada, *Nova Hedwigia* 57(3 –4):503(1993).

On shaded, weathered granite in coastal area. Pantropical. Known from East Asia & Brazil.

Zhejiang(Kalb et al., 2012, p. 35), Taiwan(Aptroot, 2003, p. 156), Hongkong(Aptroot & Seaward, 1999, p. 62).

Agonimia papillata(O. E. Erikss.) Diederich & Aptroot 乳突鳞砖孢

In Aptroot, Diederich, Sérusiaux & Sipman, *Biblthca Lichenol.* 64: 13(1997).

≡*Flakea papillata* O. E. Erikss., *Syst. Ascom.* 11(1):14(1992).

On soil and shale, and on *Diospyros kaki*. So far only known from Hong Kong, where it is common(Aptroot & Sipman 2002).

Taiwaqn(Aptroot, 2003, p. 156).

Agonimia tristicula(Nyl.) Zahlbr. 鳞砖孢

Öst. bot. Z. 5: 351(1909)

≡*Verrucaria tristicula* Nyl., *Flora*, Regensburg 48: 356(1865)

On granite, on *Populus* sp., and *Castonopsis* sp.

Taiwan(Aptroot, 2003, pp. 156 – 157).

Agonimia vouauxii (B. de Lesd.) M. Brand & Diederich 沃氏鳞砖孢

in Sérusiaux, Diederich, Brand & Boom, *Lejeunia*, n. s. 162: 13(1999).

Hongkong(Aptroot et al., 2001, p. 319).

≡*Polyblastia vouauxii* B. de Lesd. *Recherch. Lich. Dunkerque*: 259(1910).

On sheltered concrete. Known from Europe and Asia. Living culture CBS – 101362.

Hongkong (Aptroot & Seaward, 1999, p. 88).

Catapyrenium Flotow **鳞核衣属**

Bot. Ztg. 8: 361(1850).

= *Endopyrenium* Flot., in Schuchart, *Bot. Ztg.* 13: 131(1855).

= *Involucrocarpon* Servít, *Ann. Mus. Civ. Stor. Nat. Genova* 66: 244(1953).

Catapyrenium atrocinereum(H. Magn.) J. C. Wei 黑灰鳞核衣

Enum. Lich. China(Beijing): 52(1991).

≡*Dermatocarpon atrocinereum* H. Magn. *Lich. Centr. Asia*, 1: 26(1940).

Type: Gansu, alt. 3350m Bohlin, no. 56a(S!).

≡*Endopyrenium atrocinereum* (H. Magn.) Golubk. *Novit. Syst. Plant. non Vascul.* 8: 232(1971).

On calcareous rock.

Catapyrenium bohlinii(H. Magn.) J. C. Wei 包氏鳞核衣

Enum. Lich. China, 52(1991).

≡*Dermatocarpon bohlinii* H. Magn. *Lich. Centr. Asia*, 1: 27(1940).

Type: Gansu, Bohlin no. 80(holotype, not seen), nos. 5a, 22c & 33c(paratypes in S).

= *Endopyrenium bohlinii* (H. Magn.) Golubk. *Novit. Syst. Plant. non Vascul.* 8: 233(1971).

On calcareous rock.

Neimenggu, Gansu(H. Magn. 1940, p. 27 &1944, p. 15), Xinjiang(Abdulla Abbas et al., 1993, p. 76; Abdulla Abbas & Wu, 1998, p. 153; Abdulla Abbas et al., 2001, p. 361).

Catapyrenium cinereorufescens(Vain.) J. C. Wei 淡棕鳞核衣

Enum. Lich. China, 52(1991).

≡*Endocarpon cinereorufescens* Vain. in *Acta Hort. Petropolit.*, Vol. X: 561(1888).

On non calcareous rock with a thin, grayish yellow cover of calciferous earth.

Gansu(H. Magn. 1940, p. 28).

Catapyrenium crustosum(H. Magn.) Baibul. 大鳞核衣

Bot. Zh. SSSR 73(3): 352(1988).

Catapyrenium crustosum(H. Magn.) J. C. Wei *Enum. Lich. China*, 52(1991), nom. illegit., Art. 53. 1.

≡*Dermatocarpon crustosum* H. Magn., *Lich. Centr. Asia*, 1: 22(1940), Pl. II, fig. 3.

Type: Gansu, Bohlin no. 73.

≡ *Endopyrenium crustosum*(H. Magn.) Golubk., *Conspect Flori Lishainikov Mongolskoi Narodnoi Respubliki* (Conspectus of lichen flora from Peoples Republic of Mongolia), 15(1981).

Xinjiang(Abdulla Abbas & Wu, 1998, p. 153; Abdulla Abbas et al., 2001, p. 362).

Catapyrenium daedaleum(Kremp.) Stein 波纹鳞核衣

In Cohn, *Krypt. -Fl. Schlesien*(Breslau) 2(2): 312(1879).

≡*Endocarpon daedaleum* Kremp., *Flora, Regensburg* 38: 66(1855).

On soil of roadbank.

Taiwan(Aptroot 2003, p. 158).

Catapyrenium inaequale(H. Magn.) J. C. Wei 不等鳞核衣

Enum. Lich. China, 52(1991).

≡*Dermatocarpon inaequale* H. Magn. *Lich. Centr. Asia*, 1: 23(1940), Pl. VI, fig. 5.

Type: Gansu, Bohlin no. 50(S!).

≡ *Endopyrenium inaequale* (H. Magn.) Golubk. *Conspect Flori Lishainikov Mongolskoi Narodnoi Respubliki* (Conspectus of lichen flora from Peoples Republic of Mongolia), 16(1981).

Xinjiang(Abdulla Abbas et al., 1993, p. 76; Abdulla Abbas & Wu, 1998, pp. 153 – 154; Abdulla Abbas et al., 2001, p. 362), Gansu(H. Magn., 1940, 23).

Catapyrenium kansuense(H. Magn.) J. C. Wei 甘肃鳞核衣

Enum. Lich. China, 52(1991).

≡*Dermatocarpon kansuense* H. Magn. *Lich. Centr. Asia*, 1: 28(1940).

Type: Gansu, Bohlin no. 57 in S!.

≡*Endopyrenium kansuense* (H. Magn.) Golubk., *Conspect Flori Lishainikov Mongolskoi Narodnoi Respubliki* (Conspectus of lichen flora from Peoples Republic of Mongolia), 17(1981).

On calcareous and ochroceous rocks.

Catapyrenium lachneum(Ach.) R. Sant. 绵毛鳞核衣

In D. Hawksw., James & Coppins, *Lichenologist* 12(1): 106(1980).

≡*Lichen lachneus* Ach., *Lich. suec. prodr.* (Linköping): 140(1799)[1798].

Xinjiang(Abdulla Abbas et al., 1993, p. 76; Abdulla Abbas & Wu, 1998, p. 154; Abdulla Abbas et al., 2001, p. 361).

Catapyrenium minutum(H. Magn.) J. C. Wei 微片鳞核衣

Enum. Lich. China, 52(1991).

≡*Dermatocarpon minutum* H. Magn. *Lich. Centr. Asia*, 1: 29(1940).

Type: Gansu, Bohlin, no. 17 in S!.

≡*Endopyrenium minutum* (H. Magn.) Golubk., *Conspect Flori Lishainikov Mongolskoi Narodnoi Respubliki* (Conspectus of lichen flora from Peoples Republic of Mongolia), 17(1981).

Catapyrenium modestum(H. Magn.) J. C. Wei 小鳞核衣

Enum. Lich. China, 52(1991).

≡*Dermatocarpon modestum* H. Magn. *Lich. Centr. Asia*, 2: 15(1944).

Type: Neimenggu, Bohlin no. 148 in S(not seen), no. 156(paratype) in S!.

Upon other lichens on siliceous and calcareous rocks.

Catapyrenium mongolicum(H. Magn.) J. C. Wei 蒙古鳞核衣

Enum. Lich. China, 52 – 53(1991).

≡*Dermatocarpon mongolicum* H. Magn. *Lich. Centr. Asia*, 2: 14(1944).

Type: Neimenggu, Bohlin no. 163 in S!.

On calcareous rock, upon another lichen crust.

≡*Endopyrenium mongolicum* (H. Magn.) Golubk. *Conspect Flori Lishainikov Mongolskoi Narodnoi Respubliki* (Conspectus of lichen flora from Peoples Republic of Mongolia), 17(1981).

Catapyrenium perminutum(H. Magn.) J. C. Wei 大孢鳞核衣

Enum. Lich. China, 53(1991).

≡*Dermatocarpon perminutum* H. Magn. *Lich. Centr. Asia*, 1: 24(1940).

Type: Gansu, Bohlin no. 30 b(S, not seen).

On calcareous rock.

≡*Endopyrenium perminutum*(H. Magn.) Golubk. *Conspect Flori Lishainikov Mongolskoi Narodnoi Respubliki* (Conspectus of lichen flora from Peoples Republic of Mongolia), 16(1981).

Catapyrenium perumbratum(Nyl.) J. C. Wei 土生鳞核衣

Enum. Lich. China, 53(1991).

≡*Verrucaria perumbrata* Nyl. Flora 61: 343(1878).

≡*Dermatocarpon perumbratum*(Nyl.) Zahlbr., *Cat. Lich. Univ.* I: 232(1922), in H. Magn. 25(1940).

On earth.

Xinjiang(Abdulla Abbas & Wu, 1998, p. 154; Abdulla Abbas et al., 2001, p. 362), Gansu(H. Magn. 1940, p. 25).

Catapyrenium subcompactum(Stnr.) J. C. Wei 密鳞核衣

Enum. Lich. China, 53(1991).

≡*Endocarpon subcompactum* Stnr. *Eur Flechtenflora der Sahara*, 1895, p. 391.

≡*Dermatocarpon subcompactum*(Stnr.) Zahlbr., *Cat. Lich. Univ.* I: 236(1922), in H. Magn. 25(1940).

On a crustaceous lichen.

Gansu(H. Magn. 1940, p. 25).

Dermatocarpella H. Harada **小果衣属**

Nat. Hist. Res. 2(2): 137(1993).

Dermatocarpella yunnana H. Harada & Li S. Wang 云南小果衣

Bull. natn. Sci. Mus., Tokyo, N. S. 21(2): 107(1995).

Type: Yunnan, Deqin, on thin soil with mosses on rocky slope beside a road near the bottom of the valley along Lancangiang River, October 1994, Harada & wang ls 14790(holotypus in HKAS, isotypi in CBM, TNS).

On thin soil with mosses on rocky slope.

Dermatocarpon Eschw. **皮果衣属**

Syst. Lich.: 21(1824).

Dermatocarpon intestiniforme(Körb.) Hasse 肠形皮果衣

Bryologist 15: 46(1912).

≡*Endocarpon intestiniforme* Körb., *Parerga lichenol.* (Breslau) 1: 42(1859).

On rock.

Guizhou(Zhang T & J. C. Wei, 2006, p. 6).

Dermatocarpon leptophyllum(Ach.) K. G. W. Lång 薄叶皮果衣

Acta Soc. Fauna Flora fenn. 34(no. 3): 42(1912)[1910 – 1911].

≡*Lichen leptophyllus* Ach., *Lich. suec. prodr.* (Linköping): 141(1799)[1798].

≡*Endocarpon miniatum* var. *leptophyllum*(Ach.) Fr. in Jatta I: 480(1902) & Zahlbr., 17(1930).

On rocks.

Shaanxi(Jatta, 1902, P. 480; Zahlbr., 1930 b, p. 17).

Dermatocarpon luridum(Dill. ex With.) J. R. Laundon 水生皮果衣

Lichenologist 16(3): 222(1984).

≡*Lichen luridus* Dill. ex With., *Bot. arr. veg. Gr. Brit.* (London) 2: 720(1776)

= *Dermatocarpon weberi*(Ach.) W. Mann, *Lich. Bohem. Observ. Dispos.*: 66(1825).

≡*Lichen weberi* Ach., *Lich. suec. prodr.* (Linköping): 142(1799)[1798].

On rocks and stones by water.

Shaanxi, Anhui, Zhejiang(Wu J. L., 1987, p. 30; Xu B. S., 1989, p. 171).

Dermatocarpon miniatum(L.) W. Mann 皮果衣

Lich. Bohem. Observ. Dispos.: 66(1825).

≡*Lichen miniatus* L., Sp. pl. 2: 1149(1753).

≡*Endocarpon miniatum*(L.) P. Gaertn., G. Mey. & Scherb., *Ökonom. -techn. Fl Wetterau*3(2): 230(1802), Zahlbr., 17(1930).

On rocks.

Beijing(Moreau et Moreau, 1951, p. 185), Neimenggu(Chen et al. 1981, p. 134), Gansu(H. Magn. 1940, p.

30), Shaanxi(Baroni, 1894, p. 49; Zahlbr., 1930 b, p. 17; Wu J. L. 1981, p. 162; J. C. Wei et al. 1982, p. 18), Yunnan(Zahlbr., 1930 b, p. 17; Wei X. L. et al. 2007, p. 151), Shanxi(M. Sato, 1981, p. 64), North-eastern China(prov. not indicated: M. Sato, 1961, p. 42), Jiangsu(Wu J. N. & Xiang T. 1981, p. 2), Hubei(Chen, Wu & J. C. Wei, 1989, p. 400), Shandong(Zhao ZT et al., 1998, p. 28; Zhao ZT et al., 2002, p. 5; Hou YN et al., 2008, p. 68; Li y et al., 2008, p. 70), China(prov. not indicated: Asahina & M. Sato in Asahina, 1939, p. 613).

var. miniatum 原变种

xinjiang(Wang XY, 1985, p. 334; Abdulla Abbas & Wu, 1998, p. 155; Abdulla Abbas et al., 2001, p. 363 as ‘Dermatocarpo’ miniatum var. miniztum), Shaanxi(He & Chen 1995, p. 43, no specimen and locality cited).

f. miniatum 原变型

On rock.

Xinjiang(Abbas et al., 1993, p. 77; Abbas et al., 1994, p. 21.), Xizang(J. C. Wei & Y. M. Jiang, 1986, p. 10).

var. complicatum(Lightf.) Th. Fr. 重瓣变种

Nova Acta R. Soc. Scient. upsal., Ser. 3 3: 354(1861) [1860].

≡*Lichen miniatus* var. *complicatus* Lightf., Fl. Scot. 2: 858(1777).

On rocks.

Xinjiang(Wang XL, 1985, p. 334; Abudula, Abbas et al., 1993, p. 77; (Abdulla Abbas & Wu, 1998, p. 155), Abudula, Abbas et al., 2001, p. 363), Shaanxi(He & Chen 1995, p. 43, no specimen and locality cited), Xizang(J. C. Wei & Chen, 1974, p. 175).

var. crispum(A. Massal.) Zahlbr. 卷叶变种

Cat. Lich. Univ. 1: 228(1921).

≡*Endocarpon miniatum* var. *crispum* A. Massal., *Ric. auton. lich. crost.* (Verona) : 376(1852).

On rock.

Xizang(J. C. Wei & Y. M. Jiang, 1986, p. 10).

var. imbricatum(A. Massal.) Dalla Torre & Sarnth. 覆瓦变种

Die Flecht. Tirol, 502(1902).

≡*Endocarpon miniatum* var. *imbricatum* A. Massal., *Sched. critic.* (Veronae) 5: 102(1856).

On rocks.

Xinjiang(Wang XL, 1985, p. 334; Abudula, Abbas et al., 1993, p. 77; Abdulla Abbas & Wu, 1998, p. 155; Abudula, Abbas et al., 2001, p. 363), Xizang(Wei & Chen, 1974, p. 175).

var. papillosum(Anzi) Müll. Arg. 乳突变种

Bull. Trav. Soc. Murithienne du Valais 10: 58(1881).

≡*Endocarpon miniatum* var. *papillosum* Anzi, *Cat. Lich. Sondr.* : 102(1860).

Xinjiang(Wang XL, 1985, p. 334; Abudula, Abbas et al., 1993, p. 77; Abudula, Abbas et al., 2001, p. 363).

Dermatocarpon moulinsii(Mont.) Zahlbr. 长根皮果衣

In Engler & Prantl, *Nat. Pflanzenfam.*, Teil. I(Leipzig) 1 *: 60(1903).

≡*Endocarpon moulinsii* Mont., *Annls Sci. Nat.*, *Bot.*, sér. 2 20: 358(1843).

On rocks.

Xinjiang(Wang XL, 1985, p. 335;), Hebei(M. Sato, 1961, p. 42).

Dermatocarpon muhlenbergii(Ach.) Müll. Arg. [as ‘mühlenbergii’] 黑腹皮果衣

Bot. Jb. 6: 377(1885).

≡*Endocarpon muhlenbergii* Ach. [as ‘mühlenbergii’], *Syn. meth. lich.* (Lund) : 101(1814).

On rocks.

Liaoning(Shenyang [fengtien] : Moreau et Moreau, 1951, p. 185), Shaanxi(Jatta, 1902, p. 480; Zahlbr., 1930 b, p. 17).

Dermatocarpon sinense Räsänen 中华皮果衣

Arch. Soc. Zool. Bot. Fenn. Vanamo, 3: 87(1949). Lamb, Index *Nom. Lich*. 1963, p. 223.

Type: China(prov. not indicated).

Saxicolous.

Dermatocarpon vellereum Zschacke 短绒皮果衣

Rabenh. Krypt. -Fl., Edn 2(Leipzig) 9. 1(1): 638(1934).

On rocks.

Xinjiang(Wang XL, 1985, p. 335; Abudula, Abbas et al., 1993, p. 77; Abdulla Abbas & Wu, 1998, p. 155; Abudula, Abbas et al., 2001, p. 363 as '*Dermatocarpo*' *vellereum*), Shaanxi(He & Chen 1995, p. 43, no specimen and locality were cited), Xizang(J. C. Wei & Chen, 1974, p. 175; J. C. Wei, 1981, p. 81; J. C. Wei & Jiang, 1986, p. 10).

Endocarpon Hedw. **石果衣属**

Descr. micr. -anal. musc. frond. 2(3): 56(1788) [1789].

Endocarpon adscendens(Anzi) Müll. Arg. 翘石果衣

Bull. Murith. Soc. Valais. Sci. Nat. 10: 58(1881).

≡ *Dermatocarpon pusillum* var. *adscendens* Anzi, Cat. Lich. Sondr.: 103(1860).

On rock and also on mortar and dusty tree.

Taiwan(Aptroot 2003, p. 159).

Endocarpon aridum P. M. McCarthy 旱石果衣

Lichenologist 23(1): 28(1991).

Ningxia(Yang & Wei J. C., 2014, p. 1031).

Endocarpon crystallinum J. C. Wei & J. Yang 晶体石果衣

Mycotaxon 106: 446 – 448(2008).

Type: Shanxi, Yang-gao, Xiejiatun, September 23, 2004, Yang J. R Zhang E. R., SX- 28(HMAS-L. holotypus).

Endocarpon deserticola T. Zhang, X. L. Wei & J. C. Wei 荒漠石果衣

In Zhang, Liu, Wang, Wei, Wei Scientific Reports 7: 7193: 6(2017).

Type: China, Ningxia Region, Zhongwei city, Shapotou north experimental zone

On soil crust, January 2007, Zhangtao, Z07090(holotype-HMAS-L-135062).

Endocarpon globosum H. Harada & Li S. Wang 球形石果衣

Lichenologist 28(4): 303(1996).

On non-calareous stones submerged in a stream.

Yunnan(Harada & LS Wang, 1996, p. 303.).

Endocarpon nigromarginatum H. Harada 黑边石果衣

Nova Hedwigia 56(3 – 4): 342(1993).

On rock.

Hebei(Ye Jia et al., pp. 762 – 763).

Endocarpon pallidum Ach. 淡石果衣

Lich. univ.: 301(1810).

≡ *Dermatocarpon pallidum*(Ach.) Mudd, *Man. Brit. Lich.*: 269(1861), Zahlbr., 18(1930).

Shaanxi(Jatta, 1902, p. 480; Zahlbr., 1930 b, p. 18), Ningxia(Yang & Wei J. C., 2014, p. 1031).

Endocarpon pallidulum(Nyl.) Nyl. 白色石果衣

In Hue, *Nouv. Arch. Mus. Hist. Nat.*, Paris, 3 sér. 4: 106(1892).

≡ *Verrucaria pallidula* Nyl., *Flora*, Regensburg 57: 73(1874).

On concrete and sandstone.

Taiwan(Aptroot 2003, p. 159).

Endocarpon pusillum Hedw. 石果衣

Descr. micr. -anal. musc. frond. 2(3): 56(1788) [1789].

On earth. On exposed and sheltered soil, cement and granite. Cosmopolitan.

Ningxia(Yang & Wei J. C., 2014, p. 1031), Sichuan(Zahlbr., 1930 b, p. 17), Jiangsu, Shanghai(Xu B. S. 1989, p. 173, on rocks), Hongkong(Aptroot & Seaward, 1999, p. 74; Aptroot et al., 2001, p. 326).

Endocarpon rogersii P. M. McCarthy 罗杰氏石果衣

Lichenologist 23(1): 43(1991).

Ningxia(Yang & Wei J. C., 2014, p. 1031).

Endocarpon simplicatum(Nyl.) Nyl. 全缘石果衣

In Hue, *Rev. Bot. Bull. Mens.* 6: 104(1888) [1887 – 88].

≡ *Verrucaria simplicata* Nyl., *Flora*, Regensburg 67: 390(1884).

It is referable to var. bisporum P. M. McCarthy, which was so far thought to be restricted to Australia(McCarthy 2001).

On sandstone.

Taiwan(Aptroot 2003, pp. 159 – 160).

Endocarpon sinense H. Magn. 中华石果衣

Lich. Centr. Asia 1: 30(1940). var. sinense 原变种

Type: Gansu, Bohlin no. 41 b in S(!).

Xinjiang(Abdulla Abbas et al., 1994, p. 21; Abdulla Abbas & Wu, 1998, p. 156; Abdulla Abbas et al., 2001, p. 363).

var. ascendens H. Magn. 翘叶变种

Lich. Centr. Asia 1: 31(1940).

Type: Gansu, Bohlin no S(!).

Xinjiang(AbdullaAbbas et al., 1993, p. 77; Abdulla Abbas & Wu, 1998, p. 156; Abdulla Abbas et al., 2001, p. 363), Gansu(H. Magn., 1940, p. 31).

Endocarpon unifoliatum T. Zhang, X. L. wei & J. C. Wei 单鳞石果衣

Scientific Reports 7: 7193: 7(2017).

Type: China, Ningxia, Zhongwei city, Shapotou north experimental zone.

On soil crust, april 8, 2010, T. Zhang, Z10020(holotype-HMAS-L-134711).

Involucropyrenium Breussb **被核衣属**

Annln naturh. Mus. Wien, Ser. B, Bot. Zool. 98(Suppl.): 37(1996).

Involucropyrenium waltheri(Kremp.) Breuss 被核衣

Annln naturh. Mus. Wien, Ser. B, Bot. Zool.

98(Suppl.): 38(1996).

≡ *Verrucaria waltheri* Kremp., *Flora*, Regensburg 38: 69(1855).

On soil and humus.

Qinghai(Sun Z. S. et al. 2008, pp. 655 – 656).

* **Merismatium** Zopf **平裂菌属**

Nova Acta Acad. Caes. Leop. -Carol. German. Nat. Cur. 70: 263(1898).

* Merismatium decolorans(Rehm) Triebel 褪色平裂菌

Biblthca Lichenol. 35: 182(1989).

≡ *Tichothecium decolorans* Rehm, *Ascomyceten Dign.*: no. 490(1872).

In Hafellner & Obermayer, Cryptogamie, Bryol. Lichénol. 1995, 16(3): 185.

On *Arthrorhaphis alpina* var. *jungens*

Xizang(J. Hafellner & W. Obermayer 1995, pp. 185 – 186).

* **Muellerella** Hepp ex Müll. Arg. 米勒氏菌属

Mém. Soc. Phys. Hist. nat. Genève 16(2):419(1862).

* Muellerella lichenicola(Sommerf.) D. Hawksw. 地衣米勒氏菌

Bot. Notiser 132(3):289(1979).

≡*Sphaeria lichenicola* Sommerf., *Suppl. Fl. lapp.* (Oslo):218(1826).

On *Caloplaca* sp.

Yunnan (Hawksw. & M. Cole, 2003, p. 361).

* Muellerella pygmaea(Körb.) D. Hawksw. 矮小米勒氏菌

Bot. Notiser 132(3):289(1979).

≡*Tichothecium pygmaeum* Körb., *Parerga lichenol.* (Breslau) 5:467(1865).

On *Acarospora* sp., *Caloplaca* sp., *Aspicilia asiatica* (≡*Lecanora asiatica*), *Lecidea albida*, *L. tessellata* (as *L. pavimentans*), and *Protoblastenia* sp. (Magnusson 1940).

Gansu(D. Hawksw. & M. Cole, 2003, p. 361).

Neocatapyrenium H. Harada 新鳞核衣属

Nat. Hist. Res. 2(2):129(1993).

Neocatapyrenium cladonioideum(Vain.) H. Harada 石蕊新鳞核衣

Nat. Hist. Res. 2(2):129(1993).

≡*Siphula cladonioidea* Vain., *Bot. Mag.*, Tokyo 35:47(1921).

On soil and on coastal rock. Known from East Asia.

Hongkong(Aptroot & Seaward, 1999, p. 83), China(Breuss, 1998).

Parmentaria Fée 大星衣属

Essai Crypt. Exot. (Paris): xxxix(1825) [1824].

Parmentaria chungii Zahlbr. 钟氏大星衣

In Handel-Mazzetti, *Symb. Sin.* 3:29(1930).

Type: Fujian, Gu shan, Chung no. 602 a.

Notes: This lichen was cited by Zahlbr. (1932 a, p. 133) from western China("China occidentalis"). In fact, the locality of the type Fujian is located in south-eastern China rather than western China.

On bark of trees.

Parmentaria obtecta Zahlbr. 覆盖大星衣

In Handel-Mazzerri, *Symb. Sin.* 3:29 – 30(1930).

Type: Fujian, Chung no. 446 b.

Notes: This lichen was cited also by Zahlbr. (1932 a, p. 133) from western China("China occidentalis"). The type locality Fujian is located in south-eastern China rather than western China.

On bark of trees.

Placidiopsis Beltr. 类盾鳞衣属

Lich. Bassan.: 212(1858).

Placidiopsis hypothallina Aptroot 下叶类盾鳞衣

Fungal Diversity 9:20(2002).

On Trapelia sp., and on shale. New to northern hemisphere.

Taiwan(Aptroot 2003, p. 163).

Placidiopsis poronioides Aptroot 岬类盾鳞衣

In Aptroot & Seaward, *Tropical Biology* 17:87(1999).

Typus: Hong Kong, new territories, Tai Mo Shan, 800m, June 1998, Aptroot 43566(HKU(M)—holotypus, ABL—isotypus); living culture CBS – 101262.

Hongkong(Aptroot et al., 2001, p. 337).

Placidium A. Massal. **盾鳞衣属**

Symmict. Lich: 75(1855).

Placidium pilosellum(Breuss) Breuss 毛状盾鳞衣

Annln naturh. Mus. Wien, Ser. B, *Bot. Zool.* 98(Suppl.): 39(1996).

≡*Catapyrenium pilosellum* Breuss, Stapfia 23: 98(1990).

On soil.

Taiwan(Aptroot 2003, p. 163).

Placidium squamulosum(Ach.) Breuss 盾鳞衣

Annln naturh. Mus. Wien, Ser. B, Bot. Zool. 98(Suppl.): 39(1996).

≡*Endocarpon squamulosum* Ach., *Methodus*, Sectio prior: 126(1803).

On soil.

Taiwan(Aptroot 2003, p. 163).

Placopyrenium Breuss **叶核衣属**

Stud. Geobot. 7(Suppl.): 182(1987).

Placopyrenium trachyticum(Hazsl.) Breuss 粗糙叶核衣

In Nimis & Poelt, *Stud. Geobot.* 7(suppl. 1): 183(1987).

≡*Endopyrenium trachyticum* Hazsl., *Verh. Ver. Nat.*, *Heilk. Pressb.* 5: 7(1861)[1860 –61].

On coastal rock. Known from temperate and subtropical Africa, Europe and Asia.

Hongkong(Aptroot & Seaward, 1999, p. 87).

Pleurotheliopsis Zahlbr. **侧乳头衣属**

Cat. Lich. Univ. 1: 512(1922).

Pleurotheliopsis asahinae Zahlbr. 朝比氏侧乳头衣

In Fedde, *Repertorium*, 31: 204(1933).

Type: Taiwan, Asahina no. 297.

On bark of trees.

Taiwan(Zahlbr., 1940, p. 103; Wang & Lai, 1973, p. 95).

Polyblastia A. Massal. **多囊衣属**

Ric. auton. lich. crost. (Verona): 147(1852).

Polyblastia integrascens(Nyl.) Vain. 全缘多囊衣

Acta Soc. Fauna Flora fenn. 49(2): 104(1921).

≡*Verrucaria intercedens* subsp. *integrascens* Nyl., *Bull. Soc. linn. Normandie*, sér. 4 1: 237(1887).

≡*Polyblastia hyperborea* var. *integrascens* Lynge, *Lich. Nov. Zemlya*: 29(1928).

Xinjiang(Abdulla Abbas et al., 1993, p. 80 & Abdulla Abbas et al., 1994, p. 22; Abdulla Abbas & Wu, 1998, p. 157; Abdulla Abbas et al., 2001, p. 367, as *P.* "*interascens*").

Polyblastia kansuensis H. Magn. 甘肃多囊衣

Lich. Centr. Asia, 1: 19(1940).

Type locality: GANSU, Bohlin no. 40a. in S(not seen).

Xinjiang(Abdulla Abbas et al., 1993, p. 80; Abdulla Abbas & Wu, 1998, p. 157; Abdulla Abbas et al., 2001, p. 367), Gansu(H. Magn., 1940, p. 19).

Polyblastia sinoalpina Zahlbr. 高山多囊衣

In Handel-Mazzetti, *Symb. Sin.* 3: 11(1930) & Cat. Lich. Univ. 8: 52(1932).

Syntype: Yunnan, Zhongdian(as "Chung-tien"), Handel-Mazzetti no. 4677 & Sichuan, between Yenyuen & Kwapi, Handel-Mazzetti no. 2362.

On calcareous rocks.

Polyblastia subvinosa Zahlbr. 淡紫多囊衣

In Handel-Mazzerri, *Symb. Sin.* 3: 12(1930) & *Cat. Lich. Univ.* 8: 12(1932).

Type: Sichuan, Handel-Mazzetti no. 2897.

On calcareous rocks.

Psoroglaena Müll. Arg. **鳞痂衣属**

Flora, Regensburg 74(3): 381(1891).

Psoroglaena cubensis Müll. Arg. 鳞痂衣

Flora, Regensburg 74(3): 381(1891).

On tree branches, and on *Machilus thunbergii*. Pantropical species.

Taiwan(Aptroot 2003, p. 165).

Psoroglaena cubensis var. tereticola O. Eriksson 荚果变种

Syst. Ascom. 11(1): 13(1992).

Type: Yunnan, (O. Eriksson, UME – 29244, 29245) collected by O. Eriksson in 1980, and Tao Guoda in 1983 respectively.

Yunnan(O. Eriksson, 1992, p. 13).

Staurothele Norman **矮疣衣属**

Nytt Mag. Natur. 7: 28(1853)[1852].

Staurothele chlorospora Zahlbr. 绿孢矮疣衣

In Handel-Mazzetti, *Symb. Sin.* 3: 14(1930), & *Cat. Lich. Univ.* 8: 55(1932).

Syntype localities: Yunnan, at about 3825 – 4750 m, 11, 31/VI/1915, Handel-Mazzetti nos. 3559, 6762, and Sichuan, 24/VII & 6/VIII/1915, Handel-Mazzetti nos. 7207, 7460.

On rocks.

Staurothele clopima(Wahlenb.) Th. Fr. 矮疣衣

Nova Acta R. Soc. Scient. upsal., Ser. 3 3: 363(1861)[1860]; Lich. Arct.: 363(1860) in H. Magn. 20(1940).

≡ *Verrucaria clopima* Wahlenb., in Acharius, *K. Vetensk-Acad. Nya Handl.* 30: 152(1809).

On rocks(geol. simple 1246).

Xinjiang(Abdulla Abbas et al., 1993, p. 81; Abdulla Abbas & Wu, 1998, p. 158 as *S.* "*colopima*" Abdulla Abbas et al., 2001, p. 368 as *S.* "*colopima*"), Gansu, Qinghai(H. Magn. 1940, p. 20).

Staurothele desquamescens(Zahlbr.) Aptroot 无鳞矮疣衣

Symb. Bot. Ups. 34: 32 – 33(2004).

≡ *Endocarpon desquamescens* Zahlbr., in Fedde, *Repertorium*, 31: 198(1933).

Type locality: Taiwan, saxicolous, Asahina no. 302.

Taiwan(Zahlbr., 1940, p. 67; Lamb, 1963, p. 234; Wang & Lai, 1973, p. 90).

= *Staurothele yunnana* H. Harada & Li S. Wang *Lichenologist* 28(4): 298(1996).

Type: Yunnan in HKAS(holotype) and SBM, TNS(isotype).

On non-carcareous stones submerged in a stream.

Yunnan(Harada & LS Wang, 1996, p. 298).

Staurothele fauriei B. de Lesd. 法氏矮疣衣

Bull. Soc. bot. Fr. 68: 494(1921).

On siliceous rock.

Taiwan(Zahlbr., 1933 b, p. 198; Wang & Lai, 1973, p. 96).

Staurothele honghensis H. Harada & Li S. Wang 红河矮疣衣

Lichenology 5(1): 15(2006).

Type: Yunnan: Holotype CBM(FL) 16423.

On rocks alongside river.

Staurothele iwatsukii H. Harada 岩氏矮疣衣

Nat. Hist. Res. 2(1): 39(1992).

Yunnan(H. Harada & L. S. Wang, 2006, p. 17) .

Staurothele kwapiensis Zahlbr. 四川矮疣衣

In Handel-Mazzerri, *Symb. Sin*. 3: 13(1930) & Cat. Lich. Univ. 8: 57(1932) .

Type: Sichuan, at about 2125 m, 29/V/1914, Handel-Mazzetti no. 2702.

On non calcareous rock.

Staurothele microlepis Zahlbr. 网纹矮疣衣

In Handel-Mazzerri, *Symb. Sin*. 3: 15(1930) & Cat. Lich. Univ. 8: 57(1932) .

Type: Yunnan, at 2750 m, 27/VI/1914, Handel-Mazzetti no. 3247.

On calciferous rock.

Staurothele muliensis Zahlbr. 木里矮疣衣

In Handel-Mazzerri, *Symb. Sin*. 3: 16(1930) &Cat. Lich. Univ. 8: 57(1932) .

Type: Sichuan, at about 3950 m, 24/VII/1915 , Handel-Mazzetti no. 7204.

On calciferous rock.

Staurothele ochroplaca Zahlbr. 黄面矮疣衣

In Handel-Mazzetti, *Symb. Sin*. 3: 13(1930) , & *Cat. Lich. Univ*. 8: 58(1932) .

Type: Yunnan, Lijiang, at 4100 – 4300 m, 11/VII/1914 & 11/VI/1915, Handel-Mazzetti no. 3558.

On calciferous rock.

Staurothele pallidipora P. M. McCarthy [as ' pallidopora'] 珊矮疣衣

Muelleria 8(3) : 275(1995) .

On sandstone and shale. New to northern hemisphere.

Taiwan(Aptroot 2003, p. 168) .

Staurothele rufa(A. Massal.) Zschacke 淡红矮疣衣

Hedwigia 54: 190(1913) .

≡*Polyblastia rufa* A. Massal. , *Ric. auton. lich. crost.* (Verona) : 147(1852) .

On rock.

Xinjiang(Abdulla Abbas et al. , 1993, p. 81; Abdulla Abbas & Wu, 1998, p. 158; Abdulla Abbas et al. , 2001, p. 368) , Gansu(H. Magn. 1940, p. 20) .

Staurothele sinensis Zahlbr. 中国矮疣衣

In Handel-Mazzetti, *Symb. Sin*. 3: 15(1930) , & *Cat. Lich. Univ*. 8: 58(1932) .

f. obscurata Zahlbr. 暗色变型

In Handel-Mazzetti, *Symb. Sin*. 3: 15(1930) , & *Cat. Lich. Univ*. 8: 58(1932) .

On quartzite.

Type: Yunnan, Handel-Mazzetti no. 10871(pr. p.) .

f. pallescens Zahlbr. 白色变型

In Handel-Mazzetti, *Symb. Sin*. 3: 15(1930) , & *Cat. Lich. Univ*. 8: 58(1932) .

On quartzite.

Type: Yunnan , at 4200 – 4400 m, 27/VIII/1916 , Handel-Mazzetti no. 10871(pr. p.) .

Notes: The author indicated in the protologue only the same locality and specimen(no. 10871) for the new species including both the forms. So, the typification remains to be carried out.

Staurothele yunnana H. Harada & L. S. Wang 云南矮疣衣

Lichenologist 28(4) : 298(1996) .

Type in HKAS(holotype) and SBM, TNS(isotype) .

On non-carcareous stones submerged in a stream.

Yunnan(Harada & LS Wang, 1996, p. 298 & 2006, p. 19) .

Thelidium A. Massal. **乳突衣属**

Framm. Lichenogr. : 15(1855) .

Thelidium luchunense H. Harada & Li S. Wang 绿春乳突衣

Lichenology 5(1): 24(2006).

Type: Yunnan, Luchun, Holotype CBM(FL) 16576

On rocks at side of river: Yunnan

Thelidium minutulum Körb. 小乳突衣

Parerga lichenol. (Breslau) 4: 351(1863).

On brick in sheltered park. Northern temperate. In Europe.

Yunnan(H. Harada & L. S. Wang, 2006, p. 25), Taiwan(Aptroot 2003, pp. 169 – 170, on mortar and sand stone), Hongkong(Aptroot & Seaward, 1999, p. 95; Aptroot & Sipman, 2001, p. 340),

Thelidium pluvium Orange 湿地乳突衣

Lichenologist 23(2): 101(1991).

On wet concret.

Taiwan(Aptroot 2003, p. 170).

Thelidium sinense H. Harada & Li S. Wang 中华乳突衣

Lichenology 5(1): 25(2006).

On rocks periodically submerged in stream: Yunnan.

Type: Yunnan, Holotype CBM(FL) 16675.

Thelidium yunnanum H. Harada & Li S. Wang 云南乳突衣

Lichenology 3(2): 47(2004).

On rocks in small waterfall:

Yunnan: Holotype CBM(FL) 15305.

Verrucaria Schrad. **瓶口衣属**

Spicil. fl. germ. 1: 108(1794).

Sanctioning author: Fr., Nomenclatural comment: Nom. cons., see Art. 14.

Verrucaria aethiobola Wahlenb. 焦瓶口衣

In Acharius, *Methodus*, Suppl.: 17(1803).

On rock instream.

Yunnan(H. Harada &Wang L. S. 2008, pp. 3 – 8).

Verrucaria aethiobolizans Zahlbr. 黑条瓶口衣

In Handel-Mazzetti, *Symb. Sin.* 3: 6(1930) & *Cat. Lich. Univ.* 8: 6(1932).

Type: Yunnan, at 2000 m, 10/III/1914, Handel Mazzetti no 540.

On rocks.

Verrucaria applanatula H. Magn. 平坦瓶口衣

Lich. Centr. Asia 1: 14(1940).

Type: Gansu, at about 2500 m, Bohlin geol. sample no. 1436 in S(!).

On rocks.

Xinjiang(Abdulla Abbas et al., 1994, p. 23; Abdulla Abbas & Wu, 1998, pp. 158 – 159; Abdulla Abbas et al., 2001, p. 368).

Verrucaria aquatilis Mudd 水生瓶口衣

Man. Brit. Lich.: 285(1861).

On rock in stream.

Yunnan(H. Harada &Wang L. S. 2008, pp. 8 – 9).

Verrucaria arboricola Zahlbr. 树皮瓶口衣

In Handel-Mazzetti, *Symb. Sin.* 3: 11(1930) & *Cat. Lich. Univ.* 8: 8(1932).

Type: Hunan, near Changsha, at about 200 m, 10/XII/1917, Handel-Mazzitti no. 11430.

On trees.

Yunnan(Zahlbr. 1930, p. 11), Zhejiang(Xu B. S., 1989, p. 171).

Verrucaria atricolor H. Magn. 黑瓶口衣

Lich. Centr. Asia 2: 11(1944).

Type: Neimenggu, 25/X/1929, Bohlin no. 113 in S(!).

On siliceous rock.

Verrucaria aucklandica Zahlbr. 奥克兰瓶口衣

Denkschr. Kaiserl. Akad. Wiss. Wien, Math. -Naturwiss. Kl. 104: 250(1941).

On littoral volcanic rock and littoral raised coral reef. Australia and New Zealand, new to northern hemisphere.

Taiwan(Aptroot 2003, p. 171).

Verrucaria bella Zahlbr. 泡沫瓶口衣

In Handel-Mazzetti, *Symb. Sin.* 3: 9(1930) & *Cat. Lich. Univ.* 8: 9(1932).

Type: Yunnan, at about 4125 m, 24/VIII/1915, Handel-Mazzetti no. 7769.

On rocks.

Verrucaria caesiocinerata Zahlbr. 青灰瓶口衣

In Handel-Mazzetti, *Symb. Sin.* 3: 9(1930) & *Cat. Lich. Univ.* 8: 9(1932).

Type: Hunan, on sandy rock near Changsha at about 150 m, 8/III/1918, Handel-Mazzetti no. 11508.

Verrucaria caesiopsila Anzi 淡兰瓶口衣

Comm. Soc. crittog. Ital. 2(1): 23(1864).

On calcareous rock.

Neimenggu(H. Magn. 1944, p. 12).

Verrucaria cataleptoides(Nyl.) Nyl. 硬瓶口衣

Lich. Scand. (Helsinki): 272(1861).

≡ *Verrucaria margacea* var. *cataleptoides* Nyl., *Act. Soc. linn. Bordeaux* 21: 428(1856).

var. sinensis Zahlbr. 中华变种

In Handel-Mazzetti, *Symb. Sin.* 3: 7(1930) & *Cat. Lich. Univ.* 8: 10(1932).

Type: Sichuan, at about 2325 m, Handel-Mazzetti no. 2458.

On calcareous rock.

Verrucaria compaginata Zahlbr. 密集瓶口衣

In Handel-Mazzetti, *Symb. Sin.* 3: 10(1930) & *Cat. Lich. Univ.* 8: 11(1932).

Type: Yunnan, at about 1275 m, 14/VIII/1916, Handel-Mazzetti no. 9784.

On calcareous rock & quartzite.

Verrucaria contraria H. Magn. 反向瓶口衣

Lich. Centr. Asia 1: 15(1940).

Type: Gansu, 1931, Bexell at about 3500 m. in S(!).

On black, apparently non-calcareous schist.

Verrucaria cupreocervina Zahlbr. 古铜瓶口衣

In Handel-Mazzetti, *Symb. Sin.* 3: 7(1930) & *Cat. Lich. Univ.* 8: 12(1932).

Type: Yunnan, at about 4200 – 4400 m, 27/VIII/1916, Handel-Mazzetti no. 9998.

On quartzite.

Verrucaria dolosa Hepp 砖筋瓶口衣

Flecht. Europ.: no. 689(1860).

On sheltered concrete and brick. Known from Europe, North Africa, temperate Asia and North America. Living culture CBS 101360.

Taiwan(Aptroot 2003, p. 171), Hongkong(Aptroot & Seaward, 1999, p. 97; Aptroot & Sipman, 2001, p. 341).

Verrucaria evanidula Nyl. 消瓶口衣

Flora, Regensburg 70: 136(1887).

On rock.

China(prov. not indicated: Hue, 1892, p. 126; Zahlbr., 1922, p. 39; incertae sedis).

Verrucaria funckii(Spreng.) Zahlbr. 芬克氏瓶口衣

Cat. Lich. Univ. 1: 41(1921)[1922].

≡*Pyrenula funckii* Spreng., *Krypt. Gerwächse* 32: 5(1826).

On stones in stream.

Yunnan(Harada & Wang L. S. 2008, pp. 9 – 10).

Verrucaria funebris Zahlbr. 蜀瓶口衣

In Handel-Mazzerri, *Symb. Sin.* 3: 8(1930) & *Cat. Lich. Univ.* 8: 15(1932).

Type: Sichuan, at about 2850 m, 12/VI/1914, Handel-Mazzetti, no. 2898.

On calcareous rock.

Verrucaria fusconigrescens Nyl. 黑棕瓶口衣

Bull. Soc. linn. Normandie, sér. 26: 313(1872).

On littoral volcanic rock. New to Asia, cosmopolitan.

Taiwan(Aptroot 2003, p. 171).

Verrucaria glaucina Ach. 灰白瓶口衣

Lich. univ.: 675(1810).

≡*Verrucaria fuscella* var. *glaucina*(Ach.) Schaer., *Enum. critic. lich. europ.* (Bern): 215(1850).

On mortar of walls & rock.

Shaanxi(Jatta, 1902, p. 480; Zahlbr., 1930 b, p. 9), Shandong(Li y et al., 2008, pp. 70 – 71 as *V. glaucina*), Shanghai(Nyl. & Cromb. 1883, p. 66; Zahlbr., 1930 b, p. 9), Jiangxi(Xu B. S., 1989, pp. 171 – 172).

On conglomeratic rock.

Taiwan(Aptroot 2003, p. 171).

Verrucaria gongshanensis H. Harada & Wang L. S. 贡山瓶口衣

Lichenology 7(1): 10 – 13(2008).

Type: Yuannan, Gongshan co., on rock in stream, Sept. 6, 2003, Harada 20459(CBM-FL-15281-holotypus; isotypus in KUN-L).

Verrucaria halizoa Leight. 海石瓶口衣

Lich. -Fl. Great Brit.: 436(1871).

On maritime rocks. Cosmopolitan.

Hongkong(Aptroot & Seaward, 1999, p. 97), Taiwan(Aptroot 2003, p. 171).

Verrucaria handelii Zahlbr. 汉氏瓶口衣

In Handel-Mazzetti, *Symb. Sin.* 3: 9(1930) & *Cat. Lich. Univ.* 8: 17(1932).

Type: Yunnan, at 3100 – 3200 m, 23/IX/1915, Handel-Mazzetti no. 8308

On rock.

Verrucaria hochstetteri Fr. 霍氏瓶口衣

Lich. eur. reform. (Lund): 435(1831).

On raised coral reef.

Taiwan(Aptroot 2003, p. 171).

Verrucaria honghensis H. Harada & Wang L. S. 红河瓶口衣

Lichenology 7(1): 13 – 15(2008).

Type: Yunnan, Hekou co., on riverside rocks along Honghe river. Harada 21265(CBM-FL-16426- holotypus; isotypus in KUN-L).

Verrucaria hydrela Ach. 噬水瓶口衣

Syn. meth. lich. (Lund): 94(1814).

On rock in stream.

Yunnan(Harada & Wang L. S. 2008, pp. 15 – 16).

Verrucaria impressula H. Magn. 凹纹瓶口衣

Lich. Centr. Asia 1: 16(1940).

Type: Gansu, 19/XII/1931, Bohlin no. 54 b in S(!).

On calcareous rock among other lichens.

Verrucaria inaequalis H. Magn. 斜瓶口衣

Lich. Centr. Asia 1: 17(1940).

Type: Qinghai, at about 3600 m, 17/X/1931, Bohlin, geol. sample no. 1001 in S(!).

On rock.

Neimenggu(H. Magn. 1944, p. 12), Xinjiang(Abdulla Abbas et al., 1993, p. 81; Abdulla Abbas & Wu, 1998, p. 159; Abdulla Abbas et al., 2001, p. 368), Qinghai(H. Magn., 1940, p. 17).

Verrucaria kukunorensis H. Magn. 青海瓶口衣

Lich. Centr. Asia 1: 18(1940).

Type: Qinghai, at about 4000 m, 19/IV/1932, Bohlin no. B: b 2 in S(!).

Verrucaria latebrosa Körb. 广瓶口衣

Syst. lich. germ. (Breslau): 349(1855).

On rock in stream.

Yunnan(Harada & Wang L. S. 2008, pp. 16 – 17).

Verrucaria luchunensis H. Harada & Wang L. S. 绿春瓶口衣

Lichenology 7(1): 17 – 19(2008).

Type: Yunnan, Luchun-xian on riverside rock. Jan. 17, 2005, Harada 21401(CBM-FL-16560-holotypus; isotypus in KUN-L).

Verrucaria macrostoma Dufour ex DC. 巨孔瓶口衣

In Lamarck & de Candolle, *Fl. franç.*, Edn 3(Paris) 2: 319(1805).

On sheltered concrete and weathered granite. Known from Europe, Australia, temperate Asia, and North America.

Hongkong(Aptroot & Seaward, 1999, p. 97).

Verrucaria mamillana Ach. 乳头瓶口衣

Methodus, Sectio prior: 120(1803).

var. diminuens(Nyl.) Nyl. 渐狭变种

In Hue, *Nouv. Arch. Mus. Hist. Nat.*, Paris, 3 sér. 4: 117(1892).

≡ *Verrucaria marginata* var. *diminuens* Nyl., *Annls Sci. Nat.*, *Bot.*, sér. 4 20: 248(1863).

China(prov. not indicated: Hue, 1892, p. 117).

Verrucaria margacea(Wahlenb.) Wahlenb. 珍珠瓶口衣

Fl. lapp.: 465(1812).

≡ *Thelotrema margaceum* Wahlenb., in Acharius, *Methodus*, Suppl.: 30(1803).

On wet or even submerged granite along streams. Cosmopolitan

Hongkong(Aptroot & Seaward, 1999, p. 97).

Verrucaria maura Wahlenb. 暗瓶口衣

In Acharius, *Methodus*, Sectio prior: 19(1803).

On siliceous rock.

Shaanxi(Jatta, 1902, p. 480; Zahlbr., 1930 b, p. 7).

Verrucaria microsporoides Nyl. 微孢瓶口衣

Bull. Soc. bot. Fr. 8: 759(1863)[1861].

On littoral raised coral reef. New to Asia, known from Europe and Australasia.

Taiwan(Aptroot 2003, p. 171).

Verrucaria mongolica H. Magn. 蒙古瓶口衣

Lich. Centr. Asia 1: 18(1940).

Type: Gansu, at about 3500 m, 1931, Bexell no. 6 in S(!).

On blackish schist.

Verrucaria muralis Ach. 墙生瓶口衣

Methodus, Sectio prior: 115(1803).

On rocks. On sheltered concrete. Cosmopolitan.

Jiangsu, Zhejiang(Wu J. L., 1987, p. 27; Xu B. S., 1989, p. 172), Taiwan(Aptroot 2003, p. 171), Hongkong (Aptroot & Seaward, 1999, p. 97; Aptroot & Sipman, 2001, p. 341).

= *Verrucaria arisana* Zahlbr. in Fedde, Repertorium 31: 197(1933).

Type: Taiwan, Prov. Chiai, Mt. Arisan, Toroyen, December 1925, Asahina F 293(TNS holotype). Revised by Aptroot, 2004, p. 36.

Taiwan(Zahlbr., 1940, p. 7; Wang & Lai 1973, p. 98).

Verrucaria nigrescens Pers. 黑面瓶口衣

Ann. Bot. (Usteri) 15: 36(1795), var. nigrescens 原变种

On sheltered concrete. Cosmopolitan.

Hongkong(Aptroot & Seaward, 1999, p. 97).

var. devians Nyl. 淡色变种

J. Linn. Soc. London, Bot. 20: 65(1880).

On rocks.

Shanghai(Nyl. & Cromb. 1883, p. 65; Hue, 1892, p. 109; Zahlbr., 1922, p. 75 & 1930 b, p. 7).

Verrucaria nujiangensis H. Harada & Wang L. S. 怒江瓶口衣

Lichenology 7(1): 19 – 22(2008).

Type: Yunnan, Dali, Yunlong co., on stones submerged in calm stream. Sept. 4, 2003, Harada 20431(CBM-FL – 15253-holotypus; isotypus in KUN-L).

Verrucaria ochrostoma Borrer 褐孔瓶口衣

In Schaerer, *Lich. helv. spicil.* 7: 347(1836).

On sheltered or exposed concrete. Known from Europe and Asia.

Hongkong(Aptroot & Seaward, 1999, pp. 97 – 98), Taiwan(Aptroot 2003, pp. 171 – 172).

Verrucaria papillosa Ach. 多乳头瓶口衣

Lich. univ.: 286(1810).

On shale.

Taiwan(Aptroot 2003, p. 172).

Verrucaria parmigera J. Steiner 盾形瓶口衣

Verh. zool. -bot. Ges. Wien 61: 34(1911).

f. circumarata J. Steiner 围型变种

Verh. zool. -bot. Ges. Wien 61: 35(1911).

On calcareous rock.

Yunnan(Zahlbr., 1930 b, p. 6).

var. sinensis Zahlbr. 中华变种

In Handel-Mazzetti, *Symb. Sin.* 3: 6(1930) & *Cat. Lich. Univ.* 8: 26(1932).

Type: Yunnan, Handel-Mazzetti no. 6764.

On calcareous rock.

Verrucaria phaeoderma P. M. McCarthy 黑皮瓶口衣

Lichenologist 27(2) : 116(1995).

On submerged quazite. New to northern hemisphere, so far known from Australia and New Zealand.

Taiwan(Aptroot 2003, p. 172).

Verrucaria pinguicula A. Massal. 台湾瓶口衣

Lotos 6: 80(1856).

On sandstone boulders, mortar, shale, and conglomeratic rock. New to Asia, cosmopolitan.

Taiwan(Aptroot 2003, p. 172), Hongkong(Aptroot & Sipman, 2001, p. 341).

= *Verrucaria toroyensis* Zahlbr. *Repert. Spec. Nov. Regni Veg.* 31: 198(1933).

Type: Taiwan, Prov. Chiai, Mt. Arisan, Toroyen, December 1925, Asahina F 295(TNS holotype). Revised by Aptroot, 2004, p. 37.

Taiwan(Zahlbr., 1940, p. 29; Wang & Lai, 1973, p. 98).

Verrucaria praetermissa(Trevis.) Anzi 岗岩瓶口衣

Comm. Soc. crittog. Ital. 2(1): 24(1864).

≡*Leiophloea praetermissa* Trevis., *Conspect. Verruc.* : 10(1860).

On granite. Cosmopolitan.

Yunnan(Harada & Wang L. S. 2008, pp. 22 – 23), Hongkong(Aptroot & Seaward, 1999, p. 98; Aptroot & Sipman, 2001, p. 341).

Verrucaria prominula Nyl. ex Mudd 隆起瓶口衣

Man. Brit. Lich. : 29(1861).

On littoral raised coral reef. New to Asia, cosmopolitan.

Taiwan(Aptroot 2003, p. 172).

Verrucaria rheitrophila Zschacke [as ' rheithrophila'] 爱河瓶口衣

Verh. bot. Ver. Prov. Brandenb. 64: 108(1922).

On rock by small waterfall.

Yunnan(Harada & Wang L. S. 2008, pp. 23 – 24).

Verrucaria schisticola H. Magn. 片岩瓶口衣

Lich. Centr. Asia 1: 19(1940).

Type: Gansu, at about 3500 m, 1931, Bexell no. 6 in S(!).

On blackish schist.

Verrucaria subtropica Zahlbr. 亚热带瓶口衣

In Handel-Mazzetti, *Symb. Sin.* 3: 8(1930) & *Cat. Lich. Univ.* 8: 32(1932).

Type: Yunnan, at about 1700 m, 17/VIII/1916, Handel-Mazzetti no. 9829.

On rocks.

Verrucaria yunnana Zahlbr. 云南瓶口衣

In Handel-Mazzetti, *Symb. Sin.* 3: 9(1930) & *Cat. Lich. Univ.* 8: 35(1932).

Syntype: Yunnan, near Manhao, at about 200 m, 2/III/1915, Handel-Mazzetti no. 5856 & near Zhongdian, at about 4125 m, 24/VIII/1915, Handel-Mazzetti no. 7771.

On rocks.

4. Lecanoromycetes O. E. Erikss. & Winka **茶渍纲**

Myconet 1(1): 7(1997).

1) Acarosporomycetidae Reeb, Lutzoni & Cl. Roux **微孢衣亚纲**

Mol. Phylogen. Evol. 32(3): 1053(2004).

[8] Acarosporales Reeb, Lutzoni & Cl. Roux **微孢衣目**

in Hibbett et al., *Mycol. Res.* 111(5): 528(2007).

(21) Acarosporaceae Zahlbr. **微孢衣科**

in Engler & Prantl, *Nat. Pflanzenfam.*, Teil. I(Leipzig) 1 * : 150(1906).

Acarospora A. Massal. **微孢衣属**

Ric. auton. lich. crost. (Verona) : 27(1852).

Acarospora admissa(Nyl.) Kullh. 咎微孢衣

Not. Sällsk. Fauna et Fl. Fenn. Förh., Ny Ser. 11: 272(1871) [1870].

≡*Lecanora admissa* Nyl., *Flora*, Regensburg 50: 370(1867).

On rocks.

Shaanxi(Jatta, 1902, p. 476; Zahlbr., 1930 b, p. 140).

Acarospora americana H. Magn. 美洲微孢衣

K. svenska Vetensk-Akad. Handl., ser. 3 7(4) : 198(1929)

On rocks.

Xinjiang(Sahedat et al., 2015, p. 292).

Acarospora bohlinii H. Magn. 包氏微孢衣

Lichens Central Asia 1: 80(1940).

Type: Gansu, Bohlin no. 68 a in S(!).

On hard, non-calciferous stone.

Neimenggu(H. Magn. 1944, p. 39), Xinjiang(Abdulla Abbas et al., 1993, p. 74; Abdulla Abbas & Wu, 1998, p. 36; Abdulla Abbas et al., 2001, p. 360), Qinghai, Gansu(Magn., 1940, p. 69), Xizang (Obermayer W. 2004, p. 484).

Acarospora brevilobata H. Magn. 短片微孢衣

Lichens Central Asia 1: 79(1940).

Type: Gansu, at about 3,000 m, 18/XII/1931, Bohlin no. 48 in S(!).

Growing over other lichens and destroying them.

Xinjiang(Abdulla Abbas et al., 1993, p. 74; Abdulla Abbas & Wu, 1998, p. 36; Abdulla Abbas et al., 2001, p. 360), Gansu(Magn. 1940, p. 79).

Acarospora cinereoalba(Fink) H. Magn. 黑白微孢衣

Monogr. Acar. 205(1929),

≡*Acarospora cervina* var. *cinereoalba* Fink, *Contr. U. S. natnl. Herb.* 14: 171(1910).

On siliceous rock.

Neimenggu(H. Magn. 1944, p. 34, it is wrong as' *cinereoalbida*' & '1939').

Acarospora discurrens Zahlbr. 枝瓣微孢衣

In Handel-Mazzetti, *Symb. Sin.* 3: 140(1930) & *Cat. Lich. Univ.* 8: 503(1932).

Syntype: Sichuan, Yunnan, Handel-Mazzetti nos. 5188, 4297.

On rocks.

Sichuan, Yunnan(Zahlbr., 1930, p. 140), China(Zahlbr., 1932a, p503).

Acarospora fuscata(Nyl.) Th. Fr. 暗黑微孢衣

Lich. Scand. (Upsaliae) 1(1) : 215(1871).

≡*Lecanora fuscata* Nyl., *Bull. Soc. bot. Fr.* 10: 263(1863).

On granite in mountain area at 200m. Cosmopolitan.

Taiwan(Aptroot, and Sparrius, 2003, p. 2), Hongkong(Aptroot & Seaward, 1999, p. 62).

Acarospora geophila H. Magn. 地生微孢衣

Lich. Centr. Asia, 2: 38(1944).

Type: Neimenggu, 25/X/1929, Bohlin no. 103 a in S.

Acarospora glaucocarpa(Ach.) Körb. 苍果微孢衣

Parerga lichenol. (Breslau) 1: 57(1859).

≡*Parmelia glaucocarpa* Ach., Methodus, Sectio post. (Stockholmiæ) : 182(1803).

On rocks.

Xinjiang(Li Zhi-cheng et al., 2007, pp. 190 – 191).

Acarospora glypholecioides H. Magn. 聚盘微孢衣

Lich. Centr. Asia, 1: 76(1940).

Type: Gansu, at about 3,000 m, 13/XII/1931, Bohlin no. 42 a in S.

On conglomerate, calciferous stone.

Xinjiang(Abdulla Abbas & Wu, 1998, p. 37; Abdulla Abbas et al., 2001, p. 360).

Acarospora gobiensis H. Magn. 戈壁微孢衣

K. svenska Vetensk-Akad. Handl., ser. 37(4): 98(1929).

On calciferous stone.

Neimenggu(H. Magn. 1940, p. 73 & 1944, p. 33; Asahina, 1959 a, p. 65), Xinjiang(Wang XY, 1985, p. 345; Abdulla Abbas et al., 1993, p. 75; Abdulla Abbas & Wu, 1998, p. 37; Abdulla Abbas et al., 2001, p. 360), Qinghai, Gansu(H. Magn. 1940, p. 73), Shanxi, Hebei(Asahina, 1959 a, p. 65).

Acarospora heufleriana Körb. 凹盘微孢衣

Parerga lichenol. (Breslau) 1: 57(1859).

On siliceous rock.

Neimenggu(H. Magn. 1944, p. 33).

Acarospora invadens H. Magn. 侵占微孢衣

Lich. Centr. Asia, 1: 74(1940), Pl. VIII, fig. 6.

Type: Gansu, at about 4,200 m, 18/VIII/1932, Bohlin no. 88d in S(!).

On hard, non-calciferous stone.

Neimenggu(H. Magn. 1944, p. 37), Xinjiang(Abdulla Abbas et al., 1993, p. 75; Abdulla Abbas et al., 1994, p. 19; Abdulla Abbas & Wu, 1998, p. 37; Abdulla Abbas et al., 2001, p. 360), Qinghai, Gansu(Magn., 1940, p. 74).

Acarospora jenisejensis H. Magn. 亚球微孢衣

Svensk bot. Tidskr. 30: 248(1936).

On rocks.

Neimenggu(H. Magn. 1940, p. 78 & 1944, p. 38), Xinjiang(Abdulla Abbas et al., 1993, p. 75; Abdulla Abbas & Wu, 1998, p. 38; Abdulla Abbas et al., 2001, p. 360), Gansu(H. Magn. 1940, p. 78).

Acarospora macrospora(Hepp) A. Massal. ex Bagl. subsp. macrospora 小鳞微孢衣

Mém. R. Accad. Sci. Torino, Ser. 2 17: 397(1857)

≡*Myriospora macrospora* Hepp, *Flecht. Europ.*: no. 58(1853), nom. inval., Art. 38. 1(a) (Melbourne).

≡*Myriospora macrospora* Hepp ex Uloth, *Flora*, Regensburg 44: 617(1861).

=*Acarospora squamulosa*(Schrad.) Trevis., *Revta Period. Lav. Imp. Reale Acad.*, *Padova* 1(3): 263(1852) [1851 – 52]; Zahlbr. in Handel-Mazzetti, *Symb. Sin.* 3: 140(1930).

≡*Lichen squamulosus* Schrad., *Ann. Bot.* (Usteri) 22: 84(1797).

On calciferous breccia.

Xinjiang(Abdulla Abbas et al., 2001, p. 360 as *A.* '*microspora*' Bagl.), Yunnan(Zahlbr., 1930b, p. 140).

Acarospora mongolica H. Magn. 蒙古微孢衣

Lich. Centr. Asia, 2: 34(1944), Pl. IV, fig. 3.

Type: Neimenggu, 13/XI/1929, Bohlon no. 165 in S.

Acarospora nodulosa(Dufour) Hue 节微孢衣

Nouv. Arch. Mus. Hist. Nat., Paris, 5 sér. 1: 160(1909).

≡*Parmelia nodulosa* Dufour, in Fries, *Lich. eur. reform.* (Lund): 185(1831).

var. reagens(Zahlbr.) Clauzade & Cl. Roux 反应变种

Bull. Mus. Hist. Nat. Marseille 41: 61(1981).

≡*Acarospora reagens* Zahlbr., *Beih. Botan. Centralbl.* 13: 162(1902). Xizang(Obermayer, 2004, p. 484).

Notes: *Glypholecia tibetanica* H. Magn. might belong to this species(Obermayer, 2004, p. 484).

Acarospora oligospora(Nyl.) Arnold 寡微孢衣

Flora, Regensburg 53: 469(1870).

≡*Lecanora oligospora* Nyl., *Bot. Notiser:* 162(1853).

On granite in coastal and mountain area, also along streams. Northern temperate.

Taiwan(Aptroot, and Sparrius, 2003, p. 2), Hongkong(Aptroot & Seaward, 1999, p. 62; Aptroot et al., 2001, p. 318).

Acarospora pelioscypha(Wahlenb.) Th. Fr. 蓝杯微孢衣

Nova Acta R. Soc. Scient. upsal., Ser. 3 3: 189(1861)[1860].

≡*Parmelia pelioscypha* Wahlenb., in Acharius, Methodus, Suppl. (Stockholmiæ): 40(1803).

On rocks.

Xinjiang(Li Zhi-cheng et al., 2007, p. 191).

Acarospora pulvinata H. Magn. 垫微孢衣

Lich. Centr. Asia, 1: 77(1940).

Type: Gansu, at about 2, 600 m, 5/VII/1931, Bohlin no. 32 b in S(!).

Neimenggu(H. Magn. 1944, p. 38), Xinjiang(Abdulla Abbas et al., 1993, p. 75; Abdulla Abbas & Wu, 1998, p. 38; Abdulla Abbas et al., 2001, p. 360), Gansu(Magn., 1940, p. 77).

Acarospora sarcogynoides H. Magn. 网盘微孢衣

Lich. Centr. Asia, 2: 35(1944).

Type: Neimenggu, 25/X/1929, Bohlin no. 105 in S(!).

On siliceous rock among *Xanthoria elegans*.

Acarospora schleicheri(Ach.) A. Massal. 微孢衣

Ric. auton. lich. crost. (Verona): 27(1852).

≡*Urceolaria schleicheri* Ach., *Lich. univ.*: 332(1810).

On sandy rock.

Xinjiang(Wang XY, 1985, p. 345; Abdulla Abbas et al., 1993, p. 75; Abdulla Abbas & Wu, 1998, p. 38; Abdulla Abbas et al., 2001, p. 360), Xizang, Sichuan(Obermayer, 2004, p. 484), Yunnan(Zahlbr., 1930 b, p. 140).

Acarospora smaragdula(Wahlenb.) A. Massal. 翡翠微孢衣

Ric. auton. lich. crost. (Verona): 29(1852).

≡*Endocarpon smaragdulum* Wahlenb., in Acharius, *Methodus,* Suppl.: 29(1803).

On rock. Cosmopolitan.

Xinjiang(Sahedat et al., 2015, p. 292), Xizang(Paulson, 1925, p. 192), Taiwan(Aptroot, and Sparrius, 2003, p. 3), Hongkong(Aptroot & Seaward, 1999, p. 62; Aptroot et al., 2001, p. 319).

Acarospora sparsa H. Magn. 散生微孢衣

Annals Cryptog. Exot. 6: 44(1933).

On a hard, horizontal, calciferous rock.

Neimenggu(H. Magn. 1944, p. 38).

Acarospora strigata(Nyl.) Jatta 糙伏毛微孢衣

Malpighia 20: 10(1906).

≡*Lecanora strigata* Nyl., *Annls Sci. Nat., Bot.*, sér. 43: 155(1855).

On calcareous and siliceous rocks.

Neimenggu(H. Magn. 1944, p. 34), Gansu(H. Magn. 1940, p. 74).

Acarospora superans H. Magn. 缝裂微孢衣

Lich. Centr. Asia, 1: 75(1940), Pl. X, fig. 4.

Type: Gansu, at about 2,575 m, 5/VII/1931, Bohlin no. 29a in S.

Xinjiang(Abdulla Abbas & Wu, 1998, p. 38; Abdulla Abbas et al., 2001, p. 360), Qinghai, Gansu(H. Magn. 1940, pp. 75 – 76).

Acarospora suprasedens H. Magn. 类缝裂微孢衣

Lich. Centr. Asia, 2: 36(1944).

Type: Neimenggu, 19/XI/1929, Bohlin no. 187 in S(!).

Acarospora tominiana H. Magn. 托敏氏微孢衣

Monogr. Acarosp.: 216(1929).

Upon a sterile crustaceous lichen, but the substratum of the crustaceous lichen not given.

Xinjiang(Abdulla Abbas et al., 1993, p. 75; Abdulla Abbas & Wu, 1998, p. 39; Abdulla Abbas et al., 2001, p. 360), Gansu(H. Magn. 1940, p. 76), Jiangsu, Zhejiang(Xu B. S., 1989, p. 243).

Acarospora tuberculifera H. Magn. 瘤微孢衣

Lich. Centr. Asia, 1: 73(1940).

Type: Gansu, at about 3,000 m, Bohlin no. 76 in S(!).

On a non-calciferous stone with coarse granules of quartz on the surface.

Acarospora umbrina H. Magn. 茶褐微孢衣

Lich. Centr. Asia, 1: 81(1940), Pl. VIII, fig. 5.

Type: Gansu, at about 4,000 m, 5/VII/1931, Bohlin no. 34 in S(!).

On calcareous rock.

Neimenggu(H. Magn. 1944, p. 39), Gansu(Magn., 1940, p. 81), Xinjiang(Abdulla Abbas et al., 2001, p. 360).

Acarospora veronensis A. Massal. 维罗纳微孢衣

Ric. auton. lich. crost. (Verona): 29, tab. 48(1852)

On rocks.

Xinjiang(Li Zhi-cheng et al., 2007, pp. 191 – 192).

Acarospora verruculosa H. Magn. 疣微孢衣

Lich. Centr. Asia, 2: 37(1944), Pl. V, fig. 2.

Type: Neimenggu, 19/XI/1929, Bohlin, geol. sample no. 182 in S(!).

On calcareous rock with *Xanthoria elegans*.

Neimenggu(Magn., 1944, p. 37), Xinjiang(Abdulla Abbas et al., 1993, p. 75; Abdulla Abbas & Wu, 1998, p. 39; Abdulla Abbas et al., 2001, p. 360).

Acarospora xanthophana(Nyl.) Jatta 黄微孢衣

Malpighia 20: 10(1906).

≡*Lecanora xanthophana* Nyl., *Annls Sci. Nat., Bot.*, sér. 4 15: 379(1861).

On conglomerate.

Neimenggu(H. Magn. 1944, p. 33).

Glypholecia Nyl. **聚盘衣属**

Annls Sci. Nat., Bot., sér. 3 20: 317(1853).

Glypholecia scabra(Pers.) Müll. Arg. 糙聚盘衣

Hedwigia 31: 156(1892).

≡*Urceolaria scabra* Pers., *Ann. Wetter. Gesellsch. Ges. Naturk.* 2: 10(1811) [1810].

Gansu(H. Magn. 1940, p. 82), Xinjiang(Wang XY, 1985, p. 345; Abdula Abbas et al., 1993, p. 78; Abdulla Abbas et al., 1994, p. 21; Abdulla Abbas & Wu, 1998, p. 39), Ningxia(Liu M& Wei JC, 2013, p. 43), Xizang

(JC Wei & YM Jiang, 1986, p. 104).

Glypholecia tibetanica H. Magn. 藏聚盘衣

in Fedde, *Repertorium*, 31: 24(1932).

Type: Xizang, collected by Bosshard in Aug. of 1927, alt. 5500 m.

On calcareous rock.

Notes: This species might belong to *Acarospora nodulosa* var. *reagens*(Zahlbr.) Clauzade & Cl. Roux(Obermayer, 2004, p. 484). It remains to be checked in the future.

Pleopsidium Körb. **金卵石衣属**

Syst. lich. germ. (Breslau): 113(1855).

Pleopsidium chlorophanum(Wahlenb.) Zopf 黄金卵石衣

Ann. Chemie 284: 117(1895).

≡*Parmelia chlorophana* Wahlenb., in Acharius, *Methodus*, Suppl.: 44(1803).

≡*Acarospora chlorophana*(Wahlenb.) A. Massal., *Ric. auton. lich. crost.* (Verona): 27(1852).

=*Acarospora flava*(Bell.) Trev. (Zahl. br., Cat. Lich., V, 104) Zahlbr., *Lichenes* in Handel-Mazzetti, *Symb. Sin.* 3: 140(1930).

≡*Acarospora flava* Trevis., *Revta Period. Lav. Imp. Reale Acad.*, *Padova* 1(3): 262(1852) [1851 – 52]

≡*Pleopsidium flavum*(Trevis.) Körb., *Syst. lich. germ.* (Breslau): 114, tab. IV, fig. 4(1855)

On rocks.

neimenggu(M. Sato, 1940 a, p. 45), Xinjiang, Gansu(Futterer, 1911, p. 7), Xizang(Paulson, 1925, p. 192), Yunnan(Paulson, 1928, p. 317; Zahlbr., 1930 b, p. 140; Wei X. L. et al. 2007, p. 156).

Polysporina Vězda **煤尘衣属**

Folia geobot. phytotax. 13: 399(1978).

Polysporina cyclocarpa(Anzi) Vězda 圆果煤尘衣

Folia geobot. phytotax. 13(4): 399(1978).

≡*Lithographa cyclocarpa* Anzi, *Cat. Lich. Sondr.*: 97(1860).

On sheltered soft, calcareous rock. Known from Europe, North Africa and East Asia.

Cosmopolitan.

Hongkong(Aptroot & Seaward, 1999, p. 88).

Polysporina simplex(Taylor) Vězda 煤尘衣

Folia geobot. phytotax. 13: 399(1978), cited as "*Polysporina simplex*(Davies) Vězda" by Aptroot and Sparrius, 2003, p. 35.

On shale.

Taiwan(Aptroot and Sparrius, 2003, p. 35).

Sarcogyne Flot. **网盘衣属**

Bot. Ztg. 8: 381(1850).

Sarcogyne clavus(DC.) Kremp. 棍棒网盘衣

Denkschr. Kgl. Bayer. Bot. Ges., Abt. 24: 212(1861).

≡*Patellaria clavus* DC., in Lamarck & de Candolle, *Fl. franç.*, Edn 3(Paris) 2: 348(1805).

On the ground.

Sichuan(W. Obermayer, 2004, p. 507).

Sarcogyne gyrocarpa H. Magn. 脑纹网盘衣

In Meddel. *Göteborgs Bot. Trädgärd*, XII: 97 – 98(1938).

On pegmatite.

Neimenggu(H. Magn. 1940, p. 70 & 1944, p. 30), Xinjiang(Abdulla Abbas et al., 1993, p. 81; Abdulla Abbas & Wu, 1998, p. 40; Abdulla Abbas et al., 2001, p. 368), Gansu(H. Magn. 1940, p. 70).

Sarcogyne parviascifera Jiao-Hong Wang & J. C. Wei 小囊网盘衣

Mycosystema 35(11): 1344 – 1347(2016).

Type: China, Shaanxi province, Ansai county, Zhifanggou, 36°31′N, 109°12′E, on rock, 19 July 2006, Jun Yang AS019. (holotype: HMAS-L-300171).

Sarcogyne picea H. Magn. 黑网盘衣

Lich. Centr. Asia 2: 30(1944).

Type: Neimenggu, 10/XI/1920, Bohlin geol. sample no. 175 in S(!).

On hard rock.

Sarcogyne regularis Körb. 对称网盘衣

Syst. lich. germ. (Breslau): 267(1855).

= *Biatorella pruinosa* f. *nuda*(Kremp.) H. Olivier, Expo. Syst. Descr. Lich. Ouest Fr. 2: 59(1900), Zahlbr., 1933c, pp. 49 – 50.

≡ *Lecidea pruinosa* f. *nuda* Kremp., in *Rabenhorst, Flecht. Europ.* 1: no. 335(1857).

On Ca- influenced schist near ground.

Xizang(W. Obermayer, 2004, pp. 507 – 508), Taiwan(Zahlbr., 1933c, p. 49; Wang & Lai, 1973, p. 88)].

Sarcogyne saphyniana A. Abbas, Nurtai & K. Knudsen 萨氏网盘衣

In L. Nurtai, K. Knudsen & A. Abbas *Mycotaxon* 131: 136(2016).

Type: China. Xinjiang, Tianshan Grand Canyon, 43°18.89′N 87°19.51′E, alt. 2257m, sandstone, 10 Jul 2014, A. Abbas, B. Memet, L. Nurtai 20140414(Holotype, XJUNALH).

On sandstone.

Sarcogyne sinensis H. Magn. 中国网盘衣

Lich. Centr. Asia 1: 70(1940). f. sinensis

Type: Gansu, at about 3750 m, 3/IV/1932, Bohlin no. 85 a in S(!).

On siliceous or slightly calciferous stone.

Sarcogyne sinensis H. Magn. f. complicata H. Magn. 折叠变型

Lich. Centr. Asia 1: 71(1940).

Type: Gansu, at about 3000 m, 18/XII/1931, Bohlin no. 46 in S(!).

On rock.

Jiangsu(Xu B. S., 1989, p. 244).

Sarcogyne solitaria H. Magn. 单生网盘衣

Lich. Centr. Asia 2: 30(1944).

Type: Neimenggu, 20/XI/1929, Bohlin, geol. sample no. 157 in S(not seen).

On conglomerate(pegmatite).

Neimenggu(1944, p. 30), Xinjiang(Abdulla Abbas et al., 1993, p. 81; Abdulla Abbas & Wu, 1998, p. 40 as *S.* "*soliteria*"; Abdulla Abbas et al., 2001, p. 368).

2) Ostropomycetidae Reeb, Lutzoni & Cl. Roux **厚顶盘亚纲**

Mol. Phylogen. Evol. 32: 1055(2004).

[9] Agyriales Clem. & Shear **无座盘菌目**

Gen. fung., Edn 2(Minneapolis): 141(1931).

(22) Anamylopsoraceae Lumbsch & Lunke 柄盘衣科

In Lumbsch, Lunke, Feige & Huneck, *Pl. Syst. Evol.* 198(3 – 4): 285(1995).

Anamylopsora Timdal **柄盘衣属**

Mycotaxon 42: 250, 1991.

Anamylopsora pulcherrima(Vain.) Timdal 柄盘衣

Mycotaxon 42: 250(1991).

≡ *Lecidea pulcherrima* Vain., *Acta Horti Petropolit.* 10: 561(1888).

Xizang(Obermayer, 2004, p. 487) .

= *Lecidea hedinii* H. Magn. , Lich. Centr. Asia, 1: 56(1940) .

Type: Gansu, Bohlin no. 72 in S(not seen) .

On calciferous stone.

Neimenggu(H. Magn. 1944, p. 25) , GANSU(H. Magn. 1940, p. 56; Schneider, 1979, p. 194) .

[10] Baeomycetales Lumbsch, Huhndorf & Lutzoni 羊角衣目

In Hibbett et al. , *Mycol. Res*. 111(5) : 529(2007) .

(23) Baeomycetaceae Dumort. [as ' Baeomyceae'] 羊角衣科

Anal. fam. pl. (Tournay) : 71(1829)

Baeomyces Pers. 羊角衣属

Ann. Bot. (Usteri) 7: 19(1794) .

Baeomyces botryophorus Zahlbr. 聚果羊角衣

In Fedde, *Repertorium* 33: 45(1933) & *Cat. Lich. Univ*. 10: 379(1940) .

Type: Taiwan(M. Sato, 1941, p. 21; Asahina, 1943 g, p. 309; Wang & Lai, 1973, p. 88) .

Baeomyces brevis Zahlbr. 短羊角衣

Flechten der Insel Formosa in Feddes, *Lichenes* in Feddes, *Repertorium sp. nov*. 33: 45(1933) & *Cat. Lich. Univ*. 10: 379(1940) .

Type: Taiwan, Asahina no. 210.

Taiwan(M. sato, 1941a, p. 14; Asahina, 1943 g, p. 309; M. Lamb, 1963, p. 69; Wang & Lai, 1973, p. 88; Ikoma, 1983, p. 31) .

Baeomyces fungoides(Sw.) Ach. 羊角衣

Methodus, Sectio post. : 320(1803) .

≡ *Lichen fungoides* Sw. , Prodr. : 146(1788) .

= *Baeomyces roseus* var. *subcomplicatus* Sato. in T. Nakai & M. Honda's Nova Flora Japonica1941: 23.

Type from Taiwan.

= *Baeomyces roseus* Hook. , *Handbook New Zealand Flora*, 1867, p. 559(non Pers.) .

Yunnan(Hue, 1898, p. 237; Zahlbr. , 1930 b, p. 127; Sandst. 1932, p. 66& map 52) , Taiwan(Zahlbr. , 1933c, p. 45; M. Sato, 1941a, p. 15; Asahina, 1943 g, p. 311; M. Lamb, 1963, p. 70; Wang & Lai, 1973, p. 88 & 1976, p. 226; Ikoma, 1983, p. 31) .

Baeomyces pachypus Nyl. 无柄羊角衣

Syn. meth. lich. (Parisiis) 1: 182(1860) .

Terricolous.

Xizang(Sandst. 1932, map 53) , Sichuan(Zahlbr. , 1930 b, p. 127) , Yunnan(Hue, 1887a, p. 16 & 1889, p. 159 & 1898, p. 237; Zahlbr. , 1930 b, p. 127; Sandst. 1932, p. 66, map 53) .

Baeomyces placophyllus Ach. 叶羊角衣

Methodus, Sectio post. : 323(1803) .

Xizang(Sandst. 1932, p. 66 & map 52) , Yunnan(Wei X. L. et al. , 2007, pp. 147 – 148) , Zhejiang(Xu B. S. , 1989, p. 242) , Taiwan(M. Sato, 1941a, p. 16 & 1954, p. 122; Asahina, 1943 g, p. 309; Wang & Lai, 1973, p. 88) .

Baeomyces rufus(Huds.) Rebent. 淡红羊角衣

Prodr. fl. neomarch. (Berolini) : 315(1804) .

≡ *Lichen rufus* Huds. , Fl. Angl. : 443(1762) .

On exposed rock in forest area. Cosmopolitan.

Jilin(Chen et al. 1981, p. 128; Hertel & Zhao, 1982, p. 143) , HUBEI(Chen, Wu & JC Wei, 1989, p. 472) , Taiwan(Aptroot, Sparrius & Lai 2002, pp. 281 – 282) , Hongkong(Aptroot & Seaward, 1999, p. 67; Aptroot et al. , 2001, p. 321) .

Baeomyces sanguineus Asahina 血红羊角衣

Journ. Jap. Bot. 19(11):304(1943). var. sanguineus 原变种

Type locality: Taiwan

Taiwan(M. Lamb, 1963, p. 70; Wang & Lai, 1976, p. 226; Ikoma, 1983, p. 31).

var. ablutus Asahina 湿润变种

Journ. Jap. Bot. 19(11):304(1943).

Zhejiang(Xu B. S., 1989, p. 242).

Phyllobaeis Kalb & Gierl **鳞角衣属**

In Gierl & Kalb, *Herzogia* 9(3 –4):610(1993).

Phyllobaeis crustacea S. N. Cao & J. C. Wei 壳型鳞角衣

Mycotaxon 126:34-35(2013).

Type: China, Hainan: Changjiang County, Mt. Bawangling, 19°16′N 109°03′E, alt. 300 m,

On rock, 25 Nov. 2010, S. N. Cao CSN047(Holotype, HMAS-L 118095, GenBank KC414617; Isotype: HMAS-L 127984).

(24) Trapeliaceae M. Choisy ex Hertel 褐边衣科

Vortr. GesGeb. Bot., n. f. 4:181(1970).

Placopsis(Nyl.) Linds. **瘿茶渍属**

Trans. Linn. Soc. London 25:536(1866).

≡*Squamaria* subgen. *Placopsis* Nyl., *Annls Sci. Nat.*, *Bot.*, sér. 4 15:376(1861)

Placopsis asahinae I. M. Lamb 朝比氏瘿茶渍

Lilloa, 13:239(1947).

Type: Taiwan, Mt. Ali shan, Asahina 1925, in W(not seen).

On basaltic rock.

≡*Lecanora asahinae* "(Zahlbr.)" Sato in I*ndex Plantarum Nipponiarum*, IV, *Lichenes*, 1943, p. 116(nom. nud.). Lamb, *Index Nom. Lich.* 1963, p. 577.

Lecanora gelida sensu Zahlbr. in Fedde, *Repertorium*, 33:53(1933), non(L.) Nyl.

On vulcanic rock.

Taiwan(Zahlbr., 1933 c, p. 53; Asahina & M. Sato in Asahina, 1939, p. 711; Lamb, 1963, p. 577; Wang & Lai, 1973, p. 91).

Placopsis cribellans(Nyl.) Räsänen 突瘿茶渍

J. Jap. Bot. 16(2):90(1940). f. cribellans 原变型

≡*Lecanora cribellans* Nyl., *Lich. Japon.*:42(1890).

On large pebbles.

Jilin(Hertel & Zhao, 1982, p. 148).

f. tuberculifera Lamb 节瘤变型

Lilloa, 13:228(1947).

Type: Taiwan, Mt. Ali shan, Asahina no. 143(paratype cited on p. 226).

Placopsis gelida(L.) Linds. 冷瘿茶渍

Trans. Linn. Soc. London 25:536(1866).

≡*Lichen gelidus* L., *Mant. Pl.* 1:133(1767).

On bark.

Yunnan(Gao TL et al., 2012, pp. 464 –465, as *P. gelida*(L.) Linds.).

Placynthiella Elenkin **沥渍衣属**

Izv. Imp. St. -Peterburgsk. Bot. Sada 9:17(1909).

Placynthiella icmalea(Ach.) Coppins & P. James 流沥渍衣

Lichenologist 16(3):244(1984).

≡*Lecidea icmalea* Ach., K. Vetensk-Acad. *Nya Handl*. 29: 267(1808).

On *Eucalyptus* sp., *Picea morrisonicola* wood, and *Populus* sp.

Taiwan(Aptroot and Sparrius, 2003, p. 35), Hongkong(Aptroot et al., 2001, p. 337).

Placynthiella oligotropha(J. R. Laundon) Coppins & P. James 寡沥渍衣

Lichenologist 16(3): 245(1984).

≡*Lecidea oligotropha* J. R. Laundon, *Lichenologist* 1: 164(1960).

On exposed granite rock. Northern temperate.

Xinjiang(Abdulla Abbas et al., 2015, p. 424), Hongkong(Aptroot & Seaward, 1999, p. 87 – 88; Aptroot et al., 2001, p. 337).

Placynthiella uliginosa(Schrad.) Coppins & P. James 沼泽沥渍衣

Lichenologist 16(3): 245(1984).

≡*Lichen uliginosus* Schrad., *Spicil. fl. germ*. 1: 88(1794).

On *Abies kawakamii* wood.

Taiwan(Aptroot and Sparrius, 2003, p. 35).

Rimularia Nyl. **缝裂衣属**

Flora, Regensburg 51: 346(1868).

= *Mosigia* Fr., *Summa veg. Scand.*, *Sectio Prior*(Stockholm): 119(1845), Nom. illegit., Art. 53. 1.

Rimularia badioatra(Kremp.) Hertel & Rambold 棕黑缝裂衣

In Jahns, *Biblthca Lichenol*. 38: 164(1990).

≡*Aspicilia badioatra* Kremp., *Denkschr. Kgl. Bayer. Bot. Ges.*, Abt. 2 4: 285(1861).

= *Lecanora bockii* T. Rödig, in Fries, *Syst. orb. veg*. (Lundae) 1: 285(1825), Zahlbr., 1930 b, p. 162.

On clay rock.

Yunnan(Zahlbr., 1930 b, p. 162).

Rimularia gyromuscosa Aptroot 环藓缝裂衣

In Aptroot & Sparrius, *Fungal Diversity* 14: 38 – 39(2003).

Type: Taiwan, Hualien County: Taroko National Park, Hohuan Shan, mountain pass, 3300 m alt., 51RUG247705, on mosses on shale. Aptroot 52804(B-holotype, ABL, TUNGisotypes), 13 October 2001.

Trapelia M. Choisy **褐边衣属**

Bull. Soc. bot. Fr. 76: 523(1929).

Trapelia coarctata(Turner) M. Choisy 褐边衣

in Werner, *Bull. Soc. Sci. Nat phys. Maroc* 12: 160(1932).

≡*Lichen coarctatus* Turner, in Smith & Sowerby, *Engl. Bot*. 8: tab. 534(1799).

≡*Lecidea coarctata*(Turner) Nyl., *Act. Soc. Linn. Bordeaux* 21: 358(1856).

On rocks. Cosmopolitan.

Hubei(Chen, Wu & J. C. Wei, 1989, p. 425), Taiwan(Aptroot & Sparrius 2003, p. 47), Hongkong(Aptroot & Seawrd, 1999, p. 95; Aptroot & Sipman, 2001, p. 341).

Trapelia involuta(Taylor) Hertel 内卷褐边衣

Herzogia 2(4): 508(1973).

≡*Lecanora involuta* Taylor, in Mackay, *Fl. Hibern*. 2: 134(1836).

On exposed weathered granitic rock. Cosmopolitan.

Taiwan(Aptroot & Sparrius 2003, p. 47), Hongkong(Aptroot & Seaward, 1999, pp. 95 – 96; Aptroot & Sipman, 2001, p. 341).

Trapelia placodioides Coppins & P. James 叶状褐边衣

Lichenologist 16(3): 257(1984).

On exposed weathered graniticrock. Cosmopolitan.

Taiwan(Aptroot & Sparrius 2003, p. 47), Hongkong(Aptroot & Seaward, 1999, p. 96; Aptroot & Sipman, 2001,

p. 341).

Trapelia subconcolor(Anzi) Hertel 亚色褐边衣

Herzogia 2(4): 513(1973).

≡*Biatora subconcolor* Anzi, *Comm. Soc. crittog. Ital.* 1(no. 3): 151(1862).

= *Lecidea ochrolechioides* Zahlbr. in Fedde, *Repertorium*, 33: 38(1933).

Type: Taiwan, Asahina no. 155(holotype) in W.

= *Lecidea obsessa* Zahlbr., ibid. 33: 38(1933).

Type: Taiwan, Asahina no. 140(holotype) in W.

On rocks.

Taiwan(Zahlbr., 1933 c, p. 38 & 1940, p. 339; Lamb, 1963, p. 368; Wang & Lai, 1973, p. 91; Hertel, 1977, p. 349).

Trapeliopsis Hertel & Gotth. Schneid. **类褐衣属**

in Schneider, *Biblthca Lichenol.* 13: 143(1980) [1979].

Trapeliopsis flexuosa(Fr.) Coppins & P. James 曲类褐衣

Lichenologist 16(3): 258(1984).

≡*Biatora flexuosa* Fr., *Nov. Sched. Critic. Lich.*: 11(1826).

On *Cryptomeria japonica* and wood.

Taiwan(Aptroot & Sparrius 2003, p. 48).

Trapeliopsis cf. gelatinosa(Flörke) Coppins & P. James 胶类褐衣

Lichenologist 16(3): 258(1984).

≡*Lecidea gelatinosa* Flörke, *Mag. Gesell. naturf. Freunde*, Berlin 3(1 – 2): 201(1809).

On sheltered soil. Northern temperate.

Hongkong(Aptroot & Seaward, 1999, p. 96).

Trapeliopsis granulosa(Hoffm.) Lumbsch 粒类褐衣

in Hertel, Lecideaceae exsiccatae, Fascicle V(München) 5: no. 99(1983).

≡*Verrucaria granulosa* Hoffm., *Descr. Adumb. Plant. Lich.* 2(1): 21(1794).

On *Picea morrisonicola*, and *Abies kawakamii* wood.

Neimenggu, Heilongjiang(Zhao na et al., 2013, p. 1701), Taiwan(Aptroot & Sparrius 2003, p. 48).

Trapeliopsis hainanensis Hertel 海南类褐衣

Herzogia 5: 460(1981).

Type: Hainan, Hainan island, Mt. Jianfengling, at about 1000 m, 23/V/1980, Hertel no. 23067 in M(holotype) & in HMAS-L(isotype as Lecideaceae Exs. no. 59).

Hainan(J. C. Wei et al., 2013, p. 238).

Trapeliopsis wallrothii(Flörke ex Spreng.) Hertel & Gotth. Schneid. 类褐衣

in Schneider, *Biblthca Lichenol.* 13: 153(1980) [1979].

≡ *Lecidea wallrothii* Flörke ex Spreng., *Neue Entdeckungen im ganzen Umfang der Pflanzenkunde* 2: 96 (1821).

On exposed weathered granitic rock.

Hongkong(Aptroot & Seaward, 1999, p. 96).

Xylographa(Fr.) Fr. **木刻衣属**

Fl. Scan.: 344(1836).

≡*Stictis* subgen. *Xylographa* Fr., *Syst. mycol.* (Lundae) 2(1): 197(1822).

Xylographa parallela(Ach.) Fr. 平木刻衣

Summa veg. Scand., *Section Post.* (Stockholm): 372(1849).

≡*Lichen parallelus* Ach., *Lich. suec. prodr.* (Linköping): 23(1799) [1798].

On wood.

Xinjiang(Abdulla Abbas, 2014, pp. 579 – 580) , Taiwan(Aptroot & Sparrius 2003, p. 49) .

Xylographa trunciseda(Th. Fr.) Minks ex Redinger 截木刻衣

Rabh. Krypt. -Fl. , Edn 2(Leipzig) 9(2. 1) : 216(1938) .

≡*Lecidea trunciseda* Th. Fr. , *Lich. Scand.* (Upsaliae) 1(2) : 467(1874) .

On wood.

Taiwan(Aptroot & Sparrius 2003, p. 49) .

Xylographa vitiligo(Ach.) J. R. Laundon 藤木刻衣

Lichenologist 2: 147(1963) .

≡*Spiloma vitiligo* Ach. , *Methodus*, Sectio prior: 10(1803) .

On wood.

Taiwan(Aptroot & Sparrius 2003, p. 49) .

[11] Ostropales Nannf. **厚顶盘目**

Nova Acta R. Soc. Scient. upsal. , Ser. 4 8(no. 2) : 68(1932)

(25) Coenogoniaceae Stizenb. [as ' Coenogonieae'] **绒衣科**

Ber. Tät. St Gall. naturw. Ges. : 140(1862) [1861 – 62] . First spelt correctly by Zahlbruckner, *Sber. Akad. Wiss. Wien*, 111(Abt. I) : 394(1902) .

Coenogonium Ehrenb. **绒衣属**

in Nees von Esenbeck(ed.) , *Horae Phys. Berol.* : 120(1820) .

Coenogonium dilucidum(Kremp.) Kalb & Lücking 亮绒衣

in *Lücking & Kalb*, *Bot. Jb.* 122(1) : 32(2000) . R. Lücking *Foliicolous lichenized fungi. Flora Neotropica Monograph* 103: 573(2008) .

≡*Lecidea dilucida* Kremp. , *J. Mus. Godeffroy* 1(4) : 103(1873) .

≡*Dimerella dilucida*(Krempelh.) R. Sant. see Aptroot & Seaward, 1999, p. 73.

On living leaves of Pandanus, bamboo and trees in forests. Pantropical.

Hongkong(Aptroot & Seaward, 1999, p. 73; Aptroot et al. , 2001, p. 325 as *D. dilucida*) .

Coenogonium interplexum Nyl. 网绒衣

Annls Sci. Nat. , *Bot.* , sér. 4 16: 92(1862) .

Taiwan(Wang, 1972, p. 43; Wang & Lai, 1973, p. 90) .

Coenogonium isidiatum(G. Thor & Vězda) Lücking, Aptroot & Sipman 裂芽绒衣

in Rivas Plata, Lücking, Aptroot, Sipman, Chaves, Umaña & Lizano, *Fungal Diversity* 23: 297(2006) .

≡*Dimerella isidiata* G. Thor&Vězda, *Folia geobot. phytotax.* 19(1) : 72(1984) , W. Obermayer(2004) .

Xizang, Sichuan(W. Obermayer, 2004, p. 495) .

Coenogonium leprieurii(Mont.) Nyl. 勒氏绒衣

Ann. Sci. Nat. Bot. Ser. 4, 16: 89(1862) pl. 12, fig. 15 – 19.

≡*Coenogonium linkii* var. *leprieurii* Mont. , *Annls Sci. Nat.* , *Bot.* , sér. 3 16: 47(1851) .

On bark.

Fujian(Zahlbr. , 1930b, p. 73) .

Coenogonium linkii Ehrenb. 绒衣

in Nees von Esenbeck(ed.) , *Horae Phys. Berol.* : 120, pl. 27(1820) .

= *Coenogonium boninense* Sato, *J. J. B.* 8: 390(1933) ; Wang & Lai, 1973, p. 90.

"*Coenogonium subvirescens*(Nyl.) Nyl. " sensu Zahlbr. in Feede, *Repert.* 33: 26(1933) . Wang & Lai, 1973, p. 90.

On bark of tree and sheltered granite rocks in forest. Pantropical.

Taiwan(Zahlbr. , 1933c, p. 26; Asahina & Sato in Asahina, 1939, p. 631; Wang, 1972, p. 42; Wang & Lai, 1973, p. 90; Yoshimura, 1994, p. 189, as *C. boninense*) , Hongkong(Aptroot & Seaward, 1999, p. 72; Seaward & Aptroot, 2005, p. 284, as *C. cf. linkii* , Pfister 1978: 107, as *Coenogonium disjunctum.* ; Aptroot et al. , 2001, p.

324 as *C. linkii*).

Coenogonium luteum(Dicks.) Kalb & Lücking 金黄绒衣

in Lücking & Kalb, *Bot. Jb.* 122(1): 32(2000). R. Lücking *Foliicolous lichenized fungi. Flora Neotropica Monograph* 103: 570(2008).

≡*Lichen luteus* Dicks., *Fasc. pl. crypt. Brit.* (London) 1: 11(1785).

≡*Dimerella lutea*(Dicks.) Trevisan, See Aptroot & Seaward, 1999, p. 73.

On bark of trees and rock in forests. Cosmopolitan.

Hubei(Chen, Wu & Wei, 1989, p. 401), Zhejiang(Xu B. S. 1989, p. 182), Fujian(Zahlbr., 1930 b, p. 70), Hongkong(Aptroot & Seaward, 1999, p. 73; Seaward & Aptroot, 2005, p. 284, as *Dimerella lutea*, Pfister 1978: 99, as *Gyalecta lutea*; Aptroot et al., 2001, p. 325 as*D. lutea*).

Coenogonium pineti(Ach.) Lücking & Lumbsch 皮氏绒衣

in Lücking, Stuart & Lumbsch, *Mycologia* 96(2): 290(2004).

≡*Lecidea pineti* Ach., *Lich. univ.*: 195(1810).

≡*Dimerella pineti*(Ach.) Vězda, *Lichenes Selecti Exsiccati*(Průhonice) 52: no. 1279(1975).

On *Debeyraecea edulis* and *Cinnamomum* sp.

Taiwan(Aptroot and Sparrius, 2003, p. 15, as *Dimerella pineti*), Hongkong(Aptroot et al., 2001, p. 325 as *D. pineti*).

Coenogonium subluteum(Rehm) Kalb & Lücking 亚深黄绒衣

in Lücking & Kalb, *Bot. Jahrb. Syst.* 122: 34(2000). R. Lücking *Foliicolous lichenized fungi. Flora Neotropica Monograph* 103: 573(2008).

≡*Biatorina sublutea* Rehm, Philipp. *J. Sci., C, Bot.* 8(5): 404(1913).

=*Biatorinopsis epiphylla* Müll. Arg., *Flora, Regensburg* 64(7): 103(1881).

≡*Dimerella epiphylla*(Müll. Arg.) Malme, *Ark. Bot.* 26A(no. 13): 9(1935) [1934]. Sant. Symb. Bot. Upsal. 12 (1): 395(1952); Lücking, Lichenologist 31: 365(1999); Lücking & Kalb, Bot. Jahrb. Syst. 122: 34(2000) (non *Coenogonium epiphyllum* Vainio). Type, Brazil. Minas Gerais: unknown locality, s. col. (neotype, G; Santesson, 1952: 395).

≡*Microphiale epiphylla*(Müll. Arg.) Zahlbr., *Cat. Lich. Univers.* II: 696(1924), fide R. Sant. 1952, p. 399. non *Coenogonium epiphyllum* Vainio— R. Lücking *Foliicolous lichenized fungi. Flora Neotropica Monograph* 103: 573(2008).

=*Microphiale brachyspora*(Müll. Arg.) Zahlbr., in Rechinger, *Denkschr. Kaiserl. Akad. Wiss. Wien, Math.-Naturwiss. Kl.* 88: 22(1911), Zahlbr., *Cat. Lich.* Univ. II: 693(1924).

≡*Biatorinopsis brachyspora* Müll. Arg., *Lichenes Epiphylli Novi*: 16(1890).

=*Microphiale lutea* f. *foliicola* Zahlbr., in Rechinger, *Denkschr. Kaiserl. Akad. Wiss. Wien, Math.-Naturwiss. Kl.* 81: 247(1907).

Foliicolous. On *Alpinia speciosa* and *Diospyros* sp. Pantropical.

Yunnan(Zahlbr., 1930 b, p. 70, as *Microphiale brachyspora*, *M. epiphylla*, and p. 71, as *M. lutea* f. *foliicola*, fide R. Sant. 1952, p. 399; J. C. Wei & Y. M. Jiang, 1991, p. 207; Aptroot et al., 2003, pp. 44), Taiwan(Aptroot et al., 2003, pp. 42 – 43), Hongkong(Aptroot & Seaward, 1999, p. 73 as *Dimerella epiphylla*(Müll. Arg.) R. Sant.; Aptroot et al., 2001, p. 325 as *D. epiphylla*), *Coenogonium subvirescens Nyl., Flora, Regensburg 57: 72 (1874)*

Coenogonium zonatum(Müll. Arg.) Kalb & Lücking 带绒衣

in Lücking & Kalb, *Bot. Jb.* 122(1): 34(2000).

≡*Biatorinopsis zonata* Müll. Arg., *Lich. Epiphylli Novi*: 16(1890).

Yunnan(Aptroot et al., 2003, pp. 44).

(26) Gomphillaceae Walt. Watson 楔形衣科

New Phytol. 28: 31(1929)

≡*Gomphillaceae* Walt. Watson ex Hafellner, *Beih. Nova Hedwigia* 79: 280(1984).

Actinoplaca Müll. Arg. **辐射衣属**

Bull. Soc. R. Bot. Belg. 30(1): 56(1891).

Actinoplaca strigulacea Müll. Arg. 辐射衣

Bull. Soc. R. Bot. Belg. 30: 57(1891).

On *Phoebe formosana*

Yunnan(Aptroot et al., 2003, pp. 44), Hainan(Aptroot et al., 2003, p. 45).), Taiwan(Aptroot et al., 2003, pp. 42 – 43).

Asterothyrium Müll. Arg. **薄蜡衣属**

Lich. Epiph. Novi: 12(1890).

Asterothyrium microsporum R. Sant. 微孢薄蜡衣

Symb. bot. upsal. 12(no. 1): 320(1952).

On *Fastia* sp., *Hydrangea* sp., and *Tupinia* sp.

Yunnan(Aptroot et al., 2003, pp. 44), Taiwan(Aptroot et al., 2003, pp. 42 – 43).

Asterothyrium pittieri Müll. Arg. 皮氏薄蜡衣

Bull. Soc. R. Bot. Belg. 30: 71(1891).

On leaves of *Thunbergia grandiflora*.

Yunnan(J. C. Wei & Y. M. Jiang, 1991, p. 205).

Aulaxina Fée **坑盘衣属**

Essai Crypt. Exot. (Paris): lx, xciv(1825) [1824].

Aulaxina quadrangula(Stirt.) R. Sant. 四稜坑盘衣

J. Ecol. 40(1): 129(1952).

≡*Platygrapha quadrangula* Stirt., *Proc. Roy. phil. Soc. Glasgow* 11: 103(1879) [1878].

On *Alpinia speciosa* and *Asplenium nidus*.

Yunnan(Aptroot et al., 2003, pp. 44), Taiwan(Aptroot et al., 2003, pp. 42 – 43).

Aulaxina sp. 坑盘衣(未定名种)

Yunnan(Aptroot et al., 2003, pp. 44).

Bullatina Vězda & Poelt **泡衣属**

Folia geobot. phytotax. 22: 186(1987).

Bullatina aspidota(Vain.) Vězda & Poelt 泡衣

Folia geobot. phytotax. 22: 186(1987).

≡*Calenia aspidota*(Vain.) Vězda, *Folia geobot. phytotax.* 19(2): 195(1984).

≡*Ectolechia aspidota* Vain., in Hiern, *Cat. Welwitsch Afric. Pl.* 2(2): 428(1901).

Yunnan(Aptroot et al., 2003, pp. 44).

Calenia Müll. Arg. **浅盘衣属**

Lich. Epiph. Novi: 3(1890).

Calenia aspidota(Vain.) Vězda 多毛浅盘衣

Folia geobot. phytotax. 19: 195(1984).

≡*Ectolechia aspidota* Vain., in Hiern, *Cat. Welwitsch Afric. Pl.* 2(2): 428(1901).

≡*Bullatina aspidota*(Vain.) Vězda & Poelt, *Folia geobot. phytotax.* 22: 186(1987), R. Lücking *Foliicolous Lichenized Fungi* 519(2008).

Yunnan(Aptroot et al., 2003, pp. 44 as *Bullatina aspidota*).

Calenia depressa Müll. Arg. 浅盘衣未定种

Lichenes Epiphylli Novi: 4(1890).

On *Fatsia* sp.

Yunnan(Aptroot et al., 2003, pp. 44), Taiwan(Aptroot et al., 2003, pp. 42 – 43).

Calenia monospora Vězda 单孢浅盘衣

Folia geobot. phytotax. 14(1):56(1979).

On *Alpinia speciosa*.

Yunnan(Aptroot et al., 2003, pp. 44), Taiwan(Aptroot et al., 2003, pp. 42 – 43).

Calenia sp. 浅盘衣

On *Asplenium speciosa*.

Taiwan(Aptroot et al., 2003, pp. 42 – 43).

Calenia thelotremella Vain. 乳胶浅盘衣

Ann. Acad. Sci. fenn., Ser. A 15(no. 6):160(1921).

Yunnan(Aptroot et al., 2003, pp. 44).

Echinoplaca Fée **刺衣属**

Essai Crypt. Exot. (Paris):l, xciii(1825)[1824].

Echinoplaca cf. epiphylla Fée 刺衣

Essai Crypt. Exot. (Paris):xciii(1825)[1824].

On living leaves of ferns, *Pandnus* sp., *Asplenium nidus*, and trees in forests. Pantropical.

Yunnan(Aptroot et al., 2003, pp. 44), Taiwan(Aptroot et al., 2003, pp. 42 – 43), Hongkong (Aptroot & Seaward, 1999, p. 74; Aptroot et al., 2001, p. 325).

Echinoplaca heterella(Stirt.) R. Sant. 黄褐盘刺衣

Symb. bot. upsal. 12(no. 1):372(1952).

≡Arthonia heterella Stirt., *Proc. Roy. phil. Soc.* Glasgow 11:106(1879)[1878].

=*Sporopodium handelii* Zahlbr., in Handel-Mazzetti, *Symb. Sin.* 3:72(1930).

Type: Yunnan, foliicolous, Handel-Mazzetti no. 8999(lectotype) in UPS(R. Sant. 1952, p. 372).

Echinoplaca hispida Sipman 粗毛刺衣

Lichenologist 25(2):129(1993).

On *Fatsia* sp. and *Phoebe formosana*.

Taiwan(Aptroot et al., 2003, pp. 42 – 43).

Echinoplaca leucotrichoides(Vain.) R. Sant. 白毛刺衣

in Thorold, *J. Ecol.* 40:129(1952).

≡*Calenia leucotrichoides* Vain., *Ann. Acad. Sci. fenn.*, Ser. A 15(no. 6):166(1921).

On *Cryptomeria japonica* needles, *Cryptomeria japonica* bark, *Castanopsis* sp., living leaves of *Pasania kawakamii*, *Cinnamomum* sp., *Fatsia* sp., *Phoebe formosana*, *Alpinia speciosa*, *Asplenium nidus*, and wood.

Yunnan(Aptroot et al., 2003, pp. 44), Taiwan(Aptroot and Sparrius, 2003, p. 16; Aptroot et al., 2003, pp. 42 – 43).

Echinoplaca pellicula(Müll. Arg.) R. Sant. 表膜刺衣

Symb. bot. upsal. 12(no. 1):367(1952).

≡*Arthonia pellicula* Müll. Arg., *Bot. Jb.* 4(1):56(1883). (incl. f. *trichariosa* Müll. Arg.).

=*Gonolecania tetrapla* Zahlbr., in Handel-Mazzetti, *Symb. Sin.* 3:72(1930).

Type: Fujian, foliicolous, Chung no. F 330(R. Sant. 1952, p. 367).

Gyalectidium Müll. Arg. **榴果衣属**

Flora, Regensburg 64(7):100(1881).

Gyalectidium australe Lücking 南方榴果衣

in Ferraro, Lücking & Sérusiaux, J. Linn. Soc., Bot. 137(3):325(2001).

Yunnan(Aptroot et al., 2003, pp. 44).

Gyalectidium catenulatum(Cavalc. & A. A. Silva) L. I. Ferraro, Lücking & Sérus. 小链榴果衣

in Ferraro & Lücking, *Trop. Bryol.* 19:64(2000).

≡*Tauromyces catenulatus* Cavalc. & A. A. Silva, in Cavalcante, Cavalcanti & Leal, Publicações. *Instituto de Mi-*

cologia da Universidade de Pernambuco 647: 37(1972).

On *Hydrangea* sp.

Taiwan(Aptroot et al., 2003, pp. 42 – 43).

Gyalectidium caucasicum(Elenkin & Woron.) Vězda 哈萨克榴果衣

Folia geobot. phytotax. 18(1): 56(1983).

≡*Sporopodium caucasicum* Elenkin & Woron., *Jb. Pflanzenkranh.* St. Petersburg 2: 124(1908).

On *Fatsia* sp., *Hydrangea* sp., *Polystichum* sp., and *Turpinia* sp.

Yunnan(Aptroot et al., 2003, pp. 44), Taiwan(Aptroot et al., 2003, pp. 42 – 43).

Gyalectidium ciliatum Lücking, G. Thor & Tat. Matsumoto 缘毛榴果衣

in Thor, Lücking & Matsumoto, *Symb. bot. upsal.* 32(no. 3): 42(2000).

On *Machilus thunbergii* leaves. It is not obligately foliicolous.

Taiwan(Aptroot and Sparrius, 2003, p. 20).

Gyalectidium filicinum Müll. Arg. 榴果衣

Flora, Regensburg 64: 9(1881).

On leaves of *Pseuduvaria indochinensis*, *Polystichum* sp., and *Turpinia* sp.

Yunnan(Wei & Jiang, 1991, p. 206 – 207; Aptroot et al., 2003, pp. 44), Taiwan(Aptroot et al., 2003, pp. 42 – 43).

Gyalidea Lettau ex Vězda **星盘衣属**

Folia geobot. phytotax. bohemoslov. 1: 312(1966). nom. cons.

Gyalidea asteriscus(Anzi) Aptroot & Lücking 野星盘衣

Biblthca Lichenol. 86: 67(2003).

≡*Solorinella asteriscus* Anzi, *Cat. Lich. Sondr.*: 37(1860).

On soil.

Sichuan, Xizang(Obermayer W., 2004, p. 510, as *Solorinella asteriscus* Anzi).

subsp. *gracilispora* Jun Yang & J. C. Wei 狭孢亚种

Mycotaxon 109: 374(2009).

On soil.

Type: Hebei, Mt. Xiaowutai, April 15, 2005. Wang HY & Wei XL 3058(HMAS-L.).

Gyalidea japonica H. Harada & Vězda 日本星盘衣

Nat. Hist. Res. 1(2): 14(1991).

On sand stone.

Taiwan(Aptroot and Sparrius, 2003, p. 21).

Gyalidea luzonensis(Kalb & Vězda) Aptroot & Lücking 鲁宗星盘衣

Biblthca Lichenol. 86: 69(2003).

≡*Gyalidea novae-guineae* var. *luzonensis* Kalb & Vězda, in Vězda & Poelt, Nova Hedwigia, Beih. 53(1 – 2): 108(1991).

On shale.

Taiwan(Aptroot and Sparrius, 2003, p. 21).

Gyalideopsis Vězda **亚星盘衣属**

Folia geobot. phytotax. 7(2): 204(1972)

Gyalideopsis lambinonii Vězda 拉姆氏亚星盘衣

Folia geobot. phytotax. 14(1): 64(1979).

On *Cryptomeria japonica*, and *Shiia carlesii*.

Taiwan(Aptroot and Sparrius, 2003, p. 21).

Gyalideopsis muscicola P. James & Vězda 藓生亚星盘衣

in Vězda, *Folia geobot. phytotax.* 7(2): 211(1972).

On *Shiia carlesii*, and *Prunus persica*.

Taiwan(Aptroot and Sparrius, 2003, p. 21).

Gyalideopsis rostrata Kalb & Vězda 喙亚星盘衣

Biblthca Lichenol. 29: 46(1988).

On *Cryptomeria japonica*.

Taiwan(Aptroot and Sparrius, 2003, p. 21).

Gyalideopsis rubescens Vězda 红色亚星盘衣

Folia geobot. phytotax. 14(1) : 67(1979).

On *Fatsia* sp., *Alpinia speciosa*, and *Asplenium nidus*.

Taiwan(Aptroot et al., 2003, pp. 42 – 43).

Gyalideopsis vainioi Kalb & Vězda 万氏亚星盘衣

Biblthca Lichenol. 29: 51(1988).

On *Cinnamomum* sp., and *Prunus persica*.

Taiwan(Aptroot and Sparrius, 2003, p. 21).

Tricharia Fée **毛蜡衣属**

Essai Crypt. Exot. (Paris) : lxxxviii, xcviii, cii(1825) [1824].

Tricharia armata Vězda 装饰毛蜡衣

Folia geobot. phytotax. 10: 404(1975).

On *Arenga* sp.

Yunnan(Aptroot et al., 2003, p. 45), Taiwan(Aptroot et al., 2003, pp. 42 – 43).

Tricharia carnea(Müll. Arg.) R. Sant. 毛蜡衣

Symb. bot. upsal. 12(1): 385(1952).

≡*Lopadium carneum* Müll. Arg., *Flora*, Regensburg 64(7): 109(1881).

On *Hydrangea* sp.

Taiwan(Aptroot et al., 2003, pp. 42 – 43).

Tricharia kashiwadanii G. Thor, Lücking & Tat. Matsumoto 柏谷氏毛蜡衣

Symb. bot. upsal. 32(3) : 62(2000).

On *Alpinia speciosa*, and *Asplenium nidus*.

Taiwan(Aptroot et al., 2003, pp. 42 – 43).

Tricharia santessoni Hawksw. 桑氏毛蜡衣

Lichenologist 5: 321(1972).

Type locality: Hongkong, on leave of *Citrus sinensis* var. *minlgaui* CB 673(IMI 160016); isotypes in BM & UPS.

On bark of *Araucaria* sp. and *Citrus* sp. and on living leaves of *Pandunus* sp., bamboo and trees and in forests and parks, rarely on sheltered rock, from sea-level to 800m alt.

Pantropical.

Hongkong(Hawksw. 1972, p. 321 & 1973, p. 195; Thrower, 1988, pp. 17, 176; Aptroot & Seaward, 1999, p. 96).

Tricharia vainioi R. Sant. 万氏毛蜡衣

Symb. bot. upsal. 12(1) : 382(1952).

On leaves of *Polyalthia viridis*

Yunnan(Wei & Jiang, 1991, p. 207),.

On *Diospyros maritime*, *Alpinia speciosa*, and *Asplenium nidus* leaves.

Yunnan(Aptroot et al., 2003, p. 45), Taiwan(Aptroot & Sparrius 2003, p. 48; Aptroot et al., 2003, pp. 42 – 43).

(27) Graphidaceae Dumort. [as ‘ Graphineae’] **文字衣科**

Comment. bot. (Tournay): 69(1822).

Carbacanthographis Staiger & Kalb 炭刺文衣属

Biblthca Lichenol. 85: 98(2002).

Carbacanthographis marcescens(Fée) Staiger & Kalb 枯炭刺文衣

in Staiger, *Biblthca Lichenol.* 85: 109(2002).

≡*Graphis marcescens* Fée, *Essai Crypt. Exot.* (Paris): 38(1825)[1824].

On tree trunk.

Guangxi(Jia et al., 2017, Pp. 232 – 233), Hongkong(Aptroot et al., 2001, p. 328).

Chapsa A. Massal. 裂衣属

Atti Inst. Veneto Sci. lett., ed Arti, Sér. 3 5: 257(1860)[1859 – 1860].

Chapsa indica A, Massal. 裂衣

in Atti Reale Ist., Veneto Sci. Lett. Ed Arti, ser. 3, 5: 257(1860)

Hainan(Xu LL et al., 2016, p. 496).

Chapsa leprocarpa(Nyl.) A. Frish 斑果裂衣

Biblioth. Lichenol. 92: 108(2006).

≡ *Graphis leprocarpa* Nyl. *Acta Soc. Sci. Fenn.* 7: 472(1863).

Guangxi(Xu LL et al., 2016, p. 497).

Chapsa mirabilis(Zahlbr.) Lücking 殊茶盘衣

in Rivas Plata, Lücking, Sipman, Mangold, Kalb & Lumbsch, Lichenologist 42(2): 183(2010), Z. F. Jia & Lücking MycoKyes 21: 22(2017).

≡ *Phaeographina mirabilis* Zahlbr. *in* Handel-Mazzetti, *Symb. Sin.* 3: 62(1930) & *Cat. Lich. Univ.* 8: 217 (1932).

Type: Fujian, Mt. Gu shan, Chung no. 387(W).

On smooth bark of trees.

Diorygma Eschw. 霜盘衣属

Syst. Lich.: 13, 25(1824).

Diorygma erythrellum(Mont. & Bosch) Kalb, Staiger & Elix 红霜盘衣

Symb. bot. upsal. 34(no. 1): 150(2004).

≡*Ustalia erythrella* Mont. & Bosch, *in* Junghuhn, *Pl. Jungh.* 4: 478(1855).

≡*Graphina erythrella*(Mont. & Bosch) Zahlbr. *Cat. Lich. Univ.* 2: 405(1923)[1924].

On *Zelkova serrata*.

Taiwan(Aptroot and Sparrius, 2003, p. 18, as *G. erythrella*).

Diorygma fuscum Jian Li bis & Z. F. Jia 褐霜盘衣

Mycotaxon 131: 717 – 721(2016).

Type: China, Fujian, Jianou City, Fangdao Town, Wanmulin, 27°02′N 118°08′E, alt. 310 m,

On bark, 3/VI/2007, Q. F. Meng FJ1280(Holotype, HMAS-L 137193).

Diorygma hieroglyphicum(Pers.) Staiger & Kalb 象形霜盘衣

in Kalb et al., *Symb. bot. upsal.* 34(1): 151(2004), Z. F. Jia & R Lücking *MycoKeys* 21: 15(2017).

≡*Opegrapha hieroglyphica* Pers., *Ann. Wetter. Gesellsch. Ges. Naturk.* 2: 16(1811)[1810].

On bark.

Yunnan(Meng QF & Wei JC, 2008, p. 527; Meng 2008, p. 41; Z. F. Jia & J. C. Wei, 2016, p. 29), Fujian(Z. F. Jia & J. C. Wei, 2016, p. 29), Hainan(Meng QF & Wei JC, 2008, p. 527; Meng 2008, p. 41; J. C. Wei et al., 2013, p. 232; Z. F. Jia & J. C. Wei, 2016, p. 29).

=*Phaeographina callospora* Zahlbr. *Feddes Repert.* 31: 220(1933).

Type: Taiwan, Faurie no. 118(W).

Taiwan(Zahlbr., 1940, p. 188; Lamb, 1963, p. 544; Wang & Lai, 1973, p. 95 & 1976, p. 227).

Diorygma hololeucum(Mont. & Bosch) Kalb, Staiger & Elix 厚粉霜盘衣

Symb. bot. upsal. 34(no. 1) : 155(2004) .

≡ *Graphis hololeuca* Mont. & Bosch, *Pl. Jungh.* 4: 473(1855) .

Hainan(Li 2010, p. 38; J. C. Wei et al. , 2013, p. 233; Z. F. Jia & J. C. Wei, 2016, p. 30) .

Diorygma isabellinum(Zahlbr.) Z. F. Jia & Lücking 土黄霜盘衣

MycoKeys 25: 24(2017) .

≡ *Graphina isabellina* Zahlbr. , in Handel-Mazzetti, *Symb. Sinic.* 3: 58(1930) & *Cat. Lich. Univ.* 8: 213(1932) .

Type: China, Hunan, Handel-Mazzetti 11437(holotype W) .

On living trunk of *Symplocos* sp.

Diorygma junghuhnii(Mont. & Bosch) Kalb, Staiger & Elix 容氏霜盘衣

Symb. bot. upsal. 34(1) : 157(2004) .

≡ *Graphis junghuhnii* Mont. & Bosch, in Junghun, *Pl. Jungh.* 4: 471(1855) .

= *Graphina mendax*(Nyl.) Müll. Arg. *Revue mycol.* , Toulouse 10: 177(1888) .

≡ *Graphis mendax* Nyl. , *Annls Sci. Nat.* , *Bot.* , sér. 411: 244(1859) .

Misapplied name(revised by Aptroot & Seaward, 1999, p. 76) :

Graphina cf. *hologlauca* auct non Zahlbr. in Thrower, 1988, pp. 15, 97.

Graphina cf. *virginea* auct non(Eschw.) Müll. Arg. in Thrower, 1988, pp. 15, 101.

On sheltered trees. Known from tropical Asia.

Hongkong(Thrower, 1988, pp. 15, 97, 100, 101; Aptroot & Seaward, 1999, p. 76) , Taiwan (Aptroot and Sparrius, 2003, p. 18) .

On *Lonicera cumingiana*, and *Aphanamixis tripetala*

Yunnan, Guizhou, Guangdong, Guangxi(Z. F. Jia & J. C. Wei, 2016, p. 31 – 32) , Fujian(Meng QF & J. C. Wei, 2008, p. 528; Meng 2008, p. 42; Z. F. Jia & J. C. Wei, 2016, p. 31) , Hainan(Meng QF & J. C. Wei, 2008, p. 528; Meng 2008, p. 42; J. C. Wei et al. , 2013, p. 233; Z. F. Jia & J. C. Wei, 2016, p. 31) , Hongkong(Thrower 1988, p. 100 as *Graphina mendax*; Z. F. Jia & J. C. Wei, (2016, p. 32) .

Diorygma macgregorii(Vain.) Kalb, Staiger & Elix 马氏霜盘衣

Symb. bot. upsal. 34(1) : 159(2004) .

≡ *Helminthocarpon pervarians* var. *macgregorii* Vain. , *Ann. Acad. Sci. fenn.* , Ser. A 15(no. 6) : 266(1921) .

Sichuan, guizhou, Zhejiang(Meng QF & Wei Jc, 2008, pp. 528 – 529; Meng 2008, p. 43; Z. F. Jia & J. C. Wei, 2016, p. 33) , Yunnan, Hunan(Z. F. Jia & J. C. Wei, 2016, p. 33) , Hainan(J. C. Wei et al. , 2013, p. 233; Z. F. Jia & J. C. Wei, 2016, p. 33) .

= *Graphis hologlauca* Nyl. *Flora*, Regensburg 49: 133(1866) .

≡ *Graphina hologlauca*(Nyl.) Zahlbr. *Cat. Lich. Univ.* 2: 408(1923) [1924] .

On *Cryptomeria japonica*, *Michelia formosana*, *Schima superb*, and *Castanopsis* sp.

Taiwan(Aptroot and Sparrius, 2003, p. 18) .

See *Graphina mendax*(Nyl.) Müll. Arg. for Hongkong(Thrower, 1988, p. 97) .

Diorygma megasporum Kalb, Staiger & Elix 大孢霜盘衣

Symb. bot. upsal. 34(no. 1) : 160(2004)

On tree bark.

Guizhou(Meng 2008, p. 45; Z. F. Jia & J. C. Wei, 2016, p. 34) .

Diorygma pachygraphum(Nyl.) Kalb, Staiger & Elix 厚唇霜盘衣

Symb. bot. upsal. 34(no. 1) : 163(2004) .

≡ *Graphis pachygrapha* Nyl. , *Acta Soc. Sci. fenn.* 7(2) : 472(1863) .

= *Graphina roridula* Zahlbr. *in* Handel-Mazzetti, *Symb. Sin.* 3: 59(1930) & *Cat. Lich. Univ.* 8: 214(1932) .

Type: Hunan, Handel-Mazzetti no. 12302.

On tree bark.

Yunnan, Guizhou, Guanxi(Z. F. Jia & J. C. Wei, 2016, p. 36), Fujian(Meng QF & Wei JC, 2008, p. 529; Meng 2008, p. 46; Z. F. Jia & J. C. Wei, 2016, p. 35), Hainan(J. C. Wei et al., 2013, p. 233; Z. F. Jia & J. C. Wei, 2016, p. 36).

= *Graphina roridula* var. *platypoda* Zahlbr. *in* Handel-Mazzetti, *Symb. Sin.* 3: 60(1930) & *Cat. Lich. Univ.* 8: 214 (1932).

Type: Yunnan, Handel-Mazzetti no. 6554.

Corticolous.

Diorygma pruinosum(Eschw.) Kalb, Staiger & Elix 粉霜盘衣

Symb. bot. upsal. 34(1) : 166(2004).

≡*Leiogramma pruinosum* Eschw., in von Martius, *Icon. Plant. Cryptog.* 2: 12(1828) [1828 – 34].

Yunnan, Guizhou, Fujian(Z. F. Jia & J. C. Wei, 2016, p. 37), hainan(Meng QF & Wei JC, 2008, p. 530; Meng 2008, p. 48; J. C. Wei et al., 2013, p. 233; Z. F. Jia & J. C. Wei, 2016, p. 37).

Diorygma soozanum(Zahlbr.) M. Nakan. & Kashiw. 白粉霜盘衣

in Nakanishi, Kashiwadani & Moon, *Bull. natn. Sci. Mus.*, Tokyo, B 29(2) : 86(2003).

≡*Graphina soozana* Zahlbr., *Feddes Repert.* 31: 215(1933).

Type: Taiwan, Asahina no. 343.

On tree bark.

Sichuan, Yunnan, Guizhou(Meng & J. C. Wei 2008, p. 531; Meng 2008, p. 50; Z. F. Jia & J. C. Wei, 2016, p. 38), Zhejiang(Xu BS 1989, p. 177 as *Graphina soozana*), Fujian(Meng QF &J. C. Wei JC, 2008, p. 531; Meng 2008, p. 50; Z. F. Jia & J. C. Wei, 2016, p. 38), Taiwan(Zahlbr. 1940, p. 185; Lamb 1963, p. 252; Wang & Lai 1973, p. 90 as *Graphina soozana*).

Diploschistes Norman **双缘衣属**

Nytt Mag. Natur. 7: 232(1853) [1852].

Diploschistes actinostomus(Ach.) Zahlbr. 裂果双缘衣

Hedwigia 31: 34(1892).

≡*Verrucaria actinostoma* Ach., *Lich. univ.* : 288(1810).

≡*Urceolaria actinostoma* Pers., *Lich. univ.* : 288(1810).

≡*Limboria actinostoma*(Ach.) A. Massal., *Ric. auton. lich. crost.* (Verona) : 155(1852).

On exposed granite. Nearly cosmopolitan.

Jiangsu, Zhejiang(Xu B. S. 1989, p. 175), Taiwan(Aptroot and Sparrius, 2003, p. 15)

Hongkong(Rabh. 1873, p. 287; Krempelh. 1873, p. 471 & 1874 c, p. 67; Zahlbr., 1930 b, p. 67; Thrower, 1988, p. 88; Aptroot & Seaward, 1999, p. 73 – 74; Aptroot et al., 2001, p. 325).

Diploschistes actinostomus(Ach.) Zahlbr. var. anerythrinus Moreau et Moreau 非红变种

Rev. Bryol. Et Lichenol. 20: 185(1951). (nom. inval. : Art. 36).

Type: Beijing, Xishan, Tchou Y. T. nos. 102, 110(syntypes).

On rocks.

Diploschistes anactinus(Nyl.) Zahlbr. 白双缘衣

Hedwigia 31: 14(1892). f. anactinus 原变型

≡ *Urceolaria anactina* Nyl., *Lich. Japon.* : 59(1890).

On rocks.

Sichuan, Yunnan(Zahlbr., 1930 b, p. 67).

f. cinerata Zahlbr. 灰色变型

in Fedde, *Repertorium* 33: 25(1933).

Type: Taiwan, Faurie no. 81.

Taiwan(Zahlbr. r, 1940, p. 222; Wang & Lai, 1973, p. 90) .

On rocks.

Diploschistes caesioplumbeus(Nyl.) Vain. 浅黑双缘衣

Bot. Mag. , Tokyo 35: 70(1921) .

≡*Urceolaria actinostoma* var. *caesioplumbea* Nyl. , *Bull. Soc. linn. Normandie*, sér. 2 6: 264 (1872) .

On rocks.

Jiangsu(Wu J. N. & Xiang T. 1981, p. 2) . Heilongjiang(Qi GY et al. , 2015, p. 125) .

Diploschistes cinereocaesius(Sw.) Vain. 大环形双缘衣

Ann. Acad. Sci. fenn. , Ser. A 15(6) : 172(1921) .

≡*Lichen cinereocaesius* Sw. , in Acharius, *Lich. suec. prodr.* (Linköping) : 34(1799) [1798] .

On rocks covered with soil.

Guangdong(Wu J. L. , 1987, p. 47) .

Diploschistes diacapsis(Ach.) Lumbsch 双壳双缘衣

Lichenologist 20(1) : 20, 1988. Qi GY, et al. J. Fungal Research 13(2) : 125(2015) .

≡ *Urceolaria diacapsis* Ach. , *Methodus*: 339(1810) .

On rock.

Heilongjiang(Qi GY, et al. , 2015, p. 125) .

Diploschistes euganeus(A. Massal.) Zahlbr. 优格双缘衣

Verh. zool. -bot. Ges. Wien 69(Zool.) : 96(1892) .

≡ *Limboria euganea* A. Massal. , *Ric. auton. lich. crost.* (Verona) : 155(1852) .

On siliceous rock.

Liaoning(Ren Qiang & Li Shuxia, 2013, p. 67) .

Diploschistes hypoleucus(Vain.) Zahlbr. 白腹双缘衣

Hedwigia 31: 35. 1892. [non *Urceolaria hypoleuca* Ach. 1803] . Replaced synonym: *Urceolaria hypoleuca* Vain. (Nom. illegit. , Art. 53. 1) , Étud. Class. Lich. Brésil 2: 73, 1890. Wang LS *et al.* , Cryptogmie Mycologie 34 (4) : 344, 2013.

Yunnan(Wang LS et al. , 2013, p.)

Diploschistes muscorum(Scop.) R. Sant. 藓生双缘衣

in D. Hawksw. , James & Coppins, *Lichenologist* 12(1) : 106(1980) .

≡*Lichen muscorum* Scop. , *Fl. carniol.* , Edn 2(Wien) 2: 365(1772) .

= *Diploschistes bryophilus*(Ehrh.) Zahlbr. , *Hedwigia* 31: 34(1892) .

≡*Lichen bryophilus* Ehrh. , *Pl. crypt. exsicc.* : no. 236(1785) .

On soil, dead plants and *Cladonia* spp. (Aptroot & Sipman 2001; J. C. Wei 1991, p. 90) .

Xinjiang(Wang XL, 1985, p. 335, as *D. bryophilus*; Abdulla Abbas et al. , 1993, p. 77; Abdulla Abbas & Wu, 1998, p. 34; Abdulla Abbas et al. , 2001, p. 363) , Ningxia(Liu M & Wei J. C. , 2013, p. 46; Yang J. & Wei J. C. , 2014, p. 1030) , Xizang(J. C. Wei & Y. M. Jiang, 1981, p. 1147 & 1986, p. 12; Hawksw. & M. Cole, 2003, p. 360) , Hubei(Chen, Wu & J. C. Wei, 1989, p. 400) , Hongkong(Hawksw. & M. Cole, 2003, p. 360; Seaward & Aptroot, 2005, p. 284, Pfister 1978: 100, as *Urceolaria scruposa*; Aptroot et al. , 2001, p. 325, as ssp. *muscorum*) .

Diploschistes rampoddensis(Nyl.) Zahlbr. 斯里兰卡双缘衣

Cat. Lich. Univers. 2: 665(1924) .

≡Urceolaria rampoddensis Nyl. , Acta Soc. Sci. fenn. 26(no. 10) : 18(1900) .

On hard soil along the main road.

Yunnan(Samantha Fernández – Brime et al. , 2014, p. 392) .

Diploschistes scruposus(Schreb.) Norman 双缘衣

Nytt Mag. Natur. 7: 232(1853) [1852] .

≡*Lichen scruposus* Schreb., *Spic. fl. lips.* (Lipsiae): 133(1771).

≡*Urceolaria scruposa*(Schreb.) Ach., *Methodus*, Sectio prior: 147(1803).

On the earth and stones.

Xinjiang(Wang XL, 1985, p. 335; H. Magn. 1940, p. 33; Abdulla, Abbas et al., 1993, p. 77; Abdulla, Abbas et al., 1994, p. 21; Abdulla Abbas & Wu, 1998, p. 34; Abdulla Abbas et al., 2001, p. 363), Shaanxi(Jatta, 1902, p. 477; Zahlbr., 1930 b, p. 67), Taiwan(Zahlbr., 1933 c, p. 26; Wang & Lai, 1973, p. 90). Heilongjiang(Qi GY et al., 2015, p. 125).

Diploschistes scruposus f. argillosus(Ach.) Dalla Torre & Sarnth. 黏土变型

Die Flecht. Tirol: 298(1902).

≡Urceolaria scruposa f. argillosa Ach., Methodus, Sectio prior: 148(1803).

On dry earth of steppe.

Shaanxi, Yunnan, Hunan(Zahlbr., 1930 b, p. 68).

Diploschistes sinensis H. Magn. 中华双缘衣

Lich. Centr. Asia 1: 33(1940).

Type: Gansu, collected by Bohlin in Jan. of 1932(S, not seen).

On calcareous soil.

Diploschistes tianshanensis A. Abbas, S. Y. Guo & Ababaikeli 天山双缘衣

Mycotaxon 129(2): 569(2014).

Type: China. Xinjiang: Urumqi Co., Tianshan Mountain, Bayi forest farm, 43°21′N 86°49′E, alt. 1760 m, on rotten wood of Picea schrenkiana Fisch. & C. A. Mey., 27 Aug. 2011, A. Abbas 110821(Holotype, HMAS-L; isotype, XJU; GenBank KC959951, KC959952).

On rotten wood of *Picea schrenkiana*

Diploschistes xinjiangensis A. Abbas & S. Y. Guo 新疆双缘衣

Mycotaxon 129(2): 469(2014).

Type location: China, Xinjiang, Urumqi Co., South Mountain, Aketa, 3-VIII – 2011, A. Abbas 11821(Holotype, HMAS-L; isotype, XJU).

Dyplolabia A. Massal. **白唇衣属**

Neagenea Lich: 6(1854).

Dyplolabia afzelii(Ach.) A. Massal. 白唇衣

Neagenea Lich.: 6(1854).

≡*Graphis afzelii* Ach., *Syn. meth. lich.* (Lund): 85(1814).

On bark of trees.

Yunnan, Guangxi(Miao et al., 2007, pp. 495 – 496), Hainan(J. C. Wei et al., 2013, p. 233).

On exposed trees, e. g. Callitris, especially on branches. Pantropical.

Hongkong(Thrower, 1988, p. 102; Aptroot & Seaward, 1999, p. 76).

Fissurina Fée **裂隙衣属**

Essai Crypt. Exot. (Paris): 59(1825)[1824].

Fissurina adscribens(Nyl.) Z. F. Jia & Lücking 散裂隙衣

MycoKeys 25: 22(2017).

≡*Graphis adscribens* Nyl., *Bull. Soc. linn. Normandie*, sér. 22: 117(1868).

≡*Graphina adscribens* (Nyl.) Müll. Arg. in *Hedwigia* 31: 284(1892).

On bark.

Hongkong(Thrower 1988, p. 92).

= *Graphina olivascens* Zahlbr. In Handel-Mazzetti, *Symb. Sin.* 3: 57(1930) & *Cat. Lich. Univ.* 8: 214(1932).

Type: Hunan, Handel-Mazzetti no. 11220(W).

On bark of *Thea cuspidata*.

Fissurina baishanzuensis Kalb & Z. F. Jia 百山祖裂隙衣

Phytotaxa 189(1): 148 – 149(2014).

Type: China, Zhejiang: Qingyuan County; Mt. Baishanzu, Baishanzu National Nature Reserve, below the Baishanzu Protection Station; 27°45′18″N, 119°11′41″E, 1300 – 1500 m; at the edge of a mixed mountain rainforest; 11 October 2010, K. Kalb, Z. S. Sun & Z. F. Jia(holotype HMAS-L; isotype herb. Kalb 38639).

Fissurina cingalina(Nyl.) Staiger 锡兰裂隙衣

Biblioth. Lichenol. 85: 128 – 130(2002).

≡*Graphis cingalina* Nyl., *Acta Soc. Sci. Fenn.*, 26(10): 21(1900).

≡*Graphina cingalina* (Nyl.) Zahlbr., *Cat. Lich. Univ.* 2: 401(1923).

On bark of trees.

Hunan(Meng et al., 2011, P p. 160 – 162).

Fissurina columbina(Tuck.) Staiger 鸽色裂隙衣

Biblthca Lichenol. 85: 130(2002).

≡*Graphis columbina* Tuck., *Syn. N. Amer. Lich.* (Boston) 2: 123(1883).

On bark.

Guizhou, Hunan(Meng et al., 2011, Pp162 – 163), Hainan(Meng et al., 2011, p. 162 – 163; J. C. Wei et al., 2013, p. 233).

Fissurina dumastii Fée 裂隙衣

Essai Crypt. Exot. (Paris): 59(1825)[1824].

On bark.

Hainan(J. C. Wei et al., 2013, p. 233).

≡*Graphis dumastii*(Fée) Spreng. *Syst. veg.*, Edn 16 4(1): 254(1827).

Misapplied name(revised by Aptroot & Seaward, 1999, p. 76 – 77):

Graphis tachygrapha auct non Nyl. in Thrower, 1988, p. 16.

On exposed and sheltered trees. Pantropical.

Hainan(Meng et al., 2011, Pp163 – 164), Taiwan(Aptroot and Sparrius, 2003, p. 19 as *G. dumastii*), Hongkong(Thrower, 1988, pp. 16, 103, 105; Aptroot & Seaward, 1999, p. 76 – 77; Aptroot et al., 2001, p. 328).

Fissurina elaiocarpa(A. W. Archer) A. W. Archer 缘裂隙衣

Telopea 11(1): 71(2005).

≡*Graphina elaiocarpa* A. W. Archer, *Mycotaxon* 77: 165(2001).

≡*Fissurina marginata* Staiger, *Biblthca Lichenol.* 85: 144(2002).

On bark of trees.

Guizhou(Meng et al., 2011, p. 165).

Fissurina glauca(Müll. Arg.) Staiger 灰绿裂隙衣

Biblthca Lichenol. 85: 159(2002).

≡*Graphis glauca* Müll. Arg., *Bull. Herb. Boissier* 3: 58(1893).

On bark.

Hainan(Meng et al., 2011, p164; J. C. Wei et al., 2013, p. 234).

Fissurina incrustans Fée 硬壳裂隙衣

Essai Crypt. Exot. (Paris)(1825)[1824].

On bark of Rhus and other shrubs.

hongkong(Seaward & Aptroot, 2005, p. 284).

≡*Graphina incrustans*(Fée) Müll. Arg. *Mém. Soc. Phys. Hist. nat.* Genève 29(8): 47(1887).

Misapplied name(revised by Aptroot & Seward, 1999, p. 76):

Graphina adscribens auct non(Nyl.) Müll. Arg. in Thrower, 1988, pp. 15, 92.

Graphina colliculosa auct non(Mont.) Hale in Thrower, 1988, pp. 15, 94, 95.

On exposed trees and branches. Pantropical.

Hongkong (Thrower, 1988, pp. 15, 92, 94, 95, 98; Aptroot & Seward, 1999, p. 75 – 76; Aptroot et al. , 2001, p. 328, as *Graphina incrustans*) .

Fissurina isidiata Z. F. Jia 针芽裂隙衣

In Jia ZF et al. , *Mycotaxon* 121: 75 – 76(2012) .

Type: China. Hainan Island, Mountain Wuzhi, 18°46′N 109°31′E, alt. 950 m, 30. XI. 2010, Ze-Feng Jia 10 – 612(Holotype, HMAS-L 117919; isotype, LHS) .

Fissurina micromma(Zahlbr.) Aptroot 小裂隙衣

Symb. Bot. Ups. 34: 34(2004) .

≡ *Phaeographina micromma* Zahlbr. , *Repert. Spec. Nov. Regni Veg.* 31: 219(1933) .

Type: Taiwan, Asahina no. 376.

On bark of trees.

Taiwan(Zahlbr. , 1940, p. 191; Lamb, 1963, p. 545; Wang & Lai, 1973, p. 95) .

Fissurina nitidescens(Nyl.) Nyl. 小孢裂隙衣

Lich. Japon. : 108(1890) .

≡ *Graphis nitidescens* Nyl. , in Tuckerman, *Syn. N. Amer. Lich.* (Boston) 2: 123(1888) .

On bark & rocks.

Hunan, Guangxi, Fujian(Meng et al. , 2011, p. 166 – 168) , Hainan(Meng et al. , 2011, Pp166 – 167; J. C. Wei et al. , 2013, p. 234) .

Fissurina radiata Mont. 辐射裂隙衣

Annls Sci. Nat. , *Bot.* , sér. 2 18: 280(1842) .

≡ *Graphis radiata*(Mont.) Nyl. *Acta Soc. Sci. fenn.* 7: 473(1863) .

On *Mangifera indica*, and *Ginkgo biloba*,

Taiwan(Aptroot and Sparrius, 2003, p. 20) .

Fissurina subundulata Kalb & Z. F. Jia 亚裂隙衣

Phytotaxa 189(1) : 149(2014) .

Type: China, Zhejiang: Qingyuan County; Mt. Baishanzu, Baishanzu National Nature Reserve, above the Baishanzu Protection Station; 27°45′19″N, 119°12′06″E, 1500-1750 m; in a mixed mountain rainforest; 12 October 2010, K. Kalb, Z. S. Sun & Z. F. Jia(holotype HMAS-L; isotype herb. Kalb 38689)

Fissurina triticea(Nyl.) Staiger 磨裂隙衣

Biblthca Lichenol. 85: 156(2002) .

≡ *Graphis triticea* Nyl. *Acta Soc. Sci. fenn.* 7(2) : 470(1863) .

On *Pseudotsuga wilsoniana*, and *Ficus beeshiana*.

Taiwan(Aptroot and Sparrius, 2003, p. 20 as *G. triticea*) .

Glyphis Ach. **刻痕衣属**

Syn. meth. lich. (Lund) : 106(1814) .

Glyphis cicatricosa Ach. 刻痕衣

Syn. meth. lich. (Lund) : 107(1814) .

= *Glyphis favulosa* Ach. , *Syn. meth. lich.* (Lund) : 107(1814) , (Zahlbr. , 1930 b, p. 63) .

Corticolous.

Yunnan, Guangdong(Miao et al. , 2007, p. 496; Z. F. Jia & J. C. Wei, 2016, p. 40) , Shanghai(Nyl. & Cromb. 1883, p. 65, as *Glyphis favulosa* Ach. , fide Zahlbr. , 1930 b, p. 63; Zahlbr. , 1930 b, p. 63; Xu B. S. 1989, p.

176), Jiangsu(Xu B. S. 1989, p. 176; Wu JN & Wu JF 1991, p. 23; Tong LJ et al. 2001, p. 61), Zhejiang(Xu B. S. 1989, p. 176; Tong LJ et al. 2001, p. 61), Hainan(Z. F. Jia & J. C. Wei, 2016, p. 40), Taiwan(Zahlbr., 1933 b, p. 221; Wang & Lai, 1973, p. 90), Hongkong(Thrower, 1988, p. 91; Aptroot & Seaward, 1999, p. 75; Seaward & Aptroot, 2005, p. 285, Pfister 1978: 123, as *Glyphis confluens* Nyl.).

Glyphis cicatricosa var. intermedia(Müll. Arg.) Zahlbr. 间型变种

Cat. Lich. Univ. 2: 456(1923)[1924].

≡*Glyphis favulosa* var. *intermedia* Müll. Arg., Mém. Soc. Phys. Hist. nat. Genève 29(no. 8): 61(1887).

Corticolous.

Yunnan(Zahlbr., 1930 b, p. 63).

Glyphis scyphulifera(Ach.) Staiger 小杯刻痕衣

Biblthca Lichenol. 85: 175(2002).

≡*Lecidea scyphulifera* Ach., *Syn. meth. lich.* (Lund): 27(1814).

On bark of Rhus.

Hongkong(Seaward & Aptroot, 2005, p. 285, Pfister 1978: 101, as *Gyrostomum scyphuliferum*).

= *Phaeographina obfirmata* (Nyl.) Zahlbr. *Cat. Lich. Univ.* 2: 441(1923)[1924].

≡*Lecanactis obfirmata* Nyl., *J. Linn. Soc., Bot.* 20: 65(1883).

Type: Shanghai, A. C. Maingay, Confucian-temple garden.

Shanghai(Nyl. 1891, p. 12; Hue, 1891, p. 168; Zahlbr., 1930 b, p. 61),

On trees.

Graphis Adans. **文字衣属**

Fam. Pl. 2: 11(1763).

= *Graphina* Müll. Arg. *Flora*, Regensburg 63(2): 22(1880).

Graphis acharii Fée 阿瑞氏文字衣

Essai Crypt. Exot. (Paris): 39(1825)[1824].

≡*Graphina acharii*(Fée) Müll. Arg. *Mém. Soc. Phys. Hist. nat. Genève* 29(no. 8): 38(1887).

On *Acer* sp., and *Eriobotrya* sp.

Taiwan(Aptroot and Sparrius, 2003, p. 17, as *Graphina acharii*), Hongkong(Aptroot et al., 2001, p. 326, as *Graphina acharii*).

Graphis albissima Müll. Arg. 灰枝文字衣

Bull. Herb. Boissier 3: 319(1895).

Graphis albissima Vain., *Ann. Acad. Sci. fenn.*, Ser. A 6(no. 7): 159(1915), nom. illegit., Art. 53. 1.

On tree bark.

Yunnan, Guizhou, Hubei(Z. F. Jia & J. C. Wei, 2011, p. 221; Z. F. Jia & J. C. Wei, 2016, p. 53).

Graphis alpestris(Zahlbr.) Staiger 高山文字衣

Biblthca Lichenol. 85: 205(2002), Z. F. Jia & R. Lücking *MycoKyes* 21: 23(2017).

≡*Graphina*(*Eugraphina*) *alpestris* Zahlbr., in Handel-Mazzetti, *Symb. Sin.* 3: 54, 56(1930), Zahlbr. *Cat. Lich. Univ.* 8: 211(1931 – 1932).

Type locality: Yunnan, Handel-Mazzetti no. 6982, On trunk of coniferous trees.

= *Phaeographina pluviisilvarum* Zahlbr. *in* Handel-Mazzetti, *Symb. Sin.* 3: 60(1930) & *Cat. Lich. Univ.* 8: 217(1932).

Type locality: Yunnan, Handel-Mazzetti no. 9261(W).

On bark of *Betula* sp. & on tea trees.

Yunnan(Zahlbr. 1930, p. 56; Wang L. S. et al. 2008, p. 536; Z. F. Jia & J. C. Wei, 2011, p. 221; Z. F. Jia & J. C. Wei, 2016, p. 54), Hainan(Z. F. Jia & J. C. Wei 2011, p. 221; Z. F. Jia & J. C. Wei, 2016, p. 54).

Graphis analoga Nyl. 均文字衣

Annls Sci. Nat., *Bot.*, sér. 4 11: 244(1859).

≡*Graphis scripta* var. *analoga*(Nyl.) Tuck., Gen. lich. (Amherst): 207(1872).

= *Graphina analoga*(Nyl.) Zahlbr. *Denkschr. Kaiserl. Akad. Wiss.*, *Math.-Naturwiss.* Kl. 83: 107(1909).

≡ *Graphina analoga* var. *analoga*(Nyl.) Zahlbr., *Denkschr. Kaiserl. Akad. Wiss. Wien, Math.-Naturwiss.* Kl. 83: 107(1909).

On exposed trees and also on granite along stream. Pantropical.

Misapplied name(revised by Aptroot & Seaward, 1999, p. 75):

Graphina cleistoblephara auct non(Nyl.) Zahlbr. in Thrower, 1988, pp. 15, 93.

Graphina hiascens auct non Müll. Arg. in Thrower, 1988, pp. 15, 96.

= *Graphina subpublicaris* Zahlbr. in Fedde, *Repertorium*, 31: 212(1933).

Type: Taiwan, Faurie no. 127 & Asahina no. 300(syntypes).

Taiwan(Zahlbr., 1940, p. 185; Lamb, 1963, p. 252; Wang & Lai, 1973, p. 90 & 1976, p. 227; Aptroot and Sparrius, 2003, p. 17), Hongkong(Thrower, 1988, pp. 15, 93, 96; Aptroot & Seaward, 1999, p. 75; Aptroot et al., 2001, p. 326, as *Graphina analoga*; Seaward & Aptroot, 2005, p. 285, (a) Pfister 1978: 121, as *Graphis cleistoblephora* [sic]. No habitat details [bark]. n. d. *C. Wright*. Isotype [Label also as *Graphis cleistoblephora* [sic]; annotated by Nylander]. (b) Not cited in Pfister 1978(perhaps a duplicate of above)].

On bark of various bushes.

Mountains. 23. 3. 1854. *C. Wright*. See Staiger(2002), p. 222. holotype *ex* herb. Tuckerman H-Nyl 7589.), China(prov. not indicated: Hue, 1891, p. 155).

Graphis anfractuosa(Eschw.) Eschw. 螺曲文字衣

in Martius, *Fl. Bras.* 1: 86(1833)[1829 – 1833].

≡*Scaphis anfractuosa* Eschw., *Syst. Lich.*: 25(1824).

On bark of shrubs in dense thickets.

Hongkong(Seaward & Aptroot, 2005, p. 285, Pfister 1978: 121, as *Graphis assimilis*).

Graphis aperiens Müll. Arg. 顶生文字衣

Flora, Regensburg 74(1): 113(1891).

On tea trees.

Yunnan(Wang L. S. et al. 2008, p. 536; Z. F. Jia & J. C. Wei, 2011, p. 221; Z. F. Jia & J. C. Wei, 2016, p. 55).

Graphis aphanes Mont. & Bosch 隐文字衣

in Miquel, *Pl. Jungh.* 4: 474(1855).

Misapplied name(revised by Aptroot & Seaward, 1999, p. 76):

Phaeographina maxima Gro enh. (in Thrower, 1988, pp. 16, 136).

On sheltered trees and also on granite boulders along streams. Known from tropical Asia.

Hongkong(Thrower, 1988, pp. 16, 136; Aptroot & Seaward, 1999, p. 76; Aptroot et al., 2001, p. 328).

Graphis assimilis Nyl. 黑脉文字衣

Bull. Soc. Linn. Normandie, sér. 2, 2: 109(1868).

= *Graphis spodoplaca* Zahlbr., *Annls mycol.* 30(5/6): 429, 1932. Zahlbr., *Hedwigia*: 196(1934). Zahlbr., *Cat. Lich. Univ.*: 162(1940). J. C. Wei, *Enum. Lich. China*: 104, (1991).

Type: Fujian, Chung no. 569 f.

Corticolous.

Fujian(Zahlbr., 1934 b, p. 196 & 1940, p. 162, as *G. spodoplaca*), Hainan(JCWei et al., 2013, p. 234).

Yunnan, Guizhou, Hunan, Zhejiang(Z. F. Jia & J. C. Wei, (2016, p. 56 – 57), Fujian(Zahlbr. 1934, p. 196 & 1940, p. 162 as *G. spodoplaca Zahlbr.*; Z. F. Jia & J. C. Wei, 2011, p. 221), Guangdong, Hainan(Z. F. Jia & J.

C. Wei, 2011, p. 221; Z. F. Jia & J. C. Wei, 2016, p. 56 – 57), Hongkong(Guo W. Et al., 2015, p. 430; Z. F. Jia & J. C. Wei, 2016, p. 57).

Graphis asterizans Nyl. 星雀文字衣

Acta Soc. Sci. fenn. 7: 467(1863).

Corticolous.

China(prov. not indicated: Hue, 1891, p. 156; Zahlbr., 1924, p. 294).

See *Graphis rimulosa*(Mont.) Trevisan for Hongkong(Zahlbr., 1930 b, p. 42; Thrower, 1988, p. 15),

Graphis bifera Zahlbr. 双果文字衣

In Handel-Mazzetti, *Symb. Sin.* 3: 39, 47(1930) & *Cat. Lich. Univ.* 8: 201(1931 – 1932).

Type: Hunan, Handel-Mazzetti Nos. 12262, 12341(syntypes).

On bark of *Quercus glandulifera.*

Yunnan, Zhejiang, Fujian(Z. F. Jia & J. C. Wei, 2016, p. 58), Hunan(Zahlbr. 1930, p. 47), Hainan(Z. F. Jia & J. C. Wei, 2011, p. 221; Z. F. Jia & J. C. Wei, 2016, p. 58).

Graphis benguetensis Vain. 云杉文字衣

Ann. Acad. Sci. fenn., Ser. A 15(6): 222(1921).

On *Picea morrisonicola*.

Taiwan(Aptroot and Sparrius, 2003, p. 18).

Graphis caesiella Vain. 淡兰文字衣

Acta Soc. Fauna Flora fenn. 7(2): 122(1890). J. C. Wei, *Enum. Lich. China*: 100(1991).

Misapplied name(revised by Aptroot and Seaward, 1999, p. 76):

Graphis glaucescens auct. non Fée in Thrower, 1988, p. 16.

Graphis glauconigra auct. non Vainio in Thrower, 1988, p. 16.

On exposed and sheltered trees and branches, rarely on granite boulders. Pantropical.

Yunnan(Z. F. Jia & J. C. Wei, 2016, p. 59), Taiwan(Aptroot and Sparrius, 2003, p. 18),

Hainan(Z. F. Jia & J. C. Wei, 2011, p. 221; JCWei et al., 2013, p. 234; Z. F. Jia & J. C. Wei, 2016, p. 59), Hongkong(Thrower, 1988, p. 16 as *Graphis glaucescens* non Fée and *Graphis glauconigra* non Vainio; Aptroot & Seaward, 1999, p. 76; Aptroot et al., 2001, p. 328; Z. F. Jia & J. C. Wei, 2016, p. 59).

Graphis centrifuga Räsänen 离心文字衣

Suom. Elain-ja Kasvit. Seuran Van. Tiedon. Pöytäkirjat 3: 186(1949)

On bark.

Hongkong(Guo W. et al., 2015, p. 430).

Graphis cervina Müll. Arg. 鹿色文字衣

Nuovo G. bot. ital. 24: 199(1892), J. C. Wei, *Enum. Lich. China*: 100(1991). J. C. Wei, *Enum. Lich. China*: 100 (1991).

See Graphis lineola Ach. for Hongkong(Thrower, 1988, pp. 103, 104, 105).

Yunnan(Z. F. Jia & J. C. Wei, 2016, p. 61), Guizhou(Zhang T. et al. 2006, p. 7; Z. F. Jia & J. C. Wei 2011, p. 221; Z. F. Jia & J. C. Wei, 2016, p. 61), Hunan(Zahlbr. 1930, p. 40; Z. F. Jia & J. C. Wei, 2016, p. 61), Zhejiang(Xu BS 1989, p178; Wu JL 1987, p. 41; Z. F. Jia & J. C. Wei, 2016, p. 60), Fujian(Z. F. Jia & J. C. Wei 2011, p. 221; Z. F. Jia & J. C. Wei, 2016, p. 61), Hainan(Z. F. Jia & J. C. Wei, 2016, p. 61), Hongkong(Thrower 1988, pp. 103, 104, 105; Z. F. Jia & J. C. Wei, 2016, p. 61).

Graphis cervinonigra Zahlbr. 黑红文字衣

In Fedde, *Repertorium*, 31: 210(1933).

Type locality: Taiwan, Faurie no. 236.

Corticolous.

Taiwan(Zahlbr., 1940, p. 153; Lamb, 1963, p. 253; Wang & Lai, 1973, p. 90).

Graphis cincta(Pers.) Aptroot 环带文字衣

Flora of Australia(Melbourne) 57: 651(2009) .

≡*Opegrapha cincta* Pers. , *Ann. Wetter. Ges.* 2: 15(1811) . in Zahlbr. 's *Cat. Lich. Univ.* 2: 184(1923 – 1924) .

= *Graphis guimarana* Vain. , *Ann. Acad. Sci. Fenn.* , Ser. A, 7(6) : 160(1915) . Lücking et al. , *Lichenologist* 41 (4) : 402(2009) (Syn. Contr.) . Thower, *Hong Kong Lichens:* 103 – 105(1988) . J. C. Wei, *Enum. Lich. China*: 101(1991) .

= *Graphis latibasa* Zahlbr. in Fedde, *Repertorium*, 31: 207 (1933) . Lücking et al. , *Lichenologist* 41 (4) : 402 (2009) (Syn. Contr.) .

Type: Taiwan, Faurie no. 291.

Taiwan(Zahlbr. , 1940, p. 158; Lamb, 1963, p. 254; Wang & Lai, 1973, p. 90) .

Corticolous.

Zhejiang(Z. F. Jia & J. C. Wei, 2016, p. 62) , Fujian(Z. F. Jia & J. C. Wei 2011, p. 211; Z. F. Jia & J. C. Wei, 2016, p. 62) , Taiwan(Zahlbr. 1940, p. 158; Lamb 1963, p. 254; Wang & Lai 1973, p. 90 as *Graphis latibasa*) , Hainan(Z. F. Jia & J. C. Wei 2011, p. 211; J. C. Wei et al. , 2013, p. 234; Z. F. Jia & J. C. Wei, 2016, p. 62) , Hongkong(Thrower 1988, pp. 103, 104, 105 as *Graphis guimarana*) .

Graphis cleistoblephara Nyl. 闭毛文字衣

Ann. Sci. Nat. Bot. sér. 4, 20: 265(1863) .

≡*Graphina cleistoblephara*(Nyl.) Zahlbr. , *Cat. Lich. Univ*, 2: 401(1923) . J. C. Wei, *Enum.* Lich. China: 97(1991) .

Type: China(Hong Kong) , s. col. (H-Nylander 7589) .

On tree bark.

= *Graphina subpublicaris* Zahlbr. *Feddes Repert.* 31: 212(1933) .

Type: Taiwan, Faurie No. 127 & Asahina No. 300

Yunnan(Z. F. Jia & J. C. Wei 2011, p. 222; Z. F. Jia & J. C. Wei, 2016, p. 63) , Taiwan(Zahlbr. 1940, p. 185; Lamb 1963, p. 252; Wang & Lai 1973, p. 90 & 1976, p. 227) , Hainan(Z. F. Jia & J. C. Wei 2011, p. 222; J. C. Wei et al. , 2013, p. 234; Z. F. Jia & J. C. Wei, 2016, p. 63) , Hongkong(Thrower 1988, p. 93; Z. F. Jia & J. C. Wei 2011, p. 222; Z. F. Jia & J. C. Wei, 2016, p. 63) . China(prov. non in-dicated: Hue 1891, p. 155) .

Graphis cognata Müll. Arg. 钝盘文字衣

Nuovo G. bot. ital. 24: 200(1892) .

On tree bark.

Zhejiang, Hainan(Z. F. Jia & Kalb 2014, p. 964; Z. F. Jia & J. C. Wei, 2016, p. 65) .

Graphis conferta Zenker 密集文字衣

in Goebel et Kunze, *Pharmazeut. Waarenkunde* 1(3) : 166(1827 – 29) .

On tree bark.

Yunnan, Guizhou, Fujian(Z. F. Jia & J. C. Wei 2011, p. 222; Z. F. Jia & J. C. Wei, 2016, p. 66) , Zhejiang(Z. F. Jia & J. C. Wei, 2016, p. 66) , Hainan(Z. F. Jia & J. C. Wei 2011, p. 222; JCWei et al. , 2013, p. 234; Z. F. Jia & J. C. Wei, 2016, p. 66) .

Graphis dendrogramma Nyl. 树突文字衣

in Cromb. in *Journ. Linn. Soc. London, Botan.* , 16: 226(1877) .

On tree bark.

Fujian(Z. F. Jia & Kalb 2014, p. 965; Z. F. Jia & J. C. Wei, 2016, p. 67) .

Graphis descissa Müll. Arg. 裂出文字衣

Bull. Herb. Boissier 3: 318(1895) .

= *Graphis ocellata* Zahlbr. in Fedde, *Repertorium*, 31: 209 (1933) . J. C. Wei, *Enum. Lich. China*: 103, 1991. Lücking et al. , Lichenologist 41(4) : 409, 2009(Syn. Contr.) .

Type: Taiwan, Faurie no. 38.

Corticolous.

Yunnan, Hongkong(Guo W. et al. , 2015, p. 431; Z. F. Jia & J. C. Wei, 2016, p. 68), Hainan(Z. F. Jia & J. C. Wei, 2011, p. 222; Z. F. Jia & J. C. Wei, 2016, p. 68), Taiwan(Zahlbr. , 1940, p. 160; Lamb, 1963, p. 255; Wang & Lai, 1973, p. 90 as *Graphis ocellata*; Z. F. Jia & J. C. Wei 2011, p. 222), Hainan(J. C. Wei et al. , 2013, p. 234; Jia Z. F. et al. , 2016. p68).

Graphis deserpens Vain. 蜕皮文字衣

Ann. Acad. Sci. fenn. 15(6): 202(1921).

On tree bark.

Shaanxi, Fujian(Z. F. Jia & J. C. Wei 2011, p. 222; Z. F. Jia & J. C. Wei, 2016, p. 69).

Graphis desquamescens(Fée) Zahlbr. 无鳞文字衣

Denkschr. Kaiserl. Akad. Wiss. Wien, Math. -Naturwiss. Kl. 83: 108(1909).

≡*Opegrapha desquamescens* Fée, *Bull. Soc. Bot. France* 21: 24(1874).

On trees.

Zhejiang(Wu JL 1987, p. 41; Xu B. S. 1989, p. 178), Fujian, guangxi(Z. F. Jia & J. C. Wei 2011, p. 222; Z. F. Jia & J. C. Wei, 2016, p. 70), Taiwan(Wang & Lai, 1973, p. 90), Hainan(Z. F. Jia & J. C. Wei, 2016, p. 70; JCWei et al. , 2013, p. 234), Guangdong(Z. F. Jia & J. C. Wei, 2016, p. 70), See Graphis lineola Ach. for Hongkong(Thrower, 1988, pp. 103, 104, 105; Z. F. Jia & J. C. Wei, 2016, p. 70).

Graphis dupaxana Vain. 曲盘文字衣

Ann. Acad. Sci. fenn. , Ser. A 15(6): 241(1921). J. C. Wei, *Enum. Lich. China*: 101(1991).

On tree bark.

Yunnan(Wang LS 2008, p. 536; Jia & Wei 2011, p. 222; Z. F. Jia & J. C. Wei, 2016, p. 72), Zhejiang(Z. F. Jia & J. C. Wei, 2016, p. 72), Fujian(Jia & Wei 2011, p. 222), Taiwan(Wang & Lai, 1973, p. 90), Hainan(J. C. Wei et al. , 2013, p. 234; Z. F. Jia & J. C. Wei, 2016, p. 72). See *Graphis rimulosa* (Mont.) Trevisan for Hongkong(Thrower, 1988, pp. 103, 104, 105).

Graphis duplicata Ach. 双文字衣

Syn. meth. lich. (Lund): 81(1814).

= *Graphis lopingensis* Zahlbr. , *Symbol. Sinic.* , 3: 39, 48(1930) & *Cat. Lich. Univ.* 8: 203(1932).

Synonymized by Lücking et al. in *Lichenologist* 41(4): 425(2009).

Type: Yunnan, Handel-Mazzetti no. 10213.

On trunks of *Crataegus* sp.

= *Graphis lussuensis* Zahlbr. , *Symbol. Sinic.* , 3: 40, 49(1930) & *Cat. Lich. Univ.* 8: 203(1932).

Synonymized by Lücking et al. in *Lichenologist* 41(4): 425(2009).

Type: Yunnan, Handel-Mazzetti no. 9103.

On twigs of *Meliosma* sp.

Graphis duplicata Eschw. , in Martius, *Fl. Bras.* 1: 75(1833) [1829 – 1833]. Nomenclatural comment: Nom. illegit. , Art. 53. 1

On tree bark.

Sichuan, Fujian, Guangxi(Z. F. Jia & J. C. Wei 2011, p. 222; Z. F. Jia & J. C. Wei, 2016, p. 73), Yunnan (Zahlbr. 1930, p. 48 as *G. lopingensis* Zahlbr. & p. 49 as *G. lussuensis* Zahlbr. ; Zahlbr. 1930, p. 45; Z. F. Jia & J. C. Wei 2011, p. 222; Z. F. Jia & J. C. Wei, 2016, p. 73), Guizhou(Zahlbr. 1930, p. 45; Z. F. Jia & J. C. Wei 2011, p. 222; Z. F. Jia & J. C. Wei, 2016), Xizang, Hongkong(Guo W. et al. , 2015, p. 431; Z. F. Jia & J. C. Wei, 2016, p. 73).

Graphis duplicata var. peruviana(Fée) Zahlbr. 南美变种

Cat. Lich. Univ. 2: 303(1923) [1924]. J. C. Wei, *Enum. Lich. China*: 101(1991).

≡*Opegrapha peruviana* Fée, *Essai Crypt. Exot.* (Paris): 27(1825)[1824].

On bark of *Liquidambar formosana*.

Guizhou, Yunnan(Zahlbr., 1930 b, p. 45).

Graphis elegantula Zahlbr. 齐文字衣

in Handel-Mazzetti, *Symb. Sin.* 3: 50(1930) & *Cat. Lich. Univ.* 8: 202(1932). J. C. Wei, *Enum. Lich. China*: 101 (1991).

Type: Hunan, Handel-Mazzeti no. 11222.

On bark of *Thea cuspidata*.

Yunnan, Fujian(Z. F. Jia & J. C. Wei, 2016, p. 74), Guizhou(Z. F. Jia & J. C. Wei 2011, p. 222; Z. F. Jia & J. C. Wei, 2016, p. 74), Hunan(Zahlbr., 1930, p. 50).

Graphis elongata Zenker 长文字衣

in Goebel & Kunze, *Pharmaceutische Waarenkunde*(Eisenach) 1(1827).

Yunnan(Liu ZL et al., 2015, pp. 1 – 3). +

Graphis endoxantha Nyl. 内黄文字衣

Annls Sci. Nat., Bot., sér. 4 15: 50(1861). J. C. Wei, *Enum. Lich. China*: 101(1991).

On tree bark.

Guizhou(Z. F. Jia & J. C. Wei, 2011, p. 223; Z. F. Jia & J. C. Wei, 2016, p. 75), Fujian(Z. F. Jia & J. C. Wei, 2016, p. 75), Taiwan(Wang & Lai, 1973, p. 90).

Graphis epiphloea Zahlbr. 树表文字衣

in Fedde, *Repertorium*, 31: 210(1933). J. C. Wei, *Enum. Lich. China*: 101(1991).

Type: Taiwan, Faurie nos. 24, 205 & Asahina no. 354(syntypes).

Yunnan, Fujian(Z. F. Jia & J. C. Wei, 2016, p. 76), Taiwan(Zahlbr., 1940, p. 154; Lamb, 1963, p. 253; Wang & Lai, 1973, p. 90).

Corticolous.

Graphis filiformis Adaw. & Makhija 丝线文字衣

Mycotaxon 99: 314(2007).

On tree bark.

Guizhou, Fujian(Z. F. Jia & J. C. Wei, 2016, p. 77), Hainan(Z. F. Jia & J. C. Wei, 2011, p. 223 and 2016, p77: J. C. Wei et al., 2013, p. 234).

Graphis fissurata M. Nakan. & H. Harada 裂隙文字衣

Nat. Hist. Res. 5(2): 67(1999).

On bark.

Fujian(Jia ZF & J. C. Wei, 2007, pp. 187 – 188; Jia ZF & J. C. Wei, 2011, p. 223 & 2016, p. 78).

Graphis fujianenesis Jia Z. F. & J. C. Wei 福建文字衣

Mycotaxon 104: 107 – 109(2005).

Type: Fujian, Jianou, Wanmulin, alt. 580m, ad saxa, 2/VI/2007, Li Jing11182(holotype in LHS; isotype in HMAS-L).

Fujian(Z. F. Jia & J. C. Wei 2008, pp. 107-109; Z. F. Jia & J. C. Wei 2011, p. 223 & 2016, p. 79).

Graphis furcata Fée 叉形文字衣

Essai Cryptog. Écorc. Offcin., 40. 1824.

= *Graphis setschwanensis* Zahlbr. in Handel-Mazzetti, *Symb. Sin.* 3: 47(1930) & *Cat. Lich. Univ.* 8: 206(1932), J. C. Wei, *Enum. Lich. China*: 104(1991). Synonymized by Lücking et al. *in Lichenologist* 41(4): 397(2009).

Type: Sichuan, Handel-Mazzetti no. 3075.

On trunk of *Celti* sp.

Sichuan(Zahlbr., 1930, p. 47 as *Graphis setchwanensis* Zahlbr.), Gansu(Z. F. Jia & J. C. Wei 2011, p. 223; Z.

F. Jia & J. C. Wei, 2016, p. 81).

Graphis galactoderma (Zahlbr.) Lücking 乳皮文字衣

in Lücking et al., *Lichenologist* 41(4): 423(2009).

≡ *Graphina galactoderma* Zahlbr. in Handel-Mazzetti, *Symb. Sin.* 3: 57(1930) & *Cat. Lich. Univ.* 8: 213 (1932). J. C. Wei, *Enum. Lich. China*: 98(1991).

Type: China. Yunnan, Handel-Mazzetti no. 5912.

On living trunk of trees (Leguminosae).

Guizhou (Z. F. Jia & J. C. Wei, 2016, p. 82).

Graphis glaucescens Fée 灰白文字衣

Essai Crypt. Exot. (Paris): 36, pl. 8, fig. 3(1825) [1824]. J. C. Wei, *Enum. Lich. China*: 101(1991).

On *Mangifera indica*, *Castanopsis* sp., *Pinus* sp., and conglomeratic rock.

Yunnan (Z. F. Jia & J. C. Wei, 2016, p. 83), Taiwan (Aptroot and Sparrius, 2003, p. 19).

See Graphis caesiella Vainio for Hongkong (Thrower, 1988, pp. 103, 104, 105).

Graphis glauconigra Vain. 黑白文字衣

in Annal. Acad. Scient. Fennic. Ser. A, 15: 242, 1921. J. C. Wei, *Enum. Lich. China*: 101(1991).

= *Graphis chungii* Zahlbr. in Handel-Mazzetti, *Symbol. Sinic.* 3: 39, 41(1930). Zahlbr. in Handel-Mazzetti, *Symbol. Sinic.* 3: 41(1930). J. C. Wei, *Enum. Lich. China*: 100(1991).

Synonymized by Lücking et al. in *Lichenologist* 41(4): 428(2009).

Type: Fujian, Chung nos. 388, 466 e (syntypes).

Corticolous.

= *Graphis chungii* var. *oligospora* Zahlbr. in Handel-Mazzetti, *Symbol. Sinic.* 3: 39, 41(1930). J. C. Wei, *Enum. Lich. China*: 100(1991). Synonymized by Lücking et al. in *Lichenologist* 41(4): 428, 2009.

Type locality: Fujian, Chung no. 593 a.

Corticolous.

= *Graphis endophaea* Zahlbr., in *Annal. Mycolog.*, 30: 430(1932). Zahlbr., *Hedwigia* 74: 196, 1934. Zahlbr., *Cat. Lich. Univ.*, 10: 154, 1940. J. C. Wei, *Enum. Lich. China*: 101(1991).

Synonymized by Lücking et al. in *Lichenologist* 41(4): 428(2009).

Type: Fujian, Chung no. 585.

Fujian (Zahlbr., 1934, p. 196 & 1940, p. 154).

On tree bark.

Fujian (Zahlbr. 1930, p. 41 as *Graphis chungii* & *G. chungii var. ligospora*; Zahlbr. 1934, p. 196 & 1940, p. 154 as *G. endophaea*), Hainan (Z. F. Jia & J. C. Wei, 2011, 223 & 2016, p. 83; J. C. Wei et al., 2013, p. 234), Hongkong (Thrower 1988, pp. 103, 104, 105).

Graphis gonimica Zahlbr. 层藻文字衣

Annal. Mycolog. 30: 431(1932). J. C. Wei, *Enum. Lich. China*: 102(1991).

Type: Fujian, Chung nos. 591 a, 591 b (syntypes).

Fujian (Zahlbr., 1934 b, p. 196 & 1940, p. 156; Lamb, 1963, p. 253).

Corticolous.

= *Graphis sapii* Zahlbr., *Symbol. Sinic.*, 3: 39 – 40(1930). Zahlbr., *Cat. Lich. Univ.*, 3: 40(1930), J. C. Wei, *Enum. Lich. China*: 103(1991). Synonymized by Lücking et al. in *Lichenologist* 41(4): 417(2009).

Holotype: China, Hunan, Handel-Mazzetti 12819(W; isolectotype: S 2169) & Fujian, Chung no. 348(syntypes).

On bark of *Sapium japonicum*.

Hunan (Zahlbr. 1930, p. 40, as *Graphis sapii*), Fujian (Zahlbr. 1934, p. 196; Zahlbr. 1940, p. 156; Lamb 1963, p. 253, as *Graphis gonimica*; Z. F. Jia & J. C. Wei 2011, p. 223; Z. F. Jia & J. C. Wei, 2016, p. 85), Hainan (Z.

F. Jia & J. C. Wei, 2016, p. 85)., hongkong(Thrower 1988, p. 103, 105, as *Graphis sapii*; Z. F. Jia & J. C. Wei, 2016, p. 85).

See Graphis lineola Ach. for Hongkong(Thrower, 1988, pp. 103, 105).

Graphis guangdongensis Z. F. Jia & J. C. Wei 广东文字衣

Mycotaxon 110: 27 – 30(2009).

Type: Guangdong, Fengkai, Heishiding, 23°27′N, 113°30′E, alt. 250 m, on tree bark, X/28/1998. Shou-Yu Guo 2185(holotypus in HMAS-L X024581, isotypus in LHS).

Guangdong(Z. F. Jia & J. C. Wei 2009, pp. 27 – 33; Z. F. Jia & J. C. Wei 2011, p. 223; Z. F. Jia & J. C. Wei, 2016, p. 86).

Graphis handelii Zahlbr. 汉氏文字衣

in Handel-Mazzetti, *Symb. Sin.* 3: 44(1930) & *Cat. Lich. Univ.* 8: 202(1932).

Type: Hunan, Handel-Mazzetti no. 12788.

Corticolous.

Yunnan, Fujian(Z. F. Jia & J. C. Wei 2011, p. 224 & 2016, p. 87), Hunan(Zahlbr. 1930, p. 44; Wu JL 1987, p. 42; JCWei et al., 2013, p. 234), Jiangsu(Wu JL 1987, p. 42; Xu BS 1989, p. 179), Shanghai, Zhejiang(Xu B. S. 1989, p. 179), Hainan(Z. F. Jia & J. C. Wei, 2016, p. 87).

Graphis hiascens(Fée) Nyl. 裂文字衣

Annls Sci. Nat., Bot., sér. 4, 11: 226(1859).

≡ *Opegrapha hiascens* Fée, *Suppl. Assai Crypt. Écorc.* 25(1837).

≡ *Graphina hiascens*(Fée) Müll. Arg., *Mém. Soc. Phys. et Hist. Nat.* 29(8): 42(1887), J. C. Wei, *Enum. Lich. China*: 98(1991).

On *Eleocarpus silvestris* and *Podocarpus* sp.

Taiwan(Aptroot and Sparrius, 2003, p. 18), Hainan(Z. F. Jia & J. C. Wei 2011, p. 224, & 2016, p. 88; J. C. Wei et al., 2013, p. 234), Hongkong(Thrower, 1988, p. 96; Z. F. Jia & J. C. Wei 2011, p. 224 & 2016, p. 89).

Graphis hongkongensis Wei Guo & J. S. Hur 香港文字衣

Mycotaxon 130: 431(2015).

Type: China, Hong Kong, Lautau Island, the road from Lautau Peak to Nam Shan, on bark, 22°14′N 113°56′E, alt. 384 m, 15 Dec. 2011, W. Guo & J. X. Tian HK365(Holotype, HMAS-L 128226).

Graphis hossei Vain. 泰北文字衣

Ann. bot. Soc. Zool.-Bot. fenn. Vanamo 1(3): 53(1921).

= *Graphis bifera* var. *cinerea* Zahlbr. in Handel-Mazzetti, *Symbol. Sinic.*, 3: 47(1930). J. C. Wei, *Enum. Lich. China*: 100(1991). Synonymized by Lücking et al. in *Lichenologist* 41(4): 393(2009).

Holotype: China, Hunan, Handel-Mazzetti 12348(W).

On bark of *Corylpsis chinensis*.

= *Graphis cinerea*(Zahlbr.) M. Nakan., *J. Sci. Hiro. Univer.* Ser. B, 11(2): 63(1966).

= *Graphis connectens* Zahlbr., in Handel-Mazzetti, *Symbol. Sinic.* 3: 47(1930). J. C. Wei, *Enum. Lich. China*: 101 (1991). Synonymized by Lücking et al. in *Lichenologist* 41(4): 393(2009).

Holotype: China, Handel-Mazzetti 11429(W; isolectotype: S 2186).

On bark of Castanopsis fargesii.

Shaanxi, Yuannan, Guizhou, Fujian, Guangxi, Guangdong(Z. F. Jia & J. C. Wei, 2016, pp. 90 – 91), Hunan (Zahlbr. 1930, p. 47, as *Graphis bifera* var. *cinera*; Zahlbr. 1930, p. 41, as *Graphis connectens*; Z. F. Jia & J. C. Wei, 2011, p. 224 & 2016, p. 90), Zhejiang(Xu BS 1989, p. 178, as *Graphis connectens*; Z. F. Jia & J. C. Wei, 2016, p. 90), Hainan(J. C. Wei et al., 2013, p. 235; Z. F. Jia & J. C. Wei, 2016, pp. 90 – 91), Hongkong (Thrower 1988, pp. 103 – 105, as *Graphis bifera* var. *cinera*; Z. F. Jia & J. C. Wei, 2016, p. 91).

Graphis hunanensis(Zahlbr.) M. Nakan. & Kashiw. 湖南文字衣

Bull. Natn. Sci., *Tokyo*, Ser. B, 29(2) : 87, 2003.

≡ *Graphina hunanensis* Zahlbr. *in* Handel-Mazzetti, *Symb. Sin.* 3: 54 – 55(1930) & *Cat. Lich. Univ.* 8: 213 (1932) .

Type: Hunan, Handel-Mazzetti no. 11507(W) .

On rocks.

Graphis hyphosa Staiger 满菌文字衣

Bibl. Lichenol. 85: 235(2002) .

On tree bark.

Hongkong(Guo W. et al. , 2015, p. 433) .

Graphis immersella Müll. Arg. 浸鞍文字衣

Bull. Herb. Boissier 3: 319, 1895.

Lectotype(Archer 1999a) : Australia, Shirley 1793(G) .

= *Graphis manhaviensis* Zahlbr. in Handel-Mazzetti, *Symb. Sin.* 3: 46(1930) & *Cat. Lich. Univ.* 8: 203(1932) . J. C. Wei, *Enum. Lich. China*: 102(1991) . Synonymized by Lücking et al, in *Lichenologist* 41(4) : 396(2009) .

Type: Yunnan, Handel-Mazzetti no. 5831(W, holotype) .

On trunks of *Jasminum* sp.

Shaanxi, Guizhou, Zhejiang, Hainan, (Z. F. Jia & J. C. Wei, 2016, pp. 92 – 93) , Yunnan(Zahlbr. 1930, p. 46; 1932, p. 203 as *Graphis manhaviensis*; Z. F. Jia & J. C. Wei, 2016, p. 93) , Fujian(Z. F. Jia & J. C. Wei 2011, p. 224; Z. F. Jia & J. C. Wei, 2016, p. 92) , Hongkong(Guo W. et al. , 2015, p. 433; Z. F. Jia & J. C. Wei, 2016, p. 93) .

Graphis immersicans A. W. Archer 半陷文字衣

Aust. Syst. Bot. 14(2) : 262, 2001.

On tree bark.

Guizhou, Fujian(Z. F. Jia &J. C. Wei 2011, p. 224 & 2016, pp. 92 – 93) , Zhejiang(Z. F. Jia & J. C. Wei, 2016, pp. 92 – 93) , Hainan(J. C. Wei et al. , 2013, p. 235; Z. F. Jia & J. C. Wei, 2016, Pp. 92 – 93) , Hongkong (Guo W. et al. , 2015, pp. 433 – 434; Z. F. Jia & J. C. Wei, 2016, p. 93) .

Graphis intricata Eschw. 缠结文字衣

in Martius, *Fl. Bras.* 1: 79(1833) [1829 – 1833] .

Graphis intricata Fée, *Essai Crypt. Exot.* , *Suppl. Révis.* (Paris) : 42(1837) , nom. illegit. , Art. 53. 1.

= *Graphis hunana* Zahlbr. in Handel-Mazzetti, *Symb. Sin.* 3: 39 – 40(1930) . Synonymized by Lücking et al. in *Lichenologist* 41(4) : 410(2009) .

Holotype(saxicolous) : China, Handel-Mazzetti 12233(W; Zahlbr. 1930) .

On rocks and tree bark.

Zhejiang(Tong LJ 2001, p. 61; Z. F. Jia & J. C. Wei, 2016, p. 95) , Jiangsu(Xu BS 1989, p. 179; Wu JL 1987, p. 42) , Fujian(Zahlbr. 1934, p. 196; Wu JL 1987, p. 42; Jia & Wei 2011, p. 224; Z. F. Jia & J. C. Wei, 2016, p. 95) , Guangdong(Wu JL 1987, p. 42) , Hainan(Z. F. Jia & J. C. Wei, 2016, p. 95) .

Graphis japonica(Müll. Arg.) A. W. Archer & Lücking 日本文字衣

Lichenologist 41: 437(2009) .

≡ *Graphina japonica* Müll. Arg. , *Flora* 74: 113(1891) .

Type: Japan, Miyoshi 23(G) .

= *Graphina japonica* var. *major* Zahlbr. , in Fedde, *Repertor*. 31: 213(1933) . J. C. Wei, *Enum. Lich. China*: 98 (1991) . Synonymized by Lücking et al. in *Lichenologist* 41(4) : 437(2009) .

Lectotype: Taiwan, Asahina 338(W) .

Taiwan(Zahlbr. , 1940, p. 180; Wang & Lai, 1973, p. 90) .

= *Graphina verruculina* Zahlbr. in Handel-Mazzetti, *Symb. Sin.* 3: 54, 58 (1930). Zahlbr. in Handel-Mazzetti, *Symb. Sin.*: 58(1930). Zahlbr., *Cat. Lich. Univ.*, 8: 215(1932). J. C. Wei, *Enum. Lich. China*: 100(1991). Synonymized by Lücking et al. in *Lichenologist* 41(4): 437(2009).

Type: China, Fujian, Chung nos. 381, 600d(syntypes), Lectotype Chung 600d(W).

Notes: The locality of this species is in Fujian prov. of China but not in "China occidentalis" as cited by Zahlbr. in *Cat. Lich. Univ.* 8: 215(1932).

= *Graphina filiformis* Zahlbr., in Fedde, *Repertor.*, 31: 214(1933). J. C. Wei, *Enum. Lich. China*: 98(1991). Synonymized by Lücking et al. in *Lichenologist* 41(4): 437(2009).

Holotype: Taiwan, Faurie 156(W).

Taiwan(Zahlbr., 1940, p. 178; Wang & Lai, 1973, p. 90).

= *Graphina petrophila* Zahlbr., *in* Fedde, *Repertor.*, 31: 213(1933). J. C. Wei, *Enum. Lich. China*: 99(1991). Synonymized by Lücking et al. in *Lichenologist* 41(4): 437(2009).

Holotype: Taiwan, Faurie 83(W).

On granitic rocks.

Zhejiang(Z. F. Jia & J. C. Wei, 2016, p. 96), Fujian(Zahlbr. 1930b, p. 58; 1932a, p. 215 as *Graphina verruculina*; Z. F. Jia & J. C. Wei, 2011, p. 224; Z. F. Jia & J. C. Wei, 2016, p. 96), Hainan(Z. F. Jia & J. C. Wei, 2011, p. 224; JCWei et al., 2013, p. 235 as *G. japonica*; Z. F. Jia & J. C. Wei, 2016, p. 97), Hongkong(Guo W. et al., 2015, p. 434; Z. F. Jia & J. C. Wei, 2016, p. 97), Taiwan(Zahlbr. 1940, p. 178; Wang & Lai 1973, p. 90 as *Graphina filiformis*; Zahlbr. 1940, p. 180; Wang & Lai 1973, p. 90 as *Graphina japonica* var. *major*; Zahlbr. 1940, p. 183; Lamb 1963, p. 251; Wang & Lai 1973, p. 90 as *Graphina petrophila*).

Graphis jinhuana Kalb & Z. F. Jia 金华文字衣

Phytotaxa 189(1): 150(2014).

Type: China, Zhejiang: Pan'an County, in Jinhua city, Mt. Dapanshan, surroundings of Huaxi village(Huaxi Scenic Spot); 28°59′21″N, 120°29′56″E, 500 m; in a very disturbed mixed rainforest; 14 October 2010, K. Kalb, Z. S. Sun & Z. F. Jia(holotype HMAS-L; isotype herb. Kalb 38668).

Graphis kelungana Zahlbr. in Fedde 基隆文字衣

Repertorium, 31: 208(1933), J. C. Wei, *Enum. Lich. China*: 103(1991).

Type: Taiwan, Faurie no. 30.

Corticolous.

Yunnan(Z. F. Jia & J. C. Wei, 2016, p. 98), Fujian(Z. F. Jia & J. C. Wei, 2011, p. 225 & 2016, p. 98), Taiwan (Zahlbr., 1940, p. 158; Lamb, 1963, p. 254; Wang & Lai, 1973, p. 90), Hainan(J. C. Wei et al., 2013, p. 235; Z. F. Jia & J. C. Wei, 2016, p. 98).

Graphis lapidicola Fée 岩生文字衣

Bull. Soc. Bot. France, 21: 28(1874).

≡ *Graphina lapidicola*(Fée) Müll. Arg., *Flora*, 68: 513(1885). J. C. Wei, *Enum. Lich. China*: 98(1991).

On rocks or tree bark.

Hainan(Z. F. Jia & J. C. Wei 2011, p. 225 & 2016, p. 99; J. C. Wei et al., 2013, p. 235), Hong kong(Thrower 1988, p. 99).

Graphis lecanactiformis(Zahlbr.) Z. F. Jia & Lücking 碗形文字衣

MycoKeys 25: 24(2017).

≡ *Graphina lecanactiformis* Zahlbr. In Handel-Mazzetti, *Symb. Sin.* 3: 57 (1930) & *Cat. Lich. Univ.* 8: 213 (1932).

Type: Yunnan, Handel-Mazzetti no. 7147(W).

On trunks of *Pinus* sp.

Graphis leptocarpa Fée 细果文字衣

Essai Crypt. Exot. (Paris): 36, pl. 10, fig. 2(1825) [1824]. J. C. Wei, *Enum. Lich. China* 102(1991).

Corticolous.

Sichuan(Zahlbr. 1930, p. 42; Z. F. Jia & J. C. Wei 2011, p. 225 & 2016, p. 101), Guizhou(Zahlbr., 1930 b, p. 42), Yunnan, Hunan(Zahlbr., 1930 b, p. 42; Z. F. Jia & J. C. Wei, 2016, p. 101), Fujian(Zahlbr., 1934, p. 196; Z. F. Jia & J. C. Wei, 2016, p. 101), Zhejiang, Guangdong, Hainan(Z. F. Jia & J. C. Wei, 2016, pp. 100 – 101), See Graphis lineola Ach. for Hongkong(Thrower, 1988, pp. 103, 105).

Graphis librata C. Knight 梭盘文字衣

Trans. & Proc. New Zealand Inst. 16: 404(1884). J. C. Wei, *Enum. Lich. China*: 102(1991). Jilin, Liaoning, Shaanxi, Guizhou, Yuannan, Jiangsu, Zhejiang, Guangdong(Z. F. Jia & J. C. Wei, 2016), Fujian(Jia & Wei 2011, p. 225; Z. F. Jia & J. C. Wei, 2016), Hainan(J. C. Wei et al., 2013, p. 235; Z. F. Jia & J. C. Wei, 2016), Hongkong(Thrower 1988, p. 106; Z. F. Jia & J. C. Wei, 2016).

Graphis lineola Ach. 线文字衣

Lich. univ.: 264(1810).

Misapplied name(revised by Aptroot and Seaward, 1999, p. 77):

Graphis batanensis auct. non Vainio inThrower, 1988, pp. 15, 103.

Graphis cervina auct. non Müll. Arg. in Thrower, 1988, p. 16.

Graphis desquamescens auct non(Fée) Zahlbr. inThrower, 1988, p. 16.

Graphis guimarana auct non Vainio inThrower, 1988, p. 16.

Graphis leptocarpa auct non Ach. inThrower, 1988, p. 16.

Graphis librata auct non C. Knight inThrower, 1988, p. 16, 106.

Graphis sapii auct non Zahlbr. in Thrower, 1988, p. 16.

Graphis tenellula auct non Vainio in Thrower, 1980.

On exposed and sheltered trees. Pantropical.

Guizhou(Zhang T & Wei JC, 2006, p. 7), Zhejiang, Fujian(Jia Z. F. et al., 2016. p. 103), Hongkong(Thrower, 1980; Thrower, 1988, pp. 15, 16, 103, 104, 105, 106; Aptroot and Sparrius, 1999, p. 77; Aptroot et al., 2001, p. 328; Jia Z. F. et al., 2016. p. 103), Taiwan(Aptroot and Sparrius, 2003, p. 19).

Graphis longiramea Müll. Arg. 长枝文字衣

J. Linn. Soc. London, Bot. 29: 225(1892)

On bark.

Hainan(J. C. Wei et al., 2013, p. 235).

= *Graphis multibrachiata* Zahlbr. *in* Handel-Mazzetti, *Symb. Sin.* 3: 46(1930) & *Cat. Lich. Univ.* 8: 203(1932).

Type: Yunnan, Handel-Mazzetti no. 9486.

On trunk of *Prunus cornuta*.

= *Graphis zonatula* Zahlbr. in Handel-Mazzetti, *Symb. Sin.* 3: 43(1930) & *Cat. Lich. Univ.* 8: 207(1932).

Type: Hunan, Handel-Mazzetti no. 12345.

On bark of Corylopsis chinensis.

Guizhou, Yunnan, Fujian, Hainan, Guangdong(Jia Z. F. & J. C. Wei, 2016, p. 105).

Graphis marginata Raddi 金边文字衣

Atti Soc. Nat. Mat. Modena 18: 344(1820).

On *Symplocus* sp.

Taiwan(Aptroot and Sparrius, 2003, p. 19).

Graphis nanodes Vain. 矮小文字衣

Ann. Acad. Sci. fenn. 15(6): 209(1921).

On bark.

Gansu, Guizhou, Yunnan, Hainan(Jia Z. F. et al. , 2016. P. 106) .

Graphis oligospora Zahlbr. 寡孢文字衣

in Handel-Mazzetti, *Symb. Sin.* 3: 45(1930) & *Cat. Lich. Univ.* 8: 203(1932) .

Type: Sichuan, Handel-Mazzetti no. 2451(W, isotype S – 2171) .

On *Quercus* sp.

Guizhou(Jia Z. F. et al. , 2016. P. 107) .

Graphis oxyclada Müll. Arg. 骨针文字衣

Flora, Regensburg 68(28) : 512(1885) .

On bark & twigs.

Fujian(Jia Z. F. et al. , 2016. p. 108) , Hainan(J. C. Wei et al. , p. 235; Jia Z. F. et al. , 2016. p. 108) .

Graphis oxyspora(Zahlbr.) Lücking 尖孢文字衣

Lichenologist, 41(4) : 432(2009) .

≡ *Graphina oxyspora* Zahlbr. , *in* Fedde, *Repertor.* , 31: 214(1933) .

Holotype: Taiwan, Asahina 345(W; isotype: US) .

Taiwan(Zahlbr. , 1940, p. 182; Lamb, 1963, p. 252; Wang & Lai, 1973, p. 90) .

Graphis pananensis Kalb & Z. F. Jia 磐安文字衣

Phytotaxa 189(1) : 151(2014.

Type: China, Zhejiang: Pan'an County, in Jinhua city, Mt. Dapanshan, surroundings of Huaxi village(Huaxi Scenic Spot) ; 28°59′21″N, 120°29′56″E, 500 m; in a very disturbed mixed rainforest; 14# October 2010, K. Kalb, Z. S. Sun & Z. F. Jia(holotype HMAS-L; isotype herb. Kalb 38672) .

Graphis paradussii Z. F. Jia 近杜氏文字衣

The *Bryologist*, 114(2) : 389 – 390(2011) .

Holotype: Hainan, Ledong County, Mt. Jianfenglin, 18u709 N, 108u819 E, alt. 910 m, on bark, 1 october 2008, Bin Gao HN081009-c(HMAS-L 117098) , Jia Z. F. et al. p. 109.

Graphis parallela Müll. Arg. 平行文字衣

Nuovo Giorn. Bot. Ital. 29: 200(1892) .

On bark.

Hunan(Jia Z. F. et al. , 2016. p. 110) , Hongkong(Guo W. et al. 2015, 434; Jia Z. F. et al. , 2016. p. 110) .

Graphis persicina Meyen & Flot. 桃文字衣

Nova Acta Phys. -Med. Acad. Caes. Leop. -Carol. Nat. Cur. , Suppl. 1 19: 229(1843) , Zahlbr. *Cat. Lich. Univ.* 2: 320(1924) .

On *Ficus* sp. , and *Roystonea regia*.

Taiwan(Aptroot and Sparrius, 2003, p. 19) Hongkong(Aptroot et al. , 2001, p. 328) .

Graphis pinicola Zahlbr. 松皮文字衣

in Handel-Mazzetti, *Symb. Sin.* 3: 43(1930) & *Cat. Lich. Univ.* 8: 204(1932) .

Type: Sichuan, Handel-Mazzetti no. 2829.

On living twigs of *Pinus armandi.*

= *Graphis castanopsidis* Zahlbr. In Handel-Mazzetti, *Symb. Sinic.* 3: 40(1930) & *Cat. Lich. Univ.* 8: 201(1932) .

Type: Hunan, Handel-Mazzetti no. 11462.

On bark of *Castanopsis tibetana.*

Gansu, Shaanxi, Guizhou, Yunnan, Henan, Anhui, Hubei, Zhejiang, Fujian(Jia Z. F. et al. 2016, p. 112) , Hainan(J. C. Wei et al. , 2013, p. 235; Jia Z. F. et al. 2016, p. 112) .

Graphis plagiocarpa Fée 短盘文字衣

Essai Crypt. écorc. : 38(1825).

On bark.

Hainan(JCWei et al., 2013, p. 235; Jia Z. F. et al. 2016, p. 113), Hongkong(Guo W. et al. 2015, p. 435; Jia Z. F. et al. 2016, p. 113).

Graphis plumbea(Zahlbr.) Lücking 铅文字衣

Lichenologist, 41(4): 432, 2009.

≡ *Graphina plumbea* Zahlbr. In Handel-Mazzetti, *Symb. Sin.* 3: 55(1930).

Type: Fujian, Chung no. 602b(W).

Corticolous.

Graphis proserpens Vain. 多层文字衣

Bot. Tidsskr. 29: 132(1909).

Replaced synonym:

Graphis disserpens Vain., *Acta Soc. Fauna Flora fenn.* 7(2): 123(1890).

On bark of broadleaved trees.

Gansu(Jia Z. F. et al. 2016, p. 115), Xizang(J. C. Wei & Jiang, 1986, p. 13; Jia Z. F. et al. 2016, p. 115), Sichuan, Guizhou, Yunnan, Fujian(Jia Z. F. et al. 2016, p. 115), Anhui, Zhejiang(Xu B. S. 1989, p. 179; Jia Z. F. et al. 2016, p. 115), Hainan(J. C. Wei et al., 2013, p. 236; Jia Z. F. et al. 2016, p. 115).

Graphis prunicola Vain. 李生文字衣

Bot. Mag., Tokyo 35: 73(1921).

Guiahou(Jia Z. F. et al., 2016, p. 116).

Graphis renschiana(Müll. Arg.) Stizenb. 伦施文字衣

Ber. Tät. St Gall. naturw. Ges. : 184(1891)[1889 – 90].

≡ *Graphina renschiana* Müll. Arg., *Flora*, Regensburg 68(28): 512(1885).

On bark of trees.

Guizhou, Yunnan(Jia Z. F. et al., 2016, p. 117), Hainan(J. C. Wei et al., 2013, p. 236; Jia Z. F. et al., 2016, p. 117).

= *Graphina symplocorum* Zahlbr. *in* Handel-Mazzetti, *Symb. Sin.* 3: 55(1930) & *Cat. Lich. Univ.* 8: 215(1932), Lücking et al., *Lichenologist* 41(4): 399(2009)(Syn. Contr.).

Type: Hunan, Handel-Mazzetti no. 11459.

On trunk and particularly on both the earth and the root of *Symplocos* sp. at the same time.

Guizhou, Yunnan, Hainan(Jia Z. F. et al., 2016. p. 117.

Graphis rhizicola(Fée) Lücking & Chaves [as '*rhizocola*'] 根生文字衣

in Lücking, Chaves, Sipman, Umaña & Aptroot, Fieldiana, *Bot.* 46(no. 1549): 102(2008).

≡ *Opegrapha rhizicola* Fée [as '*rhizocola*'], *Essai Crypt. Exot.* (Paris): 33(1825)[1824].

On bark of trees.

Yunnan(Jia Z. F. et al., 2016, p. 118), Hainan(JCWei et al., 2013, p. 236; Jia Z. F. et al., 2016, p. 118).

Graphis rimulosa(Mont.) Trevis. 小隙文字衣

Spighe Paglie: 11(1853).

≡ *Opegrapha rimulosa* Mont., *Annls Sci. Nat., Bot.*, sér. 218: 271(1842).

Graphis dupaxana Vainio(Thrower, 1988, pp. 16, 105).

Graphis treubii Zahlbr. (Thrower, 1988, pp. 16, 105).

Graphis asterizans Nyl. (Thrower, 1988, p. 15; Zahlbr., 1930).

On Castanopsis sp., *Ficus beeshiana, Ficus microcarpa, Lonicera cumingiana,* and *Machilus thunbergii.*

Yunnan(Wang L. S. et al. 2008, p. 536, as *G. dupaxana*), Zhejiang(Jia Z. F. et al., 2016, p. 120), Taiwan(Aptroot and Sparrius, 2003, p. 20), Hainan(JCWei et al., 2013, p. 236; Jia Z. F. et al., 2016, p. 120), Hongkong (Thrower, 1988, pp. 15, 16, 103, 105; Zahlbr., 1930; Seaward & Aptroot, 2005, p. 285, Pfister 1978: 121, as *Graphis asterizans*).

Graphis rustica Kremp. 粗面文字衣

Nuovo Giorn. Bot. Ital. 7(1): 61(1875).

= *Graphis turgidula* Müll. Arg. *J. Linn. Soc., Bot.* 30: 457(1895).

Misapplied ng6ame(revised by Aptroot and Seaward, 1999, p. 77):

Graphis bifera var. *cinerea* auct non Zahlbr. in Thrower, 1988, p. 16.

On exposed and shel tered trees. Pantropical.

fujian(Jia ZF & J. C. Wei, 2007, pp. 187 – 188), Taiwan(Aptroot and Sparrius, 2003, p. 20), Hongkang(Aptroot & Seaward, 1999, p. 77).

On bark.

Yunnan, Fujian, Guangxi, Hainan(Jia Z. F. et al., 2016. pp. 120 – 121).

Graphis scripta(L.) Ach. 文字衣

K. Vetensk-Acad. Nya Handl. 28: 145(1809).

≡ *Lichen scriptus* L., *Sp. pl.* 2: 1140(1753).

On bark of *Alnus ferdinandi-coburgii* & other trees & on decaying timbers.

Neimenggu(Sun LY et al., 2000, p. 36), Sichuan(Zahlbr., 1930 b, p. 44), Yunnan(R. Paulson, 1928, p. 317; Zahlbr., 1930 b, p. 44; Wang L. S. et al. 2008, p. 536), Shanghai(Nyl. & Cromb. 1883, p. 65; Zahlbr., 1930 b, p. 44; Xu B. S. 1989, p. 180), Jiangsu, Anhui, Zhejiang(Xu B. S. 1989, p. 180), Taiwan(Wang & Lai, 1973, p. 91).

= *Graphis elongata* Zenker in Goebel & Kunze, Pharmaceutische Waarenkunde(Eisenach) 1: 165, tab. XXII, fig. 1 (1827) [1827 – 1829].

Graphis elongata Ehrh. in Arnold, Flora, Regensburg 63: 568(1880) [Nom. illegit., Art. 53. 1].

Graphis scripta var. *elongata*(Ehrh.) Arnold, *Flora*, Regensburg 64: 139(1923).

Graphis scripta f. *elongata*(Ehrh.) Malbr., *Bull. Soc. bot. Fr.* 31: 98(1884).

On bark of *Betula* sp.

Xizang(J. C. Wei & Y. M. Jiang, 1986, p. 13).

= *Graphis scripta* var. *pulverulenta*(Pers.) Ach., *Syn. meth. lich.* (Lund): 82(1814).

≡ *Opegrapha pulverulenta* Pers., *Ann. Bot.* (Usteri) 1: 29(1794).

On bark of trees.

Shanghai(Nyl. & Cromb. 1883, p. 65; Zahlbr., 1930 b, p. 44).

= *Graphis scripta* var. *serpentina*(Ach.) G. Mey., *Nebenst. Beschäft. Pflanzenk.*: 194(1825).

≡ *Lichen serpentinus* Ach., *Lich. suec. prodr.* (Linköping): 25(1799) [1798].

On bark of *Lithocarpus dealbata* & other trees.

Yunnan(Zahlbr., 1930 b, p. 44), Taiwan(Zahlbr., 1933 b, p. 210).

= *Graphis scripta* var. *tenerrima* Ach., *Lich. univ.*: 266(1810).

On bark of *Schoepfia jasminodora*.

Yunnan(Zahlbr., 1930 b, p. 44).

= *Graphis scripta* var. *typographica*(Willd.) Zahlbr., *Cat. Lich. Univ.* 2: 350(1923) [1924].

≡ *Verrucaria typographica* Willd., *Fl. berol. prodr.*: 370(1787).

= *Graphis scripta* var. *recta*(Humb.) Rabenh., *Deutschl. Krypt. -Fl.* (Leipzig) 2: 18(1845).

On Peach-trees.

Shanghai(Nyl. & Cromb. 1883, p. 65; Zahlbr., 1930 b, p. 44).

On bark.

Heilongjiang, Jilin, Neimenggu, Gansu, Shaanx, Xizang, Sichuan, Guizhou, Yunnan, Hebei, Anhui, Hubei, Hunan, Jiangxi, Fujian, Guangxi(Jia Z. F. et al. , 2016, pp. 123 – 124) .

Graphis streblocarpa(Bél.) Nyl. 条果文字衣

Flora, Regensburg 49: 133(1874) .

≡ *Opegrapha streblocarpa* Bél. , *Voy. Indes Or.* , *Bot.* 2(Cryptog.) : 134(1846) .

= *Graphina fissofurcata*(Leight.) Müll. Arg. *Flora*, Regensburg 65(24) : 385(1882) .

≡ *Graphis fissofurcata* Leight. , *Trans. Linn. Soc. London* 27: 177(1869) .

hongkong(Aptroot et al. , 2001, p. 328; Seaward & Aptroot, 2005, p. 285) .

= *Graphina cf. lapidicola* Müll. Arg. in Thrower, 1988, pp. 15, 99(Synonymy Contributor: Aptroot & Seaward, 1999, p. 75) .

On trees and also on granite boulders in shated foreat or along stream beds. Known from tropical Asia.

Zhejiang(Xu B. S. 1989, p. 177) , Fujian(Jia Z. F. et al. , 2016. p. 126) , Hongkong(Aptroot & Seaward, 1999, p. 75, 99) .

Graphis striatula(Ach.) Spreng. 皱沟文字衣

Syst. veg. , Edn 16 4(1) : 250(1827) .

≡ *Opegrapha striatula* Ach. , *Syn. meth. lich.* (Lund) : 74(1814) .

Corticolous, particularly on *Vernonia volkameriaefolia*, *Rosa roxburghii*, Crataegus cuneata, and on *Quercus glandulifera*.

Guizhou, Yunnan(Zahlbr. , 1930 b, p. 42) , Hainan(JCWei et al. , 2013, p. 236) .

= *Graphis striatula* var. *brachycarpa* Müll. Arg. , *Flora*, Regensburg 63: 21(1880) .

On *Quercus glandulifera*.

Hunan(Zahlbr. , 1930, p. 42) .

= *Graphis striatula* f. minor Kremp. , *Flora* 56: 286

On bark.

Guangdong(Rabh. 1873, p. 286; Krphbr. 1873, p. 467; Zahlbr. , 1930 b, p. 42 as *G. striatula*) .]

On bark.

Sichuan, Yunnan, Fujian, Hainan, Hongkong(Jia Z. F. et al. , 2016, p. 128) .

Graphis subassimilis Müll. Arg. 亚黑脉文字衣

Flora, Regensburg 65(21) : 333(1882) .

= *Graphis formosana* Zahlbr. *in* Fedde, *Repertorium*, 31: 208(1933) .

Type: Taiwan(, Faurie no. 243.

Taiwan(Zahlbr. , 1940, p. 155; Lamb, 1963, p. 253; Wang & Lai, 1973, p. 90) .

Corticolous.

Fujian, Hainan(Jia Z. F. et al. , 2016, pp. 128 – 129) .

Graphis subdisserpens Nyl. 亚纹皮文字衣

Bull. Soc. linn. Normandie, sér. 2 7: 175(1873) .

On bark.

Fujian, Hainan(Jia Z. F. et al. , 2016, p. 129) .

Graphis subserpentina Nyl. 亚蛇形文字衣

Acta Soc. Sci. fenn. 7: 465(1863) .

≡ *Graphina subserpentina*(Nyl.) Müll. Arg. , *Bull. Soc. R. Bot. Belg.* 32(2) : 152(1894) [1893] .

≡ *Graphina streblocarpa* var. *subserpentina*(Nyl.) Redinger, *Rev. Bryol. Lichénol.* , N. S. 9: 84(1936) .

= *Graphis adtenuans* Nyl. , *J. Linn. Soc.* , *Bot.* 20: 57 (1883) , Lücking et al. , *Lichenologist*41 (4) : 398 (2009) (Syn. Contr.) . Hue, Nouvelles Archives du Muséum d'Histoire Naturelle de Paris, 3(3) : 158, 1891. J. C. Wei, *Enum. Lich. China:* 100. 1991.

≡ *Graphina adtenuans* (Nyl.) Zahlbr. , *Cat. Liche. Univ.* 2: 395(1923) .

On bark of trees.

Fujian(Jia Z. F. et al. , 2016, p. 131) , Hainan(J. C. Wei et al. , 2013, p. 236; Jia Z. F. et al. , 2016, p. 131) .

= *Graphis adtenuans* var. *detecta* Nyl. *J. Linn. Soc. London, Bot.* 20: 57 (1883) . Zahlbr. *Cat. Lich. Univ.* 2: 393 (1924) .

China(prov. not indicated: Hue, 1891, p. 158) .

Graphis sundarbanensis Jagadeesh & G. P. Sinha 美林文字衣

Lichenologist 39: 231 – 233(2007) .

On bark.

Yunnan, Zhejiang, Guangdong(Jia Z. F. et al. , 2016. p. 132) .

Graphis tenella Ach. 细柔文字衣

Syn. meth. lich. (Lund) : 81(1814) .

Corticolous.

Sichuan, Yunnan, Hunan(Zahlbr. , 1930 b, p. 43) , Shanghai(Rabh. 1873, p. 286; Krphbr. 1873, p. 467 & 1874 a, p. 35; Zahlbr. , 1930 b, p. 43) , Fujian(Zahlbr. , 1934 b, p. 196; Jia Z. F. et al. , 2016. p. 133) , Hainan(J. C. Wei et al. , 2013, p. 236; Jia Z. F. et al. , 2016. p. 134) , Guangdong(Jia Z. F. et al. , 2016. p. 134) .

Graphis tenoriensis Lücking & Chaves 小柔文字衣

in Lücking, Chaves, Sipman, Umaña & Aptroot, Fieldiana, Bot. 46(no. 1549) : 115(2008) .

Yunnan(Joshi et al. , 2015, p. 120) .

Graphis tenuirima(Shirley) A. W. Archer 细裂文字衣

Telopea 11: 74, 2005.

≡ *Graphina tenuirima* Shirley, Bot. Bull. Dept. Agric. Qld. , botany Bulletin 5: 34(1892) .

On bark.

Shaanxi(Jia Z. F. et al. , 2016. p. 134) .

Graphis tsunodae Zahlbr. 粗皮文字衣

Annls mycol. 14: 47(1916) .

= *Graphis rockii* Redgr. *in* Zahlbr. et Redgr. *Lich. rarior. exs.* no. 324, c. icone(1934) .

Type: Yunnan, collected by Rock. see Zahlbr. et Redgr. Lich. rar. exs. no. 324(holotype) in UPS(!) .

Corticolous.

Yunnan(Zahlbr. , 1934 b, p. 196 & 1940, p. 161, as *G. rockii*; Lamb, 1963, p. 255, as *G. rockii*; Jia Z. F. et al. , 2016, p. 136) , Zhejiang, Fujian(Jia Z. F. et al. , 2016, pp. 135 – 136) .

Graphis urandrae Vain. 蕊木文字衣

in Annal. Acad. Scient. Fennic. , ser. A, 15(6) : 255(1921) .

On bark.

Guizhou, Yunnan, Fujian, Hainan(Jia Z. F. et al. , 2016. p. 137) .

Graphis verrucata Z. F. Jia & Kalb 疣体文字衣

Mycosystema, 33(5) : 962(2014) .

Holoype: China, Sichuan Province, Dujiangyan, Wei et al. 97255(HMAS-L 057736) .

On bark.

Sichuan(Jia Z. F. et al. , 2016. p. 138) .

Graphis vittata Müll. Arg. 条纹文字衣

Flora, Regensburg 65(21) : 335(1882)

On bark of trees.

Hainan(J. C. Wei et al. , 2013, p. 236.) .

= *Graphis treubii* Zahlbr. , *Annals Cryptog. Exot.* 1: 129, 1928. Lücking et al. , *Lichenologist*41 (4) : 420, 2009

(Syn. Contr.). Thower, *Hong Kong Lichens*: 103 – 105, 1988. J. C. Wei, *Enum. Lich. China*: 104, 1991.

= *Graphis theae* Zahlbr. *in* Handel-Mazzetti, *Symb. Sin*. 3: 49(1930) & *Cat. Lich. Univ*. 8: 207(1932), Lücking et al., *Lichenologist* 41(4): 420, 2009(Syn. Contr.).

Type: Hunan, Handel-Mazzetti no. 11217(W; isotype: US).

On bark of Theae cuspidata.

Jiangxi(Kashiwadani, 1995, Exs. No. 158).

= *Graphis flabellans* Zahlbr., *Feddes Repert. Spec. Nov. Regni Veg*. 31: 211, 1933. Lücking et al., *Lichenologist* 41(4): 420, 2009(Syn. Contr.). Zahlbr., Cat. Lich. Univ. 10: 161, 1940. Lamb, Index Nom. Lich. Int. Ann. 1932 et 1960 divulgatorum: 255, 1963. Wang-Yang & Lai, Taiwania 18: 90, 1973. J. C. Wei, *Enum. Lich. China*: 101, 1991. Holotype: Taiwan, Asahina 368(W).

Guizhou, Yunnan, Zhejiang, Fujian, Hainan, Guangdong, Hongkong(Jia Z. F. & J. C. Wei, 2016, pp. 139 – 140), Taiwan(Zahlbr., 1940, p. 155; Lamb, 1963, p. 253; Wang & Lai, 1973, p. 90 as *G. flabellans*).

Corticolous.

Graphis wangii Z. F. Jia 王氏文字衣

in Z. F. Jia et al., *Mycotaxon* 121: 75-79(2012).

Type: China, Yunnan province, Xinping county, Mont. Mopanshan, 23°55′N 101°58′E, alt. 2350 m.

On bark of *Quercus* sp.. 3. I. 2009. coll. Li-Song Wang 09 – 30091(Holotype, KUN-L). Yunnan(Jia Z. F. et al., 2016, p. 141).

Graphis weii Z. F. Jia & Kalb 魏氏文字衣

Mycosystema, 33(5): 963(2014).

Type: China, Hainan Province, Ledong, Mt. Jianfengling, Jing Li HN081363 (HMAS-L127465).

Haijnan(Jia Z. F. et al., 2016, p. 142).

Graphis yunnanensis S. Joshi, Upreti & Hur 云南文字衣

in Joshi S, Upreti DK, Wang XY, Hur JS. Mycobiology 42(2): 119, 2015

Type: China, Yunnan Province, Pu' er City, Pu' er National Forest Park, 22°52′08. 6″N, 100° 59′25. 5″E, alt. 1,552 m, on tree trunk, 20 Dec 2013, Hur, Wang & Liu CH130425(holotype KoLRI 020831).

Gyrostomum Fr. **圆片衣属**

Syst. orb. veg. (Lundae) 1: 268(1825).

Gyrostomum scyphuliferum(Ach.) Nyl. 圆片衣

Annls Sci. Nat., Bot., sér. 4 16: 96(1862).

≡ *Lecidea scyphulifera* Ach., *Syn. meth. lich*. (Lund): 27(1814).

Misapplied name(revised by Aptroot & Seaward, 1999, p. 77 – 78):

Phaeographina obfirmata(Nyl.) Zahlbr. (Thrower, 1988, pp. 16, 137)

= *Lecanactis obfirmata* Nyl. in Nyl. & Cromb. *Journ. Linn. Soc. London, Bot*. 20: 65(1883).

Type locality: Shanghai, A. C. Maingay, Confucian-temple garden.

Corticolous. On exposed trees, e. g. Kandelia, especially on branches. Pantropical.

Shanghai(Nyl. 1891, p. 12; Hue, 1891, p. 168; Zahlbr., 1930 b, p. 61), Guangdong(Rabh. 1873, p. 287; Krphbr. 1873, p. 471 & 1874 c, p. 66; Zahlbr., 1930 b, p. 67), Taiwan(Zahlbr., 1933 c, p. 25; Wang & Lai, 1973, p. 91), Hongkong(Thrower, 1988, pp. 16, 137; Aptroot & Seaward, 1999, p. 77 – 78; Aptroot et al., 2001, p. 328).

Hemithecium Trevis. **半实衣属**

Spighe Paglie: 11(1853).

Hemithecium alboglauca(Vain.) A. W. Archer 灰白半实衣

Syst. Biodiv. 5(1):17(2007).

≡*Graphis alboglauca* Vain. *Ann. Acad. Sci. fenn.*, Ser. A 15(6):258(1921).

On sheltered trees. Known from tropical Asia.

Hongkong(Aptroot & Seaward, 1999, p. 76).

Hemithecium balbisii(Fée) Trevis. 巴氏半实衣

Spighe Paglie: 13(1853).

≡*Graphis balbisii* Fée, *Essai Crypt. Exot.* (Paris):48(1825)[1824].

On bark of trees.

Hainan(Jia ZF et al., 2011, p. 872; JCWeiet al., 2013, p. 236), China(province not indicated: Kremp. 1868, p. 210 as *G. balbisii*).

Hemithecium canlaonense(Vain.) A. W. Archer 半实衣

Syst. Biodiv. 5: 17(2007).

≡*Graphis canlaonensis* Vain., *Ann. Acad. Sci. fenn.*, Ser. A 15(no. 6):261(1921).

≡*Fissurina canlaonensis*(Vain.) Staiger, *Biblthca Lichenol.* 85: 161(2002).

On bark

Guizhou(Meng et al., 2011, Pp. 159 – 160; Jia et al. 2016, p. 674).

Hemithecium chlorocarpum(Fée) Trevis. 绿果半实衣

Spighe e Paglie : 12(1853).

On bark

Hainan(Jia et al. 2016, p. 675).

Hemithecium duomurisporum Z. F. Jia 双砖孢半实衣

Mycosystema 30(6):871 – 872(2011).

Type: China. Hainan, Mt. Wuzhishan, 18°46′N, 109°31′E, alt. 640m, on bark. 20-Ⅶ – 2009. coll. Meng Liu HN09318(holotypus in HMAS-L 115536).

Hainan(J. C. Wei et al., 2013, p. 236).

Hemithecium hainanense Z. F. Jia 海南半实衣

in Jia Z. F. et al., *Mycotaxon* 131: 672 – 673(2016).

Type: China. Hainan Province, Changjiang County, Mt. Bawangling, Yajia, 19°14′N 109°26′E, alt. 800 m, on bark, 10/V/2008, J. Li HN081472(Holotype, HMAS-L 127460).

Hemithecium implicatum(Fée) Staiger 交织半实衣

Biblthca Lichenol. 85: 287(2002).

≡*Graphis implicata* Fée, *Bull. Soc. bot. Fr.* 21: 27(1874).

Hainan(Jia ZF et al., 2011, p. 872; JCWei et al., 2013, p. 236).

Hemithecium oshioi(M. Nakan.) M. Nakan. & Kashiw. 欧氏半实衣

in Nakanishi, Kashiwadani & Moon, *Bull. natn. Sci. Mus.*, Tokyo, B 29(2):88(2003).

≡*Graphis oshioi* M. Nakan., *Journal of Science of the Hiroshima University*, B, 2 11: 265(1967).

On trees.

Zhejiang(Xu B. S. 1989, p. 179), Hainan(Jia ZF et al., 2011, p. 872; JCWei et al., 2013, p. 236).

Leiorreuma Eschw. **厚基衣属**

Syst. Lich.: 13, 25(1824).

Leiorreuma crassimarginatum Z. F. Jia 厚缘厚基衣

Mycotaxon 130: 248(2015).

Type: China. Guizhou Province, Tongren City, Mt. Fanjing, Daling, 27°55′N 108°41′E, alt. 1570 m, on bark, 5/X/2004, coll. J. C. Wei & Tao Zhang G425(Holotype, HMAS-L 071776).

Leiorreuma dilatatum(Vain.) Staiger 膨大厚基衣

Biblioth. Lichenol. 85: 296(2002).

On bark.

Hainan(Wang X. H. et al. 2015, pp. 248 – 250).

Leiorreuma exaltata(Mont. & Bosch) Staiger 高举厚基衣

Biblthca Lichenol. 85: 298(2002).

≡*Lecanactis exaltata* Mont. & Bosch, in Junghun, *Pl. Jungh*. 4: 475(1855).

≡*Phaeographis exaltata*(Mont. & Bosch) Müll. Arg., *Flora*, Regensburg65(21): 336(1882); Aptroot & Seaward, (1999: 86).

Phaeographis balansana Müll. Arg. (Thrower 1988: 16, 139; Aptroot & Seaward, 1999: 86).

Phaeographis circumscripta(Krempelh.) Zahlbr. (Thrower, 1988: 16, 140; aptroot & Seaward, 1999: 86).

Phaeographis dendritica(Ach.) Müll. Arg. (Thrower 1988: 17, 142; aptroot & Seaward, 1999: 86).

Phaeographis inustoides Fink. (Thrower 1988: 17, 144; Aptroot & Seaward, 1999: 86).

On bark of *Rhus*. Mountainsides.

Hongkong(Aptroot & Seaward, 1999, p. 86; Seaward & Aptroot, 2005, p. 285).

On sheltered trees(on *Kandelia* sp.) in forests. Pantropical.

Hongkong(Thrower, 1988, pp. 16, 17, 139, 140, 141, 142, 144; Aptroot & Seaward, 1999, p. 86).

Notes: *Phaeographis computata* Müll. Arg. ≡Platygramme *computata* (Kremp.) A. W. Archer

Leiorreuma melanostalazans(Leight.) A. W. Archer 黑厚基衣

Telopea 11: 75(2005).

≡*Platygrapha melanostalazans* Leight., *Trans. Linn. Soc. London* 27: 180(1869).

On bark.

Hainan(Wang X. H. et al. 2015, p. 250).

Leiorreuma sericeum(Eschw.) Staiger 刚毛厚基衣

Biblioth. Lichenol. 85: 305(2002).

≡*Leiogramma sericeum* Eschw. In Martius, *Icon. Plant. Cryptog*., 2: 34(1828).

On bark.

Guangxi, Hainan(Wang X. H. et al. 2015, pp. 250 – 251).

Leiorreuma vicarians (Vain.) M. Nakan. & Kashiw. 替代厚基衣

in Nakanishi, Kashiwadani & Moon, *Bull. natn. Sci. Mus*., Tokyo, B 29(2): 88(2003).

≡*Graphis vicarians* Vain., *Bot. Mag*., Tokyo 35: 73(1921).

≡*Phaeographis vicarians*(Vain.) M. Nakan. *J. Sci. Hiroshima Univ*., ser. B, div. 2, 11: 88(1966) [1967].

On trees.

Zhejiang(Xu B. S., 1989, p. 181).

Leptotrema Mont. & Bosch **麻衣属**

in Miquel, *Pl. Jungh*. 4: 483(1855).

Leptotrema bahianum(Ach.) Müll. Arg. 麻衣

Mém. Soc. Phys. Hist. nat. Genève 29(8): 12(1887).

≡*Thelotrema lepadinum* var. *bahianum* Ach., *Methodus*, Sectio prior: 132(1803).

Leptotrema bahianum var. asiaticum Zahlbr. 亚洲变种

Feddes Repert. 33: 25(1933).

Type: Taiwan, Asahina no. 229.

Taiwan(Zahlbr., 1940, p. 219; Wang & Lai, 1973, p. 92).

On bark of trees.

Myriotrema Fée **多网衣属**

Essai Crypt. Exot. (Paris): xlix, 103(1825) [1824].

Myriotrema compunctum(Ach.) Hale 通点多网衣

Mycotaxon 11(1):133(1980).

≡*Lichen compunctus* Sm., in Acharius, *Methodus*, Sectio prior: 143(1803).

On shaded trees. Pantropical.

Hongkong(Aptroot & Seaward, 1999, p. 83).

Myriotrema microstomum(Müll. Arg.) Hale 微孔多网衣

Mycotaxon 11(1) :134(1980).

≡*Thelotrema microstomum* Müll. Arg., *Flora*, Regensburg 74*(1):* 113(1891).

= *Thelotrema microstomum* var. *formosanum* Zahlbr. Repert. Spec. Nov. Regni Veg. 33: 24(1933). Type: Taiwan, Prov. Changhua, Rengechi, 31 December 1925, Asahina F 233(Wlectotype n. v., designated by Hale 1981, TNS isolectotype). Revised by Aptroot, 2004, p. 36.

Myriotrema minutum(Hale) Hale 小多网衣

Mycotaxon 11(1):134(1980).

≡*Ocellularia minuta* Hale, *Mycotaxon* 7(2):379(1978).

On shaded trees. Known from tropical Asia.

Hongkong(Aptroot & Seaward, 1999, p. 83).

Myriotrema subcompunctum(Nyl.) Hale 亚通点多网衣

Mycotaxon 11(1):135(1980).

≡*Thelotrema subcompunctum* Nyl., *Bull. Soc. linn. Normandie*, sér. 22: 76(1868).

On shaded trees. Pantropical.

Hongkong(Aptroot & Seaward, 1999, p. 83).

Myriotrema viridialbum(*Kremp.*) Hale 绿白多网衣

Mycotaxon 11: 135 *(*1980*)*.

≡*Thelotrema viridialbum* Kremp., *Flora* 49: 221(1876)

On bark of trees.

Hainan, Fujian(Xu & Jia 2015, pp. 133 – 134).

Ocellularia G. Mey. **点衣属**

Nebenst. Beschäft. Pflanzenk. 1:327(1825).

Ocellularia alba(Fée) Müll. Arg. 白点衣

Mém. Soc. Phys. Hist. nat. Genève 29(8):6(1887).

≡*Myriotrema album* Fée, *Essai Crypt. Exot.* (Paris):104(1825)[1824].

On bark of trees.

Taiwan(Zahlbr., 1933 c, p. 23; Wang & Lai, 1973, p. 92).

Ocellularia cavata(Ach.) Müll. Arg. 洞点衣

Flora, Regensburg 65: 499(1882).

≡*Thelotrema cavatum* Ach., *K. Vetensk-Acad. Nya Handl.* 33: 92(1812).

On the barks, branches and twigs of trees.

Hainan(Jia Z. F. et al., 2016. pp. 555 – 556).

Ocellularia eumorpha(Stirt.) Hale 真形点衣

Mycotaxon 11(1):136(1980).

≡*Thelotrema eumorphum* Stirt., *Proc. Roy. phil. Soc. Glasgow* 10: 158(1876)[1877].

On bark.

Guizhou, hunan(Cui Can et al., 2014, p. 204).

Ocellularia leioplacoides(Nyl.) Frisch 平叶点衣

Biblthca Lichenol. 92: 234(2006).

≡*Thelotrema leioplacoides* Nyl., Mém. Soc. natn. Sci. nat. Cherbourg 5: 118(1891).

On bark.

Guangxi(Jia Z. F. et al. , 2016. p. 556) .

Ocellularia palaeoamplior Aptroot & Sipman 大孢点衣

J. Hattori bot. Lab. 91: 333(2001) .

Type: China, Hongkong, Sipman15125 = Aptroot48628(B-holotype; HKU(M) , ABL-isotypes) .

On smooth bark of small trees in secondary forest at 400m.

Ocellularia perforata(Leight.) Müll. Arg. 穿孔点衣

Hedwigia 31: 284(1892) .

≡ *Thelotrema perforatum* Leight. , *Trans. Linn. Soc. London* 25: 477(1866) .

On shaded trees. Pantropical.

Hongkong(Aptroot & Seaward, 1999. P. 83) .

Ocellularia postposita(Nyl.) Frisch 柱点衣

Biblthca Lichenol. 92: 250(2006) .

≡ *Ascidium postpositum* Nyl. , *Annls Sci. Nat. , Bot.* , sér. 5 7: 320(1867) .

On bark.

Guizhou, Hunan, Hainan(Cui Can et al. , 2014, p. 205) .

Ocellularia pyrenuloides Zahlbr. 类核点衣

in Magn. & Zahlbr. , *Ark. Bot.* 31A(1) : 46(1944) .

On bark.

Guizhou, Yunnan(Jia Z. F. et al. , 2016. pp. 556 – 557) .

Ocellularia subfumosa Z. F. Jia & L. S. Wang 亚烟点衣

Mycosystema 35(5) : 553 – 558(2016) .

Type: China. Yunnan Province, Xishuangbanna, Xishuangbanna Tropical Botanical Garden, Lat. : 21. 92N, Long. : 101. 25E, alt. 600m, on bark, 30-VIII-1982. coll. L. S. Wang 82 – 1302(Holotype, KUN-L 3434) .

Pallidogramme Staiger, Kalb & Lücking **灰线衣属**

in Lücking, Chaves, Sipman, Umaña & Aptroot, *Fieldiana, Bot.* 46(1549) : 9(2008) .

Replaced synonym:

Hemithecium subgen. *Leucogramma* Staiger, *Biblthca Lichenol.* 85: 277(2002) .

Pallidogramme chapadana(Redinger) Staiger, Kalb & Lücking 台地灰线衣

In Lücking, Chaves, Sipman, Umaña & Aptroot, *Fieldiana, Bot.* 46(no. 1549) : 9(2008) .

≡ *Phaeographina chapadana* Redinger, *Ark. Bot.* 26A(1) : 100(1934) .

≡ *Hemithecium chapadanum* (Redinger) Staiger, *Biblthca Lichenol.* 85: 281 (2002) – Yunnan, Guangdong, Guangxi(Miao et al. , 2007, p. 497) .

Yunnan, Guangxi, Guangdong, Hainan(Jia ZF et al. , 2011, pp. 873 – 874) .

Pallidogramme chlorocarpoides(Nyl.) Staiger, Kalb & Lücking 绿果灰线衣

in Lücking, Chaves, Sipman, Umaña & Aptroot, *Fieldiana, Bot.* 46(1549) : 9(2008) .

≡ *Graphis chlorocarpoides* Nyl. , *Flora*, Regensburg 49: 133(1866) .

≡ *Hemithecium chlorocarpoides*(Nyl.) Staiger, Biblthca Lichenol. 85: 283(2002) .

Guangxi, Hainan(Miao et al. , 2007, pp. 497 – 498) .

≡ *Phaeographina chlorocarpoides*(Nyl.) Zahlbr. , *Cat. Lich. Univers.* 2: 435(1923) .

Hunan, guangxi, Guangdong, Hainan(Jia ZF et al. , 2011, p. 874) .

Pallidogramme chrysenteron(Mont.) Staiger, Kalb & Lücking 乳黄灰线衣

in Lücking, Chaves, Sipman, Umaña & Aptroot, *Fieldiana, Bot.* 46(1549) : 9(2008) .

≡ *Graphis chrysenteron* Mont. , *Annls Sci. Nat. , Bot.* , sér. 218: 269(1842) .

≡ *Hemithecium chrysenteron*(Mont.) Trevis. , *Spighe Paglie*: 13(1853) .

Yunnan, Guangxi(Miao et al. , 2007, pp. 498 – 499) .

≡ *Leucogramma chrysenteron*(Mont.) A. Massal. , *Atti Inst. Veneto Sci. lett. , ed Arti*, Sér. 35: 320(1860) .

≡*Phaeographina chrysenteron*(Mont.) Müll. Arg. , [as ' chrysentera'] *Hedwigia* 30: 52(1891) . fide Jia Z. F & Lücking, *MycoKeys* 21: 15(2017) .

Yunnan, hunan, guangxi, Guangdong, Hainan(Jia ZF et al. , 2011, pp. 874 – 875) .

= *Phaegraphina fukiensis* Zahlbr. *in* Handel-Mazzetti, *Symb. Sin.* 3: 62(1930) & *Cat. Lich. Univ.* 8: 217(1932) . var. *fukiensis*

Type: Fujian, Mt. GU shan, Chung no. 399 a.

Fujian(Zahlbr. , 1934 b, p. 198 & 1940, p. 189) .

On smooth bark of trees.

= *Phaeographina fukiensis* var. *substriata* Zahlbr. *Annls mycol.* 30: 432(1932) .

Type: Fujian, Mt. Gu shan, Chung no. 597(pr. p.) .

On bark of trees.

Fujian(Zahlbr. , 1934 b, p. 198 & 1940, p. 189) .

Phaeographis Müll. Arg. **黑文衣属**

Flora, Regensburg 65(21) : 336(1882) .

Phaeographis dendroides(Leight.) Müll. Arg. 树生黑文衣

Flora, Regensburg 65(21) : 336(1882) .

≡*Platygrapha dendroides* Leight. , *Trans. Linn. Soc. London* 27: 179(1869) .

On tree bark.

Hainan(Li Jing et al. , 2014, p. 80, miscited as *Ph. dendroids*) .

Phaeographis dendritica(Ach.) Müll. Arg. 木黑文衣

Flora, Regensburg 65(24) : 382(1882) .

≡*Opegrapha dendritica* Ach. , *Methodus*, Sectio prior: 31(1803) .

On bark of Castanopsis fargesii.

Hunan(Zahlbr. , 1930 b, p. 52) , Zhejiang(Xu B. S. , 1989, p. 180) . , Also see Phaeographis exaltata(Mont. et v. d. Bosch.) Müll. Arg. for Hongkong. (Thrower, 1988, p. 142) .

Phaeographis exaltata(Mont. & Bosch) Müll. Arg. 极高黑文衣

Flora, Regensburg 65(21) : 336(1882) .

See *Leiorreuma exaltata*(Mont. & Bosch) Staiger

≡*Lecanactis exaltata* Mont. & Bosch, in Junghun, *Pl. Jungh.* 4: 475(1855) .

Phaeographis balansana Müll. Arg. (Thrower, 1988, pp. 16, 139; Aptroot&Seaward, 1999, p. 86) .

P. circumscripta (Kremp.) Zahlbr. (Thrower, 1988, pp. 16, 140) .

p. computata Müll. Arg. (Thrower, 1988, pp. 16, 141) .

P. dendritica(Ach.) Müll. Arg. (Thrower, 1988, pp. 17, 142) .

P. inustoides Fink(Thrower, 1988, pp. 17, 144) .

On sheltered trees(on *Kandelia* sp.) in forests. Pantropical.

Hongkong(Thrower, 1988, pp. 16, 17, 139, 140, 141, 142, 144; Aptroot & Seaward, 1999, p. 86) .

Notes: *Phaeographis computata* Müll. Arg. ≡Platygramme computata (Kremp.) A. W. Archer

Phaeographis fujianensis X. H. Wang, G. B. Shi & Z. F. Jia 福建黑文衣

Mycosystema 32(1) : 129 – 130(2013) .

Type: China, Fujian Province, Nanping, Jian' ou County, Wanmulin Nature Rerserve(Lat. 27. 36, Long. 109. 82) , alt. 420m, on the bark of the tree, June 1, 2007. Li Jing FJ1021(LHS) .

Phaeographis gracilenta Zahlbr. 细黑文衣

in Fedde, *Repertorium*, 31: 211(1933) .

Type: Taiwan, Asahina no. 301(301a for *Phaeographis gracilenta*, slected by Aptroot, 2004, p. 34) .

Taiwan(Zahlbr. , 1940, p. 168; Lamb, 1963, p. 546; Wang & Lai, 1973, p. 95) .

On bark of trees.

See also *Sarcographa labyrinthica*(Ach.) Müll. Arg.

Phaeographis haloniata(Zahlbr.) Z. F. Jia & Lücking 晕黑文衣

MycoKeys 25: 25(2017) .

≡*Graphina haloniata* Zahlbr. In Fedde, *Repertorium*, 31: 216(1933) .

Type locality: Taiwan, Asahina no. 356(W) .

Taiwan(Zahlbr. , 1940, p. 179; Lamb, 1963, p. 251; Wang & Lai, 1973, p. 90) .

= *Graphina plumbicolor* Zahlbr. *in* Fedde, *Repertorium*, 31: 217(1933) .

Type: Taiwan, Asahina no. 340.

Taiwan(Zahlbr. , 1940, p. 183; Lamb, 1963, p. 251; Wang & Lai, 1973, p. 90) .

Corticolous.

Phaeographis heterochroa Zahlbr. 杂色黑文衣

in Handel-Mazzetti, *Symb. Sin.* 3: 51(1930) & *Cat. Lich. Univ.* 8: 209(1932) f. heterochroa.

Type: Fujian, Mt. GU shan, Chung no. 402.

On smooth bark of trees.

Phaeographis heterochroa f. subunicolor Zahlbr. 亚单色变型

in Handel-Mazzetti, *Symb. Sin.* 3: 51(1930) .

Type: Fujian, Chung no. 597 b.

Phaeographis heterochroides Zahlbr. 类杂色黑文衣

in Handel-Mazzetti, *Symb. Sin.* 3: 53(1930) & *Cat. Lich. Univ.* 8: 209(1932) .

Type: Fujian, Chung nos. 368, 399, 593b. (syntypes)

On smooth bark of trees.

Phaeographis hypoglauca(Kremp.) Zahlbr. 兰底黑文衣

Cat. Lich. Univ. 2: 374(1923) [1924] .

≡*Graphis hypoglauca* Kremp. , *Flora*, Regensburg 56: 467(1873) .

Typey: Guangdong, collected by Rabh.

Guangdong(Kremp. 1874, p. 35; Rabh. 1873, p. 286) .

On porphiritic rocks.

Phaeographis intricans(Nyl.) Staiger 缠结黑文衣

Biblthca Lichenol. 85: 329(2002) .

≡*Graphis intricans* Nyl. , *Acta Soc. Sci. fenn.* 7(2) : 473(1863) .

= *Sarcographa albomaculans* Zahlbr. *Repert. Spec. Nov. Regni Veg.* 31: 221(1933) .

Type: Taiwan, Prov. Taitung, Raisha, 5 January 1926, Asahina F 352(TNS holotype) .

Taiwan(revised by Aptroot, 2004, p. 36) , Hainan(JCWei et al. , 2013, p. 237) .

Phaeographis inusta(Ach.) Müll. Arg. 焚黑文衣

Flora, Regensburg 65(24) : 383(1882) .

≡*Graphis inusta* Ach. , *Syn. meth. lich.* (Lund) : 85(1814) .

Phaeographis inusta var. simpliciuscula(Leight.) Müll. Arg. 单型变种

in Balfour, *Trans. R. Soc. Edinb.* 31: 378(1888) .

≡*Graphis smithii* var. *simpliciuscula* Leight. , *Ann. Mag. nat. Hist.* , Ser. 2 8: 278(1854) .

On twigs of *Rhamnus* sp.

Yunnan(A. Zahlbr. , 1930 b, p. 52) .

Phaeographis lecanographa(Nyl.) Staiger 皿形黑文衣

Biblthca Lichenol. 85: 334(2002) .

≡*Graphis lecanographa* Nyl. , *Flora*, Regensburg 52: 123(1869) .

≡*Phaeographina lecanographa*(Nyl.) Müll. Arg, *Flora*, Regensburg 65: 399(1882) .

≡*Graphis lecanographa* Nyl. , *Flora*, Regensburg 52: 123(1869) .

= *Phaeographina lecanographa* var. *pleiospora* Zahlbr. *Feddes Repert.* 31: 219(1933).

Type: Fujian, Chung no. 596 b.

Fujian(Zahlbr., 1933 b, p. 219 & 1934, p. 197 & 1940, p. 190).

On bark of trees.

Taiwan(Zahlbr., 1933 b, p. 219; Wang & Lai, 1973, p. 95).

Phaeographis lidjiangensis Zahlbr. 丽江黑文衣

in Handel-Mazzetti, *Symb. Sin.* 3: 51(1930) & *Cat. Lich. Univ.* 8: 209(1932).

Type: Yunnan, Handel-Mazzetti no. 4163.

On twigs of *Quercus semicarpifolia*.

Phaeographis lobata(Eschw.) Müll. Arg. 裂片黑文衣

Flora, Regensburg 65(24): 383(1882).

≡ *Lecanactis lobata* Eschw., *Syst. Lich.*: 25(1824).

Misapplied name:

Thelotrema expansum auct. non C. Knight in Thrower, 1988, pp. 17, 173, revised by Aptroot & Seaward, 1999, p. 86.

On tree in forest. Pantropical.

Hongkong(Thrower, 1988, pp. 17, 173; Aptroot & Seaward, 1999, p. 86; Aptroot et al., 2001, p. 337).

Phaeographis neotricosa Redinger 星盘黑文衣

Ark. Bot. 27A(3): 93(1935)

On bark, most records on *Cratoxylon ligustrinum*.

Hongkong(Thrower, 1988, p. 145; Aptroot et al., 2001, p. 337).

Notes. Previously listed erroneously as *Sarcographa protracta* (Krempelh.) Zahlbr.: *Sarcographa protracta* (Kremp.) Zahlbr. *Cat. Lich. Univ.* 2: 465(1923) [1924].

On trees.

Hongkong(Aptroot & Seaward, 1999, p. 94).

Phaeographis planiuscula(Mont. & Bosch) Müll. Arg. 平黑文衣

Flora, Regensburg 65: 336(1882).

≡ *Lecanactis planiuscula* Mont. & Bosch, *Pl. Jungh.* 4: 475(1856).

On *Shiia carlesii* and tree branches.

Taiwan(Aptroot and Sparrius, 2003, p. 33).

Phaeographis platycarpa Müll. Arg. 宽果黑文衣

Bot. Jb. 20: 284(1894).

On tree bark.

Hainan(Li Jing et al., 2014, pp. 80 – 81).

Phaeographis pleiospora(Zahlbr.) Z. F. Jia & Lücking 多孢黑文衣

MycoKeys 21: 20, 26(2017).

≡ *Phaeographina lecanographa* var. *pleiospora* Zahlbr. *Feddes Repert.* 31: 219(1933).

Type: Fujian, Chung no. 596 b(W).

Fujian(Zahlbr., 1933 b, p. 219 & 1934, p. 197 & 1940, p. 190).

On bark of trees.

Taiwan(Zahlbr., 1933 b, p. 219; Wang & Lai, 1973, p. 95).

Phaeographis pruinifera Zahlbr. 霜果黑文衣

in Handel-Mazzetti, *Symb. Sin.* 3: 53(1930) & *Cat. Lich. Univ.* 8: 210(1932).

Type: Yunnan, Handel-Mazzetti no. 305.

On bark of *Lithocarpus dealbata*.

Phaeographis scalpturata(Ach.) Staiger 雕型黑文衣

Biblthca Lichenol. 85: 345(2002).

≡*Graphis scalpturata* Ach., *Syn. meth. lich.* (Lund): 86(1814).

≡*Phaeographina scalpturata*(Ach.) Müll. Arg., *Flora*, Regensburg 65(25): 399(1882).

On sheltered trees in forests. Pantropical.

Hongkong(Aptroot & Seaward, 1999, p. 86; Aptroot et al., 2001, p. 337; Seaward & Aptroot, 2005, p. 285, Pfister 1978: 121 as *Graphis scalpturata*), Taiwan(Aptroot and Sparrius, 2003, p. 33).

Phaeographis sericea(Eschw.) Müll. Arg. 绢黑文衣

Flora, Regensburg 71: 523(1888).

≡*Leiogramma sericeum* Eschw., in Martius, *Icon. Plant. Cryptog.*, 2: 34(1834)[1828 - 34].

Zahlbr. *Cat. Lich. Univ.* 2: 385(1924).

= *Phaeographis heterochroa* Zahlbr. in Thrower, 1988, pp. 17, 143, mentioned by Aptroot & Seaward, 1999, p. 86.

On bark.

Hongkong(Thrower, 1988, p. 146).

= *Phaeographis heterochroa* Zahlbr. in Thrower, 1988, pp. 17, 143, mentioned by Aptroot & Seaward, 1999, p. 86.

Hongkong(Thrower, 1988, pp. 17, 143, 146; Aptroot & Seaward, 1999, p. 86).

On *Acer* sp.

Taiwan(Aptroot and Sparrius, 2003, p. 33).

Phaeographis silvicola Zahlbr. 林生黑文衣

in Handel-Mazzetti, *Symb. Sin.* 3: 53(1930) & *Cat. Lich. Univ.* 8: 210(1932).

Type: Yunnan, Handel-Mazzetti no. 7830.

On bark of *Taxus wallichiana*.

Phaeographis submaculata Zahlbr. 亚斑黑文衣

Annal. Mycol. 30: 433(1932).

Type: Fujian, Chung no. 596 c.

Fujian(Zahlbr., 1934 b, p. 197 & 1940, p. 173; Lamb, 1963, p. 548).

Phaeographis tortuosa(Ach.) Müll. Arg. 扭曲黑文衣

Mém. Soc. Phys. Hist. nat. Genève 29(8): 26(1887).

≡*Graphis tortuosa* Ach., *Syn. meth. lich.* (Lund): 85(1814).

On tree bark.

Phaeographis wukangensis Zahlbr. 武冈黑文衣

in Handel-Mazzetti, *Symb. Sin.* 3: 52(1930) & *Cat. Lich. Univ.* 8: 210(1932).

Type: Hunan, Handel-Mazzetti no. 12339.

On bark of *Pagus longipetiolata*.

Platygramme Fée **凸唇衣属**

Bull. Soc. bot. Fr. 21: 29(1874).

Platygramme computata (Kremp.) A. W. 计凸唇衣

Syst. Biodiv. 5(1): 19(2007).

≡*Phaeographis computata*(Kremp.) Müll. Arg., *Flora*, Regensburg 65: 504(1882).

≡*Graphis computata* Kremp., *Nuovo G. bot. ital.* 7(1): 36(1875).

Hongkong(Thrower, 1988, pp. 16, 17, 139, 140, 141, 142, 144; Aptroot & Seaward, 1999, p. 86).

Platygramme discurrens(Nyl.) Staiger 热带凸唇衣

Biblthca Lichenol. 85: 362(2002).

≡*Graphis discurrens* Nyl., *Annls Sci. Nat., Bot.*, sér. 419: 358(1863).

≡*Phaeographis discurrens*(Nyl.) Müll. Arg., *Flora, Regensburg* 65: 504(1882).

Yunnan, Fujian(Miao et al. , 2007, pp. 499 – 500; Jia ZF & Kalb, 2013, pp. 148 – 149), Hainan(Jia ZF & Kalb, 2013, pp. 148 – 149), Hongkong(Seaward & Aptroot, 2005, p. 285, Pfister 1978: 121, as Graphis discurrens. Ascospores mostly 5-septate, 25X 7 mm; thallus K1 red.

See Staiger(2002), p. 362: Holotype ex herb. Tuckerman H-Nyl 7166.).

China(prov. not indicated: Hue, 1891, p. 151).

Platygramme elaeoplaca(Zahlbr.) Z. F. Jia & Lücking 橄榄凸唇衣

MycoKeys 21: 16 – 17, 27(2017).

≡ *Phaeographina elaeoplaca* Zahlbr. *Annls mycol.* 30: 431(1932). fide Jia Z. F.

Type: Fujian, Chung nos 382, 399 c, 596 d(syntypes).

Fujian(Zahlbr. , 1934 b, p. 197 & 1940, p. 189).

Platygramme hainanensis Z. F. Jia & Kalb 海南凸唇衣

The Lichenologist 45(2): 146 – 147(2013).

Typus: China, Hainan Island, Mt. Wuzhishan, 18_920N, 109_680 E, alt. 680 m, 28 August 2008, coll. Jing Li HN081281(HMAS-L—holotypus).

Platygramme lückingii Z. F. Jia & Kalb 吕氏凸唇衣

The Lichenologist 45(2): 147 – 148(2013).

Typus: China, Hainan, Mt. Jianfengling, 18_710N, 108_830 E, alt. 740 m, 1 Oct. 2008, Li Jing HN081395 (HMAS-L-holotypus).

Platygramme mülleri(A. W. Archer) Staiger 米勒氏凸唇衣

Biblthca Lichenol. 85: 364(2002).

≡ *Phaeographina muelleri* A. W. Archer, *Telopea* 8(4): 473(2000).

On bark of trees.

Yunnan(Jia ZF & Kalb, 2013, pp. 148 – 149), hainan(Jia ZF & Kalb, 2013, pp. 148 – 149; J. C. Wei et al. , 2013, p. 237).

Platygramme pachyspora(Redinger) Staiger 肥孢凸唇衣

Biblthca Lichenol. 85: 364(2002).

≡ *Phaeographis pachyspora* Redinger, in Arkiv för Bot. 27A, no. 3: 70 et 77(1935).

Fujian(Jia ZF & Kalb, 2013, pp. 149 – 150).

Platygramme platyloma(Müll. Arg.) M. Nakan. & Kashiw. 宽边凸唇衣

in Nakanishi, Kashiwadani & Moon, *Bull. natn. Sci. Mus.* , Tokyo, B 29(2): 89(2003), Z. F. Jia & R. Lücking *MycoKeys* 21: 19(2017).

≡ *Phaeographina platyloma* Müll. Arg. , *Flora*, Regensburg 65(25): 398(1882).

Fujian(Jia ZF & Kalb, 2013, pp. 149 – 150).

= *Phaeographina granulans* Zahlbr. *Annls mycol.* 30: 432(1932).

Type: Fujian, Mt. Gu shan, Chung no. 404.

On bark of trees.

Fujian(Zahlbr. , 1934 b, p. 198 & 1940, p. 189; Lamb, 1963, p. 544).

Platygramme pudica(Mont. & Bosch) M. Nakan. & Kashiw. 凸唇衣

in Nakanishi, Kashiwadani & Moon, *Bull. natn. Sci. Mus.* , Tokyo, B 29(2): 89(2003).

≡ *Graphis pudica* Mont. & Bosch, *Syll. gen. sp. crypt.* (Paris): 347(1856).

Fujian(Jia ZF & Kalb, 2013, p. 151).

Platygramme taiwanensis(J. C. Wei) Z. F. Jia & Lücking 台湾凸唇衣

MycoKeys 25: 26(2017).

≡ *Graphina taiwanensis* J. C. Wei, *Enum. Lich. China*, 99(1991). f. taiwanensis

≡ *Graphina olivascens* Zahlbr. , in Feddes, *Repertorium*, 31: 215(1933), non Zahlbr. in Handel-Mazzetti, *Symb. Sin.* 3: 57(1930).

Type: Taiwan, Asahina no. 346(W).

Taiwan(Zahlbr., 1933 b, p. 215 & 1940, p. 182; Lamb, 1963, p. 251; Wang & Lai, 1973, p. 90).

On bark of a tree.

= *Graphina taiwanensis* f. *obscurata*(Zahlbr.) J. C. Wei *Enum. Lich. China*, 99(1991).

≡ *Graphina olivascens* Zahlbr. f. *obscurata* Zahlbr. in Fedde, *Repertorium*, 31: 215(1933).

Type locality: Taiwan, Asahina no. 375(W).

Taiwan(Zahlbr., 1933 b, p. 215; Lamb, 1963, p. 251; Wang & Lai, 1973, p. 90).

On bark of trees.

Platythecium Staiger **扁盘衣属**

Biblthca Lichenol. 85: 385(2002).

Platythecium colliculosum(Mont.) Staiger 小突扁盘衣

Biblthca Lichenol. 85: 380(2002).

≡ *Sclerophyton colliculosum* Mont., *Annls Sci. Nat.*, *Bot.*, sér. 316: 61(1851).

≡ *Graphina colliculosa*(Mont.) Hale *Lich. Amer. Exs.*: 156(1976).

Guizhou(Miao et al., 2007, pp. 501 – 502), Hainan(J. C. Wei et al., 2013, p. 237).

Taiwan(Aptroot and Sparrius, 2003, p. 18, as *G. colliculosa*), Hongkong(Aptroot et al., 2001, p. 326).

See also *Graphina incrustans*(Fée) Müll. Arg. for Hongkong(Thrower, 1988, pp. 94, 95).

Platythecium leiogramma(Nyl.) Staiger 光滑扁盘衣

Biblthca Lichenol. 85: 388(2002).

≡ *Graphis leiogramma* Nyl., *Acta Soc. Sci. fenn.* 7(2): 470(1863).

≡ *Phaeographis leiogramma*(Nyl.) Zahlbr., *Cat. Lich. Univers.* 2: 378(1923) [1924].

On bark of trees.

Hainan(JCWei et al., 2013, p. 237).

Platythecium maximum(Groenh.) Z. F. Jia & Lücking 巨型扁盘衣

MycoKeys 21: 21, 27(2017).

≡ *Phaeographina maxima Groenhart* in Blumea, Suppl. 5(H. J. Lam Jubilee Vol.). 107, 1958.

On rock.

Hongkong(Thrower 1988, p. 136).

Platythecium pyrrhochroa(Mont. & Bosch) Z. F. Jia & Lücking 红扁盘衣

MycoKeys 21: 24, 27 – 28(2017).

≡ *Ustalia pyrrhochroa* Mont. & Bosch, *Syll. gen. sp. crypt.* (Paris): 352(1856).

≡ *Phaeographina pyrrhochroa*(Mont. & Bosch) Zahlbr. *Cat. Lich. Univ.* 8: 218(1928).

On bark of trees, pantropical.

guizhou(Zhang T & Wei JC, 2006, p. 9), Taiwan(Zahlbr., 1933 b, p. 218; Wang & Lai, 1973, p. 95), Hongkong(Thrower, 1988, pp. 16, 135 as *Phaeographina chlorocarpoides*; Aptroot & Seaward, 1999, p. 86).

= *Platythecium dimorphodes*(Nyl.) Staiger *Biblthca Lichenol*. 85: 383(2002).

≡ *Graphis dimorphodes* Nyl., *Trans. Linn. Soc. London* 27: 176(1869).

≡ *Graphina dimorphodes* (Nyl.) Zahlbr., *Cat. Lich. Univers.* 2: 404(1923).

On bark of trees.

Guizhou(Miao et al., 2007, pp. 502 – 503), hainan(JCWei et al., 2013, p. 237).

Notes: Whether the *Phaeographina chlorocarpoides*(Nyl.) Zahlbr. in Thrower, 1988, pp. 16, 135 is a synonymy of *Phaeographina pyrrhochroa*, as mentioned in Aptroot & Seaward in 1999, p. 86, or a misapplied name for *Phaeographina chlorocarpoides*, remains to be checked in the future.

Platythecium serpentinellum(Nyl.) Staiger 蛇形扁盘衣

Biblthca Lichenol. 85: 390(2002).

≡*Graphis serpentinella* Nyl., *Acta Soc. Sci. fenn.* 7(2):469(1863).

≡*Phaeographis serpentinella*(Nyl.) Zahlbr., *Cat. Lich. Univers.* 2:386(1923)[1924].

On rocks.

Hainan(JCWei et al., 2013, p. 237).

Rhabdodiscus Vain. **棒盘衣属**

Ann. Acad. Sci. fenn., Ser. A 15(no. 6):184(1921).

Rhabdodiscus asiaticus(Vain.) Rivas Plata 亚洲棒盘衣

Lücking & Lumbsch, *Taxon* 61(6):1175(2012).

≡*Thelotrema asiaticum* Vain., *Hedwigia* 46:175(1907).

≡*Ocellularia asiatica* (Vain.) Hale, *Mycotaxon* 11(1): 136(1980).

= *Thelotrema formosanum* Zahlbr. in Fedde, *Repertorium*, 33:24(1933).

Type: Taiwan, Changhua, Rengechi, 31 December 1925, Asahina F230(TNS-holotype). Revised by Aptroot, 2004, p. 36.

Type: Taiwan, Asahina no. 330.

Taiwan(Zahlbr., 1940, p. 217; Wang & Lai, 1973, p. 97).

On bark of trees.

Sarcographa Fée **肉文衣属**

Essai Crypt. Exot. (Paris): xxxv, xc, 58(1825)[1824].

Sarcographa albo-maculans Zahlbr. 白斑肉文衣

Flechten der Insel Formosa in Feddes, *Lichenes* in Feddes, *Repertorium sp. nov.* 31:221(1933).

Type: Taiwan, Asahina no. 352.

Taiwan(Lamb, 1963, p. 655; Wang & Lai, 1973, p. 96).

On bark of trees.

Sarcographa glyphiza(Nyl.) Kr. R. Singh & G. P. Sinha 曲肉文衣

Indian Lichens, An Annotated Checklist (Kolkata):404(2010), Z. F. Jia & Lücking *MycoKeys* 21:18 – 19(2017).

≡*Graphis glyphiza* Nyl., *Annls Sci. Nat., Bot.*, sér. 419:374(1863).

Type: China, Hong Kong, Nylander 6989(H, lectotype).

≡*Phaeographina glyphiza*(Nyl.) Zahlbr. in Handel-Mazzetti, *Symb. Sinic.* 3:62, 1930. *Cat. Lich. Univers.* 8:217(1932).

≡*Phaeographis glyphiza*(Nyl.) Zahlbr. *Cat. Lich. Univers.* 2:373(1924).

Hongkong(Leighton, 1869, p. 173; Hue, 1891, p. 160; Zahlbr., 1924, p. 373 & 1930b, p. 62).

On bark of trees.

Sarcographa heteroclita(Mont.) Zahlbr. 异肉文衣

in Rechinger, *Denkschr. Kaiserl. Akad. Wiss. Wien*, Math. -Naturwiss. Kl. 88:19(1911).

≡*Glyphis heteroclita* Mont., *Annls Sci. Nat., Bot.*, sér. 2 19:83(1843).

On *Lonicera cumingiana*, *Melia azedarach*, *Roystonea regia*, *Eleocarpus silvestris*, and *Roystegia regia*.

Yunnan(Miao et al., 2007, pp. 503 – 504), Taiwan(Aptroot and Sparrius, 2003, p. 45).

Sarcographa intricans(Nyl.) Müll. Arg. 结肉文衣

Flora, Regensburg 70:77(1887).

≡*Graphis intricans* Nyl., *Acta Soc. Sci. fenn.* 7(2):473(1863).

= *Sarcographa subtricosa*(Leighton) Müll. Arg. in Thrower, 1988, p. 17, mentioned by Aptroot & Seaward, 1999, p. 94).

Misapplied name:

Sarcographa labyrinthica auct. non(Ach.) Müll. Arg. in Thrower, 1988, p. 17, revised by Aptroot & Seaward,

1999, p. 94).

On trees. Pantropical.

On *Mangifera indica*, *Roystonea regia*, *Lonicera cumingiana*, and *Podocarpus* sp.

Taiwan(Aptroot and Sparrius, 2003, p. 45), Hongkong(Aptroot & Seaward, 1999, p. 94; Aptroot & Sipman, 2001, p. 340).

Sarcographa labyrinthica(Ach.) Müll. Arg. 曲线肉文衣

Mém. Soc. Phys. Hist. nat. Genève 29(8): 62(1887)

≡*Glyphis labyrinthica* Ach., *Syn. meth. lich.* (Lund): 107(1814).

= *Sclerophyton actinoboloides Zahlbr. Repert. Spec. Nov. Regni Veg.* 31: 223(1933).

Type: Taiwan, Changhua, Rengechi, 31 December 1925, Asahina F 320(TNS holotype).

Taiwan, Asahina no. 301b for S. labyrinthica, selected by Aptroot, 2004, p. 34. Revised by Aptroot, 2004, p. 36.

Sarcographa medusulina(Nyl.) Müll. Arg. 蛇肉文衣

Flora, Regensburg 70: 77(1887).

≡*Glyphis medusulina* Nyl., *Acta Soc. Sci. fenn.* 7(2): 485(1863).

On various bushes,

Guangdong(Miao et al., 2007, pp. 504 – 505), Hongkong(Seaward & Aptroot, 2005, . 286, Pfister 1978: 123, as *Glyphis medusulina*.

Sarcographa melanocarpa Zahlbr. 黑肉文衣

Flechten der Insel Formosa in Feddes, *Repertorium sp. nov.* 31: 221(1933).

Syntype: Taiwan, Faurie no. 71 & M. Ogata(Asahina, Lich. Formos. no. 303).

Taiwan(Zahlbr., 1940, p. 196; Lamb, 1963, p. 655; Wang & Lai, 1973, p. 96).

Sarcographa tricosa(Ach.) Müll. Arg. 丝肉文衣

Mém. Soc. Phys. Hist. nat. Genève 298): 63(1887).

≡*Graphis tricosa* Ach., *Lich. univ.*: 1 – 696(1810).

On bark of numerous tree species, e. g. from the genera *Callitris* and *Kandelia*. Pantropical.

Shanghai(Xu B. S., 1989, p. 181), Guangdong(Miao et al., 2007, p. 505), Taiwan(Aptroot and Sparrius, 2003, p. 45), Hongkong(Thrower, 1988, pp. 17, 167; Aptroot & Seaward, 1999, p. 94; Aptroot & Sipman, 2001, p. 340),

Sarcographina Müll. Arg. **肉拟文衣属**

Flora, Regensburg 70: 425(1887).

Sarcographina glyphiza(Nyl.) Kr. P. Singh & D. D. Awasthi 刻画肉拟文衣

Bull. bot. Surv. India 20(1 – 4): 139(1979) [1978].

≡*Graphis glyphiza* Nyl., *Annls Sci. Nat.*, *Bot.*, sér. 4 19: 374(1863).

Type: from hongkong

Pfister 1978: 121, as *Graphis glyphiza*.

Hongkong(Seaward & Aptroot, 2005, p. 286, Pfister 1978: 121, as *Graphis glyphiza*.).

On bark of Leicher [sic] 5 Lichea. Whampoa. July 1854. C. Wright. ISOTYPE

≡*Phaeographis glyphiza*(Nyl.) Zahlbr., *Cat. Lich. Univ.* 2: 373(1924).

≡*Phaeographina glyphiza*(Nyl.) Zahlbr. in Handel-Mazzetti, Symb. Sin. 3: 62(1930) & Cat. Lich. Univ. 8: 217 (1932).

= *Phaeographis neotricosa* Redinger in Thrower 1988, pp. 17, 145, mentioned by Aptroot & Seaward, 1999, p. 94.

On bark of trees.

Hongkong (Leighton, 1869, p. 173; Hue, 1891, p. 160; Zahlbr., 1924, p. 373 & 1930 b, p. 62; Aptroot & Seaward, 1999, p. 94; Aptroot & Sipman, 2001, p. 340).

Sarcographina heterospora(Nyl.) Z. F. Jia & Lücking 异孢肉拟文衣

MycoKeys 21: 26(2017)

≡ *Graphis heterospora* Nyl. , *Annls Sci. Nat.* , Bot. , sér. 4 11: 261(1859) .

≡ *Gymnographa heterospora*(Nyl.) Staiger, *Biblthca Lichenol.* 85: 271(2002) .

≡ *Phaeographina heterospora*(Nyl.) Zahlbr. , *Cat. Lich. Univers.* 2: 439(1923) [1924] .

On *Ficus nervosa.*

Taiwan(Aptroot and Sparrius, 2003, p. 33) .

Thalloloma Trevis. **枝盘衣属**

Spighe Paglie: 13(1853) .

Thalloloma cf. anguiniforme(Vain.) Staiger [as ' anguinaeforme'] 白枝盘衣

Biblthca Lichenol. 85: 427(2002) .

Graphina virginea (Eschw.) Müll. Arg. was reported from Hong Kong(Thrower 1988, p. 101) , but according to her photograph and description, the material has norstictic acid, non-pruinose discs, 1-spored asci and hyaline muriform ascospores 50 – 60 × 15 – 25μm in size and may represent a species of *Thalloloma*, such as *T. anguinaeforme* (Vain.) Staiger. (Z. F. Jia & Lücking, 2017, p. 22) .

Thalloloma microsporum Z. F. Jia & J. C. Wei 微孢枝盘衣

Mycotaxon 107: 197 – 199(2009) .

Holotype: shaanxi, Qinling mountains, Banqiaogou, 33°88′N, 108°01′E, alt. 1520 m, on cortices of cortice *Zelkova serrata* (Thunb.) Makino. 29- VII – 2005, Ze-feng Jia SQ380(holotype in LHS; isotype in HMAS-L.) ; paratypes: ibid. , on cortices of *Pinus armandii* Franch. 29-VII – 2005, Jia Ze-feng SQ374, SQ375(LHS, HMAS-L) .

Thalloloma ochroleucum Z. F. Jia & Kalb 浅黄枝盘衣

Mycotaxon 128: 114 – 115(2014) .

Type: China, Guizhou, Tongren City, Mt. Fanjing, Daling, 27°55′N 108°41′E, alt. 1800 m, On bark of Rhododendron rufum Batalin(Ericaceae) , 22. VIII. 1963. coll. J. C. Wei 0474(Holotype, HMAS-L 047744) .

Thecaria Fée **板文衣属**

Essai Crypt. Exot. (Paris) : xlii, 97(1825) [1824] .

Thecaria montagnei(Bosch) Staiger 曼氏板文衣

Biblthca Lichenol. 85: 446(2002) , Z. F. Jia & Lücking *MycoKeys* 21: 21(2017) .

≡ *Graphis montagnei* Bosch, in Junghun, *Pl. Jungh.* 4: 472(1855) .

≡ *Phaeographina montagnei*(Bosch) Müll. Arg. *Flora*, Regensburg 65(25) : 399(1882) .

On bark of trees.

Taiwan(Zahlbr. , 1933 b, p. 218; Wang & Lai, 1973, p. 95) .

= *Phaeographina montagnei* f. *macrospora* Zahlbr. Repert. Spec. Nov. Regni Veg. 31: 219(1933) .

Type: Taiwan, Prov. Chiai, Mt. Arisan, Toroyen, 18 December 1925, Asahina F 372(TNS holotype) .

Taiwan(Aptroot, 2004, p. 34, only mentioned) , Hainan(JCWei et al. , 2013, p. 237) .

On tea trees.

Yunnan(Wang L. S. 2008, p. 538) .

≡ *Phaeographina macrospora* (Zahlbr.) M. Nakan. *J. Sci. Hiroshima Univ.* , ser. B. div. 2, 11: 101 (1966) [1967] .

On bark of trees.

Taiwan(Wang & Lai, 1973, p. 95) .

Thecaria quassiicola Fée 苦木板文衣

[as ' quassiaecola'] , *Essai Crypt. Exot.* (Paris) : xcii, tab. 1, fig. 16(1825) [1824] .

≡ *Fissurina quassiicola* Fée [as ' *quassiaecola*'] , *Essai Crypt. Exot.* (Paris) : 97(1825) [1824] .

≡*Phaeographina quassiicola*(Fée) Müll. Arg. [as ‘*quassiaecola*’], *Mém. Soc. Phys. Hist. nat. Genève* 29(8) : 47 (1887), (Aptroot et al., 2001, p. 337).

On bark of trees.

Zhejiang(Xu B. S., 1989, p. 180), Fujian(Zahlbr., 1930 b, p. 60), Hainan(J. C. Wei et al., 2013, p. 237), Taiwan(Zahlbr., 1933 b, p. 218; Wang & Lai, 1973, p. 95), Hongkong(Thrower, 1988, pp. 16, 138; Aptroot & Seaward, 1999, p. 86; Aptroot et al., 2001, p. 337).

Thecographa A. Massal. **鞘文衣属**

Atti Inst. Veneto Sci. lett., ed Arti, Sér. 3 5: 316(1860) [1859 – 1860].

Thecographa prosiliens(Mont. & Bosch) A. Massal. 壮鞘文衣

Atti Inst. Veneto Sci. Lett., ed Arti, Sér. 3 5: 316(1860), Z. F. Jia & R. Lücking *MycoKeys*21: 23, 24(2017).

≡*Opegrapha prosiliens* Mont. & Bosch, *Syll. gen. sp. crypt.* (Paris) : 349(1856).

≡*Phaeographina prosiliens*(Mont. & Bosch) Müll. Arg. *Flora*, Regensburg 65: 398(1882).

On *Abies kawakamii*, *Pseudotsuga wilsoniana* and *Abies* sp.

Taiwan(Aptroot and Sparrius, 2003, p. 33).

= *Phaeographina* valida Zahlbr. *in* Fedde, *Repertorium*, 31: 217(1933).

Type: Taiwan, Mt. Ali shan, Asahina no. 355.

Taiwan(Zahlbr., 1940, p. 193; Lamb, 1963, p. 546; Wang & Lai, 1973, p. 95).

Thelotrema Ach. **疣孔衣属**

Methodus, Sectio prior: 130(1803).

Thelotrema berkeleyanum(Mont.) Brusse 点疣孔衣

Mycotaxon 31(2) : 547(1988).

≡*Stegobolus berkeleyanus* Mont., in Hooker, *London J. Bot.* 4: 4(1845).

≡*Ocellularia berkeleyana*(Mont.) Zahlbr., Nat. Pflanzenfam., Teil. I(Leipzig) 1: 118(1905).

On conglomeratic rock.

Taiwan(Aptroot & Sparrius, 2003, p. 30).

Thelotrema expansum C. Knight 广疣孔衣

in Bailey, *Syn. Queensl. Fl.*, Suppl. 2: 86(1888).

On bark.

Hongkong(Thrower, 1988, p. 173).

Thelotrema lepadinum(Ach.) Ach. 帽贝疣孔衣

Methodus, Sectio prior: 132(1803).

≡*Lichen lepadinus* Ach., *Lich. suec. prodr.* (Linköping) : 30(1799) [1798].

On *Pseudotsuga wilsoniana*, and *Castanopsis* sp.

Taiwan(Aptroot & Sparrius 2003, p. 47).

Thelotrema microstomum Müll. Arg. 小疣孔衣

Flora, Regensburg 74(1) : 113(1891).

Thelotrema microstomum var. formosanum Zahlbr. 台湾变种

Feddes Repert. 33: 24(1933).

Type: Taiwan, Asahina no. 233.

Taiwan(Wang & Lai, 1973, p. 97).

On bark of trees.

Thelotrema murinum Zahlbr. 鼠色疣孔衣

Flechten der Insel Formosa in Feddes, Lichenes in Feddes, Repertorium sp. nov. 33: 24(1933).

Type: Taiwan, Asahina no. 228.

Taiwan(Zahlbr., 1940, p. 218; Lamb, 1963, p. 715; Wang & Lai, 1973, p. 97).

On bark of trees.

Thelotrema porinoides Mont. & Bosch 污核疣孔衣

in Miquel, *Pl. Jungh*. 4: 484(1855).

On bark. Pantropical.

Hongkong(Thrower, 1988, pp. 17, 174; Aptroot & Seaward, 1999, p. 95).

Thelotrema similans Nyl. 相似疣孔衣

Lich. Japon. : 58(1890).

On bark.

Guizhou(Zhang T & J. C. Wei, 2006, pp. 10 – 11).

Thelotrema weberi Hale 韦伯氏疣孔衣

Phytologia 27: 497(1974).

On *Abies kawakamii*.

Taiwan(Aptroot & Sparrius 2003, p. 47).

Xalocoa E. Kraichak, Lücking & Lumbsch **亚缘衣属**

Aust. Syst. Bot. 26(6): 472(2013).

Xalocoa ocellata(Fr.) E. Kraichak, R. Lücking & Lumbsch 斑眼亚缘衣

Aust. Syst. Bot. 26(6): 472(2013).

≡*Parmelia ocellata* Fr., *Lich. eur. reform.* (Lund): 190(1831).

≡*Diploschistes ocellatus*(Fr.) Norman, *Nytt Mag. Natur.* 7: 232(1853)[1852].

On calcareous rocks.

Yunnan(Zahlbr., 1930 b, p. 68).

=*Diploschistes ocellatus* f. *isabellinus* Zahlbr. *Hedwigia*, 74: 198(1934).

Type: from Sichuan.

Saxicolous.

(28) Gyalectaceae Stizenb. **凹盘衣科**

[as ‘Gyalecteae’] *Ber. Tät. St Gall. naturw. Ges.* : 158(1862)[1861 – 62].

Cryptolechia A. Massal. **隐床衣属**

Alcuni Gen. Lich. : 13(1853).

Cryptolechia saxatilis(Vězda) D. Hawksw. & Dibben 石生隐床衣

Lichenologist 14(1): 100(1982).

≡*Gyalectina saxatilis* Vězda, *Folia geobot. phytotax.* 4(4): 445(1969).

On shale.

Taiwan(Aptroot and Sparrius, 2003, p. 14).

Cryptolechia subincolorella(Nyl.) D. Hawksw. & Dibben 无色隐床衣

Lichenologist 14(1): 100(1982).

≡*Lecidea subincolorella* Nyl., *Bull. Soc. linn. Normandie*, sér. 22: 80(1868).

On *Zelkova serrata*.

Taiwan(Aptroot and Sparrius, 2003, p. 14).

Gyalecta Ach. **凹盘衣属**

K. Vetensk-Acad. Nya Handl. 29: 228(1808).

Gyalecta alutacea Zahlbr. 淡棕凹盘衣

in Handel-Mazzetti, *Symb. Sin.* 3: 71(1930) & *Cat. Lich. Univ.* 8: 256(1932).

Type: Yunnan, Handel-Mazzetti no. 9783.

Corticolous.

Gyalecta foveolaris(Ach.) Schaer. 小孔凹盘衣

Lich. helv. spicil. 7: 360(1836).

≡*Urceolaria foveolaris* Ach., *Methodus*, Sectio prior: 149(1803).

Sichuan(Obermayer W., 2004, pp. 496 – 497).

Pachyphiale Lönnr. **厚瓶衣属**

Flora, Regensburg 41: 611(1858).

Pachyphiale fagicola(Arnold) Zwackh 山毛榉厚瓶衣

Flora, Regensburg 45: 506(1862).

≡*Bacidia fagicola* Arnold, *Flora*, Regensburg 41: 504(1858).

On *Ardisia sieboldii* and *Castanopsis* sp.

Taiwan(Aptroot & Sparrius, 2003, p. 31).

(29) Phlyctidaceae Poelt ex J. C. David & D. Hawksw. **疱衣菌科**

Syst. Ascom. 10(1): 15(1991).

Phlyctis(Wallr.) Flot. **疱衣属**

Bot. Ztg. 8: 571(1850).

≡*Peltigera* sect. *Phlyctis* Wallr., *Fl. crypt. Germ.* (Norimbergae) 3: 553(1831).

Phlyctis argena(Ach.) Flot. 亮疱衣

Bot. Ztg. 8: 572(1850).

≡*Lepraria argena* Ach., *Lich. suec. prodr.* (Linköping): 8(1799)[1798].

On exposed trees. Known from Europe, North America, and temperate Asia including China(Prillinger et al. 1997, pp. 579, 582 & 583).

China(Prillinger et al. 1997, pp. 579, 582 & 583; Aptroot & Seawrd, 1999, p. 86),

Hongkong(Aptroot & Seawrd, 1999, p. 86), Taiwan(Aptroot and Sparrius, 2003, p. 33).

Phlyctis karnatana S. Joshi & Upreti 卡纳疱衣

Bryologist 113: 726(2010).

On thin-barked trees in a forest in a national park.

Yunnan(Joshi et al., 2015, p. 120).

Phlyctis schizospora Zahlbr. 裂孢疱衣

in Handel-Mazzetti, *Symb. Sin.* 3: 178(1930) & *Cat. Lich. Univ.* 8: 551(1932).

Type: Sichuan Handel-Mazzetti no. 2767.

On bark of *Myrsine africana*.

Hubei(Chen, Wu & Wei, 1989, p. 488).

Phlyctis subargena R. Ma & H. Y. Wang 亚亮疱衣

Mycotaxon 114: 362 – 365(2010).

Type: China. Gansu province, Longnan, Wenxian Co. Qiujiaba, alt. 2450m, On bark, F. Yang, 20070050, 2 August 2007. (Holotype in SDNU).

(30) Porinaceae Walt. Watson **污核衣科**

New Phytol. 28: 109(1929).

Porinaceae Rchb. [as 'Porineae'], Consp. Regni Veget. (Leipzig): 20(1828), nom. inval., Art. 32. 1(c), see Art. 18. 3(Melbourne).

Porina Ach. **污核衣属**

K. Vetensk-Acad. Nya Handl. 30: 158(1809).

Porina aenea(Wallr.) Zahlbr. 紫铜污核衣

Cat. Lich. Univ. 1: 363(1922).

≡*Verrucaria aenea* Wallr., *Fl. crypt. Germ.* (Norimbergae) 3: 299(1831).

On wood of sign post.

Ningxia(Liu M & Wei JC, 2013, p. 46), Taiwan(Aptroot 2003, p. 163).

Porina africana Müll. Arg. 非洲污核衣

Linnaea 43:41(1880).

On *Machlus thunbergii* & *Skimma superb*, and on shale. Pantropical species.

Taiwan(Aptroot 2003, p. 163).

Porina albicera(Kremp.) Overeem & D. Overeem 白污核衣

Bull. Jard. bot. Buitenz, 3 Sér. 4(1):112(1922).

≡*Verrucaria albicera* Kremp., *Lich. Foliic. Leg. Beccari*: 16(1874).

On *Pandanus* leaves.

Yunnan(Aptroot et al., 2003, p. 44), Hongkong(Aptroot et al., 2001, p. 337).

Porina applanata Vain. 扁平污核衣

Ann. Acad. Sci. fenn., Ser. A 15(6):362(1921).

On bamboo.

Taiwan(Aptroot 2003, p. 163).

Porina atrocoerulea Müll. Arg. 蓝黑污核衣

Flora, Regensburg 66(21):336(1883).

On bamboo.

Taiwan(Aptroot 2003, p. 163).

Porina austriaca(Körb.) Arnold 奥地利污核衣

Flora, Regensburg 65: 143(1882).

≡*Sagedia austriaca* Körb., *Parerga lichenol.* (Breslau) 4: 356(1863).

Zhejiang(Wu J. L. 1987, p. 32).

Porina bellendenica Müll. Arg. 钟形污核衣

Hedwigia 30:56(1891).

=*Porina farinosa* Zahlbr. *Repert. Spec. Nov. Regni Veg.* 31:200(1933).

Type: Taiwan, Prov. Nantou, Hori, 31 December 1925, Asahina F 305(W lectotype n. v., designated by McCarthy 1992, TNS isolectotype).

Taiwan(McCarthy, 2003, p. 00, as *P. farinosa*; Aptroot, 2004, p. 35).

Porina cestrensis(Tuck.) Müll. Arg. 喙污核衣

Flora, Regensburg 66(21):338(1883).

≡*Verrucaria cestrensis* Tuck., in Darlington, *Fl. Cestrica*, Edn 3:452(1853).

On *Ficus microcarpa*.

Taiwan(Aptroot 2003, p. 163).

Porina chlorotica(Ach.) Müll. Arg. 绿污核衣

Revue mycol., Toulouse 6(21):20(1884).

≡*Verrucaria chlorotica* Ach., *Lich. univ.*: 283(1810)

On sheltered granitic rock. Cosmopolitan.

Hongkong(Aptroot & Seaward, 1999, p. 88).

Porina chrysophora(Stirt.) R. Sant. 金色污核衣

Symb. bot. upsal. 12(1):222(1952).

≡*Verrucaria chrysophora* Stirt., *Proc. Roy. phil. Soc. Glasgow* 10:303(1877)[1876].

On leaves of *Diospyros maritima*, and on bamboo.

Taiwan(Aptroot 2003, p. 164; Aptroot et al., 2003, pp. 42 – 43).

Porina conica R. Sant. 圆锥污核衣

Symb. bot. upsal. 12(1):232(1952).

Yunnan(Aptroot et al., 2003, p. 44).

Porina coralloidea P. James 珊瑚污核衣

Lichenologist 5: 142(1971).

On trees in forests, and on granitic rock along stream. Known from Europe, Africa, Australia and Asia.

Hongkong(Aptroot & Seaward, 1999, p. 89).

Porina corruscans(Rehm) R. Sant. 皱污核衣

Symb. bot. upsal. 12(no. 1): 223(1952).

≡*Metasphaeria corruscans* Rehm, Leafl. of Philipp. Bot. 8: 2949(1916).

On fern fronds and leaves.

Hongkong(Aptroot et al., 2001, p. 337).

Porina cupreola(Müll. Arg.) F. Schill. 铜污核衣

Hedwigia 67: 274(1927).

≡*Phylloporina cupreola* Müll. Arg., *Hedwigia* 30: 189(1891).

Yunnan(Aptroot et al., 2003, p. 44).

Porina epiphylla Fée 叶表污核衣

Essai Crypt. Exot. (Paris): 76(1825) [1824].

≡*Phylloporina epiphylla*(Fée) Müll. Arg., *Lichenes Epiphylli Novi:* 21(1890); Zahlbr., *Cat. Lich. Univ.* 1: 533 (1922).

≡*Porina americana* Fée var. epiphylla Fée

Foliicolous. On living leaves of *Pandunus* sp. Pantropical.

Yunnan(Zahlbr., 1930 b, p. 31; Aptroot et al., 2003, p. 44), Hongkong(Aptroot & Seaward, 1999, p. 89; Aptroot et al., 2001, p. 337).

Porina epiphylloides Vězda 叶污核衣

Folia geobot. phytotax. 10: 393(1975).

Yunnan(Aptroot et al., 2003, p. 44).

Porina formosana Zahlbr. 台湾污核衣

in Feddes, *Repertorium sp. nov.* 31: 200(1933).

Syntype: Taiwan, Mt. Ali shan, Asahina nos. 305 & 308.

Taiwan(Zahlbr., 1940, p. 88; Lamb, 1963, p. 590; Wang & Lai, 1973, p. 95).

On rocks.

Porina grandispora P. M. McCarthy 大孢污核衣

Lichenologist 27(5): 338(1995).

On a tree in c. 10 m tall.

Hongkong(Aptroot et al., 2001, p. 337).

Porina guentheri(Flot.) Zahlbr. 格仑氏污核衣

Cat. Lich. Univ. 1: 384(1922).

≡*Verrucaria guentheri* Flot., *Bot. Ztg.* 8: 575(1850).

On sheltered granitic rock. Cosmopolitan.

Hongkong(Aptroot & Seaward, 1999, p. 89), Taiwan(Aptroot 2003, p. 164), on conglomeratic rock.

Porina hibernica P. James & Swinscow 冬污核衣

Lichenologist 2: 35(1962).

On *Ulmus*. New to Asia.

Taiwan(Aptroot 2003, p. 164).

Porina imitatrix Müll. Arg. 模污核衣

Flora, Regensburg 73: 196(1890).

Yunnan(Aptroot et al., 2003, p. 44).

Porina karnatakensis Makhija, Adaw. & Patw. 卡污核衣

J. Econ. Taxon. Bot. 18(3): 538(1995) [1994].

Yunnan(Aptroot et al., 2003, p. 44).

Porina leptalea(Durieu & Mont.) A. L. Sm. 纤污核衣

in Crombie & Smith, *Monogr. Brit. Lich.* 2: 333(1911) .

≡*Biatora leptalea* Durieu & Mont. , in Durieu, *Fl. d'Algérie, Cryptog.* 1: 268(1846) [1846 – 49] .

On sheltered granitic rock. Cosmopolitan.

Taiwan(Aptroot 2003, p. 164) . on pebbles and conglomeratic rocks, Hongkong(Aptroot et al. , 2001, p. 337) .

Porina leptosperma Müll. Arg. 细长污核衣

Flora, Regensburg 66(21) : 333(1883) .

On bamboo.

Taiwan(Aptroot 2003, p. 164) .

Porina limbulata(Kremp.) Vain. 檐污核衣

Ann. Acad. Sci. fenn. , Ser. A 15(no. 6) : 363(1921) .

≡*Verrucaria limbulata* Kremp. , *Lich. Foliic. Leg. Beccari*: 17(1875) .

On living leaves of trees and *Pandunus* sp.

Hongkong(Aptroot & Seaward, 1999, p. 89) .

Porina lucida R. Sant. 亮污核衣

Symb. bot. upsal. 12(1) : 240(1952) .

Yunnan(Aptroot et al. , 2003, p. 44) .

Porina mastoidea(Ach.) Müll. Arg. 乳头污核衣

Bot. Jb. 6: 399(1885) .

≡*Pyrenula mastoidea* Ach. , *Mag. Gesell. naturf. Freunde*, Berlin 6: 17(1814) .

On *Machilus thunbergii*.

Taiwan(Aptroot 2003, p. 164) .

Porina mastoidella(Nyl.) Müll. Arg. 小乳污核衣

Bot. Jb. 6: 401(1885) .

≡*Verrucaria mastoidella* Nyl. , *Flora*, Regensburg 50: 8(1867) .

On shale, brick, shale, and on *Ficus*.

Taiwan(Aptroot 2003, p. 164) .

Porina mirabilis Lücking & Vězda 奇污核衣

Willdenowia 28(1/2) : 211(1998) .

Yunnan(Aptroot et al. , 2003, p. 45) .

Porina nitidula Müll. Arg. 光污核衣

Flora, Regensburg 66(21) : 336(1883) .

On living leaves of trees and *Pandunus* sp. , *Alpinia speciosa*, *Asplenium nidus*, and *Turpinia* sp. Pantropical.

Yunnan(Aptroot et al. , 2003, p. 45) , HONGKONG(Aptroot & Seaward, 1999, p. 89; Aptoot & Sipman, 2001, p. 338) , taiwan(Aptroot et al. , 2003, pp. 42 – 43) .

Porina nucula Ach. 坚果污核衣

Syn. meth. lich. (Lund) : 112(1814) .

On bark of trees. On exposed trees, e. g. *Kandelia* sp.

Taiwan (Zahlbr. , 1933 c, p. 200; Wang & Lai, 1973, p. 95; Aptroot 2003, pp. 164 – 165 +, on shale), Hongkong(Aptroot & Seaward, 1999, p. 89) .

Porina nuculastrum(Müll. Arg.) R. C. Harris 小果污核衣

More Florida Lichens, Incl. 10 Cent Tour Pyrenol. (New York) : 174(1995) .

≡*Clathroporina nuculastrum* Müll. Arg. , *Flora*, Regensburg 67(32) : 618(1884) .

On sheltered granite and corticolous. Pantropical.

Hongkong(Aptroot & Seaward, 1999, p. 72; Aptroot et al. , 2001, p. 324 as *C. nuculastrum*) .

Porina pallescens R. Sant. 苍白污核衣

Symb. bot. upsal. 12(1):263(1952).

On *Turpinia* sp.

Taiwan(Aptroot et al., 2003, pp. 42 – 43).

Porina pariata(Nyl.) Zahlbr. 双污核衣

Cat. Lich. Univ. 1:399(1922).

≡ *Verrucaria pariata* Nyl. in Nyl. & Cromb. *Journ. Linn. Soc. London*, *Bot.* 20:66(1883).

Type: from Shanghai.

On peach trees.

Shangahi(Zahlbr., 1930 b, p. 21), Jiangsu(Xu B. S., 1989, p. 174), China(Zahlbr., 1922, p. 399).

Porina papuensis P. M. McCarthy 泡污核衣

Biblthca Lichenol. 52:86(1993).

On raised coral reef. New to northern hemisphere.

Taiwan(Aptroot 2003, p. 165).

Porina perminuta Vain. 细微污核衣

University of Calif. Publ. Bot. 12(1):14(1924).

Foliicolous. On leaves of *Diospyros maritima*.

Taiwan(Aptroot 2003, p. 165; Aptroot et al., 2003, pp. 42 – 43).

Porina phyllogena Müll. Arg. 叶生污核衣

Flora, Regensburg 66:27(no. 663)(1883).

Foliicolous.

Hongkong(Thrower, 1988, p. 150).

Porina rubentior(Stirt.) Müll. Arg. 淡红污核衣

Flora, Regensburg 66:26(no. 660)(1883).

≡ *Verrucaria rubentior* Stirt., *Proc. Roy. phil. Soc. Glasgow* 11: 107(1879)[1878].

On living leaves of *Pandanus* sp. Pantropical.

Hongkong(Aptroot & Seaward, 1999, p. 89).

Porina rufula(Kremp.) Vain. 红污核衣

Acta Soc. Fauna Flora fenn. 7(2):227(1890).

≡ *Verrucaria rufula* Kremp., *Lich. Foliic. Leg. Beccari*: 20(1874).

On *Alpinia speciosa*, and *Asplenium nidus*.

Yunnan(Aptroot et al., 2003, p. 45), Taiwan(Aptroot et al., 2003, pp. 42 – 43).

Porina sinochlorotica Zahlbr. 中华污核衣

in Handel-Mazzetti, *Symb. Sin.* 3:21(1930) & *Cat. Lich. Univ.* 8:107(1932).

Type: Sichuan, Handel-Mazzetti no. 1169.

On rocks.

Porina sphaerocephala Vain. 头状污核衣

Ann. Acad. Sci. fenn., Ser. A 15(6): 368(1921).

Foliicolous.

Yunnan(R. Sant. 1952, pp. 264 – 266).

Porina tetracerae(Ach.) Müll. Arg. 四角污核衣

Bot. Jb. 6:401(1885).

≡ *Verrucaria tetracerae* Ach., *Methodus*, Sectio prior: 121(1803).

On bark of trees. On sheltered granitic rock. Pantropical.

Taiwan(Zahlbr., 1933 c, p. 200; Wang & Lai, 1973, p. 96; Aptroot 2003, p. 165), Hongkong(Aptroot & Seaward, 1999, p. 89; Aptoot & Sipman, 2001, p. 338).

Porina trichothelioides R. Sant. 毛污核衣

Symb. bot. upsal. 12(1): 227(1952).

Yunnan(Aptroot et al., 2003, p. 45).

Porina ulceratula Zahlbr. 溃污核衣

Ark. Hydrobiol. Suppl. 12: 734(1934).

On shale. New to northern hemisphere.

Taiwan(Aptroot 2003, p. 165).

Porina vicinata Zahlbr. 亚污核衣

in Handel-Mazzetti, *Symb. Sin.* 3: 20(1930) & *Cat. Lich. Univ.* 8: 108(1932).

Type: Sichuan, Handel-Mazzetti no. 1172.

On rocks.

Porina virescens(Kremp.) Müll. Arg. 绿色污核衣

Flora, Regensburg 66(21): 331(1883).

≡*Verrucaria virescens* Kremp., *Lich. Foliic. Leg. Beccari*: 21(1874).

Yunnan(Aptroot et al., 2003, p. 45).

Trichothelium Müll. Arg. **丝果衣属**

Bot. Jb. 6: 418(1885).

Trichothelium alboatrum Vain. 黑白星果衣

Ann. Acad. Sci. fenn., Ser. A 15(6): 321(1921).

Yunnan(Aptroot et al., 2003, p. 45).

Trichothelium annulatum(P. Karst.) R. Sant. 褐斑星果衣

Symb. bot. upsal. 12(1): 275(1952).

≡*Lasiosphaeria annulata* P. Karst., *Hedwigia* 28: 193(1889).

On leaves of *Celtis giganticarpa*

Yunnan(J. C. Wei &Y. M. Jiang, 1991, p. 214 – 215).

Thelopsis Nyl. **乳果衣属**

Mém. Soc. Imp. Sci. Nat. Cherbourg 3: 194(1855).

Thelopsis isiaca Stizenb. 等乳果衣

Ber. Tät. St Gall. naturw. Ges.: 262(1895) [1893 – 94].

On coastal grnitic rocks. Known from Europe, Africa, North America, extending intoMexico, and Asia.

Hongkong(Aptroot & Seaward, 1999, 95).

[12] Pertusariales M. Choisy ex D. Hawksw. & O. E. Erikss. **鸡皮衣目**

Syst. Ascom. 5(1): 181(1986).

(31) Coccotremataceae Henssen ex J. C. David & D. Hawksw. **球孔衣科**

Syst. Ascom. 10(1): 14(1991)

Coccotrema Müll. Arg. **球孔衣属**

Miss. Sci. Cap Horn, Lich.: 31(1888).

Coccotrema cucurbitula(Mont.) Müll. Arg. 帽状球孔衣

Nuovo G. bot. ital. 21: 51(1889).

≡*Pertusaria cucurbitula* Mont., in Gay, *Hist. fis. y polit. Chile, Bot.* 8: 200(1852).

≡*Perforaria cucurbitula*(Mont.) Müll. Arg., *Nuovo G. bot. ital.* 23: 126(1891).

Taiwan(Oshio, 1968, p. 90; Wang & Lai, 1973, p. 95 & 1976, p. 227).

Parasiphula Kantvilas & Grube **准白角衣属**

in Grube & Kantvilas, *Lichenologist* 38(3): 246(2006).

Parasiphula complanata(Hook. f. & Taylor) Kantvilas & Grube 准白角衣

in Grube & Kantvilas, *Lichenologist* 38(3):246(2006).

≡*Sphaerophorus complanatus* Hook. f. & Taylor [as '*Sphaerophoron*'], *London J. Bot.* 3:654(1844).

≡*Siphula complanata* (Hook. f. & Taylor) R. Sant., *Svensk naturverenskap*: 178(1968).

= *Siphula pteruloides* Nyl., *Annls Sci. Nat.*, Bot., sér. 4 11:211(1859).

On rocks among mosses.

Neimenggu(Moreau et Moreau, 1951, p. 193), Xinjiang(Abdulla Abbas et al., 1993, p. 81; Abdulla Abbas et al., 1994, p. 23; Abdulla Abbas & Wu, 1998, p. 161; Abdulla Abbas et al., 2001, p. 368).

(32) Icmadophilaceae Triebel, in Rambold, Triebel & Hertel **霜降衣科**

Biblthca Lichenol. 53:227(1993)

Dibaeis Clem. **淡盘衣属**

Gen. fung. (Minneapolis):78, 175(1909).

Dibaeis absoluta(Tuck.) Kalb & Gierl 小淡盘衣

in Gierl & Kalb, *Herzogia* 9(3 – 4):613(1993). var. *absolutus* 原变种

≡*Baeomyces absolutus* Tuck., *Amer. J. Sci. Arts*, Ser. 2 28:201(1859).

Anhui(Xu B. S., 1989, p. 241)

var. stipitatus(Mont.) Vain. 长柄变种

Ann. Acad. Sci. fenn., Ser. A 15(no. 6):59(1921).

≡*Biatora icmadophila* var. *stipitata* Mont., *Annls Sci. Nat.*, *Bot.*, sér. 4 8:298(1857).

Saxicolous

Taiwan(Zahlbr., 1933 c, p. 44; M. Sato, 1941a, p. 20; Wang & Lai, 1973, p. 88), Hainan(J. C. Wei et al., 2013, p. 232).

Dibaeis baeomyces(L. f.) Rambold & Hertel 羊角淡盘衣

Herzogia 9(3 – 4):619(1993).

≡*Lichen baeomyces* L. f., *Suppl. Pl.*:450(1782)[1781].

On acid soil.

Yunnan(Wei X. L. et al. 2007, p. 151), Taiwan(Aptroot, Sparrius & Lai 2002, p. 283).

Dibaeis pulogensis(Vain.) Kalb & Gierl 蚕窝淡盘衣

Herzogia 9(3 – 4):628(1993).

≡*Baeomyces pulogensis* Vain., *Ann. Acad. Sci. fenn.*, Ser. A 15(6):59(1921).

On roadbanks.

Taiwan(Aptroot, Sparrius & Lai 2002, p. 283).

Dibaeis sorediata Kalb & Gierl 粉芽淡盘衣

in Gierl & Kalb, *Herzogia* 9(3 – 4):615(1993).

Misapplied name: *Baeomyces absolutes* Tuck. (in Thrower, 1988, pp. 15, 63).

On soil. Paleotropical.

Hongkong(Thrower, 1988, pp. 15, 63 as *Baeomyces absolutus* Tuck; Aptroot & Seaward, 1999, p. 73; Aptroot et al., 2001, p. 325), Taiwan(Aptroot, Sparrius & Lai 2002, p. 283).

Glossodium Nyl. **舌柱衣属**

Mém. Soc. Imp. Sci. Nat. Cherbourg 3: 169(1855).

Glossodium japonicum Zahlbr. 日本舌柱衣

Bot. Mag., Tokyo 41:336(1927).

Taiwan(Wang & Lai, 1973, p. 90).

Icmadophila Trevis. **霜降衣属**

Revta Period. Lav. Imp. Reale Acad., *Padova* 1(3):267(1852)[1851 – 52].

Icmadophila ericetorum(L.) Zahlbr. 霜降衣

Wiss. Mittellung. Bosnien und der Hercegov. 3: 605(1895).

≡*Lichen ericetorum* L., *Sp. pl.* 2: 1141(1753).

= *Icmadophila aeruginosa*(Scop.) Trevis., *Revta Period. Lav. Imp. Reale Acad., Padova* 1(3): 267(1852) [1851 –52].

On rotten logs.

Jilin(Chen et al. 1981, p. 150), Xinjiang(Abbas et al., 1997, p. 2), Xizang(Wei, 1981, p. 88; Wei & Jiang, 1981, p. 1147 & 1986, p. 95; Obermayer, 2004, pp. 499 – 500), Yunnan(Hue, 1898, p. 239; A. Zahlbr., 1930 b, p. 177; Wei X. L. et al. 2007, p. 153), Shandong(Zhao ZT et al., 1999, p. 427; Zhao ZT et al., 2002, p. 5), Taiwan(Wang & Lai, 1973, p. 91 & 1976, p. 227).

Pseudobaeomyces M. Satô **拟羊角衣属**

J. Jap. Bot. 16: 42(1940).

Pseudobaeomyces insignis(Zahlbr.) M. Satô 柄拟羊角衣

J. Jap. Bot. 16: 42(1940).

≡*Baeomyces insignis* Zahlbr., *Bot. Mag.*, Tokyo 41: 335(1932).

Syntype localities: Japan(Asahina no. 512) and China: Taiwan(Faurie no. 5858).

= *Pseudobaeomyces pachycarpus*(Müll. Arg.) M. Sato var. *stipitatus* M. Sato

On rocks.

Taiwan(Zahlbr., 1927, p. 335; Sandst. 1932, p. 66 & map 53; Sato, 1940a, p. 42 & 1941a, p. 44; Wang & Lai, 1973, p. 96).

Siphula Fr. **白角衣属**

Lich. eur. reform. (Lund): 7, 406(1831), Nom. cons., see Art. 14.

Siphula ceratites(Wahlenb.) Fr. 白角衣

Lich. eur. reform. (Lund): 406(1831).

≡*Baeomyces ceratites* Wahlenb., *Fl. lapp.*: 459(1812).

On soil or bryophytes.

Xizang(W. Obermayer, 2004, p. 508).

Siphula decumbens Nyl. 卧白角衣

Lich. Nov. Zeland. (Paris): 14(1888)

On rock and bark of *Castanopsis* sp.

Taiwan(Kantvilas et al., 2005, pp. 209 – 210).

Thamnolia Ach. ex Schaer. **地茶属**

Enum. critic. lich. europ. (Bern): 243(1850).

Thamnolia subuliformis(Ehrh.) W. L. Culb. 雪地茶

Brittonia 15: 144(1963).

≡*Lichen subuliformis* Ehrh., *Beitr. Naturk.* 3: 82(1788).

≡*Thamnolia vermicularis* subsp. *subuliformis*(Ehrh.) Schaer., Enum. critic. lich. europ. (Bern): 243(1850).

= *Thamnolia subvermicularis* Asahina, *J. Jap. Bot.* 13: 317(1937).

On the ground.

Xinjiang(Abdulla Abbas & Wu, 1998, p. 161; Abdulla Abbas et al., 2001, p. 368), Shaanxi(He & Chen 1995, p. 45, no specimen and locality were cited), Sichuan, Xizang(Obermayer 2004, p. 511).

f. subuliformis 原变型

On the ground.

Jilin(Chen et al. 1981, p. 157), Shaanxi(Wei, 1981, p. 89; Wu J. L. 1981, p. 164; Wei et al. 1982, p. 59), Xinjiang(Wu J. L. 1985, p. 77), Sichuan(Obermayer 2004, p. 512), Xizang(J. C. Wei & Chen, 1974, p. 181; J. C. Wei & Jiang, 1981, p. 1147 & 1986, p. 96; Obermayer 2004, p. 512), Hubei(Chen, Wu & J. C. Wei, 1989, p. 489), Taiwan(Wang & Lai, 1973, p. 97).

f. minor(Lamy) J. C. Wei 小枝变型

in Wei & Jiang, [*Lichens of Xizang*] (China) : 97(1986) .

≡*Thamnolia vermicularis* f. *minor* Lamy, *Bull. Soc. bot. Fr.* 25: 359(1878) .

≡*Thamnolia subvermicularis* f. *minor*(Lamy) Mot. (1960) .

On the ground among mosses.

Heilongjiang, Jilin, Xinjiang, Gansu, Shaanxi, Sichuan, Yunnan, Hubei(Yang et al. , 2015, pp. 134 – 136), Xizang(J. C. Wei & Chen, 1974, p. 181; J. C. Wei & Jiang, 1981, p. 1147 & 1986, p. 97; Yang et al. , 2015, pp. 134 – 136) .

Thamnolia vermicularis(Sw.) Schaer. 地茶

Enum. critic. lich. europ. (Bern) : 243(1850) .

≡*Lichen vermicularis* Sw. , *Method. Muscor.* : 37(1781) .

On the ground.

Shaanxi(Jatta, 1902, p. 466; Zahlbr. , 1930 b, p. 208) , Xinjiang(Abbas et al. , 1993, p. 81) , Xizang(Obermayer 2004, p. 512) , Sichuan(Elenk. 1901, p. 28, cited as eastern Tibet; Zahlbr. , 1930 b, p. 208; Stenroos et al. , 1994, p. 337; Obermayer 2004, p. 512) , Yunnan(Hue, 1887, p. 18 & 1898, p. 240; Paulson, 1928, p. 319; Zahlbr. , 1930 b, p. 208) .

f. vermicularis 原变型

On the ground.

Neimenggu(M. Sato, 1952, p. 175; Chen et al. 1981, p. 157) , Jilin(J. C. Wei, 1981, p. 88; Chen et al. 1981, p. 157) , Shaanxi(J. C. Wei et al. 1982, p. 58) , Yunnan(Wei X. L. et al. 2007, p. 158) , Xizang(J. C. Wei & Chen, 1974, p. 181; J. C. Wei & Jiang, 1981, p. 1147 & Wei & Jiang, 1986, p. 96) , Hubei(Chen, Wu & J. C. Wei, 1989, p. 489) , China(prov. not indicated: Asahina 1q& M. Sato in Asahina, 1939, p. 765) .

f. qomolangmana J. C. Wei & Y. M. Jiang 珠峰变型

Lichens of Xizang(China) : 96(1986) .

Type: Xizang, Wei & Chen no. 1230 in HMAS-L.

var. taurica(Wulfen) Schaer. 曲柄变型

Enum. critic. lich. europ. (Bern) : 244(1850) .

≡*Lichen tauricus* Wulfen, in Jacquin, *Collnea bot.* 2: 177(1791) [1788] .

On the ground.

Heilongjiang, Jilin, Neimenggu, Xizang(Yang et al. , 2015, pp. 136 – 138) , Shaanxi(Jatta, 1902, p. 466; Zahlbr. , 1930 b, p. 208; Yang et al. , 2015, p. 136) , Sichuan(Zahlbr. , 1934b, p. 213; Yang et al. , 2015, p. 136) , Yunnan(Hue, 1898, p. 241; Zahlbr. , 1930 b, p. 208; Yang et al. , 2015, p. 137) .

(33) Megasporaceae Lumbsch **巨孢衣科**

in Lumbsch, Feige & Schmitz *J. Hattori bot. Lab.* 75: 302(1994)

Aspicilia A. Massal. **平茶渍属**

Ric. auton. lich. crost. (Verona) : 36(1852) .

Aspicilia acharii(Ach.) Boistel 阿氏平茶渍

Nouv. Fl. Lich. 2: 149(1903) .

≡*Lichen acharii* Ach. , *Lich. suec. prodr.* (Linköping) : 33(1799) [1798] .

var. ochraceoferruginea(Schaer.) Kremp. 赭色变种

≡*Gyalecta acharii* var. *ochraceoferruginea* Schaer. , *Enum. critic. lich. europ.* (Bern) : 94(1850) .

≡*Lecanora lacustris*(With.) Nyl. var. *ochraceoferruginea*(Schaer.) Zahlbr. in Handel-Mazzetti, *Symb. Sin.* 3: 159(1930) .

On diabasic rock.

Sichuan(Zahlbr. , 1930 b, p. 159 & 1932 a, p. 529) , Shanghai(Kremp. 1874 c, p. 67) .

Aspicilia albocretacea(Zahlbr.) J. C, Wei 白垩平茶渍

Enum. Lich. China, 30(1991) .

≡*Lecanora albocretacea* Zahlbr. , in Handel-Mazzetti, *Symb. Sin*. 3: 152(1930) & *Cat. Lich. Univ*. 8: 525(1932) .

Type: Sichuan, at about 4750m, 6/VIII/1915, Handel-Mazzetti no. 7459.

On calciferous rock.

≡*Aspicilia albocretacea* Motyka, Porosty(Lichenes) . 1, Rodzina Lecanoraceae. Hymenelia, Aspicilia, Lecanorella, Protoplacodium, Manzonia(Lublin) : 301(1995) , nom. inval. , Art. 34. 1(Melbourne) .

Aspicilia anamyloidea(Zahlbr.) J. C. Wei 类磨石平茶渍

Enum. Lich. China, 30(1991) .

≡ *Lecanora anamyloidea* Zahlbr. , in Handel-Mazzetti, *Symb. Sin*. 3: 150 (1930) & *Cat. Lich. Univ*. 8: 526 (1932) .

Type: Sichuan, at about 2125 m, 29/V/1914, Handel-Mazzetti no. 2703.

Aspicilia aquatica(Fr.) Körb. 水生平茶渍

Syst. lich. germ. (Breslau) : 146(1855) . Zahlbr. in Handel-Mazzetti, *Symb. Sin*. 3: 152(1930) .

≡*Parmelia cinerea* var. *aquatica* Fr. , *Lich. eur. reform*. (Lund) : 144(1831) .

≡*Lecanora aquatica*(Fr.) Hepp, *Flecht. Europ*. : no. 390(1867) .

= *Lecanora amphibola*(Ach.) Vain. *Meddn Soc. Fauna Flora fenn*. 6: 167(1881) , Zahlbr. in Handel-Mazzetti, *Symb. Sin*. 3: 152(1930) .

≡*Lichen amphibolus* Ach. , *Lich. suec. prodr*. (Linköping) : 23(1799) [1798] .

Shaanxi(Jatta, 1902, p. 476; Zahlbr. , 1930 b, p. 152) .

Aspicilia asiatica(H. Magn.) Oxn. 亚洲平茶渍

Nov. sist. Niz. Rast. 9: 286(1972) .

≡*Lecanora asiatica* H. Magn. , *Lichens Central Asia* 1: 90(1940) .

var. asiatica f. asiatica 原变型

Type: Gansu, Bohlin no. 77c in S.

On calciferous rock, solitary or immixed among other lichens.

Neimenggu(H. Magn. 1940, p. 92 & 1944, p. 40) , Xinjiang(Abdulla Abbas & Wu, 1998, p. 67) , Qinghai, Gansu(H. Magn. 1940, pp. 90 – 92) .

var. asiatica f. irregularis(H. Magn.) J. C. Wei 不对称变型

≡*Lecanora asiatica* var. *asiatica* f. *irregularis* H. Magn. *Lichens Central Asia* 1: 92(1940) .

Type: Gansu, Bohlin no. 86b in S.

Xinjiang(Abdulla Abbas et al. , 1993, p. 75; Abdulla Abbas et al. , 1994, p. 20; Abdulla Abbas & Wu, 1998, p. 67; Abdulla Abbas et al. , 2001, p. 360) .

var. asiatica f. *ochracea*(H. Magn.) J. C. Wei . 黄褐变型

≡*Lecanora asiatica* var. *asiatica* f. *ochracea* H. Magn. *Lichens Central Asia* 1: 92(1940) .

Type: Gansu, Bohlin no. 62 in S.

var. subfarinosa(H. Magn.) N. S. Golubk. 亚粉变种

Nov. sist. Niz. Rast. 10: 217(1973) . f. subfarinosa

≡*Lecanora asiatica* var. *subfarinosa* H. Magn. , *Lichens Central Asia* 1: 92(1940) .

Type: Gansu, 33a

Xinjiang(Abdulla Abbas & Wu, 1998, p. 68) .

var. subfarinosa f. microcarpa(H. Magn.) J. C. Wei 微果变型

Enum. Lich. China 30(1991) .

≡*Lecanora asiatica* var. *subfarinosa* f. *microcarpa* H. Magn. *Lichens Central Asia* 1: 93(1940) .

Syntypes: Gansu, Bexell no. 1, & Bohlin nos. 29 b, 42 c in S.

Xinjiang(Abduula Abbas et al. , 1993, p. 75; Abdulla Abbas et al. , 1994, p. 20; Abdulla Abbas & Wu, 1998, p. 68; Abdulla Abbas et al. , 2001, p. 360) , Gansu(Abbas et al. , 1993, p. 75) .

Aspicilia bohlinii(H. Magn.) J. C. Wei 包氏平茶渍

Enum. Lich. China 30(1991) .

≡*Lecanora bohlinii* H. Magn. *Lichens Central Asia* 1: 93(1940) .

Type: Qinghai, Shengku, Khaltain, 12/IV/1932, collected by Bohlin, preserved in S(!) .

On calciferous rock with Lecanora kukunorensis.

Xinjiang(Abdulla Abbas et al. , 1993, p. 75; Abdulla Abbas et al. , 1994, p. 20; Abdulla Abbas & Wu, 1998, p. 68; Abdulla Abbas et al. , 2001, p. 360) , Qinghai(Magn. , 1940, p. 93) .

Aspicilia caesiororida(Zahlbr.) J. C. Wei 湿兰平茶渍

Enum. Lichens China(Beijing) : 30(1991) . f. caesiororida 原变型

≡ *Lecanora caesiororida* Zahlbr. , in Handel-Mazzetti, *Symb. Sin.* 3: 151 (1930) & *Cat. Lich. Univ.* 8: 526 (1932) .

Type: Sichuan, at about 2650 m, 26/VII/1915, Handel-Mazzetti no. 7262.

On rock.

Jiangsu(Xu B. S. , 1989, p. 207) .

f. oxydascens(Zahlbr.) J. C. Wei 锈红变型

Enum. Lichens China 30(1991) .

≡*Lecanora caesiororida* f. *oxydascens* Zahlbr. , in Handel-Mazzetti, *Symb. Sin.* 3: 151(1930) & *Cat. Lich. Univ.* 8: 526(1932) .

Type: Sichuan, at about 2950m, 8/VIII/1915, Handel-Mazzetti no. 7557.

On rock.

Aspicilia cinerea(L.) Körb. 灰平茶渍

Syst. lich. germ. (Breslau) : 164(1855) .

≡*Lichen cinereus* L. , *Mant. Pl.* 1: 132(1767) .

≡*Lecanora cinerea*(L.) Sommerf. , *Deutschl. Fl.* , *Abth.* 2(Frankfurt) 3: 90(1813) .

On rocks.

Shaanxi(Jatta, 1902, p. 476; Zahlbr. , 1930 b, p. 151) , Xizang(Mt. Qomolangma, Paulson, 1925, p. 192) , Sichuan(Zahlbr. , 1930 b, p. 151) , Shandong(Moreau et Moreau, 1951, p. 188) , Zhejiang(Nyl. & Cromb. 1883, p. 64; Hue, 1891, p. 72; Zahlbr. , 1930 b, p. 151) , See *Arthonia caesiocinerea* (Nyl. Ex Malbr.) Arnold for Hongkong(Thrower, 1988, p. 58) .

Aspicilia cinereopolita(Zahlbr.) J. C. Wei 灰光平茶渍

Enum. Lichens China(Beijing) : 31(1991) . f. cinereopolita 原变型

≡ *Lecanora cinereopolita* Zahlbr. in Handel-Mazzetti, *Symb. Sin.* 3: 152 (1930) & *Cat. Lich. Univ.* 8: 527 (1932) .

Type: Yunnan, Handel-Mazzetti no. 9999.

On rock.

f. albidior(Zahlbr.) J. C. Wei 白光变型

Enum. Lich. China 31(1991) .

≡*Lecanora cinereopolita* f. *albidior* Zahlbr. , in Handel-Mazzetti, *Symb. Sin.* 3: 152(1930) & *Cat. Lich. Univ.* 8: 527(1932) .

Type: Yunnan, in Mt. Yulong shan near Lijiang, Handel-Mazzetti no. 4296.

f. major(Zahlbr.) J. C. Wei 大光变型

Enum. Lich. China 31(1991) .

≡*Lecanora cinereopolita* f. *major* Zahlbr. , in Handel-Mazzetti, *Symb. Sin.* 3: 152(1930) & *Cat. Lich. Univ.* 8: 527(1932) .

Type: Yunnan, Handel-Mazzetti no. 9993.

Aspicilia corrugatula(Arnold) Hue 皱褶平茶渍

Nouv. Arch. Mus. Hist. Nat. , Paris, 5 sér. 2: 111(1912) [1910] .

≡*Lecidea corrugatula* Arnold, *Verh. zool. -bot. Ges.* Wien 29: 357(1879) .

≡*Lecanora corrugatula*(Arnold) Nyl. , *Flora, Regensburg* 63: 393(1880) .

On rocks.

Shaanxi(Jatta, 1902, p. 476; Zahlbr. , 1930 b, p. 151) .

Aspicilia decorticata(H. Magn.) J. C. Wei 脱皮平茶渍

Enum. Lich. China 31(1991) .

≡ *Lecanora decorticata* H. Magn. *Lichens Central Asia* 1: 95(1940) .

Lectotype: Gansu, Bohlin no. 75, selected by me in 1982 in S.

On calciferous rock.

Aspicilia desertorum(Kremp.) Mereschk. 荒漠平茶渍

Trudi naturh. Ver. ksl. Univ. Kasan 43: 36(1911) .

≡*Lecanora desertorum* Kremp. , *Verh. zool. -bot. Ges.* Wien 17: 601(1867) .

Xinjiang(Abdulla Abbas & Wu, 1998, p. 68; Abdulla Abbas et al. , 2001, p. 360) .

=*Aspicilia alpino-desertorum* f. *esculenta-alpina*(Pall.) Elenk.

Xinjiang(Mt. Tian shan, Elenk. 1901, p. 35) .

[*Aspicilia alpino-desertorum* f. *fruticuloso-foliacea* Elenk.]

Xinjiang(Aksai, Elenk. 1901, p. 36) .

"Earlier reports of both *A.* '*desertorum*' sensu Krempelhuber and *A. esculenta* need to be reviewed. "(Sohrabi et al. , 2013) .

Aspicilia determinata(H. Magn.) Golubk. 定形平茶渍

Nov. sist. Niz. Rast. 9: 236(1972) .

≡*Lecanora determinata* H. Magn. , *Lichens Central Asia* 1: 96(1940) .

Type: Gansu, Bohlin no. 7 in S(!) .

On a coarse-grained sandstone.

Aspicilia disculifera(Zahlbr.) J. C. Wei 盘状平茶渍

Enum. Lich. China 31(1991) . var. disculifera 原变种

≡ *Lecanora disculifera* Zahlbr. , in Handel-Mazzetti, *Symb. Sin.* 3: 155(1930) & *Cat. Lich. Univ.* 8: 528(1932) .

Syntype localities: Sichuan, near Huili, Handel-Mazzetti no. 1015 & Yunnan, Handel- Mazzetti no. 6378.

On rocks.

var. dealbata(Zahlbr.) J. C. Wei 白粉变种

Enum. Lich. China 32(1991) .

≡ *Lecanora disculifera* var. *dealbata* Zahlbr. , in Handel-Mazzetti, *Symb. Sin.* 3: 156(1930) & *Cat. Lich. Univ.* 8: 528(1932) .

Type: Sichuan, Handel-Mazzetti no. 1769.

On sandstone.

Aspicilia disjecta(Zahlbr.) J. C. Wei 散疣平茶渍

Enum. Lich. China 32(1991) .

≡ *Lecanora disjecta* Zahlbr. , in Handel-Mazzetti, *Symb. Sin.* . 3: 154(1930) & *Cat. Lich. Univ.* 8: 528(1932) .

Type locality: Yunnan, in Mt. Guamao shan between Yungning and Yungbei, Handel-Mazzetti no. 3319.

On rocks.

Aspicilia dschungdienensis(Zahlbr.) J. C. Wei 中甸平茶渍

Enum. Lich. China 32(1991).

≡*Lecanora dschungdienensis* Zahlbr., in Handel-Mazzetti, *Symb. Sin.* 3: 157(1930) & *Cat. Lich. Univ.* 8: 528 (1932).

Type locality: Yunnan, Handel-Mazzetti no. 6948.

On rocks.

Aspicilia exuberans(H. Magn.) J. C. Wei 彩斑平茶渍

Enum. Lich. China 32(1991).

≡*Lecanora exuberans* H. Magn. *Lich. Centr. Asia* 1:97(1940).

Type: Gansu, Bohlin no. 34 in S.

On calciferous stone with Aspicilia sublaqueata and A. decorticata, Acarospora umbrina and *Glypholecia scabra*.

Xinjiang(Abdulla Abbas et al., 1993, p. 75; Abdulla Abbas & Wu, 1998, p. 69; Abdulla Abbas et al., 2001, p. 360), Gansu(Magn., 1940, p. 97).

Aspicilia galactotera(Zahlbr.) J. C. Wei 乳平茶渍

Enum. Lich. China 32(1991).

≡*Lecanora galactotera* Zahlbr., in Handel-Mazzetti, *Symb. Sin.* 3: 156(1930) & *Cat. Lich. Univ.* 8: 529(1930).

Type: Yunnan, Yulong shan near Lijiang, Handel-Mazzetti no. 3555.

On rocks.

Aspicilia hartliana(J. Steiner) Hue 拟亚洲平茶渍

Nouv. Arch. Mus. Hist. Nat., Paris, 5 sér. 2: 113(1912) [1910].

≡*Lecanora hartliana* J. Steiner, in Halácsy, *Denkschr. Kaiserl. Akad. Wiss. Wien, Math. -Naturwiss.* Kl. 61: 264 (1894).

On calcareous rock.

Xinjiang(Abdulla Abbas et al., 1993, p. 75; Abdulla Abbas et al., 1994, p. 20; Abdulla Abbas & Wu, 1998, pp. 69 – 70; Abdulla Abbas et al., 2001, p. 360), Gansu(H. Magn. 1940, p. 98).

Aspicilia hedinii(H. Magn.) Oxner 海登氏平茶渍

Nov. sist. Niz. Rast. 9: 288(1972). f. hedinii 原变型

≡*Lecanora hedinii* H. Magn., *Lich. Centr. Asia* 1: 98(1940)

Type: Gansu, Bohlin no. 55 a in S(!).

On calciferous, dark stone.

Xinjiang(Abdulla Abbas et al., 1993, p. 75; Abdulla Abbas & Wu, 1998, p. 70; Abdulla Abbas et al., 2001, p. 360), Gansu(Magn., 1940, p. 98).

f. pruinosa(H. Magn.) J. C. Wei 霜变型

Enum. Lich. China(Beijing): 32(1991).

≡*Lecanora hedinii* f. *pruinosa* H. Magn. *Lich. Centr. Asia* 1: 100(1940).

Type: Gansu, Bohlin no. 55 b in S(!).

Aspicilia hoffmannii(Ach.) Flagey 霍夫曼平茶渍

Catal. Lich. Algérie: 15(1896).

≡*Lichen hoffmannii* Ach., *Lich. suec. prodr.* (Linköping): 31(1799) [1798].

On rocks.

Neimenggu(Abbas et al., 1993, p. 75; Abbas et al., 1994, p. 20), Xinjiang(Abdulla Abbas et al., 1993, p. 75; Abdulla Abbas et al., 1994, p. 20; Abdulla Abbas & Wu, 1998, p. 70; Abdulla Abbas et al., 2001, p. 360.), Gansu(H. Magn. 1940, p. 100), Shaaxi(Jatta, 1902, p. 476; Zahlbr., 1930 b, p. 151), Shandong(zhao ZT et al., 1999, p. 427; Zhao ZT et al., 2002, p. 5; Li Y et al., 2008, p. 71), Jiangsu(Wu J. N. & Xiang T. 1981, p. 4; Xu B. S., 1989, p. 208).

Aspicilia lesleyana Darb. 高寒平茶渍

Zahlbr. *Cat. Lich. Univ.* 5: 327(1928).

≡*Lecanora lesleyana* (Darb.) Pauls.

Xizang(Mt. Qomolangma: Paulson, 1925, p. 192).

Aspicilia maculata(H. Magn.) Oxner 斑点平茶渍

in Kopaczevskaja et al., *Opredelitel' Lishaĭnikov SSSR* Vypusk(Handbook ol the lichens of the U. S. S. R.) (Leningrad) 1: 178(1971). f. maculata 原变型

≡*Lecanora maculata* H. Magn., *Lichens Central Asia* 1: 101(1940).

Type: Gansu, Bohlin no. 66 a in S(!).

On siliceous and calcareous stones.

Neimenggu(H. Magn. 1940, p. 101 & 1944, p. 41), Xinjiang(Abdulla Abbas et al., 1993, p. 75; Abdulla Abbas & Wu, 1998, p. 71; Abdulla Abbas et al., 2001, p. 360), Qinghai, Gansu, Xizang(H. Magn. 1940, p. 101).

f. subochracea(H. Magn.) J. C. Wei 亚赭变型

Enum. Lich. China: 33(1991).

≡*Lecanora maculata* f. *subochracea* H. Magn., *Lichens Central Asia* 2: 41(1944).

Type: Neimenggu, 18/XII/1929, F. Bergman no. 154 in S(!).

Aspicilia melanaspis(Ach.) Poelt & Leuckert 黑果平茶渍

Willdenowia 7(1): 25(1973).

≡*Parmelia melanaspis* Ach., *Methodus*, Sectio post.: 196(1803).

On rocks.

Jiangsu(Xu B. S., 1989, p. 208).

Aspicilia microplaca(H. Magn.) J. C. Wei 小盘平茶渍

Enum. Lich. China: 33(1991).

≡*Lecanora microplaca* H. Magn. *Lichens Central Asia* 1: 102(1940).

Type: Gansu, Bohlin no. 13 a in S(!).

On calciferous stone.

Aspicilia ochraceoalba(H. Magn.) N. S. Golubk. 赭白平茶渍

Nov. sist. Niz. Rast. 9: 240(1972).

≡*Lecanora ochraceoalba* H. Magn. *Lich. Centr. Asia* 1: 103(1940).

Type: Gansu, Bohlin no. 30a in S(!).

On calcareous rock.

Xinjiang(Abdulla Abbas et al., 1993, p. 75; Abdulla Abbas & Wu, 1998, p. 71; Abdulla Abbas et al., 2001, p. 361), Qinghai, Gansu(H. Magn. 1940, p. 103).

Aspicilia ochromelaena(Zahlbr.) J. C. Wei 赭黑平茶渍

Enum. Lich. China: 33(1991).

≡ *Lecanora ochromelaena* Zahlbr., in Handel-Mazzetti, *Symb. Sin.* 3: 153(1930) & *Cat. Lich. Univ.* 8: 530(1932).

Type: Yunnan, Mt. Guamao shan between Yungning and Yungbei, Handel-Mazzetti no. 3320.

On rocks.

Aspicilia oleifera(H. Magn.) J. C. Wei 野油平茶渍

Enum. Lich. China 33(1991).

≡*Lecanora oleifera* H. Magn. *Lich. Centr. Asia* 1: 104(1940).

Type: Gansu, Bohlin no. 54a in S(!).

On calcareous rock with many other lichens.

Xinjiang(Abdulla Abbas et al., 1993, p. 75; Abdulla Abbas & Wu, 1998, p. 72; Abdulla Abbas et al., 2001,

p. 361), Qinghai, Gansu(H. Magn. 1940, p. 104).

Aspicilia persica(Müll. Arg.) Sohrabi 桃红平茶渍

in Seaward, Sipman & Sohrabi, *Sauteria* 15: 467(2008).

≡*Placodium persicum* Müll. Arg., *Hedwigia* 31: 154(1892).

On siliceous rock.

Xinjiang(Li Shu Xia et al., 2013, pp. 91 – 92).

Aspicilia scabridula(H. Magn.) N. S. Golubk. 微糙平茶渍

Nov. sist. Niz. Rast. 10: 219(1973).

≡*Lecanora scabridula* H. Magn. *Lich. Centr. Asia* 1: 105(1940).

Type: Gansu, Bohlin no. 15 in S(!).

On calcareous rock.

Neimenggu(H. Magn. 1944, p. 42), Qinghai, Gansu(H. Magn. 1940, p. 105).

Aspicilia schisticola(H. Magn.) Wei 片岩平茶渍

Enum. Lich. China 33(1991).

≡*Lecanora schisticola* H. Magn. *Lich. Centr. Asia* 1: 106(1940).

Type: Gansu, Bexell no. 11 in S(!).

On balckish schist without positive HCl-reaction.

Aspicilia sinensis(Zahlbr.) J. C. Wei 中华平茶渍

Enum. Lich. China 33(1991).

≡*Ionaspis sinensis* Zahlbr., in Handel-Mazzetti, *Symb. Sin.* 3: 69(1930) & *Cat. Lich. Univ.* 8: 254(1932).

Syntype: Yunnan, Handel-Mazzetti no. 6632; Sichuan, Handel-Mazzetti nos. 1010, 1017.

≡*Lecanora sinensis*(Zahlbr.) H. Magn. Act. Hort. Gothob. 8: 46(1933), non Zahlbr., 1928.

On rocks.

Sichuan, Yunnan(Zahlbr., 1930 b, p. 69 & 1932 a, p. 254; H. Magn. 1933, p. 46).

Aspicilia subalbicans(H. Magn.) J. C. Wei 亚白平茶渍

Enum. Lich. China 33(1991).

≡*Lecanora subalbicans* H. Magn. *Lich. Centr. Asia* 1: 107(1940).

Type: Gansu, Bohlin no. 40a in S(!).

On calciferous stone.

Neimenggu, Gansu(H. Magn. 1940, p. 107, & 1944, p. 42), Xinjiang(Abdulla Abbas et al., 1994, p. 20; Abdulla Abbas & Wu, 1998, p. 72; Abdulla Abbas et al., 2001, p. 361).

Aspicilia subcaesia(H. Magn.) J. C. Wei 亚兰灰平茶渍

Enum. Lich. China 34(1991).

≡*Lecanora subcaesia* H. Magn. *Lich. Centr. Asia* 1: 108(1940).

Type: Qinghai, Bohlin no. 88 e in S(!).

On calciferous rock.

Xinjiang(Abdulla Abbas et al., 1993, p. 75 as *A. subcaesis*; Abdulla Abbas & Wu, 1998, p. 72; Abdulla Abbas et al., 2001, p. 361), Qinghai(Magn., 1940, p. 108).

Aspicilia subconfluens(H. Magn.) J. C. Wei 亚汇平茶渍

Enum. Lich. China 34(1991).

≡*Lecanora subconfluens* H. Magn. *Lich. Centr. Asia* 1: 108(1940).

Type: Qinghai, Bohlin no. 88 a in S(!).

On hard, slightly calciferous, dark stone.

Gansu, Qinghai(H. Magn. 1940, p. 108).

Aspicilia subdepressa Arnold 凹平茶渍

Verh. zool. -bot. Ges. Wien 19: 611(1869).

On rock.

Heilongjiang(Li Shu Xia et al., 2013, pp. 94 – 95).

Aspicilia subdiffracta(H. Magn.) J. C. Wei 小裂平茶渍

Enum. Lich. China: 34(1991).

≡*Lecanora subdiffracta* H. Magn. *Lich. Centr. Asia* 2: 40(1944).

Type: Neimenggu, Belimiao, 1929, geol. samples no. 186 in S(!).

Aspicilia subflavida(H. Magn.) J. C. Wei 微黄平茶渍

Enum. Lich. China 34(1991).

≡*Lecanora subflavida* H. Magn. *Lich. Centr. Asia* 1: 109(1940).

Type: Gansu, Bexell no. 11 in S(!).

On black schist without positive HCl-reaction.

Aspicilia sublaqueata(H. Magn.) J. C. Wei 白边平茶渍

Enum. Lich. China: 34(1991).

≡*Lecanora sublaqueata* H. Magn. *Lich. Centr. Asia* 1: 110(1940).

Type: Gansu, Bohlin no. 35 a in S(!).

Xinjiang(Abdulla Abbas et al., 1993, p. 75; Abdulla Abbas & Wu, 1998, pp. 72 – 73; Abdulla Abbas et al., 2001, p. 361), Gansu(Magn., 1940, p. 110).

Aspicilia superposita(Zahlbr.) J. C. Wei 重叠平茶渍

Enum. Lich. China: 34(1991).

≡*Lecanora superposita* Zahlbr., in Handel-Mazzetti, *Symb. Sin.* 3: 154(1930) & *Cat. Lich. Univ.* 8: 532(1932).

Type: Sichuan, Handel-Mazzetti no. 1736.

On rocks.

Aspicilia tesselans(Zahlbr.) J. C. Wei 骰子平茶渍

Enum. Lich. China: 34(1991).

≡*Lecanora tesselans* Zahlbr., in Handel-Mazzetti, *Symb. Sin.* 3: 156(1930) & *Cat. Lich. Univ.* 8: 532(1932).

Type: Sichuan, Handel-Mazzetti no. 7259.

On rocks.

Aspicilia tibetica Sohrabi & Owe-Larss. 西藏平茶渍

Mycol. Progr. 9(4): 492(2010).

Type China. Xizang(Tibet): Himalaya Range, 135 km SSW of Lhasa, SSE of Pomo Tso(= Puma Yumco), near the pass into the Kuru valley, way from the pass-road to the glacier, 28°28′N, 090°37′E, alt. 5, 100 – 5, 300 m, Kobresiameadows and slopes covered with rock debris, on soil, 18. VII. 1994, Obermayer 04386(GZU, holotype; H, isotype).

Aspicilia tortuosa(H. Magn.) N. S. Golubk. 扭曲平茶渍

Nov. sist. Niz. Rast. 9: 238(1972). var. tortuosa 原变种

≡*Lecanora tortuosa* H. Magn., *Lich. Centr. Asia* 1: 111(1940).

Type: Gansu, Bohlin no. 3 a in S).

Xinjiang(Abdulla Abbas et al., 2001, p. 361).

var. ferruginea(H. Magn.) J. C. Wei 锈色变种

Enum. Lich. China: 34(1991).

≡*Lecanora tortuosa* var. *ferruginea* H. Magn. *Lich. Centr. Asia* 1: 113(1940).

Type: Gansu, Bohlin no. 9 in S(!).

On slightly calciferous rock.

var. simplicior(H. Magn.) J. C. Wei 单一变种

Enum. Lich. China: 34(1991) .

≡*Lecanora tortuosa* var. *simplicior* H. Magn. *Lich. Centr. Asia* 1: 112(1940) .

Type: Gansu, Bohlin no. 13b in S.

On stone, sometimes calciferous.

Xinjiang(Abdulla Abbas & Wu, 1998, p. 73; Abdulla Abbas et al. , 2001, p. 361) .

Aspicilia transbaicalica Oxner 小角平茶渍

Zhurnal Biobot. Tsiklu Bu Akad. Nauk(Journ. Cycle Bot. Acad. Sci. Ukraine) 7/8: 171(1933) .

On rocks.

Shaanxi, Heilongjiang(J. C. Wei, 1981, p. 82) , Shandong(Li Y et al. , 2008, p. 71) .

Aspicilia verrucigera Hue 疣平茶渍

Nouv. Arch. Mus. Hist. Nat. , Paris, 5 sér. 2: 48(1912) [1910] .

On siliceous rock.

Xinjiang, Qinghai(Li Shu Xia et al. , 2013, pp. 93 – 94) .

Aspicilia volcanica Ismayil, A. Abbas & S. Y. Guo 火山平茶渍

Mycotaxon 130: 545(2016) .

Type: China. Heilongjiang: Wudalianchi scenic area, 48°39. 321′N 126°09. 311′E, alt. 280 m, 15 July 2011, Gulbostan Ismayil & A. Abbas 20111154(Holotype, HMAS-L; isotype, XJU; GenBank KM609324) .

On black rock from volcanic mountain

Circinaria Link **野粮衣属**

Neues J. Bot. 3(1, 2) : 5(1809) .

Circinaria affinis(Eversm.) Sohrabi 风滚野粮衣

in Sohrabi, Stenroos, Myllys, Søchting, Ahti & Hyvönen, *Mycol. Progr*. 11: 18(2012) .

≡*Lecanora affinis* Eversm. in *Nova Acta Phys. Med. Acad. Caes. Leop. Carol. Nat. Cur*. 15: 351. 1831. Lectotype: designated by Sohrabi and Ahti(2010) , Kazakhstan. ' Desertis Kirgisorum' [1820] , *Eversmann* in Elenkin: *Lich. Fl. Ross*. No. 24e(H) .

= *Aspicilia vagans* Oxner, in Kopaczevskaja et al. , Opredelitel' Lishaĭnikov SSSR Vypusk(Handbook ol the lichens of the U. S. S. R.) (Leningrad) 1: 196(1971) , Nom. inval. , Art. 41. 5(Melbourne) .

Xinjiang(Wu J. L. , 1985, p. 73, Abdulla Abbas & Wu, 1998, p. 73; and Abdulla Abbas et al. , 2001, p. 361 as *A. vagans*; Tacheng Toli, 1232m, 2007, Abbas & Xahidin 20080364(H) (M. Sohrabi et al. 2013, p. 249) .

Circinaria alpicola(Elenkin) Sohrabi 高山野粮衣

in Sohrabi, Stenroos, Myllys, Søchting, Ahti & Hyvönen, *Mycol. Progr*. 12: 244(2013) .

≡*Aspicilia alpicola* Elenkin, *Fl. Lishaynikov Sredney Rossii* [Lichenes Florae Rossiae Mediae] 2: 222. 1907.

= *Aspicilia alpinodesertorum* f. *esculenta-alpina* Elenkin in Izv. *Imp. S. -Peterburgsk. Bot.*

Sada 1: 36. 1901(16 July) , as ' *esculenta alpina*' or ' *esculenta(alpina)* ' .

= *Aspicilia alpinodesertorum* f. *fruticulosofoliacea* Elenkin in *Izv. Imp. S. -Peterburgsk. Bot.*

Sada 1: 27, 36, 39, tab. 2, rows IX & X, figs. 1 – 7. 1901, as '*fruticuloso-foliacea*'

≡*Aspicilia fruticulosofoliacea*(Elenkin) Sohrabi in *Taxon* 59: 627. 2010. Lectotype: designated by Sohrabi and Ahti(2010) , Kyrgyzstan. ' Ad terram argillosam in regione alpina montium Tian-Shan(Kaschgariae: Werchnij Syrt 12000 ft. ped) ' , 1889, Roborowsky in Elenkin, *Lich. Fl. Ross*. No. 24d(H, isolectotypes LE 4 specimens) .

Xinjiang(Mt. Tian shan, Elenk. 1901, p. 35) .

[*Aspicilia alpino-desertorum* f. *fruticuloso-foliacea* Elenk.]

Xinjiang(Aksai, Elenk. 1901, p. 36) .

Notes: Some specimens of *Aspicilia fruticulosofoliacea*(Litterski 2002; No. Litterski 4848, 4876, and 5137; H) have been misidentified as *A. vagans*. They belong to *C. alpicola*. M. Sohrabi et al. Mycol. Progress(2013) 12:

248. Whether *Aspcilia vagans*, reported by Wu(1985, p. 73) from Xinjiang, belongs to *Circinaria affinis* or *C. alpicola, it remains to be checked in the future.*

"Earlier reports of both A. 'desertorum' sensu Krempelhuber and A. esculenta need to be reviewed."(Sohrabi et al., 2013).

Circinaria caesiocinerea(Nyl. ex Malbr.) A. Nordin, Saviü & Tibell 灰绿野粮衣

Mycologia 102(6): 1341(2010).

≡*Lecanora caesiocinerea* Nyl. ex Malbr., *Bull. Soc. Amis Sci. Nat. Rouen*, Sér. II 5: 320(1869).

≡*Aspicilia caesiocinerea*(Nyl. ex Malbr.) Arnold, *Verh. zool.-bot. Ges.* Wien 36: 67(1886).

Misapplied name:

Aspicilia cf. *cinerea* auct non(L.) Köber in Thrower, 1988, pp. 15, 58 and in Chu, 1997, p. 48, revised by Aptroot & Seaward, 1999, p. 66.

On exposed granite. Cosmopolitan.

Shaanxi(Jatta, 1902, p. 476; Zahlbr., 1930 b, p. 151), Shandong(Li y et al., 2008, p. 71 as *L. caesiocinerea*), Hongkong(Thrower, 1988, pp. 15, 58; Chu, 1997, p. 48; Aptroot & Seaward, 1999, p. 66).

Circinaria contorta(Hoffm.) A. Nordin, Saviü & Tibell 内卷野粮衣

Mycologia 102(6): 1341(2010).

≡*Verrucaria contorta* Hoffm., *Descr. Adumb. Plant. Lich.* 1(4): 97(1790).

≡*Aspicilia contorta*(Hoffm.) Körb. *Syst. lich. germ.* (Breslau): 166(1855).

On rocks.

Xinjiang(Gulbostan et al., 2015, pp. 1230 – 1231), Hebei(Ye Jia et al., 2009, pp. 763 – 764 as *A. contorta*).

Circinaria esculenta(Pall.) Sohrabi 野粮衣

in Sohrabi, Stenroos, Myllys, Søchting, Ahti & Hyvönen, *Mycol. Progr.* 11: 21(2012).

≡*Lichen esculentus* Pall., *Reise Prov. Russ. Reichs* 3: 80(1778).

≡*Aspicilia esculenta*(Pall.) Flagey, *Catal. Lich. Algérie*: 52(1896).

Xinjiang(Wu J. L., 1985, p. 73, as *A. esculenta*).

Circinaria fruticulosa(Eversm.) Sohrabi 果野粮衣

in Sohrabi, Stenroos, Myllys, Søchting, Ahti & Hyvönen, *Mycol. Progr.* 12: 253(2013).

≡*Lecanora fruticulosa* Eversm. in *Nova Acta Phys.-Med. Acad. Caes. eop.-Carol. Nat. Cur.* 15: 352, tab. 78A. 1831. Lectotype: designated by Sohrabi and Ahti(2010), Kazakhstan. 'Desertis Kirgisorum' [1820], Eversmann(H-NYL 25676).

Xinjiang, Tuo Li County, Laofeng Kou, 1994, Abbas 940001(H). Xinjiang, Tacheng Toli, 1232 m, 2008, Abbas & Xahidin 20080363-a(H).

≡*Aspicilia fruticulosa*(Eversm.) Flagey, *Catal. Lich. Algérie*: 52(1896).

On the ground.

xinjiang(Abdulla Abas et al., 1996, p. 232 as *Aspicilia fruticulosa*; Abdulla Abbas & Wu, 1998, p. 69; Abdulla Abbas et al., 2001, p. 360).

Circinaria gibbosa(Ach.) A. Nordin, Saviü & Tibell 囊野粮衣

Mycologia 102(6): 1346(2010).

≡*Lichen gibbosus* Ach., *Lich. suec. prodr.* (Linköping): 30(1799) [1798].

≡*Aspicilia gibbosa* (Ach.) Körb., *Syst. lich. germ.* (Breslau): 163(1855).

≡*Urceolaria gibbosa* Ach., *Methodus, Sectio prior*(Stockholmiæ): 144(1803).

≡*Lecanora gibbosa* (Ach.) Nyl., *Mém. Soc. Imp. Sci. Nat. Cherbourg* 5: 113(1857).

On siliceous rocks.

Shaanxi(Jatta, 1902, p. 476; Zahlbr., 1930 b, p. 152).

Circinaria lacunosa(Mereschk) Sohrabi 凹点野粮衣

in Sohrabi, M. et al., *Mycol Progress* 12: 260(2013).

≡*Aspicilia lacunosa* Mereschk. in *Trudy Obshch. Estestvoisp. Imp. Kazansk. Univ.* 43(5): 11. 1911. Lectotype: designated by Sohrabi and Ahti(2010): Kazakhstan. Semipalatinsk Region [East Kazakhstan Prov.], Zaisan District, Keller(icon seen in H, W, voucher not found).

Xinjiang, Tacheng Toli, 1050m, 1994, Abbas 94003(H) (M. Sohrabi et al. 2013, p. 260).

On the groun.

Xinjiang(Abdulla Abas et al., 1996, p. 232; Abdulla Abbas & Wu, 1998, p. 70 – 71 and Abdulla Abbas et al., 2001, p. 360 as *Aspicilia lacunosa*).

Lobothallia(Clauzade & Cl. Roux) Hafellner **瓣茶衣属**

Acta bot. Malac. 16(1): 138(1991).

≡*Aspicilia* subgen. *Lobothallia* Clauzade & Cl. Roux, *Bull. Soc. bot.* Centre-Ouest, Nouv. sér. 15: 140(1984).

Lobothallia alphoplaca(Wahlenb.) Hafellner 粉瓣茶衣

Acta bot. Malac. 16(1): 138(1991).

≡*Parmelia alphoplaca* Wahlenb., in Acharius, *Methodus*, Suppl.: 41(1803).

≡*Lecanora alphoplaca*(Wahlenb.) Ach., *Lich. univ.*: 428(1810).

≡*Aspicilia alphoplaca*(Wahlenb.) Poelt & Leuckert, *Willdenowia* 7(1): 25(1973).

On rocks.

Liaoning(Togashi, 1968, p. 358), Neimenggu(H. Magn. 1940, p. 114 & 1944, p. 43; Asahina, 1958, p. 65), Xinjiang(Abdulla Abbas et a., 1993, p. 75; Abdulla Abbas et al., 1994, p. 20; Abdulla Abbas & Wu, 1998, p. 67 and Abdulla Abbas et al., 2001, p. 360 as *Aspiclia alphoplaca*), Gansu(H. Magn. 1940, p. 114), Shaanxi (Jatta, 1902, p. 474; Zahlbr., 1930 b, p. 172), Shanxi, Hebei(Asahina, 1958, p. 65).

Lobothallia cheresina(Müll. Arg.) A. Nordin, Cl. Roux & Sohrabi 石瓣茶衣

in Roux, *Bull. Soc. linn. Provence, num. spéc.* 16: 216(2012).

≡*Lecanora cheresina* Müll. Arg., *Revue mycol., Toulouse* 2(2): 75(1880).

≡*Aspicilia cheresina*(Müll. Arg.) Hue, *Bull. Soc. linn. Provence* 27: 54(1974).

=*Lecanora cheresina* Müll. Arg. ssp. *septentrionalis* Zahlbr., in Fedde, *Repertorium* 31: 24(1933). "A typo differ lobis thalli marginalibus et areolis centralibus convexulis, turgidulis et Apotheciis semper et persistenter margine thallino albido, prominulo cinctis. Tibet: Quellgebiet des Sirigh-Jilganang-Sees, 5250m, leg. Bossihard(VIII. 1927), auf Kalksteinfelsen."

Type: Xizang, Sirigh-Jilganang-Sees, at about 5250 m, VIII/1927, collected by Bosshard.

On calcareous rocks.

Lobothallia crassimarginata X. R. Kou & Q. Ren 厚瓣茶衣

Mycotaxon 123: 243 – 245(2013).

Type: China. Inner Mongolia, Mt. Helan, on rock, alt. 1500 m, 19 Aug 2011, H. Y. Wang 20122565(Holotype, SDNU; Genbank, JX476026); Paratypes: all conserved in SDNU)

Neimenggu(Kou &Ren, 2013, pp. 243 – 245).

Lobothallia helanensis X. R. Kou & Q. Ren 贺兰瓣茶衣

Mycotaxon 123: 245 – 246(2013).

Type: Neimenggu, Mt. Helan, on rock, alt. 1500 m, 19 Aug 2011, D. B. Tong20122517(Holotype, SDNU; Genbank, JX476030). Paratypes: all conserved in SDNU.

Neimenggu(Kou &Ren, 2013, pp. 243 – 245).

Lobothallia praeradiosa(Nyl.) Hafellner 原辐瓣茶衣

Acta bot. Malac. 16(1): 138(1991).

≡*Lecanora praeradiosa* Nyl., *Flora*, Regensburg 67: 389(1884).

On rock.

Xinjiang(Kou & Ren, 2013, p. 248).

Lobothallia pruinosa X. R. Kou & Q. Ren 粉霜瓣茶衣

Mycotaxon 123: 246 – 247(2013).

Type: China. Neimenggu, Mt. Helan, on rock, alt. 1500 m, 19 Aug 2011, H. Y. Wang20123278 (Holotype, SDNU; Genbank, JX476028).

Paratypes: all conserved in SDNU).

Neimenggu(Kou & Ren, 2013, pp. 246 – 247).

Megaspora(Clauzade & Cl. Roux) Hafellner & V. Wirth **巨孢衣属**

in Wirth, *Flechten Baden-Württemb. Verbr. -Atlas*: 511(1987).

≡*Aspicilia* subgen. *Megaspora* Clauzade & Cl. Roux, *Bull. Soc. bot. Centre-Ouest, Nouv.* sér. 15: 139(1984).

Megaspora verrucosa(Ach.) Hafellner & V. Wirth 小疣巨孢衣

in Wirth, *Die Flecht. Baden-Württembergs. Verbreitungsatlas* (Stuttgart): 511(1987).

var. verrucosa 原变种.

≡*Lecanora verrucosa* Ach., *Lich. univ.*: 354(1810).

≡*Lecanora verrucosa*(Ach.) Laur. teste Nyl. in *Lich. scand.* 156(1861); H. Magn. 1940, p. 113.

≡*Pachyospora verrucosa*(Ach.) Massal. (Abbas et al., 1993, p. 79).

On old tree trunk among *Cladonia pocillum*(Ach.) O. J. Rich.

Xinjiang(H. Magn. 1940, p. 113; Abdulla Abbas et al., 1993, p. 79; Abdulla Abbas et al., 1994, p. 22; Abdulla Abbas & Wu, 1998, p. 74; Abdulla Abbas et al., 2001, p. 365), shaanxi(Ren Qiang et al., 2009, p. 104), Xizang, Sichuan(Obermayer, 2004, p. 502).

var. mutabilis(Ach.) Nimis & Cl. Roux 突变种

Lichens of Italy. An Annotated Catalogue, Monografia XII, Museo Regionale di Scienze Naturali, Torino(Torino): 430(1993).

≡*Urceolaria mutabilis* Ach., *Lich. univ.*: 335(1810).

On bark.

Neimenggu, Gansu, Ningxia, Qinghai, Xinjiang(Zhao Na et al., 2013, p. 1701).

(34) Ochrolechiaceae R. C. Harris ex Lumbsch & I. Schmitt **肉疣衣科**

in Schmitt, Yamamoto & Lumbsch, *J. Hattori Bot. Lab.* 100: 760(2006).

Ochrolechia A. Massal. **肉疣衣属**

Ric. auton. lich. crost. (Verona): 30(1852).

Ochrolechia africana Vainio 非洲肉疣衣

Annal. Univ. Fenn. Aboensis Ser. A, 2(3): 3. [May] 1926. Type: South Africa, "Durban. Trunk of tree. 1921, van der Bijl136. "(Holotype: TUR-V 34315; contains gyrophoric, lecanoric, 5-0-methylhiascic and 4-0-demethylmicrophyllinic acids.)

= *Ochrolechia africana* Zahlbr., Engler's Bot. Jahrb. 60: 504. [October] 1926; (*ICN Nom. illegit.*, *Art.* 53. 1).. Type: South Africa, Cape Province, "Zwartkops bei Port Elizabeth, auf Portulacaria afra [africana], 19. 11. 1909, Brunnthaler. "(Holotype W; contains gyrophoric, lecanoric, 5-0-methylhiascic, 4, 5-di-0- methylhiascic, and a trace of 4-0-demethylmicrophyl-linic acids.). Revised by Brodo, I. M. 1991, pp. 739 – 741.

On *Abies kawakamii*.

Yunnan(Gao TL et al., 2012, p. 462, as *O. africana* Vain. *Ann. Univ. Fenn.* Aboënsis, ser. A, 2(3): 3. 1926.), Taiwan(Aptroot & Sparrius, 2003, p. 30).

Ochrolechia akagiensis Yasuda & Vain. 颗粒肉疣衣

Bot. Mag., Tokyo 35: 54(1921).

On bark of trees.

Jilin(Ren Q. & Jia ZF, 2011, p. 172, 2011, p. 172), Gansu(Zhang Q. et al., 2008, p. 615; Ren Q. & Jia ZF, 2011, p. 172, 2011, p. 172), Hunan(Jia ZF et al., 2003, p. 31; Ren & Jia, 2011, p. 172), Jiangxi(Wu J. L. 1987, p. 135), Zhejiang(Ren & Jia, 2011, p. 172).

Ochrolechia alboflavescens(Wulfen) Zahlbr. 黄粉肉疣衣

Cat. Lich. Univ. 4: 676(1927).

≡*Lichen alboflavescens* Wulfen, in Jacquin, *Collnea bot.* 3: 111(1791) [1789].

On bark.

Gansu(Zhang Q. et al., 2008, p. 615; Ren Q. & Jia ZF, 2011, p. 172), Shaanxi, Sichuan, Yunnan(Jia ZF et al., 2008, p. 317; Ren Q. & Jia ZF, 2011, p. 172, 2011, p. 172).

Ochrolechia androgyna(Hoffm.) Arnold 阴阳肉疣衣

Flora, Regensburg 68: 236(1885).

≡*Lichen androgynus* Hoffm., *Enum. critic. lich. europ.* (Bern): 56(1784).

= *Lecanora tartarea* var. *arborea* Schaer. Jatta I, 474; Zahlbr., 1930 b, p. 176.

On bark of trees.

Gansu(Zhang Q. et al., 2008, p. 616; Ren Q. & Jia ZF, 2011, Pp. 172 – 173), Shaanxi(Jatta, 1902, p. 474; Zahlbr., 1930 b, p. 176; Zhao JZ et al., 2006, p. 180; Ren Q. & Jia ZF, 2011, Pp. 172 – 173), Sichuan(Ren Q. & Jia ZF, 2011, Pp. 172 – 173), Yunnan(Zahlbr., 1930 b, p. 176; Jia ZF et al., 2003, p. 31; Ren Q. & Jia ZF, 2011, Pp. 172 – 173).

Ochrolechia antillarum Brodo 珊瑚肉疣衣

Can. J. Bot. 69(4): 743(1991).

On moss covered rocks.

Gansu(Zhang Q. et al., 2008, p. 616; Ren Q. & Jia ZF, 2011, p. 173), Hubei(Zhao ZT & Jia ZF, 2002, p. 291; Jia ZF et al., 2003, p. 31; Ren Q. & Jia ZF, 2011, p. 173).

Ochrolechia balcanica Verseghy 瘤型肉疣衣

Beih. Nova Hedwigia 1: 85(1962).

On bark of *Pinus* sp.

Yunnan(Jia ZF et al., 2008, Pp. 317 – 318; Ren Q. & Jia ZF, 2011, p. 173).

Ochrolechia chondriocarpa Zahlbr. 粒果肉疣衣

in Handel-Mazzetti, *Symb. Sin.* 3: 176(1930) & *Cat. Lich. Univ.* 8: 548(1932).

Type: Fujian, Gushan near Fuzhou, alt. 500 – 600 m collected by Chung in 1926, ex herb. Wien(not seen).

Fujian(K. Verseghy, 1962, p. 117).

On schistose rocks.

Ochrolechia frigida(Sw.) Lynge 寒生肉疣衣

Lich. Nov. Zemlya: 182(1928).

≡*Lichen frigidus* Sw., *Method. Muscor.*: 36(1781).

Gansu(Zhang Q. et al., 2008, p. 616; Ren Q. & Jia ZF, 2011, p. 173), Xizang(Ren Q. & Jia ZF, 2011, p. 173), Sichuan(Jia ZF et al., 2003, p. 31; Ren Q. & Jia ZF, 2011, p. 173).

Ochrolechia glacialis Poelt 冰川肉疣衣

Khumbu Himal 1: 255(1966).

On plant debris over ground.

Xizang(Obermayer, 2004, p. 504).

Ochrolechia harmandii Verseghy 哈氏肉疣衣

Beih. Nova Hedwigia 1: 107(1962), as "*harmandi*", nom. nov. for *Lecanora pallescens* Nyl. ex Harm. [non(L.) Massal. 1853]. f. harmandii 原变型

On bark.

Gansu(Zhang Q. et al. , 2008, p. 616; Ren Q. & Jia ZF, 2011, P p. 173 – 174) .

f. granulosa Vers. 颗粒变型

Beih. Nova Hedwigia, 1: 108(1962) . D. Hawksw. Index of Fungi Supplement, 1972, p. 44.

Type: Yunnan, Rock JF, 1922 ex herb. in Wien.

On bark of trees.

Gansu(Zhang Q. et al. , 2008, p. 616; Ren Q. & Jia ZF, 2011, p. 174) .

f. pustulata Vers. 疱突变型

Beihefte zur *Nova Hedwigia* 1: 110(1962) . D. Hawksw. Index of Fungi Supplement, 1972, p. 44.

Type: Sichuan , Handel-Mazzetti, 1914 ex herb. in Wien. & Yunnan, Handel-Mazzetti, 1914(syntypes) .

On bark of trees.

Gansu(Zhang Q. et al. , 2008, p. 616) .

Ochrolechia isidiata(Malme) Verseghy 柱芽肉疣衣

in Ann. Mus. Nat. Hungar. n. s. VII: 298(1956) .

≡ *Ochrolechia tartarea* f. *isidiata* Malme, in *Ark. F. Bot.* XXIXA, no. 6: 33. (1937) .

On mosses.

Gansu(Zhang Q. et al. , 2008, p. 616) , Chongqing(Ren Q. & Jia ZF, 2011, p. 174) , Hunan(Jia ZF et al. , 2003, p. 31; Ren Q. & Jia ZF, 2011, p. 174) .

Ochrolechia laevigata(Räsänen) Verseghy 平滑肉疣衣

Beih. Nova Hedwigia 1: 86(1962) .

≡ *Ochrolechia pallescens* var. *laevigata* Räsänen, in *Ann. Bot. Soc. Zool. -Bot. Fenn. Vanamo*, XX, no. 3: 9 (1944) .

Yunnan(Gao TL et al. , 2012, 463 – 464) .

Ochrolechia margarita Poelt 珍珠肉疣衣

Khumbu Himal 1: 257(1966) .

On *Picea morrisonicola*, *Cryptomeria japonica*, and *Pseudotsuga wilsoniana*.

Taiwan(Aptroot & Sparrius, 2003, p. 30) .

Ochrolechia mexicana Vain. 墨西哥肉疣衣

Dansk bot. Ark. 4(11) : 9(1926) .

On bark.

Gansu(Zhang Q. et al. , 2008, pp. 616 – 617; Ren Q. & Jia ZF, 2011, p. 175) .

Ochrolechia microstictoides Räsänen 粉末肉疣衣

Lich. Fenn. Exs. : no. 226(1936) .

Sichuan, yunnan(Jia ZF et al. , 2003, p. 32; Ren Q. & Jia ZF, 2011, P p. 174 – 175) , Hunan(Ren Q. & Jia ZF, 2011, P p. 174 – 175) .

Ochrolechia montana Brodo 山地肉疣衣

Can. J. Bot. 69(4) : 753(1991) .

On bark of birch.

Gansu(Zhang Q. et al. , 2008, p. 617; Ren Q. & Jia ZF, 2011, p. 175) , Shaanxi(Zhao ZT & Jia ZF, 2002, p. 292, miscited as Shanxi; Jia ZF et al. , 2003, p. 32; Zhao JZ et al. , 2006, p. 180; Ren Q. & Jia ZF, 2011, p. 175) , Sichuan(Jia ZF et al. , 2003, p. 32; Ren Q. & Jia ZF, 2011, p. 175) .

Ochrolechia oregonensis H. Magn. 俄勒冈肉疣衣

Meddn Göteb. Bot. Trädg. 13: 251(1939) .

On bark of lignum or pine, sometimes on mosses.

Gansu(Zhang Q. et al. , 2008, pp. 617 – 618; Ren Q. & Jia ZF, 2011, p. 175) .

Ochrolechia pallentiisidiata Jia Z. F. & Ren Q. 白裂芽肉疣衣

Mycotaxon 106: 234 – 236(2008).

Type: Yunnan, Weixi, Yezhi, Baimaluo, alt. 2400m, 8/V, 1982. Su J. J. 241(Holotype: HMAS-L), 805(paratype: HMAS. L.).

On bark of trees.

Yunnan(Ren Q. & Jia Z. F., 2011, p. 176).

Ochrolechia pallescens(L.) A. Massal. 苍白肉疣衣

Nuovi Ann. Sci. nat. Bologna 7: 212(1853). f. pallescens 原变型

≡*Lichen pallescens* L., *Sp. pl.* 2: 1142(1753).

≡*Lecanora pallescens*(L.) Röhl., *Deutschl. Fl.* (Frankfurt) 3(2): 76(1813).

On bark of trees.

Jilin(Ren Q. & Jia ZF, 2011, Pp. 175 – 176), Xinjiang(Abdulla Abbas et al., 1993, p. 79; Abdulla Abbas & Wu, 1998, p. 142; Abdulla Abbas et al., 2001, p. 365; Ren Q. & Jia ZF, 2011, Pp. 175 – 176), shaanxi(Zhao JZ et al., 2006, p. 180; Ren Q. & Jia ZF, 2011, Pp. 175 – 176), Xizang, Hubei(Ren Q. & Jia ZF, 2011, Pp. 175 – 176), Sichuan(Zahlbr., 1934 b, p. 206; Jia ZF et al., 2003, p. 32; Ren Q. & Jia ZF, 2011, Pp. 175 – 176), Yunnan, Hunan(Zahlbr., 1930 b, p. 176; Ren Q. & Jia ZF, 2011, Pp. 175 – 176), Anhui(Xu B. S., 1989, p. 245; Ren Q. & Jia ZF, 2011, Pp. 175 – 176), China(prov. not indicated: Verseghy, 1962, p. 57).

On *Abies kawakamii*, and *Tsuga formosana*.

Taiwan(Aptroot & Sparrius, 2003, pp. 30 – 31).

f. corticola Arnold 树生变型

Verh. zool.-bot. Ges. Wien 23: 514(1873).

≡ *Lecanora pallescens* var. *corticola* Massal., Jatta I, 474, Zahlbr. *in* Handel-Mazzetti, *Symb. Sin.* 3: 176 (1930).

On bark of trees, particularly of *Rhododendron* sp. and *Abies* sp. Shaanxi(Jatta, 1902, p. 474; Zahlbr., 1930 b, p. 176), Sichuan, Yunnan, Fujian(Zahlbr., 1930 b, p. 176).

var. rosella(Tuck.) Zahlbr. 莲座变种

Cat. Lich. Univers. 5: 686(1928).

≡*Lecanora pallescens* var. *rosella* Tuck., *Syn. N. Amer. Lich.* (Boston) 1: 196(1882), Zahlbr. *in* Handel-Mazzetti, *Symb. Sin.* 3: 176(1930).

On bark of *Rhododendron*.

Yunnan(Zahlbr., 1930, p. 176).

Ochrolechia parella(L.) A. Massal. 肉疣衣

Ric. auton. lich. crost. (Verona): 30(1852).

≡*Lichen parellus* L., *Mant. Pl.* 1: 132(1767).

≡*Lecanora parella*(L.) Ach., *Lich. univ.*: 370(1810).

On bark of trees.

Jilin(Jia ZF et al., 2003, p. 32; Ren Q. & Jia Z. F., 2011, p. 176), Liaoning(Ren Q. & Jia Z. F., 2011, p. 176), Xinjiang(Abdulla Abbas et al., 1993, p. 79; Abdulla Abbas & Wu, 1998, p. 142; Abdulla Abbas et al., 2001, p. 365), shaanxi(Zhao JZ et al., 2006, p. 180; Ren Q. & Jia Z. F., 2011, p. 176), Yunnan(Hue, 1889, p. 173; Zahlbr., 1930 b, p. 176; Ren Q. & Jia Z. F., 2011, p. 176). chongqing, anhui(Jia ZF et al., 2003, p. 32; Ren Q. & Jia Z. F., 2011, p. 176), Sichuan, Hunan, Hubei(Ren Q. & Jia Z. F., 2011, p. 176).

Ochrolechia pseudopallescens Brodo 拟苍白肉疣衣

Syllogeus 29: 58(1981).

On bark.

Shaanxi(Zhao JZ et al., 2006, pp. 180 – 181; Ren Q. & Jia Z. F., 2011, Pp. 176 – 177).

Ochrolechia rosella(Müll. Arg.) Verseghy 莲座肉疣衣

Beih. Nova Hedwigia 1: 110(1962) .

≡*Pertusaria pallescens* var. *rosella* Müll. Arg. , *Flora*, Regensburg 62: 483(1879) .

On bark of *Rhododendron* sp.

Jilin, jiangxi, guangxi(Jia ZF et al. , 2003, p. 32; Ren Q. & Jia Z. F. , 2011, p. 177) , Gansu, Anhui(Ren Q. & Jia Z. F. , 2011, p. 177) , Yunnan(Zahlbr. , 1930 b, p. 176; Verseghy, 1962, p. 110; Ren Q. & Jia Z. F. , 2011, p. 177) , Taiwan(Aptroot & Sparrius, 2003, p. 31) .

Ochrolechia subisidiata Brodo 亚裂芽肉疣衣

Can. J. Bot. 69(4) : 759(1991) .

Gansu(Ren Q. & Jia Z. F. , 2011, p. 177) , Sichuan(Jia ZF et al. , 2003, pp. 32 – 33; Ren Q. & Jia Z. F. , 2011, p. 177) .

Ochrolechia subpallescens Verseghy 亚苍白肉疣衣

Beih. Nova Hedwigia 1: 118(1962) .

On bark of trees.

Gansu(Zhang Q. et al. , 2008, p. 618; Ren Q. & Jia Z. F. , 2011, p. 177) , Xizang(Ren Q. & Jia Z. F. , 2011, p. 177) , Sichuan(Jia ZF et al. , 2003, pp. 32 – 33; Ren Q. & Jia Z. F. , 2011, p. 177) , Yunnan(Verseghy, 1962, p. 118; Ren Q. & Jia Z. F. , 2011, p. 177) , Anhui, Hubei, Hunan(Ren Q. & Jia Z. F. , 2011, p. 177) , TAIWAN (Aptroot & Sparrius, 2003, p. 31) .

Ochrolechia subrosella Jia Z. F. & Zhao Z. T. 亚莲座肉疣衣

Mycosystema 24(2) : 162 – 163(2005) .

Type: Yunnan, Lijiang, on *Abies fabri*, Oct. 18, 2002, Wei J. C. WSY014 – 1(holotype in Liaocheng University) .

On bark of trees.

Yunnan(Ren Q. & Jia Z. F. , 2011, p. 178) .

Ochrolechia subviridis(Høeg) Erichsen 亚绿肉疣衣

Meddn Göteb. Bot. Trädg. 5: 3(1930) .

≡*Pertusaria subviridis* Høeg, *Nytt Mag. Natur.* 61: 150(1924) .

On bark.

jilin(Jia ZF et al. , 2008, p. 318; Ren Q. & Jia Z. F. , 2011, p. 178) , Anhui(Jia ZF et al. , 2008, p. 318; Jia ZF et al. , 2003, pp. 32 – 33; Ren Q. & Jia Z. F. , 2011, p. 178) , zhejiang(Jia ZF et al. , 2003, pp. 32 – 33; Ren Q. & Jia Z. F. , 2011, p. 178) , fujian(Jia ZF et al. , 2008, p. 318; Jia ZF et al. , 2003, pp. 32 – 33) .

Ochrolechia tartarea(L.) A. Massal. 酒石肉疣衣

Ric. auton. lich. crost. (Verona) : 30(1852) . var. tartarea 原变种

≡*Lichen tartareus* L. , *Sp. pl.* 2: 1141(1753) .

On shded rock. Cosmopolitan.

Jilin, Sichuan, anhui(Jia ZF et al. , 2003, pp. 32 – 33; Ren Q. & Jia Z. F. , 2011, p. 178) , shaanxi(Zhao JZ et al. , 2006, p. 181; Ren Q. & Jia Z. F. , 2011, p. 178) , Gansu(Zhang Q. et al. , 2008, p. 618; Ren Q. & Jia Z. F. , 2011, p. 178) , Shandong(Li Y et al. , 2008, p. 73) , Hongkong(Thrower, 1988, p. 16, 124 as *Ochrolechia* sp. ; Aptroot & Seaward, 1999, p. 83) Yunnan(Verseghy, 1962, p. 79) , Hubei(Chen, Wu & Wei, 1989, p. 475) .

var. frigida(Sw.) Körb. 寒生变种

Parerga lichenol. (Breslau) 1: 92(1859) .

On rocks.

Hebei(Moreau et Moreau, 1951, p. 190) , Yunnan(Zahlbr. , 1934 b, p. 206) , Sichuan(Zahlbr. , 1930 b, p. 176) .

Ochrolechia trochophora(Vain.) Oshio 轮生肉疣衣

J. Sci. Hiroshima Univ. , ser. B. div. 2, 12: 145(1968) . var. trochophora 原变种

≡*Pertusaria trochophora* Vain. , *Bot. Mag.* , Tokyo 32: 155(1918) .

On bark of trees.

Jilin, Yunnan, guangxi(Jia ZF et al. , 2003, pp. 32 – 33; Ren Q. & Jia Z. F. , 2011, p. 178) , Gansu(Zhang Q. et al. , 2008, p. 618; Ren Q. & Jia Z. F. , 2011, p. 178) , shaanxi(Zhao JZ et al. , 2006, p. 181; Ren Q. & Jia Z. F. , 2011, p. 178) , guizhou(Jia ZF et al. , 2003, pp. 32 – 33; Zhang T & J. C. Wei, 2006, p. 9; Ren Q. & Jia Z. F. , 2011, p. 178) , Anhui, Zhejiang(Xu B. S. , 1989, p. 245) , Taiwan(Wang & Lai, 1976, p. 228) .

var. pruinirosella Brodo 粉霜变种

Can. J. Bot. 69(4) : 763(1991) .

On bark of trees.

Jilin, Gansu, Guizhou, Yunnan, Anhui, Hubei, Hunan(Ren Q. & Jia Z. F. , 2011, p. 179) .

Ochrolechia upsaliensis(L.) A. Massal. 乌普萨拉肉疣衣

Ric. auton. lich. crost. (Verona) : 31(1852) .

≡*Lichen upsaliensis* L. , *Sp. pl.* 2: 1142(1753) .

On soil.

Xinjiang(Reyim Mamut et al. 2014, p. 856) .

Ochrolechia yasudae Vain. 裂芽肉疣衣

Bot. Mag. , Tokyo 32: 155(1918) .

On mosses covering rocks.

Jilin, Sichuan(Jia ZF et al. , 2003, pp. 33 – 34; Ren Q. & Jia Z. F. , 2011, p. 179) , shaanxi(Zhao JZ et al. , 2006, p. 181; Ren Q. & Jia Z. F. , 2011, p. 179) , Gansu(Zhang Q. et al. , 2008, p. 618; Ren Q. & Jia Z. F. , 2011, p. 179) , Chongqing(Ren Q. & Jia Z. F. , 2011, p. 179) , Shandong(Li Y et al. , 2008, p. 73) , Hubei(Ren Q. & Jia Z. F. , 2011, p. 179) , Hunan(Wu J. L. 1987, p. 136) , Anhui(Xu B. S. , 1989, p. 245; Jia ZF et al. , 2003, pp. 33 – 34; Ren Q. & Jia Z. F. , 2011, p. 179)) , Zhejiang(Wu J. L. 1987, p. 136; Xu B. S. , 1989, p. 245; Jia ZF et al. , 2003, pp. 33 – 34; Ren Q. & Jia Z. F. , 2011, p. 179) , Taiwan(Aptroot & Sparrius, 2003, p. 31) .

Varicellaria Nyl. **果疣衣属**

Mém. Soc. Imp. Sci. Nat. Cherbourg 5: 119(1858) .

Varicellaria hemisphaerica(Flörke) I. Schmitt & Lumbsch 半球界疣衣

in Schmitt, Otte, Parnmen, Sadowska-Deś, Lücking & Lumbsch, *MycoKeys* **4**: 29(2012) .

≡*Variolaria hemisphaerica* Flörke, *Deutsche Lich.* **2**: 6(1815) .

≡*Pertusaria hemisphaerica*(Flörke) Erichsen, *Hedwigia* **72**: 85(1932) .

On bark, rarely on rock.

Gansu(Yang F. et al. , 2008, p. 624) , Sichuan, Yunnan, Xizang, Fujian(Ren qiang & Zhao ZT, 2004, p. 330; Zhao ZT et al. , 2004. p. 535) .

Varicellaria lactea(L.) I. Schmitt & Lumbsch 乳白果疣衣

in Schmitt, Otte, Parnmen, Sadowska-Deś, Lücking & Lumbsch, *MycoKeys* 4: 31(2012) .

≡*Byssus lacteus* L. , *Sp. pl.* 2: 1169(1753) .

On siliceous rocks.

Neimenggu, jilin, Xizang(Ren Qiang & Li Shuxia, 2013, pp. 66 – 67) .

Varicellaria microsticta Nyl. 小果疣衣

Mém. Soc. Imp. Sci. Nat. Cherbourg 5: 119(1858) .

Shaanxi(Ren Q. , et al. 2003, p. 217) .

Varicellaria rhodocarpa(Körb.) Th. Fr. 粉色果疣衣

Lich. Scand. (Upsaliae) 1(1) : 322(1871) .

≡*Pertusaria rhodocarpa* Körb. , *Syst. lich. germ.* (Breslau) : 384(1855) .

Shaanxi (Zhao JZ et al. , 2006, p. 183) .

Varicellaria velata(Turner) I. Schmitt & Lumbsch 包被果疣衣

in Schmitt, Otte, Parnmen, Sadowska-ś, Lücking & Lumbsch, MycoKeys 4: 31(2012) .

≡*Parmelia velata* Turner, *Trans. Linn. Soc. London* 9: 143(1808) .

≡*Pertusaria velata*(Turner) Nyl. , *Lich. Scand.* (Helsinki) : 179(1861) .

= *Pertusaria* albovelata Zahlbr. in Handel-Mazzetti, *Symb. Sin.* 3: 146(1930) & *Cat. Lich. Univ.* 8: 519(1932) . 白被鸡皮衣

Type: Yunnan, Handel-Mazzetti no. 6504, 6793 (W – syntypes) .

On twigs & bark of trees.

Heilongjiang, Sichuan, Xizang, hubei, hunan, Jiangxi, Hainan(Zhao ZT et al. , 2004, p. 540) , Gansu(Yang F. et al. , 2008, p. 626) , Shaanxi(Jatta, 1902, p. 477; Zahlbr. , 1930 b, p. 146; Zhao ZT et al. , 2004, p. 540; Zhao JZ et al. , 2006, p. 183) , guizhou(Zhao ZT et al. , 2004, p. 540; Zhang T & Wei JC, 2006, p. 9) , Yunnan(Hue, 1889, p. 174; R. Paulson, 1928, p. 317; Zahlbr. , 1930 b, p. 146 & 1934b, p. 204; Zhao ZT et al. , 2004, p. 540) , Anhui, Zhejiang(Xu B. S. , 1989, p. 248; Zhao ZT et al. , 2004, p. 540) , Taiwan(Oshio, 1968, p. 108; Wang & Lai, 1973, p. 95; Zhao ZT et al. , 2004, p. 540) , Hainan(JCWei et al. , 2013, p. 237) .

(35) Pertusariaceae Körb. ex Körb. **鸡皮衣科**

[as ' Pertusarieae'] *Syst. lich. germ.* (Breslau) : 377(1855)

Pertusaria DC. **鸡皮衣属**

in Lamarck & de Candolle, *Fl. franç.* , Edn 3(Paris) 2: 319(1805) .

Pertusaria aberrans Müll. Arg. 畸形鸡皮衣

Bull. Herb. Boissier 1: 42(1893) .

On tree bark.

Taiwan(Zhao ZT & Ren Q, 2003, p. 669; Zhao ZT et al. , 2004, p. 533) .

Pertusaria albescens(Huds.) M. Choisy & Werner 微白鸡皮衣

Cavanillesia 5: 165(1932) var. albescens

≡*Lichen albescens* Huds. , *Fl. Angl.* : 445(1762) .

= *Pertusaria globulifera*(Turner) A. Massal. , Bull. Mie Imp. Coll. Agric. & For. : 87(1855) var. *globulifera*

≡*Pertusaria multipuncta* var. *globulifera*(Turner) H. Olivier, Rev. Bot. 8: 14(1890) .

≡*Variolaria globulifera* Turner, Trans. Linn. Soc. London 9: 139(1808) .

On twigs.

Shaanxi(Jatta, 1902, p. 477; A. Zahlbr. , 1930 b, p. 146) , Yunnan(Hue, 1887, p. 174; Zahlbr. , 1930 b, p. 146) , Fujian(Zahlbr. , 1934, p. 204) .

Pertusaria albiglobosa Q. Ren 白球鸡皮衣

Mycotaxon 124: 349 – 351(2013) .

Type: China. Shaanxi: Taibai Co. , Mt. Qinling, Baxiantai, 33°67′N 107°58′E, alt. 3760 m, 5 Aug. 2005, J. Zh. Zhao 899(Holotype, SDNU) .

Pertusaria alticola Q. Ren 高地鸡皮衣

Mycotaxon 130: 690(2015) .

Type: China, Yunnan, Yulong County, Lijiang Alpine Botanical Garden, alt. 3590 m, on *Pinus densata* Mast. , 16 Aug. 2011, Ren 2011180(holotype, SDNU) .

Pertusaria allothwaitesii Jariangpr. & A. W. Archer 阿氏鸡皮衣

in Jariangprasert, Archer & Anusarnsunthorn, *Mycotaxon* 89(1) : 124(2004) .

On tree bark.

Yunnan(Zhao Na et al. , 2014, p. 629) .

Pertusaria alpina Hepp ex Ahles 高山鸡皮衣

Pertus. et Conotr. : 12(1860) .

On bark of trees.

Gansu(Yang F. et al. , 2008, p. 623) , Shandong(Li Y et al. , 2008, p. 72) , Hunan, Zhejiang(Wu J. L. 1987, p. 123; Zhao ZT et al. , 2004, p. 533, See also *Pertusaria giberosa*) .

Pertusaria amara(Ach.) Nyl. 苦味鸡皮衣

Bull. Soc. linn. Normandie, sér. 2 6: 288(1872) .

≡*Variolaria amara* Ach. , *K. Vetensk-Acad. Nya Handl.* **30**: 163(2809)

On bark of trees.

Shaanxi(Wu J. L. 1987, p. 123; Yu S. H. et al. , 1999, p. 95; Zhao ZT et al. , 2004, p. 533; Zhao JZ et al. , 2006, p. 181) , guizhou(Zhang T & Wei JC, 2006, p. 9) , Hubei(Yu, S. H. et al. , 1997, p. 332; Yu S. H. et al. , 1999, p. 95; Zhao ZT et al. , 2004, p. 533) , Hunan, Anhui, Jiangxi, Zhejiang, Fujian(Yu S. H. et al. , 1999, pp. 95 – 96) , Taiwan(Aptroot and Sparrius, 2003, p. 32) , Hainan(JCWei et al. , 2013, p. 237) .

var. amara 原变种

Gansu(Yang F. et al. , 2008, p. 623) , Shaanxi(Zhao ZT et al. , 2004, p. 533) , Hubei(Zhao ZT et al. , 2004, p. 533) , Hubei(Zhao ZT et al. , 2004, p. 533) , Hunan, Anhui, Jiangxi, Zhejiang, Fujian(Zhao ZT et al. , 2004, p. 533) , Jilin, Heilongjiang, Sichuan, Chongqing, Guiahou, Fujian, Guangxi(Zhao ZT et al. , 2004, p. 533) .

var. pulvinata(Erichsen) Oxner 垫状变种

Flora Lishaĭnikiv Ukraïni (Kiev) 2(1) : 482(1968) .

≡*Pertusaria pulvinata* Erichsen, *Rabenh. Krypt. -Fl.* , Edn 2(Leipzig) 9(5. 1) : 524(1936) .

≡*Pertusaria amara* f. *pulvinata*(Erichsen) Almb. , Bot. Notiser, Suppl. 1(2) : 77(1948) .

On rocks or bark of trees.

Gansu(Yang F. et al. , 2008, p. 623) , Shaanxi "on rocks" (Yu et al. , 1999, 113) , Yunnan "on bark" (Yu et al. , 1999, p. 113 Zhao ZT et al. , 2004, p. 533) . Chongqing, Hunan(Zhao ZT et al. , 2004, p. 533) .

Pertusaria areolata(Ach.) A. Massal. 网纹鸡皮衣

Ric. auton. lich. crost. (Verona) : 189(1852) .

≡*Porina pertusa* var. *areolata* Ach. , *Syn. meth. lich.* (Lund) : 109(1814) .

On schistose rocks.

Shaanxi(Jatta, 1902, p. 477; Zahlbr. , 1930 b, P. 143) , Shandong(Li Y et al. , 2008, p. 72) , Jiangsu(Wu J. N. & Xiang T. 1981, p. 4, as P. *areolata*(Clem.) Nyl.) .

Pertusaria bambusetorum Zahlbr. 竹生鸡皮衣

in Handel-Mazzetti, *Symb. Sin.* 3: 147(1930) , & *Cat. Lich. Univ.* 8: 520(1932) .

Type: Yunnan, Handel-Mazzetti no. 7893(W) .

On dead bamboo and tree bark. Endemic to China.

Yunnan, Taiwan(Zhao ZT et al. , 2004, p. 534) .

Pertusaria borealis Erichsen 北方鸡皮衣

Ann. Mycol. 36: 354(1938) .

On bark of trees.

Yunnan(Ren Q. 2015, p. 692) .

Pertusaria brachyspora Erichsen 短孢鸡皮衣

Rabenh. Krypt. -Fl. , Edn 2(Leipzig) 9(5. 1) : 526(1936) .

On bark.

Gansu(Yang F. et al. , 2008, pp. 623 – 624),

Pertusaria carneopallida(Nyl.) Anzi ex Nyl. 肉白鸡皮衣

Flora, Regensburg 51: 478(1868).

≡*Lecanora carneopallida* Nyl. , *Bot. Notiser*: 183(1853).

On bark of broad-leaved trees.

Yunnan, Hunan(Zhao ZT et al. , 2004, p. 534).

Pertusaria chinensis Müll. Arg. 中国鸡皮衣

Flora. 67: 402(1884).

Type: Guangdong, Huangpu, collected by R. Rabenhorst.

Guangdong(Vain. 1900, p. 16; Zahlbr. , 1928, p. 129 & 1930 b, p. 145; Zhao ZT et al. , 2004, p. 534), Hainan (Zhao ZT et al. , 2004, p. 534; JCWei et al. , 2013, p. 237).

Pertusaria cicatricosa Müll. Arg. 疤痕鸡皮衣

Proc. R. Soc. Edinb. 11: 461(1882).

On tree bark.

Gansu(Yang F. et al. , 2008, p. 624), Yunnan(Zhao ZT et al. , 2004, p. 534), Taiwan(Zhao ZT & Ren Q, 2003, p. 670; Zhao ZT et al. , 2004, p. 534).

Pertusaria colorata Awasthi & Srivstava 赭色鸡皮衣

the *Bryologist* 96(2): 212(1993).

On tree bark.

Xizang(Zhao Na et al. , 2014, p. 629 – 630).

Notes. The name of this species is not cited in the Index Fungorum.

Pertusaria communis DC. 普通鸡皮衣

in Lamarck & de Candolle, *Fl. franç.* , Edn 3(Paris) 2: 320(1805).

Shandong(Li Y et al. , 2008, p. 72).

Pertusaria commutata Müll. Arg. 鸡皮衣

Flora, Regensburg 67(15): 269(1884).

On tree trunk.

Shaanxi(Jatta, 1902, p. 478; Zahlbr. , 1930 b, p. 146), Yunnan(Zahlbr. , 1934 b, p. 204), Shandong(Li Y et al. , 2008, p. 72), Hubei (Yu & Wu, 1997, p. 332) Anhui, Zhejiang(Xu B. S. , 1989, p. 246 on trees and rocks), TAIWAN(Oshio, 1968, p. 114; Wang & Lai, 1973, p. 95).

Pertusaria composita Zahlbr. 复合鸡皮衣

Annls mycol. 14(1/2): 58(1916).

Heilongjiang, Yunnan, Hubei, Fujian(Zhao ZT et al. , 2004, p. 534), Taiwan(Wang & Lai, 1973, p. 95).

Pertusaria copiosa Erichsen 丰鸡皮衣

Annls mycol. 39(4/6): 391(1941).

On bark or rock.

Sichuan, Yunnan(Ren Q et al. , 2008, p. 613).

Pertusaria corallina(L.) Arnold 珊瑚鸡皮衣

Flora, Regensburg 44: 658(1861).

≡*Lichen corallinus* L. , *Mant. Pl.* 1: 131(1767).

On rocks.

Yunnan(Zhao ZT et al. , 2004, p. 534), Fujian(Yu S. H. et al. , 1999, p. 112; Zhao ZT et al. , 2004, p. 534).

Pertusaria dactylina(Ach.) Nyl. 乳头鸡皮衣

Acta Soc. Sci. fenn. 7(2): 447(1863).

≡*Lichen dactylinus* Ach., *Lich. suec. prodr.* (Linköping): 89(1799)[1798].

In bogs.

Shaanxi(Jatta, 1902, p. 478; Zahlbr., 1930 b, p. 145).

Pertusaria dealbata(Ach.) Cromb. 转白鸡皮衣

Lich. Scand. (Helsinki): 180(1861).

≡*Lichen dealbatus* Ach., *Lich. suec. prodr.* (Linköping): 29(1799)[1798].

In bogs and on waste wood.

Shaanxi(Jatta, 1902, p. 478; Zahlbr., 1930 b, p. 146).

Pertusaria effigurata Zahlbr. 纹饰鸡皮衣

in Handel-Mazzetti, *Symb. Sin.* 3: 148(1930) & *Cat. Lich. Univ.* 8: 521(1932).

Type: Yunnan, Handel-Mazzetti no. 6734.

On bamboo.

Pertusaria elliptica Müll. Arg. 椭圆鸡皮衣

Bull. Herb. Boissier 3: 635(1895).

On bark.

Gansu(Yang F. et al., 2008, p. 624).

Pertusaria excludens Nyl. 外鸡皮衣

Flora, Regensburg 68: 296(1885).

Misapplied name(revised by Aptroot & Seaward, 1999, p85):

Pertusaria subvaginata var. *orientalis* auct non Räsänen inThrower, 1988, pp. 16, 134 and in Chu, 1997, p. 48.

Pertusaria leucopsora auct non Krempelh in Thrower, 1988, p. 16.

On coastal and inland, exposed granitic rocks. Northern temperate.

Yunnan(Zhao ZT et al., 2004, pp. 534 – 535), Hongkong(Rabenhorst, 1873, p. 287; Kremp. 1873, p. 469 & 1874, p. 61; Zahlbr., 1930, p. 143; Thrower, 1988, pp. 16, 134; Chu, 1997, p. 48; Aptroot & Seaward, 1999, p85; Zhao ZT et al., 2004, pp. 534 – 535).

Pertusaria fecunda Zahlbr. 育鸡皮衣

Hedwigia, 74: 203(1934).

Type: Yunnan, collected by Rock.

Yunnan(Zahlbr., 1940, p. 450; Lamb, 1963, p. 531).

On bark of trees.

Pertusaria flavicans Lamy 淡绿鸡皮衣

Bull. Soc. bot. Fr. 25: 427(1879)[1878].

= *Pertusaria amarescens* Nyl., *Flora*, Regensburg 57: 311(1874). Synonymy contributor: Aptroot & Seaward, 1999, p. 85.

On rocks.

Jiangsu(Wu J. N. & Xiang T. 1981, p. 4; Xu B. S., 1989, p. 247), Shanghai, Zhejiang(Xu B. S., 1989, p. 247), Hongkong(Thrower, 1988, pp. 16, 132 as *P. amarescens*; Chu, 1997, p. 48 as *P. amarescens*; Aptroot & Seaward, 1999, p. 67 together with *Buellia subdisformis* and p. 85, and also see *Xanthoparmelia congensis*, Aptroot & Seaward, 1999, p. 98).

Pertusaria flavosulphurea Vain. 淡黄鸡皮衣

Bot. Mag., Tokyo 35: 54(1921).

On rocks.

Jiangsu(Wu J. L. 1987, p. 125).

Pertusaria fukiensis Zahlbr. 福建鸡皮衣

Annal. Mycol. 30: 437(1932) var. fukiensis 原变种

Type: Fujian, Chung no. 606 b.

Fujian(Zahlbr., 1934 b, p. 205 & 1940, p. 450; Lamb, 1963, p. 531).

On bark of trees.

var. kushana Zahlbr. 鼓山变种

Annal. Mycol. 30: 437(1932).

Type: Fujian, Gushan, Chung no. 445 a.

Zhejiang(Xu B. S., 1989, p. 247), Fujian(Zahlbr., 1934 b, p. 205 & 1940, p. 450; Lamb, 1963, p. 531).

On bark of trees.

f. ochrascens Zahlbr. 赭色变型

Hedwigia, 74: 205(1934).

Type: Fujian, Chung no. 584.

Fujian(Zahlbr., 1940, p. 450; Lamb, 1963, p. 531).

Pertusaria gibberosa Müll. Arg. 驼峰鸡皮衣

Flora, Regensburg 65: 486(1882).

Pertusaria alpina auct. non. Hepp ex Ahles: Zhao *et al.*, *The Bryologist* 107(4): 533, 2004. (p. p.).

On bark.

Jilin, Yunnan, Sichuan(Zhao ZT et al., 2004, p. 533 as *P. alpina*), Sichuan(Ren Q et al., 2008, p. 612).

Pertusaria gracilenta Zahlbr. 疣果鸡皮衣

in Handel-Mazzetti, *Symb. Sin.* 3: 145(1930) & *Cat. Lich. Univ.* 8: 522(1932).

Type: Sichuan, Handel-Mazzetti no. 5312.

On bark of trees.

Pertusaria haematina Zahlbr. 红鸡皮衣

Hedwigia, 74: 205(1934).

Type: Sichuan, collected by F. Rock.

On bark of *Abies* sp.

Pertusaria haematommoides Zahlbr. 赤星鸡皮衣

in Fedde, *Repertorium*, 33: 50(1933).

Type: Taiwan, Asahina no. 263.

Taiwan(Zahlbr., 1940, p. 452; Lamb, 1963, p. 532; Wang & Lai, 1973, p. 95).

On smooth bark of trees.

Mycotaxon 127: 222(2014).

Type: China. Yunnan, Gongshan County, Qiqi Nature Reserve, alt. 1900 m, on tree trunk, 19 Jul. 1982, M. Zang Herb. No. 1452(holotype, KUN).

Pertusaria himalayensis D. D. Awasthi & Preeti Srivast. 喜马拉雅鸡皮衣

Bryologist 96(2): 212(1993).

On barks.

Yunnan(Yu et al., 1999, p. 114; Zhao ZT et al., 2004, p. 535).

Pertusaria huangshanensis S. Yu & J. Wu *ex* Q. Ren 黄山鸡皮衣

in Zhao Z. T., Q. Ren, A. Aptroot 2004. The *Bryologist*, 107(4): 531 – 541(2004).

Type: Anhui. Mt. Huangshan, 1982, Wu & Wang Zhen 820155(NNU, holotype)

Yunnan, Hubei, Anhui(Zhao ZT et al., 2004, p. 535).

Pertusaria lacericans A. W. Archer 撕裂鸡皮衣

Mycotaxon 41(1): 230(1991).

Yunnan(Ren Q & Zhao ZT, 2008, pp. 735 – 736) .

Pertusaria leiocarpella Müll. Arg. 平滑果鸡皮衣

Bull. Herb. Boissier 3: 636(1895) .

On bark of trees.

Yunnan(Zhao ZT et al. , 2004, p. 535) .

Literature records for China. —Yunnan(Archer 1997: 85; Yu 1997: 34 – 35) .

Pertusaria leioplaca DC. in Lam. & DC. 平台鸡皮衣

Fl. Franc ? . ed. 3, 6: 173(1815) .

On bark of trees, mainly on smooth bark of deciduous trees in shaded sites.

Heilongjiang, Yunnan, Hubei, Taiwan(Zhao ZT et al. , 2004, p. 535) . Previously reported erroneously as *P. leucostoma* (Bernh.) A. Massal. (Yu 1997: 35; Yu et al. 1999: 114), Gansu (Yang F. et al. , 2008, p. 625), shaanxi(Zhao ZT et al. , 2004, p. 535; Zhao JZ et al. , 2006, p. 181), guizhou(Zhao ZT et al. , 2004, p. 535; Zhang T & JCWei, 2006, p. 9) .

Pertusaria leioplacoides Müll. Arg. 平叶鸡皮衣

Flora, 67: 304(1884) .

Type: China(prov. not indicated) .

Pertusaria leucopsara Kremp. 白围鸡皮衣

Flora, 56: 469(1873) .

Type : Hongkong, 1871 – 1872, Rabenhorst 7568(W – type) (Zhao ZT et al. , 2004, p. 536)

Hongkong(Kremp. 1874, p. 61; Zahlbr. , 1928, p. 171 & 1930 b, p. 143) .

Pertusaria leucosora Nyl. 白孢鸡皮衣

Flora 60: 223(1877) .

On rocks.

Shaanxi, Yunnan(Q. Ren, 2015, p. 692) .

Pertusaria leucosorodes Nyl. 粗鸡皮衣

Acta Soc. Sci. Fenn. 26(10) : 16(1900) .

= *Pertusaria scaberula* A. W. Archer, Mycotaxon 41: 240(1991) .

On bark of trees.

Yunnan, Fujian, Guangdong, Guangxi, Hainan(Q. Ren, 2015, pp. 692 – 693) .

Pertusaria leucostigma Müll. Arg. 黑白鸡皮衣

Flora, Regensburg 67(24) : 462(1884) .

On bark of trees.

shaanxi(Zhao JZ et al. , 2006, p. 182) , Gansu(Yang F. et al. , 2008, p. 625) . Yunnan(Yu et al. , 1999, p. 114; Zhao ZT et al. , 2004, p. 536) .

Pertusaria leucostoma(Ach.) A. Massal. 白口鸡皮衣

Ric. auton. lich. crost. (Verona) : 188(1852) .

≡ *Thelotrema leucostomum* Ach. , *Methodus*, Sectio prior: 132(1803) .

Sphaeria leucostoma Bernh. , Bol. Geol. (Bogota) 2: 11(1832) , nom. illegit. , Art. 53. 1(later homonym)

On bark of trees.

Yunnan(Yu et al. , 1999, p. 114) .

Pertusaria lijiangensis Q. Ren 丽江鸡皮衣

Mycotaxon 127: 224(2014) .

Type: China. Yunnan, Yulong County, Lijiang Alpine Botanical Garden, alt. 3590 m, on bark of *Rhododendron* sp. , 16 Aug. 2011, Q. Ren 2011174(holotype, SDNU) .

Yunnan, Guizhou, Xizang, Zhejiang(Ren Q. 2014, p. 225) .

Pertusaria margaritacea Zahlbr. 珍珠鸡皮衣

Hedwigia, 74: 202(1934).

Type: Yunnan, collected by Rock.

Yunnan(Zahlbr., 1940, p. 457; Lamb, 1963, p. 535).

Pertusaria monogona Nyl. 四川鸡皮衣

Bull. Soc. Linn. Normand. 2(6): 289(1872).

= *Pertusaria setschwanica* Zahlbr. in Handel-Mazzetti, *Symb. Sin*. 3: 147(1930) & *Cat. Lich. Uinv*. 8: 524(1932).

Type: Sichuan, Huili Co., 1914, *Handel-Mazzetti* 1014, W) (Zhao ZT et al., 2004, p. 536)

On non calcareous rocks.

On sunny rocks.

Neimenggu, Sichuan, Yunnan(Zhao ZT et al., 2004, p. 536), shaanxi(Zhao JZ et al., 2006, p. 182). Previously reported erroneously as *P. pulvinata*(Yu 1997) and *P. stalactizoides*(Yu 1997). During the re-examination of one specimen(Sichuan, Huili Co., 1914, Handel-Mazzetti 1014, W) Zhao ZT et al. (2004, p. 536) found that *P. setschwanica* is similar to *P. monogona* in chemistry and morphology. Until the type specimen is examined, they cannot determine whether the two species are synonymous.

Pertusaria multipuncta(Turner) Nyl. 斑点鸡皮衣

Lich. Scand. (Helsinki): 179(1861). var. multipuncta 原变种

≡ *Variolaria multipuncta* Turner, *Trans. Linn. Soc. London* 9: 137(1806).

≡ *Pertusaria leptospora* Nitschke, *Jber. Westfäl. Prov. -Vereins* 21: 122(1883)

Pertusaria millepuncta Jatta, Nuov. Giorn. Bot. Italiano, 2(9): 477(1902), nom. illeg. (lapsu"*millepuncta*").

On bark of trees.

Heilongjiang, jilin, Chongqing, Xizang, guangxi(Zhao ZT et al., 2004, p. 536), Xinjiang(Abdulla Abbas et al., 1993, p. 79; Abdulla Abbas & Wu, 1998, pp. 142 – 143; Abdulla Abbas et al., 2001, p. 366), Shaanxi(Jatta, 1902, p. 477; Zahlbr., 1930 b, p. 146; Zhao ZT et al., 2004, p. 536), Sichuan(Zahlbr., 1930 b, p. 146; Zhao ZT et al., 2004, p. 536), Yunnan(Zahlbr, 1930 b, p. 146 & 1934 b, p. 204; Yu et al., 1999, p. 112; Zhao ZT et al., 2004, pp. 535 – 536), Shandong(Li Y et al., 2008, p. 72), Hubei(Yu S. H. &Wu J. N., 1997, p. 333), Anhui(Yu et al., 1999, p. 112 as *P. leptospora*; Zhao ZT et al., 2004, p. 535 as *P. leptospora*), Zhejiang(Xu B. S., 1989, p. 246 – 247; Zhao ZT et al., 2004, p. 536), Jiangxi, & Fujian(Yu et al., 1999, p. 112; Zhao ZT et al., 2004, p. 535 as *P. leptospora*), Taiwan(A. Zahlbr, 1933 c, p. 50; Wang & Lai, 1973, p. 95).

var. colorata Zahlbr. 色斑变种

in Handel-Mazzetti, *Symb. Sin*. 3: 146(1930) & *Cat. Lich. Univ*. 8: 523(1932).

Type: Sichuan, Handel-Mazzetti no. 1582.

On the rotten tree trunk of *Corylus tibetica*.

Pertusaria nakamurae(Yasuda ex Räsänen) Dibben 裂疣鸡皮衣

Publications in *Biology and Geology, Milwaukee Public Museum* 5: 19(1980).

≡ *Pertusaria tuberculifera* var. *nakamurae* Yasuda ex Räsänen, *J. Jap. Bot*. 16: 95(1940).

On bark of trees.

Anhui, Fujian(Yu 1997: 35 – 36; Yu et al., 1999, p. 114; Zhao ZT et al., 2004, p. 536),
Taiwan(Zhao ZT et al., 2004, p. 536).

Pertusaria oculata(Dicks.) Th. Fr. 眼点鸡皮衣

Lich. Scand. (Upsaliae) 1(1): 307(1871).

≡ *Lichen oculatus* Dicks., *Fasc. pl. crypt. brit*. (London) 2: 17(1790).

Muscicolous over soil or rocks, rarely corticolous.

Xinjiang(Abdulla Abbas & Wu 1998: 143; Zhao ZT et al., 2004, p. 536; Abdulla Abbas et al., 2001, p. 366), yunnan(Zhao ZT et al., 2004, p. 536).

Pertusaria ophthalmiza(Nyl.) Nyl. 睛鸡皮衣

Flora 48: 354(1865) .

≡*Pertusaria multipuncta* var. *ophthalmiza* Nyl. , *Lich. Scand.* 180(1861) .

Gansu(Yang F. et al. , 2008, p. 625) . Hubei(Yu et al. , 1999, p. 113; Zhao ZT et al. , 2004, p. 537) . HUBEI (Yu 1997: 30 – 31) . Neimenggu, Jilin, Jiangxi, Sichuan, Yunnan, Xizang, Zhejiang(Zhao ZT et al. , 2004, p. 537) , shaanxi(Zhao ZT et al. , 2004, p. 537; Zhao JZ et al. , 2006, p. 182) .

Pertusaria oshioi J. C. Wei 巨孢鸡皮衣

Enum. Lich. China, 190(1991) .

Type: The holotype is from Japan, and one of the paratypes is from China: Taiwan.

Gansu(Yang F. et al. , 2008, p. 625) . Hubei(Zhao ZT et al. , 2004, p. 537) , Taiwan(Wang & Lai, 1973, p. 95; Zhao ZT et al. , 2004, p. 537) .

Pertusaria macrospora Oshio, *J. Sci. Hiroshima Univ.* ser. B, div. 2, 12: 134(1968) , nom. illegit. , Art. 53. 1, non *Pertusaria macrospora* Naeg. ex Hepp, *Flecht Europ.* , no. 424(1857) .

Pertusaria parapycnothelia Ren Q. , Zhao Z. T. 准密生鸡皮衣

in Ren Q et al. *Mycotaxon* 108: 232 – 233(2009) .

Type: Yunnan, Lijiang, Xiangshan, alt. 2400m, on tree, J. CWei 2738(holotype: HMAS. L.) .

Pertusaria paraqilianensis Sun ZS & Zhao Z. T. in Ren Q et al. *Mycotaxon* 108: 443 – 444(2008) .

Type: Qinghai, qilian, sun zs 20072496(SDNU, holotype) .

Pertusaria pertusa(L.) Tuck. 孔鸡皮衣

Enum. N. America Lich. : 56(1845) .

≡*Lichen pertusus* L. , *Mant. Pl.* 1: 131(1767) .

On smooth to rough bark of trees, rarely on rocks.

Jilin, Sichuan, yunnan (Zhao ZT et al. , 2004, p. 537), shaanxi (Zhao JZ et al. , 2006, p. 182), Gansu(Yang F. et al. , 2008, p. 625) . Jiangsu, Shanghai, Zhejiang(Xu B. S. , 1989, pp. 247 – 248), Taiwan(Oshio, 1968, p. 128; Wang & Lai, 1973, p. 95) . Yunnan(Yu 1997: 36)

Pertusaria phaeosporina Zahlbr. 暗孢鸡皮衣

Hedwigia, 74: 204(1934) .

Type: Sichuan, collected by F. Rock.

Sichuan(Zahlbr. , 1940, p. 460; Lamb, 1963, p. 537) .

On bark of trees.

Pertusaria platycarpiza Zahlbr. 宽果鸡皮衣

in Fedde, *Repertorium*, 33: 50(1933) .

Type: Taiwan, Faurie no. 36.

Taiwan(Zahlbr. , 1940, p. 460; Wang & Lai, 1973, p. 95) .

On smooth bark of trees.

Pertusaria plittiana Erichsen 坚疣鸡皮衣

Feddes Repert. Spec. Nov. Regni Veg. 41: 77(1937) .

Saxicolous on acidic rocks.

Shaanxi(Zhao ZT et al. , 2004, p. 537; Zhao JZ et al. , 2006, p. 182) , hubei(Zhao ZT et al. , 2004, p. 537) .

Pertusaria pseudocorallina(Sw.) Arnold 拟珊瑚鸡皮衣

Verh. zool. -bot. Ges. Wien 37: 84(1887) .

≡*Lichen pseudocorallinus* Sw. , in Westring, K. *Vetensk-Acad. Handl.* : 129(1791) .

=*Pertusaria westringii*(Lilj.) Leight. , *Lich. -Fl. Great Brit.* : 236(1871) .

≡*Lichen westringii* Lilj. , *Utkast Svensk Fl.* : 418(1792) .

≡*Isidium westringii*(Lilj.) Ach. , *K. Vetensk-Acad. Nya Handl.* 15: 179(1794) .

On dead twigs. On exposed granitic rocks. Northern temperate.

Shaanxi(Zhao ZT et al. , 2004, p. 537; Zhao JZ et al. , 2006, p. 182) , Yunnan(Hue, 1889, p. 174; Zahlbr. , 1930 b, p. 144) , Hongkong(Aptroot & Seaward, 1999, p. 85) .

Pertusaria pustulata(Ach.) Duby 橄榄鸡皮衣

Bot. Gall. , Edn 2(Paris) 2: 673(1830) .

≡*Porina pustulata* Ach. , *Lich. univ.* : 1 –696(1810) .

On bark of trees.

Jiangsu(Wu J. N. & Xiang T. 1981, p. 4; Xu B. S. , 1989, p. 248) , Shanghai, Zhejiang(Xu B. S. , 1989, p. 248) .

Pertusaria pycnothelia Nyl. 密生鸡皮衣

Bull. Soc. linn. Normandie, sér. 2 2: 70(1868) .

Yunnan, Taiwan(Zhao ZT et al. , 2004, p. 537) .

Pertusaria qilianensis Ren Q & Zhao ZT 祁连鸡皮衣

in Ren Q et al. *Mycotaxon* 108: 441 –442(2008) .

Type: Qinghai, Qilian, on tree bark. Wang HY 20071461 –1(SDNU, holotype) .

Pertusaria quartans Nyl. 藓生鸡皮衣

Lich. Japon. : 54(1890) .

On mosses.

Heilongjiang, Liaoning, Yunnan, anhui(Zhao ZT et al. , 2004, p. 538) , (Zhao ZT et al. , 2004, p. 538) , Zhejiang, Fujian(Yu et al. , 1999, p. 114 –115; Zhao ZT et al. , 2004, p. 538) .

Pertusaria rigida Müll. Arg. 硬鸡皮衣

J. Linn. Soc. , *Bot.* 29: 221(1893) .

On tree bark.

Gansu, Yunnan, Fujian(Zhao Na et al. , 2014, p. 629) .

Pertusaria sanguinulenta Zahlbr. 血红鸡皮衣

in Handel-Mazzetti, *Symb. Sin.* 3: 147(1930) & *Cat. Lich. Uinv.* 8: 524(1932) .

Type: Hunan, Handel-Mazzetti no. 11855.

On bark of rotten tree trunk.

Pertusaria schizostomella Müll. Arg. 裂孔鸡皮衣

Bull. Herb. Boissier 3: 637(1895) .

Jiangsu(Yu et al. , 1999, p. 96) .

Pertusaria sommerfeltii(Sommerf.) Flörke 黑口鸡皮衣

in Fries, *Lich. eur. reform.* (Lund) : 423(1831) .

≡*Endocarpon sommerfeltii* Sommerf. , *Suppl. Fl. lapp.* (Oslo) : 135(1826) .

On bark of trees.

Heilongjiang, Sichuan, Yunnan(Zhao ZT et al. , 2004, p. 538) , Gansu(Yang F. et al. , 2008, p. 625) . Shaanxi (Wu J. L. 1987, p. 126; Zhao ZT et al. , 2004, p. 537) .

Pertusaria sphaerophora Oshio 球鸡皮衣

Journ. Sci. Hiroshima Univ. ser. B, Div. 2, 12(1) : 95(1968) .

Type: Japan(holotype) , China: Taiwan(paratype) .

Gansu(Yang F. et al. , 2008, p. 625) , Yunnan, Xizang, anhui, Fujian(Zhao ZT et al. , 2004, p. 538) , Hubei(Yu et al. , 1997, p. 333) , Taiwan(Oshio, 1968, p. 95; Wang & Lai, 1973, p. 95; Zhao ZT et al. , 2004, p. 538) .

On bark of trees, especially conifers.

Pertusaria stalactiza Nyl. 钟乳鸡皮衣

Flora, Regensburg 57: 311(1874).

On rocks.

Shaanxi(Jatta, 1902, p. 477; Zahlbr., 1930 b, p. 145).

Pertusaria stalactizoides Savicz 类钟乳鸡皮衣

Notul. syst. Inst. cryptog. Horti bot. petropol. 1: 96(1922).

On rocks.

Shaanxi(Yu et al., 1999, p. 113).

Pertusaria subcomposita Oshio 亚复合鸡皮衣

Journ. Sci. Hiroshima Univ. ser. B, Div. 2, 12(1): 97(1968).

Type: Japan(holotype), China: Taiwan, paratype).

Taiwan(Oshio, 1968, p. 97; Wang & Lai, 1973, p. 95).

Pertusaria submultipuncta Nyl. 亚多斑鸡皮衣

Lich. Japon.: 55(1890).

On bark of trees.

Heilongjiang, Jilin, Sichuan, Chongqing, Yuannan, Xizang, Anhui, Zhejiang(Zhao ZT et al., 2004, p. 538), Shaanxi(Zhao ZT et al., 2004, p. 538; Zhao JZ et al., 2006, p. 182), Hubei(Yu & Wu, 1997, p. 333; Zhao ZT et al., 2004, p. 538).

Pertusaria subobductans Nyl. 海滨鸡皮衣

Lich. Japon.: 52(1890).

On trees and rocks.

shaanxi(Zhao JZ et al., 2006, p. 182), Gansu(Yang F. et al., 2008, p. 625), Heilongjiang, Sichuan, yunnan, Chongqing, xizang, hunan, Jiangxi, zhejiang, Fujian(Zhao ZT et al., 2004, p. 538), guizhou(Zhao ZT et al., 2004, p. 538; Zhang T & JCWei, 2006, p. 9), Shandong(zhang F et al., 1999, p. 31), Hubei(Yu & Wu, 1997, p. 333 – 334; Zhao ZT et al., 2004, p. 538, miscited as Bubei), Jiangsu, Anhui(Xu B. S., 1989, p. 248; Zhao ZT et al., 2004, p. 538), Taiwan(Oshio, 1968, p. 139; Wang & Lai, 1973, p. 95; Zhao ZT et al., 2004, p. 538).

Pertusaria subochracea Stirt. 亚赭鸡皮衣

in *Proceed. Philosoph. Soc. Glasgow*, XI: 312(1928).

On bark of trees.

Yunnan, Hubei(Yu et al., 1999, p. 115; Zhao ZT et al., 2004, pp. 538 – 539).

Pertusaria subpertusa Brodo 亚孔鸡皮衣

Bull. N. Y. St. Mus. Sci. Surv. 410: 207(1968).

On bark of trees.

Gansu(Yang F. et al., 2008, p. 625), Sichuan, hubei(Zhao ZT et al., 2004, p. 539), Yunnan, Anhui, Zhejiang (Yu et al., 1999, p. 115; Zhao ZT et al., 2004, p. 539).

Pertusaria subrosacea Zahlbr. 亚玫瑰鸡皮衣

in Handel-Mazzetti, *Symb. Sin.* 3: 143(1930) & *Cat. Lich. Univ.* 8: 525(1932).

var. subrosacea 原变种

Type: Yunnan, Handel-Mazzetti nos. 133, 475, 6030, 6520, 6791 & Sichuan, no. 5539(syntypes).

On different plants, such as *Vernonia volkamenaefolia*, *Pirus pashia*, *Pinus tabulaeformis*, *Rhamnus* sp. and *Myrsine africana*.

Yunnan. Between Kunming(Yunnanfu) and Puduhe, 1914, *Handel-Mazzetti* 567 (W – type) (Zhao ZT et al., 2004, p. 539).

var. evolutior Zahlbr. 大孢变种

in Handel-Mazzetti, *Symb. Sin.* 3: 143(1930) & *Cat. Lich. Univ.* 8: 525(1932).

Type: Sichuan, Handel-Mazzetti no. 5541.

On living twigs.

var. octospora Zahlbr. 八孢变种

in Handel-Mazzetti, *Symb. Sin.* 3: 144(1930) & *Cat. Lich. Univ.* 8: 525(1932).

Type: Yunnan, Handel-Mazzetti no. 413.

On bark of *Quercus* sp.

Pertusaria substerilis Zahlbr. 亚育鸡皮衣

Hedwigia, 74: 203(1934).

Type: Yunnan, Lijiang, collected by Rock.

Yunnan(Zahlbr., 1940, p. 465).

On twigs.

Pertusaria subtruncata Müll. Arg. 亚截鸡皮衣

Flora, Regensburg 67: 397(1884).

Hongkong(Thrower, 1988, p. 133).

Pertusaria subvaginata Nyl. 亚鞘鸡皮衣

Flora, Regensburg 49: 290(1866).

var. orientalis Räsänen 东方变种

in Arch. Soc. Zool. Bot. Fenn. III: 80(1949).

On rock.

Hongkong(Thrower, 1988, p. 134).

Pertusaria subventosa Malme 亚风鸡皮衣

Ark. fo¨r Bot. 28A: 7(1936).

Taiwan(Zhao ZT et al., 2004, p. 539), Hongkong(Aptroot et al., 2001, p. 335; Zhao ZT et al., 2004, p. 539).

Pertusaria subviridis Høeg 亚翠鸡皮衣

Nytt Mag. Natur. 61: 150(1924).

On bark of trees.

Anhui, Zhejiang, Fujian(Yu et al., 1999, p. 113).

Pertusaria tetrathalamia(Fée) Nyl. 四孢鸡皮衣

Acta Soc. Sci. fenn. 7(2): 448(1863).

≡ *Trypethelium tetrathalamium* Fée, *Essai Crypt. Exot.* (Paris): 69(1825) [1824].

Misapplied name(revised by Aptroot & Seaward, 1999, p. 85):

Pertusaria cf. *subtruncata* auct. non Müll. Arg. in Thrower, 1988, pp. 16, 133.

On exposed trees, especially conifers. pantropical.

Gansu(Yang F. et al., 2008, p. 626), Hongkong(Thrower, 1988, pp. 16, 133; Aptroot & Seaward, 1999, p. 85).

var. tetrathalamia 原变种

On bark of trees.

Yunnan(Zhao ZT et al., 2004, p. 539), Taiwan(A. Zahlbr., 1933 c, p. 50; Wang & Lai, 1973, p. 95; Zhao ZT et al., 2004, p. 539).

var. octospora Müll. Arg. 八孢变种

Revue mycol., *Toulouse* 9: 84(1887).

On bark of trees.

Taiwan(Zahlbr., 1933 c, p. 50; Wang & Lai, 1973, p. 95; Zhao ZT et al., 2004, p. 539).

Pertusaria thiospoda Knight 硫点鸡皮衣

Trans. Linn. Soc. London, Bot. 2: 47(1882).

Jiangsu, Taiwan(Zhao ZT et al., 2004, p. 539).

Pertusaria thwaitesii Müll. Arg. 特韦氏鸡皮衣

Flora, Regensburg 67(24):470(1884).

On bark.

Shaanxi(Zhao JZ et al., 2006, p. 183), Gansu(Yang F. et al., 2008, p. 626).

Pertusaria trachythallina Erichsen 粗果鸡皮衣

Ark. fo¨r Bot. 30A(1):36(1940).

Corticolous on hardwoods and conifers.

jilin(Ren Qiang & Zhao ZT, 2004, p. 330), Heilongjiang, Yunnan(Ren Qiang & Zhao ZT, 2004, p. 330; Zhao ZT et al., 2004, pp. 539 – 540), Gansu(Yang F. et al., 2008, p. 626).

Pertusaria variolosa(Kremp.) Vain. 颗粒鸡皮衣

Acta Soc. Fauna Flora fenn. 7(1): 106(1890).

≡*Pertusaria subvaginata* f. *variolosa* Kremp., *Flora*, Regensburg 49:218(1896).

Sichuan, Yunnan, anhui(Ren Q & Zhao ZT, 2008, p. 735).

Pertusaria violacea Oshio 紫罗兰鸡皮衣

Journ. Sci. Hiroshima Univ. ser. B, Div. 2, 12(1):92(1968).

Type: Japan(holotype), China: Taiwan(paratype).

Taiwan(Oshio, 1968, p. 92; Wang & Lai, 1973, p. 95).

Pertusaria wangii Q. Ren 王氏鸡皮衣

Mycotaxon 127:222(2014).

Type: China. Yunnan, Lushui County, on the roadside between Fugong and Lushui Counties, 68km north of Lushui county, alt. 2950 m, on dead twigs, 7 Jun. 1981, X. Y. Wang, X. Xiao &J. J. Su 2718(holotype, HMAS-L).

Pertusaria weii Q. Ren 魏氏鸡皮衣

The Lichenologist 45(3): 337 – 339(2013).

Type: China, Xinjiang, Urumqi, TianshanMountains, Glacier No. 1 at the headwaters of Urumqi River, alt. 3500 m, on calcareous rock, 27 August 2011, Li Lin 20125913(SDNU—holotype).

Pertusaria wulingensis Z. T. Zhao & Z. S. Sun 武陵鸡皮衣

in Ren Qiang, Sun Zhongshuai, Zhao Zuntian *The Bryologist,* 112(2):394 – 396(2009).

Type: China: Hunan: Sangzhi Co., Bagongshan, 29u239N, 110u119E, 1400 m, 24 Mar 2002, Chen et al. 9866 (HMAS-L 072896, holotype).

Pertusaria xanthodes Müll. Arg. 黄鸡皮衣

Flora, Regensburg 67:286(1884).

Gansu(Yang F. et al., 2008, p. 626), Yunnan, Anhui, Jiangsu, Shanghai(Yu et al., 1999, p. 96 – 97; Zhao ZT et al., 2004, p. 540).

Corticolous on hardwoods.

Pertusaria xanthoplaca Müll. Arg. 黄台鸡皮衣

Flora, Regensburg 65:485(1882).

On volcanic rock.

Shandong, Jiangsu, Zhejiang, hongkong(Aptroot et al., 2001, p. 335; Zhao ZT et al., 2004, p. 540), Taiwan (Aptroot and Sparrius, 2003, p. 32; Zhao ZT et al., 2004, p. 540).

Pertusaria yunnana G. L. Zhou & Lu L. Zhang 云南鸡皮衣

The *Lichenologist* 46(2):169 – 172(2014).

Type: China, Yunnan Province, Kunming, Mt. Jiaozixue, on bark, 3800 m alt., 27 October 2008, Z. J. Ren 20081952(SDNU—holotype).

Variolaria Pers. **瘤果衣属**

Ann. Bot. (Usteri) 7:23(1794).

Variolaria multipunctoides(Dibben) Lendemer, B. P. Hodk. & R. C. Harris 多斑瘤果衣

Mem. N. Y. bot. Gdn 104: 88(2013) .

≡ *Pertusaria multipunctoides* Dibben, *Publications in Biology and Geology, Milwaukee Public Museum* 5: 59 (1980) .

≡ *Lepra multipunctoides(Dibben) Lendemer & R. C. Harris, Bryologist* 120(2) : 187(2017) .

On bark.

Shaanxi(Zhao JZ et al. , 2006, p. 182 as *P. multipunctoides*) .

Ostropomycetidae, families incertae sedis 厚顶盘亚纲未定位科

(36) Arctomiaceae Th. Fr. **极地衣科**

[as ' Arctomiei'] , *Nova Acta R. Soc. Scient. upsal.* , Ser. 3 3: 287(1861) [1860]

Arctomia Th. Fr. **极地衣属**

Nova Acta R. Soc. Scient. upsal. , Ser. 3 3: 287(1861) [1860] .

Arctomia teretiuscula P. M. Jorg. 极地衣

Lichenologist 35(4) : 287 – 289(2003) .

Type: Sichuan, Daxue Shan, 57 km S of Kangding, Gongga Shan, Hailougou glacier and forest park, NW of Hailougou Station, alt. 2980 – 3150 m, on mossy rocks and soil, 30 July 2000, W. Obermayer 09016(GZU—holotypus) .

(37) Arthrorhaphidaceae Poelt & Hafellner, **珠节衣科**

Phyton, Horn 17: 220(1976)

Arthrorhaphis Th. Fr. **珠节衣属**

Lich. Arctoi 3: 203(1860) .

Arthrorhaphis alpina(Schaer.) R. Sant. 高山珠节衣

in Hawksworth, James & Coppins, *Lichenologist* 12(1) : 106(1980) .

≡ *Lecidea flavovirescens* var. *alpina* Schaer. , *Lich. helv. spicil.* 4 – 5: 162(1833) .

On shale & soil.

Taiwan(Aptroot and Sparrius, 2003, p. 5) .

var. alpina 原变种

Xizang, Sichuan(W. Obermayer, 2004, p. 488) .

var. jungens Obermayer & Poelt 联合变种

in W. Obermayer *Bibliotheca Lichenologica* 88: 488(2004) .

Host of lichenicolous fungus *Cercidospora trypetheliza*, *Cercidospora soror* and *Stigmidium Arthrorhaphidis*.

Sichuan(J. Hafellner & W. Obermayer 1995, pp. 184 – 187; W. Obermayer, 2004, p. 488) .

Host of lichenicolous fungus *Cercidospora trypetheliza*, *Cercidospora soror*, and *Merismatium decolorans*.

Xizang(J. Hafellner & W. Obermayer 1995, pp. 180 & 182, 184 – 186) .

Arthrorhaphis citrinella(Ach.) Poelt 柠檬珠节衣

Best. europ. Flecht. (Vaduz) 2: 126(1969) .

≡ *Lichen citrinellus* Ach. , *K. Vetensk-Acad. Nya Handl.* : 135(1796) .

Host of *Stigmidium arthrorhaphidis*

Sichuan(J. Hafellner & W. Obermayer 1995, p. 187)

On *Tsuga* bark and soil.

Taiwan(Aptroot and Sparrius, 2003, p. 5) .

Arthrorhaphis grisea Th. Fr. 灰色珠节衣

Lich. Arctoi 3: 204(1860) .

Xizang(W. Obermayer, 2004, p. 488) , TAIWAN(Aptroot and Sparrius, 2003, pp. 5 – 6) .

Arthrorhaphis vacillans Th. Fr. & Almq. ex Th. Fr. 摆珠节衣

Bot. Notiser: 107(1867) .

Host of lichenicolous fungus *Cercidospora trypetheliza*

Xizang(J. Hafellner & W. Obermayer 1995, pp. 180 & 182; W. Obermayer, 2004, p. 488).

(38) Hymeneliaceae Körb. 膜衣科

[as ' Hymenelieae'], *Syst. lich. germ.* (Breslau): 327(1855)

Hymenelia Kremp. 膜衣属

Flora, Regensburg 35: 25(1852).

Hymenelia ceracea(Arnold) M. Choisy 蜡膜衣

Bull. mens. Soc. linn. Soc. Bot. Lyon 18: 145(1949).

≡*Aspicilia ceracea* Arnold, in Kremp., Denkschr. Kgl. Bayer. Bot. Ges. 4(2. Abth.): 180(1861).

On shale.

Taiwan(Aptroot and Sparrius, 2003, p. 22).

Hymenelia epulotica(Ach.) Lutzoni 疤膜衣

in Lutzoni & Brodo, *Syst. Bot.* 20(3): 250(1995).

≡*Gyalecta epulotica* Ach., *Lich. univ.*: 1 –696(1810).

On raised coral reef.

Taiwan(Aptroot and Sparrius, 2003, p. 22).

Hymenelia lacustris(With.) M. Choisy 湖膜衣

Bull. mens. Soc. linn. Soc. Bot. Lyon 18: 145(1949).

≡*Lichen lacustris* With., *Arr. Brit. pl.*, Edn 3(London) 4: 21(1796).

≡*Aspicilia lacustris*(With.) Th. Fr. cited by Thrower, 1988, pp. 15, 59, contributed by Aptroot & Seaward, 1999, p. 79.

On exposed rocks. Cosmopolitan.

Taiwan(Aptroot and Sparrius, 2003, p. 22), Hongkong(Thrower, 1988, pp. 15, 59; Aptroot & Seaward, 1999, p. 79; Aptroot et al., 2001, p. 328).

Ionaspis Th. Fr. 尤纳衣属

Lich. Scand. (Upsaliae) 1(1): 273(1871).

Ionaspis alpina Zahlbr. 高山尤纳衣

in Handel-Mazzetti, *Symb. Sin.* 3: 70(1930) & *Cat. Lich. Univ.* 8: 253(1932).

Type: Yunnan, Handel-Mazzetti no. 10000.

On quartzite irrigated by water from melting snow.

Yunnan(H. Magn. 1933, p. 25).

Ionaspis handelii Zahlbr. 汉氏尤纳衣

in Handel-Mazzetti, *Symb. Sin.* 3: 68(1930) & *Cat. Lich. Univ.* 8: 254(1932).

Type: Yunnan, Handel-Mazzetti no. 10001.

On quartzite irrigated by water from melting snow.

Yunnan(H. Magn. 1933, p. 28).

Ionaspis odora(Ach.) Th. Fr. 香尤纳衣

Lich. Scand. (Upsaliae) 1(1): 273(1871).

≡*Gyalecta odora* Ach., in Schaerer, *Lich. helv. spicil.* 2: 80(1826).

On schistose rocks.

Taiwan(Zahlbr., 1933 c, p. 26; Wang & Lai, 1973, p. 91).

Melanolecia Hertel 黑衣属

in Poelt & Vězda, *Biblthca Lichenol.* 16: 364(1981).

Melanolecia transitoria(Arnold) Hertel 间黑衣

in Poelt & Vězda, *Biblthca Lichenol.* 16: 365(1981).

≡*Lecidea transitoria* Arnold, *Flora*, Regensburg 53: 123, tab. III, fig. 9 – 10(1870).

= *Lecidea henrici* Zahlbr. in Handel-Mazzetti, Symb. Sin. 3: 100(1930).

Type: Yunnan, Handel-Mazzetti no. 6934(holotype) in WU.

Yunnan(Zahlbr., 1932a, p. 343; Hertel, 1977, p. 356).

Tremolecia M. Choisy **震盘衣属**

Bull. mens. Soc. linn. Soc. Bot. Lyon 22: 177(1953).

Tremolecia atrata(Ach.) Hertel 黑震盘衣

Ergebn. Forsch. Unternehmens Nepal Himal. 6(3): 351(1977).

≡ *Gyalecta atrata* Ach., *K. Vetensk-Acad. Nya Handl.* 29: 229(1808).

On siliceous rocks.

Xizang(Obermayer 2004, p. 512 – 513), Taiwan(Aptroot & Sparrius 2003, p. 48).

Tremolecia lividonigra(Zahlbr.) Hertel 肝震盘衣

Khumbu Himal 6(3): 352(1977).

≡ *Lecidea lividonigra* Zahlbr., in Handel-Mazzetti, *Symb. Sin.* 3: 91 & 98(1930).

On shale.

Taiwan(Aptroot & Sparrius 2003, p. 48).

(39) Protothelenellaceae Vězda, H. Mayrhofer & Poelt, **原乳衣科**

in Mayrhofer & Poelt, *Herzogia* 7(1 – 2): 26(1985)

Protothelenella Räsänen **原乳衣属**

Ann. bot. Soc. Zool. -Bot. fenn. Vanamo 18(1): 102(1943).

Protothelenella sphinctrinoidella(Nyl.) H. Mayrhofer & Poelt 原乳衣

Herzogia 7(1 – 2): 47(1985).

≡ *Verrucaria sphinctrinoidella* Nyl., *Flora*, Regensburg 47: 355(1864).

On brypphytes.

Xizang(W. Obermayer, 2004, p. 506).

Thrombium Wallr. **凝血衣属**

Fl. crypt. Germ. (Norimbergae) 1: 287(1831).

Thrombium cercosporum Zahlbr. 尾孢凝血衣

Ann. Mycol. 30: 427(1932) & in Fedde, *Repertorium*, 31: 23(1933).

Type: from Xizang.

Xizang(Zahlbr., 1940, p. 55).

Thrombium mongolicum H. Magn. 蒙古凝血衣

Lich. Centr. Asia 1: 21(1940).

Type: Neimenggu, about 80 km west of Bailimiao, at about 1700 m, 1931, Bexell no. 17 in S(not seen).

On pegmatit with much feldspar.

(40) Thelenellaceae O. E. Erikss. ex H. Mayrhofer **乳衣科**

Biblthca Lichenol. 26: 16(1987)

Anzina Scheid. **安乳衣属**

in Vezda, Lichenes Selecti Exsiccati(Průhonice) 73: 5(1982)

Anzina carneonivea(Anzi) Scheid. 安乳衣

in Vězda, Lichenes Selecti Exsiccati, Fascicle 73(Průhonice) 73: 5(1982)

≡ *Gyalolechia carneonivea* Anzi, Atti Soc. ital. Sci. nat. (Modena) 11: 163(1868).

Anhui(Wang YL, 2012. p. 179).

Chromatochlamys Trevis. **色衣属**

Conspect. Verruc.: 7(1860).

Chromatochlamys muscorum(Th. Fr.) H. Mayrhofer & Poelt 藓色衣

Herzogia 7(1 – 2): 28(1985).

≡*Microglaena muscorum* Th. Fr., Nova Acta R. Soc. Scient. upsal., Ser. 3 3: 362(1861) [1860]

≡*Thelenella muscorum*(Th. Fr.) Vain., *Term. Füz.* 22: 341(1899).

On soil of roadbank.

Taiwan(Aptroot 2003, p. 158).

Julella Fabre 多囊衣属

Annls Sci. Nat., Bot., sér. 69: 113(1879) [1878].

Julella vitrispora(Cooke & Harkn.) M. E. Barr 瘤多囊衣

Sydowia 38: 13(1986) [1985].

≡*Pleospora vitrispora* Cooke & Harkn., Grevillea 9(51): 86(1881).

= *Polyblastiopsis geminella* Zahlbr., in Engler & Prantl, *Nat. Pflanzenfam.*, Teil. I(Leipzig) 1: 65(1903), cited by Thrower, 1988, pp. 17, 148, mentioned by Aptroot & Seaward, 1999, p. 79.

≡*Verrucaria geminella* Nyl., (1858)

On bark.

Hongkong(Thrower, 1988, pp. 17, 148; Aptroot & Seaward, 1999, p. 79).

Thelenella Nyl. 头衣属

Mém. Soc. Imp. Sci. Nat. Cherbourg 3: 193(1855).

Thelenella brasiliensis(Müll. Arg.) Vain. 巴西头衣

J. Bot., Lond. 34: 293(1896).

≡*Microglaena brasiliensis* Müll. Arg., *Flora*, Regensburg 71: 547(1888).

= *Microglaena rosacea* Zahlbr. in Handel-Mazzetti, *Symb. Sin.* 3: 16(1930) & *Cat. Lich. Univ.* 8: 64(1932).

Type: Sichuan, Handel-Mazzetti, no. 5313.

On shale.

Taiwan(Aptroot 2003, p. 169).

Thelenella justii(Servít) H. Mayrhofer & Poelt 贾氏头衣

Herzogia 7(1 – 2): 61(1985).

≡*Microglaena justii* Servít, in Zschacke, *Rabenh. Krypt. -Fl.*, Edn 2(Leipzig) 9(1. Abt., 1. Teil): 665(1934).

On *Roystonea regia* and *Abies kawakmii*.

Taiwan(Aptroot 2003, p. 169).

Thelenella luridella(Nyl.) H. Mayrhofer 棕色头衣

Biblthca Lichenol. 26: 45(1987).

≡*Verrucaria luridella* Nyl., *Expos. Synopt. Pyrenocarp.*: 41(1858).

Thelopsis sp. (Chu, 1997, p. 48).

On granite. Pantropical.

Hongkong(Chu, 1997, p. 48; Aptroot & Seaward, 1999, p. 95).

Thelenella marginata(Groenh.) H. Mayrhofer 边缘头衣

Biblthca Lichenol. 26: 48(1987).

≡*Microglaena marginata* Groenh., *Reinwardtia* 2: 391(1954).

On sandstone.

Taiwan(Aptroot 2003, p. 169).

3) Lecanoromycetidae P. M. Kirk, P. F. Cannon, J. C. David & Stalpers ex Miadl., Lutzoni & Lumbsch in Hibbett et al., *Mycol. Res.* 111(5): 529(2007). 茶渍亚纲

[13] Lecanorales Nannf. 茶渍目

Nova Acta R. Soc. Scient. upsal., Ser. 4 8(no. 2): 68(1932).

(41) Biatorellaceae M. Choisy ex Hafellner & Casares 小蜡盘科

Nova Hedwigia 55(3 – 4): 316(1992).

Biatorellaceae M. Choisy, Bull. mens. Soc. linn. Soc. Bot. Lyon 18: 140(1949) , Nom. inval. , Art. 39. 1 (Melbourne) .

Biatorella De Not. 小蜡盘衣属

G. bot. ital. 2(1. 1) : 192(1846) .

Biatorella bambusarum Zahlbr. 竹小蜡盘衣

in Handel-Mazzetti, *Symb. Sin*. 3: 139(1930) & *Cat. Lich. Univ*. 8: 497(1932) .

Syntype: Sichuan, Handel-Mazzetti no. 1499; Yunnan, Handel-Mazzetti no. 6521.

Bamboosicolous.

Biatorella saxicola Aptroot & Sipman 岩小蜡盘衣

J. Hattori bot. Lab. 91: 321(2001) .

Type: China, Hongkong, Aptroot 48706(B-holotype: HKU(M) , ABL-isotypes) .

On granite boulders.

Biatorella torvula(Nyl.) Blomb. & Forssell 黑小蜡盘衣

Enum. Pl. Scand. : 83(1880) .

≡*Lecidea torvula* Nyl. , *Flora*, Regensburg 58: 9(1875) .

Jiangsu(Xu B. S. , 1989, p. 244) .

Piccolia A. Massal. 沥青衣属

Miscell. Lichenol. : 41(1856) .

Piccolia conspersa(Fée) Hafellner 散沥青衣

Biblthca Lichenol. 58: 109(1995) .

≡*Lecidea conspersa* Fée, *Essai Crypt. Exot*. (Paris) : 108(1825) [1824] .

On *Aralia* sp.

Taiwan(Aptroot and Sparrius, 2003, p. 35) .

Piccolia elmeri(Vain.) Hafellner 埃默氏沥青衣

Biblthca Lichenol. 58: 116(1995) .

≡*Biatorella elmeri* Vain. , *Ann. Acad. Sci. fenn*. , Ser. A 15(6) : 140(1921) .

On *Shiia* sp.

Taiwan(Aptroot and Sparrius, 2003, p. 35) .

(42) Byssolomataceae Zahlbr. 旋衣科

[as ' *Byssolomaceae*'] , in Engler & Prantl, *Nat. Pflanzenfam*. , Edn 2(Leipzig) 8: 133(1926) .

Pilocarpaceae Zahlbr. , in Engler & Prantl, *Nat. Pflanzenfam*. , Teil. I(Leipzig) 1 * : 116(1905) . Nom. illegit. , Art. 53. 1

Bapalmuia Sérus. 孢衣属

Nordic Jl Bot. 13(4) : 449(1993)

Bapalmuia palmularis(Müll. Arg.) Sérus. 掌孢衣

Nordic Jl Bot. 13(4) : 451(1993)

≡*Patellaria palmularis* Müll. Arg. , *Lichenes Epiphylli* Novi: 10(1890) .

≡*Bacidia palmularis*(Müll. Arg.) Zahlbr. , *Cat. Lich. Univers*. 4: 231(1926) [1927] .

On leaves of *Knema furfuracea*.

Yunnan(J. C. Wei & Jiang, 1991, p. 206; Wang L. S. et al. , 2008, p. 536) .

Byssolecania Vain. 絮衣属

Ann. Acad. Sci. fenn. , Ser. A 15(6) : 167(1921) .

Byssolecania deplanata(Müll. Arg.) R. Sant. 平絮衣

Symb. bot. upsal. 12(1) : 555(1952) .

≡*Patellaria deplanata* Müll. Arg. , *Lich. Epiph. Novi*: 8(1890) .

On leaves of *Cleidion brevipetiolatum*. Pantropical.

Yunnan(J. C. Wei & Jiang, 1991, p. 212) , Hongkong(Aptroot & Seaward, 1999, p. 68; Aptroot et al. , 2001, p. 322) .

Byssolecania variabilis Vězda, Kalb & Lücking 异絮衣

in Lücking & Kalb, *Bot. Jb.* 122(1) : 21(2000) .

Yunnan(Aptroot et al. , 2003, pp. 44) .

Byssoloma Trevis. **旋衣属**

Spighe Paglie: 6(1853) .

Byssoloma leucoblepharum(Nyl.) Vain. 白缘毛旋衣

Dansk bot. Ark. 4(11) : 23(1926) .

≡*Lecidea leucoblephara* Nyl. , *Annls Sci. Nat.* , *Bot.* , sér. 4 19: 337(1863) .

On *Trema orientalis* and *Alpinia speciosa*.

Taiwan(Aptroot and Sparrius, 2003, p. 10; Aptroot et al. , 2003, pp. 42 – 43) .

Byssoloma subdiscordans(Nyl.) P. James 刺旋衣

Lichenologist 5: 126(1971) .

≡*Chiodecton subdiscordans* Nyl. , *Flora*, Regensburg 62: 221(1879) .

On *Cryptomeria japonica* needles, *Polystichum fronds* and *Arenga* sp.

Yunnan(Aptroot et al. , 2003, pp. 44) , Taiwan(Aptroot and Sparrius, 2003, p. 10; Aptroot et al. , 2003, pp. 42 – 43) , Hongkong(Aptroot et al. , 2001, p. 322) .

Byssoloma tricholomum(Mont.) Zahlbr. 毛旋衣

Cat. Lich. Univers. 2: 569(1923) [1924] .

≡*Biatora tricholoma* Mont. , *Annls Sci. Nat.* , *Bot.* , sér. 3 16: 53(1851) .

On leaves and twigs of *Cryptomeria japonica*.

Fujian(Zahlbr. , 1930 b, p. 67) .

Calopadia Vězda **铃衣属**

Folia geobot. phytotax. 21(2) : 208, 215(1986) .

Calopadia phyllogena(Müll. Arg.) Vězda 叶生铃衣

Folia geobot. phytotax. 21(2) : 215(1986) .

≡*Heterothecium phyllogenum* Müll. Arg. , *Flora*, Regensburg 64(7) : 106(1881) .

On fern fronds.

Hongkong(Aptroot et al. , 2001, p. 322) .

Calopadia puiggarii(Müll. Arg.) Vězda 普氏铃衣

Folia geobot. phytotax. 21(2) : 215(1986) .

≡*Heterothecium puiggarii* Müll. Arg. , *Flora*, Regensburg 64(7) : 105(1881) .

On living leaves of trees in forests. Pantropical.

Taiwan(Aptroot and Sparrius, 2003, p. 10) , Hongkong(Aptroot & Seaward, 1999, p. 68, Living culture CBS – 101254) .

Calopadia sp. 铃衣(未定名种)

On Fatsia sp. , *Phoebe formosana*, and *Asplenium nidus*.

Taiwan(Aptroot et al. , 2003, pp. 42 – 43) .

Calopadia subcoerulescens(Zahlbr.) Vězda 天蓝铃衣

Lichenes Selecti Exsiccati, Fascicle 88(nos2176 – 2200) (Průhonice) 88: 3, no. 2185(1988) .

≡*Lopadium subcoerulescens* Zahlbr. , *Trans. Proc. N. Z. Inst.* 59: 312(1928) .

On *Cryptomeria japonica* needles, *Hydrangea* sp. , *Alpinia speciosa*, and *Arenga* sp.

Yunnan(Aptroot et al. , 2003, pp. 44) , Taiwan(Aptroot and Sparrius, 2003, p. 11; Aptroot et al. , 2003, pp. 42 – 43) , Hongkong(Aptroot et al. , 2001, p. 322) .

Fellhanera Vězda **肉盘衣属**

Folia geobot. phytotax. 21(2):200(1986).

Fellhanera bouteillei(Desm.) Vězda 布氏肉盘衣

Folia geobot. phytotax. 21(2):214(1986).

≡*Parmelia bouteillei* Desm., *Annls Sci. Nat.*, *Bot.*, sér. 3 8:191(1847).

≡*Catillaria bouteillei*(Desmaz.) Zahlbr. (Thrower, 1988, p. 73).

On leaves, e. g. of *Citrus* and *Pandanus*, *Eleagnus* leaves, *Cryptomeria japonica* Needles, *Fatsia* sp., *Hydrangea* sp., *Asplenium nidus*, and *Polystichum* sp. \ Cosmopolitan Yunnan(Aptroot et al., 2003, pp. 44), Taiwan(Aptroot and Sparrius, 2003, p. 16), Hongkong(Thrower, 1988, p. 73 ; Aptroot & Seaward, 1999, p. 75; Aptroot et al., 2001, p. 326).

Fellhanera fuscatula(Müll. Arg.) Vězda 暗色肉盘衣

Folia geobot. phytotax. 21(2):214(1986).

≡*Patellaria fuscatula* Müll. Arg., *Flora, Regensburg* 64(15):231(1881).

Yunnan(Aptroot et al., 2003, pp. 44).

Fellhanera rhaphidophylli(Rehm) Vězda [as ‘rhapidophylli’] 针晶肉盘衣

Folia geobot. phytotax. 21(2):214(1986).

≡*Bilimbia rhaphidophylli* Rehm, *Leafl. of Philipp. Bot*. 6:2237(1914).

On sandstone, and *Castanopsis* sp.

Taiwan(Aptroot and Sparrius, 2003, p. 17).

Fellhanera semecarpi(Vain.) Vězda 半果肉盘衣

Folia geobot. phytotax. 21(2):215(1986).

≡*Catillaria semecarpi* Vain., *Ann. Acad. Sci. fenn.*, Ser. A 15(6):110(1921).

On leaves of trees, *Phoebe formosana*.

Yunnan(J. C. Wei & Jiang, 1991, p. 213), Taiwan(Aptroot et al., 2003, pp. 42 – 43).

Fellhanera subfuscatula Lücking 亚暗黑肉盘衣

Trop. Bryol. 13: 162(1997).

On shale.

Taiwan(Aptroot and Sparrius, 2003, p. 17).

Fellhanera subternella(Nyl.) Vězda 肉盘衣

Folia geobot. phytotax. 21(2):215(1986).

≡*Lecidea subternella* Nyl., *Lich. Ins. Guin.*:22(1889).

On leaves of *Podocarpus* sp.

Taiwan(Aptroot and Sparrius, 2003, p. 17).

Fellhanera subtilis(Vězda) Diederich & Sérus. 精细肉盘衣

in Sérusiaux, *Mémoires de la Société Royale de Botanique de Belgique* 12:142(1990)[1988].

≡*Bacidia subtilis* Vězda, *Preslia* 33:367(1961).

On *Podocarpus phyllocladia*.

Taiwan(Aptroot and Sparrius, 2003, p. 17).

Fellhanera viridisorediata Aptroot, M. Brand & Spier 绿粉芽肉盘衣

Lichenologist 30(1):22(1998).

On roots.

Taiwan(Aptroot and Sparrius, 2003, p. 17).

Fellhaneropsis Sérus. & Coppins **类肉盘衣属**

Lichenologist 28(3):198(1996).

Fellhaneropsis kurokawana G. Thor, Lücking & Tat. Matsumoto 黑氏类肉盘衣

Symb. bot. upsal. 32(3):41(2000).

On *Cunninghamia* needles and *Dendropanax* bark.

Taiwan(Aptroot and Sparrius, 2003, p. 17) .

Lasioloma R. Sant. **绵毛衣属**

Symb. bot. upsal. 12(no. 1) : 545(1952) .

Lasioloma arachnoideum(Kremp.) R. Sant. 蛛丝绵毛衣

Symb. bot. upsal. 12(1) : 547(1952) .

≡*Phlyctis arachnoidea* Kremp. , *Lich. Foliic. Leg. Beccari:* 11(1874) .

On *Diospyros* maritima leaves.

Taiwan(Aptroot and Sparrius, 2003, p. 23) .

Lopadium Körb. **锈疣衣属**

Syst. lich. germ. (Breslau) : 210(1855) .

Lopadium disciforme(Flot.) Kullh. 盘状锈疣衣

Not. Sällsk. Fauna et Fl. Fenn. Förh. , Ny Ser. 11: 275(1871) [1870] .

≡*Heterothecium pezizoideum* var. *disciforme Flot.* , *Bot. Ztg.* 8: 553(1850) .

On *Abies kawakamii.*

Taiwan(Aptroot & Sparrius, 2003, p. 28) .

Lopadium hiroshii Kurok. 双孢囊锈疣衣

Bull . Natn. Sci. Mus. 12: 686(1969) .

On bark of trees.

Taiwan(Kurok. & Kashiwadani, 1977, p. 128) .

Lopadium pezizoideum(Ach.) Körb. 锈疣衣

Syst. lich. germ. (Breslau) : 210(1855) .

≡*Lecidea pezizoidea* Ach. , *Lich. univ.* : 182(1810) .

On bark of pine tree.

Yunnan(Zahlbr. 1930 b, p. 120) .

Micarea Fr. **亚网衣属**

Syst. orb. veg. (Lundae) 1: 256(1825) .

Micarea adnata Coppins 贴亚网衣

Bull. Br. Mus. nat. Hist. , *Bot.* 11(2) : 108(1983) .

On wood in *Pseudotsuga* forest.

Taiwan(Aptroot & Sparrius, 2003, p. 28) .

Micarea assimilata(Nyl.) Coppins 同亚网衣

Bull. Br. Mus. nat. Hist. , *Bot.* 11(2) : 114(1983) .

≡*Lecidea assimilata* Nyl. , *Lich. Scand.* (Helsinki) : 221(1861) .

Xizang(Mt. Everest: Paulson, 1952, p. 193) .

Micarea bauschiana(Körb.) V. Wirth & Vězda 巴乌亚网衣

in Vězda & Wirth, *Folia geobot. phytotax.* 11: 95(1976) .

≡*Biatora bauschiana* Körb. , *Parerga lichenol.* (Breslau) 2: 157(1860) .

On compacted soil at 200m. Northern temperate

Hongkong(Aptroot & Seaward, 1999, p. 82: living culture CBS 101363) .

Micarea botryoides(Nyl.) Coppins 葡萄亚网衣

in Hawksworth, James & Coppins, *Lichenologist* 12(1) : 107(1980) .

≡*Lecidea apochroeella* var. *botryoides* Nyl. , Flora, Regensburg 50: 373(1867)

On wood.

Taiwan(Aptroot & Sparrius, 2003, p. 28) .

Micarea cinerea(Schaer.) Hedl. 灰亚网衣

Bih. K. svenska VetenskAkad. Handl. , Afd. 3 18(3) : 81, 93(1892) .

≡*Lecidea cinerea* Schaer., *Lich. helv. spicil.* 3: 156(1828).

On wood.

Taiwan(Aptroot & Sparrius, 2003, p. 29).

Micarea denigrata(Fr.) Hedl. 变黑亚网衣

Bih. K. svenska VetenskAkad. Handl., Afd. 3 18(3): 78, 89(1892).

≡*Biatora denigrata* Fr., *K. Vetensk-Acad. Nya Handl.*: 265(1822).

On *Michelia formosana* wood.

Taiwan(Aptroot & Sparrius, 2003, p. 29).

Micarea lignaria(Ach.) Hedl. 木亚网衣

Bih. K. svenska VetenskAkad. Handl., Afd. 3 18(3): 82(1892).

≡*Lecidea lignaria* Ach., *K. Vetensk-Acad. Nya Handl.* 29: 236(1808).

On shale.

Taiwan(Aptroot & Sparrius, 2003, p. 29).

Micarea lithinella(Nyl.) Hedl. 石亚网衣

Bih. K. svenska VetenskAkad. Handl., Afd. 3, 18(3): 78(1892).

≡*Lecidea lithinella* Nyl., *Flora*, Regensburg 45: 464(1862).

On shale.

Taiwan(Aptroot & Sparrius, 2003, p. 29).

Micarea lutulata(Nyl.) Coppins 黄亚网衣

in Hawksworth, James & Coppins, *Lichenologist* 12(1): 107(1980).

≡*Lecidea lutulata* Nyl., *Flora*, Regensburg 56(19): 297(1873).

On iron-containing rock and soil. Known from temperate Europe, North America and Asia.

Hongkong(Aptroot & Seaward, 1999, p. 82).

Micarea melaena(Nyl.) Hedl. 黑亚网衣

Bih. K. svenska VetenskAkad. Handl., Afd. 3 18(3): 82(1892).

≡*Lecidea melaena* Nyl., *Bot. Notiser*: 182(1853).

≡*Bacidia melaena* (Nyl.) Zahlbr. (Abbas et al., 1993, p. 75).

On stones.

Xinjiang(Abdulla Abbas et al., 1993, p. 75; Abdulla Abbas et al., 1994, p. 22; Abdulla Abbas & Wu, 1998, p. 86; Abdulla Abbas et al., 2001, p. 365), Fujian(Wu J. L. 1987, p. 92).

Micarea micrococca(Körb.) Gams 小球亚网衣

Kl. Krypt. -Fl., Edn 3(Stuttgart) 3: 67(1967).

≡*Biatora micrococca* Körb., *Parerga lichenol.* (Breslau) 2: 155(1860).

On *Alnus japonica* and *Cryptomeria japonica*.

Taiwan(Aptroot & Sparrius, 2003, p. 29).

Micarea misella(Nyl.) Hedl. 低亚网衣

Bih. K. svenska VetenskAkad. Handl., Afd. 3 18(3): 78(1892).

≡*Lecidea anomala* var. *misella* Nyl., *Lich. Scand.* (Helsinki): 202(1861).

On wood.

Taiwan(Aptroot & Sparrius, 2003, p. 29).

Micarea neostipitata Coppins & P. F. May 新柄亚网衣

Lichenologist 33(6): 487(2001).

On *Schima superba*.

Taiwan(Aptroot & Sparrius, 2003, p. 29).

Micarea peliocarpa(Anzi) Coppins 蓝果亚网衣

Bull. Br. Mus. nat. Hist., *Bot.* 11: 169(1983).

≡Bilimbia peliocarpa Anzi, *Atti Soc. ital. Sci. nat.* (Modena) 9: 250(1866).

On shaded trees, and also on granite, moss on rock and soil. Csmopolitn.

Hongkong(Aptroot & Seaward, 1999, p. 82).

On roots, *Cryptomeria japonica* and *Tsuga* branches.

Taiwan(Aptroot & Sparrius, 2003, p. 29), Hongkong(Aptroot & Sipman 2001, p. 331).

Micarea stipitata Coppins & P. James 柄亚网衣

Lichenologist 11(2): 156(1979).

On *Cryptomeria japonica*.

Taiwan(Aptroot & Sparrius, 2003, p. 30).

Micarea sylvicola(Flot. ex Körb.) Vězda & V. Wirth 林亚网衣

Folia geobot. phytotax. 11: 99(1976).

≡*Lecidea sylvicola* Flot. ex Körb., *Syst. lich. germ.* (Breslau): 254(1855).

On shale.

Taiwan(Aptroot & Sparrius, 2003, p. 30).

Psilolechia A. Massal. **裸衣属**

Atti Inst. Veneto Sci. lett., ed Arti, Sér. 3 5: 264(1860) [1859 – 1860].

Psilolechia lucida(Ach.) M. Choisy 裸衣

Bull. mens. Soc. linn. Soc. Bot. Lyon 18: 142(1949).

≡*Lichen lucidus* Ach., *Lich. suec. prodr.* (Linköping): 39(1799) [1798].

On lignum of *Juniperus* sp.

Xizang(W. Obermayer, 2004, p. 507), Taiwan(Aptroot and Sparrius, 2003, p. 36).

* **Scutula** Tul. **芽菌属**

Annls Sci. Nat., Bot., sér. 3 17: 118(1852).

* Scutula sp. 芽菌

On *Endocarpon sinense*(Magnusson 1940).

Gansu(Hawksw. & M. Cole, 2003, p. 362).

Sporopodium Mont. **孢足衣属**

Annls Sci. Nat., Bot., sér. 3 16: 54(1851).

Sporopodium albonigrum Zahlbr. 黑白孢足衣

in Handel-Mazzetti, *Symb. Sinic.* 3: 71(1930), & *Cat. Lich. Univ.* 8: 251(1932).

Type locality: Yunnan, near Manhao, Handel-Mazzetti no. 5886 in W(not seen).

On living leaves.

R. Santesson writes about this lichen: " *Sporopodium albonigrum* A. Zahlbruckner is an undeterminable *Sporopodium* or *Lopadium*. The type specimen is in a bad condition. The accounts in the diagnosis, ' excipulum distinctum non evolutum', and ' hymenum-Jcupreolutescens', are incorrect. The excipulum is well developed, and the hymenium is I +dark blue. "(See R. Sant. 1952, p. 511).

Sporopodium flavescens(R. Sant.) Vězda 淡黄孢足衣

Lichenes Selecti Exsiccati, Fascicle 88(nos 2176 – 2200) (Průhonice): 5, no. 2193(1988).

≡*Sporopodium phyllocharis* var. *flavescens* R. Sant., *Symb. bot. upsal.* 12(1): 518(1952).

On *Aplpinia speciosa* and *Asplenium nidus*.

Taiwan(Aptroot et al., 2003, pp. 42 – 43).

Sporopodium leprieurii Mont. 孢足衣

Annls Sci. Nat., Bot., sér. 3 16: 54, pl. 16, fig. 1(1851).

On *Alpinia speciosa*.

Yunnan(Aptroot et al., 2003, p. 45), Taiwan(Aptroot et al., 2003, pp. 42 – 43).

var. leprieurii 原变种

On leaves of Knema furfuracea

Yunnan(J. C. Wei & Jiang, 1991, p. 206).

var. citrinum(Zahlbr.) R. Sant. 柠檬变种

Foliicolous lichens I; 515(1952).

≡*Lopadium citrinum* Zahlbr., ibid. 3: 121(1930) & ibid. 8: 416(1932).

Type: Yunnan, Handel-Mazzetti no. 9866(holotype) in W(not seen).

Yunnan(R. Sant. 1952, p. 515).

On living leaves of Aeschynanthus bracteatus.

Sporopodium lucidum Aptroot & Sipman 亮孢足衣

Lichenologist 25(2): 130(1993).

Hongkong(Aptroot & Sipman, 2001, p. 340).

Sporopodium phyllocharis(Mont.) A. Massal. 嗜叶孢足衣

Alcuni Gen. Lich.: 9(1853).

≡*Biatora phyllocharis* Mont., *Annls Sci. Nat.*, *Bot.*, sér. 3 10: 128(1848).

Yunnan(Aptroot et al., 2003, p. 45), Hainan(Aptroot et al., 2003, p. 45).

Sporopodium xantholeucum(Müll. Arg.) Zahlbr. 黄白孢足衣

Cat. Lich. Univers. 2: 681(1924).

≡*Gyalectidium xantholeucum* Müll. Arg., *Flora*, Regensburg 64(7): 101(1881).

On *Diospyros maritima* leaves, *Myristica* leaves, and *Ficus nervosa* leaves. *Alpinia speciosa*, and *Asplenium nidus*.

Yunnan(Aptroot et al., 2003, p. 45), Taiwan(Aptroot and Sparrius, 2003, p. 46; Aptroot et al., 2003, pp. 42 – 43).

Tapellaria Müll. Arg. **黑壳盘衣属**

Lich. Epiph. Novi: 11(1890); Santesson, *Symb. Bot. Upsal.* 12(1): 494(1952).

Tapellaria epiphylla(Müll. Arg.) R. Sant. 叶黑壳盘衣

in Thorold, *J. Ecol.* 40: 129(1952).

≡*Lopadium epiphyllum* Müll. Arg., *Flora*, Regensburg 64(7): 107(1881).

On shale.

Taiwan(Aptroot & Sparrius 2003, p. 47).

Tapellaria nana(Fée) R. Sant. 矮黑壳盘衣

Symb. bot. upsal. 12(no. 1): 507(1952).

≡*Lecanora nana* Fée, *Bull. Soc. bot.* Fr. 20: 315(1873).

= *Lopadium melaleucum* Müll. Arg. *Flora* 64: 107(1881).

Foliicolous.

Fujian(Zahlbr., 1930 b, p. 120, as *L. melaleucum*).

(43) Cladoniaceae Zenker **石蕊科**

in Goebel & Kunze, *Pharmaceutische Waarenkunde*(Eisenach) 1(3): 124(1827) [1827 – 1829].

Cladia Nyl. **筛蕊属**

Recogn. Ram. 69: (167) (1870).

Cladia aggregata(Swartz) Nyl. 筛蕊

Recogn. Ram. 69: (167) (1870).

≡*Lichen aggregatus* Swartz, *Prodr.*: 147(1788).

≡*Cladonia agreggata*(Swartz) Sprenger, Syst. veg. 4(1): 270(1827).

= *Cladonia aggregata* var. *straminea* Müll. Arg., in Vainio, Revue mycol., Toulouse 1(4): 162(1879).

Misapplied name:

Cetraria aculeata sensu Tchou YT: *Contr. Inst. Bot. Nat. Acad.* Peiping, 3(6): 311(1935), non Fr.

On the ground. Pantropical.

Xizang(J. C. Wei & Chen, 1974, p. 176; J. C. Wei & Jiang, 1981, p. 1147; J. C. Wei et al. 1985, p. 58; J. C. Wei & Jiang, 1986, p. 77) , Sichuan(Hue, 1898, p. 258; J. C. Wei et al. 1985, p. 58) , Guizhou(Zahlbr. , 1930b, p. 130; Sandst. 1932, p. 71; J. C. Wei et al. 1985, p. 58; Zhang T & J. C. Wei, 2006, p. 5) , Yunnan(Hue, 1898, P. 258; Harm. 1928, p. 322; Zahlbr. , 1930b, p. 130; Sandst. 1932, p. 71; J. C. Wei, 1981, p. 87; J. C. Wei et al. 1982, p. 22; J. C. Wei et al. 1985. p. 58; J. C. Wei X. L. et al. , 2007, p. 150.) , Anhui(J. C. Wei et al. 1985, p. 58; Xu B. S. , 1989, p. 232) , Hunan, Guangxi(J. C. Wei et al. 1985, p. 58) , Jjiangxi(Zahlbr. , 1930 b, p. 130; J. C. Wei et al. 1982, p. 22; J. C. Wei et al. 1985, p. 58) , Zhejiang(Tchou, 1935, p. 311, as Cetraria aculeata non Fr. ; J. C. Wei et al. 1985, p. 58; Xu B. S. , 1989, p. 232) , Fujian(Zahlbr. , 1930b, p. 130; Sandst. 1932, p. 71; J. C. Wei et al. 1985, p. 58) , Taiwan(A. Zahlbr. , 1933, p. 46; Asahina & M. Sato in Asahina, 1939, p. 685; Asahina, 1950 b, p. 73; M. Sato, 1958, p. 83; Wang-Yang & Lai, 1973, p. 89 & 1976, p. 227; Ahti & Lai, 1979, p. 234; Liu et al. 1980, p. 14; Lai, 2000, p. 223) , Hainan(JCWei et al. , 2013, p. 230) , Hongkong(Thrower, 1988, p. 75; Aptroot & Seaward, 1999, p. 71; Aptroot et al. , 2001, p. 324) .

Cladonia P. Browne 石蕊属

Prim. fl. holsat. (Kiliae) : 90(1756) .

Type sp. : *Cladonia subulata*(L.) Weber ex F. H. Wigg. , *Prim. fl. holsat.* (Kiliae) : 90(1780)

≡*Lichen subulatus* L. , *Sp. pl.* 2: 1153(1753) .

Cladonia aberrans(Abbayes) Stuckenb. 黄雀石蕊

in Notul. System. E Sect. Cryptog. Inst. Bot. nomine V. L. Komarovii Acad. Sci. URSS) 11: 12(1956) [*Bot. Materialy Otd. Sporov. Rast. Bot. Inst. Komarova Acad. Nauk. SSSR* 11: 12(1956)] .

≡*Cladonia alpestris* f. *aberrans* Abbayes, *Bull. Soc. sci. Bretagne* 16: 93(1939) .

≡*Cladina aberrans*(Abbayes) Hale & W. L. Culb. , *Bryologist* 73(3) : 510(1970) , J. C. Wei & Y. M. Jiang *Acta Mycologica Sinica* 5(4) : 243(1986) .

Neimenggu(J. C. Wei et al. 1986, p. 243) .

Cladonia acuminata(Ach.) Norrl. 尖石蕊

in Norlin & Nylander, *Herb. Lich. Fenn.* : 57a[label] . ante Sept. (1875) .

≡*Cenomyce pityrea* f. *acuminata* Ach. , *Syn. Lich.* 254(1814) .

Heilongjiang(Hinggan Ling: Asahina, 1943 b, p. 55; Kurok. 1959 a, p. 23) , Neimenggu(Chen et al. 1981, p. 129) , Shaanxi(He & Chen 1995, p. 42, no specimen and locality were cited) , Sichuan(Stenroos et al. , 1994, p. 325) , Hubei(Chen, Wu & J. C. Wei, 1989, p. 464) .

Cladonia amaurocraea(Flörke) Schaer. 黑穗石蕊

Lich. Helv. Spic. : 34(1823) . *Monogr. Cladon.* 1: 339(1887) .

≡*Capitularia amaurocraea* Flörke in Beitr. Naturk. 2: 334, 19 Sep. (1810) . f. *amaurocraea*

= *Cladonia amaurocraea* f. *celotea* Ach. see Vainio, *Monographia Cladoniarum Universalis. Acta Soc. Fauna Fl. Fenn.* I: 249(1887) .

= *Cladonia amaurocraea* f. *craspedia*(Ach.) Schaer. see Vainio, *Monographia Cladoniarum Universalis. Acta Soc. Fauna Fl. Fenn.* I: 250, 254(1887) .

= *Cladonia amaurocraea* f. *fruticulescens*(Norrl.) Vain. , *Acta Soc. Fauna Flora fenn.* 4(no. 1) : 253(1887) .

= *Cladonia amaurocraea* f. *oxyceras*(Ach.) Vain. , *Acta Soc. Fauna Flora fenn.* 4: 249, 254(1887) .

≡*Cenomyce oxycera* Ach. , *Lich. univ.* : 557(1810) .

= *Cladonia amaurocraea* f. *tenuisecta* Vain. , *Acta Soc. Fauna Flora fenn.* 10: 448(1894) .

On the ground.

Heilongjiang(Asahina, 1952d, p. 373; Chen et al. 1981, p. 129; J. C. Wei et al. 1982, p. 24; Luo, 1984, p. 85) , Neimenggu(Da Hinggan Ling: Sato, 1952, p. 173; Asahina, 1952d, p. 373; Chen et al. 1981, p. 129; J. C. Wei et al. 1982, p. 24; Sun LY, et al. , 2000, p. 35) , Jilin(Moreau et Moreau, 1951, p. 188; Chen et al. 1981, p. 129; J. C. Wei et al. 1982, p. 24) , Xinjiang(Abdulla Abbas & Wu, 1998, p. 52; Abdulla Abbas et al. , 2001, p. 362) ,

Shaanxi(He & Chen 1995, p. 42, no specimen & locality were cited), Xizang(J. C. Wei & Chen, 1974, p. 176; J. C. Wei & Jiang, 1981, p. 1146 & 1986, p. 85), Sichuan(Stenroos et al., 1994, pp. 325 – 326), Yunnan (Hue, 1887a, p. 18 & 1898, p. 264; Zahlbr., 1930 b, p. 130; Sandst. 1932, p. 71; Asahina & Sato in Asahina, 1939 c, p. 687), Hebei(Moreau et Moreau, 1951, p. 188).

Cladonia arbuscula(Wallr.) Flot. 林石蕊

in Wendt, *Flechten Hirschberg-Warmbrunn*: 94(1839) subsp. arbuscula. 原亚种

≡*Patellaria foliacea* var. *arbuscula* Wallr., *Naturg. Saulchen.* - Fl. : 169(1829).

≡*Cladina arbuscula* (Wallr.) Hale & Culb. -J. C. Wei& Y. M. Jiang 1986a. *Acta MycologicaSinica* 5(4) : 240 – 250.

= *Cladonia sylvatica* sensu auct. brit. ; fide Checklist of Lichens of Great Britain and Ireland(2002)

On the ground and rotten logs.

Heilongjiang(J. C. Wei et al. 1986, p. 247), Jilin(Tchou, 1935, p. 305; Moreau et Moreau, 1951, p. 187; Chen et al. 1981, p. 130; J. C. Wei et al. 1986, p. 245, 246, 247), Liaoning(Zahlbr., 1934, p. 201; Chen et al. 1981, p. 130), Neimenggu(Sato, 1952, p. 173; J. C. Wei et al. 1986, p. 246), Shaanxi(J. C. Wei et al. 1989, p. 246, 247), Xizang(J. C. Wei & Jiang, 1986, p. 80; J. C. Wei et al. 1986, p. 246, 247), Sichuan(Zahlbr., 1934b, p. 201), Guizhou(J. C. Wei et al. 1986, p. 245; Zhang T & J. C. Wei JC, 2006, p. 5), Yunnan(Hue, 1889, p. 161 & 1898, p. 257; Zahlbr., 1930 b, p. 130; Tchou, 1935, p. 305; J. C. Wei et al. 1986, p. 246), Taiwan(Zahlbr., 1933 c, p. 46; Wang & Lai, 1973, p. 89).

subsp. beringiana Ahti 黄林亚种

Ann. Bot. Soc. zool. -bot. fenn. "Vanamo" 32(1): 109(1961).

≡*Cladina adrbuscula* (Wallr.) Hale & Culb. ssp. *beringiana*(Ahti) Golubk. Jilin(J. C. Wei et al., 1986a, pp. 244 – 245 as *Cladina imshaugii*, see Ahti, 1991, p. 60), Guizhou(J. C. Wei et al., 1986a, pp. 244 – 245 as *Cladina imshaugii*, see Ahti, 1991, p. 60), Yunnan(Ahti, 1961, p. 109), Taiwan(Wang & Lai, 1976, p. 227).

subsp. *sphagnoides*(Flk.) 泥炭亚种

≡ *Cladonia rangiferina. silvatica* c. *sphagnoides* Flk., *Clad. Comm.* 168 (1828). Vain. Monogr. Clad. I, 26 (1887).

Neimenggu(Asahina, 1952, p. 373)

Cladonia bacilliformis(Nyl.) Sarnth. 类黄粉石蕊

Öst. bot. Z. 46: 264(1896).

≡*Cladonia carneola* var. *bacilliformis* Nyl. *Herb. Mus. Fenn.* 79(1859).

Misapplied name:

Cladonia cyanipes auct. non(Sommerf.) Nyl. : J. C. Wei & J. B. Chen, in *Report on the Scientific Investigations (*1966 – 1968*) in Mt. Qomolangma district*, 1974, p. 177. J. C. Wei & Jiang, in *Proccdings of Symposium on Qinghai-Xizang(Qinghai-Tibet) Plateau*, 1981, p. 1146 & *Lichens of Xizang*, 1986, p. 94.

Xizang(J. C. Wei & Chen, 1974, p. 177; J. C. Wei & Jiang, 1981, p. 1146 & 1986, p. 94; Guo & J. C. Wei, 1994, p. 30), Sichuan(Stenroos et al., 1994, p. 326). Yunnan(Guo & J. C. Wei, 1994, P. 30).

Cladonia beaumontii Vain. 比蒙氏石蕊

Acta Soc. Fauna Flora fenn. 4(1) : 411(1887).

≡*Cladonia santensis* var. *beaumontii* Tuck., Syn. N. Amer. Lich. (Boston) 1: 245(1882).

≡*Cladonia beaumontii*(Tuck.) Fink *Lich. Fl. U. S.* : 261(1935).

Xinjiang(Abdulla Abbas et al., 2000, p. 17), Yunnan(Guo & J. C. Wei, 1994, P. 30).

Cladonia bellidiflora(Ach.) Schaer. 菊花石蕊

Lich. helv. spicil. 1: 35(1823).

≡*Lichen bellidiflorus* Ach., *Lich. Suec. Prodr.* : 194(1799) ["1798"].

On the ground.

Taiwan(Ahti & Lai, 1979, p. 229; Liu et al. 1980, p. 30; Lai, 2000, p. 241).

Cladonia botrytes(K. G. Hagen) Willd. 葡萄石蕊

Fl. berol. prodr. : 365(1787).

≡*Lichen botrytes* K. G. Hagen, *Tentam. Hist. Lich. Prussic.* : 121(1782).

On rotten logs.

Heilongjiang(Asahina, 1952d, p. 374), Fujian(Zahlbr., 1930b, p. 134; Sandst. 1939, p. 96).

Cladonia cariosa(Ach.) Spreng 腐石蕊

Syst. veg., Edn 16 4(1): 272(1827).

≡*Lichen cariosus* Ach., *Lichenogr. Suec. Prodr.* : 198(1799) ["1798"].

Misapplied name:

Cladonia brevis auct. non Sandst. : Magn. *Lich. Centr. Asia*, f1: 68(1940). Heilongjiang, Liaoning, Neimenggu (Chen et al. 1981, p. 130), Xinjiang(H. Magn. 1940, p. 68; Ahti, 1976, p. 369; Wu J. L. 1986, p. 74; Abdulla Abbas et al., 1993, p. 76; Abdulla Abbas & Wu, 1998, p. 53; Abdulla Abbas et al., 2001, p. 362), Yunnan(J. C. Wei X. L. et al., 2007, p. 150, as *Cl. ceriosa*), Hubei(Chen, Wu & J. C. Wei, 1989, p. 464), Hongkong (Krphbr. 1877, p. 436; Vain. 1894, p. 49; Zahlbr., 1930 b, p. 133; Sandst. 1938, p. 89).

Cladonia carneola(Fr.) Fr. 肉石蕊

Lich. Eur. Reform. (Lund): 233(1831).

≡*Cenomyce carneola* Fr. *Sched. Crit. Lich. Suec.* 1 – 4: 23(1825).

On the ground.

Yunnan(Hue, 1898, p. 280; Zahlbr., 1930b, p. 134; Sandst. 1939, p. 97; Sato, 1952, p. 173).

Cladonia cenotea(Ach.) Schaer. 斜漏斗石蕊

Lich. helv. spicil. 1: 35(1823). f. cenotea 原变型

≡*Baeomyce cenoteus* Ach., *Methodus*: 345(1803).

On the ground and old wood.

Heilongjiang(Asahina, 1952d, p. 374; Chen et al. 1981, p. 130; J. C. Wei et al. 1982, p. 25), Jilin, Neimenggu (Chen et al. 1981, p. 130; J. C. Wei et al. 1982, p. 25), Xinjiang(Abdulla Abbas et al., 1993, p. 76; Abdulla Abbas & Wu, 1998, p. 53; Abdulla Abbas et al., 2001, p. 362), Shaanxi(Jatta, 1902, p. 465; Zahlbr., 1930b, p. 131; Sandst. 1938, p. 86), Sichuan(Stenroos et al., 1994, p. 326).

f. exaltata Nyl. 高漏斗变型

Zahlbr. *Cat. Lich. Univ.* Ⅳ: 459(1927) & in Handel-Mazzetti, *Symbolae Sinicae* 3: 131(1930).

On sandstone

Sichuan(Zahlbr., 1930b, p. 131; Sandst. 1938, p. 86).

Cladonia cervicornis(Ach.) Flot. 颈石蕊

Uebers. Arbeiten Veränd. Schles. Ges. Vaterl. Kultur 27: 31(1849), subsp. cervicornis. 原亚种

≡*Lichen cervicornis* Ach., *Lichenogr. Suec. Prodr.* : 184(1799) ["1798"].

≡*Cladonia verticillata* var. *cervicornis*(Ach.) Flörke, *De Cladonia's uit de sectie Cocciferae in België* (*morfologie, chemie, ecologie, sociologie, verspreiding en systematiek*) (Zahlbr. Cat. Lich. Ⅳ: 623)

On the ground.

Yunnan(Hue, 1887, p. 17; Zahlbr., 1930b, p. 134).

var. simplex Schaer. 单柄变种

Moreau, F. et Moreau, Mme. F. *Rev. Bryol. et Lichenol.* 20: 188(1951).

≡*Cladonia cervicornis scyphosa simplex* Schaer., *Enum. Lich. Eur.* 195(1850); Vainio. *Monographia Cladoniarum Universalis* II: 194(1894).

≡*Cladonia verticillata* var. *cervicornis* Flk. f. *simplex* Schaer.

On rocks.

Shandong(Moreau et Moreau, 1951, p. 188).

Cladonia chlorophaea(Flörke ex Sommerf.) Spreng. 喇叭粉石蕊

Syst. Veg., Edn 16 4(1):273(1827).

≡*Cenomyce chlorophaea* Flörke ex Sommerf., *Suppl. Fl. Lapp.*:130(1826).

≡*Cladonia pyxidata* subsp. *chlorophaea*(Flörke ex Sommerf.) V. Wirth, *Stittg. Beitr. Naturk.*, Ser. A(Biol.) 517:62(1994).

≡*Cladonia pyxidata* var. *chlorophaea* Flk. inVain. *Monographia Cladoniarum Universalis Acta Soc. Fauna Fl. Fenn.* II:232(1894).

= *Cladonia chlorophaea* f. *centralis*(Flot.) Asahina, *Lichens of Japan*, 1. *Genus Cladonia*: 196(1950).

≡*Cladonia chlorophaea* f. *centralis* Flot. *Lich. Fl. Siles* 30(1849) in Vain. *Monographia Cladoniarum Universalis. Acta Soc. Fauna Fl. Fenn.* II:223(1894).

= *Cladonia chlorophaea* f. *costata*(Flk.) Arn., *Lich. Fragm.* 29:7(1888) in Vain. *Monographia Cladoniarum Universalis. Acta Soc. Fauna Fl. Fenn.* II:239(1894).

≡*Cladonia pyxidata* f. *costata* Flk. Clad. Comm. 66(1828) in in Vain. *Monographia Cladoniarum Universalis. Acta Soc. Fauna Fl. Fenn.* II:238(1894).

= *Baeomyces fimbriatus* r. B. *syntheta* Ach. Method. Lich. 342(1803) in Vain. *Monographia Cladoniarum Universalis. Acta Soc. Fauna Fl. Fenn.* II:218(1894).

On the ground.

Heilongjiang(Hinggan Ling: Asahina, 1940, p. 726, & Asahina, 1952d, p. 374; Chen et al. 1981, p. 130; Luo, 1984, p. 85), Jilin, Liaoning(Chen et al. 1981, p. 130), Neimenggu(Sato, 1952, p. 173; Chen et al. 1981, p. 130; Sun LY et al., 2000, p. 35), Xinjiang(Wang XY, 1985, p. 343; Abdula Abbas et al., 1993, p. 76; Abdulla Abbas & Wu, 1998, p. 53; Abdulla Abbas et al., 2001, p. 362), Shaanxi(Jatta, 1902, p. 464; Zahlbr., 1930b, p. 134, as f. *scyphoso-prolifera* & f. *costata*; Moreau et Moreau, 1951, p. 188), Xizang(J. C. Wei & Y. M. Jiang, 1986, p. 91, as f. *chlorophaea*, f. *prolifera* & f. *simplex*), Sichuan(Stenroos et al., 1994, p. 326; Chen S. F. et al., 1997, p. 221), Guizhou(Oliv. 1904, p. 193), Shandong(Zhao ZT et al., 1998, p. 29; Zhang F et al., 1999, p. 31; Zhao ZT et al., 2002, p. 7; Hou YN et al., 2008, p. 67; Li Y et al., 2008, p. 72), AnhuiI, Zhejiang(Xu B. S. 1989, p. 238), Hubei(Müll. Arg. 1893, p. 235; Zahlbr., 1930b, p. 134; Chen, Wu & J. C. Wei, 1989, p. 465), Fujian(Zahlbr., 1930b, p. 134), Taiwan(Ahti & Lai, 1979, p. 229; Lai, 2000, p. 232), China(prov. not indicated: Sandst. 1939, p. 95; Asahina & Sato in Asahina, 1939, p. 689).

Cladonia chondrotypa Vain. 粒皮石蕊

Acta Soc. Fauna Flora fenn. 4(1): 449(1887).

Heilongjiang(Chen et al. 1981, p. 130), Taiwan(Asahina, 1941b, p. 136 & 1942, p. 677 & 1950b, p. 156; Wang & Lai, 1973, p. 89).

Cladonia ciliata Stirt. 缘毛石蕊

Scott. Natural., N. S. 3('9'):308(1888) var. ciliata.

var. tenuis(Flörke) Ahti 纤柄变种

Best. europ. Flecht., *Ergdnz.* 1(Vaduz):68(1977), & in *Biblioth. Lichenol.* 9:48(1977).

≡*Cladonia rangiferina* var. *tenuis* Flörke, De *Cladoniis difficillimo* lichenum genere commentatio nova, 164. 1828.—Lectotype(Ahti, 1961): Germany, Herb. H. G. Flörke no. 108(H).

≡*Cladonia tenuis*(Flörke) Harmand, *Lichens de France* III:228(1907), Paris.

≡*Cladina tenuis*(Flörke) Hale & W. Culb., *Bryologist* 73(3):510(1970).

≡*Cladina ciliata*(Stirt.) Trass var. *tenuis*(Flk.) Ahti & Lai, *Ann. Bot. Fennici* 16:235(1979).

= *Cladonia ciliata* f. *flavicans*(Flörke) Ahti & DePriest, *Mycotaxon* 78: 501(2001).

≡ *Cladonia rangiferina* f. *flavicans* Flörke, *De Cladoniis*, Difficillimo lichenum genere, Commentatio nova(Rostock) : 164(1828).

= *Cladina laxiuscula*(Delise) Sandst., in Zopf, *Die Flecht.* : 405(1907).

On the ground.

Jilin(Moreau et Moreau, 1951, p. 188).

= *Cladonia tenuiformis* Ahti, *Ann. Bot. Soc. Vanamo*, 32(1) : 63(1961). —Type: North Korea, 1934. 7. 28, Asahina 734(isotype in H!).

On the ground.

Heilongjiang, Liaoning(Chen et al. 1981, p. 134), Jilin(J. C. Wei et al. 1986, p. 244), Xizang(J. C. Wei & Jiang, 1986, p. 80; J. C. Wei et al. 1989, p. 244), Yunnan(Ahti, 1961 p. 57; J. C. Wei et al. 1989, p. 244), Anhui, Jiangxi(J. C. Wei et al. 1989, p. 244), Taiwan(Zahlbr., 1933 c, p. 46; Asahina, 1950, p. 64; Ahti, 1961, p. 64; Lamb, 1963, p. 179; Wang & Lai, 1973, p. 89 & 1976, p. 227; Ahti & Lai, 1979, p. 235; Liu et al. 1980, p. 15; J. C. Wei et al. 1989, p. 244; Lai, 2000, pp. 224 – 225).

This species, reported by Chen et al. (1981, p. 134) as *Cladonia tenuiformis* in Hulin county of Heilongjiang prov., is based on the specimen collected by Liou TN(no. 13400 in FPI) from Yunnan prov. (J. C. Wei et al. 1985, pp. 56, 58, 59).

Cladonia clavulifera Vain. 棒石蕊

in Robbins, *Rhodora* 26: 145(1924).

Shaanxi(He & Chen 1995, p. 42, the specimen and locality was not cited.), Xizang, Sichuan, Yunnan(Guo & J. C. Wei, 1994, P. 30).

Cladonia coccifera(L.) Willd. 红石蕊

Fl. berol. prodr. : 361(1787). f. coccifera 原变型

≡ *Lichen cocciferus* L., *Sp. pl.* 2: 1151(1753).

On the ground.

Heilongjiang, Jilin, Neimenggu(Chen et al. 1981, p. 130), Xinjiang(Abdulla Abbas & Wu, 1998, p. 54; Abdulla Abbas et al., 2001, p. 362), Shaanxi(Baroni, 1894, p. 48; Jatta, 1902, p. 466; Zahlbr., 1930b, p. 135; He & Chen 1995, p. 42 without citation of specimen and locality), Yunnan(J. C. Wei X. L. et al., 2007, p. 150), Hubei(Müll. Arg. 1893, p. 235; Zahlbr., 1930b, p. 135), China(prov. not indicated: Asahina & Sato in Asahina, 1939, p. 685).

f. phyllocoma(Flörke) Anders 鳞芽变型

≡ *Cenomyce coccifera* β *phyllocoma* Flörke see Vain. *Monographia Cladoniarum Universalis. Acta Soc. Fauna Fl. Fenn.* I: 155(1887).

On moss-covered rotten stump.

Xizang(J. C. Wei & Jiang, 1986, p. 84).

var. stemmatina(Ach.) Vain. 花冠变种

Acta Soc. Fauna Flora fenn. 4(1) : 158(1887).

≡ *Cenomyce coccifera* α *stemmatina* Ach., *Lich. univ.* : 537(1810).

Cladonia cornucopioides Hue(1889, p. 161 & 1898, p. 261), non alior.

On the ground.

Yunnan(Hue, 1889, p. 161 & 1898, p. 261; Zahlbr., 1930b, p. 135).

Cladonia coniocraea(Flörke) Spreng. 枪石蕊

Syst. veg., Edn 16 4(1) : 273(1827). f. coniocraea 原变型

≡ *Cenomyce coniocraea* Flörke, *Deutsche Lich.* 7: 11(1821).

≡ *Cladonia fimbriata* var. *coniocraea*(Flk.) Vain. (as "*phyllostrota*" in Mull. Arg. Lich. Chin. Henryan., p. 235) *Monographia Cladoniarum Universalis. Acta Soc. Fauna Fl. Fenn.* II: 311(1894).

On the ground.

Heilongjiang(Chen et al. 1981, p. 130; Luo, 1984, p. 85) , Jilin, Liaoning, Neimenggu(Chen et al. 1981, p. 130; Sun LY et al. , 2000, p. 36) , xinjiang(Wang XY, 1985, p. 343; Abdulla Abbas & Wu, 1998, p. 54) , Shaanxi (Jatta, 1902, p. 464; Zahlbr. , 1930b, p. 133) , guizhou(Zhang T & J. C. Wei, 2006, p. 5) , Hubei(Müll. Arg. 1893, p. 235; Vain. 1894, p. 311; Chen, Wu & J. C. Wei, 1989, p. 465) , Taiwan(Ahti & Lai, 1979, p. 230; Liu et al. 1980, p. 23; Lai, 2000, pp. 232 – 233) , China(prov. not indicated: Hue, 1898, p. 276; Asahina & Sato in Asahina. 1939, p. 689) .

f. phyllostrota(Flörke) Vain. 鳞芽变型

Acta Soc. Fauna Flora fenn. 53(1) : 113(1922) .

≡ *Cladonia ochrochlora* β *phyllostrota* Flörke, Clad. Comm. 79(1828) , *Monographia Cladoniarum Universalis. Acta Soc. Fauna Fl. Fenn.* II: 315(1894) .

≡ *Cladonia fimbriata* var. *coniocraea* f. *phyllostrota* (Flk.) Vain. *Monographia Cladoniarum Universalis. Acta Soc. Fauna Fl. Fenn.* II: 315(1894) .

On the ground.

Hubei(Müll. Arg. 1893, p. 235; Zahlbr. , 1930b, p. 133) , Xizang(J. C. Wei & Jiang, 1986, p. 92) , Zhejiang(Xu B. S. 1989, p. 240) .

f. truncata(Flörke) Dalla Torre & Sarnth. 截顶变型

≡ *Cladonia ochrochlora* b *truncata* Flörke, Clad. Comm. 77(1828) , Vain. *Monographia Cladoniarum Universalis. Acta Soc. Fauna Fl. Fenn.* II: 314(1894) .

≡ *Cladonia pyxidata* var. *ochrochlora* f. *truncata*(Flk.) Vain. *Monographia Cladoniarum Universalis. Acta Soc. Fauna Fl. Fenn.* II: 315(1894) .

Xinjiang(H. Magn. 1940, p. 69; Ahti, 1976, p. 369: *Cladonia coniocraea*; Abdulla Abbas et al. , 1993, p. 77; Abdulla Abbas & Wu, 1998, p. 55; Abdulla Abbas et al. , 2001, p. 362) , Xizang(J. C. Wei & Y. M. Jiang, 1986, p. 93) , Yunnan(Hue, 1898, p. 277; Zahlbr. , 1930b, p. 133) .

Cladonia corniculata Ahti & Kashiw. 钝角石蕊

in Inoue Stud. Cryptog. S. Chile: 136(1984) .

Sichuan(Stenroos et al. , 1994, p. 327) .

Cladonia cornuta(L.) Hoffm. 角石蕊

Descr. Adumb. Plant. Lich. 2: tab. 25(1794) . f. cornuta 原变型

≡ *Lichen cornutus* L. , *Sp. pl.* 2: 1152(1753) .

On the ground.

Heilongjiang(Asahina, 1952d, p. 374; Chen et al. 1981, p. 131; Luo, 1984, p. 85) , Xinjiang(Abdulla Abbas & Wu, 1998, p. 55; Abdulla Abbas et al. , 2001, p. 362) , Shaanxi(Jatta, 1902, p. 465; Zahlbr. , 1930b, p. 132) . Sichuan(Chen S. F. et al. , 1997, p. 221) , Anhui(Xu B. S. , 1989, p. 237) .

f. phyllotoca(Flörke) Arn. 尖梢鳞芽变型

Lich. Jur. 27(1884) , in Vain. *Monographia Cladoniarum Universalis. Acta Soc. Fauna Fl. Fenn.* II: 133(1894) .

≡ *Cladonia coniocraea* [unranked] *phyllotoca* Flörke, *Clad. Comm.* 87(1828) .

On the ground.

Xizang(J. C. Wei & Chen, 1974, p. 176; J. C. Wei & Jiang, 1986, p. 9) .

Cladonia corymbescens(Nyl.) Nyl. 细枝石蕊

in Leighton, *Ann. Mag. nat. Hist.* , Ser. 3 18: 407(1866) .

≡ *Cladonia degenerans* var. *corymbescens* Nyl. , *Annls Sci. Nat.* , *Bot.* , sér. 4 15: 40(1861) .

On soil-covered rock.

Shaanxi(He & Chen 1995, p. 42 without citation of specimens and their localities) , Xizang(J. C. Wei & Jiang,

1986, p. 87), Yunnan(Wei X. L. et al., 2007, p. 150).

Cladonia crispata(Ach.) Flot. 穿杯石蕊

Flechten Hirschberg-Warmbrunn: 4(1839). f. crispata 原变型

≡*Baeomyces turbinatus* var. *crispatus* Ach., *Methodus*: 341(1803).

On the ground & rotten logs.

Heilongjiang, Jilin(Chen et al. 1981, p. 131), Yunnan(Hue, 1898, p. 268; Zahlbr., 1930 b, p. 131), China (prov. not indicated: Sandst. 1938, p. 87, may probably be from Yunnan), Taiwan(Zahlbr., 1933 c, p. 46).

f. cetrariiformis(Delise) Vain. 岛衣变型

in *Acta Soc. Fauna Fl. Fenn.* 4: 392(1887). – J. W. Thomson [as '*cetrariaeformis*'], *Lich. Gen. Cladonia North America*(Toronto): 131(1968)[1967].

≡*Cenomyce gracilis* var. *cetrariiformis* Delise in Duby, *Bot. Gall.*: 625(1830).

Misapplied name:

Cladonia delessertii sensu Hue, non(Del.) Vain.

On the ground.

Yunnan(Hue, 1898, p. 268; Zahlbr., 1930b, p. 131; Sandst. 1938, p. 86), Taiwan(Wang-Yang & Lai, 1973, p. 89, as *C. crispata*; Ahti & Lai, 1979, p. 230; Liu et al. 1980, p. 19; Lai, 2000, p. 228).

f. infundibulifera(Schaer.) J. W. Thomson 漏斗变型

Lich. Gen. Cladonia North America(Toronto): 131(1968)[1967].

≡*Cladonia ceranoides* tax. vag. *infundibulifera* Schaer.

Yunnan(Hue, 1898, p. 268; Zahlbr., 1930b, p. 131), Jiangxi, Fujian(Zahlbr., 1930 b, p. 131).

f. divulsa(Delise) Vain. 撕裂变型

Monographia Cladoniarum Universalis. Acta Soc. Fauna Fl. Fenn. II: 454(1894).

≡*Cenomyce divulsa* Delise

On rocks.

Anhui(Xu B. S., 1989, p. 236).

Cladonia cryptochlorophaea Asahina 隐喇叭粉石蕊

J. Jap. Bot. 16: 711(1940).

On stonewall.

Anhui(Xu B. S. 1989, p. 239).

Cladonia cylindrica(A. Evans) A. Evans, 圆筒石蕊

Rhodora 52: 116(1950)

≡*Cladonia borbonica* f. *cylindrica* A. Evans 1930

≡*Cladonia borbonica* f. *cylindrica* A. Evans, *Trans. Conn. Acad. Arts & Sci* 30: 482(1930).

Yunnan(Guo & J. C. Wei, 1994, P. 31).

Cladonia cyanipes(Sommerf.) Nyl. 黄粉石蕊

in Mém. Soc. Sci. Nat. Cherbourg 5: 95(1858). (1858)[1857].

≡*Cenomyce cyanipes* Sommerf. in *Kongel. Norske Videnskabersselsk. Skr.* 19 *de Aarhundr*. 2: 62(1826).

On rotten logs.

Heilongjiang(Chen et al. 1981, p. 131), Shaanxi, Sichuan(J. C. Wei et al. 1982, p. 26), Hubei(Chen, Wu & J. C. Wei, 1989, p. 465).

Cladonia dactylota Tuck. 枝石蕊

Amer. J. Sci. Arts, Ser. 2 28: 200 – 206(1859).

On the ground.

Taiwan(Zahlbr., 1933c, p. 47; Sandst. 1939, p. 94; Wang & Lai, 1973, p. 89).

Cladonia decorticata(Flörke) Spreng. 脱皮石蕊

Syst. veg., Edn 16 4(1): 271(1827).

≡*Capitularia decorticata* Flörke in *Beitr. Naturk.* 2: 297(1810).

On the ground.

Shaanxi(Jatta, 1902, p. 464; Zahlbr., 1930 b, p. 132; Sandst. 1938, p. 89).

Cladonia deformis(L.) Hoffm. 正硫石蕊

Deutschl. Fl., Zweiter Theil(Erlangen): 120(1796)[1795].

≡*Lichen deformis* L., *Sp. pl.* 2: 1152(1753).

On the ground & rotten logs.

Neimenggu, Heilongjiang(Chen et al. 1981, p. 131).

Cladonia dehiscens Vain. 裂石蕊

in Hue, *Nouv. Arch. Mus. Hist. Nat.*, Paris, 3 sér. 10: 271(1898).

Guangxi(according to the distribution map 79 in Sandst. 1939; Guo 1999, p. 341, the specimen and locality was not cited.); Hainan(Guo 1999, p. 341, the specimen and locality was not cited.).

Cladonia delavayi Abbayes 戴氏石蕊

Candollea 16: 203(1958).

Type: Yunnan, R. P. Delavay no. 1558 in PC.

Misapplied name:

Cladonia sylvatica auct. non(L.) Hoffm.: Hue, *Bull. Soc. Bot. France* 34: 17(1887).

Cladonia medusina var. *luteola* auct. non(Bory) Vain.: Hue, *Nouv. Arch. du Mus.* 3(10): 263(1898).

Cladonia impexa f. *laxiuscula* auct. non(Del.) Sandst.: Zahlbr., in Handel-Mazzetti, *Symb. Sin.* 3: 130(1930).

Cladonia tenuis f. *flavicans* auct. non(Flk.) Sandst.: Zahlbr., *Hedwigia* 74: 201(1934 – 1935). (Delavay no. 1558)

Saxicolous & terricolous.

Shaanxi(He & Chen 1995, p. 42), Xizang(J. C. Wei & Y. M. Jiang, 1986, p. 86), Sichuan(Zahlbr., 1930b, p. 130, Handel-Mazzetti no. 2663; Ahti, 1962, p. 256; Stenroos et al., 1994, p. 327), Yunnan(Hue, 1887, p. 17 & 1898, p. 263; Zahlbr., 1930b, p. 130; Sandst. 1932, p. 71; Zahlbr., 1934, p. 201, reported as Cladonia tenuis f. flavicans, on the basis of specimen collected by Rock no. 18518; Ahti, 1962, p. 256).

Cladonia didyma(Fée) Vain. 小红石蕊

Acta Soc. Fauna Flora fenn. 4(1): 137(1887).

≡*Scyphophorus didymus* Fée, *Essai Crypt. Exot.* (Paris): XCVIII(1825)[1824].

Cladonia vulcanica Zollinger & Moritzi(1844). (Wang-Yang & Lai, 1973, p. 90; Lai, 2000, pp. 243 – 244).

= *Cladonia didyma* var. *vulcanica*(Zoll. & Moritzi) Vain., *Acta Soc. Fauna Flora fenn.* 4(1): 145(1887), (Wang-Yang & Lai, 1973, p. 89).

≡*Cladonia vulcanica* Zoll. & Moritzi, in Hasskarl, *Natur Geneerk. Arch. Neerl. Indie* 1: 396(1847).

On the ground.

Taiwan(Zahlbr., 1933c, p. 46; Asahina, 1950b, p. 88 & 1970a, p. 101; Yoshim. 1968, p. 236; Wang-Yang & Lai, 1973, pp. 89, 90; Ahti & Lai, 1979, p. 233; Lai, 2000, pp. 243 – 244).

Cladonia digitata(L.) Hoffm. 胀石蕊

Deutschl. Fl., Zweiter Theil(Erlangen): 124(1796)[1795]. f. digitata 原变型

≡*Lichen digitatus* L., *Sp. pl.* 2: 1152(1753).

On the ground & rotten logs.

Heilongjiang, Jilin(Chen et al. 1981, p. 131), Shandong(Li Y et al., 2008, p. 72).

f. glabrata Delise 光变型

See Vainio, *Monographia Cladoniarum Universalis. Acta Soc. Fauna Fl. Fenn.* I: 133(1887).

Heilongjiang(Asahina, 1952 d, p. 373).

Cladonia diversa Asperges ex S. Stenroos 狭杯红石蕊

in Ahti & Stenroos, Botanica Complutensis 36: 32(2012).

≡ *Cladonia diversa* Asperges, De *Cladonia*'s uit de sectie Cocciferae in België(morfologie, chemie, ecologie, sociologie, verspreiding en systematiek) 2. Ph. D. Thesis(Wilrijk): 358(1983).

On rotten wood.

Xinjiang(Abbas et al., 2000. p. 18), Neimenggu, Yunnan, Guangxi, Hunan, (Guo 1999, p. 341).

Cladonia ecmocyna(Ach.) Nyl. 长石蕊

Lich. Lapp. Or. 176(1866), Vain., *Acta Soc. Fauna Flora fenn.* 10: 125(1894)..

≡ *Cenomyce ecmocyna* Ach. in Vet. *Acad. Nya Handl.* 299 pro. p. (1810), (excl. δ.), and *Lich. Univ.*: 549(1810), in Vain. *Acta Soc. Fauna Flora fenn.* 10: 82(1894).

≡ *Cladonia ecmocyna* Leight. Not. Lichenol. XI: 406(1866), XII: 108(1867), Vain., *Acta Soc. Fauna Flora fenn.* 10: 125(1894), in Index Fungorum: *Ann. Mag. nat. Hist.*, Ser. 3 18: 406(1866).

≡ *Cladonia gracilis* var. *ecmocyna*(Ach.) Vain., *Acta Soc. Fauna Flora fenn.* 10: 125(1894).

Xinjiang(Abdulla Abbas & Wu, 1998, p. 55 & Abdulla Abbas et al., 2001, p. 362 as *Cladonia* "*ecmocyma*").

Cladonia erythrosperma Vain. 红心石蕊

Acta Soc. Fauna Flora fenn. 4(1): 374[376](1887). var. thomsoni Vain. 汤姆氏变种

Hainan(des Abb. 1956, p. 264).

Cladonia farinacea(Vain.) A. Evans 粉芽石蕊

Rhodora 52: 95(1950).

≡ *Cladonia furcata* var. *farinacea* Vain., in Hariot, *J. Bot.*, Paris **1:** 283(1887).

Sichuan(Stenroos et al., 1994, p. 328), Yunnan(Guo & J. C. Wei, 1994, P. 31).

Cladonia fenestralis Nuno 繁鳞石蕊

J. Jap. Bot. 50(10): 291(1975).

= *Cladonia squamossima*(Müll. Arg.) Ahti, i*Ann. Bot. Fenn.*, 17: 233(1980).

≡ *Cladonia gracilis* var. *squamosissima* Müll. Arg. *Flora*, 74: 372(1891).

Type: Hubei, A. Henry, no. 6959 in BM.

= *Cladonia ceratophyllina*(Nyl.) Vain., in Hue, *Nouv. Arch. Mus. Hist. Nat.*, Paris, 3 sér. 10: 273(1898).

≡ *Cladonia degenerans* f. *ceratophyllina* Nyl., *Syn. Lich.* 200(1858 – 60), Vain. *Monogr Clad.* II: 164(1894).

= *Cladonia rangiformis* var. *incurva* Müll. Arg. *Lich. Beitr.* n. 1611(1891), Vain. *Monogr Clad.* II: 452(1894).

= *Cladonia ceratophyllina* f. *subnuda* Asahina, *Lichens of Japan*, 1. *Genus Cladonia:* 184(1950).

Misapplied name:

Cladonia alinii auct. non Trass: Ahti & Lai, *Ann. Bot. Fennici*, 16: 229(1979).

Xinjiang(Abbas et al., 2000, p. 18), Shaanxi(Ahti, 1980, p. 233; He & Chen 1995, p. 42, as *Cl. squamossima*, without citation of specimens and their localities), Xizang(Ahti, 1980, p. 233; J. C. Wei & Jiang, 1986, p. 89), Sichuan(Ahti, 1980, p. 233; Stenroos et al., 1994, p. 328), Yunnan(Ahti, 1980, p. 233), Guizhou(Ahti, 1980, p. 233; Zhang T & J. C. Wei, 2006, p. 5), Anhui(Xu B. S. 1989, p. 237), Hubei(Müll. Arg., 1891, p. 372 & 1893, p. 235; Vain. 1894, p. 113; Zahlbr., 1930b, p. 133; Ahti, 1980, p. 233; Chen, Wu & J. C. Wei, 1989, p. 470), Guangxi(Guo 1999, p. 341, the specimen and locality was not cited.), Taiwan(Wang & Lai, 1976, p. 227; Ahti & Lai, 1979, p. 229; Ahti, 1980, p. 233; Liu et al. 1980, p. 27; Lai, 2000, p. 237).

Cladonia fimbriata(L.) Fr. 粉石蕊

Lich. Eur. Reform. (Lund) : 222(1831) . f. fimbriata 原变型

≡*Lichen fimbriatus* L. , *Sp. pl.* 2: 1152(1753) .

= *Cladonia fimbriata* var. *simplex* f. *minor*(Hag.) Vain. Monogr. II: 258(1894) .

≡*Lichen pyxidatus* b. *minor* Hag. , *Tent. Hist. Lich.* 113(1782) , Vain. *Monogr.* II: 262(1894) .

= *Cladonia major* (K. G. Hagen) Sandst. in *Abhandl. naturw. Verein. Bremen.* 25: 223(1921) .

≡*Lichen pyxidatus* b. *major* Hag. , *Tent. Hist. Lich.* 113(1782) , Vain. *Monogr.* II: 262(1894) .

Misapplied name:

Cladonia chlorophaea var. *pachphyllina* auct. non(Wallr.) Vain. : H. Magn. *Lich. Centr. Asia,* 1: 69(1940) .

On the ground and rotten logs.

Heilongjiang, Jilin(Chen et al. 1982, p. 132) , Hubei(Chen, Wu & J. C. Wei, 1989, p. 467) .

On the ground, rotten logs, & bases of tree trunk.

Heilongjiang(Asahina, 1952d, p. 374) , Xinjiang(H. Magn. 1940, p. 69; Ahti, 1976, p. 369; Abdulla Abbas et al. , 1993, p. 77; Abdulla Abbas & Wu, 1998, p. 55; Abdulla Abbas et al. , 2001, p. 362) , Shaanxi(Jatta, 1902, p. 464; Zahlbr. , 1930b, p. 133) , Yunnan(Hue, 1889, p. 26; Vain. 1894, p. 252) , Hubei(Müll. Arg. 1893, p. 235; Vain. 1894, p. 252; Chen, Wu & J. C. Wei, 1989, p. 466) .

var. cornuta Flörke 角形变种

Deutsch. Lich. III: 7, no. 50(1815) , Vain. *Monogr.* II: 286(1894) .

Xinjiang, Hebei(Tchou, 1935, p. 303) .

var. prolifera(Retz.) A. Massal. 重生变种

in Vainio, *Monographia Cladoniarum Universalis. Acta Soc. Fauna Fl. Fenn.* II: 270(1894) .

≡*Lichen fimbriatus* β *prolifer* Retz. , *Fl. Scand.* 232(1779) .

On the ground & rotten wood.

Shaanxi(Jatta, 1902, p. 464; Zahlbr. , 1930b, p. 133) , Yunnan(Hue, 1889, p. 159; Zahlbr. , 1930b, p. 133) , Guangxi(Guo 1999, p. 341, the specimen and locality was not cited.) , China(prov. not indicated: Hue, 1898, p. 276) .

Cladonia firma(Nyl.) Nyl. 坚石蕊

Bot. Ztg. 47: 352(1861) .

≡*Cladonia alcicornis* var. *firma* Nyl. , *Syn Meth. Lich.* 1: 191(1860) , Jatta, 1902, p. 464.

≡*Cladonia foliacea* var. *firma* (Nyl.) Vain. , *Acta Soc. Fauna Flora fenn.* 10: 400(1894) .

On the ground.

Shaanxi(Giraldi after Jatta, 1902, p. 464; Zahlbr. , 1930b, p. 134; Sandst. 1939, p. 96) , China(prov. not indicated: Sandst. 1939, p. 96) .

Cladonia floerkeana(Fr.) Flörke 红头石蕊

De Cladonia's uit de sectie Cocciferae in België(morfologie, chemie, ecologie, sociologie, verspreiding en systematiek) 2. Ph. D. Thesis(Wilrijk) : 99(1828) . f. floerkeana 原变型

≡*Cenomyce floerkeana* Fr. , Sched. Crit. Lich. Suec. 1 – 4: 18(1825) .

On the ground & rotten logs.

Heilongjiang, Jilin(J. C. Wei et al. 1982, p. 22) , Hubei(Chen, Wu & J. C. Wei, 1989, p. 462) , Guangxi(Guo 1999, p. 341, the specimen and locality was not cited.) , Taiwan(Ahti & Lai, 1979, p. 231; Liu et al. 1980, p. 31; Lai, 2000, p. 242) .

= *Cladonia floerkeana* f. *carcata*(Ach.) J. W. Thomson *Lich. Gen. Cladonia North America*(Toronto) : 71(1968) [1967] .

≡*Cenomyce carcata* Ach. see Vainio, *Monographia Cladoniarum Universalis. Acta Soc. Fauna Fl. Fenn.* I: 80 & II: 441(1894) .

Heilongjiang(Asahina, 1952d, p. 373) , Hubei(Müll. Arg. 1893, p. 235; Zahlbr. , 1930b, p. 135) .

f. intermedia(Hepp ex Vain.) J. W. Thomson 粉芽变型

Lich. Gen. Cladonia North America(Toronto) : 71(1968) [1967] .

≡*Cladonia floerkeana* var. *intermedia* Hepp ex Vain. , *Flecht. Europ.* : no. 222(1857) .

Misapplied name:

Cladonia bacillaris auct. non(Ach.) Nyl. : Hue, *Bull. Soc. Bot. France*, 34: 17(1887) .

Heilongjiang(Asahina, 1952 d, p. 373) , Yunnan(Hue, 1887a, p. 17 & 1889, p. 161 & 1898, p. 259; Zahlbr. , 1930 b, p. 135) , Taiwan(Asahina, 1950b, p. 81) .

var. suboceanica Asahina 亚海洋变种

J. J. Bot. XV: 663(1939) .

Taiwan(Wang & Lai, 1973, p. 89) .

Cladonia foliacea(Huds.) Willd. 叶石蕊

Fl. berol. prodr. : 363(1787) .

≡*Lichen foliaceus* Huds. , *Fl. Angl.* ed. 1: 457, n. 62(1762) .

var. alcicornis(Lightf.) Schaer. 茸角变种

Lich. helv. spicil. 1: 249(1823) .

≡*Cladonia alcicornis* Flk. ; F. Wils in papers and proceed. Roy. Soc. Tasmania for 1892, 1893, p. 146 in Zahlbr. *Cat. Lich. Univ.* 4: 521(1926 – 1927) .

Guizhou(Zahlbr. , 1930b, p. 134) .

Cladonia fruticulosa Kremp. 果石蕊

Verh. zool. -bot. Ges. Wien 30: 331(1880) .

Guangxi(Guo 1999, p. 341, the specimen and locality was not cited.) , Taiwan(Lai, 2000, pp. 238 – 239; Seaward & Aptroot, 2005, p. 283) , Hainan(JCWei et al. , 2013, p. 230) , Hongkong(Seaward & Aptroot, 2005, p. 283) .

Cladonia furcata(Huds.) Schrad. 分枝石蕊

Spicil. fl. germ. 1: 107(1794) . f. furcata 原变型

≡*Lichen furcatus* Huds. , *Fl. Angl.* : 458(1762) .

On the ground.

Heilongjiang(Chen et al. 1981, p. 131; Luo, 1984, p. 85) , Jilin, Liaoning, Neimenggu(Chen et al. 1981, p. 131) , xinjiang(Wang XY, 1985, p. 340; Abdulla Abbas & Wu, 1998, p. 56; Abdulla Abbas et al. , 2001, p. 362)) , Shaanxi(He & Chen 1995, p. 42) , Xizang(J. C. Wei & Chen, 1974, p. 176; J. C. Wei & Y. M. Jiang, 1981, p. 1147) , Yunnan(Hue, 1889, p. 160 <p. p. > & 1898, p. 265) , Taiwan(Ahti & Lai, 1979, p. 231; Liu et al. 1980, p. 19; Lai, 2000, p. 228) .

= *Cladonia scabriuscula* f. *cancellata*(Müll. Arg.) Sandst. , in Rabenhorst, 220(1931) .

Xinjiang(Wang XY, 1985, p. 341; Abudula Abbas et al. , 1993, p. 77) .

= *Cladonia furcata* var. *racemosa* (Hoffm.) Flörke, *De Cladonia's uit de sectie Cocciferae* in België(morfologie, chemie, ecologie, sociologie, verspreiding en systematiek) 2. Ph. D. Thesis(Wilrijk) : 152(1828) .

≡*Cladonia racemosa* Hoffm. , *Deutschl. Fl.* , *Zweiter Theil*(Erlangen) : 114(1796) [1795] .

On the ground.

Heilongjiang(Asahina, 1952d, p. 373) , Neimenggu(Sato, 1952, p. 174) , Shaanxi(Baroni, 1894, p. 48; Zahlbr. , 1930 b, p. 131) , Yunnan(Hue, 1887a, p. 17 & 1889, p. 160; Zahlbr. , 1930b, p. 131) , Hubei(Chen, Wu, J. C. Wei, 1989, p. 466) .

f. fissa(Flörke) Aigret 缝裂变型

Bull. Soc. R. Bot. Belg. 40: 109(1901) .

≡*Cladonia furcata* d *fissa* Flk. in Vainio, *Monographia Cladoniarum Universalis. Acta Soc. Fauna Fl. Fenn.* I: 329(1887) .

Guizhou(Oliv. 1904, p. 195) .

f. squamulifera Sandst. 裂芽变型

in Xu B. S. ed. 1989 Cryptogamic flora of the Yangtze Delta and adjecent regions) , p. 235.

On rocks.

Anhui(Xu B. S. , 1989, p. 235) .

var. corymbosa(Ach.) Nyl. 伞形变种

Lich. Envir. Paris: 30(1896) . Vainio, *Monogr.* , vol. I: 328(1887) & vol. III: 239(1897) .

≡*Cenomyce allotropa* n. C. *corymbosa* Ach. , see Vainio, *Monogr.* , vol. I: 328(1887) .

On the ground in forests.

Yunnan(Hue, 1889, p. 160; Zahlbr. , 1930 b, p. 131) .

= *Cladonia furcata* var. *palamaea*(Ach.) Nyl. *Acta Soc. Fauna Flora fenn.* 4(1) : 347(1887) .

≡*Baeomyces spinosus* var. *palamaeus* Ach. , *Methodus,* Sectio post. : 359(1803) .

Yunnan(Hue, 1898, p. 266; Zahlbr. , 1930 b, p. 132) .

= *Cladonia furcata* var. *pinnata*(Flörke) Vain. *Acta Soc. Fauna Flora fenn.* 4(1) : 332(1887) .

≡*Cenomyce racemosa* var. *pinnata* Flörke, in Schleicher, *Cat. pl. Helv.* : 47(1821) .

Yunnan(Hue, 1898, p. 265; Zahlbr. , 1930b. p. 131) , Jiangxi(Zahlbr. , 1930b, p. 131) , Taiwan(Sasaoka, 1919, p. 180; Zahlbr. , 1933c, p. 46; Ras. 1940, p. 149; Wang & Lai, 1973, p. 89) .

f. foliolosa(Duby) Vain. 直柄变型

Acta Soc. Fauna Flora fenn. 4(1) : 333(1887) .

≡*Cenomyce racemosa* d. *foliolosa* Del. in Duby

Xinjiang(Abdulla Abbas & Wu, 1998, p. 56) , Sichuan(Zahlbr. , 1930 b, p. 131) , Yunnan(Hue, 1898, P. 256; Zahlbr. , 1930 b, p. 131) , Taiwan(Asahina, 1950 b, p. 128) .

f. recurva(Hoffm.) Sandst. 曲枝变型

Expo. Syst. Descr. Lich. Ouest Fr. 1: 68(1897) .

≡*Cladonia furcata* C. *recurva* Hoffm. , Deutchl. Fl. II: 115(1796) , Vainio, *Monogr.* , II: 312(1894) .

On moss-covered rock.

Xizang(J. C. Wei & Jiang, 1986, p. 87) .

f. regalis(Flörke) H. Olivier 冠变型

Expo. Syst. Descr. Lich. Ouest Fr. 1: 68(1897) .

≡*Cladonia furcata* var. *pinnata* f. *regalis*(Flk.) Oliv. , Fide Asahina *J. J. B.* 18: 665(1942) .

= *Cladonia macroptera* f. *ramosa* Asahina, *J. J. B.* 18: 665(1942) . *not.*

= *Cladonia macroptera* f. *subnuda* Asahina, *J. J. B.* 18: 665(1942) . *not.*

in Lamb I. M. Index Nominum Lichenum 181*(1963)*.

≡*Cladonia furcata* var. *racemosa* b. *regalis* Flk. see Vainio, *Monogr.* , vol. I: 335(1887) .

Heilongjiang(Asahina, 1952d, p. 373) , Xinjiang(Moreau et Moreau, 1951, p. 188) , Shaanxi(Jatta, 1902, p. 465; Zahlbr. , 1930 b, p. 131) , Sichuan(Zahlbr. , 1930b, p. 131) , Taiwan(Asahina, 1950b, p. 129) .

f. truncata Vain. 截顶变型

Acta Soc. Fauna Flora fenn. 4(1) : 333(1887) .

Yunnan(Hue, 1898, p. 266; Zahlbr. , 1930b, p. 132) , Anhui, Zhejiang(Xu B. S. , 1989, p. 235) .

f. turgida Scriba ex Sandst. 肿枝变型

On rocks.

Anhui(Xu B. S. , 1989, p. 235) .

var. rigidula A. Massal 硬枝变种

in Zahlbr. *Cat. Lich. Univ.* 4: 533(1927) .

= *Cladonia racemosa* var. *rigidula* Massal.

Shaanxi(Jatta, 1902, p. 465; Zahlbr. , 1930b, p. 132) , Xizang(Wei & Jiang, 1986, p. 86) , Yunnan(Hue, 1898,

p. 266; Zahlbr. , 1930b, p. 132) , Hebei(Moreau et Moreau, 1951, p. 188) , Anhui(Xu B. S. , 1989, p. 235) .

Subsp. subrangiformis(L. Scriba ex Sandst.) Pišút, 类鹿角亚种

Acta Facultatis Rerum Naturalium Universitatis Comenianae, Botanica 6: 525(1961) .

≡*Cladonia furcata* var. *subrangiformis*(L. Scriba ex Sandst.) Hennipman, *Persoonia* 4(4) : 427(1967) .

≡*Cladonia subrangiformis* L. Scriba ex Sandst. , in Abhandl. naturw. Verein. Bremen, 25: 165(1922) [1924] .

On rotten wood, rocks, ground, soil and mosses over rocks.

Xizang, Yunnan(Guo & J. C. Wei, 1994, P. 34) .

Cladonia glauca Flörke 白粉石蕊

Clad. Comm. : 140(1828) . f. glauca 原变型

On *Pseudotsuga wilsonianan*.

Shaanxi(He & Chen 1995, p. 42 without citation of specimens and their localiyies) , Taiwan(Aptroot, Sparrius & Lai 2002, p. 282) .

f. subacuta Asahina 尖头变型

Lichens of Japan, 1. *Genus Cladonia*: 161(1950) .

Heilongjiang, Jilin, Neimenggu(Chen et al. 1981, p. 131) .

Cladonia gonggaensis S. Y. Guo & J. C. Wei 贡嘎石蕊

Mycosystema 7: 31(1995) [1994]

Type: China, Sichuan, Mt. Gongga, Lat. 29°37′, Long. 101°53′, alt. 3750 m, on the ground, VII 30, 1982, Wang XY, Xiao X & Li B. No. 9189 (holotype: HMAS – L084006) .

= *Cladonia monomorpha* Aptroot, Sipman & van Herk, Lichenologist 33(4) : 273 – 285(2001) .

On soil and rotten wood.

Sichuan, Yunnan (S. Y. Guo & J. C. Wei, 1994, p. 31) .

Cladonia gracilis(L.) Willd. 细石蕊

Fl. Berol. Prodr. : 363(1787) .

≡*Lichen gracilis* L. , *Sp. pl.* 2: 1152(1753) . var. gracilis 原变种

On the ground and rotten logs in forest.

Heilongjiang, Jilin, Neimenggu(Chen et al. 1981, p. 132) , Xinjiang(Abdulla Abbas & Wu, 1998, p. 56; Abdulla Abbas et al. , 2001, p. 362) , Shaanxi(He & Chen 1995, p. 42without citation of specimens and their localities) , Yunnan(Wei X. L. et al. , 2007, p. 150) , Hubei(Vain. 1894, p. 86) , Anhui, Zhejiang(Xu B. S. , 1989, p. 237) .

var. chordalis(Flörke) Schaer. 绿色变种

Lich. helv. spicil. 1: 32(1823) .

≡*Capitularia gracilis*βcap. (*chordalis*) Floerk. , *Beschr. Braunfr. Becherfl.* 324 pr. p. (1810) .

= *Cladonia gracilis* var. *aspera* Flörke, in Weber & Mohr, *Beitr. Naturk.* 2: 260(1810) .

Jilin(Moreau et Moreau, 1951, p. 188; Asahina, 1950b, p. 176, cited as var. chordalis f. *foliolosa* Sandst.) , Neimenggu(Sato, 1952, p. 174) , xinjiang(Wang XY, 1985, pp. 341&342 as var. *aspersa* & var. *chodalis*) , Sichuan (Zahlbr. , 1930b, p. 132) , Yunnan(Hue, 1898, p. 272; Zahlbr. , 1930b, p. 132) , Zhejiang(Moreau et Moreau, 1951, p. 188) .

subsp. nigripes(Nyl.) Ahti 侧裂亚种

Ann. bot. fenn. 17(2) : 204(1980) .

≡*Cladonia ecmocyna* f. *nigripes* Nyl. in *Norrl. öfvers. Torn. Lapp.* 319(1873) .

Jilin(Chen et al. 1981, p. 132) .

subsp. turbinata(Ach.) Ahti 陀螺亚种

Ann. bot. fenn. 17(2) : 212(1980) .

≡*Lichen turbinatus* Ach. , *Lich. Suec. Prodr.* 192(1798) .

≡*Cladonia elongata* f. *turbinata*(Ach.) Choisy in *Bull. Mens. Soc. Linn. Lyon*, 20: 17(1951) .

= *Cladonia dilatata* Hoffm. , *Deutschl. Fl.* , Zweiter Theil(Erlangen) : 126(1796) [1795] .

Heilongjiang(Asahina, 1952d, p. 374; J. C. Wei, 1981, p. 88; Ahti, 1980, p. 212, as in "Manchuria"; Ahti, 1991, p. 60) , Xinjiang(Wang XY, 1985, p. 342 as var. *dilatata*) .

Cladonia granulans Vain. 粒红石蕊

Bot. Mag. , Tokyo 35: 65(1921) . f. granulans 原变型

On the ground.

Neimenggu(Chen et al. 1981, p. 132) .

f. prolifera Sandst. 重生变型

in Evans, *Rhodora*, 40: 19(1938) , Thomson, *Lich. Gen. Clad. N. Amer.* 108(1967) .

On the base of tree-trunk & on the ground.

Xizang(J. C. Wei & Jiang, 1986, p. 92) .

Cladonia grisea (Ahti) J. C. Wei, comb. nov. 珍珠灰石蕊

FN570459

≡*Cladonia rangiferina* ssp. *grisea* Ahti, *Ann. Bot. Soc.* "Vanamo" 32(1) : 96(1961) . Holotype: Japan, Pref. Yamanashi, Lake Yamanake at foot of Mt. Fuji, Aug. 16, 1952, Y. Asahina s. n. (TNS; isotype in H!) .

≡*Cladina rangiferina* ssp. *grisea*(Ahti) Ahti & Lai, *Ann. Bot. Fenn.* 16: 234(1979) .

≡*Cladina grisea*(Ahti) Trass *Opredelitel' Lishaĭnikov SSSR*(Handbook of the Lichens of the USSR) , 5, Cladoniaceae-Acarosporaceae(Leningrad) : 76(1978) . J. C. Wei & Y. M. Jiang in *Acta Mycologica Sinica* 5(4) : 247 (1986) . Lai, 2000, 225 – 226.

On the ground.

Heilongjiang, Jilin, Neimenggu, Shaanxi, Xizang, Yunnan, Jiangxi(J. C. Wei et al. 1986, pp. 247, 248) , Xinjiang (Abdulla Abbas & Wu, 1998, p. 50; Abdulla Abbas et al. , 2001, p. 362) , Guizhou(J. C. Wei et al. 1986, pp. 247, 248; Zhang T & J. C. Wei, 2006, p. 5) , Hebei(Ahti, 1961, p. 96) , Hubei(Chen, Wu & J. C. Wei, 1989, p. 460) , Taiwan(Ahti, 1961, p. 96; Ahti & Lai, 1979, p. 234; Liu et al. 1980, p. 17; Lai, 2000, pp. 225 – 226) .

Cladonia groenlandica(Å. E. Dahl) Trass 杯角石蕊

Floristilised Mdrkmed(Eesti NSV Tead. Akad. Loodus, Selts, Tallin) 1: 2(1972) .

≡*Cladonia cornuta* var. *groenlandica* Å. E. Dahl, in *Medd. om Gröland, Cl.* , 2: 100(1950) , cum. icon. (Tab. II, fig. 1) .

≡*Cladonia cornuta subsp. groenlandica*(Å. E. Dahl) Ahti, Ann. bot. fenn. 17(2) : 221(1980)

On the ground.

Xizang(J. C. Wei & Jiang, 1986, p. 89) , Guangxi(Guo 1999, p. 342, the specimen and locality was not cited.) . Taiwan(Wang-Yang & Lai, 1973, p. 89, as *Cladonia cornuta*; Ahti & Lai, 1979, p. 230 < pr. p. > ; Ahti, 1980, p. 221; Liu et al. 1980, p. 24; Lai, 2000, pp. 233 – 234) .

Cladonia gymnopoda Vain. 裸柄石蕊

Acta Soc. Fauna Flora fenn. 10: 172(1894) .

On the ground.

Guangxi(Guo 1999, p. 342, the specimen and locality was not cited) , Taiwan(Zahlbr. , 1933c, p. 46; Sandst. 1939, map 76; Asahina, 1943e, p. 231 & 1970c, p. 257; Wang-Yang & Lai, 1973, p. 89; Ahti & Lai, 1979, p. 231; Liu et al. 1980, p. 24; Lai, 2000, p. 234) .

Cladonia hokkaidensis Asahina 北海道石蕊

J. Jap. Bot. 44: 353(1969) .

Sichuan, Yunnan(Guo & J. C. Wei, 1994, P. 32) .

Cladonia humilis(With.) J. R. Laundon 矮石蕊

Lichenologist 16(3) : 220(1984) .

≡*Lichen humilis* Withering, *Bot. Arr. Veg. Gr. Brit.* : 721(1776).

= *Cenomyce pyxidata* var. *conistea* Delise, in Duby, *Bot. Gall.*, *Edn* 2(Paris) 2: 630(1830).

≡*Cladonia conistea*(Delise) Asahina, *J. Jap. Bot.* 19: 234(1943).

= *Cladonia conoidea* Ahti, *Lichenologist* 12(1): 129(1980).

= *Cladonia fimbriata* f. *conista*(Ach.) Nyl., Addit. Lich. Boliv.: 370(1862).

≡*Cenomyce fimbriata* ß *conista* Ach., Syn. meth. lich. (Lund): 257(1814).

Cladonia conista sensu auct. brit.; fide *Checklist of Lichens of Great Britain and Ireland*(2002).

On the ground, and on rotten logs.

Heilongjiang(Asahina, 1941c, p. 436 & 1952d, p. 374), Jilin, Liaoning(Chen et al. 1981, p. 131; Ahti et al., 1996, pp. 55 – 56), Neimenggu(Chen et al. 1981, p. 131), Xinjiang(Wang XY, 1985, p. 344; Abdula Abbas et al., 1993, p. 77; Abdulla Abbas & Wu, 1998, p. 56; Abdulla Abbas et al., 2001, p. 362), Hubei(Chen, Wu & J. C. Wei, 1989, p. 466; Ahti et al., 1996, pp. 55 – 56), Jiangsu(Wu J. N. & Xiang T. 1981, p. 4; Xu B. S. 1989, p. 239), Anhui, Shanghai, Zhejiang(Xu B. S. 1989, p. 239), Fujian(Zahlbr., 1930b, p. 133), Taiwan (Ahti & Lai, 1979, p. 230; Liu et al. 1980, p. 24).

Cladonia incrassata Flörke 厚石蕊

Clad. Comm. : 21(1828).

On *Abies kawakamii* wood.

Taiwan(Aptroot, Sparrius & Lai 2002, p. 282).

Cladonia kanewskii Oxn. 东亚石蕊

in Ucraninian Botan. Review, fasc. III: 9(1926); Sandst. In Rabh. *Kryptog. -Flora* vol. IX, abt. IV, 2 Hälfte, 194 (1931).

On the ground.

Heilongjiang, Jilin, Neimenggu(Chen et al. 1981, p. 132).

Cladonia krempelhuberi(Vain.) Zahlbr. 橹石蕊

Cat. Lich. Univ. 4: 552(1927). var. krempelhuberi 原变种

≡*Cladonia verticillata* var. *krempelhuberi* Vain., *Acta Soc. Fauna Flora fenn.* 10: 187(1894).

On the ground.

Taiwan(Wang & Lai, 1973, p. 89).

var. sublepidota(Asahina) Asahina 亚鳞芽变种

J. Jap. Bot. 31: 322(1956).

≡*Cladonia verticillata* var. *sublepidota* Asahina, in *J. Jap. Bot.* 16: 465(1940) cum. icon.

On rocks.

Anhui(Xu B. S. 1989, p. 238).

Cladonia kurokawae Ahti & Stenroos 短柄石蕊

Mycosystema 8/9: 53(1996).

Paratypes: China- Hubei, Jiangxi(Ahti et al., 1996, p. 54), Taiwan(Ahti et al., 1996, p. 54; Lai, 2000, p. 233).

Cladonia laii Stenroos 赖氏石蕊

Acta Bryolichenologica Asiatica 1(1, 2): 53(1989).

Type: Taiwan, Chiayi-Nantou: Yu-shan National Park, trai from Pai-yun cottage to western peak of Yu-shan (Mt. Morrison), lower oroboreal Abies kawakamii forest on NNE slope, alt. c. 3500 m, Stenroos 3225(H, Holotype; Herb. Lai, isotype).

Yunnan(Stenroos, 1989, p. 54), Taiwan(Stenroos, 1989, p. 54; Lai, 2000, pp. 244 – 245).

Cladonia libifera Savicz 带饼石蕊

Nov. sist. Niz. Rast. 2: 167(1965).

on soil and among mosses.

Xinjiang(Yang Y & Abbas A, 2003, p. 512), Shaanxi, Guangxi(Guo 1999, p. 342, the specimen and locality was not cited.).

Cladonia macilenta Hoffm. 瘦柄红石蕊

Deutschl. Fl., *Zweiter Theil*(Erlangen): 126(1796) [1795].

On soil.

Xinjiang(Abdulla Abbas & Wu, 1998, p. 57; Abdulla Abbas et al., 2001, p. 362), Sichuan(Stenroos et al., 1994, p. 329; Chen S. F. et al., 1997, p. 221, as *C. macilenta* incl. *C. bacillaris*), Yunnan(Wei X. L. et al., 2007, p. 150), Hongkong(Aptroot et al., 2001, p. 324).

≡*Cladonia macilenta var. macilenta* = Chemotype I(= *Cl. macilenta*)

Guizhou(Zhang T & J. C. Wei, 2006, p. 6,), Hainan(JCWei et al., 2013, p. 230), Taiwan(Lai, 2000, p. 242).

= *Cladonia macilenta var. bacillaris*(Genth) Schaer. *Enum. Crit. Eur.*: 186(1850).

≡*Cladonia bacillaris* Genth, *Fl. Nssau*: 406(1835) = Chemotype II(*Cl. bacillaris*)

≡*Cladonia macilenta* f. *bacillaris* = *Cl. macilenta* chemotype II

= *Cladonia bacillaris* Nyl. & *C. bacillaris* f. *nana* Asah. (in Thrower, 1988, pp. 15, 77, 78: pr. p.

≡*Cladonia macilenta* subsp. *bacillaris*(Ach.) Boist., (1903).

≡*Baeomyces bacillaris* Ach., *Methodus*, Sectio post.: 329(1803).

= *Cladonia bacillaris* f. *clavata*(Ach.) Vain., *Acta Soc. Fauna Flora fenn.* 4(no. 1): 92(1887).

≡*Cladonia macilenta* subvar. *clavata* Boistel, *Nouv. Flore Lich.*, *Edn* 2: 25(1903).

Sichuan(Hue, 1898, p. 260; Zahlbr., 1930b, p. 135), Guizhou(Oliv. 1904, p. 193), Anhui(Xu B. S. 1989, p. 234), Jiangxi, Fujian(Zahlbr., 1930b, p. 135), Hubei(Chen, Wu & J. C. Wei, 1989, p. 462), Taiwan(Asahina, 1950b, p. 83; Ahti & Lai, 1979, p. 229), Hainan(J. C. Wei et al., 2013, p. 230), Hongkong(Thrower, 1988, p. 77).

On trees, rocks and soil in open areas and in forests. Comsoplitan.

Neimenggu(Chen et al. 1981, p. 132), Hubei(Chen, Wu & J. C. Wei, 1989, p. 463), guizhou(Zhang T & J. C. Wei, 2006, p. 5, as *Cl. bacillaris*), Shandong(Zhang F et al., 1999, p. 31, as *Cl. bacillaris*; Zhao ZT et al., 2002, p. 7 as *Cl. bacillaris*), Taiwan(Ahti & Lai, 1979, p. 231; Lai, 2000, pp. 240241), Hongkong(Thrower, 1988, pp. 15, 77, 78; Aptroot & Seaward, 1999, p. 71).

= *Cladonia macilenta* var. *squamigera* Vain., *Acta Soc. Fauna Flora fenn.* 4(1): 109(1887) – (*Cladonia macilenta f. squamigera*(Vain.) Sandst).

Jilin, Neimenggu(J. C. Wei et al. 1982, p. 23).

(*Cladonia macilenta* f. *bacillaris*(Nyl.) Jatta; *Cladonia bacilaris* f. *monstrosa* Harm.)

Heilongjiang(Asahina, 1952d, p. 373; Chen et al. 1981, p. 130; J. C. Wei et al. 1982, p. 23; Luo, 1984, p. 85), Neimenggu(J. C. Wei et al. 1982, p. 23), Shaanxi(Jatta, 1902, p. 466; Zahlbr., 1930b, p. 134), Xizang(J. C. Wei & Jiang, 1981, p. 1147 & 1986, p. 83), Hunan(Zahlbr., 1930 b, p. 134), Guangxi(Guo 1999, p. 342, the specimen and locality was not cited).

(*Cladonia bacillaris* f. *nana* Asahina *Lichens of Japan*, 1. *Genus Cladonia*: 85(1950).

Hongkong(Thrower, 1988, p. 78).

[*Cladonia bacillaris* var. *pacifica* Asahina in *Journ. Jap. Bot.* 15: 666(1939), cum. icon., Lamb. I. M. *Index Nominum Lichenum* 171(1963)].

Taiwan(Wang & Lai, 1973, p. 89).

Cladonia macroceras(Floerk.) Ahti 硬柄石蕊

Ann. bot. fenn. 15: 13(1978).

≡*Capitularia gracilis* γ. *Cap. macroceras* Floerk., *Beschr. Braunfr. Becherfl.* 330(1810) in Vainio, *Monographia Cladoniarum Universalis. Acta Soc. Fauna Fl. Fenn.* II: 116(1894).

≡*Cenomyce ecmocyna*γ. *C. macroceras* Ach., Syn. Lich. 263(1814), (herb. Ach.).

≡*Cladonia macroceras*(Delise) Ahti, *Ann. bot. fenn.* 15: 13(1978).

≡*Cenomyce gracilis* η. *macroceras* Del. in Dub. Bot. Gall. 624(1830).

= *Cladonia gracilis* var. *elongata* f. "*ceratostelis*" Anzi, in Vainio, *Monographia Cladoniarum Universalis. Acta Soc. Fauna Fl. Fenn.* II: 119(1894).

= *Cladonia gracilis* var. *hugueninii* Del. In litt. [Mass., Sched. Crit. 53(1855), in Vainio, *Monographia Cladoniarum Universalis. Acta Soc. Fauna Fl. Fenn.* II: 97(1894)].

On the ground among mosses in forests.

Neimenggu(Sato, 1952, p. 174), Xinjiang(Moreau et Moreau, 1951, p. 188; Wang XY, 1985, p. 342 *as Cladonia elongata*), Shaanxi(Baroni, 1894, p. 47; Zahlbr., 1930b, p. 132; Sandst. 1938, p. 90; Asahina & Sato in Asahina, 1939, p. 687; Ahti, 1980, p. 215 – 219; J. C. Wei, 1981, p. 88), Xizang(J. C. Wei & Chen, 1974, p. 176; J. C. Wei & Y. M. Jiang, 1981, p. 1147 & 1986, p. 88), Sichuan(Stenroos et al., 1994, p. 329), Yunnan (Hue, 1898, p. 273; Zahlbr., 1930b, p. 133).

Cladonia macrophylla(Schaer.) Stenh. 大叶石蕊

Lich. Suec. exs.: no. 3(1865).

On rotten wood and mosses over rocks.

Yunnan(Guo & J. C. Wei, 1994, P. 33); Guangxi(Guo, 1999, p. 342, the specimen and locality was not cited)

Cladonia macroptera Räsänen 大翅石蕊

J. Jap. Bot. 16: 149(1940).

On the ground among mosses.

Xinjiang(Abdulla Abbas & Wu, 1998, p. 57; Abdulla Abbas et al., 2001, p. 362), Shaanxi(J. C. Wei, 1981, p. 88; He & Chen 1995, p. 42 without citation of specimens and their localities), Xizang(J. C. Wei & Jiang, 1986, p. 87), guizhou(Zhang T & J. C. Wei, 2006, p. 6), Hubei(Chen, Wu & J. C. Wei, 1989, p. 467).

Cladonia magyarica Vain. ex Gyeln. 鳞杯石蕊

in Sched. ad *Floram Hungar. Exsicc.*, *cent.* VIII, 8(1927)

On soil.

Xinjiang, Guangxi(Guo 1999, p. 342).

Cladonia mateocyatha Robbins 丛杯石蕊

Rhodora 27: 50(1925).

On rotten wood and mosses over rocks.

Yunnan(Guo & J. C. Wei, 1994, P. 33).

Cladonia maxima(Asahina) Ahti 小葱石蕊

Acta bot. fenn. 15(1): 12(1978).

≡*Cladonia gracilis* f. *maxima* Asahina, *Atlas of Japanese Cladoniae*(Tokyo): 19(1971).

Jilin(Ahti, 1991, p. 61), Guangxi(Guo, 1999, p. 342, the specimen and locality was not cited).

Cladonia melanocaulis S. Stenroos 黑柄石蕊

Ann. bot. fenn. 25(2): 138(1988).

Guangxi(Guo, 1999, p. 343).

Cladonia merochlorophaea Asahina 类粗粉石蕊

J. Jap. Bot. 16: 713(1940).

On the ground, rotten logs & tree trunk bases.

Heilongjiang(Asahina, 1940c, p. 721 & 1950b, p. 200).

Cladonia metacorallifera Asahina 结珊瑚石蕊

J. Jap. Bot. 15: 612(1939).

On the ground rich humus.

Jilin(Chen et al. 1981, p. 132) .

Cladonia metalepta Nyl. 细长石蕊

Flora, Regensburg 59: 559(1876) .

On the ground.

Taiwan(Sandst. 1932, p. 70; Zahlbr. , 1933c, p. 46; Wang & Lai, 1973, p. 89) .

Cladonia mitis Sandst. 软鹿石蕊

Sandstede: *Clad. Exs.* : no. 55(1918) .

≡*Cladina mitis*(Sandst.) Mong. , *Bull. Soc. Agric. Sarthe*, ser. 3 7: 118(1938) .

On the ground and rotten logs.

Jilin(Ahti, 1961, p. 116; Chen et al. 1981, p. 132; J. C. Wei et al. 1986, p. 245) , Xinjiang(Abdulla Abbas & Wu, 1998, p. 50; Abdulla Abbas et al. , 2001, p. 362) , Shaanxi(J. C. Wei et al. 1982, p. 21) , Xizang(J. C. Wei & Chen, 1974, p. 175) , Taiwan(Ahti, 1961, p. 116; Wang-Yang & Lai, 1973, p. 89; Ahti & Lai, 1979, p. 234; Liu et al. 1980, p. 16; Lai, 2000, p. 225) .

Cladonia mongolica Ahti 蒙古石蕊

in Huneck, Poelt, Ahti, Vitikainen & Cogt, *Nova Hedwigia* 44(1 – 2) : 196(1987) . Heilongjiang, Neimenggu, Yunnan(Ahti, 1991, p. 61) , Sichuan(Stenroos et al. , 1994, p. 329) , Hubei(Chen, Wu & J. C. Wei, 1989, p. 467; Ahti, 1991, p. 61) , Guangxi(Guo, 1999, p. 342, the specimen and locality was not cited.) , Hainan(JCWei et al. , 2013, p. 230 – 231) .

Cladonia multiformis G. Merr. 多型石蕊

Bryologist 11: 110(1908) .

On rotten wood and ground.

Shaanxi(He & Chen 1995, p. 42 without citation of specimens and their localities) , Xizang, Yunnan(Guo & J. C. Wei, 1994, P. 33)

Cladonia nana Vain. 矮小石蕊

Acta Soc. Fauna Flora fenn. 10: 23(1894) .

On exposed iron-containing rocks and granite. Pantropical.

Hongkong(Thrower, 1988, pp. 15, 78; Aptroot & Seaward, 1999, p. 71; Aptroot et al. , 2001, p. 324) ; not *Cladonia bacillaris* f. *nana* Asah. (in Thrower, 1988, pp. 15, 78) .

Cladonia ochrochlora Flörke 黄绿石蕊

De Cladoniis, Difficillimo lichenum genere, Commentatio nova (Rostock) : 75(1828) .

≡*Cladonia coniocraea* var. *ochrochlora*(Flörke) Oxner, Flora Lishaĭnikiv Ukraïni(Kiev) 2(1) : 284(1968) .

≡*Cladonia fimbriata* var. *ochrochlora* (Flk.) Vain. , *Monographia Cladoniarum Universalis. Acta Soc. Fauna Fl. Fenn.* II: 319(1894) .

= *Cladonia coniocraea* f. *ceratodes* (Flörke) Dalla Torre & Sarnth. , Die Flecht. Tirol: 72(1902) .

≡*Cladonia ochrochlora*α. *ceratodes* Flörke, Clad. Comm. 77(1828) , Vainio, *Monographia Cladoniarum Universalis. Acta Soc. Fauna Fl. Fenn.* II: 314(1894) .

On rocks covered with mosses.

Heilongjiang(Asahina, 1952d, p. 374) , Xinjiang(Abdulla Abbas et al. , 1993, p. 77; Abdulla Abbas & Wu, 1998, p. 57 & Abdulla Abbas et al. , 2001, p. 362 as C. *coniocraea* f. *ceratodes*) , Xizang(J. C. Wei & Jiang, 1986, p. 92) , Anhui(Xu B. S. 1989, p. 240) .

(*Cladonia cornuta* f. *subdilata* Asahina)

Misapplied name:

Cladonia cornuta var. *groenlandica*(pr. p.) auct. non E. Dahl: Ahti & Lai, Ann. Bot. Fennici, 16: 230(1979) .

On the ground in forests.

Heilongjiang(Chen et al. 1981, p. 132), Shaanxi(Jatta, 1902, p. 465; Zahlbr., 1930 b, p. 133), Xizang(J. C. Wei & Jiang, 1981, p. 1147 & 1986, p. 93), Sichuan(Stenroos et al., 1994, p. 330), guizhou(Zhang T & J. C. Wei, 2006, p. 6), Hubei(Chen, Wu & J. C. Wei, 1989, p. 467), Guangxi(Guo, 1999, p. 343, the specimen and locality was not cited), Taiwan(Wang-Yang & Lai, 1973, p. 89; Ahti & Lai, 1979, p. 230 <pr. p.> & p. 231; Ahti, 1980, p. 222; Liu et al. 1980, p. 25; Lai, 2000, pp. 234 – 235).

Cladonia parasitica(Hoffm.) Hoffm. 寄生石蕊

Deutschl. Fl., Zweiter Theil (Erlangen): 127(1796) [1795].

≡*Lichen parasiticus* Hoffm., *Enum. Lich.*: 39(1784).

= *Cenomyce delicatus*(Ehrh.) Ach., *Lich. univ.*: 569(1810).

(*Lichen delicatus* Ehrh. (nom. nud.) = *Cladonia delicata* auct.)

On wood.

Shaanxi(Baroni, 1894, p. 48; Zahlbr., 1930b, p. 131; Sandst. 1938, p. 88), Anhui, Zhejiang(Xu B. S., 1989, p. 236).

Cladonia perfossa Nuno 空石蕊

Journ. Jap. Bot. 47(6): 164(1972).

Type: Taiwan, Kurok. no. 440 in TNS.

Taiwan(Nuno, 1972, p. 164; Wang-Yang & Lai, 1973, p. 89; Ahti & Lai, 1979, p. 231; Liu et al. 1980, p. 25; Lai, 2000, p. 235).

Cladonia pertricosa Kremp. 尖石蕊

Verh. zool. -bot. Ges. Wien 30: 331(1881).

= *Cladonia rangiformis* var. *pungens*(Ach.) Vain., *Acta Soc. Fauna Flora fenn.* 4(1): 361(1887).

≡*Lichen pungens* Ach., *Lich. suec. prodr.* (Linköping): 202(1799) [1798].

= *Cladonia furcata* var. *filiformis* Müll. Arg., *Flora*, Regensburg 65(19): 296(1882).

Misapplied name:

Cladonia furcata <pr. p.> sensu Hue, *Bull. Soc. Bot. France*, 36: 160(1889), non(Huds.) Schrad.

Yunnan(Hue, 1889, p. 160 <pr. p.> & 1898, p. 267; Zahlbr., 1930b, p. 132).

Cladonia peziziformis(With.) J. R. Laundon 头状石蕊

Lichenologist 16(3): 223(1984).

≡*Lichen peziziformis* With. [as '*pezizaeformis*'], *Bot. arr. veg. Gr. Brit.* (London) 2: 720(1776).

= *Cladonia capitata*(Michx.) Spreng., *Syst. veg.*, *Edn* 16 4(1): 271(1827).

≡*Helopodium capitatum* Michx., Fl. Bor. -Am. II: 329?(1803), Vainio, *Monographia Cladoniarum Universalis. Acta Soc. Fauna Fl. Fenn.* II: 29(1894).

= *Helopodium leptophyllum*(Ach.) Gray, *Nat. Arr. Brit. Pl.* (London) 1: 416(1821).

≡Cenomyce *leptophylla* Ach., *Lich. Univ.* 568(1810).

≡*Cladonia leptophylla* Floerk., Clad. Comm. 19(1828).

Cladonia squamosa var. *leptophylla*(Ach.) Rabenh., Deutschl. Krypt. – Fl. II: 102(1845). – Schaer., Enum. Lich. Eur. 199(1850).

On the ground, and among rocks on thin layers of soil.

Liaoning(Chen et al. 1981, p. 130), xinjiang(Abbas et al., 1997, p. 1), Sichuan(Stenroos et al., 1994, p. 330), Jiangsu(Wu J. N. & Xiang T. 1981, p. 4; Xu B. S., 1989, p. 237), Guangxi (Guo, 1999, p. 343, the specimen and locality was not cited), hongkong(Seaward & Aptroot, 2005, p. 283, Pfister 1978: 103, as *Cladonia mitrula* Tuck.).

Cladonia phyllophora Ehrh. ex Hoffm. 鳞叶石蕊

Deutschl. Fl., Zweiter Theil (Erlangen): 126(1796) [1795].

≡*Lichen phyllophorus* Ehrh., *Plant. Crypt.* 287(1793). in Vainio *Monogr. Cladoniarum* II: 136(1894).

Xinjiang(Abdulla Abbas & Wu, 1998, p. 57; Abdulla Abbas et al., 2001, p. 362), Guangxi (Guo, 1999, p. 343, the specimen and locality was not cited).

= *Cladonia phyllophora* f. *trachyna*(Ach.) in J. C. Wei *Enum. Lich. China*: 75(1991), Nom. inval., Art. 41. 4 (Melbourne).

≡*Baeomyces trachynus* Ach., Method. Lich. 348(1803), in Vainio *Monogr. Cladoniarum* II: 136(1894).

= *Cladonia degenerans*(Flörke) Spreng., *Linn. Syst. Veg.* IV: 273((1827), in Vainio *Monogr. Cladoniarum* II: 136 (1894).

≡*Capitularia degenerans* Flörke, Beschr. Braunfr. Becherfl. 308(1810) in Vainio *Monogr. Cladoniarum* II: 136 (1894), & in Weber & Mohr, *Beitr. Naturk.* 2: 308(1810).

On the ground.

Shaanxi(Jatta, 1902, p. 465; Zahlbr., 1930b, p. 133), Yunnan(Hue, 1887a, p. 17 & 1889, p. 27 cited as *Cladonia degenerans* var. *gracilescens* Flk.; Vain. 1894, p. 143; Sandst. 1939, p. 95), Zhejiang(Xu B. S. 1989, p. 238).

Cladonia phyllopoda(Vain.) S. Stenroos 叶足石蕊

Ann. bot. fenn. 25(2): 132(1988).

≡*Cladonia pityrea* var. *phyllopoda* Vain., *Ann. Acad. Sci. fenn.*, Ser. A 15(6): 55(1921).

On acid soil of roadbank.

Taiwan(Aptroot, Sparrius & Lai 2002, p. 282).

Cladonia pleurota(Flörke) Schaer. 粉杯红石蕊

Enum. critic. lich. europ. (Bern): 186(1850).

≡*Capitularia pleurota* Flk. In *Ges. Naturf. Freunde Berlin Mag. Neuesten Entdeck. Gesammten Naturk.* 2: 217(1808).

≡*Cladonia coccifera* var. *pleurota*(Flörke) Schaer., Lich. helv. spicil. 1: 25(1823)

On mosses.

Shaanxi(He & Chen 1995, p. 42 without citation of speciemens and their localities), Yunnan(Wei X. L. et al., 2007, pp. 150 – 151), guizhou(Zhang T & J. C. Wei, 2006, p. 6), Shandong(Zhao ZT et al., 1998, p. 29; Zhang F et al., 1999, p. 31; Zhao ZT et al., 2002, p. 7).

var. pleurota 原变种

Xizang(J. C. Wei & Jiang, 1981, p. 1147 & 1986, p. 83), Anhui, Zhejiang(Xu B. S., 1989, p. 235), Fujian (Zahlbr., 1930b, p. 135).

f. cerina(Naeg.) Anders 蜡黄变型

≡*Cladonia cerina* Naeg. ex Rabh. (1857)

On the ground & moss-covered stump.

Xizang(J. C. Wei & Jiang, 1986, p. 83).

f. dahlii C. Massal. 细粉变型

Mém. Accad. Agricolt. Arti Commerc. Verona, Ser. 3 65: 134(1889).

On the ground.

Neimenggu(J. C. Wei, 1981, P. 87).

var. esorediata Asahina 无粉芽变种

Lichens of Japan, 1. *Genus Cladonia*: 99(1950).

Neimenggu(M. Sato, 1952, p. 174).

f. extensa(Hoffm.) Mig. 伸长变型

Flecht. Deutschl. 2, 12(2): 25(1927).

≡*Cladonia extensa* Hoffm., *Deutschl. Fl.*, *Zweiter Theil(Erlangen)*: 123(1796) [1795].

On the ground & moss-covered rocks.

Xizang(J. C. Wei & Jiang, 1986, p. 83) .

Cladonia pocillum(Ach.) O. J. Rich. 莲座石蕊

Methodus, Sectio post. : 236(1803) .

≡*Baeomyces pocillum* Ach. , *Methodus*, Sectio post. : 236(1803) .

On the ground.

Heilongjiang(Asahina, 1950b, p. 192 & 1952d, p. 374) , Xinjiang(Moreau et Moreau, 1951, p. 188; Ahti, 1976, p. 370; Wang XY, 1985, p. 343 as Cladonia pyxidata var. pocillum; Abdulla Abbas et al. , 1993, p. 77; Abdulla Abbas et al. , 1994, pp. 20 – 21; Abdulla Abbas & Wu, 1998, p. 58; Abdulla Abbas et al. , 2001, p. 362) , Shaanxi(Baroni, 1894, p. 47; Jatta, 1902, p. 464; Zahlbr. , 1930b, p. 134; He & Chen 1995, p. 42) , Xizang(J. C. Wei & Jiang, 1986, p. 91) , Sichuan(Zahlbr. , 1930b, p. 134; Stenroos et al. , 1994, p. 330) , Yunnan(Hue, 1898, p. 275; Zahlbr. , 1930b, p. 134) , Hubei(Chen, Wu & J. C. Wei, 1989, p. 468) , Shandong(Zhang F et al. , 1999, p. 31; Zhao ZT et al. , 2002, p. 7; Hou YN et al. , 2008, p. 67) , Guangxi(Guo, 1999, p. 343, the specimen and locality was not cited) .

Cladonia portentosa(Dufour) Coem. 畸鹿石蕊

Bull. Acad. R. Sci. Belg. , *Cl. Sci.* , sér. 2 19: 43(1865) . f. *portentosa*

≡*Cenomyce portentosa* Dufour, *Ann. Gén. Sci. Phys*. 8: 69(1821) .

≡*Cladina portentosa* (Dufour) Follmann, *Philippia* 4: 34(1979) .

= *Cladonia impexa* Harm. , *Lich. Fr.* 3: 232(1907) .

On the ground and on rocks among mosses.

Jilin(Moreau et Moreau, 1951, p. 188) , Shaanxi(Jatta, 1902, p. 466; A. Zahlbr. , 1930b, p. 130) , Yunnan(A. Zahlbr. , 1930 b, p. 130) .

= *Cladonia spumosa*(Flörke) Sandst. , *Abh. naturw. Ver. Bremen* 21: 344, tab. II, fig. 3?4(1912)

≡*Cladina alpestris* var. *spumosa*(Flörke) Zopf, *Flechtenstoffe*: 405(1907) .

Sichuan(Zahlbr. , 1934 b, p. 201) .

This species, however, has not been found by me in China so far. The species and its infraspecific taxa, reported in literature from China, may probably be based on erroneous identification. This species remains to be revised on the basis of the original specimens cited in literature from China in the future.

Cladonia pseudodidyma Asahina 拟小红石蕊

Journ. Jap. Bot. 15(11) : 667(1939) .

Type: Taiwan(China) , Kiusiu and Hondo(Japan) .

Cladonia didyma non(Fée) Vain. : Sandst. in *Bot. Mag*. Tokyo, 61: 337(1927) .

On the ground.

Taiwan(Zahlbr. , 1933c, p. 46; Asahina, 1950b, p. 86; Lamb, 1963, p. 190; Ahti & Lai, 1979, p. 232; Liu et al. 1980, p. 32; Lai, 2000, p. 243) .

Cladonia pseudodigidata Gyeln. 拟胀石蕊

Rev. Bryol. et Lichenol. 6: 174(1933) . Lamb, *Index Nom. Lich.* 190(1963) .

= *Cladonia pleurota* var. *esorediata* Asahina, *Lichens of Japan*, 1. *Genus Cladonia*: 99 (1950) , (Lai, 2000, p. 241) .

(*Cladonia esorediata*(Asahina) Yoshim. 1968)

On earth rich in humus & on rocks among mosses.

Hubei(Chen, Wu & J. C. Wei, 1989, p. 463) , Taiwan(Ahti & Lai, 1979, p. 230; Liu et al. 1980, p. 31; Lai, 2000, pp. 241 – 242) .

Cladonia pseudoevansii Asahina 拟雀石蕊

J. Jap. Bot. 16: 187(1940) .

≡*Cladina pseudoevansii*(Asahina) Hale & W. L. Culb. , *Bryologist* 73(3) : 510(1970) .

On the ground.

Jilin(Ahti, 1961, p. 26; Chen et al. 1981, p. 133; J. C. Wei. et al. 1986 a, p. 241).

Cladonia pseudogymnopoda Asahina 拟裸柄石蕊

J. Jap. Bot. 45(9): 259(1970).

= *Cladonia pseudogymnopoda* Asahina ssp. *recurvans*(Asahina) Asahina *J. Jap. Bot.* 45(9): 260(1970).

≡ *Cladonia calycantha*(Del.) Nyl. f. *recurvans* Asahina *J. Jap. Bot.* 31: 69(1956).

Aptroot et al. (1999, pp. 71 – 72) noted: "additional Hong Kong records probably also referable to this species: *Cladonia rappii* A. Evans(in Sandstede, 1938), *Cladonia verticillaris*(Raddi) Fr. (in Sandstede, 1938), and *Cladonia verticillaris* f. *penicellata* Vainio(Hue, 1898); the latter two are both endemic to the neotropics."

On the ground. On rocks and soil in open areas and in forests. Known from tropical Asia.

Taiwan(Wang-Yang & Lai, 1973, p. 89; Ahti & Lai, 1979, p. 232; Liu et al. 1980, p. 26; Lai, 2000, p. 236), Hongkong(Sandstede, 1938, p. 90; Hue, 1898, p. 274; Thrower, 1988, pp. 15, 79; Aptroot & Seaward, 1999, p. 71 – 72; Seaward & Aptroot, 2005, p. 283, Pfister 1978: 103, as *Cladonia gracilis*; Aptroot et al., 2001, p. 324).

Cladonia pyxidata(L.) Hoffm. 喇叭石蕊

Deutschl. Fl., *Zweiter Theil*(Erlangen): 121(1796) [1795].

≡ *Lichen pyxidatus* L., *Sp. pl.* 2: 1151(1753).

On the ground.

Heilongjiang, Jilin, Liaoning(Chen et al. 1981, p. 133), Neimenggu(Chen et al. 1981, p. 133; Sun LY et al., 2000, p. 36), Xinjiang(Tchou, 1935, p. 304; J. C. Wei et al. 1982, p. 25; Wu J. L. 1985, p. 74; Abdulla Abbas et al., 1993, p. 77; Abdulla Abbas & Wu, 1998, p. 58; Abdulla Abbas et al., 2001, p. 362), Shaanxi(He & Chen 1995, p. 42 without citation of specimen and locality), Xizang(J. C. Wei & Chen, 1974, p. 176; J. C. Wei & Jiang, 1981, p. 1147 & 1986, p. 90; J. C. Wei et al. 1982, p. 25), guizhou(Zhang T & J. C. Wei, 2006, p. 6), Yunnan(Hue, 1889, p. 159 & 1898, p. 275; Paulson, 1928, p. 317; Zahlbr., 1930 b, p. 133), Hebei(Tchou, 1935, p. 304), Hubei(Müll. Arg. 1893, p. 235; Zahlbr., 1930b, p. 133), Shandong(Hou YN et al., 2008, p. 67; Li Y et al., 2008, p. 72), Anhui, Shanghai, Zhejiang(Xu B. S. 1989, p. 239), Jiangxi, Fujian(Zahlbr., 1930b, p. 133), Guangxi(Guo, 1999, p. 344, the specimen and locality was not cited), Taiwan(Ahti & Lai, 1979, p. 232; Liu et al. 1980, p. 27).

var. pyxidata 原变种

= *Cladonia pyxidata* var. *neglecta*(Flörke) A. Massal., *Symmict. Lich*: 82(1855).

= *Cladonia pyxidata* f. *staphylea*(Ach.) Harm., *Bull. Séanc. Soc. Sci. Nancy*, Sér. 2 14: 372(1896) [1895].

On the ground.

Heilongjiang(Asahina, 1950b, p. 192), xinjiang(Wang XY, 1985, pp. 342 – 343), Shaanxi(Jatta, 1902, p. 464; Z/ahlbr. 1930 b, p. 133), Xizang(J. C. Wei & Jiang, 1986, p. 90 cited as f. *pyxidata*), Yunnan(Hue, 1887a, p. 17 & 1898, p. 275; Vain. 1894, p. 226 – 229; Zahlbr., 1930b, p. 133), Hubei(Chen, Wu & J. C. Wei, 1989, p. 468).

f. lophyra(Ach.) Nyl. 鳞芽变型

≡ *Peltidea horizontalis* var. *lophyra* Ach., *Lich. univ.*: 516(1810).

Heilongjiang(Asahina, 1952d, p. 374), Xizang(J. C. Wei & Jiang, 1986, p. 90).

Cladonia ramulosa(With.) J. R. Laundon 麸皮石蕊

Lichenologist 16(3): 225(1984).

≡ *Lichen ramulosus* Withering, *Bot. Arr. Veg. Gr. Brit.*: 723(1776).

= *Cladonia adspersa* Mont. et v. d. Bosch.

= *Cladonia pityrea*(Flörk.) Fr., Nov. Sched. Crit. 21(1826) [*Fl. carniol.*, Edn 2(Wien): 21(1826)].

≡*Capitularia pityrea* Flörk. in Berl. Magaz. 2: 135(1808)[1807].

=*Cladonia anomaea* var. *scyphifera*(Flörk.) Clauzade & Cl. Roux, *Bull. Soc. bot. Centre-Ouest*, Nouv. sér., num. spec. 7: 311(1985).

=*Cenomyce pityrea* d. *scyphifera* Del. [Vainio, *Monogr. Clad. Univ.* II: 354(1894)]

Cladonia anomaea(Ach.) Ahti & P. James, in Ahti, *Lichenologist* 12(1): 128(1980), (in Thrower, 1988, pp. 15, 76).

≡*Baeomyces anomaeus* Ach., *Methodus*, Sectio post. (Stockholmiæ): 349(1803)

Misapplied name: *Cladonia cariosa*(Ach.) Sprengel(Wawra c. 1868 –71, in Kremp., 1876, p. 436).

On trees, rotten wood, rocks, and. soil in open areas and in forests. Comsopolitan.

Heilongjiang, Jilin, Liaoning, Neimenggu(Chen et al. 1981, p. 133), Shaanxi(Jatta, 1902, p. 464; Zahlbr., 1930b, p. 133; He & Chen 1995, p. 42 without citation of specimens and their localities), Xizang(J. C. Wei & Chen, 1974, p. 177; J. C. Wei & Jiang, 1981, p. 1147 & 1986, p. 93), Sichuan(Chen S. F. et al., 1997, p. 221, as cf. *C. ramulosa*), Yunnan(Wang L. S. et al., 2008, p. 536), Shandong(Zhao ZT et al., 1998, p. 30; Zhang F et al., 1999, p. 31; Zhao ZT et al., 2002, p. 7; Li Y et al., 2008, p. 72), Jiangsu(Wu J. N. & Xiang T. 1981, p. 4), Hubei(Chen, Wu & J. C. Wei, 1989, p. 469), Shanghai, Zhejiang, Anhui(Xu B. S. 1989, p. 240, 241), Fujian(Zahlbr., 1930b, p. 133), Taiwan(Zahlbr., 1933c, p. 47; Sandst. 1939, p. 94; Ras. 1940, p. 150; Asahina, 1943d, p. 196 & 1950b, pp. 211, 212, 216, 217; Wang-Yang & Lai, 1973, p. 89; Ahti & Lai, 1979, p. 232; Liu et al. 1980, p. 26; Lai, 2000, pp. 235 –236), Guangxi(Guo, 1999, p. 344, the specimen and locality was not cited.), Hainan(JCWei et al., 2013, p. 231), Hongkong(Kremp., 1876, p. 436; Thrower, 1988, pp. 15, 76 as *Cladonia anomea*; Aptroot & Seaward, 1999, p. 72; Aptroot et al., 2001, p. 324).

Cladonia rangiferina(L.) Weber ex F. H. Wigg. 鹿石蕊

in Wiggers, *Prim. fl. holsat.* (Kiliae): 90(1780).

≡*Lichen rangiferinus* L., *Sp. pl.* 2: 1153(1753).

≡*Cladina rangiferina*(L.) Nyl., *Not. Sällsk. Fauna et Fl. Fenn. Förh.*, Ny Ser. 8: 110(1866)

Heilongjiang(Asahina, 1952, p. 373; Sato, 1952, p. 173; Chen et al. 1981, p. 133; Luo, 1984, p. 85; J. C. Wei et al. 1986 a, p. 248), Jilin(Zahlbr., 1934 b, p. 202; Tchou, 1935, p. 304; Chen et al. 1981, p. 133; J. C. Wei et al. 1986a, p. 248), Liaoning(Chen et al. 1981, p. 133), Neimenggu(Asahina, 1952, p. 373; Sato, 1952, p. 173; Chen et al. 1981, p. 133; J. C. Wei et al. 1986 a, p. 248), Shaanxi(Jatta, 1902, p. 465; Zahlbr., 1930 b, p. 130; J. C. Wei et al. 1982, p. 21; He & Chen 1995, p. 42, as *Cladina rangiferina* without citation of specimens and their localities), Xizang(Ahti, 1961, p. 86; J. C. Wei & Chen, 1974, p. 175; J. C. Wei & Jiang, 1981, p. 1147; J. C. Wei et al. 1986a, p. 248; J. C. Wei & Jiang, 1986, pp. 78, 79, 80), Sichuan(Ahti, 1961, p. 86; J. C. Wei et al. 1986 a, p. 248), Yunnan(Hue, 1887, p. 17 & 1898, p. 256; Zahlbr., 1930 b, p. 130; Ahti, 1961, p. 86; J. C. Wei et al. 1986 a, p. 248: Wei X. L. et al. 2007, p. 151), Shandong(Zhao ZT et al., 2002, p. 7 as *Cladina rangiferina*), Hubei(Zahlbr., 1930b, p. 130; Chen, Wu & J. C. Wei, 1989, p. 460), Anhui, Zhejiang(Xu B. S., 1989, p. 233), Taiwan(Zahlbr., 1933c, p. 46; Wang & Lai, 1973, p. 89).

Cladonia rangiformis Hoffm. 鹿角石蕊

Deutschl. Fl., *Zweiter Theil*(Erlangen): 114(1796)[1795].

On the ground.

Shaanxi(Jatta, 1902, p. 465; Zahlbr., 1930b, p. 131), Yunnan(Hue, 1889, p. 160; Paulson, 1928, p. 317; Zahlbr., 1930b, p. 131).

=*Cladonia rangiformis* f. *foliosa*(Flörke) H. Olivier, *Expo. Syst. Descr. Lich. Ouest Fr.* 1: 70(1897).

≡*Cladonia rangiformis* var. *foliosa* Flörk., in Vainio, Monogr. Clad. Univ. 366(1887).

Yunnan(Hue, 1898, P. 267; Zahlbr., 1930b, p. 131).

Cladonia rappii A. Evans 宽杯石蕊

Trans. Ky Acad. Sci. 38: 297(1952).

= *Cladonia calycantha* var. *gracilior* Asahina, *Journ. Jap. Bot.* 31(11): 323(1956).

Type locality: Taiwan.

Cladonia calycantha non Del.: Zahlbr., 1930b, p. 134 & 1933c, p. 46. Asahina, 1940a, p. 467 & 1950b, p. 190 & 1956b, p. 323. Ahti & Lai, 1979, p. 229. Liu et al. 1980, p. 22. Wang & Lai, 1973, p. 89.

Neimenggu(Chen et al. 1981, p. 130), guizhou(Zhang T & J. C. Wei, 2006, p. 6), Anhui, Jiangsu, Zhejiang(Xu B. S. 1989, p. 238 cited as *Cladonia calycantha* f. *calycantha* and f. *foliolosa* Vain.), Fujian(Sandst. 1938, p. 90; Zahlbr., 1930b, p. 134), Taiwan(Zahlbr., 1933c, p. 46; Asahina, 1940 a, p. 467 & 1950b, p. 190 & 1956b, p. 323; Wang-Yang & Lai, 1973, p. 89; Ahti & Lai, 1979, p. 229; Liu et al. 1980, p. 22; Lai, 2000, p. 232), Guangxi(Guo, 1999, p. 344, the specimen and locality was not cited), Hainan(JCWei et al., 2013, p. 231).

See Cladonia pseudogymnopoda Asah. For Hongkong(Sandst. 1938, p. 90).

See Ahti, Taxonomic notes on some American species of the lichen genus *Cladonia in Ann. Bot. Fennici* 20: 4 (1983).

Cladonia rei Schaer. 裂杯石蕊

Lich. helv. spicil. 1: 34(1823).

= *Cladonia nemoxyna* Ach., *Lich. Nov. Zeland.* (Paris): 18(1888).

≡ *Cladonia nemoxyna* Ach., *Lich. Nov. Zeland.* (Paris): 18(1888).

Misapplied name:

Cladonia glauca f. *subacuta* auct. non Asahina: J. C. Wei & Chen in *Report on the Scientific Investigations* (1966 – 1968) in Mt. Qomolangma district, 1974, p. 176; J. C. Wei & Jiang in *Proceeding of Symposium on Qinghai-Xizang(Qinghai-Tibet) Plateau*, 1146(1981).

On the ground & rotten stump.

Heilongjiang, Neimenggu(Asahina, 1952d, p. 374; Chen et al. 1981, p. 132), Jilin(Chen et al. 1981, p. 132), Shaanxi(He & Chen 1995, p. 42, no specimen & locality were cited), Xizang(J. C. Wei & Chen, 1974, p. 176; J. C. Wei & Jiang, 1981, p. 1146 & 1986, p. 88), Hubei(Chen, Wu & J. C. Wei, 1989, p. 469), Shandong(Zhao ZT et al., 1998, p. 30; Zhang F et al., 1999, p. 31; Zhao ZT et al., 2002, p. 7). Anhui, Zhejiang(Xu B. S. 1989, p. 239), Taiwan(Zahlbr., 1933c, p. 47; Asahina, 1950b, p. 206; Wang & Lai, 1973, p. 89), Guangxi (Guo, 1999, p. 344, the specimen and locality was not cited), Hainan(J. C. Wei et al., 2013, p. 231).

Cladonia sarmentosa(Hook. f. & Taylor) C. W. Dodge 长匐石蕊

B. A. N. Z. Antarct. Res. Exped. Rep., Ser. B 7: 129(1948).

≡ *Cenomyce sarmentosa* Hook. f. & Taylor, London J. Bot. 3: 651(1844).

= *Cladonia campbelliana*(Vain.) Gyeln., *Rev. Bryol. Lichénol.*, N. S. 6: 174(1933).

≡ *Cladonia gracilis* var. *campbelliana*(Vain.) Vain., *Acta Soc. Fauna Flora fenn.* 10: 213(1894).

≡ *Cladonia gracilis* subvar. *campbelliana* Vain. *Monographia Cladoniarum Universalis.*

Acta Soc. Fauna Fl. Fenn. II: 113(1894).

Xinjiang(Abdulla Abbas, 2001, p. 361), Sichuan(Zahlbr., 1930b, p. 132; Sandst. 1938, p. 90).

Cladonia scabriuscula(Delise) Leight. 粗皮石蕊

Flora, Regensburg 58: 447(1875). f. scabriuscula 原变型

≡ *Cenomyce scabriuscula* Del. in Duby, *Bot. Gall.*: 623(1830).

≡ *Cladonia furcata* var. *scabriuscula*(Delise) Coem., in Vainio, *Acta Soc. Fauna Flora fenn.* 4(1): 338(1887).

On the ground in damp woods.

Heilongjiang(Asahina, 1952d, p. 374; J. C. Wei, 1981, p. 87; Chen et al. 1981, p. 133), Jilin, Liaoning, Neimenggu(Chen et al. 1981, P. 133), Shaanxi(Baroni, 1894, p. 48; Zahlbr., 1930b, p. 131; He & Chen 1995, p.

42 without citation of specimens and their localities), Xizang(Hue, 1898, p. 266; J. C. Wei & Y. M. Jiang, 1981, p. 1146), Sichuan(Zahlbr., 1930b, p. 132; Sandst. 1938, p. 86; Stenroos et al., 1994, p. 331), guizhou (Zhang T & J. C. Wei, 2006, p. 6), Yunnan(Hue, 1898, p. 266; Zahlbr., 1930b, p. 132; Sandst. 1938, p. 86; des Abb. 1958, p. 205), Taiwan(Zahlbr., 1933c, p. 46; Sandst. 1938, p. 86; Asahina, 1952d, p. 374; Wang & Lai, 1973, p. 89; Ahti & Lai, 1979, p. 232; Liu et al. 1980, p. 19; Lai, 2000, p. 229).

≡ *Cladonia scabriuscula* f. *scabriuscula*

Xinjiang(Wang XY, 1985, p. 341), Xizang(J. C. Wei & Jiang, 1986, p. 87), Hubei(Chen, Wu, J. C. Wei, 1989, p. 470), Jiangsu, Anhui, Zhejiang(Xu B. S., 1989, p. 236).

= *Cladonia furcata* f. *adspersa*(Flörke) Vain., *Acta Soc. Fauna Flora fenn.* 4(1): 340(1887).

≡ *Cladonia furcata* var. *asperata* Müll. Arg., *Flora*, Regensburg 65(19): 295(1882).

≡ *Cladonia furcata* f. *adspersa*(Flörke) Vain., *Acta Soc. Fauna Flora fenn.* 4(no. 1): 340(1887).

Guizhou(Oliv. 1904, p. 195).

= *Cladonia scabriuscula* f. *cancellata*(Müll. Arg.) Sandst., in Rabenhorst, 220(1931).

Xinjiang(Wang XY, 1985, p. 341; Abdula Abbas et al., 1993, p. 77; Abdulla Abbas & Wu, 1998, p. 59; Abdulla Abbas et al., 2001, p. 362).

f. elegans Rabh. 优美变型

in Sandst., Cladon. Exsicc. No. 1570(1926).

On the ground.

Xinjiang(Wang XY, 1985, p. 341; Abdulla Abbas et al., 1993, p. 77; Abdulla Abbas & Wu, 1998, p. 59; Abdulla Abbas et al., 2001, p. 362), Xizang(J. C. Wei & Jiang, 1986, p. 87), Anhui, Zhejiang(Xu B. S., 1989, p. 236).

f. robustior(Sandst.) Anders 壮枝变型

Die Strauch- und Laubflecht. Mitteleurop., 74(1927).

Xinjiang(Abdulla Abbas et al., 1993, p. 77; Abdulla Abbas & Wu, 1998, p. 59; Abdulla Abbas et al., 2001, p. 362).

f. squamulosa(Duf.) J. C. Wei 小鳞变型

Enum. Lich. China, 79(1991).

[≡ *Cenomyce furcata* β. *squamulosa* Duf. *Rev. Clad.* 31(1817)].

[≡ *Cladonia furcata* var. *scabriuscula* f. *squamulosa*(Duf.) Oliv.]

[≡ *Cladonia furcata* var. *squamulosa*(Duf.) Schaer.]

[≡ *Cladonia racemosa* var. *squamulosa*(Duf.) Schaer.]

Shaanxi(Jatta, 1902, p. 465; Zahlbr., 1930b, p. 132), Yunnan(Hue, 1889, p. 160; Zahlbr., 1930b, p. 132).

[Cladonia scabriuscula f. sublevis Sandst.] 亚光变型

Xinjiang(Wang XY, 1985, p. 341).

Cladonia sinensis S. Stenroos & J. B. Chen 中华石蕊

in Stenroos, Vitikainen & Koponen, *J. Hattori bot. Lab.* 75: 331(1994).

Type: China. Hubei: Shennongjia, 2800 m, on soil covered rock. 27 VII, 1984. J. C. Wei & J. B. Chen 10734 (holotype, HMAS-L.; isotype, H); paratype: NW Sichuan: Minshan Range, Nanping Co. Jiu-Zhai-Gou, Ze-Zha-Wa-Gou, Lake Chang-Hai. 1991. Koponen 46025/loc. 21b(H). (usnic, isousnic acids, and zeorin)

Shaanxi(He & Chen 1995, p. 42 without citation of specimens and their localities), Sichuan(Stenroos et al., 1994, p. 331).

Cladonia sobolescens Nyl. ex Vain. 匍匐石蕊

Acta Soc. Fauna Flora fenn. 8(1): 80(1892)[1890 – 1893].

On soil.

Guizhou(Zhang T & J. C. Wei, 2006, p. 6), Guangxi(Guo, 1999, p. 344, the specimen and locality was not cited).

Cladonia squamosa(Scop.) Hoffm. 鳞片石蕊

Deutschl. Fl., *Zweiter Theil* (Erlangen): 125(1796) [1795]. var. squamosa 原变种

≡*Lichen squamosus* Scop., *Fl. carniol.*, Edn 2(Wien): 368(1772).

Shaanxi(Baroni, 1894, p. 48; Zahblr., 1930b, p. 132; Ahti & Lai, 1979, p. 233), Xizang(J. C. Wei & Chen, 1974, p. 176; J. C. Wei et al. 1982, p. 25), guizhou(Zhang T & J. C. Wei, 2006, p. 6), YUNNAN(Wei X. L. et al. 2007, p. 151), Hubei(Chen, Wu & J. C. Wei, 1989, p. 470), Anhui, Zhejiang(Xu B. S., 1989, p. 236), Taiwan(Zahlbr., 1933c, p. 46; Wang-Yang & Lai, 1973, p. 89; Ahti & Lai, 1979, p. 233; Liu et al. 1980, p. 20; Lai, 2000, p. 229), Guangxi(Guo, 1999, p. 344, the specimen and locality was not cited), China(prov. not indicated: Vain. 1887, p. 417).

var. squamosa f. macrophylla Rabh. 大叶变型

Clad. Eur. 1860, tab. XXVI n. 17 – 19, XXVII n. 21 – 26.

Shaanxi(Jatta, 1902, p. 466; Zahlbr., 1930 b, p. 132).

var. microphylla Hoffm. f. cylindrica Schaer 圆柱变型

in Vainio, *Monogr. Clad. Univ.* I: 428.

Shaanxi(Jatta, 1902, p. 465; Zahlbr., 1930b, p. 132).

f. proboscidea Jatta 长角变型

I, 466, haud Fw. in Zahlbr., 132(1930) b.

Shaanxi(Jatta, 1902, p. 466; Zahlbr., 1930b, p. 132).

var. multibrachiata Flörk. 多枝变种

in Vainio, *Monogr. Clad. Univ.* I: 437(1887).

Zahlbr. *Cat. Lich. Univ.* 598(1927).

≡*Cladonia squamosa* B. *asperella* b. *multibrachiata* Flk.

Shaanxi(Jatta, 1902, p. 466; Zahlbr., 1930b, p. 132).

f. muricella Sandst. 无杯鳞变型

in Rabenhorst, Kryptog. - Flora, IX, Abt. IV, 2. Hälfte, 271(1931).

≡*Cenomyce squamosa* ? *muricella* Delise, in Duby, *Bot. Gall.*, Edn 2(Paris) 2: 626(1830).

Guizhou(Oliv. 1904, p. 195).

var. subsquamosa(Nyl. ex Leight.) Vain. 亚鳞片变种

Meddn Soc. Fauna Flora fenn. 6: 113(1881).

Guangxi(Guo, 1999, p. 344).

Cladonia stellaris(Opiz) Pouzar & Vězda 雀石蕊

Preslia 43: 196(1971). f. stellaris

≡*Cenomyce stellaris* Opiz, *Böh. phänerogam. cryptogam. Gewächse*(Prague): 141(1823).

≡*Cladina stellaris*(Opiz) Brodo, *Bryologist* 79: 363(1976) (Abbas et al., 1993, p. 76).

= *Cladonia alpestris* (L.) Rabenh., in Vainio, *Acta Soc. Fauna Flora fenn.* 4(1): 41(1887).

≡*Lichen rangiferinus* var. *alpestris* L., *Sp. pl.* 2: 1153(1753).

On the ground.

Heilongjiang(Asahina, 1952, p. 373; J. C. Wei et al. 1986, p. 242), Jilin(Zahlbr., 1934 b, p. 202; Ahti, 1961, p. 47; Chen et al. 1981, p. 133; J. C. Wei et al. 1986, p. 242), Neimenggu(Sato, 1952, p. 173; Ahti, 1961, p. 47; Chen et al. 1981, p. 133; J. C. Wei et al. 1986 a, p. 242), Xinjiang(Abdulla Abbas et al., 1993, p. 76; Abdulla Abbas & Wu, 1998, p. 50; Abdulla Abbas et al., 2001, p. 362), Shaanxi(Jatta, 1902, p. 466; Zahlbr., 1930b, p. 130; Sandst. 1932, p. 69; Wu J. L. 1981, p. 164; J. C. Wei et al. 1982, p. 21; J. C. Wei et al. 1986, p. 242; He & Chen 1995, p. 42, as *Cladina stellaris* without citation of specimens and their localities), China (prov. not indicated: Zahlbr., 1934b, p. 202).

Cladonia strepsilis(Ach.) Grognot 大叶石蕊

Pl. crypt. -cellul. Saône-et-Loire: 85(1863).

≡*Baeomyces strepsilis* Ach., Methodus, Suppl.: 52(1803).

On rotten timbers.

Heilongjiang(Chen et al. 1981, p. 134), Taiwan(Aptroot, Sparrius & Lai 2002, p. 282, on acid soil), Guangxi (Guo, 1999, p. 345, the specimen and locality was not cited).

Cladonia stricta(Nyl.) Nyl. 小鳞石蕊

Flora, Regensburg 52: 294(1869).

≡*Cladonia degenerans* var. *stricta* Nyl. in Middendorff, Reise Sibir. 6(2): 4(1867).

= *Cladonia lepidota* var. *stricta*(Nyl.) Sandst.

= *Cladonia lepidota* Nyl.

= *Cladonia cerasphora* Vain.

Jilin(Chen et al. 1981, p. 132), Taiwan(Sandst. 1938, p. 69).

Cladonia stygia(Fr.) Ruoss 白点石蕊

in Bot. Helv. 95: 241(1985).

≡*Cladonia rangiferina* f. *stygia* Fr., *Sched. Crit. Lich. Suec*. 8 – 9: 22(1826).

≡*Cladina stygia*(Fr.) Ahti, in Hertel & Oberwinkler, *Beih. Nova Hedwigia* 79: 45(1984). -Ahti T, *Mycosystema*, 4: 60(1991).

Jilin(Ahti, 1991, p. 60).

Cladonia subcariosa Nyl. 亚腐石蕊

Flora, Regensburg 59: 560(1894).

= *Cladonia polycarpoides* Nyl., in Gasilien, *Lich. Fl. St. Omer*.: 125(1894).

Heilongjiang(Asahina, 1952d, p. 374), Hebei(Moreau et Moreau, 1951, p. 188), Hubei(Chen, Wu & J. C. Wei, 1989, p. 468).

Cladonia subconistea Asahina 拟小杯石蕊

J. Jap. Bot. 17: 433(1941).

On the ground.

Heilongjiang, Liaoning(Chen et al. 1981, p. 134), Xinjiang(Abdulla Abbas & Wu, 1998, p. 59 ; Abdulla Abbas et al., 2001, p. 362), guizhou(Zhang T & J. C. Wei, 2006, p. 6), Hubei(Chen, Wu & J. C. Wei, 1989, p. 470), Anhui, Jiangsu, Zhejiang(Xu B. S. 1989, p. 240), Taiwan(Asahina, 1950b, p. 198; Ahti & Lai, 1979, p. 233; Liu et al. 1980, p. 28; Lai, 2000, pp. 237 – 238).

Cladonia submultiformis Asahina 亚多型石蕊

in J. Jap. Bot.. 18: 624(1942). f. submultiformis 原变型

On the ground.

Taiwan(Asahina, 1950b, p. 220; Lamb, 1963, p. 195; Wang & Lai, 1973, p. 89; Ahti & Lai, 1979, p. 233; Liu et al. 1980, p. 28; Lai, 2000, p. 238), Guangxi(Guo, 1999, p. 345, the specimen and locality was not cited).

f. foliolosa Asahina 小叶变型

in J. Jap. Bot. 18: 624(1942).

Taiwan(Asahina, 1950, p. 221 & 1971, fig. 26; Lamb, 1963, p. 195; Wang & Lai, 1973, p. 89).

Cladonia subpityrea Sandst. 亚麸皮石蕊

in Keissler, *Annln naturh. Mus. Wien* 42: 62(1928).

= *Cladonia formosana* Asahina, Journ. Jap. Bot. 17(9): 485(1941)

Type: Taiwan, Asahina no. 112.

= *Cladonia formosana* f. *aberans* Asahina, Journ. Jap. Bot. 17(9): 488(1941).

Type: Taiwan, Asahina no. 130.

= *Cladonia formosana* f. *sublaevigata* Asahina, Journ. Jap. Bot. 17(9): 486(1941).

Type: Taiwan, Asahina no. 127.

Xizang(J. C. Wei & Jiang, 1986, p. 94), Hubei(Chen, Wu & J. C. Wei, 1989, p. 471), Jiangsu, Anhui(Xu B. S. 1989, p. 241), Zhejiang(Sandst. 1939, p. 98; Xu B. S. 1989, p. 241), Taiwan(Asahina, 1941d, pp. 485 – 488 & 1943e, p. 240 & 1950b, pp. 218 – 220; Lamb, 1963, p. 181; Yoshim. & Sharp, 1968, p. 109; Yoshim. 1968, p. 235; Ahti & Lai, 1979, p. 233; Liu et al. 1980, p. 29).

Cladonia subradiata(Vain.) Sandst. 拟枪石蕊

Abh. naturw. Ver. Bremen 25: 230(1922).

≡ *Cladonia fimbriata* var. *subradiata* Vain., *Acta Soc. Fauna Flora fenn.* 10: 338(1894).

On exposed soil. Pantropical.

Neimenggu(Sun LY, et al., 2000, p. 36), Hubei(Chen, Wu & J. C. Wei, 1989, p. 471), Guangxi(Guo, 1999, p. 345, the specimen and locality was not cited), Hainan(J. C. Wei et al., 2013, p. 231), Hongkong(Aptroot & Seaward, 1999, p. 72).

Cladonia subsquamosa Kremp. 亚鳞石蕊

in Warming, *Vidensk. Meddel. Dansk Naturhist. Foren. Kjøbenhavn* 5: 336(1874)[1873].

≡ *Cladonia subsquamosa*(Nyl. ex Leight.) Cromb., in Warming, *J. Linn. Soc.*, *Bot.* 17: 560(1880).

≡ *Cladonia squamosa* subsp. *subsquamosa*(Nyl. ex Leight.) Hoffm., Deutschl. Fl., Zweiter Theil(Erlangen): 125(1796)[1795].

≡ *Cladonia squamosa* var. *subsquamosa* (Nyl. ex Leight.) Vain., *Meddn Soc. Fauna Flora fenn.* 6: 113(1881).

≡ *Cladonia delicata* var. *subsquamosa* Nyl. ex Leight.

On the ground in forests.

Neimenggu(Chen et al. 1981, p. 134), Xinjiang(Abdulla Abbas & Wu, 1998, p. 59; Abdulla Abbas et al., 2001, p. 362), Taiwan(Aptroot, Sparrius & Lai 2002, p. 282, as *Cl. subsquamosa* Kremp.), Guangxi(Guo, 1999, p. 345).

≡ *Cladonia subsquamosa* f. *foliosissima* Suza

Taiwan(Asahina, 1950b, p. 155).

Cladonia subulata(L.) Weber ex F. H. Wigg. 尖头石蕊

Prim. fl. holsat. (Kiliae): 90(1780). f. subulata 原变型

≡ *Lichen subulatus* L., *Sp. pl.* 2: 1153(1753).

≡ *Cladonia fimbriata* γ. *subulata*(L.) Vainio, *Monogr. Clad. Univ.* II: 282(1894)[*Cladonia fimbriata* f. *subulata* Harm.]

Xinjiang(Abdulla Abbas & Wu, 1998, p. 60; Abdulla Abbas et al., 2001, p. 362).

= *Cladonia fimbriata* var. *cornutoradiata* Coem. in Zahlbr., *Cat. Lich. Univ.* IV: 499(1927).

On the ground.

Xinjiang(Moreau et Moreau, 1951, p. 188), Shaanxi(Jatta, 1902, p. 464; Zahlbr., 1930 b, p. 133), Hubei (Chen, Wu & J. C. Wei, 1989, p. 471).

= *Cladonia fimbriata* var. *radiata*(Willd.) Cromb., *Lich. eur. reform.* (Lund): 222(1831).

≡ *Cladonia radiata* Willd., *Fl. berol. prodr.*: 363(1787).

= *Cladonia subulata* f. *radiata* (Schreb.) J. W. Thomson, *Lich. Gen. Cladonia North America*(Toronto): 112 (1968)[1967], [*Cladonia cornuto-radiata* f. *radiata*(Schreb.) Thoms.]

≡ *Lichen radiatus* Schreb., *Spic. fl. lips.* (Lipsiae)(1771).

Heilongjiang(Asahina, 1952 d, p. 374).

f. furcellata(Hoffm.) J. C. Wei 小粉叉变型

Enum. Lich. China: 81(1991).

≡*Cladonia furcellata* Hoffm. *Deutschl. Fl.* II 1796, p. 118.

≡*Cladonia cornuto-radiata* Coem. f. *furcellata* Hoffm., in Zahlbr., *Cat. Lich. Univ.* VIII: 442(19342).

Heilongjiang(Asahina, 1952d, p. 374).

Cladonia sulphurina(Michx.) Fr. 硫石蕊

Lich. eur. reform. (Lund): 237(1831).

≡*Scyphophorus sulphurinus* Michx., Fl. Bor. -Amer. 2: 328(1803).

= *Cladonia gonecha*(Ach.) Asahina, *J. Jap. Bot.* 15: 609(1939).

≡*Cladonia deformis* f. *gonecha* Ach. in Zahlbr. *Cat. Lich. Univ.* IV: 481(1927).

Jilin(Chen et al. 1981, p. 132), Xinjiang(Abdulla Abbas & Wu, 1998, p. 60; Abdulla Abbas et al., 2001, p. 362).

Cladonia symphycarpia(Flörke) Fr. 类腐石蕊

Fl. carniol., Edn 2(Wien): 20(1826).

≡*Capitularia symphycarpia* Flörke, *Trans. Am. phil. Soc.*, *New Series* 4(2): 194(1810).

Misapplied name:

Cladonia pyxidata var. *pocillum* auct. non(Ach.) Flot.: H. Magn. Lich. Centr. Asia, 1: 69(1940).

On old trunk.

Xinjiang(H. Magn. 1940, p. 69; Ahti, 1976, p. 371), Guangxi(Guo, 1999, p. 345, the specimen and locality was not cited.).

Cladonia trassii Ahti 特拉氏石蕊

Folia Cryptog. Estonica 32: 7(1998).

Jilin, Changbaishan, SW slope, 2400 m alt., 1950, S. G. Wang no. 479 [H, IFP] (Ahti, 1998, p. 16).

Cladonia umbricola Tønsberg & Ahti 荫生石蕊

Norw. J. Bot. 27: 307(1980).

On wood and *Pseudotsuga wilsoiniana*. New to Asia, known from Europe and America.

Taiwan(Aptroot, Sparrius & Lai 2002, p. 283).

Cladonia uncialis(L.) Weber ex F. H. Wigg. 寸石蕊

Prim. fl. holsat. (Kiliae): 90(1780).

≡*Lichen uncialis* L., *Sp. pl.* 2: 1153(1753).

On the ground.

Heilongjiang, Jilin, Neimenggu(Chen et al. 1981, p. 134).

Cladonia verticillata (Hoffm.) Schaer. 千层石蕊

Lich. Helv. Spic. 1(1): 31(1823).

≡*Cladonia pyxidata* [unranked] *verticillata* Hoffm., *Deutschl. Fl.* 2: 122(1796).

= *Cladonia pseudalcicornis Asahina*, J. Jap. Bot. 19: 192(1943). I. Yoshimura, *Lichen Flora of Japan in color*, 148(1994).

Type: JAPAN(not designated).

= *Cladonia schofieldii* Ahti & Brodo in Brodo & Ahti, *Canad. J. Bot.* 74: 1170(1996).

Type: CANADA. British Columbia, Queen Charlotte Islands, Moresby Island, 2.5 km SW of Tasu, 800 – 1000 m alt., 1980, Ahti no. 38989 [H – holotype; H, CANL, UBC – isotypes].

Jilin: Changbai Mts, 1977, X. L. Chen no. 4916 [H, IFP] (Ahti, 2007, pp. 14 – 15). Neimenggu(Chen et al. 1981, p. 134), Shaanxi(Baroni, 1894, p. 47; Jatta, 1902, p. 465; Zahlbr., 1930b, p. 134), Xizang(J. C. Wei & Jiang, 1981, p. 1147 & 1986, p. 90), Guizhou(Oliv. 1904, p. 194), Yunnan(Hue, 1889, p. 160; Zahlbr., 1930b, p. 134), Taiwan(Wang & Lai, 1973, p. 90), China(prov. not indicated: Kremp. 1868, p. 308; Asahina & Sato in Asahina, 1939, p. 689).

var. verticillata 原变种

On the ground.

Yunnan(Vain. 1894, p. 183; Hue, 1898, p. 274; Zahlbr. , 1930b, p. 134; Sato. 1952, p. 174) .

var. *evoluta* f. apoticta(Ach.) Vain. 离心变种展变型

Acta Soc. Fauna Flora fenn. 10: 184(1894) .

≡*Baeomyces pyxidatus* β *apotictus* Ach. , *Methodus*, Sectio post. : 338(1803) .

Yunnan(Zahlbr. , 1930b, p. 134) .

var. dilatata f. phyllocephala(Flot.) Vain. 膨大变种鳞头变型

Acta Soc. Fauna Flora fenn. 10: 185(1894) .

≡*Cladonia cervicornis* B. *verticillata* Ba. *dilatata* 1. *phyllocephala* Flot. Lich. Fl. Siles 31(1849) , in Vain. , *Acta Soc. Fauna Flora fenn.* 10: 185(1894) .

Xizang(J. C. Wei & Jiang. 1986, p. 90) , Yunnan(Hue, 1898, p. 274) .

Cladonia vulcani Savicz 多枝红石蕊

Bull. Jard. Imp. Bot. Pierre le Grand 14: 124(1914) .

≡*Cladonia polydactyla* var. *theiophila*(Asahina) Asahina, *Lichens of Japan*, 1. *Genus Cladonia*: 94(1950) [*Cladonia macilenta* ssp. *theiophila*(Asahina) Asahina]

Taiwan(Wang-Yang & Lai, 1973, p. 89; Ahti & Lai, 1979, p. 233; Liu et al. 1980, p. 33; Lai, 2000, p. 243) .

Cladonia yunnana(Vain.) des Abbayes ex J. C. Wei & Y. M. Jiang 云南石蕊

Lich. of Xizang: 84. Apr. 1986; Ahti, T. 1993. Names in current use in the Cladoniaceae(lichen – forming ascomycetes) in the ranks of genus to variety. Regnum Veg. 128: 101.

≡*Cladonia transcendens* var. *yunnana* Vain. , in Hue, *Nouv. Arch. Mus. Hist. Nat.* , sér. 3, 10: 262(1898) .

≡*Cladonia yunnana*(Vain.) des Abbayes, in *Candollea*, 16: 202(1958) , nom. inval. (ICBN-Tokyo Code: Article 33. 2.) .

Lectotype(designated here) : China, Yunnan, Mt. Tsang-shan(as “Tsang-chan”) , 1883, R. P. Delavay s. n. (137 in TUR-V.) (PC-Hue, isolectot. : TUR-V. No. 14169) .

Misapplied name:

Cladonia granulans f. *sorediascens* non Asahina: J. C. Wei & Y. M. Jiang, *Lichens of Xizang*, 1986, p. 84(No. 134) .

Xizang(J. C. Wei & Chen, 1974, p. 176; J. C. Wei &Y. M. Jiang, 1981, p. 1146 & 1986, p. 84) , Yunnan(Zahlbr. , 1930 b, p. 135; Asahina, 1939 b, p. 615; Sandst. 1939, p. 98; des Abb. 1958, p. 202; Yoshim. 1968, p. 199; C. Culb. 1969, p. 338; Wei X. L. et al. 2007, p. 151) , Hubei(Chen, Wu & J. C. Wei, 1989, p. 463) , Taiwan(Ahti & Lai, 1979, p. 233; Liu et al. 1980, p. 33; Lai, 2000, p. 244) .

Gymnoderma Nyl. **圆盘衣属**

Flora, Regensburg 43: 546(1860) .

Gymnoderma Humb. , Fl. Friberg. Spec. (Berlin) : 109(1793) , Nom. rejic. , see Arts 14. 6, Ex. 5 and 14. 7.

Gymnodermatomyces Cif. & Tomas. , Atti Ist. bot. Univ. Lab. crittog. Pavia, Ser. 4 10(1) : 40, 67(1953) .

Gymnoderma coccocarpum Nyl. 圆盘衣

in Flora 43: 546(1860) .

Taiwan(M. Sato, 1941 a, p. 95) , Guangxi(Sandst. 1932, map 54; Evans, 1947; Yoshim. & Sharp, 1968) , Xizang (Sandst. 1932, map 54) .

Gymnoderma insulare Yoshim. et Sharp 岛圆盘衣

in Amer. J. Bot. 55: 638(1968) .

Misapplied name:

Gymnoderma coccocarpum sensu Wang & Lai, *Taiwania*, 18 (1) : 91 (1973) , non Nyl. Synops. Lich. 2: 26

(1863). [non Nyl. in *Flora* 43: 546(1860)]

Taiwan(Wang & Lai, 1973, p. 91 as *G.. coccocarpum* non Nyl. & 1976, p. 227 = *G. insulare*).

Pilophorus Th. Fr. 柱衣属

Stereoc. Piloph. Comm.: 40(1857).

Pilophorus acicularis(Ach.) Th. Fr. 大柱衣

Stereoc. Piloph. Comm.: 41(1857).

≡*Baeomyces acicularis* Ach., *Methodus, Sectio prior*: 328, tab. VIII, fig. 4(1803).

On rocks.

Yunnan(M. Sato, 1941a, p. 37), Xizang(J. C. Wei & Y. M. Jiang, 1986, p. 76; W. Obermayer, 2004, p. 506, as *P.* cf. *acicularis*), Hubei(Chen, Wu & J. C. Wei, 1989, p. 459), Taiwan(Lai, 2000, pp. 263 – 264; Lai 2001, p. 176), China [prov. not indicated: Muell. Arg. 1893, p. 235; Sandst., 1932, p. 68; Asahina & Sato in Asahina, 1939, p. 681, as *Pilophoron aciculare*(Ach.) Th. Fr.]

Pilophorus cereolus(Ach.) Th. Fr. 圆头柱衣

Lich. Scand. (Upsaliae) 1(1): 55(1871).

≡*Lichen cereolus* Ach., *Lich. Suec. Prodr.* (Linköping): 89(1798).

≡*Stereocaulon cereolus*(Ach.) Ach., *Methodus, Sectio prior*: 314, tab. VII, fig. 1(1803).

On rocks.

Yunnan(Zahlbr., 1930 b, p. 128, as *Pilophoron cereolum* Th. Fr.), China(prov. not indicated: Sandst. 1932, p. 68; Asahina & Sato in Asahina, 1939, p. 681, as *Pilophoron cereolus* Th. Fr.).

Pilophorus clavatus Th. Fr. 棒柱衣

Bot. Notiser: 214(1888).

= *Pilophorus hallii*(Tuck.) Vain., *Bot. Mag.*, Tokyo 35: 59(1921).

≡*Pilophorus acicularis* f. *hallii* Tuck., *Mém. Soc. natn. Sci. nat. Cherbourg* 20: 96(1877).

= *Pilophorus japonicus* Zahlbr., *Cat. Lich. Univ.* 4: 433(1926) [1927].

Taiwan(M. Sato, 1958 a, p. 61; Kurok. 1970, p. 79; Wang-Yang & Lai, 1973, p. 95 & 1976, p. 228; Lai, 2000, p. 264; Lai 2001, p. 176).

Pilophorus fruticosus L. S. Wang & Xin Y. Wang 果柱衣

in Xin Y. Wang et al., *Lichenologist* 43(2): 7 – 140(2011).

Typus: China, Yunnan, Dali Co., Cangshan Mt., 25° 40# 47. 5 $ N, 100° 06# 02. 8 $ E, alt. 3570m, on rock, 28 July 2006, L. S. Wang 06 – 26204(KUN—holotypus; KoLRI—isotypus).

Pilophorus yunnanensis L. S. Wang & X. Y. Wang 云南柱衣

in Wang, Xin-Yu, Yogesh Joshi, Jae Seoun Hur, Soon Ok Oh, and Li Song Wang *Bryologist*, 113(2): 345 – 349 (2010).

Type: China, Yunnan: Luquan Co., Jiaozixueshan Mt., 26u059 N, 102u089 E, alt. 3600 m, On exposed rocks, 2003, L. S. Wang 03- 22641(holotype: KUN; isotype: Korean Lichen Research Institute, Sunchon National University

Thysanothecium Mont. & Berk. 扇盘衣属

London J. Bot. 5: 257(1846).

Thysanothecium scutellatum(Fr.) D. J. Galloway 扇盘衣

in Galloway & Bartlett, *Nova Hedwigia* 36(2 – 4): 390(1983) [1982].

≡*Cladonia scutellata* Fr., in Lehmann, *Pl. Preiss.* 2: 141(1846).

Guangxi(Guo, 1999, p. 345, the specimen and locality was not cited), Hainan(J. C. Wei et al., 1994, p. 24 – 26: J. C. Wei et al., 2013, p. 231).

(44) Crocyniaceae M. Choisy ex Hafellner 棉絮衣科

Beih. Nova Hedwigia 79: 274(1984).

Crocynia (Ach.) A. Massal. 棉絮衣属

Atti Inst. Veneto Sci. lett., *ed Arti*, Sér. 3 5: 251(1860)[1859 - 1860].

≡*Lecidea* sect. *Crocynia* Ach., *Lich. univ.*: 217(1810).

Crocynia faurieana B. de Lesd. apud Hue 台湾棉絮衣

Bull. Soc. Bot. France 71: 399(1924).

Type: Taiwan, collected by R. P. Faurie in 15/VI/1903.

Taiwan(Zahlbr., 1932, p. 237 & 1933c, p. 68; Wang & Lai, 1973, p. 90).

Crocynia gossypina(Sw.) A. Massal. 棉絮衣

Atti Inst. Veneto Sci. lett., *ed Arti*, Sér. 3 5: 252(1860)[1859 - 1860].

≡*Lichen gossypinus* Sw., *Prodr.*: 146(1788).

Corticolous.

Taiwan(Zahlbr., 1933 c, p. 23; Wang & Lai, 1973, p. 90).

(45) Gypsoplacaceae Timdal **鳞型衣科**

in Jahns, *Biblthca Lichenol.* 38: 423(1990)

Gypsoplaca Timdal, in Jahns **鳞型衣属**

Biblthca Lichenol. 38: 423(1990).

Gypsoplaca macrophylla(Zahlbr.) Timdal 大叶鳞型衣

in Jahns, *Biblthca Lichenol.* 38: 424(1990).

≡*Lecidea macrophylla* Zahlbr., in Handel-Mazzetti, *Symb. Sinic.* 3: 109 - 110(1930) & *Cat. Lich. Univ.* 8: 384 (1932). Timdal, E. 1990. Gypsoplacaceae and Gypsoplaca, a new family and genus of squamiform lichens. -In: H. M. Jahns(ed.): Contributions to Lichenology in Honour of A. Henssen. Bibliotheca ichenologica. No. 38. J. Cramer, Berlin-Stuttqart. Pp. 419 - 427. (Description of *G.* . *macrophylla*: p. 424: illustration: p. 421).

Type: Yunnan, Handel-Mazzetti no. 6954.

On soil.

Xinjiang(Abdulla et al., 2015, p. 510).

Sichuan(Smith, 528: UPS), (Timdal, 1984, p. 535), Yunnan(Schneider, 1979, p. 212).

On schistose rock.

On schistose rock.

Psora ochrorufa(H. Magn.) Wang XY(Abbas et al., 1993, p. 80).

≡*Lecidea ochrorufa* H. Magn. *Lichens Central Asia*, 1940, p. 57 - 58.

Type: Gansu, Jindirtetaralingin 17/7 1931 Bohlin(no. 39a, 40b: type) at about 2525m.

Lich. Centr. Asia 1: 57(1940).

Type: Gansu, no. 40 b in S.

On calcareous soil.

= *Squamarina pamirica* N. S. Golubk. Novitates Systematicae Plantarum non Vascularium 8: 239 - 242(1971).

Type: Tadzhikistania, Pamir Orientalis. N. S. Golubkova 122 in LE.

Xinjiang(Wang XY, 1985, p. 340, as *Psora ochrorufa*(H. Magn.) Wang XY; Abbas et al., 1993, p. 80, as *Psora ochrorufa*), Gansu(G. Schneider, 1979, p. 243, as *Lecidea ochrorufa*). Sichuan(Smith, 528: UPS), (Timdal, 1984, p. 535), Yunnan(Schneider, 1979, p. 212).

(46) Haematommataceae Hafellner **赤星衣科**

Beih. Nova Hedwigia 79: 281(1984).

Haematomma A. Massal. **赤星衣属**

Ric. auton. lich. crost. (Verona): 32(1852).

Haematomma africanum(J. Steiner) C. W. Dodge 非洲赤星衣

Beih. Nova Hedwigia 38: 39(1971).

≡*Haematomma puniceum* var. *africanum* J. Steiner, *Bull. Herb. Boissier*, 2 sér. 7: 641(1907).

On tree branches, and *Populus* sp.

Taiwan(Aptroot and Sparrius, 2003, p. 21) .

Haematomma fauriei Zahlbr. 佛利氏赤星衣

Annls mycol. 14: 59, 60(1916) .

On trees.

Zhejiang(Xu B. S. 1989, p. 205) .

Haematomma persoonii(Fée) A. Massal. 博松氏赤星衣

Atti Inst. Veneto Sci. lett. , ed Arti, Sér. 3 5: 253(1860) [1859 – 1860] .

≡*Lecanora persoonii* Fée, *Essai Crypt. Exot.* (Paris) : 119(1825) [1824] .

= *Haematomma similis* Bagl. , *Nuovo G. bot. ital.* 7: 248(1875) , in Thrower, 1988, pp. 16, 107, Synonymy congtributors: Aptroot & Seweaward, 1999, p. 78.

On exposed trees, especially on branches. Pantropical.

Taiwan(Aptroot and Sparrius, 2003, p. 21) , Hongkong(Thrower, 1988, pp. 16, 107; Aptroot & Seweaward, 1999, p. 78) .

Haematomma puniceum(Sw.) A. Massal. 白赤星衣

Atti Inst. Veneto Sci. lett. , ed Arti, Sér. 3 5: 253(1860) [1859 – 1860] . ssp. puniceum 原亚种

Corticolous.

Xizang(Paulson, 1928, P. 317) , Sichuan(Zahlbr. , 1930 b, p. 177) , Guizhou(Zahlbr. , 1930 b, p. 177; Zhang T & J. C. Wei, 2006, p. 7) , Yunnan(Paulson, 1928, p. 317; Zahlbr. , 1930 b, p. 177 & 1934 b, p. 206) , Taiwan (M. Sato, 1940 b, pp. 496, 500; Asahina, 1964, p. 165) ,

subsp. pacificum Asahina 太平洋亚种

J. Jap. Bot. 39: 212(1964) .

Haematomma puniceum sensu Wang & Lai in Taiwania, 18(1) : 91(1973) .

Corticolous.

Yunnan(Wang L. S. et al. 2008, p. 537) , Anhui, Shanghai, Zhejiang(Xu B. S. 1989, p. 205) , Taiwan(Wang & Lai, 1973, p. 91 & 1976, p. 227) .

Haematomma rufidulum(Fée) A. Massal. 赤星衣

Atti Inst. Veneto Sci. lett., ed Arti, Sér. 35: 253(1860) [1859 – 1860] .

≡*Lecanora rufidula* Fée, *Essai Crypt. Exot.* , *Suppl. Révis*. (Paris) : 116(1837) .

Xizang(Obermayer W. , 2004, p. 497) .

Haematomma sp. 赤星衣待定种名

On bark.

Hainan(J. C. Wei et al. , 2013, p. 231) .

Haematomma wattii(Stirt.) Zahlbr. 瓦特氏赤星衣

Cat. Lich. Univ. 5: 776(1928) .

≡*Lecanora wattii* Stirt. , *Trans. Proc. N. Z. Inst.* 30: 383(1898) [1897] .

On *Picea morrisonicola*.

Taiwan(Aptroot and Sparrius, 2003, p. 22) .

(47) Lecanoraceae Körb. [as ' Lecanoreae'] **茶渍科**

Syst. lich. germ. (Breslau) : 104(1855)

Calvitimela Hafellner **光体衣属**

Stapfia 76: 150(2001) .

Calvitimela aglaea(Sommerf.) Hafellner 藻光体衣

Stapfia 76: 151(2001) .

≡*Lecidea aglaea* Sommerf. , *Suppl. Fl. lapp.* (Oslo) : 144(1826) .

≡*Tephromela aglaea*(Sommerf.) Hertel & Rambold, Bot. Jb. 107(1 – 4) : 494(1985) .

Sichuan(W. Obermayer, 2004, p. 492).

Calvitimela armeniaca(DC.) Hafellner 美尼亚光体衣

Stapfia 76: 151(2001).

≡*Rhizocarpon armeniacum* DC., in Lamarck & de Candolle, *Fl. franç.*, Edn 3(Paris) 2: 367(1805).

≡*Tephromela armeniaca*(DC.) Hertel & Rambold(W. Obermayer, 2004, p. 492, cited as synonym)

Sichuan(W. Obermayer, 2004, p. 492).

Carbonea(Hertel) Hertel **炭盘属**

Mitt. bot. StSamml., Münch. 19: 441(1983).

≡*Lecidea* subgen. *Carbonea* Hertel, Beih. *Nova Hedwigia* 24: 101(1967).

Carbonea assimilis(Hampe ex Körb.) Hafellner & Hertel 同炭盘

in Wirth, *Die Flecht. Baden- Württembergs. Verbreitungsatlas* (Stuttgart): 511(1987).

≡*Lecidella assimilis* Hampe ex Körb., *Parerga lichenol.* (Breslau) 3: 202(1861).

≡*Lecidea assimilis*(Hampe ex Körb.) Th. Fr., *Lich. Scand.* (Upsaliae) 1(2): 556(1874).

Xizang(Paulson, 1952, p. 193; Hertel, 1977, p. 363).

Carbonea vitellinaria(Nyl.) Hertel 黄炭盘

Mitt. bot. StSamml., Münch. 19: 442(1983).

≡*Lecidea vitellinaria* Nyl., *Bot. Notiser*: 177(1852).

On stone and, on *Candelariella vitellina* (Hertel & Zhao 1982).

Jilin(Hertel & Zhao, 1982, p. 148; Hawksw. & M. Cole, 2003, p. 360 as lichenicolous fungus), Xizang(W. Obermayer, 2004, p. 492, on *Candelariella* sp.).

Carbonea vorticosa(Flörke) Hertel 轮炭盘

Mitt. bot. StSamml., Münch. 19: 442(1983).

≡*Lecidea sabuletorum* d *vorticosa* Flörke, *Mag. Gesell. naturf. Freunde*, Berlin 2: 311(1808).

On rock.

Yunnan(Zhang LL et al., 2012, p. 446).

Clauzadeana Cl. Roux **果壳藻衣属**

Bull. Soc. linn. Provence 35: 101(1984) [1983].

Clauzadeana macula(Taylor) Coppins & Rambold 网眼果壳藻衣

Biblioth. Lichenol. 34: 85(1989).

≡*Lecidea macula* Taylor, in Mackay, *Fl. Hibern.* 2: 115(1836).

On rocks.

Xinjiang, Sichuan(Zhao1 Xin et al., 2015, p. 708).

Japewia Tønsberg **厚壁孢属**

Lichenologist 22(3): 205(1990).

Japewia tornoënsis(Nyl.) Tønsberg 厚壁孢

Lichenologist 22(3): 206(1990).

≡*Lecidea tornoënsis* Nyl., Herb. Mus. Fenn.: 110(1859).

On trunk of very old Juniperus.

Sichuan(Obermayer, 2004, p. 500).

Lecanora Ach. **茶渍属**

in Luyken, *Tent. Hist. Lich.*: 90(1809).

Lecanora accumulata H. Magn. 聚茶渍

Lich. Centr. Asia, 1: 113(1940).

Type: Gansu, at about 3000 m, 1931, Bohlin no. 40a in S(!).

On calciferous rocks.

Xinjiang(Abdulla Abbas et al., 1993, p. 78; Abdulla Abbas & Wu, 1998, pp. 75 – 76; Abdulla Abbas et al.,

2001, p. 364), Gansu(H. Magn., 1940, p. 113).

Lecanora adolfii J. C. Wei 肝茶渍

Enum. Lich. China: 124(1991).

Replaced synonym:

≡ *Lecanora lividolutea* H. Magn. *Lich. Centr. Asia*, 1: 118(1940), non Räsänen in Plitt, *Bryologist*, 35: 82 (1932). [Nom. illegit., Art. 53. 1]

Type: Qinghai, Bohlin no. 88c in S(!).

On calciferous rock.

Neimenggu(H. Magn. 1944, p. 45)

Lecanora albella(Pers.) Ach. var. albella 小白茶渍

Lich. univ.: 369(1810). var. albella 原变种

≡ *Lichen albellus* Pers., *Ann. Bot.* (Usteri) 5: 18(1794).

= *Lecanora pallida* (Schreb.) Rabenh., *Deutschl. Krypt. -Fl.* (Leipzig) 2: 34(1845).

≡ *Lichen pallidus* Schreber, *Spic. Fl. Lips.* 133(1771).

Corticolous.

Gansu(Lü Lei et al., 2008, p. 100), Yunnan(Hue, 1889, p. 171; Zahlbr., 1930 b, p. 162), Taiwan(Zahlbr., 1933 c, p. 51; Wang & Lai, 1973, p. 91).

var. cinerella Flörke 灰色变种

in *Schlechtendahl, Fl. berol. prodr.* 2: 75(1824). Lecanora pallida v. cinerella(Flk.) Rabh. *Deutschl. Kryptog. -Flora*, vol. II: 34(1845).

≡ *Lecanora pallida* f. *cinerella*(Flörke) Körb., *Syst. lich. germ.* (Breslau): 145(1855).

Corticolous.

Sichuan, Yunnan, Hunan(Zahlbr., 1930 b, p. 162).

Lecanora allophana(Ach.) Nyl. 异形茶渍

Flora, Regensburg 55: 250(1872). var. allophana 原变种

≡ *Lecanora subfusca* var. *allophana* Ach., *Lich. Univ.*: 395(1810).

Corticolous.

Gansu(Lü Lei et al., 2008, p. 100), Shaanxi(Jatta, 1902, p. 474; Zahlbr., 1930 b, p. 168), Sichuan(Zahlbr., 1930 b, p. 168), Yunnan(Zahlbr., 1934 b, p. 205), Shandong(Zhang F et l., 1999, p. 30; Zhao ZT et al., 2002, p. 5), Anhui, Zhejiang(Xu B. S. 1989, p. 206).

= *Lecanora subfusca*(L.) Ach. (nom. ambig.), Corticolous.

Hunan(Zahlbr., 1930b, p. 168), Zhejiang(D. Hawksw & Weng, 1990, p. 515).

var. parisiensis J. Steiner 巴黎变种

Öst. Bot. Z. 63: 339(1913).

= *Lecanora subfusca* var. *horiza* Ach.

On tree trunk.

Shaanxi(Jatta, 1902, p. 474; Zahlbr., 1930 b, p. 168).

Lecanora alpigena(Ach.) Cl. Roux 山茶渍

in Roux, Masson, Bricaud, Coste & Poumarat, *Bull. Soc. linn. Provence, num. spéc.* 14: 108(2011).

≡ *Lecanora varia* var. *alpigena* Ach., *Lich. univ.*: 379(1810).

≡ *Lecanora polytropa* f. *alpigena*(Ach.) Schaer., *Enum. critic. lich. europ.* (Bern): 81(1850).

On rocks.

Xinjiang(Abdulla Abbas et al., 1993, p. 78; Abdulla Abbas & Wu, 1998, p. 78; Abdulla Abbas et al., 2001, p. 364), Xizang(J. C. Wei & Jiang, 1986, p. 32).

= *Lecanora polytropa*(Ehrh.) Rabh., *Deutschl. Krypt. -Fl.* (Leipzig) 2: 37(1845).

≡ *Verrucaria polytropa* Ehrh., in Hoffmann, *Deutschl. Fl.*, *Zweiter Theil* (Erlangen): 196(1796) [1795].

On siliceous rock.

Jilin(Hertel & Zhao, 1982, p. 146), Shaanxi(Jatta, 1902, p. 475; Zahlbr., 1930 b, p. 171), Sichuan(Zahlbr., 1930 b, p. 171), Jiangsu(Wu J. N. & Xiang T. 1981, p. 4; Xu B. S. 1989, p. 209), Taiwan(Aptroot and Sparrius, 2003, p. 25).

On rock.

Xizang(J. C. Wei & Chen, 1974, p. 177; J. C. Wei & Jiang, 1981, p. 1147 & 1986, p. 32).

= *Lecanora polytropa* f. *illusoria*(Ach.) Leight., *Lich. -Fl. Great Brit.*: 198(1871).

≡ *Lecanora varia* var. *illusoria* Ach., *Lich. univ.*: 380(1810).

On rock.

Xizang(J. C. Wei & Chen, 1974, p. 177; J. C. Wei & Jiang, 1986, p. 32).

Lecanora amicalis Zahlbr. 善茶渍

in Fedde, *Repertorium*, 33: 51(1933).

Type locality: Taiwan, Asahina no. 145.

Taiwan(Zahlbr., 1940, p. 476; Lamb, 1963, p. 293; Wang & Lai, 1973, p. 91).

On rocks.

Lecanora argentea Oxner & Volkova 银白茶渍

Nov. Sist. Niz. Rast. 3: 283(1966).

= *Lecanora fuliginosa* Brodo

Saxicolous.

Jilin, shaanxi(Lü Lei et al., 2009, p. 392, the name of the province miscited as Shanxi).

Lecanora argopholis(Ach.) Ach. 碎茶渍

Lich. Univ.: 346(1810).

≡ *Parmelia atra β argopholis* Ach., *Methodus*, Sectio prior: 32(1803).

≡ *Lecanora frustulosa* var. *argopholis*(Ach.) Link

= *Lecanora frustulosa* var. *thiodes* (Spreng.) Link, *Grundr. Krauterk.* 3: 195(1833).

≡ *Lecanora thiodes* Spreng

On siliceous rocks and on soil.

Neimenggu(H. Magn. 1944, p. 39), Xinjiang(Abdulla Abbas et al., 1993, p. 78 as *Lecanora frustulosa* var. *thiodes*(Spreng) Link; Abdulla Abbas & Wu, 1998, p. 76; Abdulla Abbas et al., 2001, p. 364; Obermayer, 2004, p. 500), Ningxia(Liu M & J. C. Wei, 2013, p. 44 Yang J. R Wei J. C., 2014, p. 1029), Shaanxi(Jatta, 1902, p. 475; Zahlbr., 1930 b, p. 171), Sichuan(Obermayer, 2004, p. 500).

Lecanora behringii Nyl. 贝氏茶渍

Flora, Regensburg 68: 439(1885).

≡ *Lecanora hagenii* var. *behringii*(Nyl.) Lynge, *Lich. Bear. Isl.*: 53(1926).

= *Lecanora hagenii* f. *saxifragae* Anzi, *Comm. Soc. crittog. Ital.* 2(fasc. 1): 8(1864).

On calcareous rock.

Xinjiang(Aksai-Chen: Zahlbr., 1933 a, p. 25).

Note: This species remains to be taxonomically revised.

Lecanora bruneri Imshaug & Brodo 布鲁氏茶渍

Nova Hedwigia, 12: 13(1966).

On bark.

Shaanxi(Wang CL et al., 2007, p. 47).

Lecanora byssulina Zahlbr. 粗麻茶渍

in Handel-Mazzetti, *Symb. Sin.* 3: 165(1930) & *Cat. Lich. Univ.* 8: 533(1932).

Type: Sichuan, Handel-Mazzetti no. 1806.

On rocks.

Lecanora caesioalutacea H. Magn. 兰茶渍

Lich. Centr. Asia, 1: 85(1940).

Type: Gansu, Bohlin no. 85a in S(!).

On sometimes slightly calciferous rock among different lichens.

Xinjiang(Abdulla Abbas et al., 1993, p. 78 as *L.* "*caesioalatorea*" H. Magn.; Abdulla Abbas & Wu, 1998, p. 76; Abdulla Abbas, 2001, p. 364 as *L.* "*caesioaularea*"), Gansu(H. Magn., 1940, p. 85).

Lecanora caesiorubella Ach. 蟹形茶渍

Lich. Univ.: 366(1810).

= *Lecanora cancriformis*(Hoffm.) Vain., *Hedwigia* 37: 38(1898).

≡ *Verrucaria cancriformis* Hoffm., *Deutschl. Fl.*, Zweiter Theil(Erlangen): 171(1796) [1795].

Corticolous.

Sichuan(Zahlbr., 1930 b, p. 162), Yunnan(Hue, 1889, p. 172; Zahlbr., 1930 b, p. 162).

Lecanora californica Brodo 加州茶渍

Beih. *Nova Hedwigia* 79: 107(1984).

On soil covering rock.

Ningxia(Liu M & JCWei, 2013, p. 44).

Lecanora callopizodes Nyl. 丽茶渍

Lich. Ins. Guin.: 45(1889); Hue in *Bull. Soc. Bot. France*, 26: 170(1889) et in *Nouv. Archiv. Du Museum*, ser. 3, 3: 54(1891).

Type: Yunnan. (Hue, 1891, p. 54).

Corticolous.

Lecanora campestris(Schaer.) Hue 平原茶渍

Bull. Soc. bot. Fr. 35: 47(1888).

≡ *Parmelia subfusca* var. *campestris* Schaer., *Lich. Helv. Spicil.* 8: 391(1839).

≡ *Lecanora subfusca* var. *campestris* Rabh.

On rocks.

Gansu(Lü Lei et al., 2008, p. 101), Shaanxi(Jatta, 1902, p. 474; Zahlbr., 1930 b, p. 168), Shandong(Moreau et Moreau, 1951, p. 189), Jiangsu(Wu J. N. & Xiang T. 1981, p. 4; Xu B. S. 1989, p. 207), Shanghai(Xu B. S. 1989, p. 207).

Lecanora cancriformoides Zahlbr. 拟蟹形茶渍

in Handel-Mazzetti, *Symb. Sin.* 3: 163(1930).

Type: Yunnan, Handel-Mazzetti no. 6866.

Yunnan(Zahlbr., 1932 a, p. 533).

On bark of trees.

Lecanora carnulenta Nyl. 大粒茶渍

in Nyl. & Cromb. *Journ. Linn. Soc. London, Bot.* 20: 63(1883).

Type: from Zhejiang, Ninghai.

Zhejiang(Hue, 1891, p. 66; Zahlbr., 1928, p. 402 & 1930 b, p. 164).

Lecanora carpinea(L.) Vain. 树皮茶渍

Meddn Soc. Fauna Flora fenn. 14: 23(1888). f. carpinea 原变型

≡ *Lichen carpineus* L., *Sp. pl.* 2: 1141(1753).

f. intermedia(Kremp.) Zahlbr. 中间变型

Cat. Lich. Univ. 5: 407(1928).

≡ *Lecanora intermedia* Kremp., *Denkschr. Kgl. Bayer. Bot. Ges.*, Abt. 2 4: 149(1861).

≡ *Lecanora subfusca* var. *parisiensis* f. *intermedia*(Kremp.) Oliv.

Corticolous.

Shanghai(Rabh. 1873, p. 287; Kremp. 1873, p. 470).

Lecanora cateilea(Ach.) A. Massal. 杜鹃茶渍

Ric. auton. lich. crost. (Verona): 9(1852).

≡*Lecanora subfusca* var. *cateilea* Ach., *Lich. Univ.*: 394(1810).

On rhododendra.

Gansu(Lü Lei et al., 2008, p. 101), Shaanxi(Wang CL et al., 2007, p. 47).

Lecanora cathayensis Zahlbr. 中华茶渍

in Handel-Mazzetti, *Symb. Sin.* 3: 167(1930); Zahlbr., *Cat. Lich. Univ.* 8: 533(1932).

=*Lecanora sinensis* Zahlbr., *Cat. Lich. Univ.* 5: 545(1928), non *Lecanora chinensis* Zahlbr. (orthographic variants).

Type: Shaanxi, collected by Giraldi.

On the ground.

Lecanora polycarpa sensu Jatta, *Nuov. Giorn. Bot. Ital.* 9: 475(1902), non Anzi., nec Anzi. Jiangsu, Zhejiang (Xu B. S. 1989, p. 207).

Lecanora cenisia Ach. 坚盘茶渍

Lich. Univ.: 361(1810).

=*Lecanora subfusca* var. *atrynea* Ach., *Lich. univ.*: 395(1810).

≡*Lecanora atrynea*(Ach.) Röhl., *Deutschl. Fl.* (Frankfurt) 3(2): 82(1813).

On rocks and bark of trees.

Xinjiang(Reyim Mamut et al., 2006[2005] p. 204), Gansu(Lü Lei et al., 2008, p. 101), Sichuan(Zahlbr., 1930b, p. 167), Yunnan(Hue, 1889, p. 171; Zahlbr., 1930 b, p. 167), Shandong(Zhao ZT et al., 2002, p. 5; Zhang F et al., 1999, p. 30; Li y et al., 2008, p. 71), Jiangsu(Wu J. N. & Xiang T. 1981, p. 4), Taiwan(Wang & Lai, 1973, p. 91).

Lecanora chinensis Zahlbr. 小盘茶渍

Cat. Lich. Univ. 5: 413(1928).

≡*Lecanora microcarpa* Jatta, *Nuov. Giorn. Bot. Ital.* 9: 475(1902), non Fée. (nom. illegit.)

Type: Shaanxi, collected by Giraldi.

Shaanxi(Zahlbr., 1930 b, p. 163).

On siliceous rocks.

Lecanora chlarotera Nyl. 亚丽茶渍

Bull. Soc. linn. Normandie, sér. 2 6: 274(1872).

On trees(*Populus* sp.).

Gansu(Lü Lei et al., 2008, p. 101), Shandong(Li y et al., 2008, p. 71), Jiangsu(Xu B. S. 1989, p. 207), Taiwan(Aptroot and Sparrius, 2003, p. 23).

=*Lecanora chlarona* (Ach.) Nyl., *Flora*, Regensburg 55: 250(1872).

≡*Lecanora distincta* var. *chlarona* Ach., *Lich. univ.*: 158(1810).

Sichuan, Guizhou, Hunan, Fujian(Zahlbr., 1920 b, p. 167).]

[*Lecanora chlarona* var. *chlarona* f. *geographica*(Massal.) Nyl.]

Yunnan(Zahlbr., 1930 b, p. 167).

[L. *chlarona* var. *chlarona* f. *microcarpa*(Kremp.) Zahlbr. *Cat. Lich. Univ.* 5: 416(1928).]

≡*Lecanora subfusca* var. *chlarona* f. *microcarpa* Krphbr. Flora, 56: 470(1873) & Hedwigia, 13: 65(1874).

Type: Guangdong, Hongkong(syntypes).

Guangdong(Rabh. 1873, p. 287).

Lecanora chrysocardia Zahlbr. 黄心茶渍

in Handel-Mazzetti, *Symb. Sin.* 3: 165(1930) & *Cat. Lich. Univ.* 8: 534(1932).

Type: Sichuan, Handel-Mazzetti no. 1800.

On granite.

Hongkong(Aptroot & Sewaward, 1999, p. 79; Aptroot et al. , 2001, p. 328) .

Lecanora cinereocarnea(Eschw.) Stizenb. 肉灰茶渍

Ber. Tät. St Gall. naturw. Ges. : 218(1890) [1888 – 89] .

≡*Parmelia varia* var. *cinereocarnea* Eschw. , in *Martius, Fl. Bras.* 1: 187(1833) [1829 – 1833] .

Corticolous.

Gansu(Lü Lei et al. , 2009, p. 438) .

Lecanora cinereofusca H. Magn. 棕灰茶渍

Meddn Göteb. Bot. Trädg. 7: 86(1932) .

On bark.

Gansu(Lü Lei et al. , 2009, p. 312) .

Lecanora circumborealis Brodo & Vitik. 空果茶渍

Mycotaxon 21: 288(1984) .

= *Lecanora coilocarpa* auct.

= *Lecanora subfusca* var. *coilocarpa* Ach.

On trunk of Euphorbia royleana.

Gansu(Lü Lei et al. , 2008, p. 101) , Sichuan(A. Zahlbr. , 1930 b, p. 167) .

Lecanora coccocarpiopsis Nyl. 红果茶渍

in Nyl. & Cromb. *Journ. Linn. Soc. London, Bot.* 20: 62(1883) .

Type: Zhejiang, a hill near Ninghai.

Zhejiang(Hue, 1891, p. 61; Zahlbr. , 1928, p. 610 & 1930 b, p. 172) .

On rocks.

Lecanora compendiosa Nyl. 灌木茶渍

in Nyl. & Cromb. *Journ. Linn. Soc. London, Bot.* 20: 63(1883) .

Type: from Zhejiang, Ninghai.

Zhejiang(Hue, 1891, p. 61; Zahlbr. , 1930 b, p. 168) .

On rocks associated with Lecanora coccocarpiopsis Nyl.

Lecanora crenulata(Dicks.) Hook. 钝齿茶渍

in Smith, *Engl. Fl.* (London) 5: 194(1844) .

≡*Lichen crenulatus* Dicks. , *Fasc. pl. crypt. Brit.* (London) 3: 14(1793) .

Zahlbr. , *Cat. Lich. Univ.* 5: 436(1928) | Lamb's *Index nom. lich.* : 302 Page Image in Published List

On rocks.

Xinjiang(Mamut et al. , 2006[2005] , pp. 204 – 205) , Shaanxi(Jatta, 1902, p. 475; Zahlbr. , 1930 b, p. 162) .

Lecanora dispersa(Pers.) Röhl. 散茶渍

Deutschl. Fl. (Frankfurt) 3(2) : 96(1813) .

≡*Lichen dispersus* Pers. , *Ann. Bot.* (Usteri) 1: 27(1794) .

Xinjiang(Abdulla Abbas, 1993, p. 78; Abdulla Abbas et al. , 1994, p. 21; Abdulla Abbas &Wu, 1998, p. 76; Abdulla Abbas, 2001, p. 364) , Gansu(Lü Lei et al. , 2008, p. 101) .

Lecanora dispersogranulata Szatala 散粒茶渍

Annls Mus. Natn. Hung. , n. s. 7: 46(1956) .

Corticolous.

Gansu(Lü Lei et al. , 2009, pp. 438 – 439) .

Lecanora distracta Zahlbr. 离生茶渍

in Handel-Mazzetti, *Symb. Sin.* 3: 170(1930) & *Cat. Lich. Univ.* 8: 535(1932) .

Type: Yunnan, Handel-Mazzetti no. 6947.

On calciferous rocks.

Lecanora endophaeoides Hue 内褐茶渍

in Bull. Soc. Bot. France, 36: 172(1889).

Type from Yunnan, collected by Delavay.

Yunnan(Hue, 1891, p. 66; Zahlbr., 1928, p. 447 & 1930 b, p. 168), Taiwan(Aptroot and Sparrius, 2003, p. 23).

Corticolous.

Lecanora expallens Ach. 负苍白茶渍

Lich. Univ.: 374(1810).

On *Tsuga formosana*.

Taiwan(Aptroot and Sparrius, 2003, p. 23).

Lecanora farinaria Borrer 粉末茶渍

Suppl. Engl. Bot. 2: tab. 2727(1834).

On exposed trees. Now known from temperate Europe, North Americ and Asia.

Hongkong(Aptroot & Seaward, 1999, p. 79).

Lecanora flavidorufa Hue 红黄茶渍

in Bull. Soc. Bot. France 36: 173(1889); Zahlbr. in Handel-Mazzetti, *Symb. Sin.* 3: 170(1930).

Type: Yunnan, 17/VI/1887, Delavay no. 3000.

Corticolous.

Taiwan(Aptroot and Sparrius, 2003, p. 23).

Lecanora flavovirens Fée 黄绿茶渍

Essai Crypt. Exot. (Paris): 115(1825)[1824]. var. flavovirens Fée 原变种

var. subaeruginosa(Nyl.) Vain. 亚铜绿变种

Étud. Class. Lich. Brésil 1: 83(1890); Zahlbr., 1930 b, p. 169 as "*Lecanora flavovirescens* Fée var. *subaeruginosa*(Nyl.) Vain."

≡*Lecanora granifera* var. *subaeruginosa* Nyl., *Acta Soc. Sci. fenn.* 7(2): 445(1863).

Corticolous.

Yunnan(Zahlbr., 1930 b, p. 169 & 1934, p. 206).

Notes:

The current name of *Lecanora flavovirens* Fée(1825: 115)[1824] is *Vainionora flavovirens* (Fée) Kalb(1991), and the systematic position of *Lecanora flavovirens* var. *subaeruginosa*(Nyl.) Vain. remains to be studied. As to the misspelled *Lecanora* "*flavovirescens* Fée" var. *subaeruginosa* (Nyl.) Vain. in Zahlbr. 1930: 169 is presumably a typographical error and is to be reverted to the original "flavovirens Fée"(1825), and the name of the should be "*Lecanora flavovirens* Fée var. *subaeruginosa*(Nyl.) Vain.".

Lecanora flavoviridis Krempelh. 绿茶渍

Flora, 56: 470(1873).

Type: from Hongkong.

Hongkong(Rabh. 1873, p. 287; Kremp. 1873, p. 470 & 1874 c, p. 65; Zahlbr., 1928, p. 454 & 1930 b, p. 169; Thrower, 1988, p. 16; Lumbsch, 1994, p. 157; Aptroot & Seaward, 1999, p. 79; Aptroot et al, 2001, p. 328).

On porphyritic rocks.

Lecanora fulvastra Kremp. 黄褐星茶渍

Vidensk. Meddel. Dansk Naturhist. Foren. Kjøbenhavn 5: 16, tab. I, fig. 19(1874)[1873].

On *Cryptomeria japonica*, *Diospyros kaki*, and *Populus* sp. Taiwan(Aptroot and Sparrius, 2003, p. 24).

Lecanora gangaleoides Nyl. 美盘茶渍

Flora, Regensburg 55: 354(1872).

On siliceous rock. Cosmopolitan.

Shaanxi(Jatta, 1902, p. 475; Zahlbr., 1930 b, p. 168), Taiwan(Aptroot and Sparrius, 2003, p. 24), Hongkong (Aptroot & Seaward, 1999, p. 79; Aptroot et al., 2001, p. 330).

Lecanora gansuensis L. Lü & H. Y. Wang 甘肃茶渍

in Lü Lei et al., *Mycotaxon* 123: 286 – 287(2013).

Type: Gansu, Wen County, Qiujiaba, 2300 m, on twigs, 3 Aug 2006, Yang et al. 061482 – 2(sdnu holotype).

Gansu, Guizhou(L. Lü et al., 2013, pp. 286 – 287).

Lecanora garovaglioi(Körb.) Zahlbr. [as '*garovaglii*'] 戛氏茶渍

Annln naturh. Mus. Wien 15: 208(1900).

≡*Placodium garovaglioi* Körb. [as 'garovaglii'], *Parerga lichenol.* (Breslau) 1: 54(1859).

On rock.

Gansu(Lü Lei et al., 2008, pp. 102 – 103).

Lecanora geoica H. Magn. 地茶渍

Lich. Centr. Asia 1: 86(1940).

Type: Gansu, Bohlin no. 38b in S(!).

On calcareous soil.

Lecanora glabrata(Ach.) Malme 裸茶渍

in Magnusson, *Meddn Göteb. Bot.* Trädg. 7: 79(1932).

≡ *Lecanora subfusca* var. *glabrata* Ach., *Lich. Univ.*: 393(1810).

≡*Lecanora allophana* var. *glabrata*(Ach.) J. Steiner, *Verh. zool. -bot. Ges.* Wien 65: 200(1915)

On trees.

Gansu(Lü Lei et al., 2008, p. 103), Fujian(Zahlbr., 1934 b, p. 205).

Lecanora griseomurina Zahlbr. 鼠灰茶渍

in Handel-Mazzetti, *Symb. Sin.* 3: 164(1930) & *Cat. Lich. Univ.* 8: 536(1932).

Type: Yunnan, Handel-Mazzetti no. 541.

On quartziferous rock.

Lecanora hagenii(Ach.) Ach. 小茶渍

Lich. Univ.: 367(1810).

≡*Lichen hagenii* Ach., *Lich. Suec. Prodr.* (Linköping): 57(1798).

On dead twigs with *Candelariella aurella*.

Xinjiang(H. Magn. 1940, p. 87; Abdulla Abbas et al., 1993, p. 78; Abdulla Abbas et al., 1994, p. 21; Abdulla Abbas & Wu, 1998, p. 77; Abdulla Abbas et al., 2001, p. 364), Gansu(H. Magn. 1940, p. 87; Abdulla Abbas et al., 1993, p. 78; Abdulla Abbas et al., 1994, p. 21).

Lecanora hellmichiana Poelt 希腊茶渍

in Khumbu Himal. I: 196(1966).

On granite and shale.

Taiwan(Aptroot and Sparrius, 2003, p. 24).

Lecanora helva Stizenb. 淡栗茶渍

Ber. Tät. St Gall. naturw. Ges.: 218(1890) [1888 – 89].

On *Cinnamomum* sp., *Castanopsis* sp., *Roystonea regia*, and *Populus* sp. Taiwan(Aptroot and Sparrius, 2003, p. 24).

Lecanora heterocarpina Zahlbr. 杂果茶渍

in Handel-Mazzetti, *Symb. Sin.* 3: 169(1930) & *Cat. Lich. Univ.* 8: 536(1932).

Type: Yunnan, Handel-Mazzetti no. 629.

On quartziferous rock.

Lecanora horiza(Ach.) Röhl. 平线茶渍

Deutschl. Fl. (Frankfurt) 3(2) : 82(1813) .

≡*Lecanora subfusca* var. *horiza* Ach. , *Lich. Univ.* : 394(1810) .

On *Shiia carlesii.*

Taiwan(Aptroot and Sparrius, 2003, p. 25) .

Lecanora hyaliza Zahlbr. 草茶渍

in Handel-Mazzetti, *Symb. Sin.* 3: 168(1930) & *Cat. Lich. Univ.* 8: 536(1932) .

Type: Sichuan, Handel-Mazzetti no. 2766.

On trunk of *Myrsine africana.*

Lecanora imshaugii Brodo 伊穆氏茶渍

Syllogeus 29: 48(1981) .

= *Lecanora perflexuosa* (Räsänen) H. Miyaw.

On bark.

Liaoning, Gansu(Lü Lei et al. , 2009, pp. 392 – 393) .

Lecanora insignis Degel. 突茶渍

in Ark. f. Bot. XXXA3: 53(1942) .

On *Abies kawakamii. Tsuga* and *Rhododendron* wood.

Taiwan(Aptroot and Sparrius, 2003, p. 25) .

Lecanora intricata(Ach.) Ach. 缠结茶渍

Lich. Univ. : 380(1810) .

≡ *Parmelia intricata* Ach. , *Methodus*, Sectio prior: 50(1803) .

Xijiang(Mamut et al. , 2006' 2005?' , p. 205) , Yunnan(R. Paulson, 1928, p. 317; Zahlbr. , 1939 b, p. 169) , Taiwan(Aptroot and Sparrius, 2003, p. 25) .

On shale.

Lecanora intumescens(Rebent.) Rabh. 肿茶渍

Deutschl. Krypt. -Fl. (Leipzig) 2: 34(1845) .

≡*Parmelia intumescens* Rebent. , *Prodr. fl. neomarch.* (Berolini) : 301(1804) .

On trunk of *Salix* sp.

Yunnan(Zahlbr. , 1930 b, p. 167) .

Lecanora invadens H. Magn. 侵生茶渍

Lich. Centr. Asia 1: 87(1940) .

Type: Gansu, Bohlin no. 69a in S(!) .

On non calciferous rock with other lichens, settling upon Aspicilia asiatica and destroying it.

Lecanora irridens Zahlbr. 灰盘茶渍

in Handel-Mazzetti, Symb. Sin. 3: 171(1930) & *Cat. Lich. Univ.* 8: 537(1932) .

Type: Sichuan, Handel-Mazzetti no. 1865.

On phyllite.

Lecanora isabellina H. Magn. 赭茶渍

Lich. Centr. Asia 1: 115(1940) .

Type: Gansu, Bohlin no. 67a(lectotype) in S(!) .

On non calciferous rock with other lichens.

Gansu(H. Magn. , 1940, p. 115) , Xinjiang(Abdulla Abbas et al. , 1993, p. 78; Abdulla Abbas & Wu, 1998, p. 77; Abdulla Abbas et al. , 2001, p. 364) .

Lecanora japonica Müll. Arg. 日本茶渍

in Flora, LXII: 482(1879) .

On bark. Or rock.

Shaanxi(Wang CL et al. , 2007, p. 47) .

Lecanora kelungensis Zahlbr. 基隆茶渍

in Fedde, *Repertorium*, 33: 52(1933) .

Type: Taiwan, Faurie nos. 254, 261(syntypes) .

Taiwan(Zahlbr. , 1940, p. 481; Lamb, 1963, p. 310; Wang & Lai, 1973, p. 91) .

Lecanora kukunorensis H. Magn. 青海茶渍

Lich. Centr. Asia 1: 117(1940) .

Type: Qinghai, collected by Bohlin in 1932, at about 4000 m , perserved in S(!) .

On hardly calciferous rock.

Xinjiang(Abdulla Abbas et al. , 1993, p. 78; Abdulla Abbas & Wu, 1998, p. 77; Abdulla Abbas, 2001, p. 364) , Qinghai(H. Magn. , 1940, p. 117) .

Lecanora lavidofusca Müll. Arg. 松酸茶渍

Bull. Herb. Boissier 3: 633(1895) .

On bark.

Gansu(Lü Lei et al. , 2009, p. 313) .

Lecanora lemokensis Zahlbr. 蜀茶渍

in Handel-Mazzetti, *Symb. Sin.* 3: 163(1930) . f. lemokensis 原变型

Type: Sichuan, Handel-Mazzetti no. 1583.

On non calcareous rock.

f. infumata Zahlbr. 烟熏变型

in Handel-Mazzetti, *Symb. Sin.* 3: 164(1930) & *Cat. Lich. Univ.* 8: 537(1932) .

Type: Yunnan, Handel-Mazzetti no. 12978.

On diabasic stone.

Lecanora leprosa Fée 癞屑茶渍

Essai Crypt. Exot. (Paris) : 118(1825) [1824] . var. leprosa 原变种

On exposed trees. Pantropical.

Yunnan(Wang L. S. et al. 2008, p. 537) , Taiwan(Aptroot and Sparrius, 2003, p. 23, as *Lecanora achroa* Nyl.) , Hongkong(Thrower, 1980 & 1988, pp. 16, 111; Aptroot & Seaward, 1999, p. 79; Aptroot et al. , 2001, p. 330; Seaward & Aptroot, p. 2005, p. 285, Bushes in thickets. , Pfister 1978: 95, as *Lecanora subfusca* var. *cinereocarnea* Tuck.) .

= *Lecanora achroa* Nyl. , in Crombie, *J. Bot.* , *Lond.* 14: 26(1876) .

Corticolous.

Jiangsu(Wu J. N. & Xiang T. 1981, p. 4; Xu B. S. 1989, p. 208) , Shanghai(Nyl. & Cromb. 1883, p. 63; A. Zahlbr. , 1930b, p. 162; Xu B. S. 1989, p. 208) , Zhejiang(Xu B. S. 1989, p. 208) , Fujian(A. Zahlbr. , 1930 b, p. 162) , Taiwan(Wang & Lai, 1973, p. 91) , Hongkong(Thrower, 1988, p. 111) .

var. phaeochroa(Nyl.) Zahlbr. 暗色变种

Cat. Lich. Univers. 5: 481(1928) .

≡ *Lecanora achroa* var. *phaeochroa* Nyl. in *Journal Linn. Soc. London Bot.* 20: 53(1883) .

Shanghai(Nyl. & Cromb. 1883, p. 63; Zahlbr. , 1930 b, p. 162) .

Lecanora loekoesii L. Lü, Y. Joshi & Hur 洛凯氏茶渍

in Lü, Joshi, Elix, Lumbsch, Wang, Koh & Hur, *Lichenologist* 43(4) : 324(2011) .

Heilongjiang, jilin, Liaoning(Wang HY et al. , 2013, pp. 236 – 238) .

Lecanora megalospora Hue 大孢茶渍

Ann. Mycolog. 13: 88(1915) .

Type: Yunnan, Delavay no. 21.

Yunnan(Zahlbr. , 1928, p. 485 & 1930 b, p. 168) .

Lecanora microphaea Zahlbr. 叶上茶渍

in Handel-Mazzetti, Symb. Sin. 3: 166(1930) & *Cat. Lich. Univ.* 8: 538(1932) .

Type: Guizhou, Handel-Mazzetti no. 10544.

On twigs of *Schizophragma integrifolium* and *Xylosma racemosum*.

Lecanora mikuraensis Miyaw. 米库尔茶渍

J. Hattori bot. Lab. 64: 296(1988) .

On bark.

Hebei(Han LF et al. , 2011, p. 22) .

Lecanora nipponica H. Miyaw. H. Miyaw. 瘤体茶渍

J. Hattori Bot. Lab. 64: 296(1988) .

Corticolous.

Shaanxi, Yunnan(Lü Lei et al. , 2009, pp. 393 – 394, the province name was miscited as ' Shannxi') .

Lecanora novae-hollandiae Lumbsch [as ' *novaehollandiae*'] 晶缘茶渍

J. Hattori bot. Lab. 77: 118(1994) .

Corticolous.

Gansu, anhui(Lü Lei et al. , 2009, pp. 394 – 395) .

Lecanora opiniconensis Brodo 石茶渍

Syllogeus 29: 52(1981) .

Saxicolous.

Shaanxi(Lü Lei et al. , 2009, p. 440, the province name was miscited as miscited as ' Shanxi') .

Lecanora oreinoides(Körb.) Hertel & Rambold 滇茶渍

in Rambold, Biblthca Lichenol. 34: 100(1989)

≡*Aspicilia oreinoides* Körb. , *Abh. Schles. Ges. Vaterl. Kult. Abth. Naturwiss.* 2: 32(1862) .

≡*Lecidea oreinoides*(Körb.) W. A. Weber & Hertel(in Hertel 1977, p. 297)

=*Lecidea internigrans* Krempelh. *Flora*, 56: 468 (1873) , (in Rabh. 1873, p. 286; Krempelh. 1873, p. 468 & 1874, p. 60; Thrower, 1988, p. 16) .

Type: Hongkong, R. Rabh. 1871 – 1872(lectotype) .

=*Lecidea yunnana* Zahlbr. , in Handel-Mazzetti, *Symbolae Sinicae* 3: 98(1930) .

Type: Yunnan, Handel-Mazzetti no. 127.

Yunnan(Hertel, 1977, p. 297) .

On quartziferous rock. Cosmopolitan.

Hongkong(Rabenhorst 1873, p. 286; Krempelh. 1873, p. 468 & 1874, p. 60; Zahlbr. , 1925, p. 901 & 1930 b, p. 98; Hertel, 1971, p. 238 & 1977, p. 297; Thrower, 1988, p. 16; Aptroot & Seaward, 1999, p. 79 – 80; Aptroot et al. , 2001, p. 330) .

Lecanora orosthea(Ach.) Ach. 口茶渍

Lich. univ. : 400(1810) .

≡*Lichen orostheus* Ach. , *Lich. suec. prodr.* (Linköping) : 38(1799) [1798] .

On exposed coastal granite. Known from temperate Europe, Africa and North America, and Also Hong Kong in Asia.

Hongkong(Aptroot & Seaward, 1999, p. 80) .

Lecanora pachirana Zahlbr. 厚茶渍

in Fedde, *Repertorium*, 33: 53(1933) .

Type: Taiwan, Faurie no. 279.

Taiwan(Zahlbr, 1940, p. 483; Lamb, 1963, p. 318; Wang & Lai, 1973, p. 91) .

Lecanora pachyphylla H. Magn. 黄厚叶茶渍

Lich. Centr. Asia 1: 120 – 121(1940).

Type: Gansu, Gun-tsan 20/3 1932 Bohlin(A: a, A: b = type) at about 3825m.

Xinjiang(Abbas et al., 1993, p. 78), Gansu(H. Magn., 1940, pp. 120 – 121).

Lecanora pachysperma Hue 粗子茶渍

Ann. Mycolog. 13: 97(1915).

Type: Yunnan, collected by Delavay.

Yunnan(Zahlbr., 1928, p. 496 & 1930 b, p. 168).

Corticolous.

Lecanora perflexuosa(Räsänen) H. Miyaw. 多曲茶渍

Journ. Hattori Bot. Lab. 64: 320(1988).

≡*Lecanora glabrata* var. *perflexuosa* Räsänen, J. J. B. 16: 90(1940).

Corticolous.

Shaanxi(Lü Lei et al., 2009, pp. 440 – 441, the province name was miscited as Shannxi).

Lecanora perplexa Brodo 多网茶渍

Beih. Nova Hedwigia 79: 148(1984).

On bark or rock.

Gansu(Lü Lei et al., 2009, p. 313).

Lecanora perpruinosa Fröberg 粉霜茶渍

The Calcicolous Lichens on the Great Alvar of Öland, Sweden(Lund): 50(1989).

Sichuan(Obermayer, 2004, p. 501).

Lecanora pseudistera Nyl. 拟鳞茶渍

Flora, Regensburg 55: 354(1872).

Misapplied name:

Squamaria sp. in Thrower, 1988, pp. 17, 168, revised by Aptroot & Seaward, 1999, p. 80.

On granite. Cosmopolitan.

Hongkong(Thrower, 1988, pp. 17, 168; Aptroot & Seaward, 1999, p. 80; Aptroot et al., 2001, p. 330), Taiwan (Aptroot and Sparrius, 2003, p. 25).

Lecanora pulicaris(Pers.) Ach. 丽盘茶渍

Syn. meth. lich. (Lund): 336(1814).

≡*Patellaria pulicaris* Pers., *Ann. Wetter. Gesellsch.* Ges. Naturk. 2: 13(1811)[1810].

=*Lecanora pinastri*(Schaer.) H. Magn., *Meddn Göteb. Bot. Trädg.* 7: 82(1932).

≡*Lecanora subfusca* var. *pinastri* Schaer., *Lich. helv. spicil.* 8: 390(1839).

≡*Lecanora chlarona* f. *pinastri*(Schaer.) Cromb., *Monogr. Lich. Brit.* 1: 413(1894).

Corticolous.

Gansu(Lü Lei et al., 2008, p. 103), Shaanxi(Moreau et Moreau, 1951, p. 188), Sichuan, Guizhou, Hunan, (Zahlbr., 1930 b, p. 167), Yunnan(Hue, 1889, p. 17; R. Paulson, 1928, p. 317; Zahlbr., 1930 b, p. 167).

Lecanora queenslandica C. Knight 昆士兰茶渍

in Bailey, *Syn. Queensl. Fl.*, *Suppl.* 2: 85(1888).

Shaanxi(Han LF et al., 2011, pp. 22 – 23).

Lecanora rupicola(L.) Zahlbr. 岩茶渍

Cat. Lich. Univers. 5: 525(1928).

≡*Lichen rupicola* L., *Mant. Pl.* 1: 132(1767).

Shandong(Li Y et al., 2008, p. 71).

Lecanora saligna(Schrad.) Zahlbr. 柳茶渍

Cat. Lich. Univers. 5: 536(1928).

≡*Lichen salignus* Schrad., Spicil. fl. germ. 1: 84(1794)

= *Lecanora effusa*(Hoffm.) Ach., *Lich. univ.*: 386(1810).

Xinjiang(Abdulla Abbas et al., 1993, p. 78; Abdulla Abbas & Wu, 1998, p. 78; Abdulla Abbas et al., 2001, p. 364), Gansu(Lü Lei et al., 2008, p. 103), Yunnan(Paulson, 1928, p. 317; Zahlbr., 1930 b, p. 169).

Lecanora setschwana Zahlbr. 四川茶渍

in Handel-Mazzetti, *Symb. Sin.* 3: 169(1930) & *Cat. Lich. Univ.* 8: 540(1932).

Type: Sichuan, Handel-Mazzetti no. 1373.

Shandong(Li Y et al., 2008, p. 71).

Lecanora subfusca(L.) Ach. 亚棕茶渍

Zahlbr., *Cat. Lich. Univ.* V: 551, & *in* Handel-Mazzetti, *Symb. Sin.* 3: 168(1930). var. subfusca

Corticolous. 原变种

Shanghai, Hunan(Zahlbr., 1930 b, p. 168), Hongkong(Rabenhorst, as f. *microcapa* Kremphbr in Flora, LVI, 470, 1873, nom. nud.).

var. *rugosa*(Pers.) Nyl. 皱形变种

Zahlbr., *Cat. Lich. Univ.* V: 566 & *in* Handel-Mazzetti, *Symb. Sin.* 3: 168(1930).

≡*Lichen rugosus* Pers. in Nyl.

Shaanxi(Jatta, 1902, p. 474; Zahlbr., 1930 b, p. 168).

var. subgranulata Nyl. 齿裂变种

Zahlbr., *Cat. Lich. Univ.* V: 567 & *in* Handel-Mazzetti, *Symb. Sin.* 3: 168(1930).

Corticolous.

Yunnan(Hue, 1889, p. 171; Zahlbr., 1930 b, p. 168).

Lecanora subimmergens Vain. 半埋茶渍

Bot. Mag., Tokyo 35: 51(1921).

On shale.

Taiwan(Aptroot and Sparrius, 2003, p. 25).

Lecanora subimmersa(Fée) Vain. 亚沉茶渍

Acta Soc. Fauna Flora fenn. 7(1): 98(1890) subsp. subimmersa 原亚种

≡*Lecidea subimmersa* Fée, *Bull. Soc. bot. Fr.* 20: 315(1873).

On granite in coastal areas and along mountain stream on lower slope. Pantropical.

Hongkong(Thrower, 1988, p. 16, as *Lecanora* sp.; Aptroot & Seaward, 1999, p. 80; Aptroot et al., 2001, p. 330).

≡*Aspicilia subimmersa*(Fée) Hue, *Nouv. Arch. Mus. Hist. Nat.*, Paris, 5 sér. 2: 25(1912)[1910].

On granite. Pantropical.

Misapplied name:

Lecania sp. (Thrower, 1988, p. 16).

Shandong(Li y et al., 2008, p. 71), Hongkong(Thrower, 1988, p. 16; Aptroot & Seaward, 1999, p. 80).

Taiwan(Wang & Lai, 1973, p. 91).

= *Aspicilia subimmersa* subsp. *asiatica*(Zahlbr.) J. C. Wei, *Enum. Lich. China*: 34(1991).

≡*Lecanora subimmersa* ssp. *asiatica* Zahlbr., in Handel-Mazzetti, *Symb. Sin.* 3: 158(1930) & *Cat. Lich. Univ.* 8: 531(1932).

Syntype: Yunnan, Handel-Mazzetti no. 128; Fujian, Chung no. 609(p. p.).

On rocks.

Jiangsu(Wu J. N. & Xiang T. 1981, p. 4, lapsu '*substica*'; Xu B. S., 1989, p. 209), Taiwan(Zahlbr., 1933 c, p. 51).

= *Aspicilia subimmersa* subsp. *umbrinascens*(Zahlbr.) J. C. Wei, *Enum. Lich. China*: 34(1991).

≡*Lecanora subimmersa* ssp. *umbrinascens* Zahlbr., in Handel-Mazzetti, *Symb. Sin.* 3: 158(1930) & *Cat. Lich. Univ.* 8: 531(1932).

Type: Yunnan, Handel-Mazzetti no. 5968.

On rocks.

Taiwan(Zahlbr., 1933 c, p. 51; Wang & Lai, 1973, p. 91).

Lecanora subisabellina H. Magn. 次深黄茶渍

Lich. Centr. Asia 2: 43(1944).

Type: Neimenggu, geol. sample no. 165 in S.

On hard, non calciferous stone, mixed with other lichens.

Lecanora subjaponica L. Lü & H. Y. Wang 东茶渍

in Lu Lei et al., *Lichenologist* 44(4): 466 – 467(2012).

Typus: China, Yunnan, Kunming, Mt. Jiaozi, alt. 3800 m, on bark, 27 October 2008, Wang 20083503(SDNU—holotypus). Shaanxi, Xizang, Sichuan, Yunnan(Lü Lei et al., 2012, p. 468, the province name of Shaanxi was miscited as Shanxi).

Lecanora subminuta H. Magn. 亚微茶渍

Lich. Centr. Asia 2: 44(1944).

Type: Neimenggu, Belimiao, 25/X/1929, Bohlin no. 115 in S.

On rock.

Lecanora subrugosa Nyl. 褶皱茶渍

Flora 58: 15(1875).

Corticolous.

Gansu, Shaanxi, Yunnan(Lü Lei et al., 2009, p. 395 as 'Shannxi').

Lecanora subumbrina Müll. Arg. 赭土茶渍

Bull. Herb. Boissier 3: 632(1895).

On specimen of *Lecanora leprosa*, Bushes in thickets., Pfister 1978: 95, as *Lecanora subfusca* var. *cinereocarnea* Tuck.

Hongkong(Seaward & Aptroot, 2005, p. 285; Aptroot et al., 2001, p. 330).

Lecanora sulfurescens Fée 硫茶渍

Bull. Soc. Bot. France 20: 313(1873).

On granite boulders and on granite rock in walls on lower slopes. Pantropical.

= *Lecanora subfusca* var. *chlarona* f. *microcarpa* Kremp. in Flora, LVI: 470(1873); Zahlbr., 1930, p. 168.

Hongkong(Rabh., 1873, p. 287; Kremp., 1873, p. 470 and 1874, p. 65; Zahlbr., 1930, p. 168; Thrower, 1988, p. 16; Aptroot & Seaward, 1999, p. 80; Aptroot et al., 2001, p. 330).

Lecanora sulphurea(Hoffm.) Ach. 硫茶渍

Lich. univ.: 339(1810).

≡*Parmelia sulphurea*(Hoffm.) Ach., *Methodus, Sectio post.* (Stockholmiæ): 159(1803) var. *sulphurea*.

= *Lichen sulphureus* Hoffm., *Enum. critic. lich. europ.* (Bern): 32(1784).

Lecidea sulphurea(Hoffm.) Wahlenb., Fl. lapp.: 477(1812).

On stone.

Gansu(Hertel, 1977, p. 296.)

Lecanora swartzii (Ach.) Ach 斯瓦氏茶渍

Lich. Univ. 363. (1810).

On siliceous rocks.

Xinjiang(Mamut Reyim et al., 2009, p. 155).

Lecanora symmicta(Ach.) Ach. 合茶渍

Syn. Meth. Lich.: 340(1814).

On *Picea morrisonicola*, *Alnus japonica*, *Abies kawakamii*, *Tsuga formosana*, and Rhododendron wood.

Gansu(Lü Lei et al. , 2008, pp. 103 – 104) , Taiwan(Aptroot and Sparrius, 2003, p. 25) .

Lecanora teretiuscula Zahlbr. 长园茶渍

in Handel-Mazzetti, *Symb. Sin*. 3: 173(1930) & *Cat. Lich. Univ*. 8: 547(1932) .

Type: Yunnan, Handel-Mazzetti no. 4719.

Calcicolous.

Lecanora thysanophora R. C. Harris 流苏茶渍

The Bryologist 103: 790(2000) .

Anhui(Han LF et al. , 2011, p. 24) .

Lecanora toroyensis Zahlbr. 台湾茶渍

in Fedde, *Repertorium*, 33: 52(1933) .

Type: Taiwan, Asahina no. 144.

Taiwan(Zahlbr. , 1940, p. 487; Lamb, 1963, p. 330; Wang & Lai, 1973, p. 91) .

Lecanora tropica Zahlbr. 热带茶渍

Cat. Lich. Univ. 5: 589(1928) .

On *Roystonea regia*.

Taiwan(Aptroot and Sparrius, 2003, p. 26) .

Lecanora umbrina(Ach.) A. Massal. 赭色茶渍

Ric. auton. lich. crost. (Verona) : 10(1852) . var. umbrina 原变种

≡*Lecanora hagenii* var. *umbrina* Ach. , K. Vetensk-Acad. Nya Handl. 31: 75(1810) .

Xinjiang(Abdull Abbas & Wu, 1998, p. 79) .

var. paupercula Jatta 光头山变种

Nuov. Giorn. Bot. Ital. , nov. ser. , vol. IX, 475(1902) .

Type: Shaanxi, guangtou shan, collected by Giraldi.

On smooth tree trunk.

Xinjiang(Abbas et al. , 1993, p. 78) , Shaanxi(Jatta, 1902, p. 475; Zahlbr. , 1930 b, p. 162) .

Lecanora vainioi Vänskä 韦氏茶渍

in Aptroot and Sparrius, New microlichens from Taiwan. *Fungal Diversity* 14: 26(2003) .

On shale.

Taiwan(Aptroot and Sparrius, 2003, p. 26) .

Lecanora varia(Hoffm.) Ach. 异茶渍

On *Rhododendron* wood.

Taiwan(Aptroot and Sparrius, 2003, p. 26) .

Lecanora weii L. F. Han & S. Y. Guo 魏氏茶渍

Mycotaxon 107: 157 – 161(2009) .

Type: Heilongjiang. Mt. Dailing, Liangshuilinchang(Liangshui forest farm) , alt. 350 m, On bark, 1975 V 8, Wei Jiang-Chun 2139(HMAS-L 75828 – 1, holotype) .

Heilongjiang, jilin(Han LF & Guo SY, 2009, pp. 158 – 160) .

Lecanora yenpingensis Zahlbr. 盐边茶渍

in Handel-Mazzetti, *Symb. Sin*. 3: 166(1930) & *Cat. Lich. Univ*. 8: 542(1932) .

Type: Fujian, Chung no. 162.

Corticolous.

Lecanora yunnana(Zahlbr.) H. Magn. 云南茶渍

Acta Hort. Gothob. 8: 46(1933) .

≡*Ionaspis yunnana* Zahlbr. , in Handel-Mazzetti, *Symb. Sin*. 3: 69(1930) .

Type; Yunnan, Handel-Mazzetti no. 1385.

On sandstone.

Lecidella Körb. **小网衣属**

Syst. lich. germ. (Breslau): 233(1855).

Lecidella achristotera(Nyl.) Hertel & Leuckert 无色小网衣

Willdenowia 5: 374(1969).

≡*Lecidea achristotera* Nyl., *Flora*, Regensburg 60: 223(1877).

On *Tsuga formosana* branches.

Taiwan(Aptroot and Sparrius, 2003, p. 27).

Lecidella alaiensis(Vain.) Hertel 中亚小网衣

Herzogia 2(4): 501(1973).

≡*Lecidea alaiensis* Vain., *Bot. Tidsskr.* 26: 247(1904).

On calcareous stone.

Xinjiang(Abdulla Abbas et al., 1993, p. 78; Abdulla Abbas & Wu, 1998, pp. 79 – 80; Abdulla Abbas et al., 2001, p. 364), Gansu(H. Magn. 1940, p. 48; Hertel, 1977, p. 318), Qinghai(TUR-Vain. no. 24. 635, with *Lecidea tessellata*, Hertel, 1977, p. 318).

Lecidella bullata Körb. 泡状小网衣

Parerga lichenol. (Breslau) 3: 200(1861).

On rock.

Yunnan, Xizang(Zhang LL et al., 2012, pp. 448 – 450).

Lecidella carpathica Körb. 破小网衣

Parerga lichenol. (Breslau) 3: 212(1861).

= *Lecidea loudiana* Zahlbr., in Handel-Mazzetti, *Symb. Sin.* 3: 107(1930).

Type: Hunan, Handel-Mazzetti no. 11734. (Hertel, 1977, p. 323).

= *Lecidea diffractula* H. Magn. Lich. Centr. Asia, 2: 22(1944).

Type: Neimenggu, Bohlin, geol. sample no. 187, with *Acarospora suprasedens* H. Magn. in S(fragnment of it in UPS).

Neimenggu(Hertel, 1977, p. 323).

= *Lecidea subsmaragdula* H. Magn. Lich. Centr. Asia, 1: 54(1940).

Type: Gansu, Bohlin no. 42: d in S(fragment of it in UPS). (Hertel, 1977, p. 323).

On rocks.

Neimenggu Hertel, 1977, p. 323), Xinjiang(Abdulla Abbas et al., 1993, p. 78; Abdulla Abbas et al., 1994, p. 21; Abdulla Abbas & Wu, 1998, p. 80; Abdulla Abbas et al., 2001, p. 364), Gansu(H. Magn., 1940, p. 54), Hunan(Zahlbr., 1930, p. 107), Hongkong(Aptroot et al., 2001, p. 330).

Lecidella elaeochroma(Ach.) M. Choisy 油色小网衣

Bull. mens. Soc. linn. Soc. Bot. Lyon 19: 1 9(1950).

≡ *Lecidea parasema* var. *elaeochroma* Ach., *Methodus*, Sectio prior: 36(1803).

≡*Lecidea elaeochroma*(Ach.) Ach., *Syn. meth. lich.* (Lund): 18(1814).

= *Lecidea elaeochroma* var. *hyalina*(Mart.) Zahlbr., Cat. Lich. Univers. 10: 306(1925).

≡*Lecidea hyalina* Mart., Fl. Crypt. Erl.: 248(1817).

On twigs of *Berberis sanguinea*

Sichuan(Zahlbr., 1930 b, p. 101).

= *Lecidea elaeochroma* f. *geographica*(Bagl.) Zahlbr., Cat. Lich. Univers. 3: 571(1925).

≡*Lecidea enteroleuca* var. *geographica* Bagl., Mém. R. Accad. Sci. Torino, Ser. 2 17: 421(1857).

On twigs of *Elaeagnus* sp.

Sichuan(Zahlbr., 1930 b, p. 101).

[*Lecidea parasema*(Ach.) Ach. = *Lecidea enteroleuca* Ach.]

Corticolous in forests.

Xinjiang(Moreau et Moreau, 1951, p. 187), Sichuan, Guangdong(Nyl. & Cromb. 1883, p. 65; Hue, 1891, p. 126; Zahlbr., 1930 b, p. 101), Yunnan(Hue, 1889, p. 175; Zahlbr., 1930 b, p. 101 & 1934 b, p. 199), Jiangsu (Xu B. S., 1989, p. 203).

= *Lecidea parasema* var. *tenebricosula*(Jatta) Zahlbr., Cat. Lich. Univers. 3: 664 (1925).

≡ *Lecidea enteroleuca* var. *tenebricosula* Jatta, Nuovo G. bot. ital. 9: 478(1902).

≡ *Melaspilea enteroleuca*(Ach.) Ertz & Diederich, Fungal Diversity 71: 151(2015).

Type: Shaanxi, Guangtou shan, collected by Giraldi.

Shaanxi(Zahlbr., 1930 b, p. 101).

On bark of tree.

Taiwan(Aptroot and Sparrius, 2003, p. 27). On *Michelia formosana*.

Lecidella enteroleucella(Nyl.) Hertel 小盘小网衣

Khumbu Himal 6: 330(1977).

≡ *Lecidea enteroleucella* Nyl., *J. Linn. Soc., Bot.* 20: 67(1883).

On coastal rocks and on exposed granite on mountain peak of low mountain at 250m. Paleotropical.

Hongkong(Aptroot & Seaward, 1999, p. 80; Aptroot et al., 2001, p. 330).

= *Lecidea setschwanensis* Zahlbr., in Handel-Mazzetti, *Symb. Sin.* 3: 108(1930).

Type: Sichuan, Handel-Mazzetti no. 1170 in W & isotype in US.

= *Lecidea tapetiformis* Zahlbr, in Handel-Mazzetti, *Symb. Sin.* 3: 107(1930).

Type: Sichuan, Handel-Mazzetti no. 2699(lectotype in W & isotype in US).

= *Lecidea kelungana* Zahlbr. in Fedde, *Repertorium*, 33: 36(1933).

Type: Taiwan, Abbe U. Faurie no. 257 in W.

= *Lecidea lentigerella* Zahlbr., in Fedde, *Repertorium*, 33: 36(1933).

Type: Taiwan, Abbe U. Faurie no. 277 in W.

On rocks.

Sichuan(Hertel, 1977, p. 330), Taiwan(Hertel, 1977, p. 330 – 332).

Lecidella euphorea(Flörke) Hertel 优果小网衣

in Hawksworth, James & Coppins, *Lichenologist* 12(1): 107(1980).

≡ *Lecidea sabuletorum* var. *euphorea* Flörke, *Mag. Gesell. naturf. Freunde, Berlin* 2: 311(1808).

Xinjiang(Abdulla Abbas et al., 1993, p. 78; Abdull Abbas et al., 1994, p. 21; Abdulla Abbas & Wu, 1998, p. 80; Abdulla Abbas et al., 2001, p. 364).

On *Koelreuteria formosana*.

Taiwan(Aptroot and Sparrius, 2003, p. 27).

Lecidella oceanica Lu L. Zhang & Xin Y. Wang 海洋小网衣

Bryologist 115: 330(2012).

On rock.

Liaoning(Zhao X. et al., p. 710).

Lecidella stigmatea(Ach.) Hertel & Leuckert 平小网衣

Willdenowia 5: 375(1969).

≡ *Lecidea stigmatea* Ach., *Lich. Univ.*: 161(1810).

= *Lecidea planiformis* Zahlbr. in Handel-Mazzetti, *Symb. Sin.* 3: 100(1930).

Type: Sichuan, Handel-Mazzetti, no. 2648 in W & isotype in US.

= *Lecidea vulgata* Zahlbr., *Cat. Lich. Univ.* 3: 718(1925), (ut nom. nov.).

[= *Lecidea latypea* var. *pilularia* Th. Fr.]

On rocks.

Xinjiang(Abdulla Abbas & Wu, 1998, pp. 80 – 81; Abdulla Abbas et al., 2001, p. 364), Shaanxi(Jatta, 1902,

p. 478; Zahlbr. , 1930 b, p. 10) , Sichuan(Zahlbr. , 1930 b, p. 100; Hertel, 1977, pp. 334 & 337) , Hebei(H. Smith no. 5305 in UPS, Hertel, 1977, p. 337) .

Lecidella subincongrua(Nyl.) Hertel & Leuckert 小网衣

Willdenowia 5: 375(1969) .

≡*Lecidea subincongrua* Nyl. , *Bull. Soc. linn. Normandie*, sér. 26: 291(1872) .

On large pebbles of ground.

Jilin(Hertel & Zhao, 1982, p. 148) .

Lecidella tumidula(A. Massal.) Knoph & Leuckert 肿胀小网衣

Biblioth. Lichenol. 68: 131(1997) .

≡*Lecidea tumidula* A. Massal. , Ric. auton. lich. crost. (Verona) : 71, fig. 137(1852) .

On bark.

Xinjiang(Zhao X. et al. , 2015, p. 711) .

Lecidella wulfenii(Ach.) Körb. 苔生小网衣

Parerga lichenol. (Breslau) 3: 216(1861) .

≡*Urceolaria wulfenii* Ach. , *Methodus*, Sectio prior: 152(1803) .

≡*Lecidea wulfenii*(Ach.) Ach. , *Lich. univ.* : 210(1810) .

On mosses.

Xinjiang(Abdulla Abbas et al. , 1994, p. 21; Abdulla Abbas & Wu, 1998, p. 81; Abdulla Abbas et al. , 2001, p. 364) .

Lecidella xylophila(Th. Fr.) Knoph & Leuckert 木生小网衣

Progr. Probl. Lichenol. Nineties. Proc. Third Symp. Intern. Assoc. Lichenol. , *Biblthca Lichenol.* 68: 131(1997) .

≡*Lecidea xylophila* Th. Fr. , *Oestra Blek. Lafflora*: 16(1874) .

On wood.

Taiwan(Aptroot and Sparrius, 2003, p. 27) .

Miriquidica Hertel & Rambold **奇果衣属**

Mitt. bot. StSamml. , *Münch.* 23: 378(1987) .

Miriquidica complanata(Körb.) Hertel & Rambold 奇果衣

Mitt. bot. StSamml. , *Münch.* 23: 382(1987) .

≡*Lecanora complanata* Körb. , *Parerga lichenol.* (Breslau) 1: 84(1859) .

On siliceous rocks.

Shaanxi(Jatta, 1902, p. 475; Zahlbr. , 1930 b, p. 152) .

Miriquidica obnubila(Th. Fr. & Hellb.) Hertel & Rambold 覆盖奇果衣

Mitt. Bot. Staatssamml. München 23: 389(1987) .

On rock.

Yunnan(Zhao X. et al. , 2015, p. 713) .

Miriquidica yunnanensis Lu L. Zhang & X. Zhao 云南奇果衣

in Zhao Xin et al. , *Mycotaxon* 123: 364(2013) .

Type: China. Yunnan province, Mt. Laojun, alt. 4000 m, on rock, 7 Nov. 2009, H. Y. Wang20121132(Holotype, SDNU) .

Myriolecis Clem. **多盘衣属**

Gen. fung. (Minneapolis) : 79(1909) .

=*Arctopeltis* Poelt, *Int. J. Mycol. Lichenol.* 1(2) : 147(1983) .

Myriolecis flowersiana(H. Magn.) Śliwa, Zhao Xin & Lumbsch 厚缘多盘衣

in Zhao, Leavitt, Zhao, Zhang, Arup, Grube, Pérez-Ortega, Printzen, Śliwa, Kraichak, Divakar, Crespo & Lumbsch, *Fungal Diversity* 78(1) : 301(2015) .

≡*Lecanora flowersiana* H. Magn. *Acta Hort. Gotoburg.* 19: 2(1952).

On bark or rock.

Gansu(Lü Lei et al., 2008, pp. 101 – 102).

Myriolecis percrenata(H. Magn.) Śliwa, Zhao Xin & Lumbsch 多齿多盘衣

in Zhao, Leavitt, Zhao, Zhang, Arup, Grube, Pérez-Ortega, Printzen, Śliwa, Kraichak, Divakar, Crespo & Lumbsch, *Fungal Diversity* 78(1): 301(2015).

≡*Lecanora percrenata* H. Magn. *Lich. Centr. Asia* 1: 88(1940).

Type: Gansu, Bohlin no. 45b in S(!).

On calcareous rock associated with different lichens.

Myriolecis semipallida(H. Magn.) Śliwa, Zhao Xin & Lumbsch 半苍多盘衣

in Zhao, Leavitt, Zhao, Zhang, Arup, Grube, Pérez-Ortega, Printzen, Śliwa, Kraichak, Divakar, Crespo & Lumbsch, *Fungal Diversity* 78(1): 301(2015).

≡*Lecanora semipallida* H. Magn., *Lichens Central Asia* 1: 105(1940).

Type: Gansu, Bohlin no. 42d in S.

Protoparmeliopsis M. Choisy **原类梅属**

Bull. Soc. bot. Fr. 76: 524(1929).

Protoparmeliopsis muralis(Schreb.) M. Choisy 石墙原类梅

Contr. Lichénogr. 1: tab. 1(1929).

≡*Lichen muralis* Schreb., *Spic. fl. lips.* (Lipsiae): 130(1771).

≡*Lecanora muralis*(Schreb.) Rabenh., *Deutschl. Krypt. -Fl.* (Leipzig) 2: 42(1845).

≡*Squamaria muralis*(Schreb.) Elenkin, *Acta Horti Petropolit.* 24: 95(1904).

= *Lecanora saxicola*(Pollich) Ach., *Lich. univ.*: 431(1810).

≡*Lichen saxicola* Pollich, *Hist. pl. Palat.* 3: 225(1777).

≡*Squamaria saxicola*(Pollich) Howitt, *Nottinghamsh. Fl.*: 108(1839).

≡*Lichen muralis* Schreb., *Spic. fl. lips.* (Lipsiae): 130(1771).

On granite.

Heilongjiang(Asahina, 1952d, p. 375), Neimenggu(H. Magn. 1940, p. 120), Xinjiang (Abdulla Abbas et al., 1993, p. 78; Abdulla Abbas et al., 1994, p. 21; Abdulla Abbas & Wu, 1998, p. 78; Abdulla Abbas, 2001, p. 364), Shaanxi(Jatta, 1902, p. 474; Zahlbr., 1930 b, p. 172), Hebei(Tchou, 1935, p. 321), Shandong(Zhao ZT et al., 2002, p. 5; Li Y et al., 2008, p. 71), Jiangsu(Wu J. N. & Xiang T. 1981, p. 6: here should be on p. 5 according the sequence of pages; Xu B. S. 1989, p. 208), Shanghai(Nyl. & Cromb. 1883, p. 62; Hue, 1891, p. 60; Xu B. S. 1989, p. 208), Zhejiang(Xu B. S. 1989, p. 208).

= *Lecanora muralis* var. *versicolor*(Pers.) Tuck., *Syn. N. Amer. Lich.* (Boston) 1: 185(1882).

≡*Lichen versicolor* Pers., *Ann. Bot.* (Usteri) 1: 24(1794).

Beijing(Moreau et Moreau, 1951, p. 189), Yunnan(Zahlbr., 1930 b, p. 172).

Protoparmeliopsis pseudogyrophorica S. Y. Kondr., S. O. Oh & Hur 假三苔原类梅

[as ‘pseudogyrophoricum’], in Kondratyuk, Lőkös, Tschabanenko, Moniri & Farkas, *Acta bot. hung.* 55(3 – 4): 301(2013).

Type: China: Songmudao village, Wafangdian city, Dalian county, Liaoning province, on rock. Lat.: 39°23′51.4″N; Long.: 121°48′28.4″E; Alt.: 55 m a. s. l. Coll.: Oh, S. -O., Hur, J. -S., 26. 07. 2012. Holotype: KoLRI-016651(CH – 120041).

Pyrrhospora Körb. **红蜡盘衣属**

Syst. lich. germ. (Breslau): 209(1855).

Pyrrhospora chlororphnia (Tuck.) Aptroot & Seaward 绿体红蜡盘衣

Bryologist, 108(2):285(2005).

≡*Lecidea chlororphnia* Tuck., *Proc. Amer. Acad. Arts*. 6: 275. 1866. Pfister 1978: 108, as

≡*Biatora chlororphnia*. On bark of *Ficus* sp. 23. 3. 1854. *C. Wright*. Isotype(from Hongkong) *Biatora chlororphnia* Tuck. As *Lecidea chlororphnia* Tuck. in Aptroot and Seaward(1999: 99).

Hongkong(Zahlbr., *Cat. Lich. Univ*. 3: 746(1925) & in Handel-Mazzetti, *Symb. Sin*. 3: 104(1930) as *Lecidea chlororphnia* Tuck.; Seaward & Aptroot, 2005, p. 285)

Pyrrhospora quernea(Dicks.) Körb. 栎红蜡盘衣

Syst. lich. germ. (Breslau):209, tab. 1, fig. 7(1855).

≡ *Lichen querneus* Dicks., *Fasc. pl. crypt. brit*. (London) 1: 9(1785).

On exposed and sheltered trees, e. g. Araucoria, Callitris and Kandelia, in parks forests or mangrove.

Hongkong(Aptroot & Seawrd, 1999, p. 91; Aptroot & Sipman, 2001, p. 338).

Rhizoplaca Zopf 脐鳞属

Justus Liebigs Annln Chem. 340: 291(1905).

Rhizoplaca chrysoleuca(Sm.) Zopf 红脐鳞

Justus Liebigs Annln Chem. 340: 291(1905).

≡*Lichen chrysoleucus* Sm., *Trans. Linn. Soc. London* 1: 82(1791).

≡*Lecanora chrysoleuca*(Sm.) Ach., *Lich. univ*.: 411(1810).

≡*Lecanora rubina* var. *chrysoleuca*(Sm.) Rabenh., *Deutschl. Krypt. -Fl*. (Leipzig) 2: 42(1845)

≡*Squamaria rubina* var. *chrysoleuca*(Sm.) Elenkin, *Acta Horti Petropolit*. 19: 30(1901).

[*Lichen rubinus* Vill., Hist. pl. Dauphiné 3: 977(1789), nom. illegit., Art. 53. 1.

Lecanora rubina(Hoffm.) Ach., Lich. univ.: 412(1810)

Squamaria rubina Hoffm., Descr. Adumb. Plant. Lich. 2(2):27(1794)].

On rocks.

Neimenggu(Chen et al. 1981, p. 156), Jilin(J. C. Wei, 1984, p. 211; Sun LY et al., 2000, p. 36), Xinjiang(J. C. Wei, 1984, p. 210; WangXY, 1985, p. 346 as *Squamaria rubina*; Abdula Abbas et al., 1993, p. 80; Abdulla Abbas & Wu, 1998, p. 82; Abdulla Abbas et al., 2001, p. 367),

Shaanxi(J. C. Wei, 1984, p. 211; He & Chen 1995, p. 45, no specimen and locality are cited.), ningxia(Liu M & J. C. Wei, 2013, p. 44), Xizang(Mt. Qomolangma, Paulson, 1925, p. 192; J. C. Wei & Chen, 1974, p. 178; J. C. Wei, 1984, p. 210; J. C. Wei & Jiang, 1986, p. 33), Yunnan(Paulson, 1928, p. 317; Zahlbr., 1930 b, p. 172), ShanxiI(M. Sato, 1981, p. 64), Hebei(Moreau et Moreau, 1951, p. 189; J. C. Wei, 1984, p. 210), China (prov. not indicated: Asahina & M. Sato in Asahina, 1939, p. 709).

Seven chemical strains of this species from China are reported by me in *Acta Mycologica Sinica*, 3(4):210 – 211(1984).

Rhizoplaca fumida X. Q. Gao 褐脐鳞

Acta Mycol. Sin. 6(4):233(1987).

Type: Neimenggu, Da Hinggan Ling, alt. 1800 m, no. 954 – 2 in HMAS-L.

On rocks.

Rhizoplaca haydenii(Tuck.) W. A. Weber 海氏脐鳞

Philippia 4(1):35(1979).

≡*Lecanora haydenii* Tuck., *Proc. Amer. Acad. Arts & Sci*. 6: 227(1866)[1864].

Xinjiang(Zheng X. L. et al., 2007. p. 757).

Notes. This species as a new record for China is only mentioned from Tianshan in the abstract of this paper and lack of necessary data support. It remains to be checked in the future.

Rhizoplaca huashanensis J. C. Wei 华脐鳞

Acta Mycol. Sin. 3(4):208(1984).

Type: Shaanxi, Mt. Hua shan, 15/VI/1964, JC Wei no. 59 in HMAS-L.

On rocks.

Rhizoplaca melanophthalma(DC.) Leuckert 垫脐鳞

Nova Hedwigia 28: 72(1977) .

≡*Squamaria melanophthalma* DC. , in Lamarck & de Candolle, *Fl. franç.* , Edn 3(Paris) 2: 376(1805) .

On rocks.

Xinjiang(Abdulla Abbas & Wu, 1998, p. 82; Abdulla Abbas et al. , 2001, p. 367) , Xizang(J. C. Wei, 1984, p. 211; J. C. Wei & Y. M. Jiang, 1986, p. 33) .

Rhizoplaca peltata(Ramond) Leuckert & Poelt 盾脐鳞

Nova Hedwigia 28(1) : 73(1977) . var. peltata 原变种

≡*Lecanora peltata*(Ram.) Steud. , Nomenclat. Botan. , 237(1824) .

≡*Lichen peltatus* Ramond, in Lamark & de Candolle, *Fl. franç.* , Edn 3(Paris) 2: 377(1805) .

≡*Squamaria peltata*(Ramond) DC. , in Lamark & de Candolle, *Fl. franç.* , Edn 3(Paris) 2: 377(1805) .

≡*Protoparmeliopsis peltata*(Ramond) Arup, Zhao Xin & Lumbsch, in Zhao, Leavitt, Zhao, Zhang, Arup, Grube, Pérez-Ortega, Printzen, Śliwa, Kraichak, Divakar, Crespo & Lumbsch, *Fungal Diversity* 78(1) : 301(2015) .

=*Lecanora heteromorpha*(Ach.) J. Steiner, *Öst. bot. Z.* 49: 249(1899) .

≡*Lecanora rubina* var. *heteromorpha* Ach. , *Lich. univ.* : 412(1810) .

On rocks.

Xinjiang(H. Magn. 1940, p. 121; Moreau et Moreau, 1951, p. 189; J. C. Wei, 1984, p. 209; Abdulla Abbas et al. , 1993, p. 81; Abdulla Abbas & Wu, 1998, p. 82; Abdulla Abbas et al. , 2001, p. 367) .

var. regalis(H. Magn.) J. C. Wei 黑腹变种

Acta Mycol. Sin. 3(4) : 210(1984) .

≡*Lecanora regalis* H. Magn. *Lich. Centr. Asia* 1: 122(1940) .

Type: Xinjiang, Mt. Bogdo ola, at about 2100 m, 10/VIII/1928, D. Hummel in S(!) .

On boulders.

Xinjiang(H. Magn. , 1940, p. 122; Abdulla Abbas et al. , 1993, p. 81; Abdulla Abbas & Wu, 1998, p. 83; Abdulla Abbas et al. , 2001, p. 367) .

Rhizoplaca subdiscrepans(Nyl.) R. Sant. 异脐鳞

Lichens of Sweden and Norway(Stockholm) : 278(1984) .

≡*Squamaria chrysoleuca* var. *subdiscrepans* Nyl. , *Syn. meth. lich.* (Parisiis) 2: 61(1863) .

≡*Lecanora subdiscrepans* (Nyl.) Stizenb. , *Ber. Tät. St. Gall. naturw. Ges.* : 341(1882) . [1880 – 81]

≡*Lecanora rubina* var. *subdiscrepans*(Nyl.) Zahlbr. *Symb. Sin.* 172(1930) .

=*Lecanora chrysoleuca* var. *subdiscrepans*(Nyl.) Zahlbr. *Symb. Sin.* 172(1930) .

=*Lecanora rubina*(Vill.) Ach. in Zahlbr. *Symb. Sin.* 172(1930) .

On rocks.

Shaanxi(Jatta, 1902, p. 474; Zahlbr. , 1930 b, p. 172) .

Woessia D. Hawksw. & Poelt **沃氏衣属**

Pl. Syst. Evol. 154(3 – 4) : 207(1986) .

Woessia pseudohyphophorifera Lücking & Sérus. 拟沃氏衣

in Sérusiaux, *Biblthca Lichenol.* 58: 422(1995) .

On shaded rock in park. Known from neotropics and Asia.

Hongkong(Aptroot & Seaward, 1999, p. 98; Aptroot & Sipman, 2001, p. 341) .

(48) Malmideaceae Kalb, Rivas Plata & Lumbsch **柄座衣科**

in Kalb, Rivas Plata, Lücking & Lumbsch, *Biblthca Lichenol.* 106: 150(2011) .

Malmidea Kalb, Rivas Plata & Lumbsch **柄座衣属**

in Kalb, Rivas Plata, Lücking & Lumbsch, *Biblthca Lichenol.* 106: 150(2011) .

Malmidea aurigera(Fée) Kalb, Rivas Plata & Lumbsch 带耳柄座衣

in Kalb, Rivas Plata, Lücking & Lumbsch, *Biblthca Lichenol*. 106: 153(2011).

≡*Lecidea aurigera* Fée, *Essai Crypt. Exot*. (Paris) : 106(1825) [1824].

On a tree.

Hongkong(Aptroot et al. , 2001, p. 330, as *L. aurigera*).

Malmidea granifera(Ach.) Kalb, Rivas Plata & Lumbsch 颗粒柄座衣

in Kalb, Rivas Plata, Lücking & Lumbsch, *Biblthca Lichenol*. 106: 165(2011).

≡*Lecanora granifera* Ach. , *Syn. meth. lich*. (Lund) : 163(1814).

≡*Lecidea granifera*(Ach.) Vain. , in Hiern, *Cat. Welwitsch Afric. Pl*. 2(2) : 424(1901).

Corticolous

Sichuan, Yunnan, Fujian(Zahlbr. , 1930 b, p. 109) , Taiwan(Zahlbr. , 1933, p. 39; Wang & Lai, 1973, p. 91), Hongkong(Thrower, 1988, pp. 16, 113; Aptroot & Seaward, 1999, p. 80; Aptroot et al. , 2001, p. 330, as *Lecidea granifera*).

Malmidea hypomelaena(Nyl.) Kalb & Lücking 下黑柄座衣

in Kalb, Rivas Plata, Lücking & Lumbsch, *Biblthca Lichenol*. 106: 165(2011).

On shaded trees. Pantropical.

Taiwan(Aptroot and Sparrius, 2003, p. 26) , Hongkong(Aptroot & Seaward, 1999, p. 80).

≡*Lecidea hypomela*(vel *hypomelaena*) Nyl. in *Annal. Scienc. Nat.* , *Botan.* , ser. 4. , vol. XI, 1859, p. 223 et vol. XIX, Cat. Lich. Univ. 3: 782 – 783(1925).

≡*Malcolmiella hypomelaena*(Nyl.) M. Cáceres & Lücking [as ' *hypomela*'] , in Cáceres, *Libri Botanici* 22: 106 (2007).

On shaded trees. Pantropical.

Taiwan(Aptroot and Sparrius, 2003, p. 26) , Hongkong(Aptroot & Seaward, 1999, p. 80).

Malmidea vinosa(Eschw.) Kalb, Rivas Plata & Lumbsch 紫红柄座衣

in Kalb, Rivas Plata, Lücking & Lumbsch, *Biblthca Lichenol*. 106: 166(2011).

≡*Lecidea vinosa* Eschw. , in Martius, *Fl. Bras*. 1: 251(1833) [1829 – 1833].

≡*Malcolmiella vinosa*(Eschw.) Kalb & Lücking, in Lücking & Kalb, *Bot. Jb*. 122(1) : 43(2000).

Pfister 1978: 108, as *Biatora tephraea* Tuck.

On bark of trees. Mountainsides. [Label as *Lecidea tephraea*].

≡*Biatora tephraea* Tuck. , Lichens of California(Berkeley) : 31(1866).

Corticolous.

China(prov. not indicated: Zahlbr. , 1925, p. 838).

(49) Megalariaceae Hafellner **托盘衣科**

Beih. Nova Hedwigia 79: 302(1984).

Megalaria Hafellner **大盘衣属**

Beih. Nova Hedwigia 79: 302(1984).

Megalaria laureri(Hepp ex Th. Fr.) Hafellner 劳氏大盘衣

Lichens of Italy. An Annotated Catalogue, Monografia XII, Museo Regionale di Scienze Naturali, Torino(Torino) : 429(1993).

≡*Catillaria laureri* Hepp ex Th. Fr. , *Lich. Scand*. 2: 582(1874).

≡*Catinaria laureri*(Hepp ex Th. Fr.) Degel. , *Göteborgs Kungl. Vetensk. Samhälles Handl.* , Ser. B, *Math. Naturvensk*. Skr. 1(no. 7) : 12(1941).

≡*Catillaria laureri* Hepp ex Arnold, *Lich. Exs*. : no. 353(1867).

On trees in sheltered forest. Known from temperate Europe, Australia, North America and Asia.

Taiwan(Aptroot and Sparrius, 2003, p. 13 as *C. laurei*), Hongkong(Aptroot & Seaward, 1999, p. 71 as *C. laurei*).

(50) Mycoblastaceae Hafellner 黑红衣科

[as '*Mykoblastaceae*'], Beih. Nova Hedwigia 79: 310(1984).

Mycoblastus Norman 黑红衣属

Nytt Mag. Natur. 7: 236(1853) [1852].

Mycoblastus affinis(Schaer.) T. Schauer 邻黑红衣

Lich. Alp. 12: 230(1964).

≡*Lecidea affinis* Schaer., *Enum. critic. lich. europ.* (Bern) : 132(1850).

≡*Mycoblastus alpinus* f. *affinis*(Schaer.) Räsänen, *Lichenotheca Fennica*: no. 73(1943).

≡*Mycoblastus alpinus* var. *affinis* (Schaer.) Räsänen, *Lichenotheca Fennica*: no. 73(1946).

=*Lecidea melina* Kremp. ex Nyl., *Annls Sci. Nat.*, *Bot.*, sér. 4 19: 357(1863).

≡ *Mycoblastus melinus* (Kremp. ex Nyl.) Hellb., K. svenska Vetensk-Akad. Handl., ny följd 9(11): 74 (1870).

On bark of *Pinus tabulaeformis*.

Yunnan(Hue, 1889, p. 176; Zahlbr., 1930 b, p. 112).

On bark of pine tree and *Rhododendron* sp.

Yunnan(Zahlbr., 1930 b, p. 112), China(prov. not indicated: Ras. 1940, p. 151).

On *Pseudotsuga wilsoniana*.

Taiwan(Aptroot & Sparrius, 2003, p. 30).

Mycoblastus sanguinarius(L.) Norman 黑红衣

[as 'sanguinaria'], *Cat. Lich. Univers.* 4: 5(1926). var. sanguinarius 原变种

=*Lichen sanguinarius* L., Sp. pl. 2: 1140(1753).

On bark of trees.

China(prov. not indicated: Asahina & Sato in Asahina, 1939, p. 669).

var. endorhodus(Th. Fr.) Stein 蔷薇变种

in Cohn, *Krypt. -Fl. Schlesien*(Breslau) 2(2): 256(1879).

≡*Lecidea sanguinaria* var. *endorhoda* Th. Fr., *Lich. Scand.* (Upsaliae) 1(2) : 479(1874).

On bark of pine trees.

Yunnan(Zahlbr., 1930 b, p. 111), TAIWAN(Zahlbr., 1933 c, p. 40; Wang & Lai, 1973, p. 92).

Tephromela M. Choisy 灰衣属

Bull. Soc. bot. Fr. 76: 522(1929).

Tephromela atra(Huds.) Hafellner 黑盘灰衣

in Kalb, *Lichenes Neotropici*, Fascicle VII(nos 251 – 300) (Neumarkt) 7: no. 297(1983).

≡*Lichen ater* Huds., *Fl. Angl.* : 445(1762).

≡*Lecanora atra*(Huds.) Ach., *Lich. univ.* : 344(1810).

On rocks and rarely on bark of trees.

Xingjiang(Abdulla Abbas & Wu, 1998, p. 45; Abdulla Abbas et al., 2001, p. 368), Sichuan(Zahlbr., 1930 b, p. 168), Yunnan(Wang L. S. 2008, p. 538), Shandong(zhao ZT et al., 1999, p. 427; Zhao ZT et al., 2002, p. 5; Li y et al., 2008, p. 71 as *L. atra*), Anhui, Zhejiang(Xu B. S., 1989, p. 206), Taiwan(Asahina & M. Sato in Asahina, 1939, p. 711; Wang & Lai, 1973, p. 91).

=*Lecanora atra* var. *americana* Fée, Essai Crypt. Exot., Suppl. Révis. (Paris) : 110(1837).

≡*Tephromela americana*(Fée) Kalb, *Lichenes Neotropici*, Fascicle VIII(nos 301 – 350) (Neumarkt) : no. 348 (1984).

On twigs.

Sichuan(Zahlbr., 1930 b, p. 168 & 1934, p. 206), Hunan(Zahlbr., 1930 b, p. 168).]

(51) Parmeliaceae Eschw. 梅衣科

in Goebel & Kunze, *Syst. Lich.* : 19(1824).

Alectoria Ach. **树发属**

in Luyken, *Tent. Hist. Lich.* : 95(1809) .

Alectoria acanthodes Hue 多刺树发

Nouv. Arch. Mus. Hist. Nat. 4(1) : 89(1899) .

Type: Yunnan, leg. R. P. Delavay, 31/VII/1888.

Xizang(J. C. Wei & Chen, 1874, p. 181) , Sichuan(Zahlbr. , 1930b, p. 201) , Yunnan(Du Rietz 1926, p. 17; Zahlbr. , 1930 b, p. 201 & 1934b, p. 212) , Taiwan(Asahina, 1936 c, p. 692; Asahina & M. Sato in Asahina, 1939, p. 747; Wang & Lai, 1973, p. 87) , China(Asahina, 1939, p. 747; Zahlbr. , 1930a, 376) .

Alectoria lactinea Nyl. 乳树发

Lich. Japon. : 23(1890) .

Taiwan: Hwalien county. Mt. Hohuang shan, K. Yoshida 6865, 6866.

Notes: The following specimens are identified with *A. asiatica* Du Rietz: Taiwan. Hwalien county Mt. Hohuang shan. K. Yoshida 6861, 6862.

Alectoria ochroleuca(Hoffm.) A. Massal. 金黄树发

Sched. critic. (Veronae) 2: 47(1855) .

≡*Lichen ochroleucus* Hoffm. , *Enum. critic. lich. europ.* (Bern) : 43(1784) .

≡*Usnea ochroleuca* Hoffm. , *Descr. Adumb. Plant. Lich.* 2(1) : 7(1794) .

On the ground among grasses and on rocks.

Heilongjiang(Chen et al. 1981, p. 128) , Neimenggu(Sato, 1952, p. 175; Chen et al. 1981, p. 128; J. C. Wei et al. 1982, p. 49; Wang et al. , 2015, p. 161) , Sichuan(Obermayer W. , 2004, p. 485: Chemical race 1(usnic acid, diffractaic acid) , Xizang, Yunnan(Wang et al. , 2015, p. 161) .

Alectoria sarmentosa(Ach.) Ach. 长匐树发

Lich. univ. : 595(1810) .

≡*Lichen sarmentosus* Ach. , in Liljeblad, *Utkast Svensk Fl.* : 427(1792) .

On tree trunk and twigs of Abies sp.

Jilin(Chen et al. 1981, p. 128) .

Alectoria spiculatosa Li S. Wang & Xin Y. Wang 粉刺树发

in Wang, Liu, Shi, Zhang, Ye, Chen & Wang, *Index Fungorum* 250: 1(2015) .

= *Alectoria spinosa* Li S. Wang & Xin Y. Wang, *Mycosphere* 6(2) : 161(2015) .

Type: Yunnan Prov. , Lijiang Co. , on branches of *Rhododendron*, 2014-June, Wang Li-song et al. 14 – 44046 (KUN-L 45926, Holotype) .

Yunnan(Wang et al. , 2015, p161) .

Alectoria variabilis Bystrek 变色树发

Khumbu Himal 6(1) : 22(1969) .

On bush.

Xizang(J. C. Wei & Jiang, 1986, p. 64) .

Allantoparmelia (Vain.) Essl. **实心袋属**

Mycotaxon 7(1) : 46(1978) .

Allantoparmelia almquistii(Vain.) Essl. 实心袋

Mycotaxon 7: 46(1978) .

≡*Parmelia almquistii* Vain, *Ark. Bot.* 8(4) : 32(1909) .

≡*Hypogymnia almquistii*(Vain.) Rass. , *Nov. sist. Niz. Rast.* 4: 298(1967) . Xinjiang(Abdula Abbas, 1998, fide Chen J. B. , 2016, p. 261; Abdulla Abbas & Wu, 1998, p. 89; Abdulla Abbas et al. , 2001, p. 360) .

Allocetraria Kurok. & Lai **厚枝衣属**

Bull. Natn. Sci. Mus. Tokyo, Ser. B, 17(2):60(1991).

Type: *Evernia stracheyi* Bab., Hook., *Journ. Bot.* 4:244(1852).

≡*Allocetraria stracheyi*(Bab.) Kurok. & Lai. Bull. natn. Sci. Mus., Tokyo, B17(2):62(1991).

Allocetraria ambigua(Bab.) Kurok. & Lai 黄条厚枝衣

Bull. Natn. Sci. Mus. Tokyo, Ser. B, 17(2):62(1991).

≡*Cetraria ambigua* C. Bab., in Hooker, *Hooker's J. Bot. Kew Gard. Misc.* 4:244(1852).

On wood and mosses.

Gansu(Rui-Fang Wang et al., 2015. p. 585), Shaanxi(He & Chen 1995, p. 42, specimen and locality are not cited; Rui-Fang Wang et al., 2015. p. 585), Xizang(Nyl. 1860, p. 311; J. C. Wei & Chen. 1974, p. 180; J. C. Wei & Jiang, 1981, p. 1146 & 1986, p. 56; Rui-Fang Wang et al., 2015. p. 585), Sichuan(Obermayer W. 2004, p. 486; Rui-Fang Wang et al., 2015. p. 585), Yunnan(Rui-Fang Wang et al., 2015. p. 585).

Allocetraria capitata R. F. Wang, L. S. Wang & J. C. Wei 粉头厚枝衣

Mycosystema 33(1): 21 –22. (2014).

Type: Sichuan, Dege County, Manigangge Village, Mt Que' ershan, alt. 4 510m, on the ground, N 31°55′, E 98° 56′, Wang LS et al. 07 –28259(holotype in KUN-L20001; isotype in HMAS-L 125690), Rui-Fang Wang et al., 2015. p. 585.

= *Cetraria weii* Divakar, Crespo & Lumbsch, in Divakar, Crespo, Kraichak, Leavitt, Singh, Schmitt & Lumbsch, Fungal Diversity 84: 111(2017).

Allocetraria corrugatula R. F. Wang, X. L. Wei & J. C. Wei 皱厚枝衣

Mycotaxon 130(2):585 –587(2015).

Type-China, Yunnan Prov., Deqin Co., Meli village, Meili snow mountain, on rock, 28°38′N, 98°37′E, alt. 4400 m, 10 Sep. 2012, R. F. Wang 12033(Holotype, HMAS-L).

≡*Cetraria corrugata*(R. F. Wang, X. L. Wei & J. C. Wei) Divakar, Crespo & Lumbsch, in Divakar, Crespo, Kraichak, Leavitt, Singh, Schmitt & Lumbsch, Fungal Diversity 84: 111 (2017).

Allocetraria endochrysea(Lynge) Kärnef. & Thell 杏黄厚枝衣

Nova Hedwigia 62:507(1996).

≡*Dactylina endochrysea* Lynge, *Skrift. Svalbard Ishavet*(Oslo) 59, V. Supplement: 62(1933).

Type: Yunnan, Delavay from Lijiang, alt. 4000 m in 1886, preserved in H & isotype in O.

≡ *Cetraria endochrysea*(Lynge) Divakar, Crespo & Lumbsch, in Divakar, Crespo, Kraichak, Leavitt, Singh, Schmitt & Lumbsch, Fungal Diversity 84: 111(2017).

On soil, overgrowing mosses.

Sichuan(Zahlbr., 1934 b, p. 212 & 1940, Lamb, 1963, p. 216; Follm., Huneck & Weber, 1968, p. 9; Kärnef. & Thell, 1996, p. 507 –508; Rui-Fang Wang et al., 2015. p. 587), Yunnan(Zahlbr., 1934 b, p. 212 & 1940, p. 552; Lamb, 1963, p. 216; Kärnef. & Thell, 1996, p. 507 –508; Rui-Fang Wang et al., 2015. p. 587).

Notes: Both the "Kangting" and "Tachienlu" of "Sikang" province, the locality of this species in China, collected by Harry Smith in 1934, are the synonymes of the same place Kangding(not "Kangting"), which at present belong to Sichuan province. [Kangding, Tapao shan, 4000m, Smith 1400(O) and 14052(B); Mt. Konka, 4730 m, 1928, Rock 16362(B), from Kärnef. & Thell, 1996, p. 508]

Allocetraria flavonigrescens A. Thell & Randlane 黑黄厚枝衣

in Thell, Randlane, Kärnefelt, Gao & Saag, Flechten Follmann, Contributions to Lichenology in Honour of Gerhard Follmann(Cologne):359(1995).

≡*Cetraria flavonigrescens*(A. Thell & Randlane) Divakar, Crespo & Lumbsch, in Divakar, Crespo, Kraichak, Leavitt, Singh, Schmitt & Lumbsch, Fungal Diversity 84: 111(2017).

On soil and corticolous on branches of Rhododendron.

Qinghai, shaanxi, Yunnan(Rui-Fang Wang et al. , 2015. p. 588), Xizang, Sichuan(Obermayer W. , 2004, p. 486; Rui-Fang Wang et al. , 2015. p. 588).

Allocetraria globulans(Nyl. ex Hue) A. Thell & Randlane 小球厚枝衣

in Thell, Randlane, Kärnefelt, Gao & Saag, Flechten Follmann, Contributions to Lichenology in Honour of Gerhard Follmann(Cologne): 360(1995).

≡*Platysma globulans* Nyl. ex Hue, *Nouv. Arch. Mus. Hist. Nat.* , Paris, 4 sér. 1: 213(1899).

On bark.

Qinghai, Shaanxi, Xizang, Yunnan(Rui-Fang Wang et al. , 2015. p. 588), Sichuan(Obermayer W. 2004, p. 486; Rui-Fang Wang et al. , 2015. p. 588).

Allocetraria isidiigera Kurok. & Lai 裂芽厚枝衣

Bull. Natn. Sci. Mus. Tokyo, Ser. B, 17(2): 62(1991).

Type: Xizang, Nylalam, on *Rhododendron* stem, alt. 3910 m. , J. C. Wei & J. B. Chen 1857(holotype in HMAS-L and isotype in TNS).

≡*Cetraria isidiigera*(Kurok. & M. J. Lai) Divakar, Crespo & Lumbsch, in Divakar, Crespo, Kraichak, Leavitt, Singh, Schmitt & Lumbsch, Fungal Diversity 84: 111(2017).

Corticolous on branches of *Rhododendron*.

Xizang(Kurokawa & Lai, 1991, p. 62; Rui-Fang Wang et al. , 2015. p. 588).

Allocetraria madreporiformis(Ach.) Kärnef. & Thell 小管厚枝衣

Nova Hedwigia 62: 508(1996).

≡*Dufourea madreporiformis* Ach. , *Lich. univ.* : 525(1810).

≡*Dactylina madreporiformis*(Ach.) Tuck. , *Proc. Amer. Acad. Arts & Sci.* 5: 398(1862) [1860].

On the ground.

Xinjiang(Elenk. pp. 12 – 27; H. Magn. 1940, p. 129; Moreau et Moreau, 1951, p. 191; Wu J. L. 1985, p. 75; Wang XY, 1985, p. 348 as *Dufourea madreporiformis* Ach. ; Rui-Fang Wang et al. , 2015. p. 589), Xizang (Lynge, 1933, p. 39), Yunnan(Hue, 1899, p. 60; Zahlbr. , 1930 b, p. 200).

Allocetraria sinensis X. Q. Gao 中华厚枝衣

in Thell, Randlane, Kärnefelt, Gao & Saag, Flechten Follmann, *Contributions to Lichenology* in Honour of Gerhard Follmann(Cologne): 365(1995).

Holotype HMAS, Gao 3052.

≡ *Cetraria sinensis* (X. Q. Gao) Divakar, Crespo & Lumbsch in Divakar, Crespo, Kraichak, Leavitt, Singh, Schmitt & Lumbsch, Fungal Diversity 84: 111(2017).

On ground between bryophytes.

Shaanxi, Yunnan(Rui-Fang Wang et al. , 2015. p. 589), Sichuan(Obermayer, 2004, p. 487).

Allocetraria stracheyi(Bab.) Kurok. & Lai 叉蔓厚枝衣

Bull. Natn. Sci. Mus. Tokyo, Ser. B, 17(2): 62(1991).

≡*Evernia stracheyi* Bab. , Hook. , *Journ. Bot.* 4: 244(1852). (Type from India, Kumaon, Gori River, r. Strachey & J. E. Winterbottom: lectotype in H-Nyl. Herb. 36055).

= *Cetraria everniella*(Nyl.) Kremp. , Verhandl. Zool. -bot. Gesellsch. Wien 18: 315(1068).

≡*Platysma everniella* Nyl. , Mem. Soc. Sci. Nat. Cherb. 5: 100(1857), based on *Evernia stracheyi* Bab.

= *Cetraria laii* Divakar, Crespo & Lumbsch, in Divakar, Crespo, Kraichak, Leavitt, Singh, Schmitt & Lumbsch, Fungal Diversity 84: 111(2017).

On soil and branches of shrubs.

Xinjiang(Rui-Fang Wang et al. , 2015. pp. 589 – 590), Shaanxi(Kurok. et al. , 1991, p. 64 as "Suanxi"; He & Chen 1995, p. 42 without citation of specimens and their localities; Rui-Fang Wang et al. , 2015. pp. 589 – 590), Xizang(J. C. Wei & Jiang, 1981, p. 1147 & 1986, p. 57; Rui-Fang Wang et al. , 2015. pp. 589 – 590), Sichuan(Kurok. et al. , 1991, p. 64; Obermayer, 2004, p. 487; Rui-Fang Wang et al. , 2015. pp. 589 – 590),

Yunnan(Zahlbr., 1930 b, p. 198; Kurok. et al., 1991, p. 64; Rui-Fang Wang et al., 2015. pp. 589 – 590), Taiwan(Lai, 1980, p. 216; Kurok. et al., 1991, p. 64; Lai, 2000, pp. 195 – 196).

Allocetraria subteres(Asahina) J. C. Wei, comb. nov. 亚柱厚岛衣

MB: 821260

≡ *Cetraria everniella* f. *subteres* Asahina in H. Kihara, *Fauna and Fl. of Nepal Himalaya*, 1952 – 53, 1: 58 (1955), cum icon. in Lamb, I. M. *Index Nominum Lichenum*: 155(1963); Wei Jiang-Chun, Jiang Yu-Mei, *Lichens of Xizang*, p. 57(1986).

Type collection from Nepal, Between Sangda and Kagbeni, 4300 m, H. Kihara s. n. holotype(in TNS!).

Xizang(J. C. Wei & Jiang, 1986, p. 57, as *C. everniella f. subteres*).

Specimens examined: China, Xizang, Nyanam, June 23, 1975, Chen Shu-Kun No. 7(HMAS-L. 2795), Sikkim, Herb. W. Nylander, 36054(H).

Note. *Allocetraria subteres* is different from *A. stracheyi* by having subteres and not dorsiventral lobes, which of the later one is dorsal ventral and oblate.

Allocetraria yunnanensis R. F. Wang, X. L. Wei & J. C. Wei 云南厚岛衣

The Lichenologist 47(1): 33(2015).

Type: China, Yunnan Province, Deqin County, Meli Village, Meili Snow Mountain, on soil, 28°38′1910N, 98°36′3040 E, alt. 4800 m, 10 Sept. 2012, R. F. Wang YK12012(HMAS-L128218—holotype).

= *Cetraria wangii* Divakar, Crespo & Lumbsch Divakar, Crespo, Kraichak, Leavitt, Singh, Schmitt & Lumbssh, in *Fungal Diversity*, 84: 111(2017).

On soil.

Anzia Stizenb. **绵腹衣属**

Flora, Regensburg 44: 393(1861).

Anzia colpota Vain. 霜绵腹衣

Bot. Mag., Tokyo 35: 19(1921).

On *Pinus armandi* and *Quercus* sp.

Yunnan(Wu JN, Wang LS, pp. 42 – 43; Wang X. Y. et al. 2015, pp. 102 – 103).

Anzia cristulata(Ach.) Stizenb. 小鸡冠绵腹衣

Flora, Regensburg 45: 243(1862).

≡ *Parmelia cristulata* Ach., *Syn. meth. lich.* (Lund): 218(1814).

Taiwan(Zahlbr., 1933 c, p. 58; Wang & Lai, 1973, p. 88).

Anzia formosana Asahina 台湾绵腹衣

Journ. Jap. Bot. 13(4): 221(1937).

Type: Taiwan, Mt. Alishan, leg. M. Ogata. Xizang, Sichuan, Yunnan(Wang X. Y. et al. 2015, pp. 103 – 106), Hunan(J. C. Wei et al. 1982, p. 33), Taiwan(M. Sato, 1938 b, p. 786 & 1939, p. 18; Lamb, 1963, p. 32; Wang & Lai, 1973, p. 88; Ikoma, 1983, p. 29).

Anzia hypoleucoides Müll. Arg. 淡绵腹衣

Flora, Regensburg 74(1): 111(1891).

On bark of trees.

Sichuan(Wang X. Y. et al. 2015, p. 106), Yunnan(J. C. Wei, 1981, p. 84; Wei X. L. et al., 2007, p. 147; Wang X. Y. et al. 2015, p. 106), Zhejiang(Xu B. S., 1989, p. 230), Taiwan(Sato, 1938 b, p. 788; Wang & Lai, 1973, p. 88).

A large number of specimens collected from Yunnan under the name "Anzia aff. hypoleucoides" preserved in KUN-L are cited by Wang X. Y. et al. 2015, pp. 106 – 108. It remains to be studied further in the future.

Anzia hypomelaena(Nyl.) Xin Y. Wang & Li S. Wang 黑腹绵腹衣

The Lichenologist 47(2): 108(2015).

≡ *Anzia leucobatoides*(Nyl.) Zahlbr. f. *hypomelaena* Zahlbr. in *Symb. Sin.* 3: 196(1930).

Type: China, Yunnan, Lijiang Co. , Rock 11575, 11778(Y—syntypes) .

Yunnan(Zahlbr. , 1934 b, p. 211; Wu JN, Wang LS, p. 43; Wang X. Y. et al. 2015, pp. 108 – 109) .

Anzia japonica(Tuck.) Müll. Arg. 日本绵腹衣

Flora, Regensburg 72: 507(1889) .

≡*Parmelia japonica* Tuck. , *Proc. Amer. Acad. Arts* 5: 399(1862) [1860] .

On trees.

NE China(prov. not indicated: Sato, 1939, p. 14) , Sichuan, Yunnan(Wang X. Y. et al. 2015, p. 109) , Zhejiang, Anhui, Hunan(J. C. Wei et al. 1982, p. 32) , Fujian(Wu J. N. et al. 1982, p. 9) , Taiwan(M. Sato, 1938 b, p. 787 & 1939, p. 14; Asahina & M. Sato in Asahina, 1939, p. 731; Wang & Lai, 1973, p. 88) .

f. japonica 原变型

Taiwan(Asahina, 1935 c, p. 227) .

f. saxicola Moreau et Moreau 石生变型

Rev. Bryol. et Lichenol. 20: 191(1951) , (nom. inval. : Art. 36) .

Type: Zhejiang, Liou T. N. (Fl. Chin. no. 50) .

Anzia leucobatoides(Nyl.) Zahlbr. 白绵腹衣

in Engler-Prantl, Naturl. Pflanzenfamil 1(1) : 214(1907) . f. leucobatoides 原变型

≡*Parmelia leucobatoides* Nyl. in Hue, *Bull. Soc. Bot. France*, 34: 21(1887) .

Type: Yunnan, Delavay no. 1608.

Sichuan(Wang X. Y. et al. 2015, pp. 109 – 111) , Yunnan(Hue, 1889, p. 166, 1890, p. 293 & 1899, p. 134; Zahlbr. , 1930 a, p. 278, 1930 b, p. 196 & 1934, p. 211; Wu JN, Wang LS, p. 43; Wang X. Y. et al. 2015, pp. 109 – 111) .

f. hypomelaena Zahlbr. 黑腹变型

in Handel-Mazzetti, *Symb. Sin.* 3: 196(1930) & *Cat. Lich. Univ.* 8: 570(1932) .

Syntype: Yunnan, Lijiang, Mt. Yulong shan, on rock, nos. 11575, 11778.

Yunnan(Zahlbr. , 1934 b, p. 211; Wu JN, Wang LS, p. 43) .

Anzia opuntiella Müll. Arg. 仙人掌绵腹衣

Flora, Regensburg 64(7) : 112(1881) .

Guizhou, Yunnan(Wang X. Y. et al. 2015, p. 111) , Anhui(Xu B. S. , 1989, p. 230) , Zhejiang(Xu B. S. , 1989, p. 230; Wang X. Y. et al. 2015, p. 111) , Hubei(Chen, Wu, J. C. Wei, 1989, p. 450) , Fujian(Wu J. N. et al. 1982, p. 9) , Taiwan(M. Sato, 1939, p. 12; Wang & Lai, 1973, p. 88) , China(prov. not indicated: Zahlbr. , 1930 a, p. 278) .

Anzia ornata(Zahlbr.) Asahina 瘤绵腹衣

J. Jap. Bot. 13(4) : 221(1937) .

≡*Anzia japonica* var. *ornata* Zahlbr. , *Feddes Repert.* 33: 59(1933) .

Type: Taiwan, colleceted by S. Sasaki(Asahina no. 409) .

Guizhou, Yunnan(Wang XinYu et al. 2015, pp. 111 – 112) , Anhui, Zhejiang(Xu B. S. , 1989, p. 230) , Hubei (Chen, Wu, J. C. Wei, 1989, p. 451) , Hunan(J. C. Wei et al. 1982, p. 33) , Fujian(Wu J. N. et al. 1982, p. 9) , Taiwan(M. Sato, 1938 b, p. 787 & 1939, p. 17; W. Culb. 1961, p. 381; Yoshim. & Sharp, 1968, p. 110; Wang & Lai, 1973, p. 88; Ikoma, 1983, p. 30; Wang X. Y. et al. 2015, pp. 111 – 112) .

Anzia physoidea A. L. Smith 膀果绵腹衣

Trans. Br. mycol. Soc. 16: 131(1931) .

On trees.

Sichuan, Yunnan(Wang L. S. , 1995, p. 313) .

Anzia pseudocolpota Xin Y. Wang & Li S. Wang 拟霜绵腹衣

The Lichenologist 47(2):112(2015).

Type: China, Yunnan Prov., Weixi Co., Lidiping Mt., on Loranthus bark, 3350 m, 15 June 2013, Li S. Wang 13-38274(KUN-L 22479—holotype).

Sichuan, Yunnan(Wang X. Y. et al. 2015, p. 112).

Anzia rhabdorhiza Li S. Wang & M. M. Liang 蔷薇绵腹衣

in Liang M. M. et al. *Bryologist*, 115(3):383-386(2012).

Type: Yunnan, Lijiang Co., Jiuhe village, Laojunshan Mt., 26u379N, 99u439E, elevation 3860 m. On bark of *Rhododendron* sp., 22 May 2011, Wang Li-song & Liang Meng-meng 11-32047(KUN-L 20000—holotypus; HMAS-isotypus).

Sichuan, Yunnan(Wang X. Y. et al. 2015, p. 113).

Anzia semiteres(Mont. & Bosch) Stizenb. 半圆柱绵腹衣

Flora, Regensburg 45:243(1862).

≡*Parmelia semiteres* Mont. & Bosch, *Syll. gen. sp. crypt.* (Paris):328(1856).

Taiwan(Zahlbr., 1933 c, p. 59; Wang & Lai, 1973, p. 88).

Arctoparmelia Hale **北极梅属**

Mycotaxon 25(1):251(1986).

Arctoparmelia centrifuga(L.) Hale 离心北极梅

Mycotaxon 25(1):252(1986).

≡*Lichen centrifugus* L., Sp. pl. 2: 1142(1753).

≡*Parmelia centrifuga*(L.) Ach., *Methodus*, Sectio post. (Stockholmiæ):206(1803).

≡*Xanthoparmelia centrifuga*(L.) Hale, *Phytologia* 28(5):486(1974).

On rocks and on thalli of species of Umbilicariaceae.

Heilongjiang(Gao, 1988, p. 34), Neimenggu(Gao, 1988, p. 34; Chen J. B. 2015, p. 42-43).

Arctoparmelia incurva(Pers.) Hale 曲北极梅

Mycotaxon 25(1):252(1986).

≡*Lichen incurvus* Pers., *Ann. Bot.* (Usteri) 1:24(1794).

≡*Parmelia incurva*(Pers.) Fr., *Nov. Sched. Critic. Lich.*:31(1826).

≡*Xanthoparmelia incurva* (Pers.) Hale, *Phytologia* 28(5):488(1974).

Jilin(J. C. Wei, 1983, p. 225; Chen J. B. 2015, p. 43-44).

Arctoparmelia separata(Th. Fr.) Hale 平坦北极梅

Mycotaxon 25(1):252(1986).

≡*Parmelia separata* Th. Fr., *J. Linn. Soc.*, *Bot.* 17:353(1880).

Heilongjiang(Chen J. B. 2011, pp. 882-883), Neimenggu(Chen J. B. 2011, pp. 882-883 & 2015, p. 44).

Asahinea W. L. Culb. & C. F. Culb. **裸腹叶属**

Brittonia 17:183(1965).

Asahinea chrysantha(Tuck.) W. L. Culb. & C. F. Culb. 金黄裸腹叶

Brittonia 17:183(1965). f. chrysantha 原变型

≡*Cetraria chrysantha* Tuck., *Amer. J. Sci. Arts*, Ser. 2 25:423(1858).

On the ground.

Heilongjiang, Jilin(Lai et al., 2009, pp. 366-367), Neimenggu(Sato, 1952, p. 174; Lai et al., 2009, pp. 366-367).

f. cinerascens(Asahina) J. C. Wei 灰面变型

Enum. Lich. China: 29(1991).

≡*Cetraria chrysantha* f. *cinerascens* Asahina, Journ. Jap. Bot. 10(8):481(1934).

Taiwan(Sato, 1939, p. 66; Wang-Yang & Lai, 1973, p. 88; Lai, 2000, pp. 196-197).

Austroparmelina A. Crespo, Divakar & Elix 自梅属

Syst. Biodiv. 8(2):216(2009).

Austroparmelina subtiliacea(Nyl.) A. Crespo, Divakar & Elix 亚椴自梅

Syst. Biodiv. 8(2):218(2009).

≡*Parmelia subtiliacea* Nyl., *Flora*, Regensburg 68:614(1885).

≡*Canoparmelia subtiliacea*(Nyl.) Elix & Hale in Elix, Johnston & Verdon, *Mycotaxon* 27:279(1986).

≡*Pseudoparmelia subtiliacea*(Nyl.) Hale, *Phytologia* 29:191(1974).

On rock.

Shandong(Jia ZF et al., 2008, p. 462).

Brodoa Goward 小腊肠衣属

Bryologist 89(3):222(1987)[1986].

Brodoa oroarctica(Krog) Goward 北极小腊肠衣

Bryologist 89(3):222(1987)[1986].

≡*Hypogymnia oroarctica* Krog, *Lichenologist* 6(2): 136(1974).

On rocks.

Xinjiang(J. C. Wei & Y. M. Jiang, 1999, p. 446).

Bryoria Brodo & D. Hawksw. 小孢发属

Op. Bot. 42:78(1977).

Bryoria alaskana Myllys & Goward 阿拉斯加小孢发

Lichenologist 48(5):355 – 365(2016).

Yunnan(Myllys et al., 2016, p357).

Bryoria asiatica(Du Rietz) Brodo & Hawksw. 亚洲小孢发

Op. Bot. 42:155(1977).

≡*Alectoria asiatica* Du Rietz, Ark. Bot. 20 A(11):18(1926).

Type: Sichuan, H. Smith, no. 5018(holotype) in UPS.

≡*Bryopogon asiaticus*(Du Rietz) Gyeln. in Fedde, *Repertorium* 38:235(1935).

On *Abies* sp.

Shaanxi(He & Chen 1995, p. 42 without citation of specimens and their localities), Yunnan(Wu JN, Wang LS, p. 38; Wang et Chen, 1994, p. 146), Hubei(Chen, Wu & J. C. Wei, 1989, p. 452;).

Bryoria barbata Li S. Wang & D. Liu 美髯小孢发

Phytotaxa 297(1):34(2017).

Type: China, Yunnan Prov. Lijiang Co., on bark Abies sp., 2011 – 5 – 22, Wang Lisong & Liang Mengmeng 11 – 32052(KUN-L 46008, Holotype).

Jilin, Neimenggu, Shaanxi(as Shannxi, Taibai Mt.), Sichuan, Xizang, Yunnan(Wang et al., 2017, p. 34).

Bryoria bicolor(Ehrh.) Brodo & D. Hawksw. 双色小孢发

Op. Bot. 42:99(1977).

≡*Lichen bicolor* Ehrh., *Pl. crypt. exsicc.*: no. 40(1785).

≡*Alectoria bicolor*(Ehrh.) Nyl., *Act. Soc. linn. Bordeaux* 21:291(1856).

≡*Bryopogon bicolor*(Ehrh.) Stein, in Cohn, *Krypt. -Fl. Schlesien* (Breslau) 2(2):35(1879).

Corticolous.

Heilongjiang(Chen et al. 1981, p. 128), Neimenggu(Chen et al. 1981, p. 128), Shaanxi(Baroni, 1894, p. 47), Yunnan(Hue, 1899, p. 88; Du Rietz 1926, p. 16; Paulson 1928, p. 318; Zahlbr., 1930b, p. 201; Asahina & Sato in Asahina, 1939, p. 751; Wang et Chen, 1994, p. 146; Wei X. L et al., 2007, p. 148), Taiwan(Wang & Lai, 1973, p. 87).

≡*Cornicularia bicolor*(Ehrh.) Ach., *Methodus, Sectio post.* (Stockholmiæ):304(1803)

= *Cornicularia bicolor* var. *melaneira* Ach. *Lichenogr. Univers*. 614(1810).

Corticolous.

Yunnan(Hue, 1899, p. 89; Zahlbr., 1930 b, p. 201).

Bryoria confusa(D. D. Awasthi) Brodo & D. Hawksw. 刺小孢发

Op. Bot. 42: 155(1977).

≡*Alectoria confusa* D. D. Awasthi, *Proc. Indian Acad. Sci.*, Sect. B 72(4): 152(1970).

Alectoria acanthodes auct. non Hue: J. C. Wei & Chen, in Report on the Scientific Investigations in Mt. Qomolangma district, Science Press, Beijing, 1974, p. 181.

Xinjiang(Abdulla Abbas & Wu, 1998, p. 90; Abdulla Abbas et al., 2001, 361), Shaanxi(He & Chen 1995, p. 42 without citation of specimen and locality), Yunnan(Wu JN, Wang LS, pp. 38 – 39; Wang et Chen, 1994, p. 146; Wei X. L et al., 2007, p. 148), Xizang(J. C. Wei & Chen, 1974, p. 181; J. C. Wei & Jiang, 1986, p. 63), Hubei(Chen, Wu & J. C. Wei, 1989, p. 452), Taiwan(Wang & Lai, 1976, p. 226).

Bryoria cornicularioides(P. Jorg.) Brodo & Hawksw. 类角小孢发

Op. Bot. 42: 155(1977).

≡*Alectoria cornicularioides* P. Jorg. *Bryologist* 78: 77(1975).

Alectoria bicolor auct. non(Ehrh.) Nyl.: Jatta, Nuorn. Giorn. Bot. Italiano, 2(9): 462(1902).

Type: Shaanxi, Giraldi no. 1896(holotype) in Fl.

Among mosses.

Bryoria divergescens(Nyl.) Brodo & Hawksw. 广开小孢发

Op. Bot. 42: 155(1977).

≡*Alectoria divergescens* Nyl. Flora 69: 466(1866) & in Hue, *Bull. Soc. Bot. France* 34: 20(1887).

≡*Bryopogon divergescens*(Nyl.) Gyeln. in Fedde, Repertorium, 38: 238(1935).

Type: Yunnan, R. P. Delavay, no. 1885(holotype) in H-Nyl. no. 35972. (Hue, 1890, p. 277 & 1899, p. 90; Stzbgr. 1892, p. 126; Du Rietz, 1926, p. 20; Zahlbr., 1930 a, p. 383 & 1930 b, p. 201; Gyeln. 1935, p. 238; Wang et Chen, 1994, pp. 146 – 148; Wei X. L et al., 2007, p. 148; Wang L. S. et al., 2012, p. 107 – 108).

Corticolous.

Bryoria fastigiata Li S. Wang & H. Harada 密枝小孢发

J. Hattori Bot. Lab. 100: 865(2006).

Typus: China, Yunnan prov., Zhongdian County, on branches of *Rhododendron impeditum*, 2004-6-15, Wang Li-song 04-23181(KUN-L 19023-holotypus; CBM-isotypus).

Sichuan, YunnaN(Wang et al., 2006, P 867).

Bryoria fruticulosa Li S. Wang & Myllys 卷毛小孢发

Phytotaxa 297(1): 35(2017).

Type: China, Sichuan prov., Xiangcheng Co., on bushes of *Rhododendron aganniphum*, 2002-9-12, Wang Lisong 02-23521(KUN-L 18795, Holotype in KUN-L).

Sichuan, Xizang, Yunnan(Wang et al., 2017, p36).

Bryoria furcellata(Fr.) Brodo & D. Hawksw. 叉小孢发

Op. Bot. 42: 103(1977).

≡*Cetraria furcellata* Fr., *Syst. orb. veg*. (Lundae) 1: 283(1825).

= *Alectoria nidulifera* Norrl., *Flora, Regensburg* 58: 8(1875).

Corticolous.

Heilongjiang(Chen et al. 1981, p. 128; Luo, 1984, p. 84), Neimenggu(Chen et al. 1981, p. 128), Xinjiang(Abdulla Abbas & Wu, 1998, p. 90; Abdulla Abbas et al., 2001, 361), Schuan, Yunnan(Wang L. S. etg al., 2005, pp. 174 – 176).

Bryoria fuscescens(Gyeln.) Brodo & D. Hawksw. 淡褐小孢发

Op. bot. 42: 83(1977) var. fuscescens

≡*Alectoria fuscescens* Gyeln. , *Annals Cryptog. Exot*. 4: 171(1931) .

= *Alectoria jubata*(L.) Ach. *Lich. univ*. : 592, tab. XIII, fig. 1(1810) .

≡*Lichen jubatus* L. , *Sp. pl*. 2: 1155(1753) .

= *Alectoria jubata* var. *prolixa* Ach. , *Lich. univ*. : 592(1810) .

= *Alectoria prolixa* f. *sublustris* Stizenb. , *Annln K. K. naturh. Hofmus*. Wien 7: 128(1892) .

Neimenggu (M. Sato, 1952, p. 174; Chen et al. 1981, p. 128) , Heilongjiang(Asahina, 1952 d, p. 375; Chen et al. 1981, p. 128; Luo, 1984, p. 84) , Shaanxi(Jatta, 1902, pp. 460, 462; Zahlbr. , 1930 b, p. 202; Wu J. L. , 1981, p. 162) , Yunnan(Hue, 1899, p. 86; Paulson, 1928, p. 318; Zahlbr. , 1930 b, p. 202; China(prov. not indicated: Asahina & Sato in Asahina, 1939, p. 747) .

Bryoria hengduanensis L. S. Wang & H. Harada 横断山小孢发

Acta Phytotax. Geobot. 54(2) : 100(2003) .

Type: Yunnan, Zhongdian, Tian-chi lake, alt. 3900m.

On *Abies georgei*, Aug. 1993, Wang L. S. 93 – 13673(KUN-L, 13927, holotypus, CM < B-FL: – 13390, isotypus) .

Yunnan(Wei X. L et al. , 2007, p. 148) .

Bryoria *himalayana*(Motyka) Brodo & D. Hawksw. 喜马拉雅小孢发

Op. Bot. 42: 155(1977) .

≡*Alectoria himalayana* Motyka *Fragm. Florist. Geobot*. 6: 450(1960) .

On *Rhodendron* sp.

Yunnan(Wang et Chen, 1994, p. 148; Wei X. L. et al. , 2007, p. 148, as *B. himalayensis*) .

Bryoria lactinea(Nyl.) Brodo & D. Hawksw. 乳白小孢发

Op. Bot. 42: 155(1977) .

≡*Alectoria lactinea* Nyl. , *Lich. Japon*. : 23(1890) .

On *Bambusa* sp. & *Rhododendron* sp.

Yunnan(Wu JN, Wang LS, p. 39; Wang et Chen, 1994, p. 148; Wei X. L. et al. , 2007, pp. 148 – 149) .

Bryoria lanestris(Ach.) Brodo & D. Hawksw. 绵毛小孢发

Op. Bot. 42: 88(1977) .

≡*Alectoria jubata* var. *lanestris* Ach. , *Lich. univ*. : 1 – 696(1810) .

≡*Alectoria lanestris*(Ach.) Gyeln. , *Nytt Mag. Natur*. 70: 58(1932) .

Corticolous.

Heilongjiang(Chen et al. 1981, p. 128; Luo, 1984, p. 84) , Neimenggu(Chen et al. 1981, p. 128) , Taiwan(Wang & Lai, 1973, p. 87) .

Bryoria levis D. D. Awasthi, in Awasthi & Awasthi 光滑小孢发

Candollea 40(1) : 310(1985) .

On *Bambusa* sp. , *Rhododendron* sp. & on rocks in forest.

Yunnan(Wang et Chen, 1994, p. 149) .

Bryoria nadvornikiana(Gyeln.) Brodo & D. Hawksw. 蚕丝小孢发

Op. Bot. 42: 122(1977) .

≡*Alectoria nadvornikiana* Gyeln. , *Acta Faun. Fl. Univers*. , Ser. 2, Bot. 1: 6(1932) .

On trees.

Sichuan, Yunnan(Wang L. S. et al. 2005, p. 174) .

Bryoria nepalensis D. D. Awasthi 尼泊尔小孢发

in Awasthi & Awasthi, *Candollea* 40(1) : 312(1985) .

On *Abies* sp. & *Rhodendron* sp.

Yunnan(Wang et Chen, 1994, p. 149 – 150).

Bryoria nitidula(Th. Fr.) Brodo & D. Hawksw. 光亮小孢发

Op. Bot. 42: 107(1977).

≡*Bryopogon jubatus* var. *nitidulus* Th. Fr., *Nova Acta R. Soc. Scient. upsal.*, Ser. 3 3: 125(1861) [1860].

On *Rhododendron* sp. & on rocks.

Yunnan(Wang et Chen, 1994, p. 150; Wei X. L. et al., 2007, p. 149).

Bryoria perspinosa(Bystrek) Brodo & D. Hawksw. 多叉小孢发

Op. Bot. 42: 155(1977).

≡*Alectoria perspinosa* Bystrek, *Khumbu Himal* 6(1): 21(1969).

On *Larix* sp.

Yunnan(Wang et Chen, 1994, p. 150).

Bryoria poeltii(Bystrek) Brodo & D. Hawksw. 波氏小孢发

Op. Bot. 42: 155(1977).

≡*Alectoria poeltii* Bystrek, *Khumbu Himal* 6(1): 20(1969).

On *Picea* sp. & *Larix* sp.

Yunnan(Wang et Chen, 1994, pp. 150 – 151; Wei X. L. et al., 2007, p. 149).

Bryoria rigida P. M. Jorg. & Myllys 硬质小孢发

in P. M. Jorg. Et al., *Lichenologist* 44(6): 777 – 779(2012).

Typus: China, Yunnan, Dali, Diancang Shan Mt., alt. 3570 m, on rock, 28 July 2005, L. S. Wang 06 – 26208 (KUN-holotypus; H-isotypus). [GenBank accession numbers: HQ402703, HQ402652].

Bryoria smithii(Du Rietz) Brodo & Hawksw. 珊粉小孢发

Op. Bot. 42: 152(1977).

≡*Alectoria smithii* Du Rietz, *Arkiv For Botanik.* Band 20 A no. 11: 14(1926).

Type: Sichuan, H. Smith no. 5025 b(lectotype) in UPS. (Gyeln. 1935, p. 233).

= *Alectoria bicolor* var. *berengeriana* Massal. ex Stiz.

≡*Bryopogon berengerianus*(Massal.) Gyeln.

Xizang(J. C. Wei & Jiang, 1986, p. 64), Yunnan(Wu JN, Wang LS, p. 39; Wang et Chen, 1994, p. 151).

Bryoria trichodes subsp. americana(Motyka) Brodo & D. Hawksw. 毛状小孢发，美洲亚种

Op. Bot. 42: 96(1977).

≡*Alectoria americana* Motyka, *Fragm. flor. geobot.* (Kraków) 6(3): 449(1960).

Alectoria jubata Ach. var. *lanestris* auct. non Ach.

Jilin, Neimenggu(Chen et al. 1981, p. 127), Yunnan(Wei X. L. et al., 2007, p. 149, as *B. trichodes ssp. americana*), Taiwan(Wang & Lai, 1976, p. 226).

Bryoria variabilis(Bystrek) Brodo & D. Hawksw. 多形小孢发

Op. Bot. 42: 156(1977).

≡*Alectoria variabilis* Bystrek, *Khumbu Himal* 6(1): 22(1969).

On Abies sp. & Picea sp.

Yunnan(Wu JN, Wang LS, p. 39 – 40; Wang et Chen, 1994, p. 151.).

Bryoria wuii Li S. Wang 吴氏小孢发

Phytotaxa 297(1): 37(2017).

Type: China, Yunnan Prov., Deqin Co., on bark of *Larix*, 2013 – 6 – 21, Wang Li-song & Wang Xin-yu13 – 38467(KUN-L 23999, Holotype).

Sichuan, Yunnan(Wang et al., 2017, p. 37).

Bryoria yunnana Li. S. Wang & Xin Y. Wang 云南小孢发

Phytotaxa 297(1):38(2017).

Type: Yunnan, Dali Co., Cangshan Mt., on branches of *Abies delavayi*, 2004 – 7 – 21, Wang Lisong 04 – 23414 (KUN-L 23994, Holotype).

Jilin, Shaannxi(as Shannxi, Taibai Mt.), Sichuan, Xizang, Yunnan, Taiwan(Wang et al., 2017, p39).

Bulbothrix Hale **球针叶属**

Phytologia 28:479(1974).

Bulbothrix asiatica Y. Y. Zhang & Li S. Wang 亚洲球针叶

Bryologist 117:379 – 385(2014).

Type: China, Yunnan, L. S. Wang et al. (KUN-L 46314—holotype).

Fujian(Zhang Y. Y. et al. 2016, pp. 129 – 130).

Bulbothrix goebelii(Zenker) Hale 戈氏球针叶

Smithson. Contr. Bot. 32:14(1976).

≡*Parmelia goebelii* Zenker, in Goebel & Kunze, *Pharmaceutische Waarenkunde*(Eisenach) 1(3):134(1827) [1827 – 1829].

= *Parmelia scortella* Nyl. *Flora*, Regensburg 68:615(1885), (Lai, 2000, p. 100).

Pantropical.

On calcareous rocks & on bark of trees.

Heilongjiang(Chen et al. 1981, p. 153), Yunnan(Zhao, 1964, p. 155; Zhao et al. 1982, p. 37), Fujian(Chen JB et al., 2009, pp. 93 – 94 & 2015. 46), Taiwan(Hale, 1976, p. 14; Lai, 2000, 99 – 100).

Bulbothrix isidiza(Nyl.) Hale 裂芽球针叶

Phytologia 28:480(1974).

≡*Parmelia isidiza* Nyl., *Bolm Soc. broteriana, Coimbra*, sér. 1 3:130(1884).

= *Parmelia subscortea* Asahina, *Journ. Jap. Bot.* 32(4):98(1957); Lai, 2000, p. 100.

Type: Taiwan, Asahina no. 3324 in TNS. Pantropical.

Misapplied name: *Parmelina quercina*(Willd.) Hale(Tchou, 1935, p. 311).

Pantropical.

Sichuan, Fujian, Guangxi, Hainan(Chen JB et al., 2009, p. 94; Chen JB, 2015. P. 47), Yunnan(Wang L. S. et al., 2008, p. 536; Chen JB et al., 2009, p. 94; Chen JB, 2015. P. 47; Zhang Y. Y. et al. 2016, p. 130), Taiwan (M. Lamb, 1963, p. 504; Wang & Lai, 1973, p. 93 & 94; Hale, 1976, p. 16; Lai, 2000, pp. 100 – 101; Chen JB et al., 2009, p. 94; ; Chen JB, 2015. P. 47), Hainan(JCWei et al., 2013, p. 228), Hongkong(Tchou, 1935, p. 311; Thrower, 1988, pp. 15, 67; Aptroot & Seaward, 1999, p. 68; Chen JB et al., 2009, p. 94; ; Chen JB, 2015. P. 48).

Bulbothrix lacinia Y. Y. Zhang & Li S. Wang 细长球针叶

The *Lichenologist* 48(2):126 – 127(2016).

Type: China, Yunnan Prov., Chuxiong, Chahe Village, 2013, L. S. Wang et al. 13 – 41296 (KUN-L22192—holotype).

On sandstone.

Bulbothrix mammillaria Y. Y. Zhang & Li S. Wang 乳头球针叶

The *Lichenologist* 48(2):124 – 126(2016).

Type: China, Yunnan Prov., Yongsheng Co., Dashan Village, on bark, 5 December 2013, L. S. Wang et al. 13 – 41171(KUN-L22067—holotype).

Sichuan, Yunnan, Guangxi(Zhang Y. Y. et al. 2016, pp. 124 – 126).

Bulbothrix meizospora(Nyl.) Hale 大孢球针叶

Phytologia 28(5):480(1974).

On the bark of *Cinnamomum* sp.

Xizang(Zhang Y. Y. et al. 2016, pp. 127 – 129).

Bulbothrix scortella(Nyl.) Hale 尾球针叶

Phytologia 28(5): 480(1974).

On bark.

Yunnan, Fujian(Zhang Y. Y. et al. 2016, p. 139).

Bulbothrix setschwanensis(Zahlbr.) Hale 四川球针叶

Phytologia 28: 481(1974).

≡*Parmelia setschwanensis* Zahlbr., in Handel-Mazzetti, *Symb. Sin.* 3: 184(1930) & Cat. Lich. Univ. 8: 567 (1932).

Type: Sichuan, Handel-Mazzetti no. 2739(lectotype) in WU & Handel-Maazzetti no. 2017 in W(not seen).

Sichuan(Zhang Y. Y. et al. 2016, p. 130), Yunnan(Zahlbr., 1934, p. 211; Chen JB et al., 2009, p. 94; Chen JB, 2015. P. 48; Zhang Y. Y. et al. 2016, p. 130), xizang(Chen JB et al., 2009, p. 94; Chen JB, 2015. P. 49).

Corticolous.

Bulbothrix subscortea(Asahina) Marcelli & Benatti 亚尾球针叶

Mycosphere 3: 46 – 55(2012).

Type: China, Taiwan, Asahina 3324(TNS—lectotype).

On bark.

Fujian(Zhang Y. Y. et al. 2016, p. 131).

Bulbothrix tabacina(Mont. & Bosch) Hale 烟草球针叶

Phytologia 28: 481(1974).

≡*Parmelia tabacina* Mont. & Bosch, in Miquel, *Pl. Jungh.* 4: 443(1855).

On bark. Pantropical.

Yunnan, Hunan, Guangxi, Fujian(Chen JB et al., 2009, pp. 94 – 95; Chen JB, 2015. p. 49 – 50), Hainan(Chen JB et al., 2009, pp. 94 – 95; Chen JB, 2015. P. 49 – 50; Zhang Y. Y. et al. 2016, p. 131; JCWei et al., 2013, p. 228), Taiwan(Hale, 1976, p. 24; Lai, 2000, p. 101).

Bulbothrix yunnana Wang S. L. & Chen J. B. & Elix 云南球针叶

Mycotaxon 76: 293 – 294(2000).

Type: Yunnan, Zhongdian, alt. 3700m, on bark of Acer sp., Aug. 14, 1981, Wang X. Y, Xiao X., & Su J. J. 5669(holotype: HMAS. L.; isotype: CANB), paratypes: Su J. J. 3162, Wang, Xiao, Su 3434, 5381, 5753, 5756 in HMAS. L. Yunnan(Chen JB et al., 2009, p. 95; Chen JB, 2015. P. 50).

Canomaculina Elix & Hale **灰点衣属**

Mycotaxon 29: 239(1987).

= *Rimeliella* Kurok., *Annals of the Tsukuba Botanical Garden* 10: 1(1991).

Canomaculina subsumpta(Nyl.) Elix 粉斑灰点衣

Mycotaxon 65: 477(1997).

≡*Parmelia subsumpta* Nyl., *Flora*, Regensburg 52: 117(1869).

≡*Parmotrema subsumptum*(Nyl.) Hale, *Mycotaxon* 5(2): 434(1977).

≡*Rimeliella subsumpta* (Nyl.) Kurok., *Annals of the Tsukuba Botanical Garden* 10: 9(1991).

Yunnan(Chen J. B. et al., 2003, p. 20; Chen J. B., 2015, p. 214), Taiwan(Kurok., 1991, p. 9; Lai, 2000, pp. 103 – 104).

Canomaculina subtinctoria(Zahlbr.) Elix 多色灰点衣

Mycotaxon 65: 477(1997).

≡*Parmelia subtinctoria* Zahlbr., in Handel-Mazzetti, *Symb. Sinic.* 3: 193(1930).

Type: Yunnan, Handel-Mazzetti no. 5645.

≡*Parmotrema subtinctorum*(Zahlbr.) Hale *Phytologia* 28: 339(1974).

≡*Rimeliella subtinctoria* (Zahlbr.) Kurok., *Annals of the Tsukuba Botanical Garden* 10: 10(1991).

On bark of trees & on rocks.

Jilin(Chen JB, 2015. p. 215), sichuan(Chen J. B. et al. 2003, p. 21; Chen JB, 2015. p. 216), Yunnan(Zahlbr., 1932, p. 568; Zhao, 1964, p. 162; Zhao et al. 1982, p. 51; Chen J. B. et al. 2003, p. 21; Chen JB, 2015. p. 216), Shandong(Zhang F et l., 1999, p. 30; Zhao ZT et al., 2002, p. 6 as *P. subtinctorum*), Anhui, Zhejiang(Zhao, 1964, p. 162; Zhao et al. 1982, p. 51; Chen JB, 2015. p. 215), Hubei(Chen, Wu & J. C. Wei, 1989, p. 448; Chen JB, 2015. p. 215), Guangxi(Chen J. B. et al. 2003, p. 21; Chen JB, 2015. p. 215 – 216), Taiwan(Wang & Lai, 1973, p. 94; Wang & Lai, 1976, p. 228; Lai, 2000, pp. 104 – 105), Hainan(Chen J. B. et al. 2003, p. 21; Chen JB, 2015. p. 215 – 216; JCWei et al., 2013, p. 228 as *C. subtinctoria*).

Canoparmelia Elix & Hale **灰叶属**

in Elix, Johnston & Verdon, *Mycotaxon* 27: 277(1986).

Canoparmelia amazonica(Nyl.) Elix & Hale 针牙灰叶

in Elix, Johnston & Verdon, *Mycotaxon* 27: 278(1986).

≡*Parmelia amazonica* Nyl., *Flora, Regensburg* 68: 611(1885).

≡*Pseudoparmelia amazonica*(Nyl.) Hale, *Phytologia* 29(3): 189(1974).

On trees and rocks.

Chongqing(Chen CL et al., 2008, p. 40), Yunnan(Ahti et al., 1999, p. 123), Taiwan(Hale, 1976, p. 16; Lai, 2000, pp. 106 – 107).

Canoparmelia ecaperata(Müll. Arg.) Elix & Hale 松萝酸灰叶

in Elix, Johnston & Verdon, *Mycotaxon* 27: 278(1986).

≡*Parmelia ecaperata* Müll. Arg., *Flora, Regensburg* 74(3): 378(1891).

On bark of trees, such as *Pinus yunnanensis*.

Yunnan (Liu D et al., 2014, p. 784; Chen JB, 2015. p. 52).

Parmelia conformata auct. non Vain. In Zhao JD(1964) and Zhao JD et al. (1982).

Yunnan(Zhao, 1964, p. 158; Zhao et al. 1982, p. 41).

Canoparmelia owariensis(Asahina) Elix 石芽灰叶

Mycotaxon 47: 127(1993).

≡*Parmelia owariensis* Asahina, *J. Jap. Bot.* 28: 135(1953).

On rocks.

Guangdong(Chen JB, 2015. P. 53; Chen JB, 2015. p. 53).

≡*Parmelia owariensis* Asahina, *J. Jap. Bot.* 28: 135(1953).

On rocks.

Hongkong(Hale, 1976, p. 39; Thrower, 1988, p. 154).

≡*Paraparmelia owariensis*(Asahina) Elix & J. Johnst., in Elix, Johnston & Verdon,

Mycotaxon 27: 280(1986); Aptroot & Seaward, 83(1999).

≡*Pseudoparmelia owariensis*(Asahina) Hale, *Phytologia* 29(3): 190(1974); Thrower, 1988, pp. 17, 154.

On granite boulders. Paleotropical.

Hongkong(Thrower, 1988, pp. 17, 154; Aptroot & Seaward, 1999, p83; Aptroot et al., 2001, p. 324).

Canoparmelia texana(Tuck.) Elix & Hale 粉芽灰叶

in Elix, Johnston & Verdon, *Mycotaxon* 27: 279(1986).

≡*Parmelia texana* Tuck., *Amer. J. Sci. Arts*, Ser. 2 25: 424(1858).

≡*Pseudoparmelia texana* (Tuck.) Hale

On bark of *Pinus yunnanensis*.

Yunnan(Liu D et al., 2014, p. 785), Shandong(Jia ZF et al., 2008, p. 462; Chen JB, 2015. p. 53 – 54), Guangxi(Chen JB, 2015. p. 53 – 54), Taiwan(Lai, 2000, pp. 107 – 108).

Cetraria Ach. **岛衣属**

Methodus, Sectio post. : 292(1803).

Cetraria denticulata Hue 齿岛衣

Nouv. Arch. du Museum 4(1): 85(1899).

Type: Yunnan, R. P. Delavay, 8 Aug. 1888.

Cetraria ericetorum Opiz 栗卷岛衣

Böh. phänerogam. cryptogam. Gewächse(Prague): 135(1823).

≡ *Cetraria ericetorum* Opiz, *Seznam Rostlin Kveteny Ceské*: 175(1852).

= *Cetraria crispa*(Ach.) Nyl., *Bull. Soc. linn. Normandie,* sér. 4 1: 202(1887).

≡ *Cetraria islandica* var. *crispa* Ach., *Lich. univ.*: 513(1810).

= *Cetraria islandica* var. *tenuifolia*(Retz.) Vain., *Ark. Bot.* 8(no. 4): 21(1909).

≡ *Lichen islandicus* ß *tenuifolius* Retz., *Fl. scand. prodr.*: 227(1779).

On the ground.

Neimenggu, Jilin(Chen et al. 1981, p. 129), Xinjiang(Moreau et Moreau, 1951, p. 191), Shaanxi(A. Zahlbr., 1930b, p. 198), Yunnan(Hue, 1887, p. 18 & 1899, p. 84; R. Paulson, 1928, p. 317; A. Zahlbr., 1930 b, p. 198), HUBEI(Chen, Wu & Wei, 1989, p. 433).

Cetraria hepatizon(Ach.) Vain. 肝褐岛衣

Term. Füz. 22: 278(1899).

≡ *Lichen hepatizon* Ach., *Lich. suec. prodr.* (Linköping): 110(1799)[1798].

Saxicolous.

Heilongjiang, Jilin(Lai et al., 2009, p. 367), Xizang(J. C. Wei & Jiang, 1981, p. 1146 & 1986, p. 58).

Cetraria islandica(L.) Ach. 岛衣

Methodus, Sectio post. : 293(1803). subsp. islandica 原亚种

≡ *Lichen islandicus* L., *Sp. pl.* 2: 1145(1753).

On the ground.

Jilin(Chen et al, 1981, p. 129), Xinjiang(Tchou, 1935, p. 312; J. C. Wei et al. 1982, p. 44; Wu J. L. 1985, p. 74; Wang XY, 1985, p. 348; Abdulla Abbas & Wu, 1998, p. 91; Abdulla Abbas et al., 2001, p. 361), Shaanxi (J. C. Wei, 1981, P. 85; J. C. Wei et al. 1982, p. 44; He & Chen 1995, p. 42 without citation of specimen and locality), Hubei(Chen, Wu & J. C. Wei, 1989, p. 432), Shandong(zhao zt et al., 1999, p. 427; Zhao ZT et al., 2002, p. 6).

subsp. orientalis(Asahina ex M. Satô) Kärnefelt 东方亚种

Michigan Bot. 17(4): 107(1979).

≡ *Cetraria islandica* var. *orientalis* Asahina ex M. Satô, *J. Jap. Bot.* 14: 786(1938).

China(prov. non indicated: Sato, 1939, p. 30), Xinjiang(Abdulla Abbas et al., 1993, p. 76; Abdulla Abbas & Wu, 1998, p. 91; Abdulla Abbas et al., 2001, p. 362), Taiwan(Lai, 2000, pp. 197 – 198).

f. angustifolia Asahina ex M. Satô 细叶变型

J. Jap. Bot. 14: 786(1938).

Shaanxi(Jatta, 1902, p. 467), Taiwan(Sato, 1938b, p. 786 & 1939, p. 31 & 1959, p. 42; Wang & Lai, 1973, p. 89).

var. thyreophora Ach. 聚伞变型

Lich. univ.: 512(1810).

Yunnan(Zahlbr., 1930 b, p. 198 as "f. thyrsophora").

Cetraria laevigata Rassad. 白边岛衣

Bot. Zh. SSSR 28: 79(1943).

= *Cetraria crispa* var. *japonica* Asahina ex M. Satô, *J. Jap. Bot.* 14: 787(1938).

Heilongjiang(Sato, 1939, p. 32; Asahina, 1952 d, p. 375; Lai et al., 2009, p. 368), JILIN(Lai et al., 2009, p. 368), Neimenggu(Sato, 1959, p. 40; Wei, 1981, p. 86 ; Chen et al. 1981, p. 129; Lai et al., 2009, p. 368), xingjiang(Abdulla Abbas et al. , 1997, p. 1) , Shaanxi(He & Chen 1995, p. 42 without citation of specimen and locality) , Xizang(J. C. Wei & Chen, 1974, p. 181; J. C. Wei & Jiang, 1981, p. 1146 & 1986, p. 58) , Yunnan (Wei X. L. et al. , 2007, p. 149) , Taiwan(Wang & Lai, 1973, p. 88; Lai, 1980, p. 216) .

Cetraria melaloma(Nyl.) Kremp. 黑缘岛衣

Verh. zool. -bot. Ges. Wien 18: 315(1868) .

≡*Platysma melalomum* Nyl. , *Syn. meth. lich.* (Parisiis) 1: 303(1860) .

On rocks.

Shaanxi(Jatta, 1902, p. 464; Zahlbr. 1930 b, p. 198) .

Cetraria nigricans Nyl. 黑岛衣

Musci Fennic. : 109(1859) .

On thin layer of soil covered rock.

Xizang(J. C. Wei & Chen, 1974, p. 180)

Cetraria odontella(Ach.) Ach. 刺岛衣

Syn. meth. lich. (Lund) : 230(1814) .

≡*Lichen odontellus* Ach. , *Lich. suec. prodr.* (Linköping) : 213(1799) [1798] .

≡*Cornicularia odontella*(Ach.) Röhl. , *Deutschl. Fl.* (Frankfurt) 3(2) : 141(1813) .

Xizang(J. C. Wei & Y. M. Jiang, 1981, p. 1146 & 1986, p. 63) , Taiwan(Lai, 2000, p. 198) .

Cetraria pachysperma(Hue) Zahlbr. 厚果岛衣

in Engler-Prantl. Natürl. Pflamzenfamil. I. Teil, Abt. 1 * : 215(1907) .

≡*Platysma pachyspermum* Hue, Nouv. Arch. Mus. Hist. Nat. 4(1) : 215(1899) &(2) , tab. 2, fig. 11, 1 bis et 1, ter. (1900) .

Syntype: Yunnan, collected by R. P. Delavay from Yenzihai, 17/VI/1887 & Loping shan at about 3200 m, 31/VII/1888. (also see A. Zahlbruckner, 1930 a, p. 309 & 1930 b, p. 198) .

Corticolous.

Cetraria pallescens Schaer. 皮革岛衣

in Moritzi, *Syst. Verz.* : 129(1846) .

Corticolous.

Xizang(J. C. Wei & Chen, 1974, p. 180; J. C. Wei & Jiang, 1981, p. 1147 & 1986, p. 59) , Hubei(Chen, Wu, J. C. Wei, 1989, p. 433) , Anhui, Zhejiang(Xu B. S. , 1989, p. 222) , Taiwan(Zahlbr, 1933c, p. 60; Asahina, 1934 c, p. 484; Sato, 1938 b, p. 785 & 1939, p. 50; Wang & Lai, 1973, p. 89) .

Cetraria pallida D. D. Awasthi 淡色岛衣

in Proc. Indian Acad. Sci. XLV: 130(1957) .

On moss covered rock & on bush.

Xizang(J. C. Wei & Y. M. Jiang, 1986, p. 56) .

Cetraria potaninii Oxn. 蒲氏岛衣

Journ. Cycle Bot. Acad. Sci. Ukraine 7 – 8: 168(1933) . M. Lamb, *Index Nom. Lich.* 159(1963) .

Type: Xizang, 31/V/1893 collected by G. Potanin(LE) .

Among mosses.

Cetraria reticulata Kremp. ex Räsänen 网岛衣

in Kuopion Luonnon Ystäväin Yhdistyksen Yulkaisuja, ser B, II(6) : 28, 44(1952) , (nom. inval. , in lingua Latina haud descriptum) . M. Lamb, *Index Nom. Lich.* : 159(1963) .

Syntype: Taiwan & Himalaya, substratum not indicated.

Cetraria xizangensis J. C. Wei & Y. M. Jiang 藏岛衣

Acta Phytotaxon. Sin. 18(3) : 388(1980) & *Lichens of Xizang*: 60(1986) .

Type: Xizang, J. C. Wei & Chen no. 1899 in HMAS-L.

On bark of tree.

Xizang(J. C. Wei & Y. M. Jiang, 1980, p. 388; W. Obermayer, 2004, p. 494) .

Cetrariella Kärnefelt & A. Thell 小岛衣属

in Kärnefelt, Mattson & Thell, *Bryologist* 96(3) : 402(1993) .

Cetrariella delisei(Bory ex Schaer.) Kärnefelt & A. Thell 细裂小岛衣

Bryologist 96(3) : 403(1993) .

≡*Cetraria islandica* f. *delisei* Bory ex Schaer. , *Enum. critic. lich. europ.* (Bern) : 114(1850) .

≡*Cetraria delisei*(Bory ex Schaer.) Nyl. , K. svenska Vetensk-Akad. Handl. , ny följd 7(no. 2) : 11(1867) [1866] .

On the ground among mosses and *Cladonia* spp. in larch forest.

Neimenggu(M. Sato, 1952, p. 174 & 1959, p. 45; J. C. Wei, 1981, p. 85; Wei et al. 1982, p. 44; Lai et al. , 2009, pp. 368 – 369) , Xinjiang(Mahira Muhammat et al. , 2012, pp. 6 – 7) .

Cetrariopsis Kurok. 类岛衣属

Mem. Natn Sci. Mus, Tokyo 13: 140(1980) .

Cetrariopsis wallichiana(Taylor) Kurok. 类岛衣

Mem. Natn Sci. Mus, Tokyo 13: 140(1980) .

≡*Sticta wallichiana* Taylor, *London J. Bot.* 6: 177(1847) .

≡*Platysma wallichianum*(Taylor) Nyl. , *Flora*, Regensburg 52: 443(1869) .

≡*Cetraria wallichiana*(Taylor) Müll. Arg. , *Flora*, Regensburg 71: 139(1888) .

≡*Ahtia wallichiana*(Taylor) M. J. Lai, *Quarterly Journal of the Taiwan Museum* 33(3 & 4) : 220(1980)

On bark and twigs of trees.

Xizang(Wei & Jiang, 1986, p. 60) , Sichuan(Zahlbr. , 1930 b, p. 197; Chen S. F. et al. , 1997, p. 221) , Chongqing(Chen CL et al. , 2008, p. 46) , Yunnan(Hue, 1887a, p. 19 & 1889, p. 163 & 1899, p. 211; R. Paulson, 1928, p. 317; Zahlbr. , 1930 a, p. 319 & 1930 b, p. 197) , Taiwan(M. Sato, 1938 b, p. 786 & 1939, p. 55; Kurok. 1980, p. 141; Lai, 1980b, p. 220; Lai, 2000, p. 199) .

Cetrelia W. L. Culb. & C. F. Culb. 斑叶属

Contr. U. S. natnl. Herb. 34: 490(1968) .

Cetrelia braunsiana(Müll. Arg.) W. L. Culb. & C. F. Culb. 粒芽斑叶

Contr. U. S. Natnl. Herb. 34: 493(1968) .

≡*Parmelia braunsiana* Müll. Arg. , *Flora, Regensburg* 64: 506(1881) .

Parmelia perlata auct. non Ach. : Tchou, *Contr. Inst. Bot. Nat. Acad. Peiping* 3: 308(1935) .

Parmelia pseudolivetorum auct. non Asahina: Zhao, *Acta Phytotaxon. Sin.* 9(2) : 147(1964) .

Zhao et al. Prodr. Lich. Sin. 1982, p. 45.

Corticolous.

Heilongjiang(Zhao, 1964, p. 147; Zhao et al. 1982, p. 45; Chen J. B. 1986, p. 388; Lai et al. , 2009, pp. 369 – 370) , Jilin(Chen J. B. 1986, p. 388; Lai et al. , 2009, pp. 369 – 370) , Neimenggu(Lai et al. , 2009, pp. 369 – 370) , Xizang(Wei & Jiang, 1981, p. 1147 & 1986, p. 51) , Sichuan(Zhao, 1964, p. 147; Zhao et al. 1982, p. 45; Chen J. B. 1986, p. 388; Chen S. F. et al. , 1997, p. 221) , Chongqing(Chen CL et al. , 2008, p. 40) , Guizhou(Zhang T & J. C. Wei, 2006, p. 5) , Yunnan(Zhao, 1964, p. 147; Zhao et al. 1982, p. 45; Culb. & C. Culb. 1968, p. 493; Chen J. B. , 1986, p. 388) , Hebei(Tchou, 1935, p. 308; Zhao, 1964, p. 147; Zhao et al. 1982, p. 45; Chen J. B. 1986, p. 388) , Shandong(Zhao ZT et al. , 1998, p. 29; Zhao ZT et al. , 2002, p. 6; Anhui (Zhao, 1964, p. 147; Zhao et al. 1982, p. 45; Chen J. B. 1986, p. 388; Xu B. S. , 1989, p. 223) , Hubei(Chen J. B. 1986, p. 388; Chen, Wu, J. C. Wei, 1989, p. 436) , Hunan(Chen J. B. 1986, p. 388) , Jiangxi(Zhao, 1964, p. 147; Zhao et al. 1982, p. 45; Chen J. B. 1986, p. 388) , Zhejiang(Zhao, 1964, p. 147; Zhao et al. 1982, p. 45; Chen J. B. 1986, p. 388; Xu B. S. , 1989, p. 223) , Guangxi(Chen J. B. 1986, p. 388) , Taiwan(Lai, 2000, p. 201) .

Cetrelia cetrarioides(Delise) W. L. Culb. & C. F. Culb. 粉缘斑叶

Contr. U. S. Natnl. Herb. 34: 498(1968).

≡*Parmelia cetrarioides* Delise, in Duby, *Bot. Gall.*, Edn 2(Paris) 2: 601(1830).

≡*Parmelia perlata* ß *cetrarioides*(Delise) Delise ex Duby, *Bot. Gall.*, *Edn* 2(Paris) 2: 601(1830).

Corticolous.

Heilongjiang(Chen et al. 1981, p. 152; Luo, 1984, p. 85; Chen J. B. 1986, p. 389; Lai et al., 2009, p. 370), Jilin(Zhao, 1964, p. 158; Zhao et al. 1982, p. 46; Chen et al. 1981, p. 152; Chen J. B. 1986, p. 389; Lai et al., 2009, p. 370), Liaoning(Lai et al., 2009, p. 370), Shaanxi(Zhao, 1964, p. 158; Zhao et al. 1982, p. 46; Chen J. B. 1986, p. 389; He & Chen 1995, p. 42, miscited as *Cetraria cetrarioides*), Xizang(J. C. Wei & Chen, 1974, p. 179; J. C. Wei & Jiang, 1981, p. 1147 & 1986, p. 51; J. C. Wei et al. 1982, p. 41; Chen J. B. 1986, p. 389), Sichuan(Elenk. 1904, p. 5; A. Zahlbr., 1930 b, p. 192; Zhao, 1964, p. 158; Zhao et al. 1982, p. 46; Culb. & C. Culb. 1968, p. 498; Chen J. B. 1986, p. 389), Chongqing(Chen CL et al., 2008, p. 40), Guizhou(Zhang T & J. C. Wei, 2006, p. 5), Yunnan(A. Zahlbr., 1934 b, p. 210; Culb. & C. Culb. 1968, p. 498; Zhao, 1964, p. 158; Zhao et al. 1982, p. 46; Chen J. B. 1986, p. 389), Hebei(Moreau et Moreau, 1951, p. 191), Anhui, Zhejiang (Zhao, 1964, p. 158; Zhao et al. 1982, p. 46), Hubei(Chen J. B. 1986, p. 389; Chen, Wu, & Wei, 1989, p. 437), Taiwan(A. Zahlbr., 1933 c, p. 56; Wang-Yang & Lai, 1973, p. 93 & 1976, p. 226; Lai, 2000, p. 202).

Cetrelia chicitae(W. L. Culb.) W. L. Culb. & C. F. Culb. 奇氏斑叶

Contr. U. S. Natnl. Herb. 34: 504(1968).

≡*Cetraria chicitae* W. L. Culb., *Bryologist* 68: 95(1965).

Parmelia cetrarioides auct. non(Del.) Nyl.: Zhao, Acta Phytotaxon. Sin. 9(2): 147(1964).

Zhao et al. Prodr. Lich. Sin. 1982, p. 46.

Corticolous.

Heilongjiang(Chen J. B. 1986, p. 389; Lai et al., 2009, p. 370), Jilin(Zhao, 1964, p. 147; Zhao et al. 1982, p. 46; Chen J. B. 1986, p. 389; Lai et al., 2009, p. 370), Shaanxi(He & Chen 1995, p. 42 miscited as *C. chilcitae*), Sichuan(Chen J. B. 1986, p. 389), Yunnan, Anhui, Zhejiang(Zhao, 1964, p. 147; Zhao et al. 1982, p. 46; Chen J. B. 1986, p. 389; Xu B. S., 1989, p. 223), Hubei(Chen J. B. 1986, p. 389; Chen, Wu & J. C. Wei, 1989, p. 437), Taiwan(Culb. & C. Culb. 1968, p. 504; Wang-Yang & Lai, 1976, p. 227; Lai, 2000, p. 202).

Cetrelia collata(Nyl. in Hue) W. Culb. & C. Culb. 领斑叶

Contrib. U. S. Nat. Herb. 34(7): 505(1968).

≡*Platysma collatum* Nyl. Flora 70: 134(1887) & in Hue, *Bull. Soc. Bot. France* 34: 19(1887).

Type: Yunnan, Delavay no. 1590(H-Herb. Nyl. no. 36116 a, holotype).

≡*Cetraria collata*(Nyl.) Müll. Arg. *Nouv. Giorn. Bot. Ital.* 24: 192(1892).

Corticolous.

Jilin(Lai et al., 2009, pp. 370 – 371), Sichuan(Sato, 1939, p. 57; Culb. & C. Culb. 1968, p. 505; Ikoma, 1983, p. 33; Chen J. B. 1986, p. 390), Chongqing(Chen CL et al., 2008, p. 40), Guizhou(Chen J. B. 1986, p. 390; Zhang T & J. C. Wei, 2006, p. 5), Yunnan(Hue, 1889, p. 163, & 1890, p. 274 & 1899, p. 207; Paulson, 1928, p. 317; A. Zahlbr., 1930 b, p. 197; Sato, 1939, p. 57; Culb. & C. Culb. 1968, p. 505; Ikoma, 1983, p. 33), Anhui(J. C. Wei, 1981, p. 85; Chen J. B. 1986, p. 390; Xu B. S., 1989, p. 223), Hubei(Chen J. B. 1986, p. 390; Chen, Wu, & J. C. Wei, 1989, p. 438), Zhejiang(Xu B. S., 1989, p. 223), Taiwan(Asahina & Sato in Asahina, 1939, p. 737; Lai, 2000, pp. 202 – 203).

Cetrelia davidiana W. Culb. & C. Culb. 大维氏斑叶

Contrib. U. S. Nat. Herb. 34(7): 507(1968).

Type: Yunnan, Handel-Mazzaetti no. 4253 in W.

On the bark of trees.

Xizang, Yunnan(Chen J. B. 1986, p. 390), Sichuan (Culb. & C. Culb. 1968, p. 507; Hawksw. 1972, p. 15), Chongqing(Chen CL et al., 2008, p. 40), Guizhou(Zhang T & J. C. Wei, 2006, p. 5).

Cetrelia delavayana W. Culb. & C. Culb. 戴氏斑叶

Contrib. U. S. Nat. Herb. 34(7): 509(1968).

Type: Yunnan, Delavay, 1888 in PC.

On the bark of trees and shrubs.

Shaanxi(He & Chen 1995, p. 42 without citation of specimens and their localities), Xizang, Yunnan(Chen J. B. 1986, p. 390), Sichuan(Culb. & C. Culb. 1968, p. 509; Chen J. B. 1986, p. 390; Chen S. F. et al., 1997, p. 221), Chongqing(Chen CL et al., 2008, p. 40).

Cetrelia isidiata(Asahina) W. L. Culb. & C. F. Culb. 裂芽斑叶

Contr. U. S. Natnl. Herb. 34: 510(1968).

≡*Cetraria sanguinea* f. *isidiata* Asahina, in Sato, *Parmeliales*(I) in Nakal et Honda, *Nova Flora Japonica*: 73 (1939).

On trees.

Sichuan(Chen J. B. 1986, p. 390), Chongqing(Chen CL et al., 2008, p. 40), guizhou(Zhang T & J. C. Wei, 2006, p. 5), Taiwan(Sato, 1939, p. 73; W. Culb. & C. Culb. 1968, p. 510; Hawksw. 1972, p. 15; Wang-Yang & Lai, 1973, p. 89 & 1976, p. 227; Lai, 2000, p. 203).

Cetrelia japonica(Zahlbr.) W. L. Culb. & C. F. Culb. 日本斑叶

Contr. U. S. Natnl. Herb. 34: 511(1968).

≡*Cetraria japonica* Zahlbr., *Annls mycol.* 14: 60(1916).

Liaoning(Lai et al., 2009, p. 371), Taiwan(Lai, 2000, pp. 203 – 204).

Cetrelia monachorum(Zahlbr.) W. Culb. & C. Culb. 硬膜斑叶

Syst. Bot. 1(4): 326(1976).

≡ *Parmelia monachorum* Zahlbr. in Handel-Mazzetti, *Symb. Sin.* 3: 191 (1930) & *Cat. Lich. Univ.* 8: 562 (1932).

Type: Sichuan, near Muli, alt. 4350 m, 4/VIII/1915, Handle-Mazzetti no. 7399.

Sichuan(Chen, 1986, pp. 391 – 392 as "monochorum").

On rocks.

Cetrelia nuda(Hue) W. L. Culb. & C. F. Culb. 裸斑叶

Contr. U. S. Natnl. Herb. 34: 513(1968).

≡*Platysma collatum* f. *nudum* Hue, *Nouv. Arch. Mus. Hist. Nat.*, Paris, 4 sér. 1: 208(1899).

= *Parmelia yunnana* Hue f. *subnuda* Zahlbr., *Hedwigia* 74: 210(1934).

Type: Yunnan, Lijiang, snow range, Rock, 1931(holotype in W).

Corticolous and saxicolous.

Jilin, Liaoning(Lai et al., 2009, p. 371), Sichuan(A. Zahlbr., 1930, p. 197), Yunnan(Hue, 1899, p. 208; A. Zahlbr., 1930 b, p. 197 & 1934 b, p. 210; Culb. & C. Culb. 1968, p. 513; Chen J. B. 1986, p. 391), Taiwan(A. Zahlbr., 1933 c, p. 59; M. Sato, 1938 b, p. 784 & 1939, p. 58; Culb. & C. Culb. 1968, p. 513; Wang-Yang & Lai, 1973, p. 88 & 1976, p. 227; Lai, 2000, p. 204).

Cetrelia olivetorum(Nyl.) W. L. Culb. & C. F. Culb. 橄榄斑叶

Contr. U. S. Natnl. Herb. 34: 515(1968).

≡*Parmelia olivetorum* Nyl., Not. Sällsk. *Fauna et Fl. Fenn. Förh.*, Ny Ser. 8: 180(1866).

= *Parmelia perlata* ß *olivaria* Ach., Methodus, Sectio post. (Stockholmiæ): 217(1803), Nom. rejic., Art. 56. 1.

= *Parmelia olivaria*(Ach.) Th. Fr., Lich. Scand. (Upsaliae) 1(1): 112(1871), Nom. rejic., Art. 56. 1.

Parmelia cetrarioides auct. non(Del. ex Duby) Nyl.: Zhao, *Acta Phytotaxon. Sin.* 9: 158 (1964). Zhao et al. Prodr. Lich. Sin. 1982, p. 46. (Yunnan, Wang no. 21462).

Corticolous.

Heilongjiang(Chen J. B. 1986, p. 391; Lai et al. , 2009, pp. 371 – 372) , Jilin(Chen et al. 1981, p. 152; Chen J. B. 1986, p. 391; Lai et al. , 2009, pp. 371 – 372) , Liaoning (Chen et al. 1981, p. 152; Lai et al. , 2009, pp. 371 – 372) , Shaanx(Jatta, 1902, p. 468; A. Zahlbr. , 1930 b, p. 191) ,

Xizang(J. C. Wei & Jiang, 1986, p. 52; Chen J. B. 1986, p. 391) , Sichuan(Chen J. B. 1986, p. 391; Chen S. F. et al. , 1997, p. 221) , Chongqing(Chen CL et al. , 2008, p. 40) , Yunnan(Hue, 1889, p. 164; Harm. 1928, p. 325; Culb. & C. Culb. 1968, p. 515; Zhao, 1964, p. 158, 159; Zhao et al. 1982, p. 42, 46; Chen J. B. 1986, p. 391; Wei X. L. et al. , 2007, pp. 149 – 150) , Hubei(Chen J. B. 1986, p. 391; Chen, Wu & J. C. Wei, 1989, p. 438) , Taiwan(Culb. & C. Culb. 1968, p. 519; Wang-Yang & Lai, 1976, p. 227; Lai, 2000, pp. 204 – 205) .

Cetrelia pseudolivetorum(Asahina) W. L. Culb. & C. F. Culb. 拟橄榄斑叶

Contr. U. S. Natnl. Herb. 34: 519(1968) .

≡ *Parmelia pseudolivetorum* Asahina, *J. Jap . Bot.* 27: 16(1952) .

Corticolous.

Heilongjiang(Zhao, 1964, p. 159; Zhao et al. 1982, p. 45; Chen et al. 1981, p. 153; Luo, 1984, p. 85) , Jilin (Chen et al. 1981, p. 153; Chen J. B. 1986, p. 392; Lai et al. , 2009, p. 372) , Liaoning(Chen et al. 1981, p. 153; Lai et al. , 2009, p. 372) , Xizang(J. C. Wei & Chen, 1974, p. 180; J. C. Wei & Jiang, 1981, p. 1147 & 1986, p. 52; Chen J. B. 1986, p. 392) , Sichuan(Culb. & C. Culb. 1968, p. 519; Chen J. B. 1986, p. 392) , Chongqing(Chen CL et al. , 2008, p. 40) , Guizhou(Chen J. B. 1986, p. 392; Zhang T & J. C. Wei, 2006, p. 5) , Yunnan(Zhao, 1964, p. 159; Zhao et al. 1982, p. 45; Culb. & C. Culb. 1968, p. 519; Chen J. B. 1986, p. 392) , Hebei(Zhao, 1964, p. 159; Zhao et al. 1982, p. 45) , Anhui, Jiangxi(Zhao, 1964, p. 159; Zhao et al. 1982, p. 45) , Hubei(Chen J. B. 1986, p. 392; Chen, Wu, J. C. Wei, 1989, p. 439) , Hunan(Chen J. B. 1986, p. 392) , Zhejiang, (Zhao, 1964, p. 159; Zhao et al. 1982, p. 45; Xu B. S. , 1989, p. 223) , Taiwan(Culb. & C. Culb. 1968, p. 519; Wang-Yang & Lai, 1976, p. 227; Lai, 2000, p. 205) .

Cetrelia sanguinea(Schaer.) W. L. Culb. & C. F. Culb. 血红斑叶

Contr. Nat. Herb. 34(7) : 521(1968) .

≡ *Cetraria sanguinea* Schaer. , *Syst. Verz.* : 129(1846) [1845 – 46] .

= *Imbricaria megaleia*(Nyl.) Jatta, *Licheni Cinesi in Nuovo Giornale botanico italiano. Nuova serie. Volume nono,* 469(1902) .

= *Cetraria sanguinea* var. *inactiva* Zahlbr. in Handel-Mazzetti, Symb. Sin. 3: 197(1930) .

Type: from Yunnan, Gebauer.

Corticolous.

Shaanxi(Jatta, 1902, p. 469; Zahlbr. , 1930b, p. 197; Chen J. B. 1986, p. 392; He & Chen 1995, p. 42 without citation of specimens and their localities) , Sichuan(Zahlbr. , 1930b, p. 197) , Chongqing(Chen CL et al. , 2008, p. 40) , Yunnan(Zahlbr. , 1930b, p. 197; Culb. & C. Culb. 1968, p. 521) , Hubei(Chen J. B. 1986, p. 392; Chen, Wu & J. C. Wei, 1989, p. 439) .

Cetrelia sinensis W. Culb. & C. Culb. 中华斑叶

Contr. Nat. Herb. 34(7) : 523(1968) .

Type: Yunnan, Delavay no. 30 in PC.

Saxicolous.

Jilin, Liaoning(Lai et al. , 2009, pp. 372 – 373) , Xizang(J. C. Wei & Y. M. Jiang, 1986, p. 52) , Sichuan(Chen J. B. 1986, p. 392) , Chongqing(Chen CL et al. , 2008, p. 40) , Taiwan(W. Culb. & C. Culb. 1968, p. 524; Lai, 2000, pp. 205 – 206) .

Cetreliopsis M. J. Lai **类斑叶属**

Quarterly Journal of the Taiwan Museum 33: 218(1980) .

Cetreliopsis asahinae(M. Satô) Randlane & A. Thell 朝氏类斑叶

in Randlane, Thell & Saag, *Cryptog. Bryol. -Lichénol.* 16(1) : 49(1995) .

≡*Cetraria asahinae* M. Satô, in Saito, *Ho-on Kai Mus. Res. Bul.* 11: 12(1936) .

Jilin(Lai et al. , 2009, p. 373) , Taiwan(Lai, 2000, pp. 206 – 207) .

Cetreliopsis endoxanthoides(D. D. Awasthi) Randlane & Saag 黄类斑叶

in Randlane, Thell & Saag, *Cryptog. Bryol. -Lichénol.* 16(1) : 51(1995) .

≡*Cetraria endoxanthoides* D. D. Awasthi, *Bull. bot. Surv. India* 24: 9(1982) .

On bark & branches of conifer trees.

Yunnan(ChenLH et al. , 2006, p. 503) .

Cetreliopsis laeteflava(Zahlbr.) Randlane & Saag 亮黄斑叶

Cryptog. Bryol. -Lichénol. 16(1) : 51(1995) .

≡*Cetraria laeteflava* Zahlbr. , *Feddes Repert.* 33: 60(1933) .

Type: from Taiwan.

Taiwan(Lai, 2000, p. 207) .

Cetreliopsis rhytidocarpa(Mont. & Bosch) Randlane & Saag 类斑叶

in Randlane, Thell & Saag, *Mem. Natn Sci. Mus*, Tokyo 13: 218(1980) .

≡*Cetraria rhytidocarpa* Mont. & Bosch, in Junghun, *Pl. Jungh.* 4: 430(1855) .

= *Cetraria straminea* Vain. (1909) , non Krohbr. (1860)

≡*Nephromopsis straminea*(Vain.) Räs. in *Kuopion Luonnon Ystäväin Yhdistyksen Jlkaisuja,* ser. B, II, no. 6: 50 (1952) .

On bark of trees.

Taiwan(Zahlbr. , 1940, p. 547; Asahina, 1934 c, p. 474; Sato, 1938 b, p. 785 & 1939, p. 65; M. Lamb, 1963, p. 158; Wang & Lai, 1973, p. 89; Lai, 1980, p. 218) .

Coelocaulon Link **角衣属**

Handbuck zur Erkennung der Nutzbarsten und am Häufigsten Vorkommenden Gewächse 3: 165(1833) .

Coelocaulon aculeatum(Schreb.) Link 皮刺角衣

Grundr. Krauterk. 3: 165(1833) .

≡*Lichen aculeatus* Schreb. , *Spic. fl. lips.* (Lipsiae) : 125(1771) .

≡*Cetraria aculeata*(Schreb.) Fr. , *Nov. Sched. Critic. Lich.* 4: 32(1826) .

= *Cornicularia tenuissima*(L.) Zahlbr. , *Annln K. K. naturh. Hofmus.* Wien 42: 64(1928) .

≡*Lichen tenuissimus* L. , *Sp. pl.* 2: 1145(1753) .

Xinjiang(Wang XY, 1985, p. 350; Abdulla Abbas & Wu, 1998, p. 92; Abdulla Abbas et al. , 2001, p. 363) , Shaanxi(He & Chen 1995, p. 42, no specimen and locality are cited) , Sichuan(Du Rietz 1926, p. 35; Zahlbr. , 1930b, p. 202; Wu J. L. 1986, p. 75) ; Taiwan(Lai, 2000, p. 208) .

Note: *Cetraria aculeata* sensu Tchou(1935) refers to *Cladia aggregata*(Sw.) Nyl.

Coelocaulon divergens(Ach.) R. Howe 叉角衣

Classif. Fam. Usneac. : 19(1912) .

≡*Cornicularia divergens* Ach. , *Methodus, Sectio post.* (Stockholmiæ) : 303(1803) .

≡*Bryopogon divergens*(Ach.) Schwend. , in Naegeli, *Beitr. wiss. Botan.* 2: 148(1860) .

≡*Bryocaulon divergens*(Ach.) Kärnefelt, *Op. bot.* 86: 24(1986) .

≡*Alectoria divergens*(Ach.) Nyl. , *Mém. Soc. Imp. Sci. Nat. Cherbourg* 3: 171(1855) .

Xinjiang(Abdulla Abbas & Wu, 1998, P. 89 and Abdulla Abbas et al. , 2001, 361, as “*Brycaulon*” *divergens*) , Yunnan(Paulson, 1928, p. 318) .

Crespoa (D. Hawksw.) Lendemer & B. P. Hodk. **梅叶属**

N. Amer. Fung. 7(2) : 3(2013) .

Crespoa crozalsiana(B. de Lesd. ex Harm.) Lendemer & B. P. Hodk. 粉槽梅叶

N. Amer. Fung. 7(2):3(2013).

≡*Canoparmelia crozalsiana* (B. de Lesd. ex Harm.) Elix & Hale, in Elix, Johnston & Verdon, *Mycotaxon* 27: 278(1986).

≡*Parmelia crozalsiana* B. de Lesd. ex Harm., in Harmand, *Lich. Fr.* 4:555(1910)[1909].

≡*Parmotrema crozalsianum*(B. de Lesd. ex Harm.) D. Hawksw., *Lichenologist* 43(6):648(2011).

≡*Pseudoparmelia crozalsiana*(B. de Lesd. ex Harm.) Hale, *Phytologia* 29(3):189(1974).

On rock.

Shandong(Jia ZF et al., 2008, pp. 461 –462).

Dactylina Nyl. **地指衣属**

Syn. meth. lich. (Parisiis) 1(2):286(1860).

Dactylina chinensis Follm. 中华地指衣

in Follmann, Huneck & Weber, *Willdenowia* 5(1):8(1968).

Type: Shaanxi, G. Fenzel no. 17643(holotype) in B(not seen).

≡*Dactylina arctica* subsp. *chinensis*(Follmann) Kärnefelt & A. Thell, *Nova Hedwigia*62(3 –4):504(1996), nom. inval., Art. 41.4(Melbourne).

Shaanxi(Hawksw. 1972, p. 21).

Dolichousnea(Y. Ohmura) Articus **丝萝属**

Taxon 53(4):932(2004).

≡*Usnea* subgen. *Dolichousnea* Y. Ohmura, *J. Hattori bot. Lab.* 90: 80(2001).

Dolichousnea diffracta(Vain.) Articus 环裂丝萝

Taxon 53(4):932(2004).

≡*Usnea diffracta* Vain., *Bot. Mag.*, Tokyo 35:45(1921).

On trees.

Heilongjiang(J. C. Wei, 1981, p. 87; Chen et al. 1981, p. 158; J. C. Wei et al. 1982, p. 53; Luo, 1984, p. 84), Jilin(Nuno, 1958, p. 227, cited as Manchuria frontier of Korea; Chen et al. 1981, p. 158; J. C. Wei et al. 1982, p. 53), Neimenggu(Chen et al. 1981, p. 158), Guizhou(Zhang T & J. C. Wei, 2006, p. 11), Shandong(Zhao ZT et al., 1999, p. 426; Zhao ZT et al., 2002, p. 6), north eastern China(prov. not indicated: M. Sato, 1938 a, p. 463, cited as in Manchuria), Jiangxi(Asahina, 1956 a, p. 67; Nuno, 1958, p. 227), Anhui, Zhejiang(Xu B. S., 1989, p. 225 –226), Guangdong(Ohmura, 2001, p. 82 as China Kanto), Taiwan(Zahlbr., 1933 c, p. 62; Motyka, 1936, p. 389; M. Sato, 1938 a, p. 463 & 1960, p. 53; Asahina, 1956 a, p. 67; Nuno, 1958, p. 227; Wang & Lai, 1973, p. 98; Ohmura, 2001, p. 82).

f. diffracta 原变型

On trees & occasionally on rocks.

Heilongjiang, Jilin, Liaoning, Shanxi, Shandong, Zhejiang, Jiangxi(Zhao et al. 1982, p. 80).

= *Usnea diffracta* f. *huei* Asahina, *Lichens of Japan III, Genus Usnea*: 69(1956).

Liaoning, Jiangxi(Zhao et al. 1982, p. 81).

Dolichousnea longissima(Ach.) Articus 长丝萝

Taxon 53(4):932(2004).

≡*Usnea longissima* Ach., *Lich. univ.*:626(1810).

≡*Parmelia coralloidea* var. *longissima*(Ach.) Spreng., *Syst. veg.*, Edn 16 4(1):277(1827).

≡*Usnea barbata* var. *longissima*(Ach.) Schaer., *Enum. critic. lich. europ.* (Bern):[3](1850).

On trees.

Heilongjiang(Zahlbr., 1934, p. 212, cited as eastern Manchuria, it may probanly be in eastern Heilongjiang; Chen et al. 1981, p. 158; Luo, 1984, p. 84), Jilin(Chen et al. 1981, p. 158), Shaanxi(Jatta. 1902. p. 461; Zahlbr., 1930 b, p. 207; Moreau et Moreau, 1951, p. 193; Wu J. L. 1981, p. 164; J. C. Wei et al. 1982, p. 52), Gansu(H. Magn. 1940, p. 130), Sichuan(Zahlbr., 1930 b, p. 207), Yunnan(Hue, 1887a, p. 18 & 1899, p. 51;

Paulson, 1928, p. 319; Zahlbr. , 1930 b, p. 207; Wei X. L. et al. 2007, p. 159) , Xizang(Wei & Chen, 1974, p. 181; J. C. Wei & Y. M. Jiang, 1981, p. 1146 & 1986, p. 71) , Hubei(Chen, Wu & J. C. Wei, 1989, 454) , Taiwan (Wang & Lai, 1973, p. 98; Ohmura, 2001, p. 85) .

≡*Usnea longissima* var. *longisisma*

On trees.

Jilin, Shaanxi, Sichuan, Yunnan(Zhao et al. 1982, p. 83) .

= *Usnea longissima* var. *corticata* Howe Jr. , in Proceed. Thoreau Mus. Nat. Hist. I: 18(1913) ; Motyka, Lich. Gen. Usnea Stud. Monogr. Pars system. II: 430(1937) . Yunnan(Zhao et al. 1982, p. 84) .

= *Usnea longissima* var. *elegantissima* Motyka, *Lich. Gen. Usnea Stud. Monogr. Pars system.* II: 430(1937) .

Yunnan(Zhao et al. 1982, p. 84) .

= *Usnea longissima* var. *orientalis* Motyka

Yunnan(Zahlbr. , 1934, p. 212) .

= *Usnea longissima* var. *perciliata* Motyka, *Lich. Gen. Usnea Stud. Monogr. Pars system.* II: 430(1937) .

Jilin, Sichuan, Yunnan(Zhao et al. 1982, p. 84) .

= *Usnea longissima* var. *sinica* Motyka, *Lichenum Generis Usnea Studium Monographicum* 1 – 2: 431(1936 – 38) .

Type: from Sichuan, Handel-Mazzetti

≡*Usnea sinica*(Motyka) Zhao et al. *Acta Phytotax. Sin.* 13(2) : 99(1975) .

On trees.

Yunnan(Zhao et al. 1982, p. 85) .

= *Usnea longissima* f. *tenuis* Th. Fr. in Hue, *Nouv. Arch. Mus Hist. Nat.* 4(1) : 52(1899) .

Type: Yunnan, collected by Delavay.

On bark of trees.

Jilin(Zhao et al. 1882, p. 83, as Var.) , Heilongjiang(Asahina, 1952 d, p. 375, as var.) , Yunnan(Hue, 1889, p. 52; Zahlbr. , 1930 b, p. 208, & 1934, p. 212; Motyka, 1936 – 38, p. 431, as var.) .

Dolichousnea trichodeoides(Vain. ex Motyka) Articus 毛丝萝

Taxon 53(4) : 932(2004) .

≡*Usnea trichodeoides* Vain. , *Ann. Acad. Sci. fenn.* , Ser. A 6(7) : 8(1915) .

≡*Usnea trichodeoides* Vain. ex Motyka, *Usnea* 2(1) : 405(1937) .

Taiwan(Ohmura, 2001, p. 88) .

Emodomelanelia Divakar & A. Crespo **暗点衣属**

in Crespo et al. , *Taxon* 59(6) : 1749(2010) .

Emodomelanelia masonii(Essl. & Poelt) Divakar & A. Crespo 暗点衣

in Crespo et al. , *Taxon* 59(6) : 1749(2010) .

≡*Parmelia masonii* Essl. & Poelt, *Bryologist* 94(2) : 203(1991) .

On sun-exposed rock, bryophytes over rock, ground, and on *Betula utilis*.

Yunnan(Ahti et al. , 1999, p. 124; Chen JB, 2015. p. 54 – 55) , Sichuan, Xizang(W. Obermayer, 2004, p. 505, as *Parmelia masonii*) , Taiwan(Lai, 2000, p. 143) .

Evernia Ach. **扁枝衣属**

in Luyken, *Tent. Hist. Lich.* : 90(1809) .

Evernia divaricata(L.) Ach. 柔扁枝衣

Lich. univ. : 1 – 696(1810) .

≡*Lichen divaricatus* L. , *Syst. veg.* , Edn 12: 713(1768) .

Corticolous.

Xinjiang(H. Magn. 1940, p. 129; Moreau et Moreau, 1951, p. 191; J. C. Wei et al. 1982, p. 46; Wu J. L. 1985,

p. 75; Wang XY, 1985, p. 359; Abdula Abbas et al., 1993, p. 77; Abdulla Abbas & Wu, 1998, pp. 92 – 93; Abdula Abbas et al., 2001, p. 363), Gansu(Zahlbr., 1934 b, p. 212), Shaanxi(Baroni, 1894, p. 47; Jatta, 1902, p. 462; Zahlbr., 1930 b, p. 199; Tchou, 1935, p. 302; He & Chen 1995, p. 43), Xizang(J. C. Wei & Y. M. Jiang, 1981, p. 1146 & 1986, p. 66), Sichuan(Stenroos et al., 1994, p. 334).

Evernia esorediosa(Müll. Arg.) Du Rietz 裸扁枝衣

Svensk bot. Tidskr. 18: 390(1924).

≡*Evernia mesomorpha* f. *esorediosa* Müll. Arg., *Flora, Regensburg* 74(1): 110(1891).

On twigs.

Neimenggu(M. Sato, 1952, p. 175; Chen et al. 1981, p. 134; J. C. Wei et al. 1982, p. 46), Xinjiang(Abdulla Abbas & Wu, 1998, p. 93; Abdulla Abbas et al., 2001, p. 363 as *E.* '*escorediosa*').

Evernia mesomorpha Nyl. 扁枝衣

Lich. Scand. (Helsinki): 74(1861).

= *Evernia thamnodes*(Flot.) Arnold, *Verh. zool. -bot. Ges.* Wien 23: 110(1873).

≡*Evernia prunastri* var. *thamnodes* Flot. *Die merkw. Und selten. Flecht. Hirschberg. Warmbrunn.* 5(1839).

On trees.

Neimenggu(M. Sato, 1952, p. 175; Chen et al. 1981, p. 134; J. C. Wei et al. 1982, p. 45; Sun LY et al., 2000, p. 37), Jilin(Zahlbr., 1934 b, p. 212; Chen et al. 1981, p. 134; J. C. Wei et al. 1982, p. 46), Heilongjiang(Asahina, 1952 d, p. 375; Chen et al. 1981, p. 134; J. C. Wei et al. 1982, p. 45; Luo, 1984, p. 84), Shaanxi(Jatta, 1902, p. 462; Zahlbr., 1930 b, p. 199; J. C. Wei, 1981, p. 87; J. C. Wei et al. 1982, p. 45; He & Chen 1995, p. 43, no specimen and locality are cited), Xinjiang(Wu J. L. 1985, p. 76), Yunnan(Paulson, 1928, p. 318; Zahlbr., 1930 b, p. 199; Wei X. L. et al. 2007, p. 152), Xizang(J. C. Wei & Chen, 1974, p. 181; J. C. Wei & Y. M. Jiang, 1986, p. 66), Sichuan(Stenroos et al., 1994, p. 334).

Flavocetraria Kärnefelt & A. Thell **黄岛衣属**

Acta Bot. Fenn. 150: 81(1994).

Flavocetraria cucullata(Bellardi) Kärnefelt & A. Thell 卷黄岛衣

in Kärnefelt, Thell, Randlane & Saag, *Acta Bot. Fenn.* 150: 81(1994).

≡*Lichen cucullatus* Bellardi, *Osservaz. Bot.* 1: 54(1788).

≡*Cetraria cucullata*(Bellardi) Ach., *Meth. Lich.*: 293(1803).

≡*Allocetraria cucullata*(Bellardi) Randlane & Saag, *Mycotaxon* 44: 492(1992).

≡*Nephromopsis cucullata*(Bellardi) Divakar, Crespo & Lumbsch in *Fungal Diversity*, 11April 2017, p.

On the ground among mosses.

Heilongjiang(Lai et al., 2009, pp. 373 – 374), Neimenggu(M. Sato, 1952, p. 174 & 1959, p. 44; Lai et al., 2009, pp. 373 – 374), Yunnan(Paulson, 1928, p. 317; Zahlbr., 1930 b, p. 198; M. Sato, 1939, p. 34), Xizang(J. C. Wei & Chen, 1974, p. 180; J. C. Wei & Jiang, 1981, p. 1146 & 1986, p. 56; J. C. Wei et al. 1982, p. 43).

= *Cetraria cucullata* var. *vainioi* Räsänen f. *crispata* Rass., *Bot. Mater. Gerb. bot. Inst. V. A. Komarova* 6: 11 (1949).

On the ground among mosses.

Hebei(J. C. Wei, 1981, p. 86).

Flavocetraria nivalis(L.) Kärnefelt & A. Thell 雪黄岛衣

in Kärnefelt, Thell, Randlane & Saag, *Acta Bot. Fenn.* 150: 84(1994).

≡*Lichen nivalis* L., Sp. pl. 2: 1145(1753).

≡*Cetraria nivalis*(L.) Ach., *Methodus*, Sectio post. (Stockholmiæ): 294(1803).

≡*Nephromopsis nivalis*(L.) Divakar, Crespo& Lumbsch in *Fungal Diversity*, 11 April 2017, p.

On the ground.

Heilongjiang(Lai et al. , 2009, p. 374) , Neimenggu(M. Sato, 1952, p. 174 & 1959, p. 45; Lai et al. , 2009, p. 374) , Xinjiang(Wu J. L. 1985, p. 74; Abbas et al. , 1993, p. 76; Abdulla Abbas & Wu, 1998, p. 91; Abdulla Abbas et al. , 2001, p. 362) , Xizang(Paulson, 1925, p. 190) .

Flavoparmelia Hale **皱衣属**

Mycotaxon 25(2) : 604(1986) .

Flavoparmelia baltimorensis(Gyeln. & Fóriss) Hale 巴尔迪莫皱衣

Mycotaxon 25(2) : 604(1986) .

≡*Parmelia baltimorensis* Gyeln. & Fóriss, *Annals Cryptog. Exot.* 4: 167(1931) .

On rocks.

Heilongjiang, Zhejiang(Chen J. B. 2011, pp. 884 – 885 & 2015, p. 68) .

Flavoparmelia caperata(L.) Hale 皱衣

Mycotaxon 25(2) : 604(1986) .

≡*Lichen caperatus* L. , *Sp. pl.* 2: 1147(1753) .

≡ *Parmelia caperata* (L.) Ach. Method. Lich. : 216, 1803; Zahlbr. , in Handel-Mazzetti, *Symb. Sin.* 3: 192 (1930) .

On bark of trees, on rocks.

Heilongjiang(J. C. Wei, 1981, p. 84; Chen et al. 1981, p. 151; Luo, 1984, p. 85; Chen JB, 2015, p. 70) , Jilin (Chen et al. 1981, p. 151 & 2015, p. 70) , Liaoning(Sato, 1960, p. 56) , Neimenggu(Chen et al. 1981, p. 151 & 2015, p. 70) , Xinjiang, Ningxia(Chen JB, 2015, p. 71) , Shaanxi(Jatta, 1902, p. 468; Zahlbr. , 1930b, p. 192; Tchou, 1935, p. 306; J. C. Wei et al. 1982, p. 39; He & Chen 1995, p. 43, no specimen and locality are cited; Chen JB, 2015, p. 71) , Xizang(J. C. Wei & Chen, 1974, p. 179; J. C. Wei & Jiang, 1981, p. 1147 & 1986, p. 46; Chen JB, 2015, p. 71) , Sichuan(Chen S. F. et al. , 1997, p. 221; Chen JB, 2015, p. 71) , Chongqing(Chen JB, 2015, p. 70) , Yunnan(Hue, 1889, p. 163 & 1899, p. 180; Zahlbr. , 1930 b, p. 192; Chen JB, 2015, p. 71) , Hebei(Tchou, 1935, p. 306; Moreau et Moreau, 1951, p. 191) , Shandong(Zhao ZT et al. , 1998, p. 29; Zhao ZT et al. , 1999, p. 427; Zhao ZT et al. , 2002, p. 5; Hou YN et al. , 2008, p. 67; Li Y et al. , 2008, p. 71; Chen JB, 2015, p. 70) , Anhui, Hubei(Chen JB, 2015, p. 70) , Jiangxi(Zahlbr. , 1930 b, p. 192; Chen JB, 2015, p. 70) , Zhejiang(Tchou, 1935, p. 306; Chen JB, 2015, p. 70) , Taiwan(Wang & Lai, 1973, p. 93; Lai, 2000, pp. 113 – 114) , China(prov. not indicated: Hale, 1976, p. 20) .

[*Parmelia caperata* var. *caperata* f. *caperata*]

On bark of trees & on rocks.

Liaoning, Jilin, Ningxia, Yunnan, Hebei, Jiangxi, Anhui, Zhejiang(Zhao, 1964, p. 157; Zhao et al. 1982, p. 42) , Xinjiang(Wang XY, 1985, p. 347) .

[*Parmelia caperata* f. *isidiosa* Müll. Arg. , *Revue Mycolog.* 10: 56(1888) .]

Parmelia caperata var. *caperata* f. *laevissima* auct. non Gyeln. : Zhao, *Acta Phytotax. Sin.* 9(2) : 157(1964) . Zhao et al. *Prodromus Lichenun Sinicorum,* 1982, p. 43.

Parmelia caperata var. *subglauca* auct. non(Gasilien) Nyl. : Zhao, *Acta Phytotax. Sin.* 9(2) : 158(1964) . Zhao et al. *Prodromus Lichenum Sinicorum* 1982, p. 43.

On bark of trees.

Heilongjiang, Guizhou(Zhao, 1964, p. 157 – 158; Zhao et al. 1982, p. 43) .

[*Parmelia caperata* var. *caperata* f. *laevissima* Gyeln. Folia Cryptog. 1: 587(1928) .]

Parmelia caperata var. *caperata* f. *elongata* auct. non Moreau et Moreau(nom. inval.) : Zhao, *Acta Phytotax. Sin.* 9(2) : 1964) . Zhao et al. *Prodromus Lichenum Sinicorum* 1982, p. 43.

On bark of trees.

Anhui(Zhao, 1964, p. 157; Zhao et al. 1982, p. 43) .

[*Parmelia caperata* f. *papillosa* Harm. Lich. France, 4: 574(1904) .]

Xinjiang(Wang XY, 1985, p. 347) , Shandong(Asahina, 1952a, p. 124) .

Parmelia caperata f. *ramealis* Nyl. , Annls Sci. Nat. , Bot. , sér. 4 15: 373(1861) .

= *Parmelia caperata* f. *elongtata* Moreau et Moreau, Rev. Bryol. et Lichenol. 20: 191(1951) , (nom. inval.) .

Hebei(Mt. Wuling shan, alt. 1300 m, Liou K. M. no. 35) .

Parmelia caperata var. *servediosa* F. Wilson, Victorian Nat. 6: 60(1889) .

Jilin(Moreau et Moreau, 1951, p. 191) .

Flavoparmelia caperatula(Nyl.) Elix, O. Blanco & A. Crespo 小皱衣

Australas. Lichenol. 67: 11(2010) .

≡ *Parmelia caperata* var. *caperatula* Nyl. , *Syn. meth. lich.* (Parisiis) 1(2) : 377(1860) .

≡ Imbricaria *caperatula* (Nyl.) Jatta, *Nuovo G. bot. ital.* 9: 468(1902) .

= *Flavoparmelia rutidota*(Hook. f. & Taylor) Hale, *Mycotaxon* 25(2) : 605(1986) .

≡ *Parmelia rutidota* Hook. f. & Taylor, *London J. Bot.* 3: 645(1844) .

Among mosses.

Shaanxi(Jatta, 1902, P. 468; Zahlbr. , 1930 b, p. 192) , , Guizhou, Anhui, Hunan(Chen JB, 2015, p. 72 – 73 as *Parmelia rutidota*) .

Flavopunctelia(Krog) Hale **黄星点衣属**

Mycotaxon 20(2) : 682(1984) .

≡ *Punctelia* subgen. *Flavopunctelia* Krog, *Nordic Jl Bot.* 2: 291(1982) .

Flavopunctelia flaventior(Stirt.) Hale 皱黄星点衣

Mycotaxon 20(2) : 682(1984) .

≡ *Parmelia flaventior* Stirt. , *Trans. Glasgow Soc. Fld Nat.* 5: 212(1877) .

= *Parmelia andreana* Müll. Arg. *Rev. Mycol.* 1: 169, 1879; Zhao, *Acta Phytotax. Sin.* 9: 158(1964) .

= *Parmelia himalyensis* Nyl. *Flora* 68: 605(1885) .

Type: China: Himalaya, Nyl. -Herb. no. 35729(lectotype) in H(not seen) .

= *Parmelia variata* Hue, *Nouv. Arch. Mus. Hist. Nat.* 4(4) : 154(1899) .

Type: China: Yunnan, Delavay, lectotype in PC(not seen) .

On bark of trees.

Beijing, Jilin, Shaanxi(Chen JB, 2015, p. 74 – 75) , Xizang(Hale, 1980, p. 76; J. C. Wei et al. 1982, p. 39; J. C. Wei & Y. M. Jiang, 1986, p. 46; Chen JB, 2015, p. 75) , Sichuan(Chen JB, 2015, p. 74) , Yunnan(Zahlbr. , 1930 a, p. 138 1930 b, p. 185; Zhao, 1964, p. 158; Hale, 1980, p. 76; J. C. Wei, 1981, p. 85; Zhao et al. 1982, p. 42; Wei X. L. et al. 2007, p. 152; Chen JB, 2015, p. 74) , Hebei(Chen JB, 2015, p. 74) .

Flavopunctelia soredica(Nyl.) Hale 卷叶黄星点衣

Mycotaxon 20(2) : 682(1984) .

≡ *Parmelia soredica* Nyl. , *Bull. Soc. linn. Normandie*, sér. 2 6: 316(1872) .

= *Parmelia ulophyllodes*(Vain.) Savicz, *Bull. Jard. Imp. Bot.* 15: 316(1915) , Zhao, *Acta. Phytotax. Sin.* 9: 158 (1964) .

= *Parmelia manshurica* Asahina, *Journ. Jap. Bot.* 17: 75(1941) .

Type: Heilongjiang("*Manchuria*": Asahina, 1951, 2/VIII/1940) .

Parmelia caperata sensu M. Sato, : Bot. Mag. Tokyo, 65: 174(1952) , non(L.) Ach.

On bark of trees.

Beijing(Chen JB, 2015, p. 76) , Heilongjiang(M. Sato, 1952, p. 174 & 1960, p. 58; Asahina, 1952d, p. 375; Lamb, 1963, p. 486: as "*Manchuria*"; Chen et al. 1981, p. 154; J. C. Wei et al. 1982, p. 35; Luo, 1984, p. 85;

Chen JB, 2015, p. 76), Jilin, Neimenggu(Chen et al. 1981, p. 154; J. C. Wei et al. 1982, p. 35; Sun LY et al., 2000, p. 36; Chen JB, 2015, p. 76), Xinjiang(Abbas et al., 1997, p. 2), Xizang(J. C. Wei & Chen, 1974, p. 179; J. C. Wei & Y. M. Jiang, 1981, p. 1146 & 1986, p. 47; Chen JB, 2015, p. 76), Sichuan(Chen JB, 2015, p. 76), Guizhou(Oliv. 1914, p. 35), Shanxi, Hebei(Zhao, 1964, p. 158; Zhao et al. 1982, p. 43; Chen JB, 2015, p. 76).

Hypogymnia(Nyl.) Nyl. 袋衣属

Lich. Envir. Paris: 39, 139(1896) [1895].

≡*Parmelia* subgen. *Hypogymnia* Nyl., *Flora, Regensburg* 64(34): 537(1881).

Hypogymnia alpina D. D. Awasthi 高山袋衣

Kavaka 12(2): 91(1985) [1984], Bruce McCune, *Opuscula Philolichenum*, 11: 11 – 18. 2012.

Misapplied name: *Hypogymnia delavayi* auct. non(Hue) Rassad., J. C. Wei, *Enum. Lich. China*: 114(1991).

On soil and rock, also on bark and wood; subalpine and alpine.

Sichuan, Yunnan, Xizang, (McCune, B. 2012, pp. 12 – 15).

On twigs.

Jilin(Luo, 1986, pp. 158 – 159), Heilongjiang(Wei, 1981, p. 83; Luo, 1984, p. 84; Luo, 1986, pp. 158 – 159), Anhui(Wei, 1981, p. 83; Xu B. S. 1989, p. 209), Sichuan(Zahlbr., 1934 b, p. 211), Yunnan(Hue, 1887 a, p. 21 & 1887 b, p. 135 & 1890, p. 292 & 1899, p. 127; Zahlbr., 1930 a, p. 28 & 1930 b, p. 194; Zhao, 1964, p. 142; Zhao et al. 1982, p. 13), Hubei(Chen, Wu & Wei, 1989, p. 427), Zhejiang(Xu B. S. 1989, p. 209), Fujian (Wu J. N. et al. 1982, p. 9).

Hypogymnia arcuata Tchabanenko & McCune 弓形袋衣

The Bryologist 104(1): 146(2001).

Type: Russia, Paratypes: China: Jilin, Heilongjiang, Liaoning, Sichuan, Shaanxi, Yunnan.

= *H. enteromorpha* f. *inactiva* from Shaanxi. Mt. Taibi Shan, 2, 640 – 2, 680 m, Mingxing Si et Doumu Gong, ad muscos, J. C. Wei 2496, *Lich. Sinensis Exs.* 11

Heilongjiang, Jilin, Liaoning, Shaanxi, Gansu, Sichuan, Yunnan(McCune B, Wang LS 2014, pp. 35 – 36).

Hypogymnia enteromorpha auct. non(Ach.) Nyl.

On bark of trees & on moss covered rocks.

Shaanxi(Jatta, 1902, p. 471; A. Zahlbr., 1930 b, p. 194; J. C. Wei, 1981, p. 84; J. C. Wei et al. 1982, p. 30; Zhao et al. 1982, p. 12; He & Chen 1995, p. 43, no specimen and locality are cited;), Xizang(J. C. Wei & Chen, 1974, p. 178), Yunnan(Zahlbr., 1930 b, p. 194 & 1934 b, p. 211; Zhao, 1964, p. 143; Zhao et al. 1982, p. 12), Hubei(Chen, Wu & J. C. Wei, 1989, p. 429).

Hypogymnia austerodes(Nyl.) Räsänen 硬袋衣

Ann. bot. Soc. Zool. -Bot. fenn. Vanamo 18(1): 13(1943).

≡*Parmelia austerodes* Nyl., *Flora, Regensburg* 64: 537(1881).

On moss covered substratum.

Xinjiang(Wang XY, 1985, p. 346; J. C. Wei, 1986, p. 327; Abdulla Abbas & Wu, 1998, p. 93; Abdulla Abbas et al., 2001, p. 363), Xizang(J. C. Wei & Jiang, 1986, p. 35; J. C. Wei, 1986, p. 327; McCune B, Wang LS 2014, p. 36), Sichuan(McCune B, Wang LS 2014, p. 36).

Hypogymnia bitteri(Lynge) Ahti 暗粉袋衣

Ann. Bot. Fenn. 1: 20(1964). f. bitteri 原变型

≡*Parmelia bitteri* Lynge, *Skr. VidenskSelsk. Christiania, ath. -Natur.* (15): 138(1921).

Miasapplied name:

Parmelia obscurata auct. non Ach.

On moss-covered rocks & on bark of trees.

Heilongjiang(Asahina, 1952 d, p. 375; Luo, 1986, p. 156 – 157), Neimenggu(Chen et al. 1981, p. 150; Luo, 1986, p. 156 – 157), Xizang(J. C. Wei & Chen, 1974, p. 178; J. C. Wei & Jiang, 1981, p. 1146 & 1986, p. 34).

f. erumpens(Hillmann) Rassad. 多粉变种

Nov. sist. Niz. Rast. 4: 296(1967).

≡*Parmelia obscurata* f. *erumpens* Hillmann, *Rabenh. Krypt. -Fl.*, Edn 2(Leipzig) 9(5/3): 78(1936).

Heilongjiang, Jilin, Neimenggu(Luo, 1986, p. 157).

f. obscura(Bitter) Rassad. 棕色变型

Nov. sist. Niz. Rast. 4: 296(1967).

≡*Parmelia obscurata* f. *obscura* Bitter, *Hedwigia* 40: 214(1901).

Heilongjiang, Jilin, Neimenggu(Luo, 1986, p. 157).

Hypogymnia bulbosa McCune & L. S. Wang 球叶袋衣

in B. McCune, E. Martin, L. S. Wang, *The Bryologist* 106(2): 227(2003).

Type: Yunnan, Caojian Co., Zi ben Mountain, 25844. 29 N, 9983. 59 E, on *Picea* stump, Wang L-s. 00 – 18864, 12 June 2000(holotype, KUN). Paratypes from Sichuan & Yunnan.

Taiwan(McCune & L. S. Wang, 2014, p. 39).

Hypogymnia bullata Rassad. 泡袋衣

Nov. sist. Niz. Rast. 4: 295(1967).

On tree trunk. and bushes.

Heilongjiang(Luo, 1986, p. 158).

Hypogymnia capitata McCune 球粉袋衣

in McCune B, Wang LS, *Mycosphere* 5(1): 40(2014).

Type: China. Sichuan: Upper Yalong basin, Chola Shan, Dege-Garze, Manigango, on bark. 31°52′N 99° 7′E, 4730 m, 26 Sep 1994, G. & S. Miehe 94 – 416-00/06(GZU, holotype).

Sichuan(McCune & Wang LS, 2014. p. 40).

Hypogymnia congesta McCune & C. F. Culb. **密叶袋衣**

in McCune, Martin & Wang, *Bryologist* 106(2): 227(2003).

Type: Yunnan, Wei Xi Co., Wei Den Village, Lu Ma Deng Ya Kou, 27859 N, 998109E, 3, 000 m, 26 May 1982, Wang L-s. 82 – 415(KUN). Paratype: Wei Xi Co., Wei Den Village, behind Lou Ma Deng Mt., 27859 N, 998109 E, 3, 100 m, Wang L-s. 82 – 374(KUN).

On bark and wood, including conifers and bamboo.

Hypogymnia delavayi(Hue) Rassad. 肿果袋衣

Bot. Mater. Gerb. bot. Inst. V. A. Komarova 11: 5(1956).

≡*Parmelia delavayi* Hue, in Nylander, *Flora, Regensburg* 72: 135(1887).

≡*Parmelia delavayi* Hue, *Bull. Soc. Bot. France*, 34: 21(1887).

Type: Yunnan: dans la bois de Mao-kou-tchong, au dessus Tapin-tze, 2000 m, 15. v. 1885, A. Delavay 1599 (PC).

= *Hypogymnia yunnanensis* Y. M. Jiang & J. C. Wei, Acta Mycol. Sin. 9: 293. 1990(McCune, B. 2012, p. 15).

Type: Yunnan, Lijiang, Heibaishui, alt. 3000 m, on tree trunk of Pinus sp., 11/XI/1980, Jiang Y. M., no. 286 (HMAS-L).

On bark and wood.

Sichuan, Yunnan(McCune, B. 2012, pp. 15 – 18).

Hypogymnia diffractaica McCune **环萝袋衣**

in B. McCune, E. Martin, L. S. Wang, *Bryologist* 106: 228- 231(2003).

Type: China. Sichuan: Jiulong County, Tang Gu Xiang, 3000 m, on *Rhododendron*; 11 Sep 1996, L. S. Wang 96

-16604(KUN!).

Hypogymnia duplicatoides(Oxner) Rassad. 针芽袋衣

in Bot. Materialy(Notul. System. e Sect. Cryptog. Inst. Bot. nomine V. L. Komarovii Acad. Sci. URSS) 11: 5 (1956).

≡*Parmelia duplicatoides* Oxner In *Journ. Inst. Bot. Acad. Sci. RSS Ukraine*, no. 18/19(26/27) 222(1938) USSR.

On trunk of larch tree.

Helongjiang(J. C. Wei, 1986, p. 324).

Hypogymnia farinacea Zopf 粉袋衣

Justus Liebigs Annln Chem. 352: 42(1907).

=*Ceratophyllum bitterianum* (Zahlbr.) M. Choisy, *Bull. mens. Soc. linn. Soc. Bot. Lyon* 20: 138(1951).

≡*Hypogymnia bitteriana*(Zahlbr.) Räsänen, *Lichenotheca Fennica*: no. 152(1947).

≡*Parmelia bitteriana* Zahlbr., Verh. zool. -bot. Ges. Wien 76: 95(1927).

Parmelia farinacea Bitter, *Hedwigia* 40: 174(1901)[non *P. farinacea*(L.) Ach. ex anno 1803.]

On tree trunk of Larix sp. and occasionaly on rocks.

Jilin, Heilongjiang(Luo, 1986, pp. 157 – 158).

Hypogymnia fragillima(Hillmann ex Sato) Rassad. 串孔脆袋衣

Bot. Mater. Gerb. bot. Inst. V. A. Komarova 11: 8(1956).

≡*Parmelia fragillima* Hillmann ex Sato, Feddes Repert. 45: 172(1938).

On coniferous trees.

Heilongjiang(J. C. Wei, 1981, p. 83; Chen et al. 1981, p. 150; Luo, 1984, p. 84; Luo, 1986, p. 159), Jilin(Chen et al. 1981, p. 150; Luo, 1986, p. 159), hebei(Wu Di et al., 2011, p. 36).

Hypogymnia hengduanensis J. C. Wei 横断山袋衣

Acta Mycol. Sin. 3(4): 214(1984). subsp. hengduanensis 原亚种

Type: Sichuan, H. Smith no. 14078 in UPS(holotype!) & in HMAS-L(isotype!), & paratype(no. 131 – 1 in HMAS-L) is collected by Jiang YM from Yunnan.

On bark of trees.

Sichuan(J. C. Wei, 1986, p. 327), Yunnan(J. C. Wei, 1986, p. 327; Wei X. L. et al. 2007, p. 152), Hubei (Chen, Wu, & J. C. Wei, 1989, p. 428).

subsp. kangdingensis J. C. Wei 康定亚种

Acta Mycol. Sin. 3(4): 215(1984) & *Acta Mycol. Sin. Suppl.* 1: 327(1986).

Type: Sichuan, H. Smith no. 14061 in UPS(holotype!) & in HMAS-L(isotype).

≡*Hypogymnia kangdingensis*(J. C. Wei) Chen & J. C. Wei, in *Fungi and Lichens of Shennongjia*, 1989, p. 429.

Hubei(Chen, Wu & J. C. Wei, 1989, p. 429).

≡*Hypogymnia kangdingensis*(J. C. Wei) Chen & J. C. Wei(McCune et al., 2003, p. 231).

Sichuan(McCune et al., 2003, p. 231), Taiwan(McCune, 2009, p. 823).

Hypogymnia hypotrypa (Nyl.) Rassad. 黄袋衣

Bot. Materialy, Notulae System. e Sect. Cryptogam. Inst. Bot. nomine V. L. Komarovii Acad. Sci. URSS 20: 297 (1967).

≡*Parmelia hypotrypa* Nyl. Synopsis methodica Lichenum, Tomus primus p. 403(1860).

=*Hypogymnia hypotrypa* f. *balteata*(Nyl.) J. C. Wei, *Enum. Lich. China*: 115(1991).

≡*Parmelia hypotrypa* Nyl. f. *balteata* Nyl. in Hue, *Nouv. Arch. Mus. Hist. Nat.* (Paris) 3(2): 293(1890).

Type: from Yunnan.

Shaanxi(He & Chen 1995, p. 43), Sichuan(Zahlbr., 1930 b, p. 194), Yunnan(Hue, 1890, p. 293 & 1899, p. 127;. Zahlbr., 1930 b, p. 194 & 1934 b, p. 210), Anhui(Zhao, 1964, p. 142; Zhao et al. 1982, p. 11), China (prov. not indicated: M. Nuno, 1964, p. 97).

= *Hypogymnia hypotrypella*(Asahina) Rassad. *Bot. Materialy, Notulae System. e Sect. Cryptogam. Inst. Bot. nomine V. L. Komarovii Acad. Sci. URSS* 13: 23(1960).

≡ *Parmelia hypotrypella* Asahina *Acta Phytotax. Geobot.* 14: 34(1950).

= *Hypogymnia flavida* McCune & Obermayer, *Mycotaxon* 79: 23 – 27(2001).

Remark:

A yellowish and bladderlike(fere physodi) new species without mention of soredia *Parmelia hypotrypa* was published by Nylander in 1860. Ninty years later another new species similar to *P. hypotrypa* but with soredia, named *P. hypotrypella*, was published by Asahina(1950). From then on, the *P. hypotrypa* is divided into two different species, a sorediate and esorediate. Thus, the yellowish and bladderlike species *P. hypotrypa* becomes a species group, including both the esorediate *P. hypotrypa* and sorediate *P. hypotrypella*. Both the different species were transferred to the genus *Hypogymnia*, as sorediate *H. hypotrypella*(Asahina) Rassad. (1960), and esorediate *H. hypotrypa*(Nyl.) Rassad. (1967). The syntypes of *P. hypotrypa* Nyl. (coll. Hook. et Thoms. 2014 – 2016) are preserved in H-NYL and British Museum(BM). The lectotype for esorediate *P. hypotrypa* Nyl., the basionym of *H. hypotrypa*, was annotated in one of the syntypes in BM by Awasthi(1984). Both the sorediate and esorediate thalli were found in parts of its syntypes in H-NYL(No. 34197). The esorediate thalli were conjectured as parts of sorediate thalli, and the species *H. hypotrypa* was treated as sorediate one, and *H. hypotrypella* as synonym of *H. hypotrypa*, and another esorediate new species *H. flavida* was published as well(McCune & Obermayer, 2001). Since 1950 both the sorediate and esorediate species based on the morphological characteristics have been being recognized by the lichenologists.

During the period of phynotypic taxonomy there had been a tendency in both mycology and lichenology to base new species on single characters but this approach, at least in some groups of lichens, appears to be contrary to the concepts of vascular plant taxonomist. R. Santesson(in Shibat, 1974) noted that "for me it is not an acceptable taxonomy to base a species on one character only". As pointed out by W. Culberson and C. Culberson, however, what is taxonomically important in one group of organisms may not be so in another. It is now coming to be generally accepted by systematists that concepts of particular taxa should be based on as many characters as possible(W. Culberson and C. Culberson, 1968; D. Hawksworth, 1976).

Soredia are small(25 – 100μm) balls of a few phycobiont cells wraped in mycobiont hyphae. They originate in the medulla and algal layer and are released through pores or cracks in the upper surface of the thallus(J. D. Lawrey, 1984, p. 51). What are the causes of soredial production? Shading, increased moisture, and other environmental factors have been suspected since they would tend to stimulate algal growth and, as an after effect, soredial formation. There is no experimental evidence for this, nor is there any proof that soredia have a genetic origin, a likely explanation in view of their constancy(M. E. Hale, 1983, p. 20).

The result of the combined analyses based on the morphology, anatomy, chemistry, biogeography, and molecular systematics(Xinli Wei et al., 2016, Figs. 2, 5 & 6) showed that both the sorediate *H. hypotrypa* and esorediate *H. flavida* are hard to be divided into two different species. They are distributed from Himalayas through South West of China to Russian Far East and Japan. The distribution of them is neither a sorediate nor esorediate discontinuities within the distribution pattern of Eastern Asian Mountains. In point of above mentioned facts, *Hypogymnia hypotrypa*(Nyl.) Rassad. containing both sorediate and esorediate thalli is a single species, and *H. flavida* has to be treated as synonym of *H. hypotrypa*.

Whether both the sorediate and esorediate are different populations within *H. hypotrypa* or not remains to be studied further.

On trees & moss covered rocks.

Jilin(Chen et al. 1981, p. 150; Luo, 1986, pp. 159 – 160), Shaanxi(Jatta, 1902, p. 471; Zahlbr., 1930 b, p. 194; Zhao, 1964, p. 142; J. C. Wei, 1981, p. 82; J. C. Wei et al. 1982, p. 31; Zhao et al. 1982, p. 10; He & Chen

1995, p. 43, as *H. hypotrypella*, no specimen and locality are cited), Xizang(Obermayer, 2004, p. 498, as *H. flavida*), *Sichuan*(*Elenk*. 1901, p. 22; Zahlbr.. 1930 b, p. 194 & 1934, p. 210; Zhao, 1964, p. 142; Zhao et al. 1982, p. 9; Stenroos et al., 1994, p. 334 as *H. hypotrypella*; McCune B & Obermayer, 2001, p. 26; Obermayer, 2004, p. 498, as *H. flavida* & p. 498 – 499, sometimes *H. hypotrypa* together with *H. flavida*), Yunnan(Hue, 1889, p. 166 & 1899, p. 126; Zahlbr., 1930 b, p. 194; Lai, 1980, p. 210; Zhao, 1964, p. 142; Zhao et al. 1982, pp. 8 & 10; McCune B & Obermayer, 2001, p. 26; Wei X. L. et al. 2007, p. 153, as *H. flavida*), Xizang(J. C. Wei & Chen, 1974, p. 178; J. C. Wei & Jiang, 1981, p. 1147 & 1986, p. 35; McCune B & Obermayer, 2001, p. 26; Obermayer, 2004, p. 498 – 499), Hubei(Lai, 1980, p. 210; Chen, Wu & J. C. Wei, 1989, pp. 427 – 428), Taiwan(Zahlbr., 1933 c, p. 54; Asahina, 1952 a, p. 42; Asahina & Sato in Asahina, 1939, p. 717; M. Nuno, 1964, pp. 97 & 100; Wang & Lai, 1973, p. 93 & 1976, p. 227; Lai, 1980, p. 210), China(prov. not indicated: Hue, 1890, p. 293; M. Nuno, 1964, p. 97; McCune B & Obermayer, 2001, p. 26).

Hypogymnia irregularis McCune 狭叶袋衣

Mycotaxon 115: 486 – 490(2011).

Type: China: Yunnan Province, Jiaoxi Mountain, north of Kunming, 26. 100°N 102. 867°E, 3700 m, *Abies georgei* var. *smithii* – *Rhododendron* forest on slopes near hotel, on bark of *Abies*, Sept. 2000, *McCune* 5576 (holotype: KUN; isotypes: H, HMAS-L, OSC, UPS, US).

Sichuan, Yunnan(McCune, 2011, . 488 – 490).

Hypogymnia laccata J. C. Wei & Jiang 蜡光袋衣

Acta Phytotax. Sin. 18(3): 387(1980).

Type: Xizang, alt. 4400 m, Zong Y. C. & Liao Y. Z. no. 506 in HMAS-L.

Corticolous.

Shaanxi(He & Chen 1995, p. 43, no specimen and locality are cited), Xizang(J. C. Wei & Y. M. Jiang, 1981, p. 1146 & 1986, p. 35).

Hypogymnia laxa McCune 粉唇袋衣

in B. McCune, E. Martin, L. S. Wang, *Bryologist* 106: 231 – 232(2003).

Type: Yunnan, Luquan Co., Jiao Zi Mts, N of Kunming, 26869 N, 1028529 E, 3, 750 m, September 2000, McCune 25599(OSC). Paratypes from Sichuan & Yunnan.

Hypogymnia lijiangensis Chen JB 丽江袋衣

Acta Mycol. Sinica 13(2): 109(1994).

Type: Yunnan, lijiang, wang xy et al. 6991(HMAS-L. holotype).

Xizang, Sichuan, Yunnan(Wei XL & J. C. Wei, 2012, p. 784).

Hypogymnia lugubris(Pers.) Krog 表纹袋衣

Norsk Polarinstitutt Skrifter 144: 99(1968).

≡*Parmelia lugubris* Pers., in Gaudichaud-Beaupré in Freycinet, *Voy. Uranie.*, *Bot.*: 196(1827) [1826 – 1830].

On bark of trees & on mosses between rocks.

Shaanxi(Jatta, 1902, p. 471; Zahlbr., 1930 b, p. 194).

Hypogymnia macrospora(J. D. Zhao) J. C. Wei 大孢袋衣

Enum. Lich. China: 116(1991).

≡*Parmelia macrospora* Zhao, *Acta Phytotax. Sin*. 9(2): 143(1964).

Type: Yunnan, Wang H. C. no. 1065 in HMAS-L.

Sichuan(McCune et al., 2003, p. 233), Yunnan(Hawksw. 1972, p. 49; Zhao et al. 1982, p. 12; J. C. Wei, 1991, p. 118; McCune et al., 2003, p. 233).

On bark of dead twigs.

=*Hypogymnia subvittata*(Zhao) J. C. Wei, *Enum. Lich. China*: 118(1991).

≡*Parmelia subvitata* Zhao, *Acta Phytotax. Sin.* 9(2): 14(1964).

Type: Yunnan, Wang HC no. 4827 in HMAS-L.

Yunnan(Hawksw. 1972, p. 52; Zhao et al. 1982, p. 11).

Hypogymnia magnifica X. L. Wei & McCune 背孔袋衣

in XL Wei et al. *The Bryologist*, 113(1): 120 – 123(2010).

Type: China. Yunnan: Lijiang Co., Laojuen Mt., Jiushijiulong Lake, 26.632uN 99.728uE, 4100 m, L. S. Wang 00 – 20250(KUN, holotype; isotypes: HMAS, OSC). Sichuan(localities of paratype).

Hypogymnia metaphysodes(Asahina) Rassad. 变袋衣

Nov. sist. Niz. Rast. 4: 291(1967).

≡*Parmelia metaphysodes* Asahina, *Acta phytotax. geobot.*, *Kyoto* 14: 33(1950).

Heilongjiang(Luo, 1986, pp. 160 – 161), Shaanxi(He & Chen 1995, p. 43, no specimen and locality are cited), Taiwan(Lai, 1980a, p. 210).

Hypogymnia nikkoensis(Zahlbr.) Rassad. 日光山袋衣

Nov. sist. Niz. Rast. 4: 294(1967).

≡*Parmelia nikkoensis* Zahlbr., *Bot. Mag.*, *Tokyo* 41: 364(1927).

On *Larix* sp.

Neimenggu(Chen et al. 1981, p. 150).

Hypogymnia nitida McCune & L. S. Wang 光亮袋衣

Mycosphere 5(1): 52(2014).

Type: China, Yunnan: Deqin County, Bei Ma Xue Shan, Ya Kou, 4200 m, ca. 28.38°99.0° W, 10 Aug 1993, L. S. Wang 93 – 13495(KUN).

On bark.

Sichuan, Yunnan(McCune et al., 2014, p. 52).

Hypogymnia papilliformis McCune 乳头袋衣

Tchabanenko & X. L. Wei The *Lichenologist* 47(2): 117 – 122(2015).

On bark.

Shaanxi(McCune et al. 2015, pp. 117 – 119).

Hypogymnia pendula McCune & L. S. Wang 舒展袋衣

Mycosphere 5(1): 55(2014).

Type: China, Yunnan: Jianchuan County, trailhead to Lao Juen Shan, Abies delavayi-*Rhododendron* forest near hotel, 2002. 10. 18, McCune 26711(KUN, holotype).

Hypogymnia physodes(L.) Nyl. 袋衣

Lich. Envir. Paris: 39(1896) [1895]. f. physodes 原变型

≡*Lichen physodes* L., *Sp. pl.* 2: 1144(1753).

≡*Parmelia physodes* (L.) Ach., *Methodus, Sectio post.* (Stockholmiæ): 250(1803).

≡*Imbricaria physodes* (L.) DC., in Lamarck & de Candolle, *Fl. franç.*, Edn 3(Paris) 2: 393(1805)

On twigs and bark of trees.

Heilongjiang(Chen et al. 1981, p. 150; Luo, 1984, p. 84. Luo, 1986, pp. 161 – 162), Jilin, Neimenggu(Chen et al. 1981, p. 150), Xinjiang(Wang XY, 1985, p, 346; Abudula Abbas et al., 1993, p. 78), Shaanxi(Jatta, 1902, p. 471; A. Zahlbr., 1930 b, p. 193; J. C. Wei et al. 1982, p. 30), Xizang(J. C. Wei & Chen, 1974, p. 178), Yunnan(Hue, 1887 a, p. 21 & 1899, p. 123; Paulson, 1928, p. 317), Taiwan(Wang & Lai, 1973, p. 94 & 1976, p. 227; Lai, 1980a, p. 211), China(prov. not indicated: Asahina & Sato in Asahina, 1939, p. 717).

= *Hypogymnia physodes* f. *labrosa*(Ach.) Walt. Watson, *Trans. Bot. Soc. Edinb.*, 150*th anniversary supplement* 33: 186(1942).

≡*Parmelia physodes* var. *labrosa* Ach., *Lich. univ.*: 493(1810).

On twigs of coniferous trees.

Jilin(Luo, 1986, pp. 162 – 163), Heilongjiang(Wei, 1981, p. 83; Luo, 1986, pp. 162 – 163), Xinjiang(Abdulla Abbas & Wu, 1998, p. 94; Abdulla Abbas et al., 2001, p. 364), Shaanxi(Jatta, 1902, p. 471; Zahlbr., 1920 b, p. 194).

= *Hypogymnia physodes* f. *platyphylla* (Ach.) Rassad., *Nov. sist. Niz. Rast.* 4: 291(1967).

≡ *Parmelia physodes* var. *platyphylla* Ach., *Methodus, Sectio post.*: 251(1803).

Heilongjiang, Jilin(Luo, 1986, p. 162), Xinjiang(Abdulla Abbas & Wu, 1998, p. 94; Abdulla Abbas et al., 2001, p. 364).

= *Hypogymnia physodes* f. *subtubulosa*(Anders) Rassad., in Kopachevskaya et al., *Opredelitel'Lishaĭnikov SSSR Vypusk* (*Handbook ol the lichens of the U. S. S. R.*) (Leningrad): 293(1971).

≡ *Parmelia physodes* f. *subtubulosa* Anders, *Strauch- und Laubflechten Mitteleur.*: 137(1928).

Heilongjiang(Luo, 1986, p. 162).

f. maculans(H. Olivier) Rassad. 斑点变型

Nov. sist. Niz. Rast. 4: 292(1967).

≡ *Parmelia physodes* f. *maculans* H. Olivier, *Rev. Bot.* 10: 618(1892).

Heilongjiang(Luo, 1986, p. 163).

Hypogymnia pruinoidea X. L. Wei & J. C. Wei 类霜袋衣

The Lichenologist 44(6): 784 – 787(2012).

Type: Shaanxi, Mt. Taibaishan, alt. 2800 m, on *Abies* trunk, 3 August 2005, X. L. Wei 1727(HMAS—holotype).

Hypogymnia pruinosa J. C. Wei & Jiang 霜袋衣

Acta Phytotax. Sin. 18(3): 386(1980) & Lich. of Xizang, 1986, p. 36.

Type: Xizang, 1/VI/1975, Zong Y. C. & Liao Y. Z. no. 215 in HMAS-L.

Corticolous.

Shaanxi(He & Chen 1995, p. 43, no specimen and locality are cited; X. L. Wei & J. C. Wei, 2012, p. 788), Sichuan, Yunnan(J. C. Wei, 1981, p. 84; X. L. Wei & J. C. Wei, 2012, p. 788).

Hypogymnia pseudobitteriana(D. D. Awasthi) D. D. Awasthi 拟粉袋衣

Geophytology 1: 101(1971).

≡ *Parmelia pseudobitteriana* D. D. Awasthi, *Curr. Sci.* 26: 123(1957).

Taiwan(Lai, 1980 a, p. 211).

Hypogymnia pseudocyphellata McCune & E. P. Martin 假杯点袋衣

in Bruce McCune, Erin P. Martin, Li-song Wang *The Bryologist*, 106(2): 233(2003).

Type: Yunnan, Zhong-dian Co., Tian Chi(alpine lake), 278389 N, 998399 E, 3, 750 m, Wang Ls. 94 – 14916c (KUN). Paratype: Yunnan, Zhong-dian Co., Xiozhongdian, Tianchi Lake, 278509 N, 998439 E, Wang L-s. 93 – 13683e(KUN).

Hypogymnia pseudoenteromorpha Lai 拟指袋衣

Quart. Journ. Taiwan Museum 33(3, 4): 209(1980).

Type locality: Japan: Holotype from Japan, Kurok. no. 56132; Paratypes from China: Taiwan, Ilan, Mt. Taiping shan, Wang, Feb. 13, 1976; Pingtung, Kentin Park, Wang-Yang, Jan. 1, 1974, in TAIM(not seen).

Hypogymnia pseudohypotrypa(Asahina) Ajay Singh 灰袋衣

Lichenology in the Indian Subcontinent 1966 – 77(Lucknow): 2(1980).

≡ *Parmelia pseudohypotrypa* Asahina, in Nuno, *Journ. Jap. Bot.* 39(4): 99(1964).

= *Hypogymnia sinica* J. C. Wei & Y. M. Jiang, *Acta Phytotax. Sin.* 18(3): 386(1980).

Type-China. Xizang: Nyalam, Qüxiang, DeQingtang, 3660 m, on bark of *Betula*, 21 May 1966. J. C. Wei and J. B. Chen 1110(HMAS-L!).

Xizang(J. C. Wei & Jiang, 1981, p. 1146 & 1986, p. 36) .

On bark.

Shaanxi, Yunnan, Anhui(Wei X. L. ?) ,

Hypogymnia pseudophysodes(Asahina) Rassad. 拟袋衣

Nov. Sist. Niz. Rast. 1967: 294(1967) .

≡*Parmelia pseudophysodes* Asahina, *Journ. Jap. Bot.* 26(4) : 100(1951) .

Type: China: Taiwan, Japan: Hondo, Shikoku, Russia: Sakhalin(syntypes) .

Jilin(Luo, 1986, p. 163) , xinjiang(Wang XY, 1985, p. 346) , Taiwan(Asahina, 1951 a, p. 100 & 1952 a, p. 40; Lamb, 1963, p. 495; Wang & Lai, 1973, p. 94 & 1976, p. 227) .

Hypogymnia pseudopruinosa X. L. Wei & J. C. Wei 拟霜袋衣

Mycotaxon 94: 155 – 156(2005) .

Type: Yunnan, Deqin, 1981. 8. 29. Wang, Xiao & Su7606(holotype, HMAS-L) .

Shaanxi(X. L. Wei & J. C. Wei, 2012, p. 789) , Yunnan(Wei X. L. & J. C. Wei, 2005, pp. 155 – 156) .

Hypogymnia pulverata(Nyl.) Elix 粉末袋衣

Brunonia 2: 217(1979) .

≡*Parmelia mundata* var. *pulverata* Nyl. apud Cromb. , *J. Linn. Soc. Bot.* 17: 395(1879) .

= *Hypogymnia mundata* f. *sorediosa*(Bitt.) Rassad. , *Bot. Mater. Gerb. Bot. Inst. V. A. Komarova* 11: 11, 1956. J. C. Wei, *Enum. Lich. China*: 116(1991) .

≡*Parmelia mundata* f. *sorediosa* Bitt. , *Hedwigia* 40: 255(1901) .

On bark of Larix sp.

Jilin(J. C. Wei, 1986, p. 328; Luo, 1986, p. 161) .

Hypogymnia saxicola McCune & L. S. Wang 石生袋衣

Mycosphere 5(1) : 63(2014) .

Type: China, Yunnan Province, Luquan County, Jiaozixue Mountain, north of Kunming, on top of shoulder of mountain, outcrops and Rhododendron and Juniperus(Sabina) scrub in subalpine, alt. 4200 m, rock crevices in cliff, 2000. 9, McCune 25561(holotype, KUN; isotype, OSC) .

Sichuan, Yunnan(McCune et al. , 2014, p. 63) .

Hypogymnia stricta(Hillmann) K. Yoshida 长叶袋衣

Bull. natn. Sci. Mus. , Tokyo, B 27(2, 3) : 36(2001) .

≡*Parmelia elongata* var. *stricta* Hillmann, in *Fedde, Repertorium* XLV: 171(1938) .

Yunnan, Taiwan(McCune, 2009, pp. 823 – 824) .

Hypogymnia subarticulata(Zhao et al.) J. C. Wei 节肢袋衣

Lichens of Xizang, 1986: 37(1986) .

≡ *Parmelia vittata* var. *subarticulata* Zhao et al. *Acta Phytotax. Sin.* 16(3) : 95(1978) .

Type: Yunnan, Zhao & Chen no. 4414 in HMAS-L.

Yunnan(Zhao et al. 1982, p. 9; Wei X. L. et al. 2007, p. 153) .

On bark of trees & on moss covered rocks.

Shaanxi(He & Chen 1995, p. 43, no specimen and locality are cited) , Yunnan, Taiwan(McCune, 2009, p. 825) , Xizang(Wei & Jiang, 1986, p. 37) , Hubei(Chen, Wu, & J. C. Wei, 1989, p. 430) .

Hypogymnia subcrustacea(Flot.) Kurok. 亚壳袋衣

Miscell. bryol. lichenol. (Nichinan) 5(9) : 130(1971) .

≡*Imbricaria physodes* f. *subcrustacea* Flot. , (1960) .

On *Larix* sp.

Heilongjiang(Wei, 1986, p. 325) .

Hypogymnia subduplicata(Rass.) Rassad. 腋圆袋衣

Nov. sist. Niz. Rast. 10: 1917(1973) . var. subdupicata 原变种

≡*Hypogymnia subduplicata*(Rassad.) Rassad. , *Nov. sist. Niz. Rast*. 10: 1917(1973) .

Heilongjiang(Luo, 1986, pp. 163 – 164) , Shaanxi(He & Chen 1995, p. 43, no specimen and locality are cited) , Hubei(Chen, Wu & J. C. Wei, 1989, p. 430) .

var. rugosa Luo 皱褶变种

Bull. Bot. Res. 6(3) : 164(1986) .

Type: Heilongjiang, Huzhong, Mt. Dabai, alt. 1000 m, 1980, Luo G. Y. no. 1757 – 1(NEFI) .

On tree trunk or rotten logs.

var. suberecta Luo 直立变种

Bull. Bot. Res. 6(3) : 164(1986) .

Type: Jilin, Mt. Changbai, alt. 1850 – 1900 m, 1984, Luo G. Y. no. 5759(NEFI) .

Hypogymnia subfarinacea X. L. Wei & J. C. Wei 亚粉袋衣

Mycotaxon 94: 156(2005) .

Type: Sichuan, Nanping, 1983. 6. 10. Wang & Xiao 10582(HMAS-L) .

Sichuan, Yunnan(X. L. Wei & J. C. Wei, 2005, p. 156) .

Hypogymnia submundata(Oxner) Rassad. 亚洁袋衣

Nov. sist. Niz. Rast. 4: 294(1967) . f. submundata 原变型

≡*Parmelia submundata* Oxner, In *Journ. Inst. Bot. Acad. Sci. RSS Ukraine*, no. 18/19(26/27) 221(1938) USSR (1938) .

On bark.

Guizhou(Zhang T & Wei JC, 2006, p. 8) .

f. baculosorediosa Rassad. 粉芽变型

Nov. sist. Niz. Rast. 4: 294(1967) .

≡*Parmelia submundata* auct. non Oxn. : Zhao, *Acta Phytotax. Sin*. 9(2) : 141(1964) . Zhao et al. *Prodr. Lich. Sin*. 1982, p. 10.

On twigs of coniferous trees.

Jilin, Neimenggu(Chen et al. 1981, p. 150) , Anhui(Zhao, 1964, p. 141; Zhao et al. 1982, p. 10) .

Hypogymnia subpruinosa Chen JB 亚霜袋衣

Acta Mycol. Sinica 13(2) : 107(1994) .

Type: Yunnan, xiaozhongdian, wang xy et al. 7094(hmas-l. holotype) .

Hypogymnia taibaiensis He & Chen 太白袋衣

Shaanxi(He & Chen 1995, p. 43) .

Notes: nomen nudum, no specimen and locality are cited.

Hypogymnia taiwanalpina Lai 台湾高山袋衣

Quart. Journ. Taiwan Museum, 33(2, 4) : (1980) .

Type: Taiwan, Lai no. 9300 in H! (isotype) .

Yunnan(Wei X. L. et al. 2007, p. 153) .

Hypogymnia tenuispora McCune & L. S. Wang 狭孢袋衣

Mycosphere 5(1) : 70(2014) .

Type: China, Yunnan Province, Luquan County, Jiaozixue Mt. , north of Kunming, high plateau, with outcrops and scrub Rhododendron; on Sorbus in steep, shrubby riparian gully On mountain slope, 26. 10°N 102. 87°E, alt. 4100 m, 2000. 9, McCune 25573(holotype, KUN) .

Hypogymnia tubulosa(Schaer.) Hav. 管袋衣

Bergens Mus. Årbok, Naturv. raekke no. 2: 31(1918) [1917 – 1918] . var. tubulosa 原变种

≡*Parmelia physodes* var. *tubulosa* Schaer. , *Lich. helv. spicil*. 10: 459(1840) .

f. farinosa(Hillmann) Rassad. 粉芽变型

Nov. sist. Niz. Rast. 4: 294(1967) .

≡*Parmelia tubulosa* f. *farinosa* Hillmann, *Verh. bot. Ver. Prov. Brandenb.* 65: 64(1923).

On tree trunk of *Betula* sp.

Neimenggu(Luo, 1986, p. 165).

Hypogymnia vittata(Ach.) Parrique 条袋衣

Act. Soc. linn. Bordeaux 53: 66(1898).

≡*Parmelia physodes* var. *vittata* Ach., *Methodus*, Sectio post.: 250(1803).

≡*Parmelia vittata*(Ach.) Röhl., *Deutschl. Fl.* (Frankfurt) 3(2): 109(1813).

Heilongjiang(Chen et al. 1981, p. 150), Shaanxi(Jatta, 1902, p. 471; Zahlbr., 1930 b, p. 193), Sichuan(Chen S. F. et al., 1997, p. 221), Yunnan(Hue, 1887, p. 21; Zahlbr., 1930 b, p. 193; Zhao, 1964, p. 141; McCune, 2011, p. 490), Xizang(J. C. Wei & Chen, 1974, p. 178; J. C. Wei & Jiang, 1981, p. 1147, & 1986, p. 37), Taiwan(Zahlbr., 1933 c, p. 54; Asahina, 1952 a, p. 33; Wang & Lai, 1976, p. 227; Lai, 1980a, p. 212).

f. vittata 原变型

Heilongjiang, Jilin, Neimenggu(Luo, 1986, pp. 165 – 166), Yunnan(Zhao et al. 1982, p. 9).

f. hypotropodes(Nyl.) J. C. Wei 拟小孔变型

Enum. of Lichens in China(Beijing): 118(1991).

≡*Parmelia hypotrypodes* Nyl., *Flora, Regensburg* 57: 16(1874).

≡*Imbricaria hypotropodes*(Nyl.) Jatta, *Nuovo G. bot. ital.* 9: 471(1902).

Shaanxi(Jatta, 1902, P. 471; . Zahlbr., 1930 b, p. 193), Yunnan(Hue, 1887a, p. 21 & 1889, p. 166 & 1899, p. 125), Shandong(Li Y et al., 2008, p. 71).]

f. hypotrypanea(Nyl.) Kurok. 暗尖变型

Miscell. bryol. lichenol. (Nichinan) 5(9): 130(1971).

≡*Parmelia hypotrypanea* Nyl. in *Flora* 57: 306(1874).

Heilongjiang, Jilin(Luo, 1986, p. 166), China(prov. not indicated: M. Nuno, 1964, p. 97).

f. stricta(Hillmann) Kurok. 笔直变型

Miscell. bryol. lichenol. (Nichinan) 5(9): 130(1971).

≡*Parmelia elongata* Hillm. var. *stricta* Hillm. In Fedde, *Repertorium* XLV: 171(1938).

Taiwan(M. Nuno, 1964, p. 97; Wang & Lai, 1973, p. 94).

f. physodioides Rassad. 拟袋衣变型

in Kopaczevskaja et al., *Nov. sist. Niz. Rast.* 10: 197(1973).

≡*Hypogymnia duplicata* f. *physodioides* Rassad. *Index of Fungi* 4: 113(1971 – 1980).

Heilongjiang(Luo, 1986, p. 166).

f. pinicola Rassad. 多枝变型

in Kopaczevskaja et al., *Nov. sist. Niz. Rast.* 10: 197(1973).

≡*Hypogymnia duplicata* f. *pinicola* Rassad., *Index of Fungi* 4: 113(1971 – 1980).

Heilongjiang(Luo, 1986, p. 166).

Hypotrachyna(Vain.) Hale **双歧根属**

Phytologia 28(4): 340(1974).

≡*Parmelia* sect. *Hypotrachyna* Vain., *Acta Soc. Fauna Flora fenn.* 7(no. 1): 38(1890).

= *Everniastrum* Hale, *Mycotaxon* 3(3): 345(1976) - nom. inval., Art. 39. 1(Melbourne).

≡*Everniastrum* Hale ex Sipman, *Mycotaxon* 26: 237(1986).

Hypotrachyna addita(Hale) Hale 亚覆瓦双歧根

Phytologia 28(4): 340(1974).

≡*Parmelia addita* Hale, *Phytologia* 22(6): 433(1972).

Guangxi(Chen J. B. et al., 2003, pp. 361 – 362 & 2015, p. 80).

Hypotrachyna adducta(Nyl.) Hale 丽双歧根

Phytologia 28(4):340(1974).

≡*Parmelia adducta* Nyl., *Flora*, Regensburg 68:610(1885).

=*Parmelia spectabilis* Asahina, *Journ. Jap. Bot.* 26(10):292(1951).

Type: from Taiwan

On bark of trees.

guizhou, Yunnan(ChenJ. B. et al., 2003, p. 362 & 2015, p. 81 & 2015, p. 81), Anhui(Zhao, 1964, p. 156; Zhao et al. 1982, p. 40), Taiwan(Asahina, 1951c, p. 292 & 1952 a, p. 96; Wang & Lai, 1973, pp. 93, 94 & 1976, p. 228; Lai, 2000, p. 117).

Hypotrachyna adjuncta(Hale) Hale 东方双岐根

Phytologia 28(4):340(1974).

≡*Parmelia adjuncta* Hale, *Phytologia* 22(6):434(1972).

Yunnan, hunan, Zhejiang, guangxi(ChenJ. B. et al., 2003, p. 363 & 2015, p. 81 –82), Fujian(Chen JB, 2015, p. 82).

Hypotrachyna alectorialica(W. L. Culb. & C. F. Culb.) Divakar et al. 树发双岐根

in Divakar, Crespo, Núñez-Zapata, Flakus, Sipman, Elix & Lumbsch, *Phytotaxa* 132(1):30(2013).

≡*Cetrariastrum alectorialicum* W. L. Culb. & C. F. Culb., *Bryologist* 84(3):297(1981).

Type: Yunnan, Delavay, in 1889 preserved in PC.

Xizang(Y. M. Jiang & J. C. Wei, 1989, p. 45).

Hypotrachyna asiatica Wang SL, Chen JB, & Elix 亚洲双岐根

Mycotaxon 76:294 –296(2000).

Type: Yunnan, Gongshan, alt. 2200m, on bark, Aug. 6, 1982, Su J. J. 2761 (holotype: HMAS-L.; isotype: CANB; ChenJ. B. et al., 2003, p. 363 & 2015, p. 82).

Hypotrachyna ciliata Wang SL, Chen JB, & Elix 缘毛双岐根

Mycotaxon 76:296 –297(2000).

Type: Yunnan, Yulongshan, atl. 3000m, on rocks, Aug. 4, 1981. Su J. J. 5013(holotype: HMAS. L.; isotype: CANB).

Yunnan(ChenJ. B. et al., 2003, p. 363 & 2015, p. 83).

Hypotrachyna cirrhata(Fr.) Divakar, Crespo, Sipman, Elix & Lumbsch 条双岐根

in Divakar, Crespo, Núñez-Zapata, Flakus, Sipman, Elix & Lumbsch, Phytotaxa 132(1):31(2013).

≡*Parmelia cirrhata* Fr., *Syst. orb. veg.* (Lundae) 1:283(1825).

≡*Cetrariastrum cirrhatum*(Fr.) W. L. Culb. & C. F. Culb., *Bryologist* 84(3):283(1981).

≡*Evernia cirrhata*(Fr.) M. Choisy, *Bull. Soc. bot. Fr.* 104:334(1957).

≡*Everniastrum cirrhatum*(Fr.) Hale ex Sipman, *Mycotaxon* 26:237(1986).

≡*Parmelia camtschadalis* var. *cirrhata*(Fr.) Zahlbr. [as ' kamtschadalis'], Annln K. K. naturh. Hofmus. Wien 19:43(1904).

≡*Parmelia cirrhata* Fr., *Syst. orb. veg.* (Lundae) 1:283(1825).

≡*Pseudevernia cirrhata*(Fr.) R. Schub. & Klem., *Nova Hedwigia* 11:59(1966).

On bark and on twigs of trees.

Jilin(Chen JB, 2015. p.), Gansu(H. Magn. 1940, p. 128), Shaanxi(Jatta, 1902, p. 468; Zahlbr., 1930 b, p. 195), Xizang(J. C. Wei & Chen, 1974, p. 178; J. C. Wei & Jiang, 1981, p. 1147 & 1986, p. 54; J. C. Wei, 1981, p. 85; Jiang & J. C. Wei, 1989, p. 46; Chen JB, 2015. p. 59), Sichuan(Zahlbr., 1930 b, p. 195 & 1934 b, p. 210; Zhao, 1964, p. 145; Zhao et al. 1982, p. 15; Jiang & J. C. Wei, 1989, p. 45; Chen S. F. et al., 1997, p. 221; Chen JB, 2015. p. 58), Chongqing(Chen CL et al., 2008, p. 40), Guizhou(Jiang & J. C. Wei, 1989, p. 46; Chen JB, 2015. p. 58), Yunnan(Zahlbr., 1930 b, p. 195 & 1934 b, p. 210; Zhao, 1964, p. 145; Zhao et al. 1982, p. 16; Jiang & J. C. Wei, 1989, p. 45; Wei X. L. et al. 2007, p. 152; Wang L. S. et al. 2008, p. 536; Chen

JB, 2015. p. 58), Hebei(Wu Di et al., 2011, pp. 36 – 37), Anhui, Zhejiang(Zhao, 1964, p. 145; Zhao et al. 1982, p. 15), Hubei(Chen, Wu & J. C. Wei, 1989, p. 435; Chen JB, 2015. p. 58), Fujian(Wu J. N. et al. 1982, p. 9; Jiang & J. C. Wei, 1989, p. 46; Chen JB, 2015. p. 58), Taiwan(Zahlbr., 1933 c, p. 58; Asahina, 1951 a, p. 102 & 1952 a, p. 49; Wang & Lai, 1973, p. 93; Jiang & J. C. Wei, 1989, p. 46; Lai, 2000, pp. 109 – 110), Guangxi(Chen JB, 2015. p. 58).

= *Everniastrum cirrhatum* f. *confusum*(Du Rietz) J. C. Wei, *Enum. Lich. China*: 94(1991).

≡ *Parmelia confusa* Du Rietz, *Bot. Notiser*: 342(1924).

≡ *Parmelia cirrhata* f. *confusa*(Du Rietz) Asahina

Taiwan(Wang & Lai, 1973, p. 93).]

Hypotrachyna consimilis(Vain.) Hale 同双岐根

Smithson. Contr. bot. 25: 28(1975).

≡ *Parmelia consimilis* Vainio, *étud. Lich. Bresil.*, vol. I, 1890, p. 58. (Kurok., 1979, p. 128; Lai, 2000, p. 117; J. C. Wei, 1991, p. 165).

Taiwan(Kurok., 1979, p. 128; Lai, 2000, p. 117).

Hypotrachyna crenata(Kurok.) Hale 圆齿双歧根

Phytologia 28(4): 341(1974).

≡ *Parmelia crenata* Kurok., in Hale & Kurokawa, *Contr. U. S. natnl. Herb.* 36: 168(1964).

Sichuan(Chen JB, 2015, p. 84), Taiwan(Wang & Lai, 1973, p. 93; Lai, 2000, p. 118).

Hypotrachyna croceopustulata(Kurok.) Hale 疱粉双岐根

Smithson. Contr. bot. 25: 30(1975).

≡ *Parmelia croceopustulata* Kurok. in Hale & Kurokawa, *Contr. U. S. Natn. Herb.* 36: 169(1964).

On tree bark.

Hunan(Chen J. B. et al., 2003, pp. 363 – 364 & 2015, p. 84).

Hypotrachyna diffractaica(Y. M. Jiang & J. C. Wei) Divakar et al. 环萝岐根

in Divakar, Crespo, Núñez-Zapata, Flakus, Sipman, Elix & Lumbsch, *Phytotaxa* 132(1): 31(2013).

≡ *Everniastrum diffractaicum* Y. M. Jiang & J. C. Wei, *Lichenologist* 25(1): 58(1993).

Type: Yunnan, Dali, 1941. 7. Wang H. C. 1065a: holotype, 1084b: paratype(HMAS-L).

Yunnan(Y. M. Jiang &J. C. Wei, 1993, pp. 58 – 60; Chen JB, 2015. p. 60).

Hypotrachyna endochlora(Leight.) Hale 内绿双岐根

Smithson. Contr. bot. 25: 34(1975).

≡ *Parmelia endochlora* Leight., *Lich. -Fl. Great Brit.*: 140(1871).

Taiwan(Wang & Lai, 1973, p. 93 & 1976, p. 227; Kurok. & Mineta, 1973, pp. 118 – 119).

Hypotrachyna expallida(Kurok.) Divakar, Crespo, Sipman et al. 裂芽双岐根

in Divakar, Crespo, Núñez-Zapata, Flakus, Sipman, Elix & Lumbsch, *Phytotaxa* 132(1): 33(2013).

≡ *Parmelia expallida* Kurok., *Bull. natn. Sci. Mus.*, Tokyo, B 11: 191(1968).

Type: Taiwan, Kurokawa 2930(TNS, holotypus).

Hypotrachyna exsecta(Taylor) Hale 切割双岐根

Phytologia 28: 341(1974).

≡ *Parmelia exsecta* Taylor, *London J. Bot.* 6: 166(1847).

Taiwan(Wang & Lai, 1973, p. 93; Lai, 2000, pp. 119 – 120).

= *Hypotrachyna laevigata*(Sm.) Hale, Smithson. Contr. bot. 25: 44(1975).

≡ *Parmelia laevigata* ssp. *extremi-orientalis* Asahina, *Journ. Jap. Bot.* 26(10): 289(1951).

guizhou(Chen J. B. et al., 2003, p. 364; Zhang T & J. C. Wei, 2006, p. 8; Chen JB, 2015, p. 85), Yunnan, Hunan, Zhejiang, Fujian, Guangxi(Chen J. B. et al., 2003, p. 364 & 2015, p. 85), Taiwan(Asahina, 1952 a, p. 91; Wang & Lai, 1973, p. 93 & 1976, p. 227).

Hypotrachyna exsplendens(Hale) Hale 无光双岐根

Smithson. Contr. bot. 25: 37(1975).

≡*Parmelia exsplendens* Hale in Hale & Kurokawa, *Contr. U. S. Natn. Herb.* 36: 174(1964).

Taiwan(Lai, 2000, p. 120).

Hypotrachyna flexilis(Kurok.) Hale 柔双岐根

Phytologia 28(4): 341(1974).

≡*Parmelia flexilis* Kurok., in Hara, *Flora of eastern Himalaya* (Tokyo): 607(1966).

Yunnan(Chen J. B. et al., 2003, pp. 364 – 365; Chen JB, 2015, p. 86), Taiwan(Kurok., 1966a, p. 607; Wang & Lai, 1973, p. 93; Lai, 2000, pp. 120 – 121).

Hypotrachyna granulans K. H. Moon, Kurok. & Kashiw. 颗粒双岐根

Bull. natn. Sci. Mus., Tokyo, B 26(4): 135(2000).

Yunnan(Chen J. B. et al., 2003, p. 365), Zhejiang, Guangxi(Chen J. B. et al., 2003, p. 365 & 2015, p. 87).

Hypotrachyna ikomae(Asahina) Hale 义笃双岐根

Phytologia 28(4): 341(1974).

≡*Parmelia ikomae* Asahina *J. J. B.* 28: 134(1953).

On trees.

Yunnan, Guangxi(Chen J. B. et al., 2003, p. 365 & 2015, p. 88), Zhejiang(Xu B. S., 1989, 217; Chen J. B. et al., 2003, p. 365 & 2015, p. 88), Fujian(Chen JB, 2015, p. 88), Hainan(Chen J. B. et al., 2003, p. 365 & 2015, p. 88; JCWei et al., 2013, p. 228.).

Hypotrachyna imbricatula(Zahlbr.) Hale 覆瓦双岐根

Smithson. Contr. bot. 25: 41(1975).

≡*Parmelia imbricatula* Zahlbr., *Denkschr. Kaiserl. Akad. Wiss., Math. -Naturwiss.* Kl. 83: 165(1909).

Taiwan(Lai, 2000, pp. 121 – 122), Hainan(Chen J. B. et al., 2001, p. 366 & 2015, p. 88; J. C. Wei et al., 2013, p. 228).

Hypotrachyna immaculata(Kurok.) Hale 无斑双岐根

Smithson. Contr. bot. 25: 41(1975).

≡*Parmelia immaculata* Kurok., in Hale & Kurokawa, *Contr. U. S. natnl. Herb.* 36: 178(1964).

Taiwan(Lai, 2000, p. 122).

Hypotrachyna infirma(Kurok.) Hale 弱双岐根

Phytologia 28(4): 341(1974).

≡*Parmelia infirma* Kurok., in Hale & Kurokawa, *Contr. U. S. natnl. Herb.* 36: 179(1964).

Taiwan(Kurok., 1966a, pp. 19 – 20; Lai, 2000, pp. 122 – 123).

Hypotrachyna keitauensis(Asahina) Hale 溪头双岐根

Phytologia 28(4): 341(1974).

≡*Parmelia keitauensis* Asahina, *Journ. Jap. Bot.* 26(10): 293(1951).

Type: from Taiwan.

Taiwan(Lamb, 1963, p. 482; Wang & Lai, 1973, p. 93; Lai, 2000, pp. 123 – 124).

Hypotrachyna kingii(Hale) Hale 金氏双岐根

Phytologia 28(4): 341(1974).

≡*Parmelia kingii* Hale *J. J. B.* 43: 324(1968).

Taiwan(Lai, 2000, 0. 124).

Hypotrachyna koyaensis(Asahina) Hale 双岐根

Smithson. Contr. bot. 25: 44(1975).

≡*Parmelia koyaensis* Asahina, *J. Jap. Bot.* 28: 67(1953).

Yunnan, Zhejiang, Guangxi(Chen JB, 2015, p. 89), Taiwan(Asahina, 1959 b, p. 225; Wang & Lai, p. 93 & 1976, p. 227; Lai, 2000, 124), Hainan(Chen JB, 2015, p. 89; JCWei 2013, p. 228).

Hypotrachyna laevigata ssp. extremi-orientalis(Asahina) Wang & Lai 远东亚种

Taiwania 21(2):227(1976).

≡*Parmelia laevigata* ssp. *extremi-orientalis* Asahina, *Journ. Jap. Bot*. 26(10):289(1951).

= *Parmelia laevigata* f. *esorediata* Zahlbr., *Bot. Mag*., *Tokyo* 41:351(1927).

Taiwan(Asahina, 1952 a, p. 91; Wang & Lai, 1973, p. 93 & 1976, p. 227).

Hypotrachyna lipidifera(Hale & M. Wirth) Divakar et al. 唇瓣双岐根

in Divakar, Crespo, Núñez-Zapata, Flakus, Sipman, Elix & Lumbsch, *Phytotaxa* 132(1):31(2013).

≡*Parmelia lipidifera* Hale & M. Wirth, *Phytologia* 22(1):37(1971).

On tree bark.

Hubei(Chen J. B. 2011, pp. 883 – 884 & 2015. p. 61).

Hypotrachyna majoris(Vain.) Hale 大双岐根

Phytologia 28(4): 341(1974).

≡*Parmelia majoris* Vain., *Hedwigia* 37: (33)(1898).

Taiwan(Kashiwadani, 1975, in Lai, 2000, p125).

Hypotrachyna mexicana(Egan) Divakar, Crespo et al. 墨西哥双岐根

in Divakar, Crespo, Núñez-Zapata, Flakus, Sipman, Elix & Lumbsch, *Phytotaxa* 132(1):31(2013).

≡*Cetrariastrum mexicanum* Egan, in Culberson & Culberson, *Bryologist* 84(3):287(1981).

Yunnan(Chen J. B., 2011, p. 884 & 2015. p. 61).

Hypotrachyna neostictifera Elix, Chen J. B. & L. Xu 新双岐根

Mycotaxon 86:366 – 367(2003).

Yunnan 2, Aug. 1982, Y. L. Guo 242 – 4(HMAS-L. holotypus), (Chen JB, 2015, p. 90).

Hypotrachyna nepalensis(Taylor) Divakar et al. 尼泊尔双岐根

in Divakar, Crespo, Núñez-Zapata, Flakus, Sipman, Elix & Lumbsch, *Phytotaxa* 132(1):32(2013).

≡*Parmelia nepalensis* Taylor, *London J. Bot*. 6: 172(1847).

≡*Cetrariastrum nepalense* (Taylor) W. L. Culb. & C. F. Culb., *Bryologist* 84(3):301(1981).

≡*Everniastrum nepalense*(Taylor) Hale ex Sipman, *Mycotaxon* 26:241(1986).

= *Parmelia kamtschadalis* var. *minima* Zhao, *Acta Phytotaxon. Sin*. 9(2):144(1964).

Type: Yunnan, Bao shan, Mt. Gaoligong shan, Wang Q. Z. no. 1332 in HMAS-L.

Parmelia kamtschadalis auct. non(Ach.) Eschw. var. *kamtschadalis*.

On bark of trees and twigs.

Sichuan(Zahlbr., 1930 b, p. 195), Yunnan(Hue, 1887a, p. 21 & 1889, p. 165 & 1899, p. 136; R. Paulson, 1928, p. 317; Zahlbr., 1930 b, p. 195; Zhao, 1964, p. 144; Zhao et al. 1982, p. 14 – 15; Jiang & J. C. Wei, 1989, p. 47; Wei X. L. et al. 2007, p. 152; Wang L. S. et al. 2008, p. 536; Chen JB, 2015. p. 62), Xizang(J. C. Wei & Chen, 1974, p. 178; J. C. Wei & Jiang, 1981, p. 1147 & 1986, p. 53; J. C. Wei, 1981, p. 85; Jiang & J. C. Wei, 1989, p. 47; Chen JB, 2015. p. 62), Taiwan(Asahina, 1947b, p. 85 & 1952 a, p. 48; Wang & Lai, 1973, p. 93; Jiang & J. C. Wei, 1989, p. 47; Lai, 2000, p. 110).

Hypotrachyna nodakensis(Asahina) Hale 野岳双岐根

Phytologia 28(4):341(1974).

≡*Parmelia nodakensis* Asahina, *J. Jap. Bot*. 34:226(1959).

Guizhou(Chen J. B. et al., 2003, p. 368; Zhang T & Wei JC, 2006, p. 8; Chen JB, 2015, p. 91), Sichuan, Hunan, Fujian(Chen J. B. et al. 2003, p. 368 & 2015, p. 91), Zhejiang(Xu B. S. 1989, pp. 215 – 216; Chen J. B. et al. 2003, p. 368 & 2015, p. 90), Anhui(Xu B. S. 1989, pp. 215 – 216), Taiwan(Asahina, 1959 b, p. 226; Wang & Lai, 1973, p. 94), Guangxi(Chen JB, 2015, p. 91).

Hypotrachyna novella(Vain.) Hale 荧光双岐根

Smithson. Contr. bot. 25: 49(1975).

≡*Parmelia novella* Vain., *Acta Soc. Fauna Flora fenn*. 7(1):56(1890).

Zhejiang, Fujian(Chen J. B. et al. 2003, p. 368 & 2015, p. 91).

Hypotrachyna osseoalba(Vain.) Park & Hale 骨白双歧根

Taxon 38(1):88(1989).

≡*Parmelia osseoalba* Vain., *Ann. bot. Soc. Zool. -Bot. fenn. Vanamo* 1(no. 3):39(1921).

= *Hypotrachyna formosana* (Zahlbr.) Hale, *Smithson. Contr. Bot.* 25: 38(1975).

≡*Parmelia formosana* Zahlbr., in Fedde, *Repertorium*, 33: 57(1933).

Type: Taiwan, Asahina nos. 60, 70(syntypes).

Taiwan(Asahina, 1951 e, p. 291 & 1952 a, p. 100; Lamb, 1963, p. 476; Wang & Lai, 1973, p. 93 & 1976, p. 227).

Chongqing(Chen JB et al., 2003, pp. 368 – 369; Chen CL et al., 2008, p. 40; Chen JB, 2015, p. 93), Guizhou, Yunnan(Chen JB, 2015, p. 93), Hunan, Jiangxi, Fujian, Guangxi(Chen J. B. et al. 2003, pp. 368 – 369 & 2015, p. 92 – 93), Anhui(Zhao, 1964, p. 154; Zhao et al. p. 36; Xu B. S. 1989, p. 215; Chen J. B. et al. 2003, pp. 368 – 369 & 2015, p. 92), Zhejiang(Xu B. S., 1989, p. 215; Chen J. B. et al. 2003, pp. 368 – 369 & 2015, 0. 92), Taiwan(Zahlbr., 1933, p. 57; Lai, 2000, p. 126), Hainan(Chen J. B. et al. 2003, pp. 368 – 369 & 2015, p. 92 – 93; JCWei et al., 2013, p. 228).

Hypotrachyna physcioides(Nyl.) Hale 多形双歧根

Smithson. Contr. bot. 25: 54(1975).

≡*Parmelia physcioidea* Nyl., *Syn. meth. lich.* (Parisiis) 1: 385(1860).

Yunnan(Chen J. B. et al., 2003, p. 369 & 2015, p. 94).

Hypotrachyna pseudoformosana(Asahina) J. C. Wei 拟台湾双歧根

Enum. Lich. China, 1991, p. 120.

≡*Parmelia pseudoformosana* Asahina, *Journ. Jap. Bot.* 26(10):292(1951).

Type: Taiwan.

Taiwan(Lamb, 1963, p. 495; Wang & Lai, 1973, p. 94).

This species was treated by Chen JB(2015, p. 12) as the heterotypic synonym of *Hypotrachyna majoris*, but the taxonomic revision is not seen. It remains to be checked in the future.

Hypotrachyna pseudosinuosa(Asahina) Hale 灰条双歧根

Smithson. Contr. bot. 25: 58(1975).

≡ *Parmelia pseudosinuosa* Asahina *J. J. B.* 26: 329(1951).

On bark of trees.

Yunnan(Zhao, 1964, p. 154; Zhao et al. 1982, p. 35; Chen J. B. et al., 2003, pp. 369 – 70 & 2015, p. 95), Zhejiang(Xu B. S., 1989, p. 216; Chen J. B. et al., 2003, pp. 369 – 70 & 2015, p. 94), Hunan, Fujian, Guangxi (Chen J. B. et al., 2003, pp. 369 – 70 & 2015, p. 95), Taiwan(Lai, 2000, pp. 126 – 127), Hainan(Chen J. B. et al., 2003, pp. 369 – 70 & 2015, p. 95; JCWei et al., 2013, p. 228).

Hypotrachyna rockii(Zahlbr.) Hale 洛克氏双歧根

Smithson. Contr. bot. 25: 62(1975).

≡*Parmelia rockii* Zahlbr. [as ' Rocki'], *Annls mycol.* 10(4):379(1912).

On *Picea morrisonicola*.

Taiwan(Aptroot et al., 2002, p. 285).

Hypotrachyna revoluta(Flörke) Hale 卷叶双歧根

Smithson. Contr. bot. 25: 60(1975).

≡*Parmelia revoluta* Flörke, in Sprengel, *Syst. veg.*, Edn 16 4(1):248(1827).

≡*Imbricaria revoluta*(Flörke) Flot., *Jber. schles. Ges. vaterl. Kultur* 28: 129(1850).

On bark of *Pinus* sp.

Shaanxi(Jatta, 1092, p. 569; Zahlbr., 1930 b, p. 190), Anhui(Zhao, 1964, p. 154; Zhao et al. 1982, p. 34), Zhejiang(Chen J. B. et al., 2003, p. 370 & 2015, p. 96), Taiwan(Lai, 2000, pp. 127 – 128; Aptroot et al.,

2002, p. 285 mentioned under *H. rockii.*).

Hypotrachyna rhizodendroidea(J. C. Wei & Y. M. Jiang) Divakar et al. 灌双岐根

in Divakar, Crespo, Núñez-Zapata, Flakus, Sipman, Elix & Lumbsch, *Phytotaxa* 132(1):32(2013).

≡*Cetrariastrum rhizodendroideum* J. C. Wei & Y. M. Jiang, *Acta phytotax. sin.* 20(4):496(1982).

Type: Xizang, J. C. Wei & Chen no. 762 in HMAS-L.

≡*Everniastrum rhizodendroideum*(J. C. Wei & Y. M. Jiang) Sipman, *Mycotaxon* 26: 242(1986).

On bark of trees.

Xizang(J. C. Wei & Jiang, 1986, p. 54; Jiang & J. C. Wei, 1989, p. 48; Chen JB, 2015. p. 63), Chongqing(Chen CL et al., 2008, p. 40), Guizhou(Zhang T & J. C. Wei, 2006, p. 7), Yunnan(Wang L. S. et al. 2008, p. 536; Chen JB, 2015. p. 63).

Hypotrachyna sinensis(J. B. Chen & J. C. Wei) Divakar 华双岐根

in Divakar, Crespo, Núñez-Zapata, Flakus, Sipman, Elix & Lumbsch, *Phytotaxa* 132(1):32(2013).

≡*Everniastrum sinense* J. B. Chen & J. C. Wei, in Chen, Wu & J. C. Wei, *Fungi and Lichens of Shennongjia. Mycological and Lichenological Expedition to Shennongjia* (Beijing):434(1989).

Type: Hubei, Shennongjia, Dajiuhu, alt. 2200 m, 15/VII/1984, Chen J. B. no. 10488(HMAS-L).

Chongqing(Chen CL et al., 2008, p. 40).

On tree bark.

Sichuan, Yunnan, Hubei(Chen JB, 2015, p. 64).

Hypotrachyna sinuosa(Sm.) Hale 黄条双岐根

Smithson. Contr. bot. 25: 63(1975).

≡*Lichen sinuosus* Smith, in Smith & Sowerby, *Engl. Bot.* 29: tab. 2050(1809).

≡*Parmelia sinuosa*(Sm.) Ach., *Syn. meth. lich.* (Lund):207(1814).

On bark & twigs.

Shaanxi(He & Chen 1995, p. 43, no specimen and locality are cited; Chen JB, 2015, p. 97), Xizang(J. C. Wei & Chen, 1974, p. 178; J. C. Wei & Jiang, 1981, p. 1147 & 1986, p. 43; J. C. Wei et al. 1982, p. 36; Chen JB, 2015, p. 97), Sichuan(Chen S. F. et al., 1997, p. 221; Chen J. B. et al., 2003, p. 370 & 2015, p. 96), Chongqing(Chen CL et al., 2008, p. 40), Yunnan(Zhao, 1964, p. 147; Zhao et al. 1982, p. 19; Chen J. B. et al., 2003, p. 370 & 2015, p. 96), Hubei(Chen, Wu, & J. C. Wei, 1989, p. 449; Chen J. B. et al., 2003, p. 370 & 2015, p. 96), Taiwan(Nakanishi 1964, p. 71 – 73; Wang & Lai, 1973, p. 94 & 1976, p. 227; Lai, 2000, p. 128).

Hypotrachyna sorocheila(Vain.) Divakar et al. 粉芽岐根

in Divakar, Crespo, Núñez-Zapata, Flakus, Sipman, Elix & Lumbsch, *Phytotaxa* 132(1):32(2013).

≡*Parmelia sorocheila* Vain., *Hedwigia* 38(Beibl.):(123)(1899).

≡*Cetrariastrum sorocheilum*(Vain.) W. L. Culb. & C. F. Culb., *Bryologist* 84(3):292(1981).

≡*Everniastrum sorocheilum*(Vain.) Hale ex Sipman, *Mycotaxon* 26: 242(1986).

On bark of trees.

Xizang(J. C. Wei & Jiang, 1981, p. 1147 & 1986, p. 54; Jiang & J. C. Wei, 1989, p. 49; Chen JB, 2015, p. 65), Sichuan(Chen JB, 2015, p. 65), Chongqing(Chen CL et al., 2008, p. 40), Yunnan(Jiang & J. C. Wei, 1989, p. 48; Chen JB, 2015, p. 65), Taiwan (Asahina, 1951 a, p. 102 & 1952 a, p. 50; Wang & Lai, 1973, p. 93; Ikoma, 1983, p. 79; Lai, 2000, p. 111).

Hypotrachyna sublaevigata(Nyl. ex Tuck.) Hale 亚平双岐根

Smithson. Contr. bot. 25: 66(1975).

≡*Parmelia tiliacea* var. *sublaevigata* Nyl. ex Tuck., *Syn. N. Amer. Lich.* (Boston) 1: 57(1882).

≡*Parmelia sublaeviga*(Nyl.) Nyl., *Ann. Sci. Nat. Bot.* ser. 5, 7: 306(1867); Zahlbr., in Handel- Mazzetti, *Symb. Sin.* 3: 189, 1930; Wei, *Enum. Lich. China:* 171(1991).

≡*Imbricaria sublaevigata*(Nyl.) Jatta, *Nuov. Giorn. Bot. Ital. ser.* 2, 9: 469(1902).

On trunk of trees.

Shaanxi(Jatta, 1902, p. 469; Zahlbr., 1930 b, p. 189), Taiwan(Lai, 2000, pp. 128 – 129).

Hypotrachyna subplana(Sipman) Divakar et al. 亚原双岐根

in Divakar, Crespo, Núñez-Zapata, Flakus, Sipman, Elix & Lumbsch, *Phytotaxa* 132(1): 33(2013).

≡*Everniastrum subplanum* Sipman, *Mycotaxon* 26: 244(1986).

≡*Cetrariastrum subplanum*(Sipman) Kurok., *J. Jap. Bot.* 74(4): 254(1999).

Taiwan(Kurok. 1999, p. 254; Lai, 2000, pp. 111 – 112).

Hypotrachyna subsorocheila(Y. M. Jiang & J. C. Wei) Divakar et al. 亚粉双岐根

in Divakar, Crespo, Núñez-Zapata, Flakus, Sipman, Elix & Lumbsch, *Phytotaxa* 132(1): 33(2013).

≡*Everniastrum subsorocheilum* Y. M. Jiang & J. C. Wei, *Acta Bryolichenologica Asiatica*1(1 – 2): 49(1989).

Type: Yunnan, Lijiang, JC Wei 9247(HMAS-L).

Yunnan(Chen JB, 2015, p. 66).

Hypotrachyna vexans(Zahlbr. ex W. L. Culb. & C. F. Culb.) Divakar et al. 针芽双岐根

in Divakar, Crespo, Núñez-Zapata, Flakus, Sipman, Elix & Lumbsch, *Phytotaxa* 132(1): 33(2013).

≡*Cetrariastrum vexans* Zahlbr. ex W. L. Culb. & C. F. Culb., *Bryologist* 84(3): 294(1981).

≡*Everniastrum vexans*(Zahlbr. ex W. L. Culb. & C. F. Culb.) Hale ex Sipman, *Mycotaxon*26: 242(1986).

≡*Parmelia vexans* Zahlbr., *Feddes Repert.* 33: 55(1933).

Type: Taiwan, Asahina no. 79 in W.

Parmelia americana Vain., non *Parmelia americana*(Meyen & Flot.) Mont., Annls Sci. Nat., Bot., sér. 3 18: 309(1852)

Parmelia cirrhata f. *americana* (Vain.) Asahina, non(Mont.) Asahina, J. Jap. Bot. 26: 102(1951), cited by Wang & Lai, Taiwania, 18(1): 93(1973).

On bark of trees.

Sichuan(Jiang & J. C. Wei, 1989, p. 50; Chen S. F. et al., 1997, p. 221; Chen JB, 2015, p. 67), Chongqing (Chen CL et al., 2008, p. 40), Guizhou(Jiang & J. C. Wei, 1989, p. 50; Zhang T & J. C. Wei, 2006, p. 7; Chen JB, 2015, p. 67), Yunnan (Hue, 1899, p. 136; Zhao, 1964, p. 144; Zhao et al. 1982. p. 15; Chen JB, 2015, p. 67), Hebei(Moreau et Moreau, 1951, p. 191), Anhui, Zhejiang(Jiang & J. C. Wei, 1989, p. 50; Chen JB, 2015, p. 66), Hunan, Fujian(Chen JB, 2015, p. 66 – 67), Taiwan (M. Sato. 1933, p. 272; Zahlbr., 1933 c, p. 55 & 1940, p. 540; Asahina, 1951 a, p. 102 & 1951 b, p. 193 & 1952 a, p. 49; Wang & Lai, 1973, pp. 93 & 94; Culb. & C. Culb., 1981, p. 295; Ikoma, 1983, p. 79; Jiang & J. C. Wei, 1989, p. 50; Lai, 2000, pp. 112 – 113), Hainan(Chen JB, 2015, p. 66 – 67; JCWei et al. 2013, p. 228).

Imshaugia S. L. F. Mey. **小梅衣属**

Mycologia 77(2): 337(1985).

Imshaugia aleurites(Ach.) S. L. F. Mey. 小梅衣

Mycologia 77(2): 338(1985).

≡*Lichen aleurites* Ach., *Lich. suec. prodr.* (Linköping): 117(1799) [1798].

Yunnan(Ahti et al., 1999, pp. 123 – 124), Taiwan(Lai, 2000, pp. 129 – 130).

Lethariella (Motyka) Krog **金丝属**

Norw. Jl Bot. 23(2): 88(1976).

≡*Usnea* subgen. *Lethariella* Motyka, *Lich. Gen. Usnea Monogr.* 1: 39(1936).

Lethariella cashmeriana Krog 金丝绣球

Norw. J. Bot. 23: 91(1976).

Lichen substances:

①atranorin, canarione(Canarion), and gyrophoric acid(J. C. Wei & Y. M. Jiang, 1986, p. 69)

②gyrophoric acid, trace lecanoric acid(Niu et al. 2011, p. 219)

③norstictic acid, gyrophoric acid, trace connorstictic acid(Niu et al. 2011, p. 219)

④norstictic acid, placodiolic acid(Niu et al. 2011, p. 219)

On cypress.

Gansu, Sichuan(Krog, 1976, p. 91), yunnan(Niu et al., 2011, p. 222, cited as Chemical race II for *L. cladonioides*), Xizang(Wei & Jiang, 1981, p. 1146 & 1986, p. 68; Niu et al., 2011, p. 222, cited as Chemical race II for *L. cladonioides*).

Lethariella cladonioides(Nyl.) Krog 金丝刷

Norw. Jl Bot. 23(2):93(1976).

≡*Chlorea cladonioides* Nyl., *Syn. meth. lich.* (Parisiis) 1(2):276(1860).

≡*Evernia cladonioides*(Nyl.) M. Choisy, *Bull. Soc. bot. Fr.* 104: 334(1957).

≡*Letharia cladonioides*(Nyl.) Hue, *Expédit. Antarct. Franç.*: 7(1908).

≡*Nylanderaria cladonioides*(Nyl.) Kuntze, *Revis. gen. pl.* (Leipzig) 2: 876(1891).

≡*Rhytidocaulon cladonioides*(Nyl.) Elenkin, *Bull. Jard. Imp. Bot. Pierre le Grand* 16: 269(1916).

≡*Usnea cladonioides*(Nyl.) Du Rietz, *Svensk bot. Tidskr.* 20: 91(1926).

①norstictic acid, placodiolic acid(Niu et al. 2011, p. 219, cited as chem.. type IV)

②psoromic acid, trace 2'-O-demethylpsoromic acid and gyrophoric acid(Niu et al. 2011, p. 219, cited as chem.. type V)

= *Usnea reticulata Du Rietz, Svensk. Bot. Tidsk.* 20: 93(1926).

Type: Sichuan, 2/X/1922, H. Smith no. 5007 in UPS(!).

≡*Letharia reticulata*(Du Rietz) Zahlbr.

= *Usnea sernanderi* Motyka, Lich. Gen. Usnea Stud. Monogr. Pars Syst.: 46(1936), nomen novum for *Usnea reticulata* Du Rietz, Svensk Bot. Tidskr. 20: 91(1926);

Letharia flexuosa sensu Paulson(1928), non Nyl.

= *Lethariella sernanderi* (Motyka) Obermayer

type: China, Sze-ch' uan(Sichuan), reg. Bor. -occid., mellan Tsago-gamba och Tamba, 4000 m. sm., 2 October 1922, Harry Smith, Plantae Sinenses No. 5007(UPS—holotype).

Lichen substances:

①norstictic acid, trace connorstictic acid(Niu et al. 2011, p. 219, cited as Chemotype I for *Lethariella cladonioides*).

②norstictic acid, gyrophoric acid, trace connorstictic acid(Obermayer, 1995, p7, cited for *Lethariella sernanderi*).

On trees & shrubs.

Shaanxi(Wu J. L. 1981, p. 163; J. C. Wei et al. 1982, p. 50; Wu J. L. & Zhang, 1982, p. 244; He & Chen 1995, p. 43, no specimen and locality are cited; Obermayer, 1995, p7, cited as *Lethariella sernanderi*), Sichuan (Du Rietz, 1926, p. 93; A. Zahlbr., 1930 a, p. 60 & 1930 b, p. 200; Mot. 1936 –38, p. 44; Krog, 1976, p. 93; Niu et al., 2011, p. 222, as chemical race III & IV; Obermayer, 1995, p7, cited as *Lethariella sernanderi*), Yunnan (Paulson, 1928, p. 318; A. Zahlbr., 1930 b, p. 200; Krog, 1976, p. 93; Wei X. L. et al. 2007, p. 154; Niu et al., 2011, p. 222, as chemical race III & V; Obermayer, 1995, p7, cited as *Lethariella sernanderi*), Xizang (Mot. 1936 –38, p. 44; Krog, 1976, p. 93; Wei & Jiang, 1981, p. 1146 & 1982, p. 496 & 1986, p. 67; Niu et al., 2011, p. 222, as chemical race III & V; Obermayer, 1995, p7, cited as *Lethariella sernanderi*).

Remark: This species is different from *Lethariella flexuosa* by reticulated ridges surface of non-curved fruticose thalli with soredia and containing placodiolic, gyrophoric, norstictic and connorstictic acids except psoromic acid.

Lethariella flexuosa(Nyl.) J. C. Wei 曲金丝

in J. C. Wei & Y. M. Jiang, *Acta phytotax. sin.* 20(4) : 497(1982) .

≡*Chlorea flexuosa* Nyl. , *Syn. meth. lich.* (Parisiis) 1(2) : 276(1860) .

Usnea hookeri Motyka, Lich. Gen. Usnea Monogr. 1: 44, 45(1936) .

= *Usnea flexuosa*(Nyl.) Du Rietz, (1926) ; Zahlbr. , Cat. Lich. Univ. 6: 599(non Tayl.)

On shrubs.

Remark: This species is different from *Lethariella cladonioides* by a smooth surface of curved fruticose thalli lacking soredia and containing atranorin and canarione except psoromic acid.

Gansu(Du Rietz, 1926, pp. 91, 92) , Sichuan, Yunnan(Zahlbr. , 1930 b, p. 200) , Xizang(Mot. 1936 – 38, p. 45; J. C. Wei & Y. M. Jiang, 1981, p. 1146 & 1982, p. 497 & 1986, p. 68) .

Lethariella sinensis J. C. Wei et Y. M. Jiang 中华金丝

Acta Phytotax. Sin. 20(4) : 498(1982) & Lichens of Xizang, 1986, p. 69.

Type: Xizang, Li W. H. no. 76 – 95-(1) in HMAS-L.

On twigs of Thuja sp.

Remark: This species is different from *Lethariella zahlbruckneri* by incisively reticulated ridges surface of the filamentous thalli containing psoromic and 2'-O-demethylpsoromic acids except atranorin, canarione and gyrophoric acid(trace) .

Lethariella zahlbruckneri(Du Rietz) Krog 带金丝

Norw. J. Bot. 23: 96(1976) .

≡*Usnea zahlbruckneri* Du Rietz, *Svensk Bot. Tidsk.* 20: 92(1926) .

Type: Sichuan, at about 4000 m, 2/X/1922, H. Smith no. 5007 in UPS(!) .

≡*Letharia zahlbrucneri*(Du Rietz) Zahlbr. , in Handel-Mazzetti, *Symb. Sin.* 3: 200(1930) .

= *Lethariella smithii*(Du Rietz) Obermayer, *Progr. Probl. Lichenol. Nineties. Proc. Third Symp. Intern. Assoc. Lichenol.* , *Biblthca Lichenol.* 68: 58(1997) .

≡*Usnea smithii* Du Rietz, *Svensk bot. Tidskr.* 20: 91, 92(1926) .

Type: Sichuan, at about 4000 m, 3/IX/1922, H. Smith no. 5017 in UPS(!) .

≡*Letharia smithii*(Du Rietz) Zahlbr. , in Handel-Mazzetti, *Symb. Sin.* 3: 200(1930) .

On *Juniperus* sp. & *Rhododendron* sp.

Shaanxi(Wu J. L. 1981, p. 163; J. C. Wei et al. 1982, p. 50; Wu & Zhang, 1982, p. 241) , Sichuan(Zahlbr. , 1930 a, pp. 601 & 604 & 1930 b, p. 200; Mot. 1936 – 38, p. 47 – 48; Krog, 1976, p. 96; Niu D. L. et al. , 2007, pp. 552 – 553, as chemo. type 2, 3) , Yunnan(Niu D. L. et al. , 2007, p. 553, as chemotype 3, and not mentioned chemotype) , Xizang(Niu D. L. et al. , 2007, p. 552, as chemo. type1) .

Remark: This species is different from *Lethariella sinensis* by smooth surface of the filamentous thalli containing norstictic acid except atranorin, canarione and gyrophoric acid(trace) .

Melanelia Essl. **褐衣属**

Mycotaxon 7(1) : 46(1978) .

Melanelia stygia(L.) Essl. 暗褐衣

Mycotaxon 7(1) : 47(1978) .

≡*Lichen stygius* L. , *Sp. pl.* 2: 1143(1753) .

Heilongjiang, Xinjiang(Chen JB, 2015, p. 101) .

Melanelia subargentifera(Nyl.) Essl. 银白褐衣

Mycotaxon 7(1) : 48(1978) .

≡*Parmelia subargentifera* Nyl. , *Flora*, Regensburg 58: 359(1875) .

Xinjiang(Abdulla Abbas & Wu, 1998, p. 97 as *M.* "*subargenifera*"; Abdulla Abbas et al. , 2001, p. 365) .

≡*Melanelixia subargentifera* (Nyl.) O. Blanco et al. in A. Crespo, Divakar, Essl. , D. Hawksw. & Lumbsch *Mycol. Res.* 108: 882(2004) .

On bark, and occasionally on rocks.

Xinjiang(Wang XY, 1985, p. 348; Abdulla Abbas & Wu, 1998, p. 97; Chen JB, 2015, p. 105; Nazarbek Guldan1 et al. , 2016. p. 30) .

Melanelia subaurifera(Nyl.) Essl. 假杯点褐衣

Mycotaxon 7(1) : 48(1978) .

≡*Parmelia subaurifera* Nyl. , *Flora*, Regensburg 56(2) : 22(1873) .

≡*Melanelixia subaurifera*(Nyl.) O. Blanco, A. Crespo, Divakar, Essl. , D. Hawksw. & Lumbsch, *Mycol. Res*. 108(8) : 882(2004) .

On bark of trees.

Xinjiang(Nazarbek Guldan1 et al. , 2016. pp. 30 – 31) , Xizang(Du et al. , 2010, fide Chen J. B. 2016, p. 264) , Sichuan(Du Yuan-Da et al. , 2010, pp. 284 – 285) .

Melanelixia O. Blanco et al. **伊氏叶属**

Mycol. Res. 108(8) : 881(2004) .

Melanelixia albertana(Ahti) O. Blanco et al. 唇粉芽伊氏叶

Mycol. Res. 108(8) : 881(2004) .

≡*Parmelia albertana* Ahti, *Bryologist* 72: 236(1969) .

≡*Melanelia albertana* (Ahti) Essl. , *Mycotaxon* 7: 47(1978) .

On bark of trees.

Xinjiang(Nazarbek et al. , 2016, p. 29) , Sichuan(Du Yuan-Da et al. , 2010, p. 284) .

Melanelixia fuliginosa(Fr. ex Duby) O. Blanco et al. , 光伊氏叶

Mycol. Res. 108(8) : 881(2004) subsp. *fuliginosa*

≡*Parmelia olivacea* var. *fuliginosa* Fr. ex Duby, *Bot. Gall.* , *Edn* 2(Paris) 2: 602(1830) .

≡*Melanelia fuliginosa*(Fr. ex Duby) Essl. , *Bryologist* 90: 163(1987) .

≡*Imbricaria fuliginosa* (Fr. ex Duby) Arnold, *Flora* 53: 210, (1870) .

On bark.

Neimenggu(Sun LY et al. , 2000, p. 36) , Xinjiang(Nazarbek et al. , 2016, pp. 29 – 30) , Shaanxi(Jatta, 1902, p. 471; Zahlbr. , 1930 b, p. 186) .

Melanelixia glabroides(Essl.) O. Blanco et al. 类茸伊氏叶

Mycol. Res. 108: 882(2004) .

≡*Parmelia glabroides* Essl. *J. Hattori Bot. Lab*. 42: 72(1977) .

≡*Melanelia glabroides*(Essl.) Essl. , *Mycotaxon* 7: 48(1978) .

On rocks.

Xinjiang(Omar Z. et al. , 2004, p. 386; Nazarbek Guldan1 et al. , 2016. p. 30) .

Melanelixia huei (Asahina) O. Blanco et al. 胡氏伊氏叶

Mycol. Res. 108: 882, 2004.

≡*Parmelia huei* Asahina, *J. Jap. Bot*. 26: 194(1951) .

≡*Melanelia huei* (Asahina) Essl. , *Mycotaxon* 7: 48(1978) .

On bark of trees.

Liaoning(Chen et al. 1981, p. 152) , Neimenggu(Chen et al. 1981, p. 152; Sun LY et al. , 2000, p. 36) , Shaanxi (He & Chen 1995, p. 43, no specimen and locality are cited;) , Hebei(Zhao, 1964, p. 146; Zhao et al. 1982, p. 17) , Shandong(Zhang F et al. , 1999, p. 31; Zhao ZT et al. , 2002, p. 6) , Anhui, Zhejiang(Xu B. S. , 1989, p. 212) , Hubei(Chen, Wu & J. C. Wei, 1989, p. 442) .

Melanelixia subvillosella Wang H. Y. & J. C. Wei 亚珊茸伊氏叶

Mycotaxon 104: 186 – 188(2008) .

Type: Jilin, Mt. Changbaishan, alt. 1100m, on trees, Jun. 24, 1985 Lu X. D. 211(holotype: HMAS-L. 77798) , Chen JB, 2015, p. 106.

Melanelixia villosella(Essl.) O. Blanco 珊茸伊氏叶

A. Crespo, Divakar, Essl. , D. Hawksw. & Lumbsch, *Mycol. Res*. 108(8) : 882(2004) .

≡*Parmelia villosella* Essl. *J. Hattori Bot. Lab*. 42: 95(1977) .

= *Melanelia subverruculifera*(J. C. Wei & Y. M. Jiang) J. C. Wei, *Enum. Lich. China*, 153(1991) .

≡*Parmelia subverrulifera J. C.* Wei & Y. M. Jiang, *Acta Phytotax. Sin*. 18(3) : 387(1980) .

Type: Xizang(J. C. Wei & Chen JB no. 1899 – 1(HMAS-L) .

On bark of *Betula* sp.

Shaanxi(He Q & Chen JB, 1995, p. 43 as *P. subverrulifera*; Chen J. B. et al. , 2005, p. 74, as *Melanelia villosella*) , Xizang(J. C. Wei & Jiang, 1981, p. 1146 & 1986, p. 41; Wang H. Y. , Chen J. B. & J. C. Wei 2008, p. 188) , Sichuan(Chen J. B. et al. , 2005, p. 74, as *Melanelia villosella;* Chen JB, 2015, p. 107) , Xinjiang(Chen JB, 2015, p. 107; Nazarbek Guldan1 et al. , 2016. p. 31) .

Melanohalea O. Blanco et al. **黑尔衣属**

A. Crespo, Divakar, Essl. , D. Hawksw. & Lumbsch, *Mycol. Res*. 108(8) : 882(2004) .

Melanohalea elegantula(Zahlbr.) O. Blanco et al. 长芽黑尔衣

Mycol. Res. 108: 882(2004) .

≡*Parmelia aspidota* var. *elegantula* Zahlbr. , *Verh. Ver. Nat. Heilk. Pressb*. 8: 39(1894) .

≡*Parmelia elegantula*(Zahlbr.) Szatala, *Magy. Bot. Lapok* 28: 77(1930) .

≡*Melanelia elegantula*(Zahlbr.) Essl. , *Mycotaxon* 7: 47(1978) .

= *Parmelia incolorata*(Parr.) Lett. , *Hedwigia* 61: 157(1919) , auct. non *Parmelia incolorata*(Hoffm.) Flörke, *Mag. Gesell. naturf. Freunde*, Berlin 3: 200(1809) .

≡*Parmelia fuliginosa* f. *incolorata* Parrique, *Act. Soc. linn. Bordeaux* 61: 146(1896) .

≡ *Melanelia incolorata*(Parr.) Essl. in Egan, *Bryologist* 90: 163, 1987; J. C. Wei, *Enum. Lich. China*: 153 (1991) .

On rocks & tree bark.

Neimenggu(Chen et al. , 1981, p. 152; Chen JB, 2015, p. 109) , Xinjiang(Abdulla Abbas & Wu, 1998, p. 96; Abdulla Abbas et al. , 2001, p. 365 as *Melanelia incolorata*; Chen JB, 2015, p. 109; Gulnaz et al. , 2015, p. 2334) Xizang(Chen JB, 2015, p. 109) .

Melanohalea exasperata(De Not.) O. Blanco et al. 乳突黑尔衣

Mycol. Res. 108(8) : 882(2004) .

≡*Parmelia exasperata* De Not. , *G. bot. ital*. 2(7 – 8) : 193(1847) .

≡*Melanelia exasperata*(De Not.) Essl. , *Mycotaxon* 7(1) : 47(1978) .

On trees.

Sichuan, Yunnan(Meng F. G. et al. , 2010, p. 3046) , Xinjiang(Chen J. B. et al. , 2005, p. 72, as *Melanelia exasperata* & 2015, p. 110; Gulnaz et al. , 2015, p. 2334) .

Melanohalea exasperatula(Nyl.) O. Blanco et al. 微糙黑尔衣

Mycol. Res. 108(8) : 882(2004) .

≡*Parmelia exasperatula* Nyl. , *Flora, Regensburg* 56(19) : 299(1873) .

≡*Melanelia exasperatula*(Nyl.) Essl. *Mycotaxon* 7: 47(1978) .

On stump of *Picea* sp.

Xinjiang(H. Magn. 1940, p. 128; Abdulla Abbas et al. , 1993, p. 78; Abdulla Abbas & Wu, 1998, p. 95; Abdulla Abbas et al. , 2001, p. 365; Chen JB, 2015, p. 111; Gulnaz et al. , 2015, p. 2334) , Qinghai(Meng F. G. et al. , 2010, pp. 3046 – 3047; Chen JB, 2015, p. 111) , Sichuan(Chen JB, 2015, p. 111) .

Melanohalea gomukhensis(Divakar, Upreti & Elix) O. Blanco 假裂芽黑尔衣

Mycol. Res. 108(8) : 882(2004) .

≡*Melanelia gomukhensis* Divakar, Upreti & Elix, *Mycotaxon* 80: 356(2001) .

On bark of trees.

Yunnan(Du Yuan-Da et al. , 2010, pp. 285 – 286) .

Melanohalea infumata(Nyl.) O. Blanco et al. 烟色黑尔衣

Mycol. Res. 108(8) : 882(2004) .

≡*Parmelia infumata* Nyl. , Flora, Regensburg 58: 359(1875) .

On rocks.

Xinjiang(Gulnaz et al. , 2015, p. 2332) .

Melanohalea lobulata F. G. Meng & H. Y. Wang 小裂片黑尔衣

in Zhao ZT et al. , *Mycotaxon* 108: 349 – 351(2009) .

Type: Sichuan, Litang, Kazilashan, alt. 4710m, Nov. 7, 2008, Wang H. Y. 20084049(SDNU: holotypus) .

Melanohalea olivacea(L.) O. Blanco et al. 橄榄黑尔衣

Mycol. Res. 108: 883(2004) .

≡*Lichen olivaceus* L. , *Sp. Pl.* : 1143(1753) .

≡*Melanelia olivacea*(L.) Essl. , *Mycotaxon* 7(1) : 48(1978) .

≡*Parmelia olivacea* (L.) Ach. , Method. Lich. : 213, 1803; Zahlbr. , in Handel-Mazzetti, *Symb. Sin.* 3: 186 (1930) .

On bark of trees.

Neimenggu(Chen et al. 1981, p. 152; Chen JB, 2015, p. 112) , Heilongjiang(Chen et al. 1981, p. 152; Luo, 1984, p. 85; Chen JB, 2015, p. 112) , Jilin(Zhao, 1964, p. 146; Zhao et al. 1982, p. 18; Chen JB, 2015, p. 112) , Xinjiang(Abdulla Abbas & Wu, 1998, p. 96; Abdulla Abbas et al. , 2001, p. 365; Gulnaz et al. , 2015, p. 2334) , Shaanxi(Jatta, 1902, p. 470; Zahlbr. , 1930 b, p. 186) , Hubei(Chen, Wu & J. C. Wei, 1989, p. 443; Chen JB, 2015, p. 112) .

Melanohalea olivaceoides(Krog) O. Blanco et al. 拟橄榄黑尔衣

Mycol. Res. 108(8) : 883(2004) .

≡*Parmelia olivaceoides* Krog, *Norsk Polarinst. Skr.* 144: 109(1968) .

≡*Melanelia olivaceoides*(Krog) Essl. , *Mycotaxon* 7(1) : 48(1978) .

On bark.

Xizang(Wang et al. , 2009, p. 169; Chen JB, 2015, p. 113) .

Melanohalea poeltii(Essl.) O. Blanco et al. 波氏黑尔衣

Mycol. Res. 108: 883(2004) .

≡*Melanelia poeltii* Essl. , *Mycotaxon* 28: 215(1987) .

On bark.

Sichuan(Chen & Esslinger, 2005, pp. 72 – 73; Chen JB, 2015, p. 114) .

Melanohalea septentrionalis(Lynge) O. Blanco et al. 北方黑尔衣

Mycol. Res. 108(8) : 883(2004) .

≡*Parmelia olivacea* var. septentrionalis Lynge, *Bergens Mus. Årbok* 1912(10) : 4(1912) .

≡*Parmelia septentrionalis*(Lynge) Ahti, *Acta Bot. Fenn.* 70: 22(1966) .

≡*Melanelia septentrionalis* (Lynge) Essl. , *Mycotaxon* 7(1) : 48(1978) .

On bark.

Heilongjiang(Wang et al. , 2009, p. 170) .

Melanohalea subelegantula(Essl.) O. Blanco et al. 亚长芽黑尔衣

Mycol. Res. 108(8) : 883(2004) .

≡*Parmelia subelegantula* Essl. , *J. Hattori bot. Lab.* 42: 89(1977) .

≡*Melanelia subelegantula*(Essl.) Essl. , *Mycotaxon* 7(1) : 48(1978) .

On bark.

Xizang(Wang et al. , 2009, pp. 168 – 169; Chen Jb, 2015, p. 115; Gulnaz et al. , 2015, p. 2332) .

Melanohalea subexasperata F. G. Meng & H. Y. Wang 亚乳突黑尔衣

in Sun, Meng, Li, Wang & Zhao, *Mycotaxon* 111: 66(2010).

Type: Yunnan, Shangri-la, Tianshengqiao, alt. 3500m, on twigs, 3 Nov. 2008, Wang H. Y. 20084032(holotype in SDNU).

Yunnan, Sichuan(Sun Li-Yan et al., 2010, pp. 66 – 67).

Melanohalea subolivacea(Nyl. ex Hasse) O. Blanco et al. 亚橄榄黑尔衣

Mycol. Res. 108(8): 883(2004).

≡*Parmelia subolivacea* Nyl. ex Hasse, in Hasse, *Bull. Torrey bot. Club* 24: 445(1897).

On bark and branchs.

Xinjiang(Gulnaz et al., 2015, p. 2333).

Menegazzia A. Massal. **孔叶衣属**

Neagenea Lich.: 3(1854).

Menegazzia anteforata Aptroot, M. J. Lai & Sparrius 鼓面孔叶衣

Bryologist 106(1): 158(2003).

Type: Taiwan, Miaoli co., Shei-Pa National park, 25km ENE of Tungshi, near Tahsuehshan, at km 48, 1.600m elev., 51RTG996866, on *Picea morrisonicola*, 8Oct. 2001 Aptroot 51931A(BM, holotype; ABL, isotype).

Taiwan(Moon et al., 2006, p. 128).

Menegazzia asahinae(Yasuda ex Asahina) R. Sant. 凸缘孔叶衣

in *Ark. F. Bot*. XXXA, 11: 13(1943).

≡*Parmelia asahinae* Yasuda ex Asahina, *Acta phytotax. geobot.*, *Kyoto* 14: 33(1950).

On bark of trees and on twigs.

Neimenggu, JILIN(Chen et al. 1981, p. 151), Anhui(Zhao, 1964, p. 140; Zhao et al. 1982, p. 7; Xu B. S., 1989, p. 210), Zhejiang(Xu B. S., 1989, p. 210), Fujian(Wu J. N. et al. 1982, p. 9), Taiwan(Zahlbr., 1933 c, p. 54; Lamb, 1963, p. 462; Yas. ex Asahina, 1950, p. 33 & 1952, p. 24; Asahina & M. Sato in Asahina, 1939, p. 719; Wang & Lai, 1976, p. 228; Bjerke, 2004, p. 16).

f. asahinae 原变型

Taiwan(Wang & Lai, 1973, p. 93).

f. subimpertusa(Nyl.) J. C. Wei 亚孔变型

Enum. Lich. China, 1991, p. 154.

≡*Parmelia pertusa* f. *subimpertusa* Nyl. *Lich. Japon*. 1890: 30.

≡*Parmelia asahinae* f. *subimpertusa*(Nyl.) Asahina, *Acta phytotax. geobot.*, Kyoto 14: 33(1950).

Taiwan(Asahina, 1952, p. 26; Wang & Lai, 1973, p. 93).

Menegazzia caviisidia Bjerke & P. James 穴孔叶衣

The Lichenologist 36(1): 15 – 25(2004).

On bark.

Yunnan, Taiwan(Moon et al., 2006, p. 130).

Menegazzia neotropica Bjerke 新热带孔叶衣

Lichenologist 34(6): 505(2002). subsp. neotropica 原亚种

subsp. roundicarpa Bjerke & Sipman 圆果亚种

Mycotaxon 91: 420(2005).

On bark.

Sichuan(J. W. Bjerke and W. Obermayer, 2005, p. 303).

Menegazzia primaria Aptroot, M. J. Lai, & Sparrius 裸孔叶衣

The Bryologist 106(1): 159 – 160(2003).

Type: Taiwan, Hualien co., Hohuan Shan, near Field Station, 3200m elev., 51RUG260725, On *Abies kawakmii*, Aptroot 52661(BM, holotype; ABL, isotype).

Yunnan(Bjerke et al. , 2005, p. 304) , Taiwan(Moon et al. , 2006, p. 133) .

Menegazzia pseudocyphellata Aptroot, M. J. Lai, & Sparrius 假杯点孔叶衣

The Bryologist 106(1) : 160(2003) .

Type: Taiwan, Hualien co. , Troko National park, Hohuan Shan, exposed mountain ridge, 3000m elev. , 51RUG2673, on *Pinus taiwanensis*, 12Oct. 2001, Aptroot 53687(BM, holotype; ABL, isotype) .

Menegazzia subsimilis(H. Magn.) R. Sant. 漏斗孔叶衣

Ark. Bot. 30A(no. 11) : 13(1943) .

≡*Parmelia subsimilis* H. Magn. , *Ark. Bot.* 30B(no. 3) : 5(1942) .

On soil.

Jilin(Kondratyuk et al. , 2013, p. 334, on bark of Abies sp. in Changbai shan) , Xizang, Sichuan(Bjerke et al. , 2005, p. 305) , Yunnan(Wei X. L. et al. 2007, p. 155) .

Menegazzia terebrata(Hoffm.) A. Massal. 孔叶衣

Neagenea Lich: 1(1854) . f. terebrata 原变型

≡*Lobaria terebrata* Hoffm. , *Deutschl. Fl. , Zweiter Theil(Erlangen)* : 151(1796) [1795] .

= *Menegazzia pertusa*(Schaer.) J. Steiner, in Cohn, *Krypt. -Fl. Schlesien*(Breslau) 2(2) : 78(1879) .

≡*Parmelia pertusa* Schaer. , *Lich. helv. spicil.* 10: 457(1840) .

≡*Lichen pertusus* Schrank, *Baier. Fl.* (München) 2: 519(1789) .

guizhou(Zhang T & Wei JC, 2006, p. 8) , Taiwan(Moon et al. , 2006, p. 136) .

On bark of trees.

Neimenggu(Chen et al. 1981, p. 151) , Jilin(Zhao, 1964, p. 140; Zhao et al. 1982, p. 7; Chen et al. 1981, p. 151) , Heilongjiang(Zhao, 1964, p. 140; Zhao et al. 1982, p. 7; J. C. Wei, 1981, p. 84; Chen et al. 1981, p. 151; Luo, 1984, p. 85) , Shaanxi(Zhao, 1964, p. 140; Zhao et al. 1982, p. 7 ; He & Chen 1995, p. 44, no specimen and locality are cited;) , Xizang(Bjerke et al. , 2005, p. 307) , Sichuan(Zahlbr. 1930 b, p. 194 & 1934 b, p. 211; Zhao, 1964, p. 140; Zhao et al. 1982, p. 7; Chen S. F. et al. , 1997, p. 222; Bjerke et al. , 2005, p. 307) , Yunnan(Hue, 1889, p. 166 & 1899, p. 128; Zahlbr. , 1930 b, p. 194 & 1934 b, p. 211; Zhao, 1964, p. 140; Zhao et al. 1982, p. 7) , Xizang(J. C. Wei & Jiang, 1981, p. 1147 & 1986, p. 38) , Hubei(Chen, Wu & J. C. Wei, 1989, p. 431) , Shandong(zhao ZT et al. , 1999, p. 427; Zhao ZT et al. , 2002, p. 5) , Anhui, Zhejiang(Zhao, 1964, p. 140; Zhao et al. 1982, p. 7; Xu B. S. , 1989, p. 210) , Hunan, Fujian(Zahlbr. , 1930 b, p. 194; Wu J. N. et al. 1982, p. 9) , Taiwan(Zahlbr. , 1933 c, p. 53; Asahina, 1952, p. 22; Wang & Lai, 1973, p. 94; Aptroot et al. , 2001, p. 161) , Hongkong(Thrower, 1988, pp. 16, 122, on granite boulders in mountain area; Aptroot & Seaward, 1999, p. 82) .

f. ventricosa(Hue) J. C. Wei 偏肿变型

An Enum. Lich. China, 154(1991) .

≡*Parmelia pertusa* f. *ventricosa* Hue, Nouv. Archiv. du Mus. Hist. Nat. IV, 4(1) : 129(1899) .

Zahlbr. , in Handel-Mazzetti, *Symb. Sin.* 3: 195(1930) .

Type locality: Yunnan.

On bark of trees.

Montanelia Divakar et al. **山褐衣属**

in Divakar, Del-Prado, Lumbsch, Wedin, Esslinger, Leavitt & Crespo, *Am. J. Bot.* 99(12) : 2014 – 2026(2012) .

Montanelia disjuncta(Erichsen) Divakar 假杯点山褐衣

in Divakar, Del-Prado, Lumbsch, Wedin, Esslinger, Leavitt & Crespo, *Am. J. Bot.* 99(12) : 2014 – 2026(2012) .

≡*Parmelia disjuncta* Erichsen, *Annls mycol.* 37(1/2) : 78(1939) .

≡*Melanelia disjuncta*(Erichsen) Essl. , Mycotaxon 7(1) : 46(1978) .

= *Melanelia granulosa* Essl. , in Egan, *Bryologist* 90(2) : 163(1987) .

On rocks.

Xinjiang(Abdulla Abbas et al., 1993, p. 79; Abdulla Abbas & Wu, 1998, p. 96; Abdulla Abbas et al., 2001, p. 365).

Montanelia panniformis(Nyl.) Divakar et al. 毡毛山褐衣

Am. J. Bot. 99(12):2023(2012).

≡*Parmelia olivacea* var. *panniformis* Nyl., *Herb. Mus. Fenn.*:83(1859).

≡*Melanelia panniformis*(Nyl.) Essl., *Mycotaxon* 7(1):46(1978).

Xinjiang(Abdulla Abbas et al., 1998, p. 97; Abdulla Abbas et al., 2001, p. 365), Taiwan(Lai, 2000, pp. 132–133).

Montanelia predisjuncta(Essl.) Divakar et al. 异暗山褐梅

Am. J. Bot. 99(12):2023(2012).

≡*Parmelia predisjuncta* Essl., *J. Hattori bot. Lab.* 42: 50(1977).

≡*Melanelia predisjuncta*(Essl.) Essl. *Mycotaxon* 7(1):47(1978).

Jilin(Wang et al., 2009, p. 168).

Montanelia sorediata(Ach.) Divakar et al. 粉芽山褐衣

in Divakar, Del-Prado, Lumbsch, Wedin, Esslinger, Leavitt & Crespo, *Am. J. Bot.* 99(12):2023(2012).

≡*Parmelia stygia* var. *sorediata* Ach., *Lich. univ.*:471(1810).

≡*Neofuscelia sorediosa*(Almb.) Essl. in J. C. Wei, *Enum. Lich. China*: 156(1991).

≡*Melanelia sorediosa*(Almb.) Essl., *Mycotaxon* 7(1):47(1978).

≡*Parmelia sorediosa* Almb., *Bot. Notiser, Suppl.* 1(2):134(1948).

On rocks.

Heilongjiang(Chen JB, 2015, p. 99), Xizang(J. C. Wei & Y. M. Jiang 1986, p. 41; Chen JB, 2015, p. 100).

Montanelia tominii(Oxner) Divakar et al. 茸刺山褐梅

in Divakar, Del-Prado, Lumbsch, Wedin, Esslinger, Leavitt & Crespo, *Am. J. Bot.* 99(12):2023(2012).

≡*Melanelia tominii*(Oxner) Essl., *Lichenologist* 24(1):17(1992).

≡*Parmelia tominii* Oxner, *Zhurnal Biobot. Tsiklu Bu Akad. Nauk(Journ. Cycle Bot. Acad. Sci. Ukraine)* 7/8: 171(1933).

On rocks.

Beijing, Nneimenggu, Sichuan(Chen J. B. et al., 2005, p. 73 & 2015, p. 101–102), Xinjiang(Omar Z. et al., 2004, p. 386), Xizang, Sichuan(Chen JB, 2015, p. 102), Hebei(Wu Di et al., 2011, p. 35; Chen JB, 2015, p. 101).

Myelochroa(Asahina) Elix & Hale **黄髓叶属**

Mycotaxon 29: 240(1987).

≡*Parmelia* subsect. *Myelochroa* Asahina, *Lichens of Japan*, 2. Genus *Parmelia*: 74(1952).

Myelochroa amagiensis(Asahina) Elix & Hale 圆腋黄髓叶

Mycotaxon 29: 240(1987).

≡*Parmelia amagiensis* Asahina, *J. Jap. Bot.* 26: 228(1951).

≡*Parmelina amagiensis*(Asahina) Hale, *Smithson. Contr. bot.* 33: 18(1976).

On rocks.

Anhui, Zhejiang(Zhao, 1964, p. 150; Zhao et al. 1982, p. 27; Xu B. S., 1989, p. 216), Hubei, Hunan, Fujian, Guangxi(Chen JB, 2015, p. 117).

Notes: The report of this species for Anhui & Zhejiang(Zhao et al. 1982) was proved to be based on an erroneous identification(Chen JB, 2015, p. 117).

Myelochroa aurulenta(Tuck.) Elix & Hale 金色黄髓叶

Mycotaxon 29: 240(1987) .

≡*Parmelia aurulenta* Tuck. , *Amer. J. Sci. Arts*, Ser. 2 25: 424(1858) .

≡*Parmelina aurulenta*(Tuck.) Hale(Thrower, 1988, pp. 16, 125; Chen S. F. et al. , 1997, p. 222) .

= *Parmelia hunanensis* Zahlbr. in Handel-Mazzetti, *Symb. sin*. 3: 187(1920) .

Type locality: Hunan, Handel-Mazzetti no. 11454(lectotype) in WU, not seen.

On bark of trees & on rocks in mountain area. Pantropical.

Jilin, Heilongjiang(Chen et al. 1981, p. 151; Chen JB, 2015, p. 118) , Shaanxi(He & Chen 1995, p. 44, as *Parmelina aurulenta*, no specimen and locality were cited) , Xizang(J. C. Wei & Chen, 1974, p. 179; J. C. Wei & Jiang, 1986, p. 43; Chen JB, 2015, p. 120) , Sichuan(Chen S. F. et al. , 1997, p. 222; Chen JB, 2015, p. 119) , Chongqing(Chen CL et al. , 2008, p. 40 as *Parmelia aurulenta*) , Guizhou(Zhao, 1964, p. 149; Zhao et al. 1982, p. 25) , Yunnan(Zhao, 1964, p. 149; Zhao et al. 1982, pp. 25, 35; Chen JB, 2015, p. 119) , Hebei (Chen JB, 2015, p. 118) , Hunan(Zahlbr. , 1920, p. 187; Chen JB, 2015, p. 119) , Hubei(Chen, Wu & J. C. Wei, 1989, p. 445; Chen JB, 2015, p. 119) , Jiangxi(Asahina, 1952a, p. 79; Zhao, 1964, p. 149; Zhao et al. 1982, p. 25; Chen JB, 2015, p. 119) , Anhui(Zhao, 1964, p. 149; Zhao et al. 1982, p. 25; Xu B. S. , 1989, p. 216) , Zhejiang(Zhao, 1964, p. 149; Zhao et al. 1982, p. 25; Xu B. S. , 1989, p. 216; Chen JB, 2015, p. 118) , Jiangsu, ShanghaiHANGHAI(Xu B. S. , 1989, p. 216) , Guangxi, Fujian(Zhao, 1964, p. 149; Zhao et al. 1982, p. 25; Chen JB, 2015, p. 118 – 119) , Taiwan(Zahlbr. , 1933 c, p. 56; Wang & Lai, 1973, p. 93) , Hainan(Chen JB, 2015, p. 119; JCWei et al. , 2013, p. 228) , Guangdong(Chen JB, 2015, p. 119) , Centr. China(prov. not indicated: Asahina, 1951, p. 227) , Taiwan(Zahlbr. , 1933c, p. 56; Wang & Lai, 1973, p. 93; Lai, 2000, p135) , Hongkong(Thrower, 1988, pp. 16, 125; Aptroot & Seaward, 1999, p. 82 – 83: Aptroot et al. , 2001, p. 333) ,

Myelochroa denegans(Nyl.) Elix & Hale 粉芽黄髓叶

Mycotaxon 29: 240(1987) .

≡*Parmelia denegans* Nyl. , *Acta Soc. Sci. fenn*. 26(no. 10) : 6(1900) .

Taiwan(Lai, 2000, pp. 135 – 136) .

Myelochroa entotheiochroa(Hue) Elix & Hale 皱褶黄髓叶

Mycotaxon 29: 240(1987) .

≡*Parmela entotheiochroa* Hue, *Nouv. Arch. Mus*. Paris ser. 4, 1: 161(1899) .

≡*Parmelina entotheiochroa*(Hue) Hale, *Phytologia* 28(5) : 482(1974) ; J. C. Wei, *Enum. Lich. China*: 173 (1991) .

On bark of trees.

Heilongjiang, Jilin Hebei Hunan(Chen JB, 2015, p. 121) , Liaoning(Chen et al. 1981, p. 152) , Xizang(J. C. Wei & Jiang, 1981, p. 1147 & 1986, p. 44) , Jiangsu(Xu B. S. , 1989, p. 217) , Anhui(Zhao, 1964, p. 149; Zhao et al. 1982, p. 26; Xu B. S. , 1989, p. 217; Chen JB, 2015, p. 121) , Hunan(Chen JB, 2015, p. 121) , Zhejiang (Zhao, 1964, p. 149; Zhao et al. 1982, p. 26; Xu B. S. , 1989, p. 217) , Jiangxi(Zhao, 1964, p. 149; Zhao et al. 1982, p. 26) , Taiwan(Zahlbr. , 1933 c, p. 56; Wang & Lai, 1973, p. 93) .

= *Myelochroa rhytidodes*(Hale) Elix & Hale, *Mycotaxon* 29: 241(1987) .

≡*Parmelina rhytidodes* Hale, *Smithson. Contr. bot*. 33: 43(1976) .

Shandong(Zhao ZT et al. , 1998, p. 29; zhao zt et al. , 1999, p. 427; Zhao ZT et al. , 2002, p. 5; Li Y et al. , 2008, p. 72) , Hubei(Chen, Wu & J. C. Wei, 1989, p. 446) .

Myelochroa galbina(Ach.) Elix & Hale 绿色黄髓叶

Mycotaxon 29: 240(1987) .

≡*Parmelia galbina* Ach. , *Syn. meth. lich*. (Lund) : 195(1814) .

On bark of trees.

Sichuan(Chen S. F. et al. , 1997, p. 222) .

≡*Parmelina galbina*(Ach.) Hale, *Phytologia* 28(5) : 482(1974) .

On bark of trees.

Liaoning(Chen et al. 1981, p. 152) .

Myelochroa hayachinensis(Kurok.) Elix & Hale 东亚黄髓叶

Mycotaxon 29: 240(1987) .

≡*Parmelia hayachinensis* Kurok. , *J. Jap. Bot*. 43: 350(1968) .

≡*Parmelina hayachinensis*(Kurok.) Hale, *Phytologia* 28: 482(1974) .

On tree bark.

Hunan, Jiangxi(Chen J. B. 2011, p. 885 & 2015, p. 122) .

Myelochroa leucotyliza(Nyl.) Elix & Hale 疱体黄髓叶

Mycotaxon 29: 241(1987) .

≡*Parmelia leucotyliza* Nyl. , *Lich. Jap*. : 27(1890) .

≡*Parmelina leucotyliza*(Nyl.) Hale, *Phytologia* 28: 482(1974) .

On bark of trees.

Jilin, Yunnan(Chen JB, 2015, p. 122) , Anhui(Zhao, 1964, p. 155; Zhao et al. 1982, p. 26, lapsu "*leucotylia*") .

Myelochroa metarevoluta(Asahina) Elix & Hale 反卷黄髓叶

Mycotaxon 29: 241(1987) .

≡*Parmelia metarevoluta* Asahina, *J. Jap. Bot*. 35: 97(1960) .

≡*Parmelina metarevoluta*(Asahina) Hale, *Phytologia* 28(5) : 483(1974) .

On bark.

Jilin, Sichuan, Yunnan, Hunan, Zhejiang, Fujian、(Chen JB, 2015. p. 123 – 124) , Shaanxi(He & Chen 1995, p. 44, no specimen and locality are cited;) , Guizhou(Zhang T & J. C. Wei, 2006, p. 8; Chen JB, 2015. p. 124) , Anhui(Zhao, 1964, p. 154; Zhao et al. 1982, p. 34; Xu B. S. , 1989, p. 218, on rocks) , Hubei(Chen, Wu & J. C. Wei, 1989, p. 445; Chen JB, 2015. P. 123) .

Myelochroa perisidians(Nyl.) Elix & Hale 裂芽黄髓叶

Mycotaxon 29: 241(1987) .

≡*Parmelia perisidians* Ny1. , *Acta Soc. Sci. Fenn*. 26: 6(1900) .

≡*Parmelina perisidians*(Nyl.) Hale, *Phytologia* 28: 483(1974) .

= *Parmelia subsulphurata* Asahina, *J. Jap. Bot*. 26: 228(1951) .

On trees and rocks.

Heilongjiang(Chen et al. 1981, p. 154) , Xizang(J. C. Wei & Y. M. Jiang, 1981, p. 1147 & 1986, p. 44; Chen JB, 2015. p. 124 – 125) , Chongqing, Yuannan, Fujian(Chen J. B. , 2015. p. 124 – 125) , Jiangsu, Shanghai, Anhui(Xu B. S. , 1989, p. 218) , Zhejiang(Zhao, 1964, p. 149; Zhao et al. 1982, p. 24; Xu B. S. 1989, p. 218; Chen JB, 2015. p. 124 – 125) .

Myelochroa salazinica S. L. Wang, J. B. Chen & Elix 水杨嗪黄髓叶

Mycotaxon 77: 26 – 27(2001) .

Type: Yunnan, Mt. Gongshan, on rocks, J. J. Su 3324(holotype in HMAS-L, isotype in CANB)

Myelochroa sinica S. L. Wang, J. B. Chen & Elix 中华黄髓叶

Mycotaxon 77: 28 – 29(2001) .

Type: Yunnan, Zhongdian, on rocks, X. Y. Wang, X. Xiao & J. J. Su 5642(holotype in HMAS-L, isotype in CANB) .

Myelochroa subaurulenta(Nyl.) Elix & Hale 亚黄髓叶

Mycotaxon 29: 241(1987) .

≡*Parmelia subauruleuta* Nyl. , *Flora* 68: 606(1885) . Zhao, *Acta Phytotax. Sin*. 9: 149(1964) ; Zhao, Xu &

Sun, *Prodr. Lich. Sin.* : 24(1982) ; J. C. Wei & Y. M. Jiang, *Lich. Xizang*: 45(1986).

≡*Parmelina subaurulenta*(Nyl.) Hale, *Phytologia* 28: 483(1974) ; J. C. Wei, *Enum. Lich. China*: 175(1991).

= *Parmelia conspicua* Hue, *Nouv. Arch. Mus. Hist. Nat.* 4(1) : 145(1899).

Type: from Yunnan, Delavay(lectotype in PC).

= *Parmelia homalotera* Hue, *Nouv. Arch. Mus.* Paris, 4(1) : 159(1899).

Type: Yunnan, Delavay(lectotype) in PC.

= *Parmelia fecunda* Hue. *Nonv. Arch. Mus.* Paris, 4(1) : 169(1899).

Type: Yunnan, Delavay no. 5 in PC.

= *Parmelia subcremea* Zahlbr. *Hedwigia*, 74: 208(1934).

Type: Sichuan, Rock no. 16720(pr. p., lectotype) in W(not seen).

On bark of trees & on rocks.

Heilongjiang, Jilin(J. C. Wei et al. 1982, p. 36), Yunnan(Hue, 1890, p. 19 & 1899, pp. 145, 159; Zahlbr., 1930 a, pp. 163, 166, 168 & 1930 b, pp. 183, 185, 187, 189 & 1934 b, pp. 206, 208 & 1940, p. 536; Zhao, 1964, p. 156; Zhao et al. 1982, p. 40; Hale, 1976, p. 47), Sichuan(Zahlbr., 1930 b, p. 185; Hale, 1976, p. 47), Xizang(J. C. Wei & Y. M. Jiang, 1986, p. 45), Hubei(Chen, Wu & J. C. Wei, 1989, p. 446), Hunan(Zahlbr., 1930 b, p. 183; Asahina, 1952a, p. 79), Fujian(Zahlbr., 1930 b, p. 183), Jiangxi(Asahina, 1952 a, p. 79), Anhui(Zhao, 1964, p. 149; Zhao et al. 1982, p. 25; Xu B. S., 1989, p. 219), Zhejiang(Zhao, 1964, p. 149; Zhao et al. 1982, pp. 24, 25; Xu B. S., 1989, p. 219), Taiwan(H. Sasaoka, 1919, p. 180; Asahina, 1951c, p. 226; Hale, 1976, p. 47).

[*Parmelia vicinior* Hue, Nouv. Arch. Mus. Hist. Nat. 4(1) : 156(1899).]

Type: from Yunnan, Delavay.

Yunnan(Zahlbr., 1930 a, p. 222 & 1930 b, p. 185).

On bark of trees.

= *Myelochroa irrugans* (Nyl.) Elix & Hale, *Mycotaxon* 29: 241(1987).

≡*Parmelia irrugans* Nyl., *Lich. Japon.* : 26(1890).

= *Parmelia crassata*(Hale) J. C. Wei et Y. M. Jiang, *Lichens of Xizang*, 43(1986).

guizhou(Zhang T & Wei JC, 2006, p. 8), Xizang(J. C. Wei & Jiang, 1981, p. 1147 & 1986, p. 43).

≡*Parmelina crassata* Hale, *Smiths. Contr. Bot.* 33: 22(1976) ; J. C. Wei, *Enum. Lich. China*: 173(1991).

≡*Myelochroa crassata*(Hale) Elix & Hale, *Mycotaxon* 29: 240(1987).

Taiwan(Asahina, 1951, p. 226; J. C. Wei & Y. M. Jiang, 1147(1981) & 43(1986).; Lai, 136 – 138(2000).

= *Parmelia homogenes* var. *vestita* Zahlbr. in Handel-Mazzetti, *Symb. Sin.* 3: 183(1930).

Type: from Yunnan.

On bark of trees & on rocks.

Jilin, Heilongjiang(Chen et al. 1981, p. 152), Yunnan(Zahlbr., 1930 b, p. 183 & 1932a, p. 560 & 1934b, p. 206), Shandong(Zhao ZT et al., 1998, p. 28; Zhang F et al., 1999, p. 30, as *Parmelina homogenes*; Zhao ZT et al., 2002, p. 5 as *M. subaurulenta*), Zhejiang, Anhui, Jiangxi(Zhao, 1964, p. 150; Zhao et al. 1982, p. 27), Hunan, Fujian(Zahlbr., 1930 b, p. 183).

= *Parmelia insinuata* Hue, *Nouv. Arch. Mus. Hist. Nat.* 3(1) : 158(1899), non Nyl. (1856, p. 324).

Type: Yunnan, Delavay no. 3008(lectotype) in PC.

Parmelia insinuatula Zahlbr., *Cat. Lich. Univ.* 6: 169(1930), based on *Parmelia insinuata* Hue, non Nyl.

On bark of trees.

Jilin(Chen et al. 1981, p. 154), Shaanxi, Zhejiang, Anhui(Zhao, 1964, p. 151; Zhao et al. 1982, p. 23), Yunnan (Hue, 1899, pp. 158, 178; Zahlbr., 1930 a, p. 169 & 1930 b, p. 185, 188; Hale, 1976, p. 34), Xizang(J. C. Wei

& Y. M. Jiang, 1981, p. 114 & 1986, p. 44), Hubei(Chen, Wu & J. C. Wei, 1989, p. 445), Hunan(Hale, 1976, p. 34), Anhui(Xu B. S., 1989, p. 217 – 218), China(prov. not indicated: Zahlbr., 222(1930) a,).

= *Parmelia homogenes* Nyl., *Flora* 68: 607(1885); *Cat. Lich.* VI: 168; Hue, *Nouv. Arch. Mus.* Paris, ser. 4, 1: 160, 1899; (= *P. subaurulenta* Nyl., *P. sulphurata* Hue II: 163) Zahlbr., in Handel-Mazzetti, *Symb. Sin.* 3: 183(1930); Zhao, *Acta Phytotax. Sin.* 9: 150, 1964; Chen, Zhao & Luo, *J. Northeastern Forestry Inst.* 4: 1542, 1981; Zhao, Xu & Sun, Prodr. Lich. Sin.: 27(1982).

≡ *Parmelina homogenes*(Nyl.) Hale, *Phytologia* 28: 482(1974); J. C. Wei, *Enum. Lich. China*: 173(1991).

[= *Parmelia homogenes* var. *vestita* Zahlbr., in Handel-Mazzetti, *Symb. Sin.* 3: 183(1930).

Type locality: Chna. Yunnan]

= *Parmelia subaurulenta* var. *myriocarpa* Asahina, *J. Jap. Bot*. 26: 226(1951).

≡ *Parmelia myriocarpa*(Asahina) Zhao, *Acta Phytotax. Sin.* 9: 149(1964); Zhao, Xu & Sun, *Prodr. Lich. Sin.*: 25(1982).

= *Parmelia subcremea* Zahlbr., *Hedwigia* 74: 208(1934). Type: China. Sichuan, Rock 16720(lectotype in W).

= *Parmelia irrugans* Nyl., *Lich. Jap.*: 26(1890); J. C. Wei & Y. M. Jiang, *Lich. Xizang*: 44(1986).

≡ *Parmelina irrugans*(Nyl.) Hale, *Smiths. Contr. Bot.* 33: 34(1976); J. C. Wei, *Enum. Lich. China*: 174 (1991).

≡ *Myelochroa irrugans*(Nyl.) Elix & Hale, *Mycotaxon* 29: 241(1987); Kurokawa & Lai, *Mycotaxon* 77: 250 (2001).

= *Parmelia crassata*(Hale) J. C. Wei, in J. C. Wei & Y. M. Jiang, *Lich. Xizang*: 43(1986).

≡ *Parmelina crassata* Hale, *Smiths. Contr. Bot.* 33: 22(1976); J. C. Wei, *Enum. Lich. China*: 173(1991).

= *Parmelia insinuata* Hue, *Nouv. Arch. Mus.* Paris, ser 4. 1: 158, 1899(nom. illeg.) non Nylander(1886). Type: China. Yunnan, Delavay 3008(lectotype in PC).

= *Parmelia insinuatula* Zahlbr., *Cat. Lich. Univ.* 6: 169(1930) (nom. illeg.) (fide Parmelia insinuata Hue, non Nylander).

Parmelia sulphurata auct. non Nees & Flot.: Hue, *Bull. Soc. Bot. France* 36: 163(1889) (fide Zahlbr. 1930b: 183).

On bark, rocks.

Heilongjiang, Jilin, Liaoning, Xizang, Sichuan, Guizhou, Yunnan, Anhui, Hubei, Hunan, Jiangxi, Zhejiang, Fujian, Guangxi(Chen JB, 2015, p. 127 – 129).

Myelochroa xantholepis(Mont. & Bosch) Elix & Hale 细裂黄髓叶

Mycotaxon 29: 241(1987).

≡ *Parmelia xantholepis* Mont. & Bosch, *Pl. Jungh.* 4: 446(1856).

≡ *Parmelina xantholepis*(Mont. & Bosch) Hale, *Phytologia* 28: 438(1974).

On tree bark & moss covered rocks.

Yunnan(Wang S. L. et al., 2001, pp. 135 – 136; Chen JB, 2015, p. 130), Zhejiang, Fujian(Wang S. L. et al., 2001, pp. 135 – 136), Hunan, Jiangxi(Chen JB, 2015, p. 130), Taiwan(Lai, 2000, pp. 138 – 139).

= *Parmelia crenulata* J. D. Zhao, *Acta Phytotax. Sin.* 9: 150(1964).

Type: China. Yunnan, 24 April 1957, Poliyangskii 1001(holotype in HMAS-L).

≡ *Parmelina crenulata*(J. D. Zhao) J. C. Wei, *Enum. Lich. China*: 173(1991).

≡ *Myelochroa crenulata*(J. D. Zhao) Hale ex DePriest & B. Hale, *Mycotaxon* 67: 202(1998).

On bark of trees.

Yunnan(Zhao, 1964, p. 150; Zhao et al. 1982, p. 26; Hawksw. 1972 b, p. 47; Chen JB, 2015, p. 130), Shandong (Li Y et al., 2008, p. 71 as *Parmelia crenulata*; Chen JB, 2015, p. 130), Hunan, Jiangxi(Chen JB, 2015, p.

130).

Nephromopsis Müll. Arg. **肾岛衣属**

Flora, Regensburg 74(3):374(1891).

Nephromopsis ahtii(Randlane & Saag) Randlane & Saag 艾氏肾岛衣

Mycological Progress 4(4):311(2005).

≡*Tuckneraria ahtii* Randlane & Saag, *Acta Bot. Fennica*, 159: 143 – 151(1994).

Type: Yunnan, Lijiang, Alt. 3200m, 1987. 4. 23, T. Ahtii, 陈健斌和王立松 46469 (H-holotype; TU-isotype; HMAS-L 024189, isotype).

Nephromopsis asahinae(M. Satô) Räsänen 东亚肾岛衣

Kuopion Luonnon Ystâvâin Yhdistyksen Julkaisuja, ser. B 2(6): 50(1952).

≡*Cetraria asahinae* M. Satô, in Saito, *Ho-on Kai Mus. Res. Bul.* 11: 12(1936).

On dead twigs in a broadleaf forest.

Xizang(J. C. Wei & Jiang, 1981, p. 1147 & 1986, p. 58), Taiwan(M. Sato, 1936 c, p. 23 & 1938b, p. 786 & 1939, p. 42 & 1959, p. 46; Asahina & Sato in Asahina, 1939, p. 739; Wang & Lai, 1973, p. 88).

Nephromopsis endocrocea Asahina 黄髓肾岛衣

J. Jap. Bot. 11(1):24(1935).

≡*Cetraria endocrocea*(Asahina) M. Satô, in Nakai & Honda, *Nov. fl. jap.* 5: 37(1939).

On bark of Abise sp.

Jilin(Chen et al. 1981, p. 129).

Notes: This species was removed by Lai M. J. from the list of lichens in China without any explaination(Lai M. J., 2009).

Nephromopsis globulans(Nyl.) Lai 球肾岛衣

Quart. Journ. Taiwan Museum, 33(3, 4):222(1980).

≡*Platysma globulans* Nyl. in Hue, *Bull. Soc. Bot. France*, 34: 19(1887).

Notes: The basionym "*Platysma globulans*" given by Nylander has been validly published in 1887, as above cited, but not in 1890: Hue, *Nouv. Archiv. Mus. Hist. Nat.* (Paris), ser. 3, 2: 275, as miscited by Zahlbr. in his *Cat. Lich. Univ.* 6: 301(1930), and by Lai M. J. in *Quart. Journ. Taiwan Museum*, 33(3, 4):222(1980).

Type: Yunnan, Delavay nos. 1570, 1571(syntypes).

≡*Cetraria globulans*(Nyl.) Zahlbr., Trav. de la Sous-Sect. de Troiskossawsk-Khiakta, Sect. du Pays d'Amoure de la Soc. Imp. Russe de Geogr. 12: 89(1909, 1910) (not seen) & *Cat. Lich. Univ.* 6: 301(1930).

On bark of trees.

Yunnan(Hue, 1887a, p. 19 & 1890, p. 275 & 1899, p. 213; Zahlbr., < 1909 >, 1910, p. 89 & 1930 a, p. 301 & 1930 b, p. 198; Lai, 1980b, p. 222).

Nephromopsis hengduanensis L. H. Chen 横断山肾岛衣

Mycotaxon 77: 493(2001).

Type: Yunnan, 13/VII/1982. Wang X. Y. et al. 4507(HMAS-L, holotype).

Nephromopsis komarovii(Elenk.) J. C. Wei 柯氏肾岛衣

Enum. Lich. China, 158(1991).

≡*Cetraria komarovii* Elenk., *Bull. Jard. Imp. Botan. St. -Petersbourg*, III: 51(1903).

Misapplied name:

Nephroma arcticum auct. non(L.) Torss.: Chen, Zhao & Luo, Journ. North-eastern Forestry Inst. 4: 151(1981).

Jilin(Chen, Zhao & Luo, 1981, p. 151), neimenggu(Sun LY et al., 2000, p. 36), Sichuan(Obermayer, 2004, p. 503), Hebei(Rassad. 1950, p. 229).

Notes: A specimen collected by Liou T. N. (no. 4118) from Changbai shan on rocks cited by Chen et al. in

Journ. North-eastern Forestry Inst. 4: 151(1981) under the name of *Nephroma arcticum*, auct. non(L.) Torss. As a matter of fact this lichen is similar to *Nephromopsis komarovii*, but different from that by chemistry and that the lichen from Changbai shan is tightly attached mainly to rocks and sometimes to bark of trees. So, this lichen seems to be an undescribed variety under *Nephromopsis komarovii*. Many collections of this lichen were collected by me from Changbai shan several years ago.

Nephromopsis laii(A. Thell & Randlane) Saag & A. Thell 赖氏肾岛衣

in Randlane, Saag & Thell, *Bryologist* 100(1): 111(1997).

≡ *Cetrariopsis laii* A. Thell & Randlane, in Randlane, Thell & Saag, *Cryptog. Bryol. -Lichénol.* 16(1): 46 (1995).

Jilin, Liaoning, Heilongjiang(Lai et al., 2009, p. 375), Taiwan(Lai, 2000, pp. 209 – 210).

Nephromopsis morisonicola Lai 台湾肾岛衣

Quart. Journ. Taiwan Museum, 33(3, 4): 223(1980).

Type: Taiwan, Lai no. 10438(holotype) in TAIN and(isotype) in US, DUKE, no. 10459.

On *Abies fabri* and *Salix* sp.

Sichuan(Obermayer, 2004, p. 503), Taiwan(Lai, 1980, p. 223, & 2000, p. 210).

Nephromopsis nephromoides(Nyl.) Ahti & Randlane 类肾岛衣

in Randlane & Saag, *Cryptog. Bryol. -Lichénol.* 19(2 – 3): 183(1998).

≡ *Platysma nephromoides* Nyl., *Flora*, Regensburg 52: 442(1869).

Taiwan(Lai, 2000, pp. 210 – 211).

Nephromopsis ornata(Müll. Arg.) Hue 丽肾岛衣

Nouv. Arch. Mus. Hist. Nat., Paris, 4 sér. 4(2): 90(1900).

≡ *Cetraria ornata* Müll. Arg., *Nuovo G. bot. ital.* 23: 122(1891).

= *Nephromopsis delavayi* Hue, *Nouv. Archiv. Mus. Hist. Nat.* 4(1): 219(1899).

Type locality: Yunnan.

≡ *Cetraria delavayi*(Hue) M. Sato, Parmeliales(I), in Nakai et Honda, *Nova Flora Japonica*45(1939).

On bark of trees.

Jilin(Chen et al. 1981, p. 129; Lai et al., 2009, pp. 375 – 376), Yunnan(Hue, 1899, p. 219; Zahlbr., 1930 b, p. 199 & 1934, p. 211; Lai, 1980, p. 223), Sichuan(Zahlbr., 1930 b, p. 199 & 1934 b, p. 211), Taiwan(M. Sato, 1939, p. 48; Wang-Yang & Lai, 1973, p. 89; Lai, 1980b, p. 223; Lai, 2000, p. 211).

According to Lai M. J. (1980, p. 224) the reports of *N. delavayi* made by Zahlbr. (1933 c) and M. Sato(1938, 1939) in Taiwan may probably be based on erroneous identification.

Nephromopsis pallescens(Schaer.) Y. S. Park 皮革肾岛衣

Bryologist 93(2): 122(1990). var. pallescens 原变种

≡ *Cetraria pallescens* Schaer., in Moritzi, *Syst. Verz.*: 129(1846).

Yunnan(Wei X. L. et al. 2007, p. 155; Wang L. S. et al. 2008, p. 537), Taiwan(Lai, 2000, pp. 211 – 212).

var. citrine(Taylor) Thell & Randlane 柠檬变种

Bryologist 100: 110(1997).

Taiwan(Lai, 2000, p. 212).

Nephromopsis weii X. Q. Gao & L. H. Chen 魏氏肾岛衣

in Chen & Gao, *Mycotaxon* 77: 492(2001).

Type: Fujian, Mt. Wuyishan, 24/IV/1988. X. Q. Gao 7619(HMAS-L, holotypus, & UPS, isotypus).

On bark of *Pinus* sp.

Nephromopsis yunnanensis(Nyl.) Randlane & Saag 云南肾岛衣

Mycotaxon 44(2): 488(1992).

≡*Platysma yunnanense* Nyl. Lich. Nov. Zeland. 1888: 150. Hue, *Bull. Soc. Bot. France* 36: 162(1889).

Syntype localities: Yunnan, on bark of *Quercus* sp. near Dapingzi(Tapin-tze), at about 1800m, 11/V/1885, Delavay no. 1602 & Song-pin over Dapingzi, 6/VI/1887.

≡ *Cetraria yunnanensis* (Nyl.) Zahlbr., Trav, de ka Sous-Sect. de Troitzkossawsk-Khiakta Sect. du Pays d'Amour. de la Soc. Imp. Russe de Geogr. 12: 89(1909, 1911). J. C. Wei 1991 pp. 57 – 58.

On bark of trees.

Yunnan(Hue, 1889, p. 162, & 1890, p. 275, & 1899, p. 212, t. 2: tab. 1, fig. 5, 5 bis & 5 ter.; Zahlbr., 1930 a, p. 319, & 1930 b, p. 198), Xizang(Obermayer, 2004, pp. 503 – 504).

* **Nesolechia** A. Massal. **岛菌属**

Miscell. Lichenol.: 43(1856).

* Nesolechia oxyspora(Tul.) A. Massal. 尖孢岛菌

Miscell. Lichenol.: 43(1856).

≡*Abrothallus oxysporus* Tul., *Annls Sci. Nat., Bot.*, sér. 3 17: 116(1852).

≡*Lecidea oxyspora*(Tul.) Nyl. 1855, *Phacopsis oxyspora* (Tul.) Triebel & Rambold, 1988.

On *Parmelia saxitilis*.

Yunnan (Hawksw. & M. Cole, 2003, p. 361).

Nipponoparmelia (Kurok.) K. H. Moon, Y. Ohmura & Kashiw. **缘点衣属**

in Crespo et al., *Taxon* 59(6): 1749(2010).

Nipponoparmelia isidioclada (Vain.) K. H. Moon et al. ex A. Crespo *et al.*, 枝芽缘点衣

Taxon 59: 1749(2010).

≡*Parmelia isidioclada* Vain., *Bot. Mag. Tokyo* 35: 48, 1921; Hale, Smiths. *Contr. Bot.* 66: 27(1987).

= *Parmelia yasudae* Räs., *J. Jap. Bot.* 16: 84(1940).

On bark of trees & on rocks.

Heilongjiang(Luo, 1984, p. 85), Neimenggu(Chen et al. 1981, p. 154), Yunnan(Chen J. B., 2013, p. 53 & 2015, p. 132), Zhejiang(Zhao, 1964, p. 155; Zhao et al. 1982, p. 37; Chen J. B., 2013, p. 53 & 2015, p. 132), Taiwan(Lai, 2000, 143).

Nipponoparmelia laevior (Nyl.) K. H. Moon et al. ex A. Crespo *et al.* 平缘点衣

Taxon 59: 1749(2010).

≡*Parmelia laevior* Nyl., *Lich. Jap.*: 28, 1890; Hale, Smiths. *Contr. Bot.* 66: 28, 1987.

On bark of trees.

Heilongjiang(Chen et al. 1981, p. 152), Jilin(Chen J. B., 2013, pp. 53 – 54 & 2015, p. 133), Zhejiang(Zhao, 1964, p. 154; Zhao et al. 1982, p. 34; Hale, 187, p. 28, mentioned only; Chen J. B., 2013, pp. 53 – 54 & 2015, p. 133), hunnan, Fujian, guangxi(Chen J. B., 2013, pp. 53 – 54 & 2015, p. 133), Taiwan(Zahlbr., 1933 c, p. 56; Wang & Lai, 1973, p. 93; Hale, 1987, p. 28; Lai, 2000, pp. 142 – 143).

Nipponoparmelia pseudolaevior (Asahina) K. H. Moon et al. ex A. Crespo *et al.* 拟平缘点衣

Taxon 59: 1749(2010).

≡*Parmelia pseudolaevior* Asahina, *J. Jap. Bot.* 26: 331(1951).

On rock.

Anhui(Chen J. B. 2011, p. 886, as *P. pseudolaevior*; Chen J. B., 2013, p. 54 & 2015, p. 134).

Nipponoparmelia ricasolioides (Nyl.) K. H. Moon et al. ex A. Crespo *et al*, 拟实缘点衣

Taxon 59: 1749(2010).

≡*Parmelia ricasolioides* Nyl., *Flora* 52: 135, 1887; Hue, *Bull. Soc. Bot. France* 34: 20, 1887; Hale, Smiths. Conntr. Bot. 66: 38, 1987;

Type: China. Yunnan, Delavay 1594(Nylander Herb. no. 35283, lectotype in H; isolectotypes in PC, TUR).

= *Parmelia daliensis* Zahlbr., in Handel-Mazzetti, *Symb. Sin.* 3: 183(1930).

Type: China. Yunnan, Handel-Mazzetti 6582(lectotype in W; isolectotypes in TNS, US) .

= *Parmelia daliensis* f. *tardiva* Zahlbr. , in Hadel-Mazzetti, *Symb. Sin*. 3: 184, 1930; Zahlbr. , *Hedwigia* 74: 208(1934) .

Type: China. Sichuan(Setschwan) , Muli, 3, 700m, 31 July 1915, Handel-Mazzetti 7368 (lectotype in W; isolectotype in HMAS-L) .

Sichuan, yunnan(Chen J. B. , 2013, pp. 54 – 55 & 2015, p. 135) .

Notoparmelia A. Crespo, Ferencova & Divakar **南梅属**

Lichenologist 46(1) : 59(2014) .

Notoparmelia erumpens(Kurok.) A. Crespo, Ferencova & Divakar 破裂南梅

Lichenologist 46(1) : 63(2014) .

≡ *Parmelia erumpens* Kurok. , *Lichenes Rariores et Critici Exsiccati* 2: 74(1969) .

= *Parmelia tenuirima* f. *corallina* Müll. Arg. , *Flora*, Regensburg 46: 46(1863) .

On trunk of deciduous tree.

Taiwan(Kurok. 1969, p. 225; Wang & Lai, 1973, p. 93; Hale, 1987, p. 24; Lai, 2000, p. 141) .

Notoparmelia signifera(Nyl.) A. Crespo, Ferencovά & Divakar 印纹南梅

Lichenologist 46(1) : 63(2014) .

≡ *Parmelia signifera* Nyl. , *Lich. Nov. Zeland*. (Paris) : 25(1888) .

≡ *Parmelia saxatilis* var. *signifera*(Nyl.) Müll. Arg. , *Bull. Soc. R. Bot. Belg*. 31: 30(1892) .

≡ *Imbricaria saxatilis* var. *signifera*(Nyl.) Müll. Arg.

Among mosses.

Shaanxi(Jatta, 1902, p. 469; Zahlbr. , 1930 b, p. 190) .

Oropogon Th. Fr. **砖孢发属**

Gen. Heterolich. Eur. : 49(1861) .

Oropogon asiaticus Asahina 亚洲砖孢发

in Sato, *Journ. Jap. Bot*. 13(8) : 596(1937) .

= *Oropogon loxensis* f. *endoxanthus* Zahlbr. in Hadel-Mazzetti, *Symb. Sin*. 3: 203(1930) .

Type: Sichuan, Handel-Mazzetti no. 3038.

On tree trunk.

Xizang, Sichuan(Chen JB, 1996, p. 174) , Yunnan(Wu JN, Wang LS, p. 40; Chen JB, 1996, p. 174) , Hubei (Chen, Wu & J. C. Wei, 1989, p. 452; Chen JB, 1996, p. 174) , Taiwan(Asahina & Sato in Asahina, 1939, p. 751; M. Sato, 1957, p. 63) .

Oropogon formosanus Asahina 台湾砖孢发

Journ. Jap. Bot. 27: 240(1952) .

Type: Taiwan, M. Sato, 22/I/1936, in TNS(not seen) .

Xizang(J. C. Wei & Jiang, 1981, p. 1146 & 1986, p. 65; Chen JB, 1996, p. 175) , Sichuan(Chen JB, 1996, p. 175) , Yunnan(Wu JN, Wang LS, p. 40; Chen JB, 1996, p. 175) , Taiwan(M. Sato, 1957, p. 65; Lamb, 1963, p. 451; Wang & Lai, 1973, p. 92; Ikoma, 1983, p. 75) .

Oropogon loxensis(Fée) Zukal 砖孢发

Sber. Akad. Wiss. Wien, *Math. -naturw. Kl. , Abt*. 1 104: 573(1895) .

≡ *Cornicularia loxensis* Fée, *Essai Crypt. Exot*. (Paris) : 134(1825) [1824] .

On bark of trees.

Sichuan(Du Rietz, 1926, p. 29; Zahlbr. , 1930 b, p. 202 & 1934, p. 212) , Yunnan(Zahlbr. , 1930 b, p. 202) , Zhejiang(Moreau et Moreau, 1951, p. 191) , Taiwan(Asahina, 1936, p. 698 & 1952, p. 241; M. Sato, 1937, p. 596) .

Oropogon orientalis(Gyeln.) Essl. 东方砖孢发

Systematic Botany Monographs 28: 109(1989).

Sichuan, Yunnan(Chen J. B., 1996, p. 175).

≡*Bryopogon orientalis* Gyeln., *Feddes Repert*. 38: 231, 235, 254(1935). M. Lamb, *Index Nom. Lich*.: 94 (1963).

Type: Yunnan, collected by J. F. Rock, see Zahlbr. & Reding, Lich. rar. exs. no. 321 in Herb. Mus. Budapest.

≡*Bryoria orientalis*(Gyeln.) J. C. Wei, *Enumeration of Lichens in China*(Beijing): 41(1991).

On *Abies* sp. & on rocks.

Xizang, Sichuan(Chen JB, 1996, p. 175), Yunnan(Wu JN, Wang LS, 1992, p. 41; Wei X. L. et al. 2007, p. 156; Chen JB, 1996, p. 175).

Oropogon salazinicus Essl. 水杨嗪砖孢发

Systematic Botany Monographs 28: 109(1989).

Taiwan.

Oropogon satoanus Essl. 宝岛砖孢发

Systematic Botany Monographs 28: 109(1989).

Taiwan.

On rock.

Oropogon secalonicus Essl. 黑麦酮砖孢发

Systematic Botany Monographs 28: 109(1989).

Type: Yunnan, May 12, 1881, Delaway(DUKE, holotype)

On tree of *Pinus yunnanensis*, *Quercus* sp., and on rocks.

Xizang, Sichuan(Chen JB, 1996, p. 176), Yunnan(Wu JN, Wang LS, p. 41; Chen JB, 1996, pp. 175 – 176).

Oropogon yunnanensis Essl. 云南砖孢发

Systematic Botany Monographs 28: 109(1989).

Sichuan, Yunnan(Chen JB, 1996, p. 176).

Pannoparmelia(Müll. Arg.) Darb. **海绵梅属**

Wiss. Ergebn. Schwed. Südpol. -Exp., 1901 – 1903 4(11): 11(1912).

≡*Anzia* sect. *Pannoparmelia* Müll. Arg., *Flora*, Regensburg 72: 507(1889).

Pannoparmelia angustata(Pers.) Zahlbr. 海绵梅

in Handel-Mazzetii, *Symb. Sinic*. 3: 193(1930).

≡*Parmelia angustata* Pers., in Gaudichaud-Beaupré in Freycinet, Voy. Uranie., *Bot*.: 195(1827)[1826 – 1830].

On bark of *Juniperus formosana*.

Sichuan(Hue, 1899, p. 131; Zahlbr., 1930 b, p. 195).

Parmelia Ach. **梅衣属**

Methodus, Sectio prior: xxxiii, 153(1803), *sensu strict*; Hale, *Smiths. Contr. Bot*. 66: 1(1987); Elix, *Bryologist* 96: 376(1993).

Parmelia adaugescens Nyl. 成长梅衣

Lich. Japon.: 28(1890).

On bark of trees.

Heilongjiang(Chen et al. 1981, p. 151; Luo, 1984, p. 85; Hale, 1987, p. 19, cited as in prov. Heiho, Takahashi 2928 in TNS; Chen JB, 2015, p. 137), Neimenggu(Chen et al. 1981, p. 151; Sun LY et al., 2000, p. 36), Sichuan(Chen JB, 2015, p. 137), Chongqing(Chen CL et al., 2008, p. 40), guizhou(Zhang T & J. C. Wei, 2006, p. 9), Yunnan(Chen JB, 2015, p. 137), Shandong(Li Y et al., 2008, p. 71), Anhui(Xu B. S., 1989, p. 214), Taiwan(Asahina, 1951g, p. 355 & 1952a, p. 107; Wang & Lai, 1976, p. 228, Hale, 1987, p. 19; Lai, 2000, pp. 140 – 141).

Parmelia centriasiatica Gyeln. 中亚梅衣

in Fedde, *Repertorium*, 36:300(1934).

Type: Sichuan, Kangding, collected by Potanin.

Sichuan(Lamb, 1963, p.465).

On bark of *Rhododendron* sp. & on mosses.

Parmelia cochleata Zahlbr. 螺壳梅衣

Bot. Mag., Tokyo 41:350, tab. XII, fig. 3(1927).

= *Parmelia pseudosaxatilis* Asahina *J. J. B.* 26:354(1951).

Type collection is based on *Parmelia marmariza* var. *physcioides* Zahlbr., see Hale, 1987, p. 19.

Parmelia marmariza var. *physcioides* Zahlbr., *Bot. Mag.*, Tokyo 41:352(1927).

On bark of trees & on rocks.

Jilin(Chen et al. 1981, pp. 152, 153), Taiwan(Wang & Lai, 1973, p. 94).

Parmelia fertilis Müll. Arg. 亚广开梅衣

Flora, Regensburg 70:316(1887).

= *Parmelia cochleata* Zahlbr., *Bot. Mag. Tokyo* 41:350(1927); Chen, Zhao & Luo, *J. NE Forestry Inst.* 4:152(1981).

= *Parmelia marmariza* var. *physcioids* Zahlbr., *Bot. Mag. Tokyo* 41:352(1927).

= *Parmelia pseudosaxatilis* Asahina, *J. Jap. Bot.* 26:354(1951). nom. illeg. [Type collection: based on *Parmelia marmariza* var. *physcioides* Zahlbr. (fide Hale 1987, p. 19)]; Chen, Zhao & Luo, *J. Northeastern Forestry Inst.* 4:153(1981).

On tree bark.

Heilongjiang, Jilin, Guizhou(Cehn JB, 2015, p. 138), Xinjiang(Abdulla Abbas & Wu, 1998, p. 98; Abdulla Abbas et al., 2001, p. 365), Shaanxi(He & Chen 1995, p. 44, miscited as *P. fertelis*, no specimen and locality are cited;), Chongqing(Chen CL et al., 2008, p. 40), Hubei(Chen, Wu & J. C. Wei, 1989, p. 444), Anhui, Zhejiang(Zhao, 1964, p. 153; Zhao et al., 1982, pp. 33 – 34; Hale, 1987, pp. 24 – 25; Xu B. S., 1989, p. 214), Jiangxi(Zhao, 1964, p. 153; Zhao et al. 1982, pp. 33 – 34; Hale, 1987, pp. 24 – 25), Taiwan(Zahlbr., 1933c, p. 55; Asahina, 1952a, p. 110 & 1953, p. 65; Wang & Lai, 1973, p. 94 & 1976, p. 228; Hale, 1987, pp. 24 – 25).

= *Parmelia subdivaricata* Asahina, *J. J. B.* 26:356(1951).

Type : China. Taiwan, Asahina F. 71(lectotype in TNS).

Jilin, Sichuan, Yunnan(Chen JB, 2015, p. 148), Anhui(Zhao, 1964; Zhao et al., 1982; Chen JB, 2015, p. 148), Hubei(Chen JB et al., 1989; Chen JB, 2015, p. 148), Hunan(Chen JB, 2015, p. 148), Jiangxi, Zhejiang (Zhao, 1964; Zhao et al., 1982; Chen JB, 2015, p. 148), Taiwan(Asahina 1951 & 1952; Hale 1987; Lai, 2000; Kurokawa & Lai 2001).

Parmelia glabra (Schaer.) Nyl. 茸梅衣

Flora, Regensburg 55:548(1872).

≡ *Parmelia olivacea* α*glabra* Schaer., *Lich. helv. spicil.* 10:466(1840).

≡ *Melanelia glabra*(Schaer.) Essl. *Mycotaxon* 7(1):47 [as "*glabratula*(Lamy)"] (1978) and *Mycotaxon* 7(3):526 and 533(1978) [correction].

≡ *Melanelia glabra*(Schaer.) Essl., in Egan, *Bryologist* 90(2):163(1987).

≡ *Melanelixia glabra*(Schaer.) O. Blanco et al. *Mycol. Res.* 108(8):882(2004).

Misapplied name:

Parmelia olivacea var. *corrugata* sensu Moreau et Moreau, Rev. Bryol. et Lichenol. 20:190(1951), non Hue.

On tree trunk or rocks.

Beijing(Nazarbek et al. , 2016, p. 30; Chen J. B. 2015, p. 104) , Heilongjiang, Jilin, Heimenggu, Hebei, Hubei (Chen J. B. 2015, p. 104) , Xinjiang(Nazarbek et al. , 2016, p. 30; Abdulla Abbas & Wu, 1998, p. 95; Abdulla Abbas et al. , 2001, p. 365) , Gansu(Chen J. B. 2015, p. 105) , Shaanxi(Zahlbr. , 1930 b, p. 186; Moreau et Moreau, 1951, p. 190; Zhao, 1964, p. 146; Zhao et al. 1982, p. 17; Ahti, 1966, pp. 42, 49 ; He & Chen 1995, p. 43, no specimen and locality are cited; Chen J. B. , 2015, p. 104) , Xizang(J. C. Wei & Jiang, 1981, p. 1147 & 1986, p. 40; Chen J. B. , 2015, p. 104) , Shanxi, Hebei(Chen J. B. , 2015, p. 104) , Shandong(Zhao ZT et al. , 2002, p. 6) , Hubei(Chen, Wu & J. C. Wei, 1989, p. 441) .

Parmelia infirma Kurok. 弱梅衣

in Hale & Kurokawa, *Contr. U. S. natnl. Herb.* 36: 179(1964) .

Taiwan(Wang & Lai, 1973, p. 93) .

Parmelia marmariza Nyl. 蛇纹梅衣

Lich. Japon. : 28(1890) .

On bark of trees.

Shaanxi(Zhao, 1964, p. 153; Zhao et al. 1982, p. 33; Hale, 1987, pp. 28, 30, mentioned only) , Chongqing(Chen CL et al. , 2008, p. 40) , Hubei(Chen, Wu & J. C. Wei, 1989, p. 442; Chen JB, 2015, p. 139) , Anhui(Zhao, 1964, p. 153; Zhao et al. 1982, p. 33; Hale, 1987, pp. 28, 30, mentioned only; Xu B. S. , 1989, p. 214) , Zhejiang (Xu B. S. , 1989, p. 214) , Taiwan(Zahlbr. , 1933 c, p. 55; Wang & Lai, 1973, p. 93; Hale, 1987, pp. 28, 30) .

Notes: It is difficult to distinguish between *Parmelia marmariza* Nyl. and *P. marmorophylla* Kurok. by morphology because of the intermediate forms between them. It remains to be studied by pheno-genotypic analyses in the future.

Parmelia marmorophylla Kurok. , *J. Jap. Bot.* 69(3) : 68(1994) . 海叶梅衣

Shaanxi, Sichuan, Guizhou, Yunnan, Hubei, Hunan, Anhui, Fujian(Chen J. B. 2011, p. 885 – 886 & 2015, p. 139 – 140) .

Parmelia meiophora Nyl. 稀生梅衣

in Hue, *Bull. Soc. Bot. France*, 36: 164(1889) .

Type: Yunnan, Delavay, 6, 17/IV/1887(syntypes) .

Yunnan(Hue. 1890, p. 283 & 1899, p. 159; Zahlbr. , 1930a, p. 175 & 1930b, p. 185) .

On bark of trees.

Neimenggu(Sun LY et al. , 2000, p. 36) , Sichuan(Zahlbr. , 1930b, p. 185; Hale, 1987, p. 30) .

≡ *Parmelia meiophora* var. *meiophora* Zhao, *Acta Phytotax. Sin.* 9(2) : 156(1964) .

Yunnan(Zhao, 1964, p. 156; Zhao et al. 1982, p. 39) .

= *Parmelia meiophora* var. *isidiata* Zhao, *Acta, Phytotax. Sin.* 9(2) : 156(1964) .

Type: Yunnan, poliyangskii no. 1002 in HMAS-L.

Yunnan(Zhao et al. 1982, p. 39; Hawksw. 1972, p. 49; Hale, 1987, p. 30, mentioned only) .

On bark of trees.

Guizhou(Zhang T & JC Wei, 2006, p. 9; Chen JB, 2015, p. 141) , Yunnan(Chen JB, 2015, p. 141) , Taiwan (Lai, 2000, p. 144) .

Parmelia muliensis Zahlbr. 木里梅衣

Hedwigia, 74: 209(1934) & *Cat. Lich. Univ.* 10: 526(1940) .

Type: Sichuan, collected by F. Rock.

On bark of trees.

Parmelia niitakana Asahina 高山梅衣

J. J. B. 26: 332(1951) .

= *Parmelia shinanoana* Zahlbr. f. *calvescens* Zahlbr. in Fedde, *Repertorium*, 33: 56(1933) .

Type: Taiwan, collected by S. Sasaki, Asahina no. 403.

On bark of trees.

Sichuan, Yunnan(Chen JB, 2015, p. 141), Shandong(Li Y et al., 2008, p. 71), Taiwan(Wang & Lai, 1973, p. 93, Hale, 1987, p. 30; Lai, 2000, pp. 144 – 145).

Parmelia omphalodes(L.) Ach. 北方梅衣

Method. Lich.: 204, 1803; Ahti, Lai & Qian, Fung. Sci. 14: 124, 1999.

≡*Lichen omphalodes* L., Sp. Pl.: 1143, 1753.

Heilongjiang(Chen JB, 2015, p. 143), Jilin(Ahti et al., 1999, p. 124 Chen JB, 2015, p. 143)., Neimeng, Shaanxi, Hebei(Chen JB, 2015, p. 143).

Parmelia saxatilis(L.) Ach. 石梅衣

Methodus, Sectio post.: 204(1803). f. saxatilis 原变型

≡*Lichen saxatilis* L., *Sp. pl.* 2: 1142(1753).

≡*Imbricaria saxatilis*(L.) Körber, Lichenogr. German. Specim.: 9(1846).

= *Parmelia saxatilis* var. *furfuracea* (Schaer.) Linds., *Trans. R. Soc. Edinb.* 12: 227(1859).

On bark of trees & rocks.

Heilongjiang(Chen et al. 1981, p. 153; Luo, 1984, p. 85; Chen JB, 2015, p. 144), Neimenggu, Jilin(Chen et al. 1981, p. 153; Chen JB, 2015, p. 144), North-eastern China("Machuria": Asahina, 1951g, p. 354), Xinjiang (Abbas et al., 1997, p. 3), Shaanxi(Jatta, 1902, p. 469; Zahlbr., 1930 b, p. 190; J. C. Wei et al. 1982, p. 37 ; He & Chen 1995, p. 44, no specimen and locality are cited; Chen JB, 2015, p. 144), Yunnan(Hue, 1889, p. 166; Zahlbr., 1930 b, p. 190), Shandong(Li Y et al., 2008, p. 71), Zhejiang(Tchou, 1935, p. 310), Xizang(J. C. Wei & Chen, 1974, p. 179; J. C. Wei & Jiang, 1981, p. 1147 & 1986, p. 47; Chen JB, 2015, p. 144), Taiwan (Asahina & Sato in Asahina, 1939, p. 727).

≡*Parmelia saxatilis* var. *saxatilis*

Heilongjiang, Jilin(Zhao, 1964, p. 151; Zhao et al. 1982, p. 30), Hubei(Chen, Wu & J. C. Wei, 1989, p. 443), Taiwan(Wang & Lai, 1973, p. 94).

= *Parmelia saxatilis* f. *furfuracea* Schaer. *Enum. critic. lich. europ.* (Bern): 45(1850).

≡*Parmelia saxatilis* var. *furfuracea*(Schaer.) Linds., *Trans. R. Soc. Edinb.* 12: 227(1859).

North-eastern China(as "Manchuria" cited by Asahina, 1952, p. 105).

f. munda Schaer. 平滑变型

Enum. critic. lich. europ. (Bern): 45(1850).

North-eastern China(as "Manchuria" cited by Asahina, 1952, p. 104).

var. subomphalodes Zahlbr. 亚脐变种

in Handel-Mazzetti, *Symbolae Sinicae* 3: 190(1930).

Type: Shaanxi, Mt. Guangtou shan, collected by Giraldi.

Notes: Except for the citation of this species in Xizang, it from Neimenggu, Jilin, Heilongjiang, Shaanxi, Yunnan, Zhejiang and Taiwan may probably contains both *Parmelia saxatilis* & *P. squarrosa*.

Parmelia shinanoana Zahlbr. 白边梅衣

Bot. Mag., *Tokyo* 41: 349(1927).

Heilongjiang, Neimeng(Chen JB, 2015, p. 145),

Parmelia squarrosa Hale 羽根梅衣

Phytologia 22(1): 29(1971).

= *Parmelia saxatilis* var. *divaricata* Delise ex Nyl., *Lich. Jap.*: 27(1890).

≡*Parmelia divaricata*(Delise ex Nyl.) Rassad., *Nov. sist. Niz. Rast.* 10: 199(1973) Nom. illegit., Art. 53. 1

On bark & twigs of trees.

Heilongjiang(Zhao, 1964, p. 151; Chen et al. 1981, p. 154; Zhao et al. 1982, p. 30; Hale, 1987, pp. 43 – 44 as in China: Manchuria, TNS; S. Hyvoenen, 1985, 314; Chen JB, 2015, p. 146) , Jilin(S. Hyvoenen, 1985, 314; Chen JB, 2015, p. 146) , Shaanxi(He & Chen 1995, p. 44, no specimen and locality are cited; Chen JB, 2015, p. 147) , Xizang(J. C. Wei & Jiang, 1986, p. 48; Chen JB, 2015, p. 147) , Neimenggu, Sichuan, Yunnan, Hebei, Hubei(Chen JB, 2015, p. 146 – 147) , Taiwan(Zahlbr. , 1933c, p. 56; Wang & Lai, 1973, p. 94; Wang & Lai, 1976, p. 228) .

Parmelia submutata Hue 亚变梅衣

Nouv. Arch. Mus. Hist. Nat. 4(1) : 172(1899) .

Type: Yunnan, Delavay, 27/III/1890. (lectotype in PC)

On bark of trees.

Yunnan(Zahlbr. , 1930 a, p. 215 & 1930 b, p. 189; Hale, 1987, p. 44) , Taiwan(Wang & Lai, 1973, p. 94; Hale, 1987, p44; Lai, 2000, p. 146) .

= *Parmelia leiocarpodes* Zahlbr. , *Hedwigia*, 74: 207(1934) .

Type: Sichuan, collected by F. Rock(lectotype in W) .

On bark of trees.

Yunnan(Hale, 1987, p. 44) .

= *Parmelia rhododendri* Zahlbr. in Handel-Mazzetti, *Symb. Sin.* 3: 187(1930) & *Cat. Lich. Univ.* 8: 565(1932) .

Type: Yunnan, Handel-Mazzetti no. 6518(lectotype in W; isolectotypes in TNS, US) .

On trunk of *Rhododendrom* sp.

Yunnan(Zahlbr. , 1934b, p. 208; Hale, 1987, p. 44; Chen JB, 2015, p. 149) .

Xizang, Sichuan(Chen JB, 2015, p. 149) .

Parmelia sulcata Taylor 槽梅衣

in Mackay, *Fl. Hibern.* 2: 145(1836) . f. sulcata 原变型

On bark of trees.

Heilongjiang, Neimenggu, Xizang, Sichuan, Yunnan(Chen JB, 2015, p. 150) , Xinjiang(Wang XY, 1985, p. 347; Abudula Abbas et al. , 1993, p. 79; Abdulla Abbas & Wu, 1998, p. 98; Abdulla Abbas et al. , 2001, p. 365) , Chen JB, 2015, p. 150) , Hebei(Tchou, 1935, p. 310; Zhao, 1964, p. 152; Zhao et al. 1982, p. 32) ,

f. rubescens B. de Lesd. 红叶变型

Recherch. Lich. Dunkerque, Suppl. : 69(1914) .

≡ *Parmelia rosiformis* f. *rubescens*(B. de Lesd.) Gyeln. , *Folia Cryptog.* 1(no. 6) : 503(1928) .

On rocks in birch forest.

Xizang(J. C. Wei & Chen, 1974, p. 179; J. C. Wei et al. 1982, p. 38; J. C. Wei & Y. M. Jiang, 1981, p. 1147 & 1986, p. 48, cited as *Parmelia sulcata*)

f. ulophylla B. de Lesd. 卷叶变型

Recherch. Lich. Dunkerque: 97(1910) .

On bark of trees and on rocks.

Xizang(J. C. Wei & Chen, 1974, p. 179; J. C. Wei et al. 1982, p. 38; J. C. Wei & Jiang, 1981, p. 1147 & 1986, p. 48, cited as *Parmelia sulcata*) .

Parmelina Hale **缘毛梅属**

Phytologia 28(5) : 481(1974) .

Parmelina gyrophorica Elix, Wang SL. , Chen JB. 三苔酸缘毛梅

Mycotaxon 76: 297 – 298(2000) .

Type: Yunnan, Gongshan, Cikaipuladadui, alt. 3250m, on dead wood, July 25, 1982, Su JJ, 2539(holotype: HMAS-L. ; isotype: CANB) , from the same locality based on type specimen(Chen J. B. et al. , 2003, p. 22) .

Sichuan, Yunnan, Hunan(Chen JB, 2015, p. 152).

Parmelina quercina(Willd.) Hale 栎黄髓梅

Phytologia 28:483(1974).

≡*Lichen quercinus* Willd., *Fl. berol. prodr.*:353(1787).

≡*Parmelia quercina* (Willd.) Vain., Term. Füz. 22: 279(1899).

= *Parmelia quercina* f. *furfuracea*(Schaer.) Zahlbr. in Handel-Mazzetti, Symb. Sin. 3: 185(1930), & *Cat. Lich. Univ*. VI: 190(1930).

≡*Parmelia quercifolia* var. *furfuracea* Schaer. Enumer. Critic. Lich. Europ. 44(1850).

Misapplied name:

Parmelia tiliacea sensu Hue, *Bull. Soc. Bot. Frence* 34: 16, 1887, non(Hoffm.) Ach. (fide Zahlbr. 1930, p. 185).

Imbricaria tiliacea sensu Jatta, *New. Giorn. Bot. Ital.* ser. 2. 9: 469, 1902, non(Hoffm.) Ach. (fide Zahlbr. 1930, p. 185).

Imbricaria tiliacea var. *rugulosa* sensu Jatta, Nouv. Giorn. Bot. Ital. ser. 2, 9: 469, 1902, non Leight. (fide Zahlb. 1930, p. 185).

On tree trunk.

Heilongjiang, Liaoning(Chen et al. 1981, p. 153; J. C. Wei et al. 1982, p. 38), Jilin(Chen et al. 1981, p. 153; J. C. Wei et al. 1982, p. 38; Chen JB, 2015. p. 153), Neimenggu(Chen et al. 1981, p. 153; Zhao, 1964, p. 155; Zhao et al. 1982, p. 38), North-eastern China(prov. not indicated: Hale, 1976, p. 42), Xinjiang(Abdulla Abbas & Wu, 1998, p. 98; Abdulla Abbas et al., 2001, p. 365), Shaanxi(Jatta, 1902, p. 469; Zahlbr., 1930 b, p. 185; Hale, 1976, p. 42; He & Chen 1995, p. 44, no specimen and locality are cited; chen J. B. et al., 2003, p. 22, miscited as Shanxi, taibai; Chen JB, 2015. p. 154), Xizang(J. C. Wei & Jiang, 1981, p. 1147 & 1986, p. 45; chen J. B. et al., 2003, p. 22; Chen JB, 2015. p. 154), Sichuan(Zahlbr., 1934 b, p. 206; Hale, 1976, p. 42; chen J. B. et al., 2003, p. 22; Chen JB, 2015. p. 153), Guizhou(Zhang T & Wei JC, 2006, p. 9; Chen JB, 2015. p. 153), Yunnan(Hue, 1887a, p. 21; Zahlbr., 1930 b, p. 185; chen J. B. et al., 2003, p. 22; Chen JB, 2015. p. 153), Hebei(Chen JB, 2015. p. 153), Hubei(Chen, Wu & J. C. Wei, 1989, p. 446; chen J. B. et al., 2003, p. 22), Zhejiang((Zhao, 1964, p. 155; Zhao et al. 1982, p. 38; Xu B. S., 1989, p. 218 – 219), Anhui(Zhao, 1964, p. 155; Zhao et al. 1982, p. 38; Xu B. S., 1989, p. 218 – 219; chen J. B. et al., 2003, p. 22; Chen JB, 2015. p. 153), Jiangsu(Xu B. S., 1989, p. 218 – 219), Hubei, Hunan(Chen JB, 2015. p. 153), Jiangxi(Zhao, 1964, p. 155; Zhao et al. 1982, p. 38; chen J. B. et al., 2003, p. 22; Chen JB, 2015. p. 153), Guangxi(Chen JB, 2015. p. 153).

Parmelina tiliacea(Hoffm.) Hale 皮革缘毛衣

Phytologia 28(5):481(1974).

≡*Lichen tiliaceus* Hoffm., *Enum. critic. lich. europ.* (Bern):96(1784).

≡*Parmelia tiliacea*(Hoffm.) Ach., *Meth. Lich.*:215(1803).

= *Parmelia scortea*(Ach.) Ach., *Methodus*, Sectio post.:215(1803).

≡*Lichen scorteus* Ach., *Lich. suec. prodr.* (Linköping): 119(1799) [1798].

On bark of trees.

Shaanxi(Jatta, 1902, p. 469; Zahlbr., 1930 b, p. 185), Yunnan(Zhao, 1964, p. 155; Zhao et al. 1982, p. 38; Ahti et al., 1999, p. 124).

Parmelina yalungana(Zahlbr.) P. R. Nelson & Kepler *in* P. R. Nelson et al., *The Bryologist*, 115(4):557 – 565 (2012).

≡*Parmelia yalungana* Zahlbr., *Hedwigia* 74: 206(1934).

Type: China: Sichuan. Mountains between Litang and Yalung rivers, between Muli Gomba and Baurong and

Wa-Erh-Dje, on Picea, alt. 4250 m., Jul., 1928, J. F. Rock 16720(W). Sichuan, Yunnan(P. R. Nelson et al., 2012, p. 564).

Parmelinella Elix & Hale 甲衣属

Mycotaxon 29: 241(1987).

Parmelinella chozoubae(Kr. P. Singh & G. P. Sinha) Elix & Pooprang 印度甲衣

in Pooprang, Boonpragob & Elix, *Mycotaxon* 71: 121(1999).

≡*Parmelina chozoubae* Kr. P. Singh & G. P. Sinha, *Nordic Jl Bot.* 13(4): 463(1993).

Yunnan(Chen JB, 2015. p. 155), Zhejiang(chen j. b. et al., 2003, pp. 22 – 23).

Parmelinella simplicior(Hale) Elix & Hale 小孢甲衣

Mycotaxon 29: 242(1987).

≡*Parmelia simplicior* Hale, *Bryologist* 75(1): 99(1972).

Yunnan(Wang S. L. et al., 2001, p. 135; chen j. b. et al., 2003, p. 23).

Notes: The report of this species by Chen JB et al. in 2003 is revised by Chen JB(2015. p. 266) as *Parmelinella chozoubae* (Kr. P. Singh & G. P. Sinha) Elix & Pooprang(Chen JB, 2015. p. 155).

Parmelinella wallichiana(Taylor) Elix & Hale 小裂芽甲衣

Mycotaxon 29: 242(1987).

≡*Parmelia wallichiana* Taylor, *London J. Bot.* 6: 176(1847).

≡*Parmelina wallichiana*(Tayl.) Hale(Thrower, 1988, pp. 16, 126).

On granite boulders, occasionally on trees. Paleotropical.

= *Parmelia nimandairana* Zahlbr. in Fedde, *Repertorium*, 33: 55(1933).

Type: Taiwan, Asahina no. 63(lectotype) in W &(isotype) in TNS(not seen).

On bark of trees & on rock.

= *Parmelia meiphora* var. *isidiata* Zhao, *Acta Phytotax. Sin.* 9(2): 156, 1964

Type. Yunana, Xishuangbanna, Jinghong, 29 March 1957, Poliyanskii 1002(holotype in HMAS-L).

On bark & on rock.

Jilin(J. C. Wei et al. 1982, p. 40), Liaoning(as Mandschuria australis: Asahina, 1952, p. 138; Hale, 1976, p. 51), Neimenggu(Chen et al. 1981, p. 154), Xinjiang(Abbas et al.), Xizang(J. C. Wei & Chen, 1974, p. 180; J. C. Wei & Jiang, 1981, p. 1147 & 1986, p. 45; chen j. b. et al., 2003, p. 23; Chen JB, 2015. p. 157), Guizhou (Zhang T & J. C. Wei, 2006, p. 9; Chen JB, 2015. p. 157), Yunnan(chen j. b. et al., 2003, p. 23; Chen JB, 2015. p. 157), Anhui(Zhao, 1964, p. 162; Zhao et al. 1982, p. 52; J. C. Wei et al. 1982, p. 40; Xu B. S., 1989, p. 219; chen j. b. et al., 2003, p. 23; Chen JB, 2015. p. 156), hubei(chen j. b. et al., 2003, p. 23), hunan(chen j. b. et al., 2003, p. 23; Chen JB, 2015. p. 156), guangxi, Fujian(chen j. b. et al., 2003, p. 23; Chen JB, 2015. p. 156 – 157), Jiangxi(Zhao, 1964, p. 162; Zhao et al. 1982, p. 52; chen j. b. et al., 2003, p. 23; Chen JB, 2015. p. 156), Zhejiang(Xu B. S., 1989, p. 219; chen j. b. et al., 2003, p. 23; Chen JB, 2015. p. 156), Guangdong(Hale, 1976, p. 51), Taiwan(Zahlbr., 1933 c, p. 55; Asahina, 1952 a, p. 138; Lamb, 1963, p. 489; Wang & Lai, 1973, p. 94; Hale, 1976, p. 51; Lai, 2000, pp. 147 – 148), hainan(chen j. b. et al., 2003, p. 23; Chen JB, 2015. p. 156 – 157; JCWei et al. 2013, p. 228), Hongkong(Hale, 1976, p. 51; Thrower, 1988, pp. 16, 126; Aptroot & Seaward, 1999, p. 83; Chen JB, 2015. p. 157),

Parmelinopsis Elix & Hale 狭叶衣属

Mycotaxon 29: 242(1987).

Parmelinopsis afrorevoluta(Krog & Swinscow) Elix & Hale 反卷狭叶衣

Mycotaxon 29: 242(1987).

≡*Parmelia afrorevoluta* Krog & Swinscow, *Norw. Jl Bot.* 26: 22(1979).

Hunan, Guangxi(Chen JB, 2015. p. 159), Zhejiang(Chen J. B. et al., 2003, pp. 24 – 25; Chen JB, 2015. p. 159),.

Parmelinopsis cryptochlora(Vain.) Elix & Hale 头粉狭叶衣

Mycotaxon 29: 242(1987).

≡*Parmelia cryptochlora* Vain., *J. Bot.*, *Lond.* 34: 34(1896).

≡*Parmelina cryptochlora*(Vain.) Hale, *Phytologia* 28(5): 482(1974).

On rocks and trees.

Guizhou(Zhang T & J. C. Wei, 2006, p. 9; Chen JB, 2015. p. 160), Zhejiang(Xu B. S., 1989, p. 216; Chen J. B. et al., 2003, p. 25; Chen JB, 2015. p. 160).

Parmelinopsis expallida(Kurok.) Elix & Hale 淡腹狭叶衣

Mycotaxon 29: 242(1987).

≡*Parmelia expallida* Kurok. *Bull. Natn. Sci. Mus. Tokyo*, 11: 191(1968).

Type: China. Taiwan, Kaohsiung(Gaoxiong), Nanfeng, alt. 1200 m, 8 Feb. 1965, Kurokawa2930(holotype in TNS).

≡*Parmelina expallida*(Kurok.) Hale, Phytologia 28: 482(1974).

Yunnan(Chen J. B. et al., 2003, p. 25), hunan(Chen J. B. et al., 2003, p. 25; Chen JB, 2015. P. 161), Taiwan (D. Hawksw. 1972 b, p. 48; Wang & Lai, 1973, p. 93; Lai, 2000, pp. 149 – 150).

On trees.

Parmelinopsis horrescens(Taylor) Elix & Hale 毛裂芽狭叶衣

Mycotaxon 29: 242(1987).

≡*Parmelia horrescens* Taylor, in Mackay, *Fl. Hibern.* 2: 144(1836).

≡*Hypotrachyna horrescens*(Taylor) Krog & Swinscow(Lai, 2000, p. 150).

=*Parmelia dissecta* Nyl., Flora 65: 451(1882).

≡*Parmelina dissecta*(Nyl.) Hale, *Phytologia* 28: 482(1974).

On bark.

Hunan, Zhejiang, guangxi(Chen J. B. et al., 2003, pp. 25 – 26; Chen JB, 2015. p. 162), Taiwan(Lai, 2000, p. 150; Chen JB, 2015. p. 162), Hainan(Chen JB, 2015. p. 162; JCWei et al., 2013, p. 228).

Parmelinopsis microlobulata(D. D. Awasthi) Elix & Hale 裂片狭叶衣

Mycotaxon 29: 242(1987).

≡*Parmelia microlobulata* D. D. Awasthi, *Biological Memoirs* 1(1 – 2), *Lichenology* 1: 182(1977)[1976].

Guangxi(Chen J. B. et al., 2003, p. 26; Chen JB, 2015. p. 162).

Parmelinopsis minarum(Vain.) Elix & Hale 稠芽狭叶衣

Mycotaxon 29: 243(1987).

≡*Parmelia minarum* Vain., *Acta Soc. Fauna Flora fenn.* 7(no. 1): 48(1890).

≡*Parmelina minarum*(Vain.) Skorepa, *Phytologia* 53(6): 445(1983).

=*Hypotrachyna kokiuensis*(Zhao et al.) J. C. Wei, *Enum. Lich. China*, 119(1991).

≡*Parmelia kokiuensis* Zhao et al. *Acta Phytotax. Sin.* 16(3): 95(1978).

Type: Yunnan, Zhao & Chen no. 2344 in HMAS-L.

On bark of trees.

Parmelina dissecta auct. non Nyl., *Flora*, Regensburg 65: 451(1882).

Shaanxi(He & Chen 1995, p. 44 without ; Chen JB, 2015. p. 163), Guizhou(Zhang T & Wei JC, 2006, p. 9; Chen JB, 2015. p. 163),, Yunnan(Zhao et al. 1978, p. 95 & 1982, p. 36; Chen J. B. et al., 2003, p. 366; Chen J. B. et al., 2003, pp. 26 – 27, Chen JB, 2015. p. 163), Anhui(Xu B. S., 1989, p. 217), Hunan, Jiangxi(Chen J. B. et al., 2003, pp. 26 – 27, Chen JB, 2015. p. 163),, Zhejiang(Xu B. S., 1989, p. 217; Chen JB, 2015. p. 163; Chen J. B. et al., 2003, p. 366), Fujian(Zhao et al. 1978, p. 95 & 1982, p. 36; Chen J. B. et al., 2003, pp. 26 – 27, Chen JB, 2015. p. 163), Guangxi(Chen J. B. et al., 2003, pp. 26 – 27; Chen J. B. et al., 2003, p. 366; Chen JB, 2015. p. 163), Taiwan(Hale, p. 1974, p. 482; Lai, 2000, pp. 150 – 151), Hainan(Chen J. B. et al., 2003, p. 366).

Parmelinopsis protocetrarica Elix 原岛衣酸狭叶衣

Mycotaxon 47: 119(1993).

On bark.

Hainan(Chen J. B. et al., 2003, p. 27; Chen JB, 2015. p. 164; JCWei et al., 2013, p. 228).

Parmelinopsis spumosa(Asahina) Elix & Hale 疱体狭叶衣

Mycotaxon 29: 243(1987).

≡*Parmelia spumosa* Asahina, *J. Jap. Bot.* 24: 259(1951).

≡*Parmelina spumosa* (Asah.) Hale, *Phytologia* 28: 483(1974).

≡*Hypotrachyna spumosa*(Asah.) Krog & Swinscow

On bark of trees.

Shaanxi(Wu J. L. 1987, p. 154), Jiangsu, Anhui, Zhejiang(Xu B. S. 1989, p. 219), Fujian(Chen JB, 2015. p. 165), Taiwan(Hale, 1976, p. 46; Lai, 2000, pp. 151 – 152), guangxi (Chen J. B. et al., 2003, p. 27; Chen JB, 2015. p. 165), hainan(Chen J. B. et al., 2003, p. 27; Chen JB, 2015. p. 165; JCWei et al., 2013, pp. 228 – 229).

Parmelinopsis subfatiscens(Kurok.) Elix & Hale 疱体粉芽狭叶衣

Mycotaxon 29: 243(1987).

≡*Parmelia subfatiscens* Kurok. in Hale & Kurokawa, *Contr. U. S. natnl. Herb.* 36: 134(1964).

On bark.

Hunan, zhejiang(Chen J. B. et al., 2003, pp. 27 – 28; Chen JB, 2015. p. 165), hainan(Chen J. B. et al., 2003, pp. 27 – 28; JCWei et al., 2013, p. 229).

Parmelinopsis swinscowii(Hale) Elix & Hale 水杨酸狭叶衣

Mycotaxon 29: 243(1987).

≡*Parmelia swinscowii* Hale, *Phytologia* 27(1): 4(1973).

Yunnan(Wang S. L. et al., 2001, p. 136; Chen J. B. et al., 2003, p. 28).

Parmeliopsis(Nyl.) Nyl. **小叶梅属**

Lich. Lapp. Orient. 8: 121(1866).

≡*Parmelia* subgen. *Parmeliopsis* Nyl., *Not. Sällsk. Fauna et Fl. Fenn. Förh.*, Ny Ser. 5: 130(1861).

Parmeliopsis ambigua(Wulfen) Nyl. 绿小叶梅

Syn. meth. lich. (Parisiis) 2: 54(1863).

≡*Lichen ambiguus* Wulfen, in Jacquin, *Collnea bot.* 4: 239(1791) [1790].

≡*Foraminella ambigua*(Wulfen) S. L. F. Mey., *Mycologia* 74(4): 597(1982).

Xinjiang(Abdulla Abbas & Wu, 1998, p. 99; Abdulla Abbas et al., 2001, p. 365), Taiwan(Lai, 2000, pp. 152 – 153).

Parmotrema A. Massal. **大叶梅属**

Atti Inst. Veneto Sci. lett., ed Arti, Sér. 3, 5: 248(1860) [1859 – 1860].

= *Canomaculina* Elix & Hale, *Mycotaxon* 29: 239(1987).

= *Rimelia* Hale & Fletcher, *Bryologist* 93: 23(1990).

= *Rimeliella* Kurok., *Annals of the Tsukuba Botanical Garden* 10: 1(1991).

Parmotrema andinum(Müll. Arg.) Hale 茶渍酸大叶梅

Phytologia 28(4): 334(1974).

≡*Parmelia andina* Müll. Arg., *Revue m5ycol.*, *Toulouse* 1: 169(1879).

= *Parmelia hyporysalea* Vain., *Bot. Magz. Tokyo* 35: 47, 1921; J. C. Wei, *Enum. Lich. China*: 166(1991).

≡*Parmelia olivetorum* var. *hyporysalea* Vain., *Cat. Welwitsch Afric. Pl.* 2: 399(1901).

On tree bark.

Yunnan(Chen J. B. et al., 2005, p. 97; Chen JB, 2015. p. 171), Taiwan(Asahina, 1952, p. 129).

Parmotrema argentinum(Krempelh.) Hale 阿根廷大叶梅

Phytologia 28(4): 334(1974).

≡*Parmelia argentina* Kremp., *Flora*, Regensburg 61: 476(1878).

Taiwan(Lai, 2000, 157).

Parmotrema arisani(Zahlbr.) J. C. Wei, comb. nov. 阿里山大叶梅

≡*Parmelia arisani* Zahlbr. in Fedde, *Repertorium*, 33: 57 – 58(1933).

Type: Taiwan, Mt. Arisan, Nimandaira, supra muscos ad truncos arborum(Asahina no. 85).

On the mosses covered the bark of trees.

Taiwan(Zahlbr., 1940, p. 506; Lamb, 1963, p. 462; Wang & Lai, 1973, p. 93).

Parmotrema arnoldii(Du Rietz) Hale 假缘毛大叶梅

Phytologia 28: 335(1974).

≡*Parmelia arnoldii* Du Rietz, *Nytt Mag. Natur.* 62: 80(1924).

On bark of trees.

Heilongjiang(Chen et al. 1981, p. 151), Xizang(J. C. Wei & Chen, 1974, p. 180), Guizhou(Zhang T & Wei JC, 2006, p. 9), Yunnan(Zhao, 1964, p. 159; Zhao et al. 1982, p. 46; Chen J. Bet al., 2005, p. 97; Chen JB, 2015. p. 172), Taiwan(Asahina, 1940b, p. 597 & 1952a, p. 134; Wang & Lai, 1973, p. 93), Guangxi(Zhao, 1964, p. 159; Zhao et al. 1982, p. 46 Chen JB, 2015. p. 172), Hainan(Chen J. Bet al., 2005, p. 97; Chen JB, 2015. p. 172; JCWei et al., 2013, p. 229).

Parmotrema austrosinense(Zahlbr.) Hale 华南大叶梅

Phytologia 28: 335(1974).

≡*Parmelia austrosinensis* Zahlbr. in Handel-Mazzetti, *Symb. Sin.* 3: 192(1930).

Type: Guizhou, Handel-Mazzetti no. 10580 & Yunnan, Handel-Mazzetti no. 8209(syntypes).

=*Parmelia mesotropa* var. *compactior* Zahlbr. in Handel-Mazzetti, *Symb. Sin.* 3: 192(1930).

Type: Yunnan, 13 March 1914, Handel-Mazzetti no. 581(holotype in WU; isotype in W).

On bark of trees.

Parmelia hyporysalea(Vainio) Vainio var. *cinerascens* Vainio(Asahina, 1952).

Xizang(Chen JB, 2015. p. 174), Sichuan(Zhao, 1964, p. 160; Zhao et al. 1982, p. 48; Chen J. B. et al., 2005, p. 98; Chen JB, 2015. p. 174), Yunnan(Zhao, 1964, p. 160; Zhao et al. 1982, p. 48; Hale, 1965, p. 239; Wang L. S. et al. 2008, p. 537; Chen J. B. et al., 2005, p. 98; Chen JB, 2015. p. 174), Shandong(Zhang F et al., 1999, p. 30, as *P. austrosinenes*; Zhao ZT et al., 2002, p. 6; Li Y et al., 2008, p. 71), Hunan(Chen JB, 2015. p. 173), Jiangsu, Zhejiang(Xu B. S., 1989, p. 219 – 220), Fujian(Chen JB, 2015. p. 173), Guangxi(Zhao, 1964, p. 160; Zhao et al. 1982, p. 48; Chen J. B. et al., 2005, p. 98; Chen JB, 2015. p. 173), Taiwan(Asah., 1952; Lai, 2000, pp. 157 – 158).

Parmotrema cetratum (Ach.) Hale 睫毛大叶梅

Phytologia 28: 335(1974).

≡*Parmelia cetrata* Ach., Syn. Meth. Lich.: 198(1814). - Hue, *Nouv. Arch. Mus. Hist. Nat.* ser. 4, 1: 173, 1899; Zahlbr., in Handel-Mazzetti, *Symb. Sin.* 3: 189(1930).

≡*Rimelia cetrata* (Ach.) Hale & A. Fletcher, *Bryologist* 93: 26(1990).

On bark of trees & on rocks.

Heilongjiang(Chen et al. 1981, p. 152; Luo, 1984, p. 85), shaanxi(He Q & Chen JB, 1995, p. 44 as *P. cetratum*, no speciemen & locality were cited), Yunnan(Hue, 1899, p. 173; Zahlbr., 1930 b, p. 189; Zhao, 1964, p. 153; Zhao et al. 1982. p. 30), Sichuan, Guangxi, Fujian, Jiangxi(Zhao, 1964, p. 153; Zhao et al. 1982, p. 30), Shandong(Zhang F et al., 1999, p. 30, as *Parmotrema cetratum*; Zhao ZT et al., 2002, p. 6 as *P. cetrartum*), Anhui, Zhejiang(Zhao, 1964, p. 153; Zhao et al. 1982, p. 30; Xu B. S., 1989, p. 220), Jiangsu, Shanghai(Xu B. S., 1989, p. 220), Hubei(Chen, Wu & J. C. Wei, 1989, p. 447), Taiwan(H. Sasaoka, 1919, p. 180; Asahina, 1951g, p. 356 & 1952a, p. 117; Wang-Yang & Lai, 1973, p. 93 & 1976, p. 228; Lai, 2000, pp. 184 – 185), Hainan(JCWei et al., 2013, p. 230), Hongkong(Moreau et Moreau, 1951, p. 191).

f. cetratum 原变型

Taiwan(Wang-Yang & Lai, 1973, p. 93).

f. ciliosum(Viaud-Gr. -Mar.) J. C. Wei 缘毛变型

Enum. Lich. China(Beijing): 176(1991).

≡*Parmelia cetrata* f. *ciliosa* Viaud-Grand-Marais, *Bull. Soc. Ouest France*, 2: 156(1892).

Parmelia crinita sensu Hue, *Bull. Soc. Bot. France*. 36: 164(1889), non Ach.

On bark of trees.

Yunnan(Hue, 1889, p. 164 & 1899, p. 175; Zahlbr., 1930 b, p. 189), Hunan(Zahlbr., 1930 b, p. 189), Taiwan (Zahlbr., 1933 c, p. 58; Lai, 2000, p. 185).

f. sorediiferum(Vain.) J. C. Wei 粉芽变型

Enum. Lich. China(Beijing): 177(1991).

≡*Parmelia cetrata* Ach. var. *sorediifera* Vain. *Etud. Lich. Bres.* 1: 40(1890).

= *Parmotrema Parmelia perforata* sensu Hue, *Bull. Soc. Bot. France*. 36: 164(1889), non alior.

On rocks.

Yunnan(Hue, 1889, p. 164 & 1899, p. 174; Zahlbr., 1930 b, p. 190), Sichuan, Guizhou, Hunan(Zahlbr., 1930 b, p. 190).

f. subisidiosum(Müll. Arg.) J. C. Wei 裂芽变型

Enum. Lich. China(Beijing): 177(1991).

≡*Parmelia cetrata* Ach. f. *subisidiosa* Müll. Arg. in Engler, *Botan. Jahrbuch*. 15: 256(1894). Hue, *Nouv. Arch. Mus. Hist. Nat.* 4(1): 175(1899).

= *Parmelia cetrata* Ach. f. *granularis* Asahina, *Journ. Jap. Bot*. 16(10): 593(1940).

Type: from Taiwan.

Taiwan(Asahina, 1952a, p. 118; Wang-Yang & Lai, 1973, p. 93).

On bark of trees.

Parmotrema chiapense(Hale) Hale 奇大叶梅

Phytologia 28(4): 335(1974).

≡*Parmelia chiapensis* Hale, *Contr. U. S. Natn. Herb*. 36: 323(1965).

Taiwan(Lai, 2000, p. 158).

Parmotrema corniculans(Nyl.) Hale 小角大叶梅

Phytologia 28(4): 335(1974).

≡*Parmelia corniculans* Nyl., *Flora*, Regensburg 68: 607(1885).

Taiwan(Lai, 2000, p. 160).

Parmotrema crinitoides J. C. Wei 类毛大叶梅

Enum. Lich. China, 177(1991).

= *Parmelia proboscidea* var. *eciliata* Zhao, *Acta Phytotax. Sin*. 9(2): 164(1964).

Type: Yunnan, Poliyangskii no. 1003 in HMAS-L.

On bark of trees & on dead grasses.

Yunnan(Zhao et al. 1982, p. 55; Chen J. B. et al.. 2005, p. 99; Chen JB, 2015. p. 177).

Parmotrema crinitum(Ach.) M. Choisy [as ' crinita'] 毛大叶梅

Bull. mens. Soc. linn. Soc. Bot. Lyon 21: 175(1952).

≡*Parmelia crinita* Ach., *Syn. meth. lich.* (Lund): 196(1814).

= *Parmelia proboscidea* Tayl. in Mackay, F1. Hibern. 2: 143(1836).

≡*Imbricaria proboscidea*(Taylor) Jatta, *Nuov. Giorn. Bot. Ital*. ser. 2, 9: 469(1902).

On bark of trees.

Liaoning(Chen et al. 1981, p. 152), Shaanxi(Jatta, 1902, p. 469; Zahlbr., 1930 b, pp. 190 – 191), Yunnan

(Zahlbr., 1930 b, p. 190 – 191 & 1934b, p. 210; Zhao, 1964, p. 164; Zhao et al. 1982, p. 54; Chen J. B. et al.. 2005, p. 99; Chen JB, 2015. p. 178), Sichuan, zhejiang(Chen J. B. et al.. 2005, p. 99; Chen JB, 2015. p. 178), Taiwan(Zahlbr., 1933 c, p. 57; Asahina, 1940 b, p. 600 & 1952 a, p. 136; Wang & Lai, 1973, p. 94; Lai, 2000, pp. 160 – 161).

Parmotrema cristiferum(Taylor) Hale 鸡冠大叶梅

Phytologia 28: 335(1974). f. cristiferum 原变型

≡*Parmelia cristifera* Taylor, *London J. Bot.* 6: 165(1847).

=*Parmelia mesotropa* f. *sorediosa* Müll. Arg., *Flora* 74: 377(1891).

=*Parmotrema cristiferum* f. *cineratum* (Zahlbr.) J. C. Wei, *Enum. Lich. China*: 178(1991).

≡*Parmelia cristifera* f. *cinerata* Zahlbr., *Repert. Spec. Nov. Regni Veg.* 33: 58(1933).

Type: China. Taiwan, Asahina no. 51(holotype in W; isotope in BPl).

On bark of trees.

guizhou(Chen J. B. et al.. 2005, p. 99; Chen JB, 2015. p. 179), Yunnan(Zahlbr., 1930 b, p. 192; Zhao, 1964, p. 161; Zhao et al. 1982, p. 50; Chen J. B. et al.. 2005, p. 99; Chen JB, 2015. p. 179), Guangxi(Chen JB, 2015. p. 179), Taiwan(Hale, 1965, p. 243; Wang & Lai, 1973, p. 93; Lai, 2000, p. 161), Hainan(Zhao, 1964, p. 161; Zhao et al. 1982, p. 50; Chen J. B. et al.. 2005, p. 99; Chen JB, 2015. p. 179; JCWei et al., 2013, p. 229).

f. cineratum(Zahlbr.) J. C. Wei 灰叶变型

Enum. Lich. China, 178(1991).

≡*Parmelia cristifera* f. *cinerata* Zahlbr. in Fedde, *Repertorium*, 33: 58(1933).

Type: Taiwan, Asahina no. 51 in W(holotype) and in BPI(isotype).

Taiwan(Zahlbr., 1940, p. 514; Wang & Lai, 1973, p. 93; Hale, 1965, p. 241).

On bark of trees.

Parmotrema deflectens(Kurok.) Streimann 弯曲大叶梅

Biblthca Lichenol. 22: 93(1986).

≡*Parmelia deflectens* Kurok., *Studies on Cryptogams of Papua New Guinea*(Tokyo): 130(1979).

On bark.

Yunnan(Chen J. B. et al.. 2005, p. 100; Chen JB, 2015. p. 180).

Parmotrema dilatatum(Vain.) Hale 灰黄大叶梅

Phytologia 28: 335(1974).

≡*Parmelia dilatata* Vain., *Acta Soc. Fauna Flora fenn.* 7(no. 1): 32(1890).

Yunnan(Chen et al., 2005, p. 100; Chen JB, 2015. p. 181), Taiwan(Lai, 2000, p. 162), hainan(Chen et al., 2005, p. 100; Chen JB, 2015. p. 181; JCWei et al., 2013, p. 229).

Parmotrema eciliatum(Nyl.) Hale 无毛大叶梅

Phytologia 28: 336(1974).

≡*Parmelia crinita* var. *eciliata* Nyl., *Flora*, Regensburg 52: 291(1869).

≡*Parmelia eciliata*(Nyl.) Nyl., Mexic. Pl. 1: 3(1872).

On trees and rocks.

Heilongjiang(Chen JB, 2015. p. 182), Anhui, Zhejiang(Xu B. S., 1989, p. 220; Chen J. B. et al., 2005, pp. 100 – 101; Chen JB, 2015. p. 182), hubei, hunan, Fujian(Chen J. B. et al., 2005, pp. 100 – 101; Chen JB, 2015. p. 182), Taiwan(Kurok. 1968, p. 349; Wang & Lai, 1973, p. 93; Lai, 2000, pp. 162 – 163).

Parmotrema eurysacum(Hue) Hale 正大叶梅

Phytologia 28: 336(1974).

≡*Parmelia eurysaca* Hue, *Nouv. Arch. Mus. Hist. Nat.*, *Paris*, 4 sér. 1: 194(1899).

Yunnan(Wang L. S. et al. 2008, p. 538), Taiwan(Lai, 2000, pp. 163 – 164).

Parmotrema eunetum(Stirt.) Hale 三苔酸大叶梅

Phytologia 28(4) : 336(1974) .

≡*Parmelia euneta* Stirt. , *Scott. Natural*. 4: 298(1878) [1877 – 78] .

Yunnan(Chen J. B. et al. , 2005, p. 101; Chen JB, 2015. p. 183) .

Parmotrema exquisitum(Kurok.) DePriest & B. W. Hale 精美大叶梅

Mycotaxon 67: 204(1998) .

≡*Parmelia exquisita* Kurok. , *Bull. natn. Sci. Mus.* , *Tokyo*, B 13(1) : 11(1987) .

Parmotrema ecrinitum Hale in sched. (Lai, 2000, p. 164) .

Taiwan(Lai, 2000, p. 164) .

Parmotrema gardneri(C. W. Dodge) Sérus. 假鸡冠大叶梅

Bryologist 87(1) : 5(1984) .

≡*Parmelia gardneri* C. W. Dodge, *Ann. Mo. bot. Gdn* 46(1 – 2) : 179(1959) .

On bark.

Yunnan(Chen J. B. et al. , 2005, p. 101; Chen JB, 2015. p. 183) .

Parmotrema grayanum(Hue) Hale 东方大叶梅

Phytologia 28: 336(1974) .

≡*Parmelia grayana* Hue, *Nouv. Arch. Mus. Hist. Nat.* , *Paris*, 4 sér. 1: 184(1899) .

= *Parmelia simodensis* Asahina, *J. Jap. Bot*. 17: 73(1941) .

On bark of trees.

Jilin(Chen et al. 1981, p. 153) , Yunnan, Anhui, Jiangxi(Zhao, 1964, p. 160; Zhao et al. 1982, p. 47) , Xizang (J. C. Wei & Chen, 1974, p. 180; Chen J. B. et al. , 2005, p. 101; Chen JB, 2015. p. 184) , Shandong(Hou YN et al. , 2008, p. 68) .

Parmotrema hababianum(Gyeln.) Hale [as ' hababiana'] 隐斑大叶梅

Phytologia 28(4) : 336(1974) .

≡*Parmelia hababiana* Gyeln. , *Feddes Repert*. 29: 288/416(1931) .

Misapplied name:

Parmotrema applanatum Chen, Wang & Elix, *Mycotaxon* , 1: 97(2005) . auct. non Marcelli & C. H. Ribeiro, *Mitt. Inst. Allg. Bot. Hamburg* 30 – 32: 145(2002) .

On rocks.

Xizang(J. C. Wei & Jiang, 1986, p. 48) , Yunnan(Chen J. B. et al. . 2005, p. 97; Chen JB, 2015. p. 185) .

Parmotrema immiscens(Nyl.) Kurok. 黄髓大叶梅

in Kurokawa & Arakawa, *Bulletin of the Botanic Gardens of Toyama* 2: 42(1997) .

≡*Parmelia immiscens* Nyl. , *Flora*, Regensburg 68: 606(1885) .

On bark, bamboo

Yunnan(Chen J. B. et al. , 2005, p. 102; Chen JB, 2015. p. 186) .

Parmotrema incrassatum Hale ex DePriest & B. W. Hale 厚大叶梅

Mycotaxon 67: 207(1998) .

Misapplied name(revised by Aptroot & Seaward, 1999, p. 84) :

Parmelia subrugata auct non Kremp. , Verh. zool. -bot. Ges. Wien 18: 320(1868) , in Tchou, 308(1935) .

On rocks.

Hongkong(Tchou, 1935, p. 308; Thrower, 1988, pp. 16, 128; DePriest & Hale, 1998, p. 207; Chu, 1997, p. 48; Aptroot & Seaward, 1999, p. 84; Aptroot et al. , 2001, p. 335; Chen J. B. , 2005, p. 102; Chen JB, 2015. p. 186) .

Parmotrema laeve(J. D. Zhao) J. B. Chen & Elix 光滑大叶梅

Mycotaxon 91: 102(2005) .

≡*Parmelia yunnana* var. *laevis* Zhao, *Acta Phytotax. Sin.* 9(2): 163(1964).

Type: Yunnan, Zhao & Chen no. 3648 in HMAS-L.

On bark of trees.

Yunnan(Zhao, 1964, p. 163; Chen J. B. et al., 2005, p. 102; Chen JB, 2015. p. 187).

Parmotrema latissimum(Fée) Hale 宽大叶梅

Phytologia 28: 337(1974).

≡*Parmelia latissima* Fée, *Essai Crypt. Exot.*, *Suppl. Révis.* (Paris): 119(1837).

On bark of trees.

Shaanxi(Jatta, 1902, p. 469; Zahlbr., 1930 b, p. 193).

Parmotrema leucosemothetum(Hue) Hale 白大叶梅

Phytologia 28: 337(1974).

≡*Parmelia leucosemotheta* Hue, *Nouv. Arch. Mus. Hist. Nat.*, Paris, 4 sér. 1: 192(1899).

≡*Canomaculina leucosemotheta*(Hue) Elix, *Mycotaxon* 65: 477(1997).

Taiwan(Lai, 2000, pp. 102 – 103).

Parmotrema lobulascens(J. Steiner) Hale 小裂大叶梅

Phytologia 28(4): 337(1974).

≡*Parmelia lobulascens* J. Steiner, V*erh. zool. -bot. Ges. Wien* 53: 234(1903).

Taiwan(Lai, 2000, pp. 164 – 165).

Notes: Whether this species belongs to *Parmotrema lobulascens* or *Parmotrema Pseudonilgherrensis* remains to be revised in the future.

Parmotrema louisianae(Hale) Hale 北美大叶梅

Phytologia 28(4): 337(1974).

≡*Parmelia louisianae* Hale, *Phytologia* 22(2): 92(1971).

On bark & twigs.

Guangxi(Chen J. B. et al., 2005, p. 103; Chen JB, 2015. p. 188), Hainan(Chen J. B. et al., 2005, p. 103; Chen JB, 2015. p. 188; JCWei et al., 2013, p. 229 as "*lovisianae*").

Parmotrema margaritatum(Hue) Hale 中美大叶梅

Phytologia 28(4): 337(1974).

≡*Parmelia margaritata* Hue, *Nouv. Arch. Mus. Hist. Nat.*, *Paris*, 4 sér. 1: 193(1899).

On bark.

Yunnan(Chen J. B. et al., 2005, p. 103; Chen JB, 2015. p. 189), Hainan(Chen J. B. et al., 2005, p. 103; Chen JB, 2015. p. 189; JCWei et al., 2013, p. 229).

Parmotrema mellissii(C. W. Dodge) Hale 麦氏大叶梅

Phytologia 28: 337(1974).

≡*Parmelia mellissii* C. W. Dodge, *Ann. Mo. bot. Gdn* 46: 134(1959).

Parmelia allardii Hale(Lai, 2000, p. 165).

On rocks & twigs.

Chongqing(Chen CL et al., 2008, p. 40 miscited as *P. meuissii*), Guizhou(Chen JB, 2015. p. 190), Yunnan (Hale, 1965, p. 298; Chen J. B. et al., 2005, p. 103; Chen JB, 2015. p. 190), Anhui(Xu B. S., 1989, p. 221), hunan, Zhejiang(Xu B. S., 1989, p. 221; Chen J. B. et al., 2005, pp. 103 – 104; Chen JB, 2015. p. 190), Fujian(Chen J. B. et al., 2005, pp. 103 – 104; Chen JB, 2015. p. 189, 190), Taiwan (Wang & Lai, 1973, p. 93 & 1976, p. 228; Lai, 2000, pp. 165 – 166), Guangxi, Guangdong(Chen JB, 2015. p. 190), Hainan(Chen J. B. et al., 2005, pp. 103 – 104; Chen JB, 2015. p. 189, 190 ; JCWei et 2013, p. 229).

Parmotrema merrillii(Vain.) Hale 梅氏大叶梅

Phytologia 28(4): 337(1974).

≡*Parmelia merrillii* Vain., *Philipp. J. Sci.*, *C, Bot.* 4(5):658(1909).

Taiwan(Hale, 1965, p. 299; Lai, 2000, pp. 166 – 167).

Parmotrema myriolobulatum(Zhao) J. C. Wei 密裂大叶梅

Enum. Lich. China 178 – 179(1991).

≡*Parmelia myriolobulata* Zhao, *Acta Phytotax. Sin.* 9(2):165(1964).

Type: Yunnan, Mengyang, 24Nov. 1960, J. D. Zhao &Y. B. Chen 3579(holotype in HMAS-L).

On bark of trees.

Yunnan(Zhao et al. 1982, p. 53; Hawksw. 1972, p. 50; Chen J. B. et al., 2005, p. 104; Chen JB, 2015. p. 191).

Parmotrema nanfongense(Kurok.) DePriest & B. W. Hale 矮小大叶梅

Mycotaxon 67:204(1998).

≡*Parmelia nanfongensis* Kurok., *Bull. natn. Sci. Mus.*, *Tokyo*, B 13(1):13(1987).

Type: Taiwan, Kurok., 2787(TNS, holotypus), (Lai, 2000, pp. 167 – 168).

Parmotrema neopustulatum Kurok. 新疱大叶梅

J. Jap. Bot. 81:252, 2006.

≡*Parmotrema pustulatum*(Elix & Bawingan) O. Blanco, A. Crespo, Divakar, Elix & Lumbsch, *Mycologia* 97(1):157(2005), nom illeg., non *Parmotrema pustulatum* Louwhoff & Elix, *Mycotaxon*75: 199(2000).

≡*Rimelia pustulata* Elix & Bawingan, *Mycotaxon* 81:252(2002).

On rock.

Anhui(Chen J. B. 2011, pp. 886 – 887; Chen JB, 2015. p. 191), Taiwan(Aptroot, Sparrius & Lai, 2002, pp. 290 – 291, as *Rimellia pustulata* Elix & Bawingan).

Parmotrema nilgherrense(Nyl.) Hale 尼尔山大叶梅

Phytologia 28:338(1974).

≡*Parmelia nilgherrensis* Nyl., *Flora*, Regensburg 52:291(1869).

≡*Imbricaria nilgherrensis*(Nyl.) Arn., *Verh. Zool. -Bot. Ges. Wien* 25:472(1875).

= *Parmelia yunnana* Hue, *Nouv. Arch. Mus. Hist. Nat.* (Paris) 4(1):186(1899).

Type: Yunnan, Delavay, 3/VII/1888, in P(holotype, not seen, according to Hale, 1965, p. 333).

= *Parmelia yunnana* var. *yunnana* f. *subnuda* Zahlbr., *Hedwigia*, 74:210(1934).

Type: from Yunnan, Rock.

Shaanxi(Jatta, 1902, p. 469; Zahlbr., 1930 b, p. 191), Xizang(J. C. Wei, 1981, p. 85; J. C. Wei & Jiang, 1981, p. 1147 & 1986, p. 49; Chen J. B. et al., 2005, p. 104; Chen JB, 2015. p. 193), Yunnan(Hue, 1899, p. 186; Zahlbr., 1930 b, p. 191 & 1934 b, p. 210; Zhao et al. 1982, p. 56; Hale, 1965, p. 334; Hawksw. 1972, p. 52; Chen J. B. et al., 2005, p. 104; Chen JB, 2015. p. 192), Sichuan(Zahlbr., 1934 b, p. 210), Fujian(Zahlbr., 1930 b, p. 191), Taiwan(Zahlbr., 1933 c, p. 58; Wang & Lai, 1973, p. 93), China(prov. not indicated: Zahlbr., 1930 a, p. 271).

Parmotrema overeemii(Zahlbr.) Elix 奥氏大叶梅

Australas. Lichenol. 42: 22 – 27(1998).

≡*Parmelia overeemii* Zahlbr., *Annals Cryptog. Exot.* 1(2):204(1928).

= *Parmotrema subtropicum*(Zhao) J. C. Wei, *Enum. Lich. China* 181(1991).

≡*Parmelia subtropicum* Zhao, *Acta Phytotax. Sin.* 9: 165, 1964

Type: Yunnan, Xishuang- banna, 26 April 1957, Poliyangskii 1004(holotype in HMAS-L).

On bark.

Yunnan(Chen J. B. et al., 2005, pp. 104 – 105; Chen JB, 2015. p. 194), Hainan(Chen J. B. et al., 2005, pp. 104 – 105), Taiwan(Lai, 2000, p. 168).

Parmotrema parahypotropum(W. L. Culb.) Hale 近腹大叶梅

Phytologia 28:338(1974).

≡*Parmelia parahypotropa* W. L. Culb., *Bryologist* 76: 29(1973).

Taiwan(W. Culberson, 1973, p. ?; Lai, 2000, pp. 168 – 169).

Parmotrema perforatum(Wulfen in Jacq.) A. Massal. [as ' perforata'] 穿孔大叶梅

Atti Inst. Veneto Sci. lett., ed Arti, Sér. 3 5: 248(1860) [1859 – 1860].

≡*Lichen perforatus* Wulfen in Jacq., *Coll. Bot.* 1(1): 116(1786) [1786].

≡*Parmelia perforata*(Wulfen in Jacq.) Ach., *Meth. Lich.*: 217(1803).

≡*Imbricaria perforata*(Wulfen in Jacq.) Körb., *Lichenogr. German.*: 8(1846).

On bark of trees.

Shaanxi(Jatta, 1902, P. 469; Zahlbr., 1930 b, p. 193), Yunnan(Zhao, 1964, p. 164; Zhao et al. 1982, p. 54).

Notes: The record of this species from Yunnan by Zhao in1964 and Zhao et al. in 1982 remains to be revised (Chen JB, 2015. p. 267).

Parmotrema perlatum(Huds.) M. Choisy 珠光大叶梅

Bull. mens. Soc. linn. Soc. Bot. Lyon 21: 174(1952).

≡*Lichen perlatus* Huds., *Fl. Angl.* 1: 448(1762).

≡*Parmelia perlata*(Huds.) Ach., *Method. Lich.*: 216(1803).

≡*Imbricaria perlata*(Huds.) Körb., *Lichenogr. Germ.*: 8(1846).

= *Parmelia perlata* f. *sorediata*(Schaer.) Baroni, *Bull. Soc. Bot. Ital.*: 48(1894).

≡*Parmelia perlata* var. *sorediata* Schaer., *Enum. Critic. Lich. Europ.*: 34(1850).

= *Parmotrema trichoterum*(Hue) M. Choisy, *Bull. Mens. Soc. Linn. Lyon* 21: 175(1952).

≡*Parmelia trichotera* Hue, *J. Bot.* (Paris) 12: 245(1898).

= *Parmelia coniocarpa* Laurer, Linnaea 2: 39(1827), lectotype: Sieber, Australia(M) (Hale, 1961).

= *Parmelia coriacea* var. *perlata* Eschw in Martius, *Fl. Bras.* 1(1): 206(1833).

= *Parmelia trichotera* Hue, J. Bot. (Paris) 12: 245(1898). Lectotype(New): France Vendee, Ilede Noirmoutier, Viaud-Grand-Marais(PC); see Hale, 1961, fig. 4.

Misapplied name:

Parmotrema chinense(Osbeck) Hale & Ahti, *Taxon* 35: 133(1986).

≡*Lichen chinensis* Osbeck in Ostindisk Resa: 221(1757) nom. inval. Art. 32. 1c

On bark of trees.

[*Parmelia plumbeata* Zahlbr. in Handel-Mazzetti, *Symb. Sin.* 3: 188(1930) & Cat. Lich. Univ. 8: 564(1932).

Type: Yunnan, Handel-Mazzetti no. 12977.

On diabasic rocks.]

Heilongjiang(Chen et al. 1981, p. 153; Luo, 1984, p. 85; Chen J. B. et al.. 2005, p. 98, as *Parmotrema chinense*; Chen JB, 2015. p. 196), Jilin(Chen JB, 2015. p. 195), Liaoning(Chen et al. 1981, p. 153), Jilin (Chen et al. 1981, p. 153; Chen J. B. et al.. 2005, p. 98, as *Parmotrema chinense*), Shaanxi(Baroni, 1894, p. 48; Jatta, 1902, p. 468; Zahlbr., 1930 b, p. 183; He & Chen 1995, p. 44 as *P. chinense* without citation of specimens and their localities; Chen JB, 2015. p. 196), Guizhou(Zhao, 1964, p. 160; Zhao et al. 1982, p. 49; Zhang T & J. C. Wei, 2006, p. 9, as *P. chinense*; Chen JB, 2015. p. 196), Yunnan(R. Paulson, 1928, p. 317; Zahlbr., 1930 b, p. 193; Zhao, 1964, pp. 159, 160; Zhao et al. 1982, pp. 47, 49; Wang L. S. et al. 2008, p. 538, as *P. chinense*), Sichuan, hunan(Chen J. B. et al.. 2005, p. 98, as *Parmotrema chinense*; Chen JB, 2015. p. 196), Chongqing(Chen CL et al., 2008, p. 40 as *P. chinense*, miscited as *chinese*), Hebei(Tchou, 1935, p. 308; Chen J. B. et al.. 2005, p. 98, as *Parmotrema chinense*; Chen JB, 2015. p. 196), Anhui(Chen JB, 2015. p. 196), Hubei(Chen, Wu & J. C. Wei, 1989, p. 447; Chen J. B. et al.. 2005, p. 98, as *Parmotrema chinense*; Chen JB, 2015. p. 196), Guangdong(Rabh. 1873, p. 287; Krphbr. 1873, p. 471 & 1874c, p. 66; Zahlbr., 1930 b, p. 193), Hainan(Zhao, 1964, p. 160; Zhao et al. 1982, p. 49), Guangxi(Zhao, 1964, p. 160; Zhao et al. 1982, p.

49), Anhui(Xu B. S., 1989, p. 221), Zhejiang(Tchou, 1935, p. 308; Xu B. S., 1989, p. 221), Shandong, Hongkong(Tchou, 1935, p. 308), Taiwan(Zahlbr., 1933 c, p. 56; Wang-Yang & Lai, 1973, p. 94).

Parmotrema permutatum(Stirt.) Hale 双色大叶梅

Phytologia 28: 338(1974).

≡*Parmelia permutata* Stirt., *Scott. Natural.* 4: 252(1878) [1877 – 78].

On bark.

guizhou, guangxi, hongkong(Chen J. B. et al., 2005, p. 105), Yunnan(Chen J. B. et al., 2005, p. 105; Chen JB, 2015. p. 197), Taiwan(Lai, 2000, pp. 169 – 170).

Parmotrema praesorediosum(Nyl.) Hale 类粉缘大叶梅

Phytologia 28: 338(1974).

≡*Parmelia praesorediosa* Nyl., *Sert. Lich. Trop. Labuan Singapore*: 18(1891).

= *Parmelia sanctae-crucis* Vain., Ann, Acad, Sci. Fenn. ser. A, 6(7): 4(1915).;

= *Parmelia neglecta* Asahina, *J. Jap. Bot.* 17: 71(1941).

Type : Taiwan, 5 Jan. 1926, Y. Asahina F. 51(holotype in TNS).

= *Parmelia subcetrarioides* des Abbayes, *Bull. Inst. Franc.* Afr. Noire 13: 974(1951).

Misapplied name:

Parmelia perlata auct. non(Huds) Ach. in Seemann 1852 – 57: 432; & in Zahlbr., *Repert. Spec. Nov. Regni Veg.* 33: 58(1933), (fide Asahina 1952 & Kurokawa & Lai 2001) & Tchou, *Contr. Inst. Bot. Natl. Acad. Beiping* 3: 308(1935) for Guizhou, Yunnan, Guangxi, & Hongkong(fide Aptroor & Seaward 84(1999).

On trees. Pantropical.

Jilin(Chen, 1981, p. 153), Liaoning(as "Manchuria australis" cited by Asahina, 1952 a, p. 140), Xizang(J. C. Wei & Chen, 1974, p. 180), Chongqing(Chen CL et al., 2008, p. 40), Guizhou(Zhao, 1964, p. 161; Zhao et al. 1982, p. 49), Yunnan(Chen JB, 2015. p. 197), Shandong(Zhao ZT et al., 2002, p. 6), Zhejiang(Xu B. S., 1989, p. 221), Taiwan(Zahlbr., 1933 c, p. 58; Asahina, 1941 a, p. 71 & 1952 a, p. 140 & 1955, p. 222; Hale, 1965, p. 258; Togashi, 1968, p. 315; Lamb, 1963, p. 488; Wang & Lai, 1973, pp. 93, 94; Lai, 2000, pp. 170 – 171), Hainan(JCWei et al., 2013, p. 229), Hongkong(Seemann, 1852 – 57, p. 432; Tchou, 1935, p. 308; Thrower, 1988, pp. 16, 129; Chu, 1997, p. 48; Aptroot & Seawrd, 1999, p. 84; Aptroot et al., 2001, p. 335).

Parmotrema pseudonilgherrense(Asahina) Hale 粉尼尔山大叶梅

Mycotaxon 5(2): 441(1977).

≡*Parmelia pseudonilgherrensis* Asahina, *Journ. Jap. Bot.* 29(12): 370(1954).

Type: Jilin, Mt. Hokusui-Hakusan, Kankyo-Nando, 13 Aug. 1936, U. Tsutani s. n. (lectotype in TNS). Jilin (Lamb, 1963, p. 495; Chen J. B. et al., 2005, p. 105; Chen JB, 2015. p. 200), Shaanxi((He & Chen 1995, p. 44 without citation of specimens and their localities; Chen et al., 2005, p. 105; Chen JB, 2015. p. 200), Xizang (J. C. Wei & Y. M. Jiang, 1986, p. 49; Chen J. B. et al., 2005, p. 106; Chen JB, 2015. p. 200), Sichuan, Yunnan, Hebei(Chen J. B. et al., 2005, p. 105; Chen JB, 2015. p. 200), , Henan(Chen JB, 2015. p. 200), Hubei (Chen, Wu & J. C. Wei, 1989, p. 448; Chen J. B. et al., 2005, p. 106; Chen JB, 2015. p. 200).

Parmotrema rampoddense(Nyl.) Hale 粉芽大叶梅

Phytologia 28: 338(1974).

≡*Parmelia rampoddensis* Nyl., *Acta Soc. Sci. fenn.* 26(10): 7(1900).

= *Parmelia proboscidea* var. *sorediifera* Müll. Arg., *Flora*, Regensburg 67: 616(1884).

≡*Parmelia perlata* f. *sorediifera* Müll. Arg., *Flora*, Regensburg 74(3): 382(1891).

≡*Imbricaria perlata* var. *sorediifera*(Stizenb.) Jatta, *Nuovo G. bot. ital.* 9: 468(1902).

On bark & twigs.

Shaanxi(Jatta, 1902, p. 468; Zahlbr. , 1930 b, p. 191) , Xizang, Yunnan, Zhejiang, guangxi(Chen J. C. et al. , 2005, pp. 106 – 107; Chen JB, 2015. p. 202) , Taiwan(Hale, 1965, p. 305; Wang & Lai, 1973, p. 94; Lai, 2000, p. 171) , Hainan(Chen J. C. et al. , 2005, pp. 106 – 107; Chen JB, 2015. p. 202; JCWei et al. , 2013, p. 229) .

Parmotrema reticulatum (Taylor) M. Choisy 粉网大叶梅

Bull. Mens. Soc. Linn. Lyon 21: 175(1952) .

≡*Parmelia reticulata* Taylor, in Mackay, *Fl. Hibern.* 2: 148(1836) .

≡*Rimelia reticulata*(Taylor) Hale & Fletcher , *Bryologist* 93: 28(1990) .

= *Parmelia cetrata* f. *granularis* Asahina, J. Jap. Bot. 16: 593(1940) .

Type: China. Taiwan(Formosa) , Taichu, Y. Asahina 33125(holotype in TNS) .

= *Parmelia cetrata* f. *sorediifera* Vain. , *Etud. Lich. Bres.* 1: 40(1890) . ; Zahlbr. , in Handel-Mazzetii, *Symb. Sin.* 3: 190(1930) .

≡*Parmotrema cetratum* f. *sorediiferum*(Vain.) J. C. Wei, *Enum. Lich. China*: 177(1991) .

= *Parmelia clavulifera* Räsänen, *Ann. Bot. Soc. Zool. -Bot. Fenn.* 20(3) : 4(1944) .

≡*Parmotrema clavuliferum*(Räsänen) Streimann, *Biblthca Lichenol.* 22: 93(1986) .

≡*Rimelia clavuliferum*(Räsänen) Kurok, *J. Jap. Bot.* 66: 158(1991) .

≡*Parmelia clavulifera* Räsänen, *Ann. bot. Soc. Zool. -Bot. fenn. Vanamo* 20(3) : 4(1944) .

Taiwan(Lai, 2000, pp. 185 – 186) .

= *Parmelia urceolata* var. *subcetrata* Müll. Arg. , *Flora* 66: 46(1883) .

= *Canomuculina leucosemotheta*(Hue) Elix, *Mycotaxon* 65: 477(1997) .

Misapplied name: *Parmelia cetrata* Ach. (in Moreau & Moreau, 1951) .

On bark of trees, e. g. *Kandelia* sp. , and on granite bouder, and on rocks.

Cosmopolitan.

Jilin(Chen et al. 1981, p. 153; Chen JB, 2015, p. 203) , neimenggu(Sun LY et al. , 2000, p. 36) , Xizang(J. C. Wei & Y. M. Jiang, 1986, p. 47; Chen JB, 2015, p. 206) , Sichuan, Yunnan(Du Rietz, 1924, p. 333; Chen S. F. et al. , 1997, p. 222; Chen JB, 2015, p. 205) , guizhou(Zhang T & J. C. Wei, 2006, p. 9; Chen JB, 2015, p. 205) , Hubei(Chen, Wu & J. C. Wei, 1989, p. 447; Chen JB, 2015, p. 204) , Shandong(Zhao ZT et al. , 1998, p. 29; zhao ZT et al. , 1999, p. 427, miscited as*R. triticulata*; Zhao ZT et al. , 2002, p. 6; Li Y et al. , 2008, pp. 71 – 72; Chen JB, 2015, p. 204) , Jiangxi(J. C. Wei et al. 1982, p. 40; Chen JB, 2015, p. 204) , Jiangsu(Xu B. S. , 1989, p. 220) , , Anhui, Zhejiang(Xu B. S. , 1989, p. 220; Chen JB, 2015, p. 203) , Hunan, Fujian, Guangxi (Chen JB, 2015, p. 203 – 204) , Taiwan(Asahina, 1951g, p. 357 & 1952a, p. 119; Wang & Lai, 1973, p. 94; Lai, 2000, pp. 186 – 187) , Hainan(Chen JB, 2015, p. 203 – 204; JCWei et al. , 2013, p. 230) , Hongkong(Moreau & Moreau, 1951; Thrower, 1988, pp. 16, 130 ; Aptroot & Seaward, 1999, p. 93 as *Rimelia reticulate*) .

f. reticulatum 原变型

On bark of trees & on rocks.

Yunnan, Guizhou, Henan, Shandong, Anhui, Jiangxi, Zhejiang, Guangxi, Fujian(Zhao, 1964, P. 152; Zhao et al. 1982, p. 31) , Hongkong(Thrower, 1988, p. 130) .

f. nudum(Hue) J. C. Wei 裸叶变型

Enum. Lich. China 180(1991) .

≡*Parmelia reticulata* f. *nuda* Hue, Nouv. Arch. Mus. Hist. Nat. 4(1) : 177(1899) . Guizhou, Zhejiang(Zhao, 1964, p. 153; Zhao et al. 1982, p. 32) .

Parmotrema saccatilobum(Taylor) Hale 囊瓣大叶梅

Phytologia 28: 339(1974) .

≡*Parmelia saccatiloba* Taylor, *London J. Bot.* 6: 174(1847) .

= *Parmelia tinctorum* var. *inactiva* auct. non Zahlbr. : Zhao, *Acta Phytotax. Sin.* 9(2) : 162(1964).

Zhao et al. *Prodr. Lich. Sin.* 1982, p. 52.

On bark of trees.

Chongqing(Chen CL et al., 2008, p. 40), Yunnan(Chen J. C. et al., 2005, p. 107; Chen JB, 2015. p. 207), Anhui(Zhao, 1964, p. 162; Zhao et al. 1982, p. 52), Taiwan(Zahlbr., 1933 c, p. 57; Wang & Lai, 1973, p. 94; Lai, 2000, pp. 171 – 172), Hainan(Chen J. C. et al., 2005, p. 107; Chen JB, 2015. p. 207; JCWei et al., 2013, p. 229).

Parmotrema sancti-angelii(Lynge) Hale 缘毛大叶梅

Phytologia 28: 339(1974).

≡ *Parmelia sancti-angelii* Lynge, *Ark. Bot.* 13(no. 13) : 35(1914).

Parmelia pseudohyporysalea Asah. In Kihara, Fauna Fl. Nepal Himalaya 54(1955).

On rock.

Xizang(J. C. Wei & Jiang, 1981, p. 1147 & 1986, p. 49; Chen J. C. et al., 2005, p. 108; Chen JB, 2015. p. 208), Yunnan, guangxi(Chen J. C. et al., 2005, p. 108; Chen JB, 2015. p. 208), Taiwan(Wang-Yang & Lai, 1973, p. 94; Lai, 2000, pp. 172 – 173).

Parmotrema sipmanii Louwhoff & Elix 赛普曼大叶梅

Biblioth. Lichenol. 73: 115(1999).

On bark.

Fujian(Chen JB, 2015. p. 209).

Parmotrema subarnoldii(des Abbayes) Hale 长缘毛大叶梅

Phytologia 28: 339(1974).

≡ *Parmelia subarnoldii* des Abbayes, *Mém. Inst. Sci. Madagascar*, sér. B, 10: 113(1961).

On bark.

Yunnan(Chen JB, 2015. p. 210), Hainan(Chen JB, 2015. p. 210; JCWei et al., 2013, p. 229).

Parmotrema subcorallinum(Hale) Hale 亚珊瑚大叶梅

Phytologia 28(4) : 339(1974).

≡ *Parmelia subcorallina* Hale, *Journ. Jap. Bot.* 37: 345(1962).

Type: Taiwan, Y. Asahina no. 3312.

On bark.

Chongqing(Chen CL et al., 2008, p. 40), Taiwan(Hawksw. 1972, p. 51; Hale, 1962, p. 345; Wang & Lai, 1973, p. 94; Lai, 2000, pp. 173 – 174), Hainan(Chen JB, 2015. p. 211; JCWei et al., 2013, p. 229).

Parmotrema subisidiosum(Müll. Arg.) Hale 裂芽网纹大叶梅

Phytologia 28: 339(1974).

≡ *Parmelia cetrata* var. *subisidiosa* Müll. Arg., *Engler's Bot. Jahrb.* 20: 256(1894).

≡ *Rimelia subisidiosa*(Müll. Arg.) Hale & Fletcher, *Bryologist* 93: 29(1990).

≡ *Parmotrema cetratum* f. *subisidiosa*(Müll. Arg.) J. C. Wei, *Enum. Lich. China*: 177(1991).

On rocks and on bark.

Guihou, Yunnan, Fujian(Chen JB, 2015. p. 212), Hainan(Chen JB, 2015. p. 212; JCWei et al., 2013, p. 230).

Parmotrema sublatifolium(Zhao et al.) J. C. Wei 亚宽瓣大叶梅

Enum. Lich. China 180(1991).

≡ *Parmelia sublatifolia* Zhao et al. *Acta Phytotax. Sin.* 16(3) : 96(1978).

Type: Yunnan, Zhao & Chen no. 3187 in HMAS-L.

On bark of trees & on rocks.

Yunnan(Zhao et al. 1982, p. 47; Chen J. C. et al., 2005, p. 109; Chen JB, 2015. p. 213).

Parmotrema subochraceum Hale 亚黄褐大叶梅

in Jahns, Biblthca Lichenol. 38: 117(1990).

Yunnan(Chen J. C. et al., 2005, p. 109), Hainan(J. C. Wei et al., 2013, p. 230).

Parmotrema subrugatum(Kremp.) Hale 皱纹大叶梅

Phytologia 28: 339(1974).

≡*Parmelia subrugata* Kremp., *Verh. zool. -bot. Ges. Wien* 18: 320(1868).

=*Parmelia sinensis* Hue, *Nouv. Arch. Mus. Hist. Nat.* 4(1): 187(1899).

Type: Yunnan, Delavay, 8/VIII/1888(lectotype) in P(According to Hale, 1965, p. 341).

Misapplied name:

Parmelia ciliata sensu Hue, Bull. Soc. Bot. France, 34: 20(1887) & 36: 164(1889), non DC.

Parmelia latissima sensu Hue, ibid. 36: 164(1889), non alior. (fide Zahlbr. 1930b).

On bark of trees.

Yunnan(Hue, 1887a, p. 20 & 1889, p. 164 & 1899, p. 187; Zahlbr., 1930 b, p. 191; Zhao, 1964, p. 163; Zhao et al. 1982, p. 55; Hale, 1965, p. 341; Chen J. C. et al., 2005, p. 109; Chen JB, p. 213), Hongkong(Tchou, 1935, p. 308), Taiwan(Lai, 2000, p. 174), China(prov. not indicated: Zahlbr., 1930 a, p. 265; Culb. 1969, p. 461).

Parmotrema sulphuratum(Nees & Flot.) Hale 硫大叶梅

Phytologia 28: 339(1974).

≡*Parmelia sulphurata* Nees & Flot., *Linnaea* 9: 501(1835)[1834].

On bark of trees.

Yunnan(Hue, 1889, p. 163; Zahlbr., 1930 b, p. 183 & 1934, p. 206), Hunan, Fujian(Zahlbr., 1930 b, p. 183).

Notes: Whether the species *Parmelia sulphurata* Nees & Flot. reported by Hue in 1889 from Yunnan belongs to *Parmelia homogenes*(=*Myelochroa subaurulenta*) remains to be revised in the future(Zahlbr., 1930 b; Chen JB, 2015. p. 268).

Parmotrema tinctorum(Dilese ex Nyl.) Hale 大叶梅

Phytologia 28: 339(1974).

≡*Parmelia tinctorum* Despr. ex Nyl. *Flora* 55: 547(1872).

=*Parmelia pseudotinctorum* des Abb., Bull. Inst. Fr. Afr. Noire A, 13: 973(1951).

≡*Parmotrema pseudotinctorum*(des Abb.) Hale, *Phytologia* 28: 338(1974).

=*Lichen chinensis* Osbeck, *Ostindisk resa*, 221(1757). nom. inval. (Art. 32. 1c, revised by Hawksworth 2004).

On bark of trees & on rocks. Pantropical.

Beijing, Henan, Fujian (Zhao, 1964, p. 161; Zhao et al. 1982, p. 51; Chen JB, 2015. p. 217), Liaoning(Asahina, 1952a, p. 127; Chen et al. 1981, p. 154; J. C. Wei et al. 1982, p. 41), Shaanxi(Jatta, 1902, p. 469; Zahlbr., 1930 b, p. 190), Xizang(J. C. Wei & Jiang, 1986, p. 50; Chen J. B. et al., 2005, p. 110; Chen JB, 2015. p. 219), Sichuan(Chen J. B. et al., 2005, p. 110; Chen JB, 2015. p. 218), Chongqing(Chen CL et al., 2008, p. 40), Guizhou(Zhao, 1964, p. 161; Zhao et al. 1982, p. 51; Chen J. C. et al., 2005, p. 110; Zhang T & J. C. Wei, 2006, p. 9; Chen JB, 2015. p. 218), Yunnan(Hue, 1887a, p. 20 & 1889, p. 164; R. Paulson, 1928, p. 317; Zahlbr., 1930 b, p. 190; Zhao, 1964, p. 161; Zhao et al. 1982, p. 51 ; Chen J. B. et al., 2005, p. 106; Wang L. S. et al. 2008, p. 538; Chen JB, 2015. p. 218), Hebei(Chen J. B. et al., 2005, p. 110; Chen JB, 2015. p. 217), Shandong(Moreau et Moreau, 1951, p. 191; Zhao, 964, p. 161; Zhao et al. 1982, p. 51; Zhao ZT et al., 1998, p. 29; zhao ZT et al., 1999, p. 427; Zhao ZT et al., 2002, p. 6; Chen J. B. et al., 2005, p. 110; Hou YN et al., 2008, p. 68; Li Y et al., 2008, p. 72; Chen JB, 2015. p. 217), Jiangsu(Wu J. N. & Xiang T. 1981, p. 6, here should be p. 5 according to the sequence of the pages; Xu B. S., 1989, p. 221), hubei, hunan(Chen J. B. et al., 2005, p. 110; Chen JB, 2015. p. 217 – 218), Anhui(Zhao, 1964, p. 161; Zhao et al. 1982, p. 51; Xu B. S., 1989, p. 221; Chen J. B. et al., 2005, p. 110; Chen JB, 2015. p. 217), Shanghai, Zhejiang(Xu B. S., 1989, p. 221; Chen J. B. et al., 2005, p. 110; Chen JB, 2015. p. 217), Taiwan(Zahlbr., 1933 c, p. 56; M. Sato, 1957, p. 57; Wang & Lai, 1973, p. 94; Lai, 2000, pp. 174 – 175), Guangxi(Zhao, 1964, p. 161; Zhao et al. 1982, p. 51; Chen J. B. et al., 2005, p. 110; Chen JB, 2015. p. 218), Guangdong(D. Haksw., 2004, p. 40; Chen J. B. et al., 2005, p. 110; Chen JB, 2015. p. 218), Hainan(Chen J. B. et al., 2005, p. 110; Chen JB, 2015. p. 218; JCWei et al., 2013, p. 230), Hongkong(Thrower, 1980, 1988, pp. 16, 131; Aptroot & Seaward, 1999, p. 84; Seaward & Aptroot, 2005, p. 285, Pfister 1978: 79, as *Parmelia perlata*.; Chen JB, 2015. p. 219).

Notes:

1. The species *Parmelia pseudotinctorum* des Abb. (1951) is different from *Parmelia tinctorum* Despr. ex Nyl (1872) by the thick and sometimes irregular isidia only, and the *Parmotrema tinctorum* (Dilese ex Nyl.) Hale (1974) and *Parmotrema pseudotinctorum*(des Abb.) Hale(1974) were combined as well.

2. All seven specimens(the original materials of *Lichen chinensis*, the sheet LINN1273. 152 bears a single specimen, and the sheet LINN1273. 154 with six specimens), collected by Osbeck from the village of an island in Canton(Guangdong) from China in 1751 – 1752, appear to represent the same species *Parmotrema tinctorum* (Dilese ex Nyl.) Hale(D. Hawksw., 2004).

Parmotrema ultralucens(Krog) Hale 亚毛大叶梅

Phytologia 28: 339(1974).

≡*Parmelia ultralucens* Krog, *Bryologist* 77: 253(1974).

≡*Canomaculina ultralucens*(Krog) Elix & J. B. Chen, in Chen, Wang & Elix, *Mycotaxon*86: 21(2003).

On trees.

Jiangsu, Zhejiang(Xu B. S., 1989, p. 222), Taiwan(Lai, 2000, pp. 175 – 176), Hainan(Chen J. B. et al., 2003, p. 21; Chen JB, 2015. p. 220, on rocks).

Parmotrema zollingeri(Hepp) Hale 卓氏大叶梅

Phytologia 28: 339(1974).

≡*Parmelia zollingeri* Hepp, in Zollinger, *Syst. Verz.*: 9(1854).

On bark of trees and bamboo.

Yunnan(Zhao, 1964, p. 164; Zhao et al. 1982, p. 53), Taiwan(Asahina, 1952 a, p. 130 & 1952 b, p. 15; Wang & Lai, 1973, p. 94), Hainan(JCWei et al., 2013, p. 230).

Platismatia W. L. Culb. & C. F. Culb. **宽叶衣属**

Contr. U. S. natnl. Herb. 34: 524(1968).

Platismatia erosa W. Culb. & C. Culb. 裂芽宽叶衣

Contr. US Nat. Herb. 34(7): 526(1968).

Type: Taiwan, collected by Sasaki, in W(not seen).

= *Cetraria formosana* var. *isidiata* Zahlbr. in Fedde, *Repertorium*, 33: 60(1933).

Type: Taiwan, collected by Sasaki(Asahina no. 409).

Yunnan(Wei X. L. et al. 2007, p. 156), Xizang(J. C. Wei & Y. M. Jiang, 1981, p. 1147, & 1986, p. 50), Hubei (Chen, Wu & J. C. Wei, 1989, p. 440), Taiwan(Sato, 1938b, p. 785 & 1939, p. 64; Wang-Yang & Lai, 1973, p. 89 & 1976, p. 228; Lai, 2000, pp. 213 – 214).

Platismatia formosana(Zahlbr.) Culb. & C. Culb. 台湾宽叶衣

Contr. US Nat. Herb. 34: 529(1968).

≡*Cetraria formosana* Zahlbr., in Feddes, *Repertorium sp. nov.* 33: 59(1933).

Syntype: Taiwan, Mt. Ali shan, Numanaira, 7000(Asahina no. 89) & Mt. Morrison, Sasaki(Asahina no. 411).

Taiwan(Asahina, 1934c, p. 478; M. Sato, 1938b, p. 785 & 1939, p. 63; Lamb, 1963, p. 155; Wang-Yang & Lai,

1973, p. 89; Lai, 1980, p. 225; Lai, 2000, p. 214).

Platismatia glauca(L.) W. L. Culb. & C. F. Culb. 海绿宽叶衣

Contr. U. S. natnl. Herb. 34: 530(1968). f. glauca 原变型

≡*Lichen glaucus* L., *Sp. pl.* 2: 1148(1753).

≡*Cetraria glauca*(L.) Fr., *Lich. eur. reform.* (Lund): 38(1831).

≡*Platysma glaucum*(L.) Frege, *Deutsch. Botan. Taschenb.* 2: 167(1812).

Shaanxi(Jatta, 1902, p. 467; Zahlbr., 1930 b, p. 197), Yunnan(Hue, 1887a, p. 19 & 1889, p. 163).

f. coralloidea(Wallr.) Wei 珊瑚芽变型

Enum. Lich. China 207(1991).

≡*Parmelia glauca* μ. *coralloisea* Wallr. *Flora Crypt. Germ.* 3: 522(1831).

≡*Cetraria glauca*(L.) Ach. f. *coralloidea*(Wellr.) Koerb.

On bark of trees.

Sichuan(Zahlbr., 1934, p. 211).

Platismatia lacunosa(Ach.) W. L. Culb. & C. F. Culb. 多凹宽叶衣

Contr. U. S. natnl. Herb. 34: 541(1968).

≡*Cetraria lacunosa* Ach., *Methodus*, Sectio prior: 295, tab. V, fig. 3(1803).

On bark of trees.

Taiwan(Zahlbr., 1933 c, p. 59; Wang & Lai, 1973, p. 89).

Platismatia tuckermanii(Oakes) W. L. Culb. & C. F. Culb. 涂氏宽叶衣

Contr. U. S. natnl. Herb. 34: 549(1968).

≡*Cetraria tuckermanii* Oakes apud Tuck. in *Americ. Journ. Science & Arts*, XLV: 48(1843), non Herre(1906).

= *Cetraria lacunosa* var. *atlantica* Tuck.

= *Cetraria atlantica*(Tuck.) Du Rietz

Hebei(Mt. Wuling shan, Moreau et Moreau, 1951, p. 191).

Pleurosticta Petrak **皮叶属**

Kryptog. Forsh. 2: 190(1931).

Pleurosticta acetabulum(Neck.) Elix & Lumbsch 碟形皮叶

in Lumbsch, Kothe & Elix, *Mycotaxon* 33: 453(1988).

≡*Lichen acetabulum* Neck., *Deliciae Gallo-Belgic.* 2: 506(1768).

≡*Melanelia acetabulum*(Neck.) Essl., *Mycotaxon* 7(1): 47(1978).

≡*Imbricaria acetabulum*(Neck.) DC., in Lamarck & de Candolle, *Fl. franç.*, Edn 3(Paris) 2: 392(1805).

≡*Parmelia acetabulum* (Neck.) Duby, *Bot. Gall.*, Edn 2(Paris) 2: 601(1830).

On bark of trees.

Shaanxi(Jatta, 1902, p. 470; Zahlbr., 1930 b, p. 186).

Notes: The specimens of this species have been not found in China so far. The records of it in Shaanxi seems to be based on erroneous identification. The final solution of this problem remains to check the original materials cited by Jatta in 1902 and Zahlbr. in 1930.

Pleurosticta koflerae(Clauzade & Poelt) Elix & Lumbsch 石生皮叶

in Lumbsch, Kothe & Elix, *Mycotaxon* 33: 453(1988).

≡*Parmelia koflerae* Clauzade & Poelt, *Nova Hedwigia* 3: 368(1961).

≡*Melanelia koflerae*(Clauzade & Poelt) Essl., *Mycotaxon* 7: 47(1978).

On moss covered rocks.

Xinjiang(Chen JB, 2015, p. 221).

Protoparmelia M. Choisy **原梅属**

Bull. Soc. bot. Fr. 76: 523(1929).

Protoparmelia atriseda(Fr.) R. Sant. & V. Wirth 黑原梅

in Wirth, *Die Flecht. Baden-Württembergs. Verbreitungsatlas* (Stuttgart): 511(1987).

≡*Parmelia badia* var. *atriseda* Fr., *Nov. Sched. Critic. Lich.*: 6(1827).

On shale.

Taiwan(Aptroot and Sparrius, 2003, p. 36).

Protoparmelia badia(Hoffm.) Hafellner 褐原梅

Beih. Nova Hedwigia 79: 292(1984).

≡*Verrucaria badia* Hoffm., *Deutschl. Fl., Zweiter Theil(Erlangen)*: 182(1796)[1795].

On shale.

Taiwan(Aptroot and Sparrius, 2003, p. 36).

Pseudephebe M. Choisy **拟毡衣属**

Icon. Lich. Univ., ser. 2 1: [unnumbered](1930).

Pseudephebe pubescens(L.) M. Choisy 柔毛拟毡衣

Icon. Lich. Univ. 2: sine pag. (1930).

≡*Lichen pubescens* L., *Sp. pl.* 2: 1155(1753).

On rock.

Sichuan, Xizang(Wang LS & McCune, 2010, pp. 432 – 434).

Pseudevernia Zopf **拟扁枝衣属**

Beih. Botan. Centralbl. 14: 124(1903).

Pseudevernia furfuracea(L.) Zopf 拟扁枝衣

Beih. Botan. Centralbl. 14: 124(1903).

≡*Lichen furfuraceus* L., *Sp. pl.* 2: 1146(1753).

Xinjiang(Wu JN et al., 1997, p. 14; J. C. Wei & A. Abas, 2003, pp. 27 – 28).

Punctelia Krog **星点梅属**

Nordic Jl Bot. 2(3): 290(1982).

Punctelia borreri(Sm.) Krog 粉斑星点梅

Nordic Jl Bot. 2(3): 291(1982).

≡*Lichen borreri* Sm., in Smith & Sowerby, *Engl. Bot.* 25: tab. 1780(1807).

≡*Parmelia borreri*(Sm.) Turner, *Trans. Linn. Soc. London* 9: 135(1806).

= *Parmelia pseudoborreri* Asahina, *Journ. Jap. Bot.* 26(9): 259(1951).

≡*Parmelia borreri* var. *pseudoborreri* (Asahina) Targé & Lambinon, *Bull. Soc. R. Bot. Belg.* 98: 306(1965).

Type: North-eastern China(cited by Asahina as Manchuria).

Misapplied name:

Parmelia dubia auct. non(Wulf.) Schaer.: Zhao, *Acta Phytotax. Sin.* 9: 152(1964); Zhao et al. *Prodr. Lich. Sin.* 1982, p. 32(pr. p. maj.).

On bark of trees.

Beijing(Zhao, 1964, p. 152; Zhao et al. 1982, p. 32; Wei et al. 1982, p. 38; Chen JB, 2015. p. 223), Heilongjiang(Chen et al. 1981, pp. 151, 153; Chen JB, 2015. p. 223), Liaoning(Chen JB, 2015. p. 223), Neimenggu(Chen et al. 1981, p. 153; Sun LY et al., 2000, p. 36; Chen JB, 2015. p. 223), Ningxia(Zhao, 1964, pp. 152 – 153; Zhao et al. 1982, p. 32; Chen JB, 2015. p. 224),, Gansu, Shaanxi(He & Chen 1995, p. 45 without citation of specimens and their localies; Chen JB, 2015. p. 224), Xizang(Wei & Chen, 1974, p. 179), Sichuan (Zhao, 1964, pp. 152 – 153; Zhao et al. 1982, p. 32; Chen S. F. et al., 1997, p. 222; Chen JB, 2015. p. 224), Guizhou(Zhao, 1964, pp. 152 – 153; Zhao et al. 1982, p. 32; Chen JB, 2015. p. 224), Yunnan(Hue, 1889, p.

165; Zhao, 1964, p. 152; Zhao et al. 1982, p. 32; Chen JB, 2015. p. 224), Hebei(Moreau et Moreau, 1951, p. 190; Zhao, 1964, p. 152; Zhao et al. 1985, p. 32; Chen JB, 2015. p. 223), Shandong(Zhao ZT et al., 1998, p. 29; Zhang F et al., 1999, p. 30; Zhao ZT et al., 2002, p. 6; Hou YN et al., 2008, p. 68; Li Y et al., 2008, p. 72; Chen JB, 2015. p. 224), Anhui(Zhao, 1964, pp. 152 – 153; Zhao et al. 1982, p. 32: Chen JB, 2015. p. 224), Hubei(Chen, Wu & J. C. Wei, 1989, p. 441; Chen JB, 2015. p. 224), Hunan(Chen JB, 2015. p. 224), Zhejiang(Zhao, 1964, pp. 152 – 153; Zhao et al. 1982, p. 32), Fujian(Chen JB, 2015. p. 224), Taiwan(Lai, 2000, pp. 177 – 178).

Punctelia neutralis(Hale) Krog 中星点梅

Nordic Jl Bot. 2(3): 291(1982).

≡*Parmelia neutralis* Hale, *Phytologia* 22(2): 94(1971).

Taiwan(Lai, 2000, p. 178).

Punctelia perreticulata(Räsänen) G. Wilh. & Ladd 星点梅

Mycotaxon 28(1): 249(1987).

≡*Parmelia duboscqii* var. *perreticulata* Räsänen, *Ann. bot. Soc. Zool. -Bot. fenn. Vanamo*20(3): 3(1944).

On *Cryptomeria japonica* and *Castanopsis* p.

Taiwan(Aptroot and Sparrius, 2003, p. 37).

Punctelia rudecta(Ach.) Krog 粗星点梅

Nordic Jl Bot. 2(3): 291(1982).

≡*Parmelia rudecta* Ach., *Syn. meth. lich*. (Lund): 197(1814).

≡*Imbricaria rudecta*(Ach.) Nyl.

= *Parmelia ruderata* Vain., *Bot. Mag. Tokyo* 35: 47(1921).

= *Parmelia rudecta* var. *microphyllina* Nyl.; Zahlbr., *Hedwigia* 74: 209(1934).

On bark of trees and also on rocks and on moss covering rocks.

Beijing(Zhao, 1964, p. 151; Zhao et al. 1982, p. 29; Chen JB, 2015. p. 226), Heilonhjiang(Chen et al. 1981, p. 153), Jilin(Chen JB, 2015. p. 226), Shaanxi(Jatta, 1902, p. 470; Zahlbr., 1930 b, p. 185; Chen JB, 2015. p. 226), Chongqing(Chen JB, 2015. p. 226), Sichuan(Zahlbr., 1934 b, p. 209), Guizhou(Zhao, 1964, p. 151; Zhao et al. 1982, p. 29; Moreau et Moreau, 1951, p. 190; Chen JB, 2015. p. 226), Yunnan(Zahlbr., 1934 b, p. 209; Zhao, 1964, p. 151; Zhao et al. 1982, p. 29; Moreau et Moreau, 1951, p. 190; Chen JB, 2015. p. 226), Anhui(Zhao, 1964, p. 151; Zhao et al. 1982, p. 29; Moreau et Moreau, 1951, p. 190; Chen JB, 2015. P. 226), Jiangxi(Zhao, 1964, p. 151; Zhao et al. 1982, p. 29; Moreau et Moreau, 1951, p. 190), Hebei(Zahlbr., 1930 b, p. 185; Chen JB, 2015. p. 226), Shandong(Zhao ZT et al., 1998, p. 29; Zhang F et al., 1999, p. 30; Zhao ZT et al., 2002, p. 6, the authors probably miscited "*Punctelia ruderata*(Ach.) Krog" for *Punctelia rudecta*(Ach.) Krog.; Hou YN et al., 2008, p. 68; Li Y et al., 2008, p. 72; Chen JB, 2015. p. 226), Henan(Zhao, 1964, p. 151; Zhao et al. 1982, p. 29; Chen JB, 2015. p. 226), Hubei(Chen, Wu & J. C. Wei, 1989, p. 443), Hunan (Chen JB, 2015. p. 226), Zhejiang, Guangxi(Chen JB, 2015 p. 226), Taiwan(Wang & Lai, 1973, p. 94; Lai, 2000, pp. 178 – 179), North-eastern China("Manchuria": Asahina, 1951, p. 260).

Punctelia subflava(Taylor) Elix & Johnst. 裂片星点梅

Mycotaxon 31: 50(1988).

≡*Parmelia subflava* Taylor, in W. J. Hooker, *London J. Bot*. 6: 174(1847).

On bark and on rocks.

Beijing, Jilin, Guizhou, Yunnan, Shandong, Anhui, Hubei, Hunan, Jiangxi, Zhejiang, Fujian(Chen JB, 2015. p. 227).

Punctelia subrudecta(Nyl.) Krog 亚粗星点梅

Nordic Jl Bot. 2(3): 291(1982).

≡*Parmelia subrudecta* Nyl. , *Flora*, Regensburg 69: 320(1886).

=*Parmelia dubia*(Wulfen) Schaer. , *Lich. Helv. Spicil.* 10: 453(1840).

≡*Lichen dubius* Wulfen in Jacqu. , *Collect. Bot.* 4: 275(1790).

=*Parmelia dubia* var. *coralloidea*(Müll. Arg.) Zahlbr. , in Handel-Mazzetti, *Symb. Sin.* 3: 185(1930).

≡*Parmelia hypoleuca* var. *coralloidea* Müll. Arg. , *Flora* 70: 317(1887).

Misapplied name:

Parmelia borreri auct. non(Sm.) Turner: Hue, *Bull. Soc. Bot. France* 36: 165(1889) (fide Zahlbr. 1930b, p. 185).

Imbricaria borreri auct. non(Sm.) Körb. : Jatta, *Nuov. Giorn. Bot. Ital.* ser. 2, 9: 469(1902). (fide Zahlbr. 1930b, p. 185).

On trees.

Jilin(Chen JB, 2015. p. 228), Shaanxi(Jatta, 1902, p. 469; Zahlbr. , 1930 b, p. 185), Sichuan(Zahlbr. , 1930 b, p. 185), Guizhou(Zhao et al. 1982, p. 32, Wang Q. Z. no. 2596; Chen JB, 2015. p. 228), Yunnan(Hue, 1899, p. 152; Harmand, 1928, p. 325; Zahlbr. , 1930 b, p. 185; Chen JB, 2015. p. 228), Hubei(Chen JB, 2015. p. 228), Hunan(Zahlbr. , 1930 b, p. 185), Shandong(Hou YN et al. , 2008, p. 68; Chen JB, 2015. p. 228), Taiwan(Lai, 2000, p. 179).

[=*Parmelia dubia* var. *coralloides*(Müll. Arg.) Zahlbr. in Cat. Lich. VI: 155(1930)].

[=*Imbricaria borreri* var. *coralloidea*(Müll. Arg.) Jatta I: 470].

On the ground & on rocks among mosses.

Shaanxi(Jatta, 1902, p. 470; Zahlbr. , 1930 b, p. 185).

Parmelia dubia f. *ulophylla* Sandst. , Abh. naturw. Ver. Bremen 21: 202(1912).

Parmelia caperata var. *ulophylla* Ach. , Lich. univ. : 458(1810).

[=*Imbricaria borreri* var. *ulophylla*(Ach.) Jatta I: 470].

On tree trunk & on moss-covered rocks.

Notes: According to my examination of the collections from Ningxia(no. 2009), Sichuan(nos. 242 a, 705), Yunnan(nos. 2324, 2476), Guizhou(no. 81 a), Zhejiang(nos. 6194 a, 6211), and Anhui(no. 5394), cited by Zhao as *Parmelia dubia*(Wulf.) Schaer. in 1964(p. 152) and by Zhao et al. in 1982(p. 32), all the collections contain gyrophoric acid. So they should belong to *Parmelia borreri*(Sm.) Turn. , which is placed by Krog(1982) in a new genus *Punctelia* rather than to *Parmelia dubia*. Among the specimens cited by Zhao(1964) and by Zhao et al. (1982), only one specimen, no. 259 b from Guizhou, contains lecanoric acid and therefore belongs to so called *Parmelia dubia*. However, the correct name of *Parmelia dubia* should be *Parmelia subrudecta* Nyl. , which is now acceted as *Punctelia subrudecta*(Nyl.) Krog.

The two varieties of this species mentioned above should be revised based on the original materials cited by Hue(1899), Jatta(1902), and Zahlbr. (1930) in the future.

Relicina(Hale & Kurok.) Hale 球针黄叶属

Phytologia 28: 484(1974).

≡*Parmelia* ser. *Relicinae* Hale & Kurok. , *Contr. U. S. natnl. Herb.* 36: 135(1964).

Relicina abstrusa(Vain.) Hale 玄球针黄叶

Phytologia 28: 484(1974).

≡*Parmelia abstrusa* Vain. , *Acta Soc. Fauna Flora fenn.* 7(no. 1): 64(1890).

On rocks in damp, on bark & rotten wood.

Guangxi(Chen JB et al. , 2009, p. 95; Chen JB, 2015. p. 230), Taiwan(M. Sato, 1933, p. 272 & 1951, p. 196; Asahina, 1952a, p. 70; Kurok. 1965, p. 265; Wang-Yang & Lai, 1973, p. 93 & 1976, p. 228; Lai, 2000, pp. 180 – 181), Hainan(Chen JB et al. , 2009, p. 95; Chen JB, 2015. p. 230; JCWei et al. , 2013, p. 230), Hongkong (Thrower, 1988, pp. 17, 164; Aptroot & Seawar, 1999, p. 93; Aptroot & Sipman, 2001, p. 340; Seaward & Aptroot, 2005, p. 286, Pfister 1978: 80, as *Parmelia tiliacea* var. *flavicans*.).

Relicina limbata(Laurer) Hale 镶边球针黄叶

Phytologia 28: 484(1974).

≡*Parmelia limbata* Laurer, *Linnaea* 2: 39(1827).

On the ground & rocks.

Taiwan(Asahina, 1933 c, p. 54; Asahina, 1934a, p. 300 & 1951b, p. 197; Wang & Lai, 1973, p. 93; Ikoma, 1983, p. 82).

Relicina malesiana(Hale) Hale 马来球针黄叶

Phytologia 28(5): 484(1974).

≡*Parmelia malesiana* Hale, *J. J. B.* 40: 203(1965).

Taiwan(Hale, 1965, p. 203; Kurok. 1965, p. 266; Wang-Yang & Lai, 1973, p. 93 & 1976, p. 228; Lai, 2000, pp. 181 – 182).

Relicina planiuscula(Kurok.) Hale 平球针黄叶

Phytologia 28: 484(1974).

≡*Parmelia planiuscula* Kurok. , in Hale & Kurokawa, *Contr. U. S. natnl. Herb.* 36: 144(1964).

Taiwan(Wang & Lai, 1976, p. 228).

Relicina relicinula(Müll. Arg.) Hale 小球针黄叶

Smithson. Contr. bot. 26: 27(1975).

≡*Parmelia relicinula* Müll. Arg. , *Flora*, Regensburg 65: 317(1882).

= *Parmelia relicina* (Müll. Arg.) Fr. (Hale, 1976, p. 14).

On bark of *Betula* sp.

Shaanxi(Moreau et Moreau, 1951, p. 190).

Relicina schizospatha(Kurok.) Hale 裂球针黄叶

Phytologia 28: 485(1974).

≡*Parmelia schizospatha* Kurok. , in Hale & Kurokawa, *Contr. U. S. natnl. Herb.* 36: 146(1964).

Taiwan(Kurok. 1965, p. 267; Wang & Lai, 1973, p. 94; Lai, 2000, p. 182).

Relicina subabstrusa(Gyeln.) Hale 亚玄球针黄叶

Phytologia 28: 485(1974).

≡*Parmelia subabstrusa* Gyeln. , *Reprium nov. Spec. Regni veg.* 29: 288(1931).

On tree trunk of *Actinophoeus macarthurii.*

Taiwan(Kurok. 1965, p. 267; Wang-Yang & Lai, 1973, p. 94 & 1976, p. 228; Lai, 2000, pp. 182 – 183).

Relicina sydneyensis(Gyeln.) Hale 悉尼球针黄叶

Phytologia 28: 485(1974).

≡*Parmelia sydneyensis* Gyeln. , *Annls mycol.* 36(4): 292(1938).

On bark.

Hainan(Chen JB et al. , 2009, p. 95; JCWei et al. , 2013, p. 230; Chen JB, 2015, p. 231), Taiwan(Wang-Yang & Lai, 1976, p. 228; Kashiw. 1979, p. 214; Lai, 2000, p. 183).

= *Relicina subturgida*(Kurok.) Hale, *Phytologia* 28(5): 485(1974).

≡*Parmelia subturgida* Kurok. , *J. Jap. Bot.* 40(9): 268(1965).

Taiwan(Kurok, 1965, p. 268; Wang & Lai, 1973, p. 94).

Remototrachyna Divakar & A. Crespo **糙梅属**

in Divakar, Lumbsch, Ferencová, Prado & Crespo, *Am. J. Bot.* 97: 584(2010).

Remototrachyna consimilis(Vain.) Flakus, Kukwa & Sipman 同糙梅

in Flakus, Rodríguez Saavedra & Kukwa, *Mycotaxon* 119: 161(2012).

≡*Hypotrachyna consimilis*(Vain.) Hale, *Smithson. Contr. bot.* 25: 28(1975).

≡*Parmelia consimilis* Vain. , *Étud. Class. Lich. Brésil* 1: 58(1890).

Taiwan(Kurok. 1979, p. 128).

Sulcaria Bystrek **槽枝属**

Annls Univ. Mariae Curie-Skłodowska, Sect. C, Biol. 26: 275(1971).

Sulcaria sulcata(Lév.) Bystrek ex Brodo & D. Hawksw. 槽枝衣

Op. bot. 42: 156(1977).

Sulcaria sulcata(Lév.) Bystrek, *Annls Univ. Mariae Curie-Skłodowska, Sect. C, Biol.* 26(21): 276(1971). Nom. inval., Art. 41.4(Melbourre).

≡*Cornicularia sulcata* Lév., in Jacquem., *Voyage dans l'Indie, Botan.*: 179(1844) [1841 - 44].

≡*Alectoria sulcata* (Lév.) Nyl., *Mém. Soc. Imp. Sci. Nat. Cherbourg* 5: 98(1857).

≡*Bryopogon sulcatus*(Lév.) Gyeln., *Reprium nov. Spec. Regni veg.* 38: 244(1935).

Shaanxi(He & Chen 1995, p. 45, miscited as *Sucaria sulcata*, without citation of specimens and their localities).

Chemical race 1a(atranorin /major, psoromic acid/major, methyl psoromate, 2'-O-demethyl-psoromic acid, 2-methoxypsoromic acid, unknowns):

Xizang(Obermayer & Elix, 2003, p. 35).

Chemical race 1b(atranorin /major, psoromic acid/major, methyl psoromate, vulpinic acid/major, unknowns):

Sichuan(Obermayer & Elix, 2003, p. 36).

Chemical 2a(atranorin /major, 2-methoxypsoromic acid, 2-hydroxypsoromic acid, unknowns):

Sichuan, Xizang(Tibet) (Obermayer & Elix, 2003, p. 36 - 37), Yunnan(Obermayer & Elix, 2003, p. 36 - 37; Wei X. L. et al. 2007, p. 158).

Chemicl race 2b(atranorin /major, 2-methoxypsoromic acid, vulpinic acid, unknowns):

Xizang(Tibet) (Obermayer & Elix, 2003, pp. 38 - 39).

Chemicl race 3a(atranorin /major, 2-hydroxyvirensic acid/major, 2-hydroxyconvirensic acid/major, virensic acid/minor, unknowns):

Xizang(Tibet) (Obermayer & Elix, 2003, p. 39).

Chemicl race 3b(atranorin /major, virensic acid/major, 2-hydroxyvirensic acid/minor[in apothecia], convirensic acid/minor, unknowns):

Xizang(Obermayer & Elix, 2003, p. 41).

var. sulcata 原变种

On trees.

Shaanxi(Jatta, 1902, p. 462; A. Zahlbruckner, 1930 b, p. 202), Sichuan(A. Zahlbruckner, 1930 b, p. 202 & 1934b, p. 212; M. Sato, 1957, p. 62), Yunnan(Hue, 1887a, p. 20 & 1899, p. 91; A. Zahlbruckner, 1930 a, p. 397 & 1930 b, p. 202 & 1934b, p. 212; Du Rietz, 1926, p. 19; M. Sato, 1957, p. 62; Wei, 1981, p. 87; Wei et al. 1982, p. 49; Wu JN, Wang LS, p. 42, as f.), Hubei(Chen, Wu & Wei, 1989, p. 453), Anhui, Zhejiang(Xu B. S., 1989, p. 224 - 225), Taiwan(Asahina, 1936, p. 694; M. Sato, 1937b, p. 595 & 1957, p. 62; Asahina & M. Sato in Asahina, 1939, p. 749; Wang & Lai, 1973, p. 87), China(prov. not indicated: Gyeln. 1935, p. 250).

var. barbata(D. Hawksw.) D. Hawksw. 髯毛变种

in Brodo & Hawksw. *Opera Botanica*, 42: 156(1977).

≡*Alectoria sulcata* var. *barbata* D. Hawksw. Taxon, 19: 242(1970).

Type: Taiwan, alt. 3300 - 3600 m, 1/I/1964, Kurok. no. 337(holotype) in TNS(not seen).

Taiwan(D. Hawksw. 1971, p. 21; Brodo & D. Hawksw. 1977, p. 156).

f. vulpinoides(Zahlbr.) D. Hawksw. 黄枝变型

in Brodo & Hawksw. Opera Botanica, 42: 156(1977).

≡*Alectoria sulcata* var. *vulpinoides* Zahlbr. in Handel-Mazzetti, Symb. Sin. 3: 202(1930).

Type: Yunnan, near Lijiang, Mt. Yulong shan, Handel-Mazzetti no. 3606(holotype) in W

≡*Bryopogon vulpinoides* Zahlbr.) Gyeln. in Fedde, *Repertorium*, 38: 251(1935) .

Yunnan(Du Rietz, 1926, p. 20; Zahlbr. , 1932a, p. 573; Gyeln. 1935, p. 251; Wu JN, Wang LS, p. 42), Xizang (J. C. Wei & Jiang, 1981, p. 1146 & 1986, p. 65) .

Sulcaria virens(Taylor) Bystrek ex Brodo & D. Hawksw. 绿丝槽枝

Op. bot. Soc. bot. Lund 42: 254(1977) .

≡*Alectoria virens* Taylor, in Hook. , *London Journ. of Botan.* , 6: 166(1847) . (1930) .

≡*Bryopogon virens*(Tayl.) Gyeln. , *Feddes Repert.* 38: 223, 229, 254(1935) .

On bark of trees.

Yunnan(Hue, 1899, p. 96; Du Rietz, 1926, p. 19; Zahlbr. , 1930 b, p. 202; M. Sato, 1937b, p. 596; Wei X. L. et al. 2007, p. 158) , Sichuan(Du Rietz, 1926, p. 19) , Taiwan(Zahlbr. , 1933 c, p. 61; Gyeln. 1935, p. 229; Asahina, 1936c, p. 693; M. Sato, 1937b, p. 596; Asahina & M. Sato in Asahina, 1939, p. 747; Wang & Lai, 1973, p. 87; Ikoma, 1983, p. 25) .

var. virens 原变种

Yunnan , Xizang, Taiwan(D. Hawksw. 1971, p. 339) .

var. forrestii(D. Hawksw.) D. Hawksw. 佛氏变种

in Brodo & Hawksw. *Opera Botanica*, 42: 156(1977) .

≡*Alectoria virens* var. *forrestii* D. Hawksw. *Misc. Bryol. Lichenol.* , *Nichinan* 5: 1(1969) .

Type: Yunnan, near the lake "Erhai" in Dali, G. Forrest no. 13471.

On rocks & trees in coniferous forest.

Yunnan(D. Hawksw. 1971, pp. 338, 341 & 1972, p. 3) .

Tuckermanopsis Gyeln. **土可曼衣属**

Acta Faun. Fl. Univers. , *Ser.* 2, *Bot.* 1(5 – 6) : 6(1933) .

Tuckermanopsis americana(Spreng.) Hale [as ' Tuckermannopsis'] 美洲土可曼衣

in Egan, *Bryologist* 90(2) : 164(1987) .

≡*Nephroma americanum* Spreng. , *K. svenska Vetensk-Akad. Handl.* 8: 49(1820) .

≡*Nephromopsis americana*(Spreng.) Divakar, Crespo & Lumbsch in *Fungal Diversity* 11 April 2017, p.

On tree trunk & twigs.

Neimenggu(Wei, 1981, p. 86, as *C. microphillica*, see Lai et al. , 2009, p. 376; Chen et al. 1981, p. 129, as *C. ciliaris*; Chen LH et al. , 2006, pp. 503 – 504) , Heilongjiang(Lai et al. , 2009, p. 376; Chen LH et al. , 2006, pp. 503 – 504) .

Tuckermanopsis chlorophylla(Willd.) Hale [as ' *Tuckermannopsis*'] 绿色土可曼衣

in Egan, *Bryologist* 90(2) : 164(1987) .

≡*Lichen chlorophyllus* Willd. , in Humboldt, *Fl. Friberg. Spec.* (Berlin) : 20(1793) .

≡ *Cetraria chlorophylla*(Willd.) Poetsch, in Poetsch & Schiedermayr, *Meddn Soc. Fauna Flora fenn.* 6: 121 (1872) .

≡*Nephromopsis chlorophylla*(Willd.) Divakar, Crespo & Lumbsch in *Fungal Diversity* 11 April 2017, p.

On *Abies* sp.

Xizang(Obermayer 2004, p. 513) .

Tuckermanopsis ciliaris(Ach.) Gyeln. 土可曼衣

Acta Faun. Fl. Univers. , *Ser.* 2, *Bot.* 1(5 – 6) : 6(1933) .

≡*Cetraria ciliaris* Ach. , *Lich. univ.* : 508(1810) .

≡*Nephromopsis ciliaris(Ach.) Hue, Nouv. Arch. Mus. Hist. Nat.* , *Paris,* 4 sér. 1: 216(1899) .

≡*Platysma ciliare*(Ach.) Frege, *Deutsch. Botan. Taschenb.* : 162(1812) .

Corticolous.

Yunnan(Hue, 1899, p. 216; Paulson, 1928, p. 317; Zahlbr. , 1930 b, p. 199; M. Sato, 1939, p. 52) , Taiwan

(Zahlbr., 1933 c, p. 61; M. Sato, 1938 b, p. 783 & 1939, p. 52; Wang & Lai, 1973, p. 88).

It is impossible to recognize if this species belongs to *Tukermannopsis ciliaris* or to *T. halei*(Culb. & C. Culb.) Lai without any information of chemical data in literature. So, it i remains to be revised in the future.

Tuckermanopsis gilva(Asahina) M. J. Lai [as 'Tuckermannopsis'] 黄褐土可曼衣

Quarterly Journal of theTaiwan Museum 33(3 & 4): 225(1980).

≡*Cetraria gilva* Asahina, *J. J. B.* 28: 139(1953).

On trees.

Yunnan(Wei X. L. et al. 2007, p. 158).

Tuckermanopsis oakesiana(Tuck.) Hale [as 'Tuckermannopsis'] 阿克萨土可曼衣

in Egan, *Bryologist* 90(2): 164(1987).

≡*Cetraria oakesiana* Tuck., *Boston J. Nat. Hist.* 3: 445(1841).

On bark of trees.

Hubei(Chen, Wu & J. C. Wei, 1989, p. 433).

Tuckermanopsis ulophylloides(Asahina) M. J. Lai 卷缘土可曼衣

Quarterly Journal of the Taiwan Museum 33(3 & 4): 226(1980).

≡*Cetraria ulophylloides* Asahina, *J. J. B.* 28: 138(1953).

Liaoning, Heilongjiang(Lai et al., 2009, p. 377).

Tuckermanopsis weii(X. Q. Gao & L. H. Chen) Randlane & Saag 魏氏土可曼衣

Mycotaxon 87: 479(2003).

≡*Nephromopsis weii* X. Q. Gao & L. H. Chen, in Chen & Gao, *Mycotaxon* 77: 492(2001).

Type: Fujian, Mt. Wuyishan, 24/IV/1988. X. Q. Gao7619(HMAS-L, holotypus, & PS, isotypus).

On bark of *Pinus* sp.

Tuckneraria Randlane & A. Thell **缘毛衣属**

Acta bot. fenn. 150: 144(1994).

Tuckneraria ahtii Randlane & Saag 艾氏缘毛衣

in Randlane, Saag, Thell & Kärnefelt, *Acta bot. fenn.* 150: 147(1994).

Yunnan(Wei X. L. et al. 2007, p. 158), Taiwan(Lai, 2000, p. 215).

Tuckneraria laureri(Kremp.) Randlane & A. Thell 拉氏缘毛衣

in Randlane, Saag, Thell & Kärnefelt, *Acta bot. fenn.* 150: 149(1994).

≡*Cetraria laureri* Kremp., *Flora*, Regensburg 34: 673(1851).

Terricolous and on rocks coverd with bryophytes.

Neimenggu, Jilin(Lai et al., 2009, p. 377), Shaanxi(He & Chen 1995, p. 42), Xizang (Elenk. 1904, p. 57 & 1906, p. 125; Wei & Chen, 1974, p. 180; Wei & Jiang, 1981, p. 1146 & 1986, p. 59; Obermayer 2004, pp. 513 – 514), Sichuan(Obermayer 2004, pp. 513 – 514).

Tuckneraria laxa(Zahlbr.) Randlane & A. Thell 松软缘毛衣

in Randlane, Saag, Thell & Kärnefelt, *Acta bot. fenn.* 150: 149(1994).

≡*Nephromopsis ciliaris* var. *laxa* Zahlbr. in Fedde, *Repertorium*, 33: 61(1933).

Type: from Taiwan(syntypes).

≡*Nephromopsis laxa* (Zahlbr.) M. Sato, *Journ. Jap. Bot.* 14(12): 783(1938), J. C. Wei, *Enum. Lich. China*: 158(1991).

≡*Cetraria laxa*(Zahlbr.) M. Sato in Nakai et Honda, *Nova Flora Japonica*, 1939, p. 51.

=*Nephromopsis daibuensis*(Räs.) Räs. *Kuopion Luonnon Ystavain Yhdistyksen Julkaisuja*, ser. B, 2(6): 26, 33, 47 (1952).

≡*Cetraria daibuensis* Räs. *Journ. Jap. Bot.* 16(2): 85(1940).

Type: Taiwan, Mt. Daibu(449).

On bark of trees.

Taiwan(Zahlbr. , 1933 c, p. 61; Asahina, 1935a, p. 17; M. Sato, 1938b, p. 783 & 1939, p. 51; Räs. 1940, p. 85 & 1952, pp. 26, 33, 47; Lamb, 1963, pp. 154, 158, 434, 435; Wang-Yang & Lai, 1973, pp. 89, 99; Lai, 1980b, p. 222 & 2000, pp. 215 – 216; Ikoma, 1983, p. 33) .

Tuckneraria pseudocomplicata(Asahina) Randlane & Saag 拟褶缘毛衣

in Randlane, Saag, Thell & Kärnefelt, *Acta bot. fenn.* 150: 150(1994) .

≡*Cetraria pseudocomplicata* Asahina, *J. Jap. Bot.* 12: 804(1936) .

≡*Nephromopsis pseudocomplicata* (Asahina) M. J. Lai, *Quarterly Journal of the Taiwan Museum* 33(3 & 4): 224(1980) .

On *Salix*.

Sichuan(Obermayer 2004, p. 514) , Yunnan(Wei X. L. et al. 2007, p. 158) , Taiwan(M. Sato, 1938, p. 785 & 1939, p. 48; Wang & Lai, 1973, p. 89; Lai, 1980, p. 224 & 2000, p. 216) .

Tuckneraria togashii(Asahina) Randlane & A. Thell 针芽缘毛衣

in Thell, Kärnefelt & Randlane, *J. Hattori bot. Lab.* 78: 238(1995) .

≡*Cetraria togashii* Asahina, *J. Jap. Bot.* 28: 136(1953) .

Corticolous.

Jilin, Liaoning, Heilongjiang(Lai et al. , 2009, pp. 377 – 378) , Hubei(Chen, Wu & Wei, 1989, p. 433, as *C. togashii*) , Zhejiang(Xu B. S. , 1989, p. 222 as *C. togashii*) , Taiwan(Wang & Lai, 1973, p. 89 as *C. togashii*) .

Usnea Dill. ex Adans. **松萝属**

Fam. Pl. 2: 7(1763) .

Usnea aciculifera Vain. 尖刺松萝

Bot. Mag. , Tokyo 35: 45(1921) .

On Bark of trees and occasionally on rocks.

Neimenggu(Chen et al. 1981, p. 158) , Yunnan(Wang L. S. 2008, p. 538) , Hubei(Chen, Wu & J. C. Wei, 1989, p. 454) , Jiangxi(Ohmura, 2001, p. 34, as China, Kiemgsi, Lushan 26. 9. 1942, F. Fujikawa, s. n.) , Fujian (Mot. 1936, p. 322; Zhao et al. 1982, p. 75) , Taiwan(Asahina,

Usnea angulata Ach. 角度松萝

Syn. meth. lich. (Lund) : 307(1814) .

Taiwan(Ohmura, 2001, p. 35) .

Usnea arborea Stirt. 亚灌松萝

Scott. Natural. 6: 296(1882) .

≡*Usnea thomsoni* ssp. *arborea* Mot. *Lich. Gen. Usnea Stud. Monogr. , pars system* II: 617(1938) .

On dead twigs.

Xizang(J. C. Wei & Chen, 1974, p. 181; J. C. Wei & Jiang, 1985, p. 70) .

Usnea articulata(L.) Hoffm. 关节松萝

Deutschl. Fl. , Zweiter Theil (Erlangen) : 133(1796) [1795] .

≡*Lichen articulatus* L. , *Sp. pl.* 2: 1156(1753) — *Usnea barbata* var. *articulata*(L.) Ach. —*Usnea dasypoga* var. *articulata*(L.) Harm.

Jilin, Zhejiang(Tchou, 1935, p. 301; Moreau et Moreau, 1951, p. 193) , Hebei(Tchou, 1935, p. 301) .

Usnea baileyi(Stirt.) Zahlbr. 广生松萝

Denkschr. Kaiserl. Akad. Wiss. , Math. -Naturwiss. Kl. 83: 182 – 183(1909) ; Zahlbr. *Lich. rarior. exsicc.* no. 255.

≡*Eumitria baileyi* Stirt. , *Scott. Natural.* 6: 100(1881) .

On trees and rocks in the mountains. Pantropical.

Olivier 1898, p. 82 as *Usnea scabrata* Nyl. ; Zahlbr. , 1930, p. 207 as *Usnea certain* Ach. ; Motyka, 1936 – 38, p. 548 as *Usnea orientalis* Motyka; Thrower, 1988, pp. 17, 179 as *Usnea confusa* Asah. Above citations of 4 species are given by Aptroot & Seaward, 1999, p. 97.

Taiwan(Ohmura, 2001, p. 77) .

subsp. baileyi

= *Usnea baileyi* f. *endocrocea* Zahlbr. in *Bot. Magaz*. Tokyo, XLI: 357(1927) & Lich. rarior. exsicc. no. 255.

= *Usnea implicata*(Stirt.) Zahlbr. , *Cat. Lich. Univers*. 6: 582(1930) .

On bark of trees.

Yunnan, Guangxi, Zhejiang, Jjiangxi(Zhao et al. 1982, p. 60, 61) , Fujian(Zahlbr. , 1930 b, p. 208 as f. *endocrocea* ; Motyka, 1936 – 38, p. 60) , Taiwan(Asahina, 1926 b, p. 283 & 1950 a, p. 66 & 1956 a, p. 39 & 1967 a, p. 5; Zahlbr. , 1933 c, p. 62; Motyka, 1936 – 38, p. 60; M. Sato, 1938 a, p. 466; Asahina & M. Sato in Asahina, 1939, p. 761; Wang & Lai, 1973, p. 98; Ohmura, 2001, p. 77) , Hongkong(Olivier, 1898, p. 82; Zahlbr. , 1930, p. 207; Motyka, 1936 – 38, p. 548; Thrower, 1988, pp. 17, 178 or 179; Aptroot & Seaward, 1999, p. 97; Aptroot & Sipman, 2001, p. 341) .

= *Usnea baileyi* subsp. septentrionalis Asahina, in Hara, *the flora of Eastern Himalaya, Results of the Botanical Expedition to Eastern Himalaya Organized by the University Tokyo* 1960 and 1965 (Tokyo) : 598 (1966) .

Taiwan(Asahina, 1967 a, p. 6; Wang & Lai, 1973, p. 98) .

= *Usnea formosa*(Stirt.) Zahlbr. *Cat. Lich. Univers*. 6: 575(1930) .

≡ *Eumitria formosa* Stirt. , *Scott. Natural*. 6: 297(1882) [1881 – 1882] .

On trees.

Yunnan(Zhao et al. 1982, p. 62) .

Usnea barbata(L.) F. H. Wigg. 须松萝

Brit. Fl. 1: 206(1780) .

≡ *Lichen barbatus* L. , *Sp. pl.* 2: 1155(1753) .

(Zahlbr. , 1930a, p. 550; R. Santesson, 1984, p. 316) .

On bark of trees.

Zhejiang(Tchou, 1935, p. 301) .

see *Usnea dasypoga*(Ach.) Roehl. em. Mot.

Usnea bismolliuscula Zahlbr. 柔软松萝

Cat. Lich. Univers. 6: 542(1930) .

≡ *Usnea molliuscula* Vain. , *Bot. Mag.* , *Tokyo* 35: 45(1921) , non *Usnea molliuscula* Stirt. , *Scott. Natural.* , N. S. 1(‘ 7’) : 77(1883) [1883 – 1884] .

On trees & dead twigs.

Yunnan(Wang L. S. 2008, p. 538) , Fujian(Motyka, 1936, p. 451; Zhao et al. p. 87) , Taiwan(Asahina, 1956 a, p. 63; Ohmura, 2001, p. 38) .

Usnea cavernosa Tuck. 孔松萝

Lake Superior: 171(1850) . subsp. cavernosa 原亚种

On bark.

Guizhou(Zhang T & Wei JC, 2006, p. 11) .

subsp. sibirica(Räsänen) Mot. 西伯利亚亚种

Lich. Gen. Usnea Stud. Monogr. , *Pars System*. I: 80(1936) .

≡ *Usnea sibirica* Räsänen, *Animadvers. System. ex Herb. Kryloviano Univ. Tomskensis nom. Kuibyschevi* (3) : 1 (1927) .

On coniferous tree.

Anhui(Zhao et al. 1982, p. 63) .

Usnea ceratina Ach. 角松萝

Lich. univ.: 619(1810).

≡ *Usnea barbata* var. *ceratina*(Ach.) Schaer., *Enum. critic. lich. europ.* (Bern): [3], tab. I, fig. 1(1850).

Usnea scabrosa sensu Oliv. Bull. L'Academie Internationale de Geographie Botanique, 7: 82(1898), non Ach.

On trees.

Shaanxi(Jatta, 1902, p. 461; Zahlbr., 1930 b, p. 207), Xinjiang(Tchou, 1935, p. 301), Yunnan(Hue, 1887a, p. 18 & 1889, p. 162; Zahlbr., 1930 b, p. 207), Taiwan(Sasaoka, 1919, p. 180; Wang & Lai, 1973, p. 98; Ohmura, 2001, p. 40), Hongkong(Oliv. 1898, p. 82; Zahlbr., 1930 b, p. 207).

Usnea confusa Asahina 紊松萝

Lich. Japon. 3: 97(1956). subsp. confusa 原亚种

On twigs.

Jilin(Chen et al. 1981, p. 158), Shaanxi(He & Chen 1995, p. 45 without citation of specimens and their localities), Sichuan(Zhao et al. 1982, p. 91), Taiwan(Asahina, 1968b, p. 131; Wang & Lai, 1973, p. 98; Ohmura, 2001, p. 41), hongkong(Thrower, 1988, p. 179).

subsp. pygmoidea Asahina 矮小亚种

Journ Jap. Bot. 43(5): 130(1968).

Type: fromTaiwan.

Usnea pygmaea sensu Asahina, *Lich. Jap.* 3: 98(1956), non Motyka.

Taiwan(Asahina, 1968 b, p. 131; Wang & Lai, 1973, p. 98).

subsp. subconfusa Asahina 亚紊亚种

Journ Jap. Bot. 43(5): 130(1968).

Syntype: from China(Taiwan) & Japan.

Taiwan(Asahina, 1968 b, p. 131; Hawksw. 1972, p. 75; Wang & Lai, 1973, p. 98).

Usnea cornuta Körb. 牛角松萝

Parerga lichenol. (Breslau) 1: 2(1859).

= *Usnea dasaea* Stirt., *Scott. Natural.* 6: 104(1881).

Taiwan(Ohmura, 2001, p. 44 as *U. daaea*).

Usnea crassiuscula Zhao et al. 短粗松萝

Acta Phytotax. Sin. 13(2): 93(1975).

Type: Yunnan, Chen YB no. 3327 in HMAS-L.

Yunnan(Zhao et al. 1982, p. 63).

Usnea creberrima Vain. 密松萝

Bot. Mag., *Tokyo* 35: 46(1921).

Yunnan(Motyka, 1936 – 38, p. 362).

Usnea dasopoga(Ach.) Nyl. [as 'dasypoga'] 粗毛松萝

Meddn Soc. Fauna Flora fenn. 1: 14(1876).

≡ *Usnea plicata* var. *dasopoga* Ach., *Methodus*, Sectio post.: 312(1803).

Usnea barbata sensu Paulson, *Journ. Bot.* 66: 319(1928), non Wigg.

On bark of trees.

Shaanxi(He & Chen 1995, p. 45 without citation of specimens and their localities), Yunnan(Hue, 1887a, p. 18 & 1889, p. 162; Paulson, 1928, p. 319; Zahlbr., 1930 b, p. 207; Zhao et al. 1982, p. 69), Jiangxi(Zahlbr., 1930 b, p. 207).

Usnea decumbens Zhao et al. 俯仰松萝

Acta Phytotax. Sin. 13(2): 102(1975).

Type: Yunnan, Kunming, Mt. Xi shan, Zhao & Chen no. 2292 in HMAS-L(!).

Yunnan(Zhao et al. 1982, p. 94) .

Usnea dendritica Stirt. 树松萝

Scott. Natural. 6: 296(1882) .

On trees, occasionally on rocks.

Xizang(J. C. Wei & Y. M. Jiang, 1981, p. 1146 & 1986, p. 70) , Taiwan(Ohmura, 2001, p. 45) .

Usnea dorogawensis Asahina 小塔松萝

J. J. B. 28: 228(1953) .

On bark of trees.

Sichuan(Chen S. F. et al. , 1997, p. 222) , Yunnan, Guizhou(Zhao et al. 1982, p. 79) , Xizang(J. C. Wei & Y. M. Jiang, 1986, p. 71) , Anhui, Zhejiang(Xu B. S. , 1989, p. 226) .

Usnea eumitrioides Motyka 拟轴孔松萝

Lich. Gen. Usnea Stud. Monogr. 1 – 2: 322(1936 – 38) .

Type: Fujian. alt. 1100 m, 1921, Wang T. H. 420(Holotype in W) .

On trees.

Yunnan(Zhao et al. 1982, p. 76) , Fujian(Motyka, 1936 – 38, p. 322, Awasthi, 1986, p. 370) , Taiwan(Zahlbr. , 1933 c, p. 62; Motyka, 1936 – 38, p. 322; M. Sato, 1938 a, p. 464; Wang & Lai, 1973, p. 98) , Hongkong(Thrower, 1988, p. 180) .

Usnea florida(L.) Weber ex F. H. Wigg. 松萝

Prim. fl. holsat. (Kiliae) 2: 7(1780) . var. florida 原变种

≡ *Lichen floridus* L. , *Sp. pl.* 2: 1156(1753) .

On trees & dead twigs.

Neimenggu(Chen et al. 1981, p. 158) , Sichuan(Zahlbr. , 1930 b, p. 206 & 1934b, p. 212) , Yunnan(Hue, 1887a, p. 18 & 1889, p. 162 & 1899, p. 32; Paulson, 1928, p. 319; Zahlbr. , 1930 b, p. 206; Zhao et al. 1982, p. 68) , Taiwan(Sasaoka, 1919, p. 180; Asahina, 1968 a, p. 65; Wang & Lai, 1973, p. 98; Ikoma, 1983, p. 116; Ohmura, 2001, p. 26) .

var. strigosa Ach. 糙伏毛变种

Methodus, Sectio post. : 310(1803) .

On trees.

Yunnan, Fujian(Zahlbr. , 1930 b, p. 207) .

Usnea flotowii Zahlbr. 粗轴松萝

Bot. Jb. 60: 541(1926) .

= *Usnea barbata* var. *cornuta* Flot. , *Linnaea* 17: 16(1843) , non *Usnea cornuta* Körb. 1859.

On rocks.

Fujian(Zahlbr. , 1930 b, p. 207) .

Usnea fragilescens Hav. ex Lynge 脆松萝

Stud. Lich. Fl. Norway: 230(1921) .

= *Usnea malacea* Zahlbr. Cat. Lich. Univ. Vi: 586(*Usnea mollis* Stirt.) in Zahlbr. , 1930 b, p. 207.

On trees.

Sichuan, Yunnan(Zahlbr. , 1930 b, p. 207) , Taiwan(Aptroot, Sparrius & Lai, 2002, p. 292) .

Usnea fulvoreagens(Räsänen) Räsänen 黄褐松萝

Lich. Fenn. Exs. : no. 13(1935) .

≡ *Usnea glabrescens* var. *fulvoreagens* Räsänen, Die Flecht. Estl. 1: 20(1931) .

On *Salix* sp.

Taiwan(Ohmura, 2001, Pp. 27 – 28) .

Usnea fuscorubens Motyka 红褐松萝

Lich. Gen. Usnea Stud. Monogr. 2(1) : 546(1937) .

Type: Insula Mauritius, Simony, s. n. in W.

On trees.

Yunnan(Zhao et al. 1982, p. 90) , Taiwan(Asahina, 1956 a, p. 101; C. Culb. 1969, p. 541; Ohmura, 2001, p. 48) .

Usnea galbinifera Asahina 平滑松萝

J. Jap. Bot. 38(9) : 257(1963) .

Type: Taiwan, S. Asahina no. 3601.

Taiwan(Wang & Lai, 1973, p. 98; Ikoma, 1983, p. 116) .

Usnea glabrata(Ach.) Vain. 光松萝

Ann. Acad. Sci. fenn. , Ser. A 6(7) : 7(1915) . subsp. glabrata 原亚种

≡ *Usnea plicata* var. *glabrata* Ach. , *Lich. univ.* : 1 – 696(1810) .

On twigs.

Jilin(Chen et al. 1981, p. 158) , Taiwan(Ohmura, 2001, p. 50) , Hongkong(Thrower, 1988, p. 181) .

= *Usnea sorediifera*(Arnold) Lynge *Stud. Lich. Fl. Norway*: 229(1921) .

≡ *Usnea barbata* var. *sorediifera* Arnold, *Verh. zool. -bot. Ges. Wien* 25: 471(1875) .

≡ *Usnea florida* var. *sorediifera* Hue(1930) .

= *Usnea fulvoreagens* Räsänen

Usnea dasypoga sensu Hue(1887, p. 18) , non(Ach.) Roehl. (pr. p.)

On trees.

Yunnan(Hue, 1887, p. 18 & 1899, p. 37; Zahlbr. , 1930 b, p. 207; Zhao et al. 1982, p. 73) .

subsp. pseudoglabrata Asahina 拟光亚种

Lichen of Japan III, Genus Usnea: 95(1956) .

On trees.

Yunnan(Zhao et al. 1982, p. 92) .

Usnea glabrescens(Nyl. ex Vain.) Vain. 无毛松萝

Meddn Soc. Fauna Flora fenn. 48: 173(1925) [1924] .

≡ *Usnea barbata* var. *glabrescens* Nyl. ex Vain. , *Meddn Soc. Fauna Flora fenn.* 2: 46(1878) .

On tree trunk and twigs.

Shaanxi, Sichuan, Yunnan(Wu J. L. , 1987, p. 179) , Taiwan(Ohmura, 2001, p. 29) .

= *Usnea betulina* Mot. (R. Sant. , 1984) .

On trees.

Jilin, Yunnan(Zhao et al. 1982, p. 73) .

= *Usnea compacta* Mot. (R. Sant. , 1984) .

On trees & bamboo.

Sichuan(Zhao et al. 1982, p. 72) .

Usnea hakonensis Asahina 粉刺松萝

Lich. Japon. , 3, *Gen. Usnea*: 77(1956) .

On trees.

Yunnan, Jiangxi, Fujian(Zhao et al. 1982, p. 76) .

Usnea hapalotera(Harm.) Motyka 硬光松萝

Lich. Gen. Usnea Monogr. 1: 232(1936) .

≡ *Usnea florida* var. *hapalotera* Harm. , *Lich. Fr.* 3: 377(1907) .

On trees.

Yunnan(Zhao et al. 1982, p. 68) .

Usnea hesperina Motyka 黄昏松萝

Lich. Gen. Usnea Monogr. 2: 354(1937) .

On tree bark.

Taiwan(Ohmur, 2001, p. 52).

Usnea himalayana C. Bab. 喜马拉雅松萝

Hooker's J. Bot. Kew Gard. Misc. 4: 243(1852).

Usnea flexilis auct. non Stirt.: Asahina, Lich. Jap. 3: 59(1956). Wang & Lai, *Taiwania*, 18(1): 98(1973). Zhao et al. *Prodr. Lich. Sin.* 1982: 65.

On trees.

guizhou(Zhang T & Wei JC, 2006, p. 11), Anhui(Zhao et al. 1982, p. 65; Xu B. S., 1989, p. 226), Taiwan (Motyka, 1936 – 1938, p. 126; Asahina, 1956 a, p. 59; Wang & Lai, 1973, p. 98; Ohmura, 2001, p. 54).

Usnea himantodes Stirt. 带状松萝

Scott. Natural., N. S. 1('7'): 75(1883)[1883 – 1884].

Taiwan(Ohmura, 2001, p. 78).

Usnea hirta(L.) Weber ex F. H. Wigg. 硬毛松萝

Prim. fl. holsat. (Kiliae): 91(1780).

≡*Lichen hirtus* L., *Sp. pl.* 2: 1155(1753).

≡*Usnea barbata* var. *hirta*(L.) Fr., *Lich. eur. reform.* (Lund): 18(1831).

≡*Usnea florida* var. *hirta*(L.) Ach., *Methodus, Sectio post.* (Stockholmiæ): 309(1803).

On trees & bushes.

Heilongjiang(Chen et al. 1981, p. 158; Luo, 1984, p. 84), Shaanxi(Jatta, 1902, p. 461; Zahlbr., 1930 b, p. 206), Xinjiang(Moreau et Moreau, 1951, p. 193), Sichuan(Zahlbr., 1930 b, p. 206), Yunnan, Guizhou(Zhao et al. 1982, p. 64).

Usnea iteratocarpa Zhao et al. 重果松萝

Acta Phytotax. Sin. 13(2): 95(1975).

Type: Yunnan, Wang QZ no. 227 in HMAS-L.

Yunnan(Zhao et al. Prodr. Lich. Sin. 1982: 67).

Usnea kansuensis H. Magn. 甘肃松萝

Lich. Centr. Asia, 1: 129(1940).

Type: Gansu, 29/X/1930, at about 2000 m, on a woody slope, collected by Hummel and preserved in S.

On thin twigs of coniferous trees.

Sichuan(Zhao et al. 1982, p. 66).

Usnea kirinensis Zhao et al. 吉林松萝

Acta Phytotax. Sin 13(2): 99(1975).

Type: Jilin, Yang YC et al. no. 923 in HMAS-L(!).

Usnea leprosa Motyka 麻点松萝

Lich. Gen. Usnea Monogr. 1: 106(1936).

On rocks. 23. 8. 1854. C. Wright.

Hongkong(Seaward & Aptroot, 2005, p. 286, Pfister 1978: 76, as *Usnea barbata* var. *plicata*. K1 red; salazinic acid chemotype of *Usnea hirta sensu* Clerc.)

Usnea leucospilodea Nyl. 白松萝

J. Linn. Soc., Bot. 20: 50(1883).

Taiwan(Asahina, 1969 b, pp. 36 – 38; Wang & Lai, 1973, p. 98).

Usnea luridorufa Stirt. 小刺褐松萝

Scott. Natural. 6: 104(1881).

On trees & bushes.

Neimenggu(Chen et al. 1981, p. 159), Yunnan(Motyka, 1936 – 38, p. 522; Zhao et al. 1982, p. 88), Sichuan

(Motyka, 1936 – 38, p. 522), Hubei(Chen, Wu & J. C. Wei, 1989, p. 454), Zhejiang(Xu B. S., 1989, p. 226).

Usnea macrocarpa Zhao et al. 大果松萝

Acta Phytotax. Sin. 13(2): 100(1975).

Type: Yunnan, Wang CW no. 21469 in HMAS-L(!).

Yunnan(Zhao et al. 1982, p. 90).

Usnea macrospinosa Zhao et al. 大刺松萝

Acta Phytotax. Sin. 13(2): 102(1975).

Type: Yunnan, Zhao & Chen no. 4418 in HMAS-L(!).

Yunnan(Zhao et al. 1982, p. 88).

On bushes.

Usnea maculata Stirt. 斑松萝

Scott. Natural. 6: 293(1882)[1881 – 1882].

Taiwan(M. Sato, 1938 a, p. 467; Wang & Lai, 1973, p. 98, as the synonym of *Usnea rubiginea* (Michx.) Massal.).

Usnea masudana Asahina 增田氏松萝

J. Jap. Bot. 45(5): 132(1970).

Type: Taiwan, H. Masuda no. 36001 in TNS.

Yunnan(Index of Fungi, 1971, p. 48; Wang & Lai, 1973, p. 98; Ikoma, 1983, p. 118), Taiwan(Ohmura, 2001, p. 55).

Usnea mekista(Stirt.) G. Awasthi 黑轴松萝

Curr. Sci. 54(7): 354(1985).

≡ *Usnea longissima* subsp. *mekista* Stirt., *Scott. Natural.* 6: 105(1881).

On trees.

Sichuan(Zhao et al. 1982, p. 82).

Usnea mengyangensis Zhao et al. 勐养松萝

Acta Phytotax. Sin. 13(2): 97(1975).

Type: Yunnan, Zhao & Chen no. 3723 in HMAS-L(!).

Yunnan(Zhao et al. 1982, p. 75).

Usnea misaminensis(Vain.) Motyka 米萨松萝

Lich. Gen. Usnea Monogr. 2(1): 418(1937).

≡ *Usnea longissima* var. *misaminensis* Vain., *Philipp. J. Sci., C, Bot.* 4(5): 655(1909).

Sichuan(Chen S. F. et al., 1997, p. 222), Taiwan(Kurok. & Kashiw. Lich. Rar. et Cr. Exs. XI, no. 547).

Usnea montis-fuji Motyka 粗皮松萝

Lich. Gen. Usnea Monogr. 2: 405, 420(1937).

On twigs.

Jilin(Chen et al. 1981, p. 159), Liaoning(Zhao et al. 1982, p. 85), Shaanxi(Motyka, 1936 – 38, p. 420; Zhao et al. 1982, p. 85; J. C. Wei et al. 1982, p. 51; He & Chen 1995, p. 45 without citation of specimens and their loclities), Sichuan(Zhao et al. 1982, p. 85), Hubei(Zhao et al. 1982, p. 85; Chen, Wu & J. C. Wei, 1989, p. 455), Xizang(J. C. Wei & Y. M. Jiang, 1986, p. 72).

var. montisi-fuji 原变种

Jilin, Shaanxi, Sichuan(J. C. Wei et al., 1994, p. 204.).

var. sarmentosa J. C. Wei et al. 匍匐变种

Acta Mycologica Sinica 13(3): 204(1994).

On the ground.

Jilin(J. C. Wei et al., 1994, p. 204.).

Usnea mutabilis Stirt. 多型松萝

Scott. Natural. 6: 107(1881).

On *Castanopsis* sp.

Taiwan(Aptroot, Sparrius & Lai, 2002, p. 292).

Usnea neoguinensis Asahina 新几内亚松萝

J. Jap. Bot. 43(12): 496(1968). var. neoguinensis 原变种

Taiwan(Asahina, 1968, p. 496; Wang & Lai, 1973, p. 98; Ikoma, 1983, p. 118).

var. gracilior Asahina 纤细变种

J. Jap. Bot. 43: 497(1968).

Taiwan(Asahina, 1966, p. 601 & 1972 b, p. 257; Wang & Lai, 1973, p. 98).

Usnea nidifica Taylor 栖息松萝

London J. Bot. 6: 191(1847).

Taiwan(Ohmura, 2001, p. 60).

= *Usnea japonica* Vain. *Bot. Mag., Tokyo* 32: 154(1918).

On trees.

Jilin(Chen et al. 1981, p. 158), Yunnan(Zhao et al. 1982, p. 87), Taiwan(Sasaoka, 1919, p. 181; Zahlbr., 1933 c, p. 62; Motyka, 1936 – 38, p. 459; M. Sato, 1938a, p. 466; Asahina & M. Sato in Asahina, 1939, p. 759; Asahina, 1969 a, p. 1; Wang & Lai, 1973, p. 98; Yoshim. 1974, p. 32).

= *Usnea kurokawae* Asahina, *Lichens of Japan III, Genus Usnea*: 80(1956).

On twigs.

Fujian(Zhao et al. 1982, p. 77).

Usnea niparensis Asahina 光秃松萝

Lichs of Japan III: Genus Usnea: 91(1956).

Taiwan(Asahina, 1966, p. 601, & 1972 b, p. 257; Wang & Lai, 1973, p. 98; Ohmur, 2001, p. 61).

Usnea ogatai Asahina 台湾松萝

J. Jap. Bot. 45(5): 129(1970).

Type locality: Taiwan.

Taiwan(Wang & Lai, 1973, p. 98; Ikoma, 1983, p. 119).

Usnea orientalis Motyka 东方松萝

Lich. Gen. Usnea Stud. Monogr., Paris 1 – 2: 547(1936 – 38). f. orientalis 原变型

Type: Sichuan, alt. 3300 m, 15/VI/1914, collected by Handel-Mazzetti 3035(Diar. Nr. 559), holotype in W.

[*Usnea orientalis* f. *sorediosa* Motyka]

On trunk of *Picea* sp. & *Abies* sp.

Yunnan(Zahlbr., 1934b, p. 212 & 1940, p. 597; Motyka, 1936 – 38, p. 547; Lamb, 1963, p. 756; Zhao et al. 1982, p. 91), SICHUAN(Motyka, 1936 – 38, p. 547; Zahlbr., 1940, p. 597; Lamb, 1963, p. 756; Awasthi, 1986, pp. 377 – 378), XIANGGANG(Motyka, 1936 – 38, p. 547; Zahlbr., 1940, p. 597; Lamb, 1963, p. 756), Taiwan (Wang & Lai, 1973, p. 98; Ikoma, 1983, p. 119; Ohmura, 2001, p. 62).

f. esorediosa Asahina 无粉芽变型

Lich. Jap. 3: 100(1956).

Type: Syntypes from China(Taiwan) & Japan.

Yunnan(Zhao et al. 1982, p. 91), Taiwan(Asahina, 1956 a, p. 100; Wang & Lai, 1973, p. 98).

Usnea pangiana Stirt. 环基松萝

Scott. Natural. 7: 77(1883).

On bark.

Guizhou(Zhang T & J. C. Wei, 2006, p. 11), Taiwan(Ohmura, 2001, p. 64).

= *Usnea croceorubescens* Vain., *Bot. Mag.*, *Tokyo* 35: 46(1921).

On trees.

Yunnan(Zhao et al. 1982, p. 81), Fujian(Motyka, 1936, p. 362).

subsp. pangiana 原亚种

China(prov. not indicated: Asahina, 1972 a, p. 134), Taiwan(Wang & Lai, 1973, p. 98).

subsp. hondoensis(Asahina) Asahina 环裂亚种

J. Jap. Bot. 47(5): 134(1972).

≡ *Usnea hondoensis* Asahina, *J. Jap. Bot.* 3: 87(1956).

Type: Syntypes from China(Taiwan) & Japan.

Xinjiang(Abdulla Abbas & Wu, 1998, p. 100).

= *Usnea hondoensis* ssp. *inflatula* Asahina, ibid. 3: 89(1956).

Type: Syntypes from China(Taiwan) & Japan.

Jiangxi, Guangxi(Zhao et al. 1982, p. 74), Zhejiang(Xu B. S., 1989, p. 226).

On rocks.

Usnea pectinata Taylor 拟长松萝

London J. Bot. 6: 191(1847).

≡ *Eumitria pectinata*(Taylor) Articus, *Taxon* 53(4): 932(2004).

= *Usnea hossei* Vain., *Ann. bot. Soc. Zool. -Bot. fenn. Vanamo* 1(3): 34(1921).

On bark of trees.

Jilin(Chen et al. 1981, p. 159), Sichuan, Yunnan(Zhao et al. 1982, p. 82), Xizang(Motyka, 1936 – 38, p. 422; J. C. Wei & Chen, 1974, p. 181), Taiwan(Zahlbr., 1933 c, p. 62; M. Sato, 1938 a, p. 465; Asahina, 1950 a, p. 66 & 1956 a, p. 52 & 1965 a, p. 1; Wang & Lai, 1973, p. 98; Yoshim. 1973, p. 98; Ohmura, 2001, p. 79).

Usnea perplectans Stirt. 编织松萝

Scott. Natural. 6: 103(1881).

On trees.

Yunnan(Motyka, 1936 – 38, p. 293; Zhao et al. 1982, p. 72).

Usnea perplectata Motyka 缠结松萝

Lich. Gen. Usnea Monogr. 1: 50, 55(1936).

On trees.

Yunnan(Motyka, 1936 – 38, p. 293; Zhao et al. 1982, p. 72).

Usnea pseudogatai Asahina 拟台湾松萝

Journ. Jap. Bot. 45(5): 131(1970).

Type: Taiwan, Mt. Alishan, M, Ogata no. 203 in TNS.

Yunnan(Wang L. S. 2008, p. 538), Taiwan(Index of Fungi, 1971, p. 48; Wang & Lai, 1973, p. 98; Ikoma, 1983, p. 119; Ohmura, 2001, Pp. 64 – 65).

Usnea pseudomontis-fuji Asahina 拟粗皮松萝

Lichens of Japan III, Genus Usnea: 51(1956).

Taiwan(Wang & Lai, 1973, p. 98).

Usnea pseudorientalis Asahina 拟东方松萝

J. J. B. 44: 355(1969).

Taiwan(Wang & Lai, 1973, p. 98).

Usnea pycnoclada Vain. 密枝松萝

Philipp. J. Sci., *C, Bot.* 4(5): 653(1909).

On trees.

Shaanxi, Gansu(Zhao et al. 1982, p. 92), Yunnan(Motyka, 1936 – 38, p. 498; Zhao et al. 1982, p. 92).

Usnea pygmoidea(Asahina) Y. Ohmura 矮松萝

J. Hattori bot. Lab. 90: 65(2001).

≡ *Usnea confusa* subsp. *pygmoidea* Asahina, *J. J. B.* 43: 130(1968). nom. nov. for *Usnea pygmaea* sensu Asahina(*Lich. Jap.* 3: 98(1956), non Mot.

Taiwan(Ohmura, 2001, p. 67).

Usnea recurvata Zhao et al. 下弯松萝

Acta Phytotax. Sin. 13(2): 92(1975).

Type: Yunnan, Zhao & Chen no. 3334 in HMAS-L(!).

Yunnan(Zhao et al. 1982, p. 60).

Usnea roseola Vain. 红髓松萝

Bot. Mag., Tokyo 35: 46(1921). subsp. roseola 原亚种

On trees.

Yunnan, Guizhou(Zhao et al. 1982, p. 81), Zhejiang(Xu B. S., 1989, p. 227), Fujian(J. C. Wei et al. 1982, p. 52), Taiwan(Asahina, 1965 c, p. 225; Wang & Lai, 1973, p. 98).

subsp. pseudoroseola Asahina 拟红髓亚种

Lichens of Japan III, *Genus Usnea*: 105(1956).

On trunk of *Picea* sp.

Xizang(J. C. Wei & Jiang, 1986, p. 72).

Usnea rubicunda Stirt. 深红松萝

Scott. Natural. 6: 102(1881).

= *Usnea rubicunda* Stirt. v. *primaria* Motyka, *Lich. Gen. Usnea Stud. Monogr.*, Pars System. II: 341(1937).

= *Usnea florida* var. *rubiginea* Michx., *Fl. Boreali-Americ.* 2: 332(1803).

= *Usnea lacunosa* var. *rubiginea* Jatta, Nuov. Giorn. Bot. Ital. ser. 2, IX: 461(1902), non Michx.

Type: Shaanxi, Mt. Taibai shan, collected by Giraldi.

≡ *Usnea cavernosa* var. *rubiginea*(Jatta) Zahlbr., *Cat. Lich. Univers.* 6: 544(1930).

Usnea cf. *eumitroides* Motyka in Thrower, 1988, pp. 17, 180, mentioned by Aptroot & Seaward, 1999, p. 97.

Usnea cf. *glabrata*(Ach.) Vainio in Thrower, 1988, pp. 17, 181, mentioned by Aptroot & Seaward, 1999, p. 97.

On trees. On rocks in mountains. Cosmopolitan.

Jilin(Chen et al. 1981, p. 159), Shaanxi(Jatta, 1902, p. 461; Zahlbr., 1930 a, p. 544, prov. not indicated, but it seems to be in Shaanxi & 1930 b, p. 207), Sichuan(Chen S. F. et al., 1997, p. 222), Yunnan(Zhao et al., 1982, p. 77, lapsu ' rubicund' ; Wang L. S. 2008, p. 538), Guizhou(Zhao et al. 1982, p. 77, lapsu ' rubicund'), Zhejiang(Zhao et al. 1982, p. 77, lapsu ' rubicund'; Xu B. S., 1989, p. 226 – 227), Fujian(Motyka, 1936 – 38, p. 343; Zhao et al. 1982, p. 77), Taiwan(Zahlbr., 1933 c, p. 62; Motyka, 1936 – 38, p. 343; M. Sato, 1938 a, p. 467; Asahina, 1956 a, p. 115 & 1965 b, p. 129; Wang & Lai, 1973, p. 98; Ohmura, 2001, p. 69; Ohmura, 2001, p69), Hongkong(Thrower, 1988, pp. 17, 180, 181; Aptroot & Seaward, 1999, p. 97).

Usnea rubiginea(Michx.) A. Massal. 锈松萝

Mem. Imp. Reale Ist. Veneto 10: 45(1861).

≡ *Usnea florida* var. *rubiginea* Michx., *Fl. Boreali-Americ.* 2: 332(1803).

On bark of trees.

Taiwan(Zahlbr., 1933 c, p. 62).

Usnea rubrotincta Stirt. 红皮松萝

Scott. Natural. 6: 103(1881).

≡*Usnea rubescens* Stirt. var. *rubrotincta*(Stirt.) Motyka, *Lich. Gen. Usnea Monogr.* 2: 348(1937) .

Sichuan, Guangxi, Jiangxi(Zhao et al. 1982, p. 78) , Xizang(J. C. Wei & Jiang, 1986, p. 72) , Taiwan(Asahina, 1945, p. 114; Ohmura, 2001, p. 72) .

= *Usnea rubescens* Stirt. , *Scott. Natural.* , *N. S.* 1('7') : 76(1883) [1883 – 1884] .

= *Usnea ceratinella* Vain. , *Bot. Mag.* , Tokyo 35: 45(1921) .

≡*Usnea rubicunda* var. *ceratinella* Motyka, *Lich. Gen. Usnea Stud. Monogr.* , Pars System. II: 345(1937) .

On trees, occasionally on rocks.

Yunnan, Sichuan, Guangxi, Zhejiang(Zhao et al. 1982, p. 78) , Fujian(Zhao et al. 1982, p. 78; J. C. Wei et al. 1982, p. 53) , Taiwan(Zahlbr. , 1933 c, p. 62; M. Sato, 1938 a, p. 467; Asahina & M. Sato in Asahina, 1939, p. 763; Asahina, 1956 a, p. 113 & 1965 b, p. 129; Wang & Lai, 1973, p. 98) .

= *Usnea rubescens* var. *anaemica* Asahina, *J. Jap. Bot.* 44(9) : 259(1969) .

Type localities: holotype from New Guinea and paratype from CHINA: Taiwan, at 1900 m, 23/I/1965, S. Kurokawa no. 2618.

Taiwan(Ikoma, 1983, p. 120) .

= *Usnea pseudorubescens* Asahina, *J. Jap. Bot.* 40(5) : 130(1965) .

Type locality: Taiwan, Kurok. no. 831

On trees.

Zhejiang(Xu B. S. , 1989, p. 227) , Taiwan(Wang & Lai, 1973, p. 98) .

Usnea scabrata Nyl. 疣松萝

Flora, Regensburg 58: 103(1875) .

≡*Usnea dasypoga* f. *scabrata*(Nyl.) Arn.

On tree trunk.

Shaanxi(Jatta, 1902, p. 461; Zahlbr. , 1930 b, p. 207) .

Usnea schadenbergiana Goeppert et Stein 短松萝

in 60. Jahresber. Schelesisch. Gesellsch. Für vaterl. Kultur, 229(883) .

= *Usnea dasypoga* Röhling f. *dasypogoides* Zahlbr. in Denkschrift. Math. -naturw. Klasse Kais. Akad. Wiss. Wien, LXXXIII: 185(1909) .

Taiwan(Wang & Lai, 1973, p. 98) .

Usnea sensitiva(Zhao et al.) J. C. Wei 变色松萝

Enum. Lich. China: 258(1991) .

≡*Usnea sinensis* Motyka var. *sensitiva* Zhao et al. *Acta Phytotax. Sin.* 13(2) : 96(1975) .

Type: Yunnan, Zhao & Chen no. 4003 in HMAS-L.

Yunnan(Zhao et al. 1982, p. 68) .

Usnea shimadai Asahina 岛田氏松萝

J. Jap. Bot. 45(5) : 131(1970) .

Type: Taiwan, at about 1000 m, 23/VI/1928, Y. Shimada, in TNS, not seen.

On trees.

Taiwan(Index of Fungi, 1971, p. 48; Wang & Lai, 1973, p. 98; Ikoma, 1983, p. 120; Ohmura, 2001, p. 72) .

Usnea sichowensis Zhao et al. 西畴松萝

Acta Phytotax. Sin. 13(2) : 101(1975) .

Type: Yunnan, Wang Q. Z. no. 3645 in HMAS-L.

On trees.

Yunnan(Zhao et al. 1982, p. 89) .

Usnea sinensis Motyka 中华松萝

Lich. Gen. Usnea Stud. Monogr. 1 – 2: 248(1936 – 38) .

Type: Yunnan, 1915, collected by Handel-Mazzetti.

Yunnan(Zahlbr. , 1940, p. 604; Lamb, 1963, p. 761; Zhao et al. 1982, p. 67; Ohmura, 2001, p. 73), Taiwan (Ohmura, 2001, p. 73).

= *Usnea alisani* Asahina, *Journ. Jap. Bot.* 43: 67(1968).

Type: Taiwan, Mt. Ali shan, at about 2200 m, Kurok. no. 46 in TNS.

= *Usnea alisani* f. alisani

Taiwan(Asahina, 1968 a, p. 66; Hawksw. 1972, p. 75; Wang & Lai, 1973, p. 97; Ikoma, 1983, p. 113).

= *Usnea alisani* f. *condensata* Asahina, *Journ. Jap. Bot.* 43: 67(1968).

Type: Taiwan, Mt. Ali shan, Kurok. no. 46 in TNS.

Taiwan(Hawksw. 1972, p. 75; Wang & Lai, 1973, p. 97; Ikoma, 1983, p. 113; Ohmura, 2001, Pp. 72 – 73).

Usnea sinica(Motyka) Zhao et al. 华夏松萝

Acta Phytotax. Sin. 13(2) : 98 – 99(1975).

≡ *Usnea longissima* var. *sinica* Motyka, *Lichenum Generis Usnea Studium Monographicum* 1 – 2: 431(1936 – 38).

≡ *Dolichousnea longissima* var. *sinica*(Motyka) J. C. Wei

Type: Sichuan, at 2600 – 2800 m, 1914, collected by Handel-Mazzetti.

On trees.

Yunnan(Motyka, 1936 – 38, p. 431; Zhao et al. 1975, p. 98 & 1982, p. 85).

Usnea splendens Stirt. 光滑松萝

Scott. Natural. 6: 296(1882).

On trees.

Zhejiang(Xu B. S. , 1989, p. 227).

Usnea steineri Zahlbr. 毛盘松萝

Cat. Lich. Univers. 6: 592(1930). var. steineri

On trees.

Sichaun(Zahlbr. , 1930 b, p. 206), Yunnan(Zahlbr. , 1930 b, p. 206; Zhao et al. 1982, p. 93).

var. tincta Zahlbr. 红色反应变种

Denkschr. Kaiserl. Akad. Wiss. , Math. -Naturwiss. Kl. 83: 183(1909).

On bark of trees.

Yunnan(Zahlbr. , 1930 b, p. 206; Zhao et al. 1982, p. 94).

Usnea subcavata Motyka 轴亚空松萝

Lich. Gen. Usnea Stud. Monogr. 1: 51, 57(1936).

On shrubs.

Yunnan(Zhao et al. 1982, p. 59).

Usnea subcornuta Stirt. 亚角松萝

Scott. Natural. 6: 107(1881).

On dead twigs.

Yunnan(Zhao et al. 1982, p. 65).

Usnea subfloridana Stirt. 亚花松萝

Scott. Natural. 6: 294(1882) [1881 – 1882].

= *Usnea comosa*(Ach.) Röhl. , *Deutschl. Fl.* (Frankfurt) 3(2) : 144(1813).

On trees.

Xinjiang(Wang XY, 1985, p. 350; Abdulla Abbas & Wu, 1998, p. 100; Abdulla Abbas et al. , 2001, p. 368), Guizhou(Zhang T & J. C. Wei, 2006, p. 11), Zhejiang(Xu B. S. , 1989, p. 227).

On coniferous trees.

Heilongjiang(Asahina, 1952d, p. 375; Chen et al. 1981, p. 158), Meimenggu(Chen et al. 1981, p. 158), Xinjiang(Wang XY, 1985, p. 350; Abdulla Abbas et al., 1993, p. 81; Abdulla Abbas et al., 2001, p. 368), Sichuan (Zhao et al. 1982, p. 71), guizhou(Zhang T & J. C. Wei, 2006, p. 11), Yunnan(Zahlbr., 1930b, p. 207; Zhao et al. 1982, p. 71), Xizang(J. C. Wei & Jiang, 1981, p. 1147), Zhejiang(Xu B. S., 1989, p. 227), Guangdong (Ohmura, 2001, Pp. 30 – 31 as China Kanto), Fujian(Zahlbr., 1930b, p. 207), Taiwan(M. Sato, 1938a, p. 463; Asahina & M. Sato in Asahina, 1939, p. 763; Wang & Lai, 1973, p. 98; Ohmura, 2001, p. 30).

= *Usnea comosa* subsp. *sordidula* Motyka, *Lich. Gen. Usnea Monogr.* 1: 275(1936).

On twigs.

Xizang(J. C. Wei & Y. M. Jiang, 1986, p. 71).

[*Usnea comosa* ssp. *sucomosa* Mot.]

Heilongjiang(Hinggan Ling), Liaoning(as southren Manchuria cited by Asahina, 1956a, p. 92).

[*Usnea comosa* ssp. *meianopoda* Asahina.]

Heilongjiang(Hinggan Ling), Jilin(Frontier district between China and korea) (Asahina, 1956a, p. 94).

[*Usnea comosa* ssp. *praetervisa* Asahina. Lich. Jap. 3: 95(1956).]

Syntype localities: China(Jilin) and Japan.

On *Betula ermanii*.

Jilin(Frontier district between China and Korea: Asahina, 1956a, p. 95).

Usnea sublurida Stirt. 亚褐黄松萝

Trans. Proc. N. Z. Inst. 30: 389(1898) [1897].

On trees.

Yunnan(Zhao et al. 1982, p. 79).

Usnea subrectangulata Zhao et al. 亚直角松萝

Acta Phytotax. Sin. 13(2): 92(1975).

Type: Yunnan, Zhao & Chen no. 3290 in HMAS-L.

Yunnan(Zhao et al. 1982, p. 60).

Usnea subrobusta Zhao et al. 亚粗壮松萝

Acta Phytotax. Sin. 13(2): 97(1975).

Type: Yunnan, Wang QZ, no. 3658 in HMAS-L.

On trees.

Yunnan(Zhao et al. 1982, p. 77).

Usnea subsordida Stirt. 亚污褐松萝

Proc. Roy. phil. Soc. Glasgow 11: 310(1879) [1878].

On bark of trees.

Sichuan(Chen S. F. et al., 1997, p. 222).

var. subsordida 原变种

On dead twigs.

Yunnan(Zhao et al. 1982, p. 89).

var. insensitiva Motyka 迟感变种

Lich. Gen. Usnea Stud. Monogr. 1 – 2: 533(1936 – 38).

Type locality: Yunnan, 1883, Delavay, in PC.

Substratum not indicated.

Usnea thomsonii Stirt. 灌松萝

Scott. Natural. 6: 107(1881).

On trees & shrubs.

Yunnan(Zhao et al. 1982, p. 93), Xizang(J. C. Wei & Chen, 1974, p. 181; J. C. Wei & Jiang, 1981, p. 1146 &

1986, p. 70).

Usnea torquescens Stirt. 扭曲松萝

Trans. Proc. N. Z. Inst. 30: 391(1898)[1897]. var. torquescens

var. asahinae(Motyka) Asahina 棱角变种

Journ. Jap. Bot. 46(9): 26(1971), as ' var. asahinai'.

Usnea ' asahinai' Motyka, *Lich. Gen. Usnea Stud. Monogr.* 1 – 2: 393(1936 – 38), (orthographic error).

Substratum not indicated.

Yunnan(Zhao et al. 1982, p. 86).

Usnea torulosa(Müll. Arg.) Zahlbr. 结节松萝

Cat. Lich. Univers. 6: 594(1930).

≡*Usnea dasypogoides* f. *torulosa* Müll. Arg., *Flora, Regensburg* 66: 19(1883).

On tree trunk.

Shaanxi(Jatta, 1902, p. 461; Zahlbr., 1930 b, p. 207).

Usnea trichodea Ach. 毛状松萝

Methodus, Sectio post.: 312(1803).

On tree trunk.

Shaanxi(Jatta, 1902, p. 462; Zahlbr., 1930 b, p. 208), Hubei(Müll. Arg. 1893, p. 235).

Usnea undulata Stirt. 波松萝

Scott. Natural. 6: 104(1881). f. undulata 原变型

f. fruticans Asahina 灌状变型

Journ. Jap. Bot. 42(11): 322(1967).

Type: from Taiwan.

Taiwan(Wang & Lai, 1973, p. 98).

f. perspinigera Asahina 遍刺变型

Journ. Jap. Bot. 42(11): 325(1967).

Type: from Taiwan.

Taiwan(Ikoma, 1983, p. 120).

Usnea vainioi Motyka 鳞秕松萝

Lich. Gen. Usnea Stud. Monogr. 1: 51, 67(1936).

On coniferous & broadleaf trees.

Yunnan(Zhao et al. 1982, p. 61).

Usnea wasmuthii Räsänen 圆粉松萝

Die Flecht. Estl. 1: 19(1931).

On Picea sp.

Guangdong, Taiwan(Ohmura, 2001, p. 32 as China Kanto for Guangdong).

Usnea yunnanensis J. C. Wei 云南松萝

Enum. Lich. China, 261(1991).

≡*Usnea australis* Zhao et al. *Acta Phytotax. Sin.* 13(2): 94(1975), auct. non Fries, *Syst. orb. veg.* (Lundae) 1: 282(1825).

Type: Yunnan, Simao, collected by Zhao & Xu, no. 3446 in HMAS-L.

On broadleaf trees.

Yunnan(Zhao et al. 1982, p. 64).

Vulpicida J. -E. Mattsson & M. J. Lai **黄髓衣属**

Mycotaxon 46: 427(1993).

Vulpicida juniperinus(L.) J. -E. Mattsson & M. J. Lai 桧黄髓衣

Mycotaxon 46:427(1993).

≡*Lichen juniperinus* L., *Sp. pl.* 2:1147(1753).

≡*Cetraria juniperina*(L.) Ach., *Methodus, Sectio post.* (Stockholmiæ):299(1803).

≡*Tuckermanopsis juniperina*(L.) Hale [as '*Tuckermannopsis*'], *in Egan, Bryologist* 90(2):164(1987), (J. C. Wei, 1991, pp. 245 – 246).

On bark of trees.

Neimenggu(Lai et al., 2009, p. 378), Jilin(J. C. Wei, 1981, p. 86; Chen et al. 1982, p. 42; Lai et al., 2009, p. 378), Xinjiang(Abdulla Abbas et al., 1993, p. 81; Abdulla Abbas & Wu, 1998, p. 101; Abdulla Abbas et al., 2001, p. 368 as "*Vuipicida juniperina*").

Vulpicida pinastri(Scop.) J.-E. Mattsson & M. J. Lai 花黄髓衣

Mycotaxon 46:428(1993).

≡*Lichen pinastri* Scop., *Fl. carniol.*, Edn 2(Wien) 2:382(1772).

≡*Cetraria juniperina* var. *pinastri* (Scop.) Ach., *Methodus, Sectio prior*(Stockholmiæ):299(1803).

≡*Cetraria pinastri* (Scop.) Gray, *Nat. Arr. Brit. Pl.* (London) 1:432(1821).

= *Cetraria caperata* Vain., *Acta Soc. Fauna Flora fenn.* 13(no. 6):7(1896).

Corticolous.

Neimenggu(J. C. Wei et al. 1982, p. 44; Lai et al., 2009, pp. 378 – 379), Jilin, Heilongjiang(Chen et al. 1981, p. 129), Xinjiang(Abdulla Abbas & Wu, 1998, p. 101; Abdulla Abbas et al., 2001, p. 368 as "*Vuipicida*" *pinastri*), Xizang(Wei & Chen, 1974, p. 180; J. C. Wei & Jiang, 1981, p. 1146 & 1986, p. 59; J. C. Wei et al. 1982, p. 44; Obermayer 2004, p. 515).

Xanthoparmelia(Vain.) Hale **黄梅属**

Phytologia 28(5):485(1974).

≡*Parmelia* sect. *Xanthoparmelia* Vain., *Acta Soc. Fauna Flora fenn.* 7(no. 1):60(1890).

= *Xanthomaculina* Hale, *Lichenologist* 17:262(1985).

Xanthoparmelia camtschadalis(Ach.) Hale 旱黄梅

Phytologia 28:486(1974). Wei JC. *Acta Mycol. Sin.* 2:225(1983).

≡*Borrera camtschadalis* Ach., *Syn. meth. lich.* (Lund):223(1814).

≡*Parmelia camtschadalis*(Ach.) Eschw., *in* Martius, *Fl. Bras.* 1:202(1833)[1829 – 1833].

= *Xanthoparmelia convoluta*(Kremp.) Hale, *Phytologia* 28:487(1974).

≡*Parmelia convoluta* Kremp., *Verh. zool.-bot. Ges.* Wien 30:337(1880).

Misapplied name:

Parmelia vagans auct. non Nyl.: H. Magn. *Lich. Centr. Asia*, 2:47(1944). Zhao, Acta Phytotax. Sin. 9(2):148(1964). J. C. Wei & Chen, Report on the Scientific Investigations(1966 – 1968) in Mt. Qomolangma District (Biological Section), 1974, p. 179. Zhao et al. Prodr. Lich. Sin. 1982, p. 22.

Neimenggu(Gao, 1988, p. 33; Chen JB, 2015. p. 235), Xinjiang (J. C. Wei, 1983, p. 225; Wu J. L. 1985, p. 76; Wang XY, 1985, p. 347 as *Parmelia vagans*; Abudula Abbas et al., 1993, p. 81; Abdulla Abbas et al., 1994, p. 23; Abdulla Abbas & Wu, 1998, p. 102; Chen JB, 2015. p. 235), ningxia(Liu M & J. C. Wei, 2013, p. 44; Chen JB, 2015. p. 235), Shaanxi(J. C. Wei, 1983, p. 225; He & Chen 1995, p. 45 without citation of specimen and locality), Xizang(J. C. Wei, 1983, p. 225; J. C. Wei & Y. M. Jiang, 1986, p. 42; Chen JB, 2015. p. 235), Shanxi, Hebei(Chen JB, 2015. p. 234), China(prov. not indicated: Hale, 1990, p. 81).

Notes: *Parmelia camtschadalis*(Ach.) Eschw. (≡*Xanthoparmelia camtschadalis*(Ach.) Hale) had been mistaken for *Parmelia vagans* Nyl(≡*Xanthoparmelia vagans* (Nyl.) Hale) before taxonomic revision of *Xanthoparmelia*(Vain.) Hale from China(Wei J. C. 1983).

Xanthoparmelia claviculata Kurok. 棒芽黄梅

J. Jap. Bot. 64(10):296(1989).

On rocks.

Taiwan(Lai, 2000, pp. 188 – 189) , Hainan(Chen JB, 2015. p. 236; JCWei et al. , 2013, p. 230) .

Notes: The thallus with isidia and chemistry(containing usnic acid, norlobaridone and loxodin) of this species is identical with those of X. *scabrosa* except that *X. scabrosa* is different from *X. claviculata* only by the inflated and bursted top of the matured isidia(Chen JB, 2015. p. 236) . In this case, wether the *X. claviculata* is a heterotypic synonym of *X. scabrosa* or not remains to be studied further.

Xanthoparmelia coloradoensis(Gyeln.) Hale 科罗拉多黄梅

Mycotaxon 33: 402(1988) .

≡*Parmelia conspersa* var. *coloradoensis* Gyeln. , *Repert. Spec. Nov. Regni Veg.* 29: 287(1931) .

On rocks.

Heilongjiang, Neimenggu, Xinjiang, Hebei, Anhui(Chen JB, 2015. p. 236 – 237) .

Xanthoparmelia congensis(J. Steiner) Hale 刚果黄梅

Phytologia 28(5) : 486(1974) .

≡*Parmelia congensis* J. Steiner, *Jber. schles. Ges. vaterl. Kultur* 66: 140(1889) .

Misapplied name(revised by Aptroot & Seaward, 1999, p. 98) : *Parmelia mougeotii* auct. non Schaerer in Rabh. , 1873, p. 287; in Krphbr. , 1873, p. 471 & 1874, p. 66; Zahlbr. , 1930, p. 186) .

Xanthoparmelia mougeotina auct. non(Nyl.) D. J. Galloway in Thrower, 1988, pp. 17, 183; in Chu, 1997, p. 48) .

Xanthoparmelia conspersa auct. non(Ach.) Hale in Thrower, 1988, pp. 17, 183.

On coastal and inland rocks. Pantropical.

Hongkong(Rabh. , 1873, p. 287; Krphbr. , 1873, p. 471 & 1874, p. 66; Zahlbr. , 1930, p. 186; Thrower, 1988, pp. 17, 183; Hale, 1990, p. 90 ; Chu, 1997, p. 48; Thrower, 1988, pp. 17, 183; Aptroot & Seaward, 1999, p. 98: Chu MRDS 108860 with *Pertusaria flavicans*; Aptroot & Sipman, 2001, p. 341) .

Xanthoparmelia conspersa(Ehrh. ex Ach.) Hale 散生黄梅

Phytologia 28: 485(1974) .

≡*Lichen conspersus* Ehrh. ex Ach. , *Lich. suec. prodr.* (Linköping) : 118(1799) [1798] .

≡*Parmelia conspersa*(Ach.) Ach. , *Method. Lich.* : 205(1803) .

= *Parmelia conspera* f. *isidiata* Anzi, *Cat. Lich. Sondr.* : 28(1860) .

≡*Parmelia conspera* var. *isidiata*(Anzi) Stizbg. ; Tchou, *Contr. Inst. Bot. Natl Acad. Peiping* 3: 307(1935) .

Beijing, Liaoning, (Tchou, 1935, p. 307) , Heilongjiang(Da Hinggan Ling, Ta He: Gao, 1988, p. 31; Chen JB, 2015. p. 238) , Xinjiang(Wang XY, 1985, p. 347 as *Parmelia conspersa*) , Shaanxi(Baroni, 1894, p. 48; Jatta, 1902, p. 470; Zahlbr. , 1930b, p. 186) , Shanxi(M. Sato, 1981, p. 64) , Shandong(Tchou, 1935, p. 307; Zhang F et al. , 1999, p. 30; Zhao ZT et al. , 2002, p. 5; Li Y et al. , 2008, p. 71) , Guangdong(Rabh. 1873, p. 287; Krphbr. 1873, p. 471 & 1874, p. 66) .

Notes1: The record of this species in Heilongjiang(Da Hinggan Ling) reported by Gao(1988) is undoubted, but the records of this species and its infraspecific taxa in Beijing(Tchou) , Hebei(Moreau et Moreau) , Shanxi(M. Sato) , Liaoning(Moreau, Tchou) , Shaanxi(Jatta, Zahlbr.) Shandong(Tchou) , and Guangdong(Rabh. , Krphbr.) are still obscure, and should be revised in the future.

Notes2: *Parmelia(Xanthoparmelia) conspersa* and its infraspecific taxa cited for China in literature are very dubious owing to the absence of chemical information. No one specimen belonging to *Xanthoparmelia. conspersa* had been found from Chinese collections examined before 1983(Wei J. C. 1983, p. 221) . The real *X. conspersa* was found by Gao X. Q. from Da Hinggan Ling region in 1984 and published in 1987. This finding is very important to revise the lichen genus *Xanthoparmelia* in China.

Xanthoparmelia conspersula(Nyl.) Hale 小散生黄梅

Phytologia 28(5) : 486(1974) .

≡*Parmelia conspersula* Nyl., in Crombie, *J. Bot.*, *Lond.* 14: 19(1876).

Taiwan(Zahlbr., 1933 c, p. 54; Wang & Lai, 1973, p. 93).

Xanthoparmelia constrictans(Nyl.) Hale 缩黄梅

Phytologia 28: 486(1974).

≡*Parmelia constrictans* Nyl., in Crombie, *J. Linn. Soc.*, *Bot.* 15: 168(1876)[1877].

=*Parmelia conspersa* var. *constrictans*(Nyl.) Müll. Arg., *Flora* 66: 48(1883)

≡*Imbricaria constrictans*(Nyl.) Jatta, *Nuon. Giorn. Bot. Ital.* ser. 2, 9: 470(1902).

On rocks.

Shaanxi(Jatta, 1902, p. 470; Zahlbr., 1930b, p. 186).

Xanthoparmelia coreana(Gyeln.) Hale 朝鲜黄梅

Mycotaxon 33: 402(1988).

≡*Parmelia coreana* Gyeln., *Reprium nov. Spec. Regni veg.* 29: 280(1931).

Heilongjiang("Mutankiang"), Liaoning(Kurok. 1989, p. 168), Xinjiang(Abdulla Abbas & Wu, 1998, p. 102; Abdulla Abbas et al., 2001, p. 368).

Xanthoparmelia delisei(Duby) O. Blanco et al. 棕黄梅

Taxon 53: 967(2004).

≡*Parmelia delisei*(Duby) Nyl., *Flora* 55: 426(1872).

≡*Parmelia olivacea* var. *delisei* Duby, *Bot. Gall.* 2: 602(1830).

≡*Parmelia prolixa* var. *delisei*(Duby) Nyl., Lich. Scand.: 102(1861); Zhao, *Acta Phytotax. Sin.* 9: 146(1964); Zhao, Xu & Sun, *Prodr. Lich. Sin.*: 19(1982), J. C. Wei, *Enum. Lich. China*: 169(1991).

≡*Neofuscelia delisei*(Duby) Essl., *Mycotaxon* 7: 50(1978); J. C. Wei, *Enum. Lich. China*: 156(1991).

On rocks and On bark of trees.

Xinjiang(Chen JB, 2015. p. 238), Anhui(Zhao, 1964, p. 146; Zhao et al. 1982, p. 19), Hubei(Chen, Wu & J. C. Wei, 1989, p. 442).

Xanthoparmelia desertorum(Elenkin) Hale 荒漠黄梅

Mycotaxon 33: 402(1988).

≡*Parmelia molliuscula* f. *desertorum* Elenkin, *Acta Horti Petropolit.* 19: 21(1901).

on soil.

Xinjiang(Abdulla Abbas et al., 1996, p. 233; Abdulla Abbas & Wu, 1998, pp. 102 – 103; Abdulla Abbas et al., 2001, p. 368).

Xanthoparmelia durietzii Hale 杜瑞氏黄梅

Mycotaxon 30: 322(1987).

Type: Gansu, Potanin s. n., 10/IV/1885(UPS, holotype: US, isotype).

On soil & crushed stones of the ground.

Beijing, Neimenggu(Chen JB, 2015. p. 239), Xinjiang(Abdulla Abbas & Wu, 1998, p. 103), Gansu(Hale, 1987a, p. 322 & 1990, p. 105), Hebei(Xiaowutai shan: Hale, 1987a, p. 322; Chen JB, 2015. p. 239),.

Xanthoparmelia eradicata(Nyl. ex Cromb.) Hale 除黄梅

Phytologia 28: 487(1974).

≡*Parmelia constrictans* var. *eradicata* Nyl. ex Cromb., in Crombie, *J. Linn. Soc.*, *Bot.* 15: 168(1876)[1877].

≡*Parmelia conspersa* var. *eradicata*(Nyl.) Müll. Arg. *Flora* 66: 48(1883).

≡*Imbricaria constrictans* var. *eradicata*(Nyl.) Jatta *Nuon. Giorn. Bot. Ital.* ser. 2, 9: 470(1902).

Shaanxi(Jatta, 1902, p. 470; Zahlbr., 1930b, p. 186)

Xanthoparmelia formosana Kurok. 台湾黄梅

in Kurokawa & Lai, *Mycotaxon* 77: 280(2001)

≡*Xanthoparmelia formosana* Kurok., *Annals of the Tsukuba Botanical Garden* 8: 22(1989).

Nom. inval., Art. 40. 1(Melbourne).

Holotype: from Taiwan, in TNS, Kurokawa 261, Host-Substratum/Locality: On rock:

Taiwan(Lai, 2000, p. 189).

Xanthoparmelia hypoleia(Nyl.) Hale 白纹黄梅

Phytologia 28: 487(1974).

≡*Parmelia hypoleia* Nyl., *Syn. meth. lich.* (Parisiis) 1(2): 393(1860).

On rocks.

Hongkong(Krphbr. 1873, p. 56 & 1874c, p. 66; Rabh. 1873, p. 287; Zahlbr., 1930b, p. 186).

Xanthoparmelia hypopsila(Müll. Arg.) Hale 贴生黄梅

Phytologia 28: 488(1974).

≡*Parmelia hypopsila* Müll. Arg., *Flora*, Regensburg 70: 317(1887).

On rocks.

Heilongjiang(Gao, 1988, p. 32).

Xanthoparmelia lineola(E. C. Berry) Hale 线形黄梅

Phytologia 28: 488(1974).

≡*Parmelia lineola* E. C. Berry, *Ann. Mo. bot. Gdn* 28: 77(1941).

On rocks.

Heilongjiang(Gao, 1988, p. 32; Chen JB, 2015. p. 240).

Xanthoparmelia mexicana(Gyeln.) Hale 淡腹黄梅

Phytologia 28: 488(1974).

≡*Parmelia mexicana* Gyeln., *Reprium nov. Spec. Regni veg.* 29: 281(1931).

Misapplied name:

Parmelia conspersa auct. non Ach.: H. Magn. *Lich. Centr. Asia*, 2: 46(1944).

Parmelia conspersa var. *isidiosula* auct. non Hillm.: Zhao, *Acta Phytotax. Sin.* 9(2): 147(1964). (pr. p.). Zhao et al. *Prodr. Lich. Sin.* 1982, p. 21(pr. p.).

Parmelia conspersa var. *latior* auct. non Schaer.: Zaho, *Acta Phytotax. Sin.* 9(2): 147(1964). Zhao et al. ibid. 1982, p. 21.

Parmelia subramigera auct. non Gyeln.: Chen et al. *Journ. NE Forestry Inst.* 4: 154(1981).

On rocks & on bark of trees.

Beijing(Tchou, 1935, p. 307; Chen JB, 2015. p. 241), Heilongjiang(Chen JB, 2015. p. 241), Jilin(Chen JB, 2015. p. 241), Liaoning(Tchou, 1935, p. 307; Moreau et Moreau, 1951, p. 190; Chen JB, 2015. p. 241), Neimenggu(J. C. Wei, 1983, p. 224; Gao, 1988, p. 32; Sun LY et al., 2000, p. 36; Chen JB, 2015. p. 241), Shaanxi (Jatta, 1902, p. 470; Zahlbr., 1930 b, p. 186; J. C. Wei et al. 1982, p. 35; He & Chen 1995, p. 45), Xizang(J. C. Wei & Chen, 1974, p. 179; J. C. Wei & Jiang, 1986, p. 41; Chen JB, 2015. p. 241), Xinjiang(J. C. Wei, 1983, p. 224),, Shanxi, Hebei, Anhui, Zhejiang(J. C. Wei, 1983, p. 224; Chen JB, 2015. p. 241), Shandong (Tchou, 1935, p. 307; Zhao ZT et al., 1998, p. 29; Zhao ZT et al., 2002, p. 6; Hou YN et al., 2008, p. 68; Chen JB, 2015. p. 241), China(prov. not indicated: Hale, 1990, p. 147).

Xanthoparmelia molliuscula(Ach.) Hale 柔黄梅

Phytologia 28: 488(1974).

≡*Parmelia molliuscula* Ach., *Lich. univ.*: 492(1810).

Fujian(Zahlbr., 1930b, p. 186; Chung 265).

Xanthoparmelia mongolica Kurok. 蒙古黄梅

Ann. Tsukuba Bot. Gard. 8: 22, 1989.

On rocks.

Neimenggu(Chen JB, 2015. p. 242) .

Xanthoparmelia mougeotii(Schaer.) Hale 毛氏黄梅

Phytologia 28: 488(1974) .

≡*Parmelia mougeotii* Schaer. , in Dietrich, *Deutschl. Kryptog. Gewächse*, 4 Abth. : 118(1846) .

On rocks.

Hongkong(Krphbr. 1873, p. 471 & 1874, p. 66; Rabh. 1873, p. 287; Zahlbr. , 1930 b, p. 186; Thrower, 1988, p. 183) .

Xanthoparmelia mutabilis(Taylor) Hale 易变黄梅

Smithson. Contr. bot. 74: 152(1990) .

≡*Parmelia mutabilis* Taylor, in Hooker, *J. Bot.* , *Lond*. 6: 171(1847) .

Hongkong(Krphbr. 1873, p. 471; Zahlbr. , 1930b, p. 186) .

Note: It was treated by Zahlbr. (1930b, p. 186) as *Parmelia hypoleia* Nyl.

= *Xanthoparmelia hypoleia*(Nyl.) Hale

Xanthoparmelia neotinctina(Elix) Elix & J. Johnst. 新暗腹黄梅

in Elix, Johnston & Armstrong, *Bull. Br. Mus. nat. Hist.* , *Bot*. 15(3) : 297(1986) .

≡*Parmelia neotinctina* Elix, *Aust. J. Bot*. 29(3) : 363(1981) .

= *Parmelia conspersa* f. *isidiophora* Müll. Arg. , *Flora* 66: 48(1883) .

≡*Imbricaria canstrictans* var. *eradicata* f. *isidiophora*(Müll. Arg.) Jatta, *Nuov. Giorn. Bot. Ital*. ser. 2, 9: 470(1902) .

Shaanxi(Jatta, 1902, p. 470; Zahlbr. (1930b, p. 186) as *Parmelia conspersa* var. *eradicata* Jatta) .

Xanthoparmelia novomexicana(Gyeln.) Hale 新墨西哥黄梅

Phytologia 28(5) : 488(1974)

≡*Parmelia novomexicana* Gyeln. [as ' novo-mexicana'] , *Feddes Repert*. 36: 161(1934) .

On rocks.

Heilongjiang(Gao, 1988, p. 33) .

Xanthoparmelia orientalis Kurok. 东方黄梅

J. Jap. Bot. 64: 169(1989) .

On rocks.

Beijing, Jilin, Neimenggu(Chen JB, 2015. p. 243) , Xinjiang(Abdulla Abbas et al. , 1993, p. 81; Abdulla Abbas & Wu, 1998, p. 103; Abdulla Abbas et al. , 2001, p. 368) , Chen JB, 2015. p. 243) , Xizang, Shandong, Anhui, Shanghai, Zhejing(Chen JB, 2015. p. 243) , Taiwan(Kurok. , 1989, p. 171) ; Lai, 2000, pp. 189 – 190) .

Xanthoparmelia protomatrae(Gyeln.) Hale 齿裂黄梅

Phytologia 28: 488(1974) .

≡*Parmelia protomatrae* Gyeln. , *Reprium nov. Spec. Regni veg*. 29: 155(1931) .

= *Parmelia stenophylla* f. *densata* Zhao, *Acta Phytotax. Sin*. 9(2) : 148(1964) .

Type: Anhui, Mt. Huang shan, Zhao et al. no. 5860 in HMAS-L.

≡*Xanthoparmelia dentata*(Zhao) J. C. Wei, *Acta Mycol. Sin*. 2(4) : 224(1983) .

Misapplied name:

Xanthoparmelia novomexicana auct. non(Gyeln.) Hale: Gao, *Acta Mycol. Sin*. 7: 33(1988) .

On rocks.

Heilongjiang(Gao, 1988, p. 33; Chen JB, 2015. p. 244) , Anhui(Hale, 1990, pp. 176 – 177, cited as China; Chen JB, 2015. p. 244) .

Xanthoparmelia saxeti(Stizenb.) Amo de Paz et al. 无根黄梅

Amo de Paz, A. Crespo, Elix & Lumbsch, *Aust. Syst. Bot*. 23(3) : 182(2010) .

≡*Parmelia saxeti* Stizenb. , *Ber. Tät. St Gall. naturw. Ges*. : 153(1890) [1888 – 89] .

≡*Karoowia saxeti* (Stizenb.) Hale, *Mycotaxon* 35(1): 190(1989), Lai M. J. *Illustrated macrolichens of Taiwan* (I): 131(2000).

= *Xanthoparmelia squamariiformis*(Gyeln.) Hale [as 'squamaraeformis'] *in* J. C. Wei, *Enum. Lich. China* (Beijing): 265(1991).

≡*Parmelia squamaraeformis* Gyelnik in Fedde, *Repertorium*, 36: 163(1934), Zahlbr. *Cat. Lich. Univers*. 10: 535(1938 – 1940), M. Lamb, *Index Nom. Lich*. 500(1963).

Type: from Taiwan.

Taiwan(Gyelnik, 1934, p. 163; Lai, 2000, p. 131 as *K. saxeti*).

Xanthoparmelia scabrosa(Taylor) Hale 粗黄梅

Phytologia 28: 488(1974).

≡*Parmelia scabrosa* Taylor, *London J. Bot*. 6: 162(1847).

Misapplied name(revised by Aptroot & Seaward, 1999, pp. 98 – 99):

Parmelia mutabilis auct. non Taylor in Rabenhorst, 1873, p. 287; in Krempelh., 1873, p. 471 & 1874, p. 66).

Parmelia hypoleia auct. non Nyl. in Zahlbr., 1930, p. 186.

Xanthoparmelia tinctina auct. non(Mahen & A. Gillet) Hale in Thrower, 1988, pp. 17, 184; Chu, 1997, p. 48.

On granitic and basalt rocks, mainly coastal. Known from temperate Asia, Australasia and South America.

Hongkong(Rabh., 1873, p. 287; Krphbr., 1873, p. 471 & 1874, p. 66; Zahlbr., 1930, p. 186; Thrower, 1988, pp. 17, 184; Chu, 1997, p. 48; Aptroot & Sipman, 2001, p. 341).

Xanthoparmelia squamariiformis(Gyeln.) Hale [as 'squamaraeformis'] 鳞黄梅

in J. C. Wei, *Enum. Lich. China*(Beijing): 265(1991).

≡*Parmelia squamaraeformis* Gyelnik in Fedde, *Repertorium*, 36: 163(1934), Zahlbr. *Cat. Lich. Univers*. 10: 535 (1938 – 1940), M. Lamb, *Index Nom. Lich*. 500(1963).

Type: from Taiwan.

Xanthoparmelia stenophylla(Ach.) Ahti & D. Hawksw. 菊叶黄梅

Lichenologist 37(4): 363(2005).

≡*Parmelia conspersa* ß *stenophylla* Ach., *Methodus*, Sectio post.: 206(1803).

≡*Parmelia stenophylla*(Ach.) Heugel, *Corresp. Naturf. Ver. Riga* 8: 109(1855).

≡*Parmelia stenophylla*(Ach.) Du Rietz, *Svensk. Bot. Tidskr*. 15: 176(1921).

= *Xanthoparmelia somloënsis*(Gyeln.) Hale, in Ahti, Brodo & Noble, *Mycotaxon* 28: 96(1987).

≡*Parmelia somloënsis* Gyeln., *Repert. Spec. Nov. Regni Veg*. 29: 156(1931).

= *Parmelia saxatilis* var. *subomphalodes* Zahlbr., in Handel-Mazzetti, *Symb. Sin*. 3: 190(1930).

Type: Shaanxi, Mt. Guangtou-shan, Giraldi [241] (lectotype in W).

On rocks.

Beijing(Zhao, 1964, p. 148; Zhao et al., 1982, p. 22; Chen JB, 2015. p. 246), Heilongjiang(Chen et al., 1981, p. 154; Chen JB, 2015. p. 246), Jilin(Chen JB, 2015. p. 246), Neimenggu(Chen et al., 1981, p. 154; Sun LY et al., 2000, p. 36 as *X. somloënsis*; Chen JB, 2015. p. 246), Xinjiang(Abdulla Abbas & Wu, 1998, p. 104 and Abdulla Abbas et al., 2001, p. 368 as *X.* "*somloensis*"; Chen JB, 2015. p. 246), Ningxia(Zhao, 1964, p. 148; Zhao et al., 1982, p. 22), Shaanxi, Xizang(Chen JB, 2015. p. 246), Hebei, Anhui(Chen JB, 2015. p. 246), China(prov. not indicated: Hale, 1990, p. 193).

Xanthoparmelia sublaevis(Cout.) Hale 亚平黄梅

Mycotaxon 33: 406(1988).

≡*Parmelia sublaevis* Cout., *Lich. Lusit. Catal*.: 71(1916).

= *Parmelia conspersa* var. *hypoclysta* Nyl

= *Imbricaria conspersa* var. *hypoclysta* (Nyl.) Jatta.

Shaanxi(Jatta, 1902, p. 470; Zahlbr., 1930b, p. 186).

Xanthoparmelia subramigera(Gyeln.) Hale 亚分枝黄梅

Phytologia 28: 489(1974).

≡*Parmelia subramigera* Gyeln., *Feddes Repert.* 29: 281/409(1931).

= *Parmelia conspersa* var. *hypoclysta* f. *isidiosa* Müll. Arg. *Flora*, 66: 47(1883).

≡*Imbricaria conspersa* var. *hypoclysta* f. *isidiosa*(Müll. Arg) Jatta, *Nuov. Giorn. Bot. Ital.* ser. 2, 9: 470(1902).

On rocks.

Jilin(Chen et al., 1981, p. 154), Neimenggu(Chen JB, 2015. p. 247), Shaanxi(Jatta, 1902, p. 470; Zahlbr., 1930b, p. 186), Hebei, Shandong(Chen JB, 2015. p. 247 – 248).

Xanthoparmelia taractica(Kremp.) Hale 拟菊叶黄梅

Phytologia 28: 489(1974).

≡*Parmelia taractica* Kremp., *Flora*, Regensburg 61: 439(1878).

= *Imbricaria conspersa* var. *polyphylloides* (Müll. Arg.) Jatta, *Nuov. Giorn. Bot. Ital.* Ser. 2, 9: 470(1902).

≡*Parmelia conspersa* var. *polyphylloides* Müll. Arg, *Flora* , 66: 47(1883).

Misapplied name:

Parmelia stenophylla auct. non Nyl.: Zhao, *Acta Phytotax. Sin.* 9(2): 148(1964). Zhao et al. *Prodr. Lich. Sin.* 1982, p. 21.

Parmelia conspersa var. *hypoclysta* auct. non Nyl.: Zhao, *Acta Phytotax. Sin.* 9(2): 147(1964). J. C. Wei & Chen, Report on the Scientific Investigations(1966 – 1968) in Mt. Qomolangma District(Biology Section), 1974, p. 179. Zhao et al. ibid. 1982, p. 22.

Parmelia subconspersa auct. non Nyl.: Zhao, ibid. 9(2): 147(1964). Zhao et al. *Prodr. Lich. Sin.* 1982, p. 22.

Type: from Shaanxi.

Parmelia conspersa var. *imbricaria* f. *microphylla* sensu J. C. Wei & Chen: 1974, p. 179, non Hillm.

On rocks, soil among rocks & on stems of grasses.

Beijing, Xinjiang, Sichuan(J. C. Wei, 1983, p. 224), Neimenggu(J. C. Wei, 1983, p. 224; Gao, 1988, p. 34), Heilongjiang(Gao, 1988, p. 34), Shaanxi(Jatta, 1902, p. 470; Zahlbr., 1930b, p. 186 ; He & Chen 1995, p. 45), Ningxia, Anhui(Zhao, 1964, p. 147; Zhao et al. 1982, p. 21), Xizang(J. C. Wei & Chen, 1974, p. 179; J. C. Wei, 1983, p. 224; J. C. Wei & Jiang, 1986, p. 42).

Xanthoparmelia tasmanica(Hook. f. & Taylor) Hale 黑黄梅

Phytologia 28: 489(1974).

≡*Parmelia tasmanica* Hook. f. & Taylor, *London J. Bot.* 3: 644(1844).

= *Parmelia conspersa* var. *incisa*(Taylor) Zahlbr., *Cat. Lich. Univ.* 6: 132(1930).

= *Imbricaria conspersa* var. *laxa* (Müll. Arg.) Jatta, *Nuov. Giorn. Bot. Ital.* ser. 2, 9: 470(1902).

≡*Parmelia conspersa* var. *laxa* Müll. Arg., *Flora* 66: 47(1883).

On rocks.

Neimenggu, Heilongjiang(Gao, 1988, p. 33; Chen JB, 2015. p. 248). Shaanxi(Jatta, 1902, p. 470; Zahlbr., 1930b, p. 186), Shanghai(Chen JB, 2015. p. 248).

Xanthoparmelia tinctina(Maheu & A. Gillet) Hale 暗腹黄梅

Phytologia 28: 489(1974).

≡*Parmelia tinctina* Maheu & A. Gillet, *Bull. Soc. bot. Fr.* 72: 860(1925).

= *Parmelia conspersa* var. *isidiosa* Nyl, *Flora* 64: 450(1881).

Misapplied name:

Parmelia conspersa auct. non Ach.: H. Magn. ibid. 2: 46(1944). Zhao, *Acta Phytotax. Sin.* 9(2): 147(1964). Zhao et al. *Prodr. Lich. Sin.* 1982, p. 20.

Parmelia conspersa var. *isidiosula* auct. non Hillm. : Zhao, *Acta Phytotax. Sin.* 9(2): 147(1964). Zhao et al. *Prodr. Lich. Sin.* 1982, p. 21(pr. p.).

Beijing, Xinjiang, Hebei, Zhejiang, Jiangsu, Anhui(J. C. Wei, 1983, p. 223), Neimenggu(J. C. Wei, 1983, p. 223; Gao, 1988, p. 31; Sun LY et al., 2000, p. 36), Liaoning (Tchou, 1935, p. 307; Moreau et Moreau, 1951, p. 190), Shaanxi(He & Chen 1995, p. 45 without citation of specimen and locality), Xinjiang(J. C. Wei, 1983, p. 224; Wu J. L. 1985, p. 76), Xizang(J. C. Wei & Chen, 1974, p. 179; J. C. Wei, 1983, p. 223; J. C. Wei & Jiang, 1986, p. 41), Sichuan(J. C. Wei, 1983, p. 223; Chen JB, 2015. p. 249), Chongqing(Chen CL et al., 2008, p. 40), Shandong(Tchou, 1935, p. 307; Zhao ZT et al., 1998, p. 29; Zhao ZT et al., 2002, p. 5; Hou YN et al., 2008, p. 68; Li Y et al., 2008, p. 71), Hhongkong(Thrower, 1988, p. 184).

Xanthoparmelia viriduloumbrina(Gyeln.) Lendemer 北美黄梅

Mycotaxon 92: 442(2005).

≡*Parmelia conspersa* f. *viriduloumbrina* Gyeln., *Magy. Bot. Lapok* 29: 31(1930).

On rocks.

Heilongjiang, Neimenggu, Ningxia, Shaanxi, Hebei(Chen JB, 2015. p. 250).

Xanthoparmelia xizangensis(J. C. Wei) Hale 西藏黄梅

Mycotaxon 33: 406(1988).

≡*Parmelia tinctina* var. *xizangensis* J. C. Wei, *Acta Mycol. Sin.* 2(4): 223(1983).

Type: Xizang, Zong YC, no. 218-(1) in HMAS-L.

On rocks.

Neimenggu(Chen JB, 2015. p. 251), Xizang(Hale, 1990, p. 227).

(52) Psoraceae Zahlbr. **鳞网衣科**

in Engler, *Syllabus*, Edn 2(Berlin): 44(1898).

Protoblastenia(Zahlbr.) J. Steiner **原胚衣属**

Verh. zool. -bot. Ges. Wien 61: 47(1911).

≡ *Blastenia* sect. *Protoblastenia* Zahlbr., in Engler & Prantl, *Nat. Pflanzenfam.*, Teil. I(Leipzig) 1*: 226 (1908).

Protoblastenia amagiensis Räsänen 原胚衣

J. Jap. Bot. 16(2): 97(1940).

On tea trees.

Yunnan(Wang L. S. 2008, p. 538).

Protoblastenia areolata H. Magn. 网原胚衣

Lich. Centr. Asia, 1: 130(1940).

Type locality: Gansu, Bexekk nos. 1, 6 in S(not seen).

On rocks.

Xinjiang(Abdulla Abbas et al., 1994, p. 22; Abdulla Abbas & Wu, 1998, p. 124; Abdulla Abbas et al., 2001, p. 367).

Protoblastenia formosana Zahlbr. 台湾原胚衣

in Fedde, Repertorium, 33: 62(1933).

Type: Taiwan, Asahina no. 213.

Taiwan(M. Sato, 1936b, p. 574; Wang & Lai, 1973, p. 96).

On rocks.

Protoblastenia haematommoides(Zahlbr.) Hertel 赤星原胚衣

Herzogia 3(2): 400(1975)[1973 – 4].

≡*Lecidea haematommoides* Zahlbr., in Handel-Mazzetti, *Symb. Sinic.* 3: 102 – 105(1930).

Syntype: Sichuan, Handel-Mazzetti, nos. 7553, 1006.

On non calcareous rock.

Protoblastenia incrustans(DC.) J. Steiner 皮壳原胚衣

Verh. zool. -bot. Ges. Wien 65: 203(1915) .

≡*Patellaria incrustans* DC. , in Lamarck & de Candolle, *Fl. franç.* , Edn 3(Paris) 2: 361(1805) .

On rocks.

Yunnan(Zahlbr. , 1930 b, p. 209) .

Protoblastenia rupestris(Scop.) J. Steiner 石生原胚衣

Verh. zool. -bot. Ges. Wien 61: 47(1911) .

≡*Lichen rupestris* Scop. , *Fl. carniol.* , Edn 2(Wien) 2: 363(1772) .

On rocks.

Yunnan(Zahlbr. , 1930 b, p. 209) , Taiwan(Aptroot and Sparrius, 2003, p. 36) .

Psora Hoffm. **鳞网衣属**

Deutschl. Fl. , *Zweiter Theil*(Erlangen) : 161(1796) [1795] . Nom. cons. , see Art. 14.

Psora altotibetica Timdal, Obermayer & Bendiksby 高藏鳞网衣

MycoKeys 13: 43 – 45(2016) .

Type. CHINA. Xizang: Himalaya Range, 165 km SSE of Lhasa, 40 km W of Lhünze, little village on way to Nera Tso(= Ni La Hu) , 28°23′N, 92°05′E, 4300 – 4400 m alt. , dry-valley, N-exposed dry slopes, on the ground, 1 Aug 1994, W. Obermayer 5282(holotype: GZU!) . Xizang(Einar Timdal et al. , 2016. pp. 43 – 45) .

Psora asahinae Zahlbr. ex J. C. Wei 朝比氏鳞网衣

Enum. Lich. China(Beijing) : 187(1991) .

Heilongjiang(Asahina, 1952d, p. 375) .

Psora cerebriformis W. A. Weber 脑髓鳞网衣

Mycotaxon 13(1) : 104(1981) .

On soil.

Xinjiang(Reyim Mamut et al. , 2016. p. 1483) .

Psora crenata(Taylor) Reinke 凹鳞网衣

in Pringsheim, *Jb. wiss. Bot.* 28: 97(1895) .

≡*Endocarpon crenatum* Taylor, *J. Bot.* , *Lond.* 6: 156(1847) .

= *Psora concava* B. de Lesd. , *Bull. Soc. bot. Fr.* 62: 461(1910) .

= *Lecidea undulata* H. Magn. *Lich. Centr. Asia*, 1: 60(1940) .

Type: Gansu, Bohlin no. 82 in S(!) .

On calcareous soil.

Gansu(G. Schneider, 1979, p. 101) .

Psora crystallifera(Taylor) Müll. Arg. 晶鳞网衣

Flora, Regensburg 71: 140(1888) .

≡*Lecidea crystallifera* Taylor, *London J. Bot.* 6: 148(1847) .

≡*Eremastrella crystallifera*(Taylor) Gotth. Schneid. , *Biblthca Lichenol.* 13: 76(1980) [1979] .

Hongkong(Aptroot & Sipman, 2001, p. 338) .

Psora decipiens(Hedw.) Hoffm. 红鳞网衣

Descr. Adumb. Plant. Lich. 2(4) : 68(1794) .

≡*Lichen decipiens* Hedw. , *Vidensk. Meddel. Dansk Naturhist. Foren.* Kjøbenhavn 2(1) : 7(1789) .

≡*Lecidea decipiens*(Hedw.) Ach. , *Methodus, Sectio prior*(Stockholmiæ) : 80(1803) .

On calcareous soil.

Neimenggu(Tchou, 1935, p. 321; H. Magn. 1944, p. 24) , Xinjiang(Wang XY, 1985, p. 339; Abudula Abbas et al. , 1993, p. 80; Abdulla Abbas & Wu, 1998, p. 124; Abudula Abbas et al. , 2001, p. 367; Reyim Mamut et al., 2016. p. 1484), Gansu(H. Magn. 1940, p. 56), Ningxia(Liu M & J. C. Wei, 2013, p. 44; Yang J. & Wei J. C., 2014, p. 1030), Shaanxi(He & Chen 1995, p. 45 without citation of specimen & locality), Xizang(Mt. Qomolangma: Paulson, 1925, p. 193; J. C. Wei & Chen, 1974, p. 175; J. C. Wei, 1981, p. 82; J. C. Wei & Jiang, 1981, p. 1147 & 1986, p. 30), Yunnan(Hue, 1889, p. 175; Zahlbr., 1930 b, p. 110; Wei X. L. et al. 2007, p. 156).

Psora lurida(Ach.) DC. 棕鳞网衣

in Lamarck & de Candolle, *Fl. franç.*, Edn 3(Paris) 2: 370(1805).

≡*Lecidea lurida* Ach., *Methodus*, Sectio prior: 77(1803).

Hebei, Neimenggu(Tchou, 1935, p. 322).

Notes: The species "*Psora lurida*(Dill. ex with.) DC." was reported by Tchou Y. T. (1935, p. 322) from Hebei and Neimenggu, cited by J. C. Wei in *Enum. Lich. China*: 212 – 213(1991); however, this species shares the same basionym "*Lichen luridus* Dill. ex With. (1776)" with the species *Dermatocarpon luridum*(Dill. ex With.) J. R. Laundon. This problem remains to be cleared up.

Psora luridella(Tuck.) Fink 小棕鳞网衣

Lich. Fl. U. S.: 212, 213(1935).

≡*Lecidea luridella* Tuck., *Proc. Amer. Acad. Arts & Sci.* 5: 418(1862) [1860].

On soil.

Xinjiang(Reyim Mamut et al., 2016. p. 1484).

Psora nipponica(Zahlbr.) Gotth. Schneid. 日本鳞网衣

Biblthca Lichenol. 13: 117(1980) [1979].

≡*Lecidea nipponica* Zahlbr., *Bot. Mag.*, Tokyo 41: 330(1927).

On soil.

Taiwan(Aptroot, Sparrius & Lai 2002, p. 289).

Psora tenuifolia Timdal 细叶鳞网衣

Bryologist 89(4): 272(1987) [1986].

On soil(+mosses).

Xizang, Sichuan(Einar Timdal et al., 2016. p. 46).

Psora vallesiaca(Schaer.) Timdal 墙壁鳞网衣

Nordic Jl Bot. 4(4): 538(1984).

≡*Lecidea vallesiaca* Schaer., *Lich. helv. spicil.* 2: 631(1849).

On the ground.

Xizang, Sichuan(Einar Timdal et al., 2016. p. 46).

Psorula Gotth. Schneid. **小鳞衣属**

Biblthca Lichenol. 13: 135(1980) [1979].

Psorula rufonigra(Tuck.) Gotth. Schneid. 黑红小鳞衣

Biblthca Lichenol. 13: 136(1980) [1979].

≡*Biatora rufonigra* Tuck., *Proc. Amer. Acad. Arts & Sci.* 1: 250(1848).

=*Psora inconspicua* Elenk. *Bull. Jardin. Imp. St. -Petersbourg*, 5: 81(1905).

Type: Neimenggu(Mongolia austro oriental, 1898, holotype in LE(not seen).

≡*Lecidea inconspicua*(Elenk.) Zahlbr. *Cat. Lich. Univ.* 3: 879(1925).

Neimenggu(G. Schneider, 1979, p. 136), Xinjiang(Abdulla Abbas & Wu, 1998, p. 125; Abdulla Abbas et al., 2001, p. 367).

(53) Ramalinaceae C. Agardh [as 'Ramalineae'] **树花衣科**

Aphor. bot. (Lund): 93(1821)

Bacidia De Not. **杆孢衣属**

G. bot. ital. 2(1.1): 189(1846).

Bacidia arceutina(Ach.) Rehm & Arnold 始杆孢衣

Verh. zool. -bot. Ges. Wien 19: 624(1869) .

≡*Bacidia luteola* var. *arceutina* Ach. , (1803)

On *Cinnamomum* sp. , *Podocarpus* sp. & *Castanopsis* sp.

Taiwan(Aptroot and Sparrius, 2003, p. 6) .

Bacidia arnoldiana Körb. 羔杆孢衣

Parerga lichenol. (Breslau) 2: 134(1860) .

On trees, but also on granite and concreyte in shaded forests and parks. Known from Asia, Europe, and north America.

Taiwan(Aptroot and Sparrius, 2003, p. 6) , Hongkong(Aptroot & Seaward, 1999, p. 66, Living culture CBS – 101361) .

Bacidia bagliettoana(A. Massal. & De Not.) Jatta 藓杆孢衣

Syll. Lich. Ital. : 421(1900) .

≡*Scoliciosporum bagliettoanum* A. Massal. & De Not. , in Massalongo, *Memor. Lich.* : 126(1853) .

= *Bacidia muscorum*(Sm.) Mudd. (Abbas et al. , 1993, p. 76) .

= *Lecidea muscorum* f. *terrestris* Nyl. Herb. Mus. Fenn. 1859, p. 89.

= *Bacidia muscorum* var. *terrestris*(Nyl.) Vain. (Abdulla Abbas et al. , 1993, p. 76) . Xinjiang(H. Magn. 1940, p. 67; Abdulla Abbas et al. , 1993, p. 76; Abdulla Abbas & Wu, 1998, p. 42; Abdulla Abbas et al. , 2001, p. 361) , Taiwan(Aptroot and Sparrius, 2003, p. 6) .

Bacidia celtidicola Zahlbr. 朴生杆孢衣

in *Handel-Mazzetti, Symb. Sin.* 3: 117(1930) & *Cat. Lich. Univ.* 8: 399(1932) .

Type: Sichuan, Handel-Mazzetti no. 3079.

On bark of trees.

Bacidia chloroticula(Nyl.) A. L. Sm. 绿色杆孢衣

Monogr. Brit. Lich. 2: 155(1911) .

≡*Lecidea chloroticula* Nyl. , *Flora*, Regensburg 60: 564(1877) .

On sandstone & shale.

Taiwan(Aptroot and Sparrius, 2003, p. 6) .

Bacidia circumspecta(Norrl. & Nyl.) Malme 周杆孢衣

Bot. Notiser: 140(1895) .

≡*Bacidia bacillifera* var. *circumspecta* Norrl. & Nyl. , *Meddn Soc. Fauna Flora fenn.* 10: 22(1883) .

On *Abies kawakamii.*

Taiwan(Aptroot and Sparrius, 2003, p. 6) .

Bacidia delicata(Larbal. ex Leight.) Coppins 柔杆孢衣

in Hawksworth, James & Coppins, *Lichenologist* 12(1) : 106(1980) .

≡*Lecidea effusa* var. *delicata* Larbal. ex Leight. , *Lich. -Fl. Great Brit.* , Edn 3: 371(1879) .

On trees, but also on granite in shaded forests and parks. Known from Europe and Asia.

Taiwan(Aptroot and Sparrius, 2003, p. 7) , Hongkong(Aptroot & Seaward, 1999, p. 66) .

Bacidia egenula(Nyl.) Arnold 杆孢衣

Flora, Regensburg 53: 472(1870) .

≡*Lecidea egenula* Nyl. , *Flora*, Regensburg 48: 147(1865) .

On concrete, mortar and brick.

Taiwan(Aptroot and Sparrius, 2003, p. 7) , Hongkong(Aptroot et al. , 2001, p. 321) .

Bacidia friesiana(Hepp) Körb. 针形杆孢衣

Parerga lichenol. (Breslau) 2: 133(1860) .

≡*Biatora friesiana* Hepp, *Flecht. Europ.* : no. 288(1857) .

Hubei(Chen, Wu, J. C. Wei, 1989, p. 424) .

Bacidia heterochroa(Müll. Arg.) Zahlbr. 杂绿杆孢衣

Cat. Lich. Univers. 4: 204(1926) [1927] .

≡*Patellaria heterochroa* Müll. Arg. , *Flora*, Regensburg 63: 280(1880) .

On twigs.

Ningxia(Liu M & J. C. Wei, 2013, pp. 44 – 45) .

Bacidia hunana Zahlbr. 湖南杆孢衣

in Handel-Mazzetti, *Symb. Sinic*. 3: 113(1930) & *Cat. Lich. Univ*. 8: 403(1932) .

Type: Hunan, near Changsha, in Mt. Yuelu shan, Handel-Mazzetti no. 11439.

≡*Mycobilimbia hunana*(Zahlbr.) D. D. Awasthi, in Awasthi & Mathur, *Proc. Indian Acad. Sci.* , *Pl. Sci*. 97(6) : 501(1987) .

Hongkong(Thrower, 1988, p. 60) .

See Bcidia triseptata(Hepp in Zollinger) Zahlbr. for Hongkong.

Bacidia impura Zahlbr. 污杆孢衣

in Fedde, Repertorium 33: 42(1933) .

Type: Taiwan, Asahina no. 165. (Zahlbr. , 1940, p. 362; Lamb, 1963, p. 62; Wang & Lai, 1973, p. 88) .

Bacidia inconstans Zahlbr. 多形杆孢衣

in Handel-Mazzetti, *Symb. Sin*. 3: 118(1930) & *Cat. Lich. Univ*. 8: 403(1932) .

Type: Sichuan, at about 2500 m, 25/IV/1914, Handel-Mazzetti no. 1212.

On bark of trees.

f. coerulata Zahlbr. 兰色变型

in Handel-Mazzetti, *Symb. Sin*. 3: 118(1930) & *Cat. Lich. Univ*. 8: 403(1932) .

Type: Yunnan, Handel-Mazzetti no. 6025.

Corticolous.

Bacidia inundata(Fr.) Körb. 浸水杆孢衣

Syst. lich. germ. (Breslau) : 187(1855) .

≡*Biatora inundata* Fr. , *K. Vetensk-Acad. Nya Handl.* : 270(1822) .

On stones and bricks, rarely on ligneous substrates.

Jiangsu(Wu J. L. , 1987, p. 92; Xu B. S. , 1989, p. 201) .

Bacidia jucunda Zahlbr. 快活杆孢衣

in Handel-Mazzetti, *Symb. Sin*. 3: 119(1930) & *Cat. Lich. Univ*. 8: 403(1932) .

Type: Hunan, Handel-Mazzetti no. 12157.

Corticolous.

Bacidia laurocerasi(Delise ex Duby) Zahlbr. 肉白杆孢衣

Cat. Lich. Univers. 4: 213(1926) [1927] .

≡*Patellaria laurocerasi* Delise ex Duby, *Bot. Gall.* , Edn 2(Paris) 2: 653(1830) .

= *Biatora luteola* f. *endoleuca* Nyl.

On bark of trees.

Guizhou, Yunnan(Zahlbr. , 1930b, p. 115) , Taiwan(Aptroot and Sparrius, 2003, p. 7) .

Bacidia leprophora Zahlbr. 鳞杆孢衣

in Handel-Mazzetti, *Symb. Sin*. 3: 116(1930) & *Cat. Lich. Univ*. 8: 404(1932) .

Type: Yunnan, Handel-Mazzetti no. 6028.

Corticolous.

Bacidia lopingensis Zahlbr. 罗平杆孢衣

in Handel-Mazzetti, *Symb. Sin*. 3: 117(1930) & *Cat. Lich. Univ*. 8: 404(1932) .

Type: Yunnan, Handel-Mazzetti no. 11061.

Corticolous.

Bacidia manhaviensis Zahlbr. 曼耗杆孢衣

in Handel-Mazzetti, *Symb. Sin.* 3: 117(1930) & *Cat. Lich. Univ.* 8: 405(1932).

Type: Yunnan, Handel-Mazzetti no. 5906.

Corticolous.

Bacidia medialis(Tuck.) Zahlbr. 淡盘杆孢衣

Denkschr. Kaiserl. Akad. Wiss., Math. -Naturwiss. Kl. 83: 127(1909).

≡*Lecidea medialis* Tuck., in Nylander, *Annls Sci. Nat., Bot.*, sér. 4 19: 346(1863).

On trees, but also on rock in shaded forests and parks. Pantropical.

Taiwan(Aptroot and Sparrius, 2003, p. 7), Hongkong(Thrower, 1988, p. 61; Aptroot & Seaward, 1999, p. 66; Aptroot et al., 2001, p. 321).

Bacidia melanocardia Zahlbr. 黑心杆孢衣

Hedwigia B. 74: 201(1934) & *Cat. Lich. Univ.* 10: 363(1940). M. Lamb, *Index Nom. Lich.* 1963: 63.

Type: from Sichuan, collected by F. Rock.

Corticolous.

Bacidia morosa Zahlbr. 台湾杆孢衣

in *Fedde, Repertorium* 33: 42(1933) & *Cat. Lich. Univ.* 10: 364(1940), M. Lamb, *Index Nom. Lich.* 1963: 64. Wang & Lai, *Taiwania* 18(1): 88(1973).

Type: Taiwan, Asahina no. 175.

Saxicolous.

Bacidia morula(Kremp.) Zahlbr. 黑缘杆孢衣

Cat. Lich. Univ. 4: 224(1926) [1927].

≡*Lecidea morula* Kremp., *Nuovo G. bot. ital.* 7(1): 29(1875).

Hongkong(Thrower, 1988, p. 62).

Bacidia nigra Zahlbr. 黑杆孢衣

Hedwigia B. 74: 200(1934) & *Cat. Lich. Univ.* 10: 364(1940), M. Lamb, *Index Nom. Lich.* 1963: 64.

Type: Yunnan, collected by F. Rock.

Bacidia nigrosticta Zahlbr. 黑点杆孢衣

in Handel-Mazzetti, *Symb. Sin.* 3: 116(1930) & *Cat. Lich. Univ.* 8: 406(1932).

Type: Yunnan, Handel-Mazzetti no. 13066.

Corticolous.

Bacidia olivaceorufa Vain. 红绿杆孢衣

Ann. Acad. Sci. fenn., Ser. A 15(6): 65(1921).

Yunnan(Aptroot et al., 2003, pp. 44).

Bacidia pallidocarnea(Müll. Arg.) Zahlbr. 淡肤杆孢衣

Cat. Lich. Univ. 4: 231(1926) [1927].

≡*Patellaria pallidocarnea* Müll. Arg., *Flora*, Regensburg 64(15): 232(1881).

= *Bacidia spermatophora* Zahlbr., *Cat. Lich. Univ.* 3: 118(1930) & *Cat. Lich. Univ.* 8: 408(1932). R. Santesson, *Foliicolous Lichens* I: 444(1952).

Type: Hunan, Handel-Mazzetti no. 12276.

Foliicolous.

Bacidia polychroa(Th. Fr.) Körb. 多色杆孢衣

Parerga lichenol. (Breslau) 2: 131(1860).

≡*Biatora polychroa* Th. Fr., *Öfvers. K. Svensk. Vetensk. -Akad. Förhandl.* 12: 17(1855).

= *Verrucaria fuscorubella* Hoffm.

≡*Bacidia fuscorubella*(Hoffm.) Bausch.

Corticolous.

Yunnan(Lijiang Zahlbr. , 1934 b, p. 201) , Taiwan(Aptroot and Sparrius, 2003, p. 7) .

Bacidia rubella(Hoffm.) A. Massal. 淡红杆孢衣

Ric. auton. lich. crost. (Verona) : 118, fig. 231(1852) .

≡ *Verrucaria rubella* Hoffm. , *Deutschl. Fl.* , *Zweiter Theil*(Erlangen) : 174(1796) [1795] .

= *Lichen rubellus* Ehrh. (1785, nom. nud.)

= *Lichen luteolus* Schrad.

= *Bacidia luteola*(Schrad.) Mudd.

Corticolous.

Sichuan(Zahlbr. , 1930 b, p. 119) , Yunnan(Zahlbr. , 1934 b, p. 201: lapsu ' lutcola') , Zhejiang(Xu B. S. , 1989, p. 202) .

Bacidia sabuletorum(Schreb.) Lettau 鹿乡杆孢衣

Hedwigia 52: 132(1912) .

≡ *Lichen sabuletorum* Schreb. , *Spic. fl. lips.* (Lipsiae) : 134(1771) .

Xinjiang(Abdula, Abbas et al. , 1993, p. 76) .

Bacidia subincompta(Nyl.) Arnold 黑盘杆孢衣

Flora, Regensburg 53: 472(1870) .

≡ *Lecidea subincompta* Nyl. , *Flora*, Regensburg 48: 147(1865) .

= *Bacidia affinis*(Stizenb.) Vain. , *Acta Soc. Fauna Flora fenn.* 53(no. 1) : 146(1922) .

≡ *Secoliga atrosanguinea* var. *affinis* Stizenb. , *Nova Acta Acad. Caes. Leop. -Carol. German. Nat. Cur.* 30(3) : 18(1863) .

On bark of trees.

Hunan(Zahlbr. , 1930 b, p. 115) .

Bacidia trachonopsis(Nyl.) Zahlbr. 满爱杆孢衣

Cat. Lich. Univers. 4: 157(1926) [1927] .

≡ *Lecidea trachonopsis* Nyl. , *J. Linn. Soc.* , *Bot.* 20: 64(1883) .

Type: Shanghai(Hue, 1891, p. 115; Zahlbr. , 1927, p. 157 & 1930b, p. 113.) .

Tegulicolous.

Bacidia triseptata(Hepp) Zahlbr. 三隔杆孢衣

Cat. Lich. Univers. 4: 161(1926) [1927] .

≡ *Lecidea triseptata* Hepp, in Zollinger, *Syst. Verz.* : 9(1854) .

= *Bacidia* cf. *hunana* Zahlbr. in Thrower, 1988, pp. 15, 60, synonymised by Aptroot & Seaward, 1999, p. 67.

On soil of roadbanks, but also on rock. Tropical Asia and Australia.

Taiwan(Aptroot and Sparrius, 2003, p. 8) , Hongkong(Thrower, 1988, pp. 15, 60; Aptroot & Seaward, 1999, p. 67; Aptroot et al. , 2001, p. 321) .

Bacidia wuliensis Zahlbr. 云南杆孢衣

in Handel-Mazzetti, *Symb. Sin.* 3: 114(1930) & *Cat. Lich. Univ.* 8: 411(1932) .

Type: Yunnan, Handel-Mazzetti no. 9782.

Saxicolous.

Bacidina Vězda **昵孢衣属**

Folia geobot. phytotax. 25(4) : 431(1991) [1990] .

Bacidina apiahica(Müll. Arg.) Vězda 蜜昵孢衣

Folia geobot. phytotax. 25(4) : 432(1991) [1990] .

≡ *Patellaria apiahica* Müll. Arg. , *Lichenes Epiphylli* Novi: 9(1890) .

On *Polystichum* sp.

Taiwan(Aptroot et al. , 2003, pp. 42 – 43) .

≡*Bacidia apiahica*(Müll. Arg.) Zahlbr. *Cat. Lich. Univers*. 4: 174(1926) [1927] .

On leaves of *Thunbergia grandiflora*.

Yunnan(J. C. Wei & Jiang, 1991, p. 205) , Taiwan(Aptroot and Sparrius, 2003, p. 6) .

Bacidina mirabilis(Vězda) Vězda 奇昵孢衣

Folia geobot. phytotax. 25(4) : 432(1991) [1990] .

≡*Catillaria mirabilis* Vězda, *Folia geobot. phytotax*. 15(1) : 80(1980) .

Yunnan(Aptroot et al. , 2003, pp. 44) .

Bacidina pallidocarnea(Müll. Arg.) Vězda 淡白昵孢衣

Folia geobot. phytotax. 25(4) : 432(1991) [1990] .

≡*Patellaria pallidocarnea* Müll. Arg. , *Flora*, Regensburg 64(15) : 232(1881) .

On *Fatsia* sp. and *Asplenium nidus*.

Taiwan(Aptroot et al. , 2003, pp. 42 – 43) .

Bacidina scutellifera(Vězda) Vězda 盾昵孢衣

Folia geobot. phytotax. 25(4) : 432(1991) [1990] .

≡*Bacidia scutellifera* Vězda, *Folia geobot. phytotax*. 10: 421(1975) .

Yunnan(Aptroot et al. , 2003, pp. 44) .

Bacidina sp. 昵孢衣(未定名种)

Yunnan(Aptroot et al. , 2003, pp. 44, TBG) .

Biatora Fr. **蜡盘衣属**

Lichenum Dianome Nova: 7(1817) .

Biatora sphaeroides(Dicks.) Hornem. 扁球蜡盘衣

Dansk. Oecon. Pl. 2: 559(1837) .

≡*Lichen sphaeroides* Dicks. , *Fasc. pl. crypt. Brit*. (London) 1: 9(1785) .

≡ *Bacidia sphaeroides* (Dicks.) Zahlbr. , in Engler & Prantl, *Nat. Pflanzenfam*. , Teil. I(Leipzig) 1 * : 135 (1905) .

Xinjiang(Abdulla Abbas et al. , 1993, p. 76; Abdulla Abbas & Wu, 1998, p. 42; Abdulla Abbas et al. , 2001, p. 361) .

Biatora vernalis(L.) Fr. 春蜡盘衣

K. svenska Vetensk-Akad. Handl. : 271(1822) .

≡*Lichen vernalis* L. , *Syst. Nat*. 3: 234(1768) .

Xinjiang(Abdulla Abbas & Wu, 1998, p. 43; Abdulla Abbas et al. , 2001, p. 361) .

Catinaria Vain. **托盘衣属**

Acta Soc. Fauna Flora fenn. 53(no. 1) : 143(1922) .

Catinaria kelungana Zahlbr. 台湾托盘衣

in Fedde, Repertorium 31: 223(1933) & *Cat. Lich. Univ*. 10: 205(1940) . M. Lamb, *Index Nom. Lich*. 1963: 149. Wang & Lai, Taiwania 18(1) : 88(1973) .

Type: Taiwan, Faurie no. 32.

Frutidella Kalb **小枝衣属**

Hoppea 55: 582(1994) .

Frutidella caesioatra(Schaer.) Kalb 蓝黑小枝衣

Hoppea 55: 582(1994) .

≡*Lecidea caesioatra* Schaer. , *Naturw. Anzeiger Allgem. Schweizer. Gesellsch. Naturwiss*. 2: 10(1818) .

Xizang(Mt. Qomolangma: Paulso, 1925, p. 193) .

Japewia Tønsberg **碟衣属**

Lichenologist 22(3) : 205(1990) .

Japewia tornoënsis(Nyl.) Tønsberg 碟衣

Lichenologist 22(3): 206(1990).

≡*Lecidea tornoënsis* Nyl., *Herb. Mus. Fenn.*: 110(1859).

On trunk of very old *Juniperus* sp.).

Sichuan(Obermayer, 2004, p. 500).

Lecania A. Massal. **副茶渍属**

Alcuni Gen. Lich.: 12(1853).

Lecania erysibe(Ach.) Mudd 霉副茶渍

Man. Brit. Lich.: 141(1861).

≡*Lichen erysibe* Ach., *Lich. suec. prodr.* (Linköping): 50(1799)[1798].

Xizang(Mt. Qomolangma: Paulson, 1925, p. 192).

Lecania erysibopsis(Nyl.) Zahlbr. 拟霉副茶渍

Cat. Lich. Univ. 5: 731(1928).

≡*Lecanora erysibopsis* Nyl. in Nyl. & Cromb. *Journ. Linn. Soc. London, Bot.* 20: 63(1883).

Type: Shanghai, Type not indicated.

Shanghai(Hue, 1891, p. 77; Zahlbr., 1928, p. 731 & 1930 b, p. 177).

On mortar of walls.

Lecania koerberiana J. Lahm 柯氏副茶渍

in Körber, *Parerga lichenol.* (Breslau) 1: 68(1859).

Xinjiang(Abdulla Abbas et al., 1993, p. 78; Abdulla Abbas & Wu, 1998, p. 43; Abdulla Abbas et al., 2001, p. 364).

Lecania mongolica H. Magn. 蒙古副茶渍

Lich. Centr. Asia, 2: 45(1944).

Type: Neimenggu, Bohlin, no. 116 in S.

Ningxia(Yang J. & Wei J. C., 2014, p. 1019).

Lecania rabenhorstii(Hepp) Arnold 拉本氏副茶渍

Flora, Regensburg 67: 403(1884).

≡*Patellaria rabenhorstii* Hepp, *Flecht. Europ.*: no. 75(1853).

On rock.

Liaoning(Kondratyuk et al., 2013, p. 330: Dalian county, Lüshun city, Xiaoheishi village).

Lecania toninioides Zahlbr. 泡鳞型副茶渍

Beih. Botan. Centralbl. 13: 160(1902).

On rocks.

Xinjiang(Mahira Muhammat et al., 2012, pp. 5 – 6).

Phyllopsora Müll. Arg. **树痂衣属**

Bull. Herb. Boissier 2(1): 11, 45(1894).

Phyllopsora albicans Müll. Arg. 白树痂衣

Bull. Acad. R. Sci. Belg., *Cl. Sci.*, sér. 3 32(2): 132(1893).

= *Phyllopsora formosana* Zahlbr. in *Fedde, Repertorium*, 33: 43(1933).

Type: Taiwan, 5 January 1926, Asahina F 156(TNS holotype). revised by Aptroot, 2004, p. 35).

Taiwan(Zahlbr., 1940, p. 377; Lamb, 1963, p. 551; Wang & Lai, 1973, p. 95: lapsu '*Phyllospora*' *formosana* Zahlbr.; Aptroot and Sparrius, 2003, p. 34).

Phyllopsora buettneri(Müll. Arg.) Zahlbr. 布特氏树痂衣

Cat. Lich. Univers. 4: 396(1926)[1927].

≡*Psora buettneri* Müll. Arg., *Bot. Jb.* 15: 506(1893).

On bark, *Pseudotsuga wilsoniana, Castanopsis* sp., and shale.

Taiwan(Aptroot and Sparrius, 2003, p. 34).

Phyllopsora chlorophaea(Müll. Arg.) Zahlbr. 绿色树痂衣

Denkschr. Kaiserl. Akad. Wiss. Wien, Math. - Naturwiss. Kl. 83: 133(1909) .

≡*Psora chlorophaea* Müll. Arg. , *Flora*, Regensburg 70: 320(1887) .

On *Castanopsis* sp. , and *Eriobotrya* sp.

Taiwan(Aptroot and Sparrius, 2003, p. 34) .

Phyllopsora corallina(Eschw.) Müll. Arg. 珊瑚树痂衣

Bot. Jb. 20: 264(1894) .

≡*Lecidea corallina* Eschw. , in Martius, *Fl. Bras.* 1: 256(1833) [1829 – 1833] .

On *Cryptomeria japonica*, *Picea morrisonicola*, and *Castanopsis* sp.

Taiwan(Aptroot and Sparrius, 2003, p. 34) .

Phyllopsora furfuracea Zahlbr. 鳞粉树痂衣

in Engler & Prantl, *Nat. Pflanzenfam.* , Teil. I(Leipzig) 1 * : 138(1905) .

On shale.

Taiwan(Aptroot and Sparrius, 2003, p. 34) .

Phyllopsora pannosa Müll. Arg. 毡树痂衣

Bot. Jb. 20: 265(1894) .

On *Heritiera littoralis* and *Semecarpus* sp.

Taiwan(Aptroot and Sparrius, 2003, p. 34) .

Phyllopsora parvifolia(Pers.) Müll. Arg. 树痂衣

Bull. Herb. Boissier 2(1) : 90(1894) .

≡ *Lecidea parvifolia* Pers. , in Gaudichaud-Beaupré in Freycinet, *Voy. Uranie.* , *Bot.* : 192 (1827) [1826 – 1830] .

On twigs and on calcaceous rocks.

Guizhou(Zahlbr. , 1930 b, p. 127) .

Phyllopsora stenosperma Zahlbr. 小孢树痂衣

in Feddes, *Lichenes* in Feddes, *Repertorium sp. nov.* 33: 44(1933) .

Type: Taiwan, Asahina no. 170.

Taiwan(Wang & Lai, 1973, p. 95, lapsu ' *Phyllospora*' *stenosperma* Zahlbr.) .

On bark of trees.

Physcidia Tuck. **梅蚣衣属**

Proc. *Amer. Acad. Arts & Sci.* 5: 399(1862) [1860] .

Physcidia cylindrophora(Taylor) Hue 圆梅蚣衣

Bull. Soc. linn. Normandie, sér. 6 1: 97(1908) .

≡*Parmelia cylindrophora* Taylor, *London J. Bot.* 6: 165(1847) .

On shale.

Taiwan(Aptroot and Sparrius, 2003, p. 34) .

Ramalina Ach. **树花属**

in Luyken, *Tent. Hist. Lich.* : 95(1809) .

Ramalina africana(Stein) C. W. Dodge 非洲树花

Beih. Nova Hedwigia 38: 56(1971) .

≡*Ramalina rigida* var. *africana* Stein, *Jber. schles. Ges. vaterl. Kultur* 66: 137(1888) .

On bark of trees.

Sichuan(Chen S. F. et al. , 1997, p. 222) .

Ramalina almquistii Vain. 高峰树花

Ark. Bot. 8(4) : 17(1909) .

On rocks.

Liaoning(Fu Wei et al. , 2008, p. 94) , Xinjiang(Fu Wei et al. , 2009, p. 98) , Xizang(J. C. Wei & Jiang, 1986, p. 73) , Yunnan(Oh et al. , 2014, p. 233) , Shandong(zhao ZT et al. , 1999, p. 426; Zhao ZT et al. , 2002, p. 6; Li Y et al. , 2008, p. 72) .

Ramalina americana Hale 美洲树花

Bryologist 81(4) : 599(1979) [1978] .

Gansu, Shaanxi(Fu Wei et al. , 2009, p. 98) , Yunnan(Oh et al. , 2014, p. 233) .

Ramalina aspera Räsänen 粗树花

in Ann. Bot. Soc. Zool-Bor. Fenn. Vanamo, 20(3) : 5(1944) .

shaanxi(Fu Wei et al. , 2009, p. 98) , Sichuan, Yunnan(Oh et al. , 2014, p. 233) .

Ramalina attenuata(Pers.) Tuck. 狭叶树花

U. S. Explor. Expedit. Wilkes 17: 129(1861) .

≡*Physcia attenuata* Pers. , *Ann. Wetter. Gesellsch. Ges. Naturk.* 2: 18(1811) [1810] .

= *Ramalina rigida*(Pers.) Ach. syn 294, in Jatta, *Nuovo Giornale botanico italiano. Nuova serie. Volume nono...* I: 462(1902) .

On tree trunk.

Shaanxi(Jatta, 1902, p. 462; Zahlbr. , 1930 b, p. 205) .

Ramalina australiensis Nyl. 澳树花

Bull. Soc. linn. Normandie, sér. 2 4(2) : 120(1870) .

= *Ramalina furcellata*(Mont.) Zahlbr. , *Cat. Lich. Univers*. 6: 489(1930) .

≡*Evernia furcellata* Mont. , in Sagra, *Historia física, polirica y nayturál de la islea de Cuba*: 236(1842) [1838 – 1842] .

= *Ramalina gracilenta* Fries. Nyl. in Bull. Soc. Linn. Norm. ser. 2, IV, sep. 19(1870) , Zahlbr. in Handel-Mazzetti, *Symb. Sin.* 3: 204(1930) .

Shandong(Nyl. & Cromb. 1883, p. 62; Zahlbr. , 1930 b, p. 204) , Guangdong(Nyl. & Cromb. 1870, p. 117; Zahlbr. , 1930 b, p. 204) , China(prov. not indicated: Hue, 1899, p. 65) .

Ramalina calicaris(L.) Röhl. 杯树花

Deutschl. Fl. (Frankfurt) 3(2) : 139(1813) .

≡*Lichen calicaris* L. , *Sp. pl.* 2: 1146(1753) .

On bark of trees.

Heilongjiang(Luo, 1984, p. 84; Fu Wei et al. , 2008, p. 94) , Gansu(Fu Wei et al. , 2009, p. 98) , Shaanxi(He & Chen 1995, p. 45 without citation of specimens and their localities; Fu Wei et al. , 2009, p. 98) , Sichuan(Oh et al. , 2014, p. 234, Yunnan(Hue, 1887a, p. 18 & 1889, p. 161 & 1890, p. 263 & 1899, p. 70; Zahlbr. , 1930 b, p. 204; Oh et al. , 2014, p. 234) , Taiwan(Zahlbr. , 1933 c, p. 61; M. Sato, 1937b, p. 598; Sasaoka, 1919, p. 180) .

var. calicaris 原变种

Taiwan(Wang & Lai, 1973, p. 96) .

= *Ramalina calicaris* var. *subamplicata* Nyl. *Bull. Soc. linn. Normandie*, sér. 2 4(2) : 132(1870) .

On tree trunk.

Shaanxi(Jatta, 1902, p. 463; Zahlbr. , 1930 b, p. 204) .

= *Ramalina calicaris* var. *subfastigiata* Nyl. , *Bull. Soc. linn. Normandie*, sér. 2 4(2) : 132(1870) .

On bark of trees.

Shandong(Moreau et Moreau, 1951, p. 192) .

var. japonica Hue 日本变种

Nouv. Arch. Mus. Hist. Nat. , Paris, 4 sér. 1: 71(1899) .

On bark of trees.

Shaanxi(He & Chen 1995, p. 45) , Hubei(Chen, Wu & J. C. Wei, 1989, p. 455) , Anhui(Xu B. S. , 1989, p. 228

with salazinic acid), Zhejiang(Xu B. S., 1989, p. 228 without salazinic acid), Taiwan(Zahlbr., 1933 c, p. 62; Sato, 1937b, p. 598; Asahina, 1939, p. 206; Wang & Lai, 1973, p. 96; Yoshim. 1974, p. 21).

f. papillosa Hue 小疣变型

Nouv. Arch. Mus. Hist. Nat., Paris, 4 sér. 1: 70(1899).

On bark of trees.

Sichuan, Yunnan(Zahlbr., 1930 b, p. 204).

f. subpapillosa Nyl. 亚小疣变型

Nouv. Arch. Mus. Hist. Nat., Paris, 4 sér. 1: 71(1899).

On tree trunk and twigs.

Sichuan(Zahlbr., 1930 b, p. 204), Yunnan(Hue, 1899, p. 70; Zahlbr., 1930 b, p. 204).

Ramalina celastri(Spreng.) Krog & Swinscow 线树花

Norw. Jl Bot. 23: 159(1976). subsp. celastri

≡*Parmelia celastri* Spreng., *Syst. veg.*, Edn 16 4(1): 326(1827).

=*Ramalina linearis* (Sw.) Ach., *Lich. univ.*: 598(1810).

≡*Lichen linearis* Sw., *Method. Muscor.*: 36(1781).

Hongkong(Zahlbr., 1930 b, p. 204).

Ramalina chihuahuana Kashiw. & T. H. Nash 墨西哥树花

Mycotaxon 83: 386(2002).

On rock.

Xinjiang(Fu Wei et al., 2009, pp. 98 – 99).

Ramalina commixta Asahina 假杯树花

J. J. B. 15: 207(1939).

On tree twigs.

Yunnan(Wang L. S. 2008, p. 538), Zhejiang(Xu B. S., 1989, p. 228).

Ramalina complanata(Sw.) Ach. 扁平树花

Lich. univ.: 599(1810).

≡*Lichen complanatus* Sw., in Acharius, *K. Vetensk-Acad. Nya Handl.*: 290(1797).

On tree branches.

Xinjiang(Fu Wei et al., 2009, p. 99).

Ramalina conduplicans Vain. 对折树花

Ann. bot. Soc. Zool. -Bot. fenn. Vanamo 1(3): 35(1921).

On tea trees.

Heilongjiang, Jilin(Fu Wei et al., 2008, p. 95), Xinjiang(Fu Wei et al., 2009, p. 99), Sichuan(Oh et al., 2014, p. 234), Yunnan(Wei X. L. et al. 2007, p. 157; Wang L. S. 2008, p. 538; Oh et al., 2014, p. 234), Taiwan(Ohmura et al., 2008, p. 157).

Ramalina confirmata(Nyl.) Elix 假杯点树花

Lichenes Australasici Exsiccati 5: 219(1990).

On bark of various species(Quercus, Salix) at elevations between 1,460 and 3,400 m.

Xizang, Sichuan, Yunnan(Oh et al., 2014, p. 234).

Ramalina dendriscoides Nyl. 乔木树花

Flora, Regensburg 59: 412(1876).

On bark at elevations between 1,840 and 2,450 m.

Sichuan, Yunnan(Oh et al., 2014, p. 235).

var. minor Müll. Arg. 小体变种

Proc. R. Soc. Edinb. 11: 458(1882).

On bark of trees.

Fujian(Zahlbr. , 1930 b, p. 204) .

Ramalina denticulata(Eschw.) Nyl. 细齿树花

Acta Soc. Sci. fenn. 7(2) : 434(1863) . var. denticulata 原变种

≡*Parmelia denticulata* Eschw. , in Martius, *Fl. Bras.* 1: 221(1833) [1829 – 1833] .

On tree trunk.

Shaanxi(Jatta, 1902, p. 463; Zahlbr. , 1930 b, p. 205) .

var. canalicularis Nyl. 沟槽变种

Bull. Soc. linn. Normandie, sér. 2 4(2) : 126(1870) .

On bark of trees.

Sichuan(Zahlbr. , 1934 b, p. 212) .

Ramalina digitata Meyen & Flot. 掌树花

Nova Acta Acad. Caes. Leop. -Carol. Nat. Cur. 19(Suppl.) : 212, tab. III, fig. 1(1843) .

On twigs of *Theae chinensis*.

Guangdong(Meyen & Flotow, 1843, p. 212) .

Ramalina dilacerata(Hoffm.) Hoffm. 小树花

Herb. viv. , coll. plant. sicc. , caesar. univ. Mosq. : 451(1825) .

≡*Lobaria dilacerata* Hoffm. , *Deutschl. Fl. , Zweiter Theil*(Erlangen) : 140(1796) [1795] .

= *Ramalina minuscula* Nyl. *Bull. Soc. linn. Normandie*, sér. 2 4(2) : 164(1870) .

On exposed and sheltered trees. Known from Asia, North Ameica and Europe.

Xinjiang(Zhou Chunli et al. , 2009, p. 10) , Shaanxi(Jatta, 1902, p. 462; Zahlbr. , 1930b, p. 204) , Hubei(Chen, Wu & Wei, 1989, p. 456) , Hongkong(Thrower, 1988, pp. 17, 159; Aptroot & Seaward, 1999, p. 92; Aptroot & Sipman, 2001, p. 338; Seaward & Aptroot, 2005, p. 285;) .

Ramalina ecklonii(Spreng.) Meyen & Flot. [as ' eckloni'] 艾克氏树花

Nova Acta Acad. Caes. Leop. -Carol. Nat. Cur. , Suppl. 1 19: 213(1843) .

≡*Parmelia ecklonii* Spreng. , *Syst. veg.* , Edn 16 4(1) : 328(1827) .

Shaanxi(He & Chen 1995, p. 45) .

Ramalina farinacea(L.) Ach. 粉树花

Lich. univ. : 606(1810) .

≡*Lichen farinaceus* L. , *Sp. pl.* 2: 1146(1753) .

On bark of trees.

Xinjiang(Wang XY, 1985, p. 349; Abdula Abbas et al. , 1993, p. 80; Abdulla Abbas & Wu, 1998, p. 126; Zhou Chunli et al. , 2009, pp. 10 – 11; Abdula Abbas et al. , 2001, p. 367) , Shaanxi(He & Chen 1995, p. 45) , Gansu (Fu Wei et al. , 2009, p. 99) , Sichuan(Oh et al. 2014, p. 235) , Yunnan(Hue, 1889, p. 72; Paulson, 1928, p. 318; Zahlbr. , 1930 b, p. 205: Oh et al. , 2014, p. 235) , Xizang(J. C. Wei & Jiang, 1986, p. 74) , Shandong(Hou YN et al. , 2008, p. 67) , iangsu, Zhejiang(Xu B. S. , 1989, pp. 228 – 229) .

= *Ramalina farinacea* var. *pendulina*(Ach.) Ach. , *Lich. univ.* : 607(1810) .

≡*Parmelia farinacea* var. *pendulina* Ach. , *Methodus*, Sectio post. : 264(1803) .

On bark of trees.

Taiwan(Zahlbr. , 1933 c, p. 62; M. Sato, 1937, p. 598; Wang & Lai, 1973, p. 96; Ikoma, 1983, p. 94) .

= *Ramalina farinacea* var. *phalerata*(Ach.) Ach. , *Lich. univ.* : 607(1810) .

≡*Parmelia farinacea* var. *phalerata* Ach. , *Methodus*, Sectio post. : 264(1803) .

On trees.

Shaanxi(Jatta, 1902, p. 462; Zahlbr. , 1930 b, p. 205) .

Ramalina fastigiata(Pers.) Ach. 丛生树花

Lich. univ. : 603(1810) .

≡*Lichen fastigiatus* Pers. , *Ann. Bot.* (Usteri) 1: 156(1794) .

= *Ramalina farinacea* var. *minutula* Ach. , *Lich. univ.* : 1 – 696(1810) .

Shandong(Moreau et Moreau, 1951, p. 193) .

= *Ramalina populina*(Hoffm.) Vain. , *Meddn Soc. Fauna Flora fenn.* 14: 21(1888) .

On trees.

Shaanxi(Jatta, 1902, p. 463; Zahlbr. , 1930 b, p. 205; He & Chen 1995, p. 45 without citation of specimen and locality; Fu Wei et al. , 2009, p. 99) , Gansu(Fu Wei et al. , 2009, p. 99) , Yunnan(Wei et al. 1982, p. 48) , Shandong(Tchou, 1935, p. 302) .

var. glaucodissecta Jatta 灰裂变种

Nuovo G. bot. ital. 9: 463(1902) .

On trees.

Shaanxi(Jatta, 1902, p. 463; Zahlbr. , 1930 b, p. 205) .

var. lacerata Müll. Arg. 撕裂变种

Flora, Regensburg 74: 373(1891) .

On trees.

Shaanxi(Jatta, 1902, p. 463; Zahlbr. , 1930 b, p. 205) .

Ramalina fissa(Müll. Arg.) Vain. 半裂树花

Mém. Herb. Boissier 5: 2(1900) .

≡*Ramalina inflata* var. *fissa* Müll. Arg. , *Flora*, Regensburg 71: 203(1888) .

= *Ramalina fraxinea* var. *platyna* Nyl. , *Bull. Soc. linn. Normandie*, sér. 2 4(2) : 136(1870) .

Xinjiang(Moreau et Moreau, 1951, p. 193) .

Ramalina fraxinea(L.) Ach. 白蜡树花

Lich. univ. : 622(1810) .

≡*Lichen fraxineus* L. , *Sp. pl.* 2: 1146(1753) .

On trees.

Neimenggu(Chen et al. 1981, p. 157) , xinjiang(Wang XY, 1985, p. 349; Fu Wei et al. , 2009, p. 99) , Sichuan (Oh et al. , 2014, p. 235) , Yunnan(Hue, 1887, p. 18 & 1889, p. 161 & 1899, p. 76; Zahlbr. , 1930 b, p. 205; Oh et al. , 2014, p. 235) .

= *Ramalina fraxinea* var. *calicariformis* Nyl. , *Bull. Soc. linn. Normandie*, sér. 2 4(2) : 136(1870) .

On bark of trees.

Xinjiang(Moreau et Moreau, 1951, p. 192) .

Ramalina hengduanshanensis S. O. Oh & L. S. Wang 横断山树花

in Soon-Ok Oh et al. , *Mycobiology* 42(3) : 232(2014) .

Type: Sichuan Prov. , Xiangcheng County, Mt. Daxue, 12 Sep 2002, L. S. Wang, 18823(KUN: holotype) .

Ramalina holstii Krog & Swinsc. 豪氏树花

T. D. V. *Nor J Bot* 22: 275(1975) .

On bark at elevations between 2, 000 and 3, 800 m.

Yunnan(Oh et al. , 2014, p. 236) .

Ramalina hossei Vain. 侯氏树花

Ann. bot. Soc. Zool. -Bot. fenn. Vanamo 1(no. 3) : 36(1921) .

Gansu, Shaanxi(Fu Wei et al. , 2009, p. 100) , Sichuan, Yunnan(Oh et al. , 2014, p. 236) , Taiwan(Kashiw. , 1988, p. 132) .

Ramalina inflata(Hook. f. & Taylor) Hook. f. & Taylor 肿树花

in Hooker, *Bot. Antarct. Voy. Erebus Terror* 1839 – 1843 1: 194(1845) .

≡*Cetraria inflata* Hook. f. & Taylor, *J. Bot.*, *Lond*. 3: 646(1844).

Misapplied name:

Ramalina geniculata auct. non J. D. Hook. & Taylor in Thrower, 1988, pp. 17, 160, revised by Aptroot & Seaward, 1999, p. 92.

On exposed and shelterd trees, and small bushes. Pantropical.

Hongkong(Thrower, 1988, pp. 17, 160; Aptroot & Seaward, 1999, p. 92; Seaward & Aptroot, 2005, p. 285, Pfister 1978: 73, as *Ramalina calicaris* var. *farinacea* – "a variety based on *Ramalina inflata* Hook.").

On tree trunk.

Shaanxi(Zahlbr., 1930 b, p. 205), Taiwan(Asahina, 1938, p. 727; Asahina & M. Sato in Asahina, 1939, p. 755; Wang & Lai, 1973, p. 96; Yoshim. 1974, p. 21).

var. subgeniculata Nyl 亚曲变种

On rocks.

Shandong(Moreau et Moreau, 1951, p. 193).

subsp. australis G. N. Stevens 南方亚种

Bull. Br. Mus. nat. Hist., *Bot.* 16(2): 191(1987).

On *Pinus* sp.

Yunnan(Wei X. L. et al. 2007, p. 157).

Ramalina intermedia(Delise ex Nyl.) Nyl. 间枝树花

Flora, Regensburg 56: 66(1873).

≡*Ramalina minuscula* * *intermedia* Delise ex Nyl., *Bull. Soc. linn. Normandie*, sér. 2 4(2): 166(1870).

On rock.

Jilin, Liaoning(Fu Wei et al., 2008, p. 95), Xinjiang(Wang XY, 1985, p. 349; Zhou Chunli et al., 2009, p. 11; Abdulla Abbas & Wu, 1998, p. 126; Abdulla Abbas et al., 2001, p. 367 as R. "*intermidia*"), Gansu, shaanxi (Fu Wei et al., 2009, p. 100), Sichuan, Yunnan(Oh et al., 2014, p. 236).

Ramalina intermediella Vain. 瘤枝树花

Bot. Mag., Tokyo 35: 46(1921).

On bark of trees.

Yunnan(Hue, 1899, p. 73; Zahlbr., 1930 b, p. 205), Hubei(Chen, Wu & Wei, 1989, p. 456), Shandong(Zhang F et al., 1999, p. 31; Zhao ZT et al., 2002, p. 6), Taiwan(Zahlbr., 1933 c, p. 62; M. Sato, 1937 b, p. 598; Asahina, 1939a, p. 211; Wang & Lai, 1973, p. 96; Yoshim. 1974, p. 23), Hainan(JCWei et al., 2013, p. 231).

Ramalina intricata Kremp. 缠树花

Verh. zool. -bot. Ges. in Wien 26: 438(1877).

Type: from Shandong, Djifu("Tschifie") = Zhifu(芝罘)

Shandong Shandong(Zahlbr., 1930 a, p. 494 & 1930 b, p. 204).

On rocks.

Ramalina litoralis Asahina 石生树花

in J. J. B. XV: 220(1939) cum icon.

On coastal rocks, particularly on outlying islands and areas such as Cape D'Aguilar. Rarely

On bark.

Shandong(Zhao ZT et al., 2002, p. 6), Shanghai, Zhejiang(Xu B. S., 1989, p. 229).

See *Ramalina tenella* Müll. Arg. for Hongkong(Thrower, 1988, p. 161).

Ramalina maciformis(Delise) Bory 叶树花

Dict. Class. Hist. Nat. 14: 458(1828).

≡*Parmelia maciformis* Delise, *Descript. de l'Egypte* 2: 288(1813).

On rocks.

Shaanxi(Jatta, 1902, p. 462; Zahlbr., 1930 b, p. 205).

Ramalina nervulosa(Müll. Arg.) des Abbayes 细脉树花

Bull. Inst. Franç. Afrique Noire 14: 25(1952).

≡*Ramalina farinacea* var. *nervulosa* Müll. Arg., *Flora*, Regensburg 66: 21(1883).

On rocks and on twigs. Paleotropical.

Hainan, Taiwan(Race 1 & 2: Kashiw., 1986, p. 119120), Hongkong(Thrower, 1988, pp. 17, 163; Aptroot & Seaward, 1999, p. 92; Seaward & Aptroot, 2005, p. 286, Pfister 1978: 73, as *Ramalina calicaris* var. *farinacea.*, Contains divaricatic acid, but no salazinic acid.).

Ramalina obtusata(Arnold) Bitter 钝树花

in Pringsheim, *Jb. wiss. Bot.* 36: 435(1901).

≡*Ramalina minuscula* var. *obtusata* Arnold, *Verh. zool. -bot. Ges. Wien* 25: 472(1875).

On tree bark & twigs.

Liaoning(Fu Wei et al., 2008, p. 95), Xinjiang(Zhou Chunli et al., 2009, p. 11), Gansu(Fu Wei et al., 2009, p. 100), Sichuan, Yunnan(Oh et al., 2014, p. 237).

Ramalina pacifica Asahina 太平洋树花

J. Jap. Bot. 15: 213(1939).

On bark. Paleotropical.

Sichuan, Yunnan(Oh et al., 2014, p. 237), Hongkong(Thrower, 1988, pp. 17, 162; Aptroot & Seaward, 1999, p. 92).

Ramalina pentecostii Krog & Swin. 彭氏树花

Norwegian Journ. Bot. 23(3): 167(1976).

Xinjiang(Fu Wei et al., 2009, p. 100), Sichuan(Oh et al., 2014, p. 237).

Ramalina pertusa Kashiw. 穿孔树花

Mem. Natn Sci. Mus, Tokyo 18: 102(1985).

Liaoning(Fu Wei et al., 2008, p. 95).

Ramalina peruviana Ach. 芽树花

Lich. univ.: 599(1810).

Gansu, Shaanxi(Fu Wei et al., 2009, p. 100), Sichuan, Yunnan(Oh et al., 2014, p. 237), Shandong(Li Y et al., 2008, p. 72).

= *Ramalina farinacea* var. *dendroides* Müll. Arg., *Flora*, Regensburg 66: 21(1883). Zhejiang(Moreau et Moreau, 1951, p. 193).

Ramalina pollinaria(Westr.) Ach. 粉粒树花

Lich. univ.: 608(1810).

≡*Lichen pollinarius* Westr., *K. svenska Vetensk-Akad. Handl.* 16: 56(1755).

Heilongjiang, Jilin, Liaoning(Fu Wei et al., 2008, Pp. 95 – 96), Xinjiang(Wang XY, 1985, p. 349; Abdulla Abbas & Wu, 1998, p. 126; Abdulla Abbas et al., 2001, p. 367; Zhou Chunli et al., 2009, p. 12), Shaanxi(He & Chen 1995, p. 45 without citation of specimens and their localities), Sichuan, Yunnan(Oh et al., 2014, p. 238), Shandong(Zhao ZT et al., 1998, p. 29; zhao ZT et al., 1999, p. 426; Zhao ZT et al., 2002, p. 6; Hou YN et al., 2008, p. 67; Li Y et al., 2008, p. 72).

var. pollinaria 原变种

On bark of trees.

Gansu(Fu Wei et al., 2009, p. 100), Xinjiang(Abbas et al., 1993, p. 80), Hebei(Moreau et Moreau, 1951, p. 192), Shandong(Moreau et Moreau, 1951, p. 192; Zhang F et al., 1999, p. 31), Jiangsu(Wu J. N. & Xiang T. 1981, p. 6, here should be on p. 5 according to the sequence of the pages; Xu B. S., 1989, p. 229), Shanghai,

Anhui(Xu B. S. , 1989, p. 229) , Zhejiang(Moreau et Moreau, 1951, p. 192; Xu B. S. , 1989, p. 229) , Fujian (Zahlbr. , 1930 b, p. 204) , Xizang(J. C. Wei & Jiang, 1986, p. 74) , Taiwan(Sasaoka, 1919, p. 180) , Hongkong (Thrower, 1988, p. 163) , China(prov. not indicated: Asahina & M. Sato in Asahina, 1939, p. 755) .

= *Ramalina pollinaria* f. *humilis* (Ach.) Anders

Die Strauch- und Laubflecht. Mitteleurop. : 186(1928) .

≡*Ramalina pollinaria* var. *humilis* Ach. , *Lich. univ.* : 609(1810) .

On rocks.

Beijing(Moreau et Moreau, 1951, p. 192) , Shaanxi(Jatta, 1902, p. 462; Zahlbr. , 1930 b, p. 204) , Zhejiang (Nyl. & Cromb. 1883, p. 62) .

f. rupestris Flörke 石生变型

in Schaerer, *Enum. critic. lich. europ.* (Bern) : 8(1850) .

On rocks.

Shaanxi(Jatta, 1902, p. 462; Zahlbr. , 1930 b, p. 204, as *R. pollinaria*) .

Ramalina polymorpha(Lilj.) Ach. 多形树花

Lich. univ. : 600(1810) .

≡*Lichen calicaris* var. *polymorphus* Lilj. , *Utkast Sv. Fl.* : 426(1792) .

On rocks.

Liaoning(Tchou, 1935, p. 303; Chen et al. 1981, p. 157) , Hebei(Tchou, 1935, p. 303) , Shandong(zhao ZT et al. , 1999, p. 427; Zhao ZT et al. , 2002, p. 6) .

Ramalina pseudosekika Asahina 拟石树花

J. J. B. 17: 139(1941) .

On rocks.

Shaanxi(Sato, 1981, p. 64) , Taiwan(Ikoma, 1983, p. 95) .

Ramalina pumila Mont. 矮树花

Annls Sci. Nat. , Bot. , sér. 2 20: 356(1843) .

On twigs.

Taiwan(Zahlbr. , 1933 c, p. 63; M. Sato, 1937b, p. 597; Wang & Lai, 1973, p. 96; Ikoma, 1983, p. 95) , China (prov. not indicated: Hue, 1890, p. 268) .

Ramalina roesleri(Hochst. ex Schaer.) Nyl. 肉刺树花

Bull. Soc. linn. Normandie, sér. 2 4(2) : 165(not.) (1870) .

≡*Ramalina farinacea* var. *roesleri* Hochst. ex Schaer. , *Enum. critic. lich. europ.* (Bern) : 9(1850) .

= *Ramalina minuscula* var. *pollinariella* Nyl. , *Bull. Soc. linn. Normandie*, sér. 2 4(2) : 165(1870) .

On trees.

Heilongjiang(Chen et al. 1981, p. 157; J. C. Wei et al. 1982, p. 47; Luo, 1984, p. 84; Fu Wei et al. , 2008, p. 96) , Jilin(Chen et al. 1981, p. 157; Wei et al. 1982, p. 47; Fu Wei et al. , 2008, p. 96) , Shaanxi(Jatta, 1902, p, 462; Zahlbr. , 1930 b, p. 204; He & Chen 1995, p. 45 without citation of specimens and their localities) , Yunnan(Wei X. L. et al. 2007, p. 157) , Hubei(Chen, Wu & Wei, 1989, p. 456) , Shandong(zhao ZT et al. , 1999, p. 427; Zhao ZT et al. , 2002, p. 6) , Zhejiang(Xu B. S. , 1989, p. 229) .

Ramalina seawardii Aptroot & Sipman 西瓦氏树花

J. Hattori bot. Lab. 91: 338(2001) .

Type: China, Hongkong, Lantau island, Aptroot 48528 = Sipman45087(B-holotype; HKU(M) , ABL-isotypes) .

On volcanic rock.

Taiwan(Aptroot, Sparrius & Lai, 2002, p. 290) , hongkong(Aptroot & Sipman, 2001, pp. 338 – 339; Seaward & Aptroot, 2005, p. 286, Pfister 1978: 73, as *R. scopulorum.*) .

Ramalina sekika Asahina 石树花

Journ. Jap. Bot. 17(3): 138 – 139(1941).

Type: Liaoning, Mt. Laotie shan, leg. Y. Asahina, (pro maj. p.)

Jilin(Fu Wei et al., 2008, p. 96), Liaoning(Lamb, 1963, p. 624; Ikoma, 1983, p. 95; Ohmura et al., 2008, p. 157; Fu Wei et al., 2008, p. 96).

Ramalina dilacerata var. *obtusata* sensu Nakao: Journ. Pharmaceut. Soc. Jap. 1923, no. 496, pp. 29 – 38 (1923), in Germany; pp. 423 – 497(1923), in Japanese. non Vain.

Ramalina shinanoana Kashiw. 信浓树花

Bull. natn. Sci. Mus., Tokyo, B 12(4): 122(1986).

Heilongjiang, Jilin(Fu Wei et al., 2008, p. 97), Sichuan(Oh et al., 2014, p. 238), Yunnan (Kashiw., 1986, p. 124; Oh et al., 2014, p. 238).

Ramalina sinensis Jatta 中国树花

Nuov. Giorn. Boran. Ital. nov. ser. 2, 9: 462(1902).

Type: Shaanxi, Giraldi(syntypes).

Xinjiang(Abdulla Abbas & Wu, 1998, p. 127; Abdulla Abbas et al., 2001, p. 367).

≡*Desmazieria sinensis*(Jatta) D. Hawksw. [as '*Desmaziera*'], in Hawksworth & Mahmood, *Pakist. J. scient. ind. Res.* 14(1 – 2): 114(1971).

= *Ramalina asahinana* Zahlbr., *Bot. Mag.*, *Tokyo* 41: 355, tab. XI, fig. 4(1927).

= *Ramalina asahinae* W. L. Culb. & C. F. Culb., *J. Jap. Bot.* 51(12): 374(1976).

Heilongjiang, Jilin, Liaoning(Fu Wei et al., 2008, p. 97), Neimenggu(Sun LY et al., 2000, p. 37 as *Ramaria sinensis*; Fu Wei et al., 2009, p. 100), Gansu, Qinghai(Fu Wei et al., 2009, p. 100), Xinjiang(Wang XY, 1985, p. 349; Fu Wei et al., 2009, p. 100), Shaanxi(He & Chen 1995, p. 45 without citation of specimen and locality), Shandong(Hou YN et al., 2008, p. 67).

var. sinensis 原变种

On trees.

Neimenggu, Jilin, Heilongjiang(Chen et al. 1981, p. 157), Shaanxi(Jatta, 1902, p. 462; Zahlbr., 1930 a, p. 520 & 1930 b, p. 204; Moreau et Moreau, 1951, p. 192; H. Magn. 1955, p. 305; Wei et al. 1982, p. 47; He & Chen 1995, p. 45), Xinjiang(Abbas et al., 1993, p. 80; Zhou Chunli et al., 2009, p. 12), Xizang(J. C. Wei & Chen, 1974, p. 181; J. C. Wei & Jiang, 1981, p. 1146 & 1986, p. 73), Sichuan(Oh et al., 2014, p. 238), Yunnan(Wei X. L. et al. 2007, p. 157; Oh et al., 2014, p. 238), Hubei(Chen, Wu & J. C. Wei, 1989, p. 456), Taiwan(Asahina, 1939, p. 209; Wang & Lai, 1973, p. 96 & 1976, p. 228).

var. elongata Jatta 长叶变种

Nuov. Giorn. Botan Ital. nov. ser, 2, IX: 463(1902).

Type: from Shaanxi, Giraldi(syntpes).

Shaanxi(Zahlbr., 1930 a, p. 520 & 1930 b, p. 205).

Ramalina subcomplanata(Nyl.) Zahlbr. 亚平树花

Cat. Lich. Univers. 6: 522(1930).

≡*Ramalina farinacea* var. *subcomplanata* Nyl., *Bull. Soc. linn. Normandie*, sér. 2 4(2): 134(1870).

≡*Ramalina fraxinea* subsp. *Subcomplanata* Nyl.

≡*Ramalina subcomplanata* Nyl. [Zahlbr., *Cat. Lich. Univers.* 6: 522(1930), *in* Handel-Mazzetti, *Symb. Sin.* 3: 204(1930).

On trees.

Gansu(Fu Wei et al., 2009, p. 100), Shaanxi(Jatta, 1902, p. 462; Zahlbr., 1930 a, p. 522 & 1930b, p. 204; Kashiw., 1986, p. 97; Fu Wei et al., 2009, p. 100), Yunnan(Kashiw., 1986, p. 97; Oh et al., 2014, p. 239).

Race 2: Yunnan, Taiwan(Kashiw. , 1986, p. 97) .

Race 3: Yunnan, Taiwan(Kashiw. , 1986, p. 97) .

Ramalina subfarinacea(Nyl. ex Cromb.) Nyl. 亚粉树花

Bull. Soc. linn. Normandie, sér. 2 6: 258(1872) .

≡*Ramalina scopulorum* var. *subfarinacea* Nyl. ex Cromb. , *J. Bot.* , *Lond.* 10: 74(1872) .

= *Ramalina angustissima*(Anzi) Vain. , *Meddn Soc. Fauna Flora fenn.* 14: 21(1888) .

≡*Ramalina farinacea* var. *angustissima* Anzi, *Lich. Etrur. Exsicc.* : no. 6, c(1863) .

On tree trunk.

Gansu(Wu J. L. , 1987, p. 173) .

Ramalina subgeniculata Nyl. 亚曲树花

Bull. Soc. linn. Normandie, sér. 2 4(2) : 167(1870) .

= *Ramalina geniculata* var. *olivacea* Müll. Arg.

= *Ramalina inflata* var. *gracilis* Müll. Arg.

On twigs.

Shaanxi(Jatta, 1902, p. 462) , Jiangsu, Anhui, Zhejiang(Xu B. S. , 1989, p. 229) , Taiwan(Zahlbr. , 1933c, p. 61; M. Sato, 1937b, p. 597; Asahina, 1938, p. 729; Wang-Yang & Lai, 1973, p. 96; Yoshim. 1974, p. 21) .

Ramalina subleptocarpha Rundel & Bowler 长树花

Bryologist 79(3) : 368(1976) .

On tree bark.

Xinjiang(Zhou Chunli et al. , 2009, p. 12) .

Ramalina tenella Müll. Arg. 娇嫩树花

Flora, Regensburg 62(11) : 162(1879) .

On exposed granitic rock, and rarleyon tree branches. Pantropical.

Misapplied name(revised by Aptroot & Seaward, 1999, p. 93) :

Ramalina linearis auct. non(Sw.) Ach. (in Seemann, 1852 – 57, p. 432; Zahlbr. , 1930b, p. 204,) .

Ramalina rel. litoralis auct non Asah. (Thrower, 1988, pp. 17, 161; Chu, 1997, p. 48) .

Hongkong(Seemann, 1852 – 57, p. 432; Zahlbr. , 1930b, p. 204; Thrower, 1988, pp. 17, 161; Chu, 1997, p. 48; Aptroot & Seaward, 1999, pp. 92 – 93) .

Ramalina throwerae Aptroot & Sipman 斯柔氏树花

J. Hattori bot. Lab. 91: 339 – 340(2001) .

Type: China, Hongkong, Lantau island, Sipman45106 = Aptroot48672(B-holotype; HKU(M) , Aabl-isotypes

Saxicolous.

Hongkong(Thrower, 1988, p. 163, as *R. pollinaria* with photograph; Aptroot & Seaward, 1999, p. 92 as *R. nervulosa*) ,

Misapplied name:

Ramalina pollinaria auct. non(Westr.) Ach. in Thrower, 1988, pp. 17, 163.

Ramalina nervulosa auct. non(Müll. Arg.) des Abbayes in Aptroot & Seaward, 1999, p. 92.

Ramalina yasudae Räsänen 安田氏树花

in *J. J. B.* 16: 87(1940) .

Saxicola

Jilin(Ohmura et al. , 2008, p. 157) , Liaoning(Fu Wei et al. , 2008, p. 98 as *R.* "*yasudea*") , Sichuan, Yunnan (Oh et al. , 2014, p. 239) .

Squamacidia Brako **芽鳞属**

Mycotaxon 35(1) : 6(1989) .

Squamacidia janeirensis(Müll. Arg.) Brako 芽鳞衣

Mycotaxon 35(1) : 8(1989) .

≡*Thalloidima janeirense* Müll. Arg. , *Hedwigia* 31: 280(1892) .

= *Phyllopsora stenosperma* Zahlbr. *Repert. Spec. Nov. Regni Veg*. 33: 44(1933) .

Type: Taiwan, Prov. Chiai, Mt. Arisan, Toroyen, 24 December 1925, Asahina F 170(TNS holotype) .

Taiwan(revised by Aptroot, 2004, p. 35) .

Toninia A. Massal. **泡鳞衣属**

Ric. auton. lich. crost. (Verona) : 107(1852) .

Toninia alutacea(Anzi) Jatta 淡棕泡鳞衣

Fl. ital. crypt. 1(3) : 655(1911) .

≡*Thalloidima alutaceum* Anzi, Atti Soc. ital. Sci. nat. (Modena) 9: 249(1866) .

On rocks.

Xinjiang(Reymu et al. , 2015, p. 162) , Ningxia(Yang J. & Wei J. C. , 2014, p. 1031) .

Toninia aromatica(Turner) A. Massal. 香泡鳞衣

Framm. Lichenogr. : 24(1855) .

≡*Lichen aromaticus* Turner, in Smith & Sowerby, *Engl. Bot*. 25: tab. 1777(1807) .

On calcareous rocks.

Shaanxi(Wu J. L. , 1987, p. 90) , Shanghai(Timdal, 1991, p. 39) .

= *Toninia sinensis* Zahlbr. , *Cat. Lich. Univers*. 4: 291(1926) [1927] .

= *Lecidea subaromatica* Nyl. in Nyl. & Cromb. *Journ. Linn. Soc. London, Bot*. 20: 64(1883) , nom. illeg. , (Art. 64. 1; homoname of *Lecidea subaromatica* Vainio 1883: 6) .

Type: from Shanghai.

Shanghai(Zahlbr. , 1930 b, p. 119) .

On old ramparts.

Toninia athallina(Hepp) Timdal 泡鳞衣

Op. bot. 110: 42(1991) .

≡*Biatora athallina* Hepp, *Flecht. Europ*. : no. 499(1860) .

Neimenggu(Timdal, 1991, p. 42 – 44) .

= *Catillaria mongolica* Magn. *Lichens from Central Asia* II: 25(1944) .

Type: Neimenggu Bohlin no. 108 in S(!) .

On calcareous rock.

Toninia candida(Weber) Th. Fr. 白泡鳞衣

K. svenska Vetensk-Akad. Handl. , ser. 2 7(2) : 33(1867) .

≡*Lichen candidus* Weber, *Spicil. fl. goetting*. : 193(1778) .

On ochraceous, lax stone and soil.

Xinjiang(Abdulla Abbas et al. , 1993, p. 81; Abdulla Abbas et al. , 1994, p. 23; Abdulla Abbas & Wu, 1998, p. 46; Abdulla Abbas et al. , 2001, p. 368; Reyim et al. , 2015, p. 163) , Gansu(H. Magn. 1940, p. 67; Timdal, 1991, pp. 50 – 52) .

Toninia cinereovirens(Schaer.) A. Massal. 浅绿泡鳞衣

Ric. auton. lich. crost. (Verona) : 107(1852) .

≡*Lecidea cinereovirens* Schaer. , *Lich. helv. spicil*. 3: 109(1828) .

= *Toninia olivaceoatra* H. Magn. *Lichens from Central Asia* II: 27(1944) .

Neimenggu(S: Timdal, 1991, pp. 52 – 55) .

Toninia coeruleonigricans(Lightf.) Th. Fr. 蓝黑泡鳞衣

Lich. Scand. (Upsaliae) 1(2) : 336(1874) , nom. rejic. , Art. 56. 1.

≡*Lichen coeruleonigricans* Lightf. , *Fl. Scot*. 2: 805(1777) , nom. rejic. , Art. 56. 1. Xinjiang(Abdulla Abbas &

Wu, 1998, p. 46; Abdulla Abbas et al., 2001, p. 368).

Toninia episema(Nyl.) Timdal 表记泡鳞衣

Op. bot. 110: 62(1991).

≡*Lecidea episema* Nyl., *Bot. Notiser:* 161(1853) — *Scutula episema*(Nyl.) Zopf 1896.

On *Aspicilia tortuosa*(syn, *Lecanora tortuosa*);

Gansu(Magnusson 1940; Hawksw. & M. Cole, 2003, p. 362, as lichenicolous fungus).

Toninia lutosa(Ach.) Timdal 污表泡鳞衣

Op. bot. 110: 69(1991).

≡*Lecidea lutosa* Ach., *Lich. univ.*: 182(1810).

Neimenggu(Colo, S: Timdal, 1991, pp. 69 – 70).

Toninia olivaceoatra H. Magn. 橄榄黑泡鳞衣

Lichens from Central Asia 2: 27(1944).

Type: Neimenggu, Bohlin no. 183 in S.

On calciferous stone.

Toninia philippea(Mont.) Timdal 菲泡鳞衣

Op. bot. 110: 79(1991).

≡*Lecidea philippea* Mont., *Annls Sci. Nat.*, *Bot.*, sér. 3 12: 291(1849).

Gansu(COLO, S, UPS: Timdal, 1991, pp. 79 – 82).

Toninia physaroides(Opiz) Zahlbr. 囊泡鳞衣

Cat. Lich. Univers. 4: 275(1926) [1927].

≡*Thalloidima physaroides* Opiz, Lotos 6: 158(1856).

On soil.

Xinjiang(Reyim et al., 2015, pp. 163 – 164).

Toninia sedifolia(Scop.) Timdal 叶泡鳞衣

Op. bot. 110: 93(1991).

≡*Lichen sedifolius* Scop., *Fl. carniol.*, Edn 2(Wien) 2: 395(1772).

= *Toninia coeruleonigricans*(Lightf.) Th. Fr., *Lich. Scand.* (Upsaliae) 1(2): 336(1874), Nom. rejic., Art. 56. 1.

≡*Lichen coeruleonigricans* Lightf., *Fl. Scot.* 2: 805(1777).

On the ground & rocks.

Xinjiang(Tchou, 1935, p. 322; Moreau et Moreau, 1951, p. 187; Abbas et al., 1994, p. 23; Reyim et al., 2015, pp. 162 – 163, as *T. caeruleonigricans*(Lightf.).

Notes: Due to both the *Toninia coeruleonigricans*(Lightf.) Th. Fr. and its basionym *Lichen coeruleonigricans* Lightf. are nom. rejic. according to Art. 56. 1, the taxonomic revision of its collection together with that of *Toninia sedifolia*(Scop.) Timdal[*Op. bot.* 110: 93(1991)] remains to be carried out in the future.

Toninia sinensis Zahlbr. 中国泡鳞衣

Cat. Lich. Univers. 4: 291(1926) [1927].

= *Lecidea subaromatica* Nyl. in Nyl. & Cromb. *Journ. Linn. Soc. London, Bot.* 20: 64(1883), non Vain.

Type: from Shanghai.

Shanghai(Zahlbr., 1930 b, p. 119).

On old ramparts.

Toninia sculpturata(H. Magn.) Timdal 雕泡鳞衣

Op. bot. 110: 92(1991).

≡*Catillaria sculpturata* H. Magn. *Lichens from central Asia* I: : 66(1940) & Abbas et al., 1993, p. 76.

Type: Gansu, 28/I, 1932, Bohlin no. 80 in S.

On calciferous stone, mixed with several other species, as *Verrucaria superstrata* et al. Neimenggu(H. Magn.

1944, p. 26), Gansu(Timdal, 1991, pp. 92 – 93), Xinjiang(Abdulla Abbas et al., 1993, p. 76; Abdulla Abbas & Wu, 1998, p. 49; Abdulla Abbas et al., 2001, p. 362).

Toninia tristis(Th. Fr.) Th. Fr. 淡泡鳞衣

Lich. Scand. (Upsaliae) 1(2): 341(1874).

≡*Psora tabacina* var. *tristis* Th. Fr., *Bot. Notiser*: 38(1865).

≡*Thalloidima triste*(Th. Fr.) Vain., *Acta Soc. Fauna Flora fenn.* 53(no. 1): 135, 140(1922)

= *Lecanorella tristicolor*(Th. Fr.) Motyka, *Porosty(Lichenes)*. 1, *Rodzina Lecanoraceae. Hymenelia, Aspicilia, Lecanorella, Protoplacodium, Manzonia*(Lublin): 325(1995).

≡*Lecanora tristicolor* Th. Fr., *Lich. Scand.* (Upsaliae) 1(1): 269(1871).

On soil.

Taiwan(Aptroot & Sparrius 2003, p. 47).

subsp. asiae-centralis(H. Magn.) Timdal 中亚亚种

Op. bot. 110: 112(1991).

≡*Lecidea asiaecentralis* H. Magn. *Lich. Centr. Asia*, 1: 55(1940).

Type locality: Gansu, Bohlin no. 26d in S(not seen).

On calcarous soil.

Neimenggu(H. Magn. 1944, p. 24), Xinjiang(Reyim et al., 2015, p. 164), Gansu(H. Magn. 1940, p. 55 & 1944, p. 24; G. Schneider, 1979, p. 236), Shanxi(Timdal, 1991, pp. 112 – 113).

Toninia weberi Timdal 韦伯氏泡鳞衣

Op. bot. 110: 118(1991).

≡*Thalloidima grannulosum* Szat., Annls hist. -nat. Mus. natn. hung., n. Ser. 5: 132(1954).

The new species name *Toninia weberi* is needed because *Toninia granulosa*(Michaux) Vainio has already been used(see Timdal, 1991, p. 123).

Gansu(Timdal, 1991, pp. 118 – 119).

= *Catillaria kansuensis* f. *virescens* H. Magn. *Lichens from central Asia* I: : 66(1940).

Type: Gansu, Bohlin no. 57a, in S.

On calcareous, ochraceous stone.

(54) Ramboldiaceae S. Stenroos, Miądl. & Lutzoni **果衣科**

in Miądlikowska et al., *Mol. Phylogen. Evol.* 79: 132 – 168(2014).

Ramboldia Kantvilas & Elix **果衣属**

Bryologist 97(3): 296(1994).

Ramboldia cinnabarina(Sommerf.) Kalb, Lumbsch & Elix 朱砂果衣

Nova Hedwigia 86(1 – 2): 32(2008).

≡*Lecidea cinnabarina* Sommerf., *K. svenska Vetensk-Akad. Handl.*: 114(1824) [1823].

Yunnan(Paulson, 1928, p. 317; Zahlbr., 1930 b, p. 108).

Ramboldia elabens(Fr.) Kantvilas & Elix 野果衣

Lichenologist 39(2): 139(2007).

≡*Lecidea elabens* Fr., *K. svenska Vetensk-Akad. Handl.*: 256(1822).

≡*Pyrrhospora elabens*(Fr.) Hafellner, *Herzogia* 9(3 – 4): 731(1993).

On bark of trees.

Xinjiang(Abdulla Abbas et al., 1994, p. 21; Abdulla Abbas & Wu, 1998, p. 85; Abdulla Abbas et al., 2001, p. 364).

Ramboldia heterocarpa(Fée) Kalb, Lumbsch & Elix 异果衣

Nova Hedwigia 86(1 – 2): 34(2008).

≡*Lecidea heterocarpa* Fée, *Bull. Soc. bot. Fr.* 20: 316(1873).

≡*Lecidea russula* var. *heterocarpa*(Fée) Müll. Arg., *Nuovo G. bot. ital.* 21(3): 359(1889) [1888].

Sichuan, Xizang(W. Obermayer, 2004, p. 507).

Ramboldia russula(Ach.) Kalb, Lumbsch & Elix 红果衣

Nova Hedwigia 86(1 –2): 37(2008).

≡*Lecidea russula* Ach., *Methodus*, Sectio prior(Stockholmiæ): 61(1803), Zahlbr. *Cat. Lich. Univers.* 3: 823(1925). var. russula

Shaanxi(Jatta, 1902, p. 478), Yunnan(Zahlbr., 1930b, p. 108; M. Inoue, 1988, p. 51), Fujian(Zahlbr., 1930 b, p. 108).

≡*Lecidea russula* var. *heterocarpa*(Fée) Müll. Arg. *in* Zahlbr. *Cat. Lich. Univers.* 3: 824(1925).

≡*Lecidea heterocarpa* Fée

Shaanxi(Jatta, 1902, p. 478; Zahlbr., 1930 b, p. 108).

≡*Pyrrhospora russula*(Ach.) Hafellner, in Kalb & Hafellner, *Herzogia* 9(1 –2): 86(1992).

On *Michelia formosana*, *Abies kawakamii* branches, *Tsuga formosana* branches, *Shiia* branches, and *Magnolia grandiflora*.

Taiwan(Aptroot and Sparrius, 2003, p. 37).

(55) Scoliciosporaceae Hafellner **绦孢衣科**

Beih. Nova Hedwigia 79: 340(1984)

Scoliciosporum A. Massal. **绦孢衣属**

Ric. auton. lich. crost. (Verona): 104(1852).

Scoliciosporum chlorococcum(Graewe ex Stenh.) Vězda 绿球绦孢衣

Folia geobot. phytotax. 13(4): 414(1978).

≡*Biatora hypnophila* var. *chlorococcum* Graewe ex Stenh

On *Tsuga formosana*.

Taiwan(Aptroot and Sparrius, 2003, p. 46).

Scoliciosporum pruinosum(P. James) Vězda 粉霜绦孢衣

Folia geobot. phytotax. 13: 414(1978).

≡*Bacidia pruinosa* P. James, *Lichenologist* 5(1 –2): 117(1971).

On *Cryptomeria japonica* and *Schima superba*.

Taiwan(Aptroot and Sparrius, 2003, p. 46).

Scoliciosporum umbrinum(Ach.) Arnold 赭绦孢衣

Flora, Regensburg 54: 50(1871).

≡*Lecidea umbrina* Ach., *Lich. univ.*: 183(1810).

Filed under Rhizocarpon obscuratum.

Jilin(Hertel & Zhao, 1982, p. 150).

On shale.

Taiwan(Aptroot and Sparrius, 2003, p. 46).

(56) Sphaerophoraceae Fr. **球粉衣科**

Lich. eur. reform. (Lund): 7(1831)

Bunodophoron A. Massal. **垛衣属**

Mem. Reale Ist. Veneto Sci. 10: 76(1861).

Bunodophoron diplotypum(Vain.) Wedin 双型垛衣

Pl. Syst. Evol. 187(1 –4): 232(1993).

≡*Sphaerophorus diplotypus* Vain., *Hedwigia* 37: 36(1898).

On bark of tree among mosses.

Yunnan(Chen J. B., 1996, p. 106), Taiwan(Mituno, 1938, p. 669; Wang & Lai, 1973, p. 96 & 1976 a, p. 84; Sparrius et al., 2002, 359).

Bunodophoron formosanum(Zahlbr.) Wedin 台湾垛衣

Pl. Syst. Evol. 187:233(1993).

≡*Sphaerophorus melanocarpus* ssp. *formosanus* Zahlbr. in Fedde, *Repertorium*, 31:206(1933).

Type: Taiwan, Asahina no. 274.

≡*Sphaerophorus formosanus*(Zahlbr.) Asahina, Journ. Jap. Bot. 14:667(1938).

On *Pinus* sp.

Guizhou(Chen J. B., 1996, p. 106; Zhang T & J. C. Wei, 2006, p10), Yunnan(Chen J. B., 1996, p. 106), Zhejiang(Ahti et al., 1999, pp. 124 – 125), Taiwan(M. Sato, 1934, p. 425; Mituno, 1938, p. 667; Lamb, 1963, p. 665; Wang & Lai, 1973, p. 96 & 1976 a, p. 84; Sparrius et al., 2002, 359).

Bunodophoron melanocarpum(Sw.) Wedin 黑果垛衣

Mycotaxon 55:383(1995).

≡*Lichen melanocarpus* Sw., *Prodr.*: 147(1788).

≡*Sphaerophorus melanocarpus*(Sw.) DC., in de Candolle & Lamarck, *Fl. franç.*, Edn 3(Paris) 6:178(1815).

≡*Stereocaulon melanocarpum*(Sw.) Raeusch., *Nomencl. bot.*, Edn 3:328(1797).

=*Sphaerophorus compressus* Ach., *Methodus*, Sectio prior: 135(1803).

On rocks covered with mosses.

Yunnan(Chen J. B., 1996, p. 106, as *S. melanocarpus*; Wei X. L. 2007, p. 149), , Hunan(Chen J. B., 1996, p. 106, as *S. melanocarpus*), Taiwan(Mituno, 1938, p. 666; Wang & Lai, 1973, p. 96 & 1976 a, p. 84), Hainan (JCWei et al., 2013, p. 230).

Sphaerophorus Pers. **球粉衣属**

Ann. Bot. (Usteri) 7:23(1794).

Sphaerophorus digitatus Wang & Lai 掌球粉衣

Taiwania, 21(1):84(1976).

Type: Taiwan, Ilan county, at about 1600 m, Lai, no. 8108(holotype) in TAI(not seen).

On bark.

Taiwan(Index of Fungi, 1977, p. 426).

Sphaerophorus taiwanensis Wang & Lai 台东球粉衣

Taiwania, 21(1):85(1976).

On *Abies kawakamii*

Type: Taiwan, Taidong, July 1970, Hsu K. S. (holotype) in TAI(not seen).

Taiwan(Index of Fungi, 1977, p. 426; Sparrius et al., 2002, 360).

Sphaerophorus yangii Wang & Lai 杨氏球粉衣

Taiwania, 21(1):85(1976).

Type: Taiwan, Lai no. 6712(holotype) in TAI(not seen).

Taiwan(Index of Fungi, 1977, p. 426).

(57) Stereocaulaceae Chevall. [as ‘Stereocauleae’] **珊瑚枝科**

Fl. gén. env. Paris(Paris) 1:596(1826)

Lepraria Ach. **癞屑衣属**

Methodus, Sectio prior: 3(1803).

=*Leproloma* Nyl. ex Cromb., *Monogr. Lich. Brit.* 1: 348(1883).

Lepraria albicans(Th. Fr.) Lendemer & B. P. Hodk. 白色癞屑衣

Mycologia 105(4):1005(2013).

≡*Stereocaulon albicans* Th. Fr., *Stereoc. Piloph. Comm.*: 36(1857).

≡*Leprocaulon albicans*(Th. Fr.) Nyl. ex Hue *Nouv. Arch. Mus. Hist. Nat.*, Paris, 3 sér. 2:248(1890).

≡*Stereocaulon albicans* Th. Fr., *Stereoc. Piloph. Comm.*: 36(1857).

=*Stereocaulon tenellum* Tuck.

On rocks.

Fujian(Zahlbr. , 1930 b, p. 136) , Taiwan(Wang & Lai, 1973, p. 96) .

Lepraria arbuscula(Nyl.) Lendemer & B. P. Hodk. 树癞屑衣

Mycologia 105(4) : 1005(2013) .

≡*Stereocaulon nanum* var. *arbuscula* Nyl. (1857: 97. nom. nud.)

≡*Stereocaulon arbuscula* Nyl. , Syn. meth. lich. (Parisiis) 1(2) : 253(1860) , cum descript.

≡*Leprocaulon arbuscula*(Nyl.) Nyl. , in Hue, Nouv. Arch. Mus. Hist. Nat. , Paris, 3 sér. 2: 246(1890) .

Strain I: Guangdong(Lamb & Ward, 1974, p. 518) .

Strain II: Jiangsu, Zhejiang(Xu B. S. , 1989, p. 266 with physodalic acid) , Fujian(Lamb & Ward, 1974, p. 518 –519) , Taiwan(Zahlbr. , 1933 c, p. 49; M. Sato, 1941 a, p. 90; Asahina, 1943 f, p. 280; Wang & Lai, 1973, p. 96; Lamb & Ward, 1974, p. 518 –519) .

Lepraria caesioalba(B. de Lesd.) J. R. Laundon 淡蓝癞屑衣

Lichenologist 24(4) : 324(1992) .

≡*Crocynia caesioalba* B. de Lesd. [as ‘ caesio-alba’] , *Bull. Soc. bot. Fr.* 61: 84(1914) .

On soil.

Taiwan(Aptroot and Sparrius, 2003, p. 27) .

Lepraria crassissima(Hue) Lettau 厚癞屑衣

Feddes Repert. 61(2) : 125(1958) .

≡*Crocynia crassissima* Hue, *Bull. Soc. bot.* Fr. 71: 393(1924) .

On shale.

Xinjiang(Wang XY, 1985, p, 353) , Taiwan(Aptroot and Sparrius, 2003, p. 27) .

Lepraria incana(L.) Ach. 灰白癞屑衣

Methodus, Sectio prior: 4(1803) .

≡*Byssus incana* L. , *Sp. pl.* 2: 1169(1753) .

≡*Lepra incana*(L.) F. H. Wigg. , *Prim. fl. holsat.* (Kiliae) : 97(1780) .

Lepraria aeruginosa auct. non(Weis) Sm. (James, 1965, p. 129) .

On the ground.

Xinjiang(Wang XY, 1985, p. 353 as both the *L. incana* & *L. aeruginosa*; Abdulla Abbas & Wu, 1998, p. 160) , Shaanxi(Jatta, 1902, p. 480; Zahlbr. , 1930 b, p. 244) , Yunnan(Wang L. S. 2008, p. 538, on tea trees) , Shandong(Li Y et al. , 2008, p. 73) .

[*Lepraria aeruginosa*(Wigg.) Sm.

On bark of trees, and rarely on rocks.

Xinjiang(Abudula Abbas et al. , 1993, p. 78; Abdulla Abbas et al. , 2001, p. 364) , Shaanxi, Sichuan, Jiangxi (Wu J. L. 1987, p. 209)] .

Note. *Lepraria aeruginosa*(Weis) Sm. is not a lichen. (James, 1965, p. 129) .

Lepraria lobificans Nyl. 裂片癞屑衣

Flora, Regensburg 56: 196(1873) .

On shaded soil and mossy rocks. Cosmopolitan.

Yunnan(Wang L. S. 2008, p. 538) , Hongkong(Thrower, 1988, p. 16; Aptroot & Seaward, 1999, p. 81) .

Lepraria membranacea(Dicks.) Vain. 膜癞屑衣

Acta Soc. Fauna Flora fenn. 49(no. 2) : 265(1921) .

≡*Lichen membranaceus* Dicks. , *Fasc. pl. crypt. brit.* (London) 2: 21(1790) .

≡*Leproloma membranaceum*(Dicks.) Vain. , *Term. Füz.* 22: 293(1899) ; Aptroot & Seaward, 81(1999) .

On shaded granite. Cosmopolitan.

Xinjiang(Abdulla Abbas & Wu, 1998, p. 159 as *Leproloma* "*mumbranaceum*"; Abdulla Abbas et al., 2001, p. 364 as *Leproloma membranaceum*; Abdull Abbas et al., 1993, p. 78), Hongkong(Thrower, 1988, p. 114 as *Lepraria* sp.; Aptroot & Seaward, 1999, p. 81; Aptroot et al., 2001, p. 330).

Lepraria neglecta Vain. 癞屑衣

in Lettau, (1934)

On exposed to shaded granite and soil. Cosmopolitan.

Hongkong(Chu, 1997, p. 48 as *Lepraria* sp. mentioned by Aptroot & Seaward, 1999, p. 81.

Lepraria pseudoarbuscula(Asahina) Lendemer & B. P. Hodk. 拟树癞屑衣

Mycologia 105(4): 1005(2013).

≡*Stereocaulon pseudoarbuscula* Asahina, *J. Jap. Bot.* 19: 282(1943).

≡*Leprocaulon pseudoarbuscula*(Asahina) I. M. Lamb & A. Ward, *J. Hattori bot. Lab.* 38: 533(1974).

= *Stereocaulon novo-arbuscula* Asahina, *J. J. B.* 19: 283(1943).

On shaded soil. Pantropical.

Taiwan(Lamb & Ward, 1974, p. 533), Hongkong(Lamb & Ward, 1974, p. 533; Thrower, 1988, pp. 16, 115; Aptroot & Seaward, 1999, p. 81).

Lepraria vouauxii(Hue) R. C. Harris 沃氏癞屑衣

in Egan, *Bryologist* 90(2): 163(1987).

≡*Crocynia vouauxii* Hue, *Bull. Soc. bot. Fr.* 71: 392(1924).

≡*Leproloma vouauxii*(Hue) J. R. Laundon, *Lichenologist* 21(1): 13(1989).

On shaded trees along mountain stream. Cosmopolitan.

Hongkong(Aptroot & Seaward, 1999, p. 81).

Lepraria yunnaniana(Hue) Zahlbr. 云南癞屑衣

in Handel-Mazzetti, *Symb. Sinic.* 3: 224(1930).

≡*Crocynia yunnaniana* Hue, *Bull. Soc. Bot. France*, 71: 396(1924).

Type: from Yunnan.

On rock.

Yunnan(Zahlbr., 1932a, p. 244).

Squamarina Poelt **鳞茶渍属**

Mitt. bot. StSamml., Münch. 19 – 20: 524(1958).

Squamarina callichroa(Zahlbr.) Poelt 美色鳞茶渍

Mitt. Bot. Staatssaml. Muenchen, Heft. 19 – 20: 527(1958).

≡*Lecanora callichroa* Zahlbr. in Handel-Mazzetti, *Symb. Sinic.* 3: 172(1930) & *Cat. Lich. Univ* 8: 543(1932).

Type: Yunnan, Handel-Mazzetti no. 165(WU – 2730, holotype; W – 2306/1926, isotype).

On quartzite.

Squamarina cartilaginea(With.) P. James 软骨鳞茶渍

in Hawksworth, James & Coppins, *Lichenologist* 12(1): 107(1980).

≡*Lichen cartilagineus* With., *Bot. arr. veg. Gr. Brit.* (London) 2: 708(1776).

= *Lecanora crassa*(Huds.) Ach., *Lich. univ.*: 413(1810).

≡*Squamaria crassa* (Huds.) DC., in Lamarck & de Candolle, *Fl. franç.*, Edn 3(Paris) 2: 375(1805).

≡*Lichen crassus* Huds., *Fl. Angl.*, Edn 2: 1 – 690(1778).

Hebei(Tchou, 1935, p. 320).

= *Parmelia liparia* Ach. Method. Lich. 1803: 182.

= *Lecanora crassa* var. *liparia*(Ach.) Nyl. (Zahlbr., *Cat. Lich. Univ.* V: 617)

Yunnan(Zahlbr., 1930 b, p. 174).

= *Lecanora crassa* var. *livida* Schaer. *Enumer. Critic. Lich. Europ.* 1850: 58.

On the ground.

Xinjiang(Abdulla Abbas & Wu, 1998, p. 44; Abdulla Abbas et al., 2001, p. 368), Shaanxi(Jatta, 1902, p. 474; Zahlbr., 1930 b, p. 174).

Squamarina chondroderma(Zahlbr.) J. C. Wei 硬皮鳞茶渍

Enum. Lich. China 231(1991).

≡*Lecanora chodroderma* Zahlbr. in Handel-Mazzetti, *Symb. Sinic*. 3: 174(1930).

Type: Sichuan, at 3600 – 3900 m, 27/V/1914, Handel-Mazzetti no. 2643.

var. chondroderma 原变种

On rock.

Sichuan(Zahlbr., 1930 b, p. 174 & 1932, p. 543).

var. placodizans(Zahlbr.) J. C. Wei 鳞形变种

Enum. Lich. China, 232(1991).

≡*Lecanora chondroderma* var. *placosizans* Zahlbr., in Handel-Mazzetti, *Symb. Sinic*. 3: 175(1930) & *Cat. Lich. Univ*. 8: 543(1932).

Type locality: Sichuan, at 4650 m, 6/VIII/1915, Handel-Mazzetti no. 7494.

On rock.

Squamarina gypsacea(Sm.) Poelt 石膏鳞茶渍

Mitt. bot. StSamml., Münch. 19 – 20: 524(1958).

≡*Lichen gypsaceus* Sm., *Trans. Linn. Soc. London* 1: 81(1791).

≡*Lecanora gypsacea*(Sm.) Müll. Arg., *Bull. Soc. Hallerienne* 4: 132(1856) [1854 – 56].

On calcareous soil.

Gansu(H. Magn. 1940, p. 115).

Squamarina kansuensis(H. Magn.) Poelt 甘肃鳞茶渍

Mitt. Bot. Staatssamml. Muenchen, Heft 19 – 20: 543(1958).

≡*Lecanora kansuensis* H. Magn. *Lich. Centr. Asia* 1: 116(1940).

Type: Gansu, Bohlin no. 20 in S(!).

On calcareous soil.

Xinjiang(Abdulla Abbas et al., 1993, p. 81; Abdulla Abbas & Wu, 1998, p. 44; Abdulla Abbas et al., 2001, p. 368), Gansu(H. Magn., 1940, p. 116).

Squamarina lentigera(Weber) Poelt 条斑鳞茶渍

Mitt. bot. StSamml., Münch. 19 – 20: 536(1958).

≡*Lichen lentigerus* Weber, *Spicil. fl. goetting.* : 192(1778).

≡*Lecanora lentigera* (Weber) Ach., *Lich. univ.* : 423(1810).

≡*Squamaria lentigera*(Weber) DC., in Lamarck & de Candolle, *Fl. franç.*, Edn 3(Paris) 2: 376(1805).

On soil.

Neimenggu(Tchou, 1935, p. 320; H. Magn. 1944, p. 45), Gansu(H. Magn. 1940, p. 45), Ningxia(Liu M & Wei JC, 2013, p. 46; Yang J. & Wei J. C., 2014, p. 1030).

Squamarina oleosa(Zahlbr.) Poelt 油鳞茶渍

Mitt. Bot. Staatssamml. Muenchen, Heft 19: 542(1958).

≡*Lecanora oleosa* Zahlbr., in Handel-Mazzetti, *Symb. Sinic.* 3: 175(1930) & *Cat. Lich. Univ.* 8: 545(1932).

Type: Yunnan, Handel-Mazzetti no. 3576 in W. not seen.

On calcareous rock.

Squamarina pachyphylla(H. Magn.) J. C. Wei 厚叶鳞茶渍

Enum. Lich. China(Beijing) : 232(1991).

≡*Lecanora pachyphylla* H. Magn. *Lich. Centr. Asia* 1: 120(1940).

Type: Gansu, at about 3825 m, 20/III/1932, Bohlin A: b in s(!).

On calcareous rock.

Xinjiang(Abdulla Abbas & Wu, 1998, p. 45; Abdulla Abbas et al., 2001, p. 368).

Squamarina semisterilis(H. Magn.) J. C. Wei 半育鳞茶渍

Enum. Lich. China (Beijing): 232(1991).

≡*Lecanora semisterilis* H. Magn. *Lich. Centr. Asia* 1: 123(1940).

Type: Gansu, at about 2550 m, 17/VII/1931, Bohlin no. 38 b in S(!).

On earth.

Stereocaulon Hoffm. **珊瑚枝属**

Deutschl. Fl., *Zweiter Theil* (Erlangen): 128(1796)[1795]. (Nom. cons., see Art. 14).

Stereocaulon alishanum Mineta 阿里山珊瑚枝

J. Jap. Bot. 62: 257(1987).

Taiwan(Lai, 2000, p. 267; Lai 2001, p. 177).

Stereocaulon alpestre(Flot.) Dombr. 山地珊瑚枝

Botanicheskiĭ Zhurnal 77(7): 98(1992).

≡*Stereocaulon tomentosum* var. *alpestre* Flot., *Flora*, Regensburg 19(1, Beiblatter): S. 17(1836).

Sichuan(Huang & J. C. Wei, 2006, p. 436).

Stereocaulon alpinum Laurer 高山珊瑚枝

in Funck, *Cryptog. Gewächse* 33: 6(1827). var. alpinum 原变种

On rocks among mosses.

Neimenggu(Sun LY et al., 2000, p. 37), Shaanxi(Baroni, 1894, p. 47; Jatta, 1902, p. 466; Zahlbr., 1930 b, p. 136), Gansu(Zahlbr., 1934, p. 202; Lamb, 1977, p. 203), Yunnan(Zahlbr., 1930 b, p. 136), Xizang(Elenk. 1904, p. 25; Paulson, 1925, p. 193).

"var. densum Laur." 密枝变型

(Gregory nach Pauls. I, 317 in Zahlbr., 1930 b, p. 136).

Yunnan(Paulson, 1928, p. 317; Zahlbr., 1930 b, p. 136).

Stereocaulon apocalypticum Nyl. 群生珊瑚枝

in Middendorff, *Reise Sibir.* 4(6): LV(1867).

≡*Stereocladium apocalypticum*(Nyl.) Nyl., *Bull. Soc. linn. Normandie* 1: 268(1887).

Stereocaulon wrightii auct. non Tuck.

North-eastern China(Elenk. 1904, p. 91; Lamb, 1977, p. 259), Heilongjiang(Da Hinggan Ling: Sato, 1952, p. 174).

Stereocaulon botryosum Ach. 串束珊瑚枝

in de Candolle & Lamarck, *Fl. franç.* (Paris) 6: 178(1805).

≡*Stereocaulon tomentosum* var. *botryosum*(Ach.) Nyl., *Lich. Scand.* (Helsinki): 64(1861).

=*Stereocaulon fastigiatum* Anzi, *Cat. Lich. Sondr.*: 11(1860).

Jilin(Huang & Wei, 2006, pp. 436 – 437).

Stereocaulon condensatum Hoffm. 竹杆珊瑚枝

Deutschl. Fl., *Zweiter Theil*(Erlangen): 130(1796)[1795].

Taiwan(Zahlbr., 1933 c, p. 47; Wang-Yangt & Lai, 1973, p. 96; Lamb, 1977, p. 209; Lai, 2000, pp. 267 – 268; Lai 2001, p. 177).

Stereocaulon coniophyllum I. M. Lamb 锥型珊瑚枝

Bot. Notiser 114(3): 267(1961).

Jilin, Shaanxi, Sichuan(Huang & J. C. Wei, 2006, p. 437).

Stereocaulon dactylophyllum Flörke 指叶珊瑚枝

Deutsche Lich. 4: 13(1819).

Stereocaulon myriocarpum Th. Fr., Stereoc. Piloph. Comm.: 15(1857).

China(prov. not indicated: Lamb, 1977, p. 211).

= *Stereocaulon coralloides* Th. Fr., *Sched. Crit. Lich. Suec. Exsicc.* 3: 24(1825) [1824], Zahlbr., 1930b, p. 136.

On the ground & moss covered rocks.

Jilin(Tchou, 1935, p. 305; Moreau et Moreau, 1951, p. 188), Shaanxi(Jatta, 1902, p. 466; Zahlbr., 1930b, p. 136; Tchou, 1935, p. 305), Yunnan(Hue, 1887, p. 17 & 1898, p. 254; Zahlbr., 1930b, p. 136), Sichuan(Hue, 1898, p. 254), Guizhou(Zhang T & Wei JC, 2006, p. 10), Hebei(Tchou, 1935, p. 305), Hubei(Müll. Arg. 1893, p. 235; Zahlbr., 1930b, p. 136), Fujian(Zahlbr., 1930b, p. 136).

= *Stereocaulon coralloides* var. *japonicum* Th. Fr.

China(prov. not indicated: Nyl. 1857, p. 96).

Stereocaulon depreaultii Delise ex Nyl. 迪氏珊瑚枝

Syn. meth. lich. (Parisiis) 1: 249(1860).

Jilin(Huang & J. C. Wei, 2006, pp. 437 – 438).

Stereocaulon exutum Nyl. 裸珊瑚枝

Lich. Japon.: 18(1890).

On rocks.

Jilin(Huang & J. C. Wei, 2006, p. 438), Shaanxi, Gansu(Wu J. L., 1987, p. 111).

Stereocaulon fastigiatum Anzi 帚珊瑚枝

Cat. Lich. Sondr.: 11(1860).

var. dissolutum H. Magn. 粉粒变种

On rocks.

Shaanxi(Wu, 1987, pp. 111 – 112).

Stereocaulon foliolosum Nyl. 小叶珊瑚枝

Syn. meth. lich. (Parisiis) 1: 240(1860).

var. botryophorum(Müll. Arg.) I. M. Lamb 带串变种

J. Hattori bot. Lab. 43: 267(1977).

≡ *Stereocaulon botryophorum* Müll. Arg., *Flora*, Regensburg 74: 371(1891).

Sichuan, Yunnan(Huang & J. C. Wei, 2006, p. 439).

var. strictum(C. Bab.) I. M. Lamb 密枝变种

in *Canad. Journ. Bot.* 29: 582(1951).

≡ *Stereocaulon ramulosum* var. *strictum* Bab.; Zahlbr., *Cat. Lich. Univ.* 4: 660(1927).

≡ *Stereocaulon strictum* Th. Fr., *Stereoc. Piloph. Comm.*: 24(1857).

On rocks.

Yunnan(Hue, 1889, p. 159; Zahlbr., 1930b, p. 137; Lamb, 1977, p. 267).

Stereocaulon fuliginosa(Hoffm.) Ach. 煤珊瑚枝

Shaanxi(He & Chen 1995, p. 45 without citation of specimens and their localities).

Notes: The name of this species, its basionym and original literatures have been not found.

Stereocaulon graminosum Schaer. 禾草珊瑚枝

Syst. Verz.: 127(1846) [1845 – 46].

Xizang(Huang & J. C. Wei, 2006, p. 439).

Stereocaulon himalayense D. D. Awasthi & I. M. Lamb 喜马拉雅珊瑚枝

J. Hattori bot. Lab. 43: 269(1977).

Shaanxi, Sichuan, Yunnan, Xizang(Huang & J. C. Wei, 2006, p. 440).

Stereocaulon intermedium(Savicz) H. Magn. 间型珊瑚枝

in Göteborgs Kungl. Vetensk. -och Vitterh. Samh. handl. , Ser. 4 30(7): 23(1926).

≡*Stereocaulon coralloides* f. *intermedium* Savicz, *Bot. Mater. Inst. Sporov. Rast. Glavn. Bot. Sada RSFSR* 2(11 163. 1923): 163(1923).

Shaanxi, Sichuan, Yunnan, Xizang, Hubei, Zhejiang(Huang & J. C. Wei, 2006, pp. 440 – 441).

var. gracile Huang MR & J. C. Wei 细型变种

Mycotaxon 90(2): 469 – 470(2004).

Type: Sichuan, Mt. Gongga, on sandy soil, June23, 1982. Wang XY et al. 8841(HMAS-L. holotype).

Sichuan, Yunnan, Xizang(Huang MR & J. C. Wei, 2004, p. 470).

Stereocaulon japonicum Th. Fr. 东亚珊瑚枝

De Stereoc. et Pilophor. Comment. , 18(1857).

= *Stereocaulon armatulum* Zahlbr. , in Fedde, *Repertorium*, 33: 47(1933).

Type: Taiwan, Asahina no. 215 in W(holotype).

On rocks.

Guizhou(Zhang T & Wei JC, 2006, p. 10), Shandong(Zhao ZT et al. , 2002, p. 6), Zhejiang(Lamb. 1977, p. 263), Anhui, Fujian(J. C. Wei et al. 1982, p. 27), Taiwan(Zahlbr. , 1933 c, p. 47 & 1940, p. 393; M. Sato, 1941a, p. 74; Asahina, 1960b, p. 289; Wang-Yang & Lai, 1973, p. 96 – 97; Lamb, 1977, pp. 263 & 317; Lai, 2000, p. 268), Hongkong(Thrower, 1988, pp. 17, 169; Aptroot & Seaward, 1999, p. 94), Taiwan(Lai 2001, p. 177), China(prov. not indicated: Hue, 1890, p. 247).

var. japonicum 原变种

On rocks.

Anhui, Zhejiang(Xu B. S. , 1989, p. 231).

var. subfastigiatum Asahina 亚帚变种

J. J. B. 35: 289(1960).

On rocks.

Zhejiang(Wu J. L. , 1987, p. 112; Xu B. S. , 1989, p. 231), Anhui, Jiangxi, Fujian, Taiwan(Wu J. L. , 1987, p. 112).

Stereocaulon kangdingense Huang MR & J. C. Wei 康定珊瑚枝

Mycotaxon 90(2): 470(2004).

Type: Sichuan, Kangding, Oct. 18, 1999. Chen LH 990108(HMAS-L, holotype).

Stereocaulon massartianum Hue 侧顶果珊瑚枝

Nouv. Arch. Mus. Hist. Nat. , Paris, 3 sér. 10: 252(1898).

Stereocaulon nesaeum auct. (non Nyl. 1859).

= *Stereocaulon chlorocarpoides* Zahlbr. , in Fedde, *Repertorium*, 33: 48(1933).

Type: Taiwan, Mt. Ali shan, Asahina no. 219(not seen).

≡*Stereocaulon massartianum* var. *chlorocarpoides*(Zahlbr.) Lamb, *Journ. Jap. Bot.* 40(9): 271(1965).

Guizhou(Zhang T & Wei JC, 2006, p. 10), Taiwan(Zahlbr. , 1933c, p. 48 & 1940, p. 394; M. Sato, 1941, p. 83; Lamb, 1963, p. 680 & 1965, p. 271 & 1977, p. 273; Wang-Yang & Lai, 1973, p. 97; Lai, 2000, pp. 268 – 269; Lai 2001, pp. 177 – 178).

Stereocaulon myriocarpum Th. Fr. 多果珊瑚枝

Stereoc. Piloph. Comm. : 15(1857).

Misapplied name:

Stereocaulon paschale sensu Müll. Arg. *Bull. L'herb. Bois.* 1: 235(1893), non(L.) Hoffm.

Hubei(Müll. Arg. 1893, p. 235; Zahlbr. , 1930 b, p. 136; Lamb, 1977, p. 200), Taiwan(Lai, 2000, p. 269; Lai 2001, p. 178).

var. altaicum I. M. Lamb 高地变种

J. Hattori bot. Lab. 43: 225(1977).

Sichuan, Yunnan, Xizang(Huang & Wei, 2006, pp. 441 - 442).

Stereocaulon nigrum Hue 黑珊瑚枝

in Nouv. Arch. Mus. Hist. Nat. Paris ser. 3, 10: 248(1898).

On rocks.

Hunan(Wu J. L., 1987, p. 113).

Stereocaulon octomerellum Müll. Arg. 小珊瑚枝

Flora, Regensburg 74(1): 109(1891).

On rocks.

Jilin(Mt. Changbai shan: Hertel & Zhao, 1982, p. 150), Guizhou(Zhang T & J. C. Wei, 2006, p. 10), Hainan (J. C. Wei et al., 2013, p. 231).

Stereocaulon octomerum Müll. Arg. 指珊瑚枝

Flora, Regensburg 74: 109(1891).

On rocks.

Guizhou(Zhang T & J. C. Wei, 2006, p. 10), Zhejiang(Xu B. S., 1989, p. 231 - 232), Taiwaqn(Lamb, 1977, p. 265; Lai, 2000, p. 269; Lai 2001, p. 178).

Stereocaulon paschale(L.) Hoffm. 复活节珊瑚枝

Deutschl. Fl., Zweiter Theil(Erlangen): 130(1796) [1795].

≡*Lichen paschalis* L., Sp. pl. 2: 1153(1753).

On the ground & moss covered rocks.

Neimenggu(Da Xinggan Ling, M. Sato, 1952, p. 174), Xinjiang(Abdulla Abbas & Wu, 1998, p. 129; Abdulla Abbas et al., 2001, p. 368), Shaanxi(Moreau et Moreau, 1951, p. 188), Yunnan(Hue, 1889, p. 159; Paulson, 1928, p. 317; Zahlbr., 1930 b, p. 136; M. Sato, 1941a, p. 80), Shandong(Hou YN et al., 2008, p. 67).

Stereocaulon pendulum Asahina, in Sato 下垂珊瑚枝

J. J. B. 17: 247(1941). Satala, (1941).

On rocks.

Shaanxi(Wu J. L., 1987, p. 114).

Stereocaulon pileatum Ach. 粉帽珊瑚枝

Lich. univ.: 1 - 696(1810).

On rocks. On iron-congtaining boulders.

Zhejiang(Xu B. S., 1989, p. 232); Hongkong(Aptroot & Seaward, 1999, p. 94), Taiwan(Aptroot, Sparrius & Lai, 2002, p. 291).

Stereocaulon piluliferum Th. Fr. 圆头珊瑚枝

Stereoc. Piloph. Comm.: 21(1857).

= *Stereocaulon sinense* Hue, *Nouv. Arch. Mus. Hist. Nat.* 3(10): 251(1898) et ser. 4(1), tab. 3, fig. 4(1899).

Type: Yunnan, 1888, Delavay in PC(holotype, not seen).

≡*Stereocaulon piluliferum* var. *sinense*(Hue) Lamb ex Asahina in Kihara, Fauna & Flora of Nepal Himalya, 1952 - 53(1): 50(1955). (comb. inval.)

On rocks.

Yunnan(Hue, 1898, p. 251 & 1899, tab. 3, fig. 4; Zahlbr., 1927, p. 666 & 1930 b p. 137; Dodge, 1929, p. 142; Lamb, 1977, pp. 275 - 276, 318), Sichuan(may possibly be E. Xizang, localization uncertain: Lamb, 1977, p. 275), Taiwan(Zahlbr., 1933 c, p. 47; M. Sato, 1941a, p. 84; Wang & Lai, 1973, p. 97).

Stereocaulon pomiferum P. A. Duvign. 苹果珊瑚枝

in Lejeunia, *Mém.* 14: 119(1956).

= *Stereocaulon claviceps* var. *yunnanense* Hue, *Nouv. Arch. Mus. Hist. Nat.* 3(10): 251(1898).

Type: Yunnan, summit of Mt. Cangshan, above Dali, alt. 4000 m, collected by Delavay in 1884, no. 664 in PC (holotype, not seen).

≡*Stereocaulon macrocephalum* var. *yunnanense*(Hue) Dodge, *Ann. Cryptog. Exot.* 2(2): 125(1929).

≡*Stereocaulon yunnananse*(Hue) Lamb ex Asahina in Kihara, Fauna & Flora of Nepal Himalaya, 1952 – 53 (1): 50(1955). (comb. inval.)

Stereocaulon claviceps sensu Hue, *Bull. Soc. Bot. France* 34: 17(1887), non Th. Fr.

On rocks.

Shaanxi(He & Chen 1995, p. 45 without citation of specimens and their localities), Yunnan(Hue, 1887a, p. 17 & 1898, p. 251; Dodge, 1929, p. 125; Zahlbr., 1927, p. 635 & 1930 b, p. 137; Lamb, 1963, pp. 680, 683 & 1977, pp. 276, 319), Xizang(Lamb, 1977, p. 276; Wei & Jiang, 1986, p. 75), Taiwan(Lamb, 1977, p. 276; Lai, 2000, p. 270; Lai 2001, p. 178).

Stereocaulon prostratum Zahlbr. 早熟珊瑚枝

Bot. Mag., Tokyo 41: 340, tab. XI, fig. 7(1927).

Jilin(Huang & Wei, 2006, p. 442).

Stereocaulon ramulosum Raeusch. 密珊瑚枝

Nomencl. bot., Edn 3: 328(1797).

= *Stereocaulon mixtum* Nyl., *Annls Sci. Nat.*, *Bot.*, sér. 4 11: 210(1859).

= *Stereocaulon proximum* Nyl., *Annls Sci. Nat.*, *Bot.*, sér. 4 11: 210(1859).

On the ground & rocks.

Shaanxi(Baroni, 1894, p. 47; Jatta, 1902, p. 466; Zahlbr., 1930 b, p. 136), Sichuan(Zahlbr., 1930 b, p. 136; Chen S. F. et al., 1997, p. 222), Yunnan(Zahlbr., 1930 b, p. 136 & 1934b, p. 202; Lamb, 1977, p. 280), Fujian(Zahlbr., 1930 b, p. 137), Taiwan(Sasaoka, 1919, p. 47), China(prov. not indicated: Hue, 1898, p. 244; Zahlbr., 1930 b, p. 136).

Stereocaulon rivulorum H. Magn. 细纹珊瑚枝

Göteborgs Kungl. Vetensk. -och Vitterh. Samh. handl., Ser. 4 30(7): 63(1926).

Chemotype II: Neimenggu(Huang & Wei, 2006, p. 442).

Chemotype III: Jilin, Shaanxi, Sichuan, Yunnan(Huang & J. C. Wei, 2006, pp. 442443).

Stereocaulon sasakii Zahlbr. 佐木氏珊瑚枝

in Fedde, *Repertorium*, 33: 48(1933). v. sasakii 原变种

Type: Taiwan, collected by S. Sasaki(Asahina no. 144).

Lamb(1977, p. 229) pointed out the holotype was collected by S. Sasaki, 1927(no. 414 according to label) and preserved in W.

On the ground.

Taiwan(Zahlbr., 1940, p. 397; M. Sato, 1941a, p. 88; Lamb, 1963, p. 686 & 1977, p. 229; Wang-Yang & Lai, 1973, p. 97; Lai, 2000, p. 270; Lai 2001, p. 178, as v. *sasakii*), Xizang(Lamb, 1977, p. 230), China(prov. not indicated: Lamb, 1977, p. 230).

var. simplex(Ridd.) Lamb 单型变种

J. Hattori Bot. Lab. 43: 230(1977).

Taiwan(Lamb 1977, p. 230; Lai, 2000, p. 271; Lai 2001, p. 178).

Stereocaulon saviczii Du Rietz 萨氏珊瑚枝

Ark. Bot. 22A(13): 13(1929).

Hubei(Chen, Wu & J. C. Wei, 1989, p. 458).

Stereocaulon saxatile H. Magn. 石生珊瑚枝

Göteborgs Kungl. Vetensk. -och Vitterh. Samh. handl., Ser. 4 30(7): 41(1926).

Neimenggu, Heilongjiang, Jilin, Liaopning(Huang & J. C. Wei, 2006, p. 443).

Stereocaulon sorediiferum Hue 大珊瑚枝

in Nouv. Archiv. du Muéum, ser. 3, 10: 250(1898).

= *Stereocaulon exutum* var. *sorediata* Ras. In J. J. B. 16: 88(1940).

On rocks.

Guizhou(Zhang T & Wei JC, 2006, p. 10), Hongkong(Seaward & Aptroot, 2005, p. 286, Pfister 1978: 101, as *Stereocaulon ramulosum.*).

var. sorediiferum 原变种

On rocks.

Sichuan(Lamb, 1977, p. 270), Jiangsu, Anhui, Zhejiang(Xu B. S., 1989, p. 232), Taiwan(Lamb, 1965, p. 274 & 1977, p. 270; Wang-Yang & Lai, 1973, p. 97; Lai, 2000, p. 271; Lai 2001, pp. 178 – 179), Hainan(JCWei et al., 2013, p. 231), Hongkong(Hue, 1898, p. 250; Lamb, 1965, p. 274 & 1977, p. 270; Thrower, 1988, pp. 17, 170; Aptroot & Seaward, 1999, p. 95).

var. leprosolingulatum Lamb 头状变种

Journ. Hattori Bot. Lab. 43: 271(1977).

Type: Taiwan, Kurok. no. 475(Holotype) in FH &(isotype) in TNS(not seen).

On rocks.

Taiwan(Lamb, 1977, p. 271; Lai, 2000, pp. 271 – 272; Lai 2001, p. 179).

Stereocaulon sorediiphyllum Huang MR & J. C. Wei 粉叶珊瑚枝

Mycotaxon 90(2): 470(204).

Type: Jilin, Mt. Changbai shan, on rock. Sept. 9, 1997, Guo SY 1105(HMAS-L, holotype).

Shaanxi(Huang MR & J. C. Wei, 2004, p. 472).

Stereocaulon spathuliferum Vain. 匙形珊瑚枝

Ark. Bot. 8(4): 36(1909).

= *Stereocaulon botryosum* f. *dissolutum* (H. Magn.) Frey, *Rabenh. Krypt.-Fl.*, Edn 2 (Leipzig) 9. 4(1): 125 (1932).

≡ *Stereocaulon fastigiatum* var. *dissolutum* H. Magn., *Göteborgs Kungl. Vetensk.-och Vitterh. Samh. handl.*, Ser. 4 30(7): 36(1926).

On rocks.

Shaanxi(Wu J. L., 1987, pp. 111 – 112).

Stereocaulon sterile(Savicz) Lamb in Krog 无性珊瑚枝

Occ. Pap. Farlow Herb. Crypt. Bot. 2: 1(1972).

≡ *Stereocaulon evolutum* f. *sterile* Savicz, Notul. Syst. Inst. Cryptog. Horti. Bot. Petrobol. 2: 165(1923).

Jilin, Liaoning(Huang & J. C. Wei, 2006, pp. 443 – 444).

Stereocaulon tomentosum Th. Fr. 茸珊瑚枝

Sched. Crit. Lich. Suec. Exsicc. 3: 20(1825)[1824].

On the ground & rocks.

Neimenggu(M. Sato, 1952, p. 174), Heilongjiang(Asahina, 1952d, p. 374), North-eastern China(prov. not indicated: M. Sato, 1941a, p. 78), Shaanxi(Jatta, 1902, p. 466; Zahlbr., 1930 b, p. 136; Wei et al. 1982, p. 27; He & Chen 1995, p. 45 without ciatation of specimens and their localities), Yunnan(Zahlbr., 1930 b, p. 136; M. Sato, 1941a, p. 78; Lamb, 1977, pp. 236 – 238), Sichuan(Zahlbr., 1930 b, p. 136 & 1934b, p. 202; M. Sato, 1941a, p. 78), Xizang(M. Sato, 1941a, p. 78; Wei & Jiang, 1981, p. 1147 & 1986, p. 76) Hubei(Chen, Wu & J. C. Wei, 1989, p. 457).

Stereocaulon verruculigerum Hue 疣珊瑚枝

J. Hattori bot. Lab. 43: 267(1977).

≡ *Stereocaulon verruculigerum* var. *formosanum* Asahina, *Journ. Jap. Bot.* 35: 295(1960).

Type: from Taiwan.

= *Stereocaulon formosanum* Asahina, *Journ. Jap. Bot.* 35: 294(1960). (nom. inval.).

On rocks.

Taiwan(Lamb, 1963, p. 688 & 1977, pp. 267, 417; Wang-Yang & Lai, 1973, p. 97; Lai, 2000, p. 272; Lai 2001, p. 179), Hainan(JCWei et al., 2013, p. 231).

Stereocaulon vesuvianum Pers. 多型珊瑚枝

Ann. Wetter. Gesellsch. Ges. Naturk. 2: 19(1811)[1810].

var. kilimandscharoense Stein 大型变种

Jber. schles. Ges. vaterl. Kultur 66: 134(1888).

Taiwan(Lai, 2000, pp. 272 – 273; Lai 2001, p. 179).

var. nodulosum(Wallr.) I. M. Lamb 小瘤变种

Best. europ. Flecht. (Vaduz): 633(1969).

≡ *Patellaria paschalis* var. *nodulosa* Wallr., *Fl. crypt. Germ.* (Norimbergae) 1: 441(1831).

= *Stereocaulon denudatum* Flk.

= *Stereocaulon denudatum* a. *genuimum* Th. Fr.

On rocks.

Sichuan(Elenk. 1904, p. 91), Taiwan(Zahlbr., 1933 c, p. 47; Wang-Yang & Lai, 1973, p. 97; Lamb, 1977, pp. 244 – 246; Lai, 2000, p. 273; Lai 2001, p. 179).

Lecanorales, genera incertae sedis 茶渍目未定位属

(58) Strangosporaceae S. Stenroos, Miądl. & Lutzoni 峡孢菌科

in Miądlikowska et al., Mol. Phylogen. Evol. 79: 132 – 168(2014).

* **Strangospora** Körb. 峡孢菌属

Parerga lichenol. (Breslau) 2: 173(1860).

* Strangospora moriformis(Ach.) Stein 桑葚型峡孢菌

in Cohn, *Krypt. -Fl. Schlesien*(Breslau) 2(2): 176(1879).

≡ *Arthonia moriformis* Ach., *Syn. meth. lich.* (Lund): 5(1814).

On lignum of a trunk, together with *Cyphelium tigillare* and *Acroscyphus sphaerophoroides*.

Xizang(Obermayer 2004, p. 511).

anamorphic Lecanorales 茶渍目无性型

Leprocaulon Nyl. 绒枝属

in Lamy, *Bull. Soc. bot. Fr.* 25: 352(1879)[1878].

Leprocaulon microscopicum(Vill.) Gams ex D. Hawksw. 微绒枝

in Hawksworth & Skinner, *Trans. Proc. Torquay nat. Hist. Soc.* 1972 – 3 16(3): 128(1974).

≡ *Lichen microscopicus* Vill., *Hist. pl. Dauphiné* 3(2): 946(1789).

≡ *Stereocaulon microscopicum*(Vill.) Frey, Rabenh. Krypt. -Fl., Edn 2(Leipzig) 9(4. 1): 89(1932).

= *Stereocaulon quisquiliare*(Leers) Hoffm., *Deutschl. Fl.*, *Zweiter Theil*(Erlangen): 130(1796)[1795].

≡ *Lichen quisquiliaris* Leers, *Fl. herborn.*: 264(1775).

= *Leprocaulon nanum*(Ach.) Nyl., in Lamy, *Bull. Soc. bot. Fr.* 25: 352(1879)[1878].

≡ *Lichen nanus* Ach., *Lich. suec. prodr.* (Linköping): 206(1799)[1798].

On soil covered rocks.

Guizhou(Pat. & Oliv. 1907, p. 23; Zahlbr., 1930 b, p. 136; Chen J. B. et al., 2003, p. 364; Zhang T & J. C. Wei, 2006, p. 8), Shandong(Li Y et al., 2008, p. 72 as *Stereocaulon quisquiliare*), Fujian(Zahlbr., 1930 b, p. 136).

[14] Rhizocarpales Miądl. et al. 地图衣目

Mycologia 98(6): 1097(2007)[2006], Nom. inval., Art. 39. 1(Melbourne)

(59) Catillariaceae Hafellner 腊肠衣科

Beih. Nova Hedwigia 79: 271(1984).

Catillaria A. Massal. **腊肠衣属**

Ric. auton. lich. crost. (Verona): 78(1852).

Catillaria arisana Zahlbr. 阿里山腊肠衣

in Fedde, *Repertorium* 33: 40(1933) & *Cat. Lich. Univ.* 10: 351(1940). Wang & Lai, *Taiwania* 18(1): 88(1973).

Type: Taiwan, Asahina no. 168.

Catillaria chalybeia(Borrer) A. Massal. 钢灰腊肠衣

Ric. auton. lich. crost. (Verona): 79(1852).

≡*Lecidea chalybeia* Borrer, *Suppl. Engl. Bot.* 1: tab. 2687, fig. 2(1831).

On shale and volcanic rock.

Taiwan(Aptroot and Sparrius, 2003, p. 13).

Catillaria globulosa(Flörke) Th. Fr. 小球腊肠衣

Lich. Scand. (Upsaliae) 1(2): 575(1874).

≡*Lecidea globulosa* Flörke, *Deutschl. Flecht.* 10: 1(1821).

Corticolous.

Yunnan(Zahlbr., 1930 b, p. 113).

Catillaria hospitans H. Magn. 寄生腊肠衣

Lich. Centr. Asia 1: 63(1940).

Type: Gansu, Bohlin no. 1279 in S.

On rock upon Lecidea tessellata Flk.

Catillaria imperfecta H. Magn. 无性腊肠衣

Lich. Centr. Asia 1: 64(1940).

Type: Gansu, Bohlin no. 86 b in S(!).

On calciferous stone.

Catillaria limosescens Zahlbr. 泥土腊肠衣

in Fedde, *Repertorium* 33: 41(1933) & *Cat. Lich. Univ.* 10: 355(1940). M. Lamb, *Index Nom. Lich.* 147(1963). Wang & Lai, *Taiwania* 18(1): 88(1973).

Syntype: Taiwan, Faurie nos. 174, 177; Asahina no. 160.

Saxicolous.

Catillaria melaleuca(Müll. Arg.) Zahlbr. 黑白腊肠衣

K. svenska Vetensk-Akad. Handl. 57(6): 22(1917).

≡*Patellaria melaleuca* Müll. Arg., *Flora*, Regensburg 65: 487(1882).

Yunnan(A. Zahlbr., 1934 b, p. 200).

Catillaria nigroclavata(Nyl.) J. Steiner 黑棒腊肠衣

Sber. Akad. Wiss. Wien, Math. -naturw. Kl., Abt. 1 107: 157(1898).

≡*Lecidea nigroclavata* Nyl., *Bot. Notiser*: 160(1853).

On *Populus* sp.

Taiwan(Aptroot and Sparrius, 2003, p. 13).

Catillaria picila(A. Massal.) Coppins 木腊肠衣

Lichenologist 21(3): 223(1989).

≡*Biatora picila* A. Massal., *Miscell. Lichenol.*: 38(1856).

On granire inland rock and concrete. Known from temperate Europe and Asia.

Taiwan(Aptroot and Sparrius, 2003, p. 13), Hongkong(Aptroot & Seaward, 1999, p. 71; Aptroot et al., 2001, p. 324).

Catillaria pseudopeziza(Jatta) Zahlbr. 假盘腊肠衣

in Handel-Mazzaetti, *Symb. Sin.* 3: 113(1930), & *Cat. Lich. Univ.* 8: 392(1932).

≡*Biatorina pseudopeziza* Jatta, *Nuov. Giorn. Bot. Ital.* nov. ser. 9: 179(1902).

Type: Shaanxi, Mt. Jitou shan, collected by Giraldi.

On mosses.

Catillaria rengechina Zahlbr. 台湾腊肠衣

in Fedde, *Repertorium* 33: 41(1933), & *Cat. Lich. Univ.* 10: 356(1940). M. Lamb. *Index Nom. Lich.* 148(1963). Wang & Lai, *Taiwania* 18(1): 88(1973).

Type locality: Taiwan, Asahina no. 182.

Catillaria aorediantha Zahlbr. 粉芽腊肠衣

in Handel-Mazzetti, *Symb. Sin.* 3: 112(1930) & *Cat. Lich. Uinv.* 8: 393(1932).

Type: Fujian, Chung no. 572.

Catillaria tristiopsis Zahlbr. 暗色腊肠衣

Cat. Lich. Univ. 4: 83(1927) & in Handel-Mazzetti, *Symb. Sin.* 3: 112(1930).

= *Biatorina tristis* Jatta, *Nuov. Giorn. Bot. Ital.* nov. ser. 9: 478(1902).

Type: Shaanxi, near Injiapo, collected by Giraldi.

Saxicolous.

Catillaria yunnana Zahlbr. 云南腊肠衣

in Handel-Mazzetti, *Symb. Sin.* 3: 113(1930) & *Cat. Lich. Univ.* 8: 395(1932).

Type: Yunnan, Handel-Mazzetti no. 6026.

Corticolous.

Halecania M. Mayrhofer **孢壁衣属**

Herzogia 7(3 – 4): 383(1987).

Halecania alpivaga(Th. Fr.) M. Mayrhofer 山孢壁衣

Herzogia 7(3 – 4): 391(1987).

≡*Lecania alpivaga* Th. Fr., *Lich. Scand.* (Upsaliae) 1(1): 292(1871).

On shale.

Xinjiang(Abdulla Abbas & Wu, 1998, p. 43; Abdulla Abbas et al., 2001, p. 364), Taiwan(Aptroot and Sparrius, 2003, p. 22).

Solenopsora A. Massal. **霜降鳞衣属**

Framm. Lichenogr.: 20(1855).

Solenopsora asahinae Zahlbr. ex Asahina 朝比氏霜降鳞衣

Journ. Jap. Bot. 6(3): 64(1929).

Type: Liaoning, Mt. Fenghuang shan.

Liaoning(Lamb, 1963, p. 662).

Solenopsora elixiana Verdon & Rambold 艾氏霜降鳞衣

Mycotaxon 69: 401(1998).

On sandstone bolders and shale.

Taiwan(Aptroot, Sparrius & Lai, 2002, p. 291).

Sporastatia A. Massal. **多孢衣属**

Geneac. lich. (Verona): 9(1854). (Sporastatiaceae, Index Fungorum)

Sporastatia asiatica H. Magn. 亚洲多孢衣

Lich. Centr. Asia 1: 69(1940).

Type: Qinghai, 19/IV/1932, at about 4000 m. Bohlin B: b in S(!).

On non caliciderous, hard rock.

Xinjiang(Abdulla Abbas et al., 1993, p. 81; Abdulla Abbas & Wu, 1998, p. 41; Abdulla Abbas et al., 2001, p. 368 as "*Sporasttatia*" *asiatica*), Gansu, Qinghai(H. Magn. 1940, p. 69).

Sporastatia testudinea(Ach.) A. Massal. 龟甲多孢衣

Geneac. lich. (Verona): 9(1855).

≡*Lecidea cechumena* var. *testudinea* Ach., *K. Vetensk-Acad. Nya Handl.* 29: 232(1808).

Misapplied name:

Glypholecia scabra auct. non(Pers.) Müll. Arg.: J. C. Wei & Chen, Report on the Scientific investigation(1966 –68) in Mt. Qomolangma district, Science Press, Beijing, 1974, p. 177.

On siliceous rocks.

Sichuan(Obermayer 2004, p. 510), Xizang(J. C. Wei & Chen J. B. 1974, p. 177; J. C. Wei & Jiang, 1986, p. 105; Obermayer 2004, p. 510).

(60) Rhizocarpaceae M. Choisy ex Hafellner **地图衣科**

Beih. Nova Hedwigia 79: 327(1984).

Catolechia Flot. **瘤衣属**

Bot. Ztg. 8: 367(1850).

Catolechia wahlenbergii(Ach.) Körb. 袖珍瘤衣

Syst. lich. germ. (Breslau): 181(1855).

≡*Lecidea wahlenbergii* Ach., *Methodus*, Sectio prior: 81(1803).

Catolechia wahlenbergii(Ach.) Flot., *Jber. Schles. Ges. Vaterl. Kultur* 27: 135(1849), Nom. inval., Art. 35. 1 (Melbourne).

=*Buellia pulchella*(Schaer.) Tuck., *Gen. lich.* (Amherst): 185(1872).

≡*Lecidea pulchella* Schaer., *Enum. critic. lich. europ.* (Bern): 100(1850).

Xizang(W. Obermayer, 2004, p. 493), Yunnan(Wei X. L. et al., 2007, p. 149), Taiwan(Asahina & Sato in Asahina, 1939, p. 771; Wang & Lai, 1973, p. 88).

Epilichen Clem. **表衣属**

Gen. fung. (Minneapolis): 69, 174(1909).

Epilichen cf. glauconigellus(Nyl.) Hafellner 苍表衣

Nova Hedwigia 30: 678(1979)[1978].

≡*Lecidea glauconigella* Nyl., *Lich. Scand.* (Helsinki): 238(1861).

On thalli of *Baeomyces* sp.

Sichuan, Xizang(W. Obermayer, 2004, p. 495).

Epilichen scabrosus(Ach.) Clem. 粗糙表衣

Gen. fung. (Minneapolis): 174(1909).

≡*Lecidea scabrosa* Ach., *Methodus*, Sectio prior: 48(1803).

On thalli of *Baeomyces* sp.

Sichuan, Xizang(Obermayer W. 2004, p. 495). Tibetan area(Hafelolner & Obermayer W., 1995; Obermayer W., 1998).

Rhizocarpon Ramond ex DC. **地图衣属**

in Lamarck & de Candolle, *Fl. uper.*, Edn 3(Paris) 2: 365(1805).

Rhizocarpon alpicola(Wahlenb.) Rabenh. [as '*alpicolum*'] 高山地图衣

Flecht. Europ. 22: no. 618(1861).

≡*Lecidea atrovirens* var. *alpicola* Wahlenb., *Fl. Lapp.*: 474(1812).

≡*Buellia alpicola*(Wahlenb.) Anzi, *Cat. Lich. Sondr.*: 90(1860).

≡*Lecidea alpicola*(Wahlenb.) Hepp, *Flecht. Europ.*: no. 151(1853).

On rocks.

Yunnan(Zahlbr., 1930 b, p. 124).

Rhizocarpon badioatrum(Flörke ex Spreng.) Th. Fr. 黑红地图衣

Lich. Scand. (Upsaliae) 1(2): 613(1874).

≡*Lecidea badioatra* Flörke ex Spreng., *Neue Entdeckungen im ganzen Umfang der Pflanzenkunde* 2: 95

(1821).

On shale and rocks.

Xinjiang(Mahire et al., 2015, p. 423), Taiwan(Aptroot and Sparrius, 2003, p. 38).

Rhizocarpon cinereocaesium Zahlbr. 兰灰地图衣

in Handel-Mazzetti, *Symb. Sin.* 3: 125(1930) & *Cat. Lich. Univ.* 8: 419(1932).

Type: Sichuan, alt. 1700 m, 14/IV/1914, Handel-Mazzetti no. 1379.

On rocks.

Rhizocarpon copelandii(Körb.) Th. Fr. 小孢地图衣

Lich. Scand. (Upsaliae) 1(2): 615(1874).

≡*Buellia copelandii* Körb., in Hertlaub & Lindeman, *Zweite Deutsch. Nordpolarfahrt* 2: 79(1874).

On rocks.

Shaanxi(Jatta, 1902, p. 479; Zahlbr., 1930 b, p. 124).

Rhizocarpon disporum(Nägeli ex Hepp) Müll. Arg. 灰地图衣

Revue mycol., Toulouse 1: 170(1879).

≡*Lecidea dispora* Nägeli ex Hepp, *Flecht. Europ.*: no. 28(1853).

Xinjiang(Abdulla Abbas et al., 1993, p. 80; Abdulla Abbas & Wu, 1998, p. 127; Abdulla Abbas et al., 2001, p. 367).

Rhizocarpon eupetraeoides(Nyl.) Blomb. & Forssell 类石地图衣

Enum. Pl. Scand.: 93(1880).

≡*Lecidea eupetraeoides* Nyl., *Flora*, Regensburg 58: 12(1875).

Xinjiang(Wang XY, 1985, p. 340; Abudula Abbas et al., 1993, p. 80).

Rhizocarpon expallescens Th. Fr. 淡白地图衣

Lich. Scand. (Upsaliae) 1(2): 620(1874).

On shale.

Taiwan(Aptroot and Sparrius, 2003, p. 38).

Rhizocarpon geographicum(L.) DC. 地图衣

in Lamarck & de Candolle, *Fl. uper.*, Edn 3(Paris) 2: 365(1805).

≡*Lichen geographicus* L., *Sp. pl.* 2: 1140(1753).

=*Rhizocarpon atrovirens*(L.) Chevall., *Fl. gén. env. Paris*(Paris) 1: 561(1826).

≡*Lichen atrovirens* L., *Sp. pl.* 2: 1141(1753).

≡*Lecidea atrovirens*(L.) Ach., *Methodus*, Sectio prior(Stockholmiæ): 45(1803).

=*Rhizocarpon riparium* Räsänen, *Ann. bot. Soc. Zool. -Bot. fenn. Vanamo* 19(Notul. Botan. 12): 60(1942).

On rocks.

Jilin(Chen et al. 1981, p. 157; Hertel & Zhao, 1982, p. 149), Xinjiang(Abdulla Abbas et al., 1993, p. 80; Abdulla Abbas et al., 1994, p. 22; Abdulla Abbas & Wu, 1998, p. 128; Abdulla Abbas et al., 2001, p. 367; Mahire, 2015, p. 424 as *R. riparium*), Shaanxi(Jatta, 1902, p. 479; Zahlbr., 1930 b, p. 124), Yunnan(Hue, 1889, p. 175; Paulson, 1928, p. 317; Zahlbr., 1930 b, p. 124), Guizhou(Zhang T & J. C. Wei, 2006, p. 9), Hubei (Chen, Wu & J. C. Wei, 1989, . 426), Anhui(Xu B. S., 1989, p. 204), Taiwan(Wang & Lai, 1973, p. 96), China(prov. Not ndicated: Asahina & M. Sato in Asahina, 1939, p. 673).

=*Rhizocarpon tinei* subsp. *frigidum* (Räsänen) Runemark, *Op. bot.* 2(no. 1): 125(1956).

≡*Rhizocarpon frigidum* Räsänen, *Ann. bot. Soc. Zool. -Bot. fenn. Vanamo* 19(Notul. Botan. 14): 9(1944).

On rocks.

Xizang(J. C. Wei & Y. M. Jiang, 1981, p. 1147, & 1985, p. 31).

f. (v.) contiguum(Schaer.) A. Massal. 邻变种

Ric. Auton. Lich. crost. (Verona): 100, fig. 203(1852).

≡*Lecidea geographica* var. *contigua* Schaer., *Lich. helv. uperfi.* 3: 124(1828).

≡*Diplotomma geographicum* var. *contiguum* Schaer., Jatta, I; 479(1902).

On rocks.

Shaanxi(Jatta, 1902, p. 479; Zahlbr., 1930 b, p. 125).

Rhizocarpon gracile Zahlbr. var. gracile 瘦地图衣

in Handel-Mazzetti, *Symb. Sin.* 3: 126(1930) & *Cat. Lich. Univ.* 8: 421(1932).

Type: Yunnan, Handel-Mazzetti no. 1387.

On sandstone.

var. sanguineum Zahlbr. 红色变种

in Handel-Mazzetti, *Symb. Sin.* 3: 126(1930) & *Cat. Lich. Univ.* 8: 421(1932).

Type: Sichuan, Handel-Mazzetti no. 1765.

On sandstone.

Rhizocarpon grande(Flörke ex Flot.) Arnold 巨地图衣

Flora, Regensburg 54: 149(1871).

≡*Lecidea petraea* f. *grandis* Flörke ex Flot., *Flora*, Regensburg 11(2): 690(1828).

On rocks.

Neimenggu(Zhao ZT et al., 2013, pp. 218 – 219), Xinjiang(Mahire et al., 2015, p. 423).

Rhizocarpon hochstetteri(Körb.) Vain. 豪氏地图衣

Acta Soc. Fauna Flora fenn. 53(no. 1): 280(1922).

≡*Catillaria hochstetteri* Körb., *Parerga lichenol.* (Breslau) 3: 195(1861).

On rock.

Jilin(Wang W. C. et al., 2015, p. 740).

Rhizocarpon infernulum(Nyl.) Lynge 腹地图衣

Rhodora 36: 158(1934).

≡*Lecidea infernula* Nyl., *Flora*, Regensburg 68: 440(1885).

On rock.

Guizhou(Zhao ZT et al., 2013, pp. 219 – 221).

Rhizocarpon intermediellum Räsänen 间型地图衣

Feddes Repert. 52: 132(1943).

On shale.

Taiwan(Aptroot and Sparrius, 2003, p. 38).

Rhizocarpon ischnothallum Zahlbr. 萎地图衣

in Handel-Mazzetti, *Symb. Sin.* 3: 126(1930) & *Cat. Lich. Univ.* 8: 422(1932).

Type: Yunnan, Handel-Mazzetti no. 5641.

On sandstone.

Rhizocarpon kansuense H. Magn. 甘肃地图衣

Lich. Centr. Asia, 1: 67(1940).

Type: Gansu, Bohlin no. 48 in S.

On calciferous stone.

Rhizocarpon lavatum(Ach.) Hazsl. 池地图衣

Magyar Birodalom Zuzmo-Floraja: 206(1884).

≡*Lecidea lavata* Ach., in Fries, *Nov. Sched. Critic. Lich.*: 18(1827).

On rock.

Jilin, Shaanxi, Yunnan(Wang W. C. et al. 2015, pp. 884 – 885).

Rhizocarpon lecanorinum Anders 茶渍地图衣

Hedwigia 64: 261(1923).

≡*Lecidea geographica* var. *lecanorina*(Körb.) Nyl., *Lich. Envir. Paris*: 102(1896).

On rocks.

Xinjiang(Mahire et al., 2015, p. 424).

Rhizocarpon leprosulum Zahlbr. 癞盾地图衣

in Handel-Mazzetti, *Symb. Sin.* 3: 125(1930) & *Cat. Lich. Univ.* 8: 422(1932).

Type: Yunnan, Handel-Mazzetti no. 5553.

On sandstone.

Rhizocarpon macrosporum Räsänen 大孢地图衣

Feddes Repert. 52: 131(1943).

On rocks.

Xinjiang(Mahire et al., 2015, p. 424).

Rhizocarpon nipponense Räsänen 拟地图衣

in Revist. *Sudamer. de Bot.* 7: 92(1942).

On rocks.

Xinjiang(Wu J. L., 1987, p. 94).

Rhizocarpon obscuratum(Ach.) A. Massal. 暗地图衣

Ric. Auton. Lich. crost. (Verona): 103(1852).

≡*Lecidea petraea* var. *obscurata* Ach., *Lich. univ.*: 156(1810).

On large pebbles on ground.

Jilin(Hertel & Zhao, 1982, p. 149).

Rhizocarpon parvum Runemark 小地图衣

Opera Bot. 2(1): 64(1956).

On *Tremolecia atrata*(Ach.) Hertel on rock

Xinjiang(Wang W. C. et al., 2015, pp. 744 – 746).

Rhizocarpon petraeum(Wulfen) A. Massal. 石地图衣

Ric. Auton. Lich. crost. (Verona): 102(1852).

≡*Lichen petraeus* Wulfen, *Schr. Ges. Naturf. Freunde*, Berlin 8: 89(1787).

On rock.

Xinjiang(Zhao ZT et al., 2013, pp. 221 – 223).

Rhizocarpon plicatile(Leight.) A. L. Sm. 褶地图衣

Monogr. Brit. Lich. 2: 197(1911).

≡*Lecidea plicatilis* Leight., *Ann. Mag. Nat. Hist.*, Ser. 4 4: 201(1869).

On shale.

Taiwan(Aptroot and Sparrius, 2003, p. 38).

Rhizocarpon polycarpum(Hepp) Th. Fr. 多果地图衣

Lich. Scand. 1(2): 617(1874).

≡*Lecidea confervoides* var. *polycarpa* Hepp, *Flecht. Europ.*: no. 35(1853).

On rock.

Jilin, Neimenggu, Xizang(Wang W. C. et al., 2015, pp. 741 – 743).

Rhizocarpon pusillum Runemark 微地图衣

Opera Bot. 2(1): 63(1956).

On *Sporastatia testudinea*(Ach.) A. Massal. on rock.

Xinjiang(Wan W. C. et al., 2015, p. 743).

Rhizocarpon reductum Th. Fr. 疏地图衣

Lich. Scand. (Upsaliae) 1(2): 633(1874).

On shale.

Jilin, Xinjiang, Shaanxi, Zhejiang(Wang W. C. et al. 2015, pp. 885 – 887). Taiwan(Aptroot and Sparrius, 2003, p. 38).

Rhizocarpon riparium Räsänen 岸边地图衣

Ann. Bot. Soc. Zool. -Bot. fenn. Vanamo 19(Notul. Botan. 12): 60(1942).

Hubei(Chen, Wu & J. C. Wei, 1989, p. 426).

Rhizocarpon rubescens Th. Fr. 红地图衣

Lich. Scand. (Upsaliae) 1(2): 631(1874).

On rock.

Gansu, Yunnan(Zhao ZT et al., 2013, pp. 223 – 225).

Rhizocarpon saanaënse Räsänen 芬兰地图衣

Ann. Bot. Soc. Zool. -Bot. fenn. Vanamo 16(Notul. Botan. 12): 61(1942).

On rocks.

Yunnan(Run. 1956, p. 109).

Rhizocarpon saurinum(W. A. Weber) Bungartz 蜥羽地图衣

Bryologist 107: 77(2004).

≡*Buellia saurina* W. A. Weber, Bryologist 74(2): 190(1971).

On rock.

Xinjiang, Xizang(Wang W. C. et al. 2015, pp. 887 – 888).

Rhizocarpon sinense Zahlbr. 中华地图衣

in Handel-Mazzetti, *Symb. Sin.* 3: 124(1930).

Syntype: Sichuan, Handel-Mazzetti nos. 1012, 2647; Yunnan, Handel-Mazzetti no. 8088.

On rocks.

Rhizocarpon solitarium H. Magn. 孤地图衣

Lich. Centr. Asia 2: 28(1944).

Type: Neimenggu, Amtsar no. 140 in S(!).

On siliceous rock.

Rhizocarpon sublucidum Räsänen 亚光地图衣

Ann. Bot. Soc. Zool. -Bot. fenn. Vanamo 19(Noul. Botan. 14): 3(1944).

On rocks.

Xizang(Run. 1956, p. 107).

Rhizocarpon umense(H. Magn.) A. Nordin 乌皿地图衣

Graphis Scripta 17(2): 37(2005).

≡*Buellia umensis* H. Magn. in *Ark. f. Bot.* XXXIIIA, no. 1: 137(1946).

On rock.

Xinjiang(Wang W. C. et al. 2015, pp. 889 – 890).

Rhizocarpon uperficial(Schaer.) Malme 岩表地图衣

Svensk bot. Tidskr. 8, 3: 282(1914).

≡*Lecidea superficialis* Schaer., *Lich. helv. uperfi.* 3: 124(1828).

On rocks.

Jilin(Hertrl & Zhao, 1982, p. 149), Xizang(J. C. Wei & Chen, 1974, p. 175; J. C. Wei & Y. M. Jiang, 1981, p. 1147 & 1986, p. 30: as ssp. *uperficial*), Taiwan(Aptroot and Sparrius, 2003, p. 38).

Rhizocarpon viridiatrum(Wulfen) Körb. 乌绿地图衣

Syst. Lich. germ. (Breslau): 262(1855).

≡*Lichen viridiater* Wulfen, in Jacquin, *Collnea bot.* 2: 186(1791) [1788].

Xinjiang(Abdulla Abbas et al., 1993, p. 80; Abdulla Abbas et al., 1994, p. 22; Abdulla Abbas & Wu, 1998,

p. 128; Abdulla Abbas et al. , 2001, p. 367) .

[15] Peltigerales Walt. Watson 地卷目

New Phytol. 28: 9(1929) .

①**Collematineae** Miądl. & Lutzoni 胶衣亚目

Am. J. Bot. 91(3) : 459(2004) .

(61) Coccocarpiaceae Henssen 瓦衣科

Syst. Ascom. 5: 314(1986) .

Coccocarpia Pers. 瓦衣属

in Gaudichaud-Beaupré in Freycinet, *Voy. Uranie.* , *Bot.* : 206(1827) [1826 – 1830] .

Coccocarpia erythroxyli(Spreng.) Swinscow & Krog [as ' erythroxili'] 环纹瓦衣

Norw. Jl Bot. 23: 254(1976) .

≡*Lecidea erythroxyli* Spreng. , *K. svenska Vetensk-Akad. Handl*. 46: 47(1820) .

=*Pannaria parmelioides*(Hook. f.) Colmeiro, Revista Progr. Ci. Fis. Nat. 16 – 17: 111(1867) .

≡*Lecidea parmelioides* Hook. , *Syn. pl*. (Paris) 1: 15(1822) .

≡*Coccocarpia pellita* var. *parmelioides*(Hook. f.) Müll. Arg. , Revue mycol. , Toulouse 9: 139(1887) .

Misapplied name:

Coccocarpia pellita non(Ach.) Müll. Arg. : Wei & Chen, 1974, p. 175) .

On moss covered rock & on bark of trees. Pantropical.

Ningxia(Liu Huajie et al. , 2010, p. 196) , Shaanxi(He & Chen 1995, p. 42, miscited as *C. erythtoxyli* without citation of specimens and their localities) , Xizang(Wei & Chen, 1974, p. 175; Wei & Jiang, 1986, p. 28) , Sichuan(Zahlbr. , 1930 b, p. 83; Arvidsson, 1982: 57, 62) , Guizhou(Zhang T & J. C. Wei, 2006, p. 6) , Yunnan (Hue, 1889, p. 170; Zahlbr. , 1930b, p. 83; Arvidsson, 1982: 57, 62; Wei X. L. et al. 2007, p. 151; Wang L. S. et al. , 2008, p. 536) , Hebei(Moreau et Moreau, 1951, p. 186) , Hubei(Chen, Wu & J. C. Wei, 1989, p. 422) , Jiangsu(Wu J. N. & Xiang T. 1981, p. 3; Xu B. S. 1989, p. 198) , Anhui, Zhejiang(Xu B. S. 1989, p. 198) , Fujian(Arvidsson, 1982: 57, 62; Wu J. N. et al. 1984, p. 4) , Guangdong(Vainio, 1921, p. 22; Arvidsson, 1982, p. 57, 58, 62) , Taiwan(Hue, 1908, pp. 212, 217; Zahlbr. , 1930b, p. 30 and 1933c, p. 29; Wang & Lai, 1973, p. 90; Ikoma, 1983, p. 57; Arvidsson, 1982: 57, 62) , Hongkong(Hue, 1908, p. 212; Arvidsson, 1982: 57, 62; Thrower, 1988, pp. 80, 82; Aptroot & Seaward, 1999, p. 72; Seaward & Aptroot, 2005, p. 284, Pfister 1978: 91, as *Pannaria parmelioides*.) .

Coccocarpia fenicis Vain. 小鳞瓦衣

Ann. Acad. Sci. fenn. , Ser. A 15(6) : 24(1921) .

On rocks and on tree trunk.

Fujian(Wu et al. 1984, p. 5) .

Coccocarpia palmicola(Spreng.) Arv. & D. J. Galloway 粗瓦衣

Bot. Notiser 132: 242(1979) .

≡*Lecidea palmicola* Spreng. , *K. svenska Vetensk-Akad. Handl*. 8: 46(1820) .

=*Coccocarpia cronia*(Tuck.) Vain. , *Ann. Acad. Sci. fenn.* , Ser. A 6(no. 7) : 103(1915) .

≡*Parmelia cronia* Tuck. , Proc. *Amer. Acad. Arts & Sci*. 1: 228(1848) .

=*Coccocarpia cronia* var. *isidiophylla*(Müll. Arg.) Vain.

=*Coccocarpia cronia* var. *primaria* Vain.

On bark of trees in forest and terrestrial in a coastal site. Pantropical.

Heilongjiang(Chen et al. 1981, p. 134) , Shaanxi(He & Chen 1995, p. 42, no specimen & locality were cited) , Yunnan(Wang L. S. et al. 2008, p. 536) , Hubei(Chen, Wu & J. C. Wei, 1989, p. 422) , Jiangsu(Wu J. N. & Xiang T. 1981, p. 3; Xu B. S. 1989, p. 199) , Anhui, Zhejiang(Xu B. S. 1989, p. 199) , Fujian(Zahlbr. , 1930b, p. 83; Arvidsson, 1982: 72, 76; Wu et al. 1984, p. 4 – 5) , Taiwan(Zahlbr. , 1933c, p. 30; Wang & Lai, 1973, p.

90), Hainan(JCWei et al., 2013, p. 231), Hongkong(Arvidsson, 1982: 72, 76; Thrower, 1988, p. 81; Aptroot & Seaward, 1999, p. 72; Aptroot et al., 2001, p. 324), China(prov. not indicated: Asahina & M. Sato in Asahina, 1939, p. 647).

Coccocarpia pellita(Ach.) Müll. Arg. 鳞瓦衣

Flora, Regensburg 65: 320(1882).

≡*Parmelia pellita* Ach., *Lich. univ.*: 468(1810).

On rocks and trees in forest. Pantropical.

Guizhou(Zhang T & Wei JC, 2006, p. 6), Anhui, Zhejiang(Xu B. S. 1989, p. 199), Guangdong(Arvidsson, 1982: 76, 79), Hongkong(Thrower, 1988, p. 82; Aptroot & Seaward, 1999, p. 72; Aptroot et al., 2001, p. 324), Taiwan(Arvidsson, 1982: 76, 79).

Coccocarpia smaragdina Pers. 翡翠瓦衣

in Gaudichaud-Beaupré in Freycinet, *Voy. Uranie.*, *Bot.*: 206(1827)[1826 – 1830].

Jiangxi, Fujian(Zahlbr., 1930b, p. 83), Taiwan(Zahlbr., 1933c, p. 30; Wang & Lai, 1973, p. 90).

(62) Collemataceae Zenker [as 'Collemata'] **胶衣科**

in Goebel & Kunze, *Pharmaceutische Waarenkunde* (Eisenach) 1(3): 124(1827)[1827 – 1829]

Blennothallia Trevis. **黏叶衣属**

Caratt. Tre Nuov. Gen. Collem.: 2(1853).

Blennothallia crispa(Huds.) Otálora, P. M. Jørg. & Wedin 卷曲黏叶衣

Fungal Diversity 64: 282(2014).

≡*Lichen crispus* Huds., *Fl. Angl.*: 447(1762).

≡*Collema crispum* Weber ex F. H. Wigg. *Prim. fl. holsat.* (Kiliae): 89(1780).

On rocks.

Xizang(Tibet)(Wu & Liu, 2012, pp. 188 – 189), Shandong(Li y et al., 2008, p. 71), Jiangsu(Wu J. N. & Xiang T. 1981, p. 2; Xu B. S. 1989, p. 194), Zhejiang(Xu B. S. 1989, p. 194).

Blennothallia furfureola(Müll. Arg.) Otálora, P. M. Jørg. & Wedin 小鳞黏叶衣

Fungal Diversity 64: 282(2014).

≡*Collema furfureolum* Müll. Arg., *Flora*, Regensburg 72: 142(1889).

= *Synechoblastus sublaevis* Jatta, *Nuov. Giorn. Bot. Ital.* ser. 9: 481(1902).

Type: Shaanxi, Giraldii no. 126(lectotype) in S.

≡*Collema sublaeve*(Jatta) Zahlbr., *Cat. Lich. Univers.* 3: 47(1924)[1925].

Terricolous.

Shaanxi(Jatta, 1902, p. 481; Zahlbr., 1930b, p. 76; Degel. 1974, p. 79; Wu & Liu, 2012, p. 189; He & Chen 1995, p. 42 without citation of specimens and their localities), Zhejiang(Degel. 1974, p. 79), China(prov. not indicated: Zahlbr., 1925, p. 47).

Collema Weber ex F. H. Wigg. **胶衣属**

Prim. fl. holsat. (Kiliae): 89(1780). Sanctioning author: Fr. Nom. cons., see Art. 14

Collema callibotrys Tuck. 葡萄串胶衣

Proc. Amer. Acad. Arts & Sci. 5: 386(1862)[1860].

Collema callibotrys var. coccophyllizum(Zahlbr.) Degel. 柳叶变种

Symb. bot. upsal. 20(no. 2): 68(1974).

≡*Collema callibotrys* var. *coccophyllizum*(Zahlbr.) Degel., *Symb. bot. upsal.* 20(no. 2): 68(1974).

≡*Collema coccophyllizum* Zahlbr., *Cat. Lich. Univers.* 3: 71(1924)[1925].

= *Collema coccophylloides* Nyl., *Acta Soc. Sci. fenn.* 7(2): 427(1863), nom. illegit., Art. 53. 1.

Corticolous.

Yunnan(Hue, 1898, p. 217; Zahlbr., 1930b, p. 76).

Collema callopismum A. Massal. 小丽胶衣

Miscell. Lichenol. : 23(1856).

Hebei(Jiang Z. G., 1993, p. 70).

Collema coccophorum Tuck. 球胶衣

Proc. Amer. Acad. Arts & Sci. 5: 385(1862)[1860].

On the ground.

Neimenggu(Liu & J. C. Wei, 2009, p. 14; Wu & Liu, 2012, p. 176), Ningxia(Liu Huajie et al., 2010, p. 196; Yang J. & Wei J. C., 2014, p. 1031), Hunan(Wu & Liu, 2012, p. 176).

Collema complanatum Hue 扁平胶衣

J. Bot., Paris 20: 85(1906).

Corticolous.

Heilongjiang, Jilin, Neimenggu, Shaanxi, Yunnan, Anhui, Jiangxi(Wu & Liu, 2012, pp. 209 – 210), Zhejiang (Xu B. S. 1989, p. 194), Fujian(Wu et al. 1984, p. 1), Taiwan(Zahlbr., 1933c, p. 26; Wang & Lai, 1973, p. 90; Degel. 1974, p. 161), Hainan(Wu & Liu, 2012, pp. 209 – 210; JCWei et al., 2013, p. 231), Hongkong (Thrower, 1988, p. 83).

See Collema pulcellum Ach. var. subnigrescens(Müll. Arg.) Degel. for Hongkong.

Collema cristatum(L.) Weber ex F. H. Wigg. 鸡冠胶衣

Prim. fl. holsat. (Kiliae): 89(1780).

≡*Lichen cristatus* L., *Sp. pl.* 2: 1143(1753).

On rocks.

Jiangsu, Zhejiang(Xu B. S. 1989, p. 194).

var. cristatum 原变种

On rocks.

Neimenggu, Xinjiang, Hebei(Wu & Liu, 2012, pp. 191 – 192).

var. marginale(Huds.) Degel. 边缘变种

Symb. bot. upsal. 13(2): 316(1954).

≡*Lichen marginalis* Huds., *Fl. Angl.*, Edn 2: 1 – 690(1778).

On rocks.

Beijing(Mt. Baihua shan in Beijing was mistaken by Wu & Liu for that in Hebei 2012, pp. 192 – 193), Neimenggu, Xinjiang(Wu & Liu, 2012, pp. 192 – 193).

Collema fasciculare(L.) Weber ex F. H. Wigg. 束孢胶衣

Prim. fl. holsat. (Kiliae): 89(1780).

≡*Lichen fascicularis* L., *Mant. Pl.* 1: 133(1767).

Corticolous.

Heilongjiang(Chen et al. 1981, p. 134), Jilin, Sichuan(Wu & Liu, 2012, pp. 226 – 227), Shandong(Li y et al., 2008, p. 71).

Collema flaccidum(Ach.) Ach. 石胶衣

Lich. univ.: 647(1810).

≡*Lichen flaccidus* Ach., *K. Vetensk-Acad. Nya Handl.* : 14(1795).

=*Collema rupestre*(F. Desp.) Rabenh., *Deutschl. Krypt. -Fl.* (Leipzig) 2: 50(1845).

≡*Parmelia rupestris* F. Desp., *Flore de la Sarthe*: 385(1838).

≡*Synechoblastus rupestris*(F. Desp.) Trevis., *Caratt. Tre Nuov. Gen. Collem.* : 3(1853).

On trees and stones.

Heilongjiang, Sichuan(Wu & Liu, 2012, p. 203), Xinjiang(Abdulla Abbas et al., 1993, p. 77; Abdulla Abbas & Wu, 1998, p. 61; Wu & Liu, 2012, p. 203; Abdulla Abbas et al., 2001, p. 363), Shaanxi(Jatta, 1902, p. 480; Zahlbr., 1930b, p. 76), Anhui(Xu B. S. 1989, p. 194; Wu & Liu, 2012, p. 203).

Collema furfuraceum(Schaer.) Du Rietz 粉屑胶衣

Ark. Bot. 22A(13): 5(1929), var. furfuraceum.

≡*Collema nigrescens* f. *furfuraceum* Schaer., *Enum. critic. lich. europ.* (Bern): 252(1850).

≡*Synechoblastus nigrescens* f. *furfuraceus*(Schaer.) Zahlbr., *Cat. Lich. Univers.* 3: 56(1925).

On bark of trees, occasionally on rocks.

Jilin(Liu & J. C. Wei, 2009, p. 16; Wu & Liu, 2012, p. 212), Xinjiang, Shaanxi, Sichuan, Henan, Hunan, Anhui (Wu & Liu, 2012, pp. 211 – 212), Shaanxi(He & Chen 1995, p. 42 without citation of specimens and their localities).

var. luzonense(Räsänen) Degel. 吕宋变种

Symb. bot. upsal. 20(2): 179(1974).

≡*Collema(Collemodiopsis) luzonense* Räsänen in *Arch. Soc. Zool. Bot. Fenn.* Vanamo III: 82(1949).

On bark of trees.

Hunan(Wu & Liu, 2012, p. 212).

Collema fuscovirens(With.) J. R. Laundon 棕绿胶衣

Lichenologist 16(3): 219(1984).

≡*Lichen fuscovirens* With., *Bot. arr. veg. Gr. Brit.* (London) 2: 717(1776).

On soil covering rocks.

Neimenggu, Shaanxi, Xizang, Sichuan(Wu & Liu, 2012, pp. 193 – 194), Xinjiang(Abdulla Abbas et al., 1993, p. 77; Wu & Liu, 2012, pp. 193 – 194; Abdulla Abbas & Wu, 1998, p. 62; Abdulla Abbas et al., 2001, p. 363).

Collema glebulentum(Nyl. ex Cromb.) Degel. 隆胶衣

Ark. Bot. 2(2): 88(1952).

≡*Leptogium glebulentum* Nyl. ex Cromb., *J. Bot., Lond.* 20: 272(1882).

Xinjiang(Abdulla Abbas & Wu, 1998, p. 62).

= *Collema tuniforme*(Ach.) Ach. [as 'tunaeforme'], *Lich. univ.*: 330(1810).

≡*Lichen tuniforme* Ach. [as 'tunaeformis'], *K. Vetensk-Acad. Nya Handl.* 16: 17(1795).

On rocks, ground, and base of tree trunks.

Shaanxi(Wu J. L. 1987, p. 55), Jiangsu(Wu J. L. 1987, p. 55; Xu B. S. 1989, p. 195).

= *Collema furvum* (Ach.) DC., in Lamarck & de Candolle, *Fl. franç.*, Edn 3(Paris) 2: 385(1805).

≡*Lichen furvus* Ach., Lich. suec. prodr. (Linköping): 132(1799) [1798].

On the ground among mosses.

Jilin(Wu & Liu, 2012, p. 204), Xinjiang(H. Magn. 1940, p. 40; Degel. 1974, pp. 144 – 145; Abdulla Abbas et al., 1993, p. 77; Abdulla Abbas et al., 2001, p. 363), Shaanxi(Jatta, 1902, p. 480; Zahlbr., 1930b, p. 76).

Collema japonicum(Müll. Arg.) Hue 日本胶衣

Nouv. Arch. Mus. Hist. Nat., Paris, 3 sér. 10: 220(1898).

≡*Synechoblastus japonicus* Müll. Arg., *Flora*, Regensburg 63: 17(1880).

On trunks and stones.

Beijing(Mt. Baihua shan in Beijing was mistaken for that in Hebei by Wu & Liu, 2012, pp. 204 – 205), Shaanxi, Sichuan, Henan, Hunan(Wu & Liu, 2012, pp. 204 – 205), Anhui(Xu B. S. 1989, p. 195; Wu & Liu, 2012, pp. 204 – 205), Zhejiang(Xu B. S. 1989, p. 195), Taiwan(Zahlbr., 1933 c, p. 26; Wang & Lai, 1973, p. 90).

Collema kauaiense H. Magn. 夏威夷胶衣

in Magnusson & Zahlbr., *Ark. Bot.* 31A(1): 63(1944).

On trees.

Sichuan(Wu & Liu, 2012, p. 190).

Collema latzelii Zahlbr. 拉氏胶衣

Öst. bot. Z. 59: 493(1909).

On rocks.

Hebei(Jiang Z. G., 1993, p. 71). ?

Collema leptaleum Tuck 薄胶衣

Proc. Amer. Acad. Arts & Sci. 6: 263(1866)[1864].

On trees in forest. Pantropical.

Hongkong(Thrower, 1988, pp. 15, 84; Aptroot & Seaward, 1999, p. 72).

var. leptaleum 原变种

= *Collema raishanum* Zahlbr., in Fedde, *Repertorium* 33: 26(1933).

Type: Taiwan, F-194(holotype) & F-192, F-378(paratypes) in W. (not seen)

= *Collema idzuense* ssp. *raishanum*(Zahlbr.) Asahina

= *Collema brevisporum* Z. G.. Jiang, Journal of Hebei Norman University 16(3): 83(1992).

Type: Jilin, Mt. Changbaishan, 1962. 9. 2. J. D. Zhao 6399(Hmas-L).

synonymized by Liu & Wei, 2003, p. 350.

Corticolous.

Heilongjiang, Yunnan, Zhejiang(Wu & Liu, 2012, pp. 198-199), Jilin(Z. G.. Jiang, 1992; Liu & J. C. Wei, 2003, p. 350; Wu & Liu, 2012, pp. 198-199), Taiwan(Zahlbr., 1933 c, p. 26 & 1940, p. 247; Lamb, 1963, p. 206; Wang-Yang & Lai, 1973, p. 90; Degel. 1974, pp. 102 & 107).

var. biliosum(Mont.) Degel. 裂芽变种

Symb. bot. upsal. 20(2): 105(1974).

≡ *Collema nigrescens* var. *biliosum* Mont., *Annls Sci. Nat., Bot.*, sér. 2 18: 20(1842).

On trees in forest. Pantropical.

Neimenggu, Jilin, Shaanxi, Yunnan, Jiangxi(Wu & Liu, 2012, p. 200), Hongkong(Thrower, 1988, pp. 15, 85; Aptroot & Seaward, 1999, p. 72; Wu & Liu, 2012, p. 200).

Collema lushanense Jiang 庐山胶衣

Journal of Hebei Normal University 16(3): 83(1992).

Holotype: Jiangxi, Mt. Lushan, 1960. 4. 3. Zhao J. D. et al. 577(20348).

Jiangxi(Jiang Z. G., 1992, p. 83; H. J. Liu & J. C. Wei, 2009, p. 18; Wu & Liu, 2012, pp. 187-188).

Collema multipartitum Smith. 多裂胶衣

in Smith & Sowerby, *Engl. Bot.* 36: tab. 2582(1814).

= *Collema multipartitum* var. *granulosum* Z. G.. Jiang, *Journal of Hebei Norman University*16(3): 83(1992), synonymized by Liu & J. C. Wei, 2003, p. 352.

On rocks.

Hebei(Jiang Z. G. as *Collema multipartitum* var. *granulosum* Z. G.. Jiang, 1992, p. 86; Wu & Liu, 2012, pp. 200-201).

Collema nigrescens(Huds.) DC. 黑胶衣

in Lamarck & de Candolle, *Fl. franç.*, Edn 3(Paris) 2: 384(1805).

≡ *Lichen nigrescens* Huds., *Fl. Angl.*: 450(1762).

Corticolous.

Heilongjiang, Neimenggu(Wu & Liu, 2012, pp. 214-215), Shaanxi(He & Chen 1995, p. 42 without citation of specimens and their localities), Sichuan, Yunnan, Hunan, Fujian(Zahlbr., 1930 b, p. 76), Shandong(Li y et al., 2008, p. 71).

Collema nepalense Degel. 尼泊尔胶衣

Symb. bot. upsal. 20(2): 157(1974).

On bark of trees.

Xizang, Yunnan(Wu & Liu, 2012, pp. 213-214).

Collema nipponicum Degel. 东瀛胶衣

Symb. bot. upsal. 20(2): 53(1974).

Xinjiang(Liu & Wei, 2009, p. 20; Wu & Liu, 2012, pp. 184 – 185).

Collema peregrinum Degel. 台湾胶衣

Symb. Bot. Upsal. 20(2): 109(1974).

Type: Taiwan, S. Kurok. no. 785(holotype) in TNS. (Ikoma, 1983, p. 59).

Collema poeltii Degel. 珀氏胶衣

Symb. bot. upsal. 20(2): 96(1974).

On rocks and soil.

Yunnan, Hebei, Henan, Anhui(Wu & Liu, 2012, p. 197).

Collema polycarpon Hoffm. 多果胶衣

Deutschl. Fl., Zweiter Theil (Erlangen): 102(1796) [1795].

On rocks.

Yunnan(Liu & Wei, 2009, p. 20; Wu & Liu, 2012, p. 186).

Collema pulchellum Ach. [as ' pulcellum'] 美小胶衣

Syn. meth. lich. (Lund): 321(1814). var. pulchellum

Misapplied name:

Collema pulchellum Ach. var. *multipartitum*(Z. G. Jiang, 1992, p. 86).

Collema pulchellum Ach. var. *leucopeplum*(Tuck.) Degel. (Z. G. Jiang, 1993, p. 72).

Heilongjiang, Jilin, Neimenggu, Xinjiang, Shaanxi, Hebei, Anhui, Hubei, Hunan, Yunnan, Fujian(Liu & Wei, 2003, pp. 352 – 353; Wu & Liu, 2012, pp. 217 – 218).

var. subnigrescens(Müll. Arg.) Degel. [as ' pulcellum'] 亚黑变种

Symb. bot. upsal. 20(2): 173(1974).

= *Synechoblastus flaccidus* var. *subfurvus* Müll. Arg., *Proc. R. Soc. Edinb.* 11: 457(1882).

= *Collema pustuligerum* Hue, J. Bot., Paris 20: 16(1906)

= *Collema vespertilio*(Lightf.) Hoffm., Descr. Adumb. Plant. Lich. 2: 48, tab. XXXVII, fig. 2 – 3(1794)

= *Collema corniculatum* Z. G. Jiang, *Journal of Hebei Norman University* 16(3): 85(1992),

Synonymized by H. J. Liu & J. C. Wei, 2003, p. 353.

Misapplied name:

Collema complalatum Hue(Thrower, 1988, pp. 15, 83; Z. G. Jiang, 1993, p. 72).

Collema pulcellum Ach. var. *leucopeplum*(Tuck.) Degel. (Z. G. Jiang, 1993, p. 72).

Collema pulcellum Ach. var. *pulcellum*(Z. G. Jiang, 1993, p. 72).

Corticolous, and on coastal rocks. Pantropical.

Heilongjiang(Asahina, 1952 d, p. 375; Degel. 1974, p. 176; Liu & J. C. Wei, 2003, pp. 353 – 354), Qinghai, Shaanxi, Xizang Sichuan(Liu & Wei, 2003, pp. 353 – 354). guizhou(Liu & Wei, 2003, pp. 353 – 354; Zhang T & Wei JC, 2006, p. 6), Yunnan(Degel. 1974, p. 176 ; Liu & Wei, 2003, pp. 353 – 354), Hubei(Chen, Wu & Wei, 1989, p. 420; Liu & Wei, 2003, pp. 353 – 354), Taiwan(Wang & Lai, 1973, p. 90), Hongkong(Thrower, 1988, pp. 15, 83 as *C. complalatum*, 86; Aptroot & Seaward, 1999, p. 72 – 73; Liu & J. C. Wei, 2003, pp. 353 – 354),

Collema rugosum Kremp. 皱胶衣

in Fenzl, *Reise Österr. Novara Bot.* 1: 128(1870).

= *Collema subfurvum*(Müll. Arg.) Degel., *Bot. Notiser*: 139(1948).

≡ *Synechoblastus flaccidus* var. *subfurvus* Müll. Arg., *Proc. R. Soc. Edinb.* 11: 457(1882).

On rocks and rarely on bark. Paleotrpical.

Heilongjiang(Chen et al. 1981, p. 134; Wu & Liu, 2012, p. 202), Xinjiang, Xizang, Guizhou, Yunnan, Henan, Hunan(Wu & Liu, 2012, p. 202), Shaanxi(He & Chen 1995, p. 43), Jiangsu(Wu J. N. & Xiang T. 1981, p. 2;

Xu B. S. 1989, p. 195), Anhui(Xu B. S. 1989, p. 195; Wu & Liu, 2012, p. 202), Shanghai, Zhejiang(Xu B. S. 1989, p. 195), Jiangxi(Degel. 1974, p. 140; Wu & Liu, 2012, p. 202), Fujian(Wu et al. 1984, p. 1), Hongkong (Thrower, 1988, p. 87; Aptroot & Seaward, 1999, p. 73).

Notes: *Collema subfurvum* sensu Degel. (1954) = *Collema subflaccidum* Degel.

Collema subfurvum(Müll. Arg.) Degel. = *Collema rugosum* Krphbr. (Degel. 1974)

It seems that the species reported from China under the name of *Collema subfurvum* may probably be *Collema rugosum* rather than *Collema subflaccidum* Degel.

Collema ryssoleum(Tuck.) A. Schneid. 褶胶衣

Gen. lich. (Amherst): 92(1872).

≡ *Collema nigrescens* var. *ryssoleum* Tuck., *Lichens of California*(Berkeley): 34(1866).

On rocks.

Shaanxi(Wu J. L., 1987, p. 54; Wu & Liu, 2012, p. 220), . Hebei, Sichuan, Yunnan(Wu & Liu, 2012, p. 220) Shandong(zhang F et al., 1999, p. 31; Zhao ZT et al., 2002, p. 5).

Collema shiroumanum Yasuda ex Räsänen 白山胶衣

in J. J. B. 16: 147(1940). (1940).

On bark of trees.

Hubei, Jiangxi(Wu & Liu, 2012, p. 223 – 224).

Collema sichuannese H. L. Liu & J. C. Wei 四川胶衣

Mycosystema 22(4): 532(2003).

Type: Sichuan, 1983. 6. 21. X. Y. Wang, X. Xiao10993(HMAS-L).

Sichuan(Liu & Wei, 2003, p. 532).

Collema subconveniens Nyl. 砖孢胶衣

Lich. Nov. Zeland. (Paris): 8(1888).

Xinjiang(Abdulla Abbas & Wu, 1998, p. 62; Abdulla Abbas et al., 2001, p. 363).

Collema tianmuense Z. G. Jiang, *Journal of Hebei Norman University* 16(3): 84(1992), Synonymized by Liu & J. C. Wei, 2003, pp. 354 – 456.

Type: Zhejiang, Mt. Tianmushan, 1962. 9. 1. Zhao J. D. 6265(HMAS-L).

Xinjiang(Abdulla Abbas et al., 1993, p. 77), ningxia(Liu Huajie et al., 2010, p. 197), Shaanxi, Yunnan, Zhejiang(Z. G. Jiang, 1992, p. 84; Liu & J. C. Wei, 2003, pp. 354 – 356; Wu & Liu, 2012, p. 206), guizhou (Zhang T & J. C. Wei, 2006, p. 6), Shandong(Li y et al., 2008, p. 71), Hubei(Chen, Wu & J. C. Wei, 1989, p. 421; Liu & J. C. Wei, 2003, pp. 354 – 356; Wu & Liu, 2012, p. 206).

Collema subflaccidum Degel. 亚石胶衣

Symb. bot. upsal. 20(2): 140(1974).

On bark of tees and on rocks.

Beijing, Neimenggu, Jilin, Xinjiang, Shaanxi, Yunnan, Henan, Hubei, Hunan, Anhui, Shandong, Jiangxi(Wu & Liu, 2012, pp. 207 – 208), Ningxia(Liu Huajie et al., 2010, p. 197), Guizhou(Zhang T & J. C. Wei, 2006, p. 6; Wu & Liu, 2012, pp. 207 – 208), Hainan(JCWei et al., 2013, p. 231).

Collema subnigrescens Degel. 亚黑胶衣

Symb. bot. upsal. 13(2): 413(1954).

= *Collema pulcellum* Ach. var. *multipotitum* Z. G. Jiang, *Journal of Hebei Norman University*16(3): 86(1992), synonymized by Liu & J. C. Wei, 2003, pp. 356 – 357.

Misapplied name:

Collema pulcellum Ach. var. *subnigrescsns*(Müll. Arg.) Degel. (Z. G. Jiang, 1993, pp. 72 – 73).

On tree trunks.

Shaanxi(Wu J. L., 1987, p. 55), Hubei(Z. G. Jiang, 1992, p. 86; Liu & J. C. Wei, 2003, pp. 356 – 367; Wu & Liu, 2012, p. 222), Xizang, Sichuan, Yunnan, Hunan(Liu & Wei, 2003, pp. 356 – 367; Wu & Liu, 2012, p.

222), Guizhou(Zhang T & J. C. Wei, 2006, p. 6).

f. caesium(Clemente) Degel. 淡蓝变型

Symb. bot. upsal. 13(2):417(1954).

= *Collema nigrescens* f. *caesium*(Clemente) Ach., *Syn. meth. lich.* (Lund):321(1814).

≡ *Parmelia nigrescens*(Huds.) Ach. var. *caesia* Clem., *Ensayo sobre las Variedades de la Vid Comun*: 303 (1807).

On trees

Sichuan, Guizhou(Wu & Liu, 2012, p. 223).

Collema substipitatum Zahlbr. 短柄胶衣

in Handel-Mazzetti, *Symb. Sin.* 3: 76(1930).

Type: Yunnan, Handel-Mazzetti no. 10065(lectotype) in W and in K, S(Herb. Vrang), US, W, & WU(isolectotypes).

On bark of tees.

Yunnan(Degel. 1974, p. 188), Taiwan(Degel. 1974, p. 188).

var. substipitatum 原变种

Sichuan, Guizhou, Yunnan(Wu & Liu, 2012, p. 225), Taiwan(Degelius, 1974. P188).

var. gonggashanense H. J. Liu & J. C. Wei 贡嘎山变种

Mycosystema 22(4):532(2003).

Type: Sichuan, Mt. Gonggashan, 1982. 7. 2. X. Y. Wang, X. Xiao, & B. Li 8729 – 1(HMAS-L).

Sichuan(Liu & Wei, 2003, p. 532; Wu & Liu, 2012, p. 226).

Collema tenax(Sw.) Ach. 坚韧胶衣

Lich. univ.: 635(1810). var. tenax

≡ *Lichen tenax* Swartz, *Nova Acta Acad. Upsal.* 4: 249(1784).

Heilongjiang(Chen et al. 1981, p. 150), Shaanxi(Jatta, 1902, p. 480; Zahlbr. 1930 b, p. 77), Guizhou(Zhang T & Wei JC, 2006, p. 6).

var. corallinum(A. Massal.) Degel. 珊瑚变种

Symb. bot. upsal. 13(2):165(1954).

≡ *Collema pulposum* var. *corallinum* A. Massal., *Sched. critic.* (Veronae) 10: 180(1856).

= *Collema kansuense* H. Magn. *Lich. Centr. Asia*, 1: 41(1940).

Type: Gansu, Bohlin no. 61 b in S(not seen).

= *Collema minutum* H. Magn. *Lich. Centr. Asia*, 2: 20(1944).

Type: Neimenggu, Bohlin no. 106 in S.

On calcareous soil.

Neimenggu(Degel. 1974, pp. 47, 49; Liu & J. C. Wei, 2009, p. 23), Xinjiang(Abdulla Abbas & Wu, 1998, p. 63; Abdulla Abbas et al., 2001, p. 363 as var. "*corellinum*"), Gansu(H. Magn. 1944, p. 20; Degel. 1954, pp. 155, 165, 175, 183, & 1974, p. 49), Ningxia(Liu Huajie et al., 2010, p. 197), Hebei(Wu & Liu, 2012, pp. 177 – 178).

var. crustaceum(Kremp.) Degel. 壳状变种

Symb. bot. upsal. 13(2):164(1954).

≡ *Collema crustaceum* Kremp., *Denkschr. Kgl. Bayer. Bot. Ges.*, *Abt.* 2 4: 95(1861).

On the ground.

Xinjiang(Guo, 2005, p. 47; Liu & J. C. Wei, 2009, p. 24; Wu & Liu, 2012, pp. 178 – 179), Hebei(Jiang Z. G., 1993, p. 69), Hunan(Wu & Liu, 2012, pp. 178 – 179).

var. diffractoareolatum(Schaer.) Degel. 环萝变种

Symb. bot. upsal. 13(2):164(1954).

≡ *Collema pulposum* var. *diffractoareolatum* Schaer., *Lich. helv. spicil.* 11: 539(1842).

On the ground.

Xinjiang(Wu & Liu, 2012, pp. 179 – 180).

var. expansum Degel. 展型变种

Symb. bot. upsal. 13(2): 162(1954).

On the ground.

Xinjiang, Ningxia(Wu & Liu, 2012, p. 180).

var. ogatae(Zahlbr.) Degel. 黄心变种

Symb. Bot. Upsal. 20(2): 47, 49(1974).

≡*Collema ogatae* Zahlbr. in Fedde, *Repertorium*, 33: 27(1933).

Type: Taiwan, Asahina no. 377(not seen).

On the ground.

Xinjiang(Wu & Liu, 2012, p. 181), Taiwan(Degel. 1974, pp. 47, 49; Ikoma, 1983, p. 59)

var. substellatum(H. Magn.) Degel. 亚星变种

Symb. Bot. Upsal. 20(2): 47, 49(1974).

≡*Collema substellatum* H. Magn. *Lich. Centr. Asia*, 1: 41(1940).

Type: Gansu , collected by Bohlin in 18/IX/1931 , preserved in S(not seen).

On calcareous soil.

Xinjiang(Wu & Liu, 2012, p. 182; Abdulla Abbas & Wu, 1998, p. 63; Abdulla Abbas et al., 2001, p. 363; Wu & Liu, 2012, pp. 177 – 178), Gansu(H. Magn., 1940, p. 41 *as C. substellatum*; Degel., 1974, pp. 47, 49).

var. vulgare(Schaer.) Degel. 普通变种

Symb. bot. upsal. 13(2): 163(1954). f. vulgare

≡*Arthonia vulgaris* Schaer., *Lich. helv. spicil.* 1: 8(1823).

≡*Parmelia pulposum* a. *vulgare* Schaer.

Neimenggu, Ningxia, Hebei(Wu & Liu, 2012, pp. 182 – 183), Xinjiang(Abdulla Abbas et al., 1994, p. 21; Abdulla Abbas & Wu, 1998, p. 63; Guo, 2005, p. 47; Liu & J. C. Wei, 2009, p. 25; Wu & Liu, 2012, pp. 182 – 183), Xizang(Degel. 1974, p. 49).

f. papulosum(Schaer.) Degel. 多泡变型

Symb. bot. upsal. 13(2): 163(1954).

≡*Collema pulposum* f. *papulosum* Schaer.: Zahlbr. Cat. Lich. Univ. 3: 98(1924) (ut "f. *papulosum* Mass.").

≡*C. pulposum* var. *vulgare* f. *papulosum* Schaer.

Beijing(Mt. Baihua shan in Beijing was mistaken for in Hebei by Wu & Liu, 2012, p. 184), Xinjiang(Guo, 2005, p. 47; Liu & J. C. Wei, 2009, p. 25; Wu & Liu, 2012, p. 184).

Collema texanum Tuck. 得州胶衣

Amer. J. Sci. Arts, Ser. 2 28: 200(1859). var. texanum 原变种

On the ground.

Yunnan(Wu & Liu, 2010, p. 187), Zhejiang(Degel. 1974, p. 57).

Collema thamnodes Tuck. 灌丛胶衣

in Riddle, *Bull. Torrey bot. Club* 43: 155(1916).

On rocks.

Xinjiang(Abdulla Abbas et al., 1993, p. 77; Abdulla Abbas & Wu, 1998, p. 64; Abdulla Abbas et al., 2001, p. 363).

Collema undulatum Laurer ex Flot. 波缘胶衣

Linnaea 23: 161(1850). var. undulatum 原变种

On soil.

Xinjiang(Wu & Liu, 2012, pp195 – 196).

var. granulosum Degel. 颗粒变种

Symb. bot. upsal. 13(2): 369(1954).

On rocks and soil.

Jilin, Neimenggu, Shaanxi, Guangxi(Wu & Liu, 2012, p. 196).

Collemopsidium Nyl. **类胶属**

Flora, Regensburg 64: 6(1881).

Collemopsidium halodytes(Nyl.) Grube & B. D. Ryan 盐类胶

Lichen Flora of the Greater Sonoran Desert Region(Tempe) 1: 163(2002).

≡*Verrucaria halodytes* Nyl., *Mém. Soc. Imp. Sci. Nat. Cherbourg* 5: 142(1857).

see Nash et al. (2002) for a discussion of this species.

On littoral raised coral reef.

Taiwan(Aptroot 2003, p. 159).

Enchylium(Ach.) Gray **土耳衣属**

Nat. Arr. Brit. Pl. (London) 1: 396(1821).

Enchylium limosum(Ach.) Otálora, P. M. Jørg. & Wedin 土耳衣

Fungal Diversity 64: 286(2014).

≡*Lichen limosus* Ach., *Lich. suec. prodr.* (Linköping): 126(1799)[1798].

≡*Collema limosum*(Ach.) Ach. *Lich. univ.*: 629(1810).

= *Collema glaucescens* Hoffm., *Deutschl. Fl.*, *Zweiter Theil*(Erlangen): 100(1796)[1795].

On bare earth.

Shanghai(Nyl. & Cromb. 1883, p. 62; Degel. 1974, p. 52).

Lathagrium(Ach.) Gray **胶耳衣属**

Nat. Arr. Brit. Pl. (London) 1: 399(1821).

Lathagrium auriforme(With.) Otálora 胶耳衣

P. M. Jørg. & Wedin, *Fungal Diversity* 64: 287(2014).

≡*Riccia auriformis* With., *Bot. arr. veg. Gr. Brit.* (London) 1: 704(1776).

≡*Collema auriforme*(With.) Coppins & J. R. Laundon, in Laundon, *Lichenologist* 16(3): 228(1984).

On mosses, rocks, and soil.

Neimenggu, Xinjiang, Qqinghai, Hebei, Henan(Wu & Liu, 2012, pp. 194 – 195).

Leptogium(Ach.) Gray **猫耳衣属**

Nat. Arr. Brit. Pl. (London) 1: 400(1821).

≡*Collema* subdiv. *Leptogium* Ach., *Lich. univ.*: 654(1810).

Leptogium arisanense Asahina 阿里山猫耳衣

Journ. Jap. Bot. 12(4): 252(1936).

Type: Taiwan, Asahina no. 195.

Leptogium delavayi sensu Zahlbr. in Fedde, *Repertorium*, 33: 28(1933), non Hue.

On mosses.

Shaanxi(Wang Hui-Yan et al., 2010, pp. 164 – 165.), Taiwan(Zahlbr., 1940, p. 249; Wang & Lai, 1973, p. 91; Ikoma, 1983, p. 66).

Leptogium asiaticum P. M. Jørg. 亚洲猫耳衣

Herzogia 2: 466(1973).

On *Rhus verniciflua*.

ningxia(Liu Huajie et al., 2010, p. 197), Hunan(Zahlbr., 1930 b, p. 78), Fujian(Wu J. N. et al. 1984, p. 2).

Leptogium austroamericanum(Malme) C. W. Dodge 南美猫耳衣

Ann. Mo. bot. Gdn 20: 419(1933).

≡*Leptogium cyanescens* var. *austroamericanum* Malme, *Ark. Bot*. 19(no. 8): 21(1924).

On rocks.

Yunnan, Zhejiang, Guangxi, Guangdong(Liu H. J. et al., 2012, p. 484), Taiwan(Aptroot, Sparrius & Lai 2002, p. 285).

Leptogium azureum(Sw.) Mont. 兰天猫耳衣

in Webb & Berthelot, *Hist. nat. Iles Canar*. (Paris) 3(2): 129(1840).

≡*Lichen azureus* Sw., in Acharius, *Lich. suec. prodr*. (Linköping): 137(1799) [1798].

= *Leptogium tremelloides* f. *azureum*(Sw.) Nyl.

On bark of *Rhus verniciflua*.

Sichuan(Chen S. F. et al., 1997, p. 221), guizhou(Zhang T & J. C. Wei, 2006, p. 8), Xizang(J. C. Wei & Jiang, 1986, p. 26), Hunan(Zahlbr., 1930 b, p. 77), Anhui, Zhejiang(Xu B. S., 1989, p. 196), Fujian(Wu J. N. et al. 1984, p. 4), Taiwan(Wang & Lai, 1976, p. 227), Hongkong(Thrower, 1988, p. 116).

Leptogium brebissonii Mont. 粗糙猫耳衣

in Webb & Berthelot, *Hist. nat. Iles Canar*. (Paris) 3(2): 130(1840).

Xinjiang(Wang XL, 1985, p. 336, as *Leptogium brebianonii* Mont. apud Webb.).

Leptogium burgessii(L.) Mont. 伯吉氏猫耳衣

Hist. Nat. Iles Canar. 3(2): 130(1840).

Corticolous.

Yunnan, Sichuan(Liu H. J. et al., 2015, pp. 475 & 477.).

Leptogium burnetiae C. W. Dodge 伯内氏猫耳衣

Beih. Nova Hedwigia 12: 120(1964).

On bark of trees.

ningxia(Liu Huajie et al., 2010, p. 197), Sichuan(Chen S. F. et al., 1997, p. 222; Wang Hui-Yan et al., 2010, p. 165.), Yunnan(Wang Hui-Yan et al., 2010, p. 165.), Taiwan(Aptroot, Sparrius & Lai 2002, pp. 285 – 286).

Leptogium capense P. M. Jorg. & A. K. Wallace 内含猫耳衣

in Jorgensen, *Symb. Bot. Upsal*. 32(1): 113(1997).

On tree bark.

Yunnan(Xi MQ & Liu HJ, 2014, pp. 71 – 72).

Leptogium chloromelum(Ach.) Nyl. 绿猫耳衣

Mém. Soc. Imp. Sci. Nat. Cherbourg 5: 333(1857).

≡*Lichen chloromelus* Ach

≡*Parmelia chloromela* Ach.

Taiwan(Zahlbr., 1933 c, p. 28; Wang & Lai, 1973, p. 91).

Leptogium cochleatum(Dicks.) P. M. Jørg. & P. James 螺壳猫耳衣

Lichenologist 15(2): 113(1983).

≡*Lichen cochleatus* Dicks., *Fasc. pl. crypt. Brit*. (London) 1: 13(1785).

On bark of trees.

Sichuan(Chen S. F. et al., 1997, p. 222).

Leptogium corniculatum(Hoffm.) Minks 小角猫耳衣

Flora, Regensburg 35: 353(1873).

≡*Collema corniculatum* Hoffm., *Deutschl. Fl.*, *Zweiter Theil* (Erlangen): 105(1796) [1795].

= *Collema palmatum*(Huds.) Ach.

≡*Leptogium palmatum*(Huds.) Mont. in Webb & Berth.

On moss-covered rocks.

Heilongjiang(Chen et al. 1981, p. 150) Shaanxi(Jatta, 1902, p. 480; Zahlbr. , 1930 b, p. 77) .

Leptogium corticola(Taylor) Tuck. 树皮猫耳衣

in Lea, *Cat. Pl. Cincinnati*: 47(1849) .

≡*Collema corticola* Taylor, *J. Bot.* , *Lond.* 6: 195(1847) .

Misapplied name(revised by Aptroot & Seaward, 1999, p. 81) :

Leptogium azureum auct non(Sw. ex Ach.) Mont. in Thrower, 1988, pp. 16, 116.

On shaded trees and on shaded granite rock. On *Pseudotsuga wilsoniana*. Pantropical.

Taiwan(Aptroot, Sparrius & Lai 2002, p. 286) , Hongkong(Thrower, 1988, pp. 16, 116; Aptroot & Seaward, 1999, p. 81; Aptroot et al. , 2001, p. 330) .

Leptogium cyanescens(Pers.) Körb. 变兰猫耳衣

Syst. lich. germ. (Breslau) : 420(1855) .

≡*Lichen cyanescens* Pers. , Ann. Bot. (Usteri) 1: 17(1794) .

≡Collema cyanescens Rabenh. , Deutschl. Krypt. -Fl. (Leipzig) 2: 50(1845) .

Misapplied name:

Leptogium tremeloides auct. non Weis

On bark, mosses and rotten logs.

Heilongjiang(Chen et al. 1981, p. 150) , Xinjiang(Wang XL, 1985, p. 336; Abbas et al. , 1997, p. 3) , Sichuan, Hunan(Zahlbr. , 1930b, p. 77) , Anhui, Zhejiang(Xu B. S. , 1989, p. 196) , Fujian(Zahlbr. , 1930b, p. 77; Wu J. N. et al. 1984, p. 3, on both bark and rocks) , Taiwan(Zahlbr. , 1933, p. 28; Asahina, 1932, p. 28 & 210 & 1936 b, p. 250; Asahina & Sato in Asahina, 1939, p. 639; Wang & Lai, 1973, p. 91) , Hongkong(Thrower, 1988, pp. 16, 117; Aptroot & Seaward, 1999, p. 810; Aptroot et al. , 2001, p. 331; Seaward & Aptroot, 2005, p. 285, On rocks in shady ravines. , Pfister 1978: 68, as *L. tremelloides.*) .

Note: *Leptogium tremeloides* Weis = *Leptogium lichenoides*(L.) Zahlbr. (R. Sant. , 1984, p. 196) .

Leptogium delavayi Hue 戴氏猫耳衣

Bull. Soc. Bot. France 36: 158(1889) & *Nouv. Archiv. du Mus.* 3(2) : 235(1890) . f. delavayi

Type: Yunnan, alt. 2200 m, 28/VII/1886, Delavay no. 2405 & 6/VI/1887(syntypes) .

Yunnan(Nyl. 1889, p. 45; Zahlbr. , 1925, p. 176 & 1930 b, p. 77 & 1934 b, p. 198; Wang L. S. et al. 2008, p. 537) .

Corticolous.

Shaanxi(He & Chen 1995, p. 43, as f. *delavayi* without ciatation of specimens and their localities) , Sichuan (Zahlbr. , 1930 b, p. 77 & 1934 b, p. 198) , Anhui(Xu B. S. , 1989, p. 196) .

f. fuliginosulum Zahlbr. 乌色变型

in Fedde, *Repertorium* 33: 28(1933) .

Type: Taiwan, Asahina no. 197.

Neimenggu(Sun LY et al. , 2000, p. 36) , Shaanxi(He & Chen 1995, p. 43 without citation of specimen and locality) , Xizang(J. C. Wei & Y. M. Jiang, 1986, p. 28) , Taiwan(Zahlbr. , 1940, p. 251; Asahina, 1935 d, p. 550; Wang & Lai, 1973, p. 91) .

Leptogium denticulatum Nyl. 齿裂猫耳衣

Annls Sci. Nat. , *Bot.* , sér. 5 7: 302(1867) .

On rock, occasionally on trees. On *Castanopsis* sp. Cosmopolitan.

Taiwan(Aptroot, Sparrius & Lai 2002, p. 286) , Hongkong(Thrower, 1988, pp. 16, 118; Aptroot & Seaward, 1999, p. 82; Aptroot et al. , 2001, p. 331) ,

Leptogium furfuraceum(Harm.) Sierk 皱表猫耳衣

Bryologist 67: 266(1964) .

≡*Leptogium hildenbrandii* f. *furfuraceum* Harm. , *Lich. Fr.* 1: 118(1905) .

On bark of trees.

Yunnan(Wang Hui-Yan et al. , 2010, p. 162.) .

Leptogium hibernicum M. E. Mitch. ex P. M. Jørg. 沟表猫耳衣

Herzogia 2: 462(1973) .

Leptogium hibernicum M. E. Mitch. , *Bull. Soc. sci. Bretagne* 37: 119(1964) [1962] . (Nom. inval. , [type not designated]) .

On bark of trees.

Shaanxi(Wang Hui-Yan et al. , 2010, pp. 163 – 164.) .

Leptogium hildenbrandii(Garov.) Nyl. 裸果猫耳衣

Act. Soc. linn. Bordeaux 21: 272(1856) .

≡*Collema hildenbrandii* Garov. , *Lich. Prov. Comens*. 1: 3(1837) .

On bark of tree.

Xizang(J. C. Wei & Y. M. Jiang, 1986, p. 271) , Anhui, Zhejiang(Xu B. S. , 1989, p. 196) .

Leptogium hirsutum Sierk 多毛猫耳衣

Bryologist, 67: 267(1964) .

Shaanxi(He & Chen 1995, p. 43 without ciatation of specimens and their localities) , Hubei(Chen, Wu & J. C. Wei, 1989, p. 421) .

Leptogium javanicum(Mont. & Bosch) Mont. 爪哇猫耳衣

Gatlung Asterina: 379(1856) .

≡*Stephanophorus javanicus* Mont. & Bosch, *Pl. Jungh*. 4: 492(1855) .

Cornicolous.

Yunnan(Liu Huajie et al. , 2012, p. 486) .

Leptogium laceroides B. de Lesd. 撕裂猫耳衣

Annals Cryptog. Exot. 6(2) : 112(1933) .

On *Castanopsis* and *Tsuga formosana*.

Taiwan(Aptroot, Sparrius & Lai 2002, p. 286) .

Leptogium lichenoides(L.) Zahlbr. 薄猫耳衣

Cat. Lich. Univers. 3: 136(1924) [1925] . f. lichenoides 原变型

≡*Tremella lichenoides* L. , *Sp. pl*. 2: 1157(1753) .

On mosses.

Xinjiang(H. Magn. 1940, p. 42) .

f. fimbriatum(Ach.) Zahlbr. 流苏变型

Cat. Lich. Univers. 3: 140(1924) [1925] .

≡*Parmelia lacera* var. *fimbriata* Ach. , *Methodus*, Sectio post. : 226(1803) .

≡*Leptogium lacera* f. *fimbriatum*(Ach.) Hoffm.

Shaanxi(Jatta, 1902, p. 481; Zahlbr. , 1930 b, p. 77) .

var. lophaeum(Ach.) Zahlbr. 脊梁变种

Cat. Lich. Univers. 3: 140(1924) [1925] .

≡*Parmelia scotina* var. *lophaea* Ach. , *Methodus*, Sectio post. : 238(1803) .

≡*Leptogium lacerum* var. *lophaeum* Ach.

On mosses.

Shaanxi(Jatta, 1902, p. 481; Zahlbr. , 1930 b, p. 77) .

Leptogium marginellum(Sw.) Gray 具缘猫耳衣

Nat. Arr. Brit. Pl. (London) 1: 401(1821) .

≡*Lichen marginellus* Sw. , *Prodr*. : 147(1788) .

On *Delonix regia*.

Taiwan(Aptroot, Sparrius & Lai 2002, p. 286) .

Leptogium menziesii(Sm.) Mont. 猫耳衣

Annls Sci. Nat. , *Bot.* , sér. 3 18: 313(1852) . var. menziesii 原变种

≡*Lichen menziesii* Sm. , in Acharius, *Methodus*, Sectio post. : 212(1803) .

Corticolous and saxicolous or muscicolous.

Jilin(Moreau et Moreau, 1951, p. 185) , Shaanxi(Jatta, 1902, p. 481; Zahlbr. , 1930 b, p. 78) , Gansu(H. Mang. 1940, p. 42) , Ningxia(Liu Huajie et al. , 2010, pp. 197 – 198) , Sichuan(Chen S. F. et al. , 1997, p. 222) , Yunnan(Hue, 1889, p. 158 & 1898, p. 229; Zahlbr. , 1930 b, p. 78) , Xizang(Wei & Jiang, 1986, p. 26) , Anhui, Zhejiang(Xu B. S. , 1989, p. 196 as f. menzisii) , Fujian(Wu J. N. et al. 1984, p. 2) , Taiwan(Wang & Lai, 1976, p. 227) , China(prov. not indicated: Hue, 1890, p. 235; Asahina & M. Sato in Asahina, 1939, p. 641) .

f. fuliginosum Müll. Arg. 乌色变型

Flora, Regensburg 72: 60(1889) .

= *Leptogium burnetiae* Dodge

Corticolous and muscicolous.

Shaanxi(Jatta, 1902, p. 481; Zahlbr. , 1930 b, p. 78) , Sichuan(Hue, 1898, p. 230) , Yunnan(Hue, 1889, p. 158 & 1898, p. 230; Zahlbr. , 1930 b, p. 78) , Hubei(Müll. Arg. 1893, p. 235; Zahlbr. , 1930 b, p. 78) , Anhui(Xu B. S. , 1989, p. 197 on both tree bark and rocks) , Fujian(Zahlbr. , 1930 b, p. 78; Wu J. N. et al. 1984, p. 2) .

f. saxicolum Moreau et Moreau 石生变型

Rev. Bryol. et Lichenol. 20: 185(1951) . (nom. inval. : Art. 36, in lingua Latina haud descriptum) .

Leptogium menziesii f. *saxicola* C. Moreau & M. Moreau [as ' saxicolum'] , (1951) .

Type: from Zhejiang.

On rock.

Notes: 1. "-cola" is used adjectivally in such comp. as ruricola(dwelling in the country) but then treated as a noun in apposition the same for all genders even though the genus is masculine or neuter despite the use by some authors of-colus, -cola, -colum as adjectival endings. 2. This invalid name may probably be a superfluous one of the f. *fuliginosum* Müll. Arg.

Leptogium moluccanum(Pers.) Vain. 薄刃猫耳衣

Acta Soc. Fauna Flora fenn. 7(1) : 223(1890) .

≡*Thelephora moluccana* Pers. , in Gaudichaud-Beaupré in Freycinet, *Voy. Uranie.* , *Bot.* : 175(1827) [1826 – 1830] .

≡*Collema moluccanum* Pers. apud Gaudich.

On mosses.

Fujian(Zahlbr. , 1930 b, p. 77) , Taiwan(Asahina, 1932 a, p. 210; Zahlbr. , 1933 c, p. 28) .

var. moluccanum 原变种

Taiwan(Wang & Lai, 1973, p. 91) .

var. myriophyllinum(Müll. Arg.) Asahina 多叶变种

("*mycriophyllinum*") in *J. J. B.* 8: 28(1932) .

≡*Leptogium tremelloides* var. *myriophyllinum* Müll. Arg. , *Hedwigia* 30: 181(1924) .

On mosses and bark of trees.

Anhui, Zhejiang, Shanghai(Xu B. S. , 1989, p. 197) , Taiwan(Asahina, 1932 a, p. 210 & 1932 b, p. 28; Asahina & Sato in Asahina, 1939, p. 639; Wang & Lai, 1973, p. 91) .

Leptogium papillosum(B. de Lesd.) C. W. Dodge 棒芽猫耳衣

Ann. Mo. bot. Gdn 20: 418(1933) .

≡*Leptogium hildenbrandii* var. *papillosum* B. de Lesd. , *Lich. Mexique*: 30(1914) .

On rocks.

Anhui(Wang Hui-Yan et al. 2010, p. 164.) .

Leptogium pedicellatum P. M. Jög. 花梗猫耳衣

Herzogia 3: 448(1975) .

Sichuan(Etenroos et al. , 1994, p. 334) . (Wang LS 83 – 1901: KUN(HKAS – 5633 as *L. trichophorum* was confirmed by P. M. Jög. in 1989)

Leptogium phyllocarpum(Pers.) Mont. 叶果猫耳衣

Annls Sci. Nat. , *Bot.* , sér. 3 10: 134(1848) .

≡*Collema phyllocarpum* Pers. , in Gaudichaud-Beaupré in Freycinet, *Voy. Uranie.* , *Bot.* : 204(1827) [1826 – 1830] .

On *Castanopsis* etc.

Taiwan(Aptroot, Sparrius & Lai 2002, p. 286) .

Leptogium pichneum(Ach.) Nyl. 青猫耳衣

Acta Soc. Sci. fenn. 26(10) : 3(1900) .

≡*Collema tremelloides* var. *pichneum* Ach. , *Syn. meth. lich.* (Lund) : 343(1814) .

On bark of trees.

Taiwan(Zahlbr. , 1933 c, p. 28; Wang & Lai, 1973, p. 91) .

Leptogium pseudofurfuraceum P. M. Jorg. & A. K. Wallace 拟鳞粉猫耳衣

in Jorgensen, Symb. Bot. Upsal. 32(1) : 113(1997) .

On tree bark.

Yunnan(Xi MQ & Liu HJ, 2014, pp. 71 – 72) .

Leptogium pseudopapillosum P. M. Jørg. 珊瑚猫耳衣

Symb. bot. upsal. 32(1) : 120(1997) .

Replaced synonym:

≡*Leptogium menziesii* var. *coralloideum* Jatta, *Nuovo G. bot. ital.* nov. ser. 9: 481(1902) .

Non *Leptogium coralloideum*(Meyen & Flot.) Vain. , *Ann. Acad. Sci. fenn.* , Ser. A 6(7) : 110(1915) .

Type: Shaanxi. collected by 1896, G. Giraldi. (FI, lectotype selected by Jørgensen 1973: 457) .

On mosses.

Shaanxi(Jatta, 1902, p. 481; Zahlbr. , 1930 b, p. 78; Jørg. , 1997, p. 120) , Hunan(Jørg. , 1997, p. 123: Handel-Mazetti 2515 in BM) , Taiwan(Jørg. , 1997, p. 123) .

Leptogium saturninum(Dicks.) Nyl. 土星猫耳衣

Act. Soc. linn. Bordeaux 21: 272(1856) .

≡*Lichen saturninus* Dicks. , *Fasc. pl. crypt. brit.* (London) 2: 21(1790) .

[= *Leptogium myochroum*(Bernh.) Nyl.]

= *Leptogium myochroum* var. *myochroum*(Ehrh.) Nyl. , (1879) .

≡*Lichen myochrous* Ehrht. , *Plant. Cryptog. Exsicc.* , no. 286(1793) .

On mosses.

Heilongjiang(Asahina, 1952 d, p. 375) , Neimenggu(Sun LY et al. , 2000, p. 36) , Xinjiang(H. Magn. 1940, p. 42; Moreau et Moreau, 1951, p. 185; Wang XL, 1985, p. 336; Abdula Abbas et al. , 1993, p. 78; Abdula Abbas et al. , 1994, p. 22; Abdulla Abbas et al. , 2001, p. 364) , Ningxia(Liu Huajie et al. , 2010, p. 198) , Shaanxi (Jatta, 1902, p. 481 ; He & Chen 1995, p. 43 without ciatation of specimens and their localities) , Sichuan (Chen S. F. et al. , 1997, p. 222) , Xizang(J. C. Wei & Jiang, 1986, p. 27) , Hubei(Chen, Wu & J. C. Wei, 1989, p. 421) , Anhui, Zhejiang(Xu B. S. , 1989, p. 197) .

Leptogium sessile Vain. 无柄猫耳衣

Ann. Acad. Sci. fenn. , Ser. A 6(7) : 108(1915) .

On *Cryptomeria* sp. and *Populus* sp.

Yunnan, anhui(Liu Huajie et al. , 2012, p. 487) , Taiwan(Aptroot, Sparrius & Lai 2002, p. 287) .

Leptogium splendens Asahina 闪光猫耳衣

J. Jap. Bot. 12(4) : 253(1936) .

Type: Taiwan, Asahina no. 196.

Leptogium tremelloides sensu Zahlbr. in Fedde, *Repertorium*, 33: 28(1933) , non S. F. Gray.

Fujian(Wu J. N. et al. 1984, p. 4) , Taiwan(Zahlbr. , 1933 c, p. 28; Wang & Lai, 1973, p. 91; Ikoma, 1983, 67: lapsu "sprendens") .

Leptogium taibaiense H. J. Liu & M. Q. Xi 太白猫耳衣

Mycotaxon 130: 472(2015) .

Type: China. Shaanxi Province, Mt. Taibai, Temple Ping' ansi, 34°01′N 107°48′E, elev. 2700 m, on bark, 8/VI/1963, Jiang-Chun Wei 2604 – 1(Holotype, HMAS-L 045039) .

Leptogium trichophoroides P. M. Jørg. & A. K. Wallace 类黑猫耳衣

in Jørgensen, *Symb. bot. upsal.* 32(1) : 123(1997) .

Sichuan, Yunnan, Xizang, Hubei, Fujian, Guangxi(Cao Jing et al. , 2012, pp. 214 – 215) , Taiwan(P. M. Jorgensen, 1997, p. 124) .

Leptogium trichophorum Müll. Arg. 黑猫耳衣

Flora, Regensburg 72: 505(1889) . f. trichophorum

On earth, rocks, and trunk of trees.

Yunnan (Hue, 1898, p. 230; Zahlbr. , 1930 b, p. 78) , Xizang(Wei & Jiang, 1986, p. 27) , Fujian(Wu J. N. et al. 1984, p. 2) , Taiwan(Ashina & M. Sato in Asahina, 1939, p. 641) .

f. fuliginosum Müll. Arg. 乌黑变型

Flora, Regensburg 72: 505(1889) .

= *Leptogium asiaticum* Borg.

f. microsporum Zahlbr. 微孢变型

in Feddes, *Repertorium* sp. nov. 33: 28(1933) .

Type: Taiwan, Asahina nos. 198, 203, 205(syntypes) .

(Wang & Lai, 1973, p. 92; Ikoma, 1983, p. 66) .

Leptogium wangii H. J. Liu & J. S. Hu 王氏猫耳衣

Mycotaxon 130: 473(2015) .

Type: China. Beijing City, Xiaolongmen National Forest Park, 39°57′N 115°26′E, elev. 1100 m, on soil covered rocks, VII/2013, Jian-Bin Chen & Xiao-Di Liu 2013-0714(Holotype, HMAS-L 129380) .

Leptogium weii H. J. Liu & S. Guan 魏氏猫耳衣

Mycotaxon 119: 413 – 415(2012) .

Type: China. Sichuan, Mt. Gonggashan, Zimeishan(29°37′N 101°53′E) , alt. 4000 m, on branch of *Rhododendron* sp. (*Ericaceae*) , 29/VII/1982, Xian-Ye Wang, Xie Xiao & Bin Li 9082(Holotype, HMAS-L 031713) .

Physma A. Massal. **胶囊衣属**

Neagenea Lich: 6(1854) .

Physma byrsaeum(Ach.) Tuck. 拟皮胶囊衣

Syn. N. Amer. Lich. (Boston) 1: 115(1882) .

≡ *Parmelia byrsea* Ach. , *Methodus*, Sectio post. : 222(1803) .

= *Physma byrsinum* f. *hypomelaelum* Hue in Thrower, 1988, pp. 17, 147, given by Aptroot & Seaward, 1999, p. 87) .

Hongkong(Thrower, 1988, pp. 17, 147; Aptroot & Seaward, 1999, p. 87) .

Physma byrsinum(Ach.) Müll. Arg. 皮胶囊衣

Flora, Regensburg 68(28) : 531(1885) . f. byrsinum 原变型

≡*Collema byrsinum* Ach., *Lich. univ.*: 1 – 696(1810).

On bark of trees among mosses.

Taiwan(Zahlbr., 1933 c, p. 26; Asahina & M. Sato in Asahina, 1939, p. 639; Wang & Lai, 1973, p. 95).

f. hypomelaenum(Nyl.) Hue 黑腹变型

Bull. Soc. linn. Normandie, sér. 5 9: 130(1906).

≡*Collema byrsinum* f. *hypomelaenum* Nyl., *Annls Sci. Nat.*, *Bot.*, sér. 4 12: 281(1859).

On bark of a numer of tree species.

Hongkong(Thrower, 1988, p. 147).

Physma callicarpum Hue 美果胶囊衣

Bull. Soc. linn. Normandie, sér. 5 9: 124(1906).

On bark of trees.

Taiwan(Zahlbr., 1925, p. 26 & 1933 c, p. 26; Asahina & Sato in Asahina, 1939, p. 635; Wang & Lai, 1973, p. 95; Ikoma, 1983, p. 91).

Physma hondoanum Asahina 鳞叶胶囊衣

J. J. B. 38: 65(1963).

On *Abies* sp. and *Tsuga* sp.

Taiwan(Jorgensen & Aptroot, 2002, pp. 441 – 442).

Physma radians Vain. 辐射胶囊衣

Ann. Acad. Sci. fenn., Ser. A 6(no. 6): 45(1921).

On bark of trees.

Taiwan(Zahlbr., 1933 c, p. 26; Wang & Lai, 1973, p. 95; Ikoma, 1983, p. 91).

Ramalodium Nyl. **小颖衣属**

in Crombie, *J. Linn. Soc.*, *Bot.* 17: 392(1879).

Ramalodium japonicum(Asahina) Henssen 日本小颖衣

Lichenologist 3: 39(1965).

≡*Leciophysma japonicum* Asahina, in *Miscell. Bryol. et Lichenol.* II, no. 3: 30(1960)(nomen nudum).

On conglomeratic rock.

Taiwan(Aptroot and Sparrius, 2003, p. 38).

Scytinium(Ach.) Gray [as ' Scytenium'] **颈衣属**

Nat. Arr. Brit. Pl. (London) 1: 398(1821).

Scytinium plicatile(Ach.) Otálora, P. M. Jørg. & Wedin 皱皮颈衣

Fungal Diversity 64: 291(2014)

≡*Lichen plicatilis* Ach., *K. Vetensk-Acad. Nya Handl.* 16: 11(1795).

≡*Leptogium plicatile*(Ach.) Leight. *Lich. -Fl. Great Brit.*, Edn 3: 30(1879).

≡*Collema plicatile*(Ach.) Ach., *Lich. univ.*: 635(1810).

On calcareous earth.

Shaanxi(Baroni, 1894, p. 49; Zahlbr., 1930 b, p. 77).

(63) Pannariaceae Tuck. [as ' Pannariei'] **鳞叶衣科**

Gen. lich. (Amherst): xii(1872)

Erioderma Fée **毛面衣属**

Essai Crypt. Exot. (Paris): 145(1825)[1824].

Erioderma asahinae Zahlbr. 东亚毛面衣

Bot. Mag., Tokyo 41: 322, tab. XII, fig. 7(1927).

On trees.

Zhejiang(Xu B. S. 1989, p. 199; Wu JN et al. 1997, p. 62.).

Erioderma meiocarpum Nyl. 小果毛面衣

Syn. meth. lich. (Parisiis) 2: 47, tab. IX, fig. 33(1863).

On trees.

Yunnan, Hubei(Wu et al., 1997, p. 62).

Erioderma tomentosum Hue 卷曲毛面衣

Bull. Soc. bot. Fr. 48: XLIX(1902)[1901].

On *Salix* sp.

Yunnan(Wei X. L. et al. 2007, pp. 151 – 152).

Fuscopannaria P. M. Jørg. **棕鳞衣属**

Abstracts, XV International Botanical Congress, Yokohama, Japan, August 28 - September 3, 1993(Paris): 10 (1993)

Fuscopannaria ahlneri(P. M. Jørg.) P. M. Jørg. 阿氏棕鳞衣

J. Hattori bot. Lab. 76: 205(1994).

≡*Pannaria ahlneri* P. M. Jørg., *Op. bot.* 45: 15(1978).

Hubei, Anhui, Zhejiang, Fujian(Wu JN et al., 1999, p. 88).

Fuscopannaria cheiroloba(Müll. Arg.) P. M. Jørg. 掌芽棕鳞衣

Bryologist 103: 679(2000).

Terricolous, saxicolous, corticolous or on mosses.

Qinghai, Shaanxi, Sichuan, Xizang(Hua-Jie Liu et al., 2016. pp. 458 – 461).

Fuscopannaria coralloidea P. M. Jorg. 珊瑚棕鳞衣

Bryologist 103: 681(2000).

On soil-covered rocks.

Yunnan(Hua-Jie Liu et al., 2016. pp. 461 – 462).

Fuscopannaria dispersa P. M. Jørg. 散棕鳞衣

J. Hattori Bot. Lab. 89: 249(2000).

Type: China, Yunnan, Lijiang(as Likang), Yangtze watershed, eastern slopes of Lijiang(as Likang) Snow Range, 1922, J. F. Rock 11764(US, holotype).

Fuscopannaria leucophaea(Vahl) P. M. Jørg. 暗白棕鳞衣

J. Hattori bot. Lab. 76: 205(1994).

≡*Lichen leucophaeus* Vahl, *Icon. Plant. Dan.* 6(16): 8(1787).

Xinjiang(Abbas et al., 1997, p. 2; Wu JN et al., 1999, p. 88), Xizang(Obermayer W., 2004, p. 496). see also P. M. Jorgensen 2000. Taiwan.

Fuscopannaria leucosticta(Tuck.) P. M. Jørg. 雀斑棕鳞衣

J. Hattori bot. Lab. 76: 205(1994).

≡*Pannaria leucosticta* Tuck., *Annls Sci. Nat., Bot.*, sér. 4 12: 294(1859).

On *Rhododendron* or on mosses.

Heilongjiang(as Manchuria: Jørg., 2000, p. 252), Shaanxi, Anhui, Zhejiang, Fujian, Guangdong(Wu JN et al., 1999, p. 89), Xizang(Jørg., 2000, p. 252; Obermayer W., 2004, p. 496), Guizhou(Zhang T & J. C. Wei, 2006, p. 7), Yunnan(Zahlbr., 1930 b, p. 80), Anhui, Zhejiang(Xu B. S., 1989, p. 200), Taiwan (Zahlbr., 1933 c, p. 29; Wang & Lai, 1973, p. 93).

Fuscopannaria poeltii(P. M. Jørg.) P. M. Jørg. 珀氏棕鳞衣

J. Hattori bot. Lab. 76: 205(1994).

≡*Pannaria poeltii* P. M. Jørg., *Op. bot.* 45: 115(1978).

Sichuan(Obermayer W., 2004, p. 496). see also P. M. Jorgensen 2000.

Fuscopannaria praetermissa(Nyl.) P. M. Jørg. 遗漏棕鳞衣

J. Hattori bot. Lab. 76: 205(1994).

≡*Pannaria praetermissa* Nyl., *Annls Sci. Nat., Bot.*, sér. 4 12: 295(1859).

≡*Parmeliella praetermissa*(Nyl.) P. James, *Lichenologist* 3: 98(1965) .
On mosses.
Shaanxi(Jatta, 1902, p. 474; Zahlbr. , 1930 b, p. 79) , Xizang(Jørg. , 2000, pp. 253 – 254) , Yunnan(Wu JN et al. , 1999, p. 89) .

Fuscopannaria protensa(Hue) P. M. Jorg. 长棕鳞衣
J. Hattori Bot. Lab. 76: 205(1994) .
Corticolous.
Yunnan(Hua-Jie Liu et al. , 2016. pp. 464) .

Fuscopannaria rugosa H. J. Liu & J. S. Hu 皱棕鳞衣
Mycotaxon 131: 456 – 458(2016) .
Type: China. Hubei Province, Mt. Shennongjia, 31°30′N 110°16′E, alt. 2540 m, on bark, 30/VII/1984, Jian-Bin Chen 11059(Holotype, HMAS-L 098090) .
Corticolous.
Hubei, Guangxi(Hua-Jie Liu et al. , 2016. pp. 456 – 458) .

Fuscopannaria saltuensis P. M. Jørg. 森林棕鳞衣
J. Hattori bot. Lab. 89: 255(2000) .
Type: China, Sichuan, Obermayer3344(GZU, holotype"as holtype") .
On moss-covered rock
Xizang, Sichuan(Jørg. , 2000, p. 255; Obermayer W. , 2004, p. 496) .

Fuscopannaria sorediata P. M. Jørg. 粉芽棕鳞衣
Bryologist 103(1) : 105(2000) .
Xizang(Jørg. , 2000, p. 257; Obermayer W. , 2004, p. 496) .

Kroswia P. M. Jørg. **苞衣属**
Lichenologist 34(4) : 297(2002) .

Kroswia crystallifera P. M. Jorg. 晶体苞衣
Lichenologist 34(4) : 299(2002) .
≡*Fuscopannaria crystallifera*(P. M. Jorg.) Magain & Serus. , *Lichenologist* 47(1) : 39(2015) .
On barks, shrubs, and rotten branches.
Sichuan, Guizhou, Anhui, Hubei, Zhejiang(Liu H. J. et al. 2015, p. 955) .

Kroswia epispora H. J. Liu & Chao Li 周壁孢苞衣
in Liu, Hu & Li, *Mycotaxon* 130(4) : 952(2016) [2015] .
Type: China. Zhejiang Province, Mt. Jiulongshan, 28°22 ¢ N 118°52 ¢ E, alt. 1600 m, on bark, 8/V/1987, Shu-Fan Chen Z0923(Holotype, NJNU) .
Sichuan, Zhejiang(Liu H. J. et al. 2015, p. 952) .

Kroswia gemmascens(Nyl.) P. M. Jorg. 芽苞衣
Lichenologist 34(4) : 302(2002) .
≡*Pannaria gemmascens* Nyl. , *Lich. Japon.* : 36(1890) .
≡*Physma gemmascens* (Nyl.) Asahina, *Journ. Jap. Bot.* 38: 66(1963) .
= *Physma pergranulatum* Zahlbr. , *Symb. Sin.* 3: 75(1930) .
Type: On the top of Dunghua shan between Ninghua(FUJIAN) and Shicheng(Jiangxi) .
On rocks.
Anhui(Xu B. S. , 1989, p. 198) .

Leioderma Nyl. **平皮衣属**
Lich. Nov. Zeland. (Paris) : 47(1888) .

Leioderma sorediatum D. J. Galloway & P. M. Jørg. 粉芽平皮衣
Lichenologist 19(4) : 390(1987) .

On *Shiia calesii*.

Taiwan(Aptroot, Sparrius & Lai 2002, p. 285).

Pannaria Delise ex Bory **鳞叶衣属**

Dict. Class. Hist. Nat. 13: 20(1828).

Pannaria adpressa Zahlbr. 贴鳞叶衣

in Handel-Mazzetti, *Symb. Sin.* 3: 79(1930) & *Cat. Lich. Univ.* 8: 300(1932).

Type: Yunnan, Handel-Mazzetti no. 7150.

On arenaceous rocks.

Pannaria conoplea(Ach.) Bory 绵毛鳞叶衣

Dict. Class. Hist. Nat. 13: 20(1828).

≡*Parmelia conoplea* Ach., *Lich. univ.*: 467(1810).

=*Pannaria pityrea* sensu auct.; fide Checklist of Lichens of Great Britain and Ireland(2002).

=*Pannaria lanuginosa* sensu Szatala, non(Ach.) Körb.,

=*Pannaria coeruleobadia*(Schleich.) A. Massal., *Ric. auton. lich. crost.* (Verona): 111, fig. 219(1852).

On bark of trees.

Xinjiang(Abbas et al., 1997, p. 3), Sichuan(Wu JN et al., 1999, p. 86; Chen S. F. et al., 1997, p. 222; W. Obermayer, 2004, p. 505), Xizang(W. Obermayer, 2004, p. 505), Shandong(Zhao ZT et al., 1999, p. 428), Hubei(Wu JN et al., 1999, p. 86), Hunan(Wu J. L. 1987, p. 65, On rocks or tree trunk; Wu JN et al., 1999, p. 86), Jiangsu, Fujian, Guangdong(Wu J. L. 1987, p. 65, On rocks or tree trunk), Anhui, Zhejiang(Wu J. L. 1987, p. 65, On rocks or tree trunk; Xu B. S., 1989, p. 200; Wu JN et al., 1999, p. 86), Fujian(Wu JN et al., 1999, p. 86), Taiwan(Zahlbr., 1933 c, p. 29; Wang & Lai, 1973, p. 93).

Pannaria emodii P. M. Jørg. 埃默氏鳞叶衣

Lichenologist 33(4): 298(2001).

On *Rhodo- Dendron* & *Salix*.

Sichuan, Xizang(W. Obermayer, 2004, p. 505).

Pannaria formosana P. M. Jørg. 台湾麟叶衣

*Lichenologist*33: 301(2001).

Type: Taiwan, Taidong Pref.., alt. 400 – 1200m, 11 feb. 1965, Syo Kurok. 2965

(TNS-holotype; BG-isotype)

Pannaria gemmascens Nyl. 芽鳞叶衣

Lich. Japon.: 36(1890).

≡*Physma gemmascens*(Nyl.) Asahina, *J. Jap. Bot.* 38: 66(1963).

On trees or on soil, moss covered rocks.

Anhui, Zhejiang(Xu B. S., 1989, p. 197).

Pannaria leucophaea(Vahl) P. M. Jørg. 小鳞叶衣

Op. bot. 45: 38(1978).

≡*Lichen leucophaeus* Vahl, *Icon. Plant. Dan.* 6(16): 8(1787).

=*Lichen microphyllus* Sw. (nom. nud.)

[≡*Parmeliella microphylla*(Sw.) Müll. Arg.]

[≡*Pannaria microphylla* auct.]

On arenaceous rocks.

Taiwan(Zahlbr., 1933 c, p. 29; Wang & Lai, 1973, p. 94 & 1976, p. 228).

Pannaria lurida(Mont.) Nyl. 铁色鳞叶衣

Mém. Soc. Imp. Sci. Nat. Cherbourg 5: 109(1857).

≡*Collema luridum* Mont., *Annls Sci. Nat., Bot.*, sér. 2 18: 266(1842).

On bark of trees.

Sichuan(Chen S. F. et al. , 1997, p. 222; Wu JN et al. , 1999, p. 86) , Xizang(W. Obermayer, 2004, p. 505) , Hubei(Chen, Wu & Wei, 1989, p. 423; Wu JN et al. , 1999, p. 86) , Shandong(Zhao ZT et al. , 1999, p. 428) , Anhui(Xu B. S. , 1989, p. 200; Wu JN et al. , 1999, p. 86) , Jiangsu, Zhejiang(Xu B. S. , 1989, p. 200; Wu JN et al. , 1999, p. 86) , Guangdong(Wu JN et al. , 1999, p. 86) , Taiwan(Zahlbr. , 1933 c, p. 29; Wang & Lai, 1973, p. 93) , China(prov. not indicated: Asahina & Sato in Asahina, 1939, p. 645) .

Pannaria mariana(Fr.) Müll. Arg. 玛丽鳞叶衣

Flora, Regensburg 70: 321(1887) .

≡*Parmelia mariana* Fr. , *Syst. orb. veg.* (Lundae) 1: 284(1825) .

On bark of trees.

Shaanxi(He & Chen 1995, p. 44 without citation of specimens and their localities) , Hubei(Chen, Wu & J. C. Wei, 1989, p. 423) , Zhejiang Xu B. S. , 1989, p. 200) , Guangxi, Hainan(Wu JN et al. , 1999, p. 87) , Taiwan (Räs. 1940, p. 145; Wang & Lai, 1973, p. 93) . haanxi

Pannaria pezizoides(Weber) Trevis. 类盘鳞叶衣

Lichenoth. Veneta 3 – 4: no. 98(1869) .

≡*Lichen pezizoides* Weber, *Spicil. fl. goetting.* : 200(1778) .

Xinjiang(Wang XY, 1985, p. 339; Wu JN et al. , 1999, p. 87; Abdulla Abbas & Wu, 1998, p. 87; Abdulla Abbas et al. , 2001, p. 365) .

Pannaria rubiginosa(Thunb. ex Ach.) Delise 锈红鳞叶衣

in Bory, *Dict. Class. Hist. Nat.* 13: 20(1828) .

≡*Lichen rubiginosus* Thunb. ex Ach. , *Lich. suec. prodr.* (Linköping) : 99(1799) [1798] .

On bark of trees.

Yunnan(Zahlbr. , 1930 b, p. 80) , Anhui(Xu B. S. , 1989, p. 200; Wu JN et al. , 1999, p. 87) , Guizhou, Hubei, Zhejiang, Fujian(Wu JN et al. , 1999, p. 87) .

Pannaria stylophora Vain. 柱鳞叶衣

Suomal. Tiedeakat. , *Toim.* A 6(7) : 102(1917) .

guangxi(Wu JN et al. , 1999, p. 88) .

Pannaria stylophora var. perconfluens Vain. 汇鳞变种

Ann. Acad. Sci. fenn. , Ser. A 15(6) : 10(1921) .

On bark of trees on schistose rocks.

Taiwan(Zahlbr. , 1930 b, p. 29; Wang & Lai, 1973, p. 93) .

Parmeliella Müll. Arg. **甲叶属**

Mém. Soc. Phys. Hist. nat. Genève 16(2) : 376(1862) .

Parmeliella grisea(Hue) Kurok. 灰甲叶

in Misc. Rept. Res. Inst. Nat. Resources, no. 46 – 47: 49(1958) .

≡*Placynthium griseum* Hue, in Bull. *Soc. Linn. Normand.* , ser. 5, 9: 159(1906) . (1925) .

Fujian(Wu JN et al. , 1999, p. 90) .

Parmeliella incisa Müll. Arg. 鳞甲叶

Nuovo G. bot. ital. 24: 194(1958) .

xinjiang(Wu JN et al. , 1999, p. 90; Abdulla Abbas & Wu, 1998, p. 88; Abdulla Abbas et al. , 2001, p. 365) , fujian(Wu JN et al. , 1999, p. 90) , Hubei(Chen, Wu & J. C. Wei, 1989, pp. 423 – 424) .

Parmeliella mariana(Fr.) P. M. Jørg. & D. J. Galloway 玛丽甲叶

Flora of Australia(Melbourne) 54: 316(1992) .

≡*Parmelia mariana* Fr. , *Syst. orb. veg.* (Lundae) 1: 284(1825) .

On bark.

Guizhou(Zhang T & J. C. Wei, 2006, p. 9)

Parmeliella stylophora(Vain.) P. M. Jørg. 柱甲叶

Bryologist 103(4) : 698(2000) .

≡*Pannaria stylophora* Vain. , *Suomal. Tiedeakat.* , *Toim.* A 6(no. 7) : 102(1917) .

On *Salix* sp.

Yunnan(Wei X. L. et al. 2007, p. 156) .

Protopannaria(Gyeln.) P. M. Jørg. & S. Ekman **原鳞衣属**

in Jørgensen, *Bryologist* 103(4) : 699(2000) .

≡*Pannaria* subgen. *Protopannaria* Gyeln.

Protopannaria pezizoides(Weber ex F. H. Wigg.) P. M. Jørg. & S. Ekman 类盘原鳞衣

in Jørgensen, *Bryologist* 103(4) : 699(2000) .

≡*Pannaria pezizoides*(Weber ex F. H. Wigg.) Trevis.

On mosses.

Shaanxi(Ren Q et al. , 2009, p. 104) .

Psoroma Ach. ex Michx. **鳞藓衣属**

Fl. Boreali-Americ. 2: 321(1803) .

Psoroma sinense Zahlbr. 中国鳞藓衣

in Handel-Mazzetti, *Symb. Sin.* 3: 82(1930) & *Cat. Lich. Univ.* 8: 303(1932) .

Calcicolous lichen.

Type: Yunnan, Handel-Mazzetti no. 123.

Psoroma sphinctrinum(Mont.) Nyl. 粉苔鳞藓衣

Annls Sci. Nat. , *Bot.* , sér. 4 3: 181(1855) .

≡*Parmelia sphinctrina* Mont. , *Annls Sci. Nat.* , *Bot.* , sér. 2 4: 90(1835) .

On bark of trees.

Taiwan(Zahlbr. , 1933 c, p. 29; Wang & Lai, 1973, p. 96; Ikoma, 1983, p. 92: lapsu '*sphinctrium*') .

(64) Placynthiaceae Å. E. Dahl **胎座衣科**

Meddr Grønland, Biosc. 150(no. 2) : 49(1950) .

Placynthium(Ach.) Gray **胎座衣属**

Nat. Arr. Brit. Pl. (London) 2: 395(1821) .

≡*Collema* subdiv. *Placynthium* Ach. , *Lich. univ.* : 628(1810) .

Placynthium nigrum(Huds.) Gray 黑胎座衣

Nat. Arr. Brit. Pl. (London) 2: 395(1821) .

≡*Lichen niger* Huds. , *Fl. Angl.* , Edn 2 2: 524(1778) .

On calcareous rocks.

Heilongjiang(Zhao Na et al. , 2013, p. 1701 – 1702) .

Polychidium(Ach.) Gray **多柄衣属**

Nat. Arr. Brit. Pl. (London) 1: 401(1821) .

≡*Collema* subdiv. *Polychidium* Ach. , *Lich. univ.* : 658(1810) .

Polychidium stipitatum Vězda & W. A. Weber 短小多柄衣

Mycotaxon 3(3) : 355(1976) .

On *Populus* sp. covered with bryophytes.

Xizang(W. Obermayer, 2004, p. 506) , Taiwan(Aptroot, Sparrius & Lai 2002, p. 289) .

Vestergrenopsis Gyeln. **小芽衣属**

Rabenh. Krypt. -Fl. , Edn 2(Leipzig) 9(2. 2) : 265(1940) .

Vestergrenopsis isidiata(Degel.) Å. E. Dahl 鳞芽小芽衣

Meddr Grønland, Biosc. 150(2) : 55(1950) .

≡*Pannaria isidiata* Degel., *Bot. Notiser*: 90(1943).

On supra-littoral granitic rocks, affected by freshwater downflow. Known from Europe, North Americ and Asia.

Hongkong(Aptroot, 1999, p. 98).

②**Peltigerineae** Miądl. & Lutzoni 地卷亚目

Am. J. Bot. 91(3): 449 – 464(2004)

(65) Lobariaceae Chevall. [as 'Lobarieae'] 肺衣科

Fl. gén. env. Paris(Paris) 1: 609(1826)

Dendriscocaulon Nyl. 叶上枝属

Flora, Regensburg 68: 299(1885).

Dendriscocaulon bolacinum(Ach.) Nyl. 粗叶上枝

Flora, Regensburg 68: 299(1885).

≡*Parmelia lacera* var. *bolacina* Ach., *Methodus*, Sectio prior: 220(1803).

Corticolous.

Yunnan(Zahlbr., 1930 b, p. 78).

Dendriscocaulon intricatulum(Nyl.) Henssen 缠叶上枝

apud P. James & Hessen in D. H. Brown et al. (eds.) *Lichenology Progress and Problems* (London): 31(1976).

On bark of trees, rarely on rocks and the ground.

Sichuan, Yunnan, Xizang(Wu & Liu, 2012, pp. 116 – 117).

Lobaria(Schreb.) Hoffm. 肺衣属

Deutschl. Fl., Zweiter Theil(Erlangen): 138(1796) [1795].

≡*Lichen* sect. *Lobaria* Schreb., Gen.: 768(1791).

Lobaria adscripturiens(Nyl.) Hue 齿果肺衣

Nouv. Arch. Mus. Hist. Nat., Paris, 4 sér. 2: 27, tab. VI, fig. 3, 3 bis, 4, 4 bis(1900).

≡*Ricasolia adscripturiens* Nyl., *Lich. Japon.*: 31(1890).

=*Lobaria dentata* Hue, Nouv. Arch. Mus. Hist. Nat. 4(3): 36(1901).

Type: from Yunnan.

=*Lobaria laetevirens* var. *pallescens* Vain.

Lobaria adscripta sensu Zahlbr., in Handel-Mazzetti, *Symb. Sin.* 3: 83(1930), non(Nyl.) Hue According to Yoshimura(1971, p. 307) Zahlbr. (1930 b) reported *Lobaria adscripta* from China, but his Chinese lichen proved by Yoshimura(1969 b) to be *Lobaria dentata* Hue(=*Lobaria adscripturiens*(Nyl.) Hue). Thus, *L. adscripta* is excluded from the lichen flora of China.

Lobaria laetevirens sensu Zahlbr. in Fedde, *Repertorium*, 33: 30(1933), non Lightf. (see Yoshim. 1971, p. 320).

Corticolous.

Sichuan(Zahlbr., 1934 b, p. 198; Yoshim. 1971, p. 307; Chen S. F. et al., 1997, p. 222), Yunnan(Hue, 1901, p. 36; Zahlbr., 1934 b, p. 198; Yoshim. 1971, p. 307), Guizhou(Wu & Liu, 2012, pp. 71 – 73), Hunan(Zahlbr., 1930 b, p. 83), Hubei(Chen, Wu & J. C. Wei, 1989, p. 415; Wu & Liu, 2012, pp. 71 – 73), Hunan(Wu & Liu, 2012, pp. 71 – 73), Anhui, Zhejiang(Xu B. S. 1989, p. 190; Wu & Liu, 2012, pp. 71 – 73), Fujian(Wu J. N. et al. 1982, p. 10), Taiwan(Sasaoka, 1919, p. 180; Zahlbr., 1933 c, p. 30; Yoshim. 1971, p. 307; Wang-Yang & Lai, 1973, p. 92; Lin et al., 1989, p. 50; Lai, 2000, p. 249), China(prov. not indicated: Zahlbr., 1925, p. 299).

Also see *Lobaria fuscotomentosa* Yoshim.

Lobaria chinensis Yoshim. 中华肺衣

Journ. Hattori Bot. Lab. no. 34: 281(1971), Fig. 19; Pl. 16 c-d.

Type: Taiwan, Taidong, H. Masuda s. n. (holotype) in TNS & isotype in NICH(not seen).

On bark of trees.

Sichuan(Yoshim. 1971, p. 281), Yunnan(Yoshim. 1971, p. 281; Wu & Liu, 2012, pp. 46 – 48), Xizang(J. C. Wei & Jiang, 1986, p. 23; J. C. Wei et al. 1986, p. 370; Wu & Liu, 2012, pp. 46 – 48), Hubei(J. C. Wei et al. 1986, p. 370; Chen, Wu & J. C. Wei, 1989, p. 413; Wu & Liu, 2012, pp. 46 – 48), Hunan, Anhui (Wu & Liu, 2012, pp. 46 – 48), Taiwan(Yoshim. 1971, p. 281; Wang-Yang & Lai, 1973, p. 92; Ikoma, 1983, p. 70; Lin et al., 1989, p. 50; Lai, 2000, pp. 249- 250).

Lobaria crassior Vain. 厚肺衣

Bot. Mag., Tokyo 35: 64(1921).

= *Lobaria buiensis* Räs. *Journ. Jap. Bot.* 16(3): 144(1940).

Type: Taiwan, Mt. Bui, A. Yasuda no. 679(holotype) in Rasanen's Herb. & isotype in TI, TNS.

On bark of trees.

Hubei(Chen, Wu & J. C. Wei, 1989, p. 415; Wu & Liu, 2012, p. 76), Anhui, Zhejiang(J. C. Wei et al. 1986, p. 376; Xu B. S., 1989, p. 190; Wu & Liu, 2012, p. 76), Jiangxi, Guangxi(Wu & Liu, 2012, p. 76), Taiwan (Yoshim. 1971, p. 312; Wang-Yang & Lai, 1973, p. 92; Lin et al., 1989, pp. 50 – 51; Lai, 2000, p. 250).

Lobaria dendrophora Zahlbr. 树状肺衣

Annals Cryptog. Exot. 1: 173(1928).

Taiwan(Lin et al., 1989, p. 51; Lai, 2000, pp. 250251).

Lobaria dichroa(Nyl.) Zahlbr. 二色肺衣

Cat. Lich. Univers. 3: 299(1925).

≡ *Ricasolia dichroa* Nyl., *Annls Sci. Nat., Bot.*, sér. 4 11: 254(1859).

On tree trunk.

Shaanxi(Jatta, 1902, p. 468; Zahlbr., 1930 b, p. 83; Yoshim. 1971, p. 319).

Lobaria discolor(Bory in Delise) Hue 杂色肺衣

Nouv. Arch. Mus. Hist. Nat., Paris, 4 sér. 3: 23(1901).

≡ *Sticta discolor* Bory apud Del., *Hist. Lich.*, *Sticta* 136(1822).

Yunnan(Wei X. L. et al. 2007, p. 154), Taiwan(Lin et al., 1989, p. 51).

var. discolor 原变种

Corticolous.

Fujian(Wu J. N. et al. 1982, p. 10), Taiwan(Räs. 1940, p. 144).

var. inactiva(Asahina) Yoshim 负色变种

J. Hattori bot. Lab. 34: 265(1971). f. inactiva

≡ *Lobaria adscripta* f. *inactiva* Asahina, *J. J. B.* 9: 405(1933).

Sichuan, Yunnan, Fujian, Hainan(Wu & Liu, 2012, p. 35).

f. subadscripta Yoshim. 亚黑毛变型

J. Hattori bot. Lab. 34: 266(1971).

Lobaria adscripta sensu Zahlbr. in Fedde, *Repertorium*, 33: 30(1933), non(Nyl.) Hue.

Sichuan(Chen S. F. et al., 1997, p. 222; Wu & Liu, 2012, pp. 35 – 37), Yunnan(J. C. Wei et al. 1986, P. 365), Anhui(Xu B. S., 1989, p. 190; Wu & Liu, 2012, pp. 35 – 37), Zhejiang(J. C. Wei et al. 1986, P. 365; Xu B. S., 1989, p. 190; Wu & Liu, 2012, pp. 35 – 37), Taiwan(Zahlbr., 1933 c, p. 30; Yoshim. 1971, p. 266; Wang-Yang & Lai, 1973, p. 92; Lai, 2000, p. 251).

Lobaria ferax Vain. 繁育肺衣

Philipp. J. Sci., C, Bot. 8: 132(1913).

f. stenophyllodes(Vain.) Yoshim. 狭叶变型

J. Hattori bot. Lab. 34: 271(1971).

≡*Lobaria ferax* var. *stenophyllodes* Vain., *Philipp. J. Sci.*, *C*, *Bot*. 8: 133(1913).

On trees.

Yunnan(J. C. Wei et al. 1986, p. 365).

Lobaria flava(Stizenb.) Zahlbr. 黄肺衣

Cat. Lich. Univers. 3: 301(1925).

≡*Ricasolia flava* Stizenb., *Flora*, Regensburg 81: 111(1895).

Lobaria flava var. tarokoensis Inum. 台湾变种

Acta Phytotax. Geobot. 13: 218(1943).

Type: Taiwan, Inumaru no. 527 in Herb. Hirosima Univ. (not extant, according to Yoshim. 1971, p. 321).

Lobaria fuscotomentosa Yoshim. 褐毛肺衣

J. Hattori bot. Lab. 34: 311(1971).

Misapplied name:

Lobaria adscripturiens auct. non(Nyl.) Hue: Hue, *Nouv. Arch. Mus. Hist. Nat*. ser. 4, 3: 27(1901); Zahlbr. in Handel-Mazzetti, *Symb. Sin*. 3: 83(1930).

On bark of trees, mostly in deciduous forests.

Liaoning(Wu & Liu, 2012, p. 74), Shaanxi, Sichuan(Wu & Liu, 2012, p. 74), Yunnan(Hue, 1901, p. 27; Zahlbr., 1930 b, p. 83; Yoshim. 1971, p. 311; Wu & Liu, 2012, p. 74), Hubei(Wu & Liu, 2012, p. 74), Jiangxi(Wu & Liu, 2012, p. 74), Guangxi(Wu & Liu, 2012, p. 74), Hubei(Chen, Wu & J. C. Wei, 1989, p. 415), Fujian (Wu J. N. et al. 1982, p. 10).

Yoshimura(1971, p. 312) pointed out that because of the conspicuous and dark brown tomenta, *Lobaria fuscoto mentosa* has erroneously been treated as *Lobaria adscripturiens* by many lichenologists, e. g. Hue(1901), Zahlbr. (1930), etc.

Lobaria gyrophorica Yoshim. 三苔肺衣

Journ. Hattori Bot. Lab. 34: 275(1971), fig. 18, g-l; Pl. 14, c-d.

Type: Taiwan, Kurok. no. 794 in TNS(holotype) & in NICH(isotype), (not seen). (Wang & Lai, 1973, p. 92).

= *Lobaria pulmonaria* f. *hypomelaena* sensu Zahlbr., Krypt. exs. ed. a Mus. Hist. Nat. Vind. no. 2752 in S, nom. illeg., collected from Yunnan.

On bark of trees, mostly in deciduoud in cool-temperate regions.

Gansu(Wu & Liu, 2012, p. 39), Sichuan, Yunnan(J. C. Wei et al. 1986, p. 367; Wu & Liu, 2012, p. 39), Hubei(Wei et al. 1986, p. 367; Chen, Wu & J. C. Wei, 1989, p. 412; Wu & Liu, 2012, p. 39), Taiwan(Wang-Yang& Lai, 1973, p. 92; Lin et al., 1989, p. 51; Lai, 2000, pp. 251 – 252).

Lobaria isidiophora Yoshim. 针芽肺衣

J. Hattori bot. Lab. 34: 276(1971).

= *Lobaria meridionalis* Vain. var. *minor* Räs. non L. minor(Nyl.) Vain.

Misapplied name:

Lobaria pulmonaria var. *meridionalis* sensu Wu & Liu, Acta Phytotax. 14(2): 67(1976). non(Vain.) Zahlbr.

Jilin(Chen et al. 1981, p. 150; J. C. Wei et al. 1986, pp. 367, 368: chemical race I & II), Shaanxi(Wu & Liu, 1976, p. 67; J. C. Wei et al. 1986, pp. 367, 368: chemical race I; He & Chen 1995, p. 43, no specimen & locality were cited), Sichuan(Yoshim. 1971, p. 276; J. C. Wei et al. 1986, pp. 367, 368: chemical race I; Stenroos et al., 1994, pp. 334 – 335), Yunnan(Yoshim. 1971, p. 276; J. C. Wei et al. 1986, pp. 367, 368: chemical race I; Wei X. L. et al. 2007, p. 154), Guizhou(J. C. Wei et al. 1986, p. 368: chemical race I; Zhang T & J. C. Wei, 2006, p. 8), Xizang(J. C. Wei & Jiang, 1981, p. 1146 & 1986, p. 23; J. C. Wei et al. 1986, p. 368: chemical race I), Hubei(J. C. Wei et al. 1986, p. 368: chemical race I; Chen, Wu & J. C. Wei, 1989, p. 410), Anhui(J. C. Wei et al. 1986, p. 368: chemical race I), Zhejiang(Xu B. S., 1989, p. 188), Taiwan(Yoshim. 1971, p. 276;

Wang-Yang & Lai, 1973, p. 92; Lin et al., 1989, p. 51; Lai, 2000, p. 252), Hainan(JCWei et al., 2013, p. 232).

[var. *lobulata* Wu, Xiang & Qian]

Wuyi Science Journal, 2: 10(1982). (nom. inval.)

Type: Hunan, Liu Ai-Tang no. 770072.

On moss-covered rock.

Hunan, Jiangxi, Fujian(Wu J. N. et al. 1982, p. 10).]

Lobaria isidiosa(Müll. Arg.) Vain. 裂芽肺衣

Philipp. J. Sci., C, Bot. 8(2): 129(1913).

≡*Sticta retigera* f. *isidiosa* Müll. Arg., *Flora*, Regensburg 65(19): 300(1882).

= *Lobaria retigera* var. *subisidiosa*(Asahina) Yoshim. (Wu & Liu, 2012, p. 66).

On the ground, rocks, and bark of trees.

Heilongjiang(Chen et al. 1981, 151), Jilin(J. C. Wei et al. 1986, p. 373; Wu & Liu, 2012, p. 66 as *L. retigera* var. *subisidiosa*), Shaanxi(He & Chen 1995, p. 43), Xizang(J. C. Wei & Jiang, 1986, p. 24; J. C. Wei et al. 1986, p. 373: chemical race I), Sichuan(Zahlbr., 1930 b, p. 84; Wu & Liu, 2012, p. 66 as *L. retigera* var. *subisidiosa*), Yunnan(Hue, 1901, p. 39; Zahlbr., 1930 b, p. 84; Yoshim. 1971, p. 295; J. C. Wei et al. 1986, p. 373: chemical race I; Wu &Liu, 2012, pp. 60 – 62; Wu & Liu, 2012, p. 66 as *L. retigera* var. *subisidiosa*; Wei X. L. et al. 2007, p. 154), Hebei(Moreau et Moreau, 1951, p. 186), Shandong(Zhao ZT et al., 2002, p. 5), Hubei(Müll. Arg. 1893, p. 236; J. C. Wei et al. 1986, p. 373: chemical race I; Chen, Wu & J. C. Wei, 1989, p. 413; Wu & Liu, 2012, p. 66 as *L. retigera* var. *subisidiosa*), Hunan(Wu &Liu, 2012, pp. 60 – 62), Anhui(Xu B. S., 1989, p. 188 and p. 189 as *L. retigera* var. *subisidiosa*), Zhejiang(Xu B. S., 1989, p. 188 and p. 189 as *L. retigera* var. *subisidiosa;* Wu & Liu, 2012, p. 66 as *L. retigera* var. *subisidiosa*), Guangxi(Wu & Liu, 2012, pp. 60 – 62; Wu & Liu, 2012, p. 66 as *L. retigera* var. *subisidiosa*), Fujian(Wu J. N. et al. 1982, p. 10; Wu &Liu, 2012, pp. 60 – 62), Taiwan(Sasaoka, 1919, p. 180; Zahlbr., 1933 c, p. 31; Yoshim. 1971, p. 295; Wang-Yang & Lai, 1973, p. 92; Lin et al., 1989, p. 51; Lai, 2000, pp. 252 – 253).

Lobaria japonica(Zahlbr.) Asahina 日本肺衣

J. Jap. Bot. 9(7): 450(1933). f. japonica 原变型

≡*Lobaria laciniata* subsp. *japonica* Zahlbr., *Bot. Mag.*, Tokyo 41: 324(1927).

Corticolous.

Taiwan(Ras. 1940, p. 144; Lin et al., 1989, p. 51; Lai, 2000, p. 253).

f. exsecta(Nyl.) Yoshim. 黄色变型

Journ. Hattori Bot. Lab. 34: 307(1971).

≡*Ricasolia glomulifera* var. *exsecta* Nyl. *Synop. Lich.* 369(1860).

Type: North-eastern China(cited as Manchuria), H-Nyl. 3338 in H(not seen).

On bark of trees.

Guizhou, Hunan(Wu & Liu, 2012, pp. 69 – 71), Taiwan(Yoshim. 1971, p. 307; Wang & Lai, 1973, p. 92).

Lobaria kazawaensis(Asahina) Yoshim. 拟针芽肺衣

J. Hattori bot. Lab. 34: 291(1971).

≡*Lobaria sachalinensis* var. *kazawaensis* Asahina, *J. J. B.* 23: 68(1949).

On the ground.

Shaanxi(J. C. Wei et al. 1986, p. 372: chemical race I & II; Wu &Liu, 2012, pp. 56 – 58: chemotype II), Gansu (Wu &Liu, 2012, pp. 56 – 58: chemotype II), Xizang(J. C. Wei & Jiang, 1986, p. 24; Wu &Liu, 2012, pp. 56 – 58: chemotype II), Sichuan(Stenroos et al., 1994, p. 335), Hubei(J. C. Wei et al. 1986, p. 372: chemical race II; Chen, Wu & J. C. Wei, 1989, p. 412).

Lobaria kurokawae Yoshim. 光肺衣

J. Hattori bot. Lab. 34: 297(1971).

Stictina retigera sensu Elenk. *Acta Horti Petropolitani*, 19: 39(1901), non(Borry) Müll. Arg.

Lobaria retigera sensu Wu & Liu, *Acta Phytotax. Sin*. 14(2): 66(1976), non(Bory) Trevis.

On rocks or tree trunk in forests.

Shaanxi(Wu & Liu, 1976, p. 66; Wu J. L. 1981, p. 162; J. C. Wei et al. 1982, p. 19; J. C. Wei et al. 1986, p. 374; He & Chen 1995, p. 43 without citation of specimens and their localities; Wu & Liu, 2012, p. 67), Sichuan (Elenk. 1901, p. 39, cited as Dajianlu < = Kangding in Sichuan > in eastern Tibet; Yoshim. 1971, p. 297; Wei et al. 1986, p. 374; Stenroos et al., 1994, p. 335 Chen S. F. et al., 1997, p. 222; Wu & Liu, 2012, p. 67), Yunnan(Yoshim. 1971, p. 297; Wei X. L. et al. 2007, pp. 154 – 155), guizhou(Zhang T & J. C. Wei, 2006, p. 8), Xizang(J. C. Wei & Jiang, 1981, p. 1146 & 1986, p. 22; J. C. Wei et al. 1986, p. 374; Wu & Liu, 2012, pp. 67 – 69), Hubei(Chen, Wu & J. C. Wei, 1989, p. 414), Anhui, Zhejiang(Xu B. S., 1989, p. 188; Wu & Liu, 2012, p. 67), Jiangxi(Wu & Liu, 2012, p. 67), Guangxi(Wu & Liu, 2012, p. 67), Fujian(Wu J. N. et al. 1982, p. 11; Wu & Liu, 2012, p. 67), Taiwan(Yoshim. 1971, p. 297; Wang-Yang & Lai, 1973, p. 92; Lin et al., 1989, p. 51; Lai, 2000, p. 254).

Lobaria laetevirens(Lightf.) Zahlbr. 鲜绿肺衣

in Engler & Prantl, *Nat. Pflanzenfam*., Teil. I(Leipzig) 1*: 188(1906). var. laetevirens

≡*Lichen laetevirens* Lightf., *Fl. Scot*. 2: 852(1777).

=*Ricasolia herbacea* De Notrs.

Shaanxi(Jatta, 1902, p. 468; Zahlbr., 1930 b, p. 84; Yoshim. 1971, p. 320), Taiwan(Zahlbr., 1933 c, p. 30).

var. isidiosa Inum. 裂芽变种

Acta Phytotax. Geobot. 13: 224(1943).

Type: Taiwan, Inumaru no. 6156 in Herb. Hirosima Univ. (not extant, according to Yoshim. 1971).

According to Yoshimura this variety may possibly be a synonym of one of the following isidiate species: *Lobaria crassior*, *L. lobulata*, or *L. lobulata* f. *reagens*.

Taiwan(Inum. 1943, p. 224; Yoshim. 1971, p. 322).

Lobaria linita(Ach.) Rabenh. 薄叶肺衣

Deutschl. Krypt. -Fl. (Leipzig) 2: 65(1845).

≡*Sticta linita* Ach., *Syn. meth. lich*. (Lund): 234(1814).

Jilin, Heilongjiang(Wu & Liu, 2012, pp. 54 – 56).

Lobaria lobulata Yoshim. 小裂片肺衣

Journ. Hattori Bot. Lab. 34: 314(1971), fig. 26, f-k; Pl. 26, a-c). f. lobulata

Type: Taiwan, Kurok. no. 909 in TNS(holotype) & in NICH(isotype), not seen.

On bark of trees.

Taiwan(Wang-Yang & Lai, 1973, p. 92; Ikoma, 1983, p. 71; Lai, 2000, p. 254).

f. reagens Yoshim. 正色变型

Journ. Hattori Bot. Lab. 34: 315(1971), Pl. 26, d-e.

Type: Taiwan, Kurok. no. 868 in TNS(holotype) & in NICH(isotype), not seen.

On bark of trees.

Zhejiang(Xu B. S., 1989, p. 190 – 191; Wu & Liu, 2012, p. 78), Fujian(Wu & Liu, 2012, p. 78), Taiwan (Wang-Yang & Lai, 1973, p. 92; Ikoma, 1983, p. 71; Lin et al., 1989, pp. 51 – 52; Lai, 2000, p. 255).

Lobaria meridionalis Vain. 南肺衣

Philipp. J. Sci., C, Bot. 8(2): 128(1913).

≡*Lobaria pulmunaria* var. *meridionalis*(Vain.) Zahlbr. in Lamb, I. M. Index Nominum Lichenum: 404(1963) .

= *Lobaria meridionalis* var. *subplana*(Asahina) Yoshim. Miscell. bryol. lichenol. (Nichinan) 6(8) : 135(1974) .

Lobaria pulmonaria f. *papillaris* sensu H. Magn. Lich. Centr. Asia, 1: 46(1940) , non(Del.) Hue.

Jilin, Hubei, Abhui, Jiangxi, Zhejiang(Wu & Liu, 2012, pp. 50 – 52) , Shaanxi(He & Chen 1995, p. 43 without citation of specimens and their localities) .

var. meridionalis 原变种

On tree trunk.

Gansu(H. Magn. 1940, p. 46; J. C. Wei et al. 1986, p. 369) , Sichuan(Zahlbr. , 1930 b, p. 84; J. C. Wei et al. 1986, p. 369) , Yunnan(Zahlbr. , 1930 b, p. 84 & 1934 b, p. 198; J. C. Wei et al. 1986, p. 370) , Xizang(J. C. Wei & Jiang, 1981, p. 1146) , Hubei(J. C. Wei et al. 1986, p. 370; Chen, Wu & J. C. Wei, 1989, p. 411) , Anhui, Zhejiang(Xu B. S. , 1989, p. 189) , Taiwan(Yoshim. 1971, p. 284; Wang-Yang & Lai, 1973, p. 92 & 1976, p. 227; Lin et al. , 1989, p. 52as *L. meridionalis*; Lai, 2000, p. 255) .

f. melanovillosa Inumaru 黑紫变型

Acta Phytotax. Geobot. 13: 220(1943)

On bark of trees & on moss covered rocks.

Taiwan(Inum. 1943, p. 220) .

This taxon may probably be synonymous with either *L. orientalis* or *L. gyrophorica*. (see Yoshim. 1971, p. 321)

var. isidiosa Inumaru 裂芽变种

in Acta Phytotax. et Geobot. 13: 224(1943) .

On bark of trees.

Taiwan(Inum. 1943, p. 221) .

This taxon may possibly be included in either *L. meridionalis* or *L. isidiophora*. (see Yoshim. 1971, p. 321)

var. isidiosa f. atrotomentosa Inum. 黑绒变型

Acta Phytotax. Geobot. 13: 220(1943) .

Type: Taiwan, Inum. no. 5195 in Herb. Hirosima Univ. (not extant, according to Yoshim. 1971, p. 321) .

This taxon may possibly be a synonym of either *L. meridionalis* or *L. isidiophora*. (see Yoshim. 1971, p. 321)

Lobaria orientalis(Asahina) Yoshim. 东方肺衣

J. Hattori Bot. Lab. 32: 75(1969) .

≡*Lobaria pulmonaria* var. *orientalis* Asahina *J. J. B*. 23: 67(1949) . (1949) .

≡*Lobaria pulmonaria* var. *orientalis* Asahina

Misapplied names:

Lobaria meridionalis sensu H. Magn. Lich. Centr. Asia, 1: 45(1940) , non Vain.

Sticta pulmonacea sensu Tchou, Contr. Inst. Bot. Nat. Acad. Peiping, 3: 319(1935) , non Ach.

On tree trunk.

Gansu(H. Magn. 1940, p. 45; J. C. Wei et al. 1986, p. 366: chemical race I) , Sichuan(Yoshim. 1971, p. 273; J. C. Wei et al. 1986, p. 366: chemical race I; Wu & Liu, 2012, p. 37) , Yunnan(Yoshim. 1971, p. 273; J. C. Wei et al. 1986, p. 366: chemical race I; Wu & Liu, 2012, p. 37; Wei X. L. et al. 2007, p. 155) , Xizang(Wei & Jiang, 1981, p. 1146 & 1986, p. 22) , Hubei(J. C. Wei et al. 1986, p. 366; Chen, Wu & J. C. Wei, 1989, p. 412; Wu & Liu, 2012, p. 37) , Hunan(Wu & Liu, 2012, p. 37) , Abhui(Xu B. S. , 1989, p. 189; Wu & Liu, 2012, p. 37) , Zhejiang(Tchou, 1935, p. 319; J. C. Wei et al. 1986, p. 366: chemical race I) , Taiwan(Yoshim. 1971, p. 273; Wang-Yang & Lai, 1976, p. 227; Lin et al. , 1989, p. 52; Lai, 2000, pp. 255 – 256) .

Lobaria pindarensis Räsänen 悬肺衣

in Arch. Soc. Zool. Bot. Fenn. Vanamo VI, no. 2: 84(1952) .

On decayed tree.

Jilin(Wu & Liu, 2012, p. 41: chemotype I, p. 42: chemotype II), Shaanxi(Wu & Liu, 2012, p. 41), Sichuan (Chen S. F. et al., 1997, p. 222; Wu & Liu, 2012, pp. 41 – 42), Yunnan(Wu & Liu, 2012, p. 42), Hubei(Wu & Liu, 2012, p. 41: chemotype I, p. 42: chemotype II), Hunan, Fujian(Wu & Liu, 2012, p. 41: chemotype I).

Lobaria pseudopulmonaria Gyeln. 拟肺衣

Acta Faun. Fl. Univers., Ser. 2, Bot. 2(5): 6(1933).

= *Lobaria subretigera* Inum. *Acta Phytotax. Geobot.* 10: 214(1941).

Type: Taiwan, Kurok. no. 2655(neotype) in TNS & in NICH(isoneo-type), not seen.

= *Lobaria retigera* f. *microphyllina* Asahina

= *Lobaria pulmonaria* var. *pseudopulmonaria*(Gyeln.) Gretz. (Lin et al., 1989, p. 52).

On moss covered rocks or on soil.

Shaanxi(He & Chen 1995, p. 43 without citation of specimens and their localities; Wu & Liu, 2012, p. 58), Sichuan(J. C. Wei et al. 1986, p. 373: chemical race I; Wu &Liu, 2012, p. 58), Yunnan(Yoshim. 1971, p. 291; J. C. Wei et al. 1986, p. 373: chemical race I; Chen S. F. et al., 1997, p. 222; Wu &Liu, 2012, p. 58), Xizang(J. C. Wei & Jiang, 1986, p. 22), Hubei(J. C. Wei et al. 1986, p. 373: chemical race I; Chen, Wu & J. C. Wei, 1989, p. 414; Wu &Liu, 2012, p. 58), Anhui(J. C. Wei et al. 1986, p. 373: chemical race I; Xu B. S., 1989, p. 189 – 190; Wu &Liu, 2012, p. 58), Jiangxi, Guangxi(Wu &Liu, 2012, p. 58), Taiwan(Inum. 1941, p. 214; Asahina, 1947, p. 84; Lamb, 1963, p. 405; Yoshim. 1971, p. 291; Wang-Yang & Lai, 1973, p. 92; Lin et al., 1989, p. 52; Lai, 2000, pp. 256 – 257).

Lobaria pulmonaria(L.) Hoffm. 肺衣

Deutschl. Fl., *Zweiter* Theil(Erlangen): 146(1796) [1795].

≡ *Lichen pulmonarius* L., *Sp. pl.* 2: 1145(1753).

On bark of trees.

Jilin(J. C. Wei et al. 1986, p. 371: chemical race I; Wu & Liu, 2012, p. 52), Heilongjiang(Chen et al. 1981, p. 151; Wu & Liu, 2012, p. 52).

This species has been found by me and the collaborators of mine only in Mt. Changbai shan of Jilin(J. C. Wei et al. 1986, p. 371) and may be also collected from Heilongjiang of China(Chen et al. 1981, p. 151) and even from Sakhalin and northern Korea in Asia so far(M. Sato, 1943; Asahina, 1949; Yoshimura, 1971). So the records of it under the names of *Lobaria pulmonaria* or *Sticta pulmonacea* and their infraspecific taxa, made by the different authors from most of China including Hebei(Tchou, 1935, p. 319), Shaanxi(Baroni, 1894, p. 48; Jatta, 1902, p. 468; Zahlbr., 1930 b, p. 84), Gansu(H. Magnusson, 1940, p. 46), Sichuan(Elenk. 1901, p. 38; Zahlbr., 1930 b, p. 84), Yunnan(Hue, 1901, pp. 29 – 32; Paulson, 1928, p. 317; Zahlbr., 1930 b, p. 84), Hunan, Jiangxi(Zahlbr., 1930 b, p. 84), Zhejiang(Tchou, 1935, p. 319), Taiwan(Sasaoka, 1919, p. 180; Lin et al., 1989, p. 52; Lai, 2000, pp. 257 – 258), and somewhere in China(prov. not indicated: Asahina & Sato in Asahina, 1939, p. 649), may probably be based on erroneous identification(J. C. Wei et al. 1986, pp. 370 – 371).

[var. *isidiosa* Zahlbr. f. *nigroreticulata* Inum. *Acta Phytotax.* Geobot. 13: 223(1943).

Type: Taiwan, Inum. no. 5227 in Herb. Hirosima Univ. (not extant, according to Yoshim. 1971).

This taxon may be possibly a synonym of either *Lobaria meridionalis* or *L. isidiophora*(Yoshim. 1971, p. 321).]

[f. *papillaria*(Del.) Hue]

Yunnan, Xizang(Hue, 1901, p. 31), Hubei(Müll. Arg. 1893, p. 236).

The specimens under this name from China seem to be *Lobaria meridionalis* Vain. or *L. isidiophora* Yoshim.

Lobaria quercizans Michx 栎肺衣

Fl. Boreali-Americ. 2: 324(1803).

≡*Ricasolia quercizans*(Michx.) Stiz.

On bark.

Shaanxi(Ren Q. et al., 2009, pp. 103 – 104), Taiwan(Lin et al., 1989, pp. 52 – 53; Lai, 2000, p. 258).

Lobaria retigera(Bory) Trevis. 网脊肺衣

Lichenoth. Veneta 1 – 2: no. 75(1869).

≡*Lichen retiger* Bory, Voyage dans les quatre principales îles des mers d'Afrique fait par ordre du gouvernement pendant les années IX et X de la république(1801 bis(1802) 1: 391(1804).

Misapplied names:

Lobaria pulmonaria var. *hypomela* sensu H. Magn. *Lich. Centr. Asia*, 1: 46(1940), non(Del.) Cromb.

Sticta pulmonacea sensu Tchou, *Contr. Inst. Bot. Nat. Acad.* Peiping 3: 319(1935), non Ach.

Growing over mosses on rocks, tree-bases, and bark of trees.

Hebei(Tchou, 1935, p. 319; J. C. Wei et al. 1986, p. 375), Jilin, Heilongjiang(J. C. Wei et al. 1986, p. 375; Wu & Liu, 2012, pp. 62 – 63), Liaoning(Wu & Liu, 2012, pp. 62 – 63), Shaanxi (Jatta, 1902, p. 468; Zahlbr., 1930 b, p. 84; He & Chen 1995, p. 43 without citation of specimens and their localities; Wu & Liu, 2012, pp. 62 – 64), Gansu(Zahlbr., 1934 b, p. 198; H. Magn. 1940, p. 46; J. C. Wei et al. 1986, p. 375), Sichuan(J. C. Wei et al. 1986, p. 375; Chen S. F. et al., 1997, p. 222; Wu & Liu, 2012, pp. 62 – 64), Yunnan(Hue, 1887a, p. 22 & 1889, p. 167 & 1901, p. 38; Paulson, 1928, p. 317; Zahlbr., 1930 b, p. 84; J. C. Wei et al. 1986, p. 375; Wei X. L. et al. 2007, p. 155 ; Wu & Liu, 2012, pp. 62 – 64), Guizhou(Wu & Liu, 2012, pp. 62 – 64; Zhang T & J. C. Wei, 2006, p. 8), Xizang(J. C. Wei & Jiang, 1986, p. 23; J. C. Wei et al. 1986, p. 375), Hubei(J. C. Wei et al. 1986, p. 375; Chen, Wu & J. C. Wei, 1989, p. 414; Wu & Liu, 2012, pp. 62 – 64), Hunan(Wu & Liu, 2012, pp. 62 – 64), Anhui, Zhejiang(Xu B. S., 1989, p. 189; Wu & Liu, 2012, pp. 62 – 63), Jiangxi(Yoshim. 1971, p. 298; Wu & Liu, 2012, pp. 62 – 63), Guangxi(J. C. Wei et al. 1986, p. 375; Wu & Liu, 2012, pp. 62 – 64), Guangdong, Hainan(Wu & Liu, 2012, pp. 62 – 64), Fujian(Zahlbr., 1930 b, p. 84; Wu J. N. et al. 1982, p. 11; J. C. Wei et al. 1986, p. 375; Wu & Liu, 2012, pp. 62 – 63), Taiwan(Yoshim. 1971, p. 298; Wang-Yang & Lai, 1973, p. 92; Lin et al., 1989, p. 53; Lai, 2000, pp. 258 – 259).

Lobaria sachalinensis Asahina 库页岛肺衣

J. J. B. 23: 68(1949).

On the ground.

Jilin(Chen et al. 1981, p. 151).

Lobaria scrobiculata(Scop.) P. Gaertn. 蜂巢肺衣

in Lamarck & de Candolle, *Fl. franç.*, Edn 3(Paris) 2: 402(1805).

≡*Lichen scrobiculatus* Scop., *Fl. carniol.*, Edn 2(Wien) 2: 384(1772).

≡*Sticta scrobiculata*(Acop.) Ach. (Lin et al., 1989, p. 53)

≡*Stictina scrobiculata* (Scop.) Nyl. (Lin et al., 1989, p. 53)

Xinjiang(Abdulla Abbas & Wu, 1998, pp. 133 – 134; Abdulla Abbas et al., 2001, p. 365), Taiwan(Lin et al., 1989, p. 53; Lai, 2000, p. 259).

Lobaria spathulata(Inumaru) Yoshim. 匙芽肺衣

J. Hattori bot. Lab. 34: 278(1971).

≡*Lobaria meridionalis* var. *spathulata* Inumaru, in *Acta Phytotax. et Geobot.* 13: 221(1943).

≡*Lobaria pulmonaria*(L.) Hoffm. var. *spathulata*(Inum.) Asahina(Lin et al., 1989, p. 53).

=*Lobaria pulmunaria*(L.) Hoffm. f. *microphyllina* Inum. *Acta Phytotax. Geobot.* 10(3): 216(1941).

=*Lobaria pulmonaria* f. *hypomelaena* Zahlbr., *Krypt. exs.* ed. a Mus. Hist. Nat. Vind. no. 2752 in HMAS-L. nom. illeg. collected from Yunnan.

On bark of trees.

Sichuan(J. C. Wei et al. 1986, p. 369; Stenroos et al. , 1994, p. 335; Wu & Liu, 2012, pp. 42 – 44) , Yunnan (Wei et al. 1986, p. 369; Wu & Liu, 2012, pp. 42 – 44) , Hubei(J. C. Wei et al. 1986, p. 369; Chen, Wu & J. C. Wei, 1989, p. 411; Wu & Liu, 2012, pp. 42 – 44) , Jiangxi, Zhejiang(Wu & Liu, 2012, pp. 42 – 44) , Taiwan (Inum. 1941, p. 216; Asahina, 1949, p. 67; Yoshim. 1971, p. 278; Wang-Yang & Lai, p. 92; Lin et al. , 1989, p. 53; Lai, 2000, pp. 259 – 260) .

Lobaria sublaevis(Nyl.) Yoshim. 亚平肺衣

J. Hattori bot. Lab. 34: 315(1971) .

≡*Ricasolia sublaevis* Nyl. , in Krempelhuber, *Flora*, Regensburg 51: 231(1868) .

= *Lobaria ochracea* Inum. *Phytotax. Geobot.* 10(3) : 214(1941) .

Type: Taiwan, Inum. no. 6182(neotype) in TNS.

On bark of trees.

Zhejiang(Xu B. S. , 1989, p. 191; Wu & Liu, 2012, pp. 80 – 81) , Taiwan(Inum. 1941, p. 214; Yoshim. 1971, p. 315; Wang-Yang & Lai, 1973, p. 92; Lin et al. , 1989, p. 53; Lai, 2000, p. 260) .

Lobaria subscrobiculata Vain. 亚蜂窝肺衣

Philipp. J. Sci. , C, Bot. 8: 133(1913) . f. subscrobiculata 原变型

= *Lobaria ochrotropa* Zahlbr. in Fedde, *Repertorium*, 33: 30(1933) .

Type: Taiwan, Asahina F – 1, W(herb. Mus. Hist. Natur. Vindob. 1931, no. 827, holotype) and in TNS, US(in Plitt's collection ex herb. Zahlbr.) , not seen. (according to Yoshim. 1971, p. 268)

On bark of trees.

Taiwan(Zahlbr. , 1933 c, p. 30; Inum. 1941, p. 216; Yoshim. 1971, p. 268; Wang-Yang & Lai, 1973, p. 92; Lin et al. , 1989 阿, p. 53; Lai, 2000, pp. 260 – 261) .

f. subdiscolor Yoshim. 亚异色变型

J. Hattori bot. Lab. 34: 270(1971) .

Taiwan(Lai, 2000, p. 261) .

Lobaria tuberculata Yoshim. 瘤芽肺衣

J. Hattori bot. Lab. 34: 280(1971) .

On trees.

Jilin(J. C. Wei et al. 1986, p. 369; Wu & Liu, 2012, pp. 44 – 46) , Hubei(J. C. Wei et al. 1986, p. 369; Chen, Wu & J. C. Wei, 1989, p. 411; Wu & Liu, 2012, pp. 44 – 46) .

Lobaria yoshimurae Kurok. & Kashiw. 吉村氏肺衣

Bull. natn. Sci. Mus. , Tokyo, B 4(3) : 123(1978) .

Taiwan(Lin et al. , 1989, pp. 53 – 54; Lai, 2000, pp. 261 – 262) .

Lobaria yulongenesis Chen 玉龙肺衣

Acta Mycol. Sinica 14(4) : 261 – 262(1995) .

Type: Yunnan, Mt. Yulongshan, 1987, Ahti et al. , 46339(holotype: HMAS-L) , and isotypes in HKAS & H.

Yunnan(Chen, 1995, pp. 261 – 262) .

Lobaria yunnanensis Yoshim. 云南肺衣

Journ. Hattori Bot. Lab. 34: 282(1971) , Fig. 20; Pl. 17, a, b.

Type: Yunnan, Lijiang, Handel-Mazzetti no. 2315 in W(not seen) , distributed as Kryptogam.

Exs. ed. Mus. Hist. Nat. Vindob. no. 2752. However the specimens under the same number of the Exs. (no. 2752) preserved in S, belongs to *Lobaria gyrophorica* Yoshim. , and in HMAS-L, to *Lobaria spathulata*(Inum.) Yoshim. , respectively.

Shaanxi(Wu & Liu, 2012, pp. 48 – 50) , Sichuan(Yoshim. 1971, p. 282; Wu & Liu, 2012, pp. 48 – 50) , Yunnan (Yoshim. 1971, p. 282; Wei X. L. et al. 2007, p. 155; Wu & Liu, 2012, pp. 48 – 50) , Hubei(Chen, Wu & J. C. Wei, 1989, p. 413; Wu & Liu, 2012, pp. 48 – 50) .

Pseudocyphellaria Vain. 假杯点衣属

Acta Soc. Fauna Flora fenn. 7(1): 182(1890).

Pseudocyphellaria argyracea(Delise) Vain. 银白假杯点衣

Hedwigia 37: 35(1898).

≡*Sticta argyracea* Delise, *Mém. Soc. Linn. Calvados* 2: 91(1825).

On bark of trees. On exposed granitic rock. Pantropical.

Taiwan(Zahlbr., 1933 c, p. 32; Wang & Lai, 1973, p. 97), Hainan(JCWei et al., 2013, p. 232), Hongkong (Thrower, 1988, pp. 17, 152, 153; Aptroot & Seaward, 1999, p. 89).

Pseudocyphellaria aurata(Ach.) Vain. 黄假杯点衣

Acta Soc. Fauna Flora fenn. 7(1): 183(1890).

≡*Sticta aurata* Ach., *Methodus*, Sectio post.: 277(1803).

On bark of *Abies* sp.

Jilin, Heilongjiang(Chen et al. 1981, p. 156), Yunnan(Wu & Liu, 2012, pp. 82 – 83; Wang L. S. 2008, p. 538), Guizhou(Wu & Liu, 2012, pp. 82 – 83; Zhang T & J. C. Wei, 2006, p. 9), Hubei(Chen, Wu & J. C. Wei, 1989, p. 419; Wu & Liu, 2012, pp. 82 – 83), Zhejiang(Tchou, 1935, p. 318; Xu B. S., 1989, p. 191; Wu & Liu, 2012, pp. 82 – 83), Guangdong(Wu & Liu, 2012, pp. 82 – 83), Fujian(Zahlbr., 1930 b, p. 86; Wu J. N. et al. 1982, p. 11; Wu & Liu, 2012, pp. 82 – 83), Taiwan(Asahina, 1932, p. 47; Zahlbr., 1933 c, p. 31; Asahina & M. Sato in Asahina, 1939, p. 653; Wang & Lai, 1973, p. 97).

Pseudocyphellaria cinnamomea(A. Rich.) Vain. 黄褐假杯点衣

Philipp. J. Sci., C, Bot. 8(2): 120(1913).

≡*Sticta cinnamomea* A. Rich., *Voyage de Découvert. de l'Astrolabe, Botan.* 1: 28, tab. VIII, fig. 3(1832).

Corticolous.

Guangxi(Wu & Liu, 2012, *p.* 85*), Taiwan(Räs.* 1940, *p.* 143*; Wang & Lai,* 1973, *p.* 97*; Ikoma,* 1983, *p.* 104*), Hainan(J. C. Wei et al.*, 2013, *p.* 232*).*

Pseudocyphellaria crocata(L.) Vain. 金缘假杯点衣

Hedwigia 37*:* 34*(*1898*).*

≡Lichen crocatus *L.*, Mantissa Altera*:* 310*(*1771*).*

On bark of coniferous trees & that of Betula *sp. &* Populue *sp.*

Heilongjiang(J. C. Wei, 1981, *p.* 82*; Chen et al.* 1981, *p.* 156*; Luo,* 1984, *p.* 85*), Jilin(Chen et al.* 1981, *p.* 156*; Wu & Liu,* 2012, *p.* 87*), Xizang(W. Obermayer,* 2004, *p.* 506*), Sichuan(Stenroos et al.*, 1994, *p.* 336*), Hubei (Chen, Wu & J. C. Wei,* 1989, *p.* 419*; Wu & Liu,* 2012, *pp.* 87 – 89*), Hunan(Wu & Liu,* 2012, *pp.* 87 – 89*), Jiangxi(Wu & Liu,* 2012, *pp.* 87 – 89*), Anhui, Zhejiang(Xu B. S.*, 1989, *p.* 191*; Wu & Liu,* 2012, *pp.* 87 – 89*), Guangxi(Wu & Liu,* 2012, *pp.* 87 – 89*), Fujian(Wu J. N. et al.* 1982, *p.* 11*; Wu & Liu,* 2012, *p.* 87*), Taiwan (Aptroot, Sparrius & Lai* 2002, *p.* 289*), Hainan(JCWei et al.*, 2013, *p.* 232*).*

= *Sticta mougeotiana var. aurigera Nyl.* Syn. meth. lich. *(Parisiis)* 1*:* 341(1860).

≡*Sticta aurigera* Delise, *Hist. Lich. Sticta*: 54(1822).

On bark of trees.

Hunan(Zahlbr., 1930 b, p. 86).

Pseudocyphellaria desfontainii(Delise) D. J. Galloway 戴氏假杯点衣

Lichenologist 17(3): 304(1985).

≡*Sticta desfontainii* Delise, *Hist. Lich. Sticta:* 60(1822).

On *Shiia calesii*.

Taiwan(Aptroot, Sparrius & Lai 2002, p. 289).

Pseudocyphellaria hainanensis J. N. Wu 海南假杯点衣

J*ournal of Nanjing Normal University*(Natural Science) 15(2): 62 – 65(1992).

Type: Hainan, Mt. Jianfengling, Liu Ai-Tang 810029(HNNU-L).

Pseudocyphellaria intricata(Delise) Vain. 缠结假杯点衣

Hedwigia 37: 35(1898).

≡*Sticta intricata* Delise, *Hist. Lich. Sticta*: 96(1822).

Corticolous.

Taiwan(Räs. 1940, p. 143; Wang & Lai, 1973, p. 97).

Pseudocyphellaria neglecta(Müll. Arg.) H. Magn. 裂芽杯点衣

Acta horti götoburg. 14: 30(1940).

≡*Stictina neglecta* Müll. Arg., *Flora*, Regensburg 70: 58(1887).

= *Pseudocyphellaria* isidiotyla (H. Magn.) J. N. Wu, *Flora Lichenum Sinicorum* vol. 11, Peltigerales (I); 91 (2012).

≡*Pseudocyphellaria crocata* var. *isidiotyla* H. Magn., *Acta horti götoburg*. 14: 14(1940)

≡*Cyanisticta crocata* var. *isidiotyla*(H. Magn.) Szatala, *Annls Mus. natn. Hung.*, n. s. 7: 41(1956).

Hainan(Wu & Liu, 2012, pp. 91 – 93).

Sticta(Schreb.) Ach. **牛皮叶属**

Methodus, Sectio post.: 275(1803).

≡*Lichen* sect. *Sticta* Schreb., *Gen.*: 768(1791).

Sticta cyphellulata(Müll. Arg.) Hue 杯点牛皮叶

Nouv. Arch. Mus. Hist. Nat., Paris, 4 sér. 3: 99(1901).

≡*Stictina cyphellulata* Müll. Arg., *Flora*, Regensburg 65(19): 301(1882).

On bark of tree and on rocks.

Guangxi, Hainan(Wu & Liu, 2012, p. 95).

Sticta duplolimbata(Hue) Vain. 双缘牛皮叶

Philipp. J. Sci., *C*, *Bot*. 8(2): 125(1913).

≡*Sticta ciliaris* f. *duplolimbata* Hue, *Nouv. Arch. Mus. Hist. Nat.*, Paris, 4 sér. 3: 102(1901).

On stones among mosses.

Yunnan(chen J. B. et al., 1994, p. 30; Wu & Liu, 2012, pp. 95 – 96), Hubei(Chen, Wu & J. C. Wei, 1989, p. 457; Wu & Liu, 2012, pp. 95 – 96), Zhejiang(Xu B. S., 1989, p. 192; Wu & Liu, 2012, pp. 95 – 96), Guangxi (Wu & Liu, 2012, pp. 95 – 96), Fujian(Zahlbr., 1930 b, p. 86; Wu J. N. et al. 1982, p. 12; Wu & Liu, 2012, pp. 95 – 96), Taiwan(Zahlbr., 1933 c, p. 32; M. Sato, 1935, p. 241; Asahina & M. Sato in Asahina, 1939, p. 655; Wang & Lai, 1973, p. 97).

Sticta filix(Sw.) Nyl. 蕨状牛皮叶

J. Linn. Soc., *Bot*. 9: 246(1867).

≡*Lichen filix* Sw., *Method. Muscor.*: 36(1781).

= *Sticta filicina* Ach., *Methodus*, Sectio post.: 275(1803), Nom. illegit., Art. 52. 1.

Editorial comment: Circumscription includes the type of a name(Lichen filix Sw. 1781), the eipthet of which ought to have been adopted.

On *Machilus thunbergii*, *Abies kawakamii* and *Fatsia* sp.

Yunnan, Zhejiang, Hainan, Fujian(Wu & Liu, 2012, p. 99), Taiwan(Aptroot, Sparrius & Lai, 2002, p. 291).

Sticta flabelliformis Zahlbr. 扇形牛皮叶

in Fedde, *Repertorium*, 33: 32(1933).

Type: Taiwan, Mt. Ali, Asahina no. 21.

Taiwan(Zahlbr., 1940, p. 275; Wang & Lai, 1973, p. 97; Ikoma, 1983, p. 104).

On bark of trees.

Sticta formosana Zahlbr. 台湾牛皮叶

in Feddes, *Repertorium*, 33: 33(1933).

Type: Taiwan, Asahina no. 40.

Taiwan(Zahlbr., 1940, p. 275; Lamb, 1963, p. 692; Wang & Lai, 1973, p. 97; Ikoma, 1983, p. 104).

On bark of trees.

Sticta fuliginosa(Dicks.) Ach. 黑牛皮叶

Methodus, Sectio post.: 280(1803).

≡*Lichen fuliginosus* Dicks., *Fasc. Plant. Cryptog. Brit.* I: 13(1785)[*Lichen fuliginosus* Dicks., (1775)].

= *Sticta sylvatica* var. *fuliginosa*(Hoffm.) Hepp, *Flecht. Europ.*: no. 371(1857).

≡*Lichen fuliginosus* Hoffm., *Enum. critic. lich. europ.* (Bern)(1784).

On bark of trees & on rocks.

Shaanxi(He & Chen 1995, p. 45 miscited as *S. fuligimosa* without citation of specimens and their localities; Wu & Liu, 2012, pp. 104 – 106), Sichuan(Chen S. F. et al., 1997, p. 222), Yunnan(Wu & Liu, 2012, pp. 104 – 106), Guizhou(Zhang T & J. C. Wei, 2006, p. 10), Hubei(Chen, Wu & J. C. Wei, 1989, p. 417; Wu & Liu, 2012, pp. 104 – 106), Hunan(Wu & Liu, 2012, pp. 104 – 106), Anhui, Zhejiang(Xu B. S., 1989, p. 192; Wu & Liu, 2012, p. 104), Fujian(Wu J. N. et al. 1982, p. 12; Wu & Liu, 2012, pp. 104 – 106), Taiwan(Zahlbr., 1933 c, p. 33; Asahina & M. Sato in Asahina, 1939, p. 655; Wang & Lai, 1973, p. 97).

Sticta gracilis(Müll. Arg.) Zahlbr. 柄扇牛皮叶

Cat. Lich. Univers. 3: 386(1925).

≡*Stictina gracilis* Müll. Arg., *Flora*, Regensburg 74(1): 111(1891).

= *Sticta stenoloba* Vain., *Bot. Mag.*, Tokyo 35: 64(1921).

On rocks among mosses & on trees.

Yunnan(Chen J. B. et al., 1994, pp. 30 – 31.), Guizhou(J. C. Wei, 1981, p. 82), Xizang(J. C. Wei & Jiang, 1981, p. 1147 & 1986, p. 25; Zhang T & J. C. Wei, 2006, p. 10), Hunan(Wu & Liu, 2012, p. 101), Zhejiang (Xu B. S., 1989, p. 192; Wu & Liu, 2012, p. 101), Guangdong, Guangxi(Wu & Liu, 2012, p. 101), Fujian(Wu J. N. et al. 1982, p. 12; Wu & Liu, 2012, p. 101), Taiwan(M. Sato, 1935, p. 240 & 1957, p. 66; Asahina & M. Sato in Asahina, 1939, p. 655; Wang & Lai, 1973, p. 97).

Sticta henryana Müll. Arg. 亨利牛皮叶

Flora 74: 374(1891) & Bull. L'herb. Bois. 1: 236(1893). Stzbgr. *Flora* 81: 124(1895). Zahlbr. in Handel-Mazzetti, *Symb. Sin.* 3: 85(1930).

Type: from China.

On bark of trees.

Shaanxi(He & Chen 1995, p. 45 without citation of specimens and their localities), Yunnan(Zahlbr., 1930 b, p. 85 & 1934 b, p. 198), Hubei(Müll. Arg. 1893, p. 236; Zahlbr., 1930 b, p. 85; Chen, Wu & J. C. Wei, 1989, p. 417), Anhui, Zhejiang(Xu B. S., 1989, p. 192 – 193).

Sticta limbata(Sm.) Ach. 粉缘牛皮叶

Methodus, Sectio post.: 280(1803).

≡*Lichen limbatus* Sm., in Smith & Sowerby, *Engl. Bot.* 16: tab. 1104(1803). Shaanxi(He & Chen 1995, p. 45 without citation of specimens and their localities), Gansu(Wu & Liu, 2012, pp. 106 – 107), Hubei(Chen, Wu & J. C. Wei, 1989, p. 416; Wu & Liu, 2012, pp. 106 – 107).

Sticta mougeotiana Delise 假黄牛皮叶

Hist. Lich. Sticta: 62, tab. V, fig. 13(1822) var. mougeotiana

On bark of trees.

Taiwan(Zahlbr., 1933 c, p. 33; Wang & Lai, 1973, p. 97).

Sticta neocaledonica(Müll. Arg.) Hue 南洋牛皮叶

Nouv. Arch. Mus. Hist. Nat., Paris, 4 sér. 3: 101(1901).

≡*Stictina neocaledonica* Müll. Arg., *Flora*, Regensburg 65: 303(1882).

On rocks.

Guangdong(Zahlbr., 1930 b, p. 86), Taiwan(Wang & Lai, 1973, p. 97).

Sticta nylanderiana Zahlbr. 平滑牛皮叶

Cat. Lich. Univers. 3: 356(1925). var. nylanderiana Replaced synonym: *Sticta platyphylla* Nyl. 1860 Competing homonym: non *Sticta platyphylla* A. Massal. 1853

On bark of trees among mosses in forest & on the ground.

Sticta nylanderiana var. epicoila(Hue) Zahlbr. 表生变种

Cat. Lich. Univers. 3: 357(1925).

≡*Sticta platyphylla* var. *epicoila* Hue in Nouv. *Archiv. du Muséum*, ser. 4., vol. 3: 63(1901).

On bark of trees.

Yunnan(Hue, 1901, p. 63; Zahlbr., 1930 b, p. 85).

Sticta orbicularis(A. Braun ex Meyen & Flot.) Hue 园形牛皮叶

Ann. Jard. Bot. Buitenzorg 17: 193(1901).

≡*Sticta filicina* var. *orbicularis* A. Braun ex Meyen & Flot., (1843).

On *Machilus thunbergii* and *Abies kawakamii*.

Taiwan(Aptroot, Sparrius & Lai, 2002, p. 291).

Sticta platyphylloides Nyl. 宽叶牛皮叶

in Hue, Bull. Soc. Bot. France, 34: 22(1887).

Type: Yunnan, near Dapingzi, alt. 1800 m, Delavay no. 1607.

On bark of trees.

Shaanxi(Wu & Liu, 2012, p. 113), Guizhou(Wu & Liu, 2012, p. 113; Zhang T & J. C. Wei, 2006, p. 10), Sichuan(Chen J. B., 1993, p. 458; Wu & Liu, 2012, p. 113), Yunnan(Zahlbr., 1930 b, p. 85; Chen J. B., 1993, p. 458; Chen J. B. et al., 1994, p. 31; Wei X. L. et al. 2007, pp. 157 – 158; Wu & Liu, 2012, p. 113), Xizang (Chen J. B., 1993, p. 458), Hubei(Chen J. B., 1993, p. 458), Hunan, Zhejiang(Wu & Liu, 2012, p. 113), Taiwan(Aptroot, Sparrius & Lai, 2002, p. 291), China(prov. not indicated: Hue, 1890, p. 303 & 1901, p. 63; Zahlbr., 1925, p. 360 & 1934 b, p. 198).

Sticta praetextata(Räsänen) D. D. Awasthi 缝芽牛皮叶

in Joshi & Awasthi, *Biological Memoirs* 7(2): 185(1982).

≡*Sticta platyphylla* var. *praetextata* Räsänen, in *Arch. Soc. Zool. Bot. Fenn. Vanamo*, 6(2): 84(1952).

Gansu(Wu & Liu, 2012, pp. 115 – 116), Sichuan(Chen J. B., 1993, p. 458; Wu & Liu, 2012, pp. 115 – 116), Yunnan(Chen J. B., 1993, p. 458; Chen J. B. et al., 1994, p. 31; Wu & Liu, 2012, pp. 115 – 116), Xizang (Chen J. B., 1993, p. 458).

Sticta pulvinata(Meyen & Flot.) Vain. 垫状牛皮叶

Philipp. J. Sci., C, Bot. 8(2): 123(1913).

≡*Sticta filicina* var. *pulvinata* Meyen & Flot., (1843).

On bark of trees.

Guizhou(Zhang T & J. C. Wei, 2006, p. 10), Taiwan(Wang & Lai, 1973, p. 97), Hainan(J. C. Wei et al., 2013, p. 232).

Sticta sinuosa Pers. 深波牛皮叶

in Gaudichaud-Beaupré in Freycinet, *Voy. Uranie.*, *Bot.*: 200(1827)[1826 – 1830].

On bark of trees.

Taiwan(Zahlbr., 1933 c, p. 31; Räs. 1940, p. 143; Wang & Lai, 1973, p. 97; Ikoma, 1983, p. 104).

Sticta submarginifera Zahlbr. 镶边牛皮叶

in Fedde, *Repertorium*, 33: 33(1933).

Type: Taiwan, Asahina no. 20.

Guangdong(Wu & Liu, 2012, pp. 103 – 104) , Taiwan(Zahlbr. , 1940, p. 278; Wang & Lai, 1973, p. 97; Ikoma, 1983, p. 104) .

On bark of trees.

Sticta sylvatica(Huds.) Ach. 林中牛皮叶

Methodus, Sectio post. : 231(1803) .

≡*Lichen sylvaticus* Huds. , *Fl. Angl.* : 453(1762) .

On *Populus* sp. and *Fatsia* sp.

Taiwan(Aptroot, Sparrius & Lai, 2002, p. 292) .

Sticta weigelii(Ach.) Vain. 缘裂牛皮叶

Acta Soc. Fauna Flora fenn. 7(1) : 189(1890) .

≡*Sticta damicornis* var. *weigelii* Ach. , in Acharius, *Lich. univ.* : 446(1810) .

On stream-bed rocks. Pantropical.

Yunnan(Chen J. B. et al. , 1994, p. 32; Wu & Liu, 2012, p. 107) , Hubei, Anhui, Jiangxi, Zhejiang, Fujian(Wu & Liu, 2012, p. 107) .

Sticta weigelii var. weigelii 原变种

On bark of trees.

Xizang(J. C. Wei & Jiang, 1986, p. 25) , Hubei(Chen, Wu & J. C. Wei, 1989, p. 418) , Anhui, Zhejiang(Xu B. S. , 1989, p. 193) , Fujian(Wu J. N. et al. 1982, p. 12) , Hongkong(Thrower, 1988, pp. 17, 171; Aptroot & Seaward, 1999, p. 95) .

Sticta weigelii var. enteroxanthella Zahlbr. 黄缘牛皮叶

in Feddes, *Repertorium sp. nov.* 33: 34(1933) .

Type: Taiwan, Asahina no. 44.

Taiwan(Wang & Lai, 1973, p. 97; Ikoma, 1983, p. 104) .

On bark of trees.

Sticta wrightii Tuck. 深杯牛皮叶

Amer. J. Sci. Arts, Ser. 2 28: 204(1859) .

≡*Ricasolia wrightii*(Tuck.) Nyl.

≡*Lobaria wrightii*(Tuck.) Th. Fr.

= *Sticta miyoshiana* Müll. Arg.

Sichuan(Elenk. 1901, p. 40: cited as eastern Tibet) , Taiwan(Sasaoka, 1919, p. 180; Zahlbr. , 1933c, pp. 31, 32; Wang & Lai, 1973, p. 97) .

(66) Nephromataceae Wetmore ex J. C. David & D. Hawksw. **肾盘衣科**

Syst. Ascom. 10(1) : 15(1991) .

Nephroma Ach. **肾盘衣属**

in Luyken, *Tent. Hist. Lich.* : 92(1809) .

Nephroma bellum(Spreng.) Tuck. 钟形肾盘衣

Boston J. Nat. Hist. 3: 293(1841) .

≡*Peltigera bella* Spreng. , *Syst. veg.* , Edn 16 4(1) : 306(1827) .

On the ground among mosses, and on bark of trees.

Neimenggu(Wu & Liu, 2012, pp. 119 – 120) .

Nephroma flavorhizinatum Q. Tian & H. Y. Wang 黄假根肾盘衣

Mycotaxon 115: 282 – 284(2011) .

Type: China. Sichuan province, Litang, alt. 4200 m, on ground, Z. S. Sun, 20080755, 5 Nov. 2008. (Holotype in SDNU) .

Nephroma helveticum Ach. 瑞士肾盘衣

Lich. univ. : 523(1810).

≡*Nephroma tomentosum* var. *helveticum*(Ach.) Anzi, *Atti Soc. ital. Sci. nat. Mus. Civico Storia nat. Milano* 9: 244(1866).

≡*Nephroma resupinatum* f. *helveticum* Rabenh., *Deutschl. Krypt. -Fl.* (Leipzig) 2: 68(1845).

≡*Nephromium helveticum* (Ach.) Nyl., *Lich. Nov. Zeland.* (Paris): 43(1888).

On bark of trees and over soil and mosses on rocks.

Neimenggu(Chen et al. 1981, p. 151; Wu & Liu, 2012, p. 113), Jilin(Wu & Liu, 2012, pp. 120 – 121; Wang HY et al., 2013, pp. 265 – 266), Heilongjiang(Wu & Liu, 2012, pp. 120 – 121), Shaanxi(He & Chen 1995, p. 44 without citation of specimens and their localities; Wang HY et al., 2013, pp. 265 – 266), Sichuan(Chen S. F. et al., 1997, p. 222; Wu & Liu, 2012, pp. 120 – 121; Wang HY et al., 2013, pp. 265 – 266), Yunnan(Hue, 1887a, p. 22 & 1889, p. 167 & 1900, p. 105; Zahlbr., 1930 b, p. 87 & 1934, p. 199; Wei X. L. et al. 2007, p. 155; Wu & Liu, 2012, pp. 120 – 121; Wang HY et al., 2013, pp. 265 – 266), guizhou(Zhang T & J. C. Wei, 2006, p. 8; Wu & Liu, 2012, pp. 120 – 122; Wang HY et al., 2013, pp. 265 – 266), Xizang(Wu & Liu, 2012, pp. 120 – 122; Wang HY et al., 2013, pp. 265 – 266), shanxi(Wang HY et al., 2013, pp. 265 – 266), Hubei, Jiangxi(Wu & Liu, 2012, pp. 120 – 121), Anhui, Zhejiang(Xu B. S., 1989, p. 187; Wu & Liu, 2012, pp. 120 – 121), Guangxi(Wu & Liu, 2012, pp. 120 – 121), Fujian(Wu J. N. et al. 1982, p. 6), China(prov. not indicated: Hue, 1890, p. 310).

f. caespitosum Asahina, 簇生变型

J. J. B. 37: 261(1962).

Jiangxi(M. Nuno, 1963, p. 198).

f. griscum Inumaru 灰色变型

in Acta Phytotax. et Geobot. 8: 226(1939).

Taiwan(Wang & Lai, 1973, p. 92).

Nephroma isidiosum(Nyl.) Gyeln. 裂芽肾盘衣

Annals Cryptog. Exot. 4: 126(1931).

≡*Nephromium isidiosum* Nyl., *Flora*, Regensburg 69: 417(1866).

On bark of trees, rarely on the ground and soil covered rocks.

Heilongjiang, Jilin(Wu & Liu, 2012, pp. 122 – 123), Sichuan, Yunnan, Xizang(Wu & Liu, 2012, pp. 122 – 123; Wang HY et al., 2013, pp. 269 – 270), Shaanxi(He & Chen 1995, p. 44 without citation of specimens and their localities; Wang HY et al., 2013, pp. 269 – 270), Hubei(Chen, Wu & J. C. Wei, 1989, p. 408; Wu & Liu, 2012, pp. 122 – 123).

Nephroma javanicum Gyeln. 爪哇肾盘衣

in Annal. Cryptog. Exot., 4: 136 et 148(1932). (1931).

Taiwan(Wang & Lai, 1973, p. 92).

Notes: Whether the synonym of *N. helveticum* remains to be studied further(Liu, 2012).

Nephroma moeszii Gyeln. 牟氏肾盘衣

in Annal. Cryptog. Exot., 4: 132 et 145(1932)(1930).

Taiwan(Inum. 1941, p. 70; Wang & Lai, 1973, p. 92).

Notes: Whether the synonym of *N. helveticum* remains to be studied further(Liu, 2012).

Nephroma parile(Ach.) Ach. 镶边肾盘衣

Lich. univ.: 522(1810).

≡*Lichen parilis* Ach., *Lich. suec. prodr.* (Linköping): 164(1799)[1798].

≡*Nephroma resupinatum* var. *parile*(Ach.) Mudd.

On rocks and on bark of trees.

Neimenggu, Heilongjiang(Wu & Liu, 2012, pp. 123 – 124), Xinjiang(Abdulla Abbas & Wu, 1998, p. 134; Abdulla Abbas et al., 2001, p. 365; Wu & Liu, 2012, pp. 123 – 124), Xizang(J. C. Wei & Jiang, 1986, p. 20),

Hubei(Wu & Liu, 2012, pp. 123 – 124), China(prov. not indicated: Zahlbr., 1934 b, p. 199).

Nephroma resupinatum(L.) Ach. 毛腹肾盘衣

Lich. univ.: 521(1810).

≡*Lichen resupinatus* L., *Sp. pl.* 2: 1148(1753).

On bark of trees.

Neimenggu(Wu & Liu, 2012, p. 124), Heilongjiang(Chen et al. 1981, p. 151; Luo, 1984, p. 85), Jilin(Wu & Liu, 2012, p. 124).

Nephroma sinense Zahlbr. 中国肾盘衣

in Handel-Mazzetti, *Symbolae Sinicae* 3: 87(1930).

Type: Yunnan, Rock no. 11751.

Yunnan(Zahlbr., 1934 b, p. 199; M. Nuno, 1963, p. 198).

Notes: Whether the synonym of *N. helveticum* remains to be studied further(Liu, 2012).

Nephroma subhelveticum H. Y. Wang in Wang HY et al. 亚瑞士肾盘衣

Mycotaxon 125: 270 – 272(2013).

Type-China. Jiangxi, Ji-An, Qianmocun, Mt. Nanfengmian, alt. 1300 m, 1 Nov 2010, on bark, H. Y. Wang 20106583(Holotype, SDNU; GenBank, JX867680).

Nephroma subparile Gyeln 光腹肾盘衣

in Magy. Botan. Lapok: 24(1930).

On moss covered rock.

Xizang(J. C. Wei & Y. M. Jiang, 1986, p. 20), Hubei(Chen, Wu & J. C. Wei, 1989, p. 408).

Nephroma tropicum(Müll. Arg.) Zahlbr. 热带肾盘衣

Flora, Regensburg 66: 21(1883). f. tropicum

≡*Nephromium tropicum* Müll. Arg., *Flora*, Regensburg 66: 21(1883).

= *Nephroma asahinae* Zahlbr. in Fedde, *Repertorium*, 33: 35(1933).

Type: Taiwan, Asahina no. 236.

= *Nephroma sordideluteum* Inum. in *Acta Phytotax. et Geobot.* VIII: 223(1939).

On bark of *Rhododendron* sp.

Yunnan(Zahlbr., 1930 b, p. 87; M. Nuno, 1963, p. 198), Xizang(J. C. Wei & Jiang, 1986, p. 20), Hubei(Müll. Arg. 1893, p. 236; Zahlbr., 1930 b, p. 87; Chen, Wu & J. C. Wei, 1989, p. 409), Zhejiang(Xu B. S., 1989, p. 187), Fujian(Wu J. N. et al. 1984, p. 6), Taiwan(Zahlbr., 1933 c, p. 35 & 1940, p. 281; Inum. 1941, p. 69; Lamb, 1963, p. 427; Wang & Lai, 1973, p. 92, cited as f. *tropicum*).

f. lividogriseum Gyeln. 肝色变型

in Ann. Cryptog. Exot. 4: 126, 149(1931).

Taiwan(Wang & Lai, 1973, p. 92).

Notes: Whether the synonym of N. helveticum remains to be studied further(Liu, 2012).

(67) Peltigeraceae Dumort. [as ' Peltigereae'] **地卷科**

Comment. bot. (Tournay): 68(1822).

Peltigera Willd. **地卷属**

Fl. berol. prodr.: 347(1787).

Peltigera aphthosa(L.) Willd. 绿皮地卷

Fl. berol. prodr.: 347(1787). f. aphthosa 原变型

≡*Lichen aphthosus* L., *Sp. pl.* 2: 1148(1753).

On the ground among mosses.

Heilongjiang(Chen et al. 1981, p. 155; Chen, 1986, p. 20; Wu & Liu, 2012, pp. 129 – 130), Jilin(Chen et al. 1981, p. 155; Wu & Liu, 2012, p. 129), Liaoning(as "Manchuria meridionalis" cited by Zahlbr., 1934 b, p.

199), Neimenggu(M. Sato, 1952, p. 172; Chen et al. 1981, p. 155; Chen, 1986, p. 20; Wu & Liu, 2012, p. 129), Xinjiang(Wu & Liu, 2012, pp. 129 – 130; Abdulla Abbas & Wu, 1998, p. 136; Abdulla Abbas et al., 2001, p. 365), Gansu(Wu & Liu, 2012, pp. 129 – 130), Shaanxi(Jatta, 1902, p. 467; Zahlbr., 1930 b, p. 89; He & Chen 1995, p. 44 without citation of specimens and their localities; Wu & Liu, 2012, pp. 129 – 130), Sichuan(Dajianlu < = *Kangding* > cited by Elenkin as Tibet, 1901, p. 38; Wu & Liu, 2012, pp. 129 – 130), Yunnan(Hue, 1900, p. 90; Zahlbr., 1930 b, p. 89; Wu & Liu, 2012, pp. 129 – 130), Xizang(J. C. Wei & Jiang, 1986, p. 15; Wu & Liu, 2012, pp. 129 – 130), Hubei(Müll. Arg. 1893, p. 235; Zahlbr., 1930 b, p. 89), China (prov. not indicated: Asahina & Sato in Asahina, 1939, p. 665).

f. verrucosa(Weber) Dietr. 小疣变型

Lichenographia Germanica oder Deutschlands Flechten in naturgetreuen Abbildungen nebst kurzen Beschreibungen (Jena): 27, tab. 127(1832) [1832 – 37].

≡*Lichen verrucosus* Weber, *Spicil. fl. goetting.*: 272(non Linn.) (1778).

Yunnan(Hue, 1900, p. 91; Zahlbr., 1930 b, p. 89).

Peltigera canina(L.) Willd. 犬地卷

Fl. berol. prodr.: 347(1787).

≡*Lichen caninus* L., *Sp. pl.* 2: 1149(1753).

On the ground and rotten timbers among mosses.

Beijing(Mt. Baihua shan in Beijing was mistaken for in Hebei by Wu & Liu, 2012, pp. 130 – 131), Heilongjiang(Chen et al. 1981, p. 155; Chen, 1986, p. 21; Wu & Liu, 2012. pp. 130 – 131), Neimenggu, Jilin (Chen et al. 1981, p. 155; Chen, 1986, p. 21; Wu & Liu, 2012, pp. 130 – 131), Xinjiang(Tchou. 1935, p. 317; H. Magn. 1940, p. 46; Moreau et Moreau, 1951, p. 187; Wu J. L. 1985, p. 76; Abdulla Abbas et al., 1993, p. 79; Abdulla Abbas & Wu, 1998, p. 136; Abdulla Abbas et al., 2001, p. 365; Wu & Liu, 2012, pp. 130 – 132), Shaanxi(Zahlbr., 1930 b, p. 88; He & Chen 1995, p. 44 without citation of specimens and their localities; Wu & Liu, 2012, pp. 130 – 132), Ningxia(Liu Huajie et al., 2010, p. 198), Xizang(J. C. Wei & Jiang, 1981, p. 1147; Wu & Liu, 2012, pp. 130 – 132), Sichuan(Stenroos et al., 1994, p. 331; Wu & Liu, 2012, pp. 130 – 131), Guizhou(Wu & Liu, 2012, pp. 130 – 131; Zhang T & J. C. Wei, 2006, p. 9), Yunnan(Hue, 1900, p. 92; Zahlbr., 1930 b, p. 88; Wei X. L. et al. 2007, p. 156; Wu & Liu, 2012, pp. 130 – 132), Shanxi(Wu & Liu, 2012, pp. 130 – 131), Hebei(Tchou, 1935, p. 317; Wu & Liu, 2012, pp. 130 – 131), Hubei(Chen, Wu & J. C. Wei, 1989, p. 403; Wu & Liu, 2012, pp. 130 – 131), Anhui(Lu, 1958, p. 267; Xu B. S., 1989, pp. 184 – 185; Wu & Liu, 2012, pp. 130 – 131), Zhejiang(Xu B. S., 1989, p. 184 – 185), Fujian(Wu J. N. et al. 1984, p. 6), Taiwan(Kurok. et al. 1966, p. 112).

var. canina 原变种

Heilongjiang(Asahina, 1952d, p. 374), Xizang(J. C. Wei & Jiang, 1986, p. 16).

f. spongiosa Tuck. 海绵变型

Syn. N. Amer. Lich. (Boston) 1: 109(1882).

On the ground.

Xizang(J. C. Wei & Jiang, 1986, p. 16), Taiwan(Zahlbr., 1933 c, p. 34; Inum. 1943, p. 5; Wang & Lai, 1973, p. 94).

f. ulorrhiza(Flörke) Schaer. 曲根变型

Enum. critic. lich. europ. (Bern): 20, tab. II, fig. 4(1850).

≡*Peltidea ulorrhiza* Flörke, in Hepp, *Flecht. -Fl. Würzburg*: 54(1824).

On the ground.

Shaanxi(Jatta, 1902, p. 467, as var.; Zahlbr., 1930 b, p. 88).

Peltigera cichoracea Delise ex Jatta 苣地卷

Nuovo G. bot. ital. 14(3): 170(1882).

On granite, soil, and on *Pseudotsuga wilsoniana*.

Taiwan(Aptroot, Sparrius & Lai 2002, p. 287).

Peltigera collina(Ach.) Schrad. 盾地卷

J. Bot. (Schrader) 3: 78(1801).

≡*Lichen collinus* Ach. , *Lich. suec. prodr.* (Linköping) : 162(1799) [1798].

= *Peltigera scutata*(Dicks.) Duby, *Bot. Gall.* , Edn 2(Paris) 2: 599(1830).

≡*Lichen scutatus* Dicks. , *Fasc. pl. crypt. Brit.* (London) 3: 18(1793).

On bark of trees and on the soil covered rocks.

Heilongjiang(Chen et al. 1981, p. 156; Chen, 1986, p. 21; Wu & Liu, 2012, pp. 132 – 133), Jilin(Chen, 1986, p. 21; Wu & Liu, 2012, pp. 132 – 133), Liaoning(Chen, 1986, p. 21), Neimenggu(Chen et al. 1981, p. 156; Chen, 1986, p. 21; Sun LY et al. , 2000, p. 36; Wu & Liu, 2012, pp. 132 – 133), GANSU(Wu & Liu, 2012, pp. 132 – 133), Shaanxi(Jatta, 1902, p. 468; Zahlbr. , 1930 b, p. 89; He & Chen 1995, p. 44 without citation of specimens and their localities; Wu & Liu, 2012, pp. 132 – 133), Sichuan, Yunnan(Wu & Liu, 2012, pp. 132 – 133), Hebei, Shanxi(Wu & Liu, 2012, pp. 132 – 133), Fujian(Wu J. N. et al. 1984, p. 6).

Peltigera coloradoensis Gyeln. 密茸地卷

Nytt Mag. Natur. 68: 270(1930).

Xinjiang(Wang XY, 1985, p. 338; Abudula Abbas et al. , 1993, p. 79).

Peltigera continentalis Vitik. 大陆地卷

in Stenroos S. , Vitikainen O. and Koponen T. *J. Hattori Bot. Lab.* 75: 332(1994).

Type: Mongolia. Ulan Bator City: NW slope of Mt. Bogd Uul, Dazysan Canyon, ca. 2 – 4km SW of Dazyan Hill Mountain, 1600 – 1650m, 12VII 1972 T. Ahti 29176(holotype, H) ; paratype: yunnan, lijiang, yulongshan, alt. 2600 – 2800m, 1987, T. Ahti 46381, J. B. Chen & L. S. Wang(H).

On soil covered rocks and rotten wood.

Sichuan, Yunnan(Stenroos et al. , 1994, p. 332; Wu & Liu, 2012, p. 134).

Peltigera degenii Gyeln. 裂边地卷

Magy. Bot. Lapok 25: 253(1927) [1926].

On the ground.

Heilongjiang, Liaoning(Chen, 1986, p. 21; Wu & Liu, 2012, pp. 134 – 135), Jilin(Wu & Liu, 2012, pp. 134 – 135), xinjiang(Wang XY, 1985, p. 338), Sichuan(Chen S. F. et al. , 1997, p. 222; Wu & Liu, 2012, pp. 134 – 135), Guizhou, Yunnan(Zhang T & J. C. Wei, 2006, p. 9; Wu & Liu, 2012, pp. 134 – 135), Hubei(Wu & Liu, 2012, pp. 134 – 135), Anhui(Lu, 1958, p. 263; Xu B. S. , 1989, p. 185; Wu & Liu, 2012, pp. 134 – 135), Zhejiang(Xu B. S. , 1989, p. 185; Wu & Liu, 2012, pp. 134 – 135), Taiwan(Wu & Liu, 2012, pp. 134 – 135), Guangxi(Wu & Liu, 2012, pp. 134 – 135).

Peltigera didactyla(With.) J. R. Laundon 分指地卷

Lichenologist 16(3) : 217(1984).

≡*Lichen didactylus* With. , *Bot. arr. veg. Gr. Brit.* (London) 2: 718(1776).

= *Peltigera spuria*(Ach.) DC. , in Lamarck & de Candolle, *Fl. franç.* , Edn 3(Paris) 2: 406(1805).

≡*Lichen spurius* Ach. , *Lich. suec. prodr.* (Linköping) : 159(1799) [1798].

= *Peltigera erumpens*(Taylor) Lange, *Acta Soc. Fauna Flora fenn.* 7(no. 1) : 182(1890).

≡*Peltidea erumpens* Taylor, *London J. Bot.* 6: 184(1847).

On the ground, rarely on bark of trees.

Beijing(Mt. Baihua shan in Beijing was mistaken for in Hebei by Wu & Liu, 2012, p. 136,), Heilongjiang (Asahina, 1952 d, p. 374; Chen et al. 1981, p. 155; Chen, 1986, pp. 22, 27 Wu & Liu, 2012, pp. 136 – 137), Liaoning(Chen et al. 1981, p. 155, 156; Chen, 1986, pp. 22, 27), Jilin, Neimenggu(Chen et al. 1981, p. 155,

156; Chen, 1986, pp. 22, 27; Wu & Liu, 2012, pp. 136 – 137), Xinjiang(Wang XL, 1985, p. 337 as *P. erumpens* and 338 as *P. spuria*; Abudula Abbas et al., 1993, p. 79; Abdulla Abbas & Wu, 1998, p 137; Abudula Abbas et al., 2001, p. 365; Wu & Liu, 2012, pp. 136 – 137), Shaanxi(Jatta, 1902, p. 468; Zahlbr., 1930 b, p. 89; He & Chen 1995, p. 44(no speciemen and locality were cited); Wu & Liu, 2012, pp. 136 – 137), Xizang(J. C. Wei & Jiang, 1981, p. 1147 & 1986, p. 17, 19; Wu & Liu, 2012, pp. 136 – 137), Sichuan(Chen S. F. et al., 1997, p. 222), Guizhou(Pat. et Oliv. 1907, p. 23; Zahlbr., 1930 b, p. 89; Zhang T & J. C. Wei, 2006, p. 9); Wu & Liu, 2012, pp. 136 – 137), Yunnan(Wu & Liu, 2012, pp. 136 – 137), Hebei(Wu & Liu, 2012, p. 136), Shandong (zhang F et al., 1999, p. 31; Zhao ZT et al., 2002, p. 7), Hubei(Chen, Wu & J. C. Wei, 1989, p. 407; Wu & Liu, 2012, pp. 136 – 137), Zhejiang(Xu B. S., 1989, p. 185), Fujian(Wu & Liu, 2012, pp. 136 – 137), Taiwan (Zahlbr., 1933 c, p. 35; Inum. 1943, p. 14; Wang & Lai, 1973, p. 95).

Peltigera dolichospora(Lu) Vitik. 长孢地卷

Lichenologist 18(4): 387(1986).

≡ *Peltigera polydactyla* var. *dolichospora* Lu, *Acta Phytotax. Sin.* 7(3): 264(1958).

Type: Sichuan, Lu no. 615 in NJU(holotype) and in HMAS-L(isotype). (also see D. Hawksw. 1972, p. 53).

On the ground among mosses.

Shaanxi(He & Chen 1995, p. 44(no speciemens and locality cited); Wu & Liu, 2012, p. 138), Sichuan, Yunnan(Wu & Liu, 2012, p. 138).

Peltigera elisabethae Gyeln. 平盘软地卷

Bot. Közl. 24: 135(1927). var. elisabethae 原变种

On the soil and moss covered rocks.

Beijing(Mt. Baihua shan in Beijing was mistaken for in Hebei by Wu & Liu, 2012, pp. 138 – 139), Neimenggu, Jilin, Heilongjiang(Chen, 1986, p. 21; Wu & Liu, 2012, pp. 138 – 139), Xinjiang(Abdulla Abbas & Wu, 1998, p 137; Abdulla Abbas et al., 2001, p. 365), Shaanxi(He & Chen 1995, p. 44(no speciemens and locality cited); Wu & Liu, 2012, pp. 138 – 140), Ningxia(Liu Huajie et al., 2010, p. 198), Xinjiang(Wang XY, 1985, p. 339, and p. 337 : no. 1221as *P. zahlbruckneri*, the same no. is reported as *P. elisabethae* – 2012, p. 140; Wu & Liu, 2012, pp. 138 – 140),, Qinghai, Gansu, Xizang, Yunnan(Wu & Liu, 2012, pp. 138 – 140), Sichuan (Stenroos et al., 1994, p. 332; Wu & Liu, 2012, pp. 138 – 140), Shanxi(Wu & Liu, 2012, pp. 138 – 139), Hebei(Wu & Liu, 2012, pp. 138 – 139), Anhui, Zhejiang(Xu B. S., 1989, p. 185), Fujian(Wu J. N. et al. 1984, p. 5, cited as a correct name of P. *microphylla*(Ach.) Gyeln.).

var. mauritzii(Gyeln.) J. C. Wei 无芽变种

Enum. Lich. China 183(1991). f. mauritzii 原变型

≡ *Peltigera mauritzii* Gyeln. in Hedwigia 68: 1(1928).

On the ground.

Heilongjiang, Jilin, Neimenggu(Chen, 1986, p. 24), Xinjiang(Wang XY, 1985, p. 337 as *P. mauritzii*; Abudula, Abbas et al., 1993, p. 79; Abdulla Abbas & Wu, 1998, p 137; Abdulla Abbas et al., 2001, p. 365), Shaanxi (Gyeln. 1928, p. 1 – 2; Zahlbr., 1930 b, p. 89; Dombr. 1971, p. 275), Anhui, Zhejiang(Xu B. S., 1989, p. 186).

var. mauritzii f. asiatica(Gyeln.) J. C. Wei 亚洲变型

Enum. Lich. China, 183(1991).

≡ *Peltigera mauritzii* var. *asiatica* Gyeln. in Fedde, *Repertorium*, 29: 8(1931).

China(Prov. not indicated: Zahlbr., 1932, p. 322).

var. mauritzii f. stuckenbergiae(Dombr.) J. C. Wei 卷叶变型

Enum. Lich. China, 183(1991).

≡ *Peltigera mauritzii* var. *stuckenbergiae* Dombr. Nov. Syst. Plant. non Vasc. 8: 275(1971).

On the ground and soil covered rocks.

Xinjiang(Abdulla Abbas et al. , 1993, p. 79; Abdulla Abbas & Wu, 1998, p 138; Abdulla Abbas et al. , 2001, p. 365) , Xizang(Wei & Jiang, 1986, p. 17) , Zhejiang(Xu B. S. , 1989, p. 186) .

var. mauritzii f. isidiifera(Dombr.) J. C. Wei, f. comb. nov. 裂芽变型

≡*Peltigera mauritzii* var. *stuckenbergiae* f. *isidiifera* Dombr. *Nov. Syst. Plant. non Vasc*. 8: 276(1971) .

Xinjiang(Abbas et al. , 1993, p. 79) .

var. mauritzii f. subpolydacty(Savicz.) J. C. Wei, f. comb. nov. 亚多枝变型

≡*Peltigera mauritzii* var. *stuckenbergiae* f. subpolydacty Savicz. In Dombr. *Nov. Syst. Plant. non Vasc*. 8: 278(1971) .

Xinjiang(Abbas et al. , 1993, p. 79) .

Peltigera evansiana Gyeln. 粒芽地卷

Bryologist 34: 16, 18(1931) .

Heilongjiang, Jilin, Liaoning(Chen, 1986, p. 22; Wu & Liu, 2012, pp. 140 – 141) , Xinjiang, Sichuan, Yunnan (Wu & Liu, 2012, pp. 140 – 141) .

Peltigera horizontalis(Huds.) Baumg. 平盘地卷

Fl. Lips. : 562(1790) .

≡*Lichen horizontalis* Huds. , *Fl. Angl*. : 453(1762) .

On the ground, rotten timbers and the base of tree trunks.

Neimenggu, Shanxi(Wu & Liu, 2012, pp. 141 – 142) , Heilongjiang(Asahina, 1952 d, p. 375; Chen et al. 1981, p. 155; Luo, 1984, p. 85; Chen, 1986, p. 22; Wu & Liu, 2012, pp. 141 – 142) , Jilin(Chen et al. 1981, p. 155; Chen, 1986, p. 22; Wu & Liu, 2012, pp. 141 – 142) , Xinjiang(Wu & Liu, 2012, pp. 141 – 142) , Gansu(H. Magn. 1940, p. 46) , Shaanxi(Baroni, 1894, p. 48; Jatta, 1902, p. 468; Zahlbr. , 1930 b, p. 89; Wu & Liu, 2012, pp. 141 – 142) , Sichuan, Yunnan(Wu & Liu, 2012, pp. 141 – 142) , Gguiahou(Pat. et Oliv. 1907, p. 28; Zahlbr. , 1930 b, p. 89) , Hebei(Tchou, 1935, p. 317; Moreau et Moreau, 1951, p. 187; Wu & Liu, 2012, pp. 141 – 142) , Hubei(Chen, Wu & J. C. Wei, 1989, p. 404; Wu & Liu, 2012, pp. 141 – 142) , Taiwan(Kurok. et al. 1966, p. 110; Wang & Lai, 1973, p. 94) .

Peltigera hymenina(Ach.) Delise 赭腹地卷

in Duby, *Bot. Gall*. , Edn 2(Paris) 2: 597(1830) .

≡*Peltidea hymenina* Ach. , *Methodus*, Sectio post. : 284(1803) .

≡*Peltigera polydactylon* var. *hymenina*(Ach.) Flot. , *Jber. schles. Ges. vaterl. Kultur* 28: 125(1850) .

= *Peltigera polydactylon* var. *crossoides* Gyeln. , *Magy. Bot. Lapok* 28: 61(1929) .

Neimenggu, Heilongjiang, Jilin(Chen, 1986, p. 23; Wu & Liu, 2012, p. 143) , Liaoning Chen, 1986, p. 23) , Xinjiang, Sichuan, Yunnan(Wu & Liu, 2012, p. 143) , Guizhou(Zhang T & J. C. Wei, 2006, p. 9) , Taiwan(Inum. 1943, p. 10; Wang & Lai, 1973, p. 94) .

Peltigera isidiophora L. F. Han & S. Y. Guo 穴芽地卷

The Bryologist, 118(1) : 46 – 53(2015) .

Type: China, Hebei: Xiaowutai Mountain, 40u009 N, 115u089 E, on ground, 1850 m alt. , 11 August 2013, Shou-Yu Guo, Liu-Fu Han & Wen- Xia Liu 20421. (holotype: HMAS-L) .

Peltigera kristinssonii Vitik. 克氏地卷

Ann. Bot. Fenn. 22: 291(1985) .

On the ground among mosses.

Xinjiang, Gansu(Zhao ZT & Liu HJ, 2002, p. 457; Wu & Liu, 2012, p. 144) .

Peltigera lactucifolia(With.) J. R. Laundon 乳叶地卷

Lichenologist 16(3) : 221(1984) .

≡*Lichen lactucifolius* With. , Bot. arr. veg. Gr. Brit. (London) 2: 718(1776) .

Shaanxi(He & Chen 1995, p. 44 as *P. lactucifolia* without citation of speciemens and locality).

Peltigera lepidophora(Vain.) Bitter 鳞地卷

Ber. dt. bot. Ges. 22: 251(1904).

≡*Peltigera canina* var. *lepidophora* Nyl. ex Vain., *Meddn Soc. Fauna Flora fenn.* 2: 49(1878).

On the ground and soil covered rocks.

Neimenggu, Heilongjiang(Chen et al. 1981, p. 155; Chen, 1986, p. 23; Wu & Liu, 2012, pp. 144 – 145), Jilin (Chen, 1986, p. 23; Wu & Liu, 2012, pp. 144 – 145), Xinjiang(Abdulla Abbas & Wu, 1998, p 138; Abdulla Abbas et al., 2001, p. 365), Shaanxi, Sichuan, Yunnan(Wu & Liu, 2012, pp. 144 – 145), Xizang(J. C. Wei & Jiang, 1986, p. 17; Wu & Liu, 2012, pp. 144 – 145).

Peltigera leucophlebia(Nyl.) Gyeln. 白腹地卷

Magy. Bot. Lapok 24: 79(1926) [1925].

≡*Peltigera aphthosa* var. *leucophlebia* Nyl., *Syn. meth. lich.* (Parisiis) 1(2): 323(1860).

= *Peltigera aphthosa* f. *variolosa* A. Massal., Sched. critic. (Veronae) 3: 64(1856).

≡ *Peltigera aphthosa* var. *variolosa* (A. Massal.) J. W. Thomson, *Trans. Wis. Acad. Sci. Arts Lett.* 38: 253 (1946).

≡*Peltigera leucophlebia* f. *variolosa*(A. Massal.) Gyeln., (1927).

≡*Peltigera variolosa*(A. Massal.) Gyeln., *Magy. Bot. Lapok* 28: 61(1929).

misapplied name:

Peltigera aphthosa sensu Tchou: Contr. Bot. Nat. Acad. Peiping, 3(6): 316(1935), non(L.) Willd.

On moss covered rocks and on rotten timbers.

Jilin(Tchou, 1935, p. 316; Chen et al. 1981, p. 155), Xinjiang(Abdulla Abbas & Wu, 1998, p 138; Abdulla Abbas et al., 2001, p. 365), Shaanxi(He & Chen 1995, p. 44) (no speciemens and locality cited), Xizang(J. C. Wei & Jiang, 1986, p. 16), Sichuan(Stenroos et al., 1994, p. 332), Guizhou(Zhang T & J. C. Wei, 2006, p. 9), Hubei(Chen, Wu & J. C. Wei, 1989, p. 404), Taiwan(Inum. 1943, p. 169; Wang & Lai, 1973, p. 94 & 1976, p. 228).

f. leucophlebia 原变型

Neimenggu, Heilongjiang, Jilin(Chen, 1986, p. 23).

f. vrangiana(Gyeln.) Dombr. 薄叶变型

Nov. sist. Niz. Rast. 9: 263(1972).

≡*Peltigera vrangiana* Gyeln., *Magy. Bot. Lapok* 31: 46(1932).

Heilongjiang, Jilin(Chen, 1986, p. 24).

f. angustiloba(Nikolskij & Fokin) Dombr. 狭叶变型

Nov. sist. Niz. Rast. 9: 263(1972).

≡*Peltigera aphthosa* f. *angustiloba* Nikolskij & Fokin, k lishainikovoi flore Vyatckovo kraya 6(1926) [1927] (in Russion).

Neimenggu, Heilongjiang, Jilin(Chen, 1986, p. 24).

[f. leucophlebia & f. angustiloba(Rok. et Nik..) Dombr.]

On the ground, rarely on the soil or moss covered rocks, occasionally on bark of trees.

Beijing(Mt. Baihua shan and Dongling shan in Beijing were mistaken for in Hebei byWu & Liu, 2012, pp. 145 – 146), Neimenggu, Heilongjiang, Jilin, Liaoning, Xinjiang, Qinghai, Shaanxi, Sichuan, Yunnan, Guizhou, Hebei, Henan, Hubei(Wu & Liu, 2012, pp. 145 – 147).

Peltigera malacea(Ach.) Funck 软地卷

Cryptog. Gewächse 33: 5(1827).

≡*Peltidea malacea* Ach., *Syn. meth. lich.* (Lund): 240(1814).

On the ground among mosses and grasses and on the base of tree trunks.

Neimenggu, Jilin, Heilongjiang(Chen et al. 1981, p. 155; Chen, 1986, p. 24; Wu & Liu, 2012, p. 148) , Xinjiang (H. Magn. 1940, p. 46) , Shaanxi(Jatta, 1902, p. 467; Zahlbr. , 1930 b, p. 88) .

Peltigera membranacea(Ach.) Nyl. 膜地卷

Bull. Soc. linn. Normandie, sér. 4 1: 74(1887) .

≡*Peltidea canina* var. *membranacea* Ach. , *Lich. univ.* : 518(1810) .

≡*Peltigera canina* f. *membranacea*(Ach.) Duby, *Bot. Gall.* (Paris) 2: 598(1830) .

On the ground and rotten logs.

Neimenggu, Hebei(Wu & Liu, 2012, p. 149) , Heilongjiang(Chen, 1986, pp. 24 – 25; Wu & Liu, 2012, p. 149) , Jilin(Chen et al. 1981, p. 155; Chen, 1986, p. 24; Wu & Liu, 2012, p. 149) , Liaoning(Chen, 1986, pp. 24 – 25) , Xinjiang(Abdulla Abbas et al. , 1993, p. 79; Abdulla Abbas & Wu, 1998, p 138; Abdulla Abbas et al. , 2001, p. 366; Wu & Liu, 2012, pp. 149 – 150) , Shaanxi(Jatta, 1902, p. 467; Zahlbr. , 1930 b, p. 88) , Xizang (J. C. Wei & Jiang, 1986, p. 16; Wu & Liu, 2012, pp. 149 – 150) , Sichuan(Stenroos et al. , 1994, p. 333; Wu & Liu, 2012, p. 149) , Yunnan(Wu & Liu, 2012, p. 149) , guizhou(Zhang T & J. C. Wei, 2006, p. 9) , Hubei (Müll. Arg. 1893, p. 235; Zahlbr. , 1930 b, p. 88; Chen, Wu & J. C. Wei, 1989, p. 404; Wu & Liu, 2012, p. 149) , Taiwan(Inum. 1943, p. 8) , China(Prov. not indicated: Asahina, 1939, p. 663) .

Peltigera meridiana Gyeln. 南方地卷

in Magy. Bot. Lapok, 26: 47(1927) .

f. crispoides Gyeln. 皱波变型

in Magy. Bot. Lapok, 28: 61(1929) .

On the ground among mosses.

Taiwan(Zahlbr. , 1933 c, p. 34; Wang & Lai, 1973, p. 94) .

Peltigera microphylla(Anders) Gyeln. 细裂地卷

Bryologist 34: 18, 19(1931) .

≡*Peltigera polydactylon* f. *microphylla* Anders in *Lotos*, LXXVI: 320(1928) .

On tree trunk and moss- and soil-covered substrata.

Hubei(Chen, Wu & J. C. Wei, 1989, p. 405) , Yunnan(Wang L. S. 2008, p. 538) , Xizang(J. C. Wei & Jiang, 1986, p. 17) .

Peltigera neckeri Hepp ex Müll. Arg. 光滑地卷

Mém. Soc. Phys. Hist. nat. Genève 16(2) : 370(1862) .

= *Peltigera polydactyloides* Nyl. , *Flora*, Regensburg 46: 265(1863) .

On the ground, and on the soil and moss covered rocks.

Neimenggu, Jilin(Chen, 1986, p. 25; Wu & Liu, 2012, p. 150) , Heilongjiang(Chen et al. 1981, p. 158; Chen, 1986, p. 25; Wu & Liu, 2012, p. 150) , Liaoning(Wu & Liu, 2012, p. 150) , Shaanxi(He & Chen 1995, p. 44) (no speciemens and locality cited) , Sichuan, Yunnan(Wu & Liu, 2012, pp. 150 – 151) , guizhou(Zhang T & J. C. Wei, 2006, p. 9; Wu & Liu, 2012, pp. 150 – 151) , Hubei(Chen, Wu & J. C. Wei, 1989, p. 405; Wu & Liu, 2012, pp. 150 – 151) .

Peltigera neopolydactyla(Gyeln.) Gyeln. 长根地卷

Rev. Bryol. Lichénol. , N. S. 5: 68(1932) . f. neopolydactyla

≡*Peltigera polydactylon* var. *neopolydactylis* Gyeln. in *Rev. Bryol. et Lichenol.* 5: 71(1932) .

Misapplied name:

Peltigera dolichorrhiza auct. non(Nyl.) Nyl. , *Lich. Nov. Zeland.* (Paris) : 43(1888) .

On the ground among mosses.

Heilongjiang, Jilin, Liaoning(Chen, 1986, p. 25; Wu & Liu, 2012, pp. 151 – 152) , Xinjiang(Wu & Liu, 2012, pp. 151 – 152; Abdulla Abbas & Wu, 1998, p. 139; Abdulla Abbas et al. , 2001, p. 366 as P. "*neopolydatata*") , Shaanxi(Jatta, 1902, p. 468; Zahlbr. , 1930 b, p. 89 ; He & Chen 1995, p. 44) (no speciemens and locality cited), Sichuan, Guizhou, Xizang(Wu & Liu, 2012, pp. 151 – 152), Yunnan(Hue, 1900, p. 99; Zahlbr., 1930 b, p. 89; Wu & Liu, 2012, pp. 151 – 152), Hubei(Lu, 1958, p. 265; Chen, Wu & J. C. Wei, 1989, p. 405; Wu & Liu, 2012, pp. 151 – 152), Anhui(Xu B. S., 1989, p. 185), Zhejiang(Xu B. S., 1989, p. 185; Wu & Liu, 2012, pp. 151 – 152), Guangdong, Guangxi(Wu & Liu, 2012, pp. 151 – 152), Fujian(Wu J. N. et al. 1984, p. 5), Taiwan(Zahlbr., 1933 c, p. 34; Inum. 1943, p. 6; Kurok. et al. 1966, p. 109 – 110; Wang & Lai, 1973, p. 94).

f. subincusa Gyeln. 亚霜变型

in Zahlbr. in Fedde, *Repertorium*, 33: 34(1933).

Type: Taiwan, Asahina no. 243.

On the ground.

Taiwan(Inum. 1943, p. 15; Wang & Lai, 1973, pp. 94, 95 & 1976, p. 228).

Peltigera nigripunctata Bitter 黑瘿地卷

in Bericht. Deutsch. Botan. Gesellsch. 27: 194(1909). f. nigripunctata

On the ground.

Heilongjiang(Chen et al. 1981, p. 155; Chen, 1986, p. 25), Jilin(Chen et al. 1981, p. 155; Chen, 1986, p. 25; Wu & Liu, 2012, p. 153), Neimenggu(Chen et al. 1981, p. 155; Sun LY et al., 2000, p. 36; Wu & Liu, 2012, p. 153), Xinjiang(Abdulla Abbas et al., 1993, p. 79; Abdulla Abbas & Wu, 1998, p. 139; Abdulla Abbas et al., 2001, p. 366; Wu & Liu, 2012, p. 153), Shaanxi(He & Chen 1995, p. 44)(no speciemen and locality were cited), Sichuan(Stenroos et al., 1994, p. 333; Chen S. F. et al., 1997, p. 222; Wu & Liu, 2012, p. 153), Yunnan(Gyeln. 1927, p. 45; Wu & Liu, 2012, p. 153), Xizang(J. C. Wei & Jiang, 1986, p. 16; Wu & Liu, 2012, p. 153), Hebei(Wu & Liu, 2012, p. 153), Hubei(Chen, Wu & J. C. Wei, 1989, p. 406; Wu & Liu, 2012, p. 153), Taiwan(Kurok. et al. 1966, pp. 105 – 108; Wang & Lai, 1973, p. 94).

f. hypocephalodiata Gyeln. 腹瘿变型

Magy. Bot. Lapok 45 – 47(1927).

Typ: Yunnan, Rock no. 11760.

Yunnan(Zahlbr., 1930 b, p. 88 & 1932a, p. 323).

Peltigera pindarensis D. D. Awasthi & M. Joshi 粉色地卷

Kavaka 10: 58(1982).

On rocks.

Sichuan(Chen S. F. et al., 1997, p. 222).

Peltigera polydactylon(Neck.) Hoffm. 多指地卷

Descr. Adumb. Plant. Lich. 1(1): 19(1789)[1790].

≡*Lichen polydactylus* Neck., *Method. Muscor.*: 85(1771).

On the ground among mosses; on the base of tree trunks and on rotten logs.

Heilongjiang(Chen et al. 1981, p. 155; Luo, 1984, p. 85; Wu & Liu, 2012, pp. 154 – 155), Shanxi(Wu & Liu, 2012, pp. 154 – 155), Jilin, Liaoning, Neimenggu(Chen et al. 1981, p. 155; Wu & Liu, 2012, pp. 154 – 155), Xinjiang(Moreau et Moreau, 1951, P. 187; Wang XY, 1985, p. 337; Abdula, Abbas et al., 1993, p. 79; Abdulla Abbas & Wu, 1998, p. 139; Abdulla Abbas et al., 2001, p. 366 as *P.* "*polydactyla*"; Wu & Liu, 2012, pp. 154 – 156), ningxia(Liu Huajie et al., 2010, p. 198), Shaanxi(He & Chen 1995, p. 44)(no speciemens and locality cited), Sichuan(Stenroos et al., 1994, p. 333; Chen S. F. et al., 1997, p. 222; Wu & Liu, 2012, pp. 154 – 155), Yunnan(Hue, 1887a, p. 22 & 1889, p. 167 & 1900, p. 97; Zahlbr., 1930 b, p. 89 & 1934, p. 199; Wu & Liu, 2012, pp. 154 – 155), Guizhou(Zhang T & J. C. Wei, 2006, p. 9; Wu & Liu, 2012, pp. 154 – 155), Hunan (Zahlbr., 1930 b, p. 89; Wu & Liu, 2012, pp. 154 – 155), Xizang(Wu & Liu, 2012, pp. 154 – 156), Hubei (Chen, Wu & J. C. Wei, 1989, p. 406; Wu & Liu, 2012, pp. 154 – 155), Jiangxi(Wu & Liu, 2012, pp. 154 – 155), Anhui(Xu B. S., 1989, p. 185), Fujian(Wu J. N. et al. 1984, p. 5), Taiwan(Inum. 1943, p. 9).

f. polydactylon 原变型

Heilongjiang, Jilin, Liaoning(Chen, 1986, p. 26), Taiwan(Wang & Lai, 1973, p. 94).

= *Peltidea horizontalis* var. *lophyra* Ach., *Lich. univ.*: 516(1810).

= *Peltigera polydactylon* var. *dissecta* Müll. Arg., *Flora, Regensburg* 74(3): 374(1891).

Neimenggu, Heilongjiang, Jilin, Liaoning(Chen, 1986, p. 26), Hubei(Müll. Arg. 1893, p. 236; Zahlbr., 1930 b, p. 89), Taiwan(Inum. 1943, p. 10; Wang & Lai, 1973, p. 94).]

This form may probably be a synonym of *Peltigera elisabethae* Gyeln.

f. microcarpa(Ach.) Mérat 小果变型

Nouv. Flore Lich., Edn 2 1: 199(1821).

= Peltigera neckeri Hepp ex Müll. Arg., Mém. Soc. Phys. Hist. nat. Genève 16(2): 370(1862)

≡*Peltidea polydactyla* var. *microcarpa* Ach., *Lich. univ.*: 1 – 696(1810).

Heilongjiang, Liaoning(Chen, 1986, p. 26), Hunan(Zahlbr., 1930 b, p. 89).

Peltigera ponojensis Gyeln. 白脉地卷

Memor. Soc. Fauna Flora fenn. 7: 143(1931).

On the ground.

Heilongjiang(Wu & Liu, 2012, p. 157), Liaoning, Sichuan, Yunnan(Wu & Liu, 2012, p. 157; (Zhao ZT & Liu HJ, 2002, pp. 457 – 458).

Peltigera praetextata(Flörke ex Sommerf.) Zopf 裂芽地卷

Ann. Chemie 364: 299(1909).

≡*Peltidea ulorrhiza* var. *praetextata* Flörke ex Sommerf., *Suppl. Fl. lapp.* (Oslo): 123(1826).

≡*Peltigera canina* var. *praetextata*(Flörke ex Sommerf.) Vain., *Term. Füz.* 22: 306(1899).

≡*Peltigera rufescens* var. *praetextata*(Flörke ex Sommerf.) Nyl., *Syn. meth. lich.* (Parisiis) 1(2): 324(1860).

Misapplied name:

Peltigera rufescens sensu Lu: *Acta Phytotax. Sin.* 7(3): 267(1958), non(Weiss) Humb.

On the ground and on rotten logs.

Beijing(Mt. Baihua shan in Beijing was mistaken for in Hebei by Wu & Liu, 2012, p. 158), Heilongjiang, Jilin, Neimenggu(Chen et al. 1981, p. 156; Chen, 1986, p. 26; Wu & Liu, 2012, p. 158), Xinjiang(H. Magn. 1940, p. 46; Wang XY, 1985, p. 338; Abdula Abbas et al., 1993, p. 79; Abdulla Abbas et al., 1994, p. 22; Abdulla Abbas & Wu, 1998, pp. 139 – 140; Abdulla Abbas et al., 2001, p. 366; Wu & Liu, 2012, pp158-. 159), Gansu (Wu & Liu, 2012, pp158-. 159), Shaanxi(Jatta, 1902, p. 467; Zahlbr., 1930 b, p. 89; He & Chen 1995, p. 44) (no speciemens and locality cited), Ningxia(Liu Huajie et al., 2010, p. 198), Sichuan(Lu, 1958, p. 267; Stenroos et al., 1994, p. 333; Chen S. F. et al., 1997, p. 222; Wu & Liu, 2012, p. 158), Yunnan(Hue, 1900, p. 95; Zahlbr., 1930 b, p. 89; Wu & Liu, 2012, pp158-. 159), Guizhou(Wu & Liu, 2012, pp158-. 159; Zhang T & J. C. Wei, 2006, p. 9), Xizang(J. C. Wei & Jiang, 1986, p. 18; Wu & Liu, 2012, pp158-. 159), Hebei(Wu & Liu, 2012, p. 158), Hubei(Müll. Arg. 1893, p. 235; Zahlbr., 1930 b, p. 89; Chen, Wu & J. C. Wei, 1989, p. 406), Zhejiang(Xu B. S., 1989, p. 186), Fujian(Wu J. N. et al. 1984, p. 6).

f. subglabra Gyeln. 亚光变型

in Zahlbr. in Fedde, Repertorium, 33: 35(1933), nom. nud. & Lilloa, 3: 64(1938), cum descript.

Type: Taiwan, Asahina no. 237.

On the ground.

Peltigera pruinosa(Gyeln.) Inumaru 霜地卷

in Acta Phytotax. et Geobot. 12: 11(1943).

≡*Peltigera polydactylon* var. *pruinosa* Gyeln., in Zahlbr. *Cat. Lich. Univ.* 8: 324(1932).

Misapplied name:

Peltigera polydactyla(typica) sensu Lu, *Acta Phytotax. Sin.* 7(3) : 265(1958) , non(Neck.) Hoffm.

On the ground.

Xinjiang(Abbas et al. , 1993, p. 79) , Sichuan(Lu, 1958, p. 265; Chen S. F. et al. , 1997, p. 222; Wu & Liu, 2012, p160) , Yunnan, Guizhou, Xizang(Wu & Liu, 2012, pp. 160 – 161) , Anhui(Wu & Liu, 2012, p160) , Taiwan(Inum. 1943, p. 11; Kurok. et al. 1966, p. 111; Wang & Lai, 1973, p. 94, as var. *pruinosa*) .

var. congesta Inum. 密集变种

Acta Phytotax. et Geobot. XII: 12(1943) .

On the ground.

Taiwan(Inum. 1943, p. 12; Wang & Lai, 1973, p. 94) .

f. crispata Inum. 皱波变型

Acta Phytotax. et Geobot. XII: 12(1943) .

On the ground.

Taiwan(Inum. 1943, p. 12; Wang & Lai, 1973, p. 94) .

var. spongiosa Inum. 海绵变种

Acta Phytotax. et Geobot. 12(1) : 13(1943) .

Type: Taiwan, Inum. no. 6062.

On the ground.

Taiwan(Wang & Lai, 1973, p. 94) .

Peltigera rufescens(Weiss) Humb. 地卷

Fl. Friberg. Spec. (Berlin) : 2(1793) .

≡*Lichen caninus* var. *rufescens* Weiss, (1770) .

≡*Peltigera canina* var. *rufescens*(Weiss) Mudd, *Man. Brit. Lich.* : 82(1861) .

Misapplied name:

Peltigera malacea sensu Lu, *Acta Phytotax. Sin.* 7(3) : 266(1958) , non(Ach.) Funck.

On rotten logs.

Beijing(Mt. Baihua shan in Beijing was mistaken for in Hebei by Wu & Liu, 2012, p. 161) , Heilongjiang(Asahina, 1952d, p. 375; Chen et al. 1981, p. 156; Chen, 1986, p. 27; Wu & Liu, 2012, p. 161) , Jilin(Chen, 1986, p. 27; Wu & Liu, 2012, pp. 161 – 162) , Liaoning, Neimenggu(Chen, 1986, p. 27) , Xinjiang(H. Magn. 1940, p. 46; Wang XY, 1985, p. 338; Wu J. L. 1985, p. 76; Abdula Abbas et al. , 1993, p. 79; Abdulla Abbas & Wu, 1998, p. 140; Abdulla Abbas et al. , 2001, p. 366; Wu & Liu, 2012, pp. 161 – 162) , Shaanxi(Jatta, 1902, p. 467; Zahlbr. , 1930 b, p. 89; He & Chen 1995, p44(no speciemens and locality cited) ; Wu & Liu, 2012, pp. 161 – 162) , Ningxia(Liu Huajie et al. , 2010, p. 199) , Xizang(Wu & Liu, 2012, pp. 161 – 162) , Sichuan(Wu & Liu, 2012, pp. 161 – 162) , Guizhou, Hebei(Wu & Liu, 2012, p. 161) , Yunnan(Hue, 1889, p. 167 & 1900, p. 94; Zahlbr. , 1930 b, p. 89; Wu & Liu, 2012, pp. 161 – 162) , Hubei(Chen, Wu & J. C. Wei, 1989, p. 407; Wu & Liu, 2012, pp. 161 – 162) , Zhejiang(Xu B. S. , 1989, p. 186) , Fujian(Wu J. N. et al. 1984, p. 6) , Taiwan(Kurok. et al. 1966, p. 113; Wang & Lai, 1973, p. 94) .

f. dilaceratoides Gyeln. 碎裂变型

Hedwigia 68: 3(1928) .

On moss-covered substratum.

Xizang(J. C. Wei & Y. M. Jiang, 1981, p. 1147 & 1986, p. 18) .

Peltigera scabrosa Th. Fr. 小瘤地卷

Lich. Arctoi 3: 45(1861) [1860] .

Misapplied name:

Peltigera pulverulenta auct. non(Tayl.) Nyl.

On the ground.

Neimenggu(Chen, 1986, p. 27) , Heilongjiang, Jilin(Chen et al. 1981, p. 156; Chen, 1986, p. 27) , Shaanxi(Jatta, 1902, p. 467; Zahlbr. , 1930 b, p. 89) , Xizang(Wei & Chen, 1974, p. 175; Wei & Jiang, 1981, p. 1146 & 1985, p. 18) .

Peltigera sorediata(H. Olivier) Fink(1921) 粉芽地卷

in Fink, *Lich. Flora U. S.* , 186, 189(1935) .

≡ *Peltigera rufescens* f. *sorediata* H. Olivier, *Flore analytique et dichotomique des Lichens de l'Orne et départements circonvoisins* 1: 92(1882) .

≡ *Peltigera polydactylon* f. *sorediata* Schaer. , *Enum. critic. lich. europ.* (Bern) : 21(1850) .

[≡ *Peltigera canina* var. *spuria* f. *sorediata* Schaer.]

On the ground.

Neimenggu, Heilongjiang, Jilin(Chen et al. 1981, p. 156; Wu & Liu, 2012, p. 163) , Sichuan, Yunnan, Xizang (Wu & Liu, 2012, pp. 163 – 164) .

Peltigera venosa(L.) Hoffm. 小地卷

Descr. Adumb. Plant. Lich. 1(1) : 31(1789) [1790] .

≡ *Lichen venosus* L. , *Sp. pl.* 2: 1148(1753) .

On soil.

Neimenggu(Chen, 1986, p. 28; Wu & Liu, 2012, pp. 164 – 165) , Xinjiang(Abdulla Abbas et al. , 1993, p. 79; Abdulla Abbas & Wu, 1998, p. 140; Abdulla Abbas et al. , 2001, p. 366) , Shaanxi(He & Chen 1995, p. 44(no speciemen and locality were cited) ; Wu & Liu, 2012, pp. 164 – 165) , Sichuan(Stenroos et al. , 1994, p. 334; Wu & Liu, 2012, pp. 164 – 165) , China(prov. not indicated: Asahina & Sato in Asahina, 1939, p. 665) .

var. yunnana Zahlbr. 云南变种

in Handel-Mazzetti, *Symb. Sin.* 3: 90(1930) .

Type: from Yunnan, Lijiang. Ganhaizi, alt. 3300m, 13. VI. 1915(6729) .

Peltigera wulingensis L. F. Han & S. Y. Guo 雾灵地卷

Lichenologist 45(3) : 333 – 335(2013) .

Typus: China, Hebei, Wuling Mountain, 40_330N, 117_260 E, on ground, 1820 m alt. , 14 August 2011, Liu-Fu Han 8035(HMAS-L—holotypus) .

Solorina Ach. **散盘衣属**

K. Vetensk-Acad. Nya Handl. 29: 228(1808) .

Solorina bispora Nyl. 双孢散盘衣

Syn. meth. lich. (Parisiis) 1(2) : 331(1860) .

On the ground among mosses and on soil covered rock.

Xinjiang(Wang XL, 1985, p. 336; Wu J. L. , 1985, p. 77; Abdula Abbas et al. , 1993, p. 81; Abdula, Abbas et al. , 1994, p. 23; Abdulla Abbas & Wu, 1998, p. 141; Abdulla Abbas et al. , 2001, p. 368; Wu & Liu, 2012, p. 166) , ningxia(Liu Huajie et al. , 2010, p. 199) , Shaanxi, Yunnan(Wu & Liu, 2012, p. 166) , Xizang(J. C. Wei & Jiang, 1986, p. 19; W. Obermayer, 2004, p. 509; Wu & Liu, 2012, p. 166) , Sichuan, Qinghai(W. Obermayer, 2004, p. 509) , Hebei(Wu & Liu, 2012, p. 166) .

Solorina crocea(L.) Ach. 橘黄散盘衣

K. Vetensk-Acad. Nya Handl. 29: 228(1808) .

≡ *Lichen croceus* L. , *Sp. pl.* 2: 1149(1753) .

Among large boulders(gneiss) , partly overhanging Xizang(W. Obermayer, 2004, p. 508) .

Notes: The two specimens(0680 & s06455) are different from typical European material.

Solorina octospora Arnold 八孢散盘衣

Verh. zool. -bot. Ges. Wien 26: 371(1876).

On the ground or on soil covered rocks.

Beijing(Wu & Liu, 2012, pp. 166 – 167), Xinjiang(Moreau et Moreau, 1951, p. 186; Wu & Liu, 2012, pp. 166 – 167), Shaanxi, Yunnan(Wu & Liu, 2012, pp. 166 – 167), Sichuan, Xizang(W. Obermayer, 2004, p. 509).

Solorina platycarpa Hue 宽果散盘衣

Bull. Soc. bot. Fr. 54: 419(1907).

On the ground or on moss covered rocks.

Shaanxi(Wu J. L., 1987, p. 82), Sichuan, Yunnan(Wu & Liu, 2012, pp. 167 – 168).

Solorina saccata(L.) Ach. 凹散盘衣

K. Vetensk-Acad. Nya Handl. 29: 228(1808).

≡*Lichen saccatus* L., *Fl. Suec.*, Edn 2: 419(1755).

On the ground and on soil covered rocks.

Beijing(Mt. Baihua shan in Beijing was mistaken for in Hebei by Wu & Liu, 2012, p. 168), Xinjiang(Tchou, 1935, p. 318 Wu & Liu, 2012, p. 168), Qinghai(W. Obermayer, 2004, p. 509), ningxia(Liu Huajie et al., 2010, p. 199), Shaanxi(He & Chen 1995, p. 45(no speciemens and locality cited); Wu & Liu, 2012, p. 168), Sichuan(Stenroos et al., 1994, p. 334; W. Obermayer, 2004, p. 509; Wu & Liu, 2012, p. 168), Xizang, Yunnan (Wu & Liu, 2012, p. 168), guizhou(Zhang T & J. C. Wei, 2006, p10; Wu & Liu, 2012, p. 168), Shanxi(M. Sato, 1980, p. 64), Hubei(Chen, Wu & J. C. Wei, 1989, p. 407; Wu & Liu, 2012, p. 168).

var. simensis(Hochst. ex Flot.) Nyl. 平梁变种

Mém. Soc. Sci. nat. Cherbourg 5: 101(1857).

On the ground & on bark of trees.

Sichuan(Wu & Liu, 2012, p. 169), Yunnan(Hue, 1889, p. 312 & 1900, p. 88; Zahlbr., 1930 b, p. 86; Wu & Liu, 2012, p. 169), Taiwan(Zahlbr., 1933 c, p. 34; Asahina, 1935b, p. 163; Asahina & M. Sato in Asahina, 1939, p. 659; Ikoma, 1983, p. 99), China(prov. not indicated: Hue, 1890, p. 312; Zahlbr., 1925, p. 419).

Solorina spongiosa(Ach.) Anzi 绵散盘衣

Comm. Soc. crittog. Ital. 1(3): 136(1862).

≡*Collema spongiosum* Ach., *Lich. univ.*: 661(1810).

On calcareous soil or rocks.

Jilin, Gansu(Wu J. L., 1987, p. 83). Shaanxi(Wu J. L., 1987, p. 83; Wu & Liu, 2012, pp. 169 – 170), Sichuan (W. Obermayer, 2004, p. 510; Wu & Liu, 2012, pp. 169 – 170), Xizang(W. Obermayer, 2004, p. 510).

[16] Teloschistales D. Hawksw. & O. E. Erikss. **黄枝衣目**

Syst. Ascom. 5(1): 183(1986)

(68) Letrouitiaceae Hafellner & Bellem. **多极孢衣科**

Nova Hedwigia 35: 382(1982)[1981].

Letrouitia Hafellner & Bellem. **多极孢衣属**

Nova Hedwigia 35(2 & 3): 281(1982)[1981].

Letrouitia aureola(Tuck.) Hafellner & Bellem. 金黄多极孢衣

Nova Hedwigia 35(2 & 3): 281(1982)[1981].

≡*Lecidea aureola* Tuck., *Proc. Amer. Acad. Arts* 6: 281(1866)[1864].

On *Vitex quinquefolia*.

Taiwan(Aptroot and Sparrius, 2003, p. 27).

Letrouitia domingensis(Pers.) Hafellner & Bellem. 多极孢衣

Nova Hedwigia 35(2 & 3): 281(1982).[1981].

≡*Patellaria domingensis* Pers., *Ann. Wetter. Gesellsch. Ges. Naturk.* 2: 12(1811)[1810].

≡*Bombyliospora domingensis*(Pers.) Zahlbr. var. *boninensis* Asahina, *Journ. Jap. Bot.* 10(6) : 352(1934) .

Syntype: from China(Taiwan) & Japan(Bonin) .

= *Bombyliospora domingensis* (Pers.) Zahlbr. var. *glaucarpa* (Nyl.) Vain.

Taiwan(Asahina, 1934b, p. 352; M. Sato, 1936b, p. 573; Asahina & Sato in Asahina, 1939, p. 767; Wang & Lai, 1973, p. 88) .

Letrouitia parabola(Nyl.) R. Sant. & Hafellner 线形多极孢衣

in Hafellner & Bellemère, *Nova Hedwigia* 35(2 & 3) : 281(1982)[1981] .

≡*Lecidea parabola* Nyl. , *Bull. Soc. linn. Normandie, sér.* 2 2: 90(1868) .

On *Ficus* sp.

Taiwan(Aptroot and Sparrius, 2003, p. 27) .

Letrouitia transgressa(Malme) Hafellner & Bellem. 横多极孢衣

in Hafellner, *Nova Hedwigia* 35(4) : 710(1983)[1981] .

≡*Bombyliospora domingensis* f. *transgressa* Malme, *Ark. Bot.* 18(no. 12) : 5(1923) .

On bark of Ficus

Yunnan(Shuang Ban Na, Kondratyuk et al. , 2013, p. 331) .

(69) Megalosporaceae Vězda ex Hafellner & Bellem. **大孢衣科**

Nova Hedwigia 35: 216(1982)[1981] .

Megalospora Meyen **大孢衣属**

in Meyen & Flotow, *Nova Acta Acad. Caes. Leop. -Carol. Nat. Cur.* , Suppl. 1 19: 228(1843) .

= *Bombyliospora* De Not. , in Massalongo, Ric. auton. lich. crost. (Verona) : 114(1852) .

Megalospora atrorubricans(Nyl.) Zahlbr. 黑红大孢衣

Cat. Lich. Univers. 4: 86(1926)[1927] .

≡*Lecidea marginiflexa* var. *atrorubricans* Nyl. , *Flora*, Regensburg 49: 132(1866) .

On *Abies kawakamii*, *Abies kawakamii*, *Pseudotsuga* sp.

Taiwan(Aptroot & Sparrius, 2003, p. 28) .

Megalospora sulphurata Meyen 硫大孢衣

in Meyen & Flotow, *Nova Acta Acad. Caes. Leop. -Carol. Nat. Cur.* , Suppl. 1 19: 228(1843) .

On bark of trees.

Taiwan(Zahlbr. , 1933 c, p. 42; Wang & Lai, 1973, p. 92 & 1976, p. 228) .

Megalospora tuberculosa(Fée) 结瘤大孢衣

Biblthca Lichenol. 18: 156(1983) .

≡*Lecidea tuberculosa* Fée, *Essai Crypt. Exot.* (Paris) : 107(1825)[1824] .

= *Bombylospora buelliacea* Zahlbr. in Fedde, *Repertorium* 33: 63(1933) & *Cat. Lich. Univ.* 10: 617(1940) , synonymized by Sipman, *Biblthca Lichenol.* 18: 159(1983) .

Type: Taiwan, Changhua, Rengechi, 31 December 1925, Asahina F 157(TNS lectotype) .

There is no specimen in W, but the protologue mentions two specimens. Only one was present in TNS and it is selected here as lectotype. Although it was mentioned in the protologue to be immature, it contains mature spores in accordance with the species, and TLC revealed pannarin and zeorin(Aptroot, 2004, p. 32) .

= *Bombylospora japonica* f. *purpurascens* Asahina, *Journ. Jap. Bot.* 20: 133, 134(1944) , Synonymized by Sipman, *Biblthca Lichenol.* 18: 159(1983) .

Type: Taiwan, F. no. 186.

= *Bombylospora sinensis* Zahlbr. in Handel-Mazzetti, Syb. Sin. 3: 212(1930) , synonymized by Sipman, *Biblthca Lichenol.* 18: 159(1983) .

Syntype: Hunan, Handel-Mazzetti nos. 12082, 12166, 12347; Yunnan, Handel-Mazzetti nos. 3264, 6553.

Corticolous.

Xizang, Sichuan(Obermayer, 2004, pp. 501 – 502), Yunnan, Hunan(Zahlbr., 1930b, p. 212 & 1932a, p. 584 & 1934b, p. 213), Guizhou(Zhang T & Wei JC, 2006, p. 8), Anhui(Xu B. S., 1989, p. 204), Fujian(Zahlbr., 1930 b, p. 212), Taiwan(Zahlbr., 1933 c, p. 63 & 1940, p. 617; M. Sato, 1936 b, p. 573; Asahina & M. Sato in Asahina, 1939, p. 767; Asahina, 1944a, pp. 133, 134; Wang & Lai, 1973, p. 88 & 1976, p. 226), Hongkong (Seaward & Aptroot, 2005, p. 285, Pfister 1978: 115, as *Heterothecium tuberculosum*).

Megalospora weberi Sipman 韦伯氏大孢衣

Biblthca Lichenol. 18: 112(1983).

On *Abies fabri*.

Sichuan(Obermayer, 2004, p. 502).

(70) Caliciaceae Chevall. [as ' Calicineae'] **粉衣科**

Fl. gén. env. Paris (Paris) 1: 314(1826).

Acroscyphus Lév. **顶杯衣属**

Annls Sci. Nat., Bot., sér. 3 5: 262(1846).

Acroscyphus sphaerophoroides Lév. 顶杯衣

Annls Sci. Nat., Bot., sér. 3 5: 262(1846).

Xizang(Obermayer W. 2004, p. 485), Yunnan(Zahlbr. 1930 b, p. 34; M. Sato, 1935, pp. 772 – 779; Asahina & M. Sato in Asahina, 1939, p. 625).

Amandinea M. Choisy ex Scheid. & M. Mayrhofer **喜瘤衣属**

Lichenologist 25(4): 341(1993).

Amandinea punctata(Hoffm.) Coppins & Scheid. 点喜瘤衣

Lichenologist 25(4): 343(1993).

≡ *Verrucaria punctata* Hoffm., Deutschl. Fl., Zweiter Theil(Erlangen): 192(1796) [1795].

≡ *Buellia punctata*(Hoffm.) A. Massal., Ric. auton. lich. crost. (Verona): 81, fig. 165(1852).

= *Lecidea myriocarpa* Röhl., *Deutschl. Flora*, (Frankfurt) 3(2): 35(1813).

Corticolous.

Xinjiang(Abbas et al., 1993, p. 76 as *Buellis* "*punetats*"(Hoffm.) Ricerch.; Abdulla Abbas & Wu, 1998, p. 105; Abdulla Abbas et al., 2001, p. 360), Yunnan(Hue, 1889, p. 175; Zahlbr. 1930 b, p. 223), Jiangsu(Xu B. S., 1989, p. 254), Zhejiang(Hawksworth & Weng, 1990, p. 155).

Calicium Pers. **粉衣属**

Ann. Bot. (Usteri) 7: 20(1794).

Calicium abietinum Pers. 冷杉粉衣

Tent. disp. meth. fung. (Lipsiae): 59(1797).

Xinjiang(Abdulla Abbas et al., 1993, p. 76: Abdulla Abbas & Wu, 1998, p. 32; Abdulla Abbas et al., 2001, p. 361).

Calicium adspersum Pers. 散生粉衣

Icon. Desc. Fung. Min. Cognit. (Leipzig) 1: 59(1798).

Taiwan(Sparrius et al., 2002, 357).

Calicium glaucellum Ach. 苍胞粉衣

Methodus, Sectio prior: 97(1803).

On coniferous wood.

Taiwan(Sparrius et al., 2002, 358).

Calicium hyperelloides Nyl. 腭粉衣

Syn. meth. lich. (Parisiis) 1(2): 153(1860).

On wood of Pseudotsuga wilsoniana

Taiwan(Sparrius et al., 2002, 358).

Calicium lenticulare Ach 凸镜粉衣

K. Vetensk-Acad. Nya Handl. : 262(1816).

= *Calicium lenticulare* (Hoffm.) Fr., (1835), Nom. illegit., Art. 53. 1.

≡ *Trichia lenticularis* Hoffm., Veg. Crypt. 2: 16(1790).

On rotten wood.

Xizang(J. C. Wei & Jiang, 1986, p. 11), Yunnan(Wei X. L. et al., 2007, p. 149, on *Abies* sp.).

Calicium sinense Zahlbr. 中华粉衣

in Handel-Mazzetti, *Symb. Sin.* 3: 33(1930) & *Cat. Lich. Univ.* 8: 162(1932).

Type: Yunnan, Handel-Mazzetti no. 4566.

On bark of trees.

Calicium trabinellum(Ach.) Ach. 横条粉衣

Lich. univ. : 629(1803).

≡ *Calicium xylonellum* ß *trabinellum* Ach., *Methodus*, Sectio prior: 93(1803).

On wood.

Taiwan(Sparrius et al., 2002, 358).

Cratiria Marbach **杯衣属**

Biblthca Lichenol. 74: 160(2000).

Cratiria lauri-cassiae(Fée) Marbach 月桂杯衣

Biblthca Lichenol. 74: 160(2000).

≡ *Lecidea lauri-cassiae* Fée, *Essai Crypt. Exot.*, *Suppl. Révis.* (Paris): 101(1837).

On bark of bushes in dense thickets.

Pfister 1978: 118, as *Buellia parasema* var. *triphragmia*. Hillsides. 14. 8. 1854. *C. Wright*.

[Label as *Lecidea triphragmia*; annotated by H. A. Imshaug 1952 as *Buellia lauricassiae*].

Hongkong(Seaward & Aptroot, 2005, p. 284, Pfister 1978: 118, as *Buellia parasema* var. *triphragmia*).

Cyphelium Ach. **粉果衣属**

K. Vetensk-Acad. Nya Handl. : 261(1815).

Cyphelium lucidum(Rabenh.) Th. Fr. 亮粉果衣

Gen. Heterolich. Eur. : 101(1861).

≡ *Acolium lucidum* Rabenh., *Krypt. -Fl. Sachsen, Abth.* 2(Breslau): 25(1870).

On bark of *Larix* sp.

Heilongjiang(Ren Qiang & Li Shu Xia, 2013, p. 66).

Cyphelium notarisii(Tul.) Blomb. & Forssell 诺氏粉果衣

Enum. Pl. Scand. : 95(1880).

≡ *Acolium notarisii* Tul., in Stein, in Cohn, *Annls Sci. Nat.*, *Bot.*, sér. 3 17: 81(1852).

On bark of dead *Laris* sp.

Xinjiang(Wu JN et al., 1997, p. 13).

Cyphelium tigillare(Ach.) Ach. 小梁粉果衣

K. Vetensk-Acad. Nya Handl. 3: 261(1815).

≡ *Lichen tigillaris* Ach., Lich. suec. prodr. (Linköping): 67(1799)[1798].

On dead lignum of Juniperus sp., together with *Strangospora moriformis* and *Acroscyphus sphaerophoroides*.

Sichuan, Xizang(W. Obermayer, 2004, p. 494).

Dimelaena Norman **鳞饼衣属**

Nytt Mag. Natur. 7: 231(1853)[1852].

Dimelaena oreina(Ach.) Norman 鳞饼衣

Nytt Mag. Natur. 7: 231(1853)[1852]. var. oreina

≡ *Lecanora straminea* var. *oreina* Ach., *Lich. univ.* : 432(1810).

≡*Rinodina oreina*(Ach.) A. Massal., *Ric. auton. lich. crost.* (Verona): 16(1852).

= *Rinodina altissima* H. Magn. Lich. Centr. Asia, 1: 155(1940).

Type: Gansu, Bohlin no. 35 a in S.

= *Rinodina hueana* Vain. (= *Lecanora hueana* Hue) f. *microcarpoides* Moreau et Moreau, *Rev. Bryol. et Lichenol.* 20: 194(1951). (nom. inval.)

Type: Shandong, Tchou Y. T. no. 104.

On rocks.

Neimenggu, Gansu(H. Magn. 1940, p. 155 & 1944, p. 57), Xinjiang(Abnase al., 1993, p. 77, as *D. 'oreima'*; Abdulla Abbas & Wu, 1998, p. 107; Abdulla Abbas, 2001, p. 363), Xizang(Paulson, 1925, p. 191), Shandong (Moreau & Moreau, 1951, p. 194), Jiangsu(Xu B. S., 1989, p. 255).

var. exalbescens(H. Magn.) J. C. Wei 白鳞变种

Enum. Lich. Sin. 89(1991).

≡*Rinodina altissima* var. *exalbescens* H. Magn. *Lich. Centr. Asia*, 1: 156(1940).

Type: Gansu, Bohlin no. 74 b in S.

Xinjiang(Abdulla Abbas & Wu, 1998, p. 108; Abdulla Abbas, 2001, p. 363).

Chemotype I(usnic and fumarprotcetraric acids):

Xizang(Tibet), Sichuan(Obermayer W. et al., 2004, p. 337).

Chemotype IIa(usnic and gyrophoric acids):

Jilin, Xinjiang, Xizang(Tibet) (Obermayer W. et al., 2004, p. 337).

Chemotype IIb(usnic, gyrophoric, and ovoic acids):

Xinjiang, Qinghai, Xizang(Tibet) (Obermayer W. et al., 2004, p. 337).

Chemotype III(usnic acid):

Xizang(Tibet), Sichuan(Obermayer W. et al., 2004, p. 338).

Chemotype Va(usnic, stictic, and norstictic acids):

Xizang(Tibet), Sichuan(Obermayer W. et al., 2004, p. 338).

Chemotype Vb(usnic, stictic, and hypostatic acids):

Xizang(Tibet), Sichuian(Obermayer W. et al., 2004, p. 339).

Dimelaena radiata(Tuck.) Hale & W. L. Culb. 辐鳞饼衣

Bryologist 73(3): 513(1970).

≡*Rinodina radiata* Tuck., *Proc. Amer. Acad. Arts & Sci.* 12: 173(1877).

Xinjiang(Abbas et al., 1993, p. 77; Abdulla Abbas, 2001, p. 363 as *D.* cf. *radiata*).

Dimelaena tenuis(Müll. Arg.) H. Mayrhofer & Wippel 细鳞饼衣

in Mayrhofer, Matzer, Wippel & Elix, *Mycotaxon* 58: 304(1996).

≡*Catolechia tenuis* Müll. Arg., *Flora*, Regensburg 44: 510(1881).

On a very hard, siliceous boulder on a grazed mountain slope. widespread in the tropics and subtropics.

Hongkong(Aptroot net al., 2001, p. 325).

Diploicia A. Massal. **鳞瘤衣属**

Ric. auton. lich. crost. (Verona): 86(1852).

Diploicia canescens(Dicks.) A. Massal. 灰鳞瘤衣

Ric. auton. lich. crost. (Verona): 86(1852).

≡*Lichen canescens* Dicks., *Fasc. pl. crypt. Brit.* (London) 1: 10(1785).

≡*Buellia canescens* Dicks.

On *Quercus variabilis*.

Shandong(Moreau et Moreau, 1951, p. 194).

Diplotomma Flot. **多瘤胞属**

Jber. schles. Ges. vaterl. Kultur 27: 130(1849).

Diplotomma alboatrum(Hoffm.) Flot. 黑白多瘤胞

Jber. schles. Ges. vaterl. Kultur 27: 97(1849) . var. alboatrum 原变种

≡*Lichen alboater* Hoffm. , *Enum. critic. lich. europ.* (Bern) : 30(1784) .

≡*Buellia alboatra* (Hoffm.) Th. Fr. , *Gen. Heterolich. Eur.* : 91(1861) .

On *Beberis dictyophylla*.

Xinjiang(Abdulla Abbas et al. , 1993, p. 77; Abdulla Abbas & Wu, 1998, p. 108; Abdulla Abbas et al. , 2001, p. 363) , Ningxia(Liu M & J. C. Wei, 2013, p. 46, as *Buellia alboatra*) , Yunnan(Zahlbr. , 1930 b, p. 230) .

Diplotomma epipolium(Ach.) Arnold 粉型多瘤胞

Flora, Regensburg 52: 262(1869) .

≡*Lichen epipolius* Ach. , *Lich. suec. prodr.* (Linköping) : 58(1799) [1798] .

Xinjiang(Abdulla Abbas et al. , 2001, p. 363) .

≡*Diplotomma alboatrum* ** *epipolium*(Ach.) Flot. *Jber. schles. Ges. vaterl. Kultur* 27: 130(1849) .

On rocks.

Xinjiang(Abdulla Abbas et al. , 1993, p. 77; Abdulla Abbas & Wu, 1998, pp. 108 – 109) .

Diplotomma hedinii(H. Magn.) P. Clerc & Cl. Roux 海登氏多瘤胞

[as ' hedinianum'] *in* Clerc, *Cryptogamica Helvetica* 19: 292(2004) .

≡*Buellia hedinii* H. Magn. , *Lichens Central Asia* 1: 146(1940)

≡*Diplotomma hedinii* (H. Magn.) Wu et al. , in Abbas et al. , 1993, p. 77. (name not validly published according to article 41 by ICBN, 1994) .

Gansu(H. Magn. , 1940, p. 146 – 147) , Xinjiang(Abbas et al. , 1993, p. 77; Abbas et al. , 1994, p. 21; Abdulla Abbas & Wu, 1998, p. 109; Abdulla Abbas et al. , 2001, p. 363) .

Dirinaria(Tuck.) Clem. **黑囊基衣属**

Gen. fung. (Minneapolis) : 84(1909) .

≡*ë sect. Dirinaria* Tuck. , *Proc. Amer. Acad. Arts & Sci.* 12: 166(1877) .

Dirinaria aegialita(Afzel. ex Ach.) B. J. Moore 海滩黑囊基衣

Bryologist 71: 248(1968) .

≡*Parmelia aegialita* Afzel. ex Ach. , *Methodus*, Sectio post. : 191(1803) .

On exposed trees, branches and granite. Nearly cosmopolitan.

Taiwan(Aptroot, Sparrius & Lai 2002, p. 283, on sandstone boulders. This species was incorrectly reported by Zahlbr. in 1933, p. 66) , Hongkong(Thrower, 1988, p. 15; Aptroot & Seaward, 1999, p. 74) .

Dirinaria applanata(Fée) D. D. Awasthi 扁平黑囊基衣

J. Indian bot. Soc. 49: 135(1970) .

≡*Parmelia applanata* Fée, *Essai Crypt. Exot.* (Paris) : 126(1825) [1824] .

On exposed trees and granite. Nearly cosmopolitan.

On bark of trees.

Yunnan(Wang L. S. et al. 2008, p. 536) , Shandong(zhang F et al. , 1999, p. 31; Zhao ZT et al. , 2002, p. 8; Li Y et al. , 2008, p. 73) ,) , Shanghai(Awasthi, 1975, p. 78 – 81; Xu B. S. 1989, p. 265) , Jiangsu, Zhejiang(Xu B. S. 1989, p. 265) , Guangdong, Fujian, Taiwan(Awasthi, 1975, pp. 78 – 81) , Hainan(JCWei et al. , 2013, p. 232) , Hongkong(Awasthi, 1975, p. 81; Aptroot & Seaward, 1999, p. 74; Aptroot et al. , 2001, p. 325) .

var. endochroma(H. Magn. & D. D. Awasthi) D. D. Awasthi 黄髓变种

J. Indian bot. Soc. 49: 135(1970) .

≡*Physcia picta* var. *endochroma* H. Magn. & D. D. Awasthi, in Awasthi, *J. Indian bot. Soc.* 39(1) : 8(1960) .

Yunnan(Zhao et al. 1982, p. 121, substratum is not given) .

Dirinaria aspera(H. Magn.) D. D. Awasthi 粗糙黑囊基衣

Bryologist 67: 371(1964).

≡*Physcia aspera* H. Magn., in Magnusson & Zahlbr., *Ark. Bot.* 32A(no. 2): 43(1945).

On bark of tree.

Shaanxi, Yunnan(Zhao et al. 1982, p. 121).

Dirinaria caesiopicta(Nyl.) D. D. Awasthi 浅兰黑囊基衣

Bull. Soc. bot. Fr., Let. bot. 121(7 – 8): 94(1975).

≡*Physcia caesiopicta* Nyl., *Lich. Japon.*: 34(1890).

On rocks.

Taiwan(Awasthi, 1975, pp. 94 – 96).

Dirinaria confluens(Fr.) D. D. Awasthi 无芽黑囊基衣

Biblthca Lichenol. 2: 281(1975).

≡*Parmelia confluens* Fr., *Syst. veg.*, *Edn* 16: 284(1825).

On conglomeratic rock.

Taiwan(Aptroot, Sparrius & Lai 2002, p. 284), Hainan(JCWei et al., 2013, p. 232).

Dirinaria confusa D. D. Awasthi 淆黑囊基衣

Bull. Soc. bot. Fr., Let. bot. 121(7 – 8): 56(1975).

[*Dirinaria aegialita* auct. plur., non(Afz. in Ach.) Moore: *Parmelia aegialita* Afz. in Ach.; *Physcia aegialita* Nyl. (pr. p.)].

Sichuan, Yunnan, Fujian(Zahlbr., 1930 b, p. 237), Guangxi(Zhao et al. 1982, p. 120), Taiwan(Zahlbr., 1933 c, p. 66; M. Sato, 1936 b, p. 569; Asahina, 1942, p. 631; I. Sasaki, 1942, pp. 631 – 632; Wang & Lai, 1973, p. 95; Awasthi, 1975, pp. 64 – 68), Hainan(J. C. Wei et al., 2013, p. 232 as "*confuse*").

Dirinaria papillulifera(Nyl.) D. D. Awasthi 裂芽黑囊基衣

Bryologist 67: 369(1964).

≡Physcia papillifera Nyl., *Expos. Synopt. Pyrenocarp.*: 42(1858).

≡*Physcia papillulifera* Nyl.

= *Physcia picta* f. *isidiophora* Nyl. (lapsu "isidiifera").

Substratum not given.

Hubei(Zhao et al. 1982, p. 121).

Dirinaria picta(Sw.) Clem. & Shear 有色黑囊基衣

in Clements & Shear, *Gen. Fung.*, Edn 2(Minneapolis): 323(1931).

≡*Lichen pictus* Sw., *Prodr.*: 146(1788).

≡*Physcia picta*(Sw.) Nyl., *Mém. Soc. Imp. Sci. Nat. Cherbourg* 3: 175(1855), (in Zahlbr., 1930, p. 237; Awasthi, 1975, p. 75).

[= *Physcia picta* f. *sorediifera*(Nyl. & Cromb. 1883)].

On exposed trees branches and granite. Cosmopolitan.

Guizhou, Hunan, Guangxi(Zhao et al. 1982, p. 120), Yunnan(Hue, 1900, p. 79; Zahlbr., 1930 b, p. 237; Zhao et al. 1982, p. 120), Jiangsu(Wu J. N. & Xiang T. 1981, p. 5, here should be p. 6 according to the sequence of the pages), Shanghai(Nyl. & Cromb. 1883, p. 62; Zahlbr., 1930 b, p. 237), Guangdong(Rabh. 1873, p. 471; Krphbr. 1873, p. 471 & 1874 b, p. 66; Zahlbr., 1930 b, p. 237), Fujian(Wu J. N. et al. 1985, p. 228), Taiwan (Zahlbr., 1933 c, pp. 66 – 67; M. Sato, 1936 b, p. 571; Sasaki, 1942, p. 626; Wang & Lai, 1973, p. 95; Awasthi, 1975, pp. 73 – 75), Hongkong(Hue, 1900, p. 79; Zahlbr., 1930 b, p. 237; Awasthi, 1975, pp. 73 – 75; Thrower, 1988, p. 89; Aptroot & Seweard, 1999, p. 74; Seaward & Aptroot, 2005, p. 284, Pfister 1978: 84, as *Physcia picta; Aptroot et al.*, 2001, *p.* 325), China(prov. not given: Hue, 1890, p. 322).

Hafellia Kalb, H. Mayrhofer & Scheid. **哈氏衣属**

in Kalb, *Lichenes Neotropici, Fascicle* VIII(nos 301 – 350) (Neumarkt) 9(nos. 351 – 400) : 9(1986) .

Hafellia bahiana(Malme) Sheard 贝哈氏衣

Bryologist 95(1) : 82(1992) .

≡*Buellia bahiana* Malme, *Ark. Bot.* 21 A(no. 14) : 8(1927) .

Misapplied name(revised by Aptroot & Seaward, 1999, p. 78) :

Buellia dives auct non Th. Fr. in Thrower, 1988, pp. 15, 66.

Buellia disciformis auct non(Fr.) Mudd in Thrower, 1988, pp. 15, 66.

On *Pinus* branches.

Taiwan(Aptroot and Sparrius, 2003, p. 22) , Hongkong(Thrower, 1988, pp. 15, 66; Aptroot & Seaward, 1999, p. 78; Aptroot et al. , 2001, p. 328) .

Hafellia curatellae(Malme) Marbach 枝生哈氏衣

Biblthca Lichenol. 74: 255(2000) .

≡*Buellia curatellae* Malme, *Ark. Bot.* 21 A(no. 14) : 18(1927) .

On tree branches.

Taiwan(Aptroot and Sparrius, 2003, p. 22) .

Pyxine Fr. **黑盘衣属**

Syst. orb. veg. (Lundae) 1: 267(1825) .

Pyxine berteriana(Fée) Imshaug 光面黑盘衣

Trans. Am. microsc. Soc. 76: 254(1957) .

≡*Circinaria berteriana* Fée, *Essai Crypt. Exot.* (Paris) : 128(1825) [1824] .

= *Pyxine meissneri* Tuck. , in Nylander, *Annls Sci. Nat.* , *Bot.* , sér. 4 11: 255(1859) (Nom. illegit. , Art. 53. 1) .

On twigs.

Sichuan(Hu GR &Chen JB, 2003, pp. 446 – 447, 451 – 452) , Yunnan(Hue, 1900, p. 81; Zahlbr. , 1930 b, p. 234 & 1934b, p. 213; Hu GR &Chen JB, 2003, pp. 446 – 447) , Xizang(J. C. Wei & Jiang, 1986, p. 114; Hu GR &Chen JB, 2003, pp. 446 – 447) , Shandong(Zhao ZT et al. , 2002, p. 7) , Taiwan(Aptroot, Sparrius & Lai, 2002, p. 290) , Hainan(Hu GR & Chen JB, 2003, pp. 446 – 447; JCWei et al. , 2013, p. 232) .

Pyxine cocoës(Sw.) Nyl. [as ‘ cocois’] 椰子黑盘衣

Mém. Soc. Imp. Sci. Nat. Cherbourg 5: 108(1857) .

≡*Lichen cocoës* Sw. , *Prodr.* : 146(1788) .

On bark of trees. On exposed trees, rarely on coastal rocks.

Ningxia(Liu M & J. C. Wei, 2013, p. 47; Yang J. & Wei J. C. , 2014, p. 1030) , Sichuan(Zahlbr. , 1930 b, p. 234) , Yunnan(Hue, 1889, p. 169 & 1900, p84; Zahlbr. , 1930 b, 234; Hu GR &Chen JB, 2003, p. 447) , Fujian (Wu J. N. et al. 1985, p. 228) , Taiwan(Zahlbr. , 1933 c, p. 65; M. Sato, 1936b, p. 571; Wang & Lai, 1973, p. 96) , Hainan(Hu GR &Chen JB, 2003, p. 447; JCWei et al. , 2013, p. 232) , Hongkong(Thrower, 1988, pp. 17, 157; Chu, 1997, p. 48; Aptroot & Seaward, 1999, p. 91; Aptroot & Sipman, 2001, p. 338; Hu GR &Chen JB, 2003, p. 447) .

Pyxine consocians Vain. 群聚黑盘衣

Philipp. J. Sci. , *C, Bot.* 8(2) : 109(1913) .

On rocks.

Taiwan(Aptroot, Sparrius & Lai, 2002, p. 290) . Hainan(Hu GR &Chen JB, 2003, p. 448; J. C. Wei et al. , 2013, p. 232) .

Pyxine copelandii Vain. 柯普兰氏黑盘衣

Philipp. J. Sci. , *C, Bot.* 8(2) : 110(1913) .

= *Pyxine patellaris* Kurok.

Yunnan(Hu GR &Chen JB, 2003, p. 449) , Taiwan(Kashiw. 1977, p. 142; Wang & Lai, 1973, p. 96; Hu GR &Chen JB, 2003, p. 449) , Hainan(JCWei et al. , 2013, p. 232) .

Pyxine coralligera Malme 珊瑚黑盘衣

Bih. K. svenska VetenskAkad. Handl. 23(13) : 40(1897) .

Yunnan(Hu GR &Chen JB, 2003, pp. 449 – 450) .

Pyxine cylindrica Kashiw. 圆筒黑盘衣

Bull. natn. Sci. Mus. , Tokyo, B 3(2) : 66(1977) .

On *Schefflera* sp. and on rocks.

Taiwan(Aptroot, Sparrius & Lai, 2002, p. 290) .

Pyxine endochrysina Nyl. 黄髓黑盘衣

Lich. Japon. : 34(1890) .

On coastal rocks(granite and basalt) . Paleotropical.

Chu(MRDS 108666 with Physcia albinea) .

Sichuan(Zahlbr. 1930, p. 235) , Hongkong(Thrower, 1988, pp. 17, 158; Aptroot & Seaward, 1999, p. 91) .

Pyxine himalayensis D. D. Awasthi 喜马拉雅黑盘衣

Phytomorphology 30(4) : 371(1982) [1980] .

On *Michelia formosana* and *Pisana* sp.

Taiwan(Aptroot, Sparrius & Lai, 2002, p. 290) .

Pyxine limbulata Müll. Arg. 亚橄榄黑盘衣

Flora, Regensburg 71: 112(1891) .

= *Pyxine margaritacea* Zahlbr. *Repert. Spec. Nov. Regni Veg.* 33: 66(1933) .

Type: Taiwan, Chiai, Mt. Arisan, Toroyen, 24 December 1925, Asahina F 99(TNS lectotype, selected here by Aptroot, 2004, p. 35) .

Taiwan(Zahlbr. , 1940, p. 649; M. Sato, 1936b, p. 571; Lamb, 1963, p. 612; Wang & Lai, 1973, p. 96) .

Jilin, heilongjiang, Yunnan, Xizang, Zhejiang(Hu GR &Chen JB, 2003, p. 451) , Shaanxi(Ren Q. et al. , 2009, p. 105) , Sichuan(Zahlbr. , 1931, p. 576 & 1934b, p. 213) , Taiwan(Kashiw. 1977, p. 161; Hu GR &Chen JB, 2003, p. 451) .

Kashiw. points out: "*Pyxine limbulata* seems to be endemic to eastern Asia, being known to occur in Japan (Hue, 1900; Asahina, 1959) and China(Zahlbr. , 1930) ". In fact, this specis is absent neither in Zahlbr. 's publications of 1930 a and b nor in his paper of 1933 c. but is present in Zahlbr. 's publication of 1931, p. 576 & 1934, p. 213 for Sichuan. So, this species was the first record for Taiwan prov. published by Kashiw. in 1977 (1977b) .

Pyxine meissneriana Nyl. 墨氏黑盘衣

Bull. Soc. linn. Normandie, sér. 2 7: 164(1873) .

≡ *Pyxine meissneri* Tuck. , in Nylander, *Bull. Soc. linn. Normandie*, sér. 2 2: 59(1868) .

Nom. illegit. , Art. 53.

Yunnan(Hu GR & Chen JB, 2003, p. 451 – 452) .

Pyxine microspora Vain. 小孢黑盘衣

Philipp. J. Sci. , *C, Bot.* 8(2) : 110(1913) .

On coastal granite rock. Pantropical.

Sichuan, Yunnan(Zahlbr. , 1930 b, p. 235) , Hongkong(Aptroot & Seaward, 1999, p. 91) .

Pyxine petricola Nyl. 白髓黑盘衣

in Crombie, *J. Bot.* , *Lond.* 14: 263(1876) .

=*Pyxine meissneri* var. *endoleuca* Müll. Arg., *Flora*, Regensburg 62: 290(1879).

≡*Pyxine endoleuca*(Müll. Arg.) Vain., *Hedwigia* 37(Beibl.): (42)(1898).

On bark of trees.

Yunnan(Hue, 1900, p. 82; Zahlbr., 1930 b, p. 235), Taiwan(Aptroot, Sparrius & Lai, 2002, p. 290, on shale).

Pyxine philippina Vain. 菲律宾黑盘衣

Philipp. J. Sci., C, Bot. 8(2): 110(1913).

On bark of trees.

Yunnan(Hu GR &Chen JB, 2003, p. 452), Taiwan(Kashiw. 1977, p. 163; Hu GR &Chen JB, 2003, p. 452).

Pyxine sorediata(Ach.) Mont. 粉芽黑盘衣

in Sagra, *Hist. phys. Cuba, Bot. Pl. Cell.* 2: 188(1842)[1838 – 1842].

≡*Lecidea sorediata* Ach., *Syn. meth. lich.* (Lund): 54(1814).

=*Pyxine endochrysoides*(Nyl.) Degel. in Göteborgs Kgl. Vetensk. -och. Vitterh. -Samh. Handl. ser. 6B, 1(7): 38(1941).

≡*Physcia endochrysoides* Nyl., *Flora*, Regensburg 58: 442(1875).

Misapplied name:

Pyxine cf. *copelandii* Vainio(Thrower, 1988, p17 revised by Aptroot & Seaward, 1999, p. 92).

On exposed, coastal granite. Pantropical.

jilin, Heilongjiang, Sichuan, Hunan, Anhui, Guangxi, Fujian(Hu GR &Chen JB, 2003, p. 453), Guizhou(Zhang T & J. C. Wei, 2006, p. 9), Yunnan(Wei X. L. et al. 2007, p. 156; Wang L. S. 2008, p. 538; Hu GR &Chen JB, 2003, p. 453), Shandong(Zhao ZT et al., 2002, p. 7), Hongkong(Thrower, 1988, p17; Aptroot & Seaward, 1999, pp. 91 – 92).

On bark of trees and on the ground or rocks.

Jilin(Chen et al. 1981, p. 157), Shaanxi(Jatta, 1902, p. 474; Zahlbr., 1930 b, p. 235), Sichuan(Zahlbr., 1930 b, p. 235; Zhao et al. 1982, p. 117), Yunnan(Hue, 1889, p. 169 & 1900, p. 85; Zahlbr., 1930b, p. 235; Zhao et al. 1982, p. 117), Guizhou(Zahlbr., 1930 b, p. 235), Hubei, Hunan(Zhao et al. 1982, p. 117), Shandong(Zhao ZT et al., 1999, p. 428), Jiangsu(Wu J. N. & Xiang T. 1981, p. 5, here should be on p. 6 according to the sequence of the pages; Xu B. S., 1989, p. 265), Anhui, Zhejiang(Zhao et al. 1982, p. 117; Xu B. S., 1989, p. 265 & 266), Fujian(Wu J. N. et al. 1985, p. 228, as *P. sorediata* Vain.), Taiwan(Zahlbr., 1933 c, p. 66; M. Sato, 1936b, p. 571; Wang & Lai, 1973, p. 96; Kashiw. 1977, p. 143).

Pyxine subcinerea Stirt. 淡灰黑盘衣

Trans. Proc. N. Z. Inst. 30: 397(1898)[1897].

Zhejiang, Guangxi, Hainan(Hu GR &Chen JB, 2003, pp. 453- 454), Taiwan(Kashiw. 1977, p. 166).

Pyxine subolivacea Zahlbr. 亚橄榄黑盘衣

in Handel-Mazzetti, *Symb. Sin.* 3: 235(1930).

Type: Sichuan, Handel-Mazzetti, no. 2853.

On bark of trees, particularly of *Prunus* sp.

Sculptolumina Marbach **雕纹衣属**

Biblthca Lichenol. 74: 296(2000).

Sculptolumina japonica(Tuck.) Marbach 日本雕文衣

Biblthca Lichenol. 74: 296(2000).

≡*Lecidea japonica* Tuck., *Proc. Amer. Acad. Arts & Sci.* 5: 421(1862)[1860].

On *Cryptomeria japonica*, *Michelia formosana*, *Castanopsis* sp. and bark of deciduous tree.

Taiwan(Aptroot and Sparrius, 2003, p. 46).

(71) Physciaceae Zahlbr. **蜈蚣衣科**

in Engler, Syllabus, Edn 2(Berlin) : 46(1898) .

Anaptychia Körb. **雪花衣属**

Grundriss Krypt. -Kunde: 197(1848) .

Anaptychia ciliaris(L.) Körb. ex A. Massal. 毛边雪花衣

Memor. Lich. : 35(1853) . f. ciliaris

≡*Lichen ciliaris* L. , *Sp. pl.* 2: 1144(1753) .

≡*Parmelia ciliaris*(L.) Ach. , *Methodus*, Sectio post. (Stockholmiæ) : 255(1803) .

≡*Physcia ciliaris*(L.) DC. , in Lamarck & de Candolle, *Fl. franç.* , Edn 3(Paris) 2: 396(1805) .

On dead twigs and on rocks.

Xinjiang(Zhao et al. 1982, p. 123; Abbas et al. , 1993, p. 75; Chen JB & Wang DP, 1999, pp. 336 – 337) , Gansu(Chen JB & Wang DP, 1999, pp. 336 – 337) , Shaanxi(Jatta, 1902, p. 471; Zahlbr. 1930 b, p. 243; Tchou, 1935, p. 315; Moreau et Moreau, 1951, p. 196; Zhao et al. 1982, p. 123; He & Chen 1995, p. 41 without citation of specimen and locality; Chen JB & Wang D. P. , 1999, pp. 336 – 337) , Hebei(Tchou, 1935, p. 315; Zhao et al. 1982, p. 123; Chen JB & Wang D. P. , 1999, pp. 336 – 337) , Zhejiang(Nyl. & Cromb. 1883, p. 62; Zahlbr. , 1930 b, p. 243) , China(prov. non indicated: Hue, 1890, p. 316 & 1899, p. 104) .

f. melanosticta(Ach.) Harm 黑斑变型

Lich. Fr. 3: 447(1907) .

≡*Parmelia ciliaris* var. *melanosticta* Ach. , *Methodus*, Sectio post. : 255(1803) .

= *Parmelia ciliaris* var. *saxicola* Nyl.

On rocks.

Xinjiang(Moreau et Moreau, 1951, p. 196) , Shaanxi(Jatta, 1902, p. 471; Zahlbr. , 1930 b, p. 243) .

f. nigrescens(Bory) Zahlbr. 浅黑变型

Cat. Lich. Univers. 7: 714(1931) .

≡*Borrera ciliaris* f. *nigrescens* Bory, Expedit. Scientific. Morée 3, 2. part. : 307(1832 Xinjiang(Abdulla Abbas & Wu, 1998, p. 106; Abdulla Abbas et al. , 2001, p. 360) .

Anaptychia ethiopica Swinscow & Krog 东非雪花衣

Lichenologist 8(2) : 111(1976) .

Xinjiang(Chen JB & Wang DP, 1999, p. 337) .

Anaptychia isidiza Kurok. 裂芽雪花衣

Beih, Nova Hedwigia, 6: 19(1962) , Nom. nov. for *A. parmulata* var. *isidiata* Zahlbr. [non *A. isidiata* Tomin ex anno 1926] .

On rock & on bark of trees.

Misapplied name:

Anaptychia fusca auctor non(Huds.) Vain. (Chen JB & Wang DP, 1999, p. 339) .

Heilongjiang(Chen et al. 1981, p. 128; Chen JB & Wang DP, 1999, pp. 337 – 338) , Jilin, Liaoning, Neimenggu, Shaanxi(Chen JB & Wang DP, 1999, pp. 337 – 338) , Sichuan(Chen S. F. et al. , 1997, p. 221; Chen JB & Wang DP, 1999, pp. 337 – 338) , Guizhou(Zhang T & Wei JC, 2006, p. 5) , Anhui, Hunan(Zhao et al. 1982, p. 124; Chen JB & Wang DP, 1999, pp. 337 – 338) , Hubei(Chen, Wu, Wei, 1989, p. 483; Chen JB & Wang DP, 1999, pp. 337 – 338) , Zhejiang, Anhui(Xu B. S. , 1989, p. 264) , Fujian(Wu J. N. et al. 1985, p. 223) .

Anaptychia palmulata(Michx.) Vain. 掌状雪花衣

Term. Füz. 22: 299(1899) .

≡*Psoroma palmulata* Michx. , Fl. Boreali-Americ. 2: 321(1803) .

≡*Pseudophyscia aquila* var. *palmulata*(Michx.) Hue, *Nouv. Arch. Mus. Hist. Nat.* , Paris, 4 sér. 1: 117(1899) .

= *Parmelia detonsa* Fr., *Syst. veg.*, Edn 16: 284(1825).

On bark of trees & on rocks among mosses.

Heilongjiang, Jilin(Chen JB & Wang DP, 1999, p. 338), Shaanxi(Jatta, 1902, p. 472; Zahlbr. 1930 b, p. 242; He & Chen 1995, p. 42 without citation of specimens and their localities), Xizang(Chen JB & Wang DP, 1999, p. 338), Sichuan(Zhao et al. 1982, p. 125; Chen S. F. et al., 1997, p. 221; Chen JB & Wang DP, 1999, p. 338), Yunnan(Zahlbr. 1930 b, p. 242; Kurok. 1962 a, p. 17; Chen JB & Wang DP, 1999, p. 338), Hebei(Moreau et Moreau, 1951, p. 196), Anhui, Zhejiang(Zhao et al. 1982, p. 125, Xu B. S., 1989, p. 264; Chen JB & Wang DP, 1999, p. 338), Hunan(Zhao et al. 1982, p. 125; Chen JB & Wang DP, 1999, p. 338), Hubei(Chen, Wu, Wei, 1989, p. 483; Chen JB & Wang DP, 1999, p. 338), Fujian(Wu J. N. et al. 1985, p. 223), China (prov. not indicated: Yoshim. 1974, p. 6).

Anaptychia runcinata(With.) J. R. Laundon 倒齿雪花衣

Lichenologist 16(3): 225(1984).

≡ *Lichen runcinatus* With.

= *Anaptychia fusca*(Huds.) Vain., *Term. Füz.* 22: 299(1899).

≡ *Lichen fuscus* Huds., *Fl. Angl.*, Edn 2: 1 – 690(1778).

On moss covered rock.

Hebei(Zhao et al. 1982, p. 124).

On the ground among mosses.

Shaanxi(Ren Qiang et al., 2009, p. 103).

Anaptychia sanguineus Asahina 红雪花衣

in J-R Wang Y. & M. J. Lai, *a checklist of the lichens of Taiwan in Taiwania* 18(1): 88(1973).

Taiwan(Wang & Lai, 1973, p. 88).

Anaptychia setifera(Mereschk.) Räsänen 毛盘雪花衣

Ann. Acad. Sci. fenn., Ser. A 34(4): 123(1931).

≡ *Anaptychia ciliaris* f. *setifera* Mereschk

On bark.

Xinjiang(Chen JB & Wang DP, 1999, p. 338), Shaanxi(Ren Qiang et al., 2009, p. 103).

Anaptychia tentaculata(Zahlbr.) Kurok. 腺毛雪花衣

Beih. Nova Hedwigia 6: 22(1962).

≡ *Physcia tentaculata* Zahlbr. in Fedde, *Repertorium*, 33: 37(1933). in TNS.

Type: Taiwan, 11/X/1927, S. Sasaki: Asahina F 419 in W, isotype(also see Kurok. 1962, p. 22; Wang & Lai, 1976, p. 226).

Anaptychia ulotricoides(Vain.) Vain. 污白雪花衣

Bot. Tidsskr. 26: 245(1904). f. ulotricoides

≡ *Physcia ulotricoides* Vain. [as '*ulothricoides*'], *Acta Horti Petropolit.* 10: 553(1888).

On dried twigs.

Xinjiang(Zhao et al. 1982, P. 124; Wang XY, 1985, p. 352; Abudula Abbas et al., 1993, p. 75; Abudula Abbas et al., 1994, pp. 19 – 20; Abdulla Abbas & Wu, 1998, p. 106; Chen JB & Wang D. P., 1999, p. 339; Abdulla Abbas et al., 2001, p. 360), Qinghai(Kurok. 1962 a, p. 21), Gansu(Chen JB & Wang D. P., 1999, p. 339), Xizang(Kurok. 1962 a, p. 21; J. C. Wei & Y. M. Jiang, 1986, p. 112).

f. tenuior(Vain.) Dzhur. 纤细变型

Likhenoflora Tsentral'nogo Kopetdaga (Turkmenistan) (Ashkhabad): 144(1978). Nom. inval., Art. 41. 4(Melbourne).

Basionym not indicated and reference omitted

≡ *Physcia ulotricoides* f. *tenuior* Vain. in *Acta Horti Petropolit.*, 10: 554(1888).

Xinjiang(Wang XY, 1985, p. 352; Abudula Abbas et al. , 1993, p. 75) .

Anaptychia wrightii(Tuck.) Zahlbr. 黑腹雪花衣

Cat. Lich. Univers. 7: 743(1931) .

≡*Physcia wrightii* Tuck. , *Amer. J. Sci. Arts*, Ser. 2, 28: 204(1859) .

On rock.

Jiangxi(Zhao et al. 1982, p. 132) .

This name refers to *Physcidia wrightii* and the material is probably wrongly determined.

Buellia De Not. **黑瘤衣属**

G. bot. ital. 2(1. 1): 195(1846) .

Buellia aethalea(Ach.) Th. Fr. 烟黑瘤衣

Lich. Scand. (Upsaliae) 1(2) : 604(1874) .

≡*Gyalecta aethalea* Ach. , *Lich. univ*. : 1 – 696(1810) .

On shale.

Taiwan(Aptroot and Sparrius, 2003, p. 8) .

Buellia americana(Fée) Zahlbr. 美洲黑瘤衣

Cat. Lich. Univers. 7: 334(1931) .

≡*Lecidea parasema* var. *americana* Fée, *Essai Crypt. Exot.* , *Suppl. Révis*. (Paris) : 101(1837) .

On *Tsuga formosana*, *Roystonea regia*, and *Ardisia sieboldii*.

Taiwan(Aptroot and Sparrius, 2003, p. 8) .

= *Lecidea modesta* Krempelh. , *Hedwigia*, 13(4) : 59(1874) , nom. nud.

Type: from Shanghai.

Corticolous.

Buellia atrocinerella(Nyl.) Scheid. 灰黑瘤衣

Lichenologist 25(4) : 345(1993) .

≡*Lecanora atrocinerella* Nyl. , *Flora*, Regensburg 55: 428(1872) .

Shandong(Sun JJ et al. , 2013, p. 156) .

Notes: This specimen seems has lecanorine apothecia. It remains to be checked from Shandong, Mt. Taishan.

Buellia badia(Fr.) A. Massal. 红褐黑瘤衣

Memor. Lich. : 124(1853) .

≡*Lecidea badia* Fr. , *Syst. orb. veg*. (Lundae) 1: 287(1825) .

On shale.

Liaoning(Kondratyuk et al. , 2013, p. 323) , Taiwan(Aptroot and Sparrius, 2003, p. 9) .

Buellia centralis H. Magn. 中央黑瘤衣

Lich. Centr. Asia 1: 147(1940) .

Type: Gansu, at about 2750 m, 30/I/1932, Bohlin no. 79 b in S(!) .

On the ridge of a hill.

Neimenggu(H. Magn. 1944, p. 55) , Xinjiang, Xizang(Obermayer W. et al. , 2004, p. 332) , Shandong(Sun JJ et al. , 2013, p. 156) .

Buellia cervinoplaca Zahlbr. 茶面黑瘤衣

in Handel-Mazzetti, Symb. Sin. 3: 229(1930) , *Cat. Lich. Univ*. 7: 342(1931) .

Type: Sichuan, at about 1770 m, 5/IV/1914, Handel-Mazzetti no. 1803.

Buellia disciformis(Fr.) Mudd 盘形黑瘤衣

Man. Brit. Lich. : 216(1861) .

≡*Lecidea parasema* var. *disciformis* Fr. , *Fl. carniol*. , Edn 2(Wien) : 9(1826) .

Corticolous.

Yunnan(Zahlbr. , 1930b, p. 224; Asahina & Sato in Asahina, 1939, p. 771) , Jiangsu(Wu J. N. & Xiang T.

1981, p. 6: here should be p. 5 according to the sequence of page number, on rocks), Taiwan(Aptroot and Sparrius, 2003, p. 9).

var. aeruginascens(Nyl.) Vain 铜绿变种

Étud. Class. Lich. Brésil 1: 166(1890).

≡*Lecidea disciformis* var. *aeruginascens* Nyl., *Bull. Soc. linn. Normandie*, sér. 2 2: 191(1868).

Corticolous.

Fujian(Zahlbr., 1930 b, p. 224).

var. reagens J. Steiner 反应变种

Denkschr. Kaiserl. Akad. Wiss. Wien, Math. -Naturwiss. Kl. 83: 193(1909).

Corticolous.

Yunnan(Zahlbr., 1930 b, p. 224).

Buellia disjecta Zahlbr. 裂片黑瘤衣

in Handel-Mazzetti, *Symb. Sin.* 3: 224(1930) & *Cat. Lich. Univ.* 7: 356(1931).

Type: Sichuan, at about 2100 m, 6/IV/1914, Handel-Mazzetti no. 1860.

Rupicolous.

Zhejiang(Xu B. S., 1989, p. 253).

Buellia efflorescens Müll. Arg. 开放黑瘤衣

Hedwigia 32: 129(1893).

On exposed coastal trees. Pantropical.

Taiwan(Aptroot and Sparrius, 2003, p. 9), Hongkong(Aptroot & Seaward, 1999, p. 67; Aptroot et al., 2001, p. 322).

Buellia effundens Zahlbr. 散布黑瘤衣

in Handel-Mazzetti, *Symb. Sin.* 3: 225(1930) & *Cat. Lich. Univ.* 7: 359(1931).

Type: Fujian, Fuzhou, Mt. Gu shan, at about 500 – 600 m, Chung no. 608.

On non calcareous rock.

Buellia endolateritia Zahlbr. 红髓黑瘤衣

in Handel-Mazzetti, *Symb. Sin.* 3: 229(1930) & *Cat. Lich. Univ.* 7: 360(1931).

Type: Yunnan, Handel-Mazzetti no. 786.

On quartziferous rock.

Buellia erubescens Arnold 玫瑰黑瘤衣

Verh. zool. -bot. Ges. Wien 25: 493(1875)(1874)[1873].

≡*Buellia zahlbruckneri* Stnr. var. *erubescens* Stnr. apud Zahlbr. in *Denkschr. ma th. -nat. Kl. kais. Akad. Wiss.* Wien. Lxxxiii: 193(1909). Zahlbr. *in* Handel-Mazzetti, *Symb. Sin.* 3: 224(1930).

Corticolous.

Yunnan(Zahlbr., 1930b, p. 224).

Corticolous.

Sichuan, Guizhou, Yunnan(Zahlbr. 1930b, p. 224), Taiwan(Zahlbr. 1933c, p. 64; M. Sato, 1936, p. 572; Wang & Lai, 1973, p. 88).

Buellia extenuata Müll. Arg. 伸黑瘤衣

Nuovo G. bot. ital. 23: 128(1891).

On *Arenga* sp.

Taiwan(Aptroot and Sparrius, 2003, p. 9).

Buellia handelii Zahlbr. 汉氏黑瘤衣

in Handel-Mazzetti, *Symb. Sin.* 3: 228(1930) & *Cat. Lich. Univ.* 7: 366(1931).

Type: Yunnan, Handel-Mazzetti no. 414.

Corticolous.

Buellia hedinii H. Magn. 海登氏黑瘤衣

Lich. Centr. Asia 1: 146(1940).

Type: Gansu, Bohlin no. 73 in S(!).

On calcareous rock.

Neimenggu(H. Mang. 1944, p. 55).

Buellia hilaris Zahlbr. 脐黑瘤衣

in Handel-Mazzetti, *Symb. Sin.* 3: 228(1930) & *Cat. Lich. Univ.* 7: 366(1931).

Type: Hunan, Handel-Mazzetti no. 17734.

Calcicolous.

Jiangsu(Xu B. S., 1989, p. 254).

Buellia himalayensis(S. R. Singh & D. D. Awasthi) A. Nordin 喜马拉雅黑瘤衣

≡*Diplotomma himalayense* S. R. Singh & D. D. Awasthi, *Geophytology* 19(2): 175(1990)[1989].

On *Rhododendron* wood.

Taiwan(Aptroot and Sparrius, 2003, p. 9).

Buellia keteleeriae Zahlbr. 油杉黑瘤衣

in Handel-Mazzetti, *Symb. Sin.* 3: 230(1930) & *Cat. Lich. Univ.* 7: 374(1931).

Type: Yunnan, Handel-Mazzetti no. 90.

Corticolous.

Buellia lauri-cassiae(Fée) Müll. Arg. 三隔黑瘤衣

Revue mycol., Toulouse 9: 85(1887).

≡*Lecidea lauri-cassiae* Fée, *Essai Crypt. Exot., Suppl. Révis.* (Paris): 101(1837).

Misapplied name: *Buellia disciformis* var. *triphragmia* Boistel(in Thrower, 1988, p. 15).

Corticolous and on exposed coastal trees. Pantropical.

Taiwan(Aptroot and Sparrius, 2003, p. 9), Guangdong(Wu J. L., 1987, p. 192), Hongkong(Thrower, 1988, p. 15; Aptroot & Seaward 1999, p. 67; Aptroot et al., 2001, p. 322).

Buellia leproplaca Zahlbr. 鳞盘黑瘤衣

in Handel-Mazzetti, *Symb. Sin.* 3: 229(1930) & *Cat. Lich. Univ.* 7: 375(1931).

Type: Sichuan, Handel-Mazzetti no. 7197.

Corticolous.

Buellia leptocline A. Massal. 薄黑瘤衣

Geneac. lich. (Verona): 20(1854).

=*Buellia leptoclinis*(Flot.) Körb., *Syst. lich. germ.* (Breslau): 225(1855).

≡*Lecidea leptoclinis* Flot., *Bot. Ztg.* 8: 555(1850).

var. inarimensis Jatta 无节变种

Bull. Soc. bot. Ital.: 210(1892).

On rocks.

Beijing(Moreau et Moreau, 1951, P. 194).

Buellia lindingeri Erichsen 霜盘黑瘤衣

Hedwigia 66: 281(1926).

Xizang(W. Obermayer, 2004, p. 489).

Buellia metaleptodes(Nyl.) G. Pant & D. D. Awasthi 薄交黑瘤衣

Proc. Indian Acad. Sci., Pl. Sci. 99(4): 383(1989).

≡*Lecidea metaleptodes* Nyl., *Acta Soc. Sci. fenn.* 26(no. 10): 15(1900).

On *Roystonea regia*.

Taiwn(Aptroot and Sparrius, 2003, p. 10).

Buellia mongolica H. Magn. 蒙古黑瘤衣

Lich. Centr. Asia 1: 146(1940) & 2: 55(1944).

Type: Neimenggu, Bexell no. 9 in S.

Buellia obscurior(Stirt.) Aptroot 暗黑瘤衣

in Aptroot & Sparrius, *Fungal Diversity* 14: 10(2003).

≡*Pyxine obscurior* Stirt., *Trans. & Proc. Roy. Soc. Victoria* 17: 70(1881).

On *Magnolia grandiflora*.

Taiwan(Aptroot and Sparrius, 2003, p. 10).

Buellia ocellata(Flörke ex Flot.) Körb. 单眼黑瘤衣

Syst. lich. germ. (Breslau): 224(1855).

≡*Lecidea petraea* var. *ocellata* Flörke ex Flot., *Flora, Regensburg* 11(2): 691(1828).

Shandong(Sun JJ et al., 2013, p. 156).

Notes: This specimen seems has lecanorine apothecia. It remains to be checked from Shandong, Mt. Taishan.

Buellia polita Zahlbr. 光华黑瘤衣

in Handel-Mazzetti, *Symb. Sin.* 3: 227(1930) & *Cat. Lich. Univ.* 7: 389(1931).

Type: Sichuan, at about 1770 m, 5/IV/1914, Handel-Mazzetti no. 1802.

On rock.

Buellia polospora(Leight.) Shirley 轴孢黑瘤衣

Pap. Proc. R. Soc. Tasm.: 218(1894) [1893].

≡*Lecidea polospora* Leight., *Trans. Linn. Soc. London, Bot.*, Ser. 2 1: 241(1876).

On *Podocarpus* sp.

Taiwan(Aptroot and Sparrius, 2003, p. 10).

Buellia protothallina(Krphbr.) Vain. 原体黑瘤衣

Results. Voy. S. Y. Belgica, 25(1903), nom. illegit., Art. 53. 1.

≡*Lecidea stellulata* f. *protothallina* Krphbr., in *Flora*, LIX: 267(1876).

On rocks.

Yunnan(Zahlbr. 1930 b, p. 223).

Buellia punctata(Hoffm.) A. Massal. 小点黑瘤衣

Ric. auton. lich. crost. (Verona): 81, fig. 165(1852).

≡*Verrucaria punctata* Hoffm., *Deutschl. Fl.*, *Zweiter Theil*(Erlangen): 192(1796) [1795].

Corticolous.

Yunnan(Hue, 1889, p. 175; Zahlbr., 1930 b, p. 223), Jiangsu(Xu B. S. 1989, p. 254), Zhejiang(Hawksworth & Weng, 1990, p. 155).

var. globulans Zahlbr. 小球变种

in Handel-Mazzetti, *Symb. Sin.* 3: 223(1930) & *Cat. Lich. Univ.* 7: 402(1931).

Type: Yunnan, Handel-Mazzetti no. 8212.

On bark of *Pseudotsuga wilsoniana*.

f. subpersicina Zahlbr. 桃红变型

in Handel-Mazzetti, *Symb. Sin.* 3: 223(1930) & *Cat. Lich. Univ.* 7: 399(1931).

Type: Yunnan, Handel-Mazzetti no. 474.

On bark of *Diospyros mollifolia*.

Buellia sarcogynoides H. Magn. 肉黑瘤衣

Lich. Centr. Asia 1: 145(1940).

Type: Gansu, Bohlin no. 40 a in S(!).

Buellia saxorum A. Massal. 石黑瘤衣

Ric. auton. lich. crost. (Verona): 82(1852).

On shale.

Taiwan(Aptroot and Sparrius, 2003, p. 10).

Buellia schaereri De Not. 沙氏黑瘤衣

G. bot. ital. 2(1.1): 199(1846).

= *Buellia nigritula*(Nyl.) Mudd

Corticolous.

Guangdong(R. Rabh. 1873, p. 286; Krphbr. 1874, p. 60; Zahlbr. 1930 b, p. 223).

Buellia sequax(Nyl.) Zahlbr. 承黑瘤衣

Cat. Lich. Univers. 7: 410(1931).

≡*Lecidea sequax* Nyl., *Flora*, Regensburg 58: 302(1875).

On coastal granite rocks. Known from Europe and adjacent Africa.

Hongkong(Aptroot & Seaward, 1999, p. 67).

Buellia spuria(Schaer.) Anzi 拟黑瘤衣

Cat. Lich. Sondr.: 87(1860).

≡*Lecidea spuria* Schaer., *Lich. helv. spicil.* 3: 127(1828).

On exposed , inland granite rocks. Cosmopolitan.

Shandong(Sun JJ et al., 2013, pp. 156 – 157), Taiwan(Aptroot and Sparrius, 2003, p. 10), Hongkong(Thrower, 1988, p. 15: as *Buellia* cf. *spuria*-group; Aptroot & , 1999, p. 67).

Notes: This specimen seems has lecanorine apothecia. It remains to be checked from Shandong, Mt. Taishan.

Buellia stellulata(Taylor) Mudd 小星黑瘤衣

Man. Brit. Lich.: 216(1861). var. stellulata

≡*Lecidea stellulata* Taylor, in Mackay, *Fl. Hibern.* 2: 118(1836).

On exposed , inland granite rocks. Cosmopolitan.

Hongkong(Thrower, 1988, p. 15: as *Buellia* cf. *stellulata*-group; Aptroot & Seaward, 1999, p. 67).

var. macrior Zahlbr. 大果变种

in Fedde, *Repertorium* 33: 64(1933) & *Cat. Lich. Univ.* 10: 641(1940).

Syntype: Taiwan, Faurie nos. 46 & 281.

Taiwan(M. Sato, 1936b, p. 572; Wang & Lai, 1973, p. 88).

Buellia subannulata Zahlbr. 环果黑瘤衣

in Handel-Mazzetti, *Symb. Sin.* 3: 226(1930) & *Cat. Lich. Univ.* 7: 416(1931).

Type: Sichuan, Handel-Mazzetti no. 1584.

Saxicolous.

Buellia subarmeniaca Zahlbr. 杏黄黑瘤衣

in Handel-Mazzetti, *Symb. Sin.* 3: 225(1930), var. subarmeniaca

Syntype locality: Sichuan, Handel-Mazzetti no. 1801; Yunnan, Handel-Mazzetti no. 166.

Rupicolous.

var. huiliensis Zahlbr. 会理变种

in Handel-Mazzetti, *Symb. Sin.* 3: 226(1930) & *Cat. Lich. Univ.* 7: 416(1931).

Type: Yunnan, Handel-Mazzetti no. 787.

Quartzicolous.

Buellia subdisciformis(Leight.) Jatta 亚盘黑瘤衣

Syll. Lich. Ital.: 392(1900).

≡*Lecidea subdisciformis* Leight., *Lich. -Fl. Great Brit.*: 308(1871).

See it together with Caloplaca holochracea(Nyl.) Zahlbr. for Hongkong.

Saxicolous, and on exposed , inland granite rocks. Northern temperate.

Jiangsu(Xu B. S., 1989, p. 254), Taiwan(Zahlbr. 1933c, p. 64; M. Sato, 1936b, p. 572; Wang & Lai, 1973, p. 88), Hongkong(Thrower, 1988, p. 65: *Buellia* cf. *subdisciformis* Jatta; Aptroot & Seaward, 1999, p. 67 – 68, and with a specimen: Lung Kwu Tan, 1994, Chu"MRDS 10429 with *Buellia* cf. *testacea*" , and a specimen: Beach, Lantau, 1995, Chu "MRDS 108860 with *Pertusaria flavicans*"; Wu Kai Sha, NT, 1993, Seaward & Chu "MRDS 108870 with *Caloplaca holochracea*" & Aptroot & Seawaerd, 1999, p. 69; Aptroot et al. , 2001, p. 322) .

Buellia subocculta Jatta 亚隐黑瘤衣

Nuov. Giorn. Bot. Italiano 2(9) : 479(1902) . Zahlbr. in Handel-Mazzetti, *Symb. Sin.* 3: 225(1930) , & *Cat. Lich. Univ.* 7: 420(1931) .

Type: Shaanxi, collected by Giraldi.

Saxicolous.

Buellia testacea Müll. Arg. 砖红黑瘤衣

Nuovo G. bot. ital. 21: 360(1889) .

On coastal and inland granite rocks. So far known from South America. The identity of this species remains uncertain.

Hongkong(Aptroot &Seaward, 1999, p. 67, as *Buellia* cf. *testacea* together with *Buellia Subdisformis*; Aptroot et al. , 2001, p. 322) .

Buellia trachyspora Vain. 糙孢黑瘤衣

Ann. Acad. Sci. fenn. , Ser. A 6(7) : 84(1915) .

On conglomeratic rock.

Taiwan(Aptroot and Sparrius, 2003, p. 10) .

Buellia triphragmioides Anzi [as ' *triphragmoides*'] 类三壁黑瘤衣

Atti Soc. ital. Sci. nat. (Modena) 11: 171(1868) .

On *Michelia formosana*.

Taiwan(Aptroot and Sparrius, 2003, p. 10) .

Buellia venusta(Körb.) Lettau 美丽黑瘤衣

Hedwigia 52: 244(1913) .

≡ *Diplotomma alboatrum* var. *venustum* Körb. , in Rabenhorst, *Flecht. Europ.* 13: no. 384(1858) .

On rock.

Ningxia(Liu M & J. C. Wei, 2013, p. 47) .

Buellia vernicoma(Tuck.) Tuck 漆毛黑瘤衣

Lichens of California(Berkeley) : 25(1866) .

≡ *Lecidea vernicoma* Tuck. , *Amer. J. Sci. Arts*, Ser. 2 25: 429(1858) .

Xinjiang(Abbas et al. , 1993, p. 76; Abdulla Abbas & Wu, 1998, p. 107; Abdulla Abbas et al. , 2001, p. 361 as *B.* "*vericoma*") .

Buellia yunnana Zahlbr 云南黑瘤衣

in Handel-Mazzetti, *Symb. Sin.* 3: 226(1930) & *Cat. Lich. Univ.* 7: 431(1931) .

Type: Yunnan, Handel-Mazzetti no. 5791.

Quartzicolous.

Heterodermia Trevis. **哑铃孢属**

Atti Soc. ital. Sci. nat. (Modena) 11: 613(1868) .

Heterodermia albicans(Pers.) Swinscow & Krog 白哑铃孢

Lichenologist 8(2) : 113(1976) .

≡ *Parmelia albicans* Pers. , *Ann. Wetter. Gesellsch. Ges. Naturk.* 2: 17(1811) .

≡ *Physcia albicans*(Pers.) Thoms.

On bark of trees.

Beijing, Sichuan, Hunan, Zhejiang(Zhao et al. 1982, p. 106) .

Heterodermia angustiloba(Müll. Arg.) D. D. Awasthi 狭叶哑铃孢

Geophytology 3: 113(1973).

≡*Physcia speciosa* var. *angustiloba* Müll. Arg., *Flora*, Regensburg 66: 78(1883).

≡*Anaptychia speciosa*(Wulf.) Massal. var. *angustiloba* Müll. Arg.

≡*Anaptychia angustiloba*(Müll. Arg.) Kurok. (Kurok., 1962a, p. 39; Wang-Yang & Lai, 1976, p. 226).

Sichuan(Kurok. 1962a, p. 39), Yunnan(Kurok. 1962, p. 39; Yoshim. 1974, p. 7; Chen J. B., 2001, p. 109), Guizhou(Chen J. B., 2001, p. 109; Zhang T & J. C. Wei, 2006, p. 7), Hunan(Chen J. B., 2001, p. 109), Taiwan(Kurok. 1962, P. 39; Yoshim. 1974, p. 7; Wang & Lai, 1976, p. 226).

Heterodermia barbifera(Nyl.) Kr. P. Singh 须哑铃孢

Bull. bot. Surv. India 21(1 - 4): 221(1981) [1979].

≡*Physcia barbifera* Nyl., *Syn. meth. lich.* (Parisiis) 1(2): 416(1860) (Hue, 1887a, p. 23).

≡*Anaptychia barbifera*(Nyl.) Trevis. 1861(Zahlbr., 1930b, pp. 241, 243).

On twigs.

Yunnan {Hue, 1887a, p. 23; Zahlbr., 1930b, p. 243, cited as *A. barbifera*(Nyl.) Hue(1899: 110)}.

According to Kurok. (1962, p. 99) this species is only distributed in central and southern America. Therefore I doubt whether this lichen, collected by Delavay from Mt. Caengshan near Dali in Yunnan, belongs to this species. It remains to be open to question.

Heterodermia boryi(Fée) Kr. P. Singh & S. R. Singh 卷梢哑铃孢

Geophytology 6(1): 33(1976).

≡*Borrera boryi* Fée, *Essai Crypt. Exot.* (Paris): XCII, tab. II, fig. 23(1824).

≡*Heterodermia leucomelos* subsp. *boryi*(Fée) Swinscow & Krog, *Lichenologist* 8(2): 124(1976).

≡*Anaptychia boryi*(Fée) Massal., *Memor. Lichenogr.*, 41(1853).

Heilongjiang(Chen et al. 1981, p. 128; Zhao et al. 1982, p. 141; Luo, 1984, p. 84), Jilin(Zhao et al. 1982, p. 141), Guizhou(Zhang T & Wei JC, 2006, p. 7), Yunnan(Zhao et al. 1982, p. 141; Wei X. L. et al. 2007, p. 152), Shandong(Hou Ya-nan et al., 2008, p. 66; Li Y et al., 2008, p. 73), Anhui(J. C. Wei et al. 1982, p. 56; Xu B. S. 1989, p. 259), Taiwan(Wang-Yang & Lai, 1976, p. 226).

=*Anaptychia leucomelaena* var. *multifida*(Mey. et Flot.) Müll. Arg. in Englers Botan. Jahrb., XX: 249(1894).

≡*Parmelia leucomela* var. *multifida* Mey. et Flot. in Nova Acta Acad. Carol. -Leoold., IX, Suppl., 221, tab. III, fig. 6(1847).

=*Physcia leucomela* var. *angustifolia* Nyl. in Mémoir. Soc. Scienc. Natur. Cherbourg. V: 106[1857]. Hue I: 23; II: 167.

≡*Parmelia leucomela* var. *angustifolia* Jatta in Nuov. Giorn. Bot. Ital. nov. ser., IX: 471(1902).

Misapplied name:

Anaptychia leucomelaena var. *angustifolia* sensu Sato: *Journ. Jap. Bot.* 12: 429(1936, p. p.), non (Meyen & Flot.) Müll. Arg., *Bot. Jb.* 20: 249(1894).

Shaanxi(Jatta, 1902, p. 4671; Zahlbr., 1930 b, p. 243), Xizang(J. C. Wei, 1981, p. 91; Wei & Jiang, 1986, p. 109), Yunnan(Hue, 1887a, p. 23 & 1889, p. 167; Zahlbr., 1930 b, p. 243 & 1934 b, p. 213), Hubei(Chen, Wu & J. C. Wei, 1989, p. 484), Taiwan(Zahlbr., 1933 c, p. 68; Asahina, 1934 b, 355; M. Sato, 1934 b, p. 688 & 1936 a, p. 429; Wang & Lai, 1973, p. 87).

≡*Heterodermia boryi* var. *boryi* f. *boryi*(Fée) Kr. P. Singh & S. R. Singh, *Geophytology* 6(1): 33(1976).

=*Anaptychia neoleucomelaena* var. *neoleucomelaena* f. *neoleucomelaena*

Sichuan, Taiwan(Kurok. 1962, p. 78).

=*Heterodermia boryi* var. *sorediosa*(Jatta) J. C. Wei, *Enum. Lich. China* (Beijing): 107(1991).

≡*Parmelia leucomela* var. *sorediosa* Jatta, *Annali di Botan* 6: 407(1908). Zahlbr., *Cat. Lich. Univ.* 7: 733(1931).

≡*Heterodermia boryi* var. *sorediosa* f. *sorediosa* (Jatta) J. C. Wei in J. C. Wei & Y. M. Jiang, [*Lichens of Xi-*

zang] (China) : 110(1986) .

Sichuan, Yunnan, Anhui(Zhao et al. 1982, p. 142, cited as *Anaptychia leucomelaena* f. *sorediosa*(Jatta) Kurok.) .

= *Heterodermia boryi* var. *squarrosa*(Vain.) J. C. Wei *in* J. C. Wei & Y. M. Jiang, [*Lichens of Xizang*] (China) : 110(1986) .

≡*Anptychia leucomelaena* var. *squarrosa* Vain. [as ' *Anaptychia leucomelaena* f. *squarrosa*'] , *Catal. Welwitsch Afric. Plant*, 2: 408(1901) .

≡*Anaptychia neoleucomelaena* f. *squarrosa*(Vain.) Kurok. *A monograph of the genus Anaptychia*, 78(1962) .

≡ *Heterodermia boryi* var. *squarrosa* (Vain.) J. C. Wei, in Wei & Jiang, [*Lichens of Xizang*] (China): 110 (1986) .

≡*Anaptychia boryi* var. *squarrosa*(Vain.) J. C. Wei & Y. M. Jiang, in Wei, *Bull. bot. Res.* , *Harbin* 1(3) : 91 (1981) .

Anptychia leucomelaena var. *multifida* f. *squarrosa* Vainio, *Catal. Welwitsch. Afric. Plants*, II: 408(1901) .

≡*Heterodermia neoleucomelaena* f. *squarrosa*(Vain.) D. D. Awasthi, *Geophytology* 3(1) : 114(1973) .

Misapplied name:

Anaptychia leucomelaena var. *angustifolia* sensu Sato, *Journ. Jap. Bot.* 12: 429(1936) . (pr. p.) , non(Meyen & Flot.) Müll. Arg.

Xizang(J. C. Wei & Y. M. Jiang, 1986, p. 110) , Sichuan(Zhao et al. 1982, p. 142, miscited as *A. leucomelaena* f. *squarrosa*) , Yunnan(Kurok. 1962, p. 78; J. C. Wei, 1981, p. 91; Zhao et al. 1982, p. 142, mis-cited as *Anaptychia leucomelaena* f. *squarrosa*(Vain.) Kurok.) , Taiwan(Sato, 1936 a, p. 429, pr. p; Kurok. 1962, p. 78; Wang & Lai, 1973, p. 87 & 1976, p. 226) .

= *Heterodermia boryi* var. *squarrosa* f. *circinalis* (Zahlbr.) J. C. Wei *in* J. C. Wei & Y. M. Jiang, [*Lichens of Xizang*] (China) : 110(1986) .

≡*Anaptychia leucomelos* f. *circinalis* Zahlbr. , *Beih. bot. Zbl.* , Abt. 2 19: 84(1905) .

≡*Heterodermia circinalis*(Zahlbr.) W. A. Weber, *Mycotaxon* 13(1) : 101(1981) .

On bush.

Xizang(Wei & Jiang, 1986, p. 110) .

Above cited different heterotypic synonyms remain to be revised by the authors of Flora Lichenum Sinicorum in the future.

Heterodermia comosa(Eschw.) Follmann & Redón 丛毛哑铃孢

Willdenowia 6(3) : 446(1972) .

≡*Parmelia comosa* Eschw. , *Icon. Plant. Cryptog.* 2: 25, tab. XII, fig. 1(1828) [1828 – 34] .

≡*Anaptychia comosa* (Eschw.) A. Massal. , *Memor. Lich.* : 39, fig. 41(1853) .

On bark of trees.

Xizang(J. C. Wei & Y. M. Jiang, 1981, p. 1147 & 1986, p. 110) , Yunnan(Zahlbr. , 1930 b, p. 244; Kurok. 1962, p. 103; Zhao et al. 1982, p. 142; Wei X. L. et al. 2007, pp. 152 – 153; Wang L. S. et al. 2008, p. 537) , Hunan, Guangxi(Zhao et al. 1982, p. 142) , Fujian(J. C. Wei et al. 1982, p. 58; Wu J. N. et al. 1985, p. 227) , Taiwan(Wang & Lai, 1973, p. 87) .

Heterodermia coralliphora(Taylor) Skorepa [as ' corallophora'] 珊瑚哑铃孢

Bryologist 75(4) : 490(1973) [1972] .

≡*Parmelia coralliphora* Taylor, *London J. Bot.* 6: 164(1847) .

On twigs & bark of trees.

Hubei(Zahlbr. , 1930 b, p. 243) , TAIWAN(Zahlbr. , 1933 c, p. 68; M. Sato, 1936 a, p. 426) .

Heterodermia dactyliza(Nyl.) Swinscow & Krog 指哑铃孢

Lichenologist 8: 117(1976) .

≡*Siphula dactyliza* Nyl., *Flora*, Regensburg 68: 442(1885).

≡*Physcia speciosa* var. *dactyliza* Nyl., *Syn. meth. lich.* (Parisiis) 1(2): 417(1860).

= *Anaptychia speciosa* var. *leariloba* Müll. Arg., Bot. Jb. 15: 506/508(1893).

Shaanxi(Jatta, 1902, p. 471; Zahlbr., 1930 b, p. 242).

Heterodermia dendritica(Pers.) Poelt 树哑铃孢

Nova Hedwigia 9: 31(1965).

≡ *Borrera dendritica* Pers., in Gaudichaud-Beaupré in Freycinet, *Voy. Uranie.*, *Bot.*: 207(1827)[1826 – 1830].

≡*Anaptychia dendritica*(Pers.) Vain., *Acta Soc. Fauna Flora fenn.* 7(no. 1): 134(1890) var. *dendritica*

= *Anaptychia dendritica* var. *colorata* f. *esorediosa* Kurok. *Journ. Jap. Bot.* 30: 256(1955).

Type: Taiwan, M. Ogata in TNS.

= *Anaptychia subheteochroa* Kurok. in *J. J. B.*, 35: 240(1960).

Guizhou(Zhang T & J. C. Wei, 2006, p. 7), Yunnan(Zhao et al. 1982, p. 139; Chen J. B., 2001, p. 110), Zhejiang(Zhao et al. 1982, p. 139), Hunan, Anhui, Fujian(Chen J. B., 2001, p. 110), Taiwan(Kurok. 1955, p. 256 & 1960 b, p. 248 & 1962 a, p. 54; Lamb, 1963, p. 30; Wang & Lai, 1973, p. 87, 88, & 1976, p. 226; Ikoma, 1983, p. 26).

Heterodermia diademata(Taylor) D. D. Awasthi 大哑铃孢

Geophytology 3: 113(1973).

≡*Parmelia diademata* Taylor, *London J. Bot.* 6: 165(1847).

≡*Anaptychia diademata*(Taylor) Kurok., *Beih. Nova Hedwigia* 6: 28(1962).

= *Anaptychia speciosa* var. *esorediata* Vain., *Cat. Welwitsch Afric.* Pl. 2: 409(1901).

= *Anaptychia speciosa* f. *compactior* Zahlbr., *Bot. Mag.*, *Tokyo* 41: 364(1927).

[= *Chaidhuria indica* Zahlbr.]

On trees, mosses and stones.

Guizhou(Zhang T & J. C. Wei, 2006, p. 7), Shandong(Zhao ZT et al., 2002, p. 8), Jiangsu, Zhejiang, Anhui (Xu B. S. 1989, p. 260).

f. diademata 原变型

Sichuan(Zahlbr., 1934 b, p. 74; Kurok. 1962 a, p. 29; Zhao et al. 1982, p. 134; Chen S. F. et al., 1997, p. 221), Yunnan(Kurok. 1962 a, p. 29; Zhao et al. 1982, p. 134; Wang L. S. et al. 2008, p. 537), Guangdong(Kurok. 1962 a, p. 29), Anhui, Jiangxi, Hunan, Guangxi, Guizhou(Zhao et al. 1982, p. 134), Jiangsu(Wu J. N. & Xiang T. 1981, p. 5, here should be on p. 6 according to the sequence of the pages), Fujian(Wu J. N. et al. 1985, p. 224), Hongkong(Thrower, 1988, p. 108), China(prov. not indicated: Kurok. 1959 b, p. 179).

= *Anaptychia speciosa* f. *angustata*(Räsänen) M. Satô, *Index Plantarum Nipponicarum*, IV Lichenes 4: 115(1943).

≡*Anaptychia esorediata* f. *angustata* Räs. *Journ. Jap. Bot.* 16: 139(1940).

≡*Anaptychia diademata* f. *angustata*(Räs.) Kurok. *A monograph of the genus of Anaptychia*, 30(1962).

Yunnan(Zhao et al. 1982, p. 135).

f. brachyloba(Müll. Arg.) D. D. Awasthi 短叶变型

Geophytology 3(1): 113(1973).

≡*Physcia speciosa* f. *brachyloba* Müll. Arg., *Flora*, Regensburg 73: 340(1890).

≡*Anaptychia diademata* f. *brachyloba*(Müll. Arg.) Kurok. *Beih. Nova Hedwigia* 6: 29(1962).

[= *Physcia dispersa* Nyl.]

= *Anaptychia esorediata* f. *subimbricata* Räs. in *J. J. B.*, 16: 139(1940).

≡*Anaptychia speciosa* var. *esorediata* f. *subimbricata*(Ras.) M. Sato in *Index Plantarum Nipponicarum*, IV, Lichenes, 115(1943).

= *Anaptychia speciosa* f. *subtremulans* Zahlbr., in Handel-Mazzetti, *Symb. Sin*. 3: 242(1930) & *Cat. Lich. Univ*. 8: 597(1932).

Type: Sichuan, Handel-Mazzetti no. 2762.

Jilin(Zhao et al. 1982, p. 135), Sichuan(Zahlbr., 1930b, p. 242), Yunnan, Hongkong(Kurok. 1962 a, p. 29), Fujian(Kurok. 1962 a, p. 29; Wu J. N. et al. 1985, p. 224), Taiwan(Kurok. 1962 a, p. 29; Wang & Lai, 1976, p. 226), China(prov. not indicated: Hue, 1890, p. 318; Zahlbr., 1934 b, p. 213; Kurok. 1959 b, p. 180).

= *Anaptychia esorediata* f. *condensata* Kurok. *Journ. Jap. Bot*. 34: 181(1959).

≡ *Anaptychia diademata* f. *condensata* (Kurok.) Kurok. *A monograph of the genus of Anaptychia* 30 – 31 (1962).

Yunnan(Zhao et al. 1982, p. 135), Fujian(Zhao et al. 1982, p. 135; Wu J. N. et al. 1985, p. 224), Taiwan(Kurok. 1959 b, p. 181 & 1962 a, p. 30 – 31; Wang & Lai, 1973, p. 87 & 1976, p. 226).]

Heterodermia dissecta(Kurok.) D. D. Awasthi 深裂哑铃孢

Geophytology 3: 113(1973).

≡ *Anaptychia dissecta* Kurok., *Beih. Nova Hedwigia* 6: 55(1959).

Sichuan, Hunan, Guangxi(Chen J. B., 2001, pp. 109 – 110), Guizhou(Zhang T & J. C. Wei, 2006, p. 8), Yunnan(Chen J. B., 2001, pp. 109 – 110; Wang L. S. et al. 2008, p. 537).

var. dissecta 原变种

Yunnan, Hebei, Jiangxi(Zhao et al. 1982, p. 133), Anhui, Zhejiang(Zhao et al. 1982, p. 133; Xu B. S. 1989, p. 260), Taiwan(Yoshim. 1974, p. 7).

var. koyana(Kurok.) J. C. Wei 高野变种

Enum. Lich. China 109(1991).

≡ *Anaptychia dissecta* var. *koyana* Kurok. *Journ. Jap. Bot*. 34: 183(1959).

Fujian(Wu J. N. et al. 1985, p. 225), Taiwan(Kurok. 1959 b, p. 183 & 1962 a, p. 39; Wang & Lai, 1973, p. 87).

Heterodermia erinacea(Ach.) W. A. Weber 刺哑铃孢

in Egan, *Bryologist* 90(2): 163(1987).

≡ *Lichen erinaceus* Ach., *Lich. univ*.: 499(1810).

≡ *Anaptychia erinacea*(Ach.) Trevis., *Flora*, Regensburg 44: 52(1861).

≡ *Borrera erinacea* Ach. *Lich. Univ*. 499(1810).

On stones near stream.

Guangxi(Zhao et al. 1982, p. 126).

Heterodermia firmula(Linds.) Trevis. 黄髓哑铃孢

Atti Soc. ital. Sci. nat. (Modena) 11: 615(1868).

≡ *Physcia obscura* var. *firmula* Linds., *Trans. R. Soc. Edinb*. 22: 248(1859)[1861].

≡ *Physcia firmula* (Linds.) Nyl., *Syn. meth. lich*. (Parisiis) 1(2): 418(1860).

= *Anaptychia speciosa* f. *endocroscea* Zahlbr., in Handel-Mazzetti, *Symb. Sin*. 3: 242(1930) & *Cat. Lich. Univ*. 8: 597(1932).

Type: Yunnan, on the hill at Jinjishan near Luopin shan, at about 1600 m, Handel-Mazzetti no. 10179(holotype) in W & isotype in Wu(not seen).

On *Schoepfia jasminodora*

≡ *Anaptychia firmula*(Linds.) C. W. Dodge & D. D. Awasthi, *J. Indian bot. Soc*. 39(3): 423(1960).

On bark of trees & on rocks.

Yunnan(Hue, 1889, p. 168; Kurok. 1962 fide Kurok. 1973, p. 597; Zhao et al. 1982, p. 128; Wang L. S. et al. 2008, p. 537), Hebei, Anhui(Zhao et al. 1982, p. 128), Zhejiang(Zhao et al. 1982, p. 128; Xu B. S. 1989, p. 260), Fujian(Zhao et al. 1982, p. 128; Wu J. N. et al. 1985, p. 224), Taiwan(Aptroot and Sparrius, 2003, p.

22).

Heterodermia flabellata(Fée) D. D. Awasthi 扇哑铃孢

Geophytology 3: 113(1973).

≡*Parmelia flabellata* Fée, *Essai Crypt. Exot.*, *Suppl. Révis.* (Paris) : 122, tab. XXXVIII, fig. 2(1837).

≡*Anaptychia flabellata*(Fée) A. Massal., *Memor. Lich.* : 41(1853).

On stones.

Guizhou(Zhang T & J. C. Wei, 2006, p. 8), Anhui, Zhejiang(Xu B. S. 1989, p. 260).

≡*Heterodermia flabellata* var. *flabellata*

= *Anaptychia fulvescens* (Vain.) Kurok. in J. J. B. 35: 93(1960).

≡*Anaptychia hypoleuca* var. *fulvescens* Vain., *Philipp. J. Sci.*, C, *Bot.* 8: 106(1913).

≡*Anaptychia heterochroa* Vain. var. *fulvescens*(Vain.) M. Sato, J. J. B. 12: 429(1936).

= *Anaptychia dendritica* var. *colorada* f. *esorediosa* Kurok. *J. J. B.* 30: 255(1955).

Yunnan(Zhao et al. 1982, p. 140; Wang L. S. et al. 2008, p. 537), Guangxi(Zhao et al. 1982, p. 140), Fujian (Wu J. N. et al. 1985, p. 226).

= *Heterodermia flabellata* var. *rottboellii*(Vain.) J. C. Wei *Enum. Lich. China*(Beijing) : 109(1991).

≡*Anaptychia hypoleuca* var. *rottbollii* Vain. *Phil. Journ. Sci.* 8: 106(1913).

= *Anaptychia dendritica* var. *colorata* f. *hypoflavescens* Kurok. *Journ. Jap. Bot.* 30: 255(1955).

Type: Taiwan, Mt. Ali shan, Sato 6, pr. maj. p. in TI.

= *Anaptychia fulvescens* (Vain.) Kurok. in J. J. B. 35: 93(1960).

= *Anaptychia flabellata* var. *rottboellii* (Vain.) Kurok., *Beih. Nova Hedwigia* 6: 53(1962).

On bark of tree.

Yunnan(Kurok. 1960 a, p. 94 & 1962 a, p. 53), Guangxi(Zhao et al. 1982, p. 140), Taiwan(Kurok. 1955, p. 255 & 1960 a, p. 94; Wang & Lai, 1973, p. 87)].

Heterodermia fragilissima(Kurok.) J. C. Wei & Y. M. Jiang 脆哑铃孢

Lichens of Xizang(China) : 111(1986).

≡*Anaptychia fragilissima* Kurok., *Beih. Nova Hedwigia* 6: 60(1962).

= *Anaptychia dendritica* var. *japonica* f. *microphyllina* Kurok. *Journ. Jap. Bot.* 30: 255(1955).

Xizang(J. C. Wei & Y. M. Jiang, 1981, p. 1147 & 1986, p. 111), Yunnan(Zhao et al. 1982, p. 138), Guangdong(Kurok. 1962 a, p. 60; Ikoma, 1983, p. 27).

Heterodermia granulifera(Ach.) W. L. Culb., (1967). 颗粒哑铃孢

≡*Parmelia granulifera* Ach., *Syn. meth. lich.* (Lund) : 212(1814).

≡*Anaptychia granulifera*(Ach.) A. Massal., *Memor. Lich.* : 41(1853).

Jiangxi, Guangxi(Zhao et al. 1982, p. 132).

Heterodermia hypocaesia(Yasuda ex Räsänen) D. D. Awasthi 兰腹哑铃孢

Geophytology 3: 113(1973).

≡*Anaptychia hypocaesia* Yasuda ex Räsänen, *J. Jap. Bot.* 16: 139(1940).

Sichuan, Yunnan, Anhui, Guangxi(Zhao et al. 1982, p. 138), Fujian(Wu J. N. et al. 1985, p. 226), Taiwan(Kurok. 1973, p. 603; Yoshim. 1974, p. 9; Wang & Lai, 1976, p. 226).

Heterodermia hypochraea(Vain.) Swinscow & Krog 黄腹哑铃孢

Lichenologist 8(2) : 119(1976).

≡*Anaptychia hypochroea* Vain., *Bot. Mag.*, Tokyo 35: 59(1921).

≡*Anaptychia podocarpa* var. *hypochrae*(Vain.) M. Sato in *J. J. B.* 12: 431(1936).

On trees & rocks.

Sichuan(Chen S. F. et al., 1997, p. 221), Guizhou(Zhang T & J. C. Wei, 2006, p. 8), Yunnan(Kurok. 1962 a, 95; Zhao et al. 1982, p. 143), Hubei(Chen, Wu & J. C. Wei, 1989, p. 486), Hunan(Kurok. 1962 a, p. 95), Anhui(J. C, Wei et al. 1982, p. 57; Zhao et al. 1982, p. 143; Xu B. S. 1989, p. 261), Jiangxi(J. C. Wei et al. 1982, p. 57), Zhejiang(J. C. Wei et al. 1982, p. 57; Xu B. S. 1989, p. 261), Guangxi(Zhao et al. 1982, p. 143), Fujian(J. C. Wei et al. 1982, p. 57; Wu J. N. et al. 1985, pp. 226 – 7), Taiwan(Ikoma, 1983, p. 27).

Heterodermia hypoleuca(Mühl.) Trevis. 白腹哑铃孢

Atti Soc. ital. Sci. nat. (Modena) 11: 615(1868).

≡*Parmelia hypoleuca* Mühl., *Cat. Pl. Amer. Sept.*: 105(1813).

≡*Anaptychia hypoleuca*(Muhl.) A. Massal., *Atti Inst. Veneto Sci. lett.*, *ed Arti*, Sér. 3 5: 249(1860) [1859 – 1860].

≡*Physcia speciosa* var. *hypoleuca*(Muhl.) Nyl., *Syn. meth. lich.* (Parisiis) 1(2): 417(1860).

≡*Pseudophyscia hypoleuca*(Muhl.) Hue, *Nouv. Arch. Mus. Hist. Nat.*, Paris, 4 sér. 1: 111(1899).

On bark of trees.

Heilongjiang(Chen et al. 1981, p. 128), Jilin(Chen et al. 1981, p. 128; Zhao et al. 1982, p. 127), Liaoning (Zhao et al. 1982, p. 127), Shaanxi(Zhao et al. 1982, p. 127; He & Chen 1995, p. 43 without citation of specimens and their localities), Guizhou(Oliv. 1904, p. 196; Zhang T & J. C. Wei, 2006, p. 8), Yunnan(Hue, 1899, p. 111; Zahlbr., 1934 b, p. 213; Kurok. 1962 a, p. 42; Wang L. S. et al. 2008, p. 537), Hebei, Zhejiang(Moreau et Moreau, 1951, p. 195; Zhao et al. 1982, p. 127), Hubei(Chen, Wu & J. C. Wei, 1989, p. 485), Shandong (Moreau et Moreau, 1951, p. 195; Hou Ya-nan et al., 2008, p. 66), Anhui, Zhejiang(Xu B. S. 1989, p. 261, on trees and stones), Taiwan(Kurok. 1962a, p. 42; Wang & Lai, 1973, p. 87), China(prov. not indicated: Asahina & Sato in Asahina, 1939, p. 779).

= *Anaptychia speciosa* var. *hypoleuca* f. *isidiifera* Müll. Arg. *Bull. L'Herb. Bois*. 1: 236(1893).

Central China(prov. not indicated: Müll. Arg. 1893, p. 236, no. 6718, p. p. in H, not seen).

= *Physcia schaereri* Hepp apud Zolling. System. Verzeichn. Indisch. Archip. gesamm. Pflanzen1854: 10, fig. XIV, 2.

Misapplied names:

Parmelia speciosa var. *hypoleuca* sensu Ach., non(Muhl.) Ach., *Syn. meth. lich.* (Lund): 211(1814).

Parmelia hypoleuca sensu Jatta, 1902, p. 472, non Mühl., *Cat. Pl. Amer. Sept.*: 105(1813).

Physcia hypoleuca sensu Hue, 1889, p. 168., non(Muhl.) Tuck., *Syn. N. Amer. Lich.* (Boston) 1: 68(1882).

On bark of trees & on rocks.

Shaanxi(Jatta, 1902, p. 472; Zahlbr., 1930 b, p. 242), Sichuan, Hunan, Fujian(Zahlbr., 1930 b, p. 242), Yunnan(Hue, 1889, p. 168; Zahlbr., 1930 b, p. 244 & 1934b, p. 213).

[f. lobulata f. nov. Chen JB & He Q, 1995, p. 43]. nom. nud., without both the description and citation of specimen.

Heterodermia incana(Stirt.) D. D. Awasthi 灰白哑铃孢

Geophytology 3(1): 114(1973).

≡*Physcia incana* Stirt., *Proc. Roy. phil. Soc. Glasgow* 11: 322(1879).

Taiwan(Kurok. 1962 a, p. 92; Wang & Lai, 1976, p. 226).

Heterodermia isidiophora(Nyl.) D. D. Awasthi 裂芽哑铃孢

Geophytology 3: 114(1973).

≡*Physcia speciosa* f. *isidiophora* Nyl., *Syn. meth. lich.* (Parisiis) 1(2): 417(1860).

≡*Anaptychia speciosa* f. *isidiophora*(Nyl.) Zahlbr., *Cat. Lich. Univers*. 7: 740(1931).

North eastern China(prov. not indicated: Kurok. 1959 b, p. 181 & 1962 a, p. 33), Anhui(Zhao et al. 1982, p. 132; Xu B. S. 1989, p. 262), Guangxi, Yunnan(Zhao et al. 1982, p. 132), Shandong(Zhao ZT et al., 1999, p. 428; Zhao ZT et al., 2002, p. 8), Jiangsu(Wu J. N. & Xiang T. 1981, p. 5, here should be on p. 6 according to the sequence of the pages; Xu B. S. 1989, p. 262), Zhejiang(Xu B. S. 1989, p. 262), Fujian(Wu J. N. et al. 1985, p. 225), Taiwan(Wang & Lai, 1973, p. 87).

Heterodermia japonica(Sato) Swinscow & Krog 阿里山哑铃孢

Lichenologist, 8(2): 122(1976).

≡*Anaptychia dendritica* var. *japonica* M. Sato, *Journ. Jap. Bot.* 12: 427(1936).

On bark.

Guizhou(Zhang T & Wei JC, 2006, p. 8).

= *Heterodermia propagulifera*(Vain.) J. P. Dey *in* Parker & Roane, *Distr. Hist. Biota S. Appal.* 4: 403(1977).

≡*Anaptychia dendritica* var. *propagulifera* Vain., *Philipp. J. Sci.*, *C, Bot.* 8: 107(1913).

≡*Anaptychia subheterochroa* var. *propagulifera*(Vain.) Kurok., *J. Jap. Bot.* 35: 241(1960).

Heilongjiang, Jilin, Hunan, Yunnan(Chen J. B., 2001, p. 113), Sichuan(Zhao et al. 1982, p. 139), Hubei (Chen, Wu & J. C. Wei, 1989, p. 486; Chen J. B., 2001, p. 113), Anhui(Xu B. S. 1989, p. 263, on trees), Taiwan(Wang & Lai, 1973, p. 87).

var. japonica 原变种

Type: Taiwan, Mt. Alishan, Jan. 24, 1936, Sato: Taiwan 10, in TI(not seen).

Taiwan(M. Sato, 1936 a, p. 427; Kurok. 1960 c, p. 353 & 1962 a, p. 58; Wang & Lai, 1973, p. 87; Ikoma, 1983, p. 27).

On the ground among mosses.

Heilongjiang(Shang-zhi county, Mt. Moer shan, not in Jilin prov. as miscited by Zhao et al. 1982, p. 136; Chen J. B., 2001, pp. 110 – 111), Xizang(J. C. Wei & Y. M. Jiang, 1986, p. 111), Sichuan(Zhao et al. 1982, p. 136; Chen J. B., 2001, p.), Guizhou(Chen J. B., 2001, pp. 110 – 111), Yunnan(Zhao et al. 1982, p. 136; Chen J. B., 2001, pp. 110 – 111; Wang L. S. et al. 2008, p. 537), Hunan(Zhao et al. 1982, p. 136; Chen J. B., 2001, pp. 110 – 111), Shandong(Hou Ya-nan et al., 2008, p. 66), Jiangxi, Guangxi(Zhao et al. 1982, p. 136), Anhui (Zhao et al. 1982, p. 136; Xu B. S. 1989, p. 262; Chen J. B., 2001, pp. 110 – 111), Zhejiang(Zhao et al. 1982, p. 136; Xu B. S. 1989, p. 262), Fujian (Wu J. N. et al. 1985, p. 225; Chen J. B., 2001, pp. 110 – 111).

var. reagens(Kurok.) J. N. Wu & Z. G. Qian 异反应变种

in Xu, [*Cryptogamic Flora of the Yangtze Delta and Adjacent Regions*] (Shanghai): 262(1989).

≡*Anaptychia japonica* var. *reagens* Kurok., *J. Jap. Bot.* 35: 354(1960).

Yunnan, Guangxi(Zhao et al. 1982, p. 137), Anhui, Zhejiang(Xu B. S. 1989, p. 262).

Heterodermia leucomelos(L.) Poelt [as '*leucomelaena*'] 顶直哑铃孢

Nova Hedwigia 9: 31(1965).

≡*Lichen leucomelos* L., *Sclerom. Suec.* 2: no. 1613(1763).

≡*Borrera leucomelos*(L.) Ach., *Lich. univ.*: 499(1810).

≡*Anaptychia leucomelos* (L.) A. Massal., *Acta Soc. Fauna Flora fenn.* 7(no. 1): 128(1890).

= *Anptychia leucomelaena* Vainio, *étud. Lich. Bresil,* I: 128(1890).

= *Anaptychia ophioglossa*(Taylor) Kurok., *J. Jap. Bot.* 35: 354(1960).

≡*Parmelia ophioglossa* Taylor, *London J. Bot.* 6: 172(1847).

Among mosses & Evernia divaricata

Xinjiang, Gansu(H. Magn. 1940, p. 159), Xizang(Wei & Chen, 1974, p. 181), Sichuan(Chen S. F. et al., 1997, p. 221 as *H. leucimelos*), Yunnan(Hue, 1899, p. 105; Harmand. 1928, p. 324; Zhao et al. 1982, p. 142), Taiwan(Sasaoka, 1919, p. 180; Kurok. 1960 c, p. 355 & 1962 a, p. 74 – 76; Wang & Lai, 1973, p. 87 & 1976, p. 226), Hainan(JCWei et al., 2013, p. 232 as "*leucomela*").

Heterodermia lutescens(Kurok.) Follmann 黄哑铃孢

Philippia 2(2): 3(1974).

≡*Anaptychia lutescens* Kurok., *J. J. B.* 36: 54(1961).

Misapplied name:

Anaptychia leucomelaena var. *angustifolia* sensu Sato: *Journ. Jap. Bot.* 12: 429(1936, p. p.), non

(Meyen & Flot.) Müll. Arg., *Bot. Jb.* 20: 249(1894).

Yunnan(Kurok. 1961, p. 54 & 1962 a, p. 79, cited as 'China', but it seems to be Yunnan; Zhao et al. 1982, p. 141), Taiwan(Kurok. 1961, p. 54 & 1962a, p. 79; M. Sato, 1936, p. 429, p. p.; Wang & Lai, 1973, p. 87; Ikoma, 1983, p. 27).

Heterodermia microphylla(Kurok.) Skorepa 小叶哑铃孢

Lichenologist 8(2): 132(1976).

≡*Anaptychia hypoleuca* var. *microphylla* Kurok., *J. Jap. Bot.* 34: 123(1959).

≡*Anaptychia microphylla*(Kurok.) Kurok., *Beih. Nova Hedwigia* 6: 44(1962).

On bark of trees.

Heilongjiang(Xiao Hinggan Ling, Chen et al. 1981, p. 128; Chen J. B., 2001, p. 111), Jilin(Chen J. B., 2001, p. 111), Guizhou(Zhang T & J. C. Wei, 2006, p. 8), Shandong(Zhao ZT et al., 1999, p. 428; Zhao ZT et al., 2002, p. 8), Anhui, Zhejiang(Xu B. S. 1989, p. 261, on trees and stones), Taiwan(Aptroot, Sparrius & Lai 2002, p. 284).

f. microphylla 原变型

Sichuan, Yunnan(Zhao et al. 1982, p. 126).

f. granulosa(Kurok.) J. C. Wei 粒状变型

Enum. Lich. China 111(1991).

≡*Anaptychia hypoleuca* var. *microphylla* f. *granulosa* Kurok. *Journ. Jap. Bot.* 34: 123(1959).

Jilin, Liaoning, Sichuan(Zhao et al. 1982, p. 127).

Heterodermia obscurata(Nyl.) Trevis. 暗哑铃孢

Nuovo G. bot. Ital. 1: 114(1869).

≡ *Physcia obscurata* Nyl., *Acta Soc. Sci. Fenn.* 7(2): 440(1863).

≡*Anaptychia obscurata* (Nyl.) Vain., *Acta Soc. Fauna Flora fenn.* 7(no. 1): 137(1890).

=*Anaptychia heterochroa* Vain. in *Bot. Magaz. Tokyo,* 35: 60(1921).

=*Anaptychia hypoleuca* var. *colorata* Zahlbr. *Bot. Mag.*, Tokyo 41: 363(1927).

≡*Anaptychia hypoleuca* var. *colorata*(Zahlbr.) Zahlbr.; Zahlbr. *Cat. Lich. Univ.* VII: 726(1931).

≡*Anaptychia endritica*(Pers.) Vain. var. *colorata*(Zahlbr.) Kurok. in *J. J. B.* 30: 255(1955).

[=*Anaptychia hypoleuca* var. *sorediifera*(Müll. Arg.) Vain., *Trans. Br. mycol. Soc.* 16: 132(1931).]

[≡*Anaptychia sorediifera*(Müll. Arg.) Du Rietz & Lynge, *Skr. VidenskSelsk. Christiania, Kl. I, Math. -Natur.* 16: 12(1924).]

Heilongjiang, Jilin(Chen et al. 1981, p. 128), Shaanxi(Jatta, 1902, p. 472; Zahlbr., 1930 b, p. 243), Sichuan, Jiangxi, Hunan, Guangxi(Zhao et al. 1982, p. 137), Guizhou(Zhao et al. 1982, p. 137; Zhang T & J. C. Wei, 2006, p. 8), Yunnan(Zahlbr., 1934 b, p. 213; Zhao et al. 1982, p. 137), Anhui, Zhejiang(Zhao et al. 1982, p. 137; Xu B. S. 1989, p. 262), Fujian(Zahlbr., 1930 b, p. 242; Wu J. N. et al. 1985, p. 226), Taiwan(Zahlbr., 1933 c, p. 68; M. Sato, 1936a, p. 428; Kurok. 1955, p. 255 & 1962 a, p. 49; Wang & Lai, 1973, p. 87 & 1976, p. 226).

Heterodermia orientalis J. B. Chen & D. P. Wang, in Chen 东方亚铃孢

Mycotaxon 77: 102(2001).

Type: China, Yunnan, Tengchong county, alt. 2000 m. Jun. 15, 1981. X. Y. Wang et al. 2797(Holotype in HMAS-L)

Heterodermia pacifica(Kurok.) Kurok. 太平洋哑铃孢

Folia cryptog. Estonica 32: 23(1998).

≡*Anaptychia pacifica* Kurok. *J. Hattori bot. Lab.* 37: 592(1973).

Heilongjiang, Jilin, Yunnan, Hunan, Zhejiang, Fujian(Chen J. B., 2001, pp. 112), Taiwan(Kurok. 1973, p. 592; Yoshim. 1974, p. 11; Wang & Lai, 1976, p. 226).

Heterodermia pandurata(Kurok.) J. C. Wei 琴哑铃孢

Enum. Lich. China: 112(1991).

≡*Anaptychia pandurata* Kurok. Beiheft zur *Nova Hedwigia*, heft 6: 95(1962).

Taiwan(Kurok. 1962 a, p. 95; Wang & Lai, 1976, p. 226; Ikoma, 1983, p. 28).

Heterodermia pellucida(D. D. Awasthi) D. D. Awasthi 透明哑铃孢

Geophytology 3(1) : 114(1973).

≡*Anaptychia pellucida* D. D. Awasthi, in *Proc. Indian Acad. Sci.* XLV: 136(1957).

Sichuan, Yunnan(Kurok. 1962 a, p. 93), Anhui, Guangxi(Zhao et al. 1982, p. 145).

Heterodermia podocarpa(Bél.) D. D. Awasthi 毛果哑铃孢

Geophytology 3: 114(1973).

≡*Parmelia podocarpa* Bél. , *Voy. Indes Or.* , *Bot.* 2(Cryptog.) : 122, tab. XIII, fig. 2(1846).

≡*Anaptychia podocarpa* (Bél.) A. Massal. , *Atti Inst. Veneto Sci. lett.* , *ed Arti*, Sér. 3 5: 249(1860) [1859 – 1860].

Heilongjiang(Chen et al. 1981, p. 128), Xizang(Chen J. B. , 2001, pp. 112 – 113), Sichuan(Zahlbr. , 1930 b, p. 243), Yunnan(Zahlbr. , 1930 b, p. 243; Zhao et al. 1982, p. 144; Chen J. B. , 2001, pp. 112 – 113), Jiangxi, Guangxi(Zhao et al. 1982, p. 144), Hunan(Zahlbr. , 1930 b, p. 243; Zhao et al. 1982, p. 144), Taiwan(M. Sato. 1936 a, p. 430; Kurok. 1962 a, p. 86; Wang & Lai, 1973, p. 87; Ikoma, 1983, p. 28).

Heterodermia pseudospeciosa(Kurok.) W. L. Culb. 拟哑铃孢

Bryologist 69: 484(1967).

≡*Anaptychia pseudospeciosa* Kurok. , *Beih. Nova Hedwigia* 6: 25(1962).

Anaptychia corallophora sensu Zahlbr. , in *Fedde, Repertorium*, 33: 68(1933), non(Tayl.) Vain.

Neimenggu(Sun LY et al, 2000, p. 36), Xinjiang(Abbas et al. , 1997, p. 2), Liaoning, Xizang, Sichuan, Yunnan, Hubei(Chen J. B. , 2001, p. 114), Guizhou(Zhang T & J. C. Wei, 2006, p. 8), Shandong(Zhao ZT et al. , 1998, p. 30; Zhao ZT et al. , 2002, p. 8), Anhui(Zhao et al. 1982, p. 131; Xu B. S. 1989, p. 263), Hunan, Fujian(Zhao et al. 1982, p. 131; Chen J. B. , 2001, p. 114), Jiangxi(J. C. Wei et al. 1982, p. 55; Zhao et al. 1982, p. 131), Guangxi(Chen J. B. , 2001, p. 114), Taiwan(Zahlbr. , 1933 c, p. 68; M. Sato, 1936 a, p. 426; Kurok. 1959 b, p. 176 & 1962 a, p. 25; Lamb, 1963, p. 30; Wang & Lai, 1973, p. 87), Hongkong(Thrower, 1988, p. 110).

Heterodermia rubescens(Räsänen) D. D. Awasthi 红色哑铃孢

Geophytology 3(1) : 114(1973).

≡Anaptychia hypoleuca f. rubescens Räsänen, Ann. bot. Soc. Zool. -Bot. fenn. Vanamo 5(1) : 25 – 32(1951).

Yunnan(Chen J. B. , 2001, pp. 114 – 115), Taiwan(Kurok. , 1973, p. 597).

Heterodermia sinocomosa J. B. Chen 中国丛毛哑铃孢

Mycotaxon **77**: 104(2001

Type: China, Yunnan, Mt. Yulongshan, alt. 2800 m. Dec. 6, 1960. J. D. Zhao & Y. B. Chen 3856(Holotype in HMAS-L)

Heterodermia speciosa(Wulfen) Trevis. 哑铃孢

Atti Soc. ital. Sci. nat. (Modena) 11: 614(1868).

≡*Lichen speciosus* Wulfen, in *Jacquin, Collnea bot.* 3: 119(1791) [1789].

≡*Parmelia speciosa*(Wulfen) Ach. , *Methodus, Sectio post.* : 198(1803).

≡*Physcia speciosa*(Wulfen) Nyl. , *Act. Soc. linn. Bordeaux* 21: 307(1856).

≡*Pseudophyscia speciosa*(Wulfen) Müll. Arg. , *Bull. Herb. Boissier* 2(app. 1) : 40(1894).

= *Anaptychia speciosa* var. esorediata Vain. in Catal. Welwttsch. Afric. Plants, II, 409(1901).

Above synonymies were contributed by Zahlbr. , 1930 b, p. 241.

≡*Anaptychia speciosa*(Wulfen) A. Massal. , *Memor. Lich.* : 36(1853) .

= *Anaptychia speciosa* f. *foliolosa* Moreau et Moreau, *Rev. Bryol. et Lichenol.* 20: 195(1951) . (nom. inval. Art. 36) Type: from Hebei.

= *Physcia speciosa* var. *stellata* Tuck. , *Proc. Amer. Acad. Arts & Sci.* 4: 390(1860) .

= *Physcia hypoleuca* var. *tremulans* Müll. Arg. , *Flora, Regensburg* 63: 277(1880) .

≡*Anaptychia pseudospeciosa* var. *tremulans* (Müll. Arg.) Kurok. , *Beih. Nova Hedwigia* 6: 25(1962) .

≡*Anaptychia speciosa* var. *tremulans* (Müll. Arg.) Müll. Arg. , *Bot. Jb.* 15: 505(1893) .

≡*Heterodermia tremulans* (Müll. Arg.) W. L. Culb. , *Bryologist* 69: 485(1967) .

≡*Pseudophyscia speciosa* var. *tremulans* (Müll. Arg.) Müll. Arg. , *Bull. Soc. R. Bot. Belg.* 32: 130(1893) .

= *Physcia speciosa* f. *sorediosa* Müll. Arg. , *Flora, Regensburg* 66: 78(1883) .

≡*Anaptychia speciosa* f. *sorediosa* (Müll. Arg.) Zahlbr. , *Cat. Lich. Univers.* 7: 741(1931) .

≡*Heterodermia speciosa* f. *sorediosa* (Müll. Arg.) Zahlbr. , *Cat. Lich. Univers.* 7: 741(1931) .

Misapplied names: (revised by Aptroot & Seaward, 1999, p. 78 – 79) :

Heterodermia diademata aunon(Taylor) Awasthi in Thrower, 1988, pp. 16, 108.

Heterodermia diademata f. *angustata* auct non(Rässänen) Kurok. in Thrower, 1988, pp. 16, 109) .

Heteroderma pseudospeciosa auct non(Kurok.) W. Culb. in Thrower, 1988, pp. 16, 110.

Heilongjiang(Kurok. 1959 b, p. 175 & 1962 a, p. 26) , North eastern China(as Manchuria, prov. not indicated: Ikoma, 1983, p. 28, lapsu ' termulans') , Shaanxi(Baroni, 1894, p. 48; Jatta, 1902, p. 471; Zahlbr. , 1930 b, p. 241; Zhao et al. 1982, p. 130; He & Chen 1995, p. 43) , Sichuan(Zahlbr. , 1930 b, p. 241; Zhao et al. 1982, p. 130; Chen S. F. et al. , 1997, p. 221) , Yunnan(Hue, 1889, p. 168 & 1899, pp. 114, 116; Paulson, 1928, p. 319; Zahlbr. , 1930 b, p. 241; Zhao et al. 1982, pp. 130, 131) , Hebei(Tchou, 1935, p. 131; Zhao et al. 1982, p. 130) , Shandong(Zhao ZT et al. , 1998, p. 30; Zhao ZT et al. , 1999, p. 428; Zhao ZT et al. , 2002, p. 8) , Jiangxi (Zahlbr. , 1930 b, p. 241; Zhao et al. 1982, p. 130) , Hhubei(Müll. Arg. 1893, p. 236; Zahlbr. , 1930 b, p. 242; Chen, Wu, & J. C. Wei, 1989, p. 486) , Hunan(J. C. Wei et al. 1982, p. 56; Zhao et al. 1982, p. 130) , Anhui (Zhao et al. 1982, pp. 130, 131; Xu B. S. 1989, p. 263) , Zhejiang(Moreau et Moreau, 1951, p. 195; Zhao et al. 1982, p. 131; Xu B. S. 1989, p. 263) , Fuijian(Wu J. N. et al. 1985, p. 223) , Guangxi(Zhao et al. 1982, p. 130) , Hongkong

(Seemann, 1852 – 57, p. 432; Zahlbr. , 1930, p. 241; Thrower, 1988, pp. 16, 108, 109, 110 Aptroot & Seaward, 1999, p. 78 – 79; Aptroot et al. , 2001, p. 328; Seaward & Aptroot, 2005, p. 285, Pfister 1978: 83, as *Physcia speciosa* var. *hypoleuca*. On rocks. Hillsides. 8. 4. [1854] . *C. Wright*. Medulla K2.) .

Heterodermia spinulosa(Kurok.) J. C. Wei 小刺哑铃孢

Enum. Lich. China: 113(1991) .

≡*Anaptychia spinulosa* Kurok. Beihefte zur *Nova Hegwigia*, heft 6: 101(1962) .

Type locality: Taiwan, 5/I/1926, Asahina F 105 p. p. in TNS(not seen) . (D. Hawksw. 1972, p. 5; Wang & Lai, 1976, p. 226; Ikoma, 1983, p. 28) .

Heterodermia subascendens(Asahina) Trass 翘哑铃孢

Folia cryptog. Estonica 29: [20] (1992) .

≡*Anaptychia subascendens* Asahina, *Journ. Jap. Bot.* 33: 325(1958) .

On the ground among mosses.

Yunnan(Zhao et al. 1982, p. 143; Wang L. S. et al. 2008, p. 537) , Anhui, Zhejiang(Xu B. S. 1989, p. 264) , Fujian(Wu J. N. et al. 1985, p. 227) , Taiwan(Kurok. 1962 a, p. 96; Yoshim. 1974, p. 11; Wang & Lai, 1976, p. 226; Ikoma, 1983, p. 28) .

Heterodermia szechuanensis(J. D. Zhao, L. W. Hsu & Z. M. Sun) J. C. Wei 四川哑铃孢

Enum. Lich. China: 113(1991) . f. szechuanensis

≡*Anaptychia szechuanensis* J. D. Zhao, L. W. Hsu & Z. M. Sun [as ' szechuensis'], *Acta Phytotax. Sin.* 17(2): 96(1979).

≡*Anaptychia szechuanensis* Zhao et al. *Acta Phytotax. Sin.* 17(2): 96(1979).

Type: Sichuan, Zhao & Xu no. 7728 in HMAS-L.

Sichuan, Yunnan(Zhao et al. 1979, p. 96 & 1982, p. 128).

f. albo-marginata(Zhao et al.) J. C. Wei 白边变型

Enum. Lich. China: 113(1991).

≡*Anaptychia szechuanensis* f. *albo-marginata* Zhao et al. *Acta Phytotax. Sin.* 17(2): 97(1979).

Type: Sichuan, Zhao & Xu no. 8371 in HMAS-L.

Sichuan, Hebei, Anhui(Zhao et al. 1979, p. 97 & 1982, p. 128).

Heterodermia togashii(Kurok.) D. D. Awasthi 拟白腹哑铃孢

Geophytology 3(1): 114(1973).

≡*Anaptychia togashii* Kurok., *Beih. Nova Hedwigia*6: 68(1962).

Guizhou(Zhang T & J. C. Wei, 2006, p. 8), Hubei(Chen, Wu, & J. C. Wei, 1989, p. 487).

Heterodermia undulata(J. D. Zhao, L. W. Hsu & Z. M. Sun) J. C. Wei 波圆哑铃孢

Enum. Lich. China: 113(1991).

≡*Anaptychia undulata* J. D. Zhao, L. W. Hsu & Z. M. Sun, *Acta phytotax. Sin.* 17(2): 98(1979)

Type: Anhui, Zhao & Xu no. 5613 in HMAS-L.

Anhui, Zhejiang, Guangxi(Zhao et al. 1979, p. 98 & 1982, p. 134).

Heterodermia verrucifera(Kurok.) W. A. Weber 疣哑铃孢

Mycotaxon 13(1): 102(1981).

≡ *Anaptychia leucomelos* f. *verrucifera* Kurok., *Beih. Nova Hedwigia* 6: 72(1962).

On soil and *Abies kawakamii*.

Taiwan(Aptroot, Sparrius & Lai 2002, p. 284).

Heterodermia yunnanensis(Zhao et al.) J. C. Wei 云南哑铃孢

Enum. Lich. China: 113(1991).

≡*Anaptychia yunnanensis* Zhao et al. *Acta Phytotax. Sin.* 17(2): 97(1979) & *Prodr. Lich. Sin.* 1982, p. 129.

Type: Yunnan, Zhao & Chen no. 2468 in HMAS-L.

Hyperphyscia Müll. Arg. **外蜈蚣叶属**

Bull. Herb. Boissier 2(1): 10, 41(1894).

Hyperphyscia adglutinata(Flörke) H. Mayrhofer & Poelt 胶外蜈蚣叶

Herzogia 5(1-2): 62(1979).

≡*Lecanora adglutinata* Flörke, *Deutsche Lich.* 4: 7(1819).

≡*Physciopsis adglutinata*(Flörke) M. Choisy, *Bull. mens. Soc. linn. Soc. Bot.* Lyon 19: 20(1950).

≡*Parmelia adglutinata*(Flörke) Flörke, in Mougeot & Nestler(1818)

≡*Physcia adglutinata*(Flörke) Nyl., *Mém. Soc. Imp. Sci. Nat. Cherbourg* 5: 107(1857).

On bark of trees.

Yunnan(Hue, 1900, p. 77; Zahlbr., 1930 b, p. 238), Shanghai(Nyl. & Cromb. 1883, p. 62; Zahlbr., 1930 b, p. 238), Taiwan(Aptroot, Sparrius & Lai 2002, p. 284).

Hyperphyscia cochlearis Scutari 匙外蜈蚣叶

Mycotaxon 61: 94(1997).

On *Erythrina variegata*.

Taiwan(Aptroot, Sparrius & Lai 2002, p. 284).

Hyperphyscia granulata(Poelt) Moberg 粒外蜈蚣叶

Nordic Jl Bot. 7(6): 721(1987).

≡ *Physciopsis granulata* Poelt, *Khumbu Himal* 6(2):91(1974).

On *Ardisia sieboldii*. New to Asia, pantropical.

Taiwan(Aptroot, Sparrius & Lai 2002, p. 285).

Hyperphyscia syncolla(Tuck. ex Nyl.) Kalb 颈外蜈蚣叶

Lichenes Neotropici, Fascicle VI(nos 201 – 250). (Neumarkt): no. 230(1983).

≡*Physcia syncolla* Tuck. ex Nyl., *Acta Soc. Sci. Fenn.* 7(2):441(1863).

≡*Physcia adglutinata* * *Ph. syncolla* Tuck. ex Nyl.

≡*Physciopsis syncolla*(Tuck. ex Nyl.) Poelt, *Nova Hedwigia* 9: 30(1965).

Shaanxi(Hue, 1889, p. 169 & 1900, p. 78; Zahlbr., 1930 b, p. 238 Zhao et al. 1982, p. 110;).

Phaeophyscia Moberg **黑蜈蚣叶属**

Symb. bot. upsal. 22(1):29(1977).

Phaeophyscia ciliata(Hoffm.) Moberg 睫毛黑蜈蚣叶

Symb. bot. upsal. 22(1):30(1977).

≡*Lichen ciliatus* Hoffm., *Enum. critic. lich. europ.* (Bern):69(1784).

≡*Physcia ciliata*(Hoffm.) Du Rietz, *Svensk bot. Tidskr.* 15: 168(1921).

≡*Parmelia obscura* var. *ciliata*(Hoffm.) Schaer., *Lich. helv. spicil.* 9: 442(1840).

On bark of trees.

North-eastern China(prov. not indicated), Xinjiang(Zhao et al., 1982, p. 113; Abdulla Abbas et al., 1993, p. 79; Abdulla Abbas & Wu, 1998, p. 110; Abdulla Abbas et al., 2001, p. 366), Sichuan(Zhao et al. 1982, p. 113), Shaanxi(Jatta, 1902, p. 472; Zahlbr., 1930 b, p. 240; He & Chen, 1995, p. 44, no specimen and locality were cited), Yunnan(Hue, 1889, p. 168 & 1900, p. 70; Zahlbr., 1930 b, p. 240; Zhao et al. 1982, p. 113 Moberg, 1995, p. 321), Neimenggu, Hebei(Moberg, 1995, p. 321), Hubei(Chen, Wu & J. C. Wei, 1989, p. 480), Anhui, Zhejiang(Xu B. S., 1989, p. 255).

Phaeophyscia confusa Moberg 狭黑蜈蚣叶

Nordic J Bot. 3(4):512(1983).

Corticolous. Yunnan(Moberb, 1995, p. 322)

Phaeophyscia constipata(Nyl.) Moberg 密集黑蜈蚣衣

Symb. bot. upsal. 22(1): 33(1977).

≡*Physcia muscigena* var. *constipata* Nyl., in Norrlin, *Not. Sällsk. Fauna et Fl. Fenn. Förh.*, *Ny Ser.* 13: 326 (1874)[1871 – 1874].

≡*Physcia constipata*(Nyl.) Vain., *Meddn Soc. Fauna Flora fenn.* 6: 134(1881).

= *Physcia pulverulenta* var. *tenuis* Th. Fr., *Lich. Scand.* (Upsaliae) 1(1): 137(1871).

On tile.

Xinjiang(Abdull Abbas et al., 1993, p. 79; Abdull Abbas et al., 1994, p. 22; Abdulla Abbas & Wu, 1998, p. 110; Abdulla Abbas et al., 2001, p. 366), Sichuan(Zhao et al. 1982, p. 110).

Phaeophyscia denigrata(Hue) Moberg 白腹黑蜈蚣叶

Acta bot. fenn. 150: 124(1994).

≡*Physcia constipata* Hue, *Nouv. Arch. Mus. Hist. Nat.*, Paris, 4 sér. 2: 76(1900).

Beijing, Heilongjiang, Yunnan, Hebei, Shanxi(as Shansi) (Moberg, 1995, p. 323).

Phaeophyscia endococcinea(Körb.) Moberg 红髓黑蜈蚣叶

Symb. bot. upsal. 22(1):35(1977) "endococcina". [endo-: *in Gk. Comp.*, within, inside; coccineus, a, um: deep red]

≡*Parmelia endococcinea* Körb., *Reprium nov. Spec. Regni veg.* 29: 288(1931)["*endococcina*"].

On bark of trees and on rocks.

Beijing, Hubei, Hunan, Guangxi(Zhao et al. 1982, p. 109), Sichuan(Chen S. F. et al., 1997, p. 222), Hebei

(Moberg, 1995, as var. endococcina) Fujian(Wu J. N. et al. 1985, p. 228, lapsu '*endococcina*'), Anhui, Zhejiang (Xu B. S., 1989, p. 255 – 256, lapsu '*endococcina*'), Taiwan(Zahlbr., 1933 c, p. 67; M. Sato, 1936, p. 569; Wang & Lai, 1973, p. 95).

Phaeophyscia endococcinodes(Poelt) Essl. 内赤黑蜈蚣叶

Mycotaxon 7(2): 301(1978).

≡*Physcia endococcinodes* Poelt, *Khumbu Himal* 6: 77(1977).

On shady rocks, but it also found on soil and old woad.

Yunnan(Moberg, 1995, p. 324), Taiwan(Kashiw. 1984, p. 56).

Phaeophyscia erythrocardia(Tuck.) Essl. 红心黑蜈蚣叶

Mycotaxon 7(2): 302(1978).

≡*Physcia obscura* var. *erythrocardia* Tuck. in *Proc. Americ. Acad. Arts and Sci.* 4: 399(1860).

On rocks.

Xinjiang(Abdull Abbas et al., 1993, p. 80; Abdulla Abbas & Wu, 1998, p. 110; Abdulla Abbas et al., 2001, p. 366), Shaanxi(He & Chen 1995, p. 44 without citation of specimen and locality), Hunan, Anhui, Fujian (Zhao et al. 1982, p. 113).

Phaeophyscia exornatula(Zahlbr.) Kashiw. 裂芽黑蜈蚣叶

Bull. Natn. Sci. Mus. Tokyo, ser. B, 10(3): 127(1984).

≡*Physcia setosa* var. *exornatula* Zahlbr., in Handel-Mazzetti, *Symb. Sin.* 3: 240(1930).

Type: Fujian, Chung no. B3.

Fujian(Kashiw. 1984, p. 127).

On shady rocks.

Sichuan, Yunnan(Zhao et al. 1982, p. 112), Taiwan(Kashiw. 1984, p. 129).

Phaeophyscia hirtuosa(Kremp.) Essl. 白刺毛黑蜈蚣叶

Mycotaxon 7(2): 304(1978).

≡ *Physcia hirtuosa* Kremp., *Flora*, Regensburg 56: 470(1873).

Type: from Guangdong, R. Rabh.

= *Physcia setosa* f. *japonica* Hue, *Nouv. Arch. Mus. Hist. Nat.*, *Paris*, 4 sér. 2: 73, tab. IV. fig. 2, 2, bis(1900).

≡*Physcia japonica*(Hue) Vain., *Bot. Mag.*, *Tokyo* 32: 157(1918).

Heilongjiang, Sichuan, Guizhou, Shanxi, Hebei, Hunan(Zhao et al. 1982, p. 110), neimenggu(Sun LY et al., 2000, p. 36), Xinjiang(Abdulla Abbas & Wu, 1998, p. 111; Abdulla Abbas et al., 2001, p. 366), Shaanxi(He & Chen 1995, p. 44 without citation of specimens and their localities), Hubei(Zhao et al. 1982, p. 110 ; Chen, Wu & J. C. Wei, 1989, p. 480), Shandong(Zhao ZT et al., 1998, p. 30; zhang F et al., 1999, p. 31; Zhao ZT et al., 1999, p. 428; Zhao ZT et al., 2002, p. 7), Jiangxi(Asahina, 1947a, p. 6; Zhao et al. 1982, p. 110), Shanghai(Asahina, 1947a, p. 6), Anhui(Xu B. S., 1989, p. 256), Zhejiang(Zhao et al. 1982, p. 110; Xu B. S., 1989, p. 256), Guangdong(Rabh. 1873, p. 287; Krphbr. 1874c, p. 66; Zahlbr., 1931, p. 619).

Phaeophyscia hispidula(Ach.) Essl. 毛边黑蜈蚣叶

Mycotaxon 7(2): 305(1978).

≡*Parmelia hispidula* Ach., *Lich. univ.*: 468(1810).

≡*Phaeophyscia hispidula* (Ach.) Moberg, *Bot. Notiser* 131(2): 259(1978).

≡*Parmelia setosa* Ach., Syn. *meth. lich.* (Lund): 203(1814).

≡*Physcia setosa* (Ach.) Nyl., *Syn. meth. lich.* (Parisiis) 1(2): 429(1860).

= *Physcia setosa* f. *virella* B. de Lesd., *Annals Cryptog. Exot.* 2: 232(1929).

On bark of trees & on rocks.

Heilongjiang(Chen et al. 1981, p. 156), Shaanxi(Jatta, 1902, p. 473; Zahlbr., 1930 b, p. 240; He & Chen 1995, p. 44 without citation of specimens and their localities), Xinjiang(Wu J. L, 1985, p. 77), Sichuan(Zahlbr., 1930 b, p. 240; Stenroos et al., 1994, p. 336), Yunnan(Hue, 1887a, p. 23 & 1889, p. 168 & 1900, p. 73; Zahlbr., 1930 b, p. 240; Paulson, 1928, p. 319), Hebei(Moreau et Moreau, 1951, p. 195), Hubei(Chen, Wu & J. C. Wei, 1989, p. 481), Hunan(Zahlbr., 1930 b, p. 240), Shandong(Zhao ZT et al., 1999, p. 428; Zhao ZT et al., 2002, p. 8), Shanghai(Nyl. & Cromb. 1883, p. 62; Zahlbr., 1930 b, p. 240), hongkong(Hue, 1887, p. 23), Taiwan(Aptroot, Sparrius & Lai 2002, p. 287).

f. hispidula 原变型

= *Physcia setosa* f. *virella* Bouly de Lesd., *Annals Cryptog. Exot.* 2: 232(1929).

On rocks & on bark of trees.

Beijing, Sichuan(Zhao et al. 1982, p. 112), Xizang(J. C. Wei & Y. M. Jiang, 1986, p. 113), Anhui, Zhejiang (Xu B. S., 1989, p. 256).

= *Physcia setosa* f. *japonica* Hue, *Nouv. Arch. Mus. Hist. Nat.*, *Paris*, 4 sér. 2: 73, tab. IV. fig. 2, 2, bis(1900).

On bark of *Salix* sp.

Beijing(Moreau et Moreau, 1951, p. 195).

= *Physcia setosa* f. *sulphurascens* Zahlbr., in Handel-Mazzetti, *Symb. Sinic.* 3: 240(1930).

Type: Sichuan, Handel-Mazzetti no. 10067.

On arenaceous rocks.

Sichuan(Zahlbr., 1930 b, p. 240), Yunnan(Moberg, 1995, p. 326).

Phaeophyscia hunana Hu G. R. & Chen J. B. 湖南黑蜈蚣叶

Mycosystema(4) : 534 – 535(2003).

Type: Hunan, Sanzhi, alt. 1350m, on bark, Aug. 20, 1997, Chen J. B. et al. 9876(holotype: HMAS-L.).

Phaeophyscia imbricata(Vain.) Essl. 覆瓦黑蜈蚣叶

Mycotaxon 7(2) : 308(1978).

≡ *Physcia imbricata* Vain., *Bot. Mag.*, Tokyo 35: 60(1921).

Corticolous.

Xinjiang(Abdull Abbas et al., 1993, p. 80; Abdulla Abbas & Wu, 1998, p. 111; Abdulla Abbas et al., 2001, p. 366), Taiwan(Zahlbr., 1933c, p. 67; M. Sato, 1936b, p. 570; Wang & Lai, 1973, p. 95).

Phaeophyscia limbata(Poelt) Kashiw. 粉缘黑蜈蚣叶

Bull. natn. Sci. Mus., Tokyo, B 10(3) : 129(1984).

≡ *Physcia hispidula* subsp. *limbata* Poelt, *Khumbu Himal* 6(2) : 81(1974).

Shaanxi(He & Chen 1995, p. 44 without citation of specimens and their localities), Xinjiang(Abdull Abbas et al., 1993, p. 80; Abdulla Abbas & Wu, 1998, p. 111; Abdulla Abbas et al., 2001, p. 366), Hubei(Chen, Wu & J. C. Wei, 1989, p. 481), Taiwan(Aptroot, Sparrius & Lai 2002, p. 287).

Phaeophyscia nigricans(Flörke) Moberg 黑蜈蚣叶

Symb. bot. upsal. 22(1) : 429(1977).

≡ *Lecanora nigricans* Flörke, in Sprengel, *Neue Entdeckungen im ganzen Umfang der Pflanzenkunde* 2: 97 (1821).

= Physcia nigricans var. *sciastrella*(Nyl.) Lynge, Rabenh. Krypt. -Fl., Edn 2(Leipzig) 9. 6(1) : 150(1935).

≡ *Parmelia obscura* f. *sciastrella* Nyl., in Arnold, *Flora*, Regensburg 56: 354(1873).

On rocks.

Beijing(Mt. Baihua shan, Zhao et al. 1982, p. 115), Xinjiang(H. Magn. 1940, p. 159, as var. *sciastrella*; Abdulla Abbas et al., 1993, p. 80; Abdulla Abbas & Wu, 1998, pnp. 111 – 112; Abdulla Abbas et al., 2001, p. 366), Jiangsu(Xu B. S., 1989, p. 256 – 257 on rocks and trees).

Phaeophyscia orbicularis(Neck.) Moberg 圆叶黑蜈蚣叶

Symb. bot. upsal. 22(1) : 44(1977).

≡ *Lichen orbicularis* Neck., *Method. Muscor.* : 88(1771).

≡ *Physcia orbicularis* (Baumg.) Poetsch, in Poetsch & Schiedermayr, *System. Aufzähl. samenlos. Pflanzen* (Krypt.) : 247(1872) .

≡ *Psora orbicularis* Baumg. , *Fl. Lips*. : 589(1790) .

On rocks and trees.

Beijing(Moreau et Moreau, 1951, p. 195; Zhao et al. 1982, p. 113) , Xinjiang(Abdulla Abbas et al. , 1993, p. 80; Abdulla Abbas & Wu, 1998, p. 112; Abdulla Abbas et al. , 2001, p. 366) , Shaanxi (Jatta, 1902, p. 472; Zahlbr. , 1930 b, p. 239) , Sichuan(Zhao et al. 1982, p. 113; Xu B. S. 1989, p. 257; Chen S. F. et al. , 1997, p. 222) , Guizhou, Zhejiang(Zhao et al. 1982, p. 113; Xu B. S. 1989, p. 257) , Jiangsu(Pat. & Oliv. 1907: not seen; Zahlbr. , 1930 b, p. 239; Xu B. S. 1989, p. 257) .

= *Physcia virella*(Ach.) Flagey, *Revue mycol.* , *Toulouse* 13(no. 51) : 110(1891) .

≡ *Lichen virellus* Ach. , *Lich. suec. prodr.* (Linköping) : 108(1799) [1798] .

≡ *Parmelia virella*(Ach.) Ach. , *Methodus, Sectio post.* (Stockholmiæ) : 201(1803) .

= *Physcia virella* var. *hueana*(Harm.) Lindau [as ' hueiana'] , *Die Flecht.* , edn 2: 234(1923) .

≡ *Physcia obscura* var. *virella f. hueana* Harm. Soc. Sci. Hancy, 2(31) : 262(1896) , et Lich. de France, fasc. 4: 546(1909) .

On bark of trees on rocks.

Yunnan, Guizhou(Zhao et al. 1982, p. 114)

Phaeophyscia primaria(Poelt) Trass 刺黑蜈蚣叶

Folia cryptog. Estonica 15: 2(1981) .

≡ *Physcia hispidula* subsp. *primaria* Poelt, *Khumbu Himal* 6(2) : 80(1974) .

Misapplied name:

Phaeophyscia hispidula non(Ach.) Moberg, *Bot. Notiser* 131(2) : 259(1978) .

On bark of trees & on rocks.

Heilongjiang, Shaanxi, Yunnan, Hebei, Hunan, Jiangxi(Zhao et al. 1982, p. 111) , Sichuan(Chen S. F. et al. , 1997, p. 222) , Xizang(J. C. Wei & Jiang, 1986, p. 113, cited as f. *setosa*) , Shandong(Zhao ZT et al. , 2002, p. 8) , Hubei(Chen, Wu & Wei, 1989, p. 481) , Anhui, Zhejiang(Zhao et al. 1982, p. 111; Xu B. S. , 1989, p. 256, cited as f. *hispidula*) , Taiwan(Aptroot, Sparrius & Lai 2002, p. 287) .

Phaeophyscia pyrrhophora(Poelt) D. D. Awasthi & M. Joshi 火红黑蜈蚣叶

Indian J. mycol. Res. 16(2) : 278(1978) .

≡ *Physcia pyrrhophora* Poelt, *Khumbu Himal* 6(2) : 84(1974) .

On both trunks and branches, but it has also been found on mosses on rocks.

Yunnan(Moberg, 1995, pp. 329 – 330) .

Phaeophyscia rubropulchra(Degel.) Moberg 美丽黑蜈蚣叶

Bot. Notiser 131(2) : 262(1978) .

≡ *Physcia orbicularis* f. *rubropulchra* Degel. , in *Ark. f. Bot*. XXXA, no. 1: 58(1942) .

On rocks.

Heilongjiang, Yunnan(Moberg, 1995, pp. 330 – 331) . Xinjiang(Abdulla Abbas & Wu, 1998, p. 112; Abdulla Abbas et al. , 2001, p. 366) , Zhejiang(Zhao et al. 1982, p. 114) .

Phaeophyscia sciastra(Ach.) Moberg 暗裂芽黑蜈蚣叶

Symb. bot. upsal. 22(1) : 47(1977) .

≡ *Parmelia sciastra* Ach. , *Methodus*, Suppl. : 49(1803) .

≡ *Parmelia lithotea* f. *sciastra* (Ach.) Arnold, *Flora*, Regensburg 67: 228(1884) .

≡ *Physcia lithotea* var. *sciastra*(Ach.) Nyl. , *Flora*, Regensburg 60: 354(1877) .

≡ *Physcia lithotea* var. *sciastra* (Ach.) Nyl. , *Flora*, Regensburg 60: 354(1877) .

On rocks.

Beijing(Zhao et al. 1982, p. 114; Moberg, 1995, p. 331) , Heilongjiang, Neimenggu(Moberg, 1995, p. 331) , Xinjiang(Abdulla Abbas et al. , 1993, p. 80; Abdulla Abbas & Wu, 1998, p. 112; Abdulla Abbas et al. , 2001, p. 366) , Shaanxi(Jatta, 1902, p. 472; Zahlbr. 1930b, p. 239) , Sichuan(Zahlbr. , 1930 b, p. 239) , Jiangsu(Xu B. S. , 1989, p. 257) .

Phaeophyscia squarrosa Kashiw. 羽根黑蜈蚣叶

Bull. natn. Sci. Mus. , Tokyo, B 10(1) : 47(1984) .

It grows on tree trunks in fairly open situations along forest margins and on solitary trees.

Yunnan(Moberg, 1995, pp. 331 – 332) .

Phaeophyscia trichophora(Hue) Essl. 载毛黑蜈蚣叶

Mycotaxon 7(2) : 313(1978) .

≡*Physcia trichophora* Hue, *Nouv. Arch. Mus. Hist. Nat.* 4(2) : 74(1890) .

Type: Yunnan, Delavay, near Da-Ping-Zi in 1887 & 1890(syntypes) .

On bark of trees.

Yunnan(Hue, 1900, p. 74; Zahlbr. , 1930 b, p. 241; Zhao et al. 1982, p. 102) , Taiwan(Aptroot, Sparrius & Lai 2002, p. 288) .

Physcia(Schreb.) Michx. **蜈蚣衣属**

Fl. Boreali-Americ. 2: 326(1803) .

≡*Lichen* sect. *Physcia* Schreb. , *Gen.* : 768(1791) .

Physcia adscendens(Fr.) H. Olivier 翘叶蜈蚣衣

Flore analytique et dichotomique des Lichens de l'Orne et départements circonvoisins: 79(1882) .

≡*Parmelia stellaris* var. *adscendens* Fr. , *Summa veg. Scand.* (Stockholm) : 105(1845) .

= *Physcia hispida*(Schreb.) Frege, *Deutsch. Botan. Taschenb.* 2: 169(1812) , nom. rejic. , Art. 56. 1.

≡*Lichen hispidus* Schreb. , *Spic. fl. lips.* (Lipsiae) : 126(1771) , nom. rejic. , Art. 56. 1.

On dead twigs & *Parmelia exasperatula*.

Xinjiang(Moreau et Moreau, 1951, p. 195; H. Magn. 1940, p. 159; Zhao et al. 1982, p. 97; Abdulla Abbas et al. , 1993, p. 80; Abdulla Abbas & Wu, 1998, pp. 113 – 114; Abdulla Abbas et al. , 2001, p. 366) , Shaanxi(Jatta, 1902, p. 472; Zahlbr. , 1930 b, p. 241) .

Physcia aipolia(Ehrh. ex Humb.) Fürnr. 斑面蜈蚣衣

Naturhist. Topogr. Regensburg 2: 249(1839) .

≡*Lichen aipolius* Ehrh. ex Humb. , *Fl. Friberg. Spec.* (Berlin) : 19(1793) .

≡*Parmelia aipolia*(Ehrh. ex Humb.) Ach. , *Methodus, Sectio post.* (Stockholmiæ) : 209(1803) .

On bark of trees.

Beijing(Tchou, 1935, p. 314) , Neimenggu(Sun lY et al. , 2000, p. 36) , Xinjiang(Wang XY, 1985, p. 352; Abdula Abbas et al. , 1993, p. 80; Abdulla Abbas & Wu, 1998, p. 114; Abdulla Abbas et al. , 2001, p. 366) , Shaanxi(Jatta, 1902, p. 471 – 3; Zahlbr. , 1930 b, p. 238; Moreau et Moreau, 1951, p. 194) , Yunnan(Wang L. S. 2008, p. 538) , Jiangsu(Xu B. S. , 1989, p. 257) .

≡*Physcia aipolia* f. *cercidia*(Ach.) Mig. , *Flora von Deutschl.* , Abt. II 12: 51(1924) .

≡*Parmelia aipolia* var. *cercidia* Ach. , Lich. univ. : 478(1810) .

Beijing(Tchou, 1935, p. 315; Zhao et al. 1982, p. 99) .

f. aipolia 原变型

On bark of trees.

Beijing, Heilongjiang, Shaanxi, Qinghai, Xinjiang, Sichuan, Hunan, Guangxi(Zhao et al. 1982, p. 99) .

f. angustata(Nyl.) Vain. 小叶变型

Meddn Soc. Fauna Flora fenn. 6: 136(1881) .

≡*Physcia stellaris* var. *angustata* Nyl. , *Not. Sällsk. Fauna et Fl. Fenn. Förh.* , Ny Ser. 5: 1 – 312(1861) .

Beijing(Zhao et al. 1982, p. 100).

Physcia alba(Fée) Müll. Arg. 大白蜈蚣衣

Revue mycol., *Toulouse* 9: 136(1887).

≡*Parmelia alba* Fée, *Essai Crypt. Exot.* (Paris): 125, tab. XXX, fig. 4(1824).

var. obsessa(Mont.) Lynge 普生变种

Vidensk. Skrifter, I. Math. -naturv. Klasse(no. 16): 24(1924).

≡*Parmelia obsessa* Mont., *Syll. gen. sp. crypt.* (Paris): 328(1856).

On bark of trees.

Yunnan(Zhao et al. 1982, p. 119).

Physcia albinea(Ach.) Nyl. 小白蜈蚣衣

Bull. Soc. Amis Sci. Nat. Rouen, Sér. II 3: 482(1867).

≡*Parmelia albinea* Ach., *Lich. univ.*: 1 – 696(1810).

On bark of trees, and on coastal rock. Northern temperate.

Shaanxi(Jatta, 1902, p. 472; Zahlbr., 1930b, p. 239), Sichuan(Zhao et al. 1982, p. 104), Shandong(Hou Yanan et al., 2008, p. 66), Hongkong(Aptroot & Seawaerd, 1999, pp. 87, 91: it is together with *Pyxine albinea*).

Physcia atrostriata Moberg 黑纹蜈蚣衣

Nordic Jl Bot. 6(6): 853(1986).

On sheltered trees(*Kandelia* sp.) in forests. Pantropical.

China(Prillinger etal. 1997, pp. 579, 582 & 583; Aptroot & Seaward, 1999, p. 87), Shandong(Hou Ya-nan et al., 2008, p. 66), Taiwan(Aptroot, Sparrius & Lai 2002, p. 288), Hainan(JCWei et al., 2013, p. 232), Hongkong(Aptroot & Seaward, 1999, p. 87: Aptroot & Sipman, 2001, p. 337).

Physcia biziana(A. Massal.) Zahlbr. 白粉蜈蚣衣

Öst. bot. Z. 51: 349(1901).

≡*Squamaria biziana* A. Massal., *Miscell. Lichenol.*: 35(1856).

On twigs of *Camellia* sp.

Yunnan(Zhao et al. 1982, p. 98).

Physcia caesia(Hoffm.) Hampe ex Fürnr. 兰灰蜈蚣衣

Naturhist. Topogr. Regensburg 2: 250(1839).

≡*Lichen caesius* Hoffm., *Enum. lich.*: 65(1788).

≡*Parmelia caesia* (Hoffm.) Ach., Methodus, Sectio post. (Stockholmiæ): 197(1803).

= *Physcia subalbinea* Nyl., *Flora*, Regensburg 57: 306(1874).

On rocks.

Neimenggu(Sun LY et al., 2000, p. 36), Beijing, Sichuan, Shanxi, Zhejiang(Zhao et al. 1982, pp. 101, 105), Xinjiang(H. Magn. 1940, p. 157; Wang XY, 1985, p. 353; Abdula Abbas et al., 1993, p. 80; Abdulla Abbas & Wu, 1998, p. 114; Abdulla Abbas et al., 2001, p. 366), Shaanxi(Baroni, 1894, p. 48; Jatta, 1902, p. 473; Zahlbr., 1930 b, p. 239), Yunnan(R. Paulson, 1928, p. 319; Zahlbr., 1930 b, pp. 238, 239).

[*Physcia caesia-picta* Nyl. 淡兰蜈蚣衣

China(prov. no indicated: Hue, 1890, p. 322)].

Physcia clementei(Turner) Lynge 珊瑚芽蜈蚣衣

Rabenh. Krypt. -Fl., Edn 2(Leipzig) 9. 6(1): 93(1935).

≡*Parmelia clementei* Turner [as ‘*clementi*’], in Smith & Sowerby, *Engl. Bot.* 25: tab. 1779(1807).

≡*Lichen clementei*(Turner) Sm., *Engl. Bot.* 25: tab. 1780(1807).

= *Parmelia clementiana* Ach., *Lich. univ.*: 483(1810).

≡*Physcia clementiana*(Ach.) J. Kickx f., *Fl. Crypt. Flandres*(Paris) 1: 226(1867).

≡*Parmelia astroidea* var. *clementiana*(Ach.) Rabenh. , *Deutschl. Krypt. -Fl.* (Leipzig) 2: 63(1845) .

≡*Physcia clementiana*(Ach.) J. Kickx f. , *Fl. Crypt. Flandres* (Paris) 1: 226(1867) .

On rocks. and on bark of trees.

Shaanxi(Zhao et al. 1982, p. 104, on rocks) ; Xinjiang(Abdulla Abbas et al. , 1993, p. 80 as *Physcia* "*clement*"; Abdulla Abbas & Wu, 1998, p. 115 as *Physcia* "*clementi*"; Abdulla Abbas et al. , 2001, p. 366 as *Ph.* "*clement*") , Yunnan(Hue, 1889, p. 168; Zahlbr. , 1930 b, p. 238) .

Physcia crispa Nyl. 皱波蜈蚣衣

Syn. meth. lich. (Parisiis) 1(2) : 423(1860) .

On bark of trees. On coastal rock. Pantropical.

Shanghai, Guangdong(Rabh. 1873, p. 287; Krphbr. 1873, p. 471 & 1874c, p. 66; Zahlbr. , 1930 b, p. 238) , Hongkong(Aptroot & Seaward, 1999, p. 87) .

Physcia dilatata Nyl. 膨大蜈蚣衣

Syn. meth. lich. (Parisiis) 1(2) : 423(1860) .

Shandong(Hou Ya-nan et al. , 2008, p. 66) .

Physcia dimidiata(Arnold) Nyl. 半开蜈蚣衣

in Hue, *Rev. Bot.* 5: 9(1887) [1886 – 87] .

≡*Parmelia pulverulenta* var. *dimidiata* Arnold, *Flora*, Regensburg 47: 594(1864) .

Xinjiang(Abdulla Abbas & Wu, 1998, p. 115; Abdulla Abbas et al. , 2001, p. 366) .

Physcia dubia(Hoffm.) Lettau 疑蜈蚣衣

Hedwigia 52: 254(1912) .

≡*Lobaria dubia* Hoffm. , *Deutschl. Fl.* , *Zweiter Theil* (Erlangen) : 156(1796) [1795] .

Xinjiang(Abdulla Abbas & Wu, 1998, p. 115; Abdulla Abbas et al. , 2001, p. 366 as *P.* "*dubai*") .

= *Physcia teretiuscula*(Ach.) Lynge, *Skr. VidenskSelsk. Christiania, Kl. I, Math. -Natur.* 8: 96(1916) .

≡*Parmelia caesia* var. *teretiuscula* Ach. , *Lich. univ.* : 479(1810) .

= *Physcia intermedia* Vain. , *Meddn Soc. Fauna Flora fenn.* 2: 51(1878) .

On rocks & on bark of trees.

Beijing, Sichuan, Xizang, Hebei, Hunan(Zhao et al. 1982, pp. 106, 108) , Shaanxi(Jatta, 1902, p. 472; Zahlbr. , 1930 b, p. 238) .

= *Physcia endococcina* f. *lithotodes*(Vain.) J. W. Thomson, *Beih. Nova Hedwigia* 7: 117(1963) .

On rocks.

Shaanxi(Zhao et al. 1982, p. 109 as"*endococcinea*") .

This lichen seems to be necessary to check and revise in the future.

Physcia hupehensis Zhao et al. 湖北蜈蚣衣

Acta Phytotax. Sin. 17(2) : 99(1979) .

Type: Hubei, Zhao & Xu no. 10416 in HMAS-L.

Hubei(Zhao et al. 1982, p. 101) .

On rocks.

Physcia integrata Nyl. 下黑蜈蚣衣

Syn. meth. lich. (Parisiis) 1(2) : 424(1860) .

On rocks. Pantropical.

Hongkong(Aptroot & Seaward, 1999, p. 87) .

var. integrata 原变种

On bark of trees & on rocks.

Yunnan, Henan(Zhao et al. 1982, p. 107) .

var. obsessa(Mont.) Vain. 占据变种

Acta Soc. Fauna Flora fenn. 7: 141(1890) .

≡*Parmelia obsessa* Mont., *Syll. gen.* sp. *crypt.* (Paris): 328(1856).

On bark of trees.

Taiwan(Zahlbr., 1933 c, p. 67; M. Sato, 1936b, p. 570; Asahina, 1943a, p. 4; Wang & Lai, 1973, p. 95).

Physcia leptalea(Ach.) DC. 半羽蜈蚣衣

in Lamarck & de Candolle, *Fl. franç.*, Edn 3(Paris) 2: 395(1805).

≡*Lichen leptaleus* Ach., *Lich. suec. prodr.* (Linköping): 108(1799)[1798]. (superfluous name of *Lichen semipinnatus* Leers ex Gmelin, 1791).

=*Physcia semipinnata*(Leers ex J. F. Gmel.) Moberg, *Symb. bot. upsal.* 22(1): 56(1977).

≡*Lichen semipinnatus* Leers ex J. F. Gmel., *Systema Naturae*, Edn 13 2(2): 1372(1792) (Nom. rejic., Art. 56.1).

On dead twigs of *Picea* sp.

Xinjiang(Abdulla Abbas et al., 1993, p. 80, as *Ph.* "*semiinnate*"; Abdulla Abbas & Wu, 1998, p. 116 & Abdulla Abbas et al., 2001, p. 366 as *Ph. semipinnata*), Ningxia(Zhao et al. 1982, p. 97) Shandong(Tchou, 1935, p. 314; zhang F et al., 1999, p. 31; Zhao ZT et al., 2002, p. 7; Hou Ya-nan et al., 2008, p. 66), Hebei, Zhejiang(Tchou, 1935, p. 314).

Physcia nipponica Asahina, 日本蜈蚣衣

J. J. B. 21: 6(1947).

=*Physcia obscura*(Hurh.) Hue var. *chloantha*(non Rabh.) Zahlbr.

On bark of trees.

Liaoning and other place in central China(as "Mandschuria australis" and "Central China" cited by Asahina, 1947 a, pp. 6 – 7).

Physcia obscurella Müll. Arg. 小暗蜈蚣衣

Proc. R. Soc. Edinb. 11: 459(1882).

On hard sandstone, resembling rock.

Neimenggu(H. Magn. 1944, p. 59).

Physcia orientalis Kashiw. 东方蜈蚣衣

Mem. Natn Sci. Mus, Tokyo 18: 101(1985).

Taiwan(Kashiw., 1985, p. 102).

Physcia phaea(Tuck.) J. W. Thomson 异白点蜈蚣衣

Beih. Nova Hedwigia 7: 54(1963).

≡*Parmelia phaea* Tuck., in Darlington, *Fl. Cestrica*, Edn 3: 440(1853).

=*Physcia melops* Dufour ex Nyl., *Flora*, Regensburg 57: 16(1874).

On rocks.

Beijing(Zhao et al. 1982, p. 100), Heilongjiang(Asahina, 1952d, p. 375), neimenggu(Sun LY et al., 2000, p. 37), Xinjiang(Abdulla Abbas & Wu, 1998, p. 115; Abdulla Abbas et al., 2001, p. 366), shaanxi(Ren Q. et al., 2009, p. 104), Xizang(Mt. Everest: Paulson, 1925, p. 191), Anhui, Zhejiang(Xu B. S., 1989, p. 258).

Physcia sorediosa(Vain.) Lynge 粉芽蜈蚣衣

Skr. VidenskSelsk. Christiania, ath. -Natur. 16: 27(1924).

≡*Physcia integrata* var. *sorediosa* Vain., *Acta Soc. Fauna Flora fenn.* 7(no. 1): 142(1890).

Physcia sp. (Chu, 1997, p. 48).

On coastal rocks. Pantropical.

Taiwan(Aptroot, Sparrius & Lai 2002, pp. 288 – 289), Hongkong(Chu, 1997, p. 48 as *Physcia* sp.; Aptroot & Seaward, 1999, p. 87; Aptroot et al., 2001, p. 337).

Physcia stellaris(L.) Nyl. 蜈蚣衣

Act. Soc. linn. Bordeaux 21: 307(1856).

≡*Lichen stellaris* L., *Sp. pl.* 2: 1144(1753).

On bark of trees & on rocks.

Neimenggu(Sun LY et al., 2000, p. 37), Xinjiang(Wang XY, 1985, p. 352; Abdula Abbas et al., 1993, p. 80; Abdulla Abbas & Wu, 1998, p. 116; Abdulla Abbas et al., 2001, p. 366), Shaanxi(He & Chen 1995, p. 44, no specimen and locality were cited), Yunnan, Shanghai(Zahlbr., 1930 b, p. 238), Shandong(Tchou, 1935, p. 313; Moreau et Moreau, 1951, p. 194; Li Y et al., 2008, p. 73), Jiangsu(Wu J. N. & Xiang T. 1981, p. 5, here should be p. 6 according to the sequence of pages; Xu B. S., 1989, p. 258), Shanghai, Anhui, (Xu B. S., 1989, p. 258), Zhejiang(Xu B. S. 1989, p. 258; D. Hawksw. & Weng 1990, p. 515), Fujian(Wu J. N. et al. 1985, p. 227).

f. stellaris 原变型

On bark of trees.

Jilin, Ningxia, Hebei, Shanxi, Anhui(Zhao et al. 1982, p. 103), Hubei(Zhao et al. 1982, p. 103; Chen, Wu & J. C. Wei, 1989, p. 479).

f. melanophthalma Müll. Arg. 黑体变型

China(prov. not indicated, Hue, 1900, p. 59).

f. radiata(Ach.) Nyl. 辐射变型

Not. Sällsk. Fauna et Fl. Fenn. Förh., Ny Ser. 5: 111(1861).

≡*Parmelia stellaris* var. *radiata* Ach., *Lich. univ.*: 477(1810).

On bark of trees.

Beijing(Zhao et al. 1982, p. 103).

f. tuberculata(Kernst.) Dalla Torre & Sarnth. 多瘤变型

Die Flecht. Tirol: 160(1902).

≡*Parmelia stellaris* f. *tuberculata* Kernst., *Verh. zool. -bot. Ges. Wien* 47: 295(1897).

On trees.

Xizang(Zhao et al. 1982, p. 103).

Physcia stenophyllina(Jatta) Zahlbr. 狭叶蜈蚣衣

Cat. Lich. Univers. 7: 691(1931).

≡*Parmelia stenophyllina* Jatta, *Nuov. Giorn. Botan. Ital.*, nov. ser., 9: 473(1902).

Type: Shaanxi, Giraldi from Mt. Guangtou shan.

On tree trunk.

Ningxia, Shanxi(Zhao et al. 1982, p. 100), Shaanxi(Jatta, 1902, p. 473; Zahlbr., 1930 b, p. 238 & 1931, p. 691).

Physcia tenella(Scop.) DC. 长毛蜈蚣衣

in Lamarck & de Candolle, *Fl. franç.*, Edn 3(Paris) 2: 396(1805).

≡*Lichen tenellus* Scop., *Fl. carniol.*, Edn 2(Wien) 2: 394(1772).

≡*Parmelia tenella*(Scop.) Ach., *Methodus, Sectio post.* (Stockholmiæ): 250(1803).

≡*Physcia hispida* var. *tenella*(Scop.) Walt. Watson, *Lich. Somerset:* 28(1930).

On rocks.

Xinjiang(Abdulla Abbas & Wu, 1998, p. 116; Abdulla Abbas et al., 2001, p. 366), Shaanxi(Jatta, 1902, p. 473; Zahlbr., 1930 b, p. 241), Sichuan(Zhao et al. 1982, p. 98).

Physcia tribacia(Ach.) Nyl. 糙蜈蚣衣

Flora, Regensburg 62: 48(1879).

≡*Lecanora tribacia* Ach., *Lich. univ.*: 415(1810).

Beijing, Shaanxi, Jiangsu(Wu J. L. 1987, p. 199), Xinjiang(Abdulla Abbas & Wu, 1998, p. 116; Abdulla Abbas et al., 2001, p. 366).

Physcia tribacioides Nyl. [as '*tribacoides*'] 粉唇蜈蚣衣

Flora, Regensburg 57: 307(1874).

On bark of trees.

Xinjiang(Abdulla Abbas & Wu, 1998, p. 117; Abdulla Abbas et al., 2001, p. 366), Yunnan(Wang L. S. 2008, p. 538), Sichuan, Guizhou, Hubei, Hunan, Guangxi(Zhao et al. 1982, p. 105), Shandong(zhang F et al., 1999, p. 31; Zhao ZT et al., 2002, p. 7; Hou Ya-nan et al., 2008, p. 66), Jiangsu, Shanghai, Zhejiang(Xu B. S., 1989, p. 258).

Physcia verrucosa Moberg 多疣蜈蚣衣

Nordic Jl Bot. 6(6): 862(1986).

Shandong(Hou Ya-nan et al., 2008, p. 66),

Physciella Essl. **小蜈蚣衣属**

Mycologia 78(1): 93(1986).

Physciella chloantha(Ach.) Essl. 粉小蜈蚣衣

Mycologia 78(1): 94(1986).

≡*Parmelia chloantha* Ach., *Syn. meth. lich.* (Lund): 217(1814).

≡*Phaeophyscia chloantha*(Ach.) Moberg, *Bot. Notiser* 131(2): 259(1978).

≡*Physcia obscura* f. *chloantha*(Ach.) Stein, in Cohn, Krypt. -Fl. Schlesien(Breslau) 2(2. Hälfte): 81(1879).

≡*Physcia obscura* var. *chloantha* (Ach.) Rabh.; Zahlbr, *Cat. Lich. Univers.* VII: 657(1931).

On bark of trees.

Shaanxi(Jatta, 1902, p. 472; Zahlbr., 1930 b, p. 239).

Physciella melanchra(Hue) Essl. 弹坑小蜈蚣衣

Mycologia 78(1): 95(1986).

≡*Physcia melanchra* Hue, *Nouv. Arch. Mus. Hist. Nat.*, Paris, 4 sér. 2: 75(1900).

≡*Phaeophyscia melanchra*(Hue) Hale, *Lichenologist*15(2): 158(1983).

Growing on on volcanic rock and tree trunks, branches and shrubs in open or fairly open situations at various altitudes up to 3000 m in China.

Yunnan(Moberg, 1995, pp. 328 – 329), Taiwan(Aptroot, Sparrius & Lai 2002, p. 288).

Physconia Poelt **大孢衣属**

Nova Hedwigia 9: 30(1965). Nom. cons., see Art. 14.

Physconia americana Essl. 美洲大孢衣

Mycotaxon 51: 91(1994).

On mosses.

Xinjiang(Mahira-Muhammat et al., 2012, p. 6), Shaanxi(Ren Q. et al., 2009, p. 104).

Physconia chinensis Chen JB. & Hu GR. 中华大孢衣

Mycotaxon 86: 186 – 188(2003).

Type: Jilin, Jiaohe, Sept. 2, 1991. Chen JB & Jiang YM, 1448(HMAS-L. holotype).

Jilin, Liaoning, Yunnan(Chen JB. & Hu GR., 2003, pp. 186 – 188).

Physconia detersa(Nyl.) Poelt 变色大孢衣

Nova Hedwigia 9: 30(1965).

≡*Physcia pulverulenta* var. *detersa* Nyl., *Syn. meth. lich.* (Parisiis) 1: 420(1860).

≡*Parmelia pulverulenta* var. *detersa*(Nyl.) Anzi in *Comment. Soc. Crittogamolog. Ital.*, II(1): 6(1864).

= *Physcia leucoleiptes*(Tuck.) Lettau, *Hedwigia* 52: 254(1912).

≡*Parmelia pulverulenta* ß *leucoleiptes* Tuck., *Proc. Amer. Acad. Arts & Sci.* 1: 224(1847).

On the ground.

Neimenggu, jilin(Chen JB. & Hu GR., 2003, pp. 188 – 189 as *Ph. leucoleptes*), Xinjiang(Abbas et al., 1997,

p. 4), Shaanxi(Jatta, 1902, p. 472; He & Chen 1995, p. 44 without citation of specimens and their localities; Zhao ZT et al., 2008, p. 725), Yunnan(Zhao et al. 1982, p. 115), Hubei(Chen, Wu & J. C. Wei, 1989, p. 482).

Physconia distorta(With.) J. R. Laundon 大孢衣

Lichenologist 16(3): 218(1984).

≡*Lichen distortus* With., *Bot. arr. veg. Gr. Brit.* (London) 2: 711(1776).

= *Physconia pulverulacea* Moberg, *Mycotaxon* 8(1): 310(1979).

Misapplied name:

Physcia pulverulenta f. *venusta* sensu auct. brit., non(Ach.) Nyl. Prodr. P. 62; fide *Checklist of Lichens of Great Britain and Ireland* (2002).

On moss covered rocks.

Xinjiang(Abdulla Abbas & Wu, 1998, p. 118; Abdulla Abbas et al., 2001, p. 366), Shaanxi(Tchou, 1935, p. 313).

Physconia elegantula Essl. 优美大孢衣

Mycotaxon 51: 92(1994).

On bark.

Shaanxi(Zhao ZT et al., 2008, p. 725).

Physconia enteroxantha(Nyl.) Poelt 黄髓大孢衣

Nova Hedwigia 12: 125(1966).

≡*Physcia enteroxantha* Nyl., *Flora*, Regensburg 56: 196(1873).

On bark of trees.

Zhejiang(Zhao et al. 1982, p. 117).

Physconia grisea(Lam.) Poelt 灰色大孢衣

Nova Hedwigia 9: 30(1965).

≡*Lichen griseus* Lam., *Encycl. Méth. Bot.* (Paris) 3(2): 480(1789).

≡*Physcia grisea* (Lam.) Zahlbr., *Annln K. K. naturh. Hofmus.* Wien 26: 177(1912).

On bark of trees & on rocks.

Beijing, Henan, Jiangxi(Zhao et al. 1982, p. 116), Xinjiang(Wu J. L. 1985, p. 77; Abdulla Abbas et al., 1993, p. 80; Abdulla Abbas & Wu, 1998, p. 118; Abdulla Abbas et al., 2001, p. 366), Shaanxi(Jatta, 1902, p. 472; Zahlbr., 1930 b, p. 239; Moreau et Moreau, 1951, p. 195; He & Chen 1995, p. 44 without citation of specimens and their localities), Hebei(Tchou, 1935, p. 313; Zhao et al. 1982, p. 116), Shandong(Zhao ZT et al., 1998, p. 30; Zhao ZT et al., 2002, p. 8), Hubei(Chen, Wu & J. C. Wei, 1989, p. 482), Anhui, Zhejiang(Xu B. S. 1989, p. 264 – 265).

= *Parmelia pulverulenta* var. *pityrea* Sprge. *Flora Halens.*, edit. 2, 527(1832).

≡*Physcia pulverulenta* var. *pityrea* Nyl. in *Act. Soc. Linn. Bordeaux*, 21: 308(1856).

= *Lichen pityreus* Ach., *Lich. suec. prodr.* (Linköping): 124(1799) [1798] (superfluous name of *Lobaria pulveracea* Hoffm.).

= *Physcia pulveracea*(Hoffm.) Vain., *Acta Soc. Fauna Flora fenn.* 13(no. 6): 14(1896).

≡*Lobaria pulveracea* Hoffm., *Deutschl. Fl.*, *Zweiter Theil* (Erlangen): 153(1796) [1795].

= *Physcia pulverulenta* var. *pityrea* Nyl., *Act. Soc. linn. Bordeaux* 21: 308(1856).

= *Physconia farrea* (Ach.) Poelt, *Nova Hedwigia* 9: 30(1965).

≡*Parmelia farrea* Ach., *Lich. univ.*: 475(1810).

= *Physcia farrea* f. *ornata* Hue, *Nouv. Arch. Mus. Hist. Nat.*, Paris, 4 sér. 2: 69(1900).

Shaanxi(Moreau et Moreau, 1951, p. 195).]

Physconia grumosa Kashiw. & Poelt 颗粒大孢衣

Ginkgoana 3: 56(1975).

On bark or rock.

neimenggu, jilin, sichuan, Yunnan, hebei, anhui(Chen JB. & Hu GR. , 2003, p. 189) , Shaanxi(He & Chen 1995, p. 44 without citation of specimens and their localities; Chen JB. & Hu GR. , 2003, p. 189; Zhao ZT et al. , 2008, p. 725) , Hubei(Chen, Wu & J. C. Wei, 1989, p. 482) .

Physconia hokkaidensis Kashiw. 东亚大孢衣

Ginkgoana 3: 57(1975) .

On bark or rock.

Neimenggu, Jilin, Heilongjiang, Liaoning, Xinjiang, Sichuan, Yunnan, Hubei(Chen JB. & Hu GR. , 2003, pp. 189 – 190) , shaanxi(Chen JB. & Hu GR. , 2003, pp. 189 – 190; Zhao ZT et al. , 2008, p. 726) .

Physconia kansuensis(H. Magn.) Wu, Abdulla & Jiang ex Abdulla & Wu 甘肃大孢衣

in *Lichens of Xinjiang*, 1998, p. 118.

≡*Physcia kansuensis* H. Magn. *Lich. Centr. Asia*, 1: 157(1940) .

Physconia kansuensis(H. Magn.) Wu, Abdulla & Jiang, *Journal Nanjing Nomal University*(Natur. Sci.) 16 supp. : 80(1993) , nom. inval. ICN, Art. 41. 1. in Melbourne, 2011.

Type: Gansu, Bohlin no. 59 in S.

On earth & wood.

Gansu, Qinghai(Zhao et al. 1982, p. 102) , Xinjiang(H. Magn. 1940, p. 157; Abdulla Abbas et al. , 1993, p. 80; Wu & Abdulla Abbas, 1998, p. 118; Abdulla Abbas et al. , 2001, p. 367) .

Physconia kurokawae Kashiw. 黑氏大孢衣

Ginkgoana 3: 58(1975) .

Jilin(Chen JB. & Hu GR. , 2003, p. 191) .

Physconia lobulifera Kashiw. 小裂片大孢衣

Ginkgoana 3: 60(1975) .

On bark.

Shaanxi(Zhao ZT et al. , 2008, p. 726) .

Physconia leucoleiptes(Tuck.) Essl. 唇粉大孢衣

Mycotaxon 51: 94(1994) .

≡*Parmelia pulverulenta* β *leucoleiptes* Tuck. , *Proc. Amer. Acad. Arts & Sci.* 1: 224(1847) .

≡*Physcia leucoleiptes*(Tuck.) Lettau, *Hedwigia* 52: 254(1912) .

Beijing, jilin, Heilongjiang, Liaoning, Shaanxi, Xinjiang, Yunnan, Hebei, Henan, Jiangxi(Chen JB. & Hu GR. , 2003, pp. 191 – 192) .

Physconia muscigena(Ach.) Poelt 伴藓大孢衣

Nova Hedwigia 9: 30(1965) .

≡*Parmelia muscigena* Ach. , *Lich. univ.* : 472(1810) .

≡*Physcia muscigena* (Ach.) Nyl. , *Act. Soc. linn. Bordeaux* 21: 308(1856) .

On mosses.

Shaanxi(Jatta, 1902, p. 472; Zahlbr. , 1930 b, p. 240; He & Chen 1995, p. 45 without citation of specimen and locality; Zhao ZT et al. , 2008, p. 726) , Xinjiang(Wang XY, 1985, p. 352 as *Physcia muscigena*; Abudula Abbas et al. , 1993, p. 80; Abdulla Abbas et al. , 1994, p. 22) , Shandong(Li Y et al. , 2008, p. 73) , Taiwan(Aptroot, Sparrius & Lai 2002, p. 289) .

f. muscigena 原变型

On the ground & on rocks.

Xinjiang(Zhao et al. 1982, p. 119; Abdulla Abbas & Wu, 1998, pp. 118 – 119; Abdulla Abbas et al. , 2001, p. 367) , Hunan(Zhao et al. 1982, p. 119) .

f. alpina(Nádv.) J. C. Wei & Y. M. Jiang 高山变型

Lichens of Xizang(China) : 113(1986) .

≡*Physcia muscigena* f. *alpina* Nádv. , in *Stud. Bot. Cech.* 8: 110, 124(1947) .

On the ground.

Xizang(Wei & Chen, 1974, p. 182: pr. p; J. C. Wei & Jiang, 1981, p. 1146: pr. p. & 1986, p. 113) .

f. squarrosa(Ach.) J. C. Wei & Y. M. Jiang 瘤状变型

Lichens of Xizang(China) : 112(1986) .

≡*Physcia muscigena* f. *squarrosa*(Ach.) Lynge, *Vidensk. Skrifter, I. Math. -naturv. Klasse*(no. 8) : 59(1916) .

≡*Parmelia muscigena* var. *squarrosa* Ach. , *Lich. univ.* : 473(1810) .

On the ground.

Xinjiang(Abdulla Abbas & Wu, 1998, p. 119; Abdulla Abbas et al. , 2001, p. 367) , Shaanxi, Shanxi(Zhao et al. 1982, p. 119) , Xizang(J. C. Wei & Chen, 1974, p. 182 pr. p. ; J. C. Wei & Jiang, 1981, p. 1146: pr. p. & 1986, p. 112) .

Physconia perisidiosa(Erichsen) Moberg 亚灰大孢衣

Symb. bot. upsal. 22(1) : 90(1977) .

≡*Physcia perisidiosa* Erichsen, *Verh. bot. Ver. Prov. Brandenb.* 72: 57(1930) .

≡*Physcia leucoleiptes* var. *perisidiosa*(Erichsen) Nádv. , *Stud. Bot. Čechoslav.* 8: 120(1947) .

= *Physcia farrea* sensu auct. brit. ; fide *Checklist of Lichens of Great Britain and Ireland*(2002) .

[= *Physcia farrea* f. *algeriensis* Flagey, *Catal. Lich. Algérie:* 17(1896) .]

Xinjiang(Abdulla Abbas et al. , 1993, p. 80; Abdulla Abbas & Wu, 1998, p. 119; Abdulla Abbas et al. , 2001, p. 367) , Hebei(Zhao et al. 1982, p. 116) .

Physconia pulverulenta(Schreb.) Poelt 霜大孢衣

Nova Hedwigia 9: 30(1965) .

≡*Lichen pulverulentus* Schreb. , *Spic. fl. lips.* (Lipsiae) : 128(1771) .

≡*Squamaria pulverulenta*(Schreb.) Hoffm. , *Descr. Adumb. Plant. Lich.* 1(2) : 39(1789) [1790] .

≡*Physcia pulverulenta*(Schreb.) Hampe ex Fürnr. , *Naturhist. Topogr. Regensburg* 2: 249(1839) .

≡*Physcia pulverulenta*(Hoffm.) Fuernr. (Jatta, 1902 & Zahlbr. , 1930 b) .

≡*Lichen pulverulentus* Hoffm. , *Enum. critic. lich. europ.* (Bern) : 76(1784) .

On bark of trees.

Beijing(Zhao et al. 1982, p. 118; Chen JB. & Hu GR. , 2003, p. 189) , jilin, Sichuan(Chen JB. & Hu GR. , 2003, p. 189) , Neimenggu(Sun LY et al. , 2000, p. 37; Chen JB. & Hu GR. , 2003, p. 189) , Shaanxi(Tchou, 1935, p. 313 as *Ph. venusta*; Zhao et al. 1982, p. 118; Chen JB. & Hu GR. , 2003, p. 189; Zhao ZT et al. , 2008, p. 725) , Xinjiang(Tchou, 1935, p. 312) Yunnan(Paulson, 1928, p. 319; Zahlbr. , 1930b, p. 240) , Xizang (Paulson, 1925, p. 191) .

[= *Parmelia pulverulenta* f. *epigaea* Jatta, *Sylloge Lich. Ital.* 1900, p. 143.

≡*Physcia pulverulenta* var. *epigaea* Jatta, *Flora Ital. Cryptog.* 3: 340(1909) .]

On the ground.

Shaanxi(Jatta, 1902, p. 472; Zahlbr. , 1930 b, p. 240) .

[= *Parmelia pulverulenta* var. *sorediantha* Jatta , *Nuov. Giorn. Bot. Italiano*, 2(9) : 472(1902) .]

Type: Shaanxi, Mt. Guangtou shan, collected by Giraldi.

On the ground.

Shaanxi(Jatta, 1902, p. 472; Zahlbr. , 1930 b, p. 240) .]

Beijing, Neimenggu, Jilin, Shaanxi, Sichuan(Chen JB. & Hu GR. , 2003, p. 189) .

Physconia tentaculata(Zahlbr.) Poelt, (1966) . 触丝大孢衣

≡*Physcia tentaculata* Zahlbr. , *Feddes Repert.* 33: 67(1933) .

Type: Taiwan, S. Sasaki(Asahina no. 419) .

Taiwan(Zahlbr. , 1940, p. 656; Lamb, 1963, p. 564; Wang & Lai, 1973, p. 95) .

Physconia venusta(Ach.) Poelt 雅致大孢衣

Nova Hedwigia 12: 130(1966) .

≡*Parmelia venusta* Ach. , *Methodus*, Sectio post. : 211(1803) .

On bark.

Shaanxi(Zhao ZT et al. , 2008, p. 726) .

See also Ph. distorta(With.) Laundon

Rinodina (Ach.) Gray **饼干衣属**

Nat. Arr. Brit. Pl. (London) 1: 448(1821) .

≡*Lecanora* subdiv. *Rinodina* Ach. , *Lich. univ.* : 344(1810) .

Rinodina aspersa(Borrer) J. R. Laundon 散生饼干衣

Lichenologist 18(2) : 175(1986) .

≡*Lecanora aspersa* Borrer, *Suppl. Engl. Bot.* 2: tab. 2728(1834) .

On shale.

Taiwan(Aptroot and Sparrius, 2003, p. 39) .

Rinodina bischoffii(Hepp) A. Massal. 毕氏饼干衣

Framm. Lichenogr. : 26(1855) .

≡*Psora bischoffii* Hepp, *Flecht. Europ.* : no. 81(1853) .

Ningxia(Joshi S. et al. , 2014, p. 170) .

Rinodina bohlinii H. Magn. 包氏饼干衣

Lich. Centr. Asia 1: 149(1940) .

Type: Gansu, Bohlin no. 67 in S.

Rinodina bohlinii f. bohlinii 原变型

On calcareous rocks.

Xinjiang(Abdulla Abbas et al. , 1994, p. 22; Abdulla Abbas & Wu, 1998, p. 120; Abdulla Abbas et al. , 2001, p. 367) , Gansu, Qinghai(H. Magn. 1940, p. 149) .

f. lignicola H. Magn. 木生变型

Lich. Centr. Asia 1: 150(1940) .

Type: Gansu, alt. 2750 m, 1932, Bohlin no. 78a, b, c in S(not seen) .

Xinjiang(Abdulla Abbas et al. , 1993, p. 81; Abdulla Abbas & Wu, 1998, p. 120; Abdulla Abbas et al. , 2001, p. 367) , Gansu(H. Magn. , 1940, p. 150) .

Rinodina calcarea(Hepp ex Arnold) Arnold 石灰饼干衣

Verh. zool. -bot. Ges. Wien 29: 362(1879) .

≡*Rinodina caesiella* var. *calcarea* Hepp ex Arnold, *Flora*, Regensburg 43: 69(1860) .

On rocks.

Jiangsu(Xu B. S. , 1989, p. 254) .

Rinodina capensis Hampe, in Massalongo 冠状饼干衣

Mem. Imp. Reale Ist. Veneto 10: 87(1861) .

On *Castanopsis* sp. , *Abies kawakamii*, *Pseudotsuga wilsoniana*, and deciduous tree bark.

Taiwan(Aptroot and Sparrius, 2003, p. 39) .

Rinodina colobina(Ach.) Th. Fr. 短饼干衣

Lich. Scand. (Upsaliae) 1(1) : 205(1871) .

≡*Lecanora colobina* Ach. , *Lich. univ.* : 358(1810) .

On *Castanopsis* sp.

Taiwan(Aptroot and Sparrius, 2003, p. 39) .

Rinodina conradii Körb. 康拉德饼干衣

Syst. lich. germ. (Breslau): 123(1855).

On shale.

Taiwan(Aptroot and Sparrius, 2003, p. 39).

Rinodina cornutula Zahlbr. 小角饼干衣

Lichenes in Handel-Mazzetti, *Symb. Sin.* 3: 233(1930) & *Cat. Lich. Univ.* 7: 506(1931).

Type: from Yunnan.

Hongkong(Aptroot & Seaward, 1999, p. 93; Aptroot & Sipman, 2001, p. 340).

Rinodina gennarii Bagl. 金氏饼干衣

Comm. Soc. crittog. Ital. 1(1): 17(1861).

On *Dendropanax* sp.

Taiwan(Aptroot and Sparrius, 2003, p. 39).

= *Rinodina demissa*(Flörke) Arnold, *Flora*, Regensburg 42: 68(1860).

≡ *Zeora metabolica* var. *demissa* Flörke, in Flotow, *Jber. schles. Ges. vaterl. Kultur* 27: 124(1849).

Xinjiang(Abdulla Abbas et al., 1993, p. 81; Abdulla Abbas et al., 1994, p. 22; Abdulla Abbas & Wu, 1998, pp. 120 – 121; Abdulla Abbas et al., 2001, p. 367).

Rinodina globulans Zahlbr. 球饼干衣

Lichenes in Handel-Mazzetti, *Symb. Sin.* 3: 233(1930) & *Cat. Lich. Univ.* 7: 519(1931).

Type: Sichuan, at about 3000 m, 26/IV/1914, Handel-Mazzetti no. 1771.

On sandstone.

Rinodina handelii Zahlbr. 汉氏饼干衣

Lichenes in Handel-Mazzetti, *Symb. Sin.* 3: 232(1930) & *Cat. Lich. Univ.* 7: 521(1931).

Type: Yunnan, at about 1920 m, 24/II/1914, Handel-Mazzetti no. 275.

On bark of trees.

Rinodina heterospora Zahlbr. 异孢饼干衣

Lichenes in Handel-Mazzetti, *Symb. Sin.* 3: 232(1930) & *Cat. Lich. Univ.* 7: 521(1931).

Type: Sichuan, at about 2100 m, 5/IV/1914, Handel-Mazzetti no. 1861.

On granite.

Rinodina imitatrix Zahlbr. 台湾饼干衣

in Fedde, *Repertorium*, 33: 64(1933).

Type: Taiwan, Faurie no. 184.

On rocks.

Taiwan(Zahlbr., 1933 c, p. 64 & 1940, p. 645; M. Sato, 1936b, p. 571; Lamb, 1963, p. 647; Wang & Lai, 1973, p. 96).

Rinodina kansuensis H. Magn. 甘肃饼干衣

Lich. Centr. Asia 1: 151(1940). f. kansuensis

Type: Gansu, at about 3000 m, 1931, p. Bohlin no. 41 in S(!).

On calcareous rock.

f. paupera H. Magn. 贫乏变型

Lich. Centr. Asia 2: 56(1944).

Type: Neimenggu, 13/XI/1929, Bohlin, geol. sample no. 190 in S(!).

On calcareous rock.

Rinodina lecideina H. Mayrhofer & Poelt 网盘饼干衣

Biblthca Lichenol. 12: 112(1979).

On granite. Cosmopolitan.

Hongkong(Aptroot & Seaward, 1999, p. 93).

Rinodina manshurica Räs. 东北饼干衣

Arch. soc. zool. Fenn. bot. 'Vanamo' 5(1): 27(1950).

Type: north eastern China(prov. not indicated), collected by I. Korejev.

On bark of *Tilia manshurica*.

Rinodina mniaraea(Ach.) Körb. 提灯藓饼干衣

Syst. lich. germ. (Breslau): 125(1855).

≡*Lecanora mniaraea* Ach., *Syn. meth. lich.* (Lund): 339(1814).

Xizang(Paulson, 1925, p. 191).

Rinodina mongolica H. Magn. 蒙古饼干衣

Lich. Centr. Asia 2: 56(1944).

Type: Neimenggu, 24/X/1929, Bohlin no. 109 in S(not seen).

On cortex.

Rinodina nephroidea(Vain.) Zahlbr. 肾饼干衣

Cat. Lich. Univers. 7: 534(1931).

≡*Melanaspicilia nephroidea* Vain., *Bot. Mag.*, Tokyo 32: 157(1918).

On rocks.

Shandong(Li Y et al., 2008, p. 73), Jiangsu(Wu J. N. & Xiang T. 1981, p. 6, here should be p. 5 according to the sequence of the pages).

Rinodina oxydata(A. Massal.) A. Massal. 黑色饼干衣

Geneac. lich. (Verona): 19(1854).

≡*Berengeria oxydata* A. Massal., *Spighe Paglie*: 6(1853).

=*Rinodina discolor*(Hepp) Arnold, *Flora*, Regensburg 55: 36(1872); Zahlbr. (Zahlbr., 1930b, p. 233), cited by Aptroot & Seaward, 1999, p. 93.

≡*Buellia discolor* Körber(Hepp) Anzi(Rabh., 1873, p. 286; Krempelh., 1873, p. 468 &1874, p. 60; Thrower, 1988, pp. 25, 66), cited by Aptroot & Seaward, 1999, p. 93.

On rocks.

Shandong(Moreau et Moreau, 1951, p. 194), Hongkong(Rabh., 1873, p. 286; Krempelh. 1873, p. 468 &1874b, p. 60; Zahlbr., 1930 b, p. 233; Thrower, 1988, pp. 25, 66; Aptroot & Seaward, 1999, p. 93), Taiwan (Aptroot and Sparrius, 2003, p. 40).

Note: Massalongo as the author of the basionym of this species is accepted by R. Santeson(1984) and R. Egan (1987) without any explanations.

Rinodina perminuta Groenh. ex H. Mayrhofer 海岩饼干衣

J. Hattori bot. Lab. 55: 451(1984).

On siliceous rock in coastal area.

Hongkong(Aptroot & Seaward, 1999, p. 93), Taiwan(Aptroot and Sparrius, 2003, p. 40; Aptroot & Sipman, 2001, p. 340).

Rinodina placynthielloides Aptroot 胎座饼干衣

in Aptroot & Sparrius, *Fungal Diversity* 14: 40(2003).

Type: Taiwan, Taichung County: 30 km ENE of Taichung, 7 km NW of Kukwan, along mountain trail, 1000 – 1300 m alt., 51RTG9279, on granite of wall along path. Aptroot 53451(B-holotype, BL-paratypes), 20 October 2001.

Rinodina pluriloculata Aptroot & Sparrius 多室饼干衣

in Fungal Diversity 14: 41 – 42(2003).

Type: China, Yunnan Province, 5 km W of Kunming, just outside of the city , 1750 m alt., RTN604714, on bark of treebase of living Eucalyptus globulus. Aptroot 55505(B-holotype; ABL, KUN-isotypes), 16 October 2002.

Rinodina punctosorediata Aptroot & Sparrius 点状粉芽饼干衣

in Fungal Diversity 14: 42 – 44(2003).

Type: Taiwan, Hualien County: 43 km WNW of Hualien, Meifeng, roadside with relict mature trees, 2250 m alt., 51RUG1666, on *Castanopsis*. Aptroot 52373(c. ap.) and 52378(ABL – isotype) and Sparrius 6117(B-holotype; Herb. Sparrius-isotype, c. ap.), 11 October 2001.

Yunnan(Aptroot and Sparrius, 2003, p. 43).

Rinodina pycnocarpa H. Magn. 密果饼干衣

Lich. Centr. Asia 1: 152(1940).

Type: Gansu, 22/XII/1931, at about 3100 m, Bohlin no. 63 a in S(!).

On stone.

Rinodina pyrina(Ach.) Arnold 坚果饼干衣

Flora, Regensburg 64: 196(1881).

≡*Lichen pyrinus* Ach., *Lich. suec. prodr.* (Linköping): 52(1799)[1798].

On *Populus* sp.

Taiwan(Aptroot and Sparrius, 2003, p. 43).

Rinodina roboris(Dufour ex Nyl.) Arnold 栎饼干衣

Flora, Regensburg 64: 197(1881).

≡*Lecanora sophodes* var. *roboris* Dufour ex Nyl., in Crouan & Crouan, *Florule Finistère*(Paris): 96(1867).

On *Populus* sp.

Taiwan(Aptroot and Sparrius, 2003, p. 43).

Rinodina roscida(Sommerf.) Arnold 土生饼干衣

Verh. zool. -bot. Ges. Wien 37: 133(1887).

≡*Lecanora roscida* Sommerf., *Suppl. Fl. lapp.* (Oslo): 97(1826).

≡*Rinodina orbata* var. *roscida* (Sommerf.) Zahlbr., *Cat. Lich. Univers.* 7: 540(1931).

On rocks.

Xizang(Mt. Qomolangma, Paulson, 1925, p. 191).

Rinodina setschwana Zahlbr. 四川饼干衣

in Handel-Mazzetti, *Symb. Sin.* 3: 231(1930) & *Cat. Lich. Univ.* 7: 548(1931).

Type: Sichuan, at about 1450 m, 4/IV/1914, Handel-Mazzetti no. 1171.

On sandstone.

Rinodina sophodes(Ach.) A. Massal. 饼干衣

Ric. auton. lich. crost. (Verona): 14(1852).

≡*Lichen sophodes* Ach., *Lich. suec. prodr.* (Linköping): 67(1799)[1798].

≡*Lecanora sophodes*(Ach.) Ach., *Lich. univ.*: 356(1810).

On bark of trees.

Xinjiang(Abdulla Abbas & Wu, 1998, p. 121; Abdulla Abbas et al., 2001, p. 367), Yunnan(Hue, 1889, p. 171; Paulson, 1928, p. 319; Zahlbr., 1930 b, p. 231).

Rinodina stenospora H. Magn. 小孢饼干衣

Lich. Centr. Asia 2: 55(1944).

Type: Neimenggu, 20/XI/1929, Bohlin no. 141 in S(not seen).

On non-calciferous rock.

Rinodina subleprosula Jatta 亚癞屑饼干衣

Nuov. Giorn. Bot. Italiana, ser. 2, 9: 477(1902).

Type: Shaanxi, Mt. Guangtou shan, collected by Giraldi.

Shaanxi(Zahlbru., 1930 b, p. 231).

On a young tree trunk.

Rinodina subnigra H. Magn. 亚黑饼干衣

Lich. Centr. Asia 1: 152(1940).

Type: Gansu, at about 3000 m, 13/XII/1931, Bohlin no. 40a in S(not seen).

On naked, calcareous rock.

Rinodina superposita H. Magn. 叠生饼干衣

Lich. Centr. Asia 1: 153(1940).

Type: Gansu, at about 4100 m, 12/I/1932, Bohlin no. 77 in S(!).

On sometimes slightly calciferous stone.

Rinodina teichophila(Nyl.) Arnold 砂石饼干衣

Flora, Regensburg 46: 329(1863).

≡*Lecanora teichophila* Nyl., *Flora*, Regensburg 46: 78(1863).

=*Rinodina arenaria*(Hepp) Arn.

Xizang(Mt. Qomolangma, Paulson, 1925, p. 191).

On *Tsuga formosana*, *Abies kawakamii*, and *Castanopsis* sp. Taiwan(Aptroot and Sparrius, 2003, p. 43).

Rinodina terrestris Tomin 地生饼干衣

Prirod. sel'skoe chozjajst. zasuch. -pustyn S. S. S. R. (The Nature and Agricult. in the arid Regions of the U. S. S. R.) (3): 59(1928).

Syn. *Rinodina mucronatula* H. Magn.

On soil.

Xinjiang(W. Obermayer, 2004, p. 507), Ningxia(Liu M & J. C. Wei, 2013, p. 47; Yang J. & Wei J. C., 2014, p. 1030).

Rinodina thiomela(Nyl.) Müll. Arg. 硫饼干衣

Flora, Regensburg 64(32): 515(1881).

≡*Lecanora thiomela* Nyl., *Flora*, Regensburg 48: 338(1865).

On sandstone.

Taiwan(Aptroot and Sparrius, 2003, p. 44).

Rinodina turfacea(Wahlenb.) Körb. 泥炭饼干衣

Syst. lich. germ. (Breslau): 123(1855).

≡*Lichen turfaceus* Wahlenb., *Fl. lapp.*: 408(1812).

On the ground and mosses.

Xinjiang(Abdulla Abbas et al., 1994, p. 22; Abdulla Abbas & Wu, 1998, p. 121; Abdulla Abbas et al., 2001, p. 368), Taiwan(Aptroot and Sparrius, 2003, p. 44).

Rinodina varians Zahlbr. in Fedde 变异饼干衣

Repertorium, 33: 65(1933).

Type: Taiwan, Faurie no. 175.

Taiwan(Zahlbr., 1940, p. 649; M. Sato, 1936b, p. 572; Wang & Lai, 1973, p. 96).

On rocks.

Rinodina xanthomelana Müll. Arg. 黄黑饼干衣

Nuovo G. bot. ital. 23: 390(1891).

On siliceoua rock. Known from Australia and topical Asia.

Taiwan(Aptroot and Sparrius, 2003, p. 44), Hongkong(Aptroot & Seaward, 1999, p. 930; Aptroot & Sipman, 2001, p. 340).

Rinodina zwackhiana(Kremp.) Körb. 堇紫饼干衣

Syst. lich. germ. (Breslau): 126(1855).

≡*Lecanora zwackhiana* Kremp., *Flora*, Regensburg 37: 145(1854).

=*Rinodina violascens* H. Magn. *Lich. Centr. Asia* 1: 154(1940).

Type: Gansu, at about 1475 m, 10/X/1930, Bohlin no. 24 b in S(not seen).

On calciferous rocks.

Tornabea Østh. **垫雪衣属**

in Østhagen & Sunding, *Taxon* 29(5 – 6): 688(1980).

Tornabea scutellifera(With.) J. R. Laundon 盾片饼干衣

Lichenologist 16(3): 226(1984).

≡*Lichen scutelliferus* With., *Bot. arr. veg. Gr. Brit.* (London) 2: 728(1776).

= *Tornabenia atlantica*(Ach.) Kurok., *J. Jap. Bot.* 37: 291(1962).

≡*Parmelia atlantica* Ach., *Methodus*, Suppl.: 50(1803).

NW Himalayas in China(Kurok. 1962, p. 219).

(72) Teloschistaceae Zahlbr. [as ' Theloschistaceae'] **黄枝衣科**

in Engler, *Syllabus*, Edn 2(Berlin): 45(1898).

Athallia Arup, Frödén & Søchting **全果衣属**

Nordic Jl Bot. 31(1): 36(2013).

Athallia pyracea(Ach.) Arup, Frödén & Søchting 红全果衣

Nordic Jl Bot. 31(1): 36(2013).

≡*Parmelia cerina* var. *pyracea* Ach., Methodus, Sectio post. (Stockholmiæ): 176(1803).

On twigs of Juniperus formosana.

Neimenggu(Sun LY et al., 2000, p. 37), Sichuan(Zahlbr., 1930b, p. 214).

Athallia vitellinula(Nyl.) Arup, Frödén & Søchting 蛋黄果衣

Nordic Jl Bot. 31(1): 36(2013).

≡*Lecanora vitellinula* Nyl., *Flora*, Regensburg 46: 305(1863).

≡*Caloplaca vitellinula*(Nyl.) H. Olivier, *Expo. Syst. Descr. Lich. Ouest Fr.* 1: 232(1897).

≡*Callopisma vitellinulum*(Nyl.) Arn. (1870).

≡*Placodium vitellinulum*(Nyl.) Vain., *Meddr Grønland, Biosc.* 30: 131(1905).

Saxicolous.

Neimenggu(Sun LY et al., 2000, p. 37), Shaanxi (Jatta, 1902, p. 477; Zahlbr., 1930 b, p. 214), Shanghai (Nyl. & Cromb. 1883, p. 63; Hue, 1891, p. 50; Zahlbr., 1930 b, p. 214; Xu B. S., 1989, p. 252), Zhejiang(Xu B. S., 1989, p. 252).

Austroplaca Søchting, Frödén & Arup **南衣属**

Nordic Jl Bot. 31(1): 37(2013).

Austroplaca lucens(Nyl.) Søchting, Frödén & Arup 亮南衣

Nordic Jl Bot. 31(1): 38(2013)

≡*Lecanora elegans* f. *lucens* Nyl., in Crombie, *J. Linn. Soc.*, *Bot.* 15: 184(1876) [1877].

≡*Caloplaca lucens*(Nyl.) Zahlbr., *Cat. Lich. Univers.* 7: 246(1931).

On the ground among mosses.

Shaanxi(Jatta, 1902, p. 476; Zahlbr., 1930 b, p. 221).

According to Zahlbr. (1930 b, p. 221) this lichen, reported by Jatta in 1902, p. 476, may probably be another taxon of *Caloplaca*. It is different from true *Caloplaca lucens*(Nyl.) A. Zahlbr. by geographical distribution and substratum. Actually *Caloplaca lucens* is an endemic to Antarctica, and the substrata of it are hard rocks. So, this lichen reported by Jatta from Shaanxi has to remain to be revised in the future.

Blastenia A. Massal. **胚衣属**

Flora, Regensburg 35: 573(1852).

Blastenia ammiospila(Ach.) Arup 缩柄胚衣

Søchting & Frödén, *Nordic Jl Bot.* 31(1): 67(2013).

≡ *Lecidea cinereofusca* var. *ammiospila* Ach. [as ' ammiosphila'], *K. Vetensk-Acad. Nya Handl.* 29: 268

(1808).

≡*Caloplaca ammiospila*(Ach.) H. Olivier, *Mém. Soc. natn. Sci. nat. Cherbourg* 37: 136(1909).

Saxicolous.

Sichuan(W. Obermayer, 2004, p. 489).

Blastenia amoena Zahlbr. 美胚衣

in Handel-Mazzetti, *Symb. Sin.* 3: 211(1930) & *Cat. Lich. Univ.* 8: 582(1932).

Type: Sichuan, Handel-Mazzetti no. 5316.

Saxicolous.

Blastenia ferruginea(Huds.) A. Massal. 锈胚衣

Atti Inst. Veneto Sci. lett., ed Arti, Sér. 2 3: 102(1852).

≡*Lichen ferrugineus* Huds., *Fl. Angl.*: 444(1762).

≡*Caloplaca ferruginea*(Huds.) Th. Fr., *Nova Acta R. Soc. Scient. upsal.*, Ser. 3 3: 233(1861)[1860].

≡ *Callopisma ferrugineum*(Huds.) Trevis., *Revta Period. Lav. Imp. Reale Acad., Padova*1(3): 264(1852) [1851 –52].

On twigs of *Juniperus formosana*.

Sichuan(Zahlbr., 1930 b, p. 215; Chen S. F. et al., 1997, p. 221), Anhui, Zhejiang(Xu B. S., 1989, p. 251).

Notes. Is the European species *C. ferruginea* distributed in China? So, the species from Sichuan, Anhui, and Zhejiang remains to be checked.

Blastenia handelii Zahlbr. 汉氏胚衣

in Handel-Mazzetti, *Symb. Sin.* 3: 210(1930) & *Cat. Lich. Univ.* 8: 583(1932).

Type: Yunnan, Handel-Mazzetti no. 5863.

Saxicolous.

See Caloplaca leptozona(Nyl.) Zahlbr. for Hongkong(Thrower, 1988, p. 64).

Blastenia modestula Zahlbr. 胚衣

in Handel-Mazzetti, *Symb. Sin.* 3: 209(1930) & *Cat. Lich. Univ.* 8: 583(1932).

Type: Yunnan, Handel-Mazzetti no. 5723.

Corticolous.

Blastenia polioterodes(J. Steiner) Zahlbr. 丽胚衣

Cat. Lich. Univers. 7: 39(1930)[1931].

≡*Caloplaca polioterodes* J. Steiner, *Sber. Akad. Wiss. Wien, Math. -naturw. Kl., Abt.* 1 106: 218(1897).

≡*Lecidea polioterodes*(J. Steiner) Hue, *Nouv. Arch. Mus. Hist. Nat., Paris,* 5 sér. 4: 17(1914).

On rocks.

Shandong(Moreau et Moreau, 1951, p. 193).

Blastenia setschwana Zahlbr. 四川胚衣

in Handel-Mazzetti, *Symb. Sin.* 3: 211(1930) & *Cat. Lich. Univ.* 8: 583(1932).

Type: Sichuan, Handel-Mazzetti no. 5313.

Graniticolous.

Blastenia yunnana Zahlbr. 云南胚衣

in Handel-Mazzetti, *Symb. Sin.* 3: 210(1930) & *Cat. Lich. Univ.* 8: 583(1932).

Type: Yunnan, Handel-Mazzetti no. 4161.

Brownliella S. Y. Kondr. **棕衣属**

Kärnefelt, Elix, A. Thell & Hur *in* Kondratyuk et al., *Acta bot. hung.* 55(3 –4): 271(2013).

Brownliella cinnabarina(Ach.) S. Y. Kondr., Kärnefelt, A. Thell et al. 朱砂红棕衣

in Kondratyuk et al., *Acta bot. hung.* 55(3 –4): 271(2013).

≡*Lecanora cinnabarina* Ach., *Lich. univ.*: 402(1810).

Hongkong(Aptroot et al., 2001, p. 322).

Bryoplaca Søchting, Frödén & Arup 苔衣属

Nordic Jl Bot. 31(1): 68(2013).

Bryoplaca jungermanniae(Vahl) Søchting, Frödén & Arup 欧苔衣

Nordic Jl Bot. 31(1): 68(2013)

≡*Lichen jungermanniae* Vahl, *Icon. Plant. Dan.* 6(18): 6, tab. MLXIII, 1(1792).

≡*Caloplaca jungermanniae*(Vahl) Th. Fr., *Nova Acta R. Soc. Scient. upsal.*, Ser. 3 3: 221(1861)[1860].

Xizang, Sichuan(W. Obermayer, 2004, p. 491).

Bryoplaca tetraspora(Nyl.) Søchting, Frödén & Arup 四孢苔衣

Nordic Jl Bot. 31(1): 68(2013).

≡*Lecanora tetraspora* Nyl., *Acta Soc. Sci. fenn.* 7(2): 397(1863).

≡*Caloplaca tetraspora*(Nyl.) H. Olivier *Mém. Soc. natn. Sci. nat. Cherbourg* 37: 140(1909).

On plant debris.

Xizang(W. Obermayer, 2004, p. 491).

Calogaya Arup, Frödén & Søchting **美衣属**

Nordic Jl Bot. 31(1): 38(2013).

Calogaya biatorina(A. Massal.) Arup, Frödén & Søchting 蜡美衣

Nordic Jl Bot. 31(1): 38(2013).

≡*Physcia elegans* var. *biatorina* A. Massal., *Atti Inst. Veneto Sci. lett.*, ed Arti, Sér. 2 3(App. 3): 51(1852).

≡*Caloplaca biatorina*(A. Massal.) J. Steiner, *Annls mycol.* 8(2): 239(1910).

≡*Berengeria biatorina* (A. Massal.) Trevis., *Atti Soc. ital. Sci. nat. Mus. Civico Storia nat. Milano* 11: 628 (1868).

≡*Caloplaca biatorina*(A. Massal.) J. Steiner, *Annls mycol.* 8(2): 239(1910).

= *Caloplaca callopiza*(Nyl.) Jatta, *Syll. Lich. Ital.*: 240(1900).

≡*Lecanora callopiza* Nyl., *Flora*, Regensburg 66: 98(1883).

On rocks.

Gansu, Qinghai(H. Magn. 1940, p. 137).

Calogaya decipiens(Arnold) Arup, Frödén & Søchting 莲座美衣

Nordic Jl Bot. 31(1): 38(2013).

≡*Physcia decipiens* Arnold, *Flora*, Regensburg 50: 562(1867).

≡*Caloplaca decipiens*(Arnold) Blomb. & Forssell, *Cat. Lich. Univers.* 7: 226(1931).

≡*Physcia decipiens* Arnold, *Flora*, Regensburg 50: 562(1867).

Xinjiang(Abdulla Abbas & Wu, 1998, pp. 145 – 146; Abdulla Abbas et al., 2001, p. 361), Shandong(Zhao ZT et al., 2002, p. 7), Jiangsu(Wu J. L., 1987, p. 186; Xu B. S., 1989, p. 251).

On limestone.

Calogaya ferrugineoides(H. Magn.) Arup, Frödén & Søchting 类锈美衣

Nordic Jl Bot. 31(1): 38(2013).

≡*Caloplaca ferrugineoides* H. Magn., *Lich. Central Asia*, part II: 52(1944).

Type: Gansu, Bohlin no. 78 d in S.

Misapplied name:

Caloplaca ferruginea auct. sensu H. Magn. *Lich. Central Asia*, part 1: 133 – 134(1940), non(Huds.) Th. Fr., *Nova Acta R. Soc. Scient. upsal.*, Ser. 3 3: 233(1861)[1860].

On a dry twig.

Xinjiang(Abbas et al. , 1994, p. 20; Abdulla Abbas & Wu, 1998, p. 146; Abdulla Abbas et al. , 2001, p. 361) , Gansu(H. Magn. , 1940, pp. 133 – 134, & 1944, pp. 52 – 53; Abdull Abbas et al. , 1993, p. 76) .

Calogaya lobulata(Flörke) Arup, Frödén & Søchting 裂瓣美衣

Nordic Jl Bot. 31(1) : 39(2013) .

≡ *Lecanora lobulata* Flörke, *Mag. Neuesten Entdeck. Gesammten Naturk. Ges. Naturf. Freunde Berlin* 1: 219 (1820) .

≡ *Caloplaca lobulata*(Flörke) Hellb. *Bih. K. svenska VetenskAkad. Handl.* , Afd. 3 21(13) : 67(1896) .

≡ *Xanthoria lobulata*(Flörke) B. de Lesd. , *Bull. Soc. bot. Fr.* 54: 282(1907) .

Xinjiang(Abdulla Abbas et al. , 1993, p. 76; Abdulla Abbas & Wu, 1998, p. 151; Abdulla Abbas et al. , 2001, p. 369) , Yunnan(Zahlbr. , 1930b, p. 220) , Shandong(Moreau et Moreau, 1951, p. 194; Li Y et al. , 2008, p. 72 as *Xanthoria lobulata*) .

Calogaya polycarpoides(J. Steiner) Arup, Frödén & Søchting 类富果美衣

Nordic Jl Bot. 31(1) : 39(2013) .

≡ *Xanthoria polycarpoides* J. Steiner, *Annls mycol.* 8(2) : 241(1910) .

≡ *Caloplaca polycarpoides*(J. Steiner) M. Steiner & Poelt, *Pl. Syst. Evol.* 140(2 – 3) : 168(1982) .

Xinjiang(Poelt & Hinter. , 1993, p. 178) .

Calogaya pusilla(A. Massal.) Arup, Frödén & Søchting 小美衣

Nordic Jl Bot. 31(1) : 39(2013)

≡ *Physcia pusilla* A. Massal. , *Atti Inst. Veneto Sci. lett. , ed Arti,* Sér. 2 3(App. 3) : 59(1852) .

= *Caloplaca saxicola*(Hoffm.) Nordin, *Caloplaca* sect. *Gasparrinia i Nordeuropa*(Uppsala) : 87(1972) .

≡ *Psora saxicola* Hoffm. , *Descr. Adumb. Plant. Lich.* 1(3) : 82(1790) .

Xinjiang(Abdulla Abbas & Wu, 1998, p. 147; Abdulla Abbas et al. , 2001, p. 361) .

= *Caloplaca murorum*(Ach.) Th. Fr. , *Lich. Scand.* (Upsaliae) 1(1) : 170(1871) .

≡ *Lichen murorum* Hoffm. , *Enum. critic. lich. europ.* (Bern) : 63(1784) .

≡ *Placodium murorum*(Ach.) DC. , in Lamarck & de Candolle, *Fl. franç.* , Edn 3(Paris) 2: 378(1805) .

≡ *Lecanora murorum*(Ach.) Ach. , *Lich. univ.* : 433(1810) .

Neimenggu, Xinjiang(Tchou, 1935, pp. 319 – 320; Abbas et al. , 1994, p. 20) .

= *Caloplaca murorum* var. *miniata* Th. Fr. , *Lich. Scand.* (Upsaliae) 1(1) : 170(1871) .

≡ *Lichen miniatus* Hoffm. , *Enum. critic. lich. europ.* (Bern) : 62(1784) , nom. illegit. , Art. 53. 1.

Saxicolous.

Yunnan(Zahlbr. , 1930 b, p. 220) .]

= *Caloplaca murorum* f. *arnoldi* A. L. Smith apud Paulson, *J. Bot.* , Lond. 63: 191(1925) – Himalaya.

Xizang(Paulson, 1925, p. 191) .]

Calogaya schistidii(Anzi) Arup, Frödén & Søchting 史氏美衣

Nordic Jl Bot. 31(1) : 39(2013) .

≡ *Gyalolechia schistidii* Anzi, *Cat. Lich. Sondr.* : 38(1860) .

≡ *Fulgensia schistidii* (Anzi) Poelt, *Mitt. bot. StSamml. , Münch.* 5: 595(1965) .

Xinjiang(Hurnisa Xanidin, H. et al. , 2009, pp. 109 – 110) .

Caloplaca Th. Fr. **橙衣属**

Lich. Arctoi 3: 218(1860) .

Caloplaca aegyptiaca(Müll. Arg.) Stnr. 埃及橙衣

in Sitz. Ber. Kais. Akad. Wiss. , Wien, math. -naturw. Classe 102/1: 18(1893) ; Zahlbr. , *Cat. Lich. Univers* 7: 59 (1931) ; H. Magn. *Lich. from Central Asia*, part II: 54(1944) .

≡ *Callopisma aegyptiacum* Müll. Arg. in *Revue Mycol.* 2: 73(1880) . -Pl. VII, fig. 1.

On calcareous rock.

Nneimenggu L(H. Magn. 1944, p. 54) .

Caloplaca agardhiana(Flot.) Flagey 阿氏橙衣

Mém. Soc. ému. Doubs, sér. 6 1: 247(1886) .

≡*Zeora variabilis* var. *agardhiana* Flot. , *Jber. schles. Ges. vaterl. Kultur* 27: 123(1849) .

On rocks.

Jiangsu(Wu J. N. & Xiang T. 1981, p. 6: here should be p. 5 according to the sequence of page number) .

Caloplaca albovariegata(B. de Lesd.) Wetmore 白斑橙衣

Mycologia 86(6) : 816(1995) [1994] .

≡*Pyrenodesmia albovariegata* B. de Lesd. , *Rev. Bryol. Lichénol.* , N. S. 12: 62(1942) .

On rock.

Shanxi(Zhou GL et al. , pp. 308 – 310, 2012) .

Caloplaca atrosanguinea(G. Merr.) I. M. Lamb. 黑红橙衣

In *Ann. Rept. Nat. Mus. Canada*, 1952 – 1953, *Bull.* no. 132: 305(1954) .

≡*Lecanora atrosanguinea* G. Merr. , *Ottawa Nat.* : 117(1913) .

On bark.

Xinjiang(Zhou GL et al. , 2012, pp. 310 – 312) .

Caloplaca bicolor H. Magn. 双色橙衣

Lich. Centr. Asia 1: 132(1940) .

Type: Gansu, Bohlin no. 77d in S.

Xinjiang(Poelt J, Hinteregger E, 1993, P p. 86 – 87) .

Caloplaca bogilana Y. Joshi & Hur 沼橙衣

Lichenologist 42(6) : 716(2010) .

On rock.

Liaoning, Shandong(Zhou GL et al. , 2012, pp. 313 – 315) .

Caloplaca bohlinii H. Magn. 包氏橙衣

Lich. Centr. Asia 1: 137(1940) .

Type: Gansu, Bohlin A: a in S(!) .

Caloplaca cerina(Hedw.) Th. Fr. 蜡黄橙衣

Nova Acta R. Soc. Scient. upsal. , Ser. 3 3: 218(1861) [1860] . v. cerina

≡*Lichen cerinus* Hedw. , *Descr. micr. -anal. musc. frond.* 2(3) : 62(1788) [1789] .

≡*Lecanora cerina*(Hedw.) Ach. , *Lich. univ.* : 390(1810)

≡*Callopisma cerinum*(Hedw.) De Not. [as ' cerina'] , *G. bot. ital.* 2(7 – 8) : 199(1847)

≡*Lecidea aurantiaca* var. *cerina*(Ehrh.) Schaer. , *Lich. helv. spicil.* 4 – 5: 180(1833)

Corticolous.

Neimenggu(Sun LY et al. , 2000, p. 37) , Xinjiang(Abdulla Abbas & Wu, 1998, p. 145; Abdulla Abbas et al. , 2001, p. 361) , Xizang(W. Obermayer, 2004, p. 489) , Sichuan(Hue, 1889, p. 170; Zahlbr. , 1930 b, p. 215; W. Obermayer, 2004, p. 489) , Yunnan(Hue, 1889, p. 170; Zahlbr. , 1930 b, p. 215) , Jiangsu(Xu B. S. , 1989, p. 250) , Shanghai(Xu B. S. , 1989, p. 250) , Hongkong(Thrower, 1988, p. 68) .

var. atrata Jatta 黑色变种

Nuov. Giorn. Bot. Italiano ser. 2, IX: 476(1902) . Zahlbr. , in Handel-Mazzetti, *Symb. Sin.* 3: 215(1930) & *Cat. Lich. Univ.* 7: 91(1931) .

Type: Shaanxi, collected by Giraldi.

Corticolous.

var. chloroleuca(Sm.) Th. Fr. 绿色变种

Mem. Fac. Educ. Shiga Univ. , *Nat. Sci.* 1: 170(1871)

≡*Lichen chloroleucus* Sm. , in Smith & Sowerby, *Engl. Bot.* 20: tab. 1373(1805) .

≡*Caloplaca cerina* f. *chloroleuca* Hellb. *Nerikes Lafflora*: 43(1871).

= *Caloplaca stillicidiorum*(Vahl) Lynge, *Skr. VidenskSelsk. Christiania, Kl. I, Math. -Natur*. (15): 4(1921).

Xinjiang(Poelt J, Hinteregger E, 1993, p. 95), Xizang, Sichuan(W. Obermayer, 2004, p. 490).

f. effusa(A. Massal.) Jatta 扩展变种

Syll. Lich. Ital.: 253(1900).

≡*Callopisma cerinum* var. *effusum* A. Massal., *Sched. critic.* (Veronae) 7: 131(1856).

Shaanxi(Jatta, 1902, p. 476; Zahlbr., 1930 b, p. 215).

var. muscorum(A. Massal.) Jatta 藓生变种

Syll. Lich. Ital.: 254(1900).

≡*Callopisma cerinum* var. *muscorum* A. Massal., *Symmict. Lich:* 35(1855).

On mosses or among *Cladonia* sp.

Xinjiang(H. Magn. 1940, p. 133), Xizang(W. Obermayer, 2004, p. 490).

Caloplaca cervina Zahlbr. 鹿角橙衣

in Handel-Mazzetti, *Symb. Sin.* 3: 217(1930) & *Cat. Lich. Univ.* 8: 585(1932).

Type: Sichuan, Handel-Mazzetti no. 5315.

Graniticolous.

Caloplaca chrysophora Zahlbr. 金色橙衣

in Handel-Mazzetti, *Symb. Sin.* 3: 219(1930) & *Cat. Lich. Univ.* 8: 585(1932).

Type: Sichuan, Handel-Mazzetti no. 2828.

Saxicolous.

Caloplaca cinnamomea(Th. Fr.) H. Olivier 肉桂橙衣

Mém. Soc. natn. Sci. nat. Cherbourg 37: 137(1909).

≡ *Caloplaca ferruginea* var. *cinnamomea* Th. Fr., *Nova Acta R. Soc. Scient. upsal.*, Ser. 3 3: 223(1861) [1860].

Jilin(Hertel & Zhao, 1982, p. 143).

Caloplaca cirrochroopsis Poelt & Hinter. 卷橙衣

Biblthca Lichenol. 50: 108(1993).

Sichuan(W. Obermayer, 2004, p. 490).

Caloplaca conversa(Kremp.) Jatta 转橙衣

Syll. Lich. Ital.: 254(1900).

≡*Callopisma conversum* Kremp., *Denkschr. Kgl. Bayer. Bot. Ges.*, Abt. 2 4: 132(1861).

Misapplied name: *Blastenia handelii* Zahlbr. (in Thrower, 1988, pp. 15, 64).

On granite coastal rocks. Known from northern temperate.

Neimenggu(Zhou GL et al., 2012, pp. 315 – 317), Hongkong(Chu, 1997, p. 48, as *Caloplaca conversa* and *C. carneofusca;* Aptroot & Seaward, 1999, p. 68; Aptroot et al., 2001, p. 324).

Caloplaca crenulatella(Nyl.) H. Olivier 园齿橙衣

Mém. Soc. natn. Sci. nat. Cherbourg 37: 110(1909).

≡*Lecanora crenulatella* Nyl., *Flora*, Regensburg 69: 461(1886).

On trees.

Shaanxi, hunan(zhao zt & sun ly, 2002, p. 136).

Caloplaca cupreorufa Zahlbr. 铜橙衣

in Handel-Mazzetti, *Symb. Sin.* 3: 218(1930) & *Cat. Lich. Univ.* 8: 586(1932).

Type: Sichuan, Handel-Mazzetti no. 2696.

Saxicolous.

Caloplaca delicata Zahlbr. 枝橙衣

in Handel-Mazzetti, *Symb. Sin.* 3: 215(1930) & *Cat. Lich. Univ.* 8: 586(1932).

Syntype: Yunnan, Handel-Mazzetti no. 5727; SICHUAN, Handel-Mazzetti no. 1372.

Corticolous.

Caloplaca dickoreana Poelt & Hinter. 迪克斯橙衣

Biblthca Lichenol. 50: 119(1993).

Type: China, Qinghai, Mt. Tanggula, N 33°27′, E 91°13′, 5380m., HolotypeDickoré L-08 in GZU.

On calcium-poor rock.

Caloplaca diphyodes(Nyl.) Jatta 双生橙衣

Syll. Lich. Ital.: 259(1900).

≡*Lecanora diphyodes* Nyl., *Flora*, Regensburg 55: 353(1872).

On rock.

neimenggu(Zhou GL et al., 2012, pp. 317 – 319).

Caloplaca epiphyta Lynge 附生橙衣

Skr. Svalbard Ishavet(Oslo) no. 81: 119(1940).

= *Caloplaca bryochrysion* Poelt in Fedde, *Repertorium*, LVIII: 175(1955).

On the ground.

Xinjiang(Poelt & Hinter., 1993, p. 124).

Caloplaca cf. exsecuta(Nyl.) Dalla Torre & Sarnth. 缺芽橙衣

Fl. Tirol: 191(1902).

On rock associated with *Carbonea vitellinaria* on *Candelariella*.

Xizang(W. Obermayer, 2004, p. 490).

Caloplaca gambiensis Aptroot 果橙衣

Cryptog. Mycol. 22(4): 266(2001).

On *Ficus* sp., *Roystonea regia*, *Acacia confuse*, and *Prunus* sp. Taiwan(Aptroot and Sparrius, 2003, p. 11).

Caloplaca geoica H. Magn. 土生橙衣

Lich. Centr. Asia 1: 139(1940).

Type: Gansu, Bohlin no. 44 c in S(!).

On calcareous soil among other lichens.

Caloplaca giraldii Jatta 吉拉氏橙衣

Nuov. Giorn. Bot. Ital. nov. ser. 9: 477(1902).

Type: Shaanxi, collected by Giraldi. (Zahlbr., 1930 b, p. 216 & 1931, p. 141), Sichuan(Zahlbr., 1930 b, p. 216 & 1931, p. 141).

Caloplaca grimmiae(Nyl.) H. Olivier 聚盘橙衣

Mém. Soc. natn. Sci. nat. Cherbourg 37: 119(1909).

≡*Lecanora grimmiae* Nyl., *Flora*, Regensburg 69: 97(1886).

= *Caloplaca congrediens*(Nyl.) Zahlbr., *Cat. Lich. Univers.* 7: 110(1930) [1931].

≡*Lecanora congrediens* Nyl., *Flora*, Regensburg 66: 100(1883).

Parasitic on *Candelariella vitellina* (Hertel & Zhao 1982).

Jilin(Hertel & Zhao, 1982, p. 143; Hawksw. & M. Cole, 2003, p. 360), Sichuan(W. Obermayer, 2004, p. 490).

Caloplaca hedinii H. Magn. 海登橙衣

Lich. Centr. Asia 1: 141(1940).

Type: Gansu, Bohlin no. 85b in S(!).

On sometimes very slightly calciferous rock.

Caloplaca holochracea(Nyl.) Zahlbr. 全橙衣

Cat. Lich. Univers. 7: 144(1930).

≡*Lecanora holochracea* Nyl. , in Crombie, *J. Linn. Soc. London, Bot*. 16: 17(1876) .

Misapplied name:

Caloplaca cf. *cinnabarina*(Ach.) Zahlbr. (in Thrower, 1988, pp. 15, 69; Chu, 1997, p. 48) .

On exposed, coastal or inland granitic rocks. Known from temperate Asia.

Hongkong(Thrower, 1988, pp. 15, 69; Aptroot & Seawaerd, 1999, pp. 68: Seaward & Chu, (MRDS) 108870 with *Caloplaca holochracea*, and 69: Chu "HKU[M] 10432 with *Buellia subdisformis*") .

Caloplaca infestans H. Magn. 侵害橙衣

Lich. Centr. Asia 2: 51(1944) .

Type: Neimenggu L, Bohlin no. 140 in S.

On other lichens upon calcareous rock.

Caloplaca insularis Poelt 岛橙衣

Planta 51: 300(1958) .

On sand rocks.

Xinjiang(Poelt & Hinter. , 1993, Pp. 142 – 143) .

Caloplaca ionaspoidea Zahlbr. 土果橙衣

in Handel-Mazzetti, *Symb. Sin*. 3: 214(1930) & *Cat. Lich. Univ*. 8: 587(1932) .

Type: Sichuan, Handel-Mazzetti no. 2005.

Saxicolous.

Caloplaca irrubescens(Arnold) Zahlbr. 非红橙衣

Verh. zool. -bot. Ges. Wien 48: 365(1898) .

≡*Cullopisma aurantiacum* var. *irrubescens* Arnold, *Verh. zool. -bot. Ges*. Wien 29: 353(1879) .

Beijing, Neimenggu, Yunnan, Hubei, Zhejiang(zhao zt & sun ly, 2002, p. 136) , Xizang(W. Obermayer, 2004, p. 491) .

Caloplaca kansuensis H. Magn. 甘肃橙衣

Lich. Centr. Asia 1: 134(1940) .

Type: Gansu, Bohlin no. 88a in S(!) .

On calciferous stone between other lichens.

Caloplaca leptozona(Nyl.) Zahlbr. 长带橙衣

Cat. Lich. Univers. 7: 154(1930) [1931] .

≡*Lecanora leptozona* Nyl. , *J. Linn. Soc.* , *Bot*. 20: 52(1883) .

= *Caloplaca exsecuta* (Nyl.) Dalla Torre & Sarnth. (in Chu, 1997, p. 48; W. Obermayer, 2004, p. 490 as *C.* cf. *exsecuta*) .

Misapplied name:

Blastenia handelii Zahlbr. (in Thrower, 1988, pp. 15, 64) .

On exposed rocks. Known from subtropical America, Asia, and Australia.

Xizang(W. Obermayer, 2004, p. 490) , Taiwan(Aptroot and Sparrius, 2003, p. 11) , Hongkong(Thrower, 1988, pp. 15, 64; Chu, 1997, p. 48; Aptroot & Seaward, 1999, p. 69; Aptroot et al. , 2001, p. 324) .

Caloplaca mongolica H. Magn. 蒙古橙衣

Lich. Centr. Asia 2: 51(1944) .

Type: Neimenggu L, Bohlin no. 169 in S(!) .

On calcareous rock with other lichens.

Caloplaca obscurella(J. Lahm) Th. Fr. 暗橙衣

Lich. Scand. (Upsaliae) 1(1) : 182(1871) .

≡*Blastenia obscurella* J. Lahm, in Körber, *Parerga lichenol*. (Breslau) 2: 130(1860) .

On bark of *Quercus acutissima*.

Shandong(Moreau et Moreau, 1951, p. 193) , Taiwan(Aptroot and Sparrius, 2003, p. 12) .

Caloplaca ochrotropa Zahlbr. 锈节橙衣

in Handel-Mazzetti, *Symb. Sin.* 3: 217(1930) & *Cat. Lich. Univ.* 8: 588(1932).

Type: Yunnan, Handel-Mazzetti no. 5862.

Saxicolous.

Caloplaca pachythallina Poelt & Hinter. 厚体橙衣

Biblthca Lichenol. 50: 169(1993).

Type: China, Xinjiang, N 36°03′, E76°28′; alt. 4180m, Holotype Dickoré F29 in GZU.

Caloplaca plumbeoolivacea H. Magn. 铅棕橙衣

Lich. Centr. Asia 2: 53(1944).

Type: Neimenggu, Bohlin no. 162 in S(!).

Saxicolous.

Caloplaca polytropoides Zahlbr. 拟多节橙衣

in Handel-Mazzetti, *Symb. Sin.* 3: 216(1930) & *Cat. Lich. Univ.* 8: 588(1932).

Type: Yunnan, Handel-Mazzetti no. 1021.

Saxicolous.

Caloplaca pulicarioides Aptroot 骨橙衣

in Aptroot & Seaward *Tropical Bryology* 17: 69 – 71(1999).

Typus: Hongkong, New Teriitories, Sai Kung County Park, on Kandella in mangrove near entrance area, alt. 1m, June 1998, Aptroot 43538(HKU(M)-holotypus, ABL-isotypus, Aptroot et al., 2001, p. 324.

Shandong(Zhou GL et al., 2012, pp. 319 – 321), Taiwan(Aptroot and Sparrius, 2003, p. 12).

Caloplaca scrobiculata H. Magn. 蜂窝橙衣

Lich. Centr. Asia 1: 143(940).

Type: Gansu, Bohlin no. 43 b in S(!).

On calciferous stone with *Catillaria kansuensis*.

Caloplaca annularis Clauzade & Poelt(W. Obermayer, 2004, p. 491, cited as synonym) Xinjiang(Abbas et al., 1993, p. 76; Abdulla Abbas & Wu, 1998, p. 147; Abdulla Abbas et al., 2001, p. 361), Xizang(W. Obermayer, 2004, p. 491).

Caloplaca sinensis B. de Lesd. 中华橙衣

Bull. Soc. Bot. France 79: 686(1932).

Type: N. China(prov. non indicated: Zahlbr., 1934b, p. 213 & 1940, p. 628; Lamb, 1963, p. 135).

Caloplaca sororicida M. Steiner & Poelt 孢橙衣

in Poelt & Hinteregger, *Biblthca Lichenol.* 50: 201(1993).

Neimenggu(Poelt & Hinter., 1993, p. 202 as Monogolia australis: lat. 41°40′, long. 109°20′).

Caloplaca subconcilians S. Y. Kondr., L. LŐkös & J. S. Hur 黑橙衣

in Kondratyuk, LŐkös, Tschabanenko, Moniri & Farkas, *Acta bot. hung.* 55(3 – 4): 275 – 349(2013).

Shandong(eastern coastal part of China, Kondratyuk et al., 2013, p. 284).

Caloplaca submodesta H. Magn. 白边橙衣

Bot. Notiser, 81/82(1951).

= *Caloplaca modesta* H. Magn. *Lich. Centr. Asia*, 1944, p. 50(non *Caloplaca modesta*(Zahlbr.) Fink, *Lich. Fl. of the U. S.* 1935, p. 361).

Type: Neimenggu, Bohlin no. 157 in S(!).

On non calciferous conglomerate, adjacent parts more or less calciferous.

Caloplaca tegularis(Ehrh.) Sandst. 瓦橙衣

Abh. naturw. Ver. Bremen 21(1): 220(1912).

≡ *Lichen tegularis* Ehrh., in Hoffmann, *Deutschl. Fl.*, *Zweiter Theil(Erlangen)*: 158(1796) [1795].

Saxicolous.

Xizang(Zahlbr. , 1933 a, p. 25) .

Caloplaca teicholyta(Ach.) J. Steiner 墙橙衣

Sber. Akad. Wiss. Wien, Math. -naturw. Kl. , *Abt.* 1 104: 388(1895) .

≡*Lecanora teicholyta* Ach. , *Lich. univ.* : 425(1810) .

= *Caloplaca lallavei*(Clemente ex Ach.) Flagey, *Mém. Soc. ému. Doubs*, sér. 6 1: 254(1886) .

≡*Lecidea lallavei* Clemente ex Ach. , *Syn. meth. lich.* (Lund) : 45(1814) .

Misapplied name:

Caloplaca erythrocarpa sensu Ann in Abdulla Abbas & Wu, 1998, p. 145 & Abdulla Abbas, 2001, p. 361, non (Pers.) Zwackh; fide *Checklist of Lichens of Great Britain and Ireland*(2002) .

Xinjiang(Abdulla Abbas et al. , 1994, p. 20; Abdulla Abbas & Wu, 1998, p. 146; Abdulla Abbas, 2001, p. 361) .

Caloplaca tianshanensis Xahidin, A. Abbas & J. C. Wei 天山橙衣

Mycotaxon 114: 3 – 5(2010) .

Type: Xinjiang, Nanshan in Tianshan mountain chain, Miaoergou, on limestone, alt. 1280 m, April 10, 2009, A. Abbas & H. Xahidin 20090001(holotype in XJU, isotype in HMAS-L) .

Caloplaca tominii(Savicz) Ahlner 托敏氏橙衣

Ann. bot. Soc. Zool. -Bot. fenn. Vanamo 43: 159(1949) .

≡*Placodium tominii* Savicz, *Izv. glav. bot. Sada SSSR*: 194(1930) .

Xinjiang(Abdulla Abbas & Wu, 1998, pp. 147 – 148; Abdull Abbas et al. , 2001, p. 361) .

Caloplaca transcaspica(Nyl.) Zahlbr. 杷森氏橙衣

Cat. Lich. Univers. 7: 189(1931) .

≡*Lecanora transcaspica* Nyl. , in Brotherus, *Öfvers. Finska Vetensk. -Soc. Förh*. 40: 9(1897) .

= *Caloplaca paulsenii* (Vain.) Zahlbr. , *Cat. Lich. Univers*. 7: 162(1931) , Poelt & Hinter. , 1993, p. 214.

≡*Placodium paulsenii* Vain. , *Bot. Tidsskr*. 26: 244(1904) .

Saxicolous.

Neimenggu(H. Magn. 1944, p. 54) , Xinjiang(Abdulla Abbas et al. , 1993, p. 76; Abdulla Abbas et al. , 1994, p. 20; Abdulla Abbas & Wu, 1998, p. 147; Abdulla Abbas et al. , 2001, p. 361) , Qinghai, Gansu(H. Magn. 1940, p. 135) .

= *Caloplaca paulsenii* f. *ochraceella* H. Magn. *Lich. Centr. Asia* 1: 135(1940) .

Type: Gansu, Bohlin no. 59 b in S(!) .

On rocks.

Xinjiang, Gansu(Poelt & Hinter. , 1993, p. 216) .

Caloplaca triloculans Zahlbr. 三孢橙衣

in Handel-Mazzetti, *Symb. Sin*. 3: 219(1930) & *Cat. Lich. Univ*. 8: 589(1932) .

J. Hafellner & J. Poelt, Journ. Hattori Bot. Lab. 46: 37(1979) .

Type: Yunnan, Handel-Mazzetti no. 6418.

Corticolous.

Sichuan(W. Obermayer, 2004, pp. 491 – 492) , Yunnan(Zahlbr. , 1932, p. 219) .

Caloplaca variabilis(Pers.) Müll. Arg. 多变橙衣

Mém. Soc. Phys. Hist. nat. Genève 16(2) : 387(1862) .

≡*Lichen variabilis* Pers. , *Ann. Bot*. (Usteri) 1: 26(1794) .

On rock.

Xinjiang(Poelt & Hinter. , 1993, p. 222) , Shanxi(Zhou GL et al. , 2012, pp. 321 – 323) .

Fauriea S. Y. Kondr. , LŐkös & Hur **弗雷衣属**

in Kondratyuk, LŐkös, Kim, Kondratiuk, Jeong, Jang, Oh, Wang & Hur, *Acta Biologica Hungarica* 58: 305(2016) .

Fauriea orientochinensis S. Y. Kondr., X. Y. Wang & Hur 华东弗雷衣

in Kondratyuk, LŐkös, Kim, Kondratiuk, Jeong, Jang, Oh, Wang & Hur, *Acta Biologica Hungarica* 58: 307(2016).

Type: China, Shandong Prov., Rongcheng Co., Mt Chengshantou, on Pinus bark, growing together with Amandinea punctata. Lat.: 37°25′03.0″ N; Long.: 122°40′38.3″E; Alt.: 20 m a. s. l. Coll.: Wang, X. Y. and Hur, J. -S. (CH-110011), 18.07.2011(KoLRI 013954 sub Fauriea orientochinensis-holotype); thesame locality, growing as small addition among Amandinea punctata thalli, (CH-110015), (KoLRI 013957 sub Fauriea orientochinensis-isotype); the same locality, growing with Biatora pseudosambuci, (CH-110017), (KoLRI 013959 sub Fauriea orientochinensis-isotype).

Flavoplaca Arup, Søchting & Frödén **黄绿衣属**

Nordic Jl Bot. 31(1): 44(2013).

Flavoplaca citrina(Hoffm.) Arup, Frödén & Søchting 柠檬黄绿衣

Nordic Jl Bot. 31(1): 44(2013).

≡*Verrucaria citrina* Hoffm., *Deutschl. Fl.*, *Zweiter Theil*(Erlangen): 198(1796)[1795].

≡*Caloplaca citrina* (Hoffm.) Th. Fr. *Nouv. Arch. Mus. Hist. Nat.*, Paris, 2 sér. 3: 218(1861)[1860].

≡*Lecanora citrina*(Hoffm.) Ach., *Lich. univ.*: 402(1810).

≡*Callopisma citrinum*(Hoffm.) A. Massal., *Atti Inst. Veneto Sci. lett.*, *ed Arti*, Sér. 2 3(App. 3): 97(1852).

≡*Placodium citrinum*(Hoffm.) Hepp, *Flecht. Europ.*: no. 394(1857).

On cement of wall.

Shandong(Zhang F et al., 1999, p. 31; Zhao ZT et al., 2002, p. 7), Shanghai(Nyl. & Cromb. 1883, p. 63; Hue, 1891, p. 48; Zahlbr., 1930 b, p. 215; Xu B. S., 1989, p. 250), Taiwan(Aptroot and Sparrius, 2003, p. 11).

Flavoplaca coronata(Kremp. ex Körb.) Arup, Frödén & Søchting 皇冠黄绿衣

Nordic Jl Bot. 31(1): 45(2013).

≡*Callopisma aurantiacum* var. *coronatum* Kremp. ex Körb., *Parerga lichenol.* (Breslau) 1: 66(1859).

≡*Caloplaca coronata*(Kremp. ex Körb.) J. Steiner, *Verh. zool. -bot. Ges.* Wien 69: 71(1919).

Xinjiang(Abbas et al., 1993, p. 76; Abdulla Abbas & Wu, 1998, p. 145; Abdulla Abbas et al., 2001, p. 361).

Fulgensia A. Massal. & De Not. **拟橙衣属**

in Massalongo, *Alcuni Gen. Lich.*: 10(1853).

= *Gyalolechia* A. Massal., *Ric. auton. lich. crost.* (Verona): 17(1852)

Fulgensia bracteata(Hoffm.) Räsänen 拟橙衣

Cat. Lich. Univers. 8: 585(1931)[1832]. v. bracteata

≡*Psora bracteata* Hoffm., *Deutschl. Fl.*, *Zweiter Theil*(Erlangen): 169(1796)[1795].

≡*Caloplaca bracteata*(Hoffm.) Jatta, *Syll. Lich. Ital.*: 236(1900).

Xinjiang(Abdulla Abbas et al., 1993, p. 78; Abdulla Abbas et al., 1994, p. 21; Abdulla Abbas & Wu, 1998, p. 148; Abdulla Abbas et al., 2001, p. 363), Ningxia(Liu M & Wei JC, 2013, p. 48; Yang J. & Wei J. C., 2014, p. 1030), Gansu(H. Magn. 1940, p. 136).

var. alpina(Th. Fr.) Räsänen 高山变种

Ann. bot. Soc. Zool. -Bot. fenn. Vanamo 18(1): 39(1943).

≡*Placodium fulgens* var. *alpinum* Th. Fr., *Nova Acta R. Soc. Scient. upsal.*, Ser. 3 3: 181(1861)[1860].

On the ground.

Xizang(J. C. Wei & Y. M. Jiang, 1986, p. 107).

Fulgensia bracteata subsp. deformis(Erichsen) Poelt 畸形亚种

Mitt. Bot. StSamml., *Münch.* 5: 603(1965).

≡*Caloplaca bracteata* f. *deformis* Erichsen, I Mitt. Inst. Allgem. Bot. Hamburg, 10: 417(1939).

Xinjiang(Hurnisa Xanidin, H. et al. , 2009, p. 108 as *F. bacteata* ssp. *deformis*) .

Gyalolechia Trevis. **橙果衣属**

Revta Period. Lav. Regia Accad. Sci. , Padova: 267(1851) [1851 – 52] , Nomenclatural comment: Nom. inval. , Art. 38. 1(a) (Melbourne) . The name remains to be validated.

Gyalolechia bassiae(Ach.) Søchting, Frödén & Arup 深橙果衣

Nordic Jl Bot. 31(1) : 70(2013)

≡*Isidium bassiae* Ach. , *Lich. univ.* : 579(1810) .

≡*Caloplaca bassiae* (Ach.) Zahlbr. , *Cat. Lich. Univers.* 7: 78(1930) [1931] .

On *Cinnamomum*, *Roystonea regia*, *Ficus nervosa*, and *Ardisia sieboldii*.

Yunnan(China: Gejiu County, Man Hao town, on bark. Lat. : 22°55′36. 2″N; Long. : 103°36′21. 0″E; Alt. : 598 m a. s. l. Coll. : Hur, J. -S. , 25. 12. 2006 [KoLRI-006299(CH-060449)] , Kondratyuk et al. , 2013, p. 323.) , Taiwan(Aptroot and Sparrius, 2003, p. 11) .

Gyalolechia desertorum(Tomin) Søchting, Frödén & Arup 荒漠橙果衣

Nordic Jl Bot. 31(1) : 70(2013) .

≡*Placodium desertorum* Tomin, *Ueber die Bodenfl. Halbwüst. Süd-Oust-Rußl.* : 11(1925) .

≡*Fulgensia desertorum* (Tomin) Poelt, *Mitt. bot. StSamml. , Münch.* 5: 600(1965) .

Xinjiang(Hurnisa Xanidin, H. et al. , 2009, pp. 108 – 109) , Ningxia(Liu M & Wei JC, 2013, p. 49; ; Yang J. & Wei J. C. , 2014, p. 1030, as Fulgensia desertorum; Joshi S. et al. , 2014, p. 169) .

Gyalolechia flavorubescens(Huds.) Søchting, Frödén & Arup 红橙果衣

Nordic Jl Bot. 31(1) : 70(2013) .

≡*Lichen flavorubescens* Huds. , *Fl. Angl.* : 443(1762) .

≡*Caloplaca flavorubescens* (Huds.) J. R. Laundon, *Lichenologist* 8(2) : 147(1976) .

= *Caloplaca flavorubescens* f. *microthelia*(Ach.) Verseghy, *Bot. Közl.* 74 – 75(1 – 2) : 33(1988) [1987] .

≡*Lecanora aurantiaca* var. *microthelia*(Ach.) Nyl. , *Lich. Scand.* (Helsinki) : 142(1861) .

≡*Parmelia microthelia* Ach. , *Methodus, Sectio post.* (Stockholmiæ) : 174(1803)

Corticolous.

Xinjiang(Abbas et al. , 1993, p. 76; Abdulla Abbas & Wu, 1998, p. 146; Abdulla Abbas et al. , 2001, p. 361) , Gansu(H. Magn. 1940, p. 132) , Shaansu(Jatta, 1902, p. 476; Zahlbr. , 1930b, p. 216) , Sichuan, Hunan(Zahlbr. , 1930b, p. 216) , guizhou(Zhang T & J. C. Wei, 2006, p. 5) , Yunnan(Hue, 1889, p. 170; Zahlbr. , 1930b, p. 216) , Hubei(Chen, Wu & J. C. Wei, 1989, p. 477) . Jiangsu, Zhejiang(Nyl. & Cromb. 1883, p. 63; Xu B. S. , 1989, p. 251) , Shanghai(Nyl. & Cromb. 1883, p. 63; Zahlbr. , 1930b, p. 216; Xu B. S. , 1989, p. 251) , Taiwan (Zahlbr. , 1933c, p. 64; M. Sato, 1936b, p. 573; Wang & Lai, 1973, p. 88) , China(prov. not indicated: Nyl. & Cromb. 1883, p. 63) .

Gyalolechia flavovirescens(Wulfen) Søchting, Frödén & Arup 黄绿橙果衣

Nordic Jl Bot. 31(1) : 70(2013) .

≡*Lichen flavovirescens* Wulfen, *Schr. Ges. naturf. Freunde, Berlin* 8: 122(1787) .

≡*Caloplaca flavovirescens* (Wulfen) Dalla Torre et Sarnth. *Fl. Tirol*: 180(1902) .

≡*Callopisma aurantiacum* var. *flavovirescens*(Wulfen) Massal. (in Rabh. , 1873, p. 287) .

≡*Lecidea flavovirescens*(Wulfen) Fr. , *Lich. eur. reform.* (Lund) : 291(1831) .

= *Lecanora erythrella*(Ach.) Ach. , *Lich. univ.* : 401(1810) .

≡*Lichen erythrellus* Ach. , *Lich. suec. prodr.* (Linköping) : 43(1799) [1798] .

Misapplied name: *Caloplaca aurantiaca*(Lightf.) Th. Fr. (in Krempelh. , 1873, p. 471 & 1874, p. 66) .

Saxicolous. Northern temperate.

Jilin(Hertel & Zhao, 1982, p. 144) , neimenggu(Sun LY et al. , 2000, p. 37) , Sichuan(Zahlbr. , 1930 b, p. 216;

Chen S. F. et al., 1997, p. 221), Jiangsu, Zhejiang(Xu B. S., 1989, p. 251), Shanghai(Krempelh. 1873, p. 471 & 1874, p. 66; Rabh. 1873, p. 287; Nyl. & Cromb. 1883, p. 63; Hue, 1891, p. 49; Xu B. S., 1989, p. 251), Taiwan(Zahlbr., 1933c, p. 64; M. Sato, 1936 b, p. 573; Wang & Lai, 1973, p. 88), Hongkong(Rabh. 1873, p. 287; Krempelh. 1873, p. 471 & 1874, p. 66; Zahlbr., 1930 b, p. 216; Thrower, 1988, pp. 15, 70; Aptroot & Seaward, 1999, p. 68 – 69), China(prov. not indicated: Hue, 1911, p. 165 – 167).

Gyalolechia fulgens(Sw.) Søchting, Frödén & Arup 光亮橙果衣

Nordic Jl Bot. 31(1) : 70(2013).

≡*Lichen fulgens* Sw., *Nova Acta Acad. Upsal.* 4: 246(1784).

≡*Fulgensia fulgens*(Sw.) Elenkin, *Lich. Fl. Ross. Med.* 2: no. 246(1907).

Xinjiang(Wang XY, 1985, p. 351; Xanidin, H. et al., 2009, p. 109), Ningxai(Yang J. & Wei J. C., 2014, p. 1030. as Fulgensia fulgens)

Gyalolechia subbracteata(Nyl.) Søchting, Frödén & Arup 亚鳞橙果衣

Nordic Jl Bot. 31(1) : 72(2013).

≡*Lecanora subbracteata* Nyl., *Flora*, Regensburg 66: 534(1883).

≡*Fulgensia subbracteata*(Nyl.) Poelt, Lich. Alp. 7: no. 137(1961).

xinjiang(Wang XY, 1985, p. 351; Hurnisa Xanidin, H. et al., 2009, pp. 110 – 111).

Jasonhuria S. Y. Kondr., L. LŐkös & S. O. Oh **许白衣属**

in Kondratyuk, LŐkös, Kim, Kondratiuk, Jeong, Jang, Oh & Hur, *Mycobiology* 43(3) : 198(2015).

Jasonhuria bogilana(Y. Joshi & Hur) S. Y. Kondr. Et al. 韩岛许白衣

in Kondratyuk, LŐkös, Kim, Kondratiuk, Jeong, Jang, Oh & Hur, *Mycobiology* 43(3) : 198(2015).

≡*Caloplaca bogilana* Y. Joshi & Hur, *Lichenologist* 42(6) : 716(2010).

On rock.

Liaoning, Shandong(Zhou GL et al., 2012, pp. 313 – 315).

Ioplaca Poelt 胎座衣属

Khumbu Himal 6: 443(1977).

Ioplaca pindarensis(Räsänen) Poelt & Hinter. 粉胎座衣

Biblthca Lichenol. 50: 235(1993).

≡*Callopisma pindarense* Räsänen, *Ann. Soc. Zool. -Bot. fenn. Vanamo* 6(no. 2) : 83(1952).

Sichuan(Obermayer, 2004, p. 500).

Leproplaca(Nyl.) Nyl. **黄粒属**

in Hue, *Rev. Bot. Bull. Mens.* 6: 148(1888) [1887 – 88].

≡*Lecanora* subgen. *Leproplaca* Nyl., *Flora*, Regensburg 66: 107(1883).

Leproplaca cirrochroa(Ach.) Arup, Frödén & Søchting 卷黄粒

Nordic Jl Bot. 31(1) : 72(2013).

≡*Lecanora cirrochroa* Ach., *Syn. meth. lich.* (Lund) : 181(1814).

≡*Caloplaca cirrochroa*(Ach.) Th. Fr., *Lich. Scand.* (Upsaliae) 1(1) : 171(1871).

On rock.

Xinjiang(Abas et al., 1996, p. 232; Abdulla Abbas & Wu, 1998, p. 145; Abdulla Abbas et al., 2001, p. 361).

Leproplaca xantholyta(Nyl.) Nyl. 橙黄粒

in Hue, *Rev. Bot. Bull. Mens.* 6: 148(1888) [1887 – 88].

≡*Lecanora xantholyta* Nyl., *Flora*, Regensburg 62: 361(1879).

≡*Caloplaca xantholyta*(Nyl.) Jatta, Nuovo G. bot. ital. 9: 476(1902).

On rocks.

Shaanxi(Jatt, 1902, p. 476; Zahlbr., 1930 b, p. 244).

On overhanging rock.

Xinjiang(Abdulla Abbas & Wu, 1998, p. 160; Abdulla Abbas et al. , 2001, p. 364) , Xizang, Sichuan(W. Obermayer, 2004, p. 492, as *Caloplaca xantholyta*) .

Massjukiella S. Y. Kondr. Et al. **橙鳞属**

in Fedorenko, Stenroos, Thell, Kärnefelt, Elix, Hur & Kondratyuk, Biblthca Lichenol. 108: 54(2012) .

Massjukiella polycarpa(Hoffm.) S. Y. Kondr. et al. 多果橙鳞

in Fedorenko, Stenroos, Thell, Kärnefelt, Elix, Hur & Kondratyuk, Biblthca Lichenol. 108: 60(2012) .

≡*Lobaria polycarpa* Hoffm. , *Deutschl. Fl.* , *Zweiter Theil* (Erlangen) : 159(1796) [1795] .

≡*Xanthoria polycarpa*(Hoffm.) Rieber, *Jber. vaterl. Kultur Württemberg* 47: 252(1891) .

Xinjiang(Abdulla Abbas & Wu, 1998, p. 151; Abdulla Abbas et al. , 2001, p. 269) , Shaanxi(He & Chen 1995, p. 45 without citation of specimens and their localities) .

Oxneria S. Y. Kondr. & Kärnefelt **奥克衣属**

Ukr. bot. Zh. 60(4) : 428(2003) .

Oxneria fallax(Arnold) S. Y. Kondr. & Kärnefelt 拟奥克衣

Ukr. bot. Zh. 60(4) : 431(2003) .

≡*Xanthoria fallax* Arnold, *Verh. zool. -bot. Ges*. Wien 30: 121(1880) .

≡*Placodium fallax* Hepp, *Flecht. Europ.* : no. 633(1860) .

Physcia fallax Hepp ex Arnold, *Flora*, Regensburg 41(20) : 307(1858) , nom. illegit. , Art. 53. 1.

On bark of trees.

Neimenggu(Sato, 1952, p. 175; Asahina, 1952 d, p. 375; Chen et al. 1981, p. 159; Sun LY et al. , 2000, p. 37) , Xinjiang(Abdulla Abbas et al. , 1993, p. 81; Abdulla Abbas & Wu, 1998, p. 151; Abdulla Abbas et al. , 2001, p. 369) , Shaanxi(He & Chen 1995, p. 45 as *X. fallax* without citation of specimen and locality) , Sichuan(Chen S. F. et al. , 1997, p. 222) , Xizang(J. C. Wei & Jiang, 1981, p. 1147 & 1986, p. 108) , Hubei(Chen, Wu & J. C. Wei, 1989, p. 478) , Shandong(Hou YN et al. , 2008, p. 67; Li Y et al. , 2008, p. 72 as *Xanthoria fallax*) .

Pachypeltis Søchting, Arup & Frödén **厚盾属**

Nordic Jl Bot. 31(1) : 48(2013) .

Pachypeltis intrudens(H. Magn.) Søchting, Frödén & Arup 盘厚盾

Nordic Jl Bot. 31(1) : 48(2013)

≡*Caloplaca intrudens* H. Magn. , *Lich. Centr. Asia* 1: 142(1940) .

Type: Gansu, Bohlin no. 77 b in S(!) .

On calciferous rock with different lichens.

Polycauliona Hue **多枝衣属**

Bull. Soc. linn. Normandie, sér. 6 1: 75(1908)

Polycauliona candelaria(L.) Frödén, Arup & Søchting 细片多枝衣

Nordic Jl Bot. 31(1) : 51(2013) . f. candelaria

≡*Lichen candelarius* L. , *Sp. pl*. 2: 1141(1753) .

≡*Xanthoria candelaria*(L.) Th. Fr. *Gen. Heterolich. Eur.* : 61(1861) .

= *Physcia lychnea* (Ach.) Nyl. , *Not. Sällsk. Fauna et Fl. Fenn. Förh.* , *Ny Ser*. 5: 107(1861) .

≡*Xanthoria parietina* var. *lychnea* (Ach.) J. Kickx f. , Fl. Crypt. Flandres(Paris) 1: 228(1867) .

≡*Xanthoria polycarpa* var. *lychnea*(Ach.) Vain. , Acta Soc. Fauna Flora fenn. 13(no. 6) : 12(1896) .

On tree trunk of *Acacia* sp.

Xinjiang(Abdulla Abbas & Wu, 1998, p. 150; Abdulla Abbas et al. , 2001, p. 369 as *X. candelaria*) , Shaanxi (Jatta, 1902, p. 473; Zahlbr. , 1930 b, p. 221) .

Rufoplaca Arup, Søchting & Frödén **淡平衣属**

Nordic Jl Bot. 31(1) : 74(2013)

Rufoplaca arenaria(Pers.) Arup, Søchting & Frödén 砂岩淡平衣

Nordic Jl Bot. 31(1) : 74(2013) .

≡*Lichen arenarius* Pers. , *Ann. Bot*. (Usteri) 1: 27(1794) .

= *Caloplaca erythrocarpa*(Pers.) Zwackh [as ' *erythrocarpia*'] , *Flora*, Regensburg 45: 487 (1862) .

≡*Patellaria erythrocarpa* Pers. , *Ann. Wetter. Gesellsch. Ges. Naturk*. 2: 12(1811) [1810] .

On shale.

Xinjiang(Abdulla Abbas et al. , 2001, p. 361) , Taiwan(Aptroot and Sparrius, 2003, p. 11) .

Rusavskia S. Y. Kondr. & Kärnefelt **黄鳞衣属**

Ukr. bot. Zh. 60(4) : 433(2003) .

Rusavskia sorediata(Vain.) S. Y. Kondr. & Kärnefelt 粉芽黄鳞

Ukr. bot. Zh. 60(4) : 434(2003) .

≡*Lecanora elegans* var. *sorediata* Vain. , *Meddn Soc. Fauna Flora fenn*. 6: 143(1881) .

≡*Xanthoria sorediata* (Vain.) Poelt, *Mitt. bot. StSamml. , Münch*. 11: 29(1954) , Nom. inval. , Art. 41. 5(Melbourne) .

Xinjiang(Abdulla Abbas & Wu, 1998, pp. 151 – 152) .

= *Caloplaca elegans* var. *compacta* Arn. in Nyl.

Shaanxi(Jatta, 1902, p. 479) .

Seirophora Poelt **茸枝衣属**

Flora, Regensburg 174(5/6) : 440(1983) .

Seirophora contortuplicata(Ach.) Frödén in Frödén & Lassen 缠结茸枝衣

Lichenologist 36(5) : 297(2004) .

≡*Parmelia contortuplicata* Ach. , *Syn. meth. lich*. (Lund) : 210(1814) .

≡*Xanthoria contortuplicata*(Ach.) Boistel, *Nouv. Fl. Lich*. 2: 70(1903) .

≡*Teloschistes contortuplicatus*(Ach.) Clauzade & Rondon, *Bull. Soc. linn. Provence* 22: 33(1959) .

≡*Xanthoanaptychia contortuplicata* (Ach.) S. Y. Kondr. & Kärnefelt, *Ukr. bot. Zh*. 60(4) : 435(2003) .

On lime stone or on soil.

Xinjiang, Xizang(Obermayer 2004, p. 515) .

Seirophora lacunosa(Rupr.) Frödén 凹面茸枝衣

Lichenologist 36(5) : 297(2004) .

≡*Ramalina lacunosa* Rupr. , *Mém. Acad. Imp. Sci. St. -Pétersb*. 6: 235(1845) .

≡*Teloschistes lacunosus*(Rupr.) Savicz(Wu J. L. , 1987, p. 190; J. C. Wei, 1991, p. 241; Abbas et al. , 1993, p. 81; Abbas et al. , 1994, p. 23.) .

On saline ground.

Xinjiang(Wang XY, 1985, pp. 350 – 351 as *Teloschistes brevior* var. *lacunosus*; Wu J. L. , 1987, p. 190; J. C. Wei, 1991, p. 241; Abdulla Abbas et al. , 1993, p. 81; Abdulla Abbas et al. , 1994, p. 23, as *T. lacunosus*; Abdulla Abbas & Wu, 1998, p. 149; Abdulla Abbas et al. , 2001, p. 368 as *T. brevior* var. *lacunosa*) , Ningxia(Liu M. & J. C. Wei, 2013, p. 49; Yang J. & Wei J. C. , 2014, p. 1030) .

Seirophora orientalis Frödén 东方茸枝衣

in Frödén & Litterski, *Graphis Scripta* 17(1) : 22(2005) .

Ningxia(Joshi S. et al. , 2014, p. 171 – 173) .

Seirophora villosa(Ach.) Frödén 柔毛茸枝衣

Lichenologist 36(5) : 297(2004) .

≡*Parmelia villosa* Ach. , *Methodus*, Sectio post. : 254(1803) .

= *Teloschistes brevior*(Nyl.) Vain. (H. Magn. 1940, p. 144; J. C. Wei, 1991, p. 241; Abbas et al. , 23(1994) .

On earth.

Gansu(H. Magn. 1940, p. 144; J. C. Wei, 1991, p. 241), Xinjiang(Abbas et al., 1993, p. 81; Abbas et al., 1994, p. 23, as T. brevior.), Ningxia(Liu M. & J. C. Wei, 2013, p. 49; Yang J. & Wei J. C., 2014, p. 1030).

Solitaria Arup, Søchting & Frödén **单衣属**

Nordic Jl Bot. 31(1): 55(2013).

Solitaria chrysophthalma(Degel.) Arup, Søchting & Frödén 金体单衣

Nordic Jl Bot. 31(1): 55(2013).

≡*Caloplaca chrysophthalma* Degel., *Svensk bot. Tidskr.* 38: 56(1944).

On *Magnolia grandiflora*, *Populus* sp. and other tree bark.

Beijing, Jilin, Sichuan, Hunan(zhao zt & sun ly, 2002, p. 135), neimenggu(Sun LY et al., 2000, p. 37zhao zt & sun ly, 2002, p. 135), Taiwan(Aptroot and Sparrius, 2003, p. 11)

Teloschistes Norman **黄枝衣属**

Nytt Mag. Natur. 7: 228(1853)[1852].

Teloschistes brevior(Nyl.) Hillmann 短黄枝衣

Hedwigia 69: 320(1931).

≡*Physcia villosa* f. *brevior* Nyl., *Syn. meth. lich.* (Parisiis) 1(2): 408(1860).

On bark and or rotten wood.

Xinjiang(Abdulla Abbas & Wu, 1998, pp. 148 – 149).

Teloschistes exilis(Michx.) Vain. 细黄枝衣

Acta Soc. Fauna Flora fenn. 7(1): 115(1890).

≡*Physcia exilis* Michx., *Fl. Boreali-Americ.* 2: 327(1803).

≡*Parmelia exilis*(Michx.) Spreng., *Syst. veg.*, Edn 16 4(1): 281(1827).

On twigs of Theae chinensis.

China(prov. not indicated, but it may be collected from Guangdong: Mey & Flot. 1843, p. 220).

Teloschistes flavicans(Sw.) Norman 浅黄枝衣

Nytt Mag. Natur. 7: 229(1853)[1852].

≡*Lichen flavicans* Sw., *Prodr.*: 147(1788).

≡*Physcia flavicans*(Sw.) DC., in Lamarck & de Candolle, *Fl. franç.*, Edn 3(Paris) 2: 189(1805).

On bark of trees(*Salix* & *Pinus*).

Heilongjiang(Chen et al. 1981, p. 157; Luo, 1984, p. 85), Jilin(Chen et al. 1981, p. 157), North-eastern China (prov. not indicated: Oxner, 1928, M. Sato, 1940a, p. 43), Shaanxi(Jatta, p. 473; Zahlbr., 1930b, p. 221), Sichuan(Chen S. F. et al., 1997, p. 222), Yunnan(Hue, 1887a, p. 23 & 1889, p. 167 & 1899, p. 98; Oxner, 1928, Zahlbr., 1930 b, p. 221; M. Sato, 1940a, p. 43), Xizang(Obermayer 2004, p. 511), Hubei(Chen, Wu & Wei, 1989, p. 477), Zhejiang(Xu B. S., 1989, p. 252), Taiwan(Zahlbr., 1933 c, p. 64; M. Sato, 1936b, p. 571 & 1940, p. 43; Wang & Lai, 1973, p. 97).

f. hirtellus Vain. 短毛变型

Étud. Class. Lich. Brésil 1: 114(1890).

On tree.

Yunnan(Hue, 1899, p. 99; Zahlbr., 1930 b, p. 222).

var. maximus(Meyen & Flot.) Zahlbr. 大枝变种

Cat. Lich. Univers. 7: 323(1931).

≡*Evernia flavicans* f. *maxima* Meyen & Flot., *Nova Acta Acad. Caes. Leop.-Carol. Nat. Cur.* 19(Suppl.): 210 (1843).

= *Physcia flavicans* var. *intermedia* Jatta in *Nuov. Giorn. Bot. Ital.*, *nov. ser.*; IX: 472(1902).

≡*Teloschistes flavicans* var. *intermedius* Müll. Arg., *Jb. Königl. Bot. Gart. Berlin* 2: 309(1883).

On twigs.

Shaanxi(Jatta, 1902, p. 473; Zahlbr. , 1930b, p. 222) .

Variospora Arup, Søchting & Frödén **变孢衣属**

Nordic Jl Bot. 31(1) : 75(2013) .

Variospora aurantia(Pers.) Arup, Frödén & Søchting 桔黄变孢衣

Nordic Jl Bot. 31(1) : 76(2013) .

≡*Lichen aurantius* Pers. , *Ann. Bot.* (Usteri) 5: 14(1794) .

≡*Caloplaca aurantia*(Pers.) Hellb. , *Bih. K. svenska VetenskAkad. Handl.* , Afd. 3 16(1) : 60(1890) .

= *Caloplaca callopisma*(Ach.) Th. Fr. , *Lich. Scand.* (Upsaliae) 1(1) : 169(1871) .

≡*Lecanora callopisma* Ach. , *Lich. univ.* : 437(1810) .

On rocks.

Shandong(Zhang F et al. , 1999, p. 31; Zhao ZT et al. , 2002, p. 7; Li Y et al. , 2008, p. 72) , Jiangsu(Wu J. N. & Xiang T. 1981, p. 6: here should be p. 5 according to the sequence of page number; Xu B. S. , 1989, p. 250) , Shanghai(Xu B. S. , 1989, p. 250) , Zhejiang(Nyl. & Cromb. , 1883, p. 63; Hue, 1891, p. 47; Zahlbr. , 1930 b, p. 220; Xu B. S. , 1989, p. 250) .

Variospora australis(Arnold) Arup, Søchting & Frödén 南方变孢衣

Nordic Jl Bot. 31(1) : 76(2013) .

≡*Physcia australis* Arnold, *Flora*, Regensburg 58: 154(1875) .

≡*Fulgensia australis*(Arnold) Poelt *Mitt. Bot. StSamml.* , *München.* 5: 594(1965) .

On rock.

Xinjiang(Hurnisa XAHIDIN et al. , 2009, pp. 107 – 108) , Ningxia(Liu M & J. C. Wei, 2013, p. 48) .

Variospora dolomiticola(Hue) Arup, Søchting & Frödén 拟变孢衣

Nordic Jl Bot. 31(1) : 76(2013) .

≡*Lecanora dolomiticola* Hue, *Annls mycol.* 13(2) : 83(1915) .

≡*Caloplaca dalmatica* (A. Massal.) H. Olivier, *Mém. Soc. natn. Sci. nat. Cherbourg* 37: 112(1909) .

≡*Callopisma dalmaticum* A. Massal. , *Symmict. Lich*: 30(1855) .

≡*Callopisma dalmaticum* A. Massal. , *Symmict. Lich*: 30(1855) .

On shale.

Taiwan(Aptroot and Sparrius, 2003, p. 11) .

Xanthoria(Fr.) Th. Fr. **石黄衣属**

Nova Acta R. Soc. Scient. upsal. , Ser. 3 3: 166(1861) [1860] .

≡*Parmelia* subdiv. *Xanthoria* Fr. , *Syst. orb. veg.* (Lundae) 1: 243(1825) .

Xanthoria candelaria f. stenophylla(Mass.) Hillm. 细片石黄衣狭叶变型

in Hedwigia, VXIII202(1922) .

≡*Physcia controversa* var. *stenophylla* Massal. , Schedul. Critic. , II: 43(1855) .

On tree trunk.

Shaanxi(Jatta, 1902, p. 472; Zahlbr. , 1930 b, p. 221) .

Notes. The current name of *Xanthoria candelaria* is *Polycauliona candelaria*(L.) Frödén, but what is the systematic position of *X. candelaria* f. *stenophylla* remains to be studied.

Xanthoria elegans(Link) Th. Fr. 丽石黄衣

Lich. Arctoi 3: 69(1860) .

≡*Lichen elegans* Link, *Ann. Nat. urgesch.* 1: 37(1791) .

≡*Xanthoria elegans* (Link) Th. Fr. , *Lich. Arctoi* 3: 69(1860) .

≡*Rusavskia elegans*(Link) S. Y. Kondr. & Kärnefelt, *Ukr. bot. Zh.* 60(4) : 434(2003) .

≡*Lecanora elegans* (Link) Ach. , *Lich. univ.* : 435(1810) .

≡*Caloplaca elegans*(Link) Th. Fr. , *Lich. Scand.* (Upsaliae) 1(1) : 168(1871) .

On rocks.

Neimenggu(H. Magn. 1940, p. 138 & 1944, p. 55; Sun LY et al. , 2000, p. 37, as *X. elegans*) , Jilin(Chen et al. 1981, p. 129; Hertel & Zhao, 1982, p. 150) , Xinjiang(H. Magn. 1940, p. 138; Moreau et Moreau, 1951, p. 194; Wu J. L. 1985, pp. 73 – 74; Wang XY, 1985, p. 351 as *Caloplaca elegans*; Abdula Abbas et al. , 1993, p. 81; Abdula Abbas et al. , 1994, p. 23; Abdulla Abbas & Wu, 1998, p. 150; Abdulla Abbas et al. , 2001, p. 369) , Shaanxi(He & Chen 1995, p. 45 as *X. elegans* without citation of specimen and locality) , Gansu, Qinghai(H. Magn. 1940, p. 138) , ningxia(Liu M & J. C. Wei, 2013, p. 49) , Xizang(J. C. Wei & J. B. Chen, 1974, p. 181; J. C. Wei & Jiang, 1981, p. 1147 & 1986, p. 107) , Yunnan(Hue, 1889, p. 170; Zahlbr. , 1930 b, p. 220) , Shandong(Zhao ZT et al. , 2002, p. 7) , Jiangsu(Wu J. N. & Xiang T. 1981, p. 6, here should be on p. 5 according to the sequence of the pages; Xu B. S. , 1989, p. 252) .

= *Lecanora elegans* f. *lucens* Nyl. in Cromb. *Journ. Linn. Soc. London, Bot*. 15: 184(1876) .

Shaanxi(Jatta, 1902, p. 476) .

= *Caloplaca elegans* var. *tenuis* (Wahlenb.) Th. Fr. , *Lich. Scand*. (Upsaliae) 1(1) : 168(1871) .

≡ *Lichen elegans* var. *tenuis* Wahlenb. , *Fl. lapp*. : 417(1812) .

Xizang(Sirigh-Jiganang-see: Zahlbr. , 1933 a, p. 25) .

Xanthoria parietina(L.) Th. Fr. 石黄衣

Lich. Arctoi 3: 69(1860) var. parietina

≡ *Lichen parietinus* L. , *Sp. pl*. 2: 1143(1753) .

≡ *Physcia parietina*(L.) De Not. , *G. bot. ital*. 2(2) : 297(1847) .

On bark of trees.

Beijing(Tchou, 1935, p. 315) , Jilin, Liaoning(Chen et al. 1981, p. 159) , Neimenggu(Sun LY et al. , 2000, p. 37) , Shaanxi(Jatta, 1902, p. 473; Zahlbr. , 1930 b, p. 221; J. C. Wei et al. 1982, p. 54; He & Chen 1995, p. 45 without citation of specimens and their localities) , Shandong(Hou Ya-nan et al. , 2008, p. 66; Li Y et al. , 2008, p. 72) .

Notes: This species is distributed very common in Europe, but it has not been found in China so far. The materials cited by above mentioned literatures remain to be checked further.

Xanthoria sorediata(Vain.) Poelt 粉芽石黄衣

Mitt. bot. StSamml. , Münch. 11: 29(1954) , nom. inval. , Art. 41. 5(Melbourne) .

≡ *Lecanora elegans* var. *sorediata* Vain. , *Meddn Soc. Fauna Flora fenn*. 6: 143(1881) .

Xinjiang(Abdulla Abbas et al. , 2001, p. 369) .

Zeroviella S. Y. Kondr. & Hur **假杯点黄属**

in Kondratyuk, Kim, Yu, Jeong, Jang, Kondratiuk, Zarei-Darki & Hur, *Ukr. Bot. J*. 72(6) : 578(2015) .

Zeroviella laxa(Müll. Arg.) S. Y. Kondr. & Hur 疏假杯点黄

in Kondratyuk, Kim, Yu, Jeong, Jang, Kondratiuk, Zarei-Darki & Hur, *Ukr. Bot. J*. 72(6) : 582(2015) .

≡ *Amphiloma elegans* var. *laxum* Müll. Arg. , *Flora*, Regensburg 67: 465(1884) .

≡ *Caloplaca elegans* var. *laxa*(Müll. Arg.) Jatta, *Nuovo G. bot. ital*. 9: 476(1902) .

On the ground & on rocks among mosses.

Beijing(Moreau et Moreau, 1951, p. 194) , Shaanxi(Jatta, 1902, p. 479; Zahlbr. , 1930 b, p. 221) .

Zeroviella mandschurica(A. Zahlbr.) S. Y. Kondr. & Hur 东北假杯点黄

in Kondratyuk, Kim, Yu, Jeong, Jang, Kondratiuk, Zarei-Darki & Hur, *Ukr. Bot. J*. 72(6) : 582(2015) .

≡ *Xanthoria parietina*(L.) Th. Fr. var. *mandschurica* Zahlbr. , *Ann. Mycolog*. 29: 85(1931) .

Type: Liaoning(Dalian) , Asahina no. 554.

≡ *Xanthoria mandschurica*(Zahlbr.) Asahina, *Journ. Jap. Bot*. 29(10) : 290(1954) .

≡ *Rusavskia mandschurica*(Zahlbr.) S. Y. Kondr. & Kärnefelt *Ukr. bot. Zh*. 60(4) : 434(2003) .

On rocks.

Heilongjiang, Neimenggu, Hebei, Shanxi(Asahina, 1954 a, p. 291) Shandong(Zhao ZT et al. , 1998, p. 30; zhao ZT et al. , 1999, p. 427; Zhao ZT et al. , 2002, p. 7; Hou YN et al. , 2008, p. 67; Li Y et al. , 2008, p. 72 as *Xanthoria mandschurica*) .

Lecanoromycetidae, families incertae sediss 茶渍亚纲未定位科

(73) Brigantieaceae Hafellner & Bellem. **锈疣衣科**

Nova Hedwigia 35: 246(1982) [1981] .

Brigantiaea Trevis. **疣衣属**

Spighe Paglie: 7(1853) .

Brigantiaea ferruginea(Müll. Arg.) Kashiw. & Kurok. [as ' ferrugineum'] 红疣衣

in Kashiwadani, *Mem. Natn Sci. Mus*, Tokyo 18: 96(1985) .

≡*Lopadium ferrugineum* Müll. Arg. , *Nuovo G. bot. ital*. 23: 127(1891) .

On mosses.

Anhui, Zhejiang(Xu B. S. , 1989, p. 203) , Taiwan(Kurok. & Kashiwadani, 1977, p. 128; Hafellner 1997, p. 26) .

Brigantiaea leucoxantha(Spreng.) R. Sant. & Hafellner 白黄疣衣

in Hafellner & Bellemère, *Nova Hedwigia* 35(2 & 3) : 246(1982) [1981] .

≡*Lecidea leucoxantha* Spreng. , *K. svenska Vetensk-Akad. Handl*. 8: 46(1820) .

≡*Lopadium leucoxanthum* (Spreng.) Zahlbr. , *Sber. Akad. Wiss. Wien, Math. -naturw. Kl.* , *Abt.* 1 111(1) : 398 (1902) .

On bark of trees.

Taiwan(Zahlbr. , 1933 c, p. 43; Wang & Lai, 1973, p. 92; Kurok. & Kashiwadani, 1977, p. 130) .

Brigantiaea phaeomma(Nyl.) Hafellner 黑色疣衣

Symb. bot. upsal. 32(1) : 61(1997) .

≡ *Lecidea phaeomma* Nyl. , *Lich. Nov. Zeland.* (Paris) : 90(1888) .

Over mosses on a tree trunk in a mixed mountain rainforest.

Zhejiang(Kalb et al. , 2012, p. 38) .

Brigantiaea purpurata(Zahlbr.) Hafellner & Bellem. 紫色疣衣

Nova Hedwigia 35(2 & 3) : 246(1982) [1981] .

≡*Lopadium purpuratum* Zahlbr. , *Bot. Mag.* , *Tokyo* 41: 334(1927) .

Xizang(W. Obermayer, 2004, p. 488) .

Brigantiaea tricolor(Mont.) Trevis. 三色疣衣

Spighe Paglie: 9(1853) .

≡*Biatora tricolor* Mont. , *Annls Sci. Nat.* , *Bot.* , sér. 2 18: 266(1842) .

Taiwan(Hafellner, 1997, p. 70, TNS: 6) .

(74) Coniocybaceae Rchb. [as ' Coniocybeae'] **粉头衣科**

Handb. Nat. Pfl. -Syst. (Dresden) : 132(1837) .

Chaenotheca(Th. Fr.) Th. Fr. **口果粉衣属**

Lich. Arctoi 3: 250(1860) .

≡*Calicium* b *Chaenotheca* Th. Fr. , Öfvers. *K. Svensk. Vetensk. -Akad. Förhandl.* 13: 128(1856) .

Chaenotheca brunneola(Ach.) Müll. Arg. 淡棕口果粉衣

Mém. Soc. Phys. Hist. nat. Genève 16(2) : 360(1862) .

≡*Calicium brunneolum* Ach. , *K. Vetensk-Acad. Nya Handl.* : 279(1816) .

On wood.

Taiwan(Sparrius et al. , 2002, 358) .

Chaenotheca chrysocephala(Ach.) Th. Fr. 金头口果粉衣

Lich. Arctoi 3: 250(1860) . f. chrysocephala

≡*Calicium chrysocephalum* Ach., *Methodus, Suppl.*: 15(1803).

Taiwan(Hsueh et al., 2001, p. 215).

Chaenotheca chrysocephala f. filaris(Ach.) Blomb. & Forssell 丝状变型

Enum. Pl. Scand.: 96(1880).

≡*Calicium chrysocephalum* var. *filare* Ach., *K. Vetensk-Acad. Nya Handl.* 29: 281(1808).

Yunnan(A. Zahlbruckner, 1930 b, p. 33).

Chaenotheca furfuracea(L.) Tibell 麸屑口果粉衣

Beih. Nova Hedwigia 79: 664(1984).

≡*Mucor furfuraceus* L., *Sp. pl.* 2: 1185(1753).

≡*Coniocybe furfuracea*(L.) Ach., *K. Vetensk-Acad. Nya Handl.* 4: 286(1816).

On rotten wood in forests.

Shaanxi, GANSU(Wu J. L., 1987, p. 36), Yunnan(Wang L. S. et al., 2008, p. 536, as *C.* cf. *furfuracea*), Taiwan(Sparrius et al., 2002, 358).

Chaenotheca gracilenta(Ach.) Mattsson & Middelb. 纤口果粉衣

in Middelborg & Mattsson, *Sommerfeltia* 5: 45(1987).

≡*Calicium gracilentum* Ach., *Lich. univ.*: 243(1810).

On rock.

Taiwan(Sparrius et al., 2002, 359).

Chaenotheca stemonea(Ach.) Müll. Arg. 茎口果粉衣

Mém. Soc. Phys. Hist. nat. Genève 16(2): 360(1862).

≡*Calicium trichiale* var. *stemoneum* Ach., *K. Vetensk-Acad. Nya Handl.* 29: 283(1808).

On bark of Abies sp.

Xinjiang(Abdulla Abbas et al., 1993, p. 76; Abdulla Abbas et al., 1994, p. 20; Abdulla Abbas & Wu, 1998, p. 33; Abdulla Abbas et al., 2001, p. 362), Xizang(J. C. Wei & Y. M. Jiang, 1986, p. 11).

(75) Fuscideaceae Hafellner **棕网盘科**

Beih. Nova Hedwigia 79: 278(1984).

Fuscidea V. Wirth & Vězda **棕网盘属**

Beitr. Naturk. Forsch. Südwestdeutschl. 31: 92(1972).

Fuscidea cyathoides(Ach.) V. Wirth & Vězda 杯棕网盘

Beitr. Naturk. Forsch. Südwestdeutschl. 31: 92(1972).

≡*Lichen cyathoides* Ach., *Lich. suec. prodr.* (Linköping): 62(1799) [1798].

[*Lecidea rivulosa* var. *orientalis* Zahlbr. in Handel-Mazzeti, *Symb. Sin.* 3: 108(1930).

Type: Yunnan, Handel-Mazzetti no. 6749. Corticolous.]

Fuscidea cyathoides var. *orientalis*(Zahlbr.) Mas Inoue(1981). -lectotype in WU, isolectotype in W, selected by M. Inoue.

[*Lecidea rivulosa* var. *suborientalis* Zahlbr. in Fedde, *Repertorium*, 33: 37(1933).

Type: Taiwan, Asahina no. 347.

Taiwan(Zahlbr., 1940, p. 341; Wang & Lai, 1973, p. 91; Aptroot and Sparrius, 2003, p. 17).

Corticolous.]

[*Fuscidea cyathoides* var. *orientalis*(Zahlbr.) Mas Inoue(1981, p. 178). -lectotype in WU, isolectotype in W, selected by M. Inoue.

≡*Lecidea rivulosa* var. *orientalis* Zahlbr. in Handel-Mazzeti, *Symb. Sin.* 3: 108(1930).

Type: Yunnan, Handel-Mazzetti no. 6749.

Corticolous.]

[*Lecidea rivulosa* var. *suborientalis* Zahlbr. in Fedde, *Repertorium*, 33: 37(1933).

Yunnan(Zahlbr., 1930, p. 108; M. Inoue, 1981, p. 178).

Fuscidea lygaea(Ach.) V. Wirth & Vězda 暗棕网盘

Beitr. Naturk. Forsch. Südwestdeutschl. 31: 92(1972).

≡*Lecidea lygaea* Ach., *Syn. meth. lich.* (Lund) : 34(1814).

≡*Biatora lygaea* (Ach.) W. Mann, *Lich. Bohem. Observ. Dispos.* : 48(1825).

Hongkong(Rabh. 1873, p. 286; Kremp. 1874b, p. 60 as *Lecidea* "*lygea*"; Zahlbr. 1930 b, p. 106).

Fuscidea mollis(Wahlenb.) V. Wirth & Vězda 软棕网盘

Beitr. Naturk. Forsch. Südwestdeutschl. 31: 92(1972).

≡*Lecidea rivulosa* var. *mollis* Wahlenb., *Fl. Lapp.* : 472(1812).

≡*Lecidea mollis*(Wahlenb.) Nyl., *Lich. Scand.* (Helsinki) : 223(1861).

On rocks.

Shaanxi(Jatta, 1902, p. 478; Zahlbr., 1930 b, p. 107; Hertel, 1977, p. 197), Shandong(Moreau et Moreau, 1951, p. 187; Hertel, 1977, p. 197).

Maronea A. Massal. **拟孢衣属**

Flora, Regensburg 39: 291(1856).

Maronea constans(Nyl.) Hepp 恒拟孢衣

Flecht. Europ. : no. 771(1860).

≡*Lecanora constans* Nyl., *Mém. Soc. Imp. Sci. Nat. Cherbourg* 3: 199(1855).

On *Cryptomeria japonica*.

Taiwan(Aptroot & Sparrius, 2003, p. 28).

Maronea rubra Zahlbr. 红拟孢衣

in Handel-Mazzetti, *Symb. Sin.* 3: 139(1930), & *Cat. Lich. Univ.* 8: 498(1932).

Type: Sichuan, Handel-Mazzetti no. 997 & 2854(syntypes).

On twigs.

Orphniospora Körb. **暗孢衣属**

in Hertlaub & Lindeman, *Zweite Deutsch. Nordpolarfahrt* 2: 81(1874).

Orphniospora moriopsis(A. Massal.) D. Hawksw. 桑葚暗孢衣

Lichenologist 14(2): 135(1982).

≡*Catolechia moriopsis* A. Massal., *Ric. auton. lich. crost.* (Verona) : 83(1852).

On hard, montane rock.

Shaanxi(Zhao ZT et al., 2007, p. 295 – 296).

(76) Lecideaceae Chevall. [as ' Lecideae'] **网衣科**

Fl. gén. env. Paris (Paris) 1: 549(1826).

syn. *Porpidiaceae* Hertel & Hafellner 1984

Amygdalaria Norman **扁桃盘衣属**

Nytt Mag. Natur. 7: 230(1853) [1852].

Amygdalaria aeolotera(Vain.) Hertel & Brodo 高山扁桃盘衣

in Brodo & Hertel, *Herzogia* 7(3 – 4) : 501(1987).

≡*Lecidea aeolotera* Vain., *Ann. Acad. Sci. fenn.*, Ser. A 15(no. 6) : 137(1921).

≡*Huilia aeolotera*(Vain.) Hertel, *Herzogia* 3(2) : 372(1975) [1973 – 4].

= *Huilia alpina* Zahlbr., in Handel-Mazzetti, *Symb. Sinic.* 3: 81 – 82(1930).

Type: Sichuan, Huili district, Handel-Mazzetti no. 1007.

= *Huilia insularis* Zahlbr., in Handel-Mazzetti, *Symb. Sinic.* 3: 81(1930).

Type: Sichuan, Handl-Mazzetti no. 2646.

= *Lecidea caloplacoides* Zahlbr., in Handel-Mazzetti, *Symb. Sinic.* 3: 93(1930).

Type: Yunnan, Handel-Mazzetti no. 9996.

On quartziferous rock.

Sichuan(Zahlbr., 1930 b, p. 81 & 1932a, p. 299; Hertel, 1977, p. 200).

Amygdalaria pelobotryon(Wahlenb.) Norman [as '*pelobotrya*'] 串扁桃盘衣

Nytt Mag. Natur. 7: 320(1853) [1852].

≡*Urceolaria pelobotryon* Wahlenb., in Acharius, *Methodus*, Sectio prior: 31(1803).

On large pebbles of volcanic rock on ground.

Jilin(Mt. Changbai shan, Hertel & Zhao, 1982, p. 143).

Bellemerea Hafellner & Cl. Roux **锈盘衣属**

in Clauzade & Roux, *Bull. Soc. Bot. Centre-Ouest, Nouv*. Sér. 15: 129(1984).

Bellemerea cinereorufescens(Ach.) Clauzade & Cl. Roux 淡锈盘衣

Bull. Soc. Bot. Centre-Ouest, Nouv. Sér. 15: 129(1984).

≡*Urceolaria cinereorufescens* Ach., *Lich. univ*.: 677(1810).

≡*Aspicilia cinereorufescens*(Ach.) A. Massal., *Ric. auton. lich. crost*. (Verona): 37, fig. 62(1852).

≡*Lecanora cinereorufescens*(Ach.) Hepp, *Flecht. Europ*.: no. 625(1860).

On siliceous rocks. On exposed rock in mountain area. Northern temperate.

Shaanxi(Jatta, 1902, p. 476; Zahlbr., 1930 b, p. 158), Hongkong(Aptroot & Seaward, 1999, p. 67).

Bellemerea cupreoatra(Nyl.) Clauzade & Cl. Roux 铜黑锈盘衣

Bull. Soc. bot. Centre-Ouest, Nouv. sér. 15: 129(1984).

≡*Lecanora cupreoatra* Nyl., *Not. Sällsk. Fauna et Fl. Fenn. Förh*., Ny Ser. 8: 181(1866).

≡*Aspicilia cupreoatra*(Nyl.) Arnold, *Flora, Regensburg* 53: 470(1870).

On rocks.

Shaanxi(Jatta, 1902, p. 476; Zahlbr., 1930 b, p. 150).

Clauzadea Hafellner & Bellem. **封衣属**

Beih. Nova Hedwigia 79: 319(1984).

Clauzadea monticola(Ach.) Hafellner & Bellem. 山封衣

Beih. Nova Hedwigia 79: 319(1984).

≡*Lecidea monticola* Ach., in Schaerer, *Lich. helv. spicil*. 4 – 5: 161(1833).

On rock.

Xinjiang(Zhang L. L. et al. 2015, p. 900).

Farnoldia Hertel **粒衣属**

Mitt. bot. StSamml., Münch. 19: 442(1983).

Farnoldia micropsis(A. Massal.) Hertel 微粒衣

Mitt. bot. StSamml., Münch. 19: 443(1983).

≡*Lecidea micropsis* A. Massal., *Atti Inst. Veneto Sci. lett., ed Arti*, Sér. 3 2: 368, tab. II, fig. 1 – 14(1856).

On rocks.

Xinjiang(Zhao X. X. et al., 2016. pp. 1711 – 1712).

Immersaria Rambold & Pietschm. **沉衣属**

in Rambold, *Biblthca Lichenol*. 34: 239(1989).

Immersaria athroocarpa(Ach.) Rambold & Pietschm. 密果沉衣

in Rambold, *Biblthca Lichenol*. 34: 240(1989).

≡*Lichen athroocarpus* Ach., *Lich. suec. Prodr*. (Linköping): 77(1799) [1798].

Sichuan, Xizang(Obermayer, 2004, p. 500), Taiwan(Aptroot and Sparrius, 2003, p. 23).

≡*Porpidia athroocarpa*(Ach.) Hertel & Rambold *in* Hertel, *Lecideaceae exsiccatae*, Fascicle VIII(141 – 160) (München) (141 – 160): 8(1985).

≡*Lecidea athroocarpa*(Ach.) Ach. [as 'athrocarpa'], *Methodus, Sectio prior* (Stockholmiæ): 41(1803).

=*Lecidea schitakensis* Zahlbr., in Handel-Mazzetti, *Syb. Sin*. 3: 99(1930).

Type: Yunnan, Handel-Mazzetti no. 3556.

On rock.

Jilin(Hertel & Zhao, 1982, p. 146), Xinjiang(Abdulla Abbas & Wu, 1998, p. 123 as *Porpidia athroocarpa*; Abdulla Abbas et al., 2001, p. 367), Yunnan(Hertel, 1977, p. 228).

Immersaria iranica Valadb., Sipman & Rambold 伊朗沉衣

Lichenologist 43(3): 204(2011).

On rock.

Xinjiang(Zhang L. L. et al. 2015, pp. 901 – 902).

Immersaria usbekica(Hertel) M. Barbero 乌斯沉衣

Nav.-Ros. & Cl. Roux, Bull. Soc. Linn. Provence 41: 140(1990).

On rock.

Xinjiang(Zhang L. L. et al. 2015, pp. 903 – 904).

Lecidea Ach. **网衣属**

Methodus, Sectio prior: xxx, 32(1803).

Lecidea acarocarpa Zahlbr. 小果网衣

in Handel-Mazzetti, *Symb. Sin.* 3: 104(1930) & *Cat. Lich. Univ.* 8: 365(1932).

Type: Sichuan, Handel-Mazzetti no. 5311.

Corticolous.

Lecidea albohyalina(Nyl.) Th. Fr. 白明网衣

Lich. Scand. (Upsaliae) 1(2): 431(1874).

≡*Lecidea anomala* f. *albohyalina* Nyl., *Lich. Scand.* (Helsinki): 203(1861).

Shaanxi(Jatta, 1902, p. 478), Hongkong(Thrower, 1988, p. 112).

Lecidea atrobrunnea(DC.) Schaer. 黑棕网衣

Lich. helv. spicil. 3: 134(1828).

≡*Rhizocarpon atrobrunneum* DC., in Lamarck & de Candolle, *Fl. franç.*, Edn 3(Paris) 2: 367(1805).

Xinjiang(Abdulla Abbas & Wu, 1998, p. 84; Abdulla Abbas et al., 2001, p. 364).

Lecidea auriculata Th. Fr. 耳盘网衣

Lich. Arctoi 3: 213(1861) [1860].

On rocks.

Xinjiang(Abdulla Abbas et al., 1993, p. 78; Abdulla Abbas et al., 1994, p. 21; Abdulla Abbas & Wu, 1998, p. 84; Abdulla Abbas et al., 2001, p. 364), Xizang(Paulson, 1925, p. 193; J. C. Wei & Chen, 1974, p. 175; Hertel. 1977, p. 233; J. C. Wei & Jiang, 1981, p. 1146 & 1985, p. 29).

Lecidea bacculans Zahlbr. 杆网衣

in Handel-Mazzetti, *Symb. Sin.* 3: 104(1930) & *Cat. Lich. Univ.* 8: 366(1932).

Syntype: Yunnan, Handel-Mazzetti nos. 6259, 6754.

On trunk of *Illicium yunnanense*.

Lecidea berengeriana(A. Massal.) Nyl. 海蓝网衣

Not. Sällsk. Fauna et Fl. Fenn. Förh., *Ny Ser.* 8: 144(1866).

≡*Biatora berengeriana* A. Massal., *Ric. Auton. Lich. crost.* (Verona): 128(1852).

On mosses.

Gansu, Qinghai(Zhang LL et al., 2010, p. 446).

Lecidea bohlinii H. Magn. 包氏网衣

Lich. Centr. Asia, 1: 48(1940).

Type: Gansu, Bohlin no. 62 in S(!)

Gansu(H. Magn., 1940, p48; Hertel. 1977, p. 237), Xinjiang(Abdulla Abbas et al., 1993, p. 78; Abdulla Abbas & Wu, 1998, p. 84; Abdulla Abbas et al., 2001, p. 364).

Lecidea cacaotina Zahlbr. 可可网衣

in Fedde, *Repertorium* 33: 39(1933) & *Cat. Lich. Univ.* 10: 10(1940).

Type: Taiwan, Asahina no. 147.

Taiwan(M. Lamb, 1963, p. 341; Wang & Lai, 1973, p. 91)

Lecidea confluens(Weber) Ach. 汇合网衣

Methodus, Sectio prior: 14(1803).

≡*Lichen confluens* Weber, *Spicil. Fl. uperfi.*: 180(1778).

On rock.

Qinghai(Zhang LL et al., 2010, pp. 446 – 447).

Lecidea conspersa f. sorediifera Kremp. 散生网衣,粉芽变型

in Hedwigia 13(4): 60(1874)(nom. nud.).

Guangdong(Krphbr. 1874 b, p. 60).

Notes. The current name of *Lecidea conspersa* Fée is *Piccolia conspersa* (Fée) Hafellner(1995), and the systematic position of *Lecidea conspersa* f. *sorediifera* Kremp. remains to be revised and validated in the future.

Lecidea degeliana Hertel 戴盖氏网衣

Herzogia 2: 41(1970).

On rocks.

Jilin(Hertel & Zhao, 1982, p. 146).

Lecidea diducens Nyl. 分网衣

Flora, Regensburg 48: 148(1865).

Yunnan(Zhang LL et al., 2012, pp. 446 – 448).

Lecidea djagensis Zahlbr. 川网衣

Hedwigia B. 74: 199(1934) & *Cat. Lich. Univ.* 10: 332(1940).

Type: Sichuan, collected by F. Rock.

Corticolous.

Lecidea erythrophaea Flörke ex Sommerf. 红网衣

Suppl. Fl. Lapp. (Oslo): 163(1826).

On *Tsuga formosana* branches.

Taiwan(Aptroot and Sparrius, 2003, p. 26).

Lecidea fuliginosa Taylor 煤网衣

in Mackay, *Fl. Hibern.* 2: 131(1836).

On shale.

Taiwan(Aptroot and Sparrius, 2003, p. 26).

Lecidea fuscoatra(L.) Ach. 乌棕网衣

Methodus, Sectio prior: 44(1803).

≡*Lichen fuscoater* L., *Sp. pl.* 2: 1140(1753).

=*Lecidea grisella* f. *mosigii*(Ach.) Zahlbr., *Cat. Lich. Univers.* 3: 594(1925).

On vertical site of the dust covered rock by road.

Jilin(Hertel) & Zhao, 1982, p. 147, cited as Lecidea fuscoatra(Ach.) Ach.

Lecidea granifera var. leucotropa(Nyl.) Zahlbr. 白色变型

Cat. Lich. Univers. 3: 769(1925).

≡*Lecanora granifera* var. *leucotropa* Nyl., *Bull. Soc. Linn. Normandie*, sér. 2 3: 267(1869).

Corticolous.

Yunnan(Zahlbr., 1930 b, p. 109).

Lecidea hunana Zahlbr. 湖南网衣

in Handel-Mazzetti, *Symb. Sin.* 3: 103(1930) & *Cat. Lich. Univ.* 8: 372(1932) .

Type: Hunan, Handel-Mazzetti, no. 12263.

On *Quercus glandulifera*.

Lecidea hypopta Ach. 腹网衣

Methodus, Sectio prior: 61(1803) .

On *Pseudotsuga wilsoniana* wood.

Taiwan(Aptroot and Sparrius, 2003, p. 26) .

Lecidea kansuensis H. Magn. 甘肃网衣

Lich. Centr. Asia 1: 49(1940) .

Type: Gansu, Tsagaling 29/X/1930, collected by Hummel at about 2050 m, preserved in S(!) .

On twigs of *Picea* with *Usnea longissima*.

Xinjiang(Abdulla Abbas et al. , 1993, p. 78; Abdulla Abbas & Wu, 1998, p. 85; Abdulla Abbas et al. , 2001, p. 364) , Gansu H. Magn. , 1940, p. 49) .

Lecidea lactea Flörke ex Schaer. 青网衣

Lich. helv. spicil. 3: 127(1828) .

= *Lecidea cyanea*(Ach.) Röhl. , *Deutschl. Fl.* (Frankfurt) 3(2) : 32(1813) .

≡ *Lecidea lapicida* subsp. *cyanea* Ach. , *Methodus*, Sectio prior: 38(1803) .

Hubei(Chen, Wu & J. C. Wei, 1989, p. 425)

Lecidea lapicida(Ach.) Ach. 岩网衣

Methodus, Sectio prior: 37(1803) . var. lapicida 原变种

≡ *Lichen lapicida* Ach. , *Lich. suec. Prodr.* (Linköping) : 61(1799) [1798] .

On the horizontal surface of a small boulder.

Jilin(Hertel & Zhao, 1982, p. 147) , Taiwan(Aptroot and Sparrius, 2003, p. 26) .

var. pantherina(DC.) Ach. 泛体变种

K. Vetensk-Acad. Nya Handl. 29: 232(1808) .

≡ *Patellaria lapicida* var. *pantherina* DC. , in Lamarck & de Candolle, *Fl. uper.* , Edn 3(Paris) 2: 1 – 600 (1805) .

On rock.

Sichuan(Zhang LL et al. , 2010, p. 448) .

Lecidea leprothalla Zahlbr. 屑体网衣

in handel-Mazzetti, *Symb. Sin.* 3: 103(1930) & *Cat. Lich. Univ*. 8: 372(1932) .

Type: Yunnan, Handel-Mazzetti no. 6584.

On *Keteleeria* sp.

Lecidea lithophila(Ach.) Ach. 石生网衣

Syn. meth. Lich. (Lund) : 14(1814) .

≡ *Lecidea lapicida* var. *lithophila* Ach. , *Lich. univ.* : 1 – 696(1810) .

On rocks(filed under Amygdalaria pelobotryon) .

Jilin(Hertel & Zhao, 1982, p. 148) .

Lecidea loseana Zahlbr. 大孢网衣

in handel-Mazzetti, *Symb. Sin.* 3: 102(1930) & Cat. Lich. Univ. 8: 373(1932) .

Type: Sichuan, Handel-Mazzetti no. 1424.

On *Rhododendron* sp.

Lecidea ochroleprosa Zahlbr. 赭癞屑网衣

in Fedde, *Repertorium*, 33: 37(1933) & *Cat. Lich. Univ*. 10: 339(1940) .

Type: Taiwan, Asahina, no. 163.

Corticolous.

Taiwan(Wang & Lai, 1973, p. 91)

Lecidea ochrorufa H. Magn. 赭红网衣

Lich. Centr. Asia 1: 57(1940) .

Type: Gansu, no. 40 b in S.

Gansu(G. Schneider, 1979, p. 243) .

On calcareous soil.

Lecidea opaca var. crocea(B. de Lesd.) Zahlbr. 暗网衣,番红色变种

Cat. Lich. Univers. 3: 885(1925) .

≡*Psora opaca* var. *crocea* B. de Lesd. , Bull. Soc. bot. Fr. 70: 281(1922)

On rock.

Sichuan(Zahlbr. , 1930 b, p. 110; Hertel, 1977, p. 191) .

Lecidea paratropoides Müll. Arg. 类热带网衣

Flora, Regensburg 57: 348(1874) .

= *Lecidea austromongolica* H. Magn. [as ' austro-mongolica'] , *Lichens Central Asia* 2: 23(1944) .

Type: Neimenggu, Bohlin no. 157 in S(!) .

Neimenggu(Hertel, 1977, p. 262) .

On hard siliceous rock.

Lecidea piperina Zahlbr. 胡椒网衣

in Fedde, *Repertorium*, 33: 40(1933) & *Cat. Lich. Univ*. 10: 340(1940) .

Type: Taiwan, Asahina no. 164.

Taiwan(Wang & Lai, 1973, p. 91) .

Lecidea promiscens Nyl. 杂网衣

On rock.

Yunnan(Zhang LL et al. , 2012, p. 448) .

Lecidea promixta Nyl. 原混网衣

On rock.

Shaanxi(Hu Ling et al. , 2014, p. 88) .

Lecidea sinensis H. Magn. 中华网衣

Lich. Centr. Asia 1: 58(1940) .

Type: Gansu, Bohlin, geol. Sample no. 650 in S(!) .

Gansu(Lamb, 1963, p. 379; Hertel, 1977, p. 277) .

On hard stone.

Lecidea subaenea Zahlbr. 亚铜网衣

in Handel-Mazzetti, *Symb. Sin*. 3: 92(1930) .

Type: Yunnan, Handel-Mazzetti no. 3323.

Yunnan(Zahlbr. , 1932, p. 359; Hertel, 1977, p. 314) .

On sandstone.

Lecidea subconcava H. Magn. 亚凹网衣

Lich. Centr. Asia 1: 59(1940) .

Type: Gansu, Bohlin no. 42 b(S) .

Xinjiang(Abdulla Abbas & Wu, 1998, p. 85; Abdulla Abbas et al. , 2001, p. 364) , Gansu(G. Schneider, 1979, p. 247) .

Lecidea subelevata H. Magn. 微凸网衣

Lich. Centr. Asia 1: 52(1940) .

Type: Gansu, Bohlin no. 88 d.

On both calciferous and non calciferous stone.

Xinjiang(Abdulla Abbas et al. , 1993, p. 78; Abdulla Abbas & Wu, 1998, p. 85; Abdulla Abbas et al. , 2001, p. 364) , Qinghai, Gansu(H. Magn. 1940, p. 52; Hertel, 1967, p. 123 & 1977, p. 296) .

Lecidea tephraea(Tuck.) Zahlbr. 灰白网衣

Cat. Lich. Univ. III: 838(1925) .

≡*Biatora tephraea* Tuck. , *Lichens of California*(Berkeley) : 31(1866) .

Corticolous.

China(prov. not indicated: Zahlbr. , 1925, p. 838) .

Lecidea tessellata Flörke 方斑网衣

On Ca-influenced rocks.

Qinghai(Obermayer, 2004, p. 501) , Xinjiang(Abdulla Abbas & Wu, 1998, p. 86; Abdulla Abbas et al. , 2001, p. 364; Obermayer, 2004, p. 501) .

= *Cladopycnidium sinense* H. Mang. *Lich. Centr. Asia* 1: 61(1940) .

Type: Qinghai, Bohlin, collected in 19/IV/1932, preserved in S.

= *Cladopycnidium sinense* f. *chiodectonoides* H. Magn. *Lich. Centr. Asia* 1: 62(1940) .

Type: Gansu, Bohlin no. 88 b in S.

= *Lecidea percrassata* H. Magn. *Lich. Centr. Asia* 1: 52(1940) .

Type Gansu, Bohlin no. 47f in S.

= *Lecidea pavimentans* H. Magn. *Lich. Centr. Asia* 1: 50(1940) .

Type: Gansu, Bohlin no. 86 d(holotype, not seen) ; paratypes: Bohlin nos. 35 b, 52 & Bexell no. 24 in S.

= *Lecidea pavimentans* H. Magn. var. *incrassata* H. Magn. *Lich. Centr. Asia* 1: 51(1940) .

Type: Gansu, at about 3000 m, 18/XII/1931, Bohlin no. 43 d in S.

= *Lecidea pavimentans* f. *ochracea* H. Magn. *Lich. Centr. Asia* 1: 51(1940) .

Type: Gansu, at about 3000 m, 19/XII/1931, Bohlin no. 62 in S(!) .

= *Lecidea pavimentans* f. *perfecta* H. Magn. *Lich. Centr. Asia* 1: 51(1940) .

Type: Gansu, at about 3100 m, 22/XII/1931, Bohlin no. 63 c in S(!) .

On rocks.

Xinjiang(Abdulla Abbas et al. , 1993, p. 78; Abdulla Abbas & Wu, 1998, p. 86; Obermayer, 2004, p. 501) , Qinghai(H. Magn. , 1940, p. 61; Obermayer, 2004, p. 501) , Gansu(H. Magn. 1940, pp. 50, 51, 52, 62, ; Lamb, 1963, p. 372, 373; Poelt & Wirth, 1968, p. 246; Golubk. 1970, p. 230 & 1973, p. 212; Hertel, 1977, pp. 282, 287(TUR-Vain. no. 24. 635, as *Lecidea* sp.) .

Lecidea tritula Nyl. 粉网衣

in Nyl. & Cromb. *Journ. Linn. Soc. London, Bot.* 20: 64(1883) .

Type: from Shanghai.

Shanghai(Hue, 1891, p. 109; Zahlbr. , 1925, p. 840 & 1930 b, p. 102) .

On bark of trees.

Lecidoma Gotth. Schneid. & Hertel **圆顶衣属**

in Hertel, *Herzogia* 5(3 – 4) : 460(1981) .

Lecidoma demissum(Rutstr.) Gotth. Schneid. & Hertel 垂圆顶衣

in Hertel, *Herzogia* 5(3 – 4) : 460(1981) .

≡*Lichen demissus* Rutstr. , *Spicil. Pl. cryptog. Suecicae*(Aboae) : 8(1794) .

≡*Lecidea demissa*(Rutstr.) Ach. , *Methodus, Sectio prior* (Stockholmiæ) : 81(1803) .

On soil as well as on rotten wood(e. g. *Juniperus*) .

Yunnan(Zahlbr. , 1930 b, p. 109) , Xizang, Sichuan(Obermayer, 2004, p. 501) , Jiangsu(Xu B. S. , 1989, p. 203) .

Mycobilimbia Rehm **菌盘衣属**

in Winter, *Rabenh. Krypt. -Fl.* , Edn 2(Leipzig) 1. 3(32) : 295, 327(1890) [1896] .

[*Bilimbia* De Not., *G. bot. ital.* 2(1. 1): 190(1846), nom. illegit., Art. 53. 1, non *Bilimbia* Rchb. 1837(Averrhoaceae)]

Mycobilimbia sabuletorum(Schreb.) Hafellner 沙地菌盘衣

Beih. Nova Hedwigia 79: 310(1984).

≡*Lichen sabuletorum* Schreb., *Spic. fl. lips.* (Lipsiae): 134(1771).

≡*Bilimbia sabuletorum*(Schreb.) Arnold, *Verh. zool. -bot. Ges. Wien* 19: 637(1869).

Xinjiang(Abdulla Abbas & Wu, 1998, p. 122; Abdulla Abbas et al., 2001, p. 365).

Paraporpidia Rambold & Pietschm. **准衣属**

in Rambold, *Biblthca Lichenol.* 34: 243(1989).

Paraporpidia leptocarpa(Nyl.) Rambold & Hertel 细准衣

in Rambold, *Biblthca Lichenol.* 34: 250(1989).

≡*Lecidea leptocarpa* Nyl., in Babington & Mitten in Hooker, *Bot. Antarct.* Voy., III, Fl. Tasman. 2: 352(1859)[1860].

=*Lecidea lividonigra* Zahlbr. in Handel-Mazzetti, *Symb. Sin.* 3: 98(1930).

Type: Yunnan, Handel-Mazzetti, no. 108(holotype) in WU.

≡*Tremolecia lividonigra*(Zahlbr.) Hertel, *Khumbu Himal* 6(3): 352(1977).

=*Lecidea formosae* Zahlbr. in Fedde, *Repertprium* 33: 35(1933).

Type: Taiwan, Faurie no. 48(holotype) in W(not seen).

On sand stone.

=*Lecidea tuberculans* Zahlbr. in Handel-Mazzetti, *Symb. Sin.* 3: 97(1930).

Type: Yunnan, Handel-Mazzetti no. 1386(holotype) in WU and isotype in W(not seen).

≡*Tremolecia tuberculata*(Zahlbr.) Hertel, Khumbu Himal 6(3): 357(1977).

Porpidia Körb. **假网衣属**

Syst. Lich. germ. (Breslau): 221(1855).

=*Huilia* Zahlbr., in Handel-Mazzetti, *Syb. Sin.* 3: 80(1930).

Porpidia aeolotera(Vain.) Hertel 高山假网衣

Beih. Nova Hedwigia 79: 437(1984).

≡*Lecidea aeolotera* Vain., *Ann. Acad. Sci. fenn.*, Ser. A 15(no. 6): 137(1921).

≡*Huilia aeolotera*(Vain.) Hertel, *Herzogia* 3(2): 372(1975)[1973 –4].

≡*Amygdalaria aeolotera* (Vain.) Hertel & Brodo, in Brodo & Hertel, *Herzogia* 7(3 –4): 501(1987).

=*Huilia alpina* Zahlbr., in Handel-Mazzetti, *Syb. Sin.* 3: 81(1930).

Type: Sichuan, Huili district, Handel-Mazzetti no. 1007.

=*Huilia insularis* Zahlbr., in Handel-Mazzetti, *Syb. Sin.* 3: 81(1930).

Type: SICHUAN, Handl-Mazzetti no. 2646.

=*Lecidea caloplacoides* Zahlbr., in Handel-Mazzetti, *Syb. Sin.* 3: 93(1930).

Type: YUNNAN, Handel-Mazzetti no. 9996.

On quartziferous rock.

Sichuan(Zahlbr., 1930 b, p. 81 & 1932a, p. 299; Hertel, 1977, p. 200).

Porpidia albocaerulescens(Wulfen) Hertel & Knoph 白兰假网衣

in Hertel, *Beih. Nova Hedwigia* 79: 433(1984).

≡*Lichen albocaerulescens* Wulfen [as ‘albo-cærulescens’], in Jacquin, *Collnea bot.* 2: 184(1791)[1788]

≡*Lecidea albocaerulescens*(Wulfen) Ach., *Methodus, Sectio prior*(Stockholmiæ): 52(1803).

≡*Huilia albocaerulescens*(Wulfen) Hertel, *Herzogia* 3(2): 373(1975)[1973 –74].

On rocks.

Shandong(Moreau et Moreau, 1951, p. 187), Jiangsu(Wu J. N. & Xiang T. 1981, p. 4; Xu B. S., 1989, p. 202), Anhui, Zhejiang(Xu B. S., 1989, p. 202), Yunnan, Sichuan, Fujian(Zahlbr., 1930 b, p. 94), Taiwan

(Asahina & M. Sato in Asahina, 1939, p. 667; Wang & Lai, 1973, p. 91).

var. albocaerulescens 原变种

= *Lecidea ochropolia* Zahlbr., in Handel-Mazzetti, *Syb. Sin*. 3: 96(1930).

Type locality: Yunnan, Handel-Mazzetti, no. 204.

= *Lecidea galactochrysea* Zahlbr., in Handel-Mazzetti, *Syb. Sin*. 3: 105(1930).

Type: Yunnan, Handel-Mazzetti, no. 9608.

= *Lecidea albuginosa* var. *cinereofuscescens* Hue, Bull. Soc. Bot. France, 36: 175(1889).

Type locality: Yunnan, P. Delavay no. 2897 bis.

On rocks.

Yunnan(Hue, 1889, p. 175; Zahlbr., 1930 b, pp. 96, 105 & 1932a, p. 351; Hertel, 1977, pp. 205 – 207), Guizhou(Hertel, 1977, p. 207), Hebei(Hertel, 1977, p. 207), Taiwan(Zahlbr., 1933 c, p. 35; Hertel, 1977, p. 207).

var. polycarpiza(Vain.) Rambold & Hertel 多果变种

in Rambold, *Biblthca Lichenol*. 34: 282(1989).

≡ *Lecidea polycarpiza* Vain., *Ann. Acad. Sci. fenn.*, Ser. A 15(6): 136(1921).

≡ *Huilia albocaerulescens* var. *polycarpiza*(Vain.) Hertel, Herzogia 3(2): 373(1975) [1973 – 4].

= *Lecidea dalianensis* Zahlbr., in Handel-Mazzetti, *Syb. Sin*. 3: 93(1930).

Type: Sichuan, Handel-Mazzetti, no. 1770.

= *Lecidea polyasca* Zahlbr., in Handel-Mazzetti, *Syb. Sin*. 3: 106(1930).

Type: Yunnan, Handel-Mazzetti, no. 139.

= *Lecidea rosaceocinerea* Zahlbr., in Handel-Mazzetti, *Syb. Sin*. 3: 94(1930).

Type: Sichuan, Handel-Mazzetti, no. 1585.

On rocks.

Sichuan, Yunnan, Fujian, Taiwan(Hertel, 1977, p. 209).

Porpidia cervinopungens(Zahlbr.) Hertel & Knogh 鹿斑假网衣

in J. C. Wei, *Enum. Lich. China*(Beijing): 209(1991).

≡ *Lecidea cervinopungens* Zahlbr., in Handel-Mazzetti, *Symb. Sinic*. 3: 91(1930).

Type: Yunnan, Handel-Mazzetti no. 9995.

On quartziferous rock.

≡ *Huilia cervinopungens*(Zahlbr.) Hertel, *Khumbu Himal* 6(3): 210(1977).

Yunnan(Hertel, 1977, p. 210).

Porpidia chungii(Zahlbr.) Hertel 钟氏假网衣

Beih. Nova Hedwigia 79: 437(1984).

≡ *Lecidea chubgii* Zahlbr., in Handel-Mazzetti, *Symb. Sinic*. 3: 96(1930).

Type: Fujian, Chung F 608 a.

Fujian(Hertel, 1977, p. 359).

On sandstone.

Porpidia cinereoatra(Ach.) Hertel & Knoph 黑白假网衣

in Hertel, *Beih. Nova Hedwigia* 79: 437(1984).

≡ *Lecidea cinereoatra* Ach., *Lich. univ.*: 1 – 696(1810).

On granite and shale.

Taiwan(Aptroot and Sparrius, 2003, p. 35).

Porpidia crustulata(Ach.) Hertel & Knoph 壳假网衣

in Hertel, *Beih. Nova Hedwigia* 79: 435(1984).

≡ *Lecidea parasema* d *crustulata* Ach., *Lich. univ.*: 176(1810).

≡ *Lecidea crustulata*(Ach.) Spreng., *Syst. veg.*, Edn 16 4(1): 258(1827).

≡*Huilia crustulata*(Ach.) Hertel, *Herzogia* 3(2) : 373(1975) [1973 – 74].

= *Lecidea galucosarca* Zahlbr. , in Handel-Mazzetti, *Symb. Sinic.* 3: 92(1930).

Type: Sichuan, Handel-Mazzetti, no. 7555.

On rocks.

Jilin(Hertel & Zhao, 1982, p. 144), Sichuan(Hertel, 1971, p. 240 & 1977, p. 211), Taiwan(Aptroot and Sparrius, 2003, p. 36).

Porpidia flavicunda(Ach.) Gowan 黄色假网衣

Bryologist 92(1) : 43(1989).

≡*Lecidea flavicunda* Ach. , *Lich. univ.* : 1 – 696(1810).

Sichuan, Yunnan, Xizang(Wang Xin Yu et al. , 2012, p. 623).

Porpidia grisea Gowan 灰色假网衣

Bryologist 92(1): 48(1989).

Yunnan, Xizang(Hu Ling et al, 2014, pp. 88 – 90).

Porpidia hydrophila(Fr.) Hertel & A. J. Schwab 嗜水假网衣

in Hertel, *Beih. Nova Hedwigia* 79: 437(1984).

≡*Lecidea hydrophila* Fr. , *K. Vetensk-Acad. Nya Handl.* : 256(1822).

≡*Huilia hydrophila*(Fr.) Hertel, *Herzogia* 3(2) : 374(1975) [1973].

= *Lecidea chondrospora* Zahlbr. , in Handel-Mazzetti, *Symb. Sinic.* 3: 99(1930).

Type: Yunnan, Handel-Mazzetti, no. 9994.

Yunnan(Hertel, 1977, p. 217).

Porpidia hypostictica L. Hu & Z. T. Zhao 拟大果假网衣

The *Lichenologist* 48(3) : 232(2016).

Type: China, Jilin Prov. , Changbai Co. , Mt. Changbai, 41°35′4 95″N, 127°51′51· 69″E, alt. 1300 m, on rock, 25 July 2014, Ling Hu 20141385(holotype-SDNU; GenBank: KR069081).

On rock.

Porpidia macrocarpa(DC.) Hertel & A. J. Schwab 腹斑假网衣

in Hertel, *Beih. Nova Hedwigia* 79: 437(1984).

≡*Patellaria macrocarpa* DC. , in Lamarck & de Candolle, *Fl. uper.* , Edn 3(Paris) 2: 347(1805).

≡*Huilia macrocarpa*(DC.) Hertel, *Herzogia* 3(2) : 374(1975) [1973 – 74].

≡*Lecidea aquatilis* (Vain.) Vain. , *Lichenogr. Fenn.* 4: 164(1934).

≡*Lecidea macrocarpa* f. *aquatilis* Vain. , *Meddn Soc. Fauna Flora fenn.* 10: 72(1883).

≡*Lecidea convexa* var. *hydrophila* f. *aquatilis*(Vain.) Zahlbr. *Cat. Lich. Univ.* 3: 547(1925).

Taiwan(Aptroot and Sparrius, 2003, p. 36).

f. nigrocruenta(Anzi) Fryday 黑血色变型

Lichenologist 37(1) : 16(2005).

≡*Lecidea nigrocruenta* Anzi, *Comm. Soc. Crittog. Ital.* 2(1) : 18(1864).

≡*Huilia nigrocruenta* (Anzi) Hertel, *Herzogia* 3(2) : 374(1975) [1973 – 4].

≡*Porpidia nigrocruenta*(Anzi) Diederich & Sérus. *in* Diederich, Sérusiaux, Aptroot & Rose, *Dumortiera* 42: 28 (1988).

Jilin(Hertel & Zhao, 1982, p. 145)

Porpidia platycarpoides(Bagl.) Hertel 宽果假网衣

in Nimis & Poelt, *Stud. Geobot.* 7(suppl. 1) : 187(1987).

≡*Lecidea platycarpoides* Bagl. , *Nuovo G. bot. ital.* 11: 99(1879).

Misapplied name:

Lecidea albocaerulescens auct. non(Wulfen) Ach. in Thrower, 1988, pp. 17, 112, mentioned by Aptroot & Seaward, 1999, p. 89.

On exposed or sheltered granitic rock. Northern temperate.

Hobgkong(Thrower, 1988, pp. 17, 112; Aptroot & Seaward, 1999, p. 89; Aptoot & Sipman, 2001, p. 338), Taiwan(Aptroot and Sparrius, 2003, p. 36).

Porpidia shangrila Xin Y. Wang & Lu L. Zhang 香格里拉假网衣

in Xin Y. Wang et al., *Lichenologist* 44(5): 620 – 622(2012).

Typus: China, Yunnan, Shangri-La Co., Mt. Shika alt. 3650 m, on rock, 2 November 2008, Hai Ying Wang 20081939(SDNU—holotypus; KUN—isotypus).

Porpidia soredizodes(Lamy) J. R. Laundon 粉芽假网衣

J. Linn. Soc., Bot. 101(1): 104(1989).

≡*Lecidea meiospora* var. *soredizodes* Lamy, *Bull. Soc. Bot. Fr.* 30: 410(1883).

Guizhou(Wang Xin Y. et al, 2012, p. 623).

Porpidia speirea(Ach.) Kremp. 圈型假网衣

Denkschr. Kgl. Bayer. Bot. Ges., Abt. 2 4: 210(1861).

≡*Lichen speireus* Ach., *Lich. suec. Prodr.* (Linköping): 59(1799) [1798].

On rocks.

Xinjiang(Abdulla Abbas et al., 1994, p. 22; Abdulla Abbas & Wu, 1998, p. 123; Abdulla Abbas et al., 2001, p. 367).

Porpidia squamosa Xin Y. Wang & Lu L. Zhang 鳞假网衣

in Xin Y. Wang et al., *Lichenologist* 44(5): 622 – 623(2012).

Typus: China, Qinghai, Menyuan Co., Lenglongling, alt. 3650 m, on rock, 7 August 2007, Yuan Da Du 20070991(SDNU —holotypus; KUN—isotypus).

Xinjiang, Qinghai(Xin Y. Wang et al., 2012, p. 623).

Porpidia superba(Körb.) Hertel & Knoph 超假网衣

in Hertel, *Beih. Nova Hedwigia* 79: 438(1984) f. superb

≡*Lecidea superba* Körb., *Syst. lich. germ.* (Breslau): 248(1855).

≡*Huilia superba*(Körb.) Hertel, *Mitt. bot. StSamml.*, Münch. 12: 123(1975).

= *Lecidea macrocarpa* var. *subplicata* Zahlbr., in Handel-Mazzetti, *Symb. Sinic.* 3: 96(1930).

Type: Yunnan, Handel-Mazzetti, no. 4278(lectotype).

On rocks. Northern temperate.

Yunnan(Hertel, 1977, p. 222), Hongkong(Aptroot & Seaward, 1999, p. 89).

Porpidia thomsonii Gowan 汤姆氏假网衣

Bryologist 92(1): 54(1989).

Sichuan, Yunnan(Xin Y. Wang et al., 2012, p. 623).

Romjularia Timdal **网鳞衣属**

Lichen Flora of the Greater Sonoran Desert Region(Tempe) 3: 287(2007).

Romjularia lurida(Ach.) Timdal 棕网鳞衣

Lichen Flora of the Greater Sonoran Desert Region(Tempe) 3: 288(2007).

≡*Lecidea lurida* Ach., *Methodus, Sectio prior*(Stockholmiæ): 77(1803).

Replaced synonym:

Lichen luridus Sw., Nova Acta Acad. Upsal. 4: 247(1784). Nom. illegit., Art. 53. 1

Competing homonym: non *Lichen luridus* Dill. Ex With., Bot. arr. Veg. Gr. Brit. (London) 2: 720(1776)

Hebei, Neimenggu(Tchou, 1935, p. 322).

(77) Malmideaceae Kalb, Rivas Plata & Lumbsch **网耳衣科**

in Kalb, Rivas Plata, Lücking & Lumbsch, Biblthca Lichenol. 106: 150(2011).

Malmidea Kalb, Rivas Plata & Lumbsch **网耳衣属**

in Kalb, Rivas Plata, Lücking & Lumbsch, Biblthca Lichenol. 106: 150(2011).

Malmidea aurigera(Fée) Kalb, Rivas Plata & Lumbsch 网耳衣

in Kalb, Rivas Plata, Lücking & Lumbsch, Biblthca Lichenol. 106: 153(2011).

≡*Lecidea aurigera* Fée, *Essai Crypt. Exot.* (Paris) : 106(1825) [1824].

Essai Crypt. Exot. (Paris) : 106(1825) [1824].

On *Zelkova serrata*, and bark of mature tree.

Taiwan(Aptroot and Sparrius, 2003, p. 26).

(78) Vezdaeaceae Poelt & Vězda ex J. C. David & D. Hawksw. **维氏衣科**

Syst. Ascom. 10(1) : 16(1991).

Vezdaea Tscherm. -Woess & Poelt **维氏衣属**

in Brown et al., *Lichenology: Progress & Problems*: 91(1976).

Vezdaea flava Aptroot & Sparrius 黄维氏衣

Fungal Diversity 14: 48 – 49(2003, fig. 14 in p. 44).

Type: Taiwan, Nantou County, 45 km WNW of Hualien, Meifeng, around field centre, 2000 – 2100 m alt. On soil in underhang of roadbank, 51RUG1464. Aptroot 52219 and Sparrius 5955(B-holotype, ABL, Herb. Sparrius, TUNG-isotypes).

Vezdaea stipitata Poelt & Döbbeler 柄维氏衣

Lichenologist 9: 170(1977).

On sandstone.

Taiwan(Aptroot & Sparrius 2003, p. 49).

Lecanoromycetidae, genera incertae sedis 茶渍亚纲未定位属

Bryobilimbia Fryday, Printzen & S. Ekman **藓菌衣属**

Lichenologist 46(1) : 29(2014).

Bryobilimbia ahlesii(Hepp) Fryday, Printzen & S. Ekman 阿勒氏藓菌衣

Lichenologist 46(1) : 29(2014).

Heilongjiang(Hu Ling et al., 2014, p. 84).

Bryobilimbia sanguineoatra(Wulfen) Fryday, Printzen & S. Ekman 红藓菌衣

Lichenologist 46(1) : 31(2014).

Yunnan(Hu Ling et al., 2014, pp. 85 – 87).

5. Umbilicariomycetidae Bendiksby, Hestmark & Timdal **石耳亚纲**

Taxon 62(5) : 952(2013).

[17] Umbilicariales J. C. Wei & Q. M. Zhou **石耳目**

In Q. M. Zhou & J. C. Wei *Mycosystema* 26(1) : 44(2007). Published 22 Feb. 2007. Type cited as ' *Umbilicariaceae*' , see Art. 16. 1(a).

Umbilicariales Lumbsch, Hestmark & Lutzoni, [May] 2007(this *Index* 7: 915), an isonym, see Art. 6. 3 Note 2.

(79) Elixiaceae Lumbsch **盘耳衣科**

J. Hattori bot. Lab. 83: 62(1997).

Elixia Lumbsch **盘耳衣属**

J. Hattori bot. Lab. 83: 62(1997).

Elixia flexella(Ach.) Lumbsch 盘耳衣

J. Hattori bot. Lab. 83: 62(1997).

≡*Limboria flexella* Ach., *K. Vetensk-Acad. Nya Handl.* : 258(1815).

On *Abies kawakamii* wood.

Taiwan(Aptroot and Sparrius, 2003, p. 16).

(80) Ophioparmaceae R. W. Rogers & Hafellner **盾叶衣科**

Lichenologist 20(2):172(1988). *Rhizoplacopsidaceae* J. C. Wei et al. in Q. M. Zhou & J. C. Wei *Mycosystema* 25(3):380(2006).

Boreoplaca Timdal 盾叶属

Mycotaxon 51: 503(1994).

= *Rhizoplacopsis* J. C. Wei et al. in Q. M. Zhou & J. C. Wei, *Mycosystema* 25(3):381(2006).

Boreoplaca ultrafrigida Timdal 蔚青盾叶

Mycotaxon 51: 503(1994).

= *Rhizoplacopsis weichingii* J. C. Wei & Q. M. Zhou, *Mycosystema* 25(3):381(2006).

On rocks.

Type: Jilin, Wangqing, 1996.6.8., J. C. Wei & Y. M. Jiang 45(holotype in HMAS-L).

Jilin(J. C. Wei & Q. M. Zhou, 2006, p. 381).

Ophioparma Norman 蛇孢衣属

Conat. Praem. Gen. Lich.: 18(1852).

Ophioparma handelii(Zahlbr.) Printzen & Rambold 汉德氏蛇孢衣

Herzogia 12: 24(1996).

≡ *Lecidea handelii* Zahlbr. in Handel – Mazzetti, Symb. Sin. 3: 111(1930) & Cat. Lich. Univ. 8: 383(1932).

Type: Yunnan, Handel – Mazzetti no. 4749.

= *Lecidea pseudohaematomma* Asahina in J. Jap. Bot. 29(10):225 – 226(1954), fide Asahina in J. Jap. Bot. 29(10):292(1954). Lignicolous.

Yunnan(Zahlbr., 1930 b, p. 111 & 1932, p. 383; Asahina, 1954 a, p. 292; G. Schneider, 1979, p. 191).

Ophioparma lapponica(Räsänen) Hafellner & R. W. Rogers 拉普兰蛇孢衣

in Rogers & Hafellner, *Lichenologist* 20(2):173(1988).

≡ *Haematomma lapponicum* Räsänen, *Die Flecht. Estl.* 1: 67(1931).

≡ *Haematomma ventosum* var. *lapponicum*(Räsänen) Lynge, Meddr Gr?nland, Biosc. 118(no. 8):156(1937).

On rocks.

Shaanxi(Wu J. L. 1981, p. 163), Yunnan(M. Sato, 1940 b, pp. 495, 500; Wei X. L. et al. 2007, p. 156).

Ophioparma ventosa(L.) Norman 蛇孢衣

Conat. Praem. Gen. Lich.: 19(1852).

≡ *Lichen ventosus* L., *Sp. pl.* 2: 1141(1753).

≡ *Haematomma ventosum* (L.) A. Massal., *Ric. auton. lich. crost.* (Verona):33, fig. 54(1852).

Yunnan(Zahlbr., 1930 b, p. 177; Asahina & M. Sato in Asahina, 1939c, p. 713).

(81) Umbilicariaceae Chevall. 石耳科

[as 'Umbilicarieae'] *Fl. gén. env. Paris*(Paris) 1: 640(1826)

Lasallia Mérat 疱脐衣属

Nouv. Fl. Environs Paris, Edn 2 1: 202(1821).

Lasallia asiae-orientalis Asahina 东亚疱脐衣

Journ. Jap. Bot. 35(4):99(1960). var. asiae-orientalis

Type: Taiwan, Mt. Ali shan, Y. Asahina 1925, preserved in TNS(!).

≡ *Umbilicaria asiae-orientalis*(Asahina) M. Sato *Miscnera bryol. Lichen.*, *Nihinan* 2: 136(1962).

Misapplied name:

Umbilicaria pustulata auct. non Hoffm.: Zahlbr. in Handel-Mazzetti, *Symb. Sin.* 3: 138(1930) & in Fedde, *Repertorium*, 33: 49(1933). Asahina, *Journ. Jap. Bot.* 7: 103(1931).

≡ *Lasallia asiae-orientalis* var. *asiae-orientalis*

On rocks.

Sichuan(Zahlbr. , 1930 b, p. 138; J. C. Wei & Y. M. Jiang 1993, p. 25), Xizang(J. C. Wei & Y. M. Jiang, 1981, p. 1146 & 1982, p. 20 & 1986, p. 103 & 1988, p. 93; J. C. Wei & Y. M. Jiang 1993, p. 26), Yunnan (Zahlbr. , 1930 b, p. 138: Handel-Mazzetti no. 3318 in UPS < ! >; J. C. Wei, 1982, p. 20; J. C. Wei & Y. M. Jiang 1988, p. 93; J. C. Wei & Y. M. Jiang 1993, p. 25), Anhui, Zhejiang(Xu B. S. 1989, p. 242), Taiwan (Zahlbr. , 1933 c, p. 49; M. Sato, 1937, p. 298; Lamb, 1963, p. 283; Wang & Lai, 1973, p. 91 & 1976, p. 228; J. C. Wei, 1982, p. 20; J. C. Wei & Y. M. Jiang, 1988, p. 93; J. C. Wei & Y. M. Jiang 1993, pp. 24, 26).

var. fanjingensis J. C. Wei 梵净山变种

Acta Mycol. Sin. 1(1): 20(1982).

Type: Guizhou, Mt. Fanjing shan, J. C. Wei no. 830 in HMAS-L.

Guizhou(J. C. Wei & Y. M. Jiang, 1988, p. 93; J. C. Wei & Y. M. Jiang 1993, p. 24; Zhang T & J. C. Wei, 2006, p. 8).

var. major J. C. Wei et Y. M. Jiang 大叶变种

Acta Phytotax. Sin. 20(4): 500(1982).

Type: Xizang, Wei & Chen no. 884 in HMAS-L.

On rocks.

Xizang(J. C. Wei & Y. M. Jiang, 1986, p. 103 & 1988, p. 93 & also 1993, pp. 28, 29).

Lasallia daliensis J. C. Wei 大理疱脐衣

Acta Phytotax. Sin. 1(1): 21(1982), var. daliensis.

Type: Yunnan, J. C. Wei no. 2778 in HMAS-L.

On rocks.

Sichuan(J. C. Wei & Y. M. Jiang 1993, pp. 26, 28), Yunnan(J. C. Wei & Y. M. Jiang, 1988, p. 94; J. C. Wei & Y. M. Jiang 1993, pp. 26, 28; Wei X. L. et al. 2007, pp. 153 – 154).

var. caeangshanensis(J. C. Wei) J. C. Wei & Y. M. Jiang 苍山变种

Mycosystema 1: 94(1988).

≡ *Lasallia caeangshanensis* J. C. Wei, *Acta Mycol. Sin.* 1(1): 22(1982).

Type: Yunnan, Wei no. 2758 in HMAS-L.

On rocks.

Yunnan(J. C. Wei & Jiang, 1988, p. 94 J. C. Wei & Y. M. Jiang 1993, pp. 28), HUBEI(Chen, Wu & J. C. Wei, 1989, p. 474).

Lasallia laceratula(Vain.) J. C. Wei & W. Guo 撕裂疱脐衣

Mycosystema 38(10) 1608(2019).

≡ *Umbilicaria laceratula* Vainio Dansk Bot. Arkiv4(11): 2(1926).

= *Lasallia rossica* Dombr. Nov. Syst. Plant. non vascul. 15: 180(1978).

Jilin(J. C. Wei & Y. M. Jiang, 1988, p. 95 and 1993. p. 37).

Lasallia mayebarae(M. Sato) Asahina 中华疱脐衣

Journ. Jap. Bot. 35: 101(1960). var. mayebarae

≡ *Umbilicaria mayebarae* M. Sato, *Journ. Jap. Bot.* 25(8): 167(1950).

Type: Taiwan, Mt. Ali shan, K. Mayebara 1930, preserved in TNS(!).

Misapplied name:

Umbilicaria pustulata sensu Asahina: *Journ. Jap. Bot.* 7: 102(1931), non Hoffm.

On pine trees.

Taiwan(see above mentioned type, also see Asahina, 1931, p. 102 & 1960 a, p. 101; M. Sato, 1950, p. 167; Wang & Lai, 1973, p. 91 & 1976, p. 228; Ikoma, 1983, p. 109; J. C. Wei & Y. M. Jiang, 1988, p. 96; J. C. Wei & Y. M. Jiang 1993, p. 49).

var. sinensis(J. C. Wei) J. C. Wei 鳞芽变种

Bull. Bot. Res. 1(3): 90(1981).

≡*Lasallia sinensis* J. C. Wei, *Acta Phytotax. Sin.* 11(1):1(1966).

Type: Yunnan, Lijiang, Mt. Yulong shan, J. C. Wei no. 2590 in HMAS-L.

On trunk of *Pinus densata*.

Yunnan(J. C. Wei & Y. M. Jiang, 1988, p. 96; J. C. Wei & Y. M. Jiang 1993, pp. 49 – 51).

Lasallia papulosa(Ach.) Llano 淡腹疱脐衣

Monogr. Lich. Fam. Umbilicariaceae in the Western Hemisphaere 32(1950).

≡*Gyrophora papulosa* Ach., *Lich. univ.*: 1 – 696(1810).

On rocks.

Xinjiang(J. C. Wei & Y. M. Jiang, 1992, pp. 86 – 87 & 1993, p. 29.).

Lasallia pensylvanica(Hoffm.) Llano 宾州疱脐衣

Monograph of the Lichen Family Umbilicariaceae in the Western Hemisphere: 42(1950).

≡*Umbilicaria pensylvanica* Hoffm., *Descr. Adumb. Plant. Lich.* 3(no. 4):5(1801).

On rocks.

Jilin(J. C. Wei & . M. Jiang, 1988, p. 94; J. C. Wei & Y. M. Jiang 1993, p. 35; Zhang Y & Wei J. C. 2017, p. 1101), Heilongjiang(Chen et al. 1981, p. 158), Neimenggu, Xinjiang(J. C. Wei & Y. M. Jiang 1993, p. 35; Zhang Y & Wei J. C. 2017, 跑 101),Taiwan(Wang & Lai, 1976, p. 228).

Lasallia pertusa(Rassad.) Llano 孔疱脐衣

Monograph of the Lichen Family Umbilicariaceae in theWestern Hemisphere: 48(1950), f. pertusa.

≡*Umbilicaria pertusa* Rassad., in *Compt Rend. Acad. Scienc.* de l' URSS: 348 – 350(1929).

On rocks.

Xinjiang(J. C. Wei & Y. M. Jiang, 1988, p. 94; J. C. Wei & Y. M. Jiang 1993, p. 37), Xizang(Nyalam: J. C. Wei & Chen, 1974, p. 177; J. C. Wei, 1981, p. 90; J. C. Wei & Y. M. Jiang, 1981, p. 1146 & 1986, p. 102 & 1988, p. 94; J. C. Wei & Y. M. Jiang 1993, p. 37).

f. squamulifera J. C. Wei et Y. M. Jiang 鳞芽变型

Acta Phytotax. Sin. 20(4):499(1982) & *Lichens Xizang*, 102(1986).

Type: Xizang, Lhasa, Wei & Chen no. 15 in HMAS-L.

On rocks.

Xizang(J. C. Wei & Jiang, 1988, p. 95; J. C. Wei & Y. M. Jiang 1993, p. 37).

Lasallia sinorientalis J. C. Wei 华东疱脐衣

Acta Mycol. Sin. 1(1):19(1982).

Type: Jiangxi, Mt. Lu shan, Wei no. 3246 in HMAS-L.

= *Umbilicaria fokiensis* Merril in Llano: Monogr. of Umbilicariaceae in the western hemisphere, 1950, p. 31(nom. inval.).

Type: Fujian, IV-VI/1905, collected by Dunn no. 3938 in FH(!).

Shaanxi(J. C. Wei, 1982, p. 24; J. C. Wei & Jiang, 1988, p. 95; J. C. Wei & Y. M. Jiang 1993, p. 46; He & Chen 1995, p. 43 without citation of specimens and their localities), Yunnan(Wei X. L. et al. 2007, p. 154), Jiangxi(J. C. Wei, 1982, p. 24; J. C. Wei & Jiang, 1988, p. 95; J. C. Wei & Y. M. Jiang 1993, p. 46 – 47), Anhui(Lu, 1959, p. 176; Wei & Jiang, 1988, p. 95; J. C. Wei & Y. M. Jiang 1993, p. 47), Fujian(Llano, 1950, p. 31; Lamb, 1963, p. 727; J. C. Wei, 1982, p. 23; J. C. Wei & Jiang, 1988, p. 95; J. C. Wei & Y. M. Jiang 1993, p. 47).

Lasallia xizangensis J. C. Wei et Jiang 藏疱脐衣

Acta Phytotax. Sin. 20(4):500(1982). var. xizangensis 原变种

Type: Xizang, Wei & Chen no. 881 in HMAS-L.

On rocks.

Xizang(J. C. Wei & Y. M. Jiang, 1982, p. 500 & 1986, p. 104 & 1988, p. 95; J. C. Wei & Y. M. Jiang 1993, p. 48).

var. acuta J. C. Wei et Y. M. Jiang 尖粒变种

Acta Phytotax. Sin. 20(4) : 500(1982) & *Lichens of Xizang*, 104(1986).

Type: Xizang, J. C. Wei & Chen no. 888 in HMAS-L.

Xizang(J. C. Wei &Y. M. Jiang, 1988, p. 96; J. C. Wei & Y. M. Jiang 1993, p. 48).

Umbilicaria Hoffm. **石耳属**

Descr. Adumb. Plant. Lich. 1(1) : 8(1789) [1790].

Umbilicaria altaiensis J. C. Wei & Jiang Y. M. 阿尔泰石耳

Mycosystema 5: 73 – 76(1992).

Type: Xinjiang, Mt. altai, by the Halas(Kalas) lake, on rocks, alt. 2500 m, Aug. 5, 1986, Gao XQ, 2069(HMAS-L), (J. C. Wei & Jiang Y. M. 1993, p. 70).

Umbilicaria aprina Nyl. 皱面粗根石耳

Syn. meth. lich. (Parisiis) 2: 12(1863).

On rocks.

Xinjiang(J. C. Wei & Y. M. Jiang, 1992, p. 77).

Umbilicaria badia Frey 棕色石耳

in Ber. Schweiz. Bot. Ges. 59: 453(1949).

On rocks.

Xizang(J. C. Wei & Jiang, 1986, p. 100 & 1988, p. 79; J. C. Wei & Y. M. Jiang 1993, p. 106), Yunnan(J. C. Wei & Y. M. Jiang, 1988, p. 79; J. C. Wei & Y. M. Jiang 1993, p. 106).

Umbilicaria caroliniana Tuck. 卡罗里石耳

Proc. Amer. Acad. Arts & Sci. 12: 167(1877).

≡*Lasallia caroliniana*(Tuck.) Davydov, peršoh & Rambold, *Mycol Progress*(2010) 9: 264(2010).

On rocks.

Neimenggu(J. C. Wei & Y. M. Jiang, 1988, p. 80; J. C. Wei & Y. M. Jiang 1993, p. 196; Zhang Y & Wei J. C. 2017, p. 1099), Liaoning(Chen et al. 1981, p. 157).

Umbilicaria cinerascens(Nyl.) Nyl. 灰石耳

Flora, Regensburg 52: 388(1869).

≡*Umbilicaria atropruinosa* var. *cinerascens* Nyl., *Not. Sällsk. Fauna et Fl. Fenn. Förh.*, Ny Ser. 5: 114(1861).

Shaanxi(J. C. Wei & Jiang, 1988, p. 80; J. C. Wei & Y. M. Jiang 1993, p. 109; He & Chen 1995, p. 45 without citation of specimens and their localities).

Umbilicaria cylindrica(L.) Delise, in Duby 花石耳

Bot. Gall., *Edn* 2(Paris) 2: 595(1830) var. cylindrical

≡*Lichen cylindricus* L., *Sp. pl.* 2: 1144(1753).

The basionym "*Lichen cylindricus*" was incorrectly applied for the species which had been named *Lichen proboscideus*(1753, p. 1150). In the Linnaean Herbarium there is no specimen bearing the name "*Lichen cylindricus*". Delise made the combination in *Umbilicaria* based On the specimen of Schaerer's Lich. Helvet. Nos. 143 – 147, as cited in Duby's paper(1830). After careful examination of Schaerer's specimens, it is certain that both the description furnished by Delise in Duby(1830) and Schaerer's specimens in Lich. Helvet. correspond completely with specimens bearing "*Lichen proboscideus*" in the Linnaean Herbarium. Unfortunately, "*Lichen cylindricus*" used by Delise as basionym was recombined and misapplied as "*Umbilicaria cylindrical*(L.) Delise".

I sent the manuscript of mine concerning this problem to Dr. R. Santesson to ask for criticisms by the end of 1989 from London out of respect for a friend of mine. The manuscript, however, like a stone dropped into the sea and disappeared forever. Three years later, when I wrote to him at the end of 1992 to ask about the manuscript of mine sent to him three years ago, Dr. R. Santesson, a friend of mine who had been respected by me, wrote to me that Dr. Jørgensen was working on this problem in Linnaean Herbarium and he was going to publish a paper.

Misapplied name:

Umbilicaria cylindrica sensu Delise, in Duby, *Bot. Gall.*, Edn 2(Paris) 2: 595(1830), non *Lichen cylindricus* L., *Sp. pl.* 2: 1144(1753).

= *Umbilicaria neocylindrica* J. C. Wei, in *Supplement to Mycosystema* Vol. 5. International Academic Publishers, Beijing, China. May 20, 1993, Pp. 4 – 8.

Type: Linn1273 – 204, 53(LINN-lectotype) selected by J. C. Wei.

Notes: Jørgensen indicated that a much better solution is to propose conservation of the olde name with a new type(1994).

On rocks.

Jijin(Chen et al. 1981, p. 157; J. C. Wei & Y. M. Jiang, 1988, p. 80; J. C. Wei & Y. M Jiang. 1993, pp. 114 – 115), Xinjiang(J. C. Wei & Y. M. Jiang 1993, p. 115), Shaanxi(J. C. Wei & Y. M. Jiang, 1988, p. 80; J. C. Wei & Y. M. Jiang 1993, p. 115).

Umbilicaria decussata(Vill.) Zahlbr. 网脊石耳

Cat. Lich. Univers. 8: 490(1932).

≡ *Lichen decussatus* Vill., *Hist. pl. Dauphiné* 3(2): 964(1789).

Missapplied name:

Umbilicaria leiocarpa by Chen et al.: Journ. N. E. Forestry Inst. 4: 158(1981). (Chen. no. 4825 in FPI).

On rocks.

Jilin(Chen et al. 1981, p. 158, as *U. leiocarpa*; J. C. Wei & Y. M. Jiang, 1988, p. 89; J. C. Wei & Y. M. Jiang 1993, p. 82), Xinjiang(Abdulla Abbas et al., 1993, p. 81; Abdulla Abbas & Wu, 1998, p. 130; Abdulla Abbas et al., 2001, p. 368).

Umbilicaria esculenta(Miyoshi) Minks 庐山石耳

Mém. Herb. Boissier 22: 46(1900).

≡ *Gyrophora esculenta* Miyoshi, *Botan. Zbl.* 56(6): 162(1893).

Misapplied name:

Umbilicaria spodochroa sensu Müll. Arg. in *Bull. Herb. Boissier*, I: 235(1893), auct. non Ehrh. ex Hoffm.

On rocks.

Heilongjiang, Jilin(Chen et al. 1981, p. 158; J. C. Wei & Jiang, 1988, p. 82 & 1993, p. 116), Liaoning(J. C. Wei & Y. M. Jiang, 1988, p. 82 & 1993, p. 117), Yunnan(J. C. Wei & Jiang, 1988, p. 82 & 1993, p. 117), Hubei(Müll. Arg. 1893, p. 235 cited as *Umbilicaria spodochroa*; Zahlbr., 1930 b, p. 138 cited as *Gyrophora cirrhosa*), Jiangxi(J. C. Wei, 1981, p. 89; J. C. Wei et al. 1982, p. 28; J. C. Wei & Jiang, 1988, p. 82 & 1993, p. 117), Anhui, Zhejiang(Lu, 1959, p. 174; J. C. Wei & Y. M. Jiang, 1988, p. 82 & 1993, p. 118; Xu B. S., 1989, p. 243), Hunan, Guangxi(J. C. Wei & Jiang, 1988, p. 82 & 1993, p. 1 18).

Note: After reexamination of the specimens collected by Aug. Henry from Hubei(as Hupeh), no. 6184(FH!), cited by Müll. Arg. (1893, p. 235) as *Umbilicaria spodochroa*, and by Zahlbr. (1930, p. 138) as *Gyrophora cirrhosa*, I have found that it is *Umbilicaria esculenta* rather than *Umbilicaria spodochroa*(Hoffm.) DC.

Also see *Umbilicaria thamnodes* Hue

Umbilicaria flocculosa(Wulfen) Hoffm. 焦石耳

Deutschl. Fl., *Zweiter Theil*(Erlangen): 110(1796)[1795]; J. C. Wei in *Supplement to Mycosystema* Vol. 5. International Academic Publishers, Beijing, China. May 20, 1993, Pp. 10 – 12.

≡ *Lichen flocculosus* Wulfen, in Jacquin, *Collnea bot.* 3: 99, Tab. 1, fig. 2(1789).

Iconotype: Jacq. Collect. Bot. 3: 99, tab. 1. fig. 2(1789).

≡ *Gyrophora flocculosa* (Wulfen) Körb. Syst. Lich. Germ. 95(1855).

Misapplied name:

Umbilicaria deusta sensu (L.) Baumg., *Fl. Lips.*: 571(1790).

≡*Lichen deustus* L., *Sp. pl.* 2: 1150(1753).

Gyrophora deusta sensu Ach., *Methodus, Sectio prior* (Stockholmiæ): 102(1803).

Notes: Jørgensen indicated that a much better solution is to propose conservation of the olde name with a new type(1994).

On rocks.

Jilin(J. C. Wei & Y. M. Jiang, 1988, p. 81 & 1993, p. 124), Xinjiang(J. C. Wei & Y. M. Jiang 1993, p. 124), Shaanxi(He & Chen 1995, p. 45 without citation of specimens and their localities).

Umbilicaria formosana Frey 台湾石耳

Hedwigia 71: 115(1931).

Type locality: Taiwan, S. Sasaki in herb. of Asahina.

≡*Gyrophora formosana*(Frey) M. Sato in *Index Plantarum Nipponicarum*, IV, *Lichenes*: 115(1943).

≡*Omphalodiscus formosanus*(Frey) Schol. In *Nyt. Mag. F. Naturvidensk.* LXXV: 24(1936).

≡*Omphalodiscus krascheninnikovii* var. *formosanus*(Frey) Llano, *Monograph of the Lichen Family Umbilicariaceae in the Western Hemisphere*: 88(1950).

On rocks.

Jilin(J. C. Wei & Jiang, 1988, p. 90; J. C. Wei & Jiang Y. M. 1993, p. 86), Shaanxi(J. C. Wei, 1981, p. 90; J. C. Wei & Jiang, 1988, p. 90; J. C. Wei & Jiang Y. M. 1993, p. 86), Yunnan, XIZANG(J. C. Wei & Y. M. Jiang, 1988, p. 90; J. C. Wei & Jiang Y. M. 1993, p. 86), Taiwan(Frey, 1931, p. 115; Zahlbr., 1932 a, p. 491 & 1933 c, p. 49; Sato, 1950, p. 170; Lamb. 1963, p. 727; Wang & Lai, 1973, p. 97; J. C. Wei & Y. M. Jiang 1993, p. 86).

Umbilicaria herrei Frey 薄石耳

Ber. schweiz. bot. Ges. 45: 219, 222(1936).

Jilin(J. C. Wei & Y. M. Jiang, 1988, p. 82 & 1993, p. 131), Xizang(J. C. Wei & Jiang Y. M. 1993, p. 131).

Umbilicaria hirsuta(Sw. ex Westr.) Ach. 粗根石耳

K. Vetensk-Acad. Nya Handl. 15: 97(1794).

≡*Lichen hirsutus* Sw. ex Westr., *K. Vetensk-Acad. Nya Handl.*: 47(1793).

= *Umbilicaria hirsuta* var. *vuoxaensis*(J. C. Wei) J. C. Wei in J. C. Wei & Y. M. Jiang, *Acta Phytotax. Sin.* 20(4): 499(1982).

≡*Gyrophora hirsuta* var. *vuoxaensis* J. C. Wei, *Notul. Syst. E Sect. Crypt. Inst. Bot. Nom. Komarovii Acad. Sci. URSS* 15: 8(1962).

Type: Russia, Leningrad prov. Coll. J. C. Wei(LE, holotype, HMAS-L-Isotype).

On rocks near river.

Xinjiang(J. C. Wei & Jiang Y. M. 1993, p. 133), Xizang(J. C. Wei & Chen, 1974, p. 177; J. C. Wei & Jiang, 1981, p. 1146, & 1986, pp. 99 – 100, & 1988, p. 83, & 1993, p. 133).

Umbilicaria hyperborea(Ach.) Hoffm. 北方石耳

Deutschl. Fl., *Zweiter Theil* (Erlangen): 110(1796)[1795].

≡*Lichen hyperboreus* Ach., *K. Vetensk-Acad. Nya Handl.* 15: 89(1794).

= *Umbilicaria exasperata* Hoffm., Descr. Adumb. Plant. Lich. 1(1): 7(1789)[1790], Illustration: Tabula 2, Fig. 1 – 2(1790).

Type: H. Hoffm. 8596. *Umbilicaria exasperata* Hoffm. (*Gyrophora hyperborean* Ach.) Suec.

(MW-neotype). J. C. Wei, *The lectotypification of some species in the Umbilicariaceae described by Linnaeus or Hoffman* in *Supplement to Mycosystema* Vol. 5. International Academic Publishers, Beijing, China. 1993, p. 8.

= *Umbilicaria exasperata* Hoffm. , *Descr. Adumb. Plant. Lich.* 1(1) : 7(1789) [1790] , nom. rejic. , Art. 56. 1

On rocks.

Neimenggu(Sato, 1952, p. 174; J. C. Wei & Jiang, 1988, p. 83) .

Umbilicaria hypococcinea(Jatta) Llano [as ' *hypococcina*'] 红腹石耳

Monogr. Umbilicariaceae in the West. Hemisphere, 191(1950) *in* Wei J. C. & Jiang Y. M. , *Lichens of Xizang*, 1986, p. 98(ICN: Article 60. 1, ex. 4, and 60E. 1) ; *in Mycosystema* 1: 83(1988) , and also *The Asian Umbilicariaceae(Ascomycota)* : 134(1993) .

≡ *Gyrophora hypococcinea* Jatta, *Nuov. Giorn. Botan. Ital.* nov. ser. 2, vol. 9: 473(1902) .

Nomenclatural comment:

The originally used epithet *Gyrophora* "*hypococcinea*" Jatta(Nuov. Giorn. Botan. Ital. nov. ser. 2, vol. 9: 473, 1902) is compounded of a Greek prefix "*hypo-*": ' below, under, beneath, lower' and a Latin adjective "*coccineus, a, um*": ' deep red, from scarlet to carmine and crimson' , and means ' lower surface of thallus red' . However, such a hybrid epithet should not be advocated(ICN: Recommendation 23A3(c) Not make epithets by combining words from different languages) . The misspelled *Umbilicaria* "*hypococcina*" Llano, *Monogr. Umbilicariaceae in the West. Hemisphere,* 191(1950) is presumably a typographical error and is to be reverted to the original *Gyrophora* "*hypococcinea*" Jatta(1902) . The name is to be changed accordingly and is cited as *Umbilicaria hypococcinea*(Jatta) Llano *in* Wei J. C. & Jiang Y. M. , *Lichens of Xizang*, 1986, p. 98(ICN: Article 60. 1, Ex. 4, and 60E. 1) , and *in Mycosystema* 1: 83(1988) , and also in *The Asian Umbilicariaceae (Ascomycota)*: 134 (1993) .

Type: Shaanxi, Mt. Guangtou shan, collected by Giraldi in S(!) (isotype) .

= *Gyrophora versicolor* Räs. *Arch. Soc. Zool. Bot. Fenn.* 'Vanamo' 3: 78(1949) .

Type: Shaanxi, 1913, Mary Strong Clemens no. 4021 in TUR(holotype!) & in H(isotype!) .

On rocks.

Shaanxi(Jatta, 1902, p. 473; Zahlbr. , 1927, p. 718 & 1930 b, p. 138; Ras. 1949, p. 78; Llano, 1950 p. 191; Wu J. L. 1981, p. 161; Wei, 1981, p. 89; Wei et al. 1982, p. 28; Wei & Jiang, 1988, p. 83, & 1993, p. 135; He & Chen 1995, p. 45 without citation of specimens and their localities) , Xizang(J. C. Wei & J. B. Chen, 1974, p. 177; J. C. Wei & Y. M. Jiang, 1981, p. 1146, & 1982, p. 498, & 1986, p. 98, & 1993, p. 135; Obermayer 2004, pp. 513 – 514, as *U.* ' *hypococcina*') , Sichuan(Obermayer 2004, p. 514, as *U.* ' *hypococcina*') , Yunnan(J. C. Wei & Y. M. Jiang 1993, p. 135) , Shanxi(Lu, 1959, p. 175; J. C. Wei & Y. M. Jiang 1993, p. 135) .

Umbilicaria indica Frey 印度石耳

Ber. schweiz. bot. Ges. 59: 456(1949) .

≡ *Umbilicaria papillosa* Nyl. , Syn. meth. lich. (Parisiis) 2: 11(1863) , Nom. illegit. , Art. 53. 1. auct. non DC. ex anno 1805: [*Umbilicaria papillosa* DC. , in Lamarck & de Candolle, Fl. franç. , Edn 3(Paris) 2: 411(1805)]

= *Gyrophora himalayensis* Räsänen, *Ann. bot. Soc. Zool. -Bot. fenn. Vanamo* 5(1) : 25 – 32(1951) .

Misapplied name:

Umbilicaria polyrrhiza sensu J. C. Wei & Chen: 1974, p. 177, non(L.) Ach.

On rocks.

Xizang(J. C. Wei & Chen, 1974, p. 177; J. C. Wei & Jiang, 1986, p. 100) , Yunnan(Poelt, 1977, p. 420; J. C. Wei & Jiang Y. M. 1993, p. 139; Wei X. L. et al. 2007, pp. 158 – 159) , Hubei(Chen, Wu & J. C. Wei, 1989, p. 474) .

Umbilicaria kisovana(Zahlbr. ex Asahina) Zahlbr 小石耳

Cat. Lich. Univ. 10: 405(1940) .

≡ *Gyrophora kisovana* Zahlbr. ex Asahina, *Journ. Jap. Bot.* 7: 326(1931) .

Iconotype: Figs. : A photograph of the species habit, a drawing picture of the species habit, and a drawing picture of ascus and species in *Journ. Jap. Bot.* 7: 326(1931).

On rocks.

Neimenggu(J. C. Wei & Y. M. Jiang 1993, p. 1142), Shanxi(Sato, 1981, p. 64), Jiangxi(J. C. Wei & Jiang, 1988, p. 85, & 1993, p. 142).

Umbilicaria krascheninnikovii(Savicz) Zahlbr. 淡腹石耳

Cat. Lich. Univers. 10: 405(1939).

≡ *Gyrophora krascheninnikovii* Savicz, *Izv. Imp. Bot. Sada Petra Velikago* 14: 117(1914).

Misapplied name:

Umbilicaria leiocarpa sensu Chen et al.: Journ. NE Forestry Inst. 4: 158(1981), non DC.

On rocks.

Jilin(J. C. Wei, 1981, p. 90; Chen et al. 1981, p. 158; J. C. Wei & Jiang, 1988, p. 90), Shaanxi, Xizang(J. C. Wei & Jiang, 1988, p. 90).

Umbilicaria leiocarpa DC. 平网石耳

in Lamarck & de Candolle, *Fl. franç.*, Edn 3(Paris) 2: 410(1805).

≡ *Gyrophora leiocarpa*(DC.) Steud., Nomencl. bot.: 194(1824).

Xinjiang(Abdulla Abbas & Wu, 1998, p. 130; Abdulla Abbas et al., 2001, p. 368).

Umbilicaria lyngei Schol. 网脊平盘石耳

Nytt Mag. Natur. (75): 19(1934).

Jilin, Shaanxi(Wei JC& Jiang YM, 1988, p. 92; J. C. Wei & Jiang Y. M. 1993, p. 60).

Umbilicaria microphylla(Laurer) A. Massal. 小叶石耳

Ric. auton. lich. crost. (Verona): 62(1852).

≡ *Umbilicaria atropruinosa* var. *microphylla* Laurer, *Deutschl. Fl.* (Frankfurt) 24: 13(1832).

= *Gyrophora microphylloides*(Laur.) Zahlbr. *Cat. Lich. Univers* 4: 683(1927).

On rocks.

Yunnan(Zahlbr., 1930 b, p. 138).

Umbilicaria minuta J. C. Wei & Y. M. Jiang 微石耳

Asian Umbilicariaceae 142 – 145(1993).

Type: Xizang, Mt. qomolangma, alt. 4850m, on rocks, 1966, Wei J. C. & Chen J. B. 1379(HMAS-L- 6342-holotype, 6343-isotype).

Umbilicaria muhlenbergii(Ach.) Tuck. 放射盘石耳

Enum. N. America Lich.: 55(1845).

≡ *Gyrophora mühlenbergii* Ach. [as '*mühlenbergii*'], *Lich. univ.*: 227, tab. II, fig. 11(1810).

On rocks.

Heilongjiang(J. C. Wei, 1981, p. 90; Chen et al. 1981, p. 158; J. C. Wei & Y. M. Jiang, 1988, p. 89, & 1993, p. 201), Neimenggu(Chen et al. 1981, p. 158; J. C. Wei & Y. M. Jiang, 1988, p. 89, & 1993, p. 201).

Umbilicaria nanella Frey & Poelt 小黑腹石耳

in Poelt, *Khumbu Himal* 6(3): 425(1977).

Misapplied name:

Gyrophora cylindrica var. *tornata* sensu Paulson: *Journ. Bot.* 63: 193(1925) & 66: 317(1928), non(Ach.) Nyl.

Gyrophora tornata sensu Zahlbr. in Handel-Mazzetti, *Symb. Sin.* 3: 138(1930), non Ach.

Gyrophora cylindrica var. *fimbriata* sensu Zahlbr., in Handel-Mazzetti, *Symb. Sin.* 3: 138(1930), non Ach.

On rocks.

Shaanxi(J. C. Wei & Jiang, 1988, p. 85, & 1993, p. 148), Sichuan(Poelt, 1977, p. 425; J. C. Wei & Jiang Y. M. 1993, p. 148), Yunnan(Paulson, 1928, p. 317; Zahlbr., 1930 b, p. 138; J. C. Wei & Jiang, 1988, p. 85, &

1993, p. 148), Xizang(Paulson, 1925, p. 193; J. C. Wei & Jiang, 1981, p. 1146 & 1986, p. 5).

Umbilicaria nepalensis Poelt 尼泊尔石耳

Khumbu Himal 6(3): 426(1977).

On rocks. Alt. 5530m, Mt. Qomolangma.

Xizang(J. C. Wei & Jiang Y. M. 1993, p. 149).

Umbilicaria nylanderiana(Zahlbr.) H. Magn. 皱面黑腹石耳

Lich. sel. Scand. Exs.: no. 252(1937).

≡*Gyrophora nylanderiana* Zahlbr., *Cat. Lich. Univers.* 4: 720(1927).

On rocks.

Xinjiang(J. C. Wei & Jiang Y. M., 1992, pp. 79 – 80, & 1993, p. 154.).

Umbilicaria polyphylla(L.) Baumg. 复叶石耳

Fl. Lips.: 571(1790).

≡*Lichen polyphyllus* L., *Sp. pl.* 2: 1150(1753).

Taiwan(Zahlbr., 1933 c, p. 49; Wang & Lai, 1973, p. 97).

Umbilicaria proboscidea auct. 多盘石耳

The name *Umbilicaria deusta*(L.) Baumg. [Fl. Lips. 571(1790)] may be kept on the basis of the typification [Art. 55(2)]. This name is one of the misapplied names of interlocking mistakes in nomenclature. Therefore, it is easy to cause confusion on the names. So, it may be ruled as rejected(Art. 69) and replaced by a new epithet as follows(Art. 72. 1; see *Umbilicaria cylindrical* auct.): *Lichen seustus* L., Sp. Pl. 1150(1753).

Lectotype: Linn 1273 – 206: 54, Fl. Suec. 970(LINN).

Type observation: Thallus umbilicate, orbicular, c. 22mm diam., monopyhyllous, membranceous, rigid, thin and fragile, over umbo elevated, with a reticulate of sharp edged rugi arranged circularly around umbo and becoming vermiform ridged to margins. Apothecia regularly gyrose.

Misapplied name:

Lichen proboscideus auct.

Umbilicaria proboscidea sensu Schrad., *Spicil. fl. germ.* 1: 103(1794), non Linn., Sp. Pl. 1150(1753).

= *Umbilicaria neoproboscidea* J. C. Wei in *Supplement to Mycosystema* Vol. 5. International Academic Publishers, Beijing, China. May 20, 1993, Pp. 12 – 14.

≡*Lichen deustus* L., *Sp. Pl.* 1150(1753).

Lectotype: Linn1273 – 206, 54, Fl. Suec. 970(LINN, lectotype) made by J. C. Wei, *The lectotypification of some species in the Umbilicariaceae described by Linnaeus or Hoffman* in *Supplement to Mycosystema* Vol. 5. International Academic Publishers, Beijing, China. 1993, p. 12.

Notes: Jørgensen indicated that a much better solution is to propose conservation of the olde name with a new type(1994).

On rocks.

Jilin(J. C. Wei & Y. M. Jiang, 1988, p. 86), Yunnan(Hue, 1900, p. 118; Zahlbr., 1930 b, p. 138; J. C. Wei, 1981, p. 89; J. C. Wei & Y. M. Jiang, 1988, p. 86), Xizang(J. C. Wei & Chen, 1974, p. 177; J. C. Wei & Y. M. Jiang, 1981, p. 1147 & 1986, p. 99 & 1988, p. 86).

Umbilicaria pseudocinerascens J. C. Wei &Y. M. Jiang 拟灰石耳

*Mycosystema*1: 86(1988) & 162 – 163(199 3).

Type: Yunnan, Dali, Mt. Cang shan, Zhonghefeng, 3/I/1965, alt. 3850 m, J. C. Wei, no. 2827 in HMAS-L(holotype) and H, UPS(isotypes).

On rocks.

Umbilicaria rhizinata(Frey & Poelt) Krzew. 雪根石耳

Lichenologist 42(4) : 491(2010) .

≡*Umbilicaria decussata* var. *rhizinata* Frey & Poelt, in Poelt, *Khumbu Himal* 6(3) : 419(1977) .

= *Umbilicaria aprina* var. *halei* Llano *Journ. Washington Acad. Science* 46(6) : 183(1956) .

On rocks.

Jilin(J. C. Wei & Y. M. Jiang, 1993, p. 79, as var. *halei*) .

Umbilicaria rigida(Du Rietz) Frey 硬石耳

Hedwigia 71: 117(1931) .

≡*Gyrophora rigida* Du Rietz, *Ark. Bot.* 19(12) : 3(1925) .

Shaanxi(Wei & Jiang, 1988, p. 92; J. C. Wei & Y. M. Jiang 1993, p. 61) .

Umbilicaria spodochroa Ehrh. ex Hoffm. 灰叶石耳

Deutschl. Fl. , Zweiter Theil(Erlangen) : 113(1796) [1795] .

≡*Gyrophora spodochroa*(Ehrh. ex Hoffm.) Ach. , *Methodus*, Sectio prior: 108(1803) .

= *Gyrophora cirrhosa*(Hoffm.) Vain. , *Meddn Soc. Fauna Flora fenn.* 14: 23(1888) .

≡*Umbilicaria cirrhosa* Hoffm. , *Descr. Adumb. Plant. Lich.* 1: 9, tab. II, fig. 3?4(1790) .

On rocks.

Sichuan(Elenk. 1904, p. 85 & in exs. Lich. Fl. Ross. IV, 1904, no. 151 cited as *Gyrophora spodochroa* in UPS! & FH! ; J. C. Wei & Y. M. Jiang, 1988, p. 91; J. C. Wei & Y. M. Jiang 1993, p. 100) , Yunnan(Hue, 1887, p. 23) .

Also see *Umbilicaria esculenta*(Miyoshi) Minks.

Umbilicaria squamosa Wei J. C. & Jiang Y. M. 皮芽石耳

Mycosytema 5: 80 – 81(1992) .

Type: Yunnan, Dali, Mt. Cangshan, alt. 3300m, on rocks, Dec. 29, 1964, Wei J. C. 2779(HMAS-L-holotypus) .

Xizang, Yunnan(J. C. Wei & Y. M. Jiang, 1992, pp. 80 – 81, & 1993, p. 167; Wei X. L. et al. 2007, p. 159) .

Umbilicaria subglabra(Nyl.) Harm. 光面石耳

Lich. Fr. 4: 707(1910) [1909] .

≡*Gyrophora subglabra* Nyl. , *Lich. Envir. Paris*: 135(1896) [1895] .

On rocks.

Xinjiang(JC Wei & YM Jiang, 1992, pp. 81 – 84 & 1993, p. 66) .

Umbilicaria subumbilicarioides J. C. Wei & Y. M. Jiang 栅栏皮石耳

Mycosystema 5: 84 – 85(1992) , & 167(1993) .

Type: Xizang(Tibet) , Zham, alt. 3660m, on rocks. May 11, 1966. J. C. Wei & Chen JB 470 – 1(HMAS-L. holotypus) .

Umbilicaria taibaiensis J. C. Wei & Y. M. Jiang 太白石耳

Mycosystema 5: 85 – 86(1992) , & 170 – 172(1993) .

Type: Shaanxi, Mt. Taibaishan, Baxiantai, alt. 3660m, on rocks, June 3, 1963, JC Wei(HMAS-L. , holotypus) , July 31, 1963. Ma QM & Zong YC 395 – 1(HMAS-L. paratypus) .

Misapplied name:

Umbilicaria cylindrica var. *tornata* sensu J. C. Wei & Y. M. Jiang, in *Mycosystema* 1: 80(1988) , Fig. 2 in p. 81. non(Ach.) Nyl.

Jilin(J. C. Wei & Y. M. Jiang, 1988, p. 80) , Shaanxi(J. C. Wei & Y. M. Jiang, 1988, p. 80; He & Chen 1995, p. 45 without citation of specimens and their localities) .

Umbilicaria thamnodes Hue 鳞石耳

Nouv. Arch. Mus. hist. Nat. 4(2) : 121(1900) , f. thamnodes Syntype localities: Yunnan, Delavay nos. 1571,

1573, 1574.

Misapplied names:

Gyrophora polyrrhiza sensu Hue: *Nouv. Arch. Mus. Hist. Nat.* 4(2) : 121(1900), non(L.) Ach.

Gyrophora polyrrhiza var. *luxurians* Hue: *Bull. Soc. Bot. France*, 34: 24(1887), non(Ach.) Th. Fr.

Gyrophora luxurians sensu Zahlbr. in Handel-Mazzetti, *Symb. Sin.* 3: 138(1930), non(Ach.) Roehl.

Umbilicaria esculenta sensu J. C. Wei & Chen, *Report on the Scientific Investigations* (1966 – 78) in Mt. Qomolangma district, Science Press, Beijing, 1974: 177, non(Miyoshi) Minks.

On rocks & on twigs.

Shaanxi(J. C. Wei & Y. M. Jiang, 1993, p. 175), Xizang(J. C. Wei & Chen, 1974, p. 177; J. C. Wei & Jiang, 1981, p. 1146, & 1986, p. 100, & 1993, p. 175), Yunnan(Hue, 1887a, p. 24, & 1889, p. 170, & 1900, p. 121; Zahlbr., 1930 b, p. 138; Llano, 1950, p. 197; J. C. Wei & Y. M. Jiang, 1993, p. 175; Wei X. L. et al. 2007, p. 159).

Umbilicaria thamnodes f. minor Hue 小叶变型

Nouv. Arch. Mus. Hist. Nat. 4(2) : 122(1900), tab. v, fig. 3.

Type locality: Yunnan, Delavay.

Yunnan(Zahlbr., 1930 b, p. 138).

Taxonomic revision of this species and its infraspecific taxon are still required.

Umbilicaria torrefacta(Lightf.) Schrad. 齿腐石耳

Spicil. fl. germ. 1: 104(1794).

≡*Lichen torrefactus* Lightf., *Fl. Scot.* 2: 862(1777).

Jilin(J. C. Wei & Y. M. Jiang, 1988, p. 87, & 1993, p. 180).

Umbilicaria tylorhiza(Nyl.) Nyl. 肉根石耳

Flora, Regensburg 52: 389(1869).

Basionym: *Umbilicaria vellea* * tylorhiza Nyl., Not. Sällsk. *Fauna et Fl. Fenn. F?rh.*, *Ny Ser.* 8: 122(1866)

Type: Yunnan, Delavay(H – Nyl. no. 31523, holotype !).

≡*Gyrophora tylorhiza*(Nyl.) Nyl. in Hue, evue de Botan, 5: 14 et 163(1886 – 87), & *in Bull. Soc. Bot. France* 34: 23(1887). A. Zahlbr. in Handel – Mazzetti, ymb. Sin. 3: 138(1930).

Llano, A monograph of the Umbilicariaceae 1950: 264.

= Umbilicaria trabeculataFrey &Poelt in Poelt, KhumbuHimal6(3) : 429(1977).

Type: KhumbuHimal: HohewestlichLobuche, 5050 – 5100m, SteilundUberhangflache

einesgro Ben, windverfegtenGneisblocks, L248(M, not seen).

Misapplied name:

Umbilicaria griseasensu JC Wei & JB Chen: in *Report on the Scientific Investigations* (1966 – 1968) in Mt. Qomolangma district, Science Press, Beijing, 1974: 177.

Shaanxi(J. C. Wei&Y. M. Jiang, 1988, p. 87&1993, p. 183), Xizang(J. C. Wei& Chen, 1974, p. 177; J. C. Wei&Y. M. Jiang, 1981, p. 1146 & 1986, p. 101, 1988, p. 87, &1993, p. 183), Yunnan(Hue, 1887a, p. 23 & 1900, p. 115; Zahlbr., 1930 b, p. 138; Llano, 1950, p. 264, fig. 7, asGyrophoratylorrhizaNyl.; J. C. Wei& Y. M. Jiang, 1988, p. 88).

Umbilicaria vellea(L.) Ach. 毛根石耳

K. Vetensk-Acad. Nya Handl. 15: 101(1794) f. vellea

≡*Lichen velleus* L., *Sp. pl.* 2: 1150(1753).

On rocks.

Xinjiang(J. C. Wei & Y. M. Jiang, 1993, p. 187), Xizang, Sichuan, Hebei(Wei & Jiang, 1988, p. 88, & 1993, pp. 186 – 187), JILIN(Chen et al. 1981, p. 158; J. C. Wei & Y. M. Jiang, 1988, p. 88, & 1993, pp. 186 – 187).

Umbilicaria vellea f. leprosa Schaer. 粉芽变型

Enum. critic. lich. europ. (Bern) : 25(1850) .

≡ *Gyrophora vellea* f. *leprosa*(Schaer.) Zahlbr. , *Cat. Lich. Univers*. 4: 740(1927) .

≡ *Umbilicaria vellea* var. *cinereorufescens* f. *leprosa* Schaer.

Hebei, Jilin(J. C. Wei & Y. M. Jiang, 1988, p. 88) .

Umbilicaria virginis Schrad. 淡肤根石耳

Lich. helv. spicil. 2: 564(1842) , var. virginis

xinjiang(Wang XY, 1985, p. 344; J. C. Wei & Y. M. Jiang, 1988, p. 91 & 1993, p. 94; Abdula Abbas et al. , 1993, p. 81; Abdulla Abbas et al. 2001, p. 368) , Qinghai(Obermayer 2004, p. 514) , Xizang(Poelt, 1977, p. 431) .

Umbilicaria virginis var. lecanocarpoides(Nyl.) J. C. Wei & Y. M. Jiang 茶渍果变种

Mycosystema 1: 91(1988) .

≡ *Umbilicaria lecanocarpoides* Nyl. , *Flora*, Regensburg 43: 418(1860) .

Sichuan(J. C. Wei & Y. M. Jiang, 1988, p. 91 & 1993, p. 95) .

Umbilicaria yunnana(Nyl.) Hue 云南石耳

Nouv. Arch. Mus. Hist. Nat. 4(2) : 117(1900) .

≡ *Gyrophora yunnana* Nyl. in Hue, Bull. Soc. Bot. France, 34 – 23(1887) & *Flora* 70: 135(1887) .

Type: Yunnan, Delavay no. 1600 in H(!) .

≡ *Gyrophoropsis yunnana*(Nyl.) Ras. *Arch. Soc. Zoll. Bot. Fenn.* ' Vanamo', 6(2) : 81(1952) .

On bark of *Quercus* sp.

Xizang, Sichuan(J. C. Wei & Y. M. Jiang, 1993, p. 191) , Yunnan(Hue, 1887, p. 23 & 1889, p. 170 & 1891, p. 36 & 1900, p. 117; Nyl. 1887, p. 135; Zahlbr. , 1927, p. 742, & 1930 b, p. 138 & 1932, p. 496; Llano, 1950, p. 197; Ras. 1952, p. 81; Lamb, 1963, p. 266; J. C. Wei, 1981, p. 89; J. C. Wei & Y. M. Jiang, 1988, p. 89, & 1993, p. 191; Obermayer 2004, pp. 514 – 515; Wei X. L. et al. 2007, p. 159) .

Lecanoromycetes, orders incertae sedis 茶渍纲未定位目

"Candelariomycetidae" Miadlikowska et al. **"黄烛衣亚纲"**

Mycologia 98: 1088 – 1103(2006) ; Jolanta Miadlikowska et al. , *Molecular Phylogenetics and Evolution* 79: 148 – 149(2014) .

[18] Candelariales Miadl. , Lutzoni & Lumbsch **黄烛衣目**

in Hibbett et al. *Mycol. Res.* 111(5) : 530(2007) .

(82) Candelariaceae Hakul. **黄烛衣科**

Ann. bot. Soc. Zool. -Bot. fenn. Vanamo 27(3) : 11(1954)

Candelaria A. Massal. **黄烛衣属**

Flora, Regensburg 35: 567(1852) .

Candelaria concolor(Dicks.) Arnold 同色黄烛衣

Flora, Regensburg 62: 364(1879) .

≡ *Lichen concolor* Dicks. , *Fasc. pl. crypt. Brit.* (London) 3: 18(1793) .

Corticolous.

Neimenggu(Sun LY et al. , 2000, p. 35) , Xinjiang(Wang XY, 1985, p. 348; Abdula Abbas et al. , 1993, p. 76; Abdulla Abbas & Wu, 1998, p. 47; Abdulla Abbas et al. , 2001, p. 361) , Xizang(J. C. Wei & Y. M. Jiang, 1981, p. 1147 & 1986, p. 106) , Yunnan(Wang L. S. et al. , 2008, p. 536; Wei X. L. et al. , 2007, p. 149) , Hubei(Chen, Wu & J. C. Wei, 1989, p. 476) , Shandong(Zhao ZT et al. , 1998, p. 30; Zhang F et al. , 1999, p. 31; Zhao ZT et al. , 2002, p. 7; Li Y et al. , 2008, p. 72) , Jiangsu(Wu J. N. & Xiang T. 1981, p. 6: here should be p. 5 according to the sequence of page number; Xu B. S. , 1989, p. 249) , Shanghai, Anhui, (Xu B. S. , 1989, p. 249) . Zhejiang(Xu B. S. , 1989, p. 249; D. Hawksw. & Weng, 1990, p. 515) , Taiwan(Aptroot, Sparrius & Lai 2002, p. 282) .

var. substellata(Ach.) Vain. 星芒变种

(1929;Zahlbr. *Cat. Lich. Univ.* VI: 8(1929).

≡*Arthonia polymorpha* var. *substellata* Ach., *Syn. meth. lich.* (Lund): 7(1814).

≡*Candelaria concolor* f. *substellata*(Ach.) Oxner, *Flora Lishaĭnikiv Ukraïni* (Kiev) 2(2): 366(1993), Nom. inval., Art. 41.3(Melbourne).

Corticolous.

Yunnan or Sichuan(Zahlbr., 1930 b, p. 178).

Candelaria fibrosa(Fr.) Müll. Arg. 纤黄烛衣

Flora, Regensburg 70: 319(1887).

≡*Parmelia fibrosa* Fr., *Syst. orb. veg.* (Lundae) 1: 284(1825).

Corticolous.

Yunnan(Hue, 1900, p. 51; Asahina & Sato in Asahina, 1939, p. 715; Wang L. S. et al., 2008, p. 536), Shandong(Zhao ZT et al., 2002, p. 7), Anhui, Zhejiang(Xu B. S., 1989, p. 249).

f. callopizodes(Nyl.) Hue 优美变型

Nouv. Arch. Mus. Hist. Nat. 2: 52(1929).

≡*Lecanora callopizodes* Nyl., *Lich. Ins. Guin.*: 45(1889).

Corticolous.

Yunnan(Hue, 1900, p. 52; Zahlbr., 1930 b, p. 178).

Candelariella Müll. Arg. **黄茶渍属**

Bull. Herb. Boissier 2(app. 1): 11, 47(1894).

Candelariella antennaria Räsänen 帆黄茶渍

Anal. Soc. cient. argent. 128(3): 137(1939).

On twigs.

Ningxia(M. Liu & J. C. Wei, 2013, p. 44).

Candelariella athallina(Wedd.) Du Rietz 石黄茶渍

Beih. bot. Zbl., *Abt.* 2 49: 81(1932).

≡*Lecanora vitellina* f. *athallina* Wedd., *Mém. Soc. natn. Sci. nat. Cherbourg* 19: 278(1875).

Saxicolous.

Sichuan(Zahlbr., 1930 b, p. 178).

Candelariella aurella(Hoffm.) Zahlbr. 金黄茶渍

Cat. Lich. Univers. 5: 790(1928).

≡*Verrucaria aurella* Hoffm., *Deutschl. Fl.*, *Zweiter Theil* (Erlangen): 197(1796)[1795].

On dead twigs & on rocks.

Neimenggu(H. Magn. 1944, p. 46), Xinjiang(Abdulla Abbas et al., 1993, p. 76; Abdulla Abbas & Wu, 1998, p. 48; Abdulla Abbas et al., 2001, p. 361), Qinghai, Gansu(H. Magn. 1940, p. 125), Taiwan(Aptroot and Sparrius, 2003, p. 12).

Candelariella grimmiae Poelt & Reddi 黄茶渍

Khumbu Himal 6: 7(1969).

On soil.

Taiwan(Aptroot and Sparrius, 2003, p. 12. Also from the Himalayas).

Candelariella kansuensis H. Magn. 甘肃黄茶渍

Lich. Centr. Asia 1: 127(1940). f. kansuensis

On calciferous stone.

Type: Gansu, Bohlin no. 69 b in S(!).

Neimenggu(H. Mang. 1940, p. 128).

f. frustulenta H. Magn. 硅藻变型

Lich. Centr. Asia 1: 128(1940).

Type: Gansu, Bohlin no. 5 in S(!).

Candelariella medians(Nyl.) A. L. Sm. 间黄茶渍

Monogr. Brit. Lich. 1: 228(1918).

≡*Placodium medians* Nyl., *Bull. Soc. Amis Sci. Nat. Rouen*, Sér. II 12: 50(1877).

On limestone.

Gansu, Shaanxi(Wu J. L., 1987, p. 138), Shandong(Li Y et al., 2008, p. 72 as *Candelaria medians*).

Candelariella nepalensis Poelt & Reddi 尼泊尔黄茶渍

Khumbu Himal 6(1): 8(1969).

On mosses, shale, and acid soil.

Taiwan(Aptroot and Sparrius, 2003, p. 12. Also from the Himalayas).

Candelariella oleifera H. Magn. 油黄茶渍

Lich. Centr. Asia 1: 126(1940).

Type: Gansu, Bohlin no. 88 a, e(syntypes) in S(!).

Xinjiang(Abdulla Abbas et al., 1993, p. 76; Abdulla Abbas et al., 1994, p. 20; Abdulla Abbas & Wu, 1998, p. 48; Abdulla Abbas et al., 2001, p. 361), Gansu(H. Magn., 1940, p. 126).

Candelariella reflexa(Nyl.) Lettau 折黄茶渍

Hedwigia 52: 196(1912).

≡*Lecanora vitellina* var. *reflexa* Nyl., *Bull. Soc. bot. Fr.* 13: 241(1866).

On *Picea morrisonicola*, *Populus* sp., *Castanopsis* sp., *Salix* sp., bark of *Tsuga* sp., and shale.

Taiwan(Aptroot and Sparrius, 2003, p. 12).

Candelariella rosulans(Müll. Arg.) Zahlbr. 莲座黄茶渍

Cat. Lich. Univers. 5: 802(1928).

≡*Candelariella vitellina* var. *rosulans* Müll. Arg., in Enumer. Plants. collect by E. Penard in Colorado, 1892, Lich. p. 200(1892), Zahlbr., *Cat. Lich. Univers.* 5: 802(1928).

On soil.

Ningxia(Liu M & J. C. Wei, 2013, p. 44; Yang J. & Wei J. C., 2014, p. 1029).

Candelariella sorediosa Poelt & Reddi 粉芽黄茶渍

Khumbu Himal 6: 10(1969).

On *Picea morrisonicola*, and mature *Castanopsis* sp.

Taiwan(Aptroot and Sparrius, 2003, p. 13. Also from the Himalayas).

Candelariella vitellina(Hoffm.) Müll. Arg. 蛋黄茶渍

Bull. Herb. Boissier 2(app. 1): 47(1894).

≡*Patellaria vitellina* Hoffm., *Veg. Crypt.* 2: 5(1794).

Saxicolous.

Jilin(Hertel & Zhao, 1982, p. 144), Neimenggu(H. Magn. 1944, p. 46), Xizang(Paulson, 1925, p. 191), Yunnan(Zahlbr., 1930 b, p. 178), Shanghai(Xu B. S., 1989, p. 249), Taiwan(Aptroot and Sparrius, 2003, p. 13).

f. *athallina*(Wedd.) Zahlbr. see Candelariella athallina(Wedd.) Du Rietz

Candelariella xanthostigma(Pers. ex Ach.) Lettau 柱头黄茶渍

Hedwigia 52: 196(1912).

≡*Lichen xanthostigmus* Pers. ex Ach., *Lich. univ.*: 403(1810).

Xinjiang(Abdulla Abas et al., 1996, p. 233; Abdulla Abbas & Wu, 1998, p. 48; Abdulla Abbas et al., 2001, p. 361).

Lecanoromycetes, genera incertae sedis 茶渍衣纲未定位属

Botryolepraria Canals, Hern. -Mar., Gómez-Bolea & Llimona **串屑衣属**

Lichenologist 29(4): 340(1997).

Botryolepraria lesdainii(Hue) Canals, Hern. -Mar., Gómez-Bolea & Llimona 莱氏串屑衣

Lichenologist 29(4): 340(1997)

≡*Crocynia lesdainii* Hue, *Bull. Soc. bot. Fr.* 71: 397(1924).

On raised coral reef and volcanic rock.

Taiwan(Aptroot and Sparrius, 2003, p. 8).

Buelliastrum Zahlbr. **小黑瘤衣属**

in Handel-Mazzetti, *Symb. Sin.* 3: 122(1930) &*Cat. Lich. Univ.* 8: 417(1932).

Type species non indicated.

Buelliastrum crassum Zahlbr. 厚小黑瘤衣

in Handel-Mazzetti, *Symb. Sin.* 3: 123(1930) &*Cat. Lich. Univ.* 8: 417(1932).

Syntype: Sichuan, Handel-Mazzetti no. 1020; Yunnan, Handel-Mazzetti no. 3557.

Saxicolous.

Buelliastrum tenue Zahlbr. 薄小黑瘤衣

in Handel-Mazzetti, *Symb. Sin.* 3: 122(1930) & *Cat. Lich. Univ.* 8: 417(1932).

Type: Sichuan, Handel-Mazzetti no. 1581.

Saxicolous.

Stenhammarella Hertel **大孢网盘属**

Beih. Nova Hedwigia 24: 124(1967).

Stenhammarella turgida(Ach.) Hertel 肿大孢网盘

Beih. Nova Hedwigia 24: 125(1967).

≡*Biatora turgida* Ach., *Lich. univ.*: 273(1810).

≡*Lecidea turgida*(Ach.) A. Dietr., *Deutschl. Kryptog. Gewächse*, 4 Abth.: 91(1846).

=*Lecidea habana* Zahlbr. in Handel-Mazzetti, *Symb. Sin.* 3: 95(1930) & *Cat. Lich. Univ.* 8: 343(1932).

Type locality: Yunnan, Handel-Mazzetti no. 6986 in WU(holotype) & in W(isotype).

Yunnan(Hertel, 1967, p. 125 & 1977, p. 342).

6. Leotiomycetes O. E. Erikss. & Winka **垂舌菌纲**

Myconet 1(1): 7(1997)

[19] Helotiales Nannf. ex Korf & Lizoň **柔膜菌目**

Mycotaxon 75: 501(2000).

syn. *Leotiales* Carpenter 1988

(83) Helotiaceae Rehm [as 'Helotieae'] **柔膜菌科**

in Winter, Rabenh. Krypt. -Fl., Edn 2(Leipzig) 1. 3(lief. 37): 647(1892) [1896].

* **Unguiculariopsis** Rehm **拟爪毛盘菌属**

Annls mycol. 7(5): 400(1909).

* Unguiculariopsis damingshanica W. Y. Zhuang **大明山拟爪毛盘菌**

Mycol. Res. 104(4): 508(2000).

Holotype: Guangxi: on a thin layer of lichen thalli on bamboo, Shiwandashan, Shangsi, Guangxi, alt. 300m, *S. L. Chen & W. Y. Zhuang* 1942, 27 Dec. 1997(HMAS 76137). *Paratype*: on a thin layer of lichen thalli On bamboo, Damingshan, Wuming, Guangxi, alt. 1200 m, W. P. Wu & W. Y. Zhuang 1916, 22 Dec. 1997 (HMAS 74847).

Guangxi(Zhuang W. Y., 2000, p. 508).

Helotiales, genera incertae sedis **柔膜菌目未定位属**

Rhymbocarpus Zopf **顶果衣属**

Hedwigia 35(6): 357(1896).

Rhymbocarpus neglectus(Vain.) Diederich & Etayo 疏顶果衣

Lichenologist 32(5) : 467(2000) .

≡*Lepraria neglecta* Vain. , in Lettau(1934) .

Hongkong(Aptroot et al. , 2001, p. 330) .

7. Lichinomycetes Reeb, Lutzoni & Cl. Roux, **异极衣纲**

Mol. Phylogen. Evol. 32: 1055(2004) .

[20] Lichinales Henssen & Büdel, **异极衣目**

in Eriksson & Hawksworth, Syst. Ascom. 5(1) : 138(1986)

(84) Heppiaceae Zahlbr. **蜂窝衣科**

Nat. Pflanzenfam. (Leipzig) 1(1 *) : 176(1906) .

Heppia Nägeli ex A. Massal. **蜂窝衣属**

Geneac. lich. (Verona) : 7 – 8(1854) .

Heppia conchiloba Zahlbr. 壶型蜂窝衣

Beih. Botan. Centralbl. 13: 157(1902) .

On soil.

Xizang(W. Obermayer, 2004, p. 497) .

Heppia lutosa(Ach.) Nyl. 泥蜂窝衣

Syn. meth. lich. (Parisiis) 2: 45(1869)

≡*Collema lutosum* Ach. , *Syn. meth. lich.* (Lund) : 309(1814) .

Ningxia(Yang & Wei, 2014, p. 1031) .

(85) Lichinaceae Nyl. **异极衣科**

Mém. Soc. Sci. nat. Cherbourg 2: 8(1854) .

Anema Nyl. ex Forssell **阿奈玛属**

Beitr. Gloeolich. : 40, 91(1885) .

Anema asahinae Yoshim. 朝比阿奈玛

J. Jap. Bot. 43: 357(1968) .

On rocks.

Liaoning(Yoshim. 1968 a, p. 357) .

Ephebe Fr. **毡衣属**

Syst. orb. veg. (Lundae) 1: 256(1825) .

Ephebe hispidula(Ach.) Horw. 毛毡衣

Hand-list Lich. Gr. Brit. : 6(1912) .

≡*Cornicularia hispidula* Ach. , *Lich. univ.* : 1 – 696(1810) .

On rocks.

Fujian(Wu et al. 1984, p. 1; Wu J. L. , 1987, p. 49) , JIANGSU(Wu J. L. , 1987, p. 49) .

Ephebe lanata(L.) Vain. 毡衣

Meddn Soc. Fauna Flora fenn. 14: 20(1888) .

≡*Lichen lanatus* L. , *Sp. pl.* 2: 1155(1753) .

= *Ephebe pubescens* auct. brit.

On rock.

Fujian(Zahlbr. , 1930 b, p. 73; Wu et al. 1984, p. 1) , Taiwan(Aptroot, Sparrius & Lai 2002, p. 284, on schistose rock) .

Euopsis Nyl. **真衣属**

Flora, Regensburg 58: 363(1875) .

Euopsis pulvinata(Schaer.) Vain. 垫状真衣

Meddn Soc. Fauna Flora fenn. 6: 85(1881) .

≡*Lecidea pulvinata* Schaer., *Naturw. Anzeiger Allgem. Schweizer. Gesellsch. Naturwiss.* 2: 11(1818).

On exposed, wet granite rock along streams. Northern temperate, and reported also from the mountain of South East Asia.

Xizang(Obermayer W., 2004, pp. 495 – 496), Hongkong(Aptroot & Seaward, 1999, p. 75), Taiwan(Aptroot and Sparrius, 2003, p. 16).

Lempholemma Körb. **天粘衣属**

Syst. lich. germ. (Breslau): 400(1855).

Lempholemma chalazanum(Ach.) B. de Lesd. 合点天粘衣

Recherch. Lich. Dunkerque: 261(1910).

≡*Collema chalazanum* Ach., *Lich. univ.*: 630(1810).

On brick and compacted soil. Northern temperate.

Taiwan(Aptroot and Sparrius, 2003, p. 27), Hongkong(Aptroot & Seaward, 1999, pp. 80 – 81; Aptroot et al., 2001, p. 330).

Lempholemma cladodes(Tuck.) Zahlbr. 叶枝天粘衣

Cat. Lich. Univers. 3: 23(1924)[1925].

≡*Collema cladodes* Tuck., Gen. lich. (Amherst): 89(1872).

On soil crusts & rocks.

Xinjiang(Turgunay et al., 2015, pp. 2340 – 2341).

Leptopterygium Zahlbr. **薄衣属**

in Handel-Mazzetti, *Symb. Sin.* 3: 74(1930) & *Cat. Lich. Univ.* 8: 275(1932).

Type species: *Leptopterygium gracilentum* Zahlbr.

Leptopterygium gracilentum Zahlbr. 薄衣

in Handel-Mazzetti, *Symb. Sin.* 3: 74(1930) & *Cat. Lich. Univ.* 8: 275(1932).

Type locality: Sichuan, Handel-Mazzetti no. 7206.

On calcareous rock.

Lichinella Nyl. **小极衣属**

Bull. Soc. linn. Normandie, sér. 2 6: 301(1872).

Lichinella nigritella(Lettau) P. P. Moreno & Egea 黑小极衣

Cryptog. Bryol. -Lichénol. 13(3): 246(1992).

≡*Thyrea nigritella* Lettau, Feddes Repert., Beih. 119(no. 5): 276(1942).

On rocks.

Xinjiang(Turgunay et al., 2015, p. 2342).

Peccania A. Massal. ex Arnold **同枝衣属**

Flora, Regensburg 41: 93(1858)[1859 – 1860].

Peccania polyspora H. Magn. 多孢同枝衣

Lichens Central Asia 1: 39(1940).

Type: Gansu, Bohlin no. 26e in S.

On calciferous stone.

Neimenggu(H. Magn. 1944, p. 18), Xinjiang(Abdulla Abbas et al., 1993, p. 79; Abdulla Abbas & Wu, 1998, p. 131; Abdulla Abbas et al., 2001, p. 365).

Peccania terricola H. Magn. 地生同枝衣

Lich. Centr. Asia 1: 39(1940).

Type: Gansu, Bohlin no. 1 in S(not seen).

On earth.

Xinjiang(Abdulla Abbas et al., 1993, p. 79; Abdulla Abbas & Wu, 1998, pp. 131 – 132; Abdulla Abbas et al., 2001, p. 365), Gansu(H. Magn., 1940, p. 39).

Phylliscum Nyl. **菊花衣属**

Mém. Soc. Imp. Sci. Nat. Cherbourg 3: 166(1855).

Phylliscum demangeonii(Moug. & Mont.) Nyl. 多孢菊花衣

Mém. Soc. Imp. Sci. Nat. Cherbourg 3: 166(1855).

≡*Collema demangeonii* Moug. & Mont., *Annls Sci. Nat.*, *Bot.*, sér. 3 12: 291(1849).

On rocks.

Gansu(Wu J. L. 1987, p. 50).

Phylliscum japonicum Zahlbr. 日本菊花衣

Bot. Mag., Tokyo 41: 320(1927).

On rocks.

Jiangxi(Wu J. L. 1987, p. 51).

Phylliscum testudineum Henssen 玳瑁菊花衣

Svensk Bot. Tidskr. 57: 153(1963).

Misapplied name:

Peltula coriacea Büdel, Henssen & Wessels, in Büdel & Henssen, *Int. J. Mycol. Lichenol.* 2(2 – 3): 238 (1986).

On coastal granitic rocks. Known from tropical South Africa.

Hongkong(Aptroot & Seaward, 1999, p. 84).

On siliceous rock.

Hongkong(Aptroot et al., 2001, p. 337).

Psorotichia A. Massal. **鳞壁衣属**

Framm. Lichenogr.: 15(1855).

Psorotichia dispersa H. Magn. 散鳞壁衣

Lichens Central Asia 2: 17(1944)

Type: Neimenggu, Bohlin, geol. sample no. 184 in S(!).

On calcareous stone.

Psorotichia kansuensis H. Magn. 甘肃鳞壁衣

Lichens Central Asia 1: 35(1940).

Type locality: Gansu, Bohlin no. 38 b, in S(not seen).

On calcareous soil.

Psorotichia minuta H. Magn. 小鳞壁衣

Lichens Central Asia 1: 35(1940)

Type: Gansu, Bohlin no. 25 in S(!).

Neimenggu, Gansu(H. Magn. 1940, p. 36: paratypes).

On calciferous conglomerate.

Psorotichia mongolica H. Magn. 蒙古鳞壁衣

Lichens Central Asia 1: 36(1940)

Type localities: Neimenggu(Bexell no. 6: lectotype, selected here in S, 1982), Gansu(Bohlin, geol. sample nos. 1334, 1396; no. 79 b), Qinghai(Bohlin, geol. sample no. 967).

On calciferous rocks.

Psorotichia murorum A. Massal. 鳞壁衣

Framm. Lichenogr.: 15(1855).

On limestone.

Taiwan(Aptroot and Sparrius, 2003, pp. 36 – 37).

Psorotichia nigra H. Magn. 黑鳞壁衣

Lichens Central Asia 1: 37(1940).

Type: Gansu, Bohlin no. 28 d in S.

Gansu, Qinghai(H. Magn. 1940, p. 37: paratypes) .

On the ground.

Xinjiang(Abdulla Abbas et al. , 1994, p. 22; Abdulla Abbas & Wu, 1998, p. 132; Abdulla Abbas et al. , 2001, p. 367) .

Psorotichia schaereri(A. Massal.) Arnold 莎氏鳞壁衣

Flora, Regensburg 52: 265(1869) .

≡*Pannaria schaereri* A. Massal. , *Ric. auton. lich. crost.* (Verona) : 114(1852) .

On exposed or sheltered concrete and brick. Northern temperate.

Hongkong(Aptroot & Seaward, 1999, p. 90) , Taiwan(Aptroot and Sparrius, 2003, p. 37) .

Psorotichia sinensis Zahlbr. 华鳞壁衣

in Handel-Mazzetti, *Symb. Sin*. 3: 73(1930) &Cat. Lich. 8: 27(1932) .

Type locality: Sichuan, Handel-Mazzetti no. 1772.

On arenaceous rock.

Pterygiopsis Vain. **翅衣属**

Acta Soc. Fauna Flora fenn. 7(1) : 238(1890) .

Pterygiopsis convoluta Henssen [as ‘ Pterygiopis’] 卷翅衣

Lichenes Cyanophili et Fungi Saxicolae Exsiccati, Fascicle 2(nos 26 – 50) (Marburg) : 8, no. 43(1990) .

On shale.

Taiwan(Aptroot and Sparrius, 2003, p. 37) .

Pyrenopsis(Nyl.) Nyl. **类核衣属**

Syn. meth. lich. (Parisiis) 1(1) : 97(1858) .

≡*Synalissa* sect. *Pyrenopsis* Nyl. , *Mém. Soc. Imp. Sci. Nat.* Cherbourg 3: 164(1855) .

Pyrenopsis furfurea(Nyl.) Th. Fr. 糠类核衣

Bot. Notiser: 58(1866) .

≡*Collema furfureum* Nyl. , *J. Bot.* , *Lond*. 3: 286(1865) .

On coastal rocks. Known from temperate Europe and North America. Seaward & Chu, MRDS 108673 with *Buellia* cf. *testacea*.

Hongkongot & Seaward, 1999, p. 90; Aptroot & Sipman, 2001, p. 338) .

Thyrea A. Massal. **盾链衣属**

Sched. critic. (Veronae) 4: 75(1856) .

Thyrea hondana Zahlbr. 盾链衣

Annls mycol. 29(1/2) : 79(1931) .

Beijing(Badaling, Yoshim. 1968, p. 502) .

Thyrea mongolica H. Magn. 蒙古盾链衣

Lichens Central Asia 2: 19(1944) .

Type: Neimenggu, Bohlin no. 163 in S.

On rocks.

Thyrea pulvinata(Schaer.) A. Massal. 垫盾链衣

Flora, Regensburg 39: 211(1856) .

≡*Parmelia stygia* var. *pulvinata* Schaer. , *Lich. helv. spicil*. 11: 544(1842) .

≡*Omphalaria pulvinata*(Schaer.) Nyl. , *Annls Sci. Nat.* , *Bot.* , sér. 3 20: 320(1853) .

On quartzite.

Beijing(Moreau et Moreau , 1951, p. 185) , Gansu(H. Magn. 1940, p. 40) , Xinjiang(Wu J. L. 1985, p. 78) , Yunnan(Zahlbr. 1930 b, p. 74) .

(86) Peltulaceae Büdel **盾衣科**

in Eriksson & Hawksworth *Syst. Ascom.* 5: 149(1986).

Peltula Nyl. **盾衣属**

Annls Sci. Nat., *Bot.*, sér. 3 20: 216(1853).

Peltula applanata(Zahlbr.) J. C. Wei 扁盾衣

Enum. Lich. China(Beijing): 187(1991).

≡*Heppia applanata* Zahlbr., in Handel-Mazzetti, *Symb. Sinic.* 3: 78(1930) & *Cat. Lich. Univ.* 8: 295(1932).

Type locality: Guizhou, Handel-Mazzetti no. 10804.

On rocks.

Peltula bolanderi(Tuck.) Wetmore 波氏盾衣

Ann. Mo. bot. Gdn 57: 179(1971)[1970].

≡*Pannaria bolanderi* Tuck., *Gen. lich.* (Amherst): 51(1872).

On coastal granitic rocks. Known from Africa, Australia, neotropics, and Asia.

Hongkong(Aptroot & Seaward, 1999, p. 84; Aptroot et al., 2001, p. 335), Taiwan(Aptroot and Sparrius, 2003, pp. 31 – 32).

Peltula clavata(Kremp.) Wetmore 棒盾衣

Ann. Mo. bot. Gdn 57: 181(1971)[1970].

≡*Heterina clavata* Kremp., *Flora*, Regensburg 59: 56(1876).

On coastal granitic rocks. Pantropical.

Hongkong(Aptroot & Seaward, 1999, p. 84: with *Peltula placodizans*).

Peltula cylindrica Wetmore 柱盾衣

Ann. Mo. bot. Gdn 57: 182(1971)[1970].

On rocks.

Jiangsu(Wu J. N. & Xiang T. 1981, p. 2), Shaanxi(Wu J. L. 1987, p. 60).

Peltula euploca(Ach.) Poelt ex Ozenda & Clauzade 粉芽盾衣

Les Lichens(Paris): 324(1970).

≡*Lichen euplocus* Ach., *Lich. suec. prodr.* (Linköping): 181(1799)[1798].

≡*Endocarpon euplocum*(Ach.) Ach., *Methodus, Sectio prior*: 127(1803).

=*Endocarpon guepinii* Delise, *Bot. Gall.*, Edn 2(Paris) 2: 594(1830).

≡*Heppia guepinii* (Delise) Nyl., in Hue, *Rev. Bot.* 5: 18(1887).

On rocks. Pantropical.

Xinjiang(Abdulla Abbas et al., 1993, p. 79, as "*Peltule*" *euploca*; Abdulla Abbas & Wu, 1998, pp. 132 – 133; Abdulla Abbas et al., 2001, p. 366), Shandong(Zhao ZT et al., 1998, p. 30; Zhao ZT et al., 2002, p. 8), North-eastern China(prov. not indicated: Asahina, 1934 d, p. 683; Asahina & Sato in Asahina, 1939, p. 643), Jiangsu(Xu B. S., 1989, pp. 183 – 184), Hongkong(Aptroot & Seaward, 1999, p. 84 – 85), Taiwan(Aptroot and Sparrius, 2003, p. 32).

Peltula impressula(H. Magn.) N. S. Golubk. 凹盾衣

Konspekt Flory Lishaĭnikov Mongol'skoĭ Naradnoĭ Respubliki (Leningrad): 32(1981).

≡*Heppia impressula* H. Magn. *Lich. Centr. Asia*, 1: 43(1940).

Type: Gansu, Bohlin, 16/VI/1930 in S(!).

On calcareous soil.

Peltula minuta(H. Magn.) N. S. Golubk. 小盾衣

Konspekt Flory Lishaĭnikov Mongol'skoĭ Naradnoĭ Respubliki (Leningrad): 33(1981).

≡*Heppia minuta* H. Magn. *Lich. Centr. Asia*, 1: 44(1940).

Type: Gansu, Bohlin no. 7 in S(not seen), no. 12d(paratype) in S(!).

On *Catillaria kansuensis*.

Peltula obscurans(Nyl.) Gyeln. 暗盾衣

Reprium nov. Spec. Regni veg. 38: 308(1935) .

≡*Endocarpiscum obscurans* Nyl. , *Bull. Soc. linn. Normandie*, sér. 2 6: 309(1872) .

Peltula sp. (Chu, 1997, p. 48) .

On concrete and granitic rocks. Pantropical

Hongkong(Chu, 1997, p. 48 as *Peltula* sp. ; Aptroot & Seaward, 1999, p. 85; Aptroot et al. , 2001, p. 335) , Taiwan(Aptroot and Sparrius, 2003, p. 32) .

Peltula olifera(H. Magn.) J. C. Wei 油盾衣

Enum. Lichens China(Beijing) : 187(1991) .

≡*Heppia oleifera* H. Magn. , *Lichens Central Asia* 2: 21(1944) .

Type: Neimenggu, geol. sampol. sample no. 168 in S.

On calcareous rock.

Peltula placodizans(Zahlbr.) Wetmore 台盾衣

Ann. Mo. bot. Gdn 57: 179(1971) [1970] .

≡*Heppia placodizans* Zahlbr. , *Bull. Torrey bot. Club* 35: 299(1908) .

Misapplied name:

Peltula impressa auct non(Vainio) Swinsco & Krog in Chu, 1997, p. 48, revised by Aptroot & Seaward, 1999, p. 85.

On coastal granitic rocks. Pantropical.

Hongkong(Chu, 1997, p. 48; Aptroot & Seaward, 1999, p. 85: Lamma island, 1994, Chu(107735, 108704 & HKU(M) 10467, 10448 with also *Caloplaca leptozona*) , Taiwan(Aptroot & Sparrius, 2003, p. 32) .

Peltula radicata Nyl. 根盾衣

Annls Sci. Nat. , *Bot.* , sér. 3 20: 316(1853) .

≡*Heppia radicata*(Nyl.) Vain. , *Étud. Class. Lich. Brésil* 1: 215(1890) .

On earth.

Gansu(H. Magn. 1940, p. 45) .

Peltula tortuosa(Ach.) Wetmore 多曲盾衣

Ann. Mo. bot. Gdn 57(2) : 205(1971) [1970] .

≡*Parmelia tortuosa* Ach. , *Methodus*, Sectio post. : 184(1803) .

On rock.

Xinjiang(Wu JN et al. , 1997, pp. 14 – 15) .

Peltula zabolotnoji(Elenkin) N. S. Golubk. 匝氏盾衣

Konspekt Flory Lishaĭnikov Mongol'skoĭ Naradnoĭ Respubliki(Leningrad) : 31(1981) .

≡*Heppia zabolotnoji* Elenkin, *Bull. Jard. Imp. Bot.* St. -Pétersb. 5: 86(1905) .

Type locality: Neimenggu, Zabolotnaj, 16/IX/1898.

= *Heppia kansuensis* H. Magn. Lich. Centr. Asia, 1: 43(1940) .

Type: Gansu, Bohlin no. 6a in S(!) .

On calcareous rock.

Phyllopeltula Kalb **叶盾衣属**

Biblthca Lichenol. 78: 158(2001) .

Phyllopeltula corticola(Büdel & R. Sant.) Kalb 树生叶盾衣

Biblthca Lichenol. 78: 158(2001) .

≡*Peltula corticola* Büdel & R. Sant. , in Büdel, *Biblthca Lichenol.* 23: 79(1987) .

On raised coral reef, conglomeratic rock, volcanic rock. and shale shale.

Taiwan(Aptroot and Sparrius, 2003, pp. 33 – 34).

[21] **Abrothallales** Sergio Pérez-Ortega & Ave Suija 纤柔菌目

in Pérez-Ortega, Suija, Crespo & Ríos, *Fungal Diversity* 64(1): 302(2014).

Typification Details: *Ephebe* Fr. 1825

(87) **Abrothallaceae** Sergio Pérez-Ortega & Ave Suija 纤柔菌科

in Pérez-Ortega, Suija, Crespo & Ríos, *Fungal Diversity* 64(1): 303(2014).

Ephebe Fr., Syst. orb. veg. (Lundae) 1: 256(1825).

* **Abrothallus** De Not. 纤柔菌属

Abrothallus: 1(1845).

= *Abrothallomyces* Cif. & Tomas., *Atti Ist. bot. Univ. Lab. crittog. Pavia*, Ser. 4 10(1): 49, 75(1953).

= *Phymatopsis* Tul. ex Trevis., *Linnaea* 28: 296(1857).

= *Pseudolecidea* Marchand, *Énum. Méth. Rais Fam. Genres Mycophytes*(Paris): 159(1896).

* Abrothallus peyritschii(Stein) I. Kotte 佩氏纤柔菌

Centbl. Bakt. ParasitKde, Abt. II 24: 76(1909).

≡ *Abrothallus parmeliarum* var. *peyritschii* Stein, in Cohn, Krypt. -Fl. Schlesien(Breslau) 2(2): 211(1879).

≡ *Lecidea buelliana* f. *peyritschii* (Stein) Vain. in Medd. Soc. Fauna & Flora Fenn. X: 119(1883).

Notes: This lichenicolous fungus, growing on *Vulpicida pinastri*, has been de- termined by J. Hafellner in 2003. Specimen examined: CHINA, Xizang, SE Tibet: Tibetan Himalaya N of Bhutan, Kuru Chu, Hill SW of Lhozak Vy. junction, 28°18′N, 90°57′, 4410 m alt., uppermost Abies trees in supalpine *Rhododendron* scrub, steep N-facing slope, on bark(lichenicolous fungus on *Vulpicida pinastri*), 1994-07 – 22, G. Miehe(94 – 88 – 16/06B) & U. Wündisch. Walter Obermayer Additions to the lichen flora of the Tibetan region in Contribution to Lichenology. Festschrift in Honour of Hannes Hertel. P. Döbbeler & G. Rambold(eds): *Bibliotheca Lichenologica* 88: 479 – 526. J. Cramer in der Gebrüder Borntraeger Verlagsbuchhandlung, Berlin-Stuttgart, 2004. pp. 484

8. **Sordariomycetes** O. E. Erikss. & Winka 粪壳菌纲

Myconet 1(1): 10(1997).

1) **Hypocreomycetidae** O. E. Erikss. & Winka 肉座菌亚纲

Myconet 1(1): 6(1997).

[22] **Hypocreales** Lindau 肉座菌目

in Engler & Prantl(eds), Nat. Pflanzenfam. 1: 343(1897).

(88) **Bionectriaceae** Samuels & Rossman 生赤壳菌科

Stud. Mycol. 42: 15(1999).

* **Paranectria** Sacc. 游菌属

Michelia 1(3): 317(1878).

* Paranectria oropensis subsp. parviseptata M. S. Cole & D. Hawksw. 游菌小隔亚种

Mycotaxon 77: 324(2001). Etym.: parvi-(spasely) septata(septate).

Holotypus: Taiwan, Hualien Hsien, on *Parmelia*(*Hypotrachyna*) sp., on *Pinus*, alt. 2370m, May 3, 1996, D. L. Hawksworth(K(M) 69600)

Taiwan(Cole & Hawksw. 2001, pp. 324 – 326; Hawksw. & M. Cole, 2003, p. 361).

2) **Sordariomycetidae** O. E. Erikss. & Winka 粪壳菌亚纲

Myconet 1: 10(1997).

[23] **Sordariales** genera incertae sedis 粪壳菌目未定位属

* **Roselliniella** Vain. 蔷菌属

Acta Soc. Fauna Flora fenn. 49(2): 77(1921).

* Roselliniella euparmeliicola Millanes & D. Hawksw. 梅衣蔷菌

in D. Hawksw., Millanes & Wedin, *Persoonia* 24: 13(2010).

Yunnan Province, Jade Dragon Mountain, White River Stop, on *Parmelia meiophora*, 5Oct. 2005, *M. A. Allen & B. H. Hilton* 405 – 8 – 2(*c*), holotype BM 000920346.

Host—In the thallus of *Parmelia meiophora*, apparently commensalistic as the infected areas of the thallus are not discoloured and there was no sign of necrosis. Distribution—China(Yunnan Province); known only from the holotype collection.

Sordariomycetes, families incertae sedis 粪壳菌纲未定位科

(89) Apiosporaceae K. D. Hyde, J. Fröhl., Joanne E. Taylor & M. E. Barr **梨孢假壳菌科**

in Hyde, Fröhlich & Taylor, *Sydowia* 50(1): 23(1998).

* **Apiospora** Sacc. **梨孢假壳菌属**

Atti Soc. Veneto-Trent. Sci. Nat., Padova, Sér. 4 4: 85(1875).

* Apiospora mongolica H. Magn. **蒙古梨孢假壳菌**

Lich. Centr. Asia 2: 40 – 41(1944).

On *Aspicilia subdifracta*(as *Lecanora subdifracta* H. Magn. 1944, pp. 40 – 41, Pl. VI, fig. 4.).

Neimenggu(Hawksw. & M. Cole, 2003, p. 360, noted: "presumably either a species of *Cercidospora* close to C. caudate, or a synonym of the latter species".).

(90) Obryzaceae Körb. **梨壳菌科**

[as ' Obryzeae'] *Syst. lich. germ.* (Breslau): 427(1855).

* **Obryzum** Wallr. **梨壳菌属**

Naturgesch. Flecht. 1: 253(1825).

* Obryzum striguloides Aptroot & Sipman 毛梨壳菌

J. Hattori bot. Lab. 91: 333(2001).

Type: China, Hongkong, Aptroot 48617(B-holotype; HKU(M), ABL-isotypes).

On *Agnonimia vouauxii* (Aptroot & Sipman 2001).

Hongkong(Hawksw. & M. Cole, 2003, p. 361; Aptroot et al., 2001, p. 333).

(91) Thelenellaceae O. E. Erikss. ex H. Mayrhofer **乳头菌科**

Biblthca Lichenol. 26: 16(1987).

Julella Fabre **晕孢衣属**

Annls Sci. Nat., Bot., sér. 6 9: 113(1879) [1878].

Julella lactea(A. Massal.) M. E. Barr 乳白晕孢衣

Sydowia 38: 13(1986) [1985].

≡ *Blastodesmia lactea* A. Massal., *Ric. auton. lich. crost.* (Verona): 181, fig. 369(1852).

On *Schefflera* sp.

Taiwan(Aptroot 2003, p. 160).

Julella vitrispora(Cooke & Harkn.) M. E. Barr 玻璃晕孢衣

Sydowia 38: 13(1986) [1985].

≡ *Pleospora vitrispora* Cooke & Harkn., *Grevillea* 9(no. 51): 86(1881).

Polyblastiopsis geminella(Nyl.) Zahlbr. (in Thrower, 1988, pp. 17, 148).

On exposed trees. Cosmopolitan.

Hongkong(Thrower, pp. 17, 148; Aptroot & Sewaward, 1999, p. 79).

Ascomycota, families incertae sedis 子囊菌门未定位科

(92) Epigloeaceae Zahlbr. **表粘衣科**

in Engler & Prantl, Nat. Pflanzenfam., Teil. I(Leipzig) 1 [*]: 53(1903).

Epigloea Zukal **表粘衣属**

Verh. zool.-bot. Ges. Wien 39: 78(1889).

Epigloea renitens(Grummann) Döbbeler 表粘衣

Beih. *Nova Hedwigia* 79: 228(1984).

≡ *Vorarlbergia renitens* Grummann, *Sydowia* 22(1 – 4) : 219(1969).

On acid soil.

Taiwan(Aptroot 2003, p. 160).

Epigloea sparrii Aptroot 斯巴氏表粘衣

J. Hattori bot. Lab. 93: 160(2003).

Type: Taiwan, Hulien Conty: Taroko National Park, Hohuan Shan, roadside with relict mature trees, 3200m. alt., 51RUG2572. Aptroot 52851(ABL, holotype), 13Oct. 2001.

On soil.

Taiwan(Aptroot 2003, p. 160).

(93) Thelocarpaceae Zukal [as ' Thelocarpeae'] **乳头果衣科**

Öst. bot. Z. 43: 247(1893).

Thelocarpon Nyl. **乳头果属**

Annls Sci. Nat., Bot., sér. 3 20: 318(1853).

Thelocarpon olivaceum B. de Lesd. 绿乳头果

Recherch. Lich. Dunkerque 1(Suppl.) : 149(1914).

On dead fungi on woodin the garden of the field station.

Taiwan(Aptroot 2003, p. 170).

Thelocarpon superellum Nyl. 上乳头果

Flora, Regensburg 48: 261(1865).

On vertical soil banks.

Hongkong(Aptroot & Sipman, 2001, p. 340).

Ascomycota, genera incertae sedis Pezizomycotina 子囊菌门未定位属

* **Intralichen** D. Hawksw. & M. S. Cole **衣内菌属**

Fungal Diversity 11: 88(2002).

* Intralichen christiansenii(D. Hawksw.) D. Hawksw. & M. S. Cole 衣内菌

Fungal Diversity 11: 90(2002).

On Dirinaria applanata.

Yunnan(D. Hawksw. & M. Cole, 2003, pp. 360 – 361).

* **Lichenoconium** Petr. & Syd. **衣外菌属**

Beih. Reprium nov. Spec. Regni veg. 42(1) : 432(1927).

* Lichenoconium erodens M. S. Christ. & D. Hawksw. 衣外菌

in Hawksworth, Persoonia 9(2) : 174(1977).

Yunnan(Hawksw. & M. Cole, 2003, p. 361: on *Pannoparmelia* sp. by Hawksworth DCH01 in KUN, on *Everniastrum* sp. by Hawksworth DCH 07 in KUN, on *Flavoparmelia* sp. Hawksworth DCH1a in KUN, *Heterodermia* cfr. *obscurata*, *Parmelia* sp., and *Pannoparmelia* sp. by Hawksworth in KUN), Taiwan(Hawksw. & M. Cole, 2003, p. 361: On *Hypotrachina* sp., and *Lecanora* sp. by Hawksworth in herb. Aptroot).

* Lichenoconium lecanorae(Jaap) D. Hawksw. 茶渍衣外菌

Bull. Br. Mus. nat. Hist., Bot. 6(3) : 270(1979).

≡ *Coniosporium lecanorae* Jaap, in Lindau, *Verh. bot. Ver. Prov. Brandenb.* 47: 71(1905).

On *Flavoparmelia* sp., and *Lecanora leprosa*.

Yunnan(Hawksw. & M. Cole, 2003, p. 361, on *Flavoparmelia* sp. by Hawksworth DCH1b in KUN), Hongkong (Hawksw. & M. Cole, 2003, p. 361, on *Lecanora leprosa*-Aptroot & Sipman 2001, p. 331).

Milospium D. Hawksw. **鹰衣属**

Trans. Br. mycol. Soc. 65(2) : 228(1975).

Milospium planorbis Aptroot & Sipman 平鹰衣

J. Hattori bot. Lab. 91: 331(2001).

On shaded, overhanging, dry, smooth bark in secondary mountain forest.

Hongkong(Aptroot et al., 2001, p. 331).

Normandina Nyl. **小皿叶属**

Mém. Soc. Imp. Sci. Nat. Cherbourg 3: 191(1855).

Normandina pulchella(Borrer) Nyl. 小皿叶

Annls Sci. Nat., Bot., sér. 4 15: 382(1861).

≡ *Verrucaria pulchella* Borrer, *Suppl. Engl. Bot.* 1: tab. 2602, fig. 1(1831).

On *Coccocarpia* sp.

Xijinag(Abbas et al., 1997, p. 3), Xizang(J. C. Wei & Y. M. Jiang, 1981, p. 1147 & 1986, p. 116; Obermayer, 2004, p. 504), Sichuan(Chen S. F. et al., 1997, p. 222, on *Phaeophyscia primaria*; Obermayer, 2004, p. 504), Guizhou(Zhang T & Wei JC, 2006, p. 9), Shandong(Zhao ZT et al., 2002, p. 8; Li Y et al., 2008, p. 73), Anhui, Zhejiang(Xu B. S. 1989, p. 266), Fujian(Wu J. N. et al. 1984, p. 1), Taiwan(Aptroot 2003, pp. 162 – 163, on tree branches, on *Picea morrisonicola*, on *Pinus*, *Populus*, *Dendropanax*, and *Castanopsis*, and also on soil, and shale).

* **Phaeosporobolus** D. Hawksw. & Hafellner **暗孢菌属**

Nova Hedwigia 43(3 – 4): 525(1986).

* Phaeosporobolus alpinus R. Sant., Alstrup & D. Hawksw. 高山暗孢菌

in Alstrup & Hawksworth, *Meddr Grønland, Biosc.* 31: 51(1990).

On *Pertusaria* sp.

Yunnan(D. Hawksw. & M. Cole, 2003, pp. 361 – 362).

* **Ampulifera** Deighton(1960) **细颈菌属**

anamorphic *pezizomycotina*(on folicolous lichens).

* Ampulifera foliicola Deighton 1960 叶生细颈菌

On an unidentified foliicolous lichen on *Bambusa oldhamii*.

Taiwan(Hawksw. & M. Cole, in *Mycosystema* 22(3): 360, 2003).

Koerberiella Stein **柯氏菌属**

in Cohn, Krypt. -Fl. Schlesien(Breslau) 2(2): 143(1879).

Koerberiella wimmeriana (Körber) B. Stein 柯氏菌

On shale.

Taiwan(Aptroot and Sparrius, 2003, p. 23).

(94) Carbonicolaceae Bendiksby & Timdal(incertae sedis) **炭菌科**

Taxon 62(5): 950(2013).

Carbonicola Bendiksby & Timdal **炭菌属**

Taxon 62(5): 950(2013).

Carbonicola myrmecina(Ach.) Bendiksby & Timdal **蚁炭菌**

Taxon 62(5): 951(2013).

≡ *Lecidea scalaris* var. *myrmecina* Ach., *Methodus*, Sectio prior(Stockholmiæ): 78(1803).

= *Hypocenomyce scalaris*(Ach. ex Lilj.) M. Choisy, Bull. mens. Soc. linn. Soc. Bot. Lyon 20: 133(1951).

≡ *Lichen scalaris* Ach. ex Lilj., *Utkast Svensk Fl.*: 422(1792).

On trunk of very old *Juniperus*

Sichuan(W. Obermayer, 2004, p. 497, as *Hypocenomyce* scalaris).

II: Basidiomycota R. T. Moore **担子菌门**

Bot. Mar. 23(6): 371 (1980).

Agaricomycotina Doweld **伞菌亚门**

Prosyllabus Tracheophytorum, Tentamen Systematis Plantarum Vascularium (Tracheophyta) (Moscow): LXXVIII (2001).

9. Agaricomycetes Doweld 伞菌纲

Prosyllabus Tracheophytorum, Tentamen Systematis Plantarum Vascularium (Tracheophyta) (Moscow): LXXVIII (2001).

1) Agaricomycetidae Parmasto 伞菌亚纲

Windahlia 16: 16 (1986).

[24] Atheliales Jülich 无乳头菌目

Biblthca Mycol. 85: 343 (1981).

(95) Atheliaceae Jülich 无乳头菌科

Biblthca Mycol. 85: 355 (1982).

* **Athelia Pers.** 无乳头菌属

Traité champ. (Paris): 57 (1818).

* Athelia arachnoidea (Berk.) Jülich 无乳头菌

Willdenowia, Beih. 7: 53 (1972).

On sterile crustose lichens and algae.

Yunnan (Hawksw. & M. Cole, 2003, p. 360).

[25] Agaricales Underw. 伞菌目

Moulds, mildews and mushrooms. A guide to the systematic study of the Fungiand Mycetozoa and their literature (New York): 97 (1899).

(96) Hygrophoraceae Lotsy 蜡伞菌科

Vortr. bot. Stammesgesch. 1: 705 (1907).

Dictyonema C. Agardh ex Kunth 云片衣属

Syn. pl. (Paris) 1: 1 (1822).

Dictyonema irpicinum Mont. 耙状云片衣

Annls Sci. Nat., Bot., sér. 3 10: 119 (1848).

var. scabridum Zahlbr. 粗面变种

Taiwan (Asahina, 1944b, p. 235).

Dictyonema membranaceum C. Agardh 膜云片衣

Syst. alg.: 85 (1824).

Taiwan (Asahina, 1944, p. 236).

Dictyonema thelephora (Spreng.) Zahlbr. 云片衣

Cat. Lich. Univers. 7: 748 (1931).

≡*Dematium thelephora* Spreng., *K. svenska Vetensk - Akad. Handl.* 46: 53 (1820).

=*Thelephora sericea* (Sw.) Sw., *Fl. Ind. Occid.* 3: 1928 (1806).

≡*Dictyonema sericeum* (Sw.) Berk., *London J. Bot.* 2: 639 (1843) f. *sericeum*

≡*Hydnum sericeum* Sw., *Prodr.*: 149 (1788).

Jiangxi, Fujian(Zahlbr., 1930 b, p. 244), Taiwan (Asahina, 1926 a, p. 150; Zahlbr., 1933 c, p. 68; M. Sato, 1938 c, p. 424 & 1957, p. 68; Asahina & Sato in Asahina, 1939, p. 781; Wang & Lai, 1973, p. 90 & 1976, p. 227), Hongkong (Aptroot et al., 2001, p. 325).

Dictyonema yunnanum D. Liu, X. Y. Wang & Li S. Wang, 滇云片衣

in Liu, Goffinet, Wang, Hur, Shi, Zhang, Yang, Li, Yin & Wang, *Mycosystema* 37(7): 856 (2018).

Yunnan (D. Liu et al. 2018).

Lichenomphalia Redhead, Lutzoni, Moncalvo & Vilgalys 藻伞菌属

Mycotaxon 83: 36 (2002).

Lichenomphalia hudsoniana (H. S. Jenn.) Redhead, Lutzoni, Moncalvo & Vilgalys 哈德孙藻伞菌

Mycotaxon **83**: 38 (2002).

≡*Omphalina hudsoniana* (H. S. Jenn.) H. E. Bigelow, *Mycologia* **62**(1): 15 (1970); J. C. Wei 1991, p. 161.

= *Coriscium viride* (Ach.) Vain., *Acta Soc. Fauna Flora fenn.* 7(no. 2): 189 (1890); J. C. Wei & Y. M. Jiang 1986, p. 115; J. C. Wei 1991, 39 – 40, anamorphic.

≡*Botrydina viridis* (Ach.) Redhead & Kuyper, *Arctic Alpine Mycology*, II (New York): 334 (1987); J. C. Wei 1991, 39 – 40.

≡*Endocarpon viride Ach.*, *Lich. univ.*: 300 (1810); J. C. Wei 1991, 39 – 40.

= *Normandina davidis* Hue, Bull. Soc. Bot. France, 36: 176(1889).

Type locality: Yunnan, Yanzihai, at about 3000 m.

Corticolous.

Xizang (J. C. Wei & Y. M. Jiang, 1981, p. 1146 & 1986, p. 115; J. C. Wei 1991, pp. 87, 39 – 40). Yunnan (Hue, 1889, p. 176; Zahlbr., 1930b, p. 32; Wei X. L. et al. 2007, p. 154).

Lichenomphalia umbellifera (L.) Redhead, Lutzoni, Moncalvo & Vilgalys 帽状藻伞菌

Mycotaxon 83: 38 (2002).

≡*Agaricus umbelliferus* L., *Sp. pl.* 2: 1175 (1753).

≡*Omphalia umbellifera* (L.) P. Kumm., *Führ. Pilzk.* (Zerbst): 107 (1871).

≡ *Omphalina umbellifera* (L.) Quél., *Enchir. fung.* (Paris): 44 (1886) (Obermayer, 2004, p. 504 as*Omphalina umbellifera* (L.: Fr.) Quél. s. l.).

≡*Lichenomphalia umbellifera* f. *umbellifera* (L.) Redhead, Lutzoni, Moncalvo & Vilgalys, *Mycotaxon*83: 38 (2002).

On rotten wood of *Abies* sp.

Xizang (Obermayer, 2004, p. 504, as *Omphalina umbellifera*).

Lichenomphalia velutina(Quél.) Redhead, Lutzoni, Moncalvo & Vilgalys 短绒藻伞菌

Mycotaxon 83: 43 (2002).

* **Marchandiomyces** Diederich & D. Hawksw., in Diederich **地钱菌属**

Mycotaxon 37: 311 (1990).

* Marchandiomyces corallinus (Roberge) Diederich & D. Hawksw. 珊瑚地钱菌

in Diederich, *Mycotaxon* 37: 312 (1990). (Anamorphic Marchandiobasidium)

On *Diploschistes actinostomus*.

Yunnan (Hawksw. & M. Cole, 2003, p. 361).

[26] Lepidostromatales B. P. Hodk. & Lücking **莲叶衣目**

Fungal Diversity 64(1): 174 (2014).

(97) Lepidostromataceae Ertz, Eb. Fisch., Killmann, Sérus. & Lawrey **莲叶衣科**

Am. J. Bot. 95(12): 1553 (2008).

Sulzbacheromyces B. P. Hodk. & Lücking **丽烛衣属**

Fungal Diversity 64(1): 176 (2014).

Sulzbacheromyces bicolor D. Liu, Li S. Wang & Goffinet 双色丽烛衣

Mycologia 109(5): 735 (2017) & 6 – 8 (2018).

Yunnan (D. Liu et al., 2018, pp. 6 – 8).

Sulzbacheromyces fossicolus (Corner) D. Liu & Li S. Wang 湿地丽烛衣

Mycologia 109(5): 737 (2017) & 8 – 9 (2018).

≡*Clavaria fossicola* Corner, Monograph of Clavaria and allied Genera, (Annals of Botany Memoirs No. 1): 691 (1950).

≡ *Multiclavula fossicola* (Corner) R. H. Petersen, Am. Midl. Nat. 77: 208 (1967).

Yunnan (Petersen, 1986, pp. 283 – 284; Liu D. et al., 2018, pp. 8 – 9).

Sulzbacheromyces sinensis (R. H. Petersen & M. Zang) D. Liu & Li S. Wang 中华丽烛衣

Mycologia 109(5): 740 (2017) & 11 – 13(2018).

≡ *Multiclavula sinensis* R. H. Petersen & M. Zang, *Acta Bot. Yunn.* **8**(3): 284 (1986).

Typus Yunnan, Xishuangbanna, 17. IX. 83. coll. D. K. Smith, no. 45639(holotupe, TENN); roadside 3 km south of Xishuangbanna, 17. IX. 83. no. 45644 (TENN), 10465 (isotype, HKAS).

Hainan (J. C. Wei et al., 2013, p. 238).

= *Multiclavula vernalis* (Schwein.) R. H. Petersen, *Am. Midl. Nat.* 77: 216 (1967).

≡ *Clavaria vernalis* Schwein., *Schr. naturf. Ges.* Leipzig 1: 112 (1822).

On compacted soil on road banks. Known from temperate Europe, Australia, North America and Asia.

Yunnan, Hainan, Guizhou, Taiwan (Liu D. et al., 2018, pp. 11 – 13), Fujian (Jia ZF et al., 2008, p. 620; Liu D et al. 2018, pp. 11 – 13), Hongkong (Aptroot & Seaward, 1999, p. 82; Aptroot et al., 2001, p. 333).

Sulzbacheromyces yunnanensis D. Liu, Li S. Wang & Goffinet 云南丽烛衣

Mycologia 109(5): 742 (2017) & 13 – 15(2018).

Yunnan (Liu D. et al., 2018: 13 – 15).

[27] Cantharellales Gäum. **鸡油菌目**

Vergl. Morph. Pilze (Jena): 495 (1926)

(98) Cantharellaceae J. Schröt. **鸡油菌科**

in Cohn, *Krypt. – Fl. Schlesien* (Breslau) 3. 1(25 – 32): 413 (1888) [1889]

Multiclavula R. H. Petersen **藻瑚菌属**

Am. Midl. Nat. 77: 207 (1967).

Multiclavula mucida (Pers.) R. H. Petersen 腐木藻瑚菌

Am. Midl. Nat. 77: 212 (1967).

≡ *Clavaria mucida* Pers., *Comm. fung. clav.* (Lipsiae): 55 (1797).

On rotten wood.

Xizang (Obermayer, 2004, p. 503).

10. Tremellomycetes Doweld **银耳纲**

Prosyllabus Tracheophytorum, Tentamen Systematis Plantarum Vascularium (Tracheophyta) (Moscow): LXXVIII (2001).

[28] Tremellales Fr. [as 'Tremellinae'] **银耳目**

Syst. Mycol. (Lundae) 1: 2 (1821).

(99) Tremellaceae Fr. [as 'Tremellini'] **银耳科**

Syst. Mycol. (Lundae) 1: lv (1821).

* **Tremella** Pers. **银耳属**

Neues Mag. Bot. 1: 111 (1794), Sanctioning citation: Fr., *Syst. Mycol.* 2(1): 210 (1822).

* Tremella everniae Diederich 扁枝银耳

Biblthca Lichenol. 61: 77 (1996).

On *Evernia mesomorpha* (Diederich 1996).

Yunnan (Hawksw. & M. Cole, 2003, p. 362).

* Tremella sulcariae Diederich & M. S. Christ. 树发银耳

in Diederich, *Biblthca Lichenol.* 61: 162 (1996).

On *Sulcaria sulcata* (Diederich 1996).

Yunnan (Hawksw. & M. Cole, 2003, p. 362 as lichenicolous fungus).

LITERATURE CITED

ABABAIKELI G, ABBAS A, GUO S Y, et al, 2016. *Diploschistes tianshanensis* sp. nov. , a corticolous species from Northwestern China[J] . Mycotaxon, 131: 565 – 576.

ABBAS A, JIANG Y C, WU J N, 1993. The Lichens from Mt. Tianshan Region, Xinjiang, China[J] . Journal of Nanjing Normal University – Natural Science, 16(supl.) : 74 – 82.

ABBAS A, WU J N, 1994. The lichens from Tielimaiti Pass Kuqa County, Xinjiang, China[J] . Arid Zone Research, 11(4) : 19 – 23.

ABBAS A, JIANG Y C, WU J N, 1996. Five lichen species new to china[J] . Acta Mycol. Sin, 15(3) : 232 – 233.

ABBAS A, WU J N, 1998. Lichens of Xinjiang[M] . Urumqi: Sci – Tec & Hygiene Publching House of Xinjiang, China(in Chinese with English abstract) .

ABBAS A, ABDUSALIK N, TUMUR A, et al, 2000. The lichens new to Xinjiang from Mt. Tianshan[J] . Arid Zone Research, 17(3) : 17 – 19.

ABBAS A, LITIP S, WU J N, et al, 1997. The Lichens new to Xinjiang from Kanas, Xinjiang, China[J] . Arid Zone Research, 14(4) : 1 – 4.

ABBAS A, MAMUT R, ABUDOULA D, et al, 2014. A Newly Recorded Lichen Genus Xylographa(Fr.) Fr. from China] [J] . Plant Diversity and Resources, 36(5) : 578 – 580.

ABBAS A, XANIDIN H, MAMUT R, et al, 2015. A New Record of Lichen Family from Xinjiang[J] . Arid Zone Research, 32(3) : 509 – 511.

DES ABBAYES H, 1956. Quelques Cladonia(Lichens) des regions inter – tropicales, nouveaux ou peu connus, conserves dans l'Herbarium de Kew[J] . Kew Bull. , 259 – 266.

DES ABBAYES H, 1958 a. *Cladonia*(Lichens) recoltes par la Mission zoologique suisse aux Indes[J] . Candollea, 16: 211 – 214.

DES ABBAYES H, 1958 b. Resultats des expeditions Scientifiques genevoises au Nepal en 1952 et 1954(Partie botanique) [J] . Candollea, 16: 201 – 209.

ABBAS A, GUO S Y, 2014. *Diploschistes xinjiangensis*, a new saxicolous lichen from northwest China[J] . Mycotaxon, 129(2) : 465 – 471.

AHTI T, 1961. Taxonomic Studies on reindeer Lichens(*Cladonia*, Subgenus *Cladina*) [J] . Ann. Bot. Soc. "Vanomo" 32(1) : 1 – 160.

AHTI T, 1962. Distribution of the lichen *Cladoniadelavayi* des Abb[J] . Memoranda Societatis pro Fauna et Flora Fennica , 37: 256 – 257.

AHTI T, 1966. *Parmelia olivacea* and the allied non – isidiate and non – sorediate corticolous lichens in the northern hemispherre[J] . Acta Botanica Fennica, 70: 1 – 68.

AHTI T, 1976. The lichen genus *Cladonia* in Mongolia[J] . Journ. Jap. Bot. , 51(12) : 365 – 373.

AHTI T, 1980. Taxonomic revision of *Cladonia gracilis* and its allies[J] . Ann. Bot. Fennici, 17: 195 – 243.

AHTI T & LAI M. J. 1979. The lichen genus *Cladonia*, *Cladina*, and *Cladia* in Taiwan[J] . Ann. Bot. Fennici, 16: 228 – 236.

AHTI T, 1991. Some species of Cladoniaceae(Lichenized Ascomycetes) from China and adjacent countries[J] . Mycosystema, 4: 59 – 64.

AHTI T, LAI M J, QIAN Z G, 1999. Notes on the lichen flora of China: Parmeliaceae and Sphaerophoraceae[J] . Fung. Sci. , 14: 123 – 126 .

AHTI T, DAVID L, HAWKSWORTH D L, 2005. *Xanthoparmelia stenophylla*, the correct name for *X. somloënsis*, one of the most widespread usnic acid containing species of the genus[J] . The Lichenologist, 37(4) : 363 – 366.

AHTI T, 2007. Further studies on the *Cladonia verticillata* group(Lecanorales) in East Asia and western North America[J] . Bibliotheca Lichenologica, 96: 5 – 19.

AHTI T, STENROOSE S, CHEN J B, et al, 1995 – 96. The status of Cladonia humilis in East Asia[J]. Mycosystema, 8 – 9: 53 – 58.

DE BARY A, 1866. Morphologie und Physiologie der Pilze, Flechten und Myxomyceten[M]. Leipzig.

APTROOT A, 2004. Redisposition of some, mostly pyrenocarpous, lichens described by Zahlbruckner from Taiwan [J]. Acta Univ. Ups. Symb. Bot. Ups, 34(1): 31 – 38.

APTROOT A, SEAWARD M, 1999. Annotated checklist of Hongkong Lichens[J]. Tropical Bryology, 17: 57 – 101.

APTROOT A, SIPMAN H J M, 2001. New Hong Kong lichens, ascomycets and lichenicolous fungi[J]. J. Hattori Bot. Lab, 91: 317 – 343.

APTROOT A, SPARRIUS L B, LAI M J, 2002. New Taiwan Macrolichens[J]. Mycotaxon, 84: 281 – 292.

APTROOT A, SPARRIUS L B, 2003. New microlichens from Taiwan[J]. Fungal Diversity, 14: 1 – 50.

APTROOT A, 2003. Pyrenocarpous lichens and related non – lichenized ascomycets from Taiwan[J]. J. Hattori Bot. Lab., 93: 155 – 173.

APTROOT A, LAI M J, SPARRIUS L B, 2003. The Genus *Menegazzia*(Parmeliaceae) in Taiwan[J]. The Bryologist, 106(1): 157 – 161.

APTROOT A, LAI M J, SIPAMN H, et al, 2003. Foliicolous lichens and their lichenicolous Ascomycetes from Yunnan and Taiwan[J]. Mycotaxon, (88): 41 – 47.

ARVIDSSON L, 1982. A monograph of the lichen genus *Coccocarpia*[J]. Opera Botanica, 67: 1 – 96.

ASAHINA Y, 1926 a. The Raiken's Soliquy on Botanical Science, VI[J]. J. Jap. Bot., 3(7): 150.

ASAHINA Y, 1926 b. Ibid. XI[J]. J. Jap. Bot., 3(12): 283.

ASAHINA Y, 1929. Ibid. XXVIII[J]. J. Jap. Bot., 6(3): 64.

ASAHINA Y, 1931. Ibid. XXXV[J]. J. Jap. Bot., 7(4): 102.

ASAHINA Y, 1932. Ibid[J]. 8(2): 47. —. 1932a, Ibid. XLVI[J]. J. Jap. Bot., 8(5): 210.

ASAHINA Y, 1932 b. Notes on Japanese Lichens IV[J]. J. Jap. Bot., 8(5): 27 – 28.

ASAHINA Y, 1934 a. Lichenologische Notizen(IV)[J]. J. Jap. Bot., 10(5): 299.

ASAHINA Y, 1934 b. Ibid. (V)[J]. J. Jap. Bot., 10(6): 352 – 357.

ASAHINA Y, 1934 c. Aufzahlung von Cetraria – Arten aus Japan(II)[J]. J. Jap. Bot., 10(8): 474.

ASAHINA Y, 1934 d. Lichenologische Notizen(VI)[J]. J. Jap. Bot., 10(11): 682.

ASAHINA Y, 1935 a. *Nephromopsis* – Arten aus Japan[J]. J. Jap. Bot., 11(1): 10.

ASAHINA Y, 1935 b. *Solorina* – Arten aus Japan[J]. J. Jap. Bot., 11(3): 156.

ASAHINA Y, 1935 c. *Anzia* – Arten aus Japan[J]. J. Jap. Bot., 11(4): 224.

ASAHINA Y, 1935 d. Leptogium(sect. Mallotium) – Arten aus Japan[J]. J. Jap. Bot., 11(8): 544.

ASAHINA Y, 1936 a. *Leptogium tremelloides* und verwandte Arten aus Japan[J]. J. Jap. Bot., 12(4): 247.

ASAHINA Y, 1936 b. Zwei neue Arten von Leptogium aus Formosa[J]. J. Jap. Bot., 12(4): 250 – 255, figs. 1 – 7.

ASAHINA Y, 1936 c. *Alectoria* – und *Oropogon* – Arten aus Japan[J]. J. Jap. Bot., 12(10): 690.

ASAHINA Y, 1937. *Anzia* – Arten aus Japan mit besonderer Berucksichtigung[J]. J. Jap. Bot., 13(4): 219 – 226.

ASAHINA Y, 1938. *Ramalina* – Arten aus Japan(I)[J]. J. Jap. Bot., 14(11): 721.

ASAHINA Y, 1939 a. Ibid. (II)[J]. J. Jap. Bot., 15(4): 205.

ASAHINA Y, 1939 b. Japanische Arten der*Cocciferae(Cladonia – Cenomyce)*[J]. J. Jap. Bot., 15(10): 602.

ASAHINA Y, 1939 c. Nippon Inkwasyokubutu Dukan, p. 713. see Asahina, Y. & Sato, M. M. Lichenes[M]. Tokyo.

ASAHINA Y, 1940 a. *Cladonia verticillata* Hoffm. und *Cladonia calycantha*(Del.) Nyl. aus Japan[J]. J. Jap. Bot., 16(8): 462.

ASAHINA Y, 1940 b. Lichenologische Notizen(XIV)[J]. J. Jap. Bot., 16(10): 592.

ASAHINA Y, 1940 c. Chemismus der Cladonien[J]. J. Jap. Bot., 16(12): 709 – 727.

ASAHINA Y, 1941 a. Lichenologische Notizen(XV)[J]. J. Jap. Bot., 17(2): 71 – 72.

ASAHINA Y, 1941 b. *Ibid.* (XVI)[J]. J. Jap. Bot., 17(3): 136 – 138.

ASAHINA Y, 1941 c. Chemismus der Cladonien unter besonderer Berucksichtigung der japanischen Arten[J]. J.

Jap. Bot., 17(8):431 -437.
ASAHINA Y, 1941 d. Lichenologische Notizen(XVII)[J]. J. Jap. Bot., 17(9):485 -489.
ASAHINA Y, 1942. Ibid. (XIX)[J]. J. Jap. Bot., 18(11):620.
ASAHINA Y, 1943 a. Ibid. (XX)[J]. J. Jap. Bot., 19(1,2):4.
ASAHINA Y, 1943 b. *Chemismus* der Cladonien unter besonderer Berucksichtigung der japanischen Arten(Fortsetzung)[J]. J. Jap. Bot., 19(1,2):55.
ASAHINA Y, 1943 c. Chemismus der Cladonien unter besonderer Berucksichtigung der japanischen Arten(Fortsezung)[J]. J. Jap. Bot., 19(3):55.
ASAHINA Y, 1943 d. Lichenologische Notizen(XXII)[J]. J. Jap. Bot., 19(7):189.
ASAHINA Y, 1943 e. Chemismus der Cladonien unter besonderer Berucksichtigung der japanischen Arten(Fortsetzung)[J]. J. Jap. Bot. 19(8):227.
ASAHINA Y, 1943 f. Lichenologische Notizen(XXIII)[J]. J. Jap. Bot., 19(9,10):279.
ASAHINA Y, 1943 g. Ibid. (XXIV)[J]. J. Jap. Bot., 9(11):301.
ASAHINA Y, 1944 a. Ibid. (XXV)[J]. J. Jap. Bot. 2,0(3):129(134.
ASAHINA Y, 1944 b. Ibid. (XXVI)[J]. J. Jap. Bot., 20(5):233 -237.
ASAHINA Y, 1947 a. Ibid. (61 -64)[J]. J. Jap. Bot., 21(1,2):3.
ASAHINA Y, 1947 b. Ibid. (65 -67)[J]. J. Jap. Bot., 21(5,6):83.
ASAHINA Y, 1949. Ibid. (72)[J]. J. Jap. Bot., 23(5,6):65.
ASAHINA Y, 1950 a. Ibid. (73 -74). J. Jap. Bot., 25(5):65 -68.
ASAHINA Y, 1950 b. Lichens of Japan I. Genus*Cladonia*[M]. 1 -255, Tokyo.
ASAHINA Y, 1951 a. Lichenes Japoniae novae vel minus cognitae(2)[J]. J. Jap. Bot., 26(4):97.
ASAHINA Y, 1951 b. Ibid. (3)[J]. J. Jap. Bot., 26(7):193.
ASAHINA Y, 1951 c. Ibid. (4)[J]. J. Jap. Bot., 26(8):225.
ASAHINA Y, 1951 d. Ibid. (5)[J]. J. Jap. Bot., 26(9):257.
ASAHINA Y, 1951 e. Ibid. (6)[J]. J. Jap. Bot., 26(10):289.
ASAHINA Y, 1951 f. Ibid. (7)[J]. J. Jap. Bot., 26(11):329 -334.
ASAHINA Y, 1951 g. Ibid. (8)[J]. J. Jap. Bot., 26(12):353.
ASAHINA Y, 1952 a. Lichens of Japan II. Genus *Parmelia*[M]. pp. 1 -162, Tokyo.
ASAHINA Y, 1952 b. Lichenes Japoniae novae vel minus cognitae(9)[J]. J. Jap. Bot., 27(1):15.
ASAHINA Y, 1952 c. Lichenologische Notizen(83 -84)[J]. J. Jap. Bot., 27(8):239.
ASAHINA Y, 1952 d. An addition to the Sato's Lichenes Khinganenses(Bot. Mag. Tokyo, 65: 172)[J]. J. Jap. Bot., 27(12):373.
ASAHINA Y, 1953. *Ibid*[J]. 28:65.
ASAHINA Y, 1954 a. Lichenologische Notizen(107 -109)[J]. J. Jap. Bot., 29(10):289.
ASAHINA Y, 1954 b. Ibid. (112 -113)[J]. J. Jap. Bot., 29(12):370 -372.
ASAHINA Y, 1955. Lichens collected in Aogashima and Mikurazima[J]. J. Jap. Bot., 30(7):222 -224.
ASAHINA Y, 1956 a. Lichens of Japan III. Genus *Usnea*[M]. 1 -129, Tokyo.
ASAHINA Y, 1956 b. Lichenologische Notizen(120)[J]. J. Jap. Bot., 31(11):321.
ASAHINA Y, 1957. Ibid. (124 -125)[J]. J. Jap. Bot., 32(4):97 -100.
ASAHINA Y, 1958. Ibid. (137 -139)[J]. J. Jap. Bot., 33(3):65 -69.
ASAHINA Y, 1959 a. Ibid. (149)[J]. J. Jap. Bot., 34(3):65 -66.
ASAHINA Y, 1959 b. Ibid. (150 -158)[J]. J. Jap. Bot., 34(8):225.
ASAHINA Y, 1960 a. Ibid. (160 -163)[J]. J. Jap. Bot., 35(4):97.
ASAHINA Y, 1960 b. Ibid. (165 -169)[J]. J. Jap. Bot., 35(10):289.
ASAHINA Y, 1963. Ibid. (192)[J]. J. Jap. Bot., 38(9):257.
ASAHINA Y, 1964. Ibid. (193)[J]. J. Jap. Bot., 39(6):165 -166.

ASAHINA Y, 1965 a. Ibid. (195) [J]. J. Jap. Bot., 40(1): 1.
ASAHINA Y, 1965 b. Ibid. (197) [J]. J. Jap. Bot., 40(5): 129.
ASAHINA Y, 1965 c. Ibid. (200) [J]. J. Jap. Bot., 40(8): 225.
ASAHINA Y, 1967 a. Ibid. (203) [J]. J. Jap. Bot., 42(1): 1.
ASAHINA Y, 1967 b. Ibid. (204) [J]. J. Jap. Bot., 42(9): 257.
ASAHINA Y, 1967 c. Ibid. (206) [J]. J. Jap. Bot., 42(11): 321.
ASAHINA Y, 1968 a. Ibid. (207 – 208) [J]. J. Jap. Bot., 43(3): 65 – 68.
ASAHINA Y, 1968 b. Ibid. (210) [J]. J. Jap. Bot., 43(5): 129.
ASAHINA Y, 1968 c. Ibid. (211 – 212) [J]. J. Jap. Bot., 43(12): 495.
ASAHINA Y, 1969 a. Ibid. (213 – 214) [J]. J. Jap. Bot., 44(1): 1 – 5.
ASAHINA Y, 1969 b. Ibid. (215 – 216) [J]. J. Jap. Bot. 4, 4(2): 33.
ASAHINA Y, 1969 c. Ibid. (217 – 222) [J]. J. Jap. Bot., 44(9): 257.
ASAHINA Y, 1970 a. Lichenologische Notizen(231 – 234) [J]. J. Jap. Bot., 45(4): 97.
ASAHINA Y, 1970 b. Lichenologische Notizen(235) [J]. J. Jap. Bot., 45(5): 129.
ASAHINA Y, 1970 c. Lichenologische Notizen(238 – 239). J. Jap. Bot., 45(9): 257.
ASAHINA Y, 1971. Atlas of the Japanese*Cladonia*[M]. Tokyo, 27 .
ASAHINA Y, 1972 a. Lichenologische Notizen(248 – 250) [J]. J. Jap. Bot., 47(5): 129.
ASAHINA Y, 1972 b. Ibid. (251) [J]. J. Jap. Bot., 47(9): 257.
ASAHINA &SATO M M. Lichenes in Asahina, Y, 1939, Nippon Inkwasyokubutu Dukan[M], Tokyo, 713.
AWASTHI G, 1986. Lichen genus *Usnea* in India[J]. Journ. Hattori Bot. Lab., 6: 333 – 421.
BARONI E, 1894. Sopra alcuni licheni della China racolti nella provincia dello Shan – si settentrionale [J]. Bull. Soc. Bot. Ital, 46 – 59.
BJERKE J W, 2004. Revision of the lichen genus *Menegazzia* in Japan, including two new species[J]. The Lichenologist, 36(1): 15 – 25.
BJERKE J W, OBERMAYER W, 2005. The genus *Menegazzia*(*Parmeliaceae*, lichenized Ascomycetes) in the Tibetan region [J]. Nova Hedwigia, 81(3 – 4): 301 – 309.
BREUSS O, 1998. *Catapyrenium* und verwandte Gattungen (lichenisierte Ascomyceten, Verrucariaceae) in Asien [M]. – ein erster Uberblick. Annalen der Naturhistorisches Museums Wien, 100B: 657 – 669.
BRODO I M, HAWKSWAORTH D L, 1977. *Alectoria* and allied genera in North America[J]. Opera Botanica, 42: 1 – 164.
BRODO I M, 1991. Studies in the lichen genus Ochrolechia. 2. Corticolous species of North America [J]. Can. J. Bot., 69: 733 – 772.
CAO J, LIU H J, DENG H, 2012. A hairy species of *Leptogium* to Mainland China[J]. Journal of Fungal Reseach, 10(4): 213 – 215.
CAO S N, WEI X L, ZHOU Q M, et al, 2013. *Phyllobaeis crustacea* sp. nov. from China [J]. Mycotaxon, 126: 31 – 36.
CHEN C L, LI Y, ZHAO Z T, 2008. Research on the Lichen Family Parmeliaceae from Mt. Jinfo of Chongqing [J]. Shandong Sci., 21: 39 – 40(in Chinese).
CHEN J B, 1986. A study on the lichen genus *Cetrelia* in China[J]. Acta Mycol. Sin. Supplement, 1: 386 – 396(in Chinese).
CHEN J B, 1993. Chemical Notes on three species of *Sticta* from China [J]. The Lichenologist, 25: 454 – 458.
CHEN J B, WANG Z F, WANG L S, 1994. The lichen genus *Sticta* from Yunnan of China[J]. Acta Mycol. Sinica, 13(1): 29 – 33.
CHEN J B, 1994. Two new species of *Hypogymnia*(Nyl.) Nyl. (Hypogymniaceae, Ascomycotina) [J]. Acta Mycol. Sinica, 13(2): 107 – 110.
CHEN J B, 1995. A new species of *Lobaria*[J]. Acta Mycol. Sinica, 14(4): 261 – 262.

CHEN J B, 1996. The lichens of Sphaerophorus from China [J]. Acta Mycol. Sinica, 15(2): 105 – 108.

CHEN J B, 1996. A study of the lichens of Oropogon from China[J]. Acta Mycol. Sinica, 15(3): 173 – 177.

CHEN J B, HU G R, 2003. The lichen family Physciaceae(Ascomycota) in China V. the genus *Physconia*[J]. Mycotaxon, 86: 185 – 194.

CHEN J B, LIU X J, HUANG Y, 1999. A preliminary report on corticolous lichens in Dongling mountain, Beijing, China[J]. Acta Ecol. Sin., 19: 76 – 79(in Chinese).

CHEN J B, WANG D P, 1999. The Lichen family Physciaceaed(Ascomycota) in China I. The genus *Anaptychia*[J]. Mycotaxon, 73: 335 – 342.

CHEN J B, WANG D P, 2001. The Lichen family Physciaceaed(Ascomycota) in China III. Ten species of Heterodermiaq containing depsidones[J]. Mycotaxon, 77: 107 – 116.

CHEN J B, WANG S L, ELIX J A, 2003a. Parmeliaceae(Ascomycota) lichens in China' s mainland I. The genera *Canomaculina, Parmelina, Parmelinella* and *Parmelinopsis*[J]. Mycotaxon, 86: 19 – 29.

CHEN J B, XU L, QIAN Z G, et al, 2003b. Parmeliaceae(Ascomycota) lichens in China' s mainland II. The genus *Hypotrachyna*[J]. Mycotaxon, 86: 359 – 373.

CHEN J B, WANG S L, ELIX A, 2005. Parmeliaceae(Ascomycota) lichens in China' s mainland III. The genus *Parmotrema*[J]. Mycotaxon, 91: 93 – 113.

CHEN J B, ESSLINGER T L, 2005. Parmeliaceae(Ascomycota) lichens in China' s mainland IV. Melanelia species new to China[J]. Mycotaxon, 93: 71 – 74.

CHEN J B, XU L, ELIX J A, 2009. Parmeliaceae(Ascomycota) lichens in China' s mainland V. The genera *Bulbothrix* and *Relicina*[J]. Mycosystema, 28: 92 – 96.

CHEN J B, 2011a. Parmeliaceae(Ascomycota) lichens in China' s mainland VI. Eight species new to China in parmelioid lichens[J]. Mycosystema, 30: 881 – 888(in Chinese).

CHEN J B, 2011b. Lichens. In YONG SP, XING LL, LI GL(eds.): Biodiversity Catalogue of Saihanwula Nature Reserve[M]. Inner Mongolia University Press, 489 – 510(in Chinese).

CHEN J B, 2013. Parmeliaceae(Ascomycota) lichens in China' s mainland VII. The genus Nipponoparmelia[J]. Mycosystema, 32: 51 – 55(in Chinese).

WU J N, WEI J C, 1989. Lichens of Shennongjia in Fungi and Lichens of Shenongjia[M]. Wourld Publishing Corp, Beijing, China, 386 – 493.

CHEN L H, GAO X Q, 2001. Two new species of Nephromopsis(Parmeliaceae, Ascomycota) [J]. Mycotaxon, 77: 491 – 496.

CHEN L H, ZHOU Q M, GUO S Y, 2006. Two species of cetrarioid lichens new to China[J]. Mycosystema, 25(3): 502 – 504.

CHEN S F, SANDS M K, SEAWARD M R D, 1997. Macrolichens of Wolong Nature Reserve, China[M]. Memoirs of the Hong Kong Natural History Society, 21: 221 – 222.

CHEN X L, ZHAO C F, LUO G Y, 1981. A list of lichens in northeastern China[J]. J. NE Forestry Inst., 3: 127 – 135 &4: 150 – 160.

CHEN X L, 1986. Study on *Peltigera* in Northeast China[J]. Acta Mycol. Sin., 5(1): 18 – 29.

CHENG Y L, JING N, XU H P, et al, 2012. *Herpothallon weii*, a new lichen from China[J]. Mycotaxon 119: 439 – 443.

CHIEN S S, 1932. Vegetation of the rocky ridge of Chung shan, Nanking[J]. Contr. from the biological laboratory of the science society of China, Botanical Series. 7(9): 215 – 227.

CHU F J, 1997. Ecology of supralittoral lichens on Hong Kong rocky shores. Ph. D. thesis[D]. University of Hong Kong.

COLE M S, HAWKSWORTH D L, 2001. Lichenicolous fubgi, mainly from the USA, including Patriciomyces gen. nov[J]. Mycotaxon LXXVII, (77): 305 – 338.

CROMBIE I M, 1883. (I): On a collection of Exotic Lichens made in Eastern Asia by the late Dr. A. C. Maingay, in

[J]. Journ. Linn. Soc. London, Bot., XX: 62 – 66.

CUI C, WANG X H, JIA Z F, 2014. New records of lichen genus *Ocellularia* from China[J]. Journal of Fungal Research, 12(4): 203 – 209.

CULBERSON C F, 1969. Chemical and botanical guide to lichen products[M]. The University of North Carolina Press, 1 – 628.

CULBERSON W L, 1961. A second Anzia in North America[J]. Brittonia, 13(4): 381 – 384.

CULBERSON C F, 1968. The lichen genera Cetrelia and Platismatia(Parmeliaceae)[J]. Contrb. U. S. Nat. Herb., 34: 449 – 558.

CULBERSON W, CULBERSON C, 1968. The lichen genera Cetrelia and Platismatia(Parmeliaceae)[J]. Contr. U. S. natn. Herb., 34: 449 – 558.

CULBERSON W, CULBERSON C, 1981. The Genera Cetrariastrum and Concamerella(Parmeliaceae): a chemosystematic synopsis[J]. Bryologist, 84(3): 273 – 314.

DARBISHIRE O V, 1898. Monographia Roccelleorum. Ein Beitrag zur Flechten – systematik[M]. Bibliotheca Botanica, 945: (i – iv): 1 – 102, Pl. 1 – 30, fig. 1 – 29.

DARBISHIRE O V, 1928. Roccellaceae(Massal.) Nyl.[J]. Die Pflanzenareale, 2: 1 – 4, maps1 – 5.

DARWIN C, 1872. The Origin of Species[M]. (Sixth Edition) London, John Murras, Albermarie Street, W., printed at the Edinburgh Press, 1902, Pp. 1 – 703.

DAVYDOV E A, WEI J C, 2009. Boreoplaca ultrafrigida(Umbilicariales), the correct name for Rhizoplacopsis weichingii[J]. Mycotaxon, 108: 301 – 305.

DAVYDOV E A, PERŠOH D, RAMBOLD G, 2010. The systematic position of *Lasallia caroliniana* (Tuck.) Davydov, Peršoh & Rambold comb. nova and considerations on the generic concept of *Lasallia*(Umbilicariaceae, Ascomycota)[J]. Mycological Progress, 9: 261 – 266.

DEGELIUS G, 1974. The lichen genus *Collema* with special reference to the extra – European Species[J]. Symbolae Botanicae Upsalienses, 20(2): 1 – 215.

DODGE C W, 1929. A synopsis of *Stereocaulon* with notes on some exotic species[J]. Ann. Crypt. Exot., 2(2): 93 – 153.

DU RIETZ G E, 1924. Flechtensystematischen Studien IV[J]. Bot. Not., 1924: 329 – 342.

DU RIETZ G E, 1925. Ibid. V. *Ibid*[J]. 1 – 16.

DU RIETZ G E, 1926 a. Om slaktena Evernia Ach., Letharia(Th. Fr.) A. Zahlbruckner emend. Dr. och*Usnea* Ach. subgenus *Neuropogon*(Nees et Flot.) Jatta[J]. Svensk Bot. Tidskr., 20: 89 – 93.

DU RIETZ G E, 1926 b. Vorarbeiten zu einer "Synopsis Lichenum" I. Die Gattung *Alectoria*, *Oropogon*, und *Cornicularia*[J]. Arkiv for Bot., 20 A 11: 1 – 43, Pl. 1, 2.

DU Y D, MENG F G, LI H M, et al, 2010. Three new records of parmelioid lichens from the Tibetan Plateau[J]. Mycotaxon, 111: 283 – 286.

EGAN R, 1987. A fifth checklist of the lichen – forming, lichenicolous and allied fungi of the continental United States and Canada[J]. The Bryologist, 90(2): 77 – 173. 屾鸟

ELENKIN A, 1901. Lichenes Florae Rossiae et regionum confinium orientalium Fasc. 1[J]. Acta Horti Petropolitani, 19: 1 – 52.

ELENKIN A, 1904. Ibid. Fasc. 2, 3, 4. *Ibid*[J]. 24(1): 1 – 118.

ESSLINGER T L, 1978. A new status for the brown Parmeliae [J]. Mycotaxon, 7(1): 45 – 54.

ESSLINGER T L, 1978. Studies in the lichen family Physciaceae. II The genus Phaeophyscia in north America [J]. Mycotaxon, 7(2): 283 – 320.

ESSLINGER T L, 1989: Systematics of *Oropogon*(Alectoriaceae*) in the New World*[J]. —Syst. Bot. Monographs, 28: 1 – 111.

FOLLMANN G, HUNECK S, UND WEBER W A, 1968. Mitteilungen uber Flechteninhaltsstofe LIV. Zur Chemotaxonomie des *Dactylina* /*Dufourea* – Komplexes[J]. Willdenowia, 5(1): 7 – 13.

FU W, ZHAO Z T, GUO S X, et al, 2009. An annotated checklist of the lichen genus *Ramalina* from northwestern China[J] . Mycosystema, 28(1) : 097 – 101.

FUTTERER K, 1911. Verzechnis der wahred der Reise gesammelten Blüt – enpflanzen und Flechten, Durch Asien [M] . Band, III: 34 – 37, Berlin.

GAO B, LI J, WEI J C, 2009. A new foliicolous lichen——*Enterographa hainanensis*[J] . Mycosystema, 28(2) : 175 – 177.

GAO T L, REN Q, 2012. New records of *Ochrolechia* and *Placopsis* from the Hengduan Mountains, China[J] . Mycotaxon, 122: 461 – 466.

GAO X Q, 1987. A new species of *Rhizoplaca*[J] . Acta Mycol. Sin. , 6(4) : 233 – 35.

——[——] 1988. *Xanthoparmelia* from the Da Hinggan Ling mountains in China[J] . Ibid. 7(1) : 29 – 35.

[N. S. Golubkova. 1981. Synopsis of Lichen Flora from Poelple' s Republic of Mongolia[M] . < Science >, Leningrad – 1981: 1 – 200) in Russian.] = Н. С. Голубкова 1981 Конспект флоры лишайниковМонгольской НароднойРеспублики. Ленинград[J] . НАУКА1, стр. 1 – 200.

SAHEDAT G, NASHITAY M, OMAR N, et al, 2015. Study on the lichen genus Melanhalea in Xinjiang, China[J] . Acta Bot. Boreal. – Occident. Sin. , 35(11) : 2331 – 2336.

ISMAYIL G, WEN X M, ABBAS A, 2015. New Records of the genus *Circinaria* Link and Circinaria contorta Species in Xinjiang, China[J] . Arid Zone Research, 32(6) : 1229 – 1232.

GUO S Y, 2005. Lichens. in ZHUANG WY(ed.) : Fungi of Northwestern China[M] . pp. 31 – 82. Mycotaxon Ltd. , New York, USA.

GUO W, HUR J S, 2015. *Graphis hongkongensis* sp. nov. and other *Graphis* spp. new to Hong Kong[J] . Mycotaxon, 130: 429 – 436.

GYELNIK V, 1927. Lichenologische Mittelungen 1 – 3[J] . Magyar Bot. Lapok. , 47.

GYELNIK V, 1929. *Peltigera* – Daten[J] . Hedwigia 68: 1 – 2(1928 – 29) .

GYELNIK V, 1931. Additamenta ad cognitionem *Parmeliarum* II[J] . Feddes Repert. , 29: 273 – 291.

GYELNIK V, 1934. Ibid. V, VI[J] . Ibid. , 36: 151 – 166, 299 – 302.

GYELNIK V, 1935. Conspectus *Bryopogonum*[J] . Ibid. , 38: 219 – 255.

HAFELLNER J, 1997. A world monograph of Brigantiaea(lichenized Ascomycotina, Lecanorales) [J] . Symbilae Botanicae Upsaliensis, 32(1) : 35 – 74(1997) .

HAFELLNER J, OBERMAYER W, 1995. Cercidospora trypetheliza und einige weitere lichenicole Ascomyceten auf Arthrorhaphis. Cryptogamie[J] . Bryologie et Lichénologie, 16: 177 – 190.

HAFELLNER J, UND POELT J, 1979. Die Arten der Gattung Caloplaca[J] . J. Hattori Bot. Lab. , 46: 37.

HALE M E, 1962. A new species of *Parmelia* from Asia, *P. subcorallina*[J] . J. Jap. Bot. , 37(11) : 345 – 347.

HALE M E, 1965. A Monograph of *Parmelia* Subgenus *Amphygimnia*[J] . Contr. U. S. Natl. Herb. , 36: 193 – 358.

HALE M E, 1965. Six new species of Parmelia from southeast Asia[J] . J. Jap. Bot. , 40: 199 – 205.

HALE M E, 1976 a. A monograph of the lichen genus *Pseudoparmelia* Lynge(Parmeliaceae) [J] . Smithsonian Contributions. to Botany, 31: 1 – 62.

HALE M E, 1976 b. A monograph of the lichen genus *Bulbothrix* Hale(Parmeliaceae) [J] . Ibid. 32: 1 – 28.

HALE M E, 1976 c. A monograph of the lichen genus *Parmelina* Hale(Parmeliaceae) [J] . Ibid. 33: 1 – 60.

HALE M E, 1978. A Revision of the Lichen Family Thelotremataceae in Panama[J] . Ibid. 38: 1 – 60.

HALE M E, 1980. Taxonomy and distribution of the *Parmelia flaventior* group(lichens: Parmeliaceae) [J] . J. Hattori-Bot. Lab. 47: 75 – 84.

HALE M E, 1986. *Arctoparmelia*, a new genus in the Parmeliaceae(Ascomycotina) [J] . Mycotaxon, 25: 251 – 254.

HALE M E, 1987a. New or interesting species of *Xanthoparmelia*(Vain.) Hale(Ascomycotina: Parmeliaceae) [J] . Ibid. 30: 319 – 334.

HALE M E, 1987b. A monograph of the lichen genus*Parmelia* Acharius sensu strito(Ascomycotina: Parmeliaceae) [J] . Smithsonian Contributions to Botany, 66: 1 – 55.

HALE M E, 1988. New combinations in the lichen genus *Xanthoparmelia*(Ascomycotina) [J]. Mycotaxon 33: 401 – 406

HALE M E, 1990. A synopsis of the lichen genus *Xanthoparmelia* (Vain.) Hale(Ascomyconthina, Parmeliaceae) [J]. Smithsonian Contributions to Botany, 74: 1 – 250.

HALE M E, Ahti T, 1986. An earlier name for *Parmotrema perlatum*"(Huds.) Choisy"(Ascomycotina: Parmeliaceae) [J]. Taxon, 35: 133.

HAN L F, ZHAO J C, GUO S Y, 2009. *Lecanora weii,* a new multispored species of *Lecanora* s. str. from northeastern China[J]. Mycotaxon, 107: 157 – 161.

HAN L F, GUO S Y, ZHANG H, 2011. Three corticolous species of *Lecanora* (*Lecanoraceae*) new to China[J]. Mycotaxon, 116: 21 – 25.

HAN L F, ZHANG Y Y, GUO S Y, 2013. Peltigera wulingensis, a new lichen(Ascomycota) from north China[J]. The Lichenologist, 45(3) : 329 – 336(2013).

HAN L F, ZHENG T X, GUO S Y, 2015. A new species in the lichen genus Peltigera from northern China based on morphology and DNA sequence data[J]. The Bryologist, 118(1) : 46 – 53.

HANSSEN C, MAULE A F, 1973. Pehr Osbeck' s collections and Linnaeus' s Species Plantarum(1753) [J]. Bot. J. Linn. Soc. , 67: 189 – 212.

HARMAND A, 1928. Lichen d'Indo – Chine recueillis per M. V. Demange[J]. Annal. Cryptog. Exot. , 1(4) : 319 – 337.

HARADA H, WANG L, 1995. Dermatocarpella yunnana, a new lichen species in the family Verrucariaceae from Yunnan, China[J]. Bull. Natn. Sci. Mus. , Tokyo, ser. B, 21(2) : 107 – 110.

HARADA H, WANG L, 1996. Two New Freshwater species of Verucariaceae from Yunnan, China[J]. Lichenologist, 28(4) : 297 – 305.

HARADA H, WANG L, 2004. Taxonomic study on the freshwater species of Verrucariaceae(Lichenized Ascomycota) of Yunnan, China(2. *Thelidium yunnanum* sp. nov[J]. Lichenology, 3: 47 – 50.

HARADA H, WANG L, 2006. Taxonomic study on the freshwater species of Verrucariaceae(Lichenized Ascomycota) of Yunnan, China(2). *genus Staurothele*[J]. Lichenology, 5: 13 – 22.

HARADA H, WANG L, 2006. Taxonomic study on the freshwater species of Verrucariaceae(Lichenized Ascomycota) of Yunnan, China(3). Genus *Thelidium* [J]. Lichenology, 5: 23 – 30.

HARADA H, WANG L, 2008. Taxonomic study on the freshwater species of Verrucariaceae(Lichenized Ascomycota) of Yunnan, China(4). Genus Verrucaria[J]. Lichenology, 7(1) : 1 – 24.

HARRIS R C, 1989. A sketch of the Family Pyrenulaceae(Melanommatales) in Eastern North America[J]. Memoirs of the New York Botanical Garden, 49: 98.

HAWKSWORTH D L, 1971. Chemical & nomenclatural notes on *Alectoria* (Lichenes) III. The chemistry, morphology and distribution of *Alectoria virens* Tayl [J]. J. Jap. Bot. , 46(11) : 335 – 342.

HAWKSWORTH D L, 1972. A new species of *Tricharia* Fee em. R. Sant. from Hong Kong[J]. Lichenologist, 5: 321 – 322.

HAWKSWORTH D L, 1976. Lichen Chemotaxonomy *in* Edited by D. H. Brown, D. L. Hawksworth and R. H. Bailey Lichenology [J]. Progress and Problems, 139 – 184(1976).

HAWKSWORTH D L, 1991. The fungal demention of biodiversity: magnitude, significance, and conservation[J]. Mycol. Res. , 95(6) : 641 – 655.

HAWKSWORTH D L, 2004. Rediscovery of the original material of Osbeck' s *Lichen chinensis* and the re – instatement of the name *Parmotrema perlatum*(Parmeliaceae) [J]. Herzogia, 17: 37 – 44.

HAWKSWORTH D L, COLE M S, 2003. A first checklist of lichenicolous fungi from China [J]. Mycosystema, 22 (3) : 359 – 363.

HAWKSWORTH D L, WENG Y X, 1990. Lichens on camphor trees along an air pollution gradient in Hangzhou (Zhejiang Province) [J]. Forest Research, 3(5) : 514 – 517.

HAWKSWORTH D L, MILLANES A M, WEDIN M, 2010. *Roselliniella* revealed as an overlooked genus of *Hypocreales, with the description of a second species on parmelioid lichens[J]. Persoonia* , 24: 12 – 17.

HE Q, CHEN J B, 1995. Macrolichens of Taibai Mountain. In W. Z. TAN(ed.) : Recent Research Achievements of Young Mycologists in China(in Chinese) [M] . Southwest Normal University Press, Chengdu, China, 41 – 47.

HERTEL H, 1967. Revision einiger calciphiler Formenkreise der Flechtengattung *Lecidea*[J] . Nova HedwigiaHeft, 24.

HERTEL H, 1971. Uber holarktische Krustenflechten aus den venezuelanischen Anden[J] . Willdenowia 6(2) : 225 – 272.

HERTEL H, 1977. Gesteinsbewohnende Arten der Sammelgattung *Lecidea*(Lichenes) aus Zentral – , Ost – , und Sudasien[J] . Khumbu Himal, 6(3) : 145 – 378.

HERTEL H, 1981. Beitrage zur Kenntnis der Flechtenfamilie Lecideaceae[J] . Herzogia, V: 449 – 463.

HERTEL H, ZHAO C F, 1982. Lichens from Changbai Shan – some additions to the lichen flora of Northeast China [J] . Lichenologist, 14(2) : 139 – 152.

HIBBETT D S, BINDER M, BISCHOFF J F, et al, 2007. A higher – level phylogenetic classification of the Fungi [J] . Mycological Reseach, 111: 509 – 547.

HOU Y N, ZHANG C, MA Y Z, et al, 2008. Preliminary Research on Off icinal Lichen from Mountain Tai [J] . Shandong Science, 21(2) : 65 – 68(in Chinese) .

CHEN H I, WETMORE C, LAI M J, 2001. A neglected calicioid lichen new to Taiwan[J] . Mycotaxon, 79: 215 – 216.

HU G R, CHEN J B, 2003. The lichen family Physciaceae(Ascomycota) in China. A new species of *Phaeophyscia* [J] . Mycosystema, 22(4) : 534 – 535.

HU G R, CHEN J B, 2003. The Lichen family Physciaceae(Ascomycota) in China VI. The genus Pyxine[J] . Mycotaxon, 86: 445 – 454.

HU L, WANG H Y, LIU J, et al, 2013. Two *Ropalospora* lichens new to mainland China [J] . Mycotaxon, 123: 439 – 444.

HU L, ZHAO X, SUN L Y, et al, 2014. Four lecideoid lichens new to China [J] . Mycotaxon, 128: 83 – 91.

HUANG M R, WEI J C, 2004. Three new taxa of *Stereocaulon* from China [J] . Mycotaxon, 90(2) : 469 – 472.

HUANG M R, WEI J C, 2006. Overlooked taxa of *Stereocaulon*(Stereocaulaceae, Lecanorales) in China[J] . Nova Hedwigia, 82(2 – 4) : 435 – 445.

HUE A M, 1887 a. Lichenes Yunnanenses a clar. Delavay anno 1885 collectos, et quorum novae species a celeb. W. Nylander descriptae fuerunt, exponit A. M. Hue [J] . Bull. Soc. Bot. France, 34: 16 – 24.

HUE A M, 1887 b [J] . Flora 52: 135.

HUE A M, 1889 Lichenes Yunnanenses a cl. Delavay praesertim annis 1886 – 1887, collectos exponit A. M. Hue(1) [J] . Bull. Soc. Bot. France, 36: 158 – 176.

HUE A M, 1890. Lichenes exoticos a professore W. Nylander descriptos vel recognitos et in herbario Musei Parisiensis pro maxima parte asservatos in ordine systematico disposuit[J] . Nouv. Arch. Mus. Hist. Nat. (Paris) , 3(2) : 209 – 322.

HUE A M, 1891[J] . Ibid. , 3(3) : 33 – 192.

HUE A M, 1892[J] . Ibid. , 3(4) : 103 – 210.

HUE A M, 1898. Lichenes extra – europaei[J] . Nouv. Arch. Mus. Hist. Nat. (Paris) , 3(10) : 213 – 280.

HUE A M, 1899. Lichenes extra – europaei a pluribus collectoribus ad Museum Parisiense missi[J] . Ibid. 4(4) , tome , 1: 27 – 220.

HUE A M, 1900[J] . Ibid. , 4(4) , tome , 2: 49 – 122.

HUE A M, 1901[J] . Ibid. , 4(3) : 21 – 146.

HUE A M, 1908. Lichenes morphologice et anatomice[J] . Ibid. , 4(10) : 169 – 224, fig. 17 – 30.

HUE A M, 1911[J] . Ibid. , 5(3) : 133 – 198, fig. 52 – 59.

HUE A M, 1924. Monographia *Crocyniarum*[J]. Bull. Soc. Bot. France, 71: 311 - 402.

XAHIDIN H, BAHTI P, TURSUN T, et al, 2009. The lichen genus *Fulgensia* in Xinjiang, China [J]. Mycosystema, 28(1): 106 - 111.

XAHIDIN H, ABBAS A, WEI J C, 2010 . *Caloplaca tianshanensis*(lichen - forming Ascomycota), a new species of subgenus Pyrenodesmia from China [J]. Mycotaxon, 114: 1 - 6.

HYVOENEN S, 1985. *Parmelia squarrosa*, a lichen new to Europe [J]. The Lichenologist, 17(3): 311 - 314.

IKOMA Y, 1983. Macrolichens of Japan and adjacent regions[J]. Japan, Tottori City., 120.

INDEX OF FUNGI SUPPLEMENT: Lichens 1961 - -69, CMI. Kew, Surrey.

INDEX OF FUNGI 1970 - 19 CMI. Kew, Surrey.

INOUE M, 1981. A preliminary revision of extra - Japanese speces of *Fuscidea*(lichens) [J]. Hikobia Supplement, 1: 177 - 181.

INOUE M, 1982. The genera *Lecidea*, *Lecidella* and *Huilia*(Lichens) in Japan I. *Lecidea*[J]. J. of Sci. of the Hiroshima Univ. series B, Div., 218: 1 - 56, Pl. I - IV.

INOUE M, 1988. Lecideoid Lichens of Okushiri Island, West Coast of Hokkaido, Japan[J]. Mem. Natn. Sci. Mus., Tokyo, (21): 45 - 52.

INUMARU S, 1941. Lobariae novae Japonicae[J]. Acta Phytotax. Geobot., 10: 214 - 216.

INUMARU S, 1943. Lobariae nonnullae novae Japonicae[J]. Acta Phytotax. Geobot., 13: 218 - 224.

INUMARU S, 1963. On the distribution of *Peltigera aphthosa* var. *variolosa* in Japan and the vicinities[J]. J. Jap. Bot., 38(6): 171 - 172.

GULBOSTAN I, ABBAS A, GUO S Y, 2015. *Aspicilia volcanica*, a new saxicolous lichen from Northeast China[J]. Mycotaxon, 130: 543 - 548.

JAMES P, 1965. A new checklist of British Lichens[J]. Lichenologist, 3(1): 95 - 153.

JATTA A, 1902. Lichen cinesi raccolti allo Shen - si negli anni 1894 - 1898 dal. rev. Padre Missionario G. Giraldi [J]. Nuovo Giorn. Bot. Italiano ser. 2, IX: 460 - 481.

JIA Z F, 2011. *Graphis paradussii*(Graphidaceae, Ostropales), a new lichen species to science[J]. The Bryologist, 114(2): 389 - 391.

JIA Z F, KALB K, 2013. Taxonomical studies on the lichen genus Platygramme(Graphidaceae) in China [J]. The Lichenologist, 45(2): 145 - 151(2013).

JIA Z F, KALB K, 2014. *Thalloloma ochroleucum* (*Graphidaceae*), a new species from Guizhou, China [J]. Mycotaxon, 128: 113 - 115. http://dx. doi. org/10. 5248/128. 113.

JIA Z F, LÜCKING R. Resolving the lichen genus *Graphina* Müll. Arg. in China(to be published).

JIA, Z F, LüCKING R, 2017. Resolving the genus *Phaeographina* Müll. Arg. in China[J]. MycoKeys, 21: 12 - 32.

JIA Z F, REN Q, ZHAO Z T, 2008. The lichen genus *Multiclavula* R. H. Petersen in China[J]. Mycosystema, 27 (4): 619 - 621.

JIA Z F, REN Q, SUN L Y, et al, 2008. Taxonomic studies of Pertusariaceae from China[J]. Mycosystema, 27(2): 316 - 319.

JIA Z F, WEI J C, 2005. *Graphis fujianensis*, a new species of Graphidaceae from China[J]. Mycotaxon, 104: 107 - 109.

JIA Z F, REN Q, ZHAO Z T, 2008. *Ochrolechia pallentiisidiata*, a new species from China[J]. Mycotaxon, 106: 233 - 236.

JIA Z F, LI Y L, HOU J, et al, 2008. A study of lichen genus *Canoparmelia* from Taishan Mountain[J]. Mycosystema, 27(3): 461 - 463.

JIA Z F, LI Y L, HOU J, et al, 2008. A study of lichen genus *Canoparmelia* from Taishan Moutain[J]. Mycosystema, 27: 461 - 463(in Chinese).

JIA Z F, MIAO X L, WEI J C, 2011. A preliminary study of *Hemithecium* and *Pallidogramme* (Graphidaceae, Ostropales, Ascomycota) from China[J]. Mycosystema, 30(6): 870 - 876.

JIA Z F, WEI J C, 2007. Two species new to China of Graphis(Graphidaceae, Ostropales, Ascomycota) [J]. Mycosystema, 26(2): 186 – 189.

JIA Z F, WEI J C, 2008 *Graphis fujianensis*, a new species of Graphidaceae from China[J]. Mycotaxon, 104: 107 – 109.

JIA Z F, WEI J C, 2009. A new isidiate species of Graphis(lichenised Ascomycotina) from China[J]. Mycotaxon, 110: 27 – 30.

JIA Z F, WEI J C, 2009. A new species, *Thalloloma microsporum* (*Graphidaceae, Ostropales, Ascomycota*) [J]. Mycotaxon, 107: 197 – 199.

JIA Z F, WEI J C, 2012. Two new species in the Graphidaceae(Ostropales, Ascomycota) from China[J]. Mycotaxon, 121: 75 – 79.

JIA Z F, ZHAO Z T, 2003. A preliminary study of the lichen genus *Ochrolechia* A. Massal. in China[J]. Mycosystema, 22(1): 30 – 34.

JIA Z F, ZHAO Z T, 2005. A new species of *Ochrolechia*[J]. Mycosystema, 24(2): 162 – 163.

JIA Z F, KALB K, 2014. New species and new records of Graphis(Graphidaceae) from China[J]. Mycosystema, 33 (5): 961 – 966.

JIA Z F, LI J, YANG M Z, 2017. Carbacanthographis(Graphidaceae), a lichen genus new to Guangxi[J]. Guihaia, 37(2): 231 – 233.

JIA Z F, LI M F, ZHAO X, 2016. Hemithecium hainanense sp. nov. with a checklist and key to Hemithecium species from China[J]. Mycotaxon, 131(3): 671 – 678.

JIA Z F, WANG L S, WU X L, et al, 2016. A study of Ocellularia(Graphidaceae) from China[J]. Mycosystema, 35 (5): 553 – 558.

JIANG S H, WEI X L, WEI J C, 2017. Two new species of Strigula(lichenised Dothideomycetes, Ascomycota) from China, with a key to the Chinese foliicolous species [J]. MycoKeys. 19: 31 – 42.

JIANG S H, WEI X L, WEI J C, 2016. Strigula sinoaustralis sp. nov. and three Strigula spp. new to China [J]. Mycotaxon, 131: 795 – 803(2016).

JIANG Y M, WEI J C, 1989. A Preliminary Study on *Everniastrum* from China[J]. Acta Bryolichenologica Asiatica, 1(1, 2): 43 – 52.

[——, ——] ——、——1990. A new species of *Hypogymnia* [J]. Acta Mycol. Sin. 9(4): 293 – 295.

JIANG Y M, WEI J C, 1993. A new species of *Everniastrum* containing diffractaic acid[J]. Lichenologist, 25(1): 57 – 60.

JIANG Z G, 1992. New taxa of lichen genus Collema Weber from P. R. China [J]. Journal of Hebei Norman University, 16(3): 83 – 87(in Chinese).

JIANG Z G, 1993. New recorded species of the genus Collema Weber from P. R. China and Hong Kong region[J]. Journal of Hebei Norman University, 17(3): 69 – 73(in Chinese).

JORGENSEN P M, 1975. Further notes on Asian *Alectoria*[J]. Bryologist, 78: 77.

JORGENSEN P M, 1994. Linnaean lichen names and their typification[J]. Botanical Journal of the Linnean Society, 115: 261 – 405.

JORGENSEN P M, 1997. Further notes n hairy Leptogium species [J]. Symbolae Botanicae Upsalenses, 32: 113 – 130.

JORGENSEN P M, 2000. Notes on some east – asian species of the lichen genus *Fuscopannaria*[J]. Journal of the Hattori Botanical Laboratory, 89: 247 – 259.

JORGENSEN P M, 2001. Four new Asian species in the lichen genus Pannaria[J]. The Lichenologist, 33: 297 – 302.

JORGENSEN P M, APTROOT A, 2002. The lichen Physma hondoanum Asah. discovered in Taiwan[J]. Lichenologist, 34(5): 441 – 442.

JORGENSEN P M, 2003. A new species of Arctomia from Sichuan Province, China[J]. Lichenologist, 35(4): 287 –

289.

JORGENSEN P M, MYLLYS L, VELMALA S, et al, 2012. Bryoria rigida, a new Asian lichen species from the Himalayan region [J]. The Lichenologist, 44(6): 777 – 781.

SANTOSH J, UDENI J, OH S O, et al, 2014. New records of lichens from Shapotou area in Ningxia of Northwest China [J]. Mycosystema, 33(1): 167 – 173.

JOSHI S, UPRETI D K, WANG X Y, et al, 2015. *Graphis yunnanensis*(Ostropales, Graphidaceae), a New Lichen Species from China [J]. Mycobiology, 43(2): 118 – 121.

KALB K, BUARUANG K, MONGKOLSUK P, et al, 2012. New or otherwise interesting Lichens. VI, including a lichenicolous fungus [J]. Phytotaxa, 42: 35 – 47.

KALB K, JIA Z F, 2014. New species of Graphidaceae from Zhejiang Province, China[J]. Phytotaxa, 189(1): 147 – 152.

KANTVILAS G, KASHIWADANI H, MOON K H, 2005. The lichen genus Siphula Fr. (Lecanorales) in East Asia [J]. J. Jap. Bot., 80: 208 – 213.

KASHIWADANI H, 1975. Enumeration of Anaptychiae and Parmeliae of Papua New Guinea. "The Botanical Expedition to Papua New Guinea "[M]. The National Science Museum. Tokyo.

KASHIWADANI H, 1977 a. On the Japanese species of the genus *Pyxine*(Lichens) (1) [J]. J. Jap. Bot. 52(5): 137.

KASHIWADANI H, 1977 b. Ibid. (2) [J]. Ibid., 52(6): 161.

KASHIWADANI H, 1979. Three rare species of lichens from the Kii Peninsula[J]. Mem. Natn. Sci. Mus. Tokyo, 12: 213 – 217.

KASHIWADANI H, 1984 a. A note on *Phaeophyscia*(Lichens) with orange – red medulla in Japan[J]. Ibid., 17: 55 – 59.

KASHIWADANI H, 1984 b. A revision of *Physcia ciliata*(Hoffm.) Du Rietz in Japan[J]. Bull. Natn. Sci. Mus. Tokyo ser. B(Botany), 10(1): 43 – 49.

KASHIWADANI H, 1984 c. On two Species of *Phaeophyscia* in Japan[J]. Ibid., 10(3): 127 – 132.

KASHIWADANI H, 1988. *Ramalina hossei* Vain. (Lichen) found in Bhutan and Formosa[J]. Bulletin of the National ScienceMuseum, Series B(Botany), 14(4).

KASHIWADANI H, 1995. Lich. Minus Cognitae Exs[G]. No. 158.

KASHIWADANI H, 1986. Genus *Ramalina*(Lichens) in Japan(1) On Ramalina calicaris(L.) Fr. In Span[J]. Bull. Natn. Sci. Mus., Tokyo, ser. B, 12: 89 – 98.

KASHIWADANI H, 1986. Genus *Ramalina*(Lichens) in Japan(2) On Ramalina pacifica Asah. And its allies[J]. Bull. Natn. Sci. Mus., Tokyo, ser. B, 12(4): 117 – 125.

KASHIWADANI H, S. Kurokawa 1981. Notes on Japanese and Formosan species of *Anthracothecium*(2) [J]. J. Jap. Bot., 56(11): 348 – 356.

KIRK P M, CANNON P F, MINTER D W, et al, 2008. Ainsworth & Bisby's Dictionary of the Fungi[M], 10th Edition prepared by CABI Europe – UK 2008, pp. 1 – 771.

KOU X R, LI S X, et al, 2013. Three new species and one new record of *Lobothallia* from China [J]. Mycotaxon, 123: 241 – 249.

KREMPELHUBER A VON, 1868. Exotische Flechten aus dem Herbar des K. K. Botanischen Hofkabinettes in Wien [J]. Verh. Zool. Bot. Ges. Wien, 18: 303 – 330, Pl. 3, 4.

KREMPELHUBER A VON, (I) 1873. Chinesische Flechten[J]. Flora 56: 465 – 471.

KREMPELHUBER A VON, 1874 a. Chinesische Flechten(1 – 13) [J]. Hedwigia 13(3): 33 – 35(1874).

KREMPELHUBER A VON, 1874 b. Chinesische Flechten. (14 – 22) [J]. Hedwigia, 13(4): 59 – 61. (1874).

KREMPELHUBER A VON, 1876. Aufzachlung und Beschreibung der Flechtenarten, welche Dr. Heinrich Wawra Ritter von Fernsee von zwei Reisen um die Erde mitbrachte[M]. Verhandlungen der kaiserlich – koniglichen zoologisch – botanischen Gesellschaft in Wien 26: 433 – 446.

KREMPELHUBER A VON, 1877. Aufzahlung und Beschreibung der Flechten – Arten, welche Dr. H. Wawra Ritter von Fernsee von zwei Reisen um die Erde mitbrachte[M]. *Verh. Zool. Bot. Ges. Wien*26: 433 – 446.

KROG H, 1976. *Lethariella* and *Prothusnea*, two new lichen genera in Parmeliaceae[J]. Norw. J. Bot., 23: 83 – 106.

KIRZEWICKA B, 2010. *Umbilicaria rhizinata* comb. nov. (lichenized Ascomycota)[J]. The lichenologist, 42(04): 491 – 493(2010). Short communications.

KONDRATYUK S Y, LÖKÖS L, TSCHABANENKO S, et al, 2013. New and noteworthy Lichen – forming and lichenicolous fungi [J]. Acta Botanica Hungarica 55(3 – 4), pp. 275 – 349, 2013 DOI: 10.1556/ABot.55.2013.3 – 4.9.

KONDRATYUK S Y, LÖKÖS L, KIM J A, et al, 2016. Fauriea, A New Genus of The Lecanoroid Caloplacoid Lichens(Teloschistacea, Lichen – forming Ascomycetes)[J]. Acta Botanica Hungarica 58(3 – 4): 3033 – 18, 2016 DOI: 10.1556/ABot.58.2016.3 – 4.6.

KONDRATYUK S Y, KIM J A, YU N H, et al, 2015. a new genus of xanthorioid lichens(Teloschistaceae, Ascomycota) proved by three gene phylogeny[J]. Ukr. Bot. J., 72(6): 574—584.

KUROKAWA S, 1955. Notulae miscellaneae lichenum japonicorum(1)[J]. J. Jap. Bot., 30(8): 252.

KUROKAWA S, 1959 a. Ibid. (6)[J]. Ibid., 34(1): 23.

KUROKAWA S, 1959 b. *Anaptychia*(lichens) and their allies of Japan(2)[J]. Ibid., 34(6): 174 – 176.

KUROKAWA S, 1960 a. Ibid. (3)[J]. Ibid., 35(3): 91 – 94.

KUROKAWA S, 1960 b. Ibid. (4)[J]. Ibid., 35(8): 240.

KUROKAWA S, 1960 c. Ibid. (5)[J]. Ibid., 35(12): 353.

KUROKAWA S, 1961. Ibid. (6)[J]. Ibid., 36(2): 51.

KUROKAWA S, 1962 a. A monograph of the genus *Anaptychia* zur Beiheft[J]. Nova Hedwigia, 6: 115.

KUROKAWA S, 1962 b. A note on the lichen genus*Tornabenia* Trev[J]. Ibid., 37(10): 289.

KUROKAWA S, 1965. Revision of series *Relicinae* of the genus *Parmelia* in Japan and Taiwan[J]. Ibid., 40(9): 264.

KUROKAWA S, 1966a. *Anaptychiae* and *Parmeliae*. In H. Hara(ed.)[J]. The Flora of Eastern Himalaya 1: 605 – 610.

KUROKAWA S, 1966b. Chemistry of Japanese *Peltigera* with some taxonomic notes[J]. Bull. Nat. Sci. Mus. Tokyo, 9(2): 101 – 114, Pl. 1 – 2.

KUROKAWA S, 1968 a. *Parmelia expallida*, a new lichen j species from Eastern Asia[J]. Bull. Natn. Sci. Mus. Tokyo, 11(2): 191.

KUROKAWA S, 1968 b. New or noteworthy species of *Parmelia* of Japan[J]. J. Jap. Bot., 43(10 – 11): 349.

KUROKAWA S, 1969. A note on some rare lichens of Japan[J]. Ibid., 44(8): 225.

KUROKAWA S, 1970. Notes on Japanese species of *Pilophoron*[J]. Ibid., 45(3): 73.

KUROKAWA S, 1973. Supplementary notes on the genus Anaptychia[J]. J. Hattori Bot. Lab., 37: 563 – 607.

KUROKAWA S, 1978. Noteworthy Lichens Collected in the Bonin Islands[J]. Mem. Natn. Sci. Mus. Tokyo11: 29.

KUROKAWA S, 1979. Enumeration of species of *Parmelia* in Papua New Guinea. In S. Kurokawa(ed.), Studies on Cryptogams of Papua New Guinea, 128 – 129[M]. Academia Scientific Book Inc., Tokyo.

KUROKAWA S, 1980. *Cetrariopsis*, a new genus in the Parmeliaceae, and its distribution[J]. Mem. Natn. Sci. Mus. Tokyo, 13: 139 – 142.

KUROKAWA S, 1987. New or Noteworthy Species of Parmelia, Subgenus Amphigymnia(Lichens) Producing Alectoronic and α – Collatolic Acid[J]. Bull. Nat. Sci. Mus., ser. B. (Bot.), 13(1): 11 – 15(1987).

KUROKAWA S, 1989. Studies on Japanese species of Xanthoparmelia(Parmeliaceae)[J]. J. Jap. Bot., 64(6): 165 – 175.

KUROKAWA S, 1991. Rimeliella, a new lichen genus related to Rimelia of the Parmeliaceae[J]. Ann. Tsukuba Bot. Gard. 10: 1 – 14.

KUROKAWA S, 1999. A note on postulate or sorediate species of Cetrariastrum(Parmeliaceae)[J]. J. Jap. Bot. 74(4): 251 – 255.

KUROKAWA S, Kashiwadani H, 1977. Notes on the genus Lopadium in Japan & Formosa[J]. Bull. Nat. Sci. Mus., ser. B. (Bot.), 3/4: 123 – 134.

KUROKAWA S, LAI M J, 1991. Allocetraria, a New Lichen Genus in the Parmeliaceae[J]. Bull. Nat. Sci. Mus., Ser. B, 17(2): 59 – 65a91991a0.

KUROKAWA S, MINETA M, 1973. Enumeration of Parmeliae of Ceylon[J]. Ann. Rep. Noto Mar. Lab., 13: 71 – 76.

KUROKAWA S, LAI M J, 2001. Parmelioid lichen genera and speices in Taiwan[J]. Mycotaxon, 77: 225 – 284.

LAI M J, 1980a. Notes on Some *Hypogymniae*(Parmeliaceae) from East Asia[J]. Quart. J. Taiwan Museum, 33(3, 4): 209 – 214.

LAI M J, 1980 b. Studies on the Cetrarioid Lichens in Parmeliaceae of East Asia(I)[J]. Ibid., 33(3, 4): 215 – 229.

LAI M J, 2000. Illustrated macrolichens of Taiwan [M]. The Council of Agriculture, the Executive Yuan, Taipei, Taiwan., 1 – 396.

LAI M J, 2001. The lichen family Stereocaulaceae of Taiwan[J]. Taiwan J. Forestry Science, 16: 175.

LAI M J, CHEN X L, QIAN Z G, et al, 2009. Cetrarioid lichen genera and species in N E China[J]. Ann. Bot. Fennici, 46: 365 – 380.

LAMB I M, 1947. A monograph of the lichen genus *Placopsis* Nyl[J]. Lilloa13: 151 – 288.

LAMB I M, 1963. Index Nominum Lichenum inter annos 1932 et 1960 divulgatorum[M]. 1 – 809. New York.

LAMB I M, 1965. The*Stereocaulon massartianum* assemblage in East Asia[J]. J. Jap. Bot., 40: 270.

LAMB I M, 1977. A conspectus of the lichen genus*Stereocaulon*(Schreb.) Hoffm[J]. J. Hattori Bot. Lab., 43: 191 – 355.

LAMB I M, WARD A, 1974. A preliminary conspectus of the species attributed to the imperfect lichen genus *Leprocaulon* Nyl [J]. J. Hattori Bot. Lab., 38: 499 – 553.

LANJOUW J, STAFLEU F A, 1954. Index herbariorum part, II(1)[OL]: A – D.

LANJOUW J, STAFLEU F A, 1957. *Ibid.* II(2)[OL]: E – H.

LANJOUW J, STAFLEU F A, 1972. *Ibid.* II(3)[OL]: I – L.

LAUNDON J R, 1979. Deceased lichenologists: their abbreviations and herbaria[J]. Lichenologist, 11(1): 1 – 26.

LAUNDON J R, 1984. The typification of Withering' s neglected lichens[J]. Lichenologist, 16(3): 211 – 239.

LI H, WEI J C, 2016. Functional analyses of thioredoxin from the desert fungus, *Endocarpon pusillum* Hedwig(Verrucariales, Ascomycota)[J]. Scientific Reports, 6: 27184: 1 – 10(2016). | DOI: 10.1038/srep27184.

LI H, ZHANG Y L, WANG Y Y, et al, Alternative oxidase in desert lichen is crucial for survival during oxidative stress and rehydration process[J]. Applied and Environmental Microbiology(in press).

LI J, MIAO X L, JIA Z F, 2014. Additional Materials for the genus Phaeographis from China[J]. Journal of Fungal Research, 12(2): 79 – 82.

LI J, JIA Z F, 2016. Diorygma fuscum sp. nov. from China[J]. Mycotaxon, 131(3): 717 – 721.

LI S X, KOU X R, REN Q, 2013. New records of *Aspicilia* species from China[J]. Mycotaxon, 126: 91 – 96.

LI Y, CHEN C L, ZHAO Z T, 2008. A primary study on lichens from mountain Yi [J]. Journal of fungal research, 6(2): 70 – 73.

LI Z C, WEN X M, GULBOSTAN, et al, 2007. New Chinese Records of Lichen Genus *Acarospora* in Xinjiang[J]. Journal of Fungal Research, 5(4): 191 – 192(2007).

LIANG M M, QIAN Z G, WANG X Y, et al, 2012. Contributions to the lichen flora of the Hengduan Mountains, China(5). Anzia rhabdorhiza(Parmeliaceae), a new species [J]. The Bryologist, 115(3): 382 – 387.

LIN C K, HSIAO J Y, LAI M J, 1989. Studies on the Lichen Genus Lobaria in Taiwan[J]. Proc. Natl. Sci. Counc. B. ROC, 13(1): 47 – 55.

LIU D, WANG X Y, LI J W, et al, 2014. Contributions to the lichen flora of the Hengduan Mountains, China(6): Revisional study of the genus *Canoparmelia* (Lichenized Ascomycota, Parmeliaceae) [J]. Plant Diversity and Resources, 36(6): 781 – 787.

LIU D, GOFFINET B, ERTZ D, et al, 2018. Circumscription and phylogeny of the Lepidostromatales (lichenized Basidiomycota) following discovery of new species from China and Africa. Mycologia 2018: 1 - 19. Published online 25 Jan 2018, https://doi.org/10.1080/00275514.2017.1406767

LIU H J, WEI J C, 2003. Two new taxa of the lichen genus Collema from China[J]. Mycosystema, 22(4): 531 - 533.

LIU H J, WEI J C, 2003. Taxonomic revision of six taxa in the lichen genus Collema from China[J]. Mycotaxon, 86: 349 - 358.

LIU H J, WEI J C, 2009. A brief overview of and key to species of *Collema* from China[J]. Mycotaxon, 108: 9 - 29.

LIU H J, WU Q F, CHEN Z, 2010. The lichen order Peltigerales from mountain Liupan, Ningxia, Northwest China [J]. Journal of Fungal Research, 8(4): 194 - 199.

LIU H J, CAO J, GUAN S, 2012. Three non - hairy species of *Leptogium* from China [J]. Mycotaxon, 122: 483 - 490.

LIU H J, GUAN S, 2012. A new hairy species of *Leptogium* (*Collemataceae*) from China [J]. Mycotaxon, 119: 413 - 417.

LIU H J, XI M Q, HU J S, et al, 2015. Two new species and a new record of *Leptogium* from China [J]. Mycotaxon, 130: 471 - 478.

LIU H J, HU J S, LI C, 2015. The lichen genus *Kroswia* in China[J]. Mycotaxon, 130: 951 - 959.

LIU H I, HU J S, WU Q F, 2016. New species and new records of the lichen Genus *Fuscopannaria* from China[J]. Mycotaxon, 131: 455 - 465(2016).

LIU M, WEI J C, 2013. Lichen diversity in Shapotou region of Tengger Desert, China [J]. Mycosystema, 32(1): 42 - 50.

LIU T S, LAI M J, LIN H S, 1980. The cladoniform Lichens in Taiwan[J]. Quart. J. Taiwan Mus., 33(1): 1 - 35.

LIU Z L, SUN L Y, JIA Z F, 2015. *Graphis elongata*(Graphidaceae) a new record from China [J]. Journal of Fungal Research, 13(1): 1 - 3.

LLANO G A, 1950. A monograph of the lichen family Umbilicariaceae in the western hemisphere[M]. pp. 1 - 281. Office of Naval Research, Department of the Navy - Washington, D. C.

LU B S, HYDE K D, HO W H, et al, 2000. Checklist of Hong Kong fungi[J]. Fungal Diversity Research Series, 5: 1 - 207.

LU D A, 1958. Notes on Chinese lichens, 1. *Peltigera*[J]. Acta Phytotax. Sin., 7: 263 - 269.

——. 1959. Notes on Chinese lichens, 2. Umbilicariaceae)[J]. Ibid., 8(2): 173 - 180.

LV L, WANG C L, REN Q, et al, 2008. The lichen genus *Lecanora* from Bailong river valley of Gansu province, China[J]. Mycosystema, 27(1): 99 - 104.

LV L, REN Q, SUN L Y, et al, 2009. Three species of the lichen genus *Lecanora* new to China from Bailong River Valley, Gansu Province[J]. Guihaia, 29(3): 311 - 313.

LV L, REN Q, WANG H Y, et al, 2009. New records of four *Lecanora* species from China[J]. Mycotaxon, 110: 437 - 441.

LV L, WANG H Y, ZHAO Z T, 2009. Five lichens of the genus *Lecanora* new to China[J]. Mycotaxon, 107: 391 - 396.

LV L, ZHANG L L, LIU X L, et al, 2012. *Lecanora subjaponica*, a new lichen from China[J]. The Lichenologist, 44 (4): 465 - 468.

LV L, REN Q, JIANG D F, et al, 2013. *Lecanora gansuensis* sp. nov. (*subfusca* group) from China[J]. Mycotaxon, 123: 285 - 287.

LÜCKING, RIVAS PLATAE, LÜCKING R, APTROOTA, et al, 2006. A first assessment of the Ticolichen biodiversity inventory in Costa Rica: the genus Coenogonium(Ostropales: Coenogoniaceae), with a world - wide key and checklist and a phenotype - based cladlistic analysis[J]. Fungal Diversity, 23: 297(2006).

LUMBSCH H T, 1994. Die Lecanora subfusca - Gruppe in Australasien[J]. Journal of the Hattori Botanical Laboratory, 38: 499 – 553.

LUMBSCH H T, HUHNDORF S M(eds.), 2007. Outline of Ascomycota[J]. Myconet, 13: 1 – 58, ISSN 1403 – 1418.

LUO G Y, 1984. Preliminary study on the lichen species distribution and their ecological characteristics on Dailing, Liangshui Forest Farm[J]. J. Northeastern Forestry Inst., 12(Suppl.): 84 – 88.

LUO G Y, 1986. The preliminary study of*Hypogymnia*(lichen) from Northeastern China[J]. Bull. Bot. Res., 6(3): 155 – 170.

MA J, 1981. Historical notes on studies of Chinese lichens[J]. J. of Beijing Forestry College, 2: 1.

MA J, 1981. A check list of Chinese lichens[J]. Ibid. 3: 69 – 80 & 4: 61 – 8.

MA J, 1983. A check list of Chinese lichens[J]. Ibid. 1: 107 – 110 & 2: 73 – 82 & 3: 93 – 112 & 4: 87 – 106.

MA J, 1984. A check list of Chinese lichens[J]. Ibid. 1: 95 – 114 & 2: 95 – 114 & 3: 93 – 112 & 4: 79.

MA J, 1985. A check list of Chinese lichens[J]. Ibid. 1: 101 – 119.

MA R, LI H M, WANG H Y, et al, 2010. A new species of *Phlyctis* (*Phlyctidaceae*) from China [J]. Mycotaxon, 114: 361 – 366.

MUHAMMAT M, SATTAR M, ABBAS A, et al, 2012. New Records of Lichens from Middle East of TianShan Mt. Xinjiang, China [J]. Journal of Fungal Research, 10(1): 4 – 7.

NURAHMAT M, RIHAT T, WEN X M, et al, 2015. A preliminary Study on the Lichen Genus Rhizocarpon Ramont ex DC. in Xinjiang, China[J]. Acta Bot. Boreal. – Occident. Sin., 35(2): 422 – 426.

MAGNUSSON A H, 1933. A monograph of the lichen genus *Ionaspis*[J]. Acta Horti Gothob. 8: 1 – 47.

MAGNUSSON A H, 1940. Lichens from Central Asia I [M]. Rep. Sci. Exped. N. W. China S. Hedin – The Sino – Swedish expedition – (Publ. 13). XI. BoT. 1: 1 – 168, Pl. 1 – 12, 1 folded map. f. 1 – 3.

MAGNUSSON A H, 1944. *Ibid.* II[M]. Ibid. 2: 1 – 68, Pl. 1 – 8, 1 text map.

MAGNUSSON A H, 1955. New or Otherwise Interesting Swedish Lichens XV[J]. Bot. Not., 108(2): 292 – 306.

MATTSON J E, LAI M J, 1993. *Vulpicida* genus novus[J]. Mycotaxon, 48: 427.

GULDAN1 N, GULNAZ1 S, LAZZAT1 N, et al, 2016. Study on the Lichen Genus Melanelixia in Xinjiang, China [J]. Journal of Fungal Research Mar., 14(1): 28 – 32.

MCCARTHY P M, 2001. Trichotheliales and Verrucariaceae[J]. Flora of Australia, 58A: 104 – 229.

MCCUNE B, OBERMAYER, 2001. Typification of Hypogymnia hypotrypa and H. sinica[J]. Mycotaxon LXXIX: 23 – 27.

MCCUNE B, 2009. *Hypogymnia* (Parmeliaceae) species new to Japan and Taiwan [J]. The Bryologist, 12(4): 823 – 826, doi: 10.1639/0007 – 2745 – 112.4.823.

MCCUNE B, 2011. *Hypogymnia irregularis* (*Ascomycota: Parmeliaceae*)—a new species from Asia [J]. Mycotaxon 115: 485 – 494, doi: 10.5248/115.485.

MCCUNE B, 2012. The identity of *Hypogymnia delavayi*(Parmeliaceae) and its impact on H. alpina and *H. yunnanensis* [J]. Opuscula Philolichenum, 11: 11 – 18. pdf available online 3January2012 via(http://sweetgum.nybg.org/philolichenum/).

MCCUNE B, TCHABANENKO S, 2001. *Hypogymnia arcuata* and *H. sachalinensis*, Two New Lichens from East Asia [J]. The Bryologist, 104(1): 146 150.

MCCUNE B, MARTIN E P, WANG L S, 2003. Five new species of Hypogymnia with rimmed holes from the Chinese Himalayas[J]. The Bryologist, 106(2): 226 – 234.

MCCUNE B, WANG L S, 2014. The lichen genus Hypogymnia in southwest China[J]. Mycosphere, 5(1): 27 – 76, Doi 10.5943/mycosphere/5/1/2.

MCCUNE B, TCHABNENKO S, WEI X L, 2015. *Hypogymnia papilliformis*(Parmeliaceae), a new lichen from Far East Russia and China [J]. The Lichenologist, 47(2): 117 – 122.

MENG F G, ZHAO Z T, 2010. Study on the lichen genus Melanelia Essl. in West China [J]. J. Anhui Agri. Sci., 38: 3046 – 3047(in Chinese).

MENG Q F, WEI J C, 2008. A lichen genus *Diorygma* (Graphidaceae, Ascomycota) in China [J]. Mycosystema, 27 (4): 525 –531.

MENG Q F, FU S B, WEI J C, 2011. Taxonomical Study on the lichen genus Fissurina in China in *The Present Status and Potentialities of the Lichenology in China* [M]. Pp. 157 – 168(2011), Science Press, Beijing.

MEYEN J, FLOTOW I, 1843. Lichenes. Verh. der Kaiserl. Leopold – Carol Academie der Naturforschen 19(Suppl. 1) [M]. Appendis p. 231 –232. Breslau und Bonn.

MIADLIKOWSKA J, KAUFF F, HÖGNABBA F, et al, 2014. A multigene phylogenetic synthesis for the class Lecanoromycetes(Ascomycota): 1307 fungi representing 1139 infrageneric taxa, 317 genera and 66 families[J]. Molecular Phylogenetics and Evolution, 79(2014) 132 – 168.

MIAO X L, JIA Z F, MENG Q F, et al, 2007. Some species of Graphidaceae(Ostropales, Ascomycota) rare and new to China[J]. Mycosystema, 26(4): 493 –506.

MITUNO M, 1938. *Sphaerophorus* – Arten aus Japan[J]. J. Jap. Bot., 14(10): 659 –669.

MOON K H, KUROKAWA S, KASHIWADANI H, 2006. Revision of the lichen genus *Menegazzia*(Ascomycotina: Parmeliaceae) in Eastern Asia[J]. J. Jpn. Bot., 81: 127 – 138(2006).

MOREAU F, ET MOREAU, 1951. Lichens de Chine[J]. Rev. Bryol. et Lichenol., 20: 183 – 199, 1 map.

MOTYKA J, 1936 –38. Lichenum Generis *Usnea* Studium Monographicum[M]. Pars systematica. Leopoli.

MÜLL A R G, 1884. Lichenologische Betrage(XIX) [J]. Flora, 67: 304, 402.

MÜLL A R G, 1893. Lichenes chinenses Henryani a cl. Dr. Aug. Henry, anno 1889, in China media lecti quos in herbario Kewensi determinavit[J]. Bull. Herb. Boiss., 1: 235 –236.

REYIM M, TUMUR A, XAHIDIN H, et al, 2009. A species of *Lecanora* new to China [J]. Mycosystema, 28(1): 154 – 156.

REYIM M, ABBAS A, 2016. Preliminary Study on the Lichen Gnus Psora in Xinjiang, China [J]. Act Bot. Boreal – Occident Sin., 36(7): 1482 – 1485.

NAKANISHI S, 1964. Epiphytic communites in the alpine zone of Mt. Yu Shan(Mt. Niitaka), Taiwan[J]. Misc. Bryol. Lichenol., 3(5): 71 –73.

NAKAO M, 1923. Ueber die Bestandteile der chinesischen Droge "Shi – Hoa"[J]. J. of the Pharmaceutical Society of Japan 469: 29 –38(in German) &423 –497(in Japanese).

GULDAN N, GULNAZ S, LAZZAT N, et al, 2016. Study on the Lichen Genus Melanelixia in Xinjiang, China [J]. Journal of Fungal Research, 14(1): 28 –32.

NELSON P R, KEPLER R, WALTON J, et al, 2012. Parmelina yalungana resurrected and reported from Alaska, China and Russia [J]. The Bryologist, 115(4): 557 –565.

NIU D L, WANG L S, ZHANG Y J, et al, 2007. A chemotaxonomic study of Lethariella zahlbruckneri and L. smithii (lichenized Ascomycota: Parmeliaceae) from Hnegduanshan Mountain [J]. The Lichenologist, 39(6): 549 – 553.

NIU D L, HARADA H, WANG L S, et al, 2011. Chemotaxonomic study of the Lethariella cladonioides complex(lichenized Ascomycota, Parmeliaceae) [J]. The Lichenologist, 43(3): 213 –223.

NUNO M, 1958. On the chemical ingredients of *Usneadiffracta* Vain. [J]. J. Jap. Bot., 33(8): 227.

NUNO M, 1963. On some brown *Nephromata* and chromatograms of their ingredients[J]. Ibid. 38: 196 –202.

NUNO M, 1964. Chemism of*Parmelia* subgenus *Hypogymnia* Nyl. [J]. Ibid. 39(4): 97.

NUNO M, 1972. Four new species of*Cladonia* from southeastern Asia. *Ibid.* 47: 161 – 167.

LAZZAT N, KNUDSEN K, ABBAS A, 2016. *Sarcogyne saphyniana* sp. nov., a saxicolous lichen from northwestern China[J]. Mycotaxon, 131: 135 – 139.

NYLANDER W, 1857. Enumeration generale des lichenes, avec l' indication sommaire de leur distribution geographique[J]. Mem. Soc. Sci. Cherbourg, 5: 85 – 146; Supplement pp. 332 –339.

NYLANDER W, 1887. Addenda nova ad Lichenographiam europaeam[J]. Flora, 70(9): 129 – 136.

NYLANDER W, CROMBIE J M, 1883. On a collection of exotic lichens made in Eastern Asia by the late Dr. A. C.

Maingay[J] . J. Linn. Soc. London, Bot. 20: 62 – 66.

OBERMAYER W, 1998. Dupla Graecensia Lichenum(1998) [J] . Fritschiana, 16: 7 – 14.

OBERMAYER W, 2004. Additions to the lichen flora of the Tibetan region [J] . Bibliotheca Lichenologica, 88: 479 – 526.

OBERMAYER W, BLAHA J, MAYRHOFER H, 2004. Buellia centralis and chemotypes of Dimelaena in Tibet and other Central – Asian regions[J] . Symbol. Bot. Ups. , 34(1) : 327 – 342.

OBERMAYER W, ELIXIR J A, 2003[J] . Bibiliotheca Lichenologica, 86: 33 – 46.

OHMURA Y, 2001. Taxonomic study of the genus Usnea(lichinized ascomycetes) in Japan and Taiwan[J] . J. Hattori Bot. Lab. , 90: 1 – 96.

OHMURA Y, MOON K H, KASHIWADANI H, 2008. Morphology and Molecular phylogeny of *Ramalina pollinaria*, R. sekika and R. yasudae(*Ramalinaceae*, Lichenized *Ascomycotina*) [J] . J. Jpn. Bot. , 83: 156 – 164(2008) .

OMAR Z, KEYIMU A, ABBAS A, 2004. New Chinese records of lichen genus *Melanelia* [J] . Acta Bot. Yunnan. , 26: 385 – 386(in Chinese) .

OLIVIER H, 1898. Un lichen de Hong – Kong[J] . Bull. L'Academie Internationale de Geographie Botanique, 7: 82.

OLIVIER H, 1904. Lichens dus Kouy – Tcheou[J] . Ibid. 14: 193 – 196.

OSHIO M, 1968. Taxonomical studies on the family Pertusariaceae of Japan[J] . J. of Sci. of the Hiroshima Univ. series B, Div. 2(Botany) , 12(1) : 81 – 163, Pl. I – VI.

OXNER A N, 1933. Species Lichenum novae ex Asia[J] . J. du Cycle Botanique de L'Academie des Sciences D'Ukraine, 7 – 8: 168.

PATOUILLARD N, ET OLIVIER H, 1907. (I) : Champignons et Lichens chinois[J] . in Le Monde des Plantes, ser. 2, IX, 23(1907) .

PAULSON R, 1925. Lichens of Mount Everest [J] . J. Bot. London, 63: 189 – 193.

PAULSON R, 1928. Lichens from Yunnan[J] . Ibid. , 66: 313 – 319, Pl. 587.

PFISTER D H. (ed.) , 1978. Cryptogams of the United States North Pacific Exploring Expedition, 1853 – 1856. Farlow Reference Library & Herbarium of Cryptogamic Botany, Harvard University[M] . Cambridge, MA. [Lichens by Edward Tuckerman, pp. 65 – 126.]

POELT J, 1977. Die Gattung *Umbilicaria* (Umbilicariaceae) , (Flechten des Himalaya 14) [J] . Khumbu Himal, 6 (3) : 397 – 435.

POELT J, HINTEREGGER E, 1993. Beiträge zu Kenntnis der Flechtenflora des Himalaya. Die Gattungen Ⅷ Caloplaca, Fulgensia und Ioplaca [J] . Bibliotheca Lichenologica, 50: 1 – 247.

PRILLINGER H, KRAEPELIN G, LOPANDIC K, et al, 1997. New species of *Fellomyces*isolated from epiphytic lichen species [J] . Systematic and Applied Microbiology, 20: 572 – 584.

PURVIS O W, COPPINS B J, HAWKSWORTH D L, et al, 1992. The Lichen Flora of Great Britain and Ireland[M] . Natural History Museum Publications in Association with The British Lichen Society, 1 – 710.

RABENHORST L, 1873. Chinesische Flechten in der Umgegend von Saison, Hongkong, Wampoa, Shanghay u. s. w. gesammebestimmt von Dr. v. Krempelhuber in Munchen [J] . Flora, 56: 286.

RANDLANE T, SAAG A Q, OBERMAYER W, 2001. Cetrarioid lichens containing usnic acid from the Tibetan area [J] . Mycotaxon, 80: 389 – 425.

RASSADINA K A, 1960. Notulae System. e Sect. Cryptogam. Inst. Bot. nomine V. L. Komarovii Acad. Sci. [M] . URSS, 13: 23. Leningrad.

RASSADINA K A, 1967. Notulae System. e Sect. Cryptogam. Inst. Bot. nomine V. L. Komarovii Acad. Sci. [M] . URSS, 20: 297. Leningrad.

RASANEN V, 1940. Lichenes ab A. Yasuda et aliis in Japonia collecti(II) [J] . J. Jap. Bot. , 16(3) : 139 – 153.

RASANEN V, 1949. Lichenes novi IV[J] . Arch. Soc. Zool. Bot. Fenn. "Vanamo"3: 78.

RASANEN V, 1951. Ibid. VI[J] . Ibid. 5(1) : 27.

REN Q, 2015. A new species and new records of the lichen genus *Pertusaria* from China[J] . Mycotaxon, 130: 689 –

693.

REN Q, JIA Z F, 2011. Taxonomic Study of *Ochrolechia*(Ochrolechiaceae) from China in *The Present Status and Potentialities of the Lichenology in China* [M]. Science Press, Beijing, 169 – 181.

REN Q, ZHAO Z T, WEI J C, 2003. A Lichen Genus Varicellaria Nyl. from China [J]. Mycosystema, 22(2): 216 – 218.

REN Q, ZHAO Z T, 2004. Two new records of the lichen genus *Pertusaria* from China [J]. Guihaia, 24(4): 329 – 331.

REN Q, SUN Z S, LIU H J, et al, 2008. Two new records of *Pertusaria* (lichenized ascomycetes) from China[J]. Mycosystema, 27(4): 611 – 613.

REN Q, ZHAO Z T, 2008. Two species new for China of the lichen genus *Pertusaria*[J]. Guihaia, 28(6): 735 – 736.

REN Q, SUN Z S, LI Y J, et al, 2009. Additions to the lichen flora of Mount Taibai, northwestern China[J]. Mycosystema, 28(1): 102 – 105.

REN Q, 2013. *Pertusaria albiglobosa*, a new lichen from China [J]. Mycotaxon, 124: 349 – 352.

REN Q, ZHAO Z T, WEI J C, 2003. A lichen genus *Varicellaria* Nyl. from China [J]. Mycosystema, 22(2): 216 – 218.

REN Q, SUN Z S, ZHAO Z T, 2008. Two new species of Pertusaria(Pertusariaceae) from China[J]. Mycotaxon, 106: 441 – 444.

REN Q, SUN Z S, WANG L S, et al, 2009. A new species of *Pertusaria*(Pertusariaceae) from China [J]. Mycotaxon, 108: 231 – 234.

REN Q, SUN Z S, ZHAO Z T, 2009. *Pertusaria wulingensis* (Pertusariaceae), a new lichen from China[J]. The Bryologist, 112(2): 394 – 396.

REN Q, KOU X R, 2013. A new species of Pertusaria from China [J]. The Lichenologist, 45(3): 337 – 339.

REN Q, ZHAO N, 2014. New taxa of the lichen genus Pertusaria from China [J]. Mycotaxon 127: 221 – 226.

REN Q, LI S X, 2013. New records of crustose lichens from China – 1 [J]. Mycotaxon, 125: 65 – 67.

MAMUT R, XAHIDIN H, LIU L Y, et al, 2006(2005). New Records of the Lichen Genus Lecanora from Northern Xinjiang[J]. Journal of Xinjiang University(Natural Science Edition, 23(2): 203 – 206.

MAMUT R, TUMUR A, XAHIDIN H, et al, 2009. A species of *Lecanora* new to China[J]. Mycosystema, 28(1): 154 – 156.

MAMUT R, TURSUN T, ABBAS A, 2014. Ochrolechia upsaliensis(L.) A. Massal.: A new record of lichen genus *Ochrolechia* A. Massal. from China[J]. Acta Bot. Boreal. – Occident. Sin., 34(4): 0855 – 0858.

MAMUT R, TURSUN T, ABBAS A, 2015. Study on the lichen genus Toninia A. Massal. in Xinjiang, China[J]. Guihaia, 35(2): 161 – 165.

ROGERS R W, HAFELLNER J, 1988. *Haematomma* and *Ophioparma*: two superficially similar genera of lichenized fungi[J]. Lichenologist, 20(2): 167 – 174.

RUNEMARK H, 1956. Studies in *Rhizocarpon* I. Taxonomy of the yellow species in Europe[M]. Opera Botanica, 2: 1, Stockholm.

GULNAZ S, NAZZAT N, NASHITAY M, et al, 2015. Newly – Recorded Species of Acarospora from Xinjiang [J]. Plant Science Journal, 33(3): 291 – 294.

SANDSTEDE H, 1932. Cladoniaceae A. Zahlbruckner, I. Die Pflanzenareale[M]. Dritte Reihe, Heft 6 Dr. Ludwig Diels und Dr. G. Samuelsson berausgegeben von Dr. E. Hannig und Dr. H. Winkler, pp. 63 – 71, maps 51 – 60.

SANDSTEDE H, 1938. Cladoniaceae A. Zahlbruckner, II. Die Pflanzenareale[M]. 4 Reihe, Heft 7, p. 86 – 90, maps 62. 68 and 70.

SANDSTEDE H, 1939. Cladoniaceae A. Zahlbruckner, III. Die Pflanzenareale[M]. 5 Reihe, Heft 8, pp. 94 – 98, maps 72 – 74 and 76.

SANTESSON R, 1939. Amphibious pyrenolichens I [J]. Arkiv for Bot., 29 A(10): 1 – 67, Pl. 1, 2, f. 1 – 6.

SANTESSON R, 1952. Foliicolous lichens I. A revision of the taxonomy of the obligately foliicolous, lichenized fungi [J]. Symb. Bot. Ups. 12(1): 1 – 590.

SASAOKA H, 1919. Lichens of Taiwan[J]. Trans. Nat. Hist. Soc. Formosa, 8: 179 – 181(in Japanese).

SASAKI I, 1942. Distinction between *Physcia* picta and *Physcia* aegialita[J]. J. Jap. Bot., 18(11): 626.

SATO M M, 1933. Notes on some Japanese lichens determined by Dr. Edv. A. Vain. IV[J]. Ibid., 9(4): 269.

SATO M M, 1934 a. Studies on the lichens of Japan(II)[J]. Ibid., 10(7): 424.

SATO M M, 1934 b. Ibid. (III)[J]. Ibid., 10(11): 687.

SATO M M, 1935. Ibid. (V)[J]. Ibid., 11(4): 238.

SATO M M, 1936 a. Enumeratio lichenum Ins. Formosae(I)[J]. Ibid., 12(6): 426 – 432.

SATO M M, 1936 b. Ibid. (II)[J]. Ibid., 12(8): 569 – 575.

SATO M M, 1936 c. Bull. Saito Ho – on Kai Museum11: 23.

SATO M M, 1937 Ibid. (III)[J]. J. J. B. 13(8): 595 – 599.

SATO M M, 1938 a. Ibid. (IV)[J]. J. J. B. 14(7): 463 – 469.

SATO M M, 1938 b. *Ibid.* (V)[J]. J. J. B. 14(12): 783 – 791.

SATO M M, 1938 c. The place where*Dictyonema sericeum* and D. sp. grow. (In Japanese)

SATO M M, 1939. Parmeliaceae(I)[M]. in Takenoshin Nakai et Masazi Hondo's Nova Flora Japonica vel Descriptiones et Systema Nova omnium plantarum in Imperie Japonice sponte nascentium, pp. 1 – 87. Tokyo.

SATO M M, 1940 a. East Asiatic Lichens(II)[J]. J. Jap. Bot., 16(1): 42 – 47.

SATO M M, 1940 b. *Ibid.* (IV)[J]. Ibid., 16(8): 495 – 500.

SATO M M, 1941 a. Cladoniales(I)[M]. T. Nakai et M. Hondo's Nova Flora Japonica, Tokyo., 1 – 103.

SATO M M, 1943. Notes on the Japanese lung lichens[J]. Phytotax. Geobot., 13: 238 – 241.

SATO M M, 1950. Notes on some remarkable*Umbilicariae* collected in Far Eastern Asia[J]. J. Jap. Bot., 25(8): 165.

SATO M M, 1951[J]. Ibid. 26(7): 196.

SATO M M, 1952. Lichenes Khinganenses: or a list of lichens collected by Prof. T. Kira in the Khingan Range, Manchuria[J]. Bot. Mag. Tokyo, 65(769 – 770): 172 – 175, f. 1.

SATO M M, 1954. Enumeration of Lichens collected in Tohoku District, Japan. (1) Anziaceae and Baeomycetaceae [J]. Bull. of the Yamagata Univ., Nat. Sci., 3(2): 113 – 126.

SATO M M, 1956. Range of the Japanese Lichens(I)[J]. Bull. Fac. Lib. Arts, Ibaraki Univ., Nat. Sci., 6: 29 – 31.

SATO M M, 1957. Ibid. (II)[J]. Ibid., 7: 57 – 69.

SATO M M, 1958 a. Ibid. (III)[J]. Ibid., 3: 61 – 68.

SATO M M, 1958 b. Ibid. (V)[J]. Ibid., 10: 83 – 85.

SATO M M, 1959. Ibid. (IV)[J]. Ibid., 9: 39 – 51.

SATO M M, 1960. Ibid. (VI)[J]. Ibid., 11: 53 – 62.

SATO M M, 1961. Ibid. (VII)[J]. Ibid., 12: 41 – 48.

SATO M M, 1981. Notes on the cryptogamic flora of prov. Shansi, North China(III) Lichenes[J]. Miscellanea Bryologica et Lichenologica, 9(3): 64 – 65.

SATO T, 1937. On the *Umbilicariapustulata* Hoffm[J]. J. Jap. Bot., 13(4): 298 – 300.

SCHNEIDER G, 1979. Die Flechtengattung *Psora* sensu A. Zahlbruckner[J]. Bibliotheca Lichenologica, 13.

SCHWENDENER S, 1867. Ueber die Natur der Flechten[M]. Verhandl. d. Schweiz. Naturforsch. Gesellschaft zu Rheinfelden, 51: 88 – 90.

SEAWARD M R D, APTROOT A, 2005. Hong Kong Lichens Collected on the United States North Pacific ExploringExpedition, 1853 – 1856 [J]. The Bryologist, 108(2): 282 – 286.

SEEMANN B C, 1852 – 57. (Section on 'Flora of the Island of HONG Kong' published in 1857) *The Botany of the Voyage of H. M. S. Herald under the command of Captain Henry Kellett, R. N., C. B., during the Years* 1845 – 51[M]. Lovell Reeve, London.

SHIBATA S, 1974. Some aspects of lichen chemotaxonomy. In "Chemistry in Botanical Classification"[M]. (G. Bendz and J. Santesson, eds.) Academic Press, New York and London, 241 – 249.

SIPMAN H J M, 1983. A monograph of the lichen family Megalosporaceae [J]. Bibliotheca Lichenologica., 18, 1 – 241.

MOHAMMAD S, OWE – LARSSON B, NORDIN A, et al, 2009. *Aspicilia tibetica*, a new terricolous species of the Himalayas and adjacent regions[J]. Mycol Progress, DOI 0. 1007/s11557 – 010 – 0656 – 7.

MOHAMMAD S, STENROOS S, MYLLYS L, et al, 2013. Phylogeny and taxonomy of the 'manna lichens' [J]. Mycol Progress, 12: 231 – 269, DOI 10. 1007 /sII557 – 012 – 0830 – 1.

SPARRIUS L B, APTROOT A, LAI M J, 2002. New reports of calicioid lichenized and nonlichenized ascomycetes from Taiwan[J]. Mycotaxon, 83: 357 – 360.

STENROOS S, VITIKAINEN O, KOPONEN T, 1994. Cladoniaceae, Peltigeraceae and other lichens from northwestern Sichuan, China[J]. J. Hattori Bot. Lab., 75: 319 – 344.

SUN J J, WANG X H, JIA Z F, 2013. Preliminary Study of Lichen Genus Buellia from Mountain Tai [J]. Journal of Fungal Research, 111(13): 155 – 163.

SUN L Y, ZHAO Z T, JIA Z F, 2000. An investigation of l ichens of Sai Hanwula national nature reserve zone ofInner Mongolia Autonomous Region[J]. Shandong Science, 3(4): 35 – 38.

SUN L Y, MENG F G, LI H M, et al, 2010. A new lichen, *Melanohalea subexasperata*(Parmeliaceae), from the Tibetan Plateau[J]. Mycotaxon, 111: 65 – 69.

SUN Z S, REN Q, LIU H J, et al, 2008. *Involucropyrenium* (Verrucariaceae), a lichen genus new to China[J]. Acta Botanica Yunnanica, 30(6): 655 – 656.

TCHOU Y T, 1935. Note preliminaire sur les lichens de Chine[J]. Contr. Inst. Bot. Nat. Acad. Peiping, 3: 299 – 322.

THOR GOERAN, 1990. The lichen genus Chiodecton and five allied genera[M]. Opers Bot. 103: 1 – 92. Copenhagen. G. THOR & VěZDA. 1984. [J]. Folia geobot. phytotax., 19(1): 72.

THROWER S L, 1980. Air pollution and lichens in Hong Kong[J]. Lichenologist, 12: 305 – 311.

THROWER S L, 1988. Hong Kong Lichens[M]. An Urban Council Publication, Hong Kong.

TIMDAL E, 1990. Gypsoplacaceae and Gypsoplaca, a new family and genus of squamiform lichens. – In: H. M. Jahns(ed.): Contributions to Lichenology in Honour of A. Henssen[J]. Bibliotheca Lichenologica. No. 38. J. Cramer, Berlin – Stuttqart. Pp. 419 – 427. (Description of G. macrophylla: p. 424: illustration: p. 421.

TIMDAL E, 1994. *Boreoplaca* genus novus [J]. Mycotaxon, 51: 503.

TIMDAL E, OBERMAYER W, BENDIKSBY M, 2016. *Psora altotibetica*(Psoraceae, Lecanorales), a new lichen species from the Tibetan part of the Himalayas[J]. MycoKeys, 13: 35 – 48.

TIAN Q, WANG L S, WANG H Y, et al, 2011. A new species of *Nephroma* (*Nephromataceae*) from the Tibetan Plateau[J]. Mycotaxon, 115: 281 – 285.

TOGASHI M, 1968 a. Miscellaneous notes on lichens or lichenological survey(2)[J]. J. Jap. Bot., 43: 315.

TOGASHI M, 1968 b. Ibid. (4)[J]. Ibid. 43(10 – 11): 358.

TUCKERMAN E, 1882. Synopsis N. Amer. Lich. II[M]. Boston.

TURSUN T, MAMUT R, TUMUR A, et al, 2015. Preliminar Study on the Family Lichinaceae Nyl. in Xinjiang, *China*[J]. Acta Bot. Boreal. – Occident., 35(11): 2339 – 2342.

VAINIO E D V, 1887. Monographia *Cladoniarum universalis* I[J]. Acta Soc. Fauna Fl. Fenn., 4: 1 – 509.

VAINIO E D V, 1894. Ibid. II[J]. Acta Soc. Fauna Fl. Fenn. 10: 1 – 498.

VAINIO E A, 1921. Lichenes insularum Philipinarum. III[J]. Ann. Acad. Sci. Fenn., ser. A: 15.

VERSEGHY K, 1962. Die Gattung *Ochrolechia*[J]. Beihefte zur Nova Hedwigia, 1: 1 – 146.

VITIKAINEN O, 1986. *Peltigera dolichospora*, a new Himalayan – Western Chinese lichen[J]. The Lichenologist, 18 (4): 387.

WANG C L, SUN L Y, REN Q, et al, 2007. A preliminary study of multispored *Lecanora* from Mt. Taibai[J]. Mycosystema, 26(1): 46 – 50.

WANG H Y, CHEN J B, WEI J C, 2008. A new species of Melanelixia(Parmeliaceae) from China [J]. Mycotaxon, 104: 185 – 188.

WANG H Y, CHEN J B, WEI J C, 2009. A phylogenetic analysis of *Melanelia tominii* and four new records of brown parmelioid lichens from China[J]. Mycotaxon, 107: 163 – 173.

WANG H Y, JIANG D F, HUANG Y H, et al, 2013. Study on the phylogeny of *Nephroma helveticum* and allied species [J]. Mycotaxon, 125: 263 – 275.

WANG H Y, GE A N, LI H M, et al, 2013. Additional information on *Lecanora loekoesii*[J]. Mycotaxon, 123: 235 – 239.

WANG H Y, WANG H Y, ZHAO Z T, 2010. Five lichens of *Leptogium* new to China [J]. Mycotaxon, 111: 161 – 166.

WANG L S, HARADA H, KOH Y J, et al, 2005. Two species of Bryoria(Lichenized Ascomycota, Parmeliaceae) from the Sino – Himalayas[J]. Mycobiology, 33(4): 173 – 177.

WANG L S, CHEN J B, 1994. The classification of the genus Bryoria from Yunnan[J]. Acta Botanica Yunnanica, 16 (2): 144 – 152.

WANG L S, 1995. *Anzia physoidea* A. L. Smith in China[J]. Acta Mycologica Sinica, 14(4): 313 – 314.

WANG L S, OH S O, NIU D L, et al, 2008. Diversity of Epiphytic Lichens on Tea Trees in Yunnan, China[J]. Plant Diversity, 30(5): 533 – 539.

WANG L S, HARADA H, NARUI T, et al, 2003. *Bryoria hengduanensis*(Lichenized Ascomycota, Parmeliaceae), a new species from Southern China[J]. Acta Phytotax. Geobot., 54(2): 99 – 104.

WANG L S, MCCUNE B, 2010. Contributions to the lichen flora of the Hengduan Mountains, China 1. Genus *Pseudephebe* (lichenized *Ascomycota, Parmeliaceae*) [J]. Mycotaxon, 113: 431 – 437.

WANG L S, HARADA H, WANG X Y, 2012. Contributions to the lichen flora of the Hengduan Mountains, China (3). Bryoria divergescens(Parmeliaceae), an overlooked species [J]. The Bryologist, 115(1): 101 – 108.

WANG L S, WANG X Y, LUMBSCH H T, 2013. Eight Lecanoroid Lichen Species New to China[J]. Cryptogmie Mycologie, 34(4): 343 – 348.

WANG R F, WANG L S, WEI J C, 2014. *Allocetraria capitata* sp. nov. (Parmeliaceae, Ascomycota) from China[J]. Mycosystema, 33(1): 19 – 22.

WANG R F, WEI X L, WEI J C, 2015. A new species of Allocetraria(Parmeliaceae, Ascomycota) in China [J]. The Lichenologist, 47(1): 31 – 34.

WANG R F, WEI X L, WEI J C, 2015. The genus *Allocetraria*(Parmeliaceae) in China[J]. Mycotaxon, 130: 577 – 591.

WANG S L, CHEN J B, 2000. New species of Parmeliaceae(Lichenized Ascomycotina) from China[J]. Mycotaxon, 76: 293 – 298.

WANG S L, CHEN J B, 2001a. Three species of *Parmelina* (s. lat.) new to China [J]. Mycosystema, 20: 135 – 136 (in Chinese).

WANG S L, CHEN J B, ELIX J A, 2001b. Two new species of the genus *Myelochroa*(Parmeliaceae, Ascomycota) from China[J]. Mycotaxon, 77: 25 – 30.

WANG W C, ZHAO Z T, ZHANG L L, 2015. Four *Rhizocarpon* species new to China[J]. Mycotaxon, 130: 883 – 891.

WANG W C, ZHAO Z T, ZHANG L L, 2015. Four new records of *Rhizocarpon* from China[J]. Mycotaxon, 130: 739 – 747.

WANG Y Y, LIU B, ZHANG X Y, et al, 2014. Genome characteristics reveal the impact of lichenization onlichen – forming fungus *Endocarpon pusillum* Hedwig(Verrucariales, Ascomycota) [J]. BMC genomics, 15: 34(17 January 2014).

WANG J H, LI E W, WEI J C, 2017. A preliminary study of the isoprenylated chromones from lichenized fungus

Sarcogyne asciparva [J]. Mycosystema, 36(1): 1 – 11.

WANG X Y, JOSHI Y, HUR J S, et al, 2010. Taxonomic studies on the lichen flora of southwestern China(1). *Pilophorus yunnanensis* sp. nov. (Cladoniaceae) [J]. The Bryologist, 113(2): 345 – 349.

WANG X Y, JOSHI Y, OH S O, et al, 2011. *Pilophorus fruticosus* (Cladoniaceae), a new species from south – western China[J]. The Lichenologist, 43(2): 137 – 140.

WANG X Y, ZHANG L L, JOSHI Y, et al, 2012. New species and new records of the lichen genus Porpidia(Lecideaceae) from western China [J]. The Lichenologist, 44(5): 619 – 624.

WANG X, 1985. The lichens of the Mt. Tuomuer areas in Tianshan . In Scientific Expedition of Chinese Academy of Sciences(ed.): Fauna and Flora of the Mt. Tuomuer areas in Tianshan[M]. Xinjiang people's Publishing House, Urumqi (in Chinese), 328 – 353.

WANG X Y, 1985. The Lichens in the Tormul Peak, Mt. Tianshan[M]. Biology in the Tormul Peak area of Mt. Tianshan. Xinjiang People's Publishing House, Urumqi, 328 – 353.

WANG X H, SHI G B, JIA Z F, 2013. *Phaeographis fujianensis*, a new species of lichen[J]. Mycosystema, 32(1): 128 – 130.

WANG X H, XU L L, JIA Z F, 2015. The lichen genus *Leiorreuma* in China[J]. Mycotaxon, 130: 247 – 251.

WANG X Y, GOFFINET B, LIU D, et al, 2015. Taxonomic study of the genus *Anzia* (Lecanorales, lichenized Ascomycota) from Hengduan Mountains, China [J]. The Lichenologist, 47(2): 991 – 15.

WANG Y Y, ZHANG X Y, ZHOU Q M, et al, 2015. Comparativetranscriptome analysis of the lichen – forming fungus *Endocarpon pusillum* elucidates its drought adaptation mechanisms[J]. Science China Life Sciences, 58 (1): 89 – 100.

WANG – YANG J R, 1972. The taxonomic status of *Coenogonium subvirescens* and *C. interplexum*newly found in Taiwan[J]. Taiwania, 17(1): 40 – 47.

WANG – YANG J R, LAI M J, 1973. A checklist of the lichens of Taiwan[J]. Taiwania, 18(1): 83 – 104.

WANG – YANG J R, LAI M J, 1976 a. Notes on the lichen genus*Sphaerophorus* Pers. of Taiwan, with descriptions of three new species[J]. Taiwania, 21(1): 83 – 85.

WANG – YANG J R, LAI M J, 1976 b. Additions and corrections to the lichen flora of Taiwan[J]. Taiwania, 21(2): 226.

WEI J C, 1966. A new subgenus of *Lasallia* Mer. em. Wei [J]. Acta Phytotax. Sin., 11(1): 1 – 8, Pl. 1, 2.

WEI J C, 1981. Lichenes sinenses exsiccati(Fasc. I: 1 – 50) [J]. Bull. Bot. Res., 1(3): 81 – 91.

WEI J C, 1982. Some new species and materials of *Lasallia* Mer. em. Wei[J]. Acta Mycol. Sin., 1(1): 19 – 26.

WEI J C, 1983. A taxonomic revision of lichen genus *Xanthoparmelia* (Vain.) Hale from China[J]. Acta Mycol. Sin., 2(4): 221 – 227.

WEI J C, 1984 a. A preliminary study of lichen genus *Rhizoplaca* from China[J]. Acta Mycol. Sin., 3(4): 207 – 213.

WEI J C, 1984 b. A new isidiate species of *Hypogymnia* in China[J]. Ibid., 3(4): 214 – 216.

WEI J C, 1986 Notes on some isidiate species of *Hypogymnia* in Asia [J]. Acta Mycol. Sin. Suppl., 1: 323 – 329.

WEI J C, 1991. An Enumeration of Lichens in China[M]. International Academic Publishers, Beijing, 1 – 278.

WEI J C, 1993. The lectotypification of some species in the Umbilicariaceae described by Linnaeus or Hoffman in Supplement to Mycosystema Vol. 5. [M] International Academic Publishers, Beijing, China, 1 – 17(May 20, 1993).

WEI J C, 1995. An analysis of the systematics and geography ofLichen family Umbilicariaceae from Eastern Asia (D). Abstract in Petersburg, 1995. republished in Wei Jiang – Chun's Collection of Scientific Papers [M]. Science Press, Beijing, 291 – 372.

WEI J C, 2010. The biodiversity of pan – fungi and the sustainable development of human beings[J]. Bulletin of Chinese Academy of Sciences, 25(6): 645 – 650(in Chinese).

WEI J C, ABBAS A, 2003. The lichen genus *Pseudevernia* Zopf in China [J]. Mycosystema, 22(1): 26 – 29.

WEI J C, CHEN J B, 1974. Materials for the lichen flora of the Mount Qomolangma region in Southern Xizang, China. In Report on the Scientific Investigations(1966 – 1968) in Mt. Qomolangma district(Biology and Alpine Physiology) [M] . Science Press, Beijing, 173 – 182.

WEI J C, CHEN J B, JIANG Y M, 1986 a. Studies on lichen family Cladoniaceae in China II. The lichen genus *Cladina* Nyl. [J] . Acta Mycologica Sinica, 5(4) : 240 – 250.

WEI J C, CHEN J B, JIANG Y M, 1986 b. Notes on lichen genus *Lobaria* in China[J] . Acta Mycol. Sin. , Suppl. 1: 329 – 344.

WEI J C, CHEN J B, JIANG Y M, CHEN X L, 1985. Studies on lichen family Cladoniaceae in China I. A revision of *Cladia* Nyl. [J] . Acta Mycol. Sin. , 4(1) : 55 – 59.

WEI J C, JIANG Y M, 1980. Species novae lichenum e Parmeliaceis in regione xizangensi[J] . Acta Phytotax. Sin. , 18(3) : 386 – 388.

WEI J C, JIANG Y M, 1981. A biogeographical analysis of the lichen flora of Mt. Qomolangma region in Xizang. In Proceedings of Symposium on Qinghai – Xizang(Tibet) Plateau(Beijing, China) Geological and Ecological Studies of Qinghai – Xizang Plateau Volume II Environment and Ecology of Qinghai – Xizang Plateau [M] . Science Press, Beijing, Gordon and Breach, Science publishers, Inc. New York, 1145 – 1151.

WEI J C, JIANG Y M, 1982. New materials for lichen flora from Xizang[J] . Acta Phytotax. Sin. , 20(4) : 496 – 501, Pl. 2.

WEI J C, JIANG Y M, 1986. Lichens of Xizang[M] . Science Press. Beijing, China, 1 – 130.

WEI J C, JIANG Y M, 1988. A conspectus of the lichenized ascomycetes Umbilicariaceae in China[J] . Mycosystema, 1: 73 – 106.

WEI J C, JIANG Y M, 1989. The status of the genus *Llanoa* Dodge and the delimitation of the genera in the Umbilicariaceae(Ascomycotina) [J] . Mycosystema, 2: 135 – 150.

WEI J C, JIANG Y M, 1991. Some foliicolous lichens in Xishuangbanna, China. In Tropical Lichens: Their Systematics, Conservation, and Ecology edited by D. J. Galloway, p. 201 – 216. (The Systematics Association Special Volume No. 43) [M] . he Systematics Association by Clarendon Press, Oxford.

WEI J C, JIANG Y M, 1992. Some species new to science and distribution in Umbilicariceae(Ascomycota) [J] . Mycosystema, 5: 73 – 88.

WEI J C, JIANG Y M, 1993. The Asian Umbilicariaceae(Ascomycota) [M] . International Academic Publishers. , pp. 1 – 217.

WEI J C, JIANG Y M, GUO S Y, 1994. Studies on the lichen family Cladoniaceae in China III. A new genus to China: *Thysanothecium* [J] . Mycosystema, 7: 23 – 27.

WEI J C, HU Y C, JIANG Y M, et al, 1994. A study on divergent characters of populations in *Usnea montis – fuji*. Mot. [J] . Acta Mycologica Sinica, 13(3) : 199 – 204.

WEI J C, JIANG Y M, 1999. A lichen genus Brodoa, new to China in Parmeliaceae [J] . Mycosystema, 18(4) : 445 – 448.

WEI J C, WANG X Y, CHEN X L, et al, 1982. Lichenes officinales sinenses[M] . Science Press, Beijing. pp. 1 – 65.

WEI J C, JIA Z F, WU X L, 2013. An Investigation of Lichen Diversity from Hainan Island of China and Prospect of the R. & D of their Resources[J] . Journal of Fungal Research, 11(4) : 224 – 238.

WEI X L, WEI J C, 2005. Two new species of *Hypogymnia*(Lecanorales, Ascomycota) with pruinose lobe tips from China[J] . Mycotaxon, 94: 155 – 158.

WEI X L, WEI J C, 2012. A study of the pruinose species of *Hypogymnia* (Parmeliaceae, Ascomycota) from China [J] . The Lichenologist , 44(6) : 783 – 793.

WEI X L, WANG L S, HUR J S, 2007. Lichen flora of western part of Yunnan province, China [J] . Journal of Fungal Research, 5(3) : 146 – 160.

WEI X L, MCCUNE B, WANG L S, et al, 2010. *Hypogymnia magnifica*(Parmeliaceae) , a new lichen from southwest China [J] . The Bryologist, 113(1) : 120 – 123.

WEI X L, MCCUNE B, LUMBSCH H T, et al, 2016. Limitations of Species Delimitation Based on Phylogenetic Analyses: A Case Study in the *Hypogymnia hypotrypa* Group(Parmeliaceae, Ascomycota)[J]. PLoS ONE, 11(11):1 - 20.

WIRTH V, 1987, Die Flechten Baden - Worttembergs: Verbreitungsatlas[M]. Ulmer, 1 - 528.

WU D, MA J, HUA Z L, et al, 2011. Study on the lichens of Hebei Wuling Mount IV[J]. Journal of Capital Normal University(Natural Science Edition), 32(5):34 - 38(in Chinese).

WU J N, LIU H J, 2012. Flora Lichenum Sinicorum Vol. 11, Peltigerales(I)[M]. Plates 1 - 25. Science Press, Beijing, 1 - 262.

WU J L, 1981. Medicinal lichens in Qin Ling mountain[J]. Acta Pharmaceutica Sinica, 16(3):161 - 167.

WU J L, 1985a. A brief introduction of literature on the studies of Chinese lichens[J]. Act. Bot. Bor. - Occ. Sin. 5(1):101 - 107.

WU J L, 1985b. The Lichens collected from the steppe of Xinjiang[J]. Acta Phytotax. Sin. 23(1):73 - 78.

WU J L, 1987. Lichen Iconography of China[M]. China Prospect Publishing House(CPPH), Beijing.

WU J L, ZHANG Z J, 1982. A study on *Lethariella* subgen. *Chlorea* from Qinling mountain, Shaanxi province[J]. Acta Phytotax. Sin. 20(2):241 - 246.

WU J N, KANG R C, ABBAS A, 1997. The lichen genera *Cyphelium* and *Pseudevernia* and lichen species *Peltula tortuosa* first recorded in China from Hanas of Xinjiang[J]. Arid Zone Research, 14(3):13 - 15.

WU J N, LIU A T. What is the Chinese medicinal Herb: "Lao Long Pi"[J]. Acta Phytotax. Sin., 14(2):66 - 68.

WU J N, LIU H J(ed.), 2012. Flora Lichenum Sinicorum vol. 11, Peltigerales(I)[M]. Science Press, Beijing pp. 1 - 261, with plate I - XXV(in the Chinese language with the keys to different taxa in English).

WU J N, QIAN Z G, 1989. Xu B S ed.: Cryptogamic flora of the Yangtze Delta and adjecent regions), pp. 158 - 266[M]. Shanghai Scientific & Technical Publishers, Shanghai.

WU J N, QIAN Z G, YU S, 1997. The lichen genus *Erioderma* in China[J]. Journal of Nanjing Normal University(Natural Science Edition), 20(4):61 - 62, 67.

WU J N, QIAN Z G, 1999. Lichen Genus *Pannaria*, *Fuscopannia*, and *Parmeliella* in China[J]. Journal of Nanjing Normal University(Natural Science Edition), 22(3):85 - 90.

WU J N, XIANG T, 1981. A preliminary study of the lichens from Yuntai mountain in Lianyungang, Jiangsu[J]. Journal of Nanjing College(Natural Science Edition), 3: 1 - 11.

WU J N, XIANG T, QIAN Z G, 1982. Notes on Wuyi mountain lichens(I)[J]. Wuyi Science Journal, 2:9 - 13.

WU J N, XIANG T, QIAN Z G, 1984. Notes on Wuyi mountain lichens(II)[J]. Wuyi Science Journal, 4: 1 - 7.

WU J N, XIANG T, QIAN Z G, 1985. Notes on Wuyi mountain lichens(III)[J]. Wuyi Science Journal, 5: 223 - 230.

WU J N, WANG L S, 1992. The lichen families Alectoriaceae and Anziaceae in Lijiang Prefecture, Yunnan[J]. Acta Botanica Yunnanica, 14(1):37 - 44.

XANIDIN H, ABBAS A, WEI J C, 2010. *Caloplaca tianshanensis*(lichen - forming Ascomycota), a new species of subgenus Pyrenodesmia from China[J]. Mycotaxon, 114: 1 - 6.

XANIDIN H, BAHTI P, TURSUN T, et al, 2009. The lichen genus *Fulgensia* in Xinjiang, China[J]. Mycosystema, 28(1):106 - 111.

XI M Q, LIU H J, 2014. Two new species of *Leptogium* to Asia[J]. Journal of Fungal Research, 12(2):71 - 74.

XU B S(ed.), 1989. Cryptogamic flora of the Yangtze Delta and adjecent regions)[M]. Shanghai Scientific & Technical Publishers, 158 - 266.

XU L L, JIA Z F. 2015. Lichen genus *Myriotrema* and species *M. viridialbum* new to China[J]. Journal of Fungal Research, 13(3):132 - 135.

XU L L, WU Q H, WANG Q D, et al, 2016. *Chapsa*(Graphidaceae, Ostropales), a lichen grnus new to China[J]. Journal of Tropical and Subtropical Botany, 24(5):495 - 498.

YANG F, REN Q, LI S X, et al, 2008 The lichen genus *Pertusaria* from Bailong River valley of Gansu, China[J].

Mycosystema, 27(4): 622 – 626.

YANG J, WEI J C, 2008. The new lichen species *Endocarpon crystallinum* from semiarid deserts in China[J]. Mycotaxon, 106: 445 – 448.

YANG J, WEI J C, 2009. A new subspecies of Gyalidea asteriscus from China[J]. Mycotaxon, 109: 373 – 377, 2009.

YANG Y, ABBAS A, 2003. *Cladonia libifera*, a lichen species new to China[J]. Mycosystema, 22(3): 512.

YE J, ZHENG H Y, ZHANG H, et al, 2009. Two species of lichens new to China[J]. Mycosystema, 28(5): 762 – 764.

YOSHIMURA I, 1968. The phytogeographical relationships between the Japanese and North American species of Cladonia[J]. J. Hattori Bot. Lab., 31: 227 – 246.

YOSHIMURA I, 1968 a. Japanese species of *Anema*[J]. J. Jap. Bot., 43(10 – 11): 354 – 358.

YOSHIMURA I, 1968 b. Japanese species of *Thyrea*[J]. Ibid. 43(12): 500 – 502.

YOSHIMURA I, 1971. The genus *Lobaria* of Eastern Asia[J]. J. Hattori Bot. Lab. 34: 231 – 364.

YOSHIMURA I, 1994. Lichen flora of Japan in colour[M]. Hoikusha Publiching Co., Ltd. Japan, 1 – 349.

YOSHIMURA I, SHARP A J, 1968. Some Lichens from the Southern Appalachians and Mexico[J]. Bryologist, 71 (2): 108 – 113.

YU S H, WU J N, 1997. A preliminary study of *Pertusaria* D. C. in Hubei Shennongjia region[J]. Journal of Wuhan Botanical Research, 15(4): 331 – 335.

YU S H, WU J N, LI P, 1999. Some Lichen Species of *Pertusaria* New to China)[J]. Mycosystema, 18(1): 112.

YU S H, WU J N, CHEN Q X, 1999. Critical Notes on the Lichen Genus Pertusaria of China[J]. Journal of Nanjing Normal University(Natural Science), 1(2): 94 – 97.

ZAHLBRUCKNER A, 1922. Cat. Lich. Univ. 1[M]. Leipzig. reprinted by Johnson Reprint Corporation, New York, N. Y., 1951.

ZAHLBRUCKNER A, 1924. *Ibid.* 2[M]. Ibid.

ZAHLBRUCKNER A, 1925. *Ibid.* 3[M]. Ibid.

ZAHLBRUCKNER A, 1927. *Ibid.* 4[M]. Ibid.

ZAHLBRUCKNER A, 1928. *Ibid.* 5[M]. Ibid.

ZAHLBRUCKNER A, 1930 a. *Ibid.* 6[M]. Ibid.

ZAHLBRUCKNER A, 1931. *Ibid.* 7[M]. Ibid.

ZAHLBRUCKNER A, 1932 a. *Ibid.* 8[M]. Ibid.

ZAHLBRUCKNER A, 1934 a. *Ibid.* 9[M]. Ibid.

ZAHLBRUCKNER A, 1940. *Ibid.* 10[M]. Ibid.

ZAHLBRUCKNER A, 1930 b. Lichenes in Handel – Mazzetti, Symbolae Sinicae[M]. Jiulius Springer in Vienna, printed in Austria, 3: 1 – 254, Pl. 1.,

ZAHLBRUCKNER A, 1932 b. Neue Flechten XI[J]. Ann. Mycologici, 30: 427 – 441.

ZAHLBRUCKNER A, 1933 a. Lichenes in Feddes[J]. Repertorium sp. nov., 31: 23 – 25.

ZAHLBRUCKNER A, 1933 b. Flechten der Insel Formosa in Feddes, Lichenes in Feddes[J]. Repertorium sp. nov. 31: 194 – 224.

ZAHLBRUCKNER A, 1933 c. Flechten der Insel Formosa in Feddes, Lichenes in Feddes[J]. Repertorium sp. nov. 33: 22 – 68.

ZAHLBRUCKNER A, 1934 b. Nachtrage zur Flechtenflora Chinas[J]. Hedwigia, 74: 195 – 213.

ZHANG T, LI H M, WEI J C, 2006. The lichens of Mts. Fanjingshan in Guizhou province[J]. Mycosystema, 4(1): 1 – 13.

ZHANG F, ZHAO Z T, LIU H J, et al, 1999. A preliminary report on lichens from Mt. Meng[J]. Shandong forestry Science and technology, (2): 30 – 31.

ZHANG L L, WANG H Y, SUN L Y, et al, 2010. Four lichens of the genus *Lecidea* from China[J]. Mycotaxon,

112: 445 – 450.

ZHANG L L, WANG L S, WANG H Y, et al, 2012. Four new records of lecideoid lichens from China[J]. Mycotaxon, 119: 445 – 451.

ZHANG L L, HU L, ZHAO X X, et al, 2015. New records of *Clauzadea* and *Immersaria* from China[J]. Mycotaxon, 130: 899 – 905.

ZHANG Q, YANG F, REN Q, et al, 2008. The lichen family Ochrolechiaceae from Bailong River Valley in Gansu Province, China[J]. Mycosystema, 27(4): 614 – 618.

ZHANG Y Y, WANG X Y, LIU D, et al, 2016. The genus *Bulbothrix* (Parmeliaceae) in China[J]. The Lichenologist, 48(2): 1211 – 33.

ZHANG Y L, LI H, et al, 2017. A calcium – binding protein EpANN from the lichenized fungus Endocarpon pusillum enhances heat – shock tolerancein yeast[J]. Fungal Genetics and Biology, 108: 36 – 43.

ZHANG Y, WEI J C, 2017. Researches on the generic classification based on symplesiomorphy of genotype and phynotype in the family Umbilicariaceae(Ascomycota)[J]. Mycosystema, 36(8): 1089 – 1103.

ZHANG Y J, 2000. Molecular Genetics[M]. Science Press, Beijing(in Chinese), 1 – 486.

ZHAO J D, 1964. A preliminary study on Chinese *Parmelia*[J]. Acta Phytotax. Sin., 9: 166, Pl. 10 – 15.

ZHAO J D, HSU L W, SUN Z M, 1975. Species novae *Usneae* Sinicae[J]. Ibid., 13(2): 90 – 107.

[——, —— & ——] ——, ——, —— 1978. Species novae *Parmeliae* Sinicae)[J]. Ibid., 16(3): 95 – 97.

[——, —— & ——] ——, ——, —— 1979. Species novae *Anaptychiae* et *Physciae* sinicae[J]. Acta Phytotax. Sin. 17(2): 96 – 100.

[——, —— & ——) ——, ——, ——1982. Prodromus Lichenum Sinicorum[M]. Science PressBeijing, 1 – 156.

ZHAO J Z, GUO S Y, ZHAO Z T, 2006. An annotated checklist of the lichen faminly Pertusariaceae from Qinling mountains[J]. Mycosystema, 25(2): 179 – 183.

ZHAO N, SU Q J, REN Q, 2013. Three new records of crustose lichens from mainland China[J]. Acta Bot. Boreal. – Occident. Sin., 33(8): 1700 – 1702.

ZHAO N, SUN Z S, ZHOU M, et al, 2014. Three new records of the lichen genus Pertusaria from China[J]. Acta Bot. Boreal. – Occident. Sin., 34(3): 0628 – 0630.

ZHAO X, ZHANG L L, ZHAO Z T, 2013. A new species of *Miriquidica* from China[J]. Mycotaxon, 123: 363 – 367.

ZHAO X X, ZHANG L L, MIAO C C, et al, 2016. A new species of Porpidia from China[J]. The Lichenologist, 48 (3): 229 – 235 .

ZHAO X X, HU L, ZHAO Z T, 2016. Farnoldia Hertel – a new record genus for China with description of new record species[J]. Acta Bot. Boreal. – occident. Sin., 36(8): 1710 – 1712.

ZHAO X, ZHANG L L, SUN L Y, et al, 2015. Four new records ofLecanoraceae in China[J]. Mycotaxon 130: 707 – 715.

ZHAO Z T, LIU H J, JIANG C L, 1999. Study on lichens from mount Lao in Shandong province[J]. Journal of Shandong Normal University(natural science), 14(4): 426 – 428.

ZHAO Z T, MENG F G, LI H M, et al, 2009. A new species of Melanohalea(Parmeliaceae) from the Tibetan Plateau [J]. Mycotaxon, 108: 347 – 352.

ZHAO Z T, JIA Z F, 2002. Two new records of Ochrolechia from China[J]. Mycosystema, 21(2): 291 – 292.

ZHAO Z T, LIU H J, LI J T, 1998. The Lichens from Mount. Culai of Shandong province[J]. Shandong Science, 11 (4): 28 – 31(in Chinese).

ZHAO Z T, LIU H J, 2002. Two additions to to a lichen genus Peltigera in China[J]. Mycosystema , 21(3): 457 – 458.

ZHAO Z T, LI K F, WANG H, 2002. A study on lichens of Shandong province, East China[J]. Shandong Sci., 15 (3): 4 – 8(in Chinese).

ZHAO Z T, MENG F G, LI H M, et al, 2009. A new species of *Melanohalea*(Parmeliaceae) from the Tibetan Plateau

[J]. Mycotaxon, 108: 347 - 352.

ZHAO Z T, REN Q, 2003. Two new records of the lichen genus Pertusaria from China[J]. Mycosystema, 22(4): 669 - 670.

ZHAO Z T, SUN L Y, 2002. Three new records of the lichen genus Caloplaca fromChina[J]. Mycosystema, 21(1): 135 - 136.

ZHAO Z T, REN Q, APTROOT A, 2004. An Annotated Key to the Lichen Genus *Pertusaria* in China[J]. The Bryologist, 107(4): 531 - 541.

ZHAO Z T, WANG C L, SUN L Y, et al, 2007. *Orphniospora* (Fuscidiaceae), a lichen genus new to China[J]. Mycosystema, 26(2): 195 - 196.

ZHAO Z T, LI Y J, REN Q, et al, 2008. Studies on the genus *Physconia* from Qinling Mountains of Shaanxi in China [J]. Guihaia, 28(6): 724 - 727.

ZHAO Z T, LI C, ZHAO X, et al, 2013. New records of *Rhizocarpon* from China[J]. Mycotaxon, 125: 217 - 226.

ZHAO Z T, ZHAO X, GAO W, et al, 2014. *Pertusaria yunnana*, a new species from south - west China[J]. The Lichenologist, 46(2): 169 - 173.

XIAO Z G, SHEN H M, ZHAO A N, 2007. Phylogenetic Analysis of Lichen - Forming Fungi *Rhizoplaca* Zopf from China Based on ITS Data and Morphology[J]. Z. Naturforsch. 62c, 757 - 764.

ZHOU C L, WEN X M, XANIDIN H, et al, 2009. Classification of the *Ramalina* from Xinjiang[J]. Journal of Fungal Research, 7(1): 9 - 13.

ZHOU G L, ZHAO Z T, LV L, et al, 2012. Seven dark fruiting lichens of *Caloplaca* from China[J]. Mycotaxon, 122: 307 - 324.

ZHOU Q M, WEI J C, 2006. A new genus and species *Rhizoplacopsis weichingii* in a new family Rhizoplacopsitaceae (Ascomycota)[J]. Mycosystema, 25(3): 376 - 385.

ZHOU Q M, WEI J C, 2007. A new order Umbilicariales J. C. Wei & Q. M. Zhou(Ascomycota)[J]. Mycosystema, 26(1): 40 - 45.

ZHUANG W Y, 2000. Two new species of Unguiculariopsis(Helotiaceae, Encoelioideae) from China[J]. Mycol. Res., 104(4): 507 - 509.

INDEX OF SCIENTIFIC NAMES IN CHINESE
(中文名称索引)

C

D

E

F

G

H

J

K

L

M

N

O

P

Q

R

X

Y

Z

INDEX OF SCIENTIFIC NAMES
(学名索引)

A

B

C

D

E

F

G

H

I

J

K

L

M

N

R

S

T

U

V

W

X

Z